AF821832

Advances in Analysis of Flavors and Fragrances: Chemistry, Properties and Applications in Food Quality Improvement

Advances in Analysis of Flavors and Fragrances: Chemistry, Properties and Applications in Food Quality Improvement

Topic Editors

Ana Leahu
Maria Soledad Prats Moya
Cristina Ghinea

MDPI

Basel • Beijing • Wuhan • Barcelona • Belgrade • Novi Sad • Cluj • Manchester

Topic Editors
Ana Leahu
Faculty of Food Engineering
Stefan cel Mare University of Suceava
Suceava
Romania

Maria Soledad Prats Moya
Analytical Chemistry
Nutrition and Food Sciences Department
University of Alicante
Alicante
Spain

Cristina Ghinea
Faculty of Food Engineering
Stefan cel Mare University of Suceava
Suceava
Romania

Editorial Office
MDPI AG
Grosspeteranlage 5
4052 Basel, Switzerland

This is a reprint of the Topic, published open access by the journals *Beverages* (ISSN 2306-5710), *Fermentation* (ISSN 2311-5637), *Foods* (ISSN 2304-8158), *Molecules* (ISSN 1420-3049) and *Chemosensors* (ISSN 2227-9040), freely accessible at: https://www.mdpi.com/topics/K95102U32T.

For citation purposes, cite each article independently as indicated on the article page online and as indicated below:

Lastname, A.A.; Lastname, B.B. Article Title. *Journal Name* **Year**, *Volume Number*, Page Range.

ISBN 978-3-7258-4499-9 (Hbk)
ISBN 978-3-7258-4500-2 (PDF)
https://doi.org/10.3390/books978-3-7258-4500-2

© 2025 by the authors. Articles in this book are Open Access and distributed under the Creative Commons Attribution (CC BY) license. The book as a whole is distributed by MDPI under the terms and conditions of the Creative Commons Attribution-NonCommercial-NoDerivs (CC BY-NC-ND) license (https://creativecommons.org/licenses/by-nc-nd/4.0/).

Contents

About the Editors

Ana Leahu

Ana Leahu, with a PhD in Science and Agroforestry (2005), is an Associate Professor PhD. Eng at the Faculty of Food Engineering, "Stefan cel Mare" University of Suceava. Her doctoral thesis was conceived and carried out during her work at the Office of Pedological and Agrochemical Studies in Suceava, where she worked as a research chemical engineer with responsibilities in soil and plant sample analysis, thus benefiting from access to experimental fields of pasture and forage plants and the agro-phytotechnical laboratories of the Agricultural Research Center in the same locality. Ana Leahu has the ability to analyze functional products and focuses on the research and development of functional foods fortified with food waste as raw materials. She is also good at analyzing bioactive ingredients, covering phenolic compounds, antioxidant activity and anthocyanins. She focuses on food chemistry and food testing technology, and is proficient in using spectroscopy (such as nuclear magnetic resonance, Fourier-transform infrared spectroscopy, and ultraviolet–visible spectroscopy) and chromatography (gas chromatography–mass spectrometry) to carry out related research; at the same time, she pays attention to waste management and waste value-added treatment, and is committed to exploring innovative paths for resource recycling. Ana Leahu has fruitful academic achievements, publishing 3 books, 48 papers in ISI-indexed journals, 74 papers in Romanian and international peer-reviewed journals included in the database, and 8 papers in peer-reviewed conference proceedings (4 of which are included in ISI conference proceedings). She also has two patents.

Maria Soledad Prats Moya

Maria Soledad Prats Moya, with a PhD in Analytical Chemistry, is an Associate Professor at the University of Alicante in the Analytical Chemistry, Nutrition and Food Science Department. Her teaching experience includes courses in Food Science, Culinary Chemistry, and Quantitative Food Analysis. Her research areas include culinary science and food chemical and physical characterization. She is an expert in nut chemical characterization and the application of chemometrics for cultivar classification. Other research areas include nutritional food studies and the risks and benefits of ingredients. She is a co-author of 46 manuscripts in JCR journals and 2 books with more than 10 chapters. She also has two patents. She has participated in more than 80 poster or oral presentations at congresses, and she is the director of the food analysis, culinary chemistry, and nutrition research group.

Cristina Ghinea

Cristina Ghinea is a Lecturer at the Faculty of Food Engineering, "Stefan cel Mare" University of Suceava, and holds a PhD in Environmental Engineering (2012), co-supervised by "Gheorghe Asachi" Technical University of Iasi, Romania, and Twente University, The Netherlands. Her teaching experience includes courses in Organic Chemistry, Life Cycle Assessment of Products, and Techniques for Investigating Environmental Factors. Her research interests are waste generation, food waste, waste management and waste valorization processes, life cycle assessment, waste management modeling and optimization, and statistical analysis. She has authored or co-authored 11 books and chapters, and 80 articles, of which 41 articles were published in ISI journals, 25 in international database journals, 7 in ISI proceedings, and 7 in IEEE conference publications. Her Hirsch factor is 17/19/20 (WoS/Scopus/Google Scholar)

foods

MDPI

Editorial

Advances in Analysis of Flavors and Fragrances: Chemistry, Properties and Applications in Food Quality Improvement

Ana Leahu [1], Marìa Soledad Prats Moya [2,*] and Cristina Ghinea [1,*]

[1] Faculty of Food Engineering, Stefan cel Mare University of Suceava, 720229 Suceava, Romania; analeahu@fia.usv.ro

[2] Analytical Chemistry, Nutrition and Food Sciences Department, University of Alicante, San Vicente del Raspeig, 03690 Alicante, Spain

* Correspondence: maria.prats@ua.es (M.S.P.M.); cristina.ghinea@fia.usv.ro (C.G.)

Natural flavors and fragrances are an exciting area for both academic and industrial research as consumer preference for natural and sustainable products continues to grow. Flavor influences the well-being of humans and is considered a multisensory experience that encompasses taste, smell, and texture [1]. Human chemical senses are stimulated by organic compounds that can be found in fruits, spices, teas, coffee, mushrooms, milk, meat, fermented products, and others [2].

The profiles of these compounds in foods are influenced by various factors, such as species, geographical origin, growing conditions, and processing methods [3]. Raw materials and production processes (malting, crushing, fermentation, distillation, and maturation) shape the flavor of spirits. Esters are formed during alcoholic fermentation and give these alcoholic beverages floral and fruity notes. Additionally, the barrel maturation process is another crucial phase in which the spirit develops depth, character, and a harmonious blend of flavors. Fruit variety, agronomic factors related to the orchard, processing technologies, fermentation, and maturation conditions, influences the sensory profiles and chemical quality of ciders. Wheat variety, geographical location, and harvest year can affect the grain composition and volatile composition of whisky. The cultivar, growing conditions, and post-harvest processing technologies (baked, roasted, steamed, and sundried tea) influence the final sensorial perception of the teas. Pollination, bagging, harvesting period, storage conditions, and calcium regulation influence the fruit's aroma.

Flavor compounds in fruit are mainly organic acid esters, aldehydes, and volatile phenols, while alcohols, aldehydes, ketones, esters, and acids were identified in teas. Maillard reaction is responsible for meat products' flavor, while fatty acids, carbonyl compounds, and methyl sulfide can contribute to the flavor of dairy products [4]. In alcoholic beverages, alcohols, acids, aldehydes, ketones, esters, ethers, hydrocarbons, pyrrolines, pyridines, furans, and sulfur compounds were identified [5]. Aldehydes, alcohols, ketones, furans, and pyrazines are key aroma compounds determined in soy, and their identification is crucial for improving the overall sensory acceptability of soy products [6].

The food industry and scientists are actively investigating the identification, isolation, and concentration of organic flavor compounds by various methods [7]. Aroma components can be isolated from foods mainly through stir bar sorptive extraction (SBSE) and solid-phase microextraction (SPME) techniques. Additionally, headspace (HS) is the best technical mode for visualizing volatile compounds' profile [8]. The detection of volatile organic compounds (VOCs) can be performed with Gas Chromatography–Mass Spectrometry (GC-MS), Gas Chromatography–Ion Mobility Spectrometry (GC-IMS), Headspace

Solid-Phase Microextraction Gas Chromatography–Mass Spectrometry (HS-SPME-GC-MS) techniques, electronic nose (EN), electronic tongues (ETs), colorimetric sensor array (CSA), and others [9]. The main drawback of some techniques is still volatile quantification, as there is a lack of commercial volatile standards.

The final assessment of a food's flavor quality largely depends on human emotions and preferences when selecting and consuming it. Therefore, flavoring agents are commonly used to mask unpleasant flavors in food. Consumers associate health with natural additives, but artificial additives show higher sensory efficiency and flavor acceptance if the consumer has not consulted the product label. Therefore, it is crucial to understand the influence of neurocognitive mechanisms and perceptions on consumption decisions.

The primary aim of this Special Topic (ST), "Advances in Analysis of Flavors and Fragrances: Chemistry, Properties and Applications in Food Quality Improvement", is to investigate recent advances and current knowledge in the study of flavor and fragrance in different types of foods and beverages. This ST comprises 36 papers which address subject-like investigations of the aromatic components of wines (Grenache and Huangjiu), liquors (Baijiu, Luzhou-flavored liquor, Hunan Light-Flavor Baijiu), plum brandies, apple juice and cider, Jerez vinegars, Irish whiskey, sun-dried green tea, raw Pu-erh tea, and baked green tea, green coffee, traditional fermented koumiss, Hurood cheese, cow, donkey, camel, and horse milk powder, chicken carcasses, fermented soybeans, soy sauce, and others products. The usability of electronic noise and tongue data and volatile compounds' determination and finding the relationship between key metabolites and taste attributes are analyzed in different matrices and with the help of multivariate statistical methods (please see the list of contributions for further details). We believe that further investigations in this scientific field, including artificial intelligence (AI) and nanotechnology, will provide solutions for the food industry in terms of flavor and fragrance applications, as well as controlled release from food products.

Author Contributions: Conceptualization, A.L. and C.G.; writing—original draft preparation, A.L., M.S.P.M. and C.G.; review and editing, M.S.P.M. and C.G. All authors have read and agreed to the published version of the manuscript.

Acknowledgments: As the Guest Editors, we would like to thank all the authors who submitted their work on this Topic. Special thanks to all anonymous reviewers involved in the Topic who helped the authors to improve their manuscripts. Thanks also to the editorial staff of *Foods, Fermentation, Molecules, Beverage,* and *Chemosensors* for supporting the idea of this Topic: *Advances in Analysis of Flavors and Fragrances: Chemistry, Properties and Applications in Food Quality Improvement*.

Conflicts of Interest: The authors declare no conflicts of interest.

List of Contributions:

1. Stoffel, E.S.; Robertson, T.M.; Catania, A.A.; Casassa, L.F. The Impact of Fermentation Temperature and Cap Management on Selected Volatile Compounds and Temporal Sensory Characteristics of Grenache Wines from the Central Coast of California. *Molecules* **2023**, *28*, 4230.
2. Chang, R.; Zhou, Z.; Dong, Y.; Xu, Y.; Ji, Z.; Liu, S.; Mao, J. Sensory-Guided Isolation, Identification, and Active Site Calculation of Novel Umami Peptides from Ethanol Precipitation Fractions of Fermented Grain Wine (Huangjiu). *Foods* **2023**, *12*, 3398.
3. Li, L.; Fan, M.; Xu, Y.; Zhang, L.; Qian, Y.; Tang, Y.; Li, J.; Zhao, J.; Yuan, S.; Liu, J. Comparative Analysis of Volatile Flavor Compounds in Strongly Flavored Baijiu under Two Different Pit Cap Sealing Processes. *Foods* **2023**, *12*, 2579.
4. Jiao, K.; Deng, B.; Song, P.; Ding, H.; Liu, H.; Lian, B. Difference Analysis of the Composition of Iron (Hydr)Oxides and Dissolved Organic Matter in Pit Mud of Different Pit Ages in Luzhou Laojiao and Its Implications for the Ripening Process of Pit Mud. *Foods* **2023**, *12*, 3962.

5. Xu, J.; Zhang, T.; Chen, H.; Dai, Y.; Li, Z.; He, J.; Ju, R.; Hou, A. Study on the Fermented Grain Characteristics and Volatile Flavor Substances during the Tuqu Fermentation of Hunan Light-Flavor Baijiu. *Foods* **2024**, *13*, 899.
6. Balak, J.; Drábová, L.; Maťátková, O.; Doležal, M.; Marsík, D.; Jarosova Kolouchova, I. Differences in Volatile Profiles and Sensory Characteristics in Plum Spirits on a Production Scale. *Fermentation* **2024**, *10*, 235.
7. Øvsthus, I.; Martelanc, M.; Albreht, A.; Radovanović Vukajlović, T.; Česnik, U.; Mozetič Vodopivec, B. Chemical Characterization of Cider Produced in Hardanger—From Juice to Finished Cider. *Beverages* **2024**, *10*, 73.
8. Uysal, R.S. Effects of Aging in Wood Casks on Anthocyanins Compositions, Volatile Compounds, Colorimetric Properties, and Sensory Profile of *Jerez* Vinegars. *Fermentation* **2024**, *10*, 469.
9. Lan, W.; Cheng, W.; Li, R.; Zhang, M.; Li, M.; Zhang, Y.; Zhou, Y. Comparison of Flavor Differences between the Juices and Wines of Four Strawberry Cultivars Using Two-Dimensional Gas Chromatography-Time-of-Flight Mass Spectrometry and Sensory Evaluation. *Molecules* **2024**, *29*, 4691.
10. Vashishtha, A.; Kilcawley, K.N.; Skibinska, I.; Whelan, S.; Byrne, J.L.; Garcia-Cabellos, G.; Morris, S. Influence of Terroir on the Grain Composition, and Volatile Profile of Irish Grain (Wheat) New Make Spirit. *Beverages* **2024**, *10*, 106.
11. Tsitlakidou, P.; Kamplioni, D.; Kyriakoudi, A.; Irakli, M.; Biliaderis, C.G.; Mourtzinos, I. Antioxidant-Enhanced Alginate Beads for Stabilizing Rapeseed Oil: Utilizing Extracts from Post-Distillation Waste Residues of Rosemary. *Foods* **2024**, *13*, 2142.
12. Ma, C.; Gao, C.; Li, Y.; Zhou, X.; Fan, G.; Tian, D.; Huang, Y.; Li, Y.; Zhou, H. The Characteristic Aroma Compounds of GABA Sun-Dried Green Tea and Raw Pu-Erh Tea Determined by Headspace Solid-Phase Microextraction Gas Chromatography–Mass Spectrometry and Relative Odor Activity Value. *Foods* **2023**, *12*, 4512.
13. Niu, X.; Ao, C.; Yu, J.; Zhao, Y.; Huang, H. GC-MS Combined with Proteomic Analysis of Volatile Compounds and Formation Mechanisms in Green Teas with Different Aroma Types. *Foods* **2024**, *13*, 1848.
14. Li, S.; Du, D.; Cheng, S.; Wei, Z. Evaluating the Intensity of Bitter–Astringent Interactive Perception in Green Tea Based on the Weber–Fechner Law. *Chemosensors* **2025**, *13*, 137.
15. Cascos, G.; Lozano, J.; Montero-Fernández, I.; Marcía-Fuentes, J.A.; Aleman, R.S.; Ruiz-Canales, A.; Martín-Vertedor, D. Electronic Nose and Gas Chromatograph Devices for the Evaluation of the Sensory Quality of Green Coffee Beans. *Foods* **2024**, *13*, 87.
16. Latief, M.; Muntasir, R.W.; Wijaya, D.E.; Tarigan, I.L.; Sutrisno, S. Synergetic Effects of *Coffea liberica* and *Curcuma zanthorrhiza*: Study of Sensory Profile, Proximate, and Chemical Compound. *Beverages* **2025**, *11*, 9.
17. Gou, Y.; Han, Y.; Li, J.; Niu, X.; Ma, G.; Xu, Q. Discriminant Analysis of Aroma Differences between Cow Milk Powder and Special Milk Powder (Donkey, Camel, and Horse Milk Powder) in Xinjiang Based on GC-IMS and Multivariate Statistical Methods. *Foods* **2023**, *12*, 4036.
18. Gou, Y.; Ma, X.; Niu, X.; Ren, X.; Muhatai, G.; Xu, Q. Exploring the Characteristic Aroma Components of Traditional Fermented Koumiss of Kazakh Ethnicity in Different Regions of Xinjiang by Combining Modern Instrumental Detection Technology with Multivariate Statistical Analysis Methods for Odor Activity Value and Sensory Analysis. *Foods* **2023**, *12*, 2223.
19. Wang, Y.; Zeng, H.; Cao, Y.; Wang, S.; Wang, B. A Study of Key Aroma Compounds in Hurood Cheese and Their Potential Correlations with Lactic Acid Bacteria. *Fermentation* **2023**, *9*, 670.
20. Cao, L.; Hunt, C.J.; Meyer, A.S.; Lametsch, R. New Insight into the Substrate Selectivity of Bovine Milk γ-glutamyl Transferase via Structural and Molecular Dynamics Predictions. *Molecules* **2023**, *28*, 4657.
21. Zhang, X.; Liu, S.-Q. Effects of Reducing Sugars on Colour, Amino Acids, and Volatile Flavour Compounds in Thermally Treated Minced Chicken Carcass Hydrolysate. *Foods* **2024**, *13*, 991.
22. Ciobanu, M.-M.; Flocea, E.-I.; Boișteanu, P.-C. The Impact of Artificial and Natural Additives in Meat Products on Neurocognitive Food Perception: A Narrative Review. *Foods* **2024**, *13*, 3908.
23. Peng, J.; Lu, X.; Fan, R.; Ren, Y.; Sun, H.; Yu, Y.; Cheng, B. Analysis and Evaluation of Muscle Quality in Different Parts of the Bighead Carp (*Aristichthys nobilis*). *Foods* **2023**, *12*, 4430.

24. Wang, S.; Luo, L.; Zhang, R.; Guo, K.; Zhao, Z. The Biochemical Composition and Quality of Adult Chinese Mitten Crab *Eriocheir sinensis* Reared in Carbonate-Alkalinity Water. *Foods* **2024**, *13*, 362.
25. Luo, J.; Frank, D.; Arcot, J. Forging Prawn and Salmon Flavours with Non-Animal-Based Ingredients. *Foods* **2025**, *14*, 820.
26. Zeng, X.; Zhang, S.; Geng, W.; Jin, J.; Liao, K.; Tang, Z.; Wang, S.; Zhou, W. Headspace Solid-Phase Microextraction and Gas Chromatography–Mass Spectrometry Combined with Sensory Evaluation for the Analysis of Volatile Aromatic Compounds in Apricot (*Prunus armeniaca* L.) Germplasm Resources Cultivated in Xinjiang, China. *Foods* **2024**, *13*, 3912.
27. Zhang, W.; Yan, M.; Zheng, X.; Chen, Z.; Li, H.; Mao, J.; Qin, H.; Zhu, C.; Du, H.; Abd El-Aty, A.M. Exploring the Aroma Fingerprint of Various Chinese Pear Cultivars through Qualitative and Quantitative Analysis of Volatile Compounds Using HS-SPME and GC×GC-TOFMS. *Molecules* **2023**, *28*, 4794.
28. Ding, B.; Wang, F.; Zhang, B.; Feng, M.; Chang, L.; Shao, Y.; Sun, Y.; Jiang, Y.; Wang, R.; Wang, L.; et al. Flavor Characteristics of Ten Peanut Varieties from China. *Foods* **2023**, *12*, 4380.
29. He, Z.; Huan, J.; Ye, M.; Liang, D.; Wu, Y.; Li, W.; Gong, X.; Jiang, L. Based on CiteSpace Insights into *Illicium verum* Hook. f. Current Hotspots and Emerging Trends and China Resources Distribution. *Foods* **2024**, *13*, 1510.
30. Chao, K.; Qi, T.; Wan, Q.; Li, T. Insights into the Flavor Differentiation between Two Wild Edible *Boletus* Species through Metabolomic and Transcriptomic Analyses. *Foods* **2023**, *12*, 2728.
31. Xu, X.; Long, Z.; Du, W.; Chen, Q.; Zhang, Y.; Hu, S. Dynamics of Physicochemical Properties, Flavor, and Microbial Communities of Salt-Free Bamboo Shoots during Natural Fermentation: Correlation between Microorganisms and Metabolites. *Fermentation* **2023**, *9*, 733.
32. Gong, J.; Wang, X.; Ni, H.; Wang, Y. The Volatile Compounds Change during Fermentation of *Saccharina japonica* Seedling. *Foods* **2024**, *13*, 1992.
33. Masuda, M.; Terada, Y.; Tsuji, R.; Nakano, S.; Ito, K. Time-Series Sensory Analysis Provided Important TI Parameters for Masking the Beany Flavor of Soymilk. *Foods* **2023**, *12*, 2752.
34. Chen, Z.; Wu, Y.; Tong, S.; Jin, J.; Zhang, L.; Li, C.; Tan, Q.; Wen, F.; Tao, Y. Characterization of the Key Aroma Compounds of Soybean Flavor in Fermented Soybeans with *Bacillus subtilis* BJ3-2 by Gene Knockout, Gas Chromatography–Olfactometry–Mass Spectrometry, and Aroma Addition Experiments. *Fermentation* **2024**, *10*, 409.
35. Tan, Q.; Wu, Y.; Li, C.; Jin, J.; Zhang, L.; Tong, S.; Chen, Z.; Ran, L.; Huang, L.; Zuo, Z. Characterization of Key Aroma Compounds of Soy Sauce-like Aroma Produced in Ferment of Soybeans by *Bacillus subtilis* BJ3-2. *Foods* **2024**, *13*, 2731.
36. Xian, F.; Yang, L.; Ye, H.; Xu, J.; Yue, X.; Wang, X. Revealing the Mechanism of Aroma Production Driven by High Salt Stress in *Trichomonascus ciferrii* WLW. *Foods* **2024**, *13*, 1593.

References

1. Hossain, M.S.; Wazed, M.A.; Asha, S.; Hossen, M.A.; Fime, S.N.M.; Teeya, S.T.; Jenny, L.Y.; Dash, D.; Shimul, I.M. Flavor and Well-Being: A Comprehensive Review of Food Choices, Nutrition, and Health Interactions. *Food Sci. Nutr.* **2025**, *13*, e70276. [CrossRef] [PubMed]
2. Reshna, K.R.; Gopi, S.; Balakrishnan, P. Introduction to flavor and fragrance in food processing. In *Flavors and Fragrances in Food Processing: Preparation and Characterization Methods*; Balakrishnan, P., Gopi, S., Eds.; American Chemical Society: Washington, DC, USA, 2022; Volume 1433, pp. 1–19.
3. Yang, H.; Wu, M.; Shen, X.; Lai, Y.; Wang, D.; Ma, C.; Ye, X.; Sun, C.; Cao, J.; Sun, C.; et al. Comprehensive Review of VOCs as a Key Indicator in Food Authentication. *eFood* **2025**, *6*, e70057. [CrossRef]
4. Wei, G.; Dan, M.; Zhao, G.; Wang, D. Recent advances in chromatography-mass spectrometry and electronic nose technology in food flavor analysis and detection. *Food Chem.* **2023**, *405*, 134814. [CrossRef] [PubMed]
5. Wu, J.; Liu, Y.; Zhao, H.; Huang, M.; Sun, Y.; Zhang, J.; Sun, B. Recent advances in the understanding of off-flavors in alcoholic beverages: Generation, regulation, and challenges. *J. Food Compos. Anal.* **2021**, *103*, 104117. [CrossRef]
6. Xu, J.; Zeng, M.; Wang, Z.; Chen, Q.; Qin, F.; Chen, J.; He, Z. Characteristic flavor compounds in soy protein isolate and their correlation with sensory attributes. *Food Res. Int.* **2025**, *206*, 116038. [CrossRef] [PubMed]
7. Tananaki, C.; Liolios, V.; Kanelis, D.; Rodopoulou, M.A. Investigation of Volatile Compounds in Combination with Multivariate Analysis for the Characterization of Monofloral Honeys. *Appl. Sci.* **2022**, *12*, 264. [CrossRef]

8. Starowicz, M. Analysis of Volatiles in Food Products. *Separations* **2021**, *8*, 157. [CrossRef]
9. Feng, Y.; Wang, Y.; Beykal, B.; Qiao, M.; Xiao, Z.; Luo, Y. A mechanistic review on machine learning-supported detection and analysis of volatile organic compounds for food quality and safety. *Trends Food Sci. Technol.* **2024**, *143*, 104297. [CrossRef]

Disclaimer/Publisher's Note: The statements, opinions and data contained in all publications are solely those of the individual author(s) and contributor(s) and not of MDPI and/or the editor(s). MDPI and/or the editor(s) disclaim responsibility for any injury to people or property resulting from any ideas, methods, instructions or products referred to in the content.

molecules

MDPI

Article

The Impact of Fermentation Temperature and Cap Management on Selected Volatile Compounds and Temporal Sensory Characteristics of Grenache Wines from the Central Coast of California

Emily S. Stoffel [1,2], Taylor M. Robertson [2], Anibal A. Catania [3] and L. Federico Casassa [2,*]

1 Food Science & Nutrition Department, California Polytechnic State University, San Luis Obispo, CA 93407, USA; esstoffe@calpoly.edu

2 Wine & Viticulture Department, California Polytechnic State University, San Luis Obispo, CA 93407, USA; trober24@calpoly.edu

3 Centro de Estudios de Enología, Estación Experimental Agropecuaria Mendoza, Instituto Nacional de Tecnología Agro-pecuaria (INTA), San Martín 3853, Mendoza 5507, Argentina; acatania@fca.uncu.edu.ar

* Correspondence: lcasassa@calpoly.edu

Abstract: Grenache wines from the Central Coast of California were subjected to different alcoholic fermentation temperature regimes (Cold, Cold/Hot, Hot) and cap management protocols, namely, punch down (PD), or no punch down (No PD), to determine the effect of these practices on the color, aroma, and the retronasal and mouthfeel sensory characteristics of the resulting wines. Descriptive analysis (n = 8, line scale rating 0–15) results indicated that the combination of a hot fermentation temperature and no punch downs led to a significantly higher intensity in perceived color saturation (7.89) and purple hue (8.62). A two-way analysis of variance (ANOVA) showed that cap management was significantly more impactful on the perception of orthonasal aromas than fermentation temperature. The reduction aroma was significantly higher in No PD wines (5.02) compared to PD wines (3.50), while rose and hot aromas had significantly higher intensity perception for PD wines (5.18, 6.80) than for No PD wines (6.80, 6.14). Conversely, analysis of selected volatile compounds indicated that fermentation temperature was more impactful than cap management regime. Cold/Hot wines had higher concentrations of important esters such as ethyl hexanoate (650 μg/L) and isoamyl acetate (992 μg/L). Cold wines had a higher concentration of β-damascenone (0.719 μg/L). TCATA evaluation (n = 8) indicated that Cold/Hot PD wines had a significantly higher citation proportion of fruit flavor (1.0) and velvet astringency perception (0.80) without significant reduction flavors. Finally, the present study represents a contribution with the main volatile compounds (e.g., β-damascenone and esters in the Cold and Cold/Hot fermented wines, respectively; hexanol in PD wines, which may be potentially responsible for a hot mouthfeel), and sensory characteristics (red fruit, tropical fruit, white pepper, and rose) of Grenache wines grown in the Mediterranean climate of the Central Coast of California.

Keywords: red wine; fermentation temperature; cap management; punch downs; descriptive analysis; temporal check-all-that-apply; astringency; retronasal aroma; volatile chemistry

1. Introduction

Grenache has proven to be extremely versatile from a winemaking viewpoint as it can be used in rosé, blended red wines (i.e., GSM blends), and monovarietal wines [1]. Grenache is considered a Mediterranean variety that produces wine of low color, tannins, and perceivable acidity [2,3]. However, its wines have a complex aromatic composition based on esters, terpenes, and even thiols [4]. Because of this, Grenache wines have been the subject of extensive research in volatile chemistry through gas chromatography-mass spectroscopy (GC-MS) and gas chromatography-olfactometry (GC-O) analysis as well as sensory assessment, typically through descriptive analysis [5–7]. However, there have been no studies to the authors' knowledge that have utilized a dynamic sensory method to assess the retronasal profile of Grenache wines.

Time-based sensory methods have become popular in wine research as they are able to capture how a wine's retronasal and mouthfeel sensations change over time. One of the newer temporal sensory methods is a rapid method known as Temporal Check-All-That-Apply (TCATA). TCATA is a time-based version of check-all-that-apply (CATA) [8]. CATA is a rapid method that requires panelists to check all the attributes they are perceiving in a sample [9]. CATA can be used to assess both orthonasal and retronasal aromas, but it is limited to assessment at a single time point. Hence, TCATA was developed utilizing the principles of CATA but allowing for a continuous and dynamic assessment of sensory properties.

TCATA shares similarities to an earlier sensory method known as Temporal Dominance of Sensations (TDS) and the two methods have been compared extensively [8–11]. Briefly, TDS relies on the notion of dominance or "*the most attention-grabbing sensation at each moment*" which limits the panelist to selecting one attribute at a time [12]. The limitation of one dominant attribute can lead to data loss for other attributes present and of interest to the researcher. Additionally, the dominant attribute decision process can increase the possibility of dithering or, inability to decide on an attribute. Dithering can lead to a lag time in the sensory profile data [9]. These limitations are made up for in TCATA, as the panelist can select multiple attributes at a time and are not restricted by the definition of dominance [8]. However, the two methods are similar in that the ideal number of attributes is between six and ten which lessens the probability of dumping [13]. TCATA is a rapid method, meaning it requires little training (less than five hours total), and has been used in consumer panels [14]. Additionally, TCATA data are converted to curves that show the citation proportion of each attribute over time which are similar to TDS curves. However, TCATA curves do not contain calculated chance and significance lines showing statistical significance in the curves. The data can be subjected to difference curves as done with TDS to determine any statistical significance between treatments [15]. While TCATA does make up for some of the main limitations of TDS, it does have its own limitations. The main limitation of TCATA is that panelists can be slower in checking and unchecking all the applicable attributes causing a lag time in the resulting TCATA curves [16].

Although TCATA is a newer method, it has been used consistently with retronasal aroma and mouthfeel studies in wine [17–19]. For example, Syrah wine was made with two levels of ethanol, high (15%) and low (10%), and analyzed using TCATA for in-mouth sensory properties [17]. Results indicated that the high ethanol Syrah had a significantly longer finish compared to the low ethanol Syrah from the TCATA evaluation. Additionally, a principal component analysis found that the high ethanol Syrah was associated with bitterness, astringency, and dark fruit, whereas low ethanol Syrah was strongly associated with acidity, red fruit, and green character. Another study focused only on the evaluation of mouthfeel characteristics in Pinot noir, Cabernet Franc, and Cabernet Sauvignon [19]. Four mouthfeel characteristics were evaluated by two sensory panels using TCATA. Results showed that Pinot noir had the lowest citation proportion in perceived astringency compared to the other wines. However, none of these studies focused on understanding the temporal impact of a particular winemaking technique using TCATA. Only one study, to the authors' knowledge, has used TCATA to understand changes in a winemaking technique. This study looked at the effect of carbonation level due to the addition of different levels of dextrose in the *liqueur de tirage* in Chardonnay-based sparkling wines from the Columbia Valley, WA (USA) [18,20]. TCATA results indicated that an increase in carbonation (CO_2 g/L) led to a longer retronasal and mouthfeel finish [18]. However, in this study, TCATA was used to evaluate a specialized form of winemaking, as such, the results are not applicable to still red winemaking.

Understanding the impact of winemaking techniques on the retronasal and mouthfeel characteristics is important especially as consumers primarily focus on this aspect of wine when making a hedonic judgment about wine [21]. From a consumer perspective, sensory characterization is the first step in understanding how a product will be perceived by the public [22]. Two of the most important technical decisions prior to the onset of winemaking for red varietals are the fermentation temperature during alcoholic fermentation and cap management regime during skin contact (i.e., maceration) time.

The choice of fermentation temperature affects the rate of alcoholic fermentation, the evolution and diversity of yeast populations, phenolic extraction, and the sensory characteristics of the resulting wine [23–25]. Phenolic extraction during red winemaking is particularly sensitive to fermentation temperature which in turn influences the color, taste, and mouthfeel of the resulting wine [26]. A review of winemaking techniques on phenolic extraction reported that hot fermentation temperatures (25 °C to 30 °C) favored higher phenolic extraction thus resulting in wines with higher astringency and color saturation [23]. Similarly, a study was conducted on three different clones of Pinot noir and three different fermentation temperatures, including cold (10 °C), hot (25 °C), and variable (seven days at 10 °C, and seven days at 25 °C). Results reported that a hot fermentation temperature led to wines with higher total phenolics and higher red hue according to International Commission on Illuminations (CIE) Lab results [27]. In Petit Verdot wines, fermentation at three different temperatures (17 °C, 21 °C, and 25 °C) led to significant increases in total anthocyanins and flavonols whereby color intensity increased significantly when wines were fermented at 25 °C compared to 17 °C [28].

In terms of sensory implications, a study in Syrah examined the impact of three fermentation temperature regimes (15 °C, 20 °C then 30 °C, and 30 °C then 20 °C) as well as the use of cold-soak or extended maceration [29]. Results from a pair-comparison sensory test showed that a warmer fermentation temperature (30 °C) led to a higher perceived red color and black currant flavor compared to a cold fermentation temperature (15 °C). These results suggested that a hot fermentation temperature led to a higher red hue and overripe fruit flavors. Furthermore, another study examined the impact of cold (15 °C) and control (25 °C) fermentation temperatures and cold-tolerant yeast strains in Merlot. It was reported that the cold treatments regardless of yeast strains had significantly higher concentrations of total esters, but lower total terpenes compared to the control (25 °C) [30]. The results seen in the volatile chemistry corroborated with descriptive analysis data, whereby control treatments were inversely related to the fruit aromas (red fruits and banana) according to a principal component analysis, suggesting a colder fermentation temperature led to more fresh fruit aromas.

Cap management techniques during alcoholic fermentation also have critical implications for the chemical and sensory aspects of red wines. The cap refers to the mass of skins and seeds that form at the top of the fermentation vessel due to the release of carbon dioxide. The resulting cap is typically integrated into the fermenting juice by various cap management practices. The most prevalent cap management methods applied during red winemaking are pump-overs, submerged caps, punch downs, and rack-and-return (i.e., délestage). There is extensive literature on the chemical impact of cap management techniques in red wines [31,32]. However, research on the sensory impact of these practices remains limited. One such study examined the chemical and sensory impact of punch downs, submerged caps, and an eight-week extended maceration in Merlot wines [33,34]. Descriptive analysis results showed that the punch-down wines had the lowest astringency intensity score [34]. Additionally, TDS was utilized to understand the temporal sensory impact of cap management practices [33]. Results indicated that while descriptive analysis showed differences in astringency intensity; the dominance of astringency in the TDS curves was similar among all treatments except for the submerged cap at week 0 of the treatment. However, the TDS evaluation results were limited as they did not include retronasal aromas.

When utilizing time-based sensory methods to evaluate wines, it is important to consider the confounding factor of salivary flow rate. This is because the temporal perception of astringency, retronasal aromas, and tastants (bitterness) can be affected by the salivary flow rate. Previous research has indicated that individuals with a low salivary flow rate perceive astringency later and more intensely than high salivary flow rate individuals [35–37]. The same pattern has been seen with bitterness perception [36].

The relationship between retronasal aromas and salivary flow rate appears to be based on the aroma type and its corresponding volatile compounds. A study was conducted on Syrah with spiked volatile compounds (isobutyl methoxypyrazine, phenylethyl alcohol, and lactone) corresponding to vegetal, floral, and coconut aromas, respectively. It was determined that individuals with high salivary flow rates experienced the intensity of these compounds significantly later and more intensely compared to low salivary flow rate individuals [38]. As for fruit aromas, which are associated with esters, a study that examined commercial Tempranillo and Verdejo rosé wines that were spiked with four esters (isoamyl acetate, ethyl butanoate, ethyl hexanoate, and ethyl decanoate) determined that the short chain esters were perceived significantly more intensely by high salivary flow rate panelists than low salivary flow rate panelists [39].

Common to all these studies was the use of the temporal sensory method known as Time-Intensity (TI). This method is like descriptive analysis (DA) in that it uses a 10 or 15 cm scale to assess the intensity of an attribute. However, TI assesses the intensity of one or two attributes over time typically after the wine has been expectorated [40]. A recent study examined the effect of salivary flow rate and the perception of astringency, bitterness, and mineral retronasal aromas utilizing TDS in red wines that had been co-fermented or blended with a combination of two or three varietals (Merlot, Malbec, and Petite Sirah) [41]. The results of the study found the same trends between salivary flow rate and the time to onset of dominance for astringency and bitterness as seen in past studies with TI and intensity assessment. Individuals that were considered to have low salivary flow rates experienced the time to onset of dominance for astringency and bitterness later than the high salivary flow rate panelists. However, no studies to the authors' knowledge have utilized TCATA to further understand salivary flow rate and time of perception.

Overall, there is a scarcity of research examining the sensory and volatile composition of Grenache wines that have undergone different alcoholic fermentation temperature and cap management regimes. Therefore, in the present study, Grenache wines from the Central Coast of California (USA), were produced in triplicate fermentations by a combination of three contrasting fermenting temperature regimes (Cold: 12 °C; Cold/Hot: seven days 12 °C, and seven days at 28 °C; Hot: 28 °C) and two contrasting cap management regimes, punch down (PD) and no punch down (No PD). The objective of this study was to understand the impact of fermentation temperature and cap management on the detailed sensory and volatile composition of Grenache wines. A combination of descriptive analysis and TCATA was used to assess the temporal retronasal impact. Selected volatile compounds were analyzed in the wines to assess the relationship between volatile chemistry and sensory perception. An additional purpose of the study was to expand the understanding of the relationship between salivary flow rate and time of perception for selected retronasal and mouthfeel sensations.

2. Results and Discussion

2.1. Perceived Color

Two color properties (saturation and purple hue) were assessed to understand the effects of fermentation temperature (Cold, Cold/Hot, and Hot) and of two cap management styles (PD and No PD) in Grenache wines. Results for color and aroma attributes are shown in Table 1. Both color attributes, saturation ($p = <0.0001$) and purple hue ($p = <0.0001$), were statistically significant. Hot No PD wines had significantly higher color intensity compared to all other wines, including the Hot PD wines. Cold No PD wines were significantly lower in saturation compared to all other wines. According to a two-way ANOVA, fermentation temperature was statistically significant for saturation ($p = <0.0001$) with Hot wines significantly higher than all other fermentation temperatures (Supplemental Table S1). The factor of cap management on its own was not significant but there was a significant interaction between the two factors for saturation (p = <0.0001).

Table 1. One-way analysis of variance (ANOVA) showing mean separation and p-values of descriptive sensory attributes of Grenache wines assessed by a trained panel (n = 8).

Treatments	Saturation	Purple Hue	Overall Aroma Intensity	Reduction	Rose	Red Fruit	Tropical Fruit	White Pepper	Mushroom	Hot
Cold PD	3.49 c [1]	5.02 c	7.61 ab	3.53 bc	5.67 a	5.57 a	5.06 ab	3.23 ab	2.70 ab	6.55 a
Cold No PD	2.42 d	4.36 c	7.81 ab	5.10 a	4.58 b	5.03 a	4.42 b	3.30 ab	2.99 ab	6.07 a
Cold/Hot PD	6.29 b	7.00 b	7.99 a	4.20 ab	4.83 b	5.41 a	4.96 ab	3.49 a	2.84 ab	6.94 a
Cold/Hot No PD	6.65 b	6.84 b	8.03 a	4.88 a	4.95 ab	5.44 a	5.26 a	3.51 a	3.09 a	6.10 a
Hot PD	6.53 b	8.21 a	7.17 b	2.77 c	5.05 ab	5.71 a	4.72 ab	2.80 b	2.35 b	6.90 a
Hot No PD	7.98 a	8.67 a	7.83 ab	5.08 a	4.54 b	5.43 a	4.88 ab	3.22 ab	3.30 a	6.25 a
p-value [2]	**<0.0001**	**<0.0001**	0.291	**<0.0001**	0.058	0.744	0.386	0.326	0.144	0.261

[1] Different letters within columns indicate a significant difference for Fisher's least significant differences test (p < 0.05). [2] Significant p-values are shown in bold fonts.

To date, there have not been many studies that have examined the sensory impact of fermentation temperature on perceived color saturation. Results of past studies have shown that an increase in fermentation temperature led to increased color properties from a chemical perspective [24]. For example, a study conducted on Pinot noir which was fermented at 20 °C, 30 °C, and using a high intensity/short time heat treatment (90 °C to 95 °C) found that concentration of polymeric pigments had a strong, inverse correlation to the lightness (L*) parameter in CIE Lab measurements, meaning the wines were darker in saturation [42]. Another study examined the impact of flash release (heating fermenting musts 95 °C or more) on the resulting phenolic composition of Grenache, Mourvèdre, and Carignan wines over two vintages [43]. The application of flash release led to higher levels of anthocyanins during fermentation. However, after alcoholic fermentation, flash-release wines saw a decrease in anthocyanins and an increase in tannin to anthocyanin ratio. It is important to note that this study was limited in that it examined anthocyanin, tannins, and total polyphenol index, but the relationship between heat during fermentation and polymeric pigment formation was not examined. Findings from the previous literature align with the chemical results of the present study. Herein, Hot wines regardless of cap management procedure had significantly higher total polymeric pigments (p = <0.0001) compared to the other fermentation temperatures (Table 2). For anthocyanins (p = 0.001), both Cold wines and the Hot PD wines were significantly lower compared to Hot No PD wines and both Cold/Hot wines which suggested a potential influence of cap management. Both total anthocyanins (p = 0.001) and total polymeric pigments (p = <0.0001), were statistically significant for the factor fermentation temperature, but not for the factor cap management (Supplemental Table S2). Additionally, total polymeric pigment concentration showed a positive correlation to perceived color saturation whereby Hot wines had significantly higher saturation. However, there was a significant interaction between cap management and fermentation temperature for both parameters, therefore it was likely cap management had some influence on color extraction. A potential explanation for the interaction between fermentation temperature and cap management could be a mixing effect proper of alcoholic fermentation, whereby carbon dioxide naturally mixes the bottom of the fermenting cap and promotes phenolics extraction [31]. This explanation was suggested by a previous experiment that examined the impact of pump-over volume and frequency in Cabernet Sauvignon [31].

Purple hue intensity also increased with fermentation temperature. Significant differences were apparent for purple hue (p = <0.0001) between fermentation temperature regimes but not cap management (Table 1). Hot wines had a significantly higher purple hue compared to both Cold/Hot and Hot wines. Cold wines had a significantly lower purple hue compared to the other fermentation temperatures. The same trend was seen in the two-way ANOVA but the interaction between fermentation temperature and cap management was statistically significant (p = <0.0001) (Supplemental Table S2). Previous research on color as affected by alcoholic fermentation temperature primarily focused on the red hue, likely because the red hue can be directly assessed through the red/green chroma (a*) parameter in CIE Lab analysis. For example, Pinot noir wines were fermented at Cold (10 °C), Hot (25 °C), and Variable (10 °C then 25 °C) temperature regimes; resulting CIE Lab analysis showed that the Hot wines led to significantly higher a* levels or red hue [27]. The use of red hue has also been seen in sensory tests in the literature. A pair-comparison test on Syrah wines that were fermented at 15 °C and 30 °C and received cold soak, extended maceration, or neither, determined that wines with increased fermentation temperature led to more red color [29]. However, it appears that in terms of fermentation temperature purple hue in the present study showed similar behavior to that seen in red hue from past research. Overall, in the present study, a hot fermentation temperature and No PD led to wines with higher color saturation and purple hue.

Table 2. One-way analysis of variance (ANOVA) of the phenolic composition of Grenache wines. Values represent the mean of three replicates followed by the standard error of the mean (n = 3).

Treatment	Anthocyanins (mg/L Malvidin-3-Glucoside)	SPP	LPP	TPP	Total Tannins (mg/L CE)	Total Phenolics (mg/L CE)
Cold PD	187 ± 9.23 b [1]	0.480 ± 0.03 c	0.026 ± 0.02 bc	0.505 ± 0.05 c	26.4 ± 3.70 b	487 ± 36.4 b
Cold No PD	178 ± 6.98 b	0.432 ± 0.01 c	0.004 ± 0.01 c	0.436 ± 0.01 c	22.8 ± 4.11 b	509 ± 17.1 b
Cold/Hot PD	280 ± 4.61 a	0.656 ± 0.02 b	0.070 ± 0.02 bc	0.727 ± 0.01 b	94.8 ± 7.55 a	800 ± 19.4 a
Cold/Hot No PD	309 ± 8.27 a	0.680 ± 0.01 b	0.022 ± 0.02 bc	0.702 ± 0.03 b	87.7 ± 14.7 a	862 ± 44.4 a
Hot PD	220 ± 5.39 b	0.718 ± 0.02 b	0.209 ± 0.01 a	0.927 ± 0.02 a	96.8 ± 13.0 a	735 ± 19.6 a
Hot No PD	299 ± 41.2 a	0.793 ± 0.04 a	0.085 ± 0.04 b	0.878 ± 0.07 a	81.0 ± 18.8 a	893 ± 112 a
p-value [2]	**0.001**	**<0.0001**	**0.001**	**<0.0001**	**0.001**	**0.000**

[1] Different letters within columns indicate a significant difference for Fisher's least significant differences test (p < 0.05). [2] Significant p-values are shown in bold fonts. SPP: Small polymeric pigments; LPP: Large polymeric pigments; TPP: total polymeric pigments; CE: catechin equivalents.

2.2. Orthonasal Aroma Attributes

A principal component analysis (PCA) was conducted on the descriptive analysis data and calculated using Dimension 1 (eigenvalue: 6.27; variability: 0.627) and Dimension 2 (eigenvalue: 2.78; variability: 0.278) (Figure 1A,B). The resulting PCA explained 90.6% of the variation within the dataset. Dimension 1 separated wines by cap management regime while Dimension 2 separated the wines by fermentation temperature whereby Cold/Hot and Hot wines were distinguished from Cold wines. Additionally, the No PD wines appeared to be more correlated with most of the sensory loadings compared to the PD wines (Figure 1A,B). The stronger correlation between most of the sensory characteristics and No PD wines may be in part due to the sensitivity of Grenache wines to oxidation [7]. The latter has been attributed to the high concentration of hydroxycinnamic acids (270–460 mg/L caftaric acid) in Grenache wines compared to other varieties (50–60 mg/L caftaric acid) as these compounds assist in the oxidation process of wine [1,43]. Therefore, it seems plausible that excessive cap management may lead to lower aroma retention in Grenache wines due to oxidation. Indeed, a previous study examined the effect of oxygen exposure on the sensory profile of bottled Sangiovese, which is another red variety that is highly susceptible to oxidation [44]. Results indicated that after 10 months of oxygen exposure, wines had significantly higher balsamic aromas (indicative of oxidation) over other aromatic attributes.

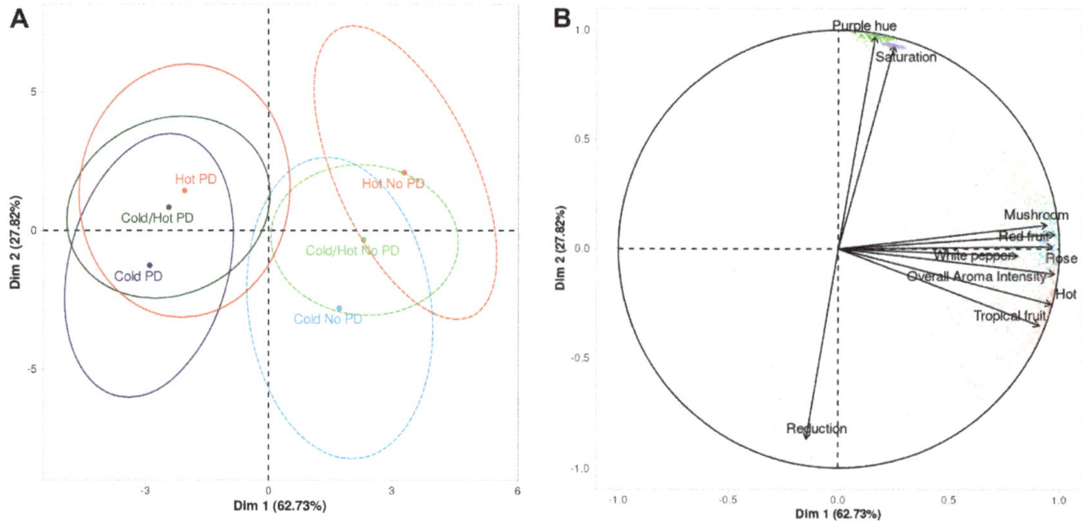

Figure 1. Principal component analysis of descriptive analysis data Grenache wines evaluated by a trained sensory panel ($n = 8$). (**A**) Treatment confidence ellipses constructed using Hotelling's T2 test ($p < 0.05$) for each pair of products indicate 95% confidence intervals. (**B**) Sensory attribute loadings.

Interestingly, according to the one-way ANOVA, the reduction aroma ($p = <0.0001$) was the only statistically significant attribute (Table 1). For reduction ($p = <0.0001$), both the Cold and Hot wines showed statistically significant variation between cap management regimes (Table 1). Cold No PD wines were significantly higher in reduction aroma than Cold PD wines; the same trend was observed between the Hot No PD wines and Hot PD wines. In the two-way ANOVA, the No PD wines regardless of fermentation temperature were significantly higher in reduction aromas (Supplemental Table S1). The No PD wines also had a significantly greater mushroom aroma ($p = 0.0019$). The aromas of reduction and mushroom are associated with hydrogen sulfide, methanethiol, and other volatile sulfur compounds [45,46]. The formation of reduction-associated aromas, particularly during alcoholic fermentation, can be attributed to a lack of, or suboptimal oxygen exposure to the fermenting must, among other factors. For example, a study conducted on Shiraz to understand the effect of oxygen exposure on volatile sulfur compounds production showed that wines that were sparged with air (20% O_2) or a 40% oxygen-nitrogen mixture had significantly less hydrogen sulfide compared to a control (no gas sparging) or nitrogen sparging post-inoculation [47]. Sensory results corroborated with the volatile chemistry, whereby the control and nitrogen sparging treatments were strongly associated with the aromas attributes rotten egg, sewage, rubber, and non-fruit persistence which were inversely related to the air (20% O_2) and 40% oxygen-nitrogen mixture wines.

During alcoholic fermentation, oxygen exposure to the wine can occur during cap management procedures such as punch downs or pump-overs. Therefore, it stands to reason that the No PD wines of the present study had significantly higher reduction aromas compared to the PD wines as no cap management occurred in the No PD wines. It is important to note that a lack of cap management is not the only factor that can contribute to the reduction of related aromas in red wines. Other factors such as yeast strain, and improper yeast nutrition can exacerbate the reduction of aromas as previously shown [48]. Another important factor related to reduction aromas is fermentation temperature, whereby a warmer fermentation temperature can lead to increased reduction aromas [49]. In the present study, however, fermentation temperature did not have a significant impact on the reduction of aromas seen in the wines (Supplemental Table S1). Nonetheless, there was a significant interaction between cap management and fermentation temperature

(Supplemental Table S1). This was also seen in the one-way ANOVA of reduction aromas whereby Cold No PD, Cold/Hot No PD, and Hot No PD wines had the highest reduction aroma intensity and were not significantly different from each other (Table 1). Therefore, fermentation temperature likely had some influence on the resulting sensory profile of these wines. It is also important to note that sulfur-based compounds were not measured during volatile compound analysis of the present study (Table 3), as sulfur compounds require very specific methods of analysis for detection [50,51].

Table 3. One-way analysis of variance (ANOVA) of the concentration of volatile compounds (μg/L) in Grenache wines. The average odor activity value (OAV) for each wine is shown in parentheses for each volatile compound. Values represent the mean of three replicates ($n = 3$).

Compounds	Cold PD	Cold No PD	Cold/Hot PD	Cold/Hot No PD	Hot PD	Hot No PD	*p*-Value	Source [1]
Esters								
Isobutyl acetate	n.d. [2] (0.00)	34.1 (0.016) bc [3]	n.d. (0.00)	n.d. (0.00)	69.5 (0.033) b	141 (0.067) a	**0.002** [4]	[52]
Ethyl butyrate	176 (3.42) bc	156 (3.02) cd	197 (3.83) ab	229 (4.46) a	137 (2.67) d	191 (3.71) bc	**0.002**	[53]
Hexyl acetate	6.44 (2.22) a	2.77 (0.955) b	4.79 (1.65) ab	5.20 (1.79) a	n.d. (0.00)	n.d. (0.00)	**0.000**	[54]
Isoamyl acetate	727 (19.7) cd	752 (20.3) cd	846 (22.9) bc	1138 (30.7) a	611 (16.5) d	998 (27.0) ab	**0.001**	[53]
Ethyl hexanoate	631 (3.16) a	523 (2.62) bc	613 (3.06) ab	687 (3.44) a	440 (2.20) c	490 (2.45) c	**0.002**	[55]
Ethyl lactate	22,125 (0.143) a	26,695 (0.173) a	26,704 (0.173) a	33,023 (0.214) a	35,740 (0.231) a	26,006 (0.168) a	0.365	[56]
Ethyl heptanoate	0.994 (-) a [5]	n.d. (-)	0.464 (-) a	n.d. (-)	0.855 (-) a	n.d. (-)	0.165	- [5]
Ethyl octanoate	356 (2.21) b	289 (1.80) b	418 (2.60) ab	525 (3.26) a	336 (2.09) b	294 (1.83) b	**0.021**	[53]
Ethyl decanoate	10.7 (0.010) a	8.33 (0.007) a	10.8 (0.010) a	12.3 (0.011) a	9.49 (0.009) a	5.49 (0.005) a	0.771	[57]
Diethyl succinate	337 (0.002) b	316 (0.002) b	354 (0.002) b	419 (0.002) b	892 (0.004) a	546 (0.003) ab	**0.043**	[56]
Ethyl hexadecanoate	396 (-) ab	614 (-) ab	163 (-) b	439 (-) ab	871 (-) a	520 (-) ab	0.234	-
Phenylethyl acetate	8.15 (0.033) bc	11.0 (0.044) abc	4.65 (0.019) c	16.7 (0.067) a	8.20 (0.033) bc	14.3 (0.057) ab	0.081	[57]
Ethyl cinnamate	1.12 (1.02) b	1.89 (1.72) ab	1.49 (1.35) ab	1.98 (1.80) ab	2.01 (1.82) ab	2.27 (2.07) a	0.142	[58]
Total Esters	24,776 a	29,403 a	29,317 a	36,497 a	39,118 a	29,209 a	0.340	
Nor-isoprenoids								
β-damascenone	0.757 (15.1) a	0.680 (13.6) a	n.d. (0.00)	n.d. (0.00)	n.d. (0.00)	n.d. (0.00)	**0.050**	[59]
Terpenes								
Citronellol	15.1 (0.838) a	14.5 (0.805) a	11.5 (0.636) b	11.6 (0.647) b	13.0 (0.723) ab	14.6 (0.809) a	0.057	[59]
Alcohols								
1-hexanol	2293 (2866) a	1462 (1828) b	2550 (3188) a	1664 (2080) b	2519 (3145) a	1592 (1990) b	**0.000**	[59]
1-octanol	n.d. (0.00)	n.d. (0.00)	1.74 (0.016) a	n.d. (0.00)	1.98 (0.018) a	n.d. (0.00)	0.570	[60]
1-nonanol	0.972 a (-)	n.d. (-)	0.677 a (-)	0.916 a (-)	n.d. (-)	n.d. (-)	0.692	-
Isobutanol	7982 (0.200) bcd	11,602 (0.290) abc	7710 (0.193) cd	7013 (0.175) d	12,682 (0.317) a	12,451 (0.311) ab	**0.047**	[56]
Isoamyl alcohol	49,280 (1.64) a	50,106 (1.67) a	40,457 (1.35) a	55,908 (1.86) a	71,490 (2.38) a	44,475 (1.48) a	0.494	[58]
Phenylethyl alcohol	15,660 (1.12) de	15,223 (1.09) e	19,369 (1.38) cd	20,646 (1.47) c	25,698 (1.84) b	32,479 (2.32) a	**<0.0001**	[58]
Total Alcohols	75,216 ab	78,395 ab	70,088 b	85,230 ab	112,387 a	90,997 ab	0.251	
Aldehydes								
Benzaldehyde	20.0 (0.001) a	18.0 (0.001) a	13.6 (0.001) a	32.2 (0.002) a	12.4 (0.001) a	n.d. (0.00)	0.767	[58]

[1] Source of detection threshold used to calculate OAV. [2] Volatile compound not detected during analysis. [3] Different letters within columns indicate a significant difference for Fisher's least significant differences test ($p < 0.05$). [4] Significant *p*-values are shown in bold fonts. [5] (-) indicates that there was no literature detection threshold found.

Overall aroma intensity and three aroma attributes, namely rose, tropical fruit, and white pepper did not have statistically significant *p*-values, but these attributes showed variation in the *post-hoc* test in the one-way ANOVA (Table 1). In the two-way ANOVA, the PD wines were significantly higher in rose ($p = 0.031$) and hot ($p = 0.018$) aromas (Supplemental Table S1). Previous reports suggest that the rose aroma in red wines and particularly Grenache wines can be circumscribed to three different classes of volatile compounds including esters (phenylethyl acetate), terpenes (citronellol), and alcohols (phenylethyl alcohol) or a combination of them [61,62]. Indeed, a previous study, which examined the sensory and volatile chemistry of Cannonau (Grenache) from different regions in Sardinia (Italy), found that the floral aromas were strongly associated with citronellol and isoamyl acetate while floral flavors were associated with phenylethyl alcohol and limonene [5]. This was not the case in the present study. Both phenylethyl acetate and citronellol had odor activity values (OAVs) lower than one and, therefore, were unlikely to be of sensory relevance (Table 3). Phenylethyl alcohol did have OAVs higher than one for all wines and had its concentration significantly effected by fermentation temperature with a significant interaction with cap management (p = <0.0001) (Table 3, Supplemental Table S3). A Partial Least Square Regression (PLSR) analysis was run to relate volatile chemistry with sensory descriptors. The PLSR model, which was calculated using Factor 1 (R^2 X: 0.33; R^2Y: 0.44; Q^2: −0.07) and Factor 2 (R^2 X: 0.65; R^2Y:0.71; Q^2: 0.13), explained 71.2% of the variation in the dataset, showing that the rose aroma was not strongly associated with phenylethyl alcohol (Figure 2A,B). It is then possible that the rose aroma could have been attributed to a volatile compound that was not detected in the GC-MS methods used. Alternatively, this aroma could be due to the wine aroma buffer which is a combination of several volatile compounds in wine that contribute to its "vinous" aroma character [61].

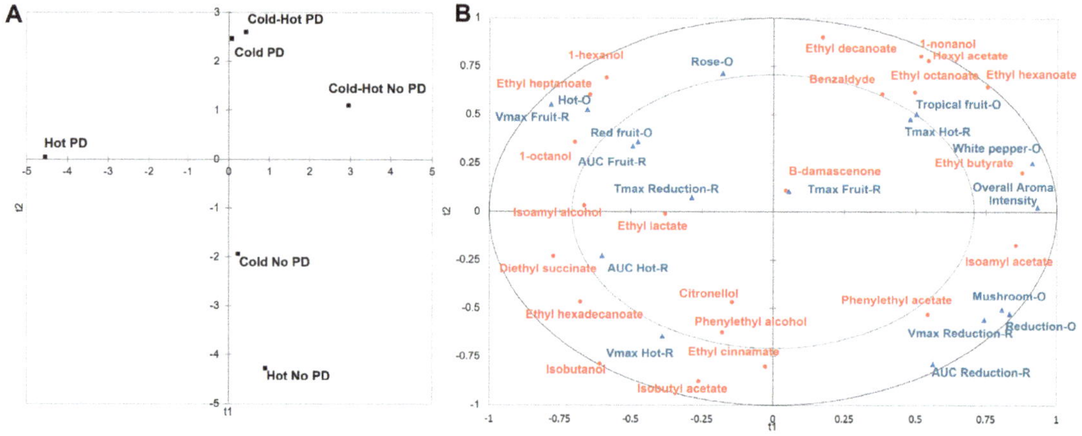

Figure 2. Partial least square regression analysis of volatile chemistry and sensory attributes. (**A**) Wine treatments (■). (**B**) Correlation loading of the relationship between the volatile chemistry (•) and sensory attributes (▲). O: indicates orthonasal aroma; R: indicates retronasal aroma; V_{max}: maximum citation proportion; T_{max}: time of maximum citation proportion; AUC: area under the curve, total intensity response.

As for the hot aroma, in the present study, it was defined as *"the aromatic burn of ethanol"*. Hot aromas in wine are typically associated with higher levels of ethanol. Empirically, monovarietal Premium or Ultra-premium Grenache wines produced commercially tend to be wines with 14.5 to 15.5% ethanol (v/v). However, this was not the case for the wines of the present experiment, whereby ethanol in the finished wines ranged from 12.8% to 13.3% ethanol (v/v) (Table 4). While the one-way ANOVA showed variation in the *post-hoc* test, the two-way ANOVA indicated that ethanol was not significantly impacted by fermentation temperature, cap management, or their interaction (Supplemental Table S4). Three factors might explain this discrepancy. First, even though panelists did detect the burn of ethanol in all the wines, they did so at a moderate level (based upon an average rating of 6.5 sensory units over 15). Second, it is possible that through cross-modal association [63] panelists might have ascribed red fruit and tropical fruit aromas to heat which is otherwise a character sometimes associated with Grenache wines from Mediterranean climates. Third, and most likely, the perception of the hot aroma could have been caused by the volatile compound 1-hexanol. 1-hexanol is a C6 alcohol attributed with herbaceous and resinous aroma characteristics [64]. 1-hexanol showed extremely high OAV values, of three orders of magnitude, in all wines indicating the compound had high sensory relevance (Table 3). In the two-way ANOVA, 1-hexanol (p = <0.0001) was significantly higher in PD wines compared to No PD wines (Supplemental Table S3). As well, 1-hexanol and the hot aroma were closely related in the PLSR (Figure 2). This aligns with the results of the previous literature, whereby the volatile and phenolic chemistry impact of a floating cap treatment compared to submerged cap treatment in Barbera wines were examined [65]. Results indicated the floating cap treatments combined with pump overs had a significantly higher concentration of hexanol than the submerged cap wines, which was thought to be due to enzymatic oxidation.

Table 4. One-way analysis of variance (ANOVA) of the basic chemical composition of Grenache wines. Values represent the mean of three replicates followed by the standard error of the mean (n = 3).

Treatments	Ethanol (v/v%)	pH	Titratable Acidity (g/L)	Acetic Acid (g/L)	Glucose + Fructose (g/L)	Lactic Acid (g/L)	Malic Acid (g/L)
Cold PD	13.3 ± 0.17 a [1]	3.61 ± 0.04 ab	5.52 ± 0.12 d	0.267 ± 0.04 b	0.140 ± 0.03 ab	1.24 ± 0.03 ab	0.060 ± 0.01 a
Cold No PD	13.1 ± 0.23 ab	3.69 ± 0.01 a	5.67 ± 0.06 cd	0.380 ± 0.03 a	0.127 ± 0.01 ab	1.18 ± 0.02 ab	0.060 ± 0.01 a
Cold/Hot PD	13.1 ± 0.11 ab	3.54 ± 0.01 c	5.74 ± 0.04 bcd	0.180 ± 0.01 c	0.113 ± 0.02 ab	1.25 ± 0.01 a	0.050 ± 0.01 a
Cold/Hot No PD	12.8 ± 0.04 b	3.55 ± 0.01 bc	5.97 ± 0.02 ab	0.177 ± 0.01 c	0.150 ± 0.01 a	1.26 ± 0.01 a	0.040 ± 0.01 a
Hot PD	13.1 ± 0.16 ab	3.53 ± 0.02 c	5.87 ± 0.06 bc	0.287 ± 0.02 b	0.097 ± 0.01 b	1.15 ± 0.03 bc	0.043 ± 0.01 a
Hot No PD	12.8 ± 0.17 ab	3.62 ± 0.03 ab	6.19 ± 0.12 a	0.277 ± 0.03 b	0.120 ± 0.01 ab	1.09 ± 0.05 c	0.047 ± 0.01 a
p-value [2]	0.300	**0.004**	**0.001**	**0.001**	0.194	**0.011**	0.318

[1] Different letters within columns indicate a significant difference for Fisher's least significant differences test ($p < 0.05$). [2] Significant p-values are shown in bold fonts.

The only aroma attribute influenced by fermentation temperature was white pepper which was higher in the Cold and Cold/Hot wines. However, intensity ratings of white pepper aroma were low, ranging from 2.80 to 3.51 in scores on a 15 cm line scale (Table 1), so this specific attribute was unlikely to have a strong sensory impact.

All the wines were similar in overall aroma intensity as well as red fruit, and tropical fruit aromas regardless of treatment. An explanation for the overall lack of differences seen with aroma intensities between the wines may have been due to a masking effect caused by reduction aromas and their associated chemical compounds. Two compounds responsible for reduction aromas in wines are hydrogen sulfide and methanethiol [46]. A previous study examined the aromatic impact of these two compounds as well as their interaction in model white, red, and red oak wines based on typical Spanish wines [66]. Wines were evaluated by a sensory method known as Rate-All-That-Apply (RATA). The results indicated that the presence of hydrogen sulfide and methanethiol significantly repressed fruit and floral aromas in the white and red wines. Based on these findings, it is likely that a similar masking effect occurred with the Grenache wines of the present study particularly as the aroma descriptors included attributes like rose, tropical fruit, and red fruit. However, the volatile chemistry results showed differences among the wines even if they may have not been perceived from a sensory perspective. Indeed, Cold wines had high concentrations of β-damascenone and Cold/Hot wines tended to have higher concentrations of the esters ethyl butyrate, isoamyl acetate, ethyl octanoate, and ethyl hexanoate (Table 3). Hot wines also had high concentrations of isoamyl acetate (Table 3). All of these compounds with high OAVs are associated with fruit aromas and therefore they were likely masked by reduction aromas.

2.3. *Taste, Mouthfeel, and Retronasal Aroma*

Overall, the most cited attributes during TCATA evaluations were bitterness, the astringency subquality suede, and acidity as indicated by the highest maximum citation proportion (V_{max}) values compared to all other attributes (Table 5). Additionally, bitterness, suede, and acidity had the highest AUC (total intensity response) values. These three attributes also had the longest length (T_{first}–T_{last}) with the addition of the astringency subquality velvet. Interestingly, the retronasal aromas of fruit, hot, and reduction consistently had the lowest time parameter values most notably, AUC, V_{max}, and length. An exception to this trend was that the astringency subquality velvet had the lowest V_{max} value (maximum citation proportion) (Table 5). The most cited attributes of the present study reflected results seen in a past experiment that utilized TDS to examine the retronasal and mouthfeel sensations in blended and cofermented Malbec, Merlot, and Petite Sirah, whereby astringency, bitterness, acidity, and the retronasal aroma mineral were the most dominant attributes [41]. Similarly, another study examined the retronasal and mouthfeel sensations in six different red varietal wines using TDS and TCATA [67]. The TDS results indicated that acidity and astringency were the most dominant attributes, while TCATA results showed a similar pattern but with the addition of bitterness as the most cited attribute. Both past studies included retronasal aromas, but, for the most part, these retronasal aromas were not as cited or dominant depending on their respective sensory methods [41,67]. The same trend has been observed in the present study. The similarities between the results of TDS and TCATA may be an indication that panelists interpreted dominance and the presence of an attribute in the same way. Additionally, astringency, bitterness, and acidity are typically constant attributes in red wines, while retronasal aromas can change depending on the varietal and winemaking style. This means that panelists may be more confident in their perception of acidity, astringency, and bitterness. Hence, these attributes had higher citation proportions in the temporal retronasal sensory studies.

Table 5. One-way analysis of variance (ANOVA) of sensory attributes and time extracted from TCATA curves parameters evaluated by the trained panel (n = 8).

Attribute	AUC [1]	T_{max} (Seconds)	V_{max} (Citation Proportion)	T_{first} (Seconds)	T_{last} (Seconds)	Length (Seconds)
Acidity	24.2 b [2]	24.1 d	0.875 a	14.8 d	69.4 b	54.7 bc
Bitter	28.6 a	28.4 bc	0.861 ab	15.3 cd	76.7 a	61.4 a
Fruit	19.4 c	23.9 d	0.781 c	14.6 d	66.1 b	51.5 c
Hot	14.5 d	22.4 d	0.778 c	15.2 d	53.1 d	37.9 d
Reduction	15.3 d	25.2 cd	0.660 d	16.1 c	58.6 c	42.5 d
Suede	28.9 a	30.1 ab	0.785 bc	17.6 b	78.9 a	61.3 a
Velvet	19.2 c	32.9 a	0.576 e	18.6 a	75.8 a	57.2 ab
p-value [3]	**<0.0001**	**<0.0001**	**<0.0001**	**<0.0001**	**<0.0001**	**<0.0001**

[1] AUC: total intensity response; T_{max}: time of maximum citation proportion; V_{max}: maximum citation proportion; T_{first}: time of first citation proportion per attribute; T_{last}: time of last citation proportion per attribute; Length: total time attribute was perceived. [2] Different letters within columns indicate a significant difference for Fisher's least significant differences test, $p < 0.05$. [3] Significant p-values are shown in bold fonts.

A PCA was conducted to show the product trajectory of each treatment and the order of attribute perception during evaluation (Figure 3). Dimension 1 (eigenvalue: 5.79; variability: 0.827) and Dimension 2 (eigenvalue: 0.76; variability: 0.108) were used to calculate the PCA and explained 95% of the variation within the dataset. The lower right quadrant was dominated by the attributes reduction, acidity, hot, and fruit, which appeared early in the evaluation at around 10 s. The upper right quadrant was dominated by the suede, velvet, and bitter attributes indicating that these attributes were perceived in the middle of the evaluation between 30- and 40-s. This order of perception has been seen previously in wine research with time-based sensory analysis [33,41,67]. However, the attribute lines are close to one another, suggesting that the retronasal and mouthfeel profiles were not largely different from each other. The exception to this trend was the Cold fermentation temperature wines which separated and were lower than the other wines from 20 to 40 s. While Figure 3 can show broad differences among wines and the order of attribute perception, it is not able to show significant differences among treatments.

The TCATA curves for each wine are shown in Figure 4A–F with highlighted lines indicating significant differences of an attribute compared to all treatments. Interestingly, while acidity and bitterness were the most cited attributes, most of the wines did not have statistically significant differences comparatively. The exception was the Cold No PD wines which had a significantly lower citation proportion of bitterness from 11 to 46 s (Figure 4B). This was expected as acidity perception is linked to titratable acidity (TA), and although there were significant differences in TA between wines, these are unlikely to be of sensory relevance (Table 4).

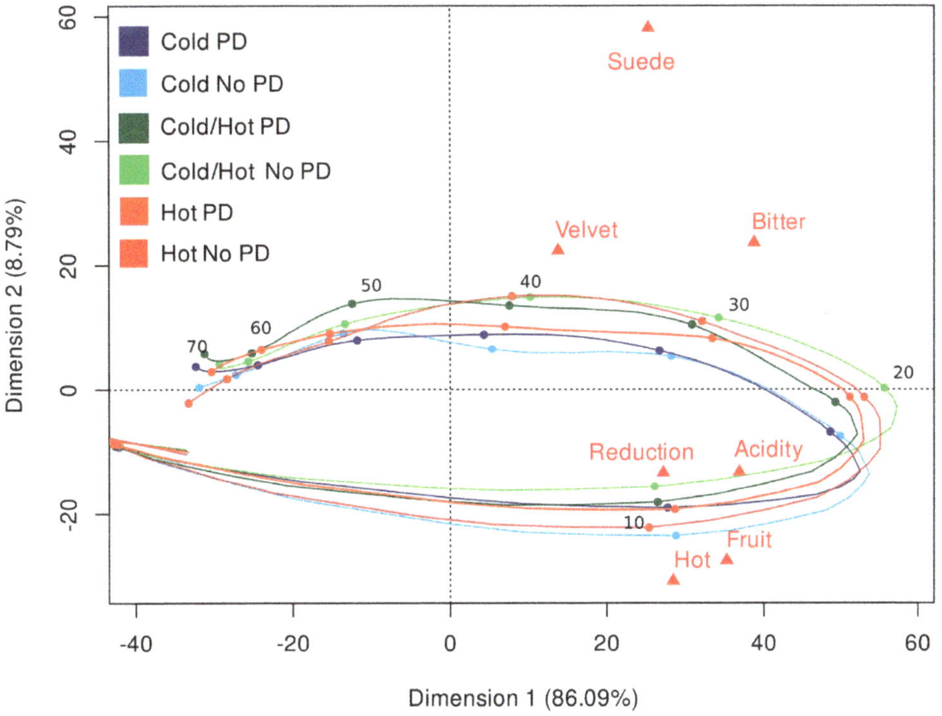

Figure 3. Principal component analysis for the trajectory timeline (seconds) of TCATA attributes in Grenache wines evaluated by a trained sensory panel (n = 8).

Similarly, suede, which also had a high V_{max} as shown in Table 5, was only significantly higher in Cold PD and Cold/Hot PD wines for a brief period (five seconds) (Figure 4A,C). Otherwise, the wines were either not statistically significant for suede or in the case of Cold No PD had a significantly lower citation proportion of suede for 5 s. As for the velvet sub-quality of astringency, both Cold wines had significantly lower citation proportions during the middle of the evaluation while Cold/Hot PD wines had significantly higher citation proportions during the middle of the evaluation. Differences in TCATA curves were reflected in phenolic composition among winemaking treatments. The Cold/Hot and Hot wines regardless of cap management had higher phenolic composition compared to the Cold wines (Table 2). The two-way ANOVA of phenolic composition indicated that Cold/Hot and Hot wines were not significantly different in terms of anthocyanins, total tannins, and phenolics (Supplemental Table S2). These results were expected as the Cold/Hot wines of the present study underwent a treatment similar to a cold soak in which the must is held at a cold temperature prior to fermentation, followed by a warm fermentation temperature [23]. However, the Cold/Hot wines were fermenting during the temperature changes. The application of cold-soak during winemaking has been shown to have no effect on the phenolic composition of wines specifically tannins, total phenolics, and anthocyanins [68]. Interestingly, the highest citation proportion and longest persistence of velvet were seen in the Cold/Hot PD wines (Figure 4C). According to a previous study that examined the chemical and sensory effect of cold soaking compared to a control in six different varietals, the use of cold soak did not significantly affect the perception of astringency [68]. However, the differences seen in the present study may be due to the definitions of astringency used which were the sub-qualities suede and velvet. These two subqualities were defined and understood by the panel as medium and low levels of astringency for suede and velvet, respectively.

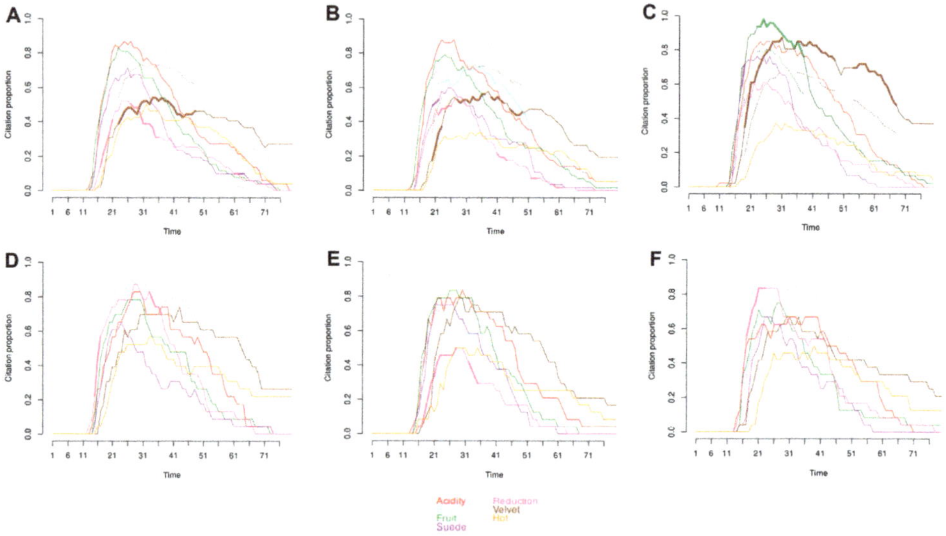

Figure 4. Temporal check-all-that-apply (TCATA) curves. (**A**) Cold PD; (**B**) Cold No PD; (**C**) Cold/Hot PD; (**D**) Cold/Hot No PD; (**E**) Hot PD; (**F**) Hot No PD. When attributes within each treatment are significantly different from all other wines the line is highlighted, and a dotted reference line of the same attribute indicates where the average citation proportion of the other wines is for that attribute.

The attribute that showed the most statistical variation in all the wines was reduction. Indeed, all wines except for Cold/Hot PD, had significantly higher citation proportions of reduction at some point during evaluation (Figure 4). For Cold/Hot No PD and both Hot wines, the only significantly different attribute was the reduction retronasal aroma (Figure 4D–F). Like the results seen in the orthonasal assessment of the wines, reduction was influenced by the cap management regime. This was probably due to the lack of oxygen exposure in the No PD wines during winemaking as well as the potential factors of fermentation temperature, and yeast nutrition as explained by previous research [49,69]. Although there has been research conducted on the reduction aroma in wines using gas chromatography [50,51,70], and static sensory methods [66,69,71], to the authors' knowledge, the temporal sensory pattern of perceived retronasal reduction has not been reported before. Based on the winemaking treatments and varietal choice of the present study, the retronasal perception of reduction appears to be first perceived at a similar timepoint as bitterness which was later compared to the other retronasal aromas (Table 5; Figure 3). However, on average, the retronasal perception of reduction did not have as long of a length as the other attributes assessed; in fact, it had the second shortest length, with the shortest length belonging to hot retronasal aroma (Table 5).

Cold/Hot PD wines were the only wines that did not have a significant occurrence of reduction flavor. They also showed a significantly higher citation proportion of fruit retronasal aromas from 11 to 31 s post-expectoration which was not seen in any of the other wines. The Cold/Hot wines were similar to that of a cold-soak treatment [68]. Previous research on the application of cold-soak to Syrah wines reported an increase in black currant flavor compared to the control according to a pair-comparison test [29]. Therefore, the Cold/Hot PD treatment may have enhanced fruit flavors due to its similarity to a prefermentative cold-soak treatment. However, it must be noted that unlike traditional cold-soak regimes whereby usually no yeast is inoculated, in the present study the wines were inoculated with a selected yeast strain at the onset of skin contact time immediately post-crushing. Overall, the TCATA evaluation method provided extensive information about the wines and selected evaluation attributes.

2.4. Salivary Flow Rate and Temporal Sensory Perception

A one-way ANOVA conducted on the salivary flow rate for each panelist is shown in Supplemental Table S5. The mean salivary flow rate was 2.50 mL/min and there was a two-fold difference between the maximum (3.18 mL/min) and minimum (1.39 mL/min) salivary flow rate within panelists. Categorical salivary flow rates were designated based on the value's position above or below the mean. Values above the mean were designated as high salivary flow rate (HF); while values below the mean were designated as low salivary flow rate (LF). Within the sensory panel ($n = 8$), there were six individuals with high salivary flow rate (2.53 mL/min to 3.18 mL/min; two panelists were considered low salivary flow rate (1.39 mL/min to 1.58 mL/min). Supplemental Table S6 shows a one-way ANOVA for the first time of perception by the two astringency subqualities and fruit retronasal aroma for salivary flow rate. The astringency subquality suede ($p = 0.001$) was the only statistically significant attribute, whereby LF panelists perceived it significantly later post-expectoration compared to HF panelists. Based on the present results, it appears the suede subquality may reflect the established pattern where LF individuals perceived astringency later than HF individuals which aligns with the results of previous research [35,37]. However, this panel was small in size ($n = 8$) and had many HF panelists ($n = 6$) compared to LF panelists ($n = 2$). Therefore, more assertive conclusions are not possible in light of these results. Further research with a more balanced panel in terms of salivary flow rate would be required for unequivocal conclusions.

3. Materials and Methods

3.1. Winemaking

Grenache grapes were sourced from the Rancho Real vineyard in the Santa Maria Valley AVA (Santa Barbara County, CA, USA) during the 2021 vintage. The winemaking protocol followed that of a previously published study [27], with the following modifications. 1260 kg of fruit were manually harvested and transported to Cal Poly Pilot Winery for processing the same day. Clusters were crushed/destemmed using a crusher–destemmer (Bucher Vaslin, Niederweningen, Switzerland), and resulting musts went immediately into 60 L food-grade fermenters (Speidel, Swabia, Germany), at a rate of 50 kg per vessel, measured on an analytical scale (Cardinal Detecto, Model 204, Webb City, MO, USA). Fifty mg/L of sulfur dioxide (SO_2) was added to each fermenter immediately post-crushing. Three different fermentation temperature regimes, combined with two contrasting cap management regimes, were established: Cold (targeting an average fermentation temperature of 12 °C), Cold/Hot (fermentation temperatures of 12 °C from day 0 to 7, and 28 °C from day 8 to 14), and Hot (targeting an average fermentation temperature of 28 °C throughout). In addition, two contrasting cap management regimes per fermentation temperature regime were established, namely no punch down (No PD), whereby the fermenters received no cap management during the fermentation/maceration period; and punch down (PD), whereby fermenters received two daily manual punch downs (9 a.m. and 5 p.m.), each lasting one-minute, performed with a must plunger. All individual treatments were established in triplicate ($n = 3$), affording a total of 18 individual fermentations, of the following treatments: Cold PD; Cold No PD; Cold/Hot PD; Cold/Hot No PD; Hot PD; Hot No PD. Musts were inoculated with dry yeast (Enartis D20, 153 Windsor, CA, USA), at a rate of 40 g/hL, 6 h after crush, and fermenters were taken to their respective fermentation conditions. The selected yeast was chosen for its ability to ferment at a wide temperature range, as previously shown [27]. Total maceration time was set to 14 days.

Cold fermentation temperature treatments were established by placing the fermenters in a temperature-controlled cold room set at 10 °C for the duration of the 14-day maceration period. The Cold/Hot fermentation regime treatments were established by placing the fermenters for 7 days in a temperature-controlled cold room set at 12 °C and were then subsequently transferred to a warmer environment in a temperature-controlled room set at 26 °C, for the last 7 days of the maceration period. The Hot fermentation regime treatments were established by placing the fermenters in a temperature-controlled room set at 26 °C, for the duration of the 14-day maceration period. Temperature probes with integrated data logging software (Elitech RC-5, 8 Elitech, Milpitas, CA, USA), were placed inside each fermenter approximately 8 cm from the bottom and the probes were set at one-minute collection intervals. Plastic tubing with holes for free liquid flow was inserted vertically into the fermenters to allow for optimal probe placement in the must, below the cap, and above the bottom of the tanks, as previously described [27]. Temperature probe data was extracted using the ElitechLog software (Elitech, Milpitas, CA, USA). Diammonium phosphate (DAP) at a rate of 200 mg/L (Fermaid K, Scott Laboratories, Petaluma, CA, USA) was added on day two of alcoholic fermentation to all fermenters. Total soluble solids (°Brix), and daily temperature were monitored via a densitometer (Anton Paar DMA 35 Basic, Graz, Austria) during the duration of the 14-day maceration period, after which wines were pressed off the fermentation solids. Press fractions were discarded, with only free-run wines transferred to 18 L glass carboys. Wines were subsequently inoculated with *Oenococcus oenii* (VP41, Scott Laboratories, Paso Robles, CA, USA), at a rate of 20 g/hL to undergo malolactic fermentation. Malolactic fermentation was monitored via enzymatic analysis of L-malic acid and L-lactic acid with an enzymatic analyzer (Admeo Y15, Angwin, CA, USA), using commercial enzymatic analysis kits (Biosystems, Barcelona, Spain). Malolactic fermentation was completed in all the wines (malic acid < 0.1 g/L) (Supplemental Table S4). Cold stabilization occurred at 10 °C for 30 days, after which all wines were racked and adjusted to 30 mg/L SO_2. The wines were filtered through 8 μm cellulose filter pads (Vintner's Vault, Paso Robles, CA, USA), and readjusted to 30 mg/L SO_2. The wines were bottled in early March 2022, using DIAM 5 micro-agglomerated cork closure (G3 Enterprises, Modesto, CA, USA; oxygen transmission rate; 0.4 ng/bottle/year; oxygen initial release: 1.3 mg), stored in a vertical position and kept in cellar-like conditions (approximately 12 to 14 °C) until analyses. Wines received a six-month bottle aging period prior to sensory analysis.

3.2. Basic Chemistry and Phenolic Measurements

pH (Thermo Scientific Orion Star A211) was measured manually. Titratable acidity (TA) was measured by titrating a known quantity of wine (5 mL) in a deionized water solution against 0.067 N NaOH (Fisher Scientific, Waltham, MA, USA) to a pH endpoint of 8.2. Ethanol (% *v*/*v*), was measured by near-infrared spectroscopy using an Alcolyzer Wine M/ME analysis system (Anton Paar, Graz, Austria). Acetic acid, glucose, fructose, malic acid, and lactic acid were determined enzymatically using commercial enzymatic analysis kits (Admeo, Biosystems Group, Hollister, CA, USA).

Phenolic compounds were analyzed using spectrophotometric methods. Wine samples were centrifuged in a 15,000× *g* microfuge (model 5415D; Eppendorf, Hamburg, Germany), and the supernatant was transferred into clean 1 mL Eppendorf tubes prior to analysis. Anthocyanins and total polymeric pigments (TPP) [the sum of small polymeric pigments (SPP) and large polymeric pigments (LPP)], were determined as previously reported [72]. Tannins were analyzed by protein precipitation [73]. Total phenolics (expressed as mg/L (+)-catechin equivalents, CE), and anthocyanins (expressed as mg/L malvidin-3-glucoside), were measured following a previously published method [72]. Non-tannin phenolics were calculated as total phenolics minus protein precipitable tannins. All spectrophotometric measurements were made with a Cary 60 UV-Vis spectrophotometer equipped with an 18-sample cell auto-sampler (Agilent Technologies, Santa Clara, CA, USA).

3.3. Volatile Compound Analysis

Volatile compounds for analysis were selected based on previous reports on Grenache [1,4,5]. Standards of the volatile compounds to be analyzed were purchased from Acro's Organics (Thermo Fisher Scientific, Fair Lawn, NJ, USA), Alfa Aesar (Thermo Fisher Scientific, Fair Lawn, NJ, USA), Fischer Chemical (Thermo Fisher Scientific, Fair Lawn, NJ, USA), Sigma Aldrich (Sigma Aldrich, Sigma Aldrich Inc., St. Louis, MO, USA), Spectrum (Spectrum Chemical MFG. Corp., Gardena, CA, USA), and TCI (Tokyo Chemical Industry Co., Ltd.,

Portland, OR, USA). Reference compound information including CAS number, manufacturer, and purity are shown in Supplemental Table S7.

Stock solutions of volatile standards to optimize compound separation and calculate the response factor for quantification were prepared in 95% ethanol. Reference compounds were grouped according to typical concentration in wines to create thirteen stock solutions with no more than five reference compounds per stock solution and 50 µg/L of 2-undecanone (Sigma Aldrich, Sigma Aldrich Inc., St. Louis, MO, USA) which was used as the internal standard for all compounds (Supplemental Table S8) [74]. The creation of stock solutions and internal standard (ISTD) solutions were based on a previously described procedure [75]. However, stocks 11–13 required higher concentrations therefore the stock solution process was modified. For stock solution 11, ethyl hexadecanoate was added at 50 µL in 250 mL and then 115 µL was pipetted into the model wine for analysis (Supplemental Table S8). A model wine solution was utilized to optimize the separation of the volatile standards and calculate the response factor for each compound which was based on a previously described procedure [74]. The volume of each stock solution pipetted into the sample was in the middle range of the typical concentration of each compound found in wine, which varied depending on the compound, and compounds were verified using their respective qualifying ions (Supplemental Table S8). Model wine solutions were analyzed in triplicate following the analysis parameters described below for solid-phase microxextraction and stir-bar sorptive extraction procedures.

3.3.1. Solid-Phase Microextraction

Solid-phase microextraction (SPME) was used to analyze and quantify the compounds isoamyl alcohol, isobutanol, ethyl lactate, and ethyl hexadecanoate. Sample preparation, sampling, and instrument analysis followed a previously outlined procedure with modifications [74]. For each sample, 10 mL aliquot (model wine solutions or wine samples) was pipetted into 20 mL round-bottomed glass vials (Gerstel, Linthicum, MD, USA) containing 3 g of NaCl (ACS reagent, Sigma Aldrich, Sigma Aldrich Inc., St. Louis, MO, USA) and 6.18 µL of 2-undecanone (Sigma Aldrich, Sigma Aldrich Inc., St. Louis, MO, USA) was used as the internal standard. Vials were topped with metal screw caps with 1.3 mm septa, blue silicone, and PTFE (Gerstel, Linthicum, MD, USA). Samples were analyzed in triplicate and the run order was randomized.

For instrument analysis the SPME fiber DVB/C-WR/PDMS, 50/30 µm, 10 mm (Restek, Restek Corporation, Bellefonte, PA, USA) was used. Samples were heated to 40 °C and agitated for 10 min at 250 rpm before sampling by the SPME fiber. The fiber was exposed to the sample for 30 min at 40 °C with an agitation speed of 250 rpm. The SPME fiber was desorbed in split mode with a 20:1 split ratio, the fiber was kept in the inlet for 10 min and the inlet temperature was held at 240 °C. The samples were analyzed using an 8890 gas chromatography system (Agilent, Agilent Technologies Inc., Santa Clara, CA, USA) as well as an MPS autosampler (Gerstel, Linthicum, MD, USA). A DB-Wax capillary column, 30 m, 0.250 mm, 0.25 µm (Agilent, Agilent Technologies Inc., Santa Clara, CA, USA) was used. Helium was used as the carrier gas at a constant flow or 1 mL/min. For analysis the oven parameters were as follows: 40 °C for 5 min, followed by an increase of 3 °C/min to 180 °C, then 30 °C/min to 240 °C where the temperature was held. The MSD interface was also held at 240 °C. Compounds were analyzed using deconvolution ranked by height and 50 retention time (RT) window size. Compounds were identified by the NIST 20 MS Search Version 2.4 using their respective qualifying ions shown in Supplemental Table S8.

3.3.2. Stir Bar Sorptive Extraction

The stir-bar sorptive extraction (SBSE) method was based on a previously described procedure [76] with modifications for a DB-Wax capillary column. A volume of 5 mL sample was pipetted into 20 mL round-bottomed glass vials (Gerstel, Linthicum, MD, USA) containing 5 mL of ultra-pure water (Millipore, Sigma Aldrich Inc., St. Louis, MO, USA). 6.18 µL of 2-undecanone (Sigma Aldrich, Sigma Aldrich Inc., St. Louis, MO, USA) was used as the internal standard. All samples were twisted using a PDMS coated Twister^{TM} stir bar, 10 mm, 0.5 mm (Gerstel, Linthicum, MD, USA) for 1 h on a Twister^{TM} 20-position magnetic stirrer at 1000 rpm (Gerstel, Linthicum, MD, USA). Wine samples were analyzed in triplicate and the run order was randomized.

The initial thermal desorption unit (TDU) temperature was set at 40 °C then heated to 60 °C at a rate of 60 °C/min and a hold time of one minute for ramp 1. The TDU temperature was then raised to 280 °C at a rate of 270 °C/min and a 3-min hold time. Desorption mode was set at solvent venting for 1.33 min. The transfer temperature was 300 °C. Cyro cooling was utilized for 15 min, and the initial cooling injection system (CIS) temperature was −120 °C with an equilibration time of 0.20 min. The CIS temperature was raised to 280 °C at a rate of 12 °C/min and a hold time of 5 min. The samples were analyzed using an 8890 gas chromatography system (Agilent, Agilent Technologies Inc., Santa Clara, CA, USA) as well as a MPS autosampler (Gerstel, Linthicum, MD, USA). A DB-Wax capillary column, 30 m, 0.250 mm, 0.25 µm (Agilent, Agilent Technologies Inc., Santa Clara, CA, USA) was used. Helium was used as the carrier gas at a constant flow of 1.33 mL/min. For GC parameters, the inlet pressure was set at 15.63 psi with a septum purge flow of 3 mL/min. The mode was set at the solvent vent with a purge flow of 25.6 mL/min. Column flow rate was set at 1.28 mL/min with a pressure of 15.63 psi and a hold-up time of 1.82 min. The column oven started at 40 °C with a hold time of 1 min, followed by a ramp of 10 °C/min to 250 °C for a hold time of 3 min and a total run time of 25 min. The MSD interface was also held at 240 °C. Compounds were analyzed using deconvolution ranked by height and a 50 RT window size. Compounds were identified using NIST 20 MS Search Version 2.4 using their respective qualifying ions shown in Supplemental Table S8.

3.4. *Descriptive Analysis and TCATA Sensory Panel*

Members of the sensory panel (n = 8) were recruited by electronic mail (e-mail) and the project received California Polytechnic State University International Review Board (IRB) approval (IRB protocol #2020-058). The panel was composed of 50% females and 50% males with an age range of 21 to 60 years old. Panelists were screened for PROP sensitivity and salivary flow rate. Sensitivity to PROP was assessed using 6-*n*-propylthiouracil (PROP) (Fluka, Lot #: 1129436, Steinheim, Germany) following a previously described procedure [77]. The panel was composed of 50% super-tasters, 38% medium-tasters, and 13% non-tasters. The salivary flow rate collection procedure and calculation were performed as previously reported [41]. Additionally, panelists were screened for visual disorders and color perception deficiencies through pseudo-isochromatic color testing plates (Ishihara maps). None of the panelists showed color deficiencies based on this test.

3.4.1. Panel Training

Wines were evaluated using general descriptive analysis (DA) and TCATA following previous procedures [8,40]. Panel training was conducted over seven sessions, each lasting from one hour to one and a half hours from August to September 2022. Descriptive analysis training focused on the evaluation procedure, attribute definition, standard review, and assessment of intensity. Nine attributes were selected by the panel which encompassed color and aroma. Attributes with standard compositions used during training are shown in Table 6. Initial training sessions focused on the identification of standards and the detection of the attributes in the research wines. The remaining training sessions incorporated the use of intensity assessment of attributes with a 15 cm, unstructured line scale. Finally, the last two training sessions had the panelists practice the use of the sensory software prior to formal evaluation. During training, the evaluation of color was conducted separately in clear, International Organization for Standardization (ISO) glasses while all other attributes were assessed using black glasses to mask the variation in color among treatments and avoid bias due to color.

Table 6. Color, aroma, taste, flavor, and mouthfeel attribute standard composition.

Attribute		Standard Composition
Color	Purple Hue	High: L* = 39.93, C* = 61.92, a* = 60.82, b* = 11.63 [1]
		Low: L* = 86.40, C* = 15.57, a* = 15.13, b* = 3.66 [2]
	Saturation	High: L* = 39.93, C* = 61.92, h = 10.83 [1]
		Low: L* = 86.40, C* = 15.57, h = 13.60 [2]
Aroma [3]	Reduction	100 mL of base reduction solution; 116.94 g raw white onion (Signature Farms, Better Living Brands LLC, Pleasanton, CA, USA), 112.02 g cooked, white onion (Signature Farms, Better Living Brands LLC, Pleasanton, CA, USA) on medium until browned and soaked in 1000 mL of base wine overnight.
	Rose	80 mL rose water (Fee Brothers, Fee Brothers, Rochester, NY, USA), 99.2 mL rose syrup (Monin, Monin Inc., Clearwater, FL, USA)
	Red Fruit	51.54 g mashed, fresh raspberries (Fresh Kampo, Meridian Fruits, Periban Los Reyes, SN 60440, Mexico), 89.08 g cut, fresh strawberries (Signature Farms, Better Living Brands LLC, Pleasanton, CA, USA)
	Tropical Fruit	23.62 g crème de banana syrup (Torani, R. Torre & Company, San Leandro, CA, USA), 12.98 g mango syrup (Torani, R. Torre & Company, San Leandro, CA, USA)
	White Pepper	6.60 g white pepper (First Street, Amerifoods Trading Co, Los Angeles, CA, USA)
	Mushroom	52.34 g chopped, raw Baby Bello mushrooms (Signature Farms, Better Living Brands LLC, Pleasanton, CA, USA)
	Hot	444 mL of vodka (New Amsterdam, New Amsterdam Spirits Company, Modesto, CA, USA)
Retronasal Aromas	Reduction	100 mL of base reduction solution; 116.94 g raw white onion (Signature Farms, Better Living Brands LLC, Pleasanton, CA, USA), 112.02 g cooked, white onion (Signature Farms, Better Living Brands LLC, Pleasanton, CA, USA) on medium until browned and soaked in 1000 mL of base wine overnight.
	Fruit	61.26 g cut, fresh strawberries (Signature Farms, Better Living Brands LLC, Pleasanton, CA, USA), 13.26 g strawberry preserves (Bonne Maman, Andros, France), 16.31 g wild blueberry preserves (Bonne Maman, Andros, France), 18.39 g cherry preserves (Bonne Maman, Andros, France), 13.95 g blackberry syrup (Torani, R. Torre & Company, San Leandro, CA, USA), 15.29 g cherry syrup (Torani, R. Torre & Company, San Leandro, CA, USA)
	Hot	444 mL of vodka (New Amsterdam, New Amsterdam Spirits Company, Modesto, CA, USA)
Taste	Acidity [4]	1.23 g tartaric acid (LD Carlson Company, LD Carlson Company, Kent, OH, USA)
	Bitter [4]	1.32 g caffeine, anhydrous (Sigma Aldrich, 1003363509, Sigma-Aldrich Co., St. Louis, MO, USA)
Mouthfeel	Suede [5]	Suede dress (Willow Ridge)
	Velvet [5]	Velvet jacket (Chico's)

[1] 2021 Graciano research wine, CIE Lab Measurements. [2] 2021 Graciano research and 50 mL of water, CIE Lab Measurements. [3] Prepared to 1000 mL of Franzia Merlot (Franzia, Franzia Vineyards, Ripon, CA, USA). [4] Prepared in 750 mL of Franzia Merlot (Franzia, Franzia Vineyards, Ripon, CA, USA). [5] Cut two 5.10 cm × 12.7 cm pieces. One was glued to the canvas board; another was pulled through the canvas hole and allowed to dangle.

TCATA training focused on the evaluation procedure, attribute definition, and retronasal aroma, taste, and mouthfeel standard review. The attribute list consisted of six categories selected by the panel. During training, panelists were asked to independently identify each standard followed by a group discussion where missed standards were reviewed again. For the panelists to understand the difference between retronasal aroma attributes and taste and mouthfeel attributes, standards were evaluated separately from each other. Taste and mouthfeel standards were evaluated using nose clips to block retronasal aromas. The TCATA evaluation procedure was practiced starting halfway through training. Panelists were instructed to take the full sample in their mouth and swish for 10 s. At the 10 s mark they were prompted to expectorate their sample and begin evaluation (Figure 5). Like TDS, the order of attributes was balanced across panelists [78]. A dual-stop feature was utilized when panelists no longer perceived any attributes before the 70 s was finished. The TCATA task lasted seventy seconds with results recorded at 0.5 s intervals by the software. Panelists were instructed to palate cleanse for fifteen seconds between samples.

Figure 5. TCATA evaluation procedure.

3.4.2. Formal Evaluations

Formal evaluations were conducted over five sessions in individual sensory cabinets at the J. Lohr Sensory Lab of the Wine & Viticulture department at California Polytechnic State University, San Luis Obispo, in September 2022. All replicates were evaluated three times. Descriptive analysis and TCATA were assessed in separate sessions. For descriptive analysis testing, 30 mL aliquots of samples were served. All samples regardless of the sensory method were served in clear, ISO glasses with four-digit random codes, and an aluminum foil top at room temperature. Wines were served monadically either under red lighting (Luna 3AO, 18:18W, Zaniboni Lighting, Clearwater, FL, USA) or for color assessment under daylight setting (Luna 3, 26:26W, Zaniboni Lighting, Clearwater, FL, USA). Order of serving was randomized according to William's Latin Square Design. The assessment used a 15 cm line scale anchored at 1 cm and 14 cm with the terms "low" and "high" respectively.

To lower the bias of panelists when assessing the different modalities of the wine, color, and aroma were assessed separately. This resulted in six separate tests with 9 wines per test. The run order of the wines was randomized per test per session. Panelists assessed wines following the evaluation procedures described during training. Following each sample there was a forced palate cleansing of 15 s. A 10 min break was given halfway through each testing session.

TCATA formal evaluations were conducted over three sessions separate from DA and evaluated in individual sensory cabinets. All replicates were evaluated three times. 15 mL aliquots of sample were served at laboratory temperature (18 °C) in ISO glasses with four-digit random codes, and an aluminum foil top. Wines were served monadically under red lighting (Luna 3AO, 18:18W, Zaniboni Lighting, Clearwater, FL, USA). Order of serving was randomized according to William's Latin Square Design. Panelists assessed eighteen wines per session, following the evaluation procedure described during training, with a 5 min break after nine samples. Following each sample there was a required palate cleansing of 15 s. Unsalted crackers (Nabisco unsalted tops, premium saltine crackers, East Hanover, NJ, USA) and water (Evian natural spring water, Evian, France) were provided for palate cleansing. All TCATA data collection was done in RedJade Sensory Software (RedJade, Silicon Valley, CA, USA). Panel performance was analyzed using the panel analysis feature for descriptive analysis and the TCATA function in the sensory package of XL STAT 2022.1.2 (Addinsoft, Paris, France) was utilized for agreement and repeatability during retronasal and mouthfeel assessment.

3.5. Statistical Analysis

The basic chemistry and phenolic data were analyzed by a one-way ANOVA ($p < 0.05$). Additionally, to understand the relationship between each main effect and the interaction in chemical and phenolic data, a two-way ANOVA was conducted ($p < 0.05$). Fisher's Least Significant Differences (LSD) was used as the *post-hoc* comparison.

Quantification of the volatile compounds in wine samples was conducted by the internal standard method outlined in a previous procedure [75]. One-way and two-way ANOVAs were conducted on the SPME and SBSE data with Fisher's LSD as the *post-hoc* test. To understand the potential relationship between volatile chemistry and sensory results, a Partial Least Square Regression was conducted whereby sensory attributes were the dependent variable and volatile compounds were the independent variable [79]. All statistical analyses were run in XLSTAT 2022.1.2 (Addinsoft, Paris, France).

Descriptive analysis data were analyzed by a one-way ANOVA ($p < 0.05$) on the wines. To understand the effect of each factor on the sensory responses, a two-way ANOVA ($p < 0.05$) was run on the descriptive analysis data. The *post-hoc* comparison for all statistical analyses was Fisher's LSD test in XLSTAT 2022.1.2 (Addinsoft, Paris, France). Principal component analysis (PCA) using the correlation matrix with no rotation was applied to the wine sensory data, including the replicates, using R software (R Foundation for Statistical Computing, Vienna, Austria). Confidence ellipses indicating 95% confidence intervals were based on the multivariate distribution of Hotelling's test for $p < 0.05$ and were constructed using the SensoMineR panellipse function on R [80].

TCATA curves were obtained using the tempR package [81] with the R-software version 4.2.2 (R Core Team, 2022). The tempR package follows a previously published protocol [10], by calculating the average citation proportions for each attribute (y-axis) over time (x-axis; 70 s) for each wine. The citation proportion of each sample and attributes at TCATA were recorded and grouped in contingency tables, where products and times were arranged in rows and attributes in columns. Data tables were submitted to Principal Component Analysis (PCA) using the approach described [82] but with attribute standardization. Trajectories were smoothed along each dimension independently and PCA biplots were visualized using the tempR package [81]. Selected time parameters (Table 7) were extracted from TCATA curves and analyzed by attribute using a one-way ANOVA ($p < 0.05$) and Fisher's LSD test in XLSTAT 2022.1.2 (Addinsoft, Paris, France).

Panelist salivary flow rate data were analyzed by a one-way ANOVA ($p < 0.05$) to determine the spread of variance in the panel. These values were converted to categorical data by designating values above the overall mean as a high flow rate and below the mean as a low flow rate [41]. A separate one-way ANOVA ($p < 0.05$) was run on the categorical flow rates to compare these values with the perception of the astringency sub-qualities suede and velvet as well as the fruit aroma. Statistical analyses were conducted in XLSTAT 2022.1.2 (Addinsoft, Paris, France).

Table 7. Definitions of time parameters extracted from TCATA curves.

Time Parameter	Definition
V_{max}	Maximum citation proportion
T_{max}	Time of maximum citation proportion
Area Under Curve	Total intensity response
T_{first}	Time of first citation proportion per attribute
T_{last}	Time of last citation proportion per attribute
Length	Total time attribute was perceived (T_{first}–T_{last})

4. Conclusions

Based on the TCATA assessment, the most optimal treatment of this experiment was the Cold/Hot PD wines as the wines had a significantly higher citation proportion of fruit flavor and velvet astringency perception relative to the other winemaking treatments. These two terms bear potentially positive hedonic connotations for consumers. Therefore, Grenache winemaking may benefit from a combination of contrasting fermentation temperatures during maceration. For instance, the practice of controlled cold soak at low temperatures, ideally inoculated with a selected yeast strain, followed by a relatively warm to hot alcoholic fermentation temperature, could lead to results comparable with those of the present study. Generally, gentler cap management regimes that minimize overexposure to oxygen may lead to more aromatic retention in Grenache wines. However, this minimalist intervention, if applied, will have to be carefully managed to avoid or minimize the production of reduction aromas.

In summary, for Grenache wines, it is recommended to utilize a Cold/Hot fermentation temperature as it will lead to more retronasal and mouthfeel and a balance of fruity and reduced aromas. Both fermentation temperature and cap management appear to be variety-specific, thus further research on both these factors on popular wine varietals needs to be carried out. Finally, the present study represents a contribution with the main volatile compounds (e.g., β-damascenone and esters in the Cold fermented wines; hexanol in PD wines, which may be potentially responsible for a hot aroma and mouthfeel), and sensory characteristics (red fruit, tropical fruit, white pepper, and rose) of Grenache wines grown in the Mediterranean climate of the Central Coast of California.

Supplementary Materials: The following supporting information can be downloaded at: https://www.mdpi.com/article/10.3390/molecules28104230/s1, Table S1: Two-way analysis of variance (ANOVA) for the main effects fermentation temperature and cap management with interaction showing the mean separation and *p*-values of descriptive sensory attributes of Grenache wines assessed by a trained panel (n = 8); Table S2: Two-way analysis of variance (ANOVA) for the main effects of fermentation temperature and cap management with interaction showing the mean separation and *p*-values of the phenolic composition of Grenache wines. Values represent the mean of three replicates followed by the standard error of the mean (n = 3); Table S3: Two-way analysis of variance (ANOVA) for the main effects of fermentation temperature and cap management with interaction showing the mean separation and *p*-values for the concentration of volatile compounds (μg/L) in Grenache wines. Values represent the mean of three replicates (n = 3); Table S4: Two-way analysis of variance (ANOVA) for the main effects of fermentation temperature and cap management with interaction showing the mean separation and *p*-values of the basic chemical composition of Grenache wines. Values represent the mean of three replicates followed by the standard error of the mean (n = 3); Table S5: One-way analysis of variance (ANOVA) of the saliva (mL/min) for each panelist. Values represent the mean of three salivary flow rate replicates followed by the standard error of the mean (n = 3); Table S6: One-way analysis of variance (ANOVA) of time to onset of perception for astringency subqualities (suede, velvet) and retronasal fruit compared to salivary flow rate (mL/min); Table S7: Compounds evaluated using SPME and SBSE on the GC-MS, CAS numbers, manufacturers, and purity. Table S8: Selected volatile compounds for stir bar sorptive extraction (SBSE) and solid-phase microextraction (SPME) analysis with a concentration in wine according to the literature (μg/L), volume added to the wine (μL), and concentration in model wine (μg/L). References [83–93] are cited in the supplementary materials.

Author Contributions: Conceptualization, L.F.C. and E.S.S.; methodology, L.F.C. and E.S.S.; software, E.S.S. and A.A.C.; validation, L.F.C., E.S.S., A.A.C. and T.M.R.; formal analysis, E.S.S. and A.A.C.; investigation, E.S.S. and T.M.R.; resources, E.S.S. and L.F.C.; data curation, E.S.S., A.A.C. and T.M.R.; writing—original draft preparation, E.S.S.; writing—review and editing, L.F.C.; visualization, E.S.S., A.A.C. and L.F.C.; supervision, L.F.C.; project administration, E.S.S.; funding acquisition, L.F.C. All authors have read and agreed to the published version of the manuscript.

Funding: This research received no external funding.

Institutional Review Board Statement: The study was conducted in accordance with the Declaration of Helsinki and approved by the Institutional Review Board California Polytechnic State University (058, 2020).

Informed Consent Statement: Informed consent was obtained from all subjects involved in the study.

Data Availability Statement: Not applicable.

Acknowledgments: The authors would like to thank the sensory panelists for their hard work and dedication to this project. Also, we would like to thank E&J Gallo Winery (Modesto, CA, USA) for their fruit donation for the making of the wines.

Conflicts of Interest: The authors declare no conflict of interest. The funders had no role in the design of the study; in the collection, analyses, or interpretation of data; in the writing of the manuscript; or in the decision to publish the results.

Sample Availability: Not applicable.

References

1. Rentzsch, M.; Schwarz, M.; Winterhalter, P.; Hermosín-Gutiérrez, I. Formation of hydroxyphenyl-pyranoanthocyanins in Grenache wines: Precursor levels and evolution during aging. *J. Agric. Food. Chem.* **2007**, *55*, 4883–4888. [CrossRef] [PubMed]
2. Leriche, C.; Molinier, C.; Caillé, S.; Razungles, A.; Symoneaux, R.; Coulon-Leroy, C. Development of a methodology to study typicity of PDO wines with professionals of the wine sector. *J. Sci. Food. Agric.* **2020**, *100*, 3866–3877. [CrossRef] [PubMed]
3. López-Giral, N.; González-Arenzana, L.; González-Ferrero, C.; López, R.; Santamaría, P.; López-Alfaro, I.; Garde-Cerdán, T. Pulsed electric field treatment to improve the phenolic compound extraction from Graciano, Tempranillo and Grenache grape varieties during two vintages. *Innov. Food Sci. Emerg. Technol.* **2015**, *28*, 31–39. [CrossRef]
4. Ferreira, V.; Ortín, N.; Escudero, A.; López, R.; Cacho, J. Chemical characterization of the aroma of Grenache rosé wines: Aroma extract dilution analysis, quantitative determination, and sensory reconstitution studies. *J. Agric. Food Chem.* **2002**, *50*, 4048–4054. [CrossRef]
5. Petretto, G.L.; Mercenaro, L.; Urgeghe, P.P.; Fadda, C.; Valentoni, A.; Del Caro, A. Grape and wine composition in *Vitis vinifera* l. cv. Cannonau explored by GC-MS and sensory analysis. *Foods* **2021**, *10*, 101. [CrossRef] [PubMed]
6. Segurel, M.A.; Razungles, A.J.; Riou, C.; Salles, M.; Baumes, R.L. Contribution of dimethyl sulfide to the aroma of Syrah and Grenache noir wines and estimation of its potential in grapes of these varieties. *J. Agric. Food Chem.* **2004**, *52*, 7084–7093. [CrossRef] [PubMed]
7. Caillé, S.; Samson, A.; Wirth, J.; Diéval, J.B.; Vidal, S.; Cheynier, V. Sensory characteristics changes of red Grenache wines submitted to different oxygen exposures pre and post bottling. *Anal. Chim. Acta* **2010**, *660*, 35–42. [CrossRef]
8. Castura, J.C.; Antúnez, L.; Giménez, A.; Ares, G. Temporal Check-All-That-Apply (TCATA): A novel dynamic method for characterizing products. *Food Qual. Prefer.* **2016**, *47*, 79–90. [CrossRef]
9. Ares, G.; Jaeger, S.R.; Antúnez, L.; Vidal, L.; Giménez, A.; Coste, B.; Picallo, A.; Castura, J.C. Comparison of TCATA and TDS for dynamic sensory characterization of food products. *Food Res. Int.* **2015**, *78*, 148–158. [CrossRef]
10. Meyners, M.; Castura, J.C. The analysis of Temporal Check-All-That-Apply (TCATA) data. *Food Qual. Prefer.* **2018**, *67*, 67–76. [CrossRef]
11. Berget, I.; Castura, J.C.; Ares, G.; Næs, T.; Varela, P. Exploring the common and unique variability in TDS and TCATA data—A comparison using canonical correlation and orthogonalization. *Food Qual. Prefer.* **2020**, *79*, 103790. [CrossRef]
12. Varela, P.; Antúnez, L.; Carlehög, M.; Alcaire, F.; Castura, J.C.; Berget, I.; Giménez, A.; Næs, T.; Ares, G. What Is dominance? An exploration of the concept in TDS Tests with trained assessors and consumers. *Food Qual. Prefer.* **2018**, *64*, 72–81. [CrossRef]
13. Jaeger, S.R.; Alcaire, F.; Hunter, D.C.; Jin, D.; Castura, J.C.; Ares, G. Number of terms to use in Temporal Check-All-That-Apply Studies (TCATA and TCATA fading) for sensory product characterization by consumers. *Food Qual. Prefer.* **2018**, *64*, 154–159. [CrossRef]
14. Jaeger, S.R.; Beresford, M.K.; Hunter, D.C.; Alcaire, F.; Castura, J.C.; Ares, G. Does a familiarization step influence results from a TCATA task? *Food Qual. Prefer.* **2017**, *55*, 91–97. [CrossRef]
15. Pineau, N.; Schlich, P.; Cordelle, S.; Mathonnière, C.; Issanchou, S.; Imbert, A.; Rogeaux, M.; Etiévant, P.; Köster, E. Temporal Dominance of Sensations: Construction of the TDS curves and comparison with Time-Intensity. *Food Qual. Prefer.* **2009**, *20*, 450–455. [CrossRef]
16. Ares, G.; Castura, J.C.; Antúnez, L.; Vidal, L.; Giménez, A.; Coste, B.; Picallo, A.; Beresford, M.K.; Cheang, S.L.; Jaeger, S.R. Comparison of two TCATA variants for dynamic sensory characterization of food products. *Food Qual. Prefer.* **2016**, *54*, 160–172. [CrossRef]
17. Baker, A.K.; Castura, J.C.; Ross, C.F. Temporal Check-All-That-Apply characterization of Syrah wine. *J. Food Sci.* **2016**, *81*, S1521–S1529. [CrossRef]
18. McMahon, K.M.; Culver, C.; Castura, J.C.; Ross, C.F. Perception of carbonation in sparkling wines using Descriptive Analysis (DA) and Temporal Check-All-That-Apply (TCATA). *Food Qual. Prefer.* **2017**, *59*, 14–26. [CrossRef]
19. Kemp, B.; Trussler, S.; Willwerth, J.; Inglis, D. Applying Temporal Check-All-That-Apply (TCATA) to mouthfeel and texture properties of red wines. *J. Sens. Stud.* **2019**, *34*, e12503. [CrossRef]
20. McMahon, K.M.; Culver, C.; Ross, C.F. The production and consumer perception of sparkling wines of different carbonation levels. *J. Wine Res.* **2017**, *28*, 123–134. [CrossRef]
21. Charters, S.; Pettigrew, S. The dimensions of wine quality. *Food Qual. Prefer.* **2007**, *18*, 997–1007. [CrossRef]
22. Francis, I.L.; Williamson, P.O. Application of consumer sensory science in wine research. *Aust. J. Grape Wine Res.* **2015**, *21*, 554–567. [CrossRef]
23. Sacchi, K.L.; Bisson, L.F.; Adams, D.O. A review of the effect of winemaking techniques on phenolic extraction in red wines. *Am. J. Enol. Vitic.* **2005**, *56*, 197–206. [CrossRef]

24. Şener, H. Effect of temperature and duration of maceration on colour and sensory properties of red wine: A review. *S. Afr. J. Enol. Vitic.* **2018**, *39*, 227–234. [CrossRef]
25. Casassa, L.F. Flavonoid Phenolics in Red Winemaking. In *Phenolic Compounds—Natural Sources, Importance and Applications*; InTech: London, UK, 2017; pp. 153–196.
26. Casassa, L.F.; Harbertson, J.F. Extraction, evolution, and sensory impact of phenolic compounds during red wine maceration. *Annu. Rev. Food Sci. Technol.* **2014**, *5*, 83–109. [CrossRef] [PubMed]
27. Reynolds, J.C.; Meusel, R.C.; Catania, A.A.; Casassa, L.F. Chemical and chromatic effects of fermentation temperature on three clones of Pinot noir over two consecutive vintages. *Am. J. Enol. Vitic.* **2022**, *73*, 75–92. [CrossRef]
28. Pérez-Navarro, J.; García Romero, E.; Gómez-Alonso, S.; Izquierdo Cañas, P.M. Comparison between the phenolic composition of Petit Verdot wines elaborated at different maceration/fermentation temperatures. *Int. J. Food Prop.* **2018**, *21*, 996–1007. [CrossRef]
29. Reynolds, A.; Cliff, M.; Girard, B.; Kopp, T. Influence of fermentation on properties of Semillon and Shiraz wines. *Am. J. Enol. Vitic.* **2001**, *52*, 235–240. [CrossRef]
30. Massera, A.; Assof, M.; Sari, S.; Ciklic, I.; Mercado, L.; Jofré, V.; Combina, M. Effect of low temperature fermentation on the yeast-derived volatile aroma composition and sensory profile in Merlot wines. *LWT* **2021**, *142*, 111069. [CrossRef]
31. Lerno, L.A.; Panprivech, S.; Ponangi, R.; Hearne, L.; Blair, T.; Oberholster, A.; Block, D.E. Effect of pump-over conditions on the extraction of phenolic compounds during Cabernet Sauvignon fermentation. *Am. J. Enol. Vitic.* **2018**, *69*, 295–301. [CrossRef]
32. Wimalasiri, P.M.; Zhan, J.; Tian, B. Characterisation of tannin and aroma profiles of Pinot noir wines made with or without grape pomace. *Fermentation* **2022**, *8*, 718. [CrossRef]
33. Frost, S.C.; Blackman, J.W.; Ebeler, S.E.; Heymann, H. Analysis of Temporal Dominance of Sensation data using correspondence analysis on Merlot wine with differing maceration and cap management regimes. *Food Qual. Prefer.* **2018**, *64*, 245–252. [CrossRef]
34. Frost, S.C.; Blackman, J.W.; Hjelmeland, A.K.; Ebeler, S.E.; Heymann, H. Extended maceration and cap management impacts on the phenolic, volatile, and sensory profiles of Merlot wine. *Am. J. Enol. Vitic.* **2018**, *69*, 360–370. [CrossRef]
35. Lesschaeve, I.; Noble, A.C. Polyphenols: Factors influencing their sensory properties and their effects on food and beverage preferences. *Am. J. Clin. Nutr.* **2005**, *81*, 330S–335S. [CrossRef] [PubMed]
36. Fischer, U.; Boulton, R.B.; Noble, A. Physiological factors contributing to the variability of sensory assessment: Relationship between salivary flow rate and temporal perception of gustatory stimuli. *Food Qual. Prefer.* **1994**, *5*, 55–64. [CrossRef]
37. Ishikawa, T.; Noble, A.C. Temporal perception of astringency and sweetness in red wine. *Food Qual. Prefer.* **1995**, *6*, 27–33. [CrossRef]
38. Baker, A.K.; Ross, C.F. Wine finish in red wine: The effect of ethanol and tannin concentration. *Food Qual. Prefer.* **2014**, *38*, 65–74. [CrossRef]
39. Criado, C.; Chaya, C.; Fernández-Ruíz, V.; Álvarez, M.D.; Herranz, B.; Pozo-Bayón, M.Á. Effect of saliva composition and flow on inter-individual differences in the temporal perception of retronasal aroma during wine tasting. *Food Res. Int.* **2019**, *126*, 108677. [CrossRef]
40. Meilgaard, M.C.; Civille, G.V.; Carr, T.B. *Sensory Evaluation Techniques*, 5th ed.; CRC Press: Boca Raton, FL, USA, 2016.
41. Stoffel, E.S.; Lesniauskas, R.O.; Anderson, S.R.; Krystoff, C.T.; Casassa, L.F. Temporal evaluation of retronasal and mouthfeel sensations in cofermented and blended red wines from California. *Am. J. Enol. Vitic.* **2023**, *74*, 1–15. [CrossRef]
42. Gao, L.; Girard, B.; Mazza, G.; Reynolds, A.G. Changes in anthocyanins and color characteristics of Pinot noir wines during different vinification processes. *J. Agric. Food Chem.* **1997**, *45*, 2003–2008. [CrossRef]
43. Morel-Salmi, C.; Souquet, J.M.; Bes, M.; Cheynier, V. Effect of flash release treatment on phenolic extraction and wine composition. *J. Agric. Food Chem.* **2006**, *54*, 4270–4276. [CrossRef] [PubMed]
44. Rinaldi, A.; Picariello, L.; Soares, S.; Brandão, E.; de Freitas, V.; Moio, L.; Gambuti, A. Effect of oxidation on color parameters, tannins, and sensory characteristics of Sangiovese wines. *Eur. Food Res. Technol.* **2021**, *247*, 2977–2991. [CrossRef]
45. Franco-Luesma, E.; Ferreira, V. Reductive off-odors in wines: Formation and release of H2S and methanethiol during the accelerated anoxic storage of wines. *Food Chem.* **2016**, *199*, 42–50. [CrossRef] [PubMed]
46. Smith, M.E.; Bekker, M.Z.; Smith, P.A.; Wilkes, E.N. Sources of volatile sulfur compounds in wine. *Aust. J. Grape Wine Res.* **2015**, *21*, 705–712. [CrossRef]
47. Bekker, M.Z.; Day, M.P.; Holt, H.; Wilkes, E.; Smith, P.A. Effect of oxygen exposure during fermentation on volatile sulfur compounds in Shiraz wine and a comparison of strategies for remediation of reductive character. *Aust. J. Grape Wine Res.* **2016**, *22*, 24–35. [CrossRef]
48. Ugliano, M.; Kolouchova, R.; Henschke, P.A. Occurrence of hydrogen sulfide in wine and in fermentation: Influence of yeast strain and supplementation of yeast available nitrogen. *J. Ind. Microbiol. Biotechnol.* **2011**, *38*, 423–429. [CrossRef] [PubMed]
49. Bohlscheid, J.C.; Osborne, J.P.; Ross, C.F.; Edwards, C.G. Interactive effects of selected nutrients and fermentation temperature on H2S production by wine strains of *Saccharomyces*. *J. Food Qual.* **2011**, *34*, 51–55. [CrossRef]
50. Fang, Y.; Qian, M.C. Sensitive quantification of sulfur compounds in wine by headspace solid-phase microextraction technique. *J. Chromatogr. A* **2005**, *1080*, 177–185. [CrossRef]
51. Siebert, T.E.; Solomon, M.R.; Pollnitz, A.P.; Jeffery, D.W. Selective determination of volatile sulfur compounds in wine by gas chromatography with sulfur chemiluminescence detection. *J. Agric. Food Chem.* **2010**, *58*, 9454–9462. [CrossRef]
52. Pineau, B.; Barbe, J.C.; Van Leeuwen, C.; Dubourdieu, D. Examples of perceptive interactions involved in specific "red-" and "black-berry" aromas in red wines. *J. Agric. Food Chem.* **2009**, *57*, 3702–3708. [CrossRef]

53. Lorrain, B.; Tempere, S.; Iturmendi, N.; Moine, V.; De Revel, G.; Teissedre, P.L. Influence of phenolic compounds on the sensorial perception and volatility of red wine esters in model solution: An insight at the molecular level. *Food Chem.* **2013**, *140*, 76–82. [CrossRef] [PubMed]
54. Cometto-Muñiz, J.E.; Cain, W.S.; Abraham, M.H.; Gil-Lostes, J. Concentration-detection functions for the odor of homologous n-acetate esters. *Physiol. Behav.* **2008**, *95*, 658–667. [CrossRef] [PubMed]
55. Cameleyre, M.; Lytra, G.; Tempere, S.; Barbe, J.C. Olfactory Impact of higher alcohols on red wine fruity ester aroma expression in model solution. *J. Agric. Food Chem.* **2015**, *63*, 9777–9788. [CrossRef] [PubMed]
56. Gómez-Míguez, M.J.; Cacho, J.F.; Ferreira, V.; Vicario, I.M.; Heredia, F.J. Volatile components of Zalema white wines. *Food Chem.* **2007**, *100*, 1464–1473. [CrossRef]
57. Rocha, S.M.; Rodrigues, F.; Coutinho, P.; Delgadillo, I.; Coimbra, M.A. Volatile composition of Baga red wine: Assessment of the identification of the would-be impact odourants. *Anal. Chim. Acta* **2004**, *513*, 257–262. [CrossRef]
58. Arias-Pérez, I.; Sáenz-Navajas, M.P.; de-la-Fuente-Blanco, A.; Ferreira, V.; Escudero, A. Insights on the role of acetaldehyde and other aldehydes in the odour and tactile nasal perception of red wine. *Food Chem.* **2021**, *361*, 130081. [CrossRef]
59. Zalacain, A.; Marín, J.; Alonso, G.L.; Salinas, M.R. Analysis of wine primary aroma compounds by stir bar sorptive extraction. *Talanta* **2007**, *71*, 1610–1615. [CrossRef]
60. Bowen, A.J.; Reynolds, A.G. Odor potency of aroma compounds in Riesling and Vidal Blanc table wines and icewines by gas chromatography-olfactometry-mass spectrometry. *J. Agric. Food Chem.* **2012**, *60*, 2874–2883. [CrossRef]
61. Ferreira, V.; De la Fuente, A.; Sáenz-Navajas, M.P. Wine Aroma Vectors and Sensory Attributes. In *Managing Wine Quality: Volume One: Viticulture and Wine Quality*; Elsevier: Amsterdam, The Netherlands, 2021; pp. 3–39, ISBN 9780081020678.
62. Sabon, I.; De Revel, G.; Kotseridis, Y.; Bertrand, A. Determination of volatile compounds in Grenache wines in relation with different terroirs in the Rhone Valley. *J. Agric. Food Chem.* **2002**, *50*, 6341–6345. [CrossRef]
63. Delwiche, J. The impact of perceptual interactions on perceived flavor. *Food Qual. Prefer.* **2004**, *15*, 137–146. [CrossRef]
64. Welke, J.E.; Zanus, M.; Lazzarotto, M.; Alcaraz Zini, C. Quantitative analysis of headspace volatile compounds using comprehensive two-dimensional gas chromatography and their contribution to the aroma of Chardonnay wine. *Food Res. Int.* **2014**, *59*, 85–99. [CrossRef]
65. Bosso, A.; Panero, L.; Petrozziello, M.; Follis, R.; Motta, S.; Guaita, M. Influence of submerged-cap vinification on polyphenolic composition and volatile compounds of Barbera wines. *Am. J. Enol. Vitic.* **2011**, *62*, 503–511. [CrossRef]
66. Franco-Luesma, E.; Sáenz-Navajas, M.P.; Valentin, D.; Ballester, J.; Rodrigues, H.; Ferreira, V. Study of the effect of H2S, MeSH and DMS on the sensory profile of wine model solutions by Rate-All-That-Apply (RATA). *Food Res. Int.* **2016**, *87*, 152–160. [CrossRef]
67. Medel-Maraboli, M.; López-Solís, R.; Valenzuela-Prieto, D.; Vargas-Silva, S.; Obreque-Slier, E. Limited relationship between temporality of sensory perception and phenolic composition of red wines. *LWT* **2021**, *142*, 111028. [CrossRef]
68. Casassa, L.F.; Bolcato, E.A.; Sari, S.E. Chemical, chromatic, and sensory attributes of 6 red wines produced with prefermentative cold soak. *Food Chem.* **2015**, *174*, 110–118. [CrossRef]
69. Ugliano, M.; Travis, B.; Francis, I.L.; Henschke, P.A. Volatile composition and sensory properties of shiraz wines as affected by nitrogen supplementation and yeast species: Rationalizing nitrogen modulation of wine aroma. *J. Agric. Food Chem.* **2010**, *58*, 12417–12425. [CrossRef]
70. Franco-Luesma, E.; Ferreira, V. Quantitative analysis of free and bonded forms of volatile sulfur compounds in wine. basic methodologies and evidences showing the existence of reversible cation-complexed forms. *J. Chromatogr. A* **2014**, *1359*, 8–15. [CrossRef] [PubMed]
71. Ugliano, M.; Fedrizzi, B.; Siebert, T.; Travis, B.; Magno, F.; Versini, G.; Henschke, P.A. Effect of nitrogen supplementation and saccharomyces species on hydrogen sulfide and other volatile sulfur compounds in Shiraz fermentation and wine. *J. Agric. Food Chem.* **2009**, *57*, 4948–4955. [CrossRef] [PubMed]
72. Harbertson, J.F.; Picciotto, E.A.; Adams, D.O. Measurement of polymeric pigments in grape berry extracts and wines using a protein precipitation assay combined with bisulfite bleaching. *Am. J. Enol. Vitic.* **2003**, *54*, 301–306. [CrossRef]
73. Harbertson, J.F.; Kennedy, J.A.; Adams, D.O. Tannin in skins and seeds of Cabernet Sauvignon, Syrah, and Pinot noir berries during ripening. *Am. J. Enol. Vitic.* **2002**, *53*, 54–59. [CrossRef]
74. Hjelmeland, A.K.; King, E.S.; Ebeler, S.E.; Heymann, H. Characterizing the chemical and sensory profiles of United States Cabernet Sauvignon wines and blends. *Am. J. Enol. Vitic.* **2013**, *64*, 169–179. [CrossRef]
75. Castro, L.F.; Ross, C.F. Determination of flavour compounds in beer using stir-bar sorptive extraction and solid-phase microextraction. *J. Inst. Brew.* **2015**, *121*, 197–203. [CrossRef]
76. Alves, R.F.; Nascimento, A.M.D.; Nogueira, J.M.F. Characterization of the aroma profile of Madeira wine by sorptive extraction techniques. *Anal. Chim. Acta* **2005**, *546*, 11–21. [CrossRef]
77. Tepper, B.J.; Christensen, C.M.; Cao, J. Development of brief methods to classify individuals by PROP taster status. *Physiol. Behav.* **2001**, *73*, 571–577. [CrossRef]
78. Pineau, N.; de Bouillé, A.G.; Lepage, M.; Lenfant, F.; Schlich, P.; Martin, N.; Rytz, A. Temporal Dominance of Sensations: What is a good attribute list? *Food Qual. Prefer.* **2012**, *26*, 159–165. [CrossRef]
79. Noble, A.C.; Ebeler, S.E. Use of multivariate statistics in understanding wine flavor. *Food Rev. Int.* **2002**, *18*, 1–21. [CrossRef]
80. Husson, F.; Lê, S.; Pagès, J. Confidence ellipse for the sensory profiles obtained by principal component analysis. *Food Qual. Prefer.* **2005**, *16*, 245–250. [CrossRef]

81. Castura, J.C. tempR: Temporal Sensory Data Analysis 2022. Available online: https://cran.r-project.org/web/packages/tempR/tempR.pdf (accessed on 15 April 2023).
82. Castura, J.C.; Baker, A.K.; Ross, C.F. Using contrails and animated sequences to visualize uncertainty in dynamic sensory profiles obtained from Temporal Check-All-That-Apply (TCATA) data. *Food Qual. Prefer.* **2016**, *54*, 90–100. [CrossRef]
83. Carrau, F.M.; Medina, K.; Boido, E.; Farina, L.; Gaggero, C.; Dellacassa, E.; Versini, G.; Henschke, P.A. De novo synthesis of monoterpenes by *Saccharomyces cerevisiae* wine yeasts. *FEMS Microbiol. Lett.* **2005**, *243*, 107–115. [CrossRef] [PubMed]
84. Yilmaztekin, M.; Kocabey, N.; Hayaloglu, A.A. Effect of maceration time on free and bound volatiles of red wines from Cv. Karaoğlan (*Vitis vinifera* l.) grapes grown in Arapgir, Turkey. *J. Food Sci.* **2015**, *80*, C556–C563. [CrossRef]
85. Dennis, E.G.; Keyzers, R.A.; Kalua, C.M.; Maffei, S.M.; Nicholson, E.L.; Boss, P.K. Grape contribution to wine aroma: Production of hexyl acetate, octyl acetate, and benzyl acetate during yeast fermentation is dependent upon precursors in the must. *J. Agric. Food Chem.* **2012**, *60*, 2638–2646. [CrossRef]
86. Pittari, E.; Moio, L.; Piombino, P. Interactions between polyphenols and volatile compounds in wine: A literature review on physicochemical and sensory insights. *Appl. Sci.* **2021**, *11*, 1157. [CrossRef]
87. Girard, B.; Yuksel, D.; Cliff, M.A.; Delaquis, P.; Reynolds, A.G. Vinification effects on the sensory, colour and gc profiles of Pinot noir wines from British Columbia. *Food Res. Int.* **2001**, *34*, 483–499. [CrossRef]
88. Rodríguez-Bencomo, J.J.; Muñoz-González, C.; Andújar-Ortiz, I.; Martín-Álvarez, P.J.; Moreno-Arribas, M.V.; Pozo-Bayón, M.Á. Assessment of the effect of the non-volatile wine matrix on the volatility of typical wine aroma compounds by headspace solid phase microextraction/gas chromatography analysis. *J. Sci. Food Agric.* **2011**, *91*, 2484–2494. [CrossRef] [PubMed]
89. Oliva, J.; Martínez-Gil, A.M.; Lorenzo, C.; Cámara, M.A.; Salinas, M.R.; Barba, A.; Garde-Cerdán, T. Influence of the use of fungicides on the volatile composition of Monastrell red wines obtained from inoculated fermentation. *Food Chem.* **2015**, *170*, 401–406. [CrossRef] [PubMed]
90. Tomasino, E.; Harrison, R.; Breitmeyer, J.; Sedcole, R.; Sherlock, R.; Frost, A. Aroma composition of 2-year-old New Zealand Pinot noir wine and its relationship to sensory characteristics using canonical correlation analysis and addition/omission tests. *Aust. J. Grape Wine Res.* **2015**, *21*, 376–388. [CrossRef]
91. Furdíková, K.; Machyňáková, A.; Drtilová, T.; Špánik, I. Comparison of different categories of Slovak Tokaj wines in terms of profiles of volatile organic compounds. *Molecules* **2020**, *25*, 669. [CrossRef]
92. Ferreira, V.; López, R.; Cacho, J.F. Quantitative determination of the odorants of young red wines from different grape varieties. *J. Sci. Food Agric.* **2000**, *80*, 1659–1667. [CrossRef]
93. Uthurry, C.A.; Varela, F.; Colomo, B.; Suárez Lepe, J.A.; Lombardero, J.; García Del Hierro, J.R. Ethyl carbamate concentrations of typical Spanish red wines. *Food Chem.* **2004**, *88*, 329–336. [CrossRef]

Disclaimer/Publisher's Note: The statements, opinions and data contained in all publications are solely those of the individual author(s) and contributor(s) and not of MDPI and/or the editor(s). MDPI and/or the editor(s) disclaim responsibility for any injury to people or property resulting from any ideas, methods, instructions or products referred to in the content.

foods

MDPI

Article

Sensory-Guided Isolation, Identification, and Active Site Calculation of Novel Umami Peptides from Ethanol Precipitation Fractions of Fermented Grain Wine (Huangjiu)

Rui Chang [1], Zhilei Zhou [1,2,3], Yong Dong [1], Yuezheng Xu [4], Zhongwei Ji [1,2,3], Shuangping Liu [1,2,3] and Jian Mao [1,2,4,*]

1 National Engineering Research Center of Cereal Fermentation and Food Biomanufacturing, School of Food Science and Technology, Jiangnan University, Wuxi 214122, China; 7200112052@stu.jiangnan.edu.cn (R.C.)
2 Jiangnan University (Shaoxing) Industrial Technology Research Institute, Shaoxing 312000, China
3 Jiangsu Provincial Engineering Research Center for Bioactive Product Processing, Jiangnan University, Wuxi 214122, China
4 National Engineering Research Center for Huangjiu, Shaoxing 312000, China
* Correspondence: maojian@jiangnan.edu.cn; Tel.: +86-510-85329062

Abstract: Huangjiu is rich in low-molecular-weight peptides and has an umami taste. In order for its umami peptides to be discovered, huangjiu was subjected to ultrafiltration, ethanol precipitation, and macroporous resin purification processes. The target fractions were gathered according to sensory evaluation. Subsequently, we used peptidomics to identify the sum of 4158 peptides in most umami fractions. Finally, six novel umami peptides (DTYNPR, TYNPR, SYNPR, RFRQGD, NFHHGD, and FHHGD) and five umami-enhancing peptides (TYNPR, SYNPR, NFHHGD, FHHGD, and TVDGPSH) were filtered via virtual screening, molecular docking, and sensory verification. Moreover, the structure–activity relationship was discussed using computational approaches. Docking analysis showed that all umami peptides tend to bind with T1R1 through hydrogen bonds and hydrophobic forces, which involve key residues HIS71, ASP147, ARG151, TYR220, SER276, and ALA302. The active site calculation revealed that the positions of the key umami residues D and R in the terminal may cause taste differences in identified peptides.

Keywords: huangjiu; umami peptide; molecular docking; peptidomics; density functional theory

1. Introduction

Umami is the fifth basic taste. Many studies have shown that not only can tasting molecules such as amino acids and nucleotides cause an umami taste, but low-molecular-weight polypeptides can also provide an umami taste [1]. The umami threshold values of most identified umami peptides are even lower than those of umami aspartate and glutamic acid [2], indicating the umami potential of peptides. In 1978, Yamasaki and Maekawa first discovered the umami-tasting octapeptide KGDEESLA from beef papain hydrolyzate [3]. Decades later, about 301 umami peptides have been identified [4], ranging in length from 2 to 15 amino acids, and many are derived from fermented foods. As a kind of typical fermented food, some types of grain-fermented wine have an obvious umami taste and are abundant with peptides [5,6]. Umami peptides are important nonvolatile flavor components in fermented wines, and they can affect overall sensory perception through their intrinsic taste profile and flavor interaction with the matrix's components. In addition, several earlier studies revealed the umami potential of peptides in fermented wine [7–9]. However, due to the complex matrix and extraction process, there is still little known about umami peptides in fermented grain wine.

In order to collect the taste peptide fractions in food, preparative or semipreparative high-performance liquid chromatography separation has frequently been used, but it is very time-consuming [10]. Guided by sensory evaluation, the separation process can be effective even with a macroporous resin medium [11,12]. For peptide identification, the traditional instrument gives limited results due to accuracy issues. Recently, widely used peptidomics technology based on nano-liquid chromatography with tandem mass spectrometry (Nano-HPLC-MS/MS) and advanced data collection modes have been able to achieve high-throughput peptide sequencing results in crude fractions. Dozens of novel umami peptides have been discovered via peptidomics in chicken [13], fish [14], and bacon [15], and even the bioactive peptides in fermented rice wines have been discovered [16]. Moreover, peptidomics coupled with a series of in silico methods (machine learning, virtual screening, and molecular docking) has been an effective strategy for umami peptide prediction [17]. As of now, this strategy has successfully screened 16 umami peptides from 508 peptides in clams [18], and narrowed 208 chicken soup peptides down to 20 target peptides [13]. Therefore, peptidomics and virtual screening can contribute to the exploration of novel umami peptides in fermented wines.

Huangjiu is a traditional Chinese fermented rice wine that has been consumed for over 2000 years and has a distinct umami taste. During the fermentation, the raw protein is degraded by microbial metabolism and produces a wealth of peptides [6]. The peptide content (1770 to 4210 mg/L) of huangjiu is significantly higher than that of Japanese rice wine, also known as sake (156 to 242 mg/L) [5]. To date, 19 umami-enhancing glutamine peptides have been identified in sake [19,20], but only one potential umami peptide, EEE, has been reported in huangjiu [21]. Hence, the umami peptides in huangjiu warrant further exploration.

This study aims to enrich, identify, and characterize the huangjiu umami peptides through a combined isolation and screening strategy. The target fractions were separated via ultrafiltration, ethanol precipitation, and macroporous resin, and then evaluated by sensory panelists. After peptidomics sequencing and virtual screening, the candidate peptides were further verified through a sensory evaluation test. Finally, the taste mechanism of the identified umami peptides was explored through molecular docking and theoretical active site calculations. These results will provide a reference for taste peptide identification in other fermented wines.

2. Materials and Methods

2.1. Materials and Chemicals

The 20 L mechanical craft semidry (total sugar ranged from 15 to 40 g/L) huangjiu used in this study was sourced from a local winery in Shaoxing, Zhejiang province, China. All analytically pure reagents used, including absolute ethanol, sodium hydroxide, hydrochloric acid, phenol, concentrated sulfuric acid, anhydrous glucose, Folin phenol, and gallic acid, were procured from Sinopharm Reagent Co., Ltd., Shanghai, China. For the sensory test, food-grade sodium chloride, citric acid, sucrose, quinine, glutathione, and monosodium glutamate (MSG) were purchased from Gino Biotech Co., Ltd., Quzhou, China. The purification equipment, specifically the membrane filtration module, was acquired from Shaoxing Haina Co., Ltd., Shaoxing, China.

2.2. Separation of Huangjiu Umami Peptide Fractions

The target umami peptide fractions were obtained through sensory evaluation of the isolate separated via ultrafiltration, ethanol precipitation, and macroporous resin processes (Figure 1). First, the raw huangjiu was subjected to centrifuge pretreatment (8000 rpm, 20 min, 4 °C) to remove insoluble impurities, and then it was passed through an ultrafiltration membrane with a molecular weight cut-off of 3000 Da (Suez, Wuxi, China). The permeate fraction (HJUF) with peptide molecular weight lower than 3000 Da was concentrated using rotary evaporation (IKA RV8, Staufen, Germany) at 50 °C and a vacuum of <0.098 MPa, followed by freeze-drying. The HJUF, with a peptide concentration of 40 mg/mL, was sequentially subjected to 60%, 75%, and 90% ethanol gradient precipitation (4 °C, 12 h). The precipitates were collected after centrifugation and named 60% ETP,

75% ETP, and 90% ETP, respectively. The final supernatant was termed 90% EST. The concentrations of peptides, total sugars, and total phenols in all fractions were quantified using the Pierce™ BCA Peptide Quantification Kit (Thermo Fisher Scientific, San Jose, CA, USA), phenol sulfate colorimetry, and Folin phenol colorimetry, respectively. The peptide content (g/100 g) was calculated as its purity in the sum of the three above components.

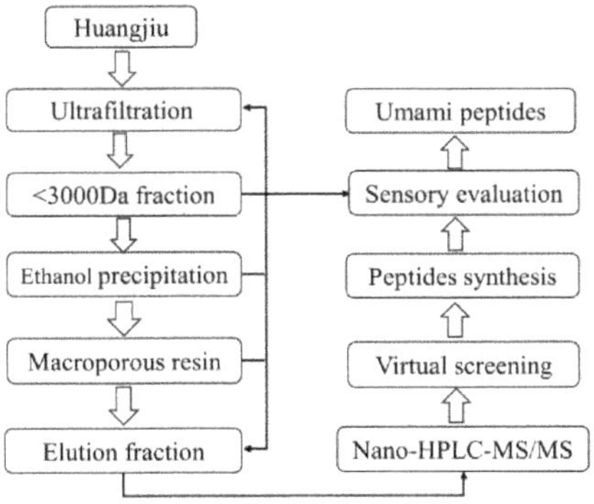

Figure 1. The flowchart of the huangjiu umami peptide separation and identification.

The preliminary fractions 60% ETP, 75% ETP, and 90% ETP were dissolved in pure water at a peptide concentration of 20 mg/mL, then loaded (15 mL) onto a chromatography column (Φ2.6 × 18 cm) equipped with nonpolar XAD16 macroporous resin (Shanghai Yuanye Biology, Shanghai, China). After reaching adsorption equilibrium, the column was sequentially eluted with 0%, 20%, 40%, and 60% ethanol at a flow rate of 3 mL/min. The elution process was monitored at 220 nm ultraviolet absorption value. The eluted samples corresponding to the same absorption peaks were collected together, concentrated, and then freeze-dried for subsequent sensory analysis.

2.3. Identification of Peptides via Nano-HPLC-MS/MS

The peptide fraction with the highest sensory umami intensity was identified using Nano-HPLC-MS/MS. The samples were desalted using Pierce C18 Spin Tips and then redissolved in solvent A (0.1% formic acid in water), followed by analysis using Q-Exactive Plus coupled to an EASY-nanoLC 1200 system (Thermo Fisher Scientific, San Jose, CA, USA). The sample (6 μL) was loaded onto an analytical column of 25 cm with 75 μm inner diameter and 1.9 μm resin (Dr Maisch) at a flow rate of 300 nL/min, then separated at 2% buffer B (80% ACN with 0.1% FA) with a 60 min gradient at the start, increasing stepwise to 35% within 47 min and then to 100% in 1 min, which was then maintained for 12 min. The column temperature was 40 °C and the electrospray voltage was set to 2 kV. The mass spectrometer was run under the data-dependent acquisition (DDA) mode. The full-scan MS spectra (m/z 200–1800) survey was conducted with a resolution of 7×10^4. The automatic gain control (AGC) target was 3×10^6, and the maximum injection time was 50 ms. Next, the precursor ions were selected in the collision cell for fragmentation via higher-energy collision dissociation (HCD). The MS/MS resolution was set to 1.75×10^4, and the AGC target was 1×10^5, the maximum injection time was 45 ms, and the dynamic exclusion was 30 s.

Data were processed using PEAKS Studio version 10.6 (Bioinformatics Solutions Inc., Waterloo, ON, Canada) and then matched with the proteomes of Oryza sativa subsp. Japonica, Triticum aestivum, and Saccharomyces cerevisiae (strain ATCC 204508/S288c) in the UniProt database. No digestion enzyme was selected. The peptides with a −logP score of ≥20 were filtered.

2.4. Virtual Screening and Molecular Docking

Considering the length of most reported umami peptides, the identified peptides with lengths no longer than ten were kept for further analysis. Machine learning models embedded in online discrimination tools were used to predict the umami potential of the peptide sequences. Potential umami peptides were retained through BIOPEP-UWM [22] (https://biochemia.uwm.edu.pl/biopep-uwm/ accessed on 15 June 2023), UMPred-FRL [23] (http://pmlabstack.pythonanywhere.com/UMPred-FRL accessed on 15 June 2023), and Umami_YYDS [4] (http://www.tastepeptides-meta.com/cal, accessed on 15 June 2023). The filtered peptides were kept for screening.

Molecular docking of filtered umami peptides was conducted with the T1R1/T1R3 homology model, which was built at SWISS-MODEL (https://swissmodel.expasy.org/, accessed on 16 June 2023). The sweet receptor (PDB code: 5X2M) was chosen as a template, and the sequences of human umami receptors Q7RTX1 (TSIR1) and Q7RTX0 (TSIR3) were retrieved from the UniProt database [24]. The model stereochemistry was assessed with the MolProbity server (http://molprobity.biochem.duke.edu/, accessed on 16 June 2023). The peptide structures were generated with Chemoffice software after MMFF94 force field energy optimization. Docking was performed with Auto Dock Vina1.1.2. The docking active sites of the receptor were determined through alignment with the original ligand of the template protein, T1R3 site x, y, z = 51.31, 40.37, 12.71 and T1R1 site x, y, z = 39.05, 25.33, 40.88, with the box region being 25 × 25 × 25 Å. The ratio of hydrophobic, acidic, basic, bitter, and sweet amino acids in all candidate synthetic peptide sequences was counted, and the taste activity profile was also calculated using BIOPEP-UWM sensory analysis.

2.5. Solid-Phase Synthesis of Umami Peptides

The ten candidate umami peptides screened from docking score were synthesized with the solid-phase synthesis method from Shanghai Gill Biochemical Co., Ltd. (Shanghai, China), with purity not lower than 95%.

2.6. Sensory Evaluation

2.6.1. Descriptive Sensory Analysis

Referring to the quantitative descriptive analysis method reported by Zhuang et al. [11] with slight modifications, the sensory evaluation was conducted inside a room at a temperature of 24 °C, in accordance with ISO-8586-1, 2012. Ten sensory evaluation panelists (five men and five women, aged from 25 to 30) who have no history of known taste impairment or disorders were recruited from Jiangnan University. The panelists could accurately identify and describe the characteristic differences in taste standard solutions, including sour, sweet, bitter, salty, umami, and kokumi, after undergoing sensory training. All panelists gave their informed consent before every sensory session, which was approved by the Jiangnan University Medical Ethics Committee (JNU20230601IRB21). The panelists were asked to describe the samples' taste and assign intensity scores ranging from 0 to 9 points, with 0 for pure water and 5 representing standard solutions. The standard taste solutions for sour, sweet, bitter, salty, umami, and kokumi were 0.08% citric acid, 1% sucrose, 0.08% quinine, 0.35% sodium chloride, 0.35% monosodium glutamate (MSG), and 0.15% glutathione, respectively. All of the peptide fractions obtained via ultrafiltration, ethanol precipitation, and resin separation were dissolved in water at a concentration of 10 mg/mL. The candidate synthetic peptide was dissolved in water at a concentration of 4 mg/mL. The panel members sipped each sample for 10 s prior to spitting it out, washed their mouths with pure water, and rested for 3 min to eliminate fatigue. The final sensory intensity score was the average of all panelists' scores.

2.6.2. Taste Thresholds of Synthetic Peptides

According to the methods of Liu et al. [25] with slight modifications, the synthetic peptides were diluted with water at a 1:1 (v/v) ratio, and judgment started as the concentration increased. The possible concentration range of threshold was preliminarily determined based on the descriptive sensory trial. Namely, 0.125 mg/mL was used as the initial evaluation concentration, followed by 0.25, 0.5, 1, and 2 mg/mL. Triangle tests were performed on one diluted peptide sample group and two water blank control groups. The individual thresholds of all the panelists were averaged to determine the overall threshold.

2.6.3. Umami Enhancement Effects of Synthetic Peptides

The peptides were dissolved in 0.35% MSG solutions to reach 1 mg/mL and then given to the panelists for umami intensity evaluation, separately, for which the intensity score range was 0 to 9. The blank solution 0.35% MSG was scored as 5. For the thresholds of umami enhancement, a comparative taste dilution analysis was adopted [13].

2.7. Theoretical Calculation of Active Sites

All ten candidate peptides' active sites were calculated. Geometry optimization of peptide structure was performed at the B3LYP level, considering the D3 dispersion correction for weak interaction. Given that the construction of the Pople basis set is not ideal, the higher-precision def series 3-zeta basis set TVZP was used. In addition, the environment was applied with gas-phase and implicit solvation (water with SMD model). All calculations were performed using the ORCA package (version 5.03), and the hybrid functional RI acceleration technology was fully utilized to reduce calculation time. Molecular orbital analysis (HOMO and LUMO) was conducted with the Multiwfn program (version 3.7) [26].

2.8. Data Analysis

All assays were conducted in triplicate. Data analysis of variance (ANOVA) and Duncan's test were performed on pairs using SPSS (version 23.0, SPSS Inc., Chicago, IL, USA). Results were deemed statistically significant at $p < 0.05$. Data visualization was carried out using Origin software (Pro 2016, Origin Lab Corp., Northampton, MA, USA). The interactions between the peptides and the receptor during docking were visualized using Ligplot+, and the interaction bonds' heatmap was drawn using Hiplot (https://hiplot.com.cn/, accessed on 1 August 2023).

3. Results and Discussion

3.1. Sensory-Guided Separation of Huangjiu Peptides

3.1.1. Ultrafiltration and Ethanol Precipitation Fractions

Compared with the raw amount (7.68 ± 0.36 g/100 g), the peptide content of the permeate fraction (HJUF) was increased to 12.88 ± 0.19 g/100 g after ultrafiltration. However, the umami intensity of HJUF did not improve very much, indicating that the umami fraction needed further enrichment (Figure 2B). The peptide content and recovery rate of ethanol precipitation (60% ETP, 75% ETP, 90% ETP) and the supernatant fraction (90% EST) are shown in Figure 2A. The results showed that a total of 68.81% of the peptides were recovered through ethanol precipitation, suggesting that most huangjiu peptides have a certain degree of hydrophobicity. Hydrophobicity is an important characteristic of umami peptides [27]. In addition, the precipitation process increased the peptide content of the 75% ETP fractions (20.36 ± 0.26 g/100 g), as well as the 60% ETP fractions (14.22 ± 0.56 g/100 g). Although the peptide purity of all precipitated fractions was not very high, the sensory panelists could still distinguish the taste difference. The sensory results showed that 60% ETP was tasteless, and 75% ETP had the highest umami intensity (score: 4.6), followed by 90% ETP (score: 3.38). Meanwhile, the 90% EST fraction had a strong bitter taste and high acidity.

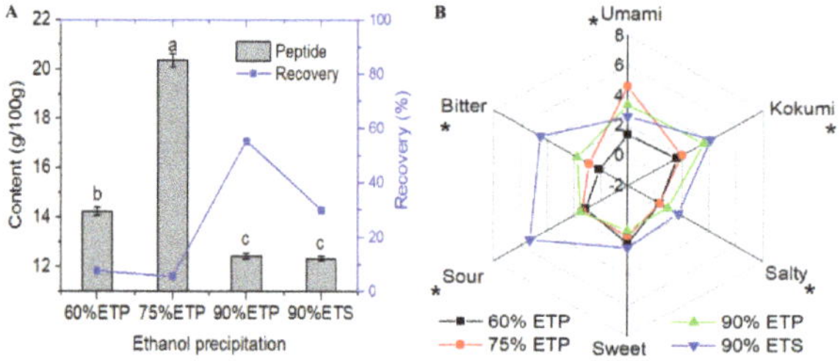

Figure 2. Sensory characteristics of precipitated huangjiu umami peptide fractions ((**A**): recovery rate and peptide content, (**B**): sensory intensity of peptide fractions). Note that letters in the same column and asterisks in sensory radar chart represent significant differences ($p < 0.05$).

3.1.2. Macroporous Resin Separation Fractions

Resin was used for further enrichment. Macroporous resin, with its high loading capacity, safe eluent, and simple purification steps, has been applied for taste peptide fraction enrichment in soy sauce [28,29] and bean paste [30]. The nonpolar resin XAD-16 has a large specific surface area (800 m^2/g) and a small pore size (0.7 nm), which have led to it being commonly used in complex food material matrices. In particular, it performed well in terms of grain protein hydrolyzate peptide enrichment [31], and was thus applied to fermented grain wine, or huangjiu. In this study, the highest umami fractions, 75% ETP and 90% ETP, were kept separated via macroporous resin separation, and different hydrophobicity samples were obtained through ethanol gradient elution (Figure 3C,D).

With elution, the peptide content of the collected samples gradually increased (Figure 3A). Most peptides in 75% ETP and 90% ETP were eluted with water (0% ethanol), with recovery rates of 54.71 $\pm$ 0.93% and 37.15 $\pm$ 0.52%, respectively (Figure 3B). Notably, the water-eluted fractions had higher umami intensity than the 20~60% ethanol-eluted samples (Figure 3E,F), even though the peptide content was much lower (Figure 3A). Meanwhile, the 20–60% ethanol-eluted fractions had a taste dominated by bitterness. Similarly, in the soy sauce umami peptide macroporous resin separation study, the author also observed that the water-eluted resin fractions had a higher umami intensity than other fractions [11]. Comparing the water-eluted umami fractions of 75% ETP and 90% ETP, the umami intensity of the former (3.67) was almost twice that of the latter. Furthermore, the water-eluted fractions of 75% ETP had a strong sour taste, which may be due to the presence of more acidic umami amino acid residues in the peptides. For instance, the umami taste of Korean soybean paste [32] and stinky mandarin fish [14] peptides was correlated to a high proportion of acidic amino acids (glutamic acid and aspartic acid). Unlike soy sauce and soybean paste, stinky mandarin fish is a fermented meat product, but its umami peptide composition also relies on hydrophilic amino acids. This suggests that for umami peptides in fermented foods, moderate hydrophilicity is a necessary key factor. Therefore, the huangjiu umami peptides may be concentrated in the 75% water-eluted resin fractions.

3.2. Virtual Screening of Umami Peptides

Nano-HPLC-MS/MS identified 4158 peptides from the 75% ETP water-eluted fractions. There are 22 peptides with the greatest potential umami taste according to the machine learning prediction tools (BIOPEP-UWM, Umami_PREL, Umami_YYDS). The MolProbity Ramachandran plot (Figure S1) showed that the stereochemistry of the built receptor model residues was reasonable, with 93.7% in the favored area and 98.7% in the allowed area, similar to the results of Cheng et al. [33]. From the docking results (Table S1), ten peptides with intensities larger than 5×10^6 and binding energies lower than -8.9 kcal/mol were selected for solid-phase synthesis and sensory evaluation. The protein accession matching analysis showed that DTYNPR, TYNPR, SYNPR, FRDEHQ, FRDEH, and RFRQGD were derived from the rice glutenin subunit. Among them, DTYNPR was reported as a lipolysis inhibitory peptide [34]. TYNPK came from the wheat helicase

MAGATAMA 3. NFHHGD, FHHGD, and TVDGPSH were mainly released from the yeasts Enolase 2 and Glyceraldehyde-3-phosphate dehydrogenase 1. These findings demonstrated that enzymatic hydrolysis of raw proteins (rice, wheat, yeast, etc.) and yeast autolysis are important sources of huangjiu umami peptides [5].

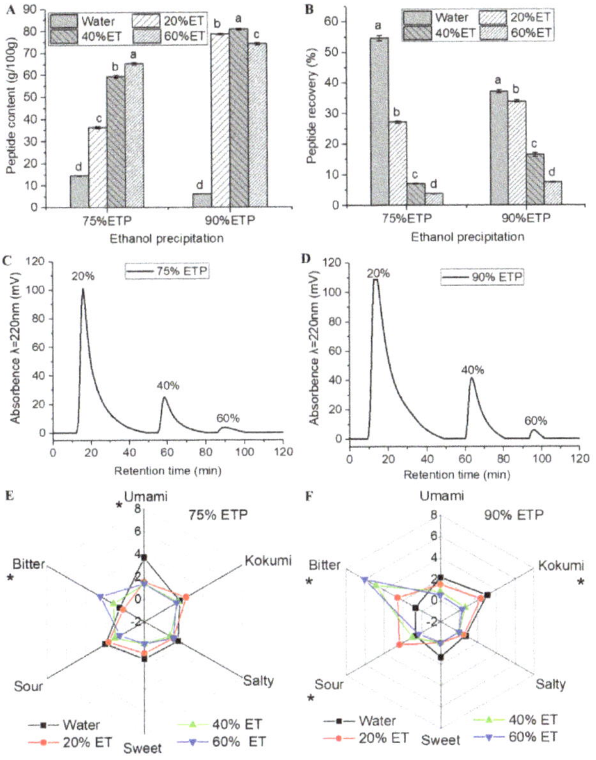

Figure 3. Sensory characteristics of resin−purified huangjiu umami peptide fractions ((**A**): peptide content, (**B**): recovery rate, (**C**,**D**): elution response curve, (**E**,**F**): sensory intensity of each elution fraction). The asterisks in sensory radar chart represent significant differences ($p < 0.05$). Different letters above bars indicate significant differences at the 0.05 level.

3.3. Synthetic Peptide Umami Characteristics and Composition Analysis

Sensory evaluation confirmed that DTYNPR, TYNPR, SYNPR, RFRQGD, NFHHGD, and FHHGD had a prominent umami taste (Figure 4A). The taste thresholds of these peptides in water were 1.63 ± 0.60, 2.11 ± 0.80, 2.36 ± 0.84, 2.41 ± 1.11, 2.75 ± 1.28, and 2.86 ± 0.75 mmol/L, respectively (Table 1). Although the thresholds of these peptides in water were higher than that of MSG (1.6 mmol/L), most of the peptides had umami-enhancing effects in 0.35% MSG solutions, and the threshold values ranged from 0.44 ± 0.15 to 0.65 ± 0.19 mmol/L. Notably, the weakly umami FRDEHQ had the best umami-enhancing effect in the 0.35% MSG solution, improving umami intensity up to 6.75 at 1 mg/mL, with a corresponding threshold value of 0.44 ± 0.15 mmol/L. Similarly, the TYNPK, FRDEH, and TVDGPSH peptides also significantly increased umami intensity in 0.35% MSG (Figure 4B). This enhancement of umami may be due to the synergistic effect of MSG with umami peptides. In a relevant study [35], the results showed that the addition of MSG increased the size of the receptor binding cavity and allowed peptides to bind to more key residues, thus inducing a stronger umami flavor. Therefore, the docking screening effectively identified peptides associated with umami tastes, but further analysis is needed to understand the differences in detail.

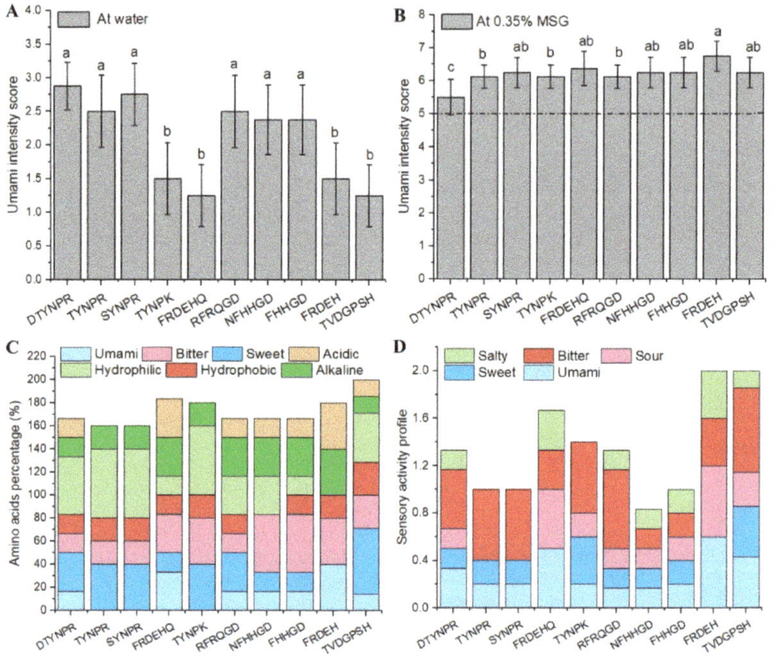

Figure 4. Sensory and sequence features of synthetic umami peptides ((**A**): umami intensity in water, (**B**): umami-enhancing effect with 0.35% MSG, (**C**): amino acid composition, (**D**): taste activity profile). Different letters in the same column represent significant differences ($p < 0.05$).

Table 1. Umami thresholds of target synthetic peptides.

Peptide	Sensory Description	Umami Threshold (mol/L) in Water	Umami Threshold (mol/L) in MSG	Protein Accession
DTYNPR	umami, slightly sweet, sour	1.63 ± 0.60 b	0.81 ± 0.28 a	P07728
TYNPR	umami, slightly bitter	2.11 ± 0.80 ab	0.63 ± 0.18 b	P14614
SYNPR	umami, slightly bitter	2.36 ± 0.84 ab	0.64 ± 0.19 ab	Q02897
FRDEHQ	weakly umami, sour	3.02 ± 0.56 a	0.65 ± 0.19 ab	P14614
TYNPK	weakly umami, slightly bitter	2.41 ± 1.11 ab	0.53 ± 0.13 b	A0A3B6AZR2
RFRQGD	umami, sour	2.41 ± 0.45 ab	0.53 ± 0.16 b	P07728
NFHHGD	umami, weakly sour	2.75 ± 1.28 a	0.52 ± 0.17 b	P00925
FHHGD	umami, weakly sour	2.86 ± 0.75 a	0.61 ± 0.20 ab	P00925
FRDEH	weakly umami, sour	3.02 ± 1.18 a	0.44 ± 0.15 b	P14614
TVDGPSH	weakly umami, weakly sour	3.16 ± 0.99 a	0.61 ± 0.16 ab	P00360

Results are mean ± SD (n = 3); different letters in the same column represent significant differences ($p < 0.05$).

The amino acid composition analysis revealed that most peptides shared umami residues D or E (Figure 4C). However, the taste activity profile indicated that umami amino acids are not necessary conditions for peptides' umami taste potential (Figure 4D) [36]. TYNPR and SYNPR lack umami amino acids but have an umami taste, which can be attributed to the umami dipeptide fragment NP [37], the C-terminal amino acids R [38], and the geometry of binding with the T1R1-T1R3 receptor's key residues. Umami amino acids are also acidic amino acids. Due to the peptides FRDEHQ and FRDEH both having the sour character of residues D and E and the fragment DE, they provide a stronger sour sensory taste than others. This aligns with the acidic dipeptides DD, DE, EE, and EEE found in soy sauce [39].

In addition, basic (R, K) and bitter amino acids (F, H, Y) also appear frequently. R and K have often been found in reported umami peptides, such as clam water extract umami peptides [40]. All of the peptides with a high taste activity profile for bitterness were not too bitter, this may be due to the masking effect of umami, sweetness, and basic fragments on bitterness [41]. Like TVDGPSH, the bitterness profile was high (0.71), but the umami fragments (VD, D) and sweet amino acids (V, G, P) contribute to the total profile score up to 0.8572 (Figure 3D), and the sensory evaluation showed that the bitterness was very weak. In a metabolomics study of taste components in pufferfish [42] and salted duck eggs [43], the author found that some non-umami amino acids can enhance umami intensity. This suggests that the synergistic effect of sweet and umami amino acids is beneficial for reducing bitterness. Similarly, the results obtained by Wang et al. also showed that sweet amino acids (A, S, G, etc.) and acidic residues D/E can significantly improve umami taste [44]. Therefore, amino acid composition is closely related to the umami taste of peptides.

3.4. Molecular Docking Interactions of Peptides with T1R1-T1R3

In order to investigate the interaction between peptides and T1R1-T1R3 receptors, a docking analysis was performed. The 2D residue interactions are shown in Figure S2 and the corresponding contact bond number heatmap is depicted in Figure 5.

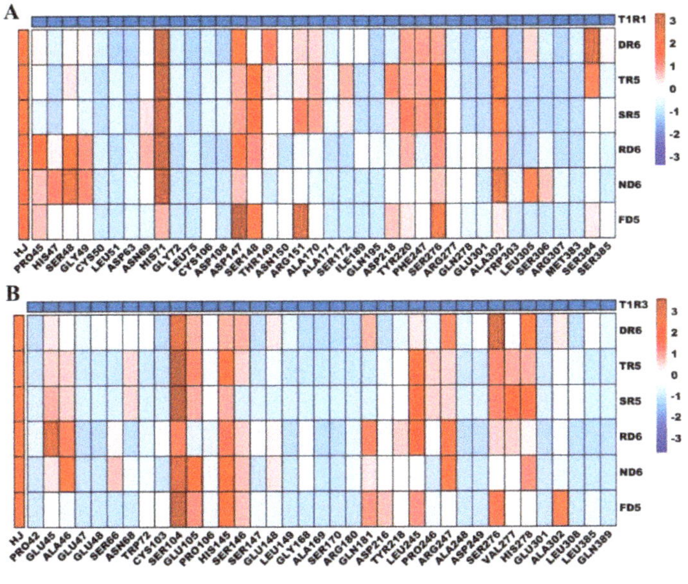

Figure 5. The interaction bond heatmap between umami peptides and T1R1−T1R3 receptors ((**A**) for T1R1 and (**B**) for T1R3. DR6, TR5, SR5, RD6, ND6, and FD5 are short for DTYNPR, TYNPR, SYNPR, RFRQGD, NFHHGD, and FHHGD, respectively).

3.4.1. Molecular Docking Interactions of Peptides with T1R1

As seen from the docking energy results in Table S1, the six umami peptides tended to interact with T1R1 due to its lower binding energy (−9.6 to −11.0 kcal/mol) compared with T1R3 (−8.9 to −10.1 kcal/mol). This was consistent with the reported studies showing that T1R1 was the main umami peptide binding domain [27,44]. And the affinity of most of the umami peptides, DTYNPR, TYNPR, and SYNPR, to T1R1 was also lower than that of RFRQGD, NFHHGD, and FHHGD. This indicated that binding affinity can be used for evaluating peptides. The results of the residue interaction (Figure S2) revealed that these peptides formed 14, 15, 16, 16, 11, and 14 hydrogen bonds, respectively, accounting for 10.18% to 20.25% of the total contact bonds, while the hydrophobic bonds occupied 86, 69, 63, 105, 97, and 95, respectively. Therefore, hydrophobic forces were a crucial factor for peptide–receptor binding. This process involved a number of high contact frequency residues including SER48, HIS71, ASP147, SER148, ARG151, TYR220, PHE247, SER276, ALA302, and SER384. Among these, SER48, ARG151, TYR220, and SER276 were key residues for the umami molecules GMP and IMP and the beef peptide KGDESSLA with T1R1 [24]. The residues HIS71, ASP147, and ALA302 also contributed the most to the chicken soup and pufferfish umami peptides with T1R1 [13,45]. In addition, when compared with RFRQGD, NFHHGD, and FHHGD, three other umami peptides (DTYNPR, TYNPR, and SYNPR) can still bind with some residues (THR149, ALA170, and SER172) that also play important roles for monosodium glutamate and sodium succinate [24].

3.4.2. Molecular Docking Interactions of Peptides with T1R3

For T1R3, the hydrogen bond numbers of the above six peptides dropped to 10, 6, 6, 9, 7, and 5, respectively (Figure S2). Meanwhile, the hydrophobic interactions were also reduced to 90, 62, 58, 82, 72, and 58, respectively, amounting to 90.11% to 92.06% of the total contact bonds. From the heatmap results, all peptides mostly shared the residues GLU45, SER104, HIS145, LEU245, ARG247, SER276, VAL277, and HIS278.

The residues GLU45, HIS145, VAL277, and HIS278 were also crucial for the chicken breast umami peptides GIQKELQF, FTERVQ, and AEINKILGN binding to T1R3 [46]. In the docking results for porcine bone soup [47] and fermented fish (Chouguiyu) umami peptides [14] with T1R3, the residues SER104, HIS145, and SER276 were important. Beyond that, the residues GLU45, SER104, LEU245, and HIS278 may have contributed to the umami-enhancing effect, as suggested by the results of the umami-enhancing peptide RFPHADF-ARP docking with T1R3 [48]. This suggests that T1R3 was also necessary in umami recognition. Moreover, it was noted that DTYNPR, TYNPR, and SYNPR interacted more with key residues such as ASN68, LEU245, SER276, VAL277, and HIS278 than others (RFRQGD, NFHHGD, and FHHGD). Furthermore, TYNPR and SNPR interacted with exactly the same residues with T1R3, which may explain the similar umami intensities.

3.5. Theoretical Calculation of Umami Peptides' Active Sites

The ten identified peptides above (Table 1) have similar binding energy ranges but differ significantly in umami intensity, indicating that the affinity energy is just one of the necessary conditions for umami peptides. To further investigate the differences between these umami peptides, active site calculations were performed in both the gas and water phases, and the results are shown in Figure 6. In general, the active sites of molecules are regions that are prone to gaining and losing electrons, and the electron density distribution in these regions is different from that in others. Density functional theory (DFT) has been widely used to describe the structure of atoms, molecules, crystals, etc., based on the electron density distribution, and the molecular orbital is directly related to the electron density. In general, the active sites can be reflected by the highest occupied molecular orbital (HOMO) and the lowest unoccupied molecular orbital (LUMO), which refer to the regions that tend to donate and accept electrons, respectively [49].

The results indicate (Figure 6) that most of the positions of HOMO and LUMO on huangjiu umami peptides in the gas and water phases were not the same. This revealed that either HOMO or LUMO could be the active sites, and the solvent environment had a great influence on them. According to Wang et al.'s study [44], the residues D, E, R, K, and H in medium-length (4–7) umami peptides are prone to being the active sites. Our results show that the terminal residues (D or R) in the DTYNPR, TYNPR, SYNPR, FHHGD, NFHHGD, and RFRQGD peptides tend to be active sites in the gas phase. This confirmed the positional distribution of umami peptide active sites proposed by Wang et al. However, the results were different in the water phase; only FHHGD and RFRQGD still had the active sites D and R, which may be related to the protonation or deprotonation of peptide amino and carboxyl groups in the water environment. Since the protonation process of peptides can be influenced by varying pH conditions, many studies have calculated the active sites of small peptides under gas-phase conditions for convenient comparison [44,48,50].

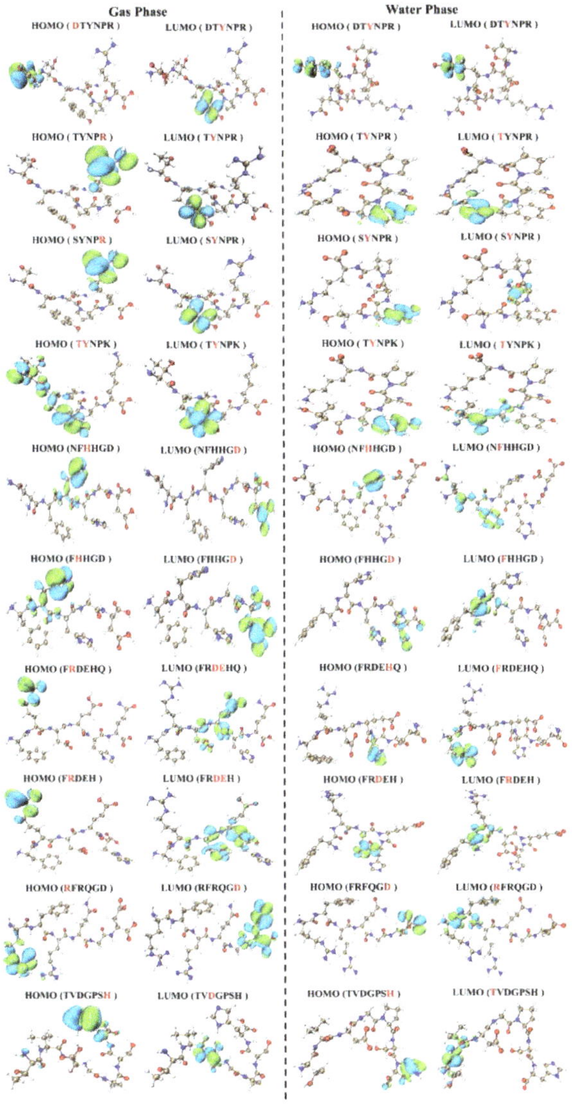

Figure 6. The calculated active sites of umami peptides (active sites are colored red in sequence, and the positive and negative phases of the orbital wave function are shown in green and blue, respectively).

As we take the sensory evaluation and DFT calculation results together, it appears that the position and types of active sites (D, E, R, and K) in the ten peptides may have the most impact on the umami taste, which is consistent with the results of Wang et al.'s study [44]. DTYNPR, TYNPR, and SYNPR all had the terminal residues R or D, and they act as active sites in the gas phase, but the C-terminal residue K at TYNPK was not one of the active sites (Y and T), which could have caused the weaker umami intensity of TYNPK. Furthermore, it is known that residues D and E at the length of 4–7 umami peptides are most involved in docking with T1R1 receptors [40]. The peptides FRDEHQ, FRDEH, and TVDGPSH, which have a weaker umami taste, had the key active sites of D or E, but these sites were also not located at the terminal position. When key active sites (D, E, and R) are in the terminal position, it may be beneficial to reduce the steric hindrance when interacting with the T1R1 receptor pocket. In addition, for peptides with multiple active sites (D, E, H, and R), the umami intensity may be affected by whether the key binding residues D or E are preferred at the terminal position. This inference was confirmed with the ten umami peptides identified above, and this offered a possible explanation for the ten peptides' umami taste intensity. These results further proved the relationship between the peptide surface charge and taste activity [51], because the most active sites (charged amino acids or groups) tend to change position according to different electron densities. In a recent study, when the terminal residues of the bitter peptide RFPHADF were modified with glucose to increase electron density through the Maillard reaction, the original bitter peptide could be transformed into a non-bitter umami-enhancing peptide [48]. Therefore, the above analysis demonstrated that the position of key active sites in small peptides has a significant effect on the overall taste.

4. Conclusions

In this study, six novel umami and five umami-enhancing peptides derived from huangjiu were identified through multistep purification, virtual screening, and sensory evaluation. All identified umami peptides could bind to the T1R1-T1R3 receptor's key residues through hydrogen bonds and van der Waals forces. Furthermore, the active site calculations revealed that the key umami residues (D and R), whether at the sequence's terminal position or not, had a significant impact on the taste features of the above huangjiu umami peptides. In the future, studying peptide stability and regulation during the brewing process will provide a better understanding of the fermented grain wine's taste. Overall, peptitomics combined with sensory-guided isolation could contribute to the discovery of taste peptides in other fermented wines and complex food matrices.

Supplementary Materials: The following supporting information can be downloaded at https://www.mdpi.com/article/10.3390/foods12183398/s1: Table S1: Umami prediction and docking screening results of 75% ETP water-eluted fraction peptides with macroporous resin; Figure S1: Ramachandran plot of the modeled umami receptor T1R1/T1R3; Figure S2: The docking residue interaction mode between umami peptides and T1R1-T1R3 receptor.

Author Contributions: Conceptualization and methodology, R.C. and Z.Z.; validation, Y.D. and Z.Z.; software and formal analysis, R.C. and Y.D.; resources, Y.X.; data curation and writing—original draft preparation, R.C.; writing—review and editing, S.L. and Z.J.; visualization, R.C.; supervision, J.M., Z.Z., S.L. and Z.J.; funding acquisition, J.M. All authors have read and agreed to the published version of the manuscript.

Funding: This research was funded by the National Key Research and Development Program of China (no. 2022YFD2101204) and the National Natural Science Foundation of China (22138004 and 32001828).

Data Availability Statement: Data will be made available on request.

Conflicts of Interest: The authors declare no conflict of interest.

References

1. Kurihara, K. Glutamate: From discovery as a food flavor to role as a basic taste (umami). *Am. J. Clin. Nutr.* **2009**, *90*, 719S–722S. [CrossRef] [PubMed]

2. Huang, Z.; Feng, Y.; Zeng, J.; Zhao, M. Six categories of amino acid derivatives with potential taste contributions: A review of studies on soy sauce. *Crit. Rev. Food Sci. Nutr.* **2023**, 1–12. [CrossRef]
3. Yamasaki, Y.; Maekawa, K. A peptide with delicious taste. *Agric. Biol. Chem.* **1978**, *42*, 1761–1765.
4. Cui, Z.; Zhang, Z.; Zhou, T.; Zhou, X.; Zhang, Y.; Meng, H.; Wang, W.; Liu, Y. A TastePeptides-Meta system including an umami/bitter classification model Umami_YYDS, a TastePeptidesDB database and an open-source package Auto_Taste_ML. *Food Chem.* **2023**, *405*, 134812. [CrossRef]
5. Zhou, M.; Bu, T.; Zheng, J.; Liu, L.; Yu, S.; Li, S.; Wu, J. Peptides in brewed wines: Formation, structure, and function. *J. Agric. Food Chem.* **2021**, *69*, 2647–2657. [CrossRef] [PubMed]
6. Shi, Y.; Feng, R.; Mao, J.; Liu, S.; Zhou, Z.; Ji, Z.; Chen, S.; Mao, J. Structural characterization of peptides from huangjiu and their regulation of hepatic steatosis and gut microbiota dysbiosis in hyperlipidemia mice. *Front. Pharmacol.* **2021**, *12*, 1517. [CrossRef]
7. Schmidt, C.V.; Olsen, K.; Mouritsen, O.G. Umami potential of fermented beverages: Sake, wine, champagne, and beer. *Food Chem.* **2021**, *360*, 128971. [CrossRef]
8. Klosse, P. Umami in wine. *Res. Hosp. Manag.* **2013**, *2*, 25–28. [CrossRef]
9. Yamamoto, T.; Watanabe, U.; Fujimoto, M.; Sako, N. Taste preference and nerve response to 5′-inosine monophosphate are enhanced by glutathione in mice. *Chem. Senses* **2009**, *34*, 809–818. [CrossRef]
10. Ding, Y.; Li, X.; Kan, J. Isolation and identification of flavor peptides from douchi (traditional Chinese soybean food). *Int. J. Food Prop.* **2017**, *20*, 1982–1994. [CrossRef]
11. Zhuang, M.; Zhao, M.; Lin, L.; Dong, Y.; Chen, H.; Feng, M.; Sun-Waterhouse, D.; Su, G. Macroporous resin purification of peptides with umami taste from soy sauce. *Food Chem.* **2016**, *190*, 338–344. [PubMed]
12. Fan, S.H.; Liu, T.T.; Wan, P.; Zhu, Q.; Xia, N.; Wang, Q.Z.; Chen, D.W. Enrichment of the umami-taste-active amino acids and peptides from crab sauce using ethanol precipitation and anion-exchange resin. *J. Food Process. Preserv.* **2021**, *45*, e15390.
13. Zhang, J.; Zhang, J.; Liang, L.; Sun, B.; Zhang, Y. Identification and virtual screening of novel umami peptides from chicken soup by molecular docking. *Food Chem.* **2023**, *404*, 134414.
14. Yang, D.; Li, C.; Li, L.; Chen, S.; Hu, X.; Xiang, H. Taste mechanism of umami peptides from Chinese traditional fermented fish (Chouguiyu) based on molecular docking using umami receptor T1R1/T1R3. *Food Chem.* **2022**, *389*, 133019. [PubMed]
15. Zhang, J.; Toldrá, F.; Zhang, W.; Yin, Y.; Zhu, Z. Study on the effects and mechanisms of ultrasound on the peptide profile and taste of unsmoked bacon using peptidomics and bioinformatics. *Food Chem.* **2023**, *414*, 135764.
16. Zheng, X.; Chi, H.; Ma, S.; Zhao, L.; Cai, S. Identification of novel α-glucosidase inhibitory peptides in rice wine and their antioxidant activities using in silico and in vitro analyses. *LWT* **2023**, *178*, 114629.
17. Qi, L.; Gao, X.; Pan, D.; Sun, Y.; Cai, Z.; Xiong, Y.; Dang, Y. Research progress in the screening and evaluation of umami peptides. *Compr. Rev. Food Sci. Food Saf.* **2022**, *21*, 1462–1490.
18. Zhang, T.; Hua, Y.; Zhou, C.; Xiong, Y.; Pan, D.; Liu, Z.; Dang, Y. Umami peptides screened based on peptidomics and virtual screening from *Ruditapes philippinarum* and *Mactra veneriformis* clams. *Food Chem.* **2022**, *394*, 133504.
19. Kiyono, T.; Hirooka, K.; Yamamoto, Y.; Kuniishi, S.; Ohtsuka, M.; Kimura, S.; Park, E.Y.; Nakamura, Y.; Sato, K. Identification of pyroglutamyl peptides in Japanese rice wine (sake): Presence of hepatoprotective pyroGlu-Leu. *J. Agric. Food Chem.* **2013**, *61*, 11660–11667.
20. Hashizume, K.; Ito, T.; Nagae, Y.; Tokiwano, T. Quantitation and sensory properties of three newly identified pyroglutamyl oligopeptides in sake. *Biosci. Biotechnol. Biochem.* **2019**, *83*, 357–364. [PubMed]
21. Han, F.L.; Xu, Y. Identification of low molecular weight peptides in Chinese rice wine (Huang Jiu) by UPLC-ESI-MS/MS. *J. Inst. Brew.* **2011**, *117*, 238–250.
22. Minkiewicz, P.; Iwaniak, A.; Darewicz, M. BIOPEP-UWM database of bioactive peptides: Current opportunities. *Int. J. Mol. Sci.* **2019**, *20*, 5978. [CrossRef] [PubMed]
23. Charoenkwan, P.; Nantasenamat, C.; Hasan, M.M.; Moni, M.A.; Manavalan, B.; Shoombuatong, W. UMPred-FRL: A new approach for accurate prediction of umami peptides using feature representation learning. *Int. J. Mol. Sci.* **2021**, *22*, 13124.
24. Liu, H.; Da, L.-T.; Liu, Y. Understanding the molecular mechanism of umami recognition by T1R1-T1R3 using molecular dynamics simulations. *Biochem. Biophys. Res. Commun.* **2019**, *514*, 967–973.
25. Liu, Z.; Zhu, Y.; Wang, W.; Zhou, X.; Chen, G.; Liu, Y. Seven novel umami peptides from Takifugu rubripes and their taste characteristics. *Food Chem.* **2020**, *330*, 127204.
26. Lu, T.; Chen, F. Multiwfn: A multifunctional wavefunction analyzer. *J. Comput. Chem.* **2012**, *33*, 580–592. [PubMed]
27. Zhang, C.; Miao, Y.; Feng, Y.; Wang, J.; Tian, Z.; Dong, J.; Gao, B.; Zhang, L. Umami polypeptide detection system targeting the human T1R1 receptor and its taste-presenting mechanism. *Biomaterials* **2022**, *287*, 121660. [PubMed]
28. Zhu, X.; Sun-Waterhouse, D.; Chen, J.; Cui, C.; Wang, W. Comparative study on the novel umami-active peptides of the whole soybeans and the defatted soybeans fermented soy sauce. *J. Sci. Food Agric.* **2021**, *101*, 158–166. [CrossRef] [PubMed]
29. Zhu, X.; Sun-Waterhouse, D.; Chen, J.; Cui, C.; Wang, W. Bitter-tasting hydrophobic peptides prepared from soy sauce using aqueous ethanol solutions influence taste sensation. *Int. J. Food Sci. Technol.* **2020**, *55*, 146–156.
30. Zhao, J.; Liao, S.; Bi, X.; Zhao, J.; Liu, P.; Ding, W.; Che, Z.; Wang, Q.; Lin, H. Isolation, identification and characterization of taste peptides from fermented broad bean paste. *Food Funct.* **2022**, *13*, 8730–8740. [PubMed]
31. Yang, H.; Zong, X.; Cui, C.; Mu, L.; Zhao, H. Wheat gluten hydrolysates separated by macroporous resins enhance the stress tolerance in brewer's yeast. *Food Chem.* **2018**, *268*, 162–170. [CrossRef]

32. Rhyu, M.-R.; Kim, E.-Y. Umami taste characteristics of water extract of Doenjang, a Korean soybean paste: Low-molecular acidic peptides may be a possible clue to the taste. *Food Chem.* **2011**, *127*, 1210–1215. [CrossRef] [PubMed]
33. Cheng, K.; Wang, S.; Wang, Y.; Bao, Y.; Gao, P.; Lei, L.; Liang, H.; Zhang, S.; Dong, L. Modification of a Novel Umami Octapeptide with Trypsin Hydrolysis Sites via Homology Modeling and Molecular Docking. *J. Agric. Food Chem.* **2023**, *71*, 5326–5336. [CrossRef] [PubMed]
34. Zhang, Y.; Tang, X.; Li, F.; Zhang, J.; Zhang, B.; Yang, X.; Tang, Y.; Zhang, Y.; Fan, J.; Zhang, B. Inhibitory effects of oat peptides on lipolysis: A physicochemical perspective. *Food Chem.* **2022**, *396*, 133621. [PubMed]
35. Dang, Y.; Hao, L.; Cao, J.; Sun, Y.; Zeng, X.; Wu, Z.; Pan, D. Molecular docking and simulation of the synergistic effect between umami peptides, monosodium glutamate and taste receptor T1R1/T1R3. *Food Chem.* **2019**, *271*, 697–706. [CrossRef]
36. Yu, Z.; Jiang, H.; Guo, R.; Yang, B.; You, G.; Zhao, M.; Liu, X. Taste, umami-enhance effect and amino acid sequence of peptides separated from silkworm pupa hydrolysate. *Food Res. Int.* **2018**, *108*, 144–150. [CrossRef]
37. Zhao, Y.; Zhao, X.; Sun-Waterhouse, D.; Waterhouse, G.I.N.; Zhao, M.; Zhang, J.; Wang, F.; Su, G. Two-stage selective enzymatic hydrolysis generates protein hydrolysates rich in Asn-Pro and Ala-His for enhancing taste attributes of soy sauce. *Food Chem.* **2021**, *345*, 128803. [CrossRef]
38. Chen, M.; Gao, X.; Pan, D.; Xu, S.; Zhang, H.; Sun, Y.; He, J.; Dang, Y. Taste characteristics and umami mechanism of novel umami peptides and umami-enhancing peptides isolated from the hydrolysates of Sanhuang Chicken. *Eur. Food Res. Technol.* **2021**, *247*, 1633–1644.
39. Oka, S.; Nagata, K. Isolation and characterization of acidic peptides in soy sauce. *Agric. Biol. Chem.* **1974**, *38*, 1195–1202. [CrossRef]
40. Li, X.; Xie, X.; Wang, J.; Xu, Y.; Yi, S.; Zhu, W.; Mi, H.; Li, T.; Li, J. Identification, taste characteristics and molecular docking study of novel umami peptides derived from the aqueous extract of the clam Meretrix meretrix Linnaeus. *Food Chem.* **2020**, *312*, 126053. [CrossRef]
41. Kim, M.J.; Son, H.J.; Kim, Y.; Misaka, T.; Rhyu, M.-R. Umami–bitter interactions: The suppression of bitterness by umami peptides via human bitter taste receptor. *Biochem. Biophys. Res. Commun.* **2015**, *456*, 586–590. [CrossRef] [PubMed]
42. Yang, L.; Dai, B.; Ayed, C.; Liu, Y. Comparing the metabolic profiles of raw and cooked pufferfish (*Takifugu flavidus*) meat by NMR assessment. *Food Chem.* **2019**, *290*, 107–113. [CrossRef] [PubMed]
43. Gao, B.; Hu, X.; Li, R.; Zhao, Y.; Tu, Y.; Zhao, Y. Screening of characteristic umami substances in preserved egg yolk based on the electronic tongue and UHPLC-MS/MS. *LWT* **2021**, *152*, 112396. [CrossRef]
44. Wang, W.; Cui, Z.; Ning, M.; Zhou, T.; Liu, Y. In-silico investigation of umami peptides with receptor T1R1/T1R3 for the discovering potential targets: A combined modeling approach. *Biomaterials* **2022**, *281*, 121338. [CrossRef]
45. Wang, W.; Yang, L.; Ning, M.; Liu, Z.; Liu, Y. A rational tool for the umami evaluation of peptides based on multi-techniques. *Food Chem.* **2022**, *371*, 131105. [CrossRef]
46. Zhang, L.; Pu, D.; Zhang, J.; Hao, Z.; Zhao, X.; Sun, B.; Zhang, Y. Identification of novel umami peptides in chicken breast soup through a sensory-guided approach and molecular docking to the T1R1/T1R3 taste receptor. *J. Agric. Food Chem.* **2023**, *71*, 7803–7811. [CrossRef] [PubMed]
47. Liang, L.; Zhou, C.; Zhang, J.; Huang, Y.; Zhao, J.; Sun, B.; Zhang, Y. Characteristics of umami peptides identified from porcine bone soup and molecular docking to the taste receptor T1R1/T1R3. *Food Chem.* **2022**, *387*, 132870. [CrossRef]
48. Zhang, J.; Su, G.; Zhao, T.; Fan, J.; Ho, C.-T.; Zhao, M. Preparation, Sensory Characterization, and Umami-Enhancing Mechanism of Novel Peptide Glycoconjugates. *J. Agric. Food Chem.* **2022**, *70*, 8043–8051. [CrossRef] [PubMed]
49. Chakraborty, D.; Chattaraj, P.K. Conceptual density functional theory based electronic structure principles. *Chem. Sci.* **2021**, *12*, 6264–6279. [CrossRef]
50. Shi, C.; Liu, M.; Zhao, H.; Lv, Z.; Liang, L.; Zhang, B. A novel insight into screening for antioxidant peptides from hazelnut protein: Based on the properties of amino acid residues. *Antioxidants* **2022**, *11*, 127. [CrossRef]
51. Yu, X.; Zhang, L.; Miao, X.; Li, Y.; Liu, Y. The structure features of umami hexapeptides for the T1R1/T1R3 receptor. *Food Chem.* **2017**, *221*, 599–605. [CrossRef] [PubMed]

Disclaimer/Publisher's Note: The statements, opinions and data contained in all publications are solely those of the individual author(s) and contributor(s) and not of MDPI and/or the editor(s). MDPI and/or the editor(s) disclaim responsibility for any injury to people or property resulting from any ideas, methods, instructions or products referred to in the content.

foods

MDPI

Article

Comparative Analysis of Volatile Flavor Compounds in Strongly Flavored Baijiu under Two Different Pit Cap Sealing Processes

Lingshan Li [1,†], Mei Fan [1,†], Yan Xu [2], Liang Zhang [3], Yu Qian [4], Yongqing Tang [3], Jinsong Li [3], Jinsong Zhao [5], Siqi Yuan [2,3,6,*] and Jun Liu [1,7,*]

1 Bioengineering College, Sichuan University of Science & Engineering, Yibin 644000, China; lilingshan@stu.suse.edu.cn (L.L.); fanmei@stu.suse.edu.cn (M.F.)
2 School of Biotechnology, Jiangnan University, Wuxi 214122, China; yxu@jiangnan.edu.cn
3 Luzhou Laojiao Group Co., Ltd., Luzhou 646000, China; zhangl@lzlj.com (L.Z.)
4 Analysis and Testing Center, Sichuan University of Science & Engineering, Zigong 643000, China; qianyu1001@sina.com
5 Sichuan Liquor Group, Luzhou Tianfu 1st Street (Liangjiang International), Wuhou District, Chengdu 610000, China; zhaojinsong@suse.edu.cn
6 Science and Technology Department, Sichuan University of Science & Engineering, Zigong 643000, China
7 Key Laboratory of Liquor-Making and Application, Sichuan University of Science & Engineering, Yibin 644000, China
* Correspondence: yuansiqi@suse.edu.cn (S.Y.); liujunbio@suse.edu.cn (J.L.)
† These authors contributed equally to the work.

Abstract: The solid-state fermentation process of strongly flavored Baijiu is complicated by the co-fermentation of many different microorganisms in the fermentation pools. The traditional fermentation pools of strong flavor Baijiu are sealed with mud, and this sealed-pit mud is not easy to maintain; therefore, the pit cap is prone to cracks and to caving in. The destruction of the sealed-pit mud may lead to instability in the composition and an abundance of microorganisms in the fermentation process that results in fluctuations of product quality. Thus, the production method of replacing the mud cap with a new steel cap is gradually attracting the attention of scientific and technical workers in the industry. However, so far, there have been relatively few reports on the use of steel lids for sealing pits for fermentation and brewing. In this study, the volatile flavor components of 270 Baijiu samples from mud-sealing and steel-sealing pits of a Chinese Baijiu distillery were studied qualitatively and quantitatively using Gas Chromatography–Mass Spectrometry (Abbreviated as GC-MS). Our statistical methods included Hierarchical Cluster Analysis (Abbreviated as HCA), Principal Component Analysis (Abbreviated as PCA), and Discriminant Analysis (Abbreviated as DA). A statistical analysis was carried out on the yield of strongly flavored Baijiu, and we made a comprehensive evaluation of the Baijiu produced under the two pit-sealing modes with regard to flavor and economic efficiency. The yield of strong flavored Baijiu was 6.7% higher with steel-sealing pits compared with mud-sealing pits. Cluster analysis categorized the strongly flavored Baijiu samples into two categories initially: (1) samples produced using mud-sealing pits and (2) samples using steel-sealing pits. Our analysis also indicated that the 28 compounds used for quantification were selected correctly. Surprising to the experimental staff, the overall score for the steel-sealing pits was greater than that of the mud-sealing pits based on PCA. Using DA, the prediction results were 100% accurate. In summary, through a comparative analysis of the flavor and yield, which are the two main factors that affect the quality of Baijiu in a distillery, and systematic combination at both experimental and theoretical levels, the differences between the Baijiu production by steel-sealing and the traditional mud-sealing were clear. Regardless of the impact of age, the detectable flavor components of Baijiu from the mud-steeling pits were very consistent with those of the steel-sealing pits in terms of richness or concentration. However, steel-sealing pits were significantly superior to mud-sealing pits with respect to output, consistency in quality, and cost (human and economic) savings.

Keywords: flavor compounds; strong flavor Baijiu; pit cap; mud-sealing pits; steel-sealing pits; Gas Chromatography-Mass Spectrometry

1. Introduction

By the end of 2022, the cumulative export value of alcoholic beverages in China was approximately 21.53 billion RMB, an increase of 25.1% year-on-year (China Industrial Economic Information Network). Topping the list of alcoholic beverages in China was Chinese Baijiu, which includes strong flavor, Maotai flavor, light flavor, and others [1–4]. The unique brewing process, which involves mixed steaming and boiling, and continuous ingredients for fermented grains of Chinese Baijiu endow itself with characteristic flavor [3]. Chinese Baijiu is highly sought after by consumers and is also renowned overseas. The fermentation of Baijiu is the result of the joint action of microbial clusters. Good fermentation technology will lead to the production of a large number of flavor substances and metabolites, which not only have rich nutritional value but also have certain medicinal value, such as cyclic dipeptide, prochemic acid, etc. [5,6].

Strongly flavored Baijiu is produced from grains with medium-high temperature, with Daqu being the flavor-producing agent, and involves continuous distillation of grain ingredients, mixed steaming and mixed burning, solid-state fermentation, distillation, aging, and blending [7]. Aromatic compounds in Baijiu mainly come from protein or amino acid metabolism. It has a strong aromatic odor and has a significant fragrance forming effect [8–10]. In fact, the aromatic compounds in Baijiu come from fermentation during the brewing process and decomposition or synthesis during distillation. Among them, esters, aldehydes, and ketones produced during fermentation have aromatic properties and are the main source of aromatic compounds in Baijiu. In the distillation process, with the evaporation of alcohol, some volatile aromatic compounds will also be distilled out and become the aromatic substances in Baijiu. Their aromatic compounds are dominated by ethyl caproate and are produced without the addition of edible alcohols, nonfermented aromatic compounds, or taste-producing substances [11,12]. The fermentation of strongly flavored Baijiu is carried out in mud pits (usually called "NiJiao" in Chinese) [13], which have an important influence on the formation of flavor compounds in strongly flavored Baijiu. With the increase in demand for Baijiu, the number and scale of Baijiu enterprises has expanded. In the face of growing market demand, traditional manual production is unable to meet the production demands for Baijiu products; the intelligent production and automation in distilleries is gradually being put on the agenda by researchers in the industry. Traditional strongly flavored Baijiu is brewed by sealing the pit with mud [14] (Figure 1). This process isolates oxygen in the early stage of production and prevents the microbial flora in the environment from invading the fermentation system. However, as fermentation time passes, this method of sealing the pits has shortcomings, such as cracks in the pit cap, maintenance of the pits, and contamination with environmental bacteria [15,16]. These factors affect the quality and taste of strongly flavored Baijiu, which causes some unnecessary, but significant, economic losses.

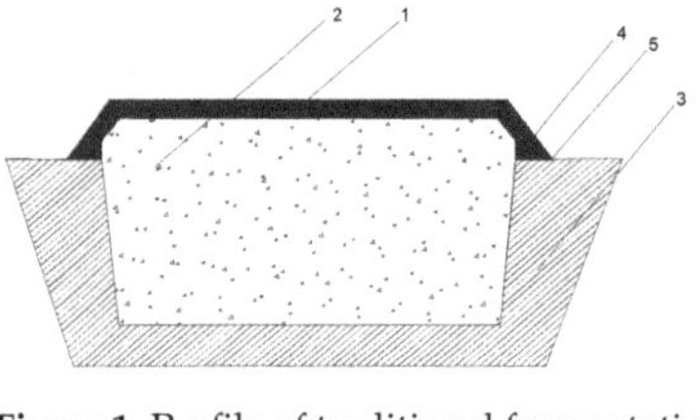

Figure 1. Profile of traditional fermentation device (pit cap: mud) for producing strongly flavored Baijiu in China. (1—mud-sealing cap; 2—Grains of Baijiu; 3—Brewing pit; 4—Sealing the pit opening; 5—Included angle of prism clamp, Modified from Figure 1 in Lu et al., 2012 [14]).

The sealing technology of pit is one of the key points in the production of Chinese Baijiu possessing varied and diversified flavors (12 types including Maotai flavor, strong flavor, Qingxiang flavor, etc.). The key fermentation elements concerned such as the types of fermentation microorganisms and the corresponding fermentation time are all related to the sealing of the pit. Poor sealing technology can lead to a series of devastating consequences such as fermentation rancidity, heavy odor, and low alcohol production rate. However, better sealing technology can make the aroma of the Baijiu stronger and more unique.

In response to these problems, different forms of pit sealing have been developed by Baijiu practitioners. For example, steel-sealing methods have been improved from the simple steel-sealing lid (Figure 2) to those that are detectable from the outside (Figure 3) [17], and plastic film sealing has also started to be applied in the fermentation sealing process of strongly flavored Baijiu. The plastic film sealing pit has disadvantages, such as difficult operation, poor insulation, and easy breakage. On the other hand, the steel-sealing pit solves the problems of entering and exiting the traditional pit fermented grains, high labor intensity, poor control of the internal temperature and humidity of the pit, and the problems of automating the production of the solid fermentation type of strongly flavored Baijiu. It also allows real-time monitoring of the physical and chemical indicators related to fermentation in the pit. Therefore, the production method of replacing the mud-sealing cap with a new steel-sealing cap has gradually received the attention of scientific and technical workers in the industry, but relatively little research has been reported on the fermentation and brewing method of steel-sealing caps.

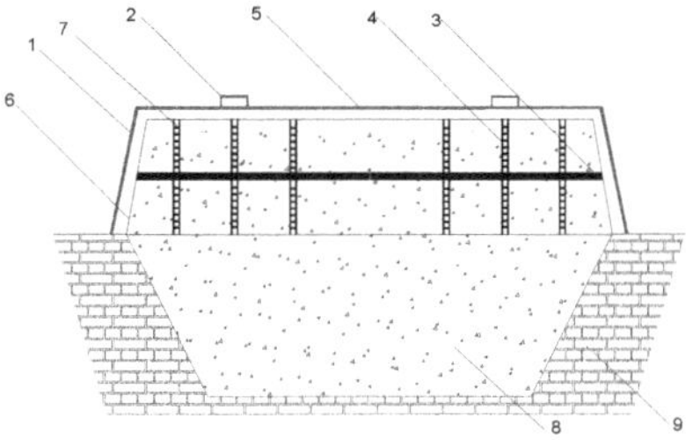

Figure 2. Profile of pit cap improved fermentation device A(pit cap: steel) for producing strongly flavored Baijiu in China. (1—steel-sealing cap; 2—Lifting ring; 3—Fixing brackets; 4—Pit mud board; 5—Cover top; 6—Lower box; 7—Pit mud; 8—Grains of Baijiu; 9—Brewing pit, Modified from Figure 2 in Zhang, 2015 [11]).

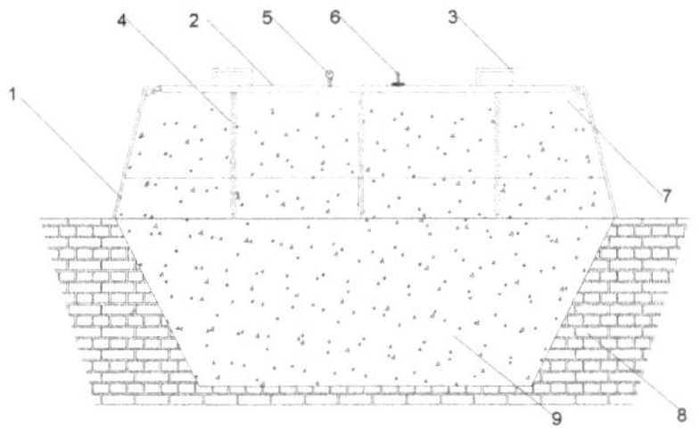

Figure 3. Profile of pit cap improved fermentation device B(pit cap: steel) for producing strongly flavored Baijiu in China. (1—Lower box; 2—Upper cover body; 3—Lifting ring; 4—Lower box side reinforcements; 5—Pressure Gauges; 6—Material handling ports; 7—Outer side wall of the lower box; 8-Brewing pit; 9—Grains of Baijiu, Modified from Figure 3 in Zhang and Zhao et al., 2015 [17]).

The quality of Baijiu products is closely related to the composition of flavor substances. For Chinese Baijiu with large export volume, the method of steel-sealing pits can be dynamically monitored from the outside, which can not only reduce a series of problems caused by mud-sealing pits but also reduce the production of harmful substances through monitoring. At present, the research on the flavor substances of strongly flavored Baijiu mainly focuses on volatile substances, which play an important role in Baijiu aroma. Compared with other analytical techniques, Gas Chromatography–Mass Spectrometry (GC-MS) has obvious advantages in flavor analysis [18–25]. GC-MS can be used to analyze qualitatively and quantitatively most of the volatile flavor components in strongly flavored Baijiu; for this reason, GC-MS has become one of the most powerful tools for the separation and detection of the larger number of flavor compounds in Baijiu [26,27].

To identify the differences between the quality of Baijiu products produced in mud-sealing and steel-sealing pits systematically, we collected strongly flavored Baijiu samples from mud-sealing and steel-sealing pits from a distillery using GC-MS to analyze flavor, studied the volatile flavor components using Hierarchical Cluster Analysis (Abbreviated as HCA), Principal Component Analysis (Abbreviated as PCA), and Discriminant Analysis (Abbreviated as DA), and analyzed the results statistically. The ultimate goal was to find a way to achieve both artificial intelligence and intensive production to save time, effort, and money. We hoped to determine whether the quality and taste of the product (Baijiu) was affected, compared with the traditional method of production, to solve the increasingly serious challenge between supply and demand faced by the Baijiu industry.

2. Materials and Methods

2.1. Materials

Materials: Baijiu samples were taken from the same vintage and same fermentation pits at a famous Baijiu enterprise in southern Sichuan. Samples were collected from steel-sealing and traditional mud-sealing pits.

Distillation: For ease of operation, this facility distilled Baijiu in upper, middle, and lower layers of grain.

Strongly flavored Baijiu picking: Because Chinese traditional Baijiu uses solid-state, retort barrel distillation technology, the quality and alcohol concentration of distilled Baijiu from different time periods are different, and the Baijiu is usually picked and stored separately in stages during the production process. In this study, the distilled strongly flavored Baijiu (i.e., first segment Baijiu, second segment Baijiu, and third segment Baijiu) in the corresponding Baijiu period was picked by a Baijiu picker.

Sample collection time: The brewing workshop for the strong flavor Baijiu distillery was divided into five times every year: the first time was in September; the second time was in November; the third time was in February of the next year; the fourth time was in April of the next year; and the fifth time was in June of the next year.

2.2. Instruments and Equipment

We used a TSQ8000 triple quadrupole GC-MS, an AI1310 multifunctional autosampler (Thermo Fisher Scientific, Waltham, MA, USA), a pipette gun (1000 μL, Dalong Xingchuang Experimental (Instruments) Co., Ltd., Beijing, China), a volumetric flask (100 mL), an injection bottle (2 mL), a TG-WAX column (30 m × 0.25 mm × 0.25 μm, Aglient Corporation, Santa Clara, CA, USA), an ultrasonic cleaner, and a microinjector 1 μL.

2.3. Chemicals

Flavor substance standards were chromatographic grade or above. *n*-propanol, ethyl valerate, *n*-butanol, isobutanol, acetic acid, hexyl butyrate, isoamyl alcohol, *n*-amyl alcohol, ethyl butyrate, *n*-hexanol, hexanoic acid, ethyl formate, ethyl acetate, ethyl caproate, ethyl heptanoate, butyl caproate, ethyl octanoate, propionic acid, butanoic acid, heptanoic acid, octanoic acid, and *tert*-amyl alcohol were purchased from Tianjin Guangfu Chemical Reagent Co., Tianjin, China. Isoamyl acetate, *n*-amyl acetate, and anhydrous ethanol were purchased from Beijing Tanmo Quality Control Standard Material Centre., Beijing, China. Ethyl phenylacetate, ethyl tetradecanoate, and ethyl hexadecanoate were purchased from Bailingway Technology Co., Ltd., Beijing, China. Propyl hexanoate and ethyl linoleate were purchased from Aladdin Reagent (Shanghai) Co., Ltd., Shanghai, China. Anhydrous ethanol was purchased from Shanghai Aladdin Biochemical Technology Co., Ltd., Shanghai, China. Methanol was purchased from Adamas-beta Co., Shanghai, China.

2.4. Methods

2.4.1. Calculations of Yield of Strongly Flavored Baijiu

Yield rate = brewing strong flavor Baijiu yield/amount of grain × 100%. The mud-sealing pit contained 130 kg of grain per retort, and the steel-sealing pit contained 225 kg of grain per retort.

2.4.2. Identification of Volatile Flavor Compounds Using GC-MS

Qualitative analysis was carried out by combining the retention index (RI) of the flavor substances with a NIST 12 mass spectrometry library. The method for the quantitative analysis of flavor components in strong flavor Baijiu was based on the national standard GB/T 10345-2007 [28] "Methods for the analysis of Baijiu". The specific processing steps of the sample were as follows: We measured the standard substance of flavor components and used 60% (*v*/*v*) ethanol solution to volume with a 100 mL volumetric flask. The measured value of all flavor components was 0.02% *v*/*v*. Before the samples were analyzed, the mixed standard solutions were analyzed, and the correction factors for each flavor compound were calculated using the peak areas and the concentration of the standard substances. Amyl acetate, 2-ethylbutyrate, and tert-pentyl alcohol were used as internal standards for esters, acids, and alcohols of flavor components, respectively.

2.4.3. Statistical Analysis

HCA, PCA, and DA were applied to extract and to analyze the data. Wherein, the specific methods for PCA are as follows:

The formula for each principal component score is $Fi = w_{i1\times1} + w_{i2}X_2 + \ldots + w_{in}X_n$

Among them, $W_{ij} = \theta_i/\sqrt{\lambda}$, which denotes the weight of each variable in the principal component. θ_i is the coefficient that corresponds to each variable in the component matrix, and $\sqrt{\lambda}$ is the open root value of the eigenvalue that corresponds to the i-th principal component.

We calculated the comprehensive score by using the contribution rate of the Baijiu on seven main components in different pit sealing methods as the weight coefficient of the comprehensive score; the formula was $F = \alpha_1F1 + \alpha_2F2 + \ldots + \alpha_nFn$, where α_i denotes the percentage of variance of the ith principal component.

ChemPattern software V2017.3 (Comayne Beijing Technology Co., Ltd., Beijing, China) and SPSS 24.0 were also used to analyze the data chemometrically. Excel 2016 and Origin 2018 software were used to collate statistics and to draw graphs for analysis of the assay data.

3. Results and Discussion

3.1. Sample Collection

According to the principle of parallel sampling, three pits were selected for mud and steel sealing, and samples were taken from each of the three pits selected. We sampled the upper, middle, and lower layers of fermented grains from each pit, and the 1st, 2nd, and 3rd segment samples of strong flavor Baijiu were obtained from distillation of each layer of fermented grains, for a total of 270 samples. For example, we use mud-sealing pits for sampling: 3 pits × 3 (upper, middle, and lower layers of fermented grains) × 3 (each layer has 1st, 2nd, and 3rd segment) × 5 (a total of five samples were taken) = 135.

3.2. Yield of Strongly Flavored Baijiu

The yield in strongly flavored Baijiu from the same pit at different sampling times showed that the yield decreased from spring to autumn to winter to summer when we used either traditional or new sealed pits (Figure 4). The steel-sealing pit cap exhibited less variation in yield than the mud-sealing cap, which provided preliminary evidence of the stability of sealing with steel. In a fast-moving society, the more stable the Baijiu yield is for roughly the same quality of Baijiu, the better this is. This result indicated that the environment had little influence on the outcome. The greater stability of sealing with steel should improve the economy of the distillery operation, and this provides more theoretical relevant data for the intelligent operation of the distillery.

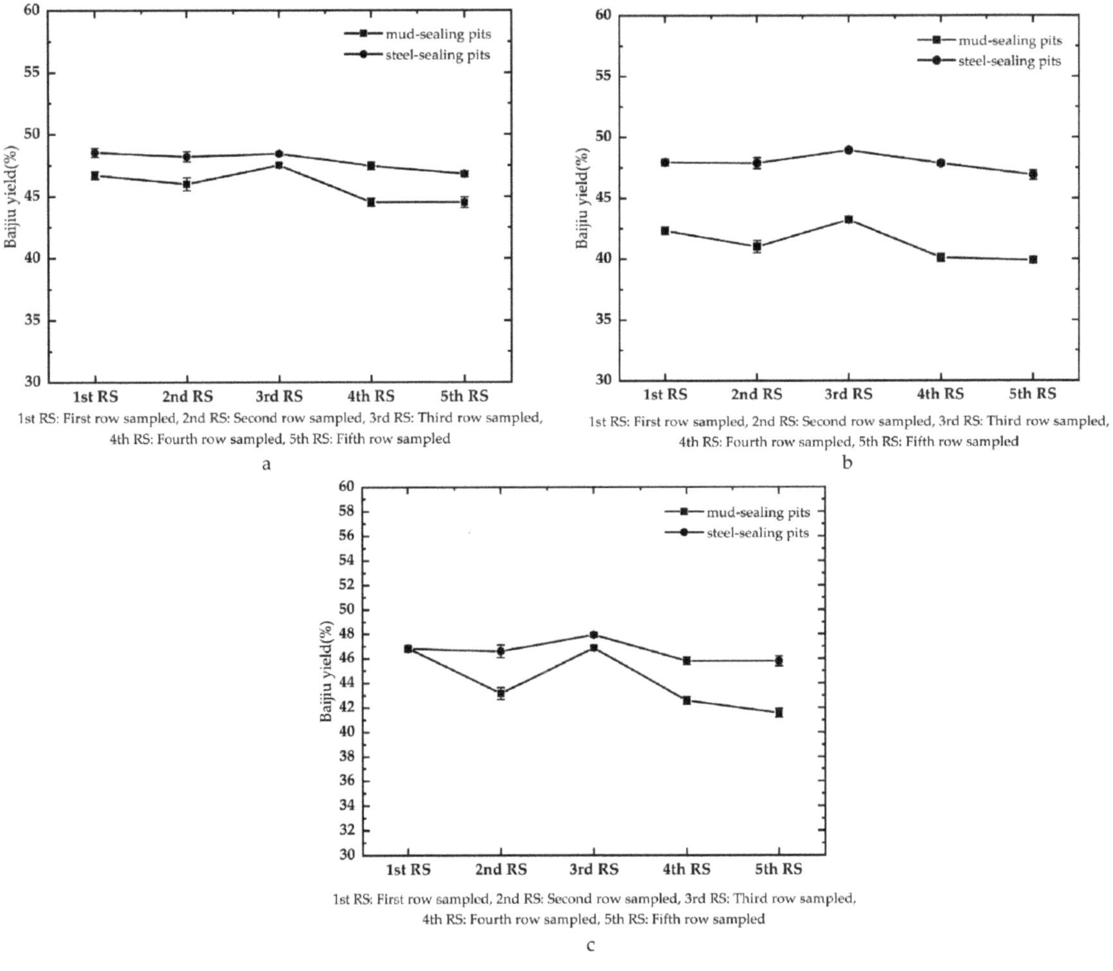

Figure 4. Variation in the yield of strongly flavored Baijiu from the same pit using mud sealing or steel sealing under five sampling batches. ((**a**)—Pit No. 1; (**b**)—Pit No. 2; (**c**)—Pit No. 3).

The yield from the two different forms of pit sealing (Figure 5) showed that for the first segment of Baijiu, the upper, middle, and lower lees yielded 20.00–24.10% (average 21.31%) in the mud-sealing pit, while the steel-sealing pit yielded 19.07–24.3% (average 21.61%). For the second segment, yield was 19.90–21.74% (average 20.72%) at each level of the lees in the mud-sealing pits, while 20.22–29.27% (average 24.31%) was produced in the steel-sealing pits. For the third segment, the yield of each layer of lees in the mud-sealed pit was 9.70–14.31% (average 11.58%), while the steel-sealing pit yielded 8.76–11.81% (average 10.04%). For the pits as a whole, the rate of strongly flavored Baijiu yield from the upper, middle, and lower lees in the mud-sealing pits was 37.03–45.21% (average 41.42%), while the rate of strong flavor Baijiu yield in the steel-sealing pits was 44.35–51.23% (average 48.12%).

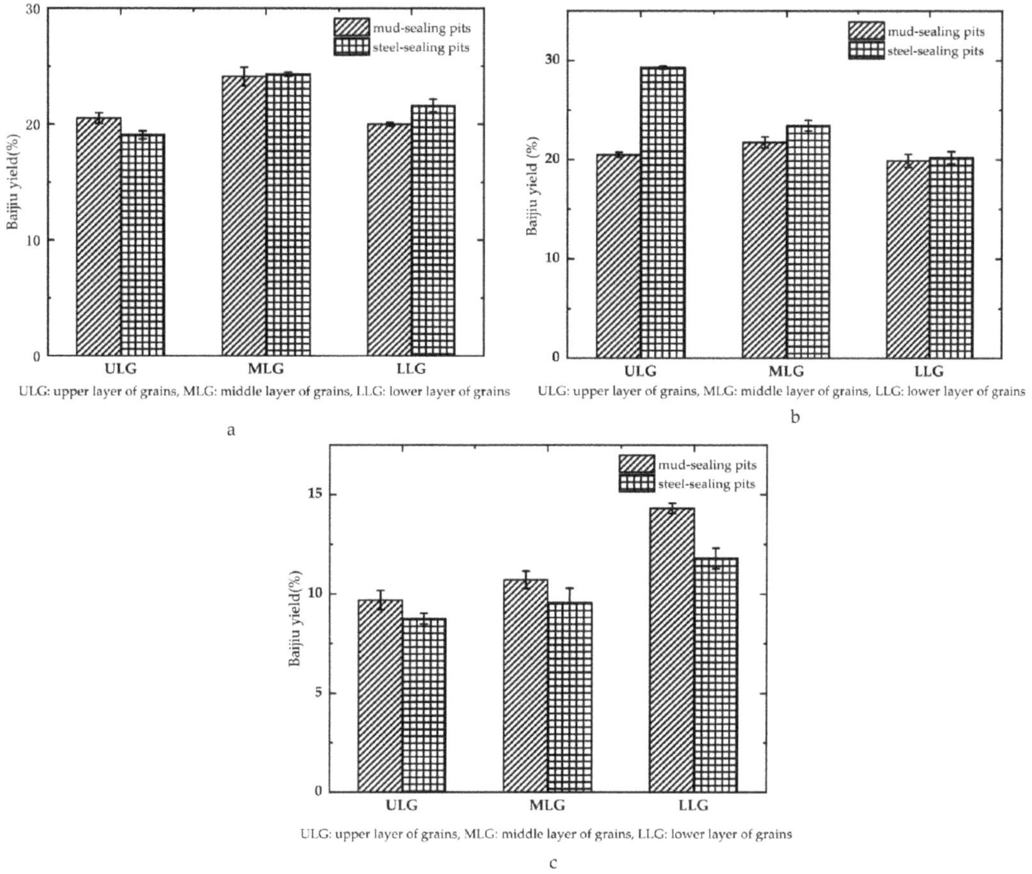

Figure 5. Yield of strongly flavored Baijiu from two different pit sealing methods. ((**a**)—1 segment of strong flavor Baijiu; (**b**)—2 segment of strong flavor Baijiu; (**c**)—3 segment of strong flavor Baijiu).

The second segment Baijiu was characterized by its high alcohol concentration and high ester concentration, and their aromas were rich, pure, and harmonious [29]. In addition, the quality of the second segment Baijiu was excellent, and for the corresponding pits with a small difference between the first segment Baijiu (0.30% difference between mud-sealing and steel-sealing pits) and the third segment Baijiu (1.54% difference), the yield of the second segment Baijiu in the steel-sealing pits (24.31%) increased compared with the second segment in the mud-sealing pits (20.72%). Overall, Baijiu quality improved.

From a comprehensive point of view, the rate of strongly flavored Baijiu production in steel-sealing pits increased in the upper and middle tiers of lees. The application of steel lids improved both quality and production of Baijiu significantly.

3.3. Analysis of Volatile Flavor Components in Strongly Flavored Baijiu Produced under Two Types of Pit Sealing Methods

3.3.1. Qualitative Analysis

The 270 strongly flavored Baijiu samples were studied qualitatively and quantitatively using GC-MS with a direct injection method. A total of 112 compounds were detected within 32 min, which included fifty-eight esters, fourteen acids, twenty-two alcohols, six aldehydes, seven ketones, and other compounds (Table 1). The analysis of the flavor compounds in the steel-sealing and the mud-sealing pits showed that there were 104 volatile flavor compounds in the mud-sealing pit and 102 volatile flavor compounds in the steel-sealing pit.

Further collation of the common and characteristic volatile compounds in the upper, middle, and lower lees in mud and steel sealing pits (Table 1) showed that the types of volatile compounds in the strongly flavored Baijiu of the mud cap pit increased from top to bottom with the fermented grain layers, and the lower layer of fermented grains had more types of compounds (Figure 6). The lower layer of fermented grains was in contact with the pit wall and the bottom of the fermentation pit. The surface of the mud pit of the cellar had more microorganisms, and more metabolites were produced in this complex microbial system. In addition, the lower layer of fermented grains was immersed under the yellow water line, and the composition of the yellow water was very complex. This composition consisted mainly of starch, polysaccharides, dextrins, proteins, some minerals, alcohols, aldehydes, acids, and esters, which can provide reasonable nutritional and material conditions for the growth of microorganisms, thus, causing the lower layer of grains to contain more flavor substances. There was no significant difference in the types and quantities of flavor compounds from different layers of fermented grains in the steel-sealing pit, probably due to the steel-sealing method, which allowed controlled internal fermentation that we observed through external monitoring.

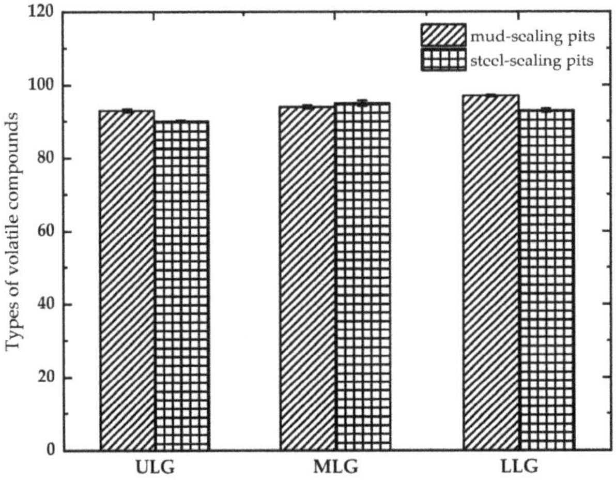

Figure 6. Amount of volatile flavor compounds in the different layers of fermented grains in the process of making strongly flavored Baijiu in China.

In addition, the following volatile compounds with antibacterial, analgesic, and anti-inflammatory activity were found in the middle fermented layer of grains in the steel-sealing pit: L-pyroglutamic acid methyl ester, and Cyclo (Phe-Pro) [30]. These compounds are beneficial when consumed and provide a basis for the direction of subsequent research on beneficial flavor compounds in this brand of Chinese Baijiu.

3.3.2. Further Analysis of Esters, Acids, and Alcohols in the Volatile Flavor Compounds

Esters are the main aromatic substances in strongly flavored Baijiu, with a pleasant fruity aroma, and they were dominated by ethyl caproate, ethyl acetate, ethyl lactate, and ethyl butyrate in our samples [31,32]. Esters are important substances that affect the flavor of Baijiu. Esters included ethyl acetate, ethyl caproate, ethyl lactate, ethyl butyrate, ethyl palmitate, ethyl anti oleate, ethyl valerate, ethyl heptanoate, and ethyl caproate in descending order of percentage, which conforms to the basic characteristics of strongly flavored Baijiu. The area percentage of ethyl acetate was the highest (3.66–18.53%), followed by ethyl caproate (4.34–16.57%), ethyl lactate (1.52–8.79%), ethyl butyrate (0.59–4.65%), ethyl palmitate (0.29–7.07%), ethyl trans oleate (0.03–5.73%), ethyl valerate (0.25–1.57%), ethyl heptanoate (0.11–0.96%), and ethyl octanoate (0.10–1.26%).

The following four types of alcohols were detected in all samples and accounted for a large proportion of all alcohols in this test: ethanol, n-butanol, propanol, and isobutanol. As the main component of Baijiu, ethanol concentration has an important impact on the yield of Baijiu. The following acids were detected at high levels: acetic acid, caproic acid, and butyric acid. Lactic acid, which is the precursor of ethyl lactate, was detected in a small amount in this test. It may be related to the weak volatility of lactic acid through analysis. In a future study, appropriate pretreatment methods can be selected to treat the samples to enrich the substances with weak volatility to obtain a more comprehensive detection of the flavor compounds of Baijiu.

3.3.3. Quantitative Analysis

Compared with the samples with more quantitative substances and considering that the Chinese Baijiu distillery uses single grain brewing, the volatile flavor components from single grain brewing contained fewer total esters compared with multi-grain brewing. We screened a total of 28 flavor components with higher concentration for quantitative analysis after determining the ethanol peaks; a total of 28 flavor components with a peak height >0.5% were selected for quantitative analysis, of which sixteen were esters, six were alcohols, and six were acids. The concentrations of flavor compounds in the steel-sealing pits were compared with the corresponding stratified (upper, middle, and lower) and segmented (1st, 2nd, and 3rd segments) pits in mud-sealing pits (Tables 2–4).

Alcohols are produced under aerobic conditions through the deamination of amino acids or the decarboxylation of sugars [33–35], most of which display a distinctive fruity flavor. In this study, six alcohols, which included n-propanol, isobutanol, n-butanol, isoamyl alcohol, n-amyl alcohol, and n-hexanol, were analyzed quantitatively in all samples. The quantitative situation is detailed in Tables 2–4.

Organic acids are the important flavor substances of strongly flavored Baijiu, and they are also the precursors of the flavor components of Baijiu. An appropriate amount of organic acids can make the Baijiu plump, harmonious, and with a long aftertaste. Among them, caproic acid, butyric acid, and acetic acid are the chromatographic skeletal components, which have obvious flavor-fixing effects in Baijiu. The corresponding esters generated by the interaction of different kinds of organic acids and alcohols constitute the main flavor substances of strongly flavored Baijiu [34]. When the concentration of these esters in all Baijiu samples was low, they exhibited a certain cheesy aroma, which increased the complexity of the aroma of strong flavor Baijiu.

Table 1. Identification of flavor components of strongly flavored Baijiu under two pit sealing methods.

No.	Compounds Specific to Mud-Sealing Pits, Steel-Sealing Pits	Volatile Flavor Compounds	Descriptions [36]
1		Acetaldehyde	Pungent, ether-like odor, fruity, coffee, wine and green aromas when diluted
2		2-Propanone	Pungent, sweet and slightly aromatic
3		Ethyl methanoate	Fruity aroma
4		Ethyl acetate	Pineapple scent
5		Methanol	Pungent odor
6		2-Methylbutyraldehyde	Asphyxiating odor
7		Isovaleraldehyde	Aroma of apple at high dilution, peach at concentration below 10 ppm
8		2-Hydroxy-3-pentanone	
9		Ethyl 2-methylpropanoate	
10		Propane,1,1-diethoxy-2-methyl-	
11		1,1-Diethoxy-pentan	

Table 1. *Cont.*

No.	Compounds Specific to Mud-Sealing Pits, Steel-Sealing Pits	Volatile Flavor Compounds	Descriptions [36]
12		2-Butanol	The smell of wine
13		Ethyl butyrate	Sweet and fruity, with notes of pineapple, banana and apple
14		1-Propanol	Strong aromas of meat at low concentrations
15		Acetaldehyde butyl ethyl acetal	
16		Ethyl isovalerate	Apple, mulberry aroma
17		Isovaleraldehyde diethyl acetal	
18		Butyl acetate	Fruity aroma, diluted with a pineapple, banana-like aroma
19		Isobutanol	Alcoholic, irritating odor
20		Isoamyl acetate	
21		2-Butanol, 3-methyl	
22		Ethyl valerate	Fruity, sourness
23		1-Butanol	Jasmine, spicy flavor
24		Amyl acetate	Fruity scent
25		Methyl hexanoate	Volatile, etheric aroma, pineapple-like
26		3-Methyl-1-butanol	Mixed alcoholic and spicy notes with mellow, etheric and banana aromas.
27		Butyl butyrate	Apple scent
28		2-Hexanol	
29		Ethyl caproate	Fruity aroma
30		1-Pentanol	
31		Isoamyl isobutyrate	
32		3-Hydroxy-2-butanone	Sweet, dairy aroma with a fatty, oily note
33		Imidazole-4-acetic acid	
34		1,1,3-Triethoxypropane	
35		Pentanoic acid, butyl ester	Sweet fruity, green aromas
36		Caproic acid propyl ester	Elegant aromas of pineapple and blackberry undertones
37		(S)-(+)-2-Heptanol	
38		Ethyl heptanoate	Fruity, green, waxy, Cornish
39		Ethyl L(-)-lactate	Sweet, tart and fruity aroma
40		1-Hexanol	
41		Methyl 2-hydroxyisobutyrate	
42		Hexanoic acid butyl ester	Elegant aromas of pineapple and apple
43		Hexyl butyrate	Green, waxy, and fruity, with a characteristic aroma of almond fruit
44		3-Methyl-2-butanol	
45		Ethyl caprylate	Waxy aroma, musty aroma, fruity apricot-like aroma, creamy aroma, milk aroma, sweet wine aroma
46		Hexyl acetate	Green, sweet and fruity aromas with hints of appleand banana peel
47		Ethylidene diacetate	
48		1-Hydroxy	Fresh, light, oily aroma with wine notes and a spicy flavor
49		Acetic acid glacial	Irritating odor
50		Furfural	Special scent
51		DL leucine ethyl ester	
52		Butyl lactate	Slightly smelly
53		Ethyl nonanoate	Fruity, waxy, estery and green aromas
54		Isoamyl lactate	

Table 1. *Cont.*

No.	Compounds Specific to Mud-Sealing Pits, Steel-Sealing Pits	Volatile Flavor Compounds	Descriptions [36]
55		Isobutyric acid	Pungent odor
56		(S)-(+)-1,2-Propanediol	
57		Hexyl hexanoate	Green, waxy, herbal and tropical fruit and berry aromas.
58		Octanoic acid, butylester	
59		Propionic acid	Irritating odor
60		Butyric acid	Putrid sour smell
61		Ethyl caprate	Aromas of grapes, Cornish wine
62		Phenylacetaldehyde	Hyacinth aroma
63		Furfuryl alcohol	Special smell and bitter spicy taste
64		Isovaleric acid	Pungent sour odor, with the aroma of cheese, dairy products and fruits after dilution
65		Heptyl formate	Floral and fruity aromas
66		(2,2-Diethoxyethyl)-Benzene	
67		2-Ethylbutyric acid	Sour, musty odor
68		Ethyl phenylacetate	Floral, fruity, powdery, woody, animal, cocoa
69		Hexyl caprylate	Fruity, green, waxy, ester
70		Phenethyl acetate	Rose aroma with dense sweet undertones
71		Ethyl dodecanoate	Gentle fragrance
72		Hexanoic acid	Sweat, cheese, sourness
73		Heptanoic acid	Fermented, waxy and fruity aromas
74		Octanoic acid	Sweat odor
75		Valeric acid	Sweat, cheese, sourness
76		Cyclopentadecanolide	
77		9-Ethyl oxynicotinate	
78		Pentadecanoic acid,ethyl ester	
79		2-Pentadecanone,6,10,14-trimethyl	
80		9-Hexadecenoic acid,ethyl ester	
81		9,9-Diethoxynonanoic acid ethyl ester	
82		Hexadecanoic acid ethyl ester	
83		Ethyl 3-phenylpropionate	
84		Ethyl 12-oxododecanoate	
85		2-Phenylethanol	Rose fragrance
86		9-Hexadecenoic acid	
87		2-[[(9Z,12Z)-9,12-Octadecadienyl]oxy]ethanol	
88		Heptadecanoic acid,ethyl ester	
89		Methyl (7Z)-7-hexadecenoate	
90		Ethyl myristate	
91		Octadecanoic acid,ethyl ester	
92		Ethyl oleate	Floral, fruity and oily aromas
93		Ethyl linoleate	
94		5,8,11,14-Eicosatetraenoicacid	
95		Ethyl alpha-linolenate	
96		Methyl linolenate	
97		Hexadecanoic acid	
98		2-Butanol	
99		2-Heptanol	
100		Heptyl heptanoate	
101		Methyl L-pyroglutamate	
102		3-Isobutyl-2,3,6,7,8,8a-hexahydrOpyrrolopyrazine-1,4-dione	

Table 1. *Cont.*

No.	Compounds Specific to Mud-Sealing Pits, Steel-Sealing Pits	Volatile Flavor Compounds	Descriptions [36]
103		Diethyl succinate	
104		Isobutyl hexanoate	Sweet fruity, green and waxy aromas
105		Heptanol	Fresh, lightly oily aroma with hints of wine and a spicy flavor
106		2-Hexadecanol	
107		2-Thiapropane	
108		Hexadecanoic acid, butyl ester	
109		Cyclo(Phe-Pro)	
110		5-Methyl-2-heptanone	
111		Methyl trans linoleic acid ester	
112		L(+)-Lactic acid	

The compounds in the table are sorted in the order of peak appearance. The colors in the table above each represent the following: Blue: Specific compounds of mud-sealing pit in the upper layer of fermented grains. Green: Specific compounds of steel-sealing pit in the upper layer of fermented grains. Brown: Specific compounds of mud-sealing pit in the middle layer of fermented grains. Grey: Specific compounds of steel-sealing pit in the middle layer of fermented grains. Purple: Specific compounds of mud-sealing pit in the lower layer of fermented grains. Orange: Specific compounds of steel-sealing pit in the lower layer of fermented grains. White: The compounds contained in the upper, middle, and lower fermented grains that corresponded to the mud-sealing pit and the steel-sealing pit.

Table 2. The concentration of flavor compounds in the upper, middle, and lower layers of the first segment of strongly flavored Baijiu in the pit under two pit sealing methods.

Flavor Substances	Flavor Substance Concentration/(mg·L^{-1})					
	Upper Fermented Layers of Grain		Middle Fermented Layers of Grain		Lower Fermented Layers of Grain	
	MSP-U	SSP-U	MSP-M	SSP-M	MSP-L	SSP-L
Ethyl methanoate	235.08 ± 94.66 ^{b}	459.76 ± 160.27 ^{a}	210.18 ± 114.29 ^{a}	402.28 ± 157.81 ^{a}	207.08 ± 122.88 ^{a}	224.87 ± 108.87 ^{a}
Ethyl acetate	1812.1 ± 223.84 ^{b}	2460.59 ± 223.35 ^{a}	2682.33 ± 1107.97^{b}	2840.74 ± 295.97 ^{a}	2774.64 ± 1005.81 ^{a}	2795.56 ± 585.39 ^{a}
Ethyl butyrate	487.92 ± 262.35 ^{a}	593.7 ± 154.07 ^{a}	530.99 ± 440.44 ^{a}	614.29 ± 228.29 ^{a}	620.86 ± 531.88 ^{a}	685.86 ± 251.58 ^{a}
1-Propanol	487.67 ± 255.75 ^{a}	695.46 ± 330.64 ^{a}	467.59 ± 284.17 ^{a}	1232.77 ± 559.95 ^{a}	612.69 ± 300.90 ^{a}	1482.69 ± 720.01 ^{a}
Isobutanol	184.2 ± 123.96 ^{a}	175.86 ± 94.34 ^{a}	166.15 ± 121.18 ^{a}	206.52 ± 94.86 ^{a}	179.86 ± 111.97 ^{a}	185.29 ± 72.20 ^{a}
Isoamyl acetate	15.9 ± 13.64 ^{a}	16.73 ± 15.12 ^{a}	19.91 ± 14.79 ^{a}	23.79 ± 14.68 ^{a}	20.51 ± 13.93 ^{a}	19.25 ± 17.82 ^{a}
Ethyl valerate	179.43 ± 89.42 ^{a}	235.58 ± 80.32 ^{a}	215.82 ± 136.95 ^{a}	297.52 ± 114.66 ^{a}	333.7 ± 280.75 ^{b}	408.28 ± 143.84 ^{a}
1-Butanol	496.08 ± 365.32 ^{a}	387.54 ± 250.84 ^{a}	443.94 ± 252.79 ^{a}	647.92 ± 193.42 ^{a}	528.92 ± 194.01 ^{a}	970.99 ± 429.79 ^{a}
3-Methyl-1-butanol	457.33 ± 230.13 ^{a}	301.37 ± 20.68 ^{a}	418.17 ± 179.38 ^{a}	457.01 ± 109.57 ^{a}	459.55 ± 178.81 ^{a}	451.23 ± 236.07 ^{b}
Ethyl caproate	2806.73 ± 790.85 ^{a}	2091.37 ± 678.98 ^{a}	2311.41 ± 826.22 ^{a}	2230.49 ± 537.40 ^{a}	2250.15 ± 1245.53 ^{a}	3350.5 ± 706.54 ^{a}
1-Pentanol	26.92 ± 22.77 ^{a}	14.7 ± 11.53 ^{a}	24.78 ± 15.11 ^{b}	28.79 ± 10.15 ^{a}	30.3 ± 12.57 ^{a}	57.9 ± 23.43 ^{a}
Caproic acid propyl ester	4.49 ± 3.25 ^{a}	2.72 ± 2.07 ^{a}	4.37 ± 3.27 ^{a}	6.45 ± 3.64 ^{a}	6.88 ± 5.68 ^{a}	11.66 ± 4.72 ^{a}
Ethyl heptanoate	133.01 ± 70.94 ^{a}	104.32 ± 37.16 ^{a}	106.9 ± 35.15 ^{a}	129.11 ± 49.92 ^{a}	166.33 ± 127.38 ^{a}	250.86 ± 63.04 ^{a}
Ethyl L(-)-lactate	698.37 ± 166.02 ^{b}	889.84 ± 262.05 ^{a}	821.9 ± 270.35 ^{a}	1041.07 ± 240.47 ^{a}	800.14 ± 262.51 ^{a}	1159.62 ± 572.36 ^{a}
1-Hexanol	115.9 ± 92.00 ^{a}	48.07 ± 27.13 ^{a}	102.79 ± 66.56 ^{a}	109.19 ± 27.98 ^{a}	117.42 ± 57.81 ^{a}	222.91 ± 32.37 ^{a}
Hexanoic acid butyl ester	46.65 ± 43.72 ^{a}	13.05 ± 7.96 ^{a}	33.22 ± 19.95 ^{a}	34.55 ± 18.97 ^{a}	55.29 ± 60.12 ^{a}	72.08 ± 27.32 ^{a}
Hexyl butyrate	20.86 ± 51.43 ^{a}	2.26 ± 2.13 ^{a}	13 ± 30.19 ^{a}	6.12 ± 4.24 ^{a}	10.52 ± 27.23 ^{a}	9.38 ± 4.15 ^{a}
Ethyl caprylate	70.19 ± 50.17 ^{a}	46.78 ± 24.12 ^{a}	49.82 ± 16.21 ^{b}	65.58 ± 26.25 ^{a}	56.82 ± 19.46 ^{a}	101.9 ± 30.50 ^{a}
Acetic acid glacial	613.8 ± 253.58 ^{a}	393.59 ± 133.25 ^{a}	569.14 ± 260.46 ^{a}	514.43 ± 140.11 ^{a}	574.06 ± 209.52 ^{a}	723.9 ± 333.80 ^{a}
Propionic acid	3.61 ± 6.46 ^{a}	2.01 ± 5.01 ^{a}	4.57 ± 5.43 ^{a}	7.83 ± 11.86 ^{a}	6.79 ± 6.36 ^{a}	11.28 ± 16.79 ^{a}
Butyric acid	197.26 ± 137.90 ^{a}	94.28 ± 48.82 ^{a}	238.03 ± 229.72 ^{a}	166.7 ± 144.17 ^{a}	227.31 ± 180.59 ^{a}	538.13 ± 315.81 ^{a}
Ethyl phenylacetate	2.66 ± 7.69 ^{a}	0.25 ± 0.32 ^{a}	5.09 ± 16.06 ^{a}	0.22 ± 0.59 ^{a}	1.63 ± 3.07 ^{a}	1.1 ± 1.11 ^{a}
Hexanoic acid	121.47 ± 76.72 ^{a}	69.2 ± 54.61 ^{a}	115.84 ± 86.22 ^{a}	116.15 ± 115.78 ^{a}	118.64 ± 76.84 ^{a}	508.84 ± 298.93 ^{a}
Heptanoic acid	3.77 ± 7.54 ^{a}	1.43 ± 2.70 ^{a}	1.79 ± 2.30 ^{a}	2.63 ± 4.66 ^{a}	2.09 ± 2.20 ^{a}	10.35 ± 11.70 ^{a}
Ethyl myristate	62.89 ± 127.36 ^{a}	8.72 ± 7.65 ^{a}	114.87 ± 301.70 ^{a}	12.06 ± 9.52 ^{a}	64.62 ± 130.19 ^{a}	7.32 ± 6.27 ^{a}
Octanoic acid	14.36 ± 24.04 ^{a}	6.73 ± 11.15 ^{a}	4.9 ± 6.65 ^{b}	10.62 ± 15.82 ^{a}	5.89 ± 5.70 ^{a}	14.63 ± 15.21 ^{a}
Ethyl hexadecanoate	94.04 ± 135.03 ^{a}	100.18 ± 103.25 ^{a}	72.32 ± 117.70 ^{a}	291.71 ± 324.12 ^{a}	74.6 ± 109.84 ^{a}	122.58 ± 178.81 ^{a}
Ethyl linoleate	282.67 ± 219.41 ^{a}	322.93 ± 238.47 ^{a}	255.41 ± 222.77 ^{a}	637.44 ± 467.96 ^{a}	359.73 ± 324.58 ^{a}	312.84 ± 238.30 ^{a}

Values are means ± SD. Within the same fermented layers of grain, the different letters $^{(a,b)}$ in the same row indicate the values are significantly different ($p \leq 0.05$) (MSP-U and SSP-U; MSP-M and SSP-M; MSP-L and SSP-L). MSP: mud-sealing pits, SSP: steel-sealing pits; U: upper layer of fermentation grains, M: middle layer of fermentation grains, L: lower layer of fermentation grains.

Table 3. The concentration of flavor compounds in the upper, middle, and lower layers of the second segment of strongly flavored Baijiu in the pit under two pit sealing methods.

Flavor Substances	Flavor Substance Concentration ($mg \cdot L^{-1}$)					
	Upper Fermented Layers of Grain		Middle Fermented Layers of Grain		Lower Fermented Layers of Grain	
	MSP-U	SSP-U	MSP-M	SSP-M	MSP-L	SSP-L
Ethyl methanoate	71.03 ± 36.90 [a]	153.65 ± 78.64 [a]	70.44 ± 33.68 [a]	114.28 ± 26.39 [a]	71.23 ± 29.80 [a]	123.05 ± 35.46 [a]
Ethyl acetate	999.44 ± 375.79 [b]	2330.36 ± 399.78 [a]	1410.46 ± 268.49 [a]	1711.45 ± 163.05 [a]	1625.14 ± 391.32 [a]	2082.55 ± 287.88 [a]
Ethyl butyrate	153.53 ± 108.87 [a]	241.96 ± 105.75 [a]	151.85 ± 68.44 [a]	243 ± 87.13 [a]	196.89 ± 93.52 [a]	402.86 ± 112.21 [a]
1-Propanol	341.22 ± 193.66 [a]	659.21 ± 369.21 [a]	340.63 ± 191.69 [b]	1016.33 ± 481.07 [a]	469.38 ± 177.48 [a]	1202.15 ± 432.57 [a]
Isobutanol	96.5 ± 75.83 [a]	97.4 ± 51.01 [a]	80.28 ± 55.70 [a]	111.07 ± 51.01 [a]	99.41 ± 61.61 [a]	127.57 ± 50.69 [a]
Isoamyl acetate	8.21 ± 6.81 [a]	10.46 ± 7.14 [a]	7.79 ± 6.11 [a]	6.71 ± 4.37 [a]	11.69 ± 12.40 [a]	7.62 ± 5.19 [a]
Ethyl valerate	74.92 ± 57.86 [b]	139.75 ± 60.19 [a]	88.31 ± 47.28 [a]	144.5 ± 63.79 [a]	134.61 ± 64.47 [a]	286.59 ± 92.00 [a]
1-Butanol	400.72 ± 320.88 [a]	382.45 ± 206.92 [a]	387.6 ± 286.48 [a]	579.2 ± 155.23 [a]	483.57 ± 247.42 [a]	855.96 ± 238.19 [a]
3-Methyl-1-butanol	362.9 ± 205.83 [a]	312.22 ± 157.90 [a]	340.95 ± 192.15 [a]	376.7 ± 93.57 [a]	368.99 ± 181.08 [a]	413.29 ± 136.30 [a]
Ethyl caproate	1666.63 ± 551.24 [a]	1345.67 ± 662.87 [a]	1875.86 ± 681.75 [a]	1269.1 ± 319.82 [a]	2231.09 ± 1083.01 [a]	2651.09 ± 649.99 [a]
1-Pentanol	21.35 ± 16.66 [a]	14.92 ± 10.17 [a]	21.18 ± 14.72 [a]	29.27 ± 10.18 [a]	28.99 ± 13.60 [a]	51.33 ± 9.22 [a]
Caproic acid propyl ester	1.51 ± 1.73 [a]	1.39 ± 1.12 [a]	1.61 ± 1.35 [a]	3.73 ± 3.22 [a]	3.44 ± 2.56 [a]	7.71 ± 2.37 [a]
Ethyl heptanoate	57.95 ± 47.71 [a]	56.22 ± 24.48 [a]	51.82 ± 21.13 [a]	65.99 ± 31.50 [a]	92.61 ± 64.03 [a]	183.83 ± 72.35 [a]
Ethyl L(-)-lactate	1359.5 ± 452.66[b]	1462.16 ± 490.25 [a]	1495.8 ± 410.03 [a]	1844.52 ± 338.16 [a]	1504.32 ± 489.43 [b]	2008.2 ± 256.74 [a]
1-Hexanol	117.51 ± 103.84 [a]	73.23 ± 43.14 [a]	117.26 ± 93.38 [a]	144.95 ± 52.89 [a]	139.59 ± 94.40 [a]	244.64 ± 39.02 [a]
Hexanoic acid butyl ester	15.19 ± 14.05 [a]	6.4 ± 3.21 [a]	14.02 ± 8.89 [a]	16.62 ± 11.87 [a]	30.89 ± 31.05 [a]	57.74 ± 21.92 [a]
Hexyl butyrate	7.79 ± 15.21 [a]	0.57 ± 0.35 [a]	5.93 ± 2.31 [a]	2.95 ± 2.61 [a]	4.76 ± 10.67 [a]	3.4 ± 3.29 [a]
Ethyl caprylate	27.34 ± 14.92 [a]	24.26 ± 10.92 [a]	27.16 ± 16.75 [a]	31.31 ± 18.78 [a]	43.78 ± 30.29 [a]	74.38 ± 28.54 [a]
Acetic acid glacial	554.36 ± 165.33 [a]	459.66 ± 149.95 [a]	572.94 ± 153.13 [b]	650.8 ± 132.11 [a]	573.73 ± 114.71 [a]	789.84 ± 206.79 [a]
Propionic acid	15.04 ± 16.58 [a]	4.65 ± 3.11 [a]	19.63 ± 18.06 [a]	12.57 ± 15.23 [a]	22.35 ± 18.44 [a]	10.16 ± 10.78 [a]
Butyric acid	171.91 ± 127.64 [a]	70.78 ± 35.55 [a]	210.31 ± 166.83 [a]	246.54 ± 214.47 [a]	335.86 ± 252.56 [a]	606.57 ± 211.58 [a]
Ethyl phenylacetate	0.86 ± 1.34 [a]	1.06 ± 2.01 [a]	1.09 ± 1.65 [a]	0.67 ± 1.19 [a]	1.3 ± 1.70 [a]	0.88 ± 1.30 [a]
Hexanoic acid	131.32 ± 137.83 [a]	27.86 ± 36.37 [a]	118.3 ± 86.30 [a]	166.11 ± 142.20 [a]	225.5 ± 213.02 [a]	546.31 ± 160.21 [a]
Heptanoic acid	1.94 ± 3.19 [a]	0.15 ± 0.52 [a]	1.4 ± 1.21 [a]	1.91 ± 3.20 [a]	4.11 ± 2.65 [a]	8.91 ± 7.78 [a]
Ethyl myristate	5.14 ± 3.21 [a]	1.8 ± 1.18 [a]	6.12 ± 11.60 [a]	1.38 ± 1.02 [a]	5.03 ± 3.25 [a]	1.05 ± 1.16 [a]
Octanoic acid	4.4 ± 63.22 [a]	0.74 ± 1.28 [a]	2.92 ± 3.27 [a]	2.79 ± 3.36 [a]	5.04 ± 2.22 [a]	10.39 ± 9.08 [a]
Ethyl hexadecanoate	15.65 ± 22.58 [a]	25.64 ± 25.66 [a]	14.19 ± 18.79 [a]	23.22 ± 22.10 [a]	16.18 ± 22.82 [b]	17.29 ± 18.80 [a]
Ethyl linoleate	37.32 ± 24.67 [a]	41.37 ± 23.08 [a]	31.42 ± 21.40 [a]	45.15 ± 16.51 [a]	35.46 ± 23.85 [a]	40.24 ± 16.88 [a]

Values are means ± SD. Within the same fermented layers of grain, the different letters [(a,b)] in the same row indicate the values are significantly different ($p \leq 0.05$) (MSP-U and SSP-U; MSP-M and SSP-M; MSP-L and SSP-L). MSP: mud-sealing pits, SSP: steel-sealing pits; U: upper layer of fermentation grains, M: middle layer of fermentation grains, L: lower layer of fermentation grains.

The comparison of the four major esters and three major acids of Baijiu in the 1st, 2nd, and 3rd segment of the upper, middle, and lower layers of fermented grains (Tables 5 and 6) is as follows:

For the distilling segment of the upper layer of fermented grains, the concentration of each of hexanoic acid, butyric acid, and acetic acid in the traditional mud pit sealing method was higher than that under the steel pit sealing method. Except for ethyl hexanoate, the total concentration of each of ethyl lactate, ethyl butyrate, and ethyl acetate was lower than that of the strongly flavored Baijiu under the steel pit sealing method. The possible reason for these results was that the upper layer of fermented grains had greater contact with the sealing mud under the mud sealing method, which resulted in a higher concentration of acid substances required for the later synthesis of esters compared with the steel sealing method.

For the middle layer of fermented grains, the concentration of caproic acid and acetic acid under the steel pit sealing method was higher than that under the traditional mud pit sealing method, and the total concentration of ethyl lactate, ethyl caproate, ethyl butyrate, and ethyl acetate was higher than that under the traditional pit sealing method. The fermentation of the middle layer of fermented grains in the pit was better than that of the upper layer of fermented grains, which may have been due to the better sealing performance under the steel pit sealing method. For the distilling segment of the lower fermented grains, the concentration of acetic acid, butyric acid, caproic acid, ethyl acetate, ethyl caproate, ethyl lactate, and ethyl butyrate under the steel pit sealing method was each higher than those under the traditional mud pit sealing method. Over time, the advantages of the steel pit sealing method became apparent because it had better sealing performance inside the pit and was less prone to pollution, which resulted in a relatively higher concentration of esters.

Table 4. The concentration of flavor compounds in the upper, middle, and lower layers of the third segment of strongly flavored Baijiu in the pit under two pit sealing methods.

Flavor Substances	Flavor Substance Concentration ($mg \cdot L^{-1}$)					
	Upper Fermented Layers of Grain		Middle Fermented Layers of Grain		Lower Fermented Layers of Grain	
	MSP-U	SSP-U	MSP-M	SSP-M	MSP-L	SSP-L
Ethyl methanoate	16.09 ± 16.37 ^{a}	15.73 ± 12.02 ^{a}	16.75 ± 12.32 ^{a}	21.62 ± 19.52 ^{a}	19.31 ± 16.24 ^{a}	23.79 ± 21.03 ^{a}
Ethyl acetate	226.64 ± 99.49 ^{b}	421.03 ± 139.23 ^{a}	522 ± 193.52 ^{a}	481.92 ± 96.07 ^{a}	645.09 ± 202.73 ^{a}	672.14 ± 131.17 ^{a}
Ethyl butyrate	48.52 ± 28.48 ^{a}	42.49 ± 27.37 ^{a}	55 ± 26.82 ^{a}	77.64 ± 41.53 ^{a}	75.72 ± 37.57 ^{a}	156.69 ± 64.26 ^{a}
1-Propanol	213.58 ± 113.31 ^{a}	404.09 ± 193.22 ^{a}	193.76 ± 115.14 ^{a}	635.73 ± 320.90 ^{a}	285.78 ± 117.20 ^{a}	877.9 ± 404.97 ^{a}
Isobutanol	39.38 ± 35.98 ^{a}	35.55 ± 22.26 ^{a}	26.29 ± 9.03 ^{a}	40.5 ± 22.06 ^{a}	43.27 ± 34.75 ^{a}	59.81 ± 30.84 ^{a}
Isoamyl acetate	2.65 ± 2.81 ^{a}	2.77 ± 3.22 ^{a}	2.62 ± 1.32 ^{b}	3.94 ± 3.00 ^{a}	3.04 ± 2.65 ^{a}	4.38 ± 3.44 ^{a}
Ethyl valerate	32.36 ± 24.54 ^{a}	38.41 ± 24.32 ^{a}	44.17 ± 26.56 ^{a}	55.62 ± 27.94 ^{a}	60.19 ± 30.55 ^{b}	122.29 ± 47.98 ^{a}
1-Butanol	249.67 ± 209.21 ^{a}	250.4 ± 130.12 ^{a}	238.58 ± 190.40 ^{a}	356.75 ± 113.81 ^{a}	306.03 ± 188.72 ^{a}	541.3 ± 184.70 ^{a}
3-Methyl-1-butanol	199.22 ± 125.42 ^{a}	195.4 ± 57.62 ^{a}	190.19 ± 128.97 ^{a}	208.92 ± 63.11 ^{a}	214.53 ± 125.64 ^{a}	247.12 ± 73.97 ^{a}
Ethyl caproate	905.53 ± 346.18 ^{a}	484.1 ± 149.54 ^{a}	1012.77 ± 281.10 ^{a}	648.4 ± 220.26 ^{a}	1265.07 ± 541.02 ^{a}	1570.82 ± 365.85 ^{a}
1-Pentanol	13.74 ± 10.48 ^{a}	13.18 ± 5.02 ^{a}	13.79 ± 10.08 ^{b}	19.28 ± 6.20 ^{a}	19.5 ± 10.88 ^{a}	37.15 ± 7.31 ^{a}
Caproic acid propyl ester	0.73 ± 1.08 ^{a}	0.53 ± 0.24 ^{a}	0.74 ± 0.54 ^{a}	1.49 ± 1.33 ^{a}	1.66 ± 1.33 ^{a}	4.17 ± 1.75 ^{a}
Ethyl heptanoate	32.51 ± 27.14 ^{a}	26.38 ± 11.54 ^{a}	30.95 ± 13.82 ^{a}	34.45 ± 12.33 ^{a}	59.17 ± 45.12 ^{a}	115.49 ± 47.20 ^{a}
Ethyl L(-)-lactate	2648.81 ± 1229.76 ^{a}	2940.37 ± 628.09 ^{a}	2842.17 ± 933.37 ^{a}	2920.39 ± 552.66 ^{a}	2521.66 ± 864.35 ^{a}	2616.47 ± 383.57 ^{a}
1-Hexanol	92.78 ± 81.72 ^{a}	91.24 ± 33.48 ^{b}	93.33 ± 77.25 ^{b}	116.73 ± 31.10 ^{a}	117.48 ± 92.38 ^{a}	192.08 ± 65.62 ^{a}
Hexanoic acid butyl ester	13.12 ± 13.55 ^{a}	4.32 ± 2.46 ^{a}	11.95 ± 7.32 ^{a}	11.38 ± 5.00 ^{a}	22.94 ± 20.95 ^{a}	41.43 ± 12.75 ^{a}
Hexyl butyrate	23.36 ± 16.25 ^{a}	0 ± 0.00 ^{a}	19.11 ± 8.65 ^{a}	0.43 ± 1.15 ^{a}	17.32 ± 10.25 ^{a}	1.82 ± 1.58 ^{a}
Ethyl caprylate	29.04 ± 26.32 ^{a}	17.01 ± 6.20 ^{a}	27.73 ± 21.94 ^{a}	20.69 ± 8.05 ^{a}	56.31 ± 36.21 ^{a}	55.46 ± 20.42 ^{a}
Acetic acid glacial	734.36 ± 241.57 ^{a}	693.69 ± 142.03 ^{a}	808.87 ± 257.65 ^{a}	899.4 ± 228.82 ^{a}	745.66 ± 218.01 ^{a}	958.74 ± 222.63 ^{a}
Propionic acid	24.52 ± 12.32 ^{a}	11.03 ± 25.44 ^{a}	24.99 ± 20.21 ^{a}	17.77 ± 35.31 ^{a}	26.54 ± 21.02 ^{a}	24.74 ± 19.67 ^{a}
Butyric acid	347.55 ± 272.00 ^{a}	172.21 ± 59.24 ^{a}	454.97 ± 437.84 ^{a}	401.59 ± 319.99 ^{a}	625.4 ± 236.27 ^{a}	848.57 ± 237.99 ^{a}
Ethyl phenylacetate	2.3 ± 2.95 ^{a}	1.51 ± 1.20 ^{a}	2.51 ± 1.35 ^{a}	1.29 ± 1.02 ^{a}	2.31 ± 2.01 ^{a}	1.67 ± 1.06 ^{a}
Hexanoic acid	235.35 ± 210.98 ^{a}	112.99 ± 41.17 ^{a}	253.99 ± 175.17 ^{a}	259.1 ± 177.62 ^{a}	345.06 ± 269.05 ^{a}	779.06 ± 172.83 ^{a}
Heptanoic acid	4.85 ± 4.25 ^{a}	1.35 ± 1.20 ^{a}	3.84 ± 3.59 ^{a}	2.89 ± 2.70 ^{a}	7.51 ± 11.84 ^{a}	13.04 ± 9.55 ^{a}
Ethyl myristate	6.88 ± 2.52 ^{a}	2.32 ± 2.01 ^{a}	8.38 ± 2.68 ^{a}	2.32 ± 1.12 ^{a}	6.95 ± 2.68 ^{a}	1.74 ± 0.89 ^{a}
Octanoic acid	9.12 ± 5.21 ^{a}	1.63 ± 1.00 ^{a}	7.02 ± 7.28 ^{a}	3.59 ± 2.21 ^{a}	9.91 ± 8.47 ^{a}	15.97 ± 12.25 ^{a}
Ethyl hexadecanoate	22.01 ± 2.23 ^{b}	30.1 ± 25.61 ^{a}	28.14 ± 33.73 ^{a}	29.49 ± 25.03 ^{a}	25.69 ± 19.66 ^{b}	33.51 ± 30.27 ^{a}
Ethyl linoleate	37.18 ± 21.02 ^{a}	42.76 ± 6.72 ^{a}	42.7 ± 30.88 ^{a}	44.43 ± 7.46 ^{a}	40.53 ± 28.81 ^{a}	47.06 ± 17.54 ^{a}

Values are means ± SD. Within the same fermented layers of grain, the different letters $^{(a,b)}$ in the same row indicate the values are significantly different ($p \leq 0.05$) (MSP-U and SSP-U; MSP-M and SSP-M; MSP-L and SSP-L). MSP: mud-sealing pits, SSP: steel-sealing pits; U: upper layer of fermentation grains, M: middle layer of fermentation grains, L: lower layer of fermentation grains.

Table 5. Concentration of four major esters in different fermented layers of grain under two pit sealing methods.

Fermented Grains Layer	Pit Sealing Method	Ethyl Acetate ($mg \cdot L^{-1}$)	Ethyl Caproate ($mg \cdot L^{-1}$)	Ethyl Lactate ($mg \cdot L^{-1}$)	Ethyl Butyrate ($mg \cdot L^{-1}$)
Upper fermented grains	mud	1012.73 ± 5.86 ^{b}	1792.96 ± 6.95 ^{a}	1568.89 ± 9.31 ^{a}	229.99 ± 14.32 ^{b}
	steel	1737.33 ± 9.55 ^{a}	1307.05 ± 8.20 ^{a}	1764.12 ± 10.11 ^{a}	292.72 ± 12.01 ^{a}
Middle fermented grains	mud	1538.26 ± 10.21 ^{a}	1733.35 ± 19.22 ^{a}	1719.96 ± 15.20 ^{a}	245.95 ± 8.95 ^{a}
	steel	1678.04 ± 15.36 ^{a}	1382.66 ± 14.32 ^{b}	1935.33 ± 13.01 ^{a}	311.64 ± 6.34 ^{a}
Lower fermented grains	mud	1681.62 ± 7.21 ^{a}	1915.44 ± 7.21 ^{a}	1608.71 ± 16.62 ^{a}	297.82 ± 7.41 ^{a}
	steel	1850.08 ± 10.22 ^{a}	2524.14 ± 6.74 ^{a}	1928.10 ± 9.89 ^{a}	415.14 ± 8.16 ^{a}

Values are means ± SD. Within the same fermented layers of grain, the different letters $^{(a,b)}$ in the same column indicate the values are significantly different ($p \leq 0.05$). (Comparing the upper/middle/lower layer of the mud- and steel-sealing pits, respectively).

Table 6. Concentration of three major acids in different layers of fermented grain under two pit sealing methods.

Fermented Grains Layer	Pit Sealing Method	Acetic Acid ($mg \cdot L^{-1}$)	Butyrate ($mg \cdot L^{-1}$)	Caproic Acid ($mg \cdot L^{-1}$)
Upper fermented grains	mud	634.17 ± 7.25 [a]	238.91 ± 3.21 [a]	162.71 ± 7.20 [a]
	steel	515.65 ± 6.68 [a]	112.42 ± 10.35 [a]	70.02 ± 9.51 [b]
Middle fermented grains	mud	650.32 ± 8.69 [a]	301.10 ± 12.27 [a]	162.71 ± 4.32 [a]
	steel	688.21 ± 10.21 [a]	271.61 ± 6.63 [a]	180.45 ± 5.88 [a]
Lower fermented grains	mud	631.15 ± 15.64 [a]	396.19 ± 7.86 [a]	229.73 ± 11.02 [a]
	steel	824.16 ± 9.44 [a]	664.42 ± 3.28 [a]	611.40 ± 12.11 [a]

Values are means ± SD. Within the same fermented layers of grain, the different letters (a,b) in the same column indicate the values are significantly different ($p \leq 0.05$). (Comparing the upper/middle/lower layer of the mud- and steel-sealing pits, respectively).

3.4. Differentiation of Strongly Flavored Baijiu with Two Different Pit Sealing Methods by HCA

HCA is a technique for natural aggregation of research objects by similarity, an unsupervised pattern recognition method, and a statistical analysis technique for classifying research objects into relatively homogeneous groups [37,38]. In this experiment, 28 volatile compounds with an area percentage > 0.5% were selected, and their concentrations were used as a variable to standardize the data. The strongly flavored Baijiu with the 1st, 2nd, and 3rd segment of the two sealing methods in the upper, middle, and lower strong flavor Baijiu were clustered into two categories, with the mud-sealing Baijiu samples divided into one category and the steel-sealing Baijiu samples clustered into another category (Figure 7). The differences between the sealing methods indicated that the clustering analysis differentiated the strongly flavored Baijiu produced by the different sealing methods and that the 28 quantitative substances selected were correct. However, the HCA method did not fully cluster the same batch of Baijiu together. For example, MGT 141, MGT 241, and MGT 341 were divided into two clusters in the top lees of the Baijiu sample. A possible reason for this is that the mud-sealing pit was not stable due to external influences during the brewing process.

3.5. Analysis of PCA Scores for Two Types of Pit Sealing Methods

PCA is a multivariate statistical analysis method in which multiple variables are transformed linearly in an unsupervised mode to select a smaller number of significant variables [39,40]. Using SPSS 23.0, PCA was performed on 28 volatile compounds of strong flavor Baijiu to extract data with eigenvalues > 1, and the original information was reduced to seven principal components. The variance contributions of F1, F2, F3, F4, F5, F6, and F7 were 30.433%, 15.441%, 11.520%, 5.452%, 4.984%, 4.413%, and 3.737%, respectively, with a cumulative variance contribution of 75.979% (Table 7). This indicated that these seven principal components explained 75.979% of the 28 variables. Taking the first seven principal components for analysis was feasible due to the large sample size and the cumulative contribution that was close to 80%, which ensured that the condensed composite variables were representative of most of the data. The 28 original variables of strong flavor base Baijiu with different pit sealing methods (ethyl valerate, ethyl heptate, propyl caproate … ethyl lactate, and propionic acid are denoted as X_1, X_2, X_3 … X_{27}, X_{28}, respectively) had a linear relationship with each principal component. Table 8 shows the coefficients of the relationships between each variable and the principal components, and the linear equations for the seven principal components and variables are as follows:

$$F1 = 0.299X_1 + 0.293X_2 + 0.287X_3 + 0.282X_4 + 0.268X_5 + 0.237X_6 + 0.234X_7 + 0.234X_8 + 0.233X_9 + 0.226X_{10} + 0.215X_{11} + 0.212X_{12} + 0.202X_{13} + 0.202X_{14} + 0.190X_{15} + 0.086X_{16} - 0.006X_{17} + 0.123X_{18} - 0.052X_{19} - 0.044X_{20} + 0.087X_{21} - 0.022X_{22} + 0.137X_{23} + 0.119X_{24} + 0.161X_{25} + 0.068X_{26} - 0.090X_{27} - 0.060X_{28}$$

$$F2 = 0.009X_1 + 0.092X_2 + 0.147X_3 - 0.077X_4 - 0.194X_5 - 0.220X_6 + 0.167X_7 - 0.022X_8 + 0.120X_9 + 0.206X_{10} + 0.049X_{11} - 0.197X_{12} + 0.083X_{13} - 0.195X_{14} - 0.172X_{15} + 0.399X_{16} + 0.314X_{17} + 0.293X_{18} + 0.283X_{19} + 0.096X_{20} + 0.023X_{21} + 0.028X_{22} + 0.261X_{23} - 0.095X_{24} - 0.169X_{25} + 0.227X_{26} + 0.236X_{27} + 0.175X_{28}$$

$$F3 = 0.145X_1 + 0.097X_2 + 0.051X_3 + 0.143X_4 + 0.070X_5 + 0.096X_6 + 0.213X_7 + 0.132X_8 - 0.238X_9 - 0.203X_{10} - 0.108X_{11} - 0.050X_{12} - 0.007X_{13} + 0.041X_{14} - 0.261X_{15} - 0.063X_{16} + 0.140X_{17} + 0.051X_{18} - 0.021X_{19} + 0.445X_{20} + 0.430X_{21} + 0.382X_{22} - 0.309X_{23} - 0.142X_{24} - 0.066X_{25} - 0.100X_{26} + 0.038X_{27} + 0.017X_{28}$$

$$F4 = -0.138X_1 - 0.007X_2 - 0.040X_3 - 0.174X_4 - 0.070X_5 + 0.008X_6 + 0.037X_7 + 0.133X_8 + 0.018X_9 - 0.029X_{10} + 0.018X11 - 0.345X_{12} + 0.096X_{13} - 0.126X_{14} - 0.079X_{15} - 0.059X_{16} + 0.121X_{17} - 0.030X_{18} - 0.227X_{19} + 0.088X_{20} + 0.030X_{21} + 0.202X_{22} + 0.040X_{23} + 0.584X_{24} + 0.431X_{25} + 0.284X_{26} - 0.200X_{27} + 0.073X_{28}$$

$$F5 = -0.005X_1 + 0.030X_2 - 0.005X_3 - 0.132X_4 - 0.074X_5 - 0.113X_6 + 0.127X_7 - 0.169X_8 + 0.321X_9 + 0.253X_{10} + 0.060X_{11} - 0.085X_{12} + 0.131X_{13} - 0.090X_{14} + 0.188X_{15} - 0.229X_{16} - 0.001X_{17} - 0.267X_{18} - 0.313X_{19} + 0.154X_{20} + 0.103X_{21} + 0.293X_{22} + 0.252X_{23} + 0.054X_{24} - 0.226X_{25} - 0.42X_{26} + 0.203X_{27} + 0.007X_{28}$$

$$F6 = 0.065X_1 + 0.006X_2 + 0.002X_3 + 0.012X_4 + 0.066X_5 + 0.019X_6 - 0.058X_7 - 0.275X_8 - 0.060X_9 - 0.072X_{10} + 0.078X_{11} + 0.001X_{12} - 0.154X_{13} + 0.273X_{14} + 0.087X_{15} - 0.009X_{16} - 0.174X_{17} + 0.186X_{18} + 0.120X_{19} + 0.030X_{20} + 0.077X_{21} - 0.040X_{22} - 0.063X_{23} + 0.280X_{24} + 0.266X_{25} - 0.160X_{26} + 0.514X_{27} + 0.511X_{28}$$

$$F7 = 0.001X_1 - 0.141X_2 - 0.014X_3 + 0.157X_4 - 0.005X_5 + 0.098X_6 - 0.229X_7 + 0.084X_8 + 0.040X_9 + 0.067X_{10} + 0.297X_{11} + 0.181X_{12} + 0.168X_{13} - 0.032X_{14} - 0.066X_{15} - 0.066X_{16} + 0.368X_{17} - 0.354X_{18} + 0.376X_{19} + 0.082X_{20} - 0.383X_{21} + 0.312X_{22} - 0.145X_{23} - 0.011X_{24} + 0.154X_{25} - 0.071X_{26} + 0.094X_{27} + 0.095X_{28}$$

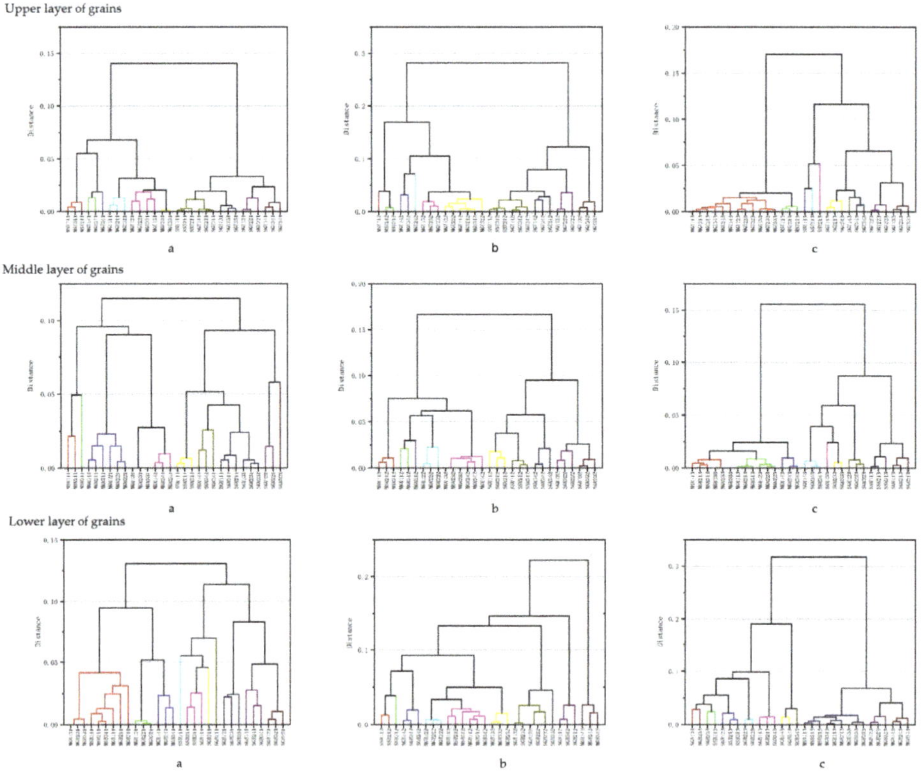

Figure 7. Clustering analysis of strongly flavored Baijiu with different pit sealing methods. ((**a**)—1 segment strong flavor Baijiu; (**b**)—2 segment strong flavor Baijiu; (**c**)—3 segment strong flavor Baijiu; The names of the letters in the above picture mean: initial "M" for mud-sealing pit, initial "S" for steel-sealing pit).

Table 7. Eigenvalues and variance contribution of principal components for the analysis of the scores of two pit sealing methods.

Components	Eigenvalue	Percent Variance	Cumulative Contribution (%)
PC1	8.521	30.433	30.433
PC2	4.323	15.441	45.874
PC3	3.226	11.52	57.394
PC4	1.526	5.452	62.845
PC5	1.396	4.984	67.829
PC6	1.236	4.413	72.242
PC7	1.046	3.737	75.979

Table 8. Composition matrix of 28 volatile compounds in strong flavor Baijiu based on principal component analysis.

Quantitative Compound	F1	F2	F3	F4	F5	F6	F7
Ethyl valerate (X_1)	0.299	0.009	0.145	−0.138	−0.005	0.065	0.001
Ethyl heptanoate (X_2)	0.293	0.092	0.097	−0.007	0.03	0.006	−0.141
Caproic acid propyl ester (X_3)	0.287	0.147	0.051	−0.04	−0.005	0.002	−0.014
Ethyl butyrate (X_4)	0.282	−0.077	0.143	−0.174	−0.132	0.012	0.157
Ethyl acetate (X_5)	0.268	−0.194	0.07	−0.07	−0.074	0.066	−0.005
Ethyl methanoate (X_6)	0.237	−0.22	0.096	0.008	−0.113	0.019	0.098
Hexanoic acid butyl ester (X_7)	0.234	0.167	0.213	0.037	0.127	−0.058	−0.229
Ethyl caproate (X_8)	0.234	−0.022	0.132	0.133	−0.169	−0.275	0.084
1-Butanol (X_9)	0.233	0.12	−0.238	0.018	0.321	−0.06	0.04
1-Pentanol (X_{10})	0.226	0.206	−0.203	−0.029	0.253	−0.072	0.067
1-Propanol (X_{11})	0.215	0.049	−0.108	0.018	0.06	0.078	0.297
Isobutanol (X_{12})	0.212	−0.197	−0.05	−0.345	−0.085	0.001	0.181
Ethyl caprylate (X_{13})	0.202	0.083	−0.007	0.096	0.131	−0.154	0.168
Isoamyl acetate (X_{14})	0.202	−0.195	0.041	−0.126	−0.09	0.273	−0.032
3-Methyl-1-butanol (X_{15})	0.19	−0.172	−0.261	−0.079	0.188	0.087	−0.066
Hexanoic acid (X_{16})	0.086	0.399	−0.063	−0.059	−0.229	−0.009	−0.066
Butyric acid (X_{17})	−0.006	0.314	0.14	0.121	−0.001	−0.174	0.368
Heptanoic acid (X_{18})	0.123	0.293	0.051	−0.03	−0.267	0.186	−0.354
Acetic acid glacial (X_{19})	−0.052	0.283	−0.021	−0.227	−0.313	0.12	0.376
Ethyl phenylacetate (X_{20})	−0.044	0.096	0.445	0.088	0.154	0.03	0.082
Ethyl myristate (X_{21})	0.087	0.023	0.43	0.03	0.103	0.077	−0.383
Hexyl butyrate (X_{22})	−0.022	0.028	0.382	0.202	0.293	−0.04	0.312
1-Hexanol (X_{23})	0.137	0.261	−0.309	0.04	0.252	−0.063	−0.145
Ethyl palmitate (X_{24})	0.119	−0.095	−0.142	0.584	0.054	0.28	−0.011
Ethyl linoleate (X_{25})	0.161	−0.169	−0.066	0.431	−0.226	0.266	0.154
Octanoic acid (X_{26})	0.068	0.227	−0.1	0.284	−0.424	−0.16	−0.071
Ethyl L(-)-lactate (X_{27})	−0.09	0.236	0.038	−0.2	0.203	0.514	0.094
Propionic acid (X_{28})	−0.06	0.175	0.017	0.073	0.007	0.511	0.095

The scoring model was $F = 30.433F1 + 15.441F2 + 11.52F3 + 5.452F4 + 4.984F5 + 4.413F6 + 3.737F7$, which we used to calculate the comprehensive score of strongly flavored Baijiu for different pit sealing methods (Table 9). In general, the average score of steel-sealing pits was higher than that of mud-sealing pits. The higher score of the strong flavor Baijiu with the steel cap pits may be attributed to the improvement in the sealing of the steel caps, which ensured the anaerobic environment in the pit to enhance the anaerobic respiration of microorganisms and metabolism of more flavor substances.

The PCA method can be applied to the discriminant function of product type and variety. We selected the samples of strongly flavored Baijiu produced from the upper, middle, and lower levels of fermented grains as the scatter diagram (Figure 8), and the results are as follows:

Table 9. Composition matrix of 28 volatile compounds in strongly flavored Baijiu based on principal component analysis.

Samples	Score
Strong flavor Baijiu made from the upper grains of the mud-sealing pit	
Strong flavor Baijiu made from the middle grains of the mud-sealing pit	0.624 ± 0.117^{b}
Strong flavor Baijiu made from the lower grains of the mud-sealing pit	
Strong flavor Baijiu made from the upper grains of the steel-sealing pit	
Strong flavor Baijiu made from the middle grains of the steel-sealing pit	0.632 ± 0.052^{a}
Strong flavor Baijiu made from the lower grains of the steel-sealing pit	

Values are means ± SD. The different letters [(ab)] in the same column indicate the values are significantly different ($p \leq 0.05$).

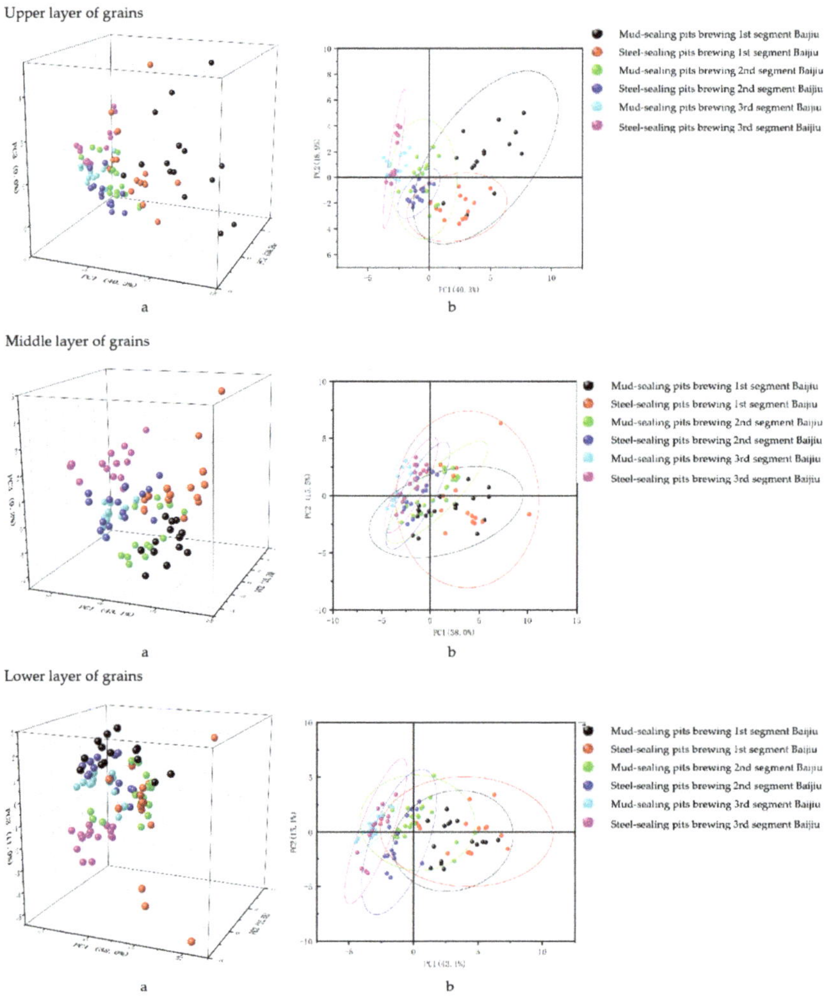

Figure 8. Scatter plot of the main components of the flavor substances of Baijiu. ((**a**)—three-dimensional scatter plot of PC1, PC2, and PC3; (**b**)—plan view of PC1 and PC2).

From the figure, the strongly flavored Baijiu produced from different pit sealing forms overlapped and could not be separated clearly. The reason may be that the sample size was too large and the fermentation process was not significantly different, which resulted in incomplete separation. Among them, the scatter diagram of the first segment Baijiu of the upper, middle, and lower levels of fermented grains was relatively scattered, which may be because the distillery usually relied on the experience of the master when picking Baijiu, and there was the possibility of picking Baijiu too early or too later. However, each layer of segmented Baijiu was obviously gathered together, which was consistent with the results obtained by HCA. This indicated that PCA and HCA were consistent with each other when classifying and identifying strongly flavored Baijiu with different pit sealing methods.

3.6. Discriminant Analysis

Five typical discriminant functions were obtained by discriminant analysis for each of the different layers of grains. The upper layer of grains with eigenvalues of 25.19, 11.00, 2.70, 1.37, and 0.44 explained 61.90%, 27.00%, 6.60%, 3.40%, and 1.10%, respectively, of the variation in the model. The first two typical discriminant analysis functions explained 88.90% of the variance. The middle layer of grains with eigenvalues of 18.03, 7.32, 5.04, 2.20, and 0.67 explained 54.20%, 22.00%, 15.10%, 6.60%, and 2.00%, respectively, of the variation in the model. The first two typical discriminant analysis functions explained 76.20% of the variation. The lower layer of grains with eigenvalues of 12.03, 5.32, 2.38, 1.91, and 0.29 explained 54.90%, 24.20%, 10.90%, 8.70%, and 1.30%, respectively, of the variation in the model. The first two typical discriminant analysis functions explained 79.10% of the variance.

It is possible to describe the differences and links between the flavor components in the top lees 1st, 2nd, and 3rd segments of strongly flavored Baijiu in mud-sealing pits and steel-sealing pits. The first two typical discriminant functions were used to make scatter plots for the different segment distillations of the same layer of fermented grains (Figure 9). For the corresponding 1st, 2nd, and 3rd segments Baijiu of the upper, middle, and lower layers of fermented grains, the centroids of the six types of Baijiu samples were separated from each other without any overlap. The group mass centers of the 1st, 2nd, and 3rd segments of the mud-sealing pits and steel-sealing pits were closer together, and the samples were more similar. However, the mud-sealing pit samples were separated approximately from the steel-sealing pit samples, but this result was in general agreement with the results of the principal component analysis, which indicated that both principal component analysis and discriminant analysis were very effective tools for the two different sealing methods. These statistical methods were useful for classification and prediction.

We analyzed 108 strongly flavored Baijiu samples from the upper, middle, and lower layers of fermented grains using DA. We randomly selected one sample from the 1st, 2nd, and 3rd segments of steel and mud sealing pit brewing as a "test sample", which totaled six samples. We used the remaining 102 samples as "training samples" to establish a discriminant function and then discriminated the test samples to determine their classification to compare them with the actual classification of the samples. In this study, six typical variables were obtained, and a discriminant function was established based on these variables to classify two different types of strongly flavored Baijiu with different pit sealing methods, with an accurate discrimination rate of 100.0% (Tables 10–12).

Based on the predicted results of the discriminant analysis, we used the GC-MS analysis of the total ion flow chromatogram of strongly flavored Baijiu combined with the discriminant analysis to classify and to identify the model for the two different types of strongly flavored Baijiu with different pit sealing methods. The correct identification rate of the samples was 100.0%.

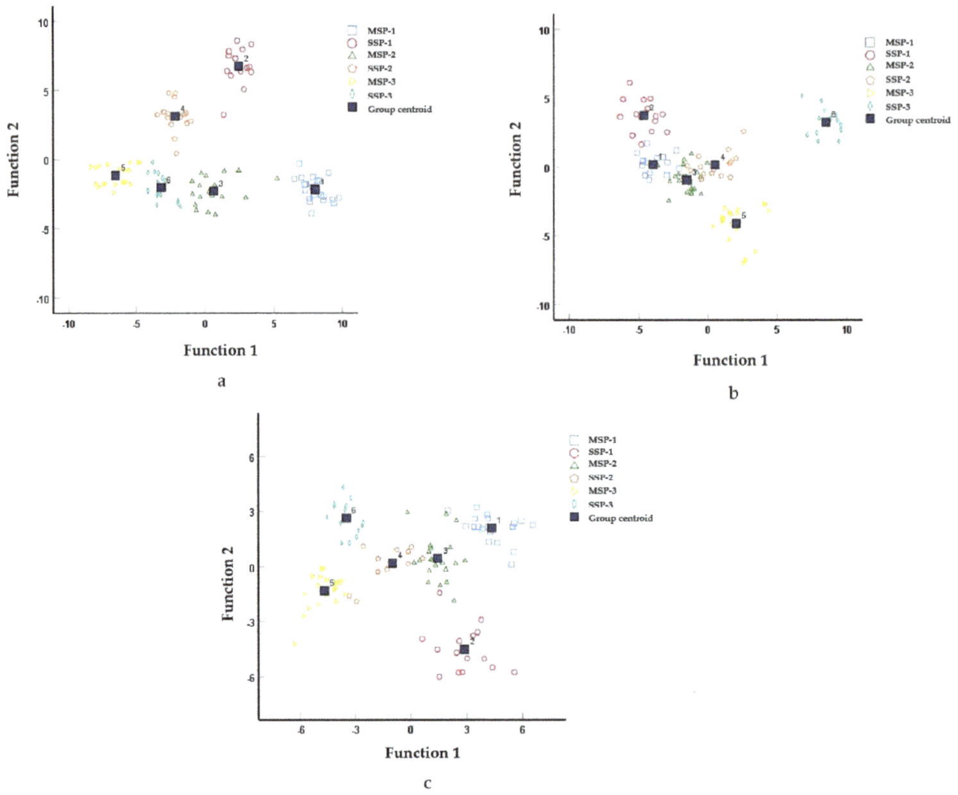

Figure 9. Scatter diagram of a typical discriminant analysis of the two sealing methods for strongly flavored Baijiu. ((**a**)—upper layer of grain; (**b**)—middle layer of grain; (**c**)—lower layer of grain; The names of the letters in the above picture mean: MSP for mud-sealing pits, SSP for steel-sealing pits; 1—1 segment strong flavor Baijiu; 2—2 segment strong flavor Baijiu; 3—3 segment strong flavor Baijiu).

Table 10. Predicted results of the typical discriminant analysis of the strongly flavored Baijiu of the upper layer of grains for the two methods for sealing the pits.

	Samples	Results of Prediction						
		MSP-U1st	SSP-U1st	MSP-U2nd	SSP-U2nd	MSP-U3rd	SSP-U3rd	Total
Original count	MSP-U1st	21						21
	SSP-U1st		15					15
	MSP-U2nd			21				21
	SSP-U2nd				15			15
	MSP-U3rd					21		21
	SSP-U3rd						15	15
Proportion (%)	MSP-U1st	100						100
	SSP-U1st		100					100
	MSP-U2nd			100				100
	SSP-U2nd				100			100
	MSP-U3rd					100		100
	SSP-U3rd						100	100

100.0% of original grouped cases were classified correctly. 1st, 2nd, 3rd represent for the first, second, and third segment strongly flavored Baijiu, respectively.

Table 11. Predicted results of the typical discriminant analysis of the strongly flavored Baijiu of the middle layer of grains for the two methods for sealing the pits.

	Samples	Results of Prediction						
		MSP-M1st	SSP-M1st	MSP-M2nd	SSP-M2nd	MSP-M3rd	SSP-M3rd	Total
Original count	MSP-M1st	18						18
	SSP-M1st		15					15
	MSP-M2nd			24				24
	SSP-M2nd				15			15
	MSP-M3rd					21		21
	SSP-M3rd						15	15
Proportion (%)	MSP-M1st	100						100
	SSP-M1st		100					100
	MSP-M2nd			100				100
	SSP-M2nd				100			100
	MSP-M3rd					100		100
	SSP-M3rd						100	100

100.0% of original grouped cases were classified correctly. 1st, 2nd, 3rd represent for the first, second, and third segment strong flavor Baijiu, respectively.

Table 12. Predicted results of the typical discriminant analysis of the strongly flavored Baijiu of the lower layer of grains for the two methods of sealing the pits.

	Samples	Results of Prediction						
		MSP-L1st	SSP-L1st	MSP-L2nd	SSP-L2nd	MSP-L3rd	SSP-L3rd	Total
Original count	MSP-L1st	21						21
	SSP-L1st		15					15
	MSP-L2nd			21				21
	SSP-L2nd				15			15
	MSP-L3rd					15		12
	SSP-L3rd						21	21
Proportion (%)	MSP-L1st	100						100
	SSP-L1st		100					100
	MSP-L2nd			100				100
	SSP-L2nd				100			100
	MSP-L3rd					100		100
	SSP-L3rd						100	100

100.0% of original grouped cases were classified correctly. 1st, 2nd, 3rd represent for the first, second, and third segment strong flavor Baijiu, respectively.

4. Conclusions

The types and the number of flavor substances produced by fermentation in mud-sealing and steel-sealing pit were relatively similar. In the PCA experiment, steel-sealing pits were more significant than traditional mud-sealing pits. HCA can divide samples into two categories, and using DA can also separate samples without aggregation. At the same time, two types of samples (one is mud-sealing and the other is steel-sealing) were successfully identified. At first, the relationship between the two methods is mutually verified. By this way, it can result in a more robust conclusion. Economically, the results of statistical analysis showed that the method of steel-sealing the pits increased the yield of Baijiu by 6.7% compared with the mud-sealing pit in the brewing process, with no significant difference in taste quality. Based on flavor and yield, both at the experimental and theoretical levels, we found no significant difference between the steel-sealing and the traditional mud-sealing final product (Baijiu) for Chinese Baijiu brewing. Compared with the disadvantages of the mud-sealing pits, where the yield was greatly influenced by the environment and was unstable, the process of steel sealing pits exhibited better controllability and stability, and it was more resistant to adverse conditions. In the future, the selection of pits cap sealing for intensive solid state fermentation process based on artificial intelligence and mechanization will present the trend of gradually replacing mud sealing with steel sealing.

Author Contributions: L.L., M.F., S.Y., Y.Q., Y.T. and J.L. (Jun Liu): Conceptualization, documentation, and data curation. M.F., S.Y., J.L. (Jun Liu), J.Z. and L.L.: writing—original draft preparation, M.F., Y.X., L.Z. and J.L. (Jun Liu): writing—review and editing. S.Y. and J.L. (Jun Liu): supervision. Y.X., L.Z. and J.L. (Jun Liu): project administration and funding acquisition. All authors have read and agreed to the published version of the manuscript.

Funding: This research was funded by the Postdoctoral Research Funding of joint station of Jiangnan University and Luzhou Laojiao Group Co. Ltd. (grant no. 206644), Key Research and Development Program of Sichuan Province (grant no. 2021YFS0341, 2020YJ0155), Key Program of Key Laboratory of Brewing Biotechnology and Application of Sichuan Province (grant no. NJ2017-01), Industry-University-Research Collaboration project of Wuliangye Co., Ltd. (grant no. CXY2020ZR02) and Talent introduction Program of SUSE (grant no. 2017RCL25).

Data Availability Statement: Data is contained within the article.

Acknowledgments: We would like to thank Thomas A. Gavin, Emeritus, Cornell University, for help with editing this paper.

Conflicts of Interest: Liang Zhang, Yongqing Tang, Jinsong Li were employed by the company Luzhou Laojiao Group Co. Ltd. Authors Jinsong Zhao was employed by the company Sichuan Liquor Group. The authors declare that this study received funding from Postdoctoral Research Funding of Luzhou Laojiao Group Co. Ltd. (grant no.206644). The funder had the following involvement with the study: This company provides materials and equipment for this project. Three authors from this company participated in program and experimental design, field experiment guidance and data analysis. We declare that we have no financial and personal relationships with other people or organizations that can inappropriately influence our work, there is no professional or other personal interest of any nature or kind in any product, service and/or company that could be construed as influencing the position presented in this article.

References

1. Cheng, P.; Fan, W.; Xu, Y. Determination of Chinese Baijiu from different geographic origins by combination of mass spectrometry and chemometric technique. *Food Control.* **2014**, *35*, 153–158. [CrossRef]
2. McGovern, P.E.; Zhang, J.; Tang, J.; Zhang, Z.; Hall, G.R.; Moreau, R.A.; Nũnez, A.; Butrym, E.D.; Richards, M.P.; Wang, C.; et al. Fermented beverages of pre- and proto-historic China. *Proc. Natl. Acad. Sci. USA* **2004**, *101*, 17593–17598. [CrossRef] [PubMed]
3. Zheng, X.W.; Han, B.Z. Baijiu, Chinese liquor: History, classification and manufacture. *J. Ethn. Foods* **2016**, *3*, 19–25. [CrossRef]
4. Fan, W.; Qian, M.C. Identification of aroma compounds in Chinese 'Yanghe Daqu' Baijiu by normal phase chromatography fractionation followed by gas chromatography olfactometry. *Flavor Fragr. J.* **2006**, *21*, 333–342. [CrossRef]
5. He, G.; Huang, J.; Wu, C.; Jin, Y.; Zhou, R. Bioturbation effect of fortified Daqu on microbial community and flavor metabolite in Chinese strong-flavor liquor brewing microecosystem. *Food Res. Int.* **2019**, *129*, 108851. [CrossRef]
6. Avalos, J.L.; Fink, G.R.; Stephanopoulos, G. Compartmentalization of metabolic pathways in yeast mitochondria improves the production of branched-chain alcohols. *Nat. Biotechnol.* **2013**, *31*, 335–341. [CrossRef]
7. Guo, M.; Hou, C.; Bian, M.; Shen, C.; Zhang, S.; Huo, D.; Ma, Y. Characterization of microbial community profiles associated with quality of Chinese strong-aromatic liquor through metagenomics. *J. Appl. Microbiol.* **2020**, *127*, 750–762. [CrossRef]
8. Belitz, H.D.; Grosch, W.; Schieberle, P. *Food Chemistry*; Springer: Berlin/Heidelberg, Germany, 2009.
9. Marilley, L.; Casey, M.G. Flavours of cheese prod-ucts: Metabolic pathways, analytical tools and identification of producing strains. *Int. J. Food Microbiol.* **2004**, *90*, 139–159. [CrossRef]
10. Wu, J.; Liu, P.; Fan, Y.; Bao, H.; Du, G.; Zhou, J.; Chen, J. Multivariate modular metabolic engineering of Escherichia colito produce resveratrol from L-tyrosine. *J. Bio-Technol.* **2013**, *167*, 404–411.
11. Liu, S. Spatial heterogeneity of prokaryotic microorganism communities in Luzhou- flavour liquor pit mud. In *Food Science*; Xihua University: Chengdu, China, 2013; p. 41.
12. Shi, S.; Zhang, L.; Wu, Z.Y.; Zhang, W.X.; Deng, Y.; Zhong, F.D.; Li, J.M. Analysis of the fungi community in multiple-and single grains Zaopei from a Luzhou-flavor liquor distillery in western China. *World J. Microbiol. Biotechnol.* **2011**, *27*, 1869–1874. [CrossRef]
13. He, G.; Huang, J.; Zhou, R.; Wu, C.; Jin, Y. Effect of fortified Daqu on the microbial community and flavor in Chinese strong flavor Baijiu brewing process. *Front. Microbiol.* **2019**, *10*, 56. [CrossRef]

14. Lu, Z.; Liu, X.; Tu, F.; Wu, X.; Chen, Y.; Ren, J. A Method of Sealing Cellar with Solid Brewing with Sealing Clay. CN 102757877A, 31 October 2012. (In Chinese)
15. Wang, W.; Fan, G.; Li, X.; Fu, Z.; Liang, X.; Sun, B. Application of wickerhamomyces anomalus in simulated solid-state fermentation for Baijiu production: Changes of microbial community structure and flavor metabolism. *Front. Microbiol.* **2020**, *11*, 598758. [CrossRef]
16. Wang, H.; Li, G.; Chen, X.; Chu, H. A Multi-Layer Composite Sealing Device for Northern Cellars. CN 210458116U, 5 May 2020. (In Chinese)
17. Zhang, L. Sealed Fermentation Device and Sealing Method for Cubic Cellar. Luzhou Pinchuang Technology, Co.: Luzhou, China, 2015. (In Chinese)
18. He, F.; Duan, J.; Zhao, J.; Li, H.; Sun, J.; Huang, M.; Sun, B. Different distillation stages Baijiu classification by temperature-programmed headspace-gas chromatography-ion mobility spectrometry and gas chromatography-olfactometrymass spectrometry combined with chemometric strategies. *Food Chem.* **2021**, *365*, 130430. [CrossRef] [PubMed]
19. Jia, W.; Fan, Z.; Du, A.; Li, Y.; Zhang, R.; Shi, Q.; Shi, L.; Chu, X. Recent advances in Baijiu analysis by chromatography based technology—A review. *Food Chem.* **2020**, *324*, 126899. [CrossRef]
20. Li, H.; Qin, D.; Wu, Z.; Sun, B.; Sun, X.; Huang, M.; Sun, J.; Zheng, F. Characterization of key aroma compounds in Chinese Guojing sesame-flavor Baijiu by means of molecular sensory science. *Food Chem.* **2019**, *284*, 100–107. [CrossRef]
21. Song, X.; Zhu, L.; Wang, X.; Zheng, F.; Zhao, M.; Liu, Y.; Li, H.; Zhang, F.; Zhang, Y.; Chen, F. Characterization of key aroma-active sulfur-containing compounds in Chinese Laobaigan Baijiu by gas chromatography-olfactometry and comprehensive twodimensional gas chromatography coupled with sulfur chemiluminescence detection. *Food Chem.* **2019**, *297*, 124959. [CrossRef] [PubMed]
22. Song, X.; Jing, S.; Zhu, L.; Ma, C.; Song, T.; Wu, J.; Zhao, Q.; Zheng, F.; Zhao, M.; Chen, F. Untargeted and targeted metabolomics strategy for the classification of strong aroma-type baijiu (liquor) according to geographical origin using comprehensive two-dimensional gas chromatography-time-of-flight mass spectrometry. *Food Chem.* **2020**, *314*, 126098. [CrossRef] [PubMed]
23. Song, X.; Wang, G.; Zhu, L.; Zheng, F.; Ji, J.; Sun, J.; Li, H.; Huang, M.; Zhao, Q.; Zhao, M.; et al. Comparison of two cooked vegetable aroma compounds, dimethyl disulfide and methional, in Chinese Baijiu by a sensory-guided approach and chemometrics. *LWT* **2021**, *146*, 111427. [CrossRef]
24. Wu, Z.; Qin, D.; Duan, J.; Li, H.; Sun, J.; Huang, M.; Sun, B. Characterization of benzenemethanethiol in sesame-flavor baijiu by high-performance liquid chromatography-mass spectrometry and sensory science. *Food Chem.* **2021**, *364*, 130345. [CrossRef] [PubMed]
25. Zhu, L.; Wang, X.; Song, X.; Zheng, F.; Li, H.; Chen, F.; Zhang, Y.; Zhang, F. Evolution of the key odorants and aroma profiles in traditional Laowuzeng baijiu during its oneyear ageing. *Food Chem.* **2020**, *310*, 125898. [CrossRef]
26. Fan, Q.; Wang, X.; Zhao, Y.; Zheng, F.; Li, H.; Zhang, F.; Chen, F. Characterization of key aroma compounds in Laobaigan Chinese Baijiu by GC× GC-TOF/MS and means of molecular sensory science. *Flavor Fragr. J.* **2019**, *34*, 514–525. [CrossRef]
27. Liu, H.; Sun, B. Effect of fermentation processing on the flavor of Baijiu. *J. Agric. Food Chem.* **2018**, *66*, 5425–5432. [CrossRef]
28. China Foods Limited Fermentation Industry Research Institute; Luzhou Laojiao Group Co., Ltd. *Method of Analysis for Chinese Spirits*, 2nd ed.; General Administration of Quality Supervision, Inspection and Quarantine: Beijing, China, 2007; pp. 1–34.
29. Chemicalbook. Available online: https://www.chemicalbook.com/ProductIndex.aspx (accessed on 6 May 2023).
30. Xiao, X.; Xu, K.; Gao, Z.-H.; Zhu, Z.-H.; Ye, C.; Zhao, B. Kuiling Ding Biomimetic asymmetric catalysis. *Sci. China Chem.* **2023**, *66*, 1553–1633. [CrossRef]
31. Mo, X.; Xu, Y.; Fan, W. Characterization of aroma compounds in Chinese rice wine Qu by solvent-assisted flavor evaporation and headspace solid-phase microextraction. *J. Agric. Food Chem.* **2010**, *58*, 2462–2469. [CrossRef]
32. Xu, M.L.; Yu, Y.; Ramaswamy, H.S.; Zhu, S.M. Characterization of Chinese liquor aroma components during aging process and liquor age discrimination using gas chromatography combined with multivariable statistics. *Sci. Rep.* **2017**, *7*, 1–9. [CrossRef]
33. Cabaroglu, T.; Yilmaztekin, M. Methanol and major volatile compounds of Turkish Raki and effect of distillate source. *J. Inst. Brew.* **2011**, *117*, 98–105. [CrossRef]
34. Garde-Cerdán, T.; Ancín-Azpilicueta, C. Effect of the addition of different quantities of amino acids to nitrogen-deficient must on the formation of esters, alcohols, and acids during wine alcoholic fermentation. *LWT-Food Sci. Technol.* **2008**, *41*, 501–510. [CrossRef]
35. Zhuang, M. The production mechanism of the fragrance in the Chinese spirit & the operation of the spirit brewing. *Liquor. Mak.* **2007**, *34*, 109–113.
36. Niu, Y.; Wang, P.; Xiao, Z.; Zhu, J.; Sun, X.; Wang, R. Evaluation of the perceptual interaction among ester aroma compounds in cherry wines by GC-MS, GC-O, odor threshold and sensory analysis: An insight at the molecular level. *Food Chem.* **2019**, *275*, 143–153. [CrossRef]
37. Li, J.; Song, C.; Hou, C.; Huo, D.; Shen, C.; Luo, X.; Yang, M.; Fa, H. Development of a colorimetric sensor array for the discrimination of Chinese Baijiu based on selected volatile markers determined by GC-MS. *J. Agric. Food Chem.* **2014**, *62*, 10422–10430. [CrossRef]
38. Versari, A.; Laurie, V.F.; Ricci, A.; Laghi, L.; Parpinello, G.P. Progress in authentication, typification and traceability of grapes and wines by chemometric approaches. *Food Res. Int.* **2014**, *60*, 2–18. [CrossRef]

39. Ma, Y.; Huo, D.Q.; Qin, H.; Shen, C.H.; Hou, C.J. Classification of aroma styles and geographic origins of Chinese Baijiu using chemometrics based on fluorescence spectroscopy. *J. Appl. Spectrosc.* **2017**, *84*, 361–368. [CrossRef]
40. Zheng, J.; Liang, R.; Zhang, L.Q. Characterization of microbial communities in strong aromatic liquor fermentation pit muds of different ages assessed by combined DGGE and PLFA analyzes. *Food Res. Int.* **2013**, *54*, 660–666. [CrossRef]

Disclaimer/Publisher's Note: The statements, opinions and data contained in all publications are solely those of the individual author(s) and contributor(s) and not of MDPI and/or the editor(s). MDPI and/or the editor(s) disclaim responsibility for any injury to people or property resulting from any ideas, methods, instructions or products referred to in the content.

foods

MDPI

Article

Difference Analysis of the Composition of Iron (Hydr)Oxides and Dissolved Organic Matter in Pit Mud of Different Pit Ages in Luzhou Laojiao and Its Implications for the Ripening Process of Pit Mud

Kairui Jiao [1], Bo Deng [2], Ping Song [3], Hailong Ding [2], Hailong Liu [1] and Bin Lian [4,*]

1 College of Life Sciences, Nanjing Normal University, Nanjing 210023, China; jkr08061999@163.com (K.J.); liuhailong@njnu.edu.cn (H.L.)
2 National Engineering Research Center of Solid State Brewing, Luzhou 646000, China; dengbo@lzlj.com (B.D.); dinghl@lzlj.com (H.D.)
3 School of Food Science and Pharmaceutical Engineering, Nanjing Normal University, Nanjing 210023, China; songping@njnu.edu.cn
4 College of Marine Science and Engineering, Nanjing Normal University, Nanjing 210023, China
* Correspondence: lianbin@njnu.edu.cn or bin2368@vip.163.com; Tel.: +86-025-85898551

Abstract: Long-term production practice proves that good liquor comes out of the old cellar, and the aged pit mud is very important to the quality of Luzhou-flavor liquor. X-ray diffraction, Fourier transform ion cyclotron resonance mass spectrometry, and infrared spectroscopy were used to investigate the composition characteristics of iron-bearing minerals and dissolved organic matter (DOM) in 2-year, 40-year, and 100-year pit mud and yellow soil (raw materials for making pit mud) of Luzhou Laojiao distillery. The results showed that the contents of total iron and crystalline iron minerals decreased significantly, while the ratio of Fe(II)/Fe(III) and the content of amorphous iron (hydr)oxides increased significantly with increasing cellar age. DOM richness, unsaturation, and aromaticity, as well as lignin/phenolics, polyphenols, and polycyclic aromatics ratios, were enhanced in pit mud. The results of the principal component analysis indicate that changes in the morphology and content of iron-bearing minerals in pit mud were significantly correlated with the changes in DOM molecular components, which is mainly attributed to the different affinities of amorphous iron (hydr)oxides and crystalline iron minerals for the DOM components. The study is important for understanding the evolution pattern of iron-bearing minerals and DOM and their interactions during the aging of pit mud and provides a new way to further understand the influence of aged pit mud on Luzhou-flavor liquor production.

Keywords: Luzhou-flavor liquor; pit mud; DOM; iron (hydr)oxides; FT-ICR-MS

1. Introduction

Chinese liquor is one of the oldest distilled spirits in the world, of which strong-flavor baijiu (SFB), also known as Luzhou-flavor liquor, is the best-selling in China, accounting for over 70% of total consumption in the Chinese baijiu market [1,2]. SFB production entails fermentation of the treated raw grains and Daqu(a kind of saccharifying and fermenting agent for Chinese liquor) in a cellar (approximate dimensions: length, 5.0 m; width, 3.0 m; and depth, 2.0 m) for 2 to 4 months [3]. The four walls and bottom of the cellar are covered with a layer of pit mud made of yellow soil [4]. Long-term production practice proves that the longer the pit mud is used, the better its quality and the higher the quality of the SFB produced by the cellar. This finding indicates that the aging of pit mud plays an important role in SFB fermentation [5]. As natural pit mud takes at least 20 years to mature, it is necessary to find ways to accelerate the maturation of new cellar and thus improve the quality of SFB. Therefore, the research into the maturation mechanism of pit mud has been a hot topic among researchers working in the domestic brewing industry.

The aging mechanism of pit mud is complicated. Previous studies focus on microbial diversity and functional analysis in pit mud. For example, Liu et al. explored the bacterial and fungal communities in pit mud of different ages and revealed the changes in microbial community structure and phylogenetic novelty therein [6,7]. Zhang et al. found significant compositional differences in bacterial community structures at different vertical depths of bottom pit mud [8]. Chai et al. studied clostridia with butyrate production potential in pit mud and elucidated the species-specific butyrate synthesis characteristics of *clostridium* [9]. These studies suggest that the microbial diversity and community structure of pit mud are closely related to the aging of pit mud. However, the study of dissolved organic matter (DOM) and mineral composition in pit mud has not been reported.

Organic matter (OM) is an important component of pit mud. Previous studies mainly focused on the determination of changes in reducing sugar, organic acid, and ammonia nitrogen in pit mud [3,10], which failed to expound all of the changes in organic components during the aging of pit mud. In addition, in terms of research methods, previous studies mainly used gas chromatography, high-performance liquid chromatography, and headspace-solid phase microextraction/gas chromatography-mass spectrometry to analyze aroma components in pit mud [3,8] and did not analyze complex organic matter components from the molecular level for pit mud. DOM is widespread in terrestrial ecosystems and is a major source of carbon and energy in soil microbial communities [11]. The pit mud contains a large amount of DOM. As the most dynamic and microbially accessible part of OM pool in pit mud, research of the variation of DOM components in pit mud will deepen the understanding of the maturation mechanism of pit mud. Fourier transform ion cyclotron resonance mass spectrometry (FT-ICR-MS) is widely used to explore the composition characteristics of complex DOM samples at the molecular level because of its ultrahigh mass accuracy and mass-resolving power [12]. Therefore, FT-ICR-MS was used in this study to analyze the composition characteristics of DOM in pit mud.

As a special form of fermented soil, the composition and structure of the minerals in pit mud have a profound influence on its physicochemical properties and microbial life activities. In the process of production practice, the periodic opening and closing of the cellar often cause a change in the redox environment in pit mud [8]. With the increase in cellar age, the color of pit mud changes from reddish to grey [10]. The active redox changes of iron were found to cause changes in the color of the soil [13]. Iron (hydr)oxides are the most important chromogenic component in soil [14]. The changes in the color of pit mud with different ages indicate that there are differences in the contents of iron and iron (hydr)oxides in pit mud, but there are no relevant reports.

In conclusion, the changes in the composition and content of DOM and iron minerals in pit mud may be an important factor affecting the aging of pit mud; therefore, in the present study, pit mud was sampled from Luzhou Laojiao distillery, a famous SFB manufacturer in China, as the research object. The chemical diversity characteristics of DOM and the composition of iron (hydr)oxides in different samples were analyzed, and the influence of their interaction on the aging of pit mud was revealed. This study helps to enrich the understanding of the maturation process of pit mud.

2. Materials and Methods

2.1. Sample Collection and Environmental Description

Samples were obtained from Luzhou Laojiao distillery, Luzhou City, Sichuan Province, China (105°29′26″ E, 28°53′51″ N, Figure 1). Raw material for making pit mud- yellow soil (unfermented, marked CK) and pit mud aged about 2, 40, and 100 years (marked 2-PM, 40-PM, and 100-PM) were selected as objects of the research. Pit mud sampling points were located at the diagonal intersection of the four walls and the bottom of the cellar. At each point, samples with a mass of 200 g were collected to a total sample mass of 1 kg. Pit mud was mixed and used as a representative sample of a cellar. Three cellars with similar ages were collected as replicates. Therefore, a total of 12 samples (4 groups × 3 replicates per group) were collected. Samples were collected using sterile sampling bags and transported to the laboratory on ice boxes. The sample was divided into two parts. One part of the fresh sample was stored in a 4 °C refrigerator for the determination of the moisture content (MC) and Fe(II), and the other part of the sample was air-dried and passed through a 100-mesh screen and then used for XRD, ATR-FTIR, FT-ICR-MS, and measurement of other physicochemical indicators.

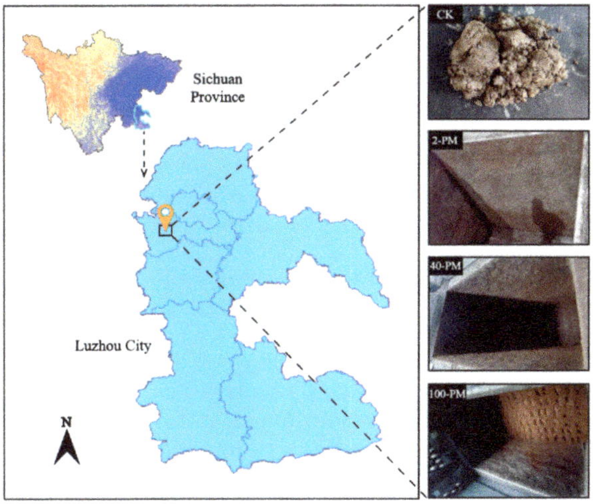

Figure 1. Location map of sampling sites in Luzhou, Sichuan. CK, yellow soil; 2-PM, 2-year pit mud; 40-PM, 40-year pit mud; 100-PM, 100-year pit mud.

2.2. Determination and Characterization of Physicochemical Properties of Samples

The MC of fresh samples was determined by drying them at 105 °C for 24 h. The pH of the samples was determined using a pH meter (PHSJ-4F, Shanghai Lei Ci, Shanghai, China) according to the soil-water ratio of 1:2.5 (*w*/*v*). The OM content of the samples was determined by the sulfuric acid-potassium dichromate oxidation method according to the Chinese National Standard NY/T 1121.6-2006 [15]. According to Chinese National Standard NY/T 1121.7-2014 [16], NH_4F-HCl (0.03 mol/L) was used to extract available phosphorus (AP) according to the soil-water ratio of 1:10 (*w*/*v*), and then AP was determined using ion chromatograph (ICS-900, Dionex, Sunnyvale, CA, USA). Dissolved organic carbon (DOC) was determined using a total organic carbon analyzer (TOC-VCPH, Shimadzu Corporation, Kyoto, Japan). Carbonate content was determined using acid-base titration [17].

Determination of iron-related indices in samples: Fe(II) contents in fresh samples were analyzed using the ferrozine method [18]; total iron (TFe) contents were digested by mixed acid (HNO_3-HF-$HClO_4$) [19]; Fe(III) contents were calculated by subtracting Fe(II) from TFe [20]; free iron oxides (Fe_d) were extracted using a dithionite-citrate-bicarbonate solution (0.1 mol/L) [21]; Amorphous iron oxides (Fe_o) were extracted from ammonium oxalate solution (0.2 mol/L, pH = 3) [22]; crystalline iron oxides (Fe_c) were calculated by subtracting Fe_o from Fe_d [23]; iron ion concentrations were measured using a flame atomic absorption spectrometer (AA-6300C, Shimadzu Corporation, Kyoto, Japan).

The mineral composition of the samples was detected using an X-ray diffractometer (XRD; XPert Pro MPD, Malvern Panalytical Corporation, Almelo, The Netherlands). The scanning angle ranged from 10 to 65°. The scanning speed was 10°/min. Organic functional groups on the sample surface were detected by attenuated total reflection Fourier transform infrared spectroscopy (ATR-FTIR; Thermo Nicolet iS5, Waltham, MA, USA). Spectral data was collected between 4000 to 400 cm^{-1}. The resolution was 4000 cm^{-1}. The number of scans was 32.

2.3. Extraction and Determination of DOM in Samples

2.3.1. DOM Extraction

The air-dried samples (3.00 g) were mixed with 30 mL of Milli-Q Water (GREEN-Q2-10T, EPED, Nanjing, China) in a 50 mL centrifuge tube and then agitated in an oscillator (25 °C, 180 rpm) for 12 h. The DOM solution was obtained by centrifuging the suspension at 8000 rpm for 8 min and filtering the supernatant through a 0.45 μm filter membrane. Acidified samples were obtained by adjusting the pH of the DOM solution to 2 with concentrated hydrochloric acid (37%, ultrapure) [12].

DOM samples were enriched and desalted using a Bond Elut PPL column (500 mg, 6 mL, Agilent, Santa Clara, CA, USA). The specific procedure was described as follows [24]: 18 mL methanol (HPLC grade) and 18 mL acidified water (pH = 2.0) were used to activate the filter element. Acidified samples were passed through the filter element at a rate of about 2 mL/min under gravity. The filter element was rinsed with 18 mL of acidified water, and then the moisture in the column was dried with N_2 (99.999% purity). Finally, 6 mL methanol (HPLC grade) was added to elute the sample. The eluent was diluted with 1:1 (v/v) methanol: water solution and then used for FT-ICR-MS analysis.

2.3.2. FT-ICR-MS Analysis

The molecular composition of DOM was resolved using a Bruker solariX 94 FT-ICR-MS (Billerica, MA, USA) equipped with a 9.4-T superconducting magnet and a negative electron spray ionization source. All formulas were assigned according to the criterion of elemental combinations of C_{1-100}, H_{1-200}, O_{0-50}, N_{0-4}, S_{0-2}, P. Only peaks with a signal-to-noise ratio (S/N) ≥ 6 and error threshold of ± 0.5 ppm were considered for further analysis; $H/C < 2.4$ and $O/C < 1.2$ were set as further restrictions on the formula and subsequent computations. The SRNOM standard was used for calibration. The operating conditions were as follows: a 4-kV spray shield voltage, a 4.5-kV capillary column voltage, and a 320-V capillary terminal voltage. To reveal the detailed characteristics of DOM molecules, the following intensity-weighted average parameters were calculated [25]: "AI" (aromaticity index), "DBE" (double bond equivalent), "NOSC" (nominal oxidation state of carbon), "CRAMs" (carboxyl-rich alicyclic molecules), "O/C" (oxygen-to-carbon element ratios), and "H/C" (hydrogen-to-carbon element ratios).

DOM compounds were classified into six types according to AI value and ratios of H/C and O/C [26–28]: (1) Combustion-derived polycyclic aromatics (CDPA): $AI > 0.66$; (2) Vascular plant-derived polyphenols (VPDP): $0.66 \geq AI > 0.50$; (3) Highly unsaturated lignin/phenolics (HULP): $AI \leq 0.50$ and $H/C < 1.5$; (4) Unsaturated aliphatic compounds (UAC): $2.2 \geq H/C \geq 1.5$; (5) Saturated fatty acid (SFA): $H/C \geq 2.0$ and $O/C < 0.9$; (6) Carbohydrates (CH): O/C: 0.67–1.2 and H/C: 1.5–2.2.

2.4. Statistical Analysis

The SPSS software package (Version 16.0, IBM, Armonk, NY, USA) was used for one-way analysis of variance (ANOVA) (Tukey test). Principal component analysis (PCA) was performed using Origin 2022b software (Origin Lab., Northampton, MA, USA). All results were expressed as ±SD; $p < 0.05$ was considered statistically significant.

3. Results and Discussion

3.1. Mineral Composition and Physicochemical Properties of Different Samples

The composition of iron-bearing minerals in yellow soil and pit mud samples was analyzed using XRD (Figure 2). The results showed that the mineral composition of the different samples was similar, but there were some differences in the intensity and location of the peaks. Compared with the yellow soil used for making pit mud, the peak intensity of crystalline iron-bearing minerals (such as goethite, mahlmoodite, and ferrorichterite, expressed as Fe_c) in different pit mud samples exhibited a trend of decreasing gradually with increasing cellar age.

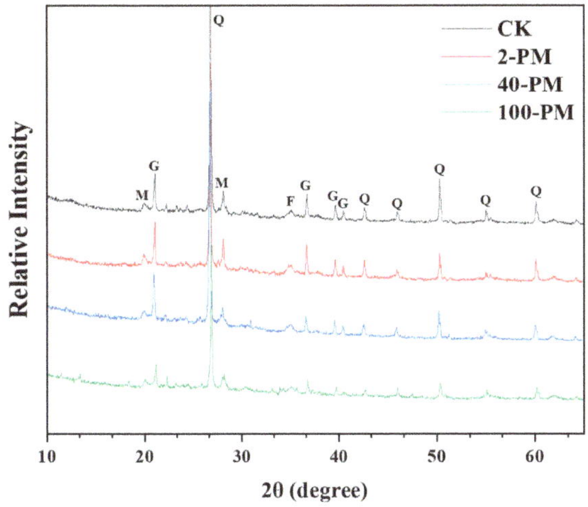

Figure 2. The mineral composition varies in different samples. CK, yellow soil; 2-PM, 2-year pit mud; 40-PM, 40-year pit mud; 100-PM, 100-year pit mud; Q, quartz; G, goethite; M, mahlmoodite; F, ferrorichterite.

The results were explained as follows: (1) Iron is a sensitive redox element in geochemistry, and changes in the surrounding redox environment can alter its chemical properties [29]. In the process of SFB fermentation, with the continuous consumption of oxygen inside the cellar, the pit mud environment gradually presents a reduced state, which promotes the reduction in Fe_c and increases the solubility of iron, so the peak intensity of Fe_c gradually decreases with the increase in cellar age; (2) The weakened peak intensity of iron-bearing minerals in the XRD pattern may also be due to the partial conversion of Fe_c to amorphous iron (hydr)oxides. In general, long-term fermentation promotes the transformation of iron-bearing crystalline minerals and the loss of iron from minerals in pit mud. To clarify the specific changes of iron-bearing minerals, the contents of iron and iron minerals and other physical and chemical indexes in yellow soil and pit mud were determined (Table 1). The results showed that the physical and chemical properties of pit mud changed significantly with the increase in cellar age.

Table 1. Physicochemical properties of yellow soil, 2-year, 40-year, and 100-year pit mud.

Parameter	CK	2-PM	40-PM	100-PM
pH	6.16 ± 0.04 [b]	4.04 ± 0.03 [d]	5.10 ± 0.22 [c]	6.62 ± 0.11 [a]
MC (%)	7.57 ± 0.33 [d]	23.71 ± 1.53 [c]	32.60 ± 3.28 [b]	43.87 ± 1.68 [a]
OM (g/kg)	22.62 ± 1.74 [d]	89.14 ± 4.72 [c]	168.99 ± 26.21 [b]	223.78 ± 6.95 [a]
AP (g/kg)	0.05 ± 0.00 [c]	0.30 ± 0.02 [c]	10.54 ± 0.40 [b]	13.20 ± 0.36 [a]
DOC (mg/L)	66.84 ± 3.38 [d]	381.58 ± 18.07 [c]	828.23 ± 91.26 [b]	1202.88 ± 78.97 [a]
Carbonate (%)	3.55 ± 0.28 [a]	1.41 ± 0.13 [c]	2.33 ± 0.11 [b]	2.71 ± 0.30 [b]
TFe (g/kg)	31.99 ± 0.50 [a]	24.26 ± 1.55 [b]	20.21 ± 0.18 [c]	15.24 ± 0.28 [d]
Fe(II) (g/kg)	0.45 ± 0.03 [d]	0.74 ± 0.05 [c]	2.04 ± 0.07 [b]	3.79 ± 0.13 [a]
Fe(III) (g/kg)	31.54 ± 0.53 [a]	23.52 ± 1.53 [b]	18.17 ± 0.19 [c]	11.44 ± 0.33 [d]
Fe(II)/Fe(III)	0.01 ± 0.00 [c]	0.03 ± 0.01 [c]	0.11 ± 0.01 [b]	0.33 ± 0.02 [a]
Fe_d (g/kg)	12.44 ± 0.22 [a]	10.52 ± 0.32 [b]	9.20 ± 0.25 [c]	8.83 ± 0.05 [c]
Fe_o (g/kg)	1.51 ± 0.22 [d]	3.69 ± 0.04 [c]	6.16 ± 0.07 [b]	7.54 ± 0.51 [a]
Fe_c (g/kg)	10.93 ± 0.31 [a]	6.83 ± 0.29 [b]	3.03 ± 0.27 [c]	1.29 ± 0.51 [d]

Note: MC, the moisture content; OM, organic matter; AP, available phosphorus; DOC, dissolved organic carbon; TFe, total iron; Fe_d, free iron oxides; Fe_o, amorphous iron oxides; Fe_c, crystalline iron oxides; CK, yellow soil; 2-PM, 2-year pit mud; 40-PM, 40-year pit mud; 100-PM, 100-year pit mud. Values are means ± SDs of three replicates. Different small letters in the same column represent significant differences at the 0.05 level.

As shown in Table 1, the pH of the unfermented yellow soil was 6.16, which was weakly acidic, while the pH of 2-year pit mud decreased to 4.04. This was due to the large amount of lactic acid secreted by the dominant lactic acid bacteria in the low-aged pit mud [8]. As the cellar age increased, the pH of pit mud increased from 4.04 in 2-year pit mud to 6.62 in 100-year pit mud, which was mainly attributed to the degradation of organic acids such as lactic acid in pit mud and the accumulation of alkaline metabolites such as NH_4^+ in the long-term fermentation process [3,30]. In addition, MC increased significantly from 7.57% to 43.87%; OM increased significantly from 22.62 g/kg to 223.78 g/kg. The reason for this change was, on the one hand, the infiltration of yellow water from fermented grains into pit mud, which brought a large amount of water and OM [31], and on the other hand, the relative proportion of crystalline mineral content in pit mud decreased during the fermentation process, thus increasing the proportion of OM and MC.

During SFB fermentation, periodic cellar sealing openings can cause changes in the redox environment in pit mud, which can lead to changes in the relative contents of Fe(II) and Fe(III) (Table 1). After sealing the cellar, the environment of pit mud gradually shows a reduced state. At this time, Fe(III) in Fe_c was continuously decreased and dissolved to produce Fe(II) in anoxic environment [32]. As a result, the content of Fe_c decreased significantly from 10.93 g/kg to 1.29 g/kg, the content of Fe(II) increased significantly from 0.45 g/kg to 3.79 g/kg, and the Fe(II)/Fe(III) increased significantly from 0.01 to 0.33. In addition, the dissolved Fe(II) can be oxidized to Fe_o because of the increase in the dissolved oxygen level in pit mud after opening the cellar [33], so the content of Fe_o increased significantly from 1.51 g/kg to 7.54 g/kg. Fe_d included Fe_c and Fe_o [23], and the content of Fe_d significantly decreased with increasing cellar age (Table 1). As the cellar age increased, although the content of Fe_c decreased significantly and the content of Fe_o increased significantly, the content of Fe_d, the sum of the two, decreased significantly (from 12.44 g/kg to 8.83 g/kg), which caused a significant decrease in the content of TFe (from 31.99 g/kg to 15.24 g/kg). This result also coincided with the result of the weakening of the peak intensity of iron minerals in the XRD pattern (Figure 2).

Duan et al. found that Fe_c can be tightly bound to organic carbon (OC)-derived DOC and AP, rendering this part of the OC difficult to use by microorganisms [34]. As the cellar age increased, the reductive dissolution of Fe_c in multiple rounds of fermentation and reduction environment resulted in the release of DOC and AP bound to Fe_c. Therefore, from yellow soil to 100-year pit mud, the content of DOC increased significantly from 66.84 mg/L to 1202.88 mg/L, and the content of AP increased significantly from 0.05 g/kg to 13.20 g/kg. The higher content of DOC and AP in aged pit mud may improve the metabolic activity of microorganisms, which may have a positive impact on the quality of SFB products.

The carbonate content of pit mud increased significantly from 1.41% to 2.71% with increasing cellar age, which may inhibit the conversion of Fe_o to Fe_c [23]. In addition, a significant increase in DOC concentration (from 66.84 mg/L to 1202.88 mg/L) could retard or inhibit the conversion of Fe_o to Fe_c [35]. In general, these results led to lower amounts of crystalline iron minerals and more amorphous iron (hydr)oxides in aged pit mud.

3.2. *Changes in DOM Molecular Composition during Aging of Pit Mud*

To compare the differences in chemical properties of DOM molecules in different samples, the intensity-weighted average parameters of DOM molecules were calculated (Table 2). The results showed that: with the increase in cellar age, (1) The DOM molecular formula number increased from 3636 to 7858, indicating that the richness of DOM molecule increased; (2) The DBE value increased from 3.20 to 6.65, indicating that unsaturated molecules rich in double bonds and/or rings gradually accumulated in DOM molecules, so the unsaturation of DOM molecules increased and the H/C ratios decreased correspondingly (from 1.94 to 1.43); (3) The AI value increased from 0.05 to 0.16, indicating the accumulation of aromatic compounds in DOM molecules; (4) The NOSC value changed from −1.44 to −0.55, implying the increase in oxidized forms of carbons in DOM molecules [36]. (5) CRAMs increased from 572 to 2616, showing the accumulation of difficult-to-degrade carboxyl-rich alicyclic compounds in DOM molecules [36]; (6) The O/C ratios increased from 0.22 to 0.28, indicating that the degree of oxidation of DOM molecules increased, which was also consistent with increasing NOSC values in Table 2.

Table 2. Intensity-weighted average parameters of DOM.

Sample	Formula Number	O/C	H/C	DBE	AI	NOSC	CRAMs
CK	3636 ± 227 [c]	0.22 ± 0.01 [c]	1.94 ± 0.08 [a]	3.20 ± 0.13 [c]	0.05 ± 0.03 [c]	−1.44 ± 0.09 [d]	572 ± 79 [c]
2-PM	5459 ± 901 [b]	0.25 ± 0.01 [bc]	1.67 ± 0.04 [b]	4.45 ± 0.14 [b]	0.10 ± 0.01 [b]	−0.88 ± 0.04 [c]	1419 ± 532 [b]
40-PM	6762 ± 294 [ab]	0.27 ± 0.01 [ab]	1.53 ± 0.02 [bc]	5.38 ± 0.32 [b]	0.13 ± 0.01 [ab]	−0.69 ± 0.01 [b]	2514 ± 142 [a]
100-PM	7858 ± 224 [a]	0.28 ± 0.16 [a]	1.43 ± 0.11 [c]	6.65 ± 0.87 [a]	0.16 ± 0.01 [a]	−0.55 ± 0.01 [a]	2616 ± 164 [a]

Note: Formula number, the number of DOM molecular formulas identified; DBE, double bond equivalence; AI, aromaticity index; NOSC, the nominal oxidation state of carbon; CRAMs, carboxyl-rich alicyclic molecules. CK, yellow soil; 2-PM, 2-year pit mud; 40-PM, 40-year pit mud; 100-PM, 100-year pit mud. Values are means ± SDs of three replicates. Different small letters in the same column represent significant differences at the 0.05 level.

The data in Table 2 were also supported by the results of ATR-FTIR spectral analysis. As shown in Figure 3, the ATR-FTIR spectra of all samples exhibited similar absorption peak positions, but the relative intensity of the peaks differed. The absorption peaks of the main organic functional groups of the samples were presented as follows: (1) The characteristic peaks at 1430 and 1535 cm^{-1} were assigned to the C-O symmetric stretch in the COO- of carboxylic acids [37]; (2) The characteristic peak at 1630 cm^{-1} was assigned to the C=C stretching in the aromatic group and the COO- asymmetric stretching vibration in the carboxyl group [38]; (3) The characteristic peaks at 2800 to 3500 cm^{-1} were assigned to aromatic, carboxylic and phenolic groups [39]. The enhancement of these peaks further showed that compounds rich in aromatic, carboxyl, and phenolic groups were present in increasing concentrations during the aging of pit mud.

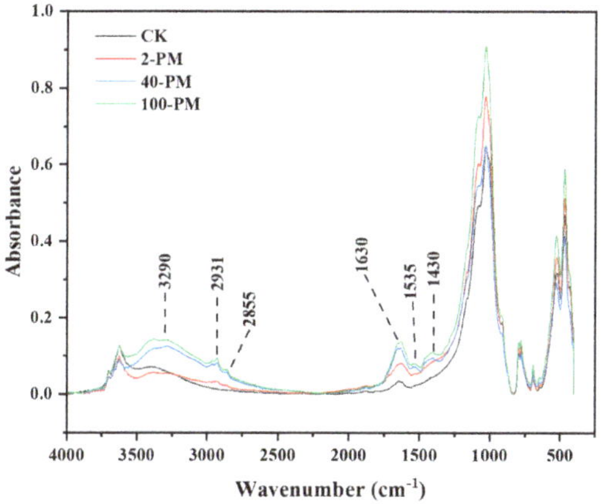

Figure 3. ATR-FTIR spectra of different samples. CK, yellow soil; 2-PM, 2-year pit mud; 40-PM, 40-year pit mud; 100-PM, 100-year pit mud.

In general, increasing cellar age significantly changed the molecular diversity and chemical properties of DOM. DOM in aged pit mud had higher DBE values, AI values, and more CRAMs (Table 2). This indicates that compounds with high unsaturation and aromaticity, as well as alicyclic compounds rich in carboxyl groups, gradually dominate as fermentation proceeds. The accumulation of these compounds may result from changes in microbial metabolites during the aging process and the accumulation of stubborn organic compounds such as highly unsaturated and aromatic compounds [40].

The classification of DOM compounds in different samples was based on H/C, O/C ratios, and AI values (Figure 4A) [26]. The results showed that there were significant differences in the types and molecular compositions of DOM compounds in yellow soil and pit mud. Saturated fatty acids (60.38%) accounted for the largest proportion in yellow soil, followed by unsaturated aliphatic compounds (31.42%). Unsaturated aliphatic compounds (56.69–68.53%) dominated in pit mud, followed by lignin (14.32–22.27%) and saturated fatty acids (15.06–18.47%). As shown in Figure 4A, the proportions of lignin, polyphenols, and polycyclic aromatics gradually increased with cellar age due to the stubborn nature of these compounds, which were not easily degraded by microorganisms [37]. In addition, these stubborn compounds may also be derived from exogenous plant residues (such as sorghum hulls and wheat hulls) during fermentation [40]. Compared with yellow soil, the proportion of unsaturated aliphatic compounds increased significantly, while the proportions of saturated fatty acids and carbohydrates decreased significantly. Unsaturated aliphatic compounds are usually derived from bacterial metabolites [39]. This suggests that, as fermentation proceeds, the bacterial diversity in pit mud has become significantly different from that of the yellow soil, and the bacteria in pit mud may be more inclined to produce more unsaturated aliphatic compounds. As carbohydrates are more bio-unstable and most easily utilized by microorganisms [40], the proportion of carbohydrates in pit mud was reduced from 0.19% to 0.06% (Figure 4A). Carbohydrates are more biologically unstable and most easily utilized by microorganisms [37], so the reduction in the proportion of carbohydrates in pit mud from 0.19% to 0.06% can be ascribed to the consumption of microbial metabolic activities (Figure 4A).

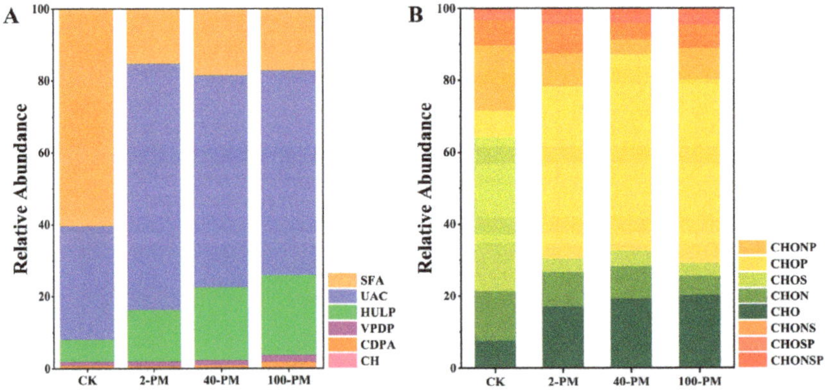

Figure 4. Type of DOM compounds (**A**) and element composition (**B**). SFA, saturated fatty acid; UAC, unsaturated aliphatic compounds; HULP, highly unsaturated lignin/phenolics; VPDP, vascular plant-derived polyphenols; CDPA, combustion-derived polycyclic aromatics; CH, carbohydrates. CHO, any formulae without heteroatoms; CHON, any formulae with N only; CHOS, any formulae with S only; CHOP, any formulae with P only; CHONS, any formulae with N and S; CHONP, any formulae with N and P; CHOSP, any formulae with S and P; CHONSP, any formulae with N, S, and P. CK, yellow soil; 2-PM, 2-year pit mud; 40-PM, 40-year pit mud; 100-PM, 100-year pit mud.

DOM compounds can also be further divided into different molecular types based on their element composition (Figure 4B) [26]. The results indicated that yellow soil had the largest proportion of CHOS (42.83%) molecules, followed by CHONP (18.07%) molecules and CHON (13.79%) molecules. Compared with yellow soil, CHOP (47.77–54.60%) molecules and CHO (17.03–20.28%) molecules increased greatly and dominated in pit mud, while CHON (5.33–9.65%), CHONP (3.98–9.0%) and CHOS (3.58–4.28%) molecules decreased greatly. Studies have shown that sulfur-containing compounds are widely applied in the production of detergents and personal care products [41]. Therefore, the high proportion of CHOS molecules in yellow soil indicated that yellow soil may be affected by human activities to some extent.

DOM is widespread in terrestrial ecosystems and plays a critical role in the evolution of soil biogeochemical processes and structure [11]. The large differences in compound type and elemental composition of DOM in yellow soil and pit mud reflect the complex biochemical transformation processes that occur in pit mud during SFB fermentation. This process may have a potentially beneficial effect on the quality of SFB. Further research will provide a theoretical basis for promoting the aging of pit mud and selecting suitable raw materials for yellow soil.

3.3. Association of DOM Molecules with Iron (Hydr)Oxides

PCA was used to evaluate the correlation between iron (hydr)oxides and DOM compounds (Figure 5). The different samples were clearly separated in the PC1 and PC2 axes. The PC1 and PC2 axes explained 75.3% and 10.3% of the total variance, respectively.

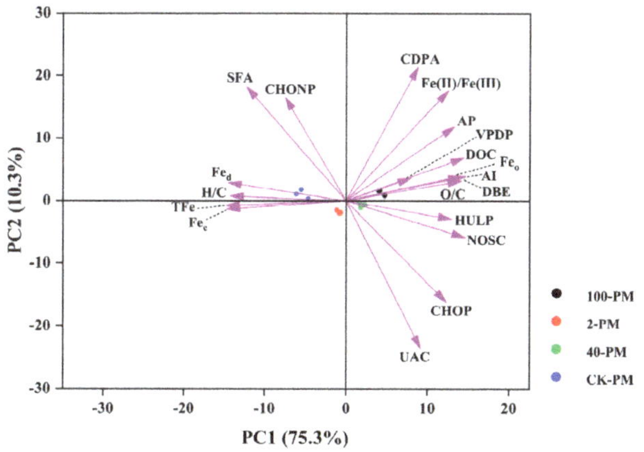

Figure 5. DOM molecular characteristics of different samples and PCA analysis of iron (hydr)oxides. CK, yellow soil; 2-PM, 2-year pit mud; 40-PM, 40-year pit mud; 100-PM, 100-year pit mud. AP, available phosphorus; DOC, dissolved organic carbon; TFe, total iron; Fe_d, free iron oxides; Fe_o, amorphous iron oxides; Fe_c, crystalline iron oxides; Fe(II)/Fe(III), ratio of divalent to trivalent iron ions; H/C, hydrogen-to-carbon element ratios; O/C, oxygen-to-carbon element ratios; DBE, double bond equivalence; AI, aromaticity index; NOSC, the nominal oxidation state of carbon; HULP, highly unsaturated lignin/phenolics; VPDP, vascular plant-derived polyphenols; CDPA, combustion-derived polycyclic aromatics; UAC, unsaturated aliphatic compounds; SFA, saturated fatty acid; CHONP, any formulae with N and P; CHOP, any formulae with P only.

The carboxyl groups of aromatic DOM molecules can undergo ligand exchange with the hydroxyl groups on the surface of iron hydroxide, and thus DOM with aromatic structures (such as CDPA, VPDP, and HULP) are more susceptible to bonding with iron (hydr)oxides to form Fe-OM complexes [42,43]. In addition, Fe_o has a larger specific surface area and more binding sites compared to Fe_c, resulting in a stronger adsorption capacity to preferentially bind to CDPA, VPDP, and HULP [44,45]. Thus, CDPA, VPDP, and HULP were negatively correlated with Fe_c and positively correlated with Fe_o (Figure 5, Table S1). In addition, OM is highly chemically reactive towards iron (hydr)oxides and can affect the transformation of their morphology [46]. CDPA, VPDP, and HULP can block the reduction in Fe_o upon binding to Fe_o, thereby delaying or inhibiting the conversion of Fe_o to Fe_c [19]. Thus, there were more Fe_o and less Fe_c in aged pit mud (Table 1).

Correlation analysis of NOSC values, O/C ratios, and Fe(II)/Fe(III) can also further explain the relationship between DOM and iron (hydr)oxides. The NOSC value and O/C ratios can represent the oxidation state of DOM molecules [36]. The continuous oxidation of DOM molecules (supplying energy to the fermentation process) led to an increase in the degree of oxidation of DOM molecules, and therefore, the NOSC values and O/C ratios of DOM molecules increased (Table 1). The electrons generated due to the oxidation of DOM molecules can be transferred to Fe(III) on the surface of iron (hydr)oxides to form Fe(II) [44], leading to an increase in Fe(II)/Fe(III). Thus, the NOSC values and O/C ratios were positively correlated with Fe(II)/Fe(III) (Figure 5, Table S1).

AP was negatively correlated with Fe_c and CHONP molecules (Figure 5, Table S1), indicating that with the increase in cellar age, the increase in the content of AP in pit mud may be due to the release of originally adsorbed phosphoric acid from the reduction in total Fe_c on the one hand, and the transformation of CHONP compounds on the other hand. In addition, CHOP compounds were negatively correlated with CHONP compounds (Figure 5, Table S1), suggesting that the increase in CHOP compounds may be attributed to the conversion of CHONP compounds.

To sum up, as the cellar age increased, the compositional characteristics of the iron-bearing minerals and DOM in pit mud changed significantly. Crystalline iron minerals predominate in young pit mud, while amorphous iron (hydr)oxides are more dominant in aged pit mud. The richness, unsaturation, and aromaticity of DOM increased significantly during the aging of pit mud. DOM from aged pit mud contained more lignin/phenolics, polyphenols, and polycyclic aromatics. Furthermore, there is a significant correlation between the changes in the morphology and content of iron-bearing minerals in pit mud and the chemical characterization of the DOM molecules. Therefore, the use of microbial fermentation and other biotechnological methods to reduce the iron content of local yellow soil (raw material for pit mud production) may help to promote the aging of new pit mud.

4. Conclusions

In summary, the contents of total iron and crystalline iron minerals decreased significantly with the increase in cellar age, and the ratio of Fe(II)/Fe(III) and the content of amorphous iron (hydr)oxides increased significantly. In connection with this, the richness, unsaturation, and aromaticity of DOM and the ratio of lignin/phenolics, polyphenols, and polycyclic aromatics in pit mud were increased. The research showed that molecular diversity and chemical properties of DOM compounds were closely correlated with the content and morphological changes of iron (hydr)oxides during the aging process of pit mud, and the interaction between DOM and iron (hydr)oxides changes the inorganic and organic components in pit mud, which further changes the physicochemical properties of pit mud and affects the aging of pit mud. The changes in the content and composition of iron-bearing minerals and DOM are important potential factors affecting the aging of pit mud, which not only provides a new idea to improve the aging of pit mud by using biotechnology but also helps to understand the process and mechanism of aging of pit mud from the perspective of mineral and DOM composition.

Supplementary Materials: The following supporting information can be downloaded at: https://www.mdpi.com/article/10.3390/foods12213962/s1, Table S1: Correlation analysis-Pearson.

Author Contributions: Investigation, methodology, data curation, writing—original draft, writing—review and editing, K.J.; investigation, writing and review, B.D.; investigation, writing and review, funding acquisition, P.S.; investigation, writing and review, H.D.; investigation, methodology, writing and review, H.L.; conceptualization, investigation, writing—review and editing, supervision, funding acquisition, B.L. All authors have read and agreed to the published version of the manuscript.

Funding: This work was funded by the Independent Research Project of Culture and Technology Innovation Centre Co-established by Luzhou Laojiao and Nanjing Normal University (grant number CY2X202200201) and the National Natural Science Foundation of China (grant number 41772360).

Data Availability Statement: The data used to support the findings of this study can be made available by the corresponding author upon request.

Acknowledgments: The authors would also like to thank Huang He of Nanjing Normal University for his guidance and support in this research project.

Conflicts of Interest: The authors declare that they have no known competing financial interests or personal relationships that could have appeared to influence the work reported in this paper.

References

1. Fu, J.; Chen, L.; Yang, S.; Li, Y.; Jin, L.; He, X.; He, L.; Ao, X.; Liu, S.; Liu, A.; et al. Metagenome and analysis of metabolic potential of the microbial community in pit mud used for Chinese strong-flavor liquor production. *Food Res. Int.* **2021**, *143*, 110294. [CrossRef]
2. Liu, H.; Sun, B. Effect of Fermentation Processing on the Flavor of Baijiu. *J. Agric. Food. Chem.* **2018**, *66*, 5425–5432. [CrossRef]
3. Chai, L.J.; Qian, W.; Zhong, X.Z.; Zhang, X.J.; Lu, Z.M.; Zhang, S.Y.; Wang, S.T.; Shen, C.H.; Shi, J.S.; Xu, Z.H. Mining the Factors Driving the Evolution of the Pit Mud Microbiome under the Impact of Long-Term Production of Strong-Flavor Baijiu. *Appl. Environ. Microbiol.* **2021**, *87*, e88521. [CrossRef]
4. Xu, Y.; Sun, B.; Fan, G.; Teng, C.; Xiong, K.; Zhu, Y.; Li, J.; Li, X. The brewing process and microbial diversity of strong flavour Chinese spirits: A review. *J. Inst. Brew.* **2017**, *123*, 5–12. [CrossRef]
5. Guan, T.; Lin, Y.; Chen, K.; Ou, M.; Zhang, J. Physicochemical Factors Affecting Microbiota Dynamics During Traditional Solid-State Fermentation of Chinese Strong-Flavor Baijiu. *Front. Microbiol.* **2020**, *11*, 2090. [CrossRef]
6. Liu, M.; Tang, Y.; Zhao, K.; Liu, Y.; Guo, X.; Ren, D.; Yao, W.; Tian, X.; Gu, Y.; Yi, B.; et al. Determination of the fungal community of pit mud in fermentation cellars for Chinese strong-flavor liquor, using DGGE and Illumina MiSeq sequencing. *Food Res. Int.* **2017**, *91*, 80–87. [CrossRef]
7. Liu, M.; Tang, Y.; Guo, X.; Zhao, K.; Tian, X.; Liu, Y.; Yao, W.; Deng, B.; Ren, D.; Zhang, X. Deep sequencing reveals high bacterial diversity and phylogenetic novelty in pit mud from Luzhou Laojiao cellars for Chinese strong-flavor Baijiu. *Food Res. Int.* **2017**, *102*, 68–76.
8. Zhang, H.; Meng, Y.; Wang, Y.; Zhou, Q.; Li, A.; Liu, G.; Li, J.; Xing, X. Prokaryotic communities in multidimensional bottom-pit-mud from old and young pits used for the production of Chinese Strong-Flavor Baijiu. *Food Chem.* **2020**, *312*, 126084.
9. Chai, L.; Xu, P.; Qian, W.; Zhang, X.; Ma, J.; Lu, Z.; Wang, S.; Shen, C.; Shi, J.; Xu, Z. Profiling the Clostridia with butyrate-producing potential in the mud of Chinese liquor fermentation cellar. *Int. J. Food Microbiol.* **2019**, *297*, 41–50.
10. Hu, X.; Du, H.; Ren, C.; Xu, Y. Illuminating Anaerobic Microbial Community and Cooccurrence Patterns across a Quality Gradient in Chinese Liquor Fermentation Pit Muds. *Appl. Environ. Microbiol.* **2016**, *82*, 2506–2515. [CrossRef]
11. Bahureksa, W.; Tfaily, M.M.; Boiteau, R.M.; Young, R.B.; Logan, M.N.; McKenna, A.M.; Borch, T. Soil Organic Matter Characterization by Fourier Transform Ion Cyclotron Resonance Mass Spectrometry (FTICR MS): A Critical Review of Sample Preparation, Analysis, and Data Interpretation. *Environ. Sci. Technol.* **2021**, *55*, 9637–9656. [CrossRef]
12. Han, H.; Feng, Y.; Chen, J.; Xie, Q.; Chen, S.; Sheng, M.; Zhong, S.; Wei, W.; Su, S.; Fu, P. Acidification impacts on the molecular composition of dissolved organic matter revealed by FT-ICR MS. *Sci. Total Environ.* **2022**, *805*, 150284. [CrossRef]
13. Torrent, J.; Barrón, V.; Liu, Q. Magnetic enhancement is linked to and precedes hematite formation in aerobic soil. *Geophys. Res. Lett.* **2006**, *33*, 2. [CrossRef]
14. Vodyanitskii, Y.N.; Kirillova, N.P. Application of the CIE-L*a*b* System to Characterize Soil Color. *Soil Phys.* **2016**, *49*, 1337–1346. [CrossRef]
15. *NY/T 1121.6-2006*; Soil Testing-Part 6: Method for Determination of Soil Organic Matter. Ministry of Agriculture of the China: Beijing, China, 2006.
16. *NY/T 1121.7-2014*; Soil Testing-Part 7: Method for Determination of Available Phosphorus in Soil. Ministry of Agriculture of the China: Beijing, China, 2014.
17. Lu, M.; Wang, X.; Li, Y.; Liu, H.; An, X.; Lian, B. Soil microbial community structure and environmental effects of serpentine weathering under different vegetative covers in the serpentine mining area of Donghai County, China. *Sci. Total Environ.* **2022**, *835*, 155452. [CrossRef]
18. Stookey, L.L. Ferrozine—A new spectrophotometric reagent for iron. *Anal. Chem.* **1970**, *42*, 779–781. [CrossRef]
19. Xue, J.; Deng, Y.; Luo, Y.; Du, Y.; Yang, Y.; Cheng, Y.; Xie, X.; Gan, Y.; Wang, Y. Unraveling the impact of iron oxides-organic matter complexes on iodine mobilization in alluvial-lacustrine aquifers from central Yangtze River Basin. *Sci. Total Environ.* **2022**, *814*, 151930. [CrossRef]
20. Li, Y.; Chen, Z.; Chen, J.; Castellano, M.J.; Ye, C.; Zhang, N.; Miao, Y.; Zheng, H.; Li, J.; Ding, W. Oxygen availability regulates the quality of soil dissolved organic matter by mediating microbial metabolism and iron oxidation. *Glob. Chang. Biol.* **2022**, *28*, 7410–7427. [CrossRef]
21. Lalonde, K.; Mucci, A.; Ouellet, A.; Gélinas, Y. Preservation of organic matter in sediments promoted by iron. *Nature* **2012**, *483*, 198–200. [CrossRef]

22. Ohno, T.; Parr, T.B.; Gruselle, M.C.I.; Fernandez, I.J.; Sleighter, R.L.; Hatcher, P.G. Molecular Composition and Biodegradability of Soil Organic Matter: A Case Study Comparing Two New England Forest Types. *Environ. Sci. Technol.* **2014**, *48*, 7229–7236. [CrossRef]
23. Rezapour, S.; Azhah, H.; Momtaz, H.R.; Ghaemian, N. Changes in forms and distribution pattern of soil iron oxides due to long-term cropping in the Northwest of Iran. *Environ. Earth Sci.* **2015**, *73*, 7275–7286. [CrossRef]
24. Li, X.; Sun, G.; Chen, S.; Fang, Z.; Yuan, H.; Shi, Q.; Zhu, Y. Molecular Chemodiversity of Dissolved Organic Matter in Paddy Soils. *Environ. Sci. Technol.* **2018**, *52*, 963–971. [CrossRef] [PubMed]
25. Koch, B.P.; Dittmar, T. From mass to structure: An aromaticity index for high-resolution mass data of natural organic matter. *Rapid Commun. Mass Spectrom.* **2006**, *20*, 926–932. [CrossRef]
26. Zhou, Y.; Zhao, C.; He, C.; Li, P.; Wang, Y.; Pang, Y.; Shi, Q.; He, D. Characterization of dissolved organic matter processing between surface sediment porewater and overlying bottom water in the Yangtze River Estuary. *Water Res.* **2022**, *215*, 118260. [CrossRef]
27. Luo, J.; Zhou, Q.; Hu, X.; Zeng, H.; Deng, P.; He, C.; Shi, Q. Lake Chemodiversity Driven by Natural and Anthropogenic Factors. *Environ. Sci. Technol.* **2022**, *56*, 5910–5919. [CrossRef]
28. Sun, Y.; Li, X.; Li, X.; Wang, J. Deciphering the Fingerprint of Dissolved Organic Matter in the Soil Amended with Biodegradable and Conventional Microplastics Based on Optical and Molecular Signatures. *Environ. Sci. Technol.* **2022**, *56*, 15746–15759. [CrossRef]
29. Torrent, J.; Liu, Q.; Bloemendal, J.; Barron, V. Magnetic enhancement and iron oxides in the upper Luochuan loess-paleosol sequence, Chinese Loess Plateau. *Soil Sci. Soc. Am. J.* **2007**, *71*, 1570–1578. [CrossRef]
30. Zhang, Q.Y.; Yuan, Y.J.; Liao, Z.M.; Zhang, W.X. Use of microbial indicators combined with environmental factors coupled with metrology tools for discrimination and classification of Luzhou-flavoured pit muds. *J. Appl. Microbiol.* **2017**, *123*, 933–943. [CrossRef]
31. Gao, Z.; Wu, Z.; Zhang, W. Effect of Pit Mud on Bacterial Community and Aroma Components in Yellow Water and Their Changes during the Fermentation of Chinese Strong-Flavor Liquor. *Foods* **2020**, *9*, 372. [CrossRef]
32. Li, J.; Wang, Y.; Xue, X.; Xie, X.; Siebecker, M.G.; Sparks, D.L.; Wang, Y. Mechanistic insights into iodine enrichment in groundwater during the transformation of iron minerals in aquifer sediments. *Sci. Total Environ.* **2020**, *745*, 140922. [CrossRef]
33. Frommer, J.; Voegelin, A.; Dittmar, J.; Marcus, M.A.; Kretzschmar, R. Biogeochemical processes and arsenic enrichment around rice roots in paddy soil: Results from micro-focused X-ray spectroscopy. *Eur. J. Soil Sci.* **2011**, *62*, 305–317. [CrossRef]
34. Duan, X.; Li, Z.; Li, Y.; Yuan, H.; Gao, W.; Chen, X.; Ge, T.; Wu, J.; Zhu, Z. Iron–organic carbon associations stimulate carbon accumulation in paddy soils by decreasing soil organic carbon priming. *Soil Biol. Biochem.* **2023**, *179*, 108972. [CrossRef]
35. Zhao, Y.; Moore, O.W.; Xiao, K.; Curti, L.; Fariña, A.O.; Banwart, S.A.; Peacock, C.L. The role and fate of organic carbon during aging of ferrihydrite. *Geochim. Cosmochim. Acta* **2022**, *335*, 339–355. [CrossRef]
36. Keiluweit, M.; Nico, P.S.; Kleber, M.; Fendorf, S. Are oxygen limitations under recognized regulators of organic carbon turnover in upland soils? *Biogeochemistry* **2016**, *127*, 157–171. [CrossRef]
37. Wu, S.; You, F.; Boughton, B.; Liu, Y.; Nguyen, T.A.H.; Wykes, J.; Southam, G.; Robertson, L.M.; Chan, T.; Lu, Y.; et al. Chemodiversity of Dissolved Organic Matter and Its Molecular Changes Driven by Rhizosphere Activities in Fe Ore Tailings Undergoing Eco-Engineered Pedogenesis. *Environ. Sci. Technol.* **2021**, *55*, 13045–13060. [CrossRef]
38. Chapman, S.J.; Campbell, C.D.; Fraser, A.R.; Puri, G. FTIR spectroscopy of peat in and bordering Scots pine woodland; relationship with chemical and biological properties. *Soil Biol. Biochem.* **2001**, *33*, 1193–1200. [CrossRef]
39. Artz, R.R.E.; Chapman, S.J.; Jean Robertson, A.H.; Potts, J.M.; Laggoun-Défarge, F.; Gogo, S.; Comont, L.; Disnar, J.; Francez, A. FTIR spectroscopy can be used as a screening tool for organic matter quality in regenerating cutover peatlands. *Soil Biol. Biochem.* **2008**, *40*, 515–527. [CrossRef]
40. Li, Y.; Heal, K.; Wang, S.; Cao, S.; Zhou, C. Chemodiversity of Soil Dissolved Organic Matter and Its Association with Soil Microbial Communities Along a Chronosequence of Chinese Fir Monoculture Plantations. *Front. Microbiol.* **2021**, *12*, 729344. [CrossRef]
41. Xu, L.; Hu, Q.; Liao, L.; Duan, Z.; Liu, S.; Chen, L.; Zhu, Q.; Zhong, A. Hydrological isolation affected the chemo-diversity of dissolved organic matter in a large river-connected lake (Poyang Lake, China). *Sci. Total Environ.* **2022**, *851*, 158047. [CrossRef]
42. Ye, Y.; Wang, Z.; Liu, L.; Qi, K.; Xie, X. Novel insights into the temporal molecular fractionation of dissolved black carbon at the iron oxyhydroxide-water interface. *Water Res.* **2023**, *229*, 119410. [CrossRef]
43. Lv, J.; Zhang, S.; Wang, S.; Luo, L.; Cao, D.; Christie, P. Molecular-Scale Investigation with ESI-FT-ICR-MS on Fractionation of Dissolved Organic Matter Induced by Adsorption on Iron Oxyhydroxides. *Environ. Sci. Technol.* **2016**, *50*, 2328–2336. [CrossRef] [PubMed]
44. Bao, Y.; Bolan, N.S.; Lai, J.; Wang, Y.; Jin, X.; Kirkham, M.B.; Wu, X.; Fang, Z.; Zhang, Y.; Wang, H. Interactions between organic matter and Fe (hydr)oxides and their influences on immobilization and remobilization of metal(loid)s: A review. *Crit. Rev. Environ. Sci. Technol.* **2022**, *52*, 4016–4037. [CrossRef]

45. Huang, Z.; Lv, J.; Cao, D.; Zhang, S. Iron plays an important role in molecular fractionation of dissolved organic matter at soil-water interface. *Sci. Total Environ.* **2019**, *670*, 300–307. [CrossRef] [PubMed]
46. Karimian, N.; Burton, E.D.; Johnston, S.G.; Hockmann, K.; Choppala, G. Humic acid impacts antimony partitioning and speciation during iron(II)-induced ferrihydrite transformation. *Sci. Total Environ.* **2019**, *683*, 399–410. [CrossRef]

Disclaimer/Publisher's Note: The statements, opinions and data contained in all publications are solely those of the individual author(s) and contributor(s) and not of MDPI and/or the editor(s). MDPI and/or the editor(s) disclaim responsibility for any injury to people or property resulting from any ideas, methods, instructions or products referred to in the content.

foods

MDPI

Article

Study on the Fermented Grain Characteristics and Volatile Flavor Substances during the Tuqu Fermentation of Hunan Light-Flavor Baijiu

Jie Xu [1,2], Ting Zhang [1,2], Huitai Chen [3], Yijie Dai [4], Zongjun Li [4], Jia He [1,2], Rongfang Ju [1,2] and Aixiang Hou [1,2,*]

1 College of Food Science and Technology, Hunan Agricultural University, Changsha 410128, China; 18798312569@163.com (J.X.); zhangting9909@163.com (T.Z.); rehjia189@163.com (J.H.); 15284676937@163.com (R.J.)
2 Key Laboratory of Food Science and Biotechnology of Hunan Province, Changsha 410128, China
3 Hunan Guoyuan Liquor Co., Ltd., Miluo 414400, China; 18673187905@163.com
4 School of Biology and Environmental Engineering, Guiyang University, Guiyang 550005, China; daiyijie866@163.com (Y.D.); hnlizongjun@163.com (Z.L.)
* Correspondence: aixianghou@163.com

Abstract: The present study employed Hunan local Tuqu for fermentation and investigated the physicochemical properties, microbial community composition, and volatile flavor compounds of the fermented grains, as well as the correlation between the physicochemical indicators and the microbial community. The findings reveal that the activities of α-amylase and glucoamylase were highest during the initial stages of the fermentation process. The acid protease activity increased to 30.6 U/g on the second day and then decreased. Cellulose and lipase activities both showed an increasing trend. The moisture content increased sharply to 73.41% and then remained relatively stable. The acidity was highest on the eighth day. Fifty genera of bacteria and twenty-two genera of fungi were detected. *Lactobacillus* was dominant among bacteria, and *Saccharomyces* was dominant among fungi. A correlation analysis showed that there were positive correlations between moisture, acidity, cellulose, lipase activities and *Lactobacillus*, and there were positive correlations between moisture content, acidity, cellulase activity, acidic protease activity and *Saccharomyces*. A total of 46 volatile flavor compounds were detected, of which 6 alcohols and 14 esters constituted the major portion, and 9 key flavor compounds with an ROAV > 1 were identified throughout the fermentation process. Isoamyl acetate had the highest ROAV and made the greatest contribution to the flavor.

Keywords: light-flavor Baijiu; fermented grains; bacteria; fungi; volatile flavor substances

1. Introduction

As one of the oldest distilled liquors in the world, Chinese Baijiu has a history of thousands of years [1]. In the process of development, due to the brewing of raw materials, fermentation strains, brewing techniques, and regional differences, 12 types of Baijiu, such as sauce-flavor, strong-flavor, light-flavor, and rice-flavor Baijiu, have gradually developed [2]. The liquor market in Hunan province is highly inclusive, open, and of high consumption grade, but the competitiveness of real estate liquor is weak and the market share is low. The annual sales of liquor are nearly 30 billion yuan, but the consumption of real estate liquor is only more than 5 billion yuan, accounting for about one sixth of the market share. Therefore, in order to improve the competitiveness and market value of local liquor, it is necessary to study a liquor with local Hunan characteristics.

As one of the oldest and most popular types of Baijiu, light-flavor Baijiu, is pure, sweet, mild, natural, and has a fresh aftertaste [3]. It can be divided into Daqu, Xiaoqu, and Fuqu light-flavor Baijiu according to the type of starter culture [4]. Microbial starter cultures play a very important role in the fermentation process of Baijiu. It has been shown that bacteria have an important function in the production of enzymes and aromas during the fermentation of Baijiu. The flavor substances produced by the metabolism of bacteria play a decisive role in the aroma and quality of the liquor [5]. Fungi play a key regulatory role in the formation of flavor compounds and in ensuring the consistent food quality of fermented foods [6]. Microorganisms of fermented grains interact with moisture, acidity, and enzyme activity during fermentation, so it was necessary to investigate their interactions during fermentation. In recent years, research on light-flavor Baijiu has focused on the influence of microbial flora, the structure of functional strains, and flavor substances on the style of Baijiu [7–9]. However, the physicochemical properties, microbial communities, and dynamics of the flavor compounds in fermented grains are still not fully understood.

During the fermentation of Baijiu, the microbial composition and environmental factors can have a great influence on the community succession during the fermentation of Baijiu. Microbial interactions affect the metabolic behavior of microorganisms involved in food fermentation and, consequently, changes in volatile flavor substances. Therefore, exploring the physicochemical properties, microbial communities, and volatile flavor compounds of fermentation grains could not only reveal the changing rules of the physicochemical, microbial community, and volatile flavor substances of the fermentation process, but also provide an important reference for controlling the quality of clear-flavored original wine from the perspective of fermentation. In this study, the microbial flora composition of different fermentation stages was analyzed by using Hunan Yueyang and Liuyang Tuqu in the mixed fermentation of light-flavor Baijiu. The moisture content, acidity, associated enzyme activity, and major volatile flavor substances were determined, and correlations between microorganisms and the moisture, acidity, and enzyme activity during fermentation were analyzed. The aim was to investigate the evolution of the microbial flora, the variations in moisture, acidity, enzymatic activity, and other physicochemical characteristics, and the formation of volatile flavor substances in the Tuqu fermentation process of Hunan light-flavored Baijiu, in order to provide data reference and theoretical guidance for the development of Hunan characteristic light-flavor Baijiu.

2. Materials and Methods

2.1. Sample Collection

All samples for this study were collected from the top of the cellar at 0 days, 2 days, 4 days, 6 days, 8 days, and 10 days of fermentation. The five-point sampling method was used for sampling. The specific sampling points are shown in Figure 1. The samples collected at each point were mixed and stored in a liquid nitrogen jar.

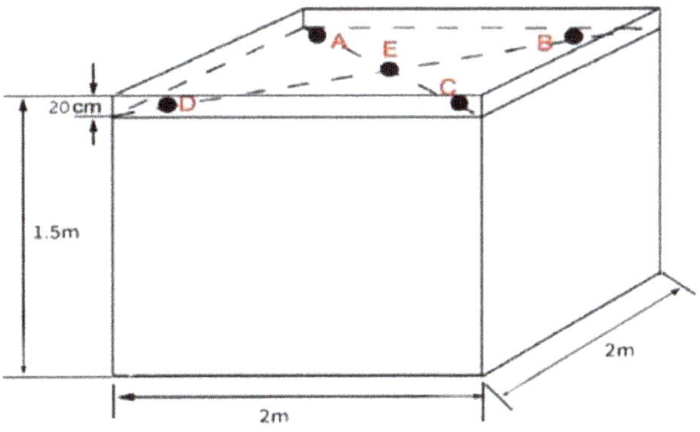

Figure 1. Sampling sites of the fermented grain samples.

2.2. The Brewing Process of the Tuqu Fermentation of Hunan Light-Flavor Baijiu

The technological process of the Tuqu fermentation of Hunan light-flavor Baijiu in this study is shown in Figure 2.

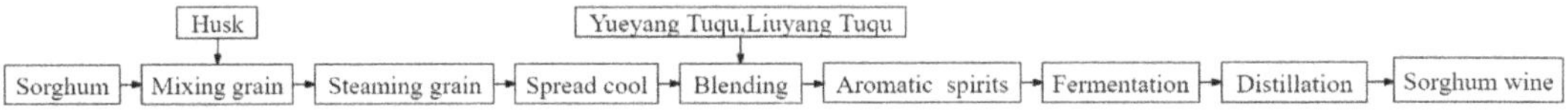

Figure 2. The brewing process of Hunan light-flavor Baijiu.

2.3. Major Reagents and Instruments

Concentrated sulfuric acid (analytical and pure): Nanjing Sheng Qing He Chemical Co., Ltd. (Nanjing, China); toluene (analytical purity): Wuxi Yashang Chemical Co., Ltd. (Yizing, China); A-amylase (A-AL) kit, acid protease (ACP) kit, glucoamylase (Gluco) kit, cellulase (CL) kit, and lipase (LPS) kit: Suzhou Kemin Biotechnology Co., Ltd. (Suzhou, China); constant temperature oscillation water bath: Hunan Jinnan Instrument Manufacturing Co., Ltd. (Jinan, China); MD enzyme standard instrument: Meigu Molecular Instruments (Shanghai, China) Co., Ltd.; CJJ78-1 Magnetic Stirrer: Wuhan Glamour Testing Equipment Co., Ltd. (Wuhan, China); H1200R high-speed frozen centrifuge: Shanghai Precision Instrument Company (Shanghai, China); F6/10 handheld high-speed homogenizer: Shanghai Jingxin Industrial Development Co., Ltd. (Shanghai, China); XUEKE ice-making machine: Beijing Zhongxing Baihui Technology Co., Ltd. (Beijing, China); ZYS-50B wise copy energy measuring water meter: XiangYi Balance Instrument Co., Ltd. (Xiangtan, China).

2.4. Determination of Physical and Chemical Indicators

The enzyme activity was determined according to the kit instructions. The moisture was determined using an intelligent moisture tester. The acidity was determined according to the "DB34/T2264-2014" [10] fermentation method for solid fermented grains, and the acid–base neutralization titration method was used for the determination.

2.5. DNA Extraction, Polymerase Chain Reaction (PCR) Amplification, and Sequencing

2.5.1. DNA Extraction

The total microbial genomic DNA samples were extracted using the DNeasy PowerSoil Kit (QIAGEN, Inc., Venlo, The Netherlands) following the manufacturer's instructions and stored at −20 °C prior to further analysis. The quantity and quality of the extracted DNAs were measured using a NanoDrop ND-1000 spectrophotometer (Thermo Fisher Scientific, Waltham, MA, USA) and agarose gel electrophoresis, respectively.

2.5.2. 16S rRNA and ITS Gene Amplicon Sequencing

PCR amplification of the bacterial 16S rRNA gene V3-V4 region was performed using the forwards primer 338F (ACTCCTACGGGAGGCAGCA) and the reverse primer 806R (GGACTACHVGGGTWTCTAAT). PCR amplification of fungal internal transcribed spacer (ITS) regions was performed with the primers ITS3 (GCATCGATGAAGAACGCAGC) andITS4 (TCCTCCGCTTATTGATATGC). Sample-specific 7-bp barcodes were incorporated into the primers for multiplex sequencing. The PCR components contained 5 μL of Q5 reaction buffer (5×), 5 μL of Q5 HighFidelity GC buffer (5×), 0.25 μL of Q5 High-Fidelity DNA Polymerase (5 U/μL), 2 μL (2.5 mM) of dNTPs, 1 μL (10 μM) of each forward and reverse primer, 2 μL of DNA template, and 8.75 μL of ddH2O. Thermal cycling consisted of initial denaturation at 98 °C for 2 min, followed by 25 cycles consisting of denaturation at 98 °C for 15 s, annealing at 55 °C for 30 s, and extension at 72 °C for 30 s, with a final extension of 5 min at 72 °C. PCR amplicons were purified with Agencourt AMPure Beads (Beckman Coulter, Indianapolis, IN, USA) and quantified using the PicoGreen dsDNA Assay Kit (Invitrogen, Carlsbad, CA, USA). After the individual quantification step, amplicons were pooled in equal amounts, and paired-end 2 × 300 bp sequencing was performed using the Illumina MiSeq platform with MiSeq Reagent Kit v3 at Shanghai Personal Biotechnology Co., Ltd. (Shanghai, China).

2.6. Determination of Volatile Flavor Substances

Headspace solid phase microextraction was employed to extract VOCs from the fermented grain samples. The detailed method was carried out as follows: in a headspace vial with a capacity of 20 mL equipped with a magnetic stirrer, 3 g of the fermented grain sample, 2 g of sodium chloride, 20 μL of amyl acetate (0.5275 mg/mL) as an internal standard solution, and 4 mL of boiled ultra-pure water were combined. The top of the glass bottle was sealed with a silicon diaphragm to create a sealed headspace environment. The headspace glass bottle was then heated to 50 °C, and the tea sample was stirred at a speed of 1100 rpm using the magnetic stirrer. After a 10 min pre-balance period, a 50/30 μm DVB/CAR/PDMS fiber from American Supelco was inserted into the headspace vial to extract the VOCs. The extraction process lasted for 45 min. Subsequently, the fiber was desorbed for five minutes at the injection port of the GC–MS, operating in splitless injection mode at 250 °C.

Head space solid phase microextraction (HS-SPME):

The gas chromatography conditions were as follows: column, Hp-5 ms, ultra-inert (−60~325 °C) 30 m × 250 mm × 0.25 μm; injection temperature, 250 °C in splitless mode; heating procedure, a starting temperature of 50 °C, maintained for 2 min, increased to 230 °C at 4 °C/min and maintained for 5 min; and column carrier gas, helium at a constant flow rate of 1 mL/min.

The mass spectrometric conditions were as follows: ionization source, electron impact (EI); ion source temperature, 230 °C; EI energy, 70 eV; quadrupole temperature, 150 °C; scan mode, constant pressure full scan; and mass spectrometric scan range, 30–600 amu. We compared the mass spectrometry information of each chromatographic peak obtained from GC-MS analysis with the reference compounds in the NIST 17 library to identify unknown compounds [11]. To calculate the retention index (RI) for each peak, the retention time (RT) of *n*-alkanes (C7~C21) under the same GC–MS conditions was utilized. The identification of VOCs was accomplished by comparing the calculated RI values and the mass spectra of all detected metabolites with those present in the NIST 17 library. For the quantification of VOCs, the peak area ratio of the VOCs to the internal standard ethyl caprate was multiplied by the concentration of ethyl caprate.

2.7. Analysis of Relative Odor Activity Value

The relative odor activity value (ROAV) is a method used to identify key volatile compounds in samples, and VOCs with an ROAV value greater than or equal to 1 are considered key volatile compounds. VOCs with an ROAV value ranging from 0.1 to <1 have a modifying effect on the overall flavor of the sample. The ROAV calculation method was based on the ratio of a volatile component's odor activity value (OAV) to the greatest odor activity value (OAVmax) among the volatile compounds in the Tuqu fermentation of Hunan light-flavor Baijiu [12], and the thresholds were obtained by finding the threshold of perception of each volatile compound in the water based on [13].

2.8. Statistical Analysis

Experimental data were analyzed using the Origin 2018 statistical program. Correlation analysis was performed using SPSS v. 22.0 software. The raw data generated by the Illumina NovaSeq250 sequencing platform were used for sequence data analysis using the QIIME and R package (v3.2.0). The community composition and heatmaps based on the R language were generated using the Illumina NovaSeq250 sequencing platform.

3. Results and Discussion

3.1. Variation Rules of Enzyme Activity during the Tuqu Fermentation of Hunan Light-Flavor Baijiu

In the Baijiu brewing process, enzymes, as key substances in Baijiu fermentation, have various functions, such as degrading fermented raw material particles, promoting microbial reproduction and catalyzing biochemical reactions to produce aroma substances. The variations in enzyme activities during the Tuqu fermentation of Hunan light-flavor Baijiu are shown in Table 1.

Table 1. Variations in enzyme activity during the Tuqu fermentation of Hunan light-flavor Baijiu.

Fermentation Days (d)	Alpha Amylase Activity/(U/g)	Acid Protease Activity/(U/g)	Glucoamylase Activity/(U/g)	Cellulase Activity/(U/g)	Lipase Activity/(U/g)
0	28.1	7.6	1027.1	118.9	31.1
2	6.5	30.6	809.8	473.8	34.3
4	4.2	25.7	786.2	607.7	34.8
6	3.5	20.5	691.8	603.1	33.5
8	4.9	14.7	732.5	607.1	35.1
10	3.8	16.7	766.7	668.7	36.4

As shown in Table 1, the α-amylase and glycolytic enzyme activities decreased during the fermentation process, and the α-amylase enzyme activity decreased from 28.1 U/g to 3.8 U/g. The acidity and alcoholic content increase during the fermentation process, which inhibits α-amylase activity [14]. The glucoamylase activity decreased from 1027.1 U/g to 766.7 U/g. *Aspergillus* can produce glucoamylase at the beginning of fermentation, which makes raw starch material decompose rapidly. As fermentation progresses, fungi such as *Aspergillus* decline gradually, and glucoamylase activity also decreases gradually [15]. Acid protease activity was highest on the second day of fermentation. Fungi such as *Aspergillus, Penicillium,* and *Rhizopus* can produce acid proteases. In the early stages of fermentation, there was sufficient oxygen and nutrients in the fermented grains, which promoted the proliferation of microorganisms; therefore, the activity of acid protease was increased. Cellulase activity and lipase activity showed an overall increasing trend, reaching peaks of 668.7 U/g and 36.4 U/g on day 10, respectively. Cellulase can act on fibrous raw materials, while lipase can act on ester bonds, making full use of the raw materials.

3.2. Variation in Moisture and Acidity during the Tuqu Fermentation Process of Hunan Light-Flavor Baijiu

As shown in Figure 3A, from the initial stage of fermentation, the moisture content increased sharply from 65.04% to 73.41% and remained stable. During the prefermentation period, the source of moisture is mainly the water added during production as well as the water produced by the interaction between microorganisms [3]. During the fermentation process, volatile flavor substances such as ethanol, acids, and esters are produced along with water, thus increasing the moisture content of the fermented grains [16]. There was a slight decrease in moisture content at 2–10 days, and then it slowly increased again. This was due to the effect of gravity; the water flowed to the bottom of the cellar, becoming the main component of the yellow water, resulting in a decrease in moisture. During the fermentation process, with the progress of fermentation, yeast and other microorganisms conducted ethanol fermentation to produce free water, so the water of the fermented grains increased.

The acid in the fermented grains is mainly produced by microbial metabolism and is an important flavor substance in the Baijiu body, which can play a role in flavor enhancement, and the right amount of acid can increase the after-taste of liquor and make the Baijiu body fuller [17]. As seen in Figure 3B, the overall acidity showed an upward trend. As fermentation proceeded, anaerobic and partly anaerobic bacteria proliferated to produce a variety of organic acids, such as lactic acid and acetic acid, to increase the acidity.

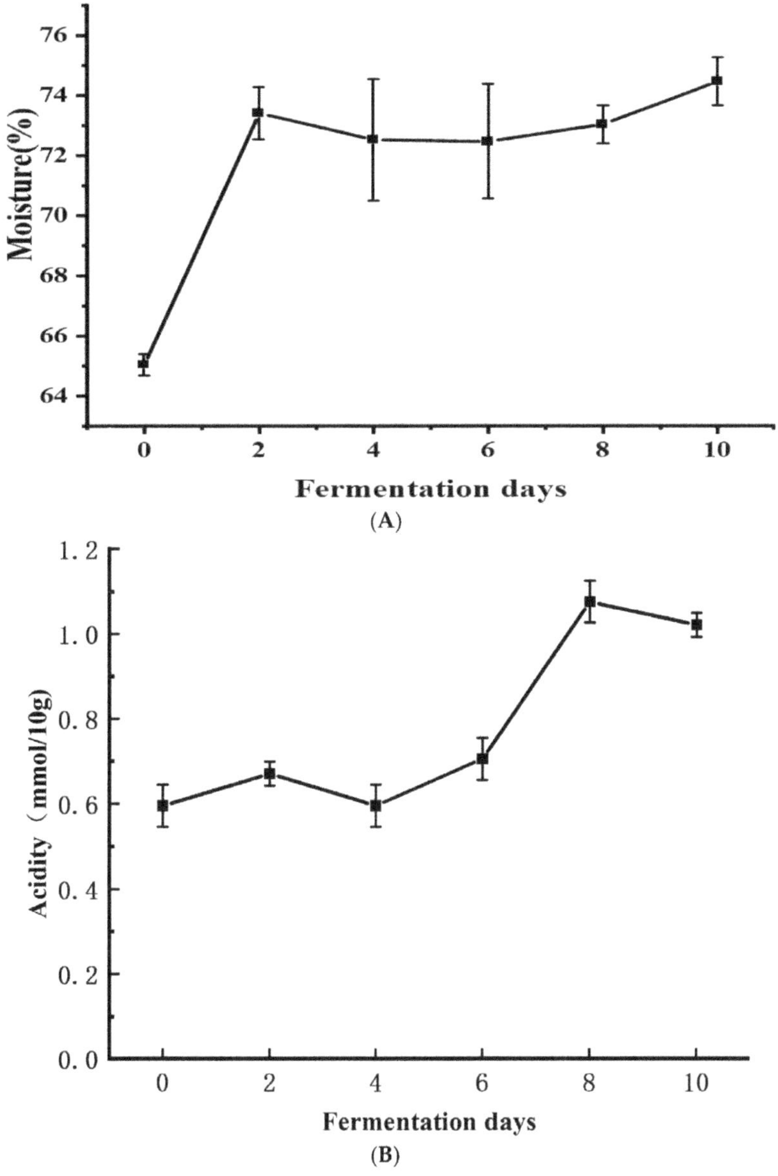

Figure 3. Variation of moisture and acidity with fermentation days during the Toqu fermentation of Hunan light-flavor Baijiu ((**A**): moisture; (**B**): acidity).

3.3. Variations in the Structure of Bacteria during the Tuqu Fermentation of Hunan Light-Flavor Baijiu

3.3.1. Bacterial Alpha Diversity Analysis of the Fermented Grains

The complexity of the species diversity of the studied samples was analyzed by calculating α-diversity indices. In α-diversity analysis, the Chao 1 index is mainly related to the abundance of the samples, and the larger the Chao 1 index is, the higher the abundance of the samples. The Shannon index is mainly related to the diversity of the sample, which reflects not only the richness but also the evenness of the species; the larger the Shannon index is, the higher the diversity [7]. As shown in Figure 4, the Chao1 index of F4 was higher than that of the other sample groups, which indicates that the total number of bacterial species in the F4 group was higher than that in the other sample groups. The Shannon index was higher in the F0 group than in the other sample groups, which indicates that the diversity of bacterial species was higher in the F0 group than in the other sample groups. The Shannon index decreased continuously as the fermentation proceeded, which indicates that the unfavorable conditions of low oxygen, low pH values, and high ethanol caused many bacteria to be inhibited. The coverage of all samples was greater than 0.996, and the *p* value of the Chao1 index was 0.069, which indicates that the library coverage of each sample was high, the sequences were basically completely sequenced, and the sequencing results were reliable.

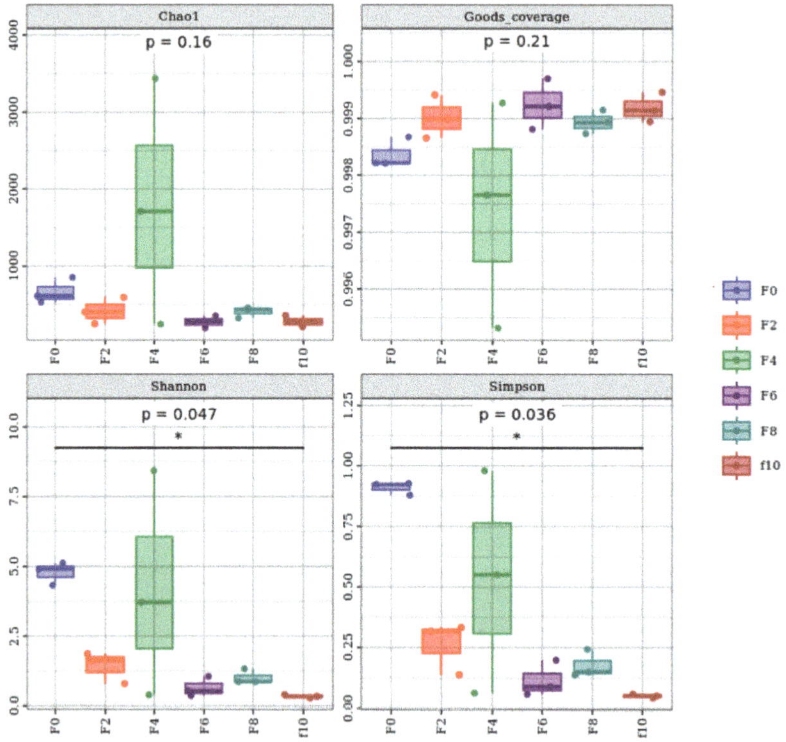

Figure 4. Alpha diversity analysis of bacteria in the Tuqu fermentation process of Hunan light-flavor Baijiu. * is a marker indicating significance.

3.3.2. Analysis of the Bacterial Community Structure of the Fermented Grains

Figure 5 shows the relative abundance of species in the top 20. As seen from the figure, the bacterial flora in the fermented grains at the genus level consisted mainly of *Lactobacillus*, *Geobacillus*, *Pediococcus*, and *Acetobacter*. As the fermentation progressed, *Lactobacillus* accounted for over 60% of the total sequence readings at 2–10 days, and a peak value of 97.97% was observed at the tenth day. The acidity and alcoholic strength increased, which caused the extinction of many bacteria; therefore, *Lactobacillus* became the dominant bacteria. *Lactobacillus* could affect the growth of other microorganisms by producing antagonistic substances such as bacteriocins and by competing with other microorganisms for substrates [18,19]. These were the dominant bacterial genera in the fermented grains. In addition, *Lactobacillus* were identified as the most abundant bacteria during the fermentation of the fermented grains, which was consistent with the results revealed in previous studies [20]. The abundances of *Geobacillus* and *Acetobacter* were highest, reaching peaks of 23.75% and 17.58% at 0 days, respectively, and then decreasing gradually. During the fermentation of the liquor, *Bacillus* can produce some flavor synthesis precursors in addition to the direct flavor substances obtained from their own fermentation. Complex flavor substances are further synthesized by a number of chemical or biological interactions [21]. *Acetobacter* are specialized aerobic bacteria that play an important role in the aroma presentation of Baijiu. There are abundant bacteria, such as *Acetobacter*, in the environment that can break down different types of sugars and alcohols into organic acids in oxygen [22].

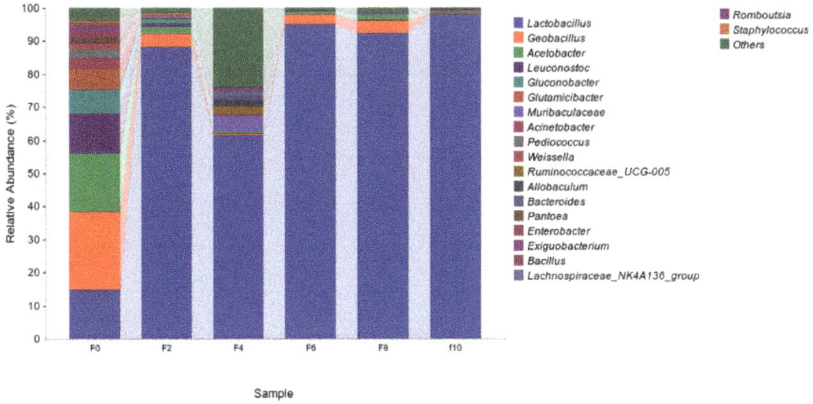

Figure 5. Community structure distribution of bacteria based on genus level during the Tuqu fermentation of Hunan light-flavor Baijiu.

3.3.3. Heat Map Analysis of the Fermented Grain Bacterial Community Succession

Figure 6 shows a heat map of the bacterial composition during the Tuqu fermentation of Hunan light-flavor Baijiu, where a color closer to green indicates a greater relative abundance of the species, and a color closer to brown indicates a smaller relative abundance of the genus [23]. As shown in Figure 6, the bacterial diversity and abundance in the F0 and F4 stages were more diverse, including bacteria such as *Bacillus* and *Weissella*. As fermentation progressed, the diversity and abundance in the fermented grains decreased rapidly. It is possible that the increase in acidity as well as the ethanol content inhibited the growth of these bacteria. *Lactobacillus* became the dominant bacterium in the fermented grains. *Lactobacillus* is a class of anaerobic or partly anaerobic bacteria that can ferment sugars to produce lactic acid and provide amino acids and various vitamins and bacteriocins for the growth and reproduction of other microorganisms. *Lactobacillus* plays an important role in the microecological environment of brewing [24].

3.4. Variations in the Structure of Fungi during the Tuqu Fermentation of Hunan Light-Flavor Baijiu

3.4.1. Fungi Alpha Diversity Analysis

From the α-diversity index in Figure 7, it can be seen that the coverage of each group of samples was greater than 0.999, and the p value of the Chao1 index was 0.014, which indicates that the sequencing depth was sufficient. The sequences in the samples were basically detected completely, and the results were true and reliable for subsequent analysis. The Chao1 index in F4 was the highest, which indicates that the total number of fungal species was the highest in the sample groups. The Shannon index of the F6 group was the highest, which indicates that the diversity of fungal species in the F6 group was the highest in the samples.

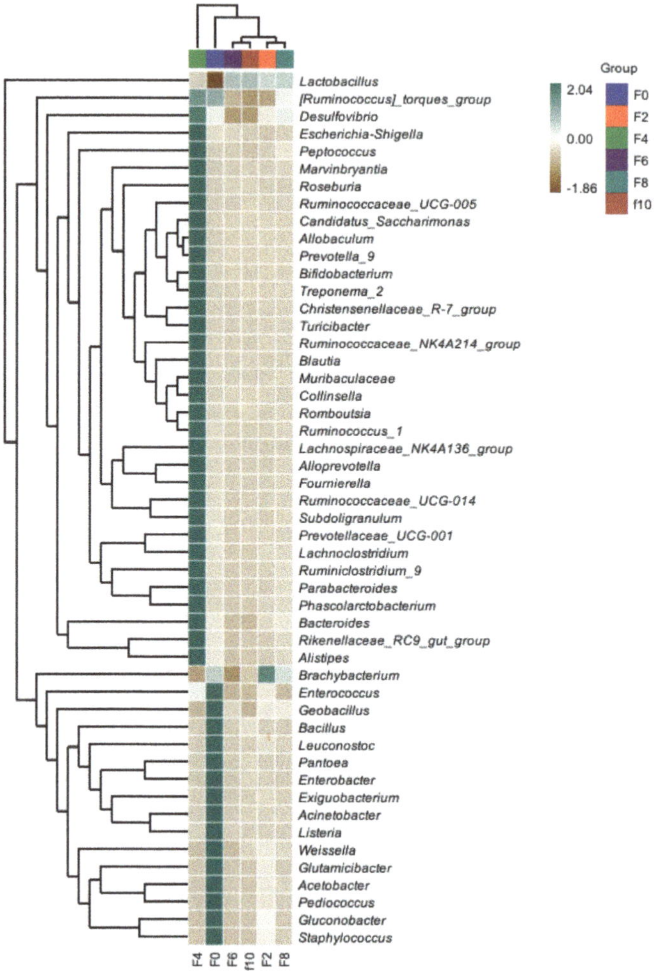

Figure 6. Heat map of the genus-based level of bacteria in the Tuqu fermentation process of Hunan light-flavor Baijiu.

3.4.2. Analysis of the Fermented Grain Fungi Community Structure

Figure 8 shows the top 20 species in terms of relative abundance. As seen from the figure, *Saccharomyces*, *Saccharomycopsis*, *Rhizopus*, and *Pichia* were the four main fungal genera. Among them, *Pichia* had the highest abundance on the second day, which gradually decreased as fermentation progressed. It has been shown that *Pichia* plays a major role in the production of aromas in Baijiu brewing and has the ability to produce liquor metabolism and at this stage, yeasts and bacteria proliferate and produce flavor substances from the available sugars and other nutrients [25]. This is consistent with previous research [26]. *Saccharomyces cerevisiae* is the main functional bacterium in the fermentation process of liquor, which not only has a high fermentation capacity but also produces esters and plays an important role in the fullness of the liquor [27]. During the prefermentation period, *Saccharomycopsis* and *Rhizopus* were the main fungi, and the relative abundance decreased as fermentation progressed. It might be that the increased concentration of organic acids and alcohols disrupted the growth environment. As fermentation proceeded, *Saccharomyces* was the main fungus, reaching a peak of 66.20% on the second day. *Saccharomyces cerevisiae* can grow and reproduce in an anaerobic environment and is also tolerant to ethanol [2].

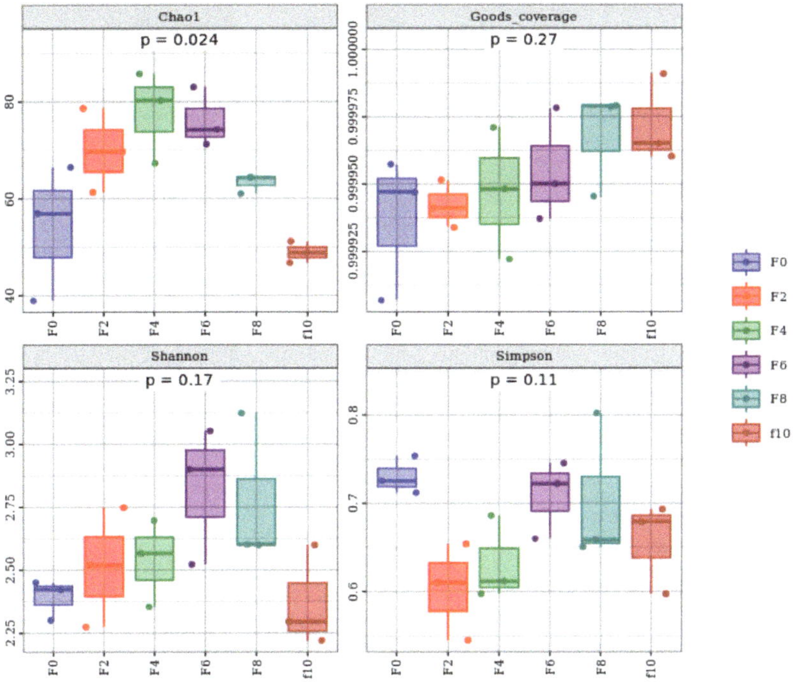

Figure 7. Alpha diversity analysis of fungi in the Tuqu fermentation process of Hunan light-flavor Baijiu.

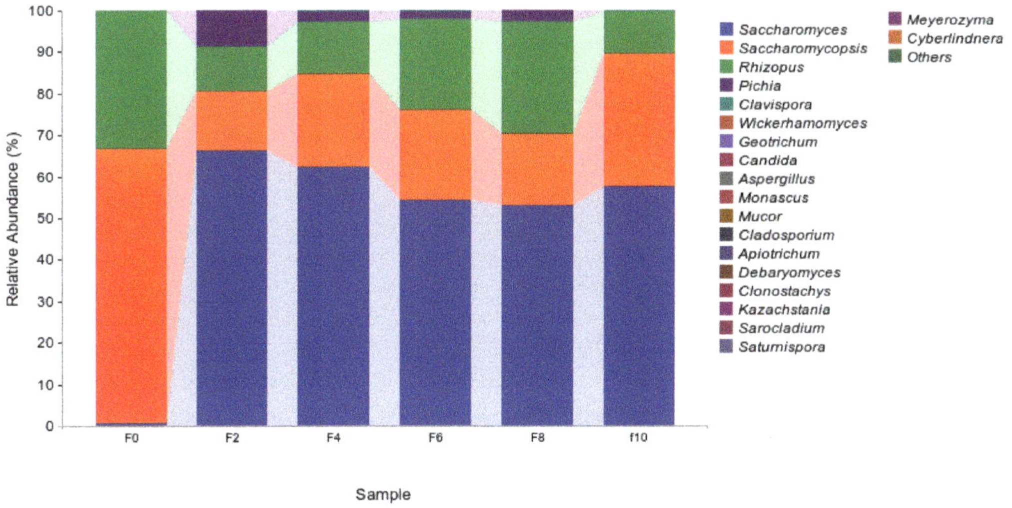

Figure 8. Community structure distribution of Fungi based on genus level during the Tuqu fermentation of Hunan light-flavor Baijiu.

3.4.3. Heat Map Analysis of Fungal Community Succession Based on the Species Taxonomic Genus Level

Figure 9 shows a heatmap of the fungal composition during the fermentation of Tuqu-fermented grains of Hunan light-flavor Baijiu. A color closer to green represents a higher abundance, and a color closer to brown represents a lower abundance. As shown in the graph, as fermentation progressed, there were different dominant fungi at different fermentation stages, indicating a rich microbial structure. The main dominant fungi at the F0 stage were *Trichosporon*, *Saccharomycopsis* and *Rhizopus*. The main dominant fungi at the F2 stage were *Pichia*, *Debaryomyces*, and *Cyberlindnera*. The main fungal genera at the F4 stage were *Mucor*, *Aspergillus*, *Kazachstania*, *Monascus*, and *Sarociadium*. The main dominant fungi at the F6 stage were *Meyerozyma*, *Cladosporium*, and *Clonostachys*. The main dominant fungi at the F8 stage were *Rhizopus* and *Satumispora*, and the main dominant fungi at the F10 stage were *Clavispora* and *Wickerhamomyces*. These fungi are important fungal genera in the fermentation process of Baijiu. Among them, *Wickerhamomyces* and *Aspergillus* play a key role in fermentation. *Aspergillus* can produce a variety of enzymes and plays a positive role in regulating the *saccharification*, *esterification*, and liquefaction power of Daqu and fermented grains. In addition, it can metabolize organic acids and fatty acids, which can lead to the production of aromatic esters that enhance and improve the flavor of Baijiu [28]. *Wickerhamomyces* is an important aromatic yeast with a high acidity tolerance and a high ester production capacity, which can impart a rich aroma to the liquor [29].

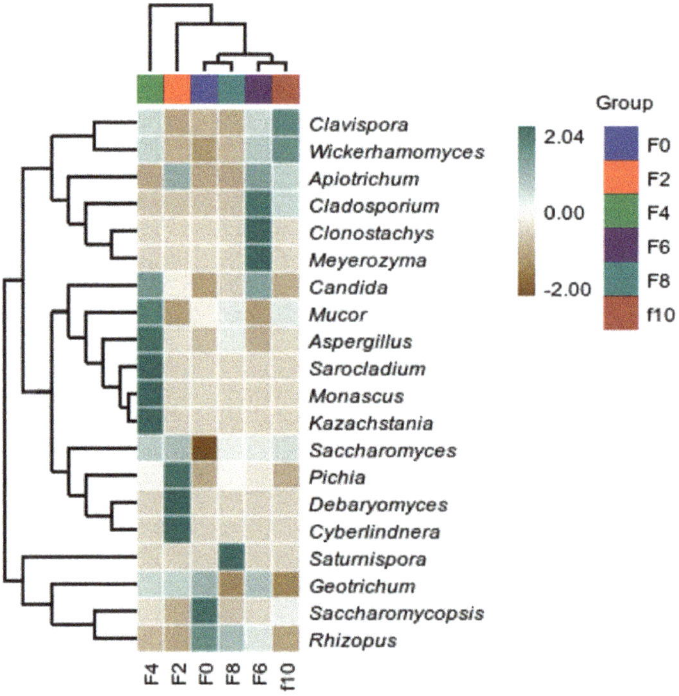

Figure 9. Heat map of the genus-based level of Fungi in the Tuqu fermentation process of Hunan light-flavor Baijiu.

3.5. Correlation Analysis of Microbial and Physicochemical System Indicators in the Tuqu Fermentation Process of Hunan Light-Flavor Baijiu

3.5.1. Correlation Analysis of Bacteria at the Genus Level and Physicochemical Indicators

Spearman correlation analysis of physicochemical indicators and bacteria that were the top 10 bacterial genera in the fermented grains was performed using SPSS software. As shown in Figure 10 The closer the color was to green, the stronger the correlation was, and the closer it was to brown, the weaker the correlation was. As shown in the graph, *Lactobacillus* was positively correlated with moisture content, acidity, cellulase activity, and lipase activity, which suggests that the acidity and moisture were the main factors affecting microbial succession, which is consistent with previous research on the light-flavor liquor [30] indicating that *Lactobacillus* might produce cellulase and lipase [31]. *Lactobacillus* has a strong ability to metabolize carbohydrates and produce acid, providing a source of carbon for other microorganisms. It is a precursor to the formation of flavor during the fermentation of liquor [32]. The other three genera, *Geobacillus*, *Pediococcus*, and *Acetobaceter*, which were ranked in the top four relative abundances, were all negatively correlated with the moisture content, acidity, cellulase activity, and lipase activity. This was due to the presence of diacetyl in the metabolites of *Lactobacillus*, which inhibits many spoilage and pathogenic bacteria. Moreover, it could inhibit the growth of gram-negative bacteria by interfering with the utilization of arginine through the binding protein reaction of gram-negative bacteria with arginine [33].

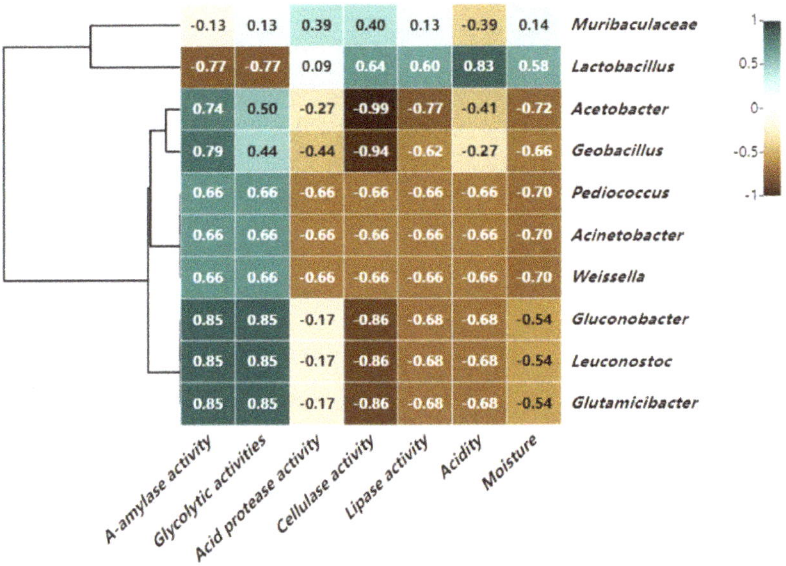

Figure 10. Heat map of correlation between bacterial microorganisms and physicochemical indicators based on genus level.

3.5.2. Correlation Analysis of Fungi at the Genus Level and Physicochemical Indicators

As shown in Figure 11, there were significant interactions between physicochemical factors and most fungal microorganisms, indicating the complexity of the fungal microbial metabolism. Moisture content and acidity were positively correlated with *Pichia* and *Wickerhamomyces*; moisture content, acidity, lipase activity, acid protease activity, and A-amylase activity were negatively correlated with *Saccharomycopsis* and *Rhizopus*; cellulase activity was positively correlated with *Clavwaspora* and *Wickerhamomyces*; and glucoamylase activity was positively correlated with *Rhizopus*. Traditionally, Rhizoctonia solani and others were considered to be the main producers of the saccharification power of macrophytes, which is consistent with the results in this study. The strong positive correlation between acid protease activity and the first dominant bacterium *Saccharomyces* suggests that acid protease production might be associated with yeasts, and *Saccharomycopsis* has been shown to produce acidic proteases capable of degrading large proteins [34].

The microbial composition and environmental factors can have a great influence on the community succession during the fermentation of Baijiu [35], indicating that those two factors are the main drivers that induce the succession of the fermentation from the initial to the end stage.

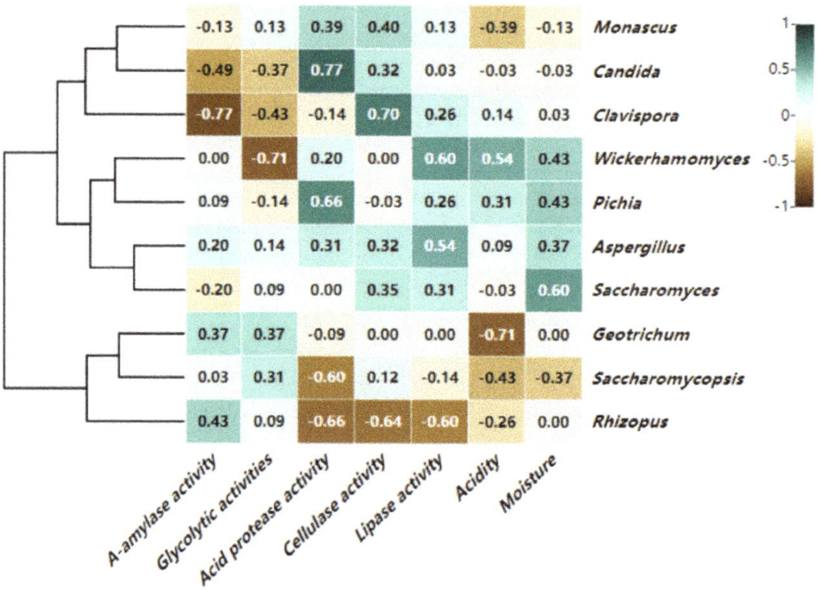

Figure 11. Heat map of correlation between fungi microorganisms and physicochemical indicators based on genus level.

3.6. Qualitative and Quantitative Analysis of VOCs by GC–MS

In this study, Table 2 provides qualitative and quantitative analysis of the identified VOCs. As shown in Figure 12A, a total of 46 VOCs were identified using GC–MS, including aldehydes (5), esters (14), alcohols (6), ketones (1), alkanes (4), phenols (2), acids (2), alkenes (9), and others (3). The relative proportion of these categorized VOCs is illustrated in Figure 12B, with esters (30.4%), alkenes (19.6%) and alcohols (13%) constituting a significant portion, and 63% of the total overall.

Esters were the most abundant class of flavor components in Baijiu, giving it fruity and floral aromas [36]. It was also the most abundant flavor substance detected in the fermented grains of Hunan light-flavor Baijiu, including ethyl lactate, isoamyl acetate, ethyl caprylate, diethyl succinate, ethyl caprate, ethyl phenylacetate, phenethyl acetate, etc. The esters accumulated during fermentation and reached a peak of 2130.6 (μg/kg) on the fourth day of fermentation, increasing from 95.68 (μg/kg) at the beginning of fermentation to 1780.7 (μg/kg) at the end of fermentation, which was 18.6 times higher than that at the beginning of fermentation. Isoamyl acetate, ethyl caprylate, diethyl succinate, and 2-methylbutyl acetate were the most abundant in the fermentation process of the fermented grains, accounting for 79.5% of the esters, with a fruity aroma, and are important aroma and flavor presenting substances in the fermented grains of Hunan light-flavor Baijiu [37].

Table 2. VOCs during the Tuqu fermentation of Hunan light-flavor Baijiu.

NO	Compounds	CAS	RI a/RI b	Identification c	F0	F2	F4	F6	F8	F10	Thresholds (μg/kg)	F0	F2	F4	F6	F8	F10
Acids																	
a1	2-Hydroxypropan	50–21-5	838/835	MS, RI	n.d.	5.86 ± 0.46 C	15.90 ± 1.19 A	11.81 ± 0.94 B	n.d.	n.d.	n.f.	-	-	-	-	-	-
a2	2-amino-4-methylbenzoic acid	2305-36-4	949/925	MS, RI	159.79 ± 19.92	240.52 ± 26.93 C	244.87 ± 16.50 AB	224.45 ± 9.42 AB	266.98 ± 10.69 A	189.91 ± 58.27 BC	n.f.	-	-	-	-	-	-
Alcohols																	
a3	3-Methyl-1-butanol	123-51-3	736/740	MS, RI	698.71 ± 23.34 B	296.73 ± 50.96 D	472.78 ± 20.98 C	459.49 ± 17.67 C	1219.52 ± 136.97 A	502.99 ± 53.32 C	4	35.49	2.44	5.29	6.85	19.32	3.52
a4	1-Octen-3-ol	3391-86-4	980/981	MS, RI	25.72 ± 8.43 A	15.91 ± 3.06 B	24.61 ± 2.56 A	15.65 ± 0.50 B	18.56 ± 2.31 B	15.30 ± 1.24 B	1.5	3.48	0.35	0.73	0.62	0.78	0.29
a5	2-Ethylhexan-1-ol	104-76-7	1030/1032	MS, RI	3.00 ± 0.40 A	n.d.	n.d.	n.d.	n.d.	n.d.		-	-	-	-	-	-
a6	Phenethyl alcohol	60-12.8	1116/1117	MS, RI	273.40 ± 1.71 E	637.84 ± 52.93 D	802.88 ± 84.86 B	1031.92 ± 99.92 A	485.74 ± 19.76 D	736.40 ± 71.82 BC	564.23	0.10	0.04	0.06	0.11	0.05	0.04
a7	2-Methyl-1-butanol	137-32-6	739/738	MS, RI	n.d.	n.d.	n.d.	1052.98 ± 17.40 B	328.97 ± 7.45 C	1513.12 ± 99.28 A	15.9	0.00	0.00	0.00	3.95	1.31	2.66
a8	1-Octanol	111-87-5	1071/1074	MS, RI	n.d.	n.d.	n.d.	20.65 ± 2.36 B	12.74 ± 0.82 C	32.64 ± 2.16 A	125.8	0.00	0.00	0.00	0.01	0.01	0.01
Esters																	
a9	Ethyl lactate	97-64-3	815/812	MS, RI	7.25 ± 0.17 E	13.30 ± 0.96 D	54.60 ± 0.90 B	69.38 ± 3.46 A	44.29 ± 1.99 B	34.78 ± 4.87 C	n.f.	-	-	-	-	-	-
a10	Isoamyl acetate	123-92-2	876/876	MS, RI	73.83 ± 6.55 D	455.61 ± 38.73 AB	335.35 ± 22.99 B	251.52 ± 19.00 C	236.75 ± 7.32 C	536.37 ± 26.61 A	0.15	100.00	100.00	100.00	100.00	100.00	100.00
a11	Ethyl caprylate	106-32-1	1196/1200	MS, RI	7.15 ± 0.91 C	352.63 ± 11.79 A	369.77 ± 33.46 A	401.26 ± 50.43 A	239.56 ± 31.40 B	369.52 ± 17.22 A	19.3	0.08	0.60	0.86	1.24	0.79	0.54
a12	Diethyl succinate	123-25-1	1182/1186	MS, RI	2.75 ± 0.46 E	513.20 ± 47.83 B	586.85 ± 23.91 A	500.78 ± 25.07 B	248.46 ± 62.24 D	379.88 ± 10.56 C	n.f.	-	-	-	-	-	-
a13	Ethyl caprate	110-38-3	1396/1401	MS, RI	4.70 ± 0.49 D	85.34 ± 3.71 C	126.78 ± 2.47 b	192.61 ± 15.70 A	97.32 ± 5.20 C	92.56 ± 5.18 C	5	0.19	0.56	1.13	2.30	1.23	0.52
a14	Ethyl phenylacetate	101-97-3	1246/1248	MS, RI	n.d.	5.99 ± 0.39 A	5.33 ± 0.41 B	5.93 ± 1.76 A	6.70 ± 0.16 A	6.87 ± 0.19 A	n.f.	-	-	-	-	-	-
a15	Phenethyl acetate	103-45-7	1258/1259	MS, RI	n.d.	58.08 ± 9.51 B	71.15 ± 3.07 A	79.47 ± 3.86 A	72.91 ± 4.07 A	52.40 ± 1.42 A	249.59	0.00	0.01	0.01	0.02	0.02	0.01
a16	Ethyl caproate	123-66-0	1000/1004	MS, RI	n.d.	311.90 ± 23.51 A	308.75 ± 27.45 A	n.d.	n.d.	n.d.	5	0.00	2.05	2.76	0.00	0.00	0.00
a17	Ethyl nonanoate	123-29-5	1296/1299	MS, RI	n.d.	7.24 ± 1.25 B	9.38 ± 1.21 A	n.d.	n.d.	n.d.	n.f.	-	-	-	-	-	-

Table 2. *Cont.*

NO	Compounds	CAS	RI a/RI b	Identification c	F0	F2	F4	F6	F8	F10	Thresholds (μg/kg)	F0	F2	F4	F6	F8	F10
a18	Ethyl laurate	106-33-2	1595/1605	MS, RI	n.d.	4.08 ± 0.82 BC	5.03 ± 0.90 ABC	6.65 ± 1.30 A	5.28 ± 1.55 AB	3.38 ± 0.17 C	n.f.	-	-	-	-	-	-
a19	2-Methylbutyl acetate	624-41-9	880/880	MS, RI	n.d.	n.d.	233.73 ± 12.48 D	422.28 ± 18.13 B	659.90 ± 28.35 A	292.66 ± 20.03 C	5	0.00	0.00	2.09	5.04	8.36	1.64
a20	2-hydroxy-4-methylvalerate ethyl ester	10348-47-7	1060/1061	MS, RI	n.d.	n.d.	16.76 ± 0.53 B	20.74 ± 1.93 A	12.74 ± 0.82 C	12.28 ± 2.09 C	n.f.	-	-	-	-	-	-
a21	Ethyl heptanoate	106-30-9	1097/1105	MS, RI	n.d.	n.d.	7.12 ± 0.31 a	n.d.	n.d.	n.d.	1.9	0.00	0.00	0.17	0.00	0.00	0.00
a22	Hexyl acetate	142-92-7	1011/1016	MS, RI	n.d.	n.d.	n.d.	n.d.	8.46 ± 0.20 A	n.d.	n.f.	-	-	-	-	-	-
Alkenes																	
a23	1-Nonene	124-11-8	889/879	MS, RI	42.09 ± 2.52 A	n.d.	n.d.	n.d.	n.d.	n.d.	n.f.	-	-	-	-	-	-
a24	3-Methyl-1-heptene	4810/9/7	1075/1071	MS, RI	14.92 ± 0.50 B	13.85 ± 0.39 B	10.50 ± 0.35 C	16.04 ± 1.50 B	23.50 ± 3.01 B	440.40 ± 12.97 A	n.f.	-	-	-	-	-	-
a25	1-Dodecene	112-41-4	1190/1192	MS, RI	6.79 ± 0.41 B	4.94 ± 0.71 B	n.d.	n.d.	11.52 ± 3.59 A	10.18 ± 0.77 A	n.f.	-	-	-	-	-	-
a26	1, 2, 3, 3 cyclopentadiene	65372-78-3	1386/1387	MS, RI	2.27 ± 0.99 C	3.21 ± 0.29 C	5.53 ± 0.32 B	7.88 ± 1.29 A	n.d.	n.d.	n.f.	-	-	-	-	-	-
a27	1-Tetradecene	1120-36-1	1392/1394	MS, RI	3.49 ± 0.78 B	n.d.	n.d.	4.83 ± 0.17 A	4.16 ± 0.71 AB	2.45 ± 0.67 C	n.f.	-	-	-	-	-	-
a28	(Z)-3-Un.d.ecene	821-97-6	1086/1074	MS, RI	n.d.	20.78 ± 5.27 A	n.d.	n.d.	n.d.	n.d.	n.f.	-	-	-	-	-	-
a29	l-Caryophyllene	87-44-5	1419/1421	MS, RI	11.40 ± 0.42 BC	11.83 ± 1.19 BC	14.53 ± 1.70 B	25.64 ± 3.57 A	8.89 ± 1.16 C	9.30 ± 1.01 C	64	0.04	0.01	0.01	0.02	0.01	0.00
a30	4-Methoxystyrene	637-69-4	1156/1151	MS, RI	n.d.	n.d.	n.d.	6.34 ± 1.15 B	23.36 ± 0.25 A	3.53 ± 1.09 C	n.f.	-	-	-	-	-	-
a31	1, 2, 3, 3 cyclopentadiene	65372-78-3	1386/1386	MS, RI	n.d.	n.d.	n.d.	n.d.	3.45 ± 0.82 A	3.01 ± 0.78 A	n.f.	-	-	-	-	-	-
Alkanes																	
a32	Cis-1, 2-dimethylcyclopentane	1192-18-3	723/724	MS, RI	715.98 ± 33.87 C	937 ± 83.22 A	814.94 ± 58.83 B	664.03 ± 58.58 C	656.27 ± 39.86 C	826.49 ± 36.56 B	n.f.	-	-	-	-	-	-
a33	2, 6, 11-Trimethyldodecane	31295-56-4	1275/1281	MS, RI	4.04 ± 0.59 A	n.d.	n.d.	n.d.	n.d.	n.d.	n.f.	-	-	-	-	-	-
a34	Tridecane	629-50-5	1300/1301	MS, RI	4.94 ± 0.77 A	n.d.	n.d.	n.d.	n.d.	n.d.	n.f.	-	-	-	-	-	-
a35	4, 6-Dimethyldodecane	61141-72-8	1325/1327	MS, RI	n.d.	n.d.	n.d.	n.d.	6.88 ± 0.59 A	5.63 ± 0.79 B	n.f.	-	-	-	-	-	-

Table 2. *Cont.*

NO	Compounds	CAS	RI a/RI b	Identification c	F0	F2	F4	F6	F8	F10	Thresholds (μg/kg)	F0	F2	F4	F6	F8	F10
Aldehydes											n.f.						
a36	Phenylacetaldehyde	122-78-1	1045/1046	MS, RI	4.63 ± 0.23 C	6.75 ± 0.64 B	7.99 ± 1.25 A	8.14 ± 0.88 A	n.d.	n.d.	6.3	0.15	0.04	0.06	0.08	0.00	0.00
a37	Decanal	112-31-2	1206/1208	MS, RI	1.34 ± 0.13 CD	4.23 ± 0.34 C	3.29 ± 0.96 CD	3.11 ± 0.34 CD	10.04 ± 2.27 A	7.81 ± 0.56 B	3	0.09	0.05	0.05	0.06	0.21	0.07
a38	Hexanal	66-25-1	800/803	MS, RI	n.d.	49.69 ± 4.45 A	n.d.	n.d.	n.d.	n.d.	5	0.00	0.33	0.00	0.00	0.00	0.00
a39	1-Nonanal	124-19-6	1104/1106	MS, RI	7.51 ± 1.75 C	51.91 ± 11.88 B	92.99 ± 7.91 A	n.d.	n.d.	n.d.	1.1	1.39	1.55	3.78	0.00	0.00	0.00
a40	2, 5-Dimethylbenzaldehyde	5779-94-2	1208/1217	MS, RI	n.d.	n.d.	6.58 ± 0.38 A	n.d.	n.d.	n.d.	n.f.	-	-	-	-	-	-
Phenols																	
a41	2, 4-Di-tert-butylphenol	96-76-4	1519/1520	MS, RI	3.55 ± 0.40 A	n.d.	n.d.	n.d.	n.d.	n.d.	n.f.	-	-	-	-	-	-
a42	4-Ethylguaiacol	2785-89-9	1282/1281	MS, RI	n.d.	13.13 ± 1.55 A	12.90 ± 0.75 A	12.39 ± 0.62 A	13.46 ± 0.62 A	n.d.	89.25	0.00	0.00	0.01	0.01	0.01	0.00
Ketones																	
a43	3-Octanone	106-68-3	986/989	MS, RI	n.d.	n.d.	n.d.	n.d.	35.02 ± 1.69 A	n.d.	21.4	0.00	0.00	0.00	0.00	0.10	0.00
Other																	
a44	Naphthalene	91-20-3	1182/1182	MS, RI	8.46 ± 0.45 A	n.d.	n.d.	n.d.	n.d.	n.d.	n.f.	-	-	-	-	-	-
a45	Benzylhydrazine	555-96-4	1118/1122	MS, RI	69.00 ± 3.06 C	319.93 ± 13.71 B	37.13 ± 2.48 D	46.18 ± 3.53 D	352.31 ± 23.38 A	46.57 ± 4.17 D	n.f.	-	-	-	-	-	-
a46	2-Pentylfuran	3777-69-3	993/992	MS, RI	n.d.	n.d.	n.d.	n.d.	n.d.	44.27 ± 5.13 A	5.8	0.00	0.00	0.00	0.00	0.00	0.21

[a]: retention index of compounds on HP-5MS; [b]: retention index of compounds in references; [c]: "MS": mass spectrum comparison using NIST17 library; A–E: different letters represent significant differences. "RI": retention index in agreement with the literature value; 'n.f.': threshold was not found in the literature; 'n.d.': data were not detected in the sample. The threshold of volatile compounds in water refers to the literatures [13].

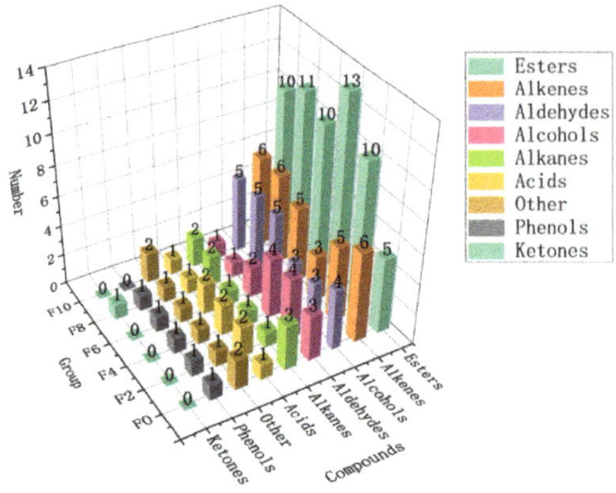

(A)

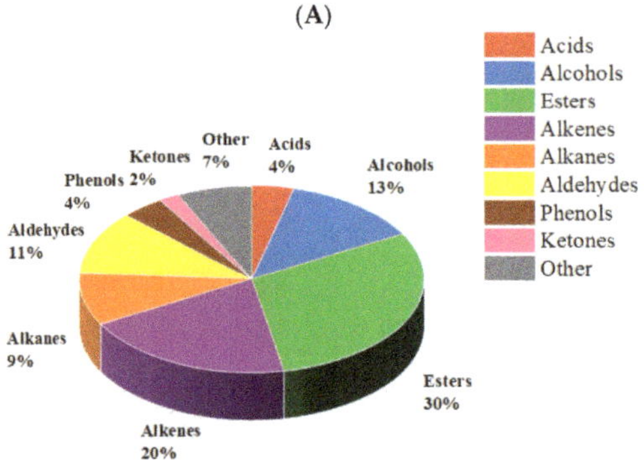

(B)

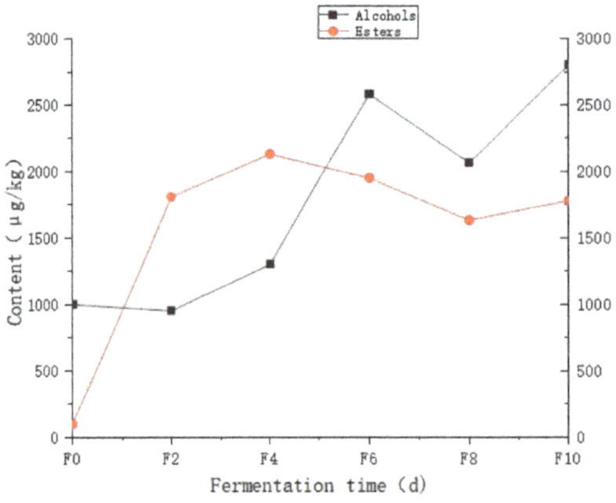

(C)

Figure 12. Numbers and percentages of volatile compounds identified in the Tuqu fermentation process of Hunan light-flavor Baijiu: (**A**) numbers of different types of volatile compounds in different steps; (**B**) the percentages of different volatile compounds in the Tuqu fermentation process of Hunan light-flavor Baijiu; (**C**) the changes in the content of alcohols and esters of Hunan Tuqu light-flavor Baijiu.

3.7. Critical Volatile Compounds

It is worth noting that not all VOCs detected in the fermented grain samples fermented at different times contribute significantly to the flavor of the fermented grains. An ROAV analysis was conducted to assess the contribution of different volatiles to the fermented grains' aroma during the fermentation of the fermented grains. The ROAV value indicates the importance of key aroma compounds, with higher scores indicating a greater contribution to the aroma quality of the fermented grains [38]. Among the VOCs analyzed (Table 2), nine key VOCs with ROAV > 1 were identified in the fermented grains: 3-methyl-1-butanol (ROAV = 35.49), 1-octen-3-ol (ROAV = 3.48), 2-methyl-1-butanol (ROAV = 3.95), isoamyl acetate (ROAV = 100), ethyl caprylate (ROAV = 1.24), ethyl caprate (ROAV = 2.30), ethyl caproate (ROAV = 2.76), 2-methylbutyl acetate (ROAV = 8.36), and 1-nonanal (ROAV = 3.78). Isoamyl acetate had the highest ROAV and contributed the most to the flavor of the fermented grains. Its content decreased and then increased, giving the fermented grains a fruity aroma and a banana flavor. 3-methyl-1-butanol was the second most important odor-active substance, contributing to the floral and cocoa flavors of the fermented grains. 1-octen-3-ol, decanal, despite being present in low concentrations, had a high ROAV due to its low odor threshold, and it contributes a mushroom flavor to the fermented grains. 2-methylbutyl acetate provides a fruity and pear aroma to the fermented grains, ethyl caprylate provides a sweet and soft floral and brandy aroma, ethyl caproate provides a brandy and pineapple aroma to the fermented grains, and ethyl caprate provides a brandy and pear aroma to the fermented grains. All these volatile organic compounds enhance the floral and fruity aromas of the fermented grains. VOCs with $0.1 \leq$ ROAV < 1 included phenethyl alcohol, ethyl heptanoate, decanal, hexanal, 3-octanone, and 2-pentylfuran, which have a moderating effect on the aroma characteristics of the fermented grains. Phenethyl alcohol increased during fermentation, giving the fermented grains its honey and rose aromas.

4. Conclusions

The results demonstrate that the microbial metabolism exhibited significant activity during the pre-fermentation and mid-fermentation stages, which were characterized by a higher water content and the efficient utilization of starch for saccharification. α-amylase and glucoamylase activities displayed a sharp decline, while acid protease, cellulase activity, and lipase activity showed an increase. Additionally, the acidity gradually increased as the fermentation progressed. The microbial structure of Hunan light-flavor Baijiu at different fermentation periods was investigated using high-throughput sequencing technology. A total of fifty bacterial genera and twenty-two fungal genera were identified in this study. Lactobacillus and Saccharomyces were found to be the predominant microorganisms among bacteria and fungi, respectively. Notably, there were substantial variations in the microbial composition and abundance observed across different fermentation periods, highlighting the complexity of the microbial structure during the fermentation stage. The results of a Spearman correlation analysis of physicochemical indicators with fungal and bacterial microorganisms show that *Lactobacillus* and *Saccharomyces* were the dominant genera for bacteria and fungi, respectively, which suggests that these physicochemical indicators drive changes in the flora of *Saccharomyces* and *Lactobacillus*. At the same time, the growth and metabolism of *Saccharomyces* and *Lactobacillus* could also have an important effect on the enzyme activity, ethanol content, and acidity, which are conducive to the formation of the Baijiu flavor. Additionally, a total of 46 volatile flavor compounds were detected via GC-MS in the fermented grains samples, of which esters constituted the major portion, and nine key flavor compounds with an ROAV > 1 were identified throughout the fermentation process. Isoamyl acetate had the highest ROAV and made the greatest contribution to the flavor, enhancing the floral and fruity aroma of the fermented grains. These results provide a data reference and theoretical guidance for the industrial development of light-flavor Baijiu.

Author Contributions: A.H.: methodology, writing—original draft preparation, writing—reviewing and editing; J.X.: data curation, writing—original draft preparation; T.Z.: investigation, validation; J.H.: software; R.J.: investigation; Y.D.: validation; H.C.: project administration, Z.L.: methodology, supervision, funding acquisition. All authors have read and agreed to the published version of the manuscript.

Funding: This research was funded by the Doctoral Research Start-up Fund of Guiyang University, Micro-ecology of Wine-making and Quality Analysis of the Rounds of Wine in the Core Region of Sauce-type Baijiu (Project No. GYU-KY-[2022]).

Institutional Review Board Statement: Not applicable.

Informed Consent Statement: Not applicable.

Data Availability Statement: The original contributions presented in the study are included in the article; further inquiries can be directed to the corresponding author.

Conflicts of Interest: Author Huitai Chen was employed by Hunan Guoyuan Liquor Co., Ltd. Huitai Chen has contributed toward project administration. The objectivity and authenticity of the experimental results and discussion were not affected by the company. The remaining authors declare that the research was conducted in the absence of any commercial or financial relationships that could be construed as a potential conflict of interest.

References

1. Yao, F.; Yi, B.; Shen, C.; Tao, F.; Liu, Y.; Lin, Z.; Xu, P. Chemical Analysis of the Chinese Liquor Luzhoulaojiao by Comprehensive Two-Dimensional Gas Chromatography/Time-of-Flight Mass Spectrometry. *Sci. Rep.* **2015**, *5*, 9553. [CrossRef]
2. Li, R.-R.; Xu, M.; Zheng, J.; Liu, Y.-J.; Sun, C.-H.; Wang, H.; Guo, X.-W.; Xiao, D.-G.; Wu, X.-L.; Chen, Y.-F. Application Potential of Baijiu Non-Saccharomyces Yeast in Winemaking through Sequential Fermentation with *Saccharomyces cerevisiae*. *Front. Microbiol.* **2022**, *13*, 902597. [CrossRef] [PubMed]
3. Huang, X.; Fan, Y.; Lu, T.; Kang, J.; Pang, X.; Han, B.; Chen, J. Composition and Metabolic Functions of the Microbiome in Fermented Grain during Light-Flavor Baijiu Fermentation. *Microorganisms* **2020**, *8*, 1281. [CrossRef] [PubMed]
4. Tang, J.; Liu, Y.; Lin, B.; Zhu, H.; Jiang, W.; Yang, Q.; Chen, S. Effects of Ultra-Long Fermentation Time on the Microbial Community and Flavor Components of Light-Flavor *Xiaoqu Baijiu* Based on Fermentation Tanks. *World J. Microbiol. Biotechnol.* **2022**, *38*, 3. [CrossRef] [PubMed]
5. Wang, L.; Wang, Y.Y.; Wang, D.Q.; Xu, J.; Yang, F.; Liu, G.; Zhang, D.Y.; Feng, Q.; Xiao, L.; Xue, W.B.; et al. Dynamic Changes in the Bacterial Community in Moutai Liquor Fermentation Process Characterized by Deep Sequencing: Dynamic Changes in the Bacterial Community in Moutai Liquor. *J. Inst. Brew.* **2015**, *121*, 603–608. [CrossRef]
6. Ji, X.; Yu, X.; Wu, Q.; Xu, Y. Initial Fungal Diversity Impacts Flavor Compounds Formation in the Spontaneous Fermentation of Chinese Liquor. *Food Res. Int.* **2022**, *155*, 110995. [CrossRef] [PubMed]
7. Wang, Q.; Wang, C.; Xiang, X.; Xu, H.; Han, G. Analysis of Microbial Diversity and Succession during *Xiaoqu Baijiu* Fermentation Using High-throughput Sequencing Technology. *Eng. Life Sci.* **2022**, *22*, 495–504. [CrossRef] [PubMed]
8. Wang, Z.; Sun, X.; Liu, Y.; Yang, H. Characterization of Key Aroma Compounds in Xiaoqu Liquor and Their Contributions to the Sensory Flavor. *Beverages* **2020**, *6*, 42. [CrossRef]
9. Pan, Y.; Wang, Y.; Hao, W.; Duan, C.; Wang, S.; Wei, J.; Liu, G. Metatranscriptomics Unravel Composition, Drivers, and Functions of the Active Microorganisms in Light-Flavor Liquor Fermentation. *Microbiol. Spectr.* **2022**, *10*, e02151-21. [CrossRef]
10. *DB34/T2264-2014*; The Analysis Method of Solid-State Fermented Grains. Anhui Provincial Bureau: Hefei, China, 2014.
11. Xiao, Y.; Huang, Y.; Chen, Y.; Zhu, M.; He, C.; Li, Z.; Wang, Y.; Liu, Z. Characteristic Fingerprints and Change of Volatile Organic Compounds of Dark Teas during Solid-State Fermentation with Eurotium Cristatum by Using HS-GC-IMS, HS-SPME-GC-MS, E-Nose and Sensory Evaluation. *LWT* **2022**, *169*, 113925. [CrossRef]
12. Xiao, L.; Yang, C.; Zhang, X.; Wang, Y.; Li, Z.; Chen, Y.; Liu, Z.; Zhu, M.; Xiao, Y. Effects of Solid-State Fermentation with Bacillus Subtilis LK-1 on the Volatile Profile, Catechins Composition and Antioxidant Activity of Dark Teas. *Food Chem. X* **2023**, *19*, 100811. [CrossRef] [PubMed]
13. Van Gemert, L.J. *Compilations of Flavour Threshold Values in Water and Other Media*, 2nd Enlarged and Revised ed.; Boelens Aroma Chemical Information Service: Huizen, The Netherlands, 2018.
14. Zeng, Y.; Wang, Y.; Chen, Q.; Xia, X.; Liu, Q.; Chen, X.; Wang, D.; Zhu, B. Dynamics of Microbial Community Structure and Enzyme Activities during the Solid-State Fermentation of Forgood Daqu: A Starter of Chinese Strong Flavour Baijiu. *Arch. Microbiol.* **2022**, *204*, 577. [CrossRef]
15. Wang, B.; Wu, Q.; Xu, Y.; Sun, B. Specific Volumetric Weight-Driven Shift in Microbiota Compositions with Saccharifying Activity Change in Starter for Chinese Baijiu Fermentation. *Front. Microbiol.* **2018**, *9*, 2349. [CrossRef]

16. Luo, A.; Yang, N.; Yang, J.; Hao, J.; Zhao, J.; Shi, S.; Hu, B. Effects of Microbial Interspecies Relationships and Physicochemical Parameters on Volatile Flavors in Sorghum-based Fermented Grains during the Fermentation of Shanxi Light-flavored Liquor. *Food Sci. Nutr.* **2023**, *11*, 1452–1462. [CrossRef]
17. Wu, Y.; Hou, Y.; Chen, H.; Wang, J.; Zhang, C.; Zhao, Z.; Ao, R.; Huang, H.; Hong, J.; Zhao, D.; et al. 'Key Factor' for Baijiu Quality: Research Progress on Acid Substances in Baijiu. *Foods* **2022**, *11*, 2959. [CrossRef]
18. Le Lay, C.; Coton, E.; Le Blay, G.; Chobert, J.M.; Haertlé, T.; Choiset, Y.; Van Long, N.N.; Meslet-Cladière, L.; Mounier, J. Identification and Quantification of Antifungal Compounds Produced by Lactic Acid Bacteria and Propionibacteria. *Int. J. Food Microbiol.* **2016**, *239*, 79–85. [CrossRef] [PubMed]
19. Kouitcheu Mabeku, L.B.; Ngue, S.; Bonsou Nguemo, I.; Leundji, H. Potential of Selected Lactic Acid Bacteria from *Theobroma cacao* Fermented Fruit Juice and Cell-Free Supernatants from Cultures as Inhibitors of *Helicobacter pylori* and as Good Probiotic. *BMC Res. Notes* **2020**, *13*, 64. [CrossRef] [PubMed]
20. Pang, X.N.; Han, B.Z.; Huang, X.N.; Zhang, X.; Hou, L.F.; Cao, M.; Gao, L.J.; Hu, G.H.; Chen, J.Y. Effect of the Environment Microbiota on the Flavour of Light-Flavour Baijiu during Spontaneous Fermentation. *Sci. Rep.* **2018**, *8*, 3396. [CrossRef]
21. Shen, T.; Liu, J.; Wu, Q.; Xu, Y. Increasing 2-Furfurylthiol Content in Chinese Sesame-Flavored Baijiu via Inoculating the Producer of Precursor l-Cysteine in Baijiu Fermentation. *Food Res. Int.* **2020**, *138*, 109757. [CrossRef]
22. Sun, W.; Xiao, H.; Peng, Q.; Zhang, Q.; Li, X.; Han, Y. Analysis of Bacterial Diversity of Chinese Luzhou-Flavor Liquor Brewed in Different Seasons by Illumina Miseq Sequencing. *Ann. Microbiol.* **2016**, *66*, 1293–1301. [CrossRef]
23. Wang, H.-Y.; Zhang, X.-J.; Zhao, L.-P.; Xu, Y. Analysis and Comparison of the Bacterial Community in Fermented Grains during the Fermentation for Two Different Styles of Chinese Liquor. *J. Ind. Microbiol. Biotechnol.* **2008**, *35*, 603–609. [CrossRef]
24. Hao, H.; Yan, R.; Miao, Z.; Wang, B.; Sun, J.; Sun, B. Volatile Organic Compounds Mediated Endogenous Microbial Interactions in Chinese Baijiu Fermentation. *Int. J. Food Microbiol.* **2022**, *383*, 109955. [CrossRef]
25. Lin, B.; Tang, J.; Yang, Q.; Su, Z.; Zhu, L.; Li, Q.; Jiang, W.; Zhang, L.; Liu, Y.; Chen, S. Microbial Succession and Its Effect on Key Aroma Components during Light-Aroma-Type *Xiaoqu Baijiu* Brewing Process. *World J. Microbiol. Biotechnol.* **2022**, *38*, 166. [CrossRef]
26. Zhang, H.; Wang, L.; Tan, Y.; Wang, H.; Yang, F.; Chen, L.; Hao, F.; Lv, X.; Du, H.; Xu, Y. Effect of Pichia on Shaping the Fermentation Microbial Community of Sauce-Flavor Baijiu. *Int. J. Food Microbiol.* **2021**, *336*, 108898. [CrossRef]
27. Schloss, P.D.; Gevers, D.; Westcott, S.L. Reducing the Effects of PCR Amplification and Sequencing Artifacts on 16S rRNA-Based Studies. Edited by Jack Anthony Gilbert. *PLoS ONE* **2011**, *6*, e27310. [CrossRef]
28. Dorn-In, S.; Hölzel, C.S.; Janke, T.; Schwaiger, K.; Balsliemke, J.; Bauer, J. PCR-SSCP-Based Reconstruction of the Original Fungal Flora of Heat-Processed Meat Products. *Int. J. Food Microbiol.* **2013**, *162*, 71–81. [CrossRef] [PubMed]
29. Wang, W.; Fan, G.; Li, X.; Fu, Z.; Liang, X.; Sun, B. Application of Wickerhamomyces Anomalus in Simulated Solid-State Fermentation for Baijiu Production: Changes of Microbial Community Structure and Flavor Metabolism. *Front. Microbiol.* **2020**, *11*, 598758. [CrossRef]
30. Cappello, M.S.; Zapparoli, G.; Logrieco, A.; Bartowsky, E.J. Linking Wine Lactic Acid Bacteria Diversity with Wine Aroma and Flavour. *Int. J. Food Microbiol.* **2017**, *243*, 16–27. [CrossRef]
31. Hao, F.; Tan, Y.; Lv, X.; Chen, L.; Yang, F.; Wang, H.; Du, H.; Wang, L.; Xu, Y. Microbial Community Succession and Its Environment Driving Factors During Initial Fermentation of Maotai-Flavor Baijiu. *Front. Microbiol.* **2021**, *12*, 669201. [CrossRef] [PubMed]
32. Song, Z.; Du, H.; Zhang, Y.; Xu, Y. Unraveling Core Functional Microbiota in Traditional Solid-State Fermentation by High-Throughput Amplicons and Metatranscriptomics Sequencing. *Front. Microbiol.* **2017**, *8*, 1294. [CrossRef] [PubMed]
33. Bartowsky, E.J.; Henschke, P.A. The 'Buttery' Attribute of Wine—Diacetyl—Desirability, Spoilage and Beyond. *Int. J. Food Microbiol.* **2004**, *96*, 235–252. [CrossRef] [PubMed]
34. Yamada, T.; Ogrydziak, D.M. Extracellular Acid Proteases Produced by *Saccharomycopsis lipolytica*. *J. Bacteriol.* **1983**, *154*, 23–31. [CrossRef] [PubMed]
35. Cheng, W.; Chen, X.; Lan, W.; Liu, G.; Xue, X.; Li, R.; Pan, T.; Li, N.; Zhou, D.; Chen, X. Insights into the Influence of Physicochemical Parameters on the Microbial Community and Volatile Compounds during the Ultra-Long Fermentation of Compound-Flavor Baijiu. *Front. Microbiol.* **2023**, *14*, 1272559. [CrossRef] [PubMed]
36. He, Q.; Zhou, W.; Chen, X.; Zhang, Q. Chemical and Bacterial Composition of *Broussonetia papyrifera* Leaves Ensiled at Two Ensiling Densities with or without *Lactobacillus plantarum*. *J. Clean. Prod.* **2021**, *329*, 129792. [CrossRef]
37. Li, W.; Fan, G.; Fu, Z.; Wang, W.; Xu, Y.; Teng, C.; Zhang, C.; Yang, R.; Sun, B.; Li, X. Effects of Fortification of Daqu with Various Yeasts on Microbial Community Structure and Flavor Metabolism. *Food Res. Int.* **2020**, *129*, 108837. [CrossRef]
38. Xia, D.; Tan, X.; Wang, L.; Li, Z.; Hou, A.; Zhu, Y.; Lai, L.; Wang, Y. GC-MS Coupled with Rate-All-That-Apply (RATA) to Analyse the Volatile Flavor Substances of Yellow Wine during Fermentation. *Foods* **2023**, *12*, 1992. [CrossRef]

Disclaimer/Publisher's Note: The statements, opinions and data contained in all publications are solely those of the individual author(s) and contributor(s) and not of MDPI and/or the editor(s). MDPI and/or the editor(s) disclaim responsibility for any injury to people or property resulting from any ideas, methods, instructions or products referred to in the content.

fermentation

Article

Differences in Volatile Profiles and Sensory Characteristics in Plum Spirits on a Production Scale

Josef Balak [1], Lucie Drábová [2], Olga Maťátková [1], Marek Doležal [2], Dominik Marsík [1] and Irena Jarosova Kolouchova [1,*]

[1] Department of Biotechnology, University of Chemistry and Technology, Technická 5, 166 28 Prague, Czech Republic; josef.balak@vscht.cz (J.B.); olga.matatkova@vscht.cz (O.M.); dominik.marsik@vscht.cz (D.M.)

[2] Department of of Food Analysis and Nutrition, University of Chemistry and Technology, Technická 5, 166 28 Prague, Czech Republic; lucie.drabova@vscht.cz (L.D.); marek.dolezal@vscht.cz (M.D.)

* Correspondence: irena.kolouchova@vscht.cz; Tel.: +420-220-444-367

Abstract: The specific sensory properties attributed to distillates from different plum varieties are intricately linked to aromatic substances, fruit quality, and technology employed during production. This study compares four plum brandies, each made from a renowned plum variety: Presenta, Valjevka, Čačanská lepotica, and Čačanská rodná on a production scale. Analytical and sensory profiles were assessed using GC-FID, an available analytical method advantageous for monitoring industrial fruit distillate production. Between 71 and 85 analytes were detected in the distillates, with the Presenta plum distillate containing the highest number of substances. Statistically significant differences in analyte concentration between plum varieties ($p < 0.05$) were observed for 11 analytes. The comparison of analytical profiles and sensory evaluation revealed that a high concentration of 1-propanol, despite its negative sensory perception, significantly impacts the overall perception of a distillate, contrasting with substances like acetaldehyde and propyl acetate, which have positive sensory evaluations but lesser significance in content. Our work identified key compounds and procedures that can be used as benchmarks for production of plum brandy with positive sensory evaluation. These findings demonstrate the broad application potential of GC-FID in fruit distillate production as an independent tool for aromatic profile assessment and quality control.

Keywords: volatile compounds; plum spirit; sensory characteristic; gas chromatography

1. Introduction

Fruit spirits production relies on the conversion of fermentable sugars into ethanol by microorganisms, primarily yeast, followed by distillation. The concentrations of volatile substances, aroma attributes, and olfactory thresholds are critical factors in determining the quality and character of each distillate [1]. The composition and condition of the source fruit play a pivotal role in the sensory attributes and the organoleptic characteristics of the final product, as most compounds are formed during fermentation through microbial activity [2–4]. Technologically important substances are mainly present in the pulp and peel of the fruit, while some sensory active and potentially harmful substances originate in the fruit stones [5,6]. Key groups of substances include saccharides, organic acids, nitrogenous compounds, polyphenols, vitamins, minerals, and volatile fragrances [2]. Pectin breakdown during fermentation, particularly in pectin-rich raw materials, contributes to the presence of methanol in fruit spirits [7]. Methanol and its metabolites present toxicity risks, prompting regulatory limits on their concentration (Regulation (EU) 2019/787 of the European Parliament and of the Council) [8]. The profile of higher alcohols produced is influenced by various factors, notably the range of available amino acids [9–11].

Aromatic compounds, also known as fragrances, constitute a diverse group of volatile compounds that are sensorially active. Many of these compounds traverse the entire production process and contribute to the desirable fruity aroma of the final distillate. Typically, aromatic volatiles from fruit are found in distillates at lower concentrations, compared to many fermentation products, such as higher alcohols [12,13]. However, their low olfactory thresholds and synergistic effects significantly influence the organoleptic properties of distillates [14–16]. Odorants are found in fruit, either in free or bound form, and include mainly various alcohols, carbonyl compounds, organic acids, esters, and terpene compounds, as well as lactones, phenols, alkanes, and others [3,17,18]. The precursors for the biosynthesis of these aromatic compounds are terpenoids, lipids, amino acids, and carotenoids, from which their primary carbon skeletons are formed. These are subsequently modified by various metabolic routes, mainly by oxidation, reduction, acylation, methylation, or cyclization, resulting in a diverse range of analytical profiles of fragrant volatiles from various fruits [3,19].

Detailed information on the composition of the selected fruit source and the organoleptic properties of individual compounds can help in predicting the quality of the final distillate [1,5,6,20,21]. Although individual varieties of a given fruit species may vary in composition, these differences are generally less significant than those observed between different fruit types [18,21].

Distillates are most often produced from locally abundant raw materials. In Central Europe, distillates are often made from plums, apricots, apples, pears, and grapes (cognac, brandy, etc.). The plum is a popular fruit crop, especially in Europe, where it is grown on 500,000 ha, with a production of around 3 million tons. In the past 20 years, 170 plum cultivars have been developed to adapt to changing climatic conditions, exhibiting characteristics such as disease resistance, a shorter growing season, late flowering, and frost resistance [22,23]. Europe's largest plum producers include Romania, Serbia, France, and Poland. Plums are cultivated for various purposes, including fresh consumption, dried fruit production, and a significant portion of the plum crop is utilized by the distillery industry [24]. In terms of area, plum orchards rank second only to apple trees among all pome and stone fruit trees [25]. The pre-tax turnover for the distillation industry in the EU is estimated to be 26.5 billion euros [26]. The production of plum spirits has a long tradition in many areas of Central, Eastern, and South-Eastern Europe, among other regions. The compounds 1-hexanol, *cis*-3-hexenol, hexanal, nonanal, benzaldehyde, linalool, hexyl acetate, ethyl 2-methylbutanoate, ethyl nonanoate, methyl cinnamate, ethyl cinnamate, γ-octalactone, γ-decalactone, and others are usually associated with plum aromas [5,6,18,27,28].

Studies have primarily focused on the differences in the composition of volatile compounds in plum brandies, but mostly on a laboratory scale, with total sample sizes typically not exceeding 100 L [13,23,25,29–33]. Other works focused on advanced analytical methods such as GC–HRMS (gas chromatography/high-resolution mass spectrometry), HS-SPME (Headspace solid-phase microextraction), GCxGC–TOF-MS (comprehensive gas chromatography time of flight mass spectrometry), SPME–GC–MS (solid-phase microextraction coupled to gas chromatography–mass spectrometry), and in-line conductivity measurements [13,25,34]. To our knowledge, our study stands out for its unique production scale data source and the utilization of available instruments, such as GC-FID. This method can help the producer verify the quality of the distillate, even during the process. We conducted a comparison between the analytical profiles of four plum brandies fermented and distilled on a production scale, alongside their sensory analysis. We identified compounds that occurred frequently and which were correlated with less favorable sensory outcomes. Moreover, compounds preferred by consumers as characteristic of plum brandies were identified. The results show that this method can be used to identify plum brandies with high commercial potential during production scale distillation.

2. Materials and Methods

2.1. Samples of Plum Brandy

The distillates compared were made from plums of varieties Čačanská lepotica, Čačanská rodná, Valjevka, and Presenta. The fruit of these four varieties were grown in Moravia Region of Czech Republic under similar climatic conditions. The fermentation of each variety was conducted in a 4 t vertical stainless steel tank, with a total content of 3.5 t of plums disintegrated by toothed roller crusher (without crushing the fruit stones). Alcoholic fermentation was carried out in closed vessels at 20 ± 2 °C under anaerobic conditions in an air-conditioned fermentation hall. Fermentation was carried out by the natural autochthonous microorganisms. The duration of alcoholic fermentation was 2 months. The distillation of all the plum varieties was conducted using a single pot still with a distillation column (volume 600 L), heated by a gas burner. The process took place one month after the end of fermentation. Distillate samples were collected as average samples from the collected distillate without standardization of the ethanol content. For analysis, samples were standardized to 40% (v/v).

2.2. Chemicals and Reagents

All compounds were purchased from Sigma-Aldrich Ltd. (Taufkirchen, Germany) and were of at least p.a. purity or higher. This included the following: methyl acetate, ethyl acetate, ethyl butyrate, octyl acetate, ethyl benzoate, propyl benzoate, hexyl hexanoate, isoamyl benzoate, isoamyl decanoate, ethyl hexanoate, 2-phenylethyl octanoate, methyl cinnamate, ethyl cinnamate, ethyl undecanoate, ethyl (E)-2-decenoate, ethyl myristate, methyl palmitate, ethyl oleate, ethyl linoleate, ethyl salicylate, diethyl succinate, methanol, ethanol, 1-propanol, isopropyl alcohol, 1-butanol, 2-butanol, isobutyl alcohol, 2-methyl butanol, isoamyl alcohol, 1-pentanol, 1-hexanol, 1-heptanol, 1-octanol, 1-nonanol, 1-decanol, benzyl alcohol, 2-phenylethanol, 3-ethoxypropionaldehyde diethyl acetal, diethyl acetal, eugenol, geraniol, linalool, citronellol, α-terpineol, acetone, hexanal, heptanal, farnesene, nonanal, isovaleraldehyde, furfural, benzaldehyde, 2,4-butanedione, damascenone, and ethyl carbamate. Water purified by a Milli-Q® Integral system supplied by Merck (Darmstadt, Germany) was used throughout the study.

2.3. Instruments

For the identification of the distillate components, the gas chromatograph Agilent 6890N, coupled to the 5975 mass spectrometer (GC-MS, Agilent Technologies, Palo Alto, CA, USA), was employed. For the separation of the target analytes, an HP-INNOWAX column (30 m × 0.250 mm i.d., 0.25 μm film thickness) obtained from Agilent Technologies (Palo Alto, CA, USA) was used. The GC conditions were as follows: oven temperature program: 40 °C (3 min); 5 $°C \cdot min^{-1}$ to 240 °C (5 min); carrier gas helium with a flow rate of 1 $mL \cdot min^{-1}$; injection mode—split (50:1); injection volume 1 μL; inlet temperature 250 °C. The MSD interface temperature was set to 230 °C, the quadrupole temperature to 150 °C and the ion source temperature to 230 °C. The mass spectrometer was operated in the selected ion-monitoring mode, detecting at least two ions per analyte. The quantification of individual compounds was carried out using an Agilent 7890B gas chromatograph, coupled to a flame ionization detector (GC-FID, Agilent Technologies, Palo Alto, CA, USA). Target analytes were separated using an HP-INNOWAX column (60 m × 0.250 mm i.d., 0.25 μm film thickness) under the following conditions: 35 °C (8 min); 5 $°C \cdot min^{-1}$ to 250 °C (5 min). The carrier gas was helium, with a flow rate of 1.7 $mL \cdot min^{-1}$; injection mode: split 1:50 with an injection volume of 1 μL. Analyses were carried out in triplicate and their averages were used as a single data point. Calibration was performed and repeated with a standard solution, according to Spaho et al. [11].

2.4. Sensory Evaluation

The results of the chemical analyses were supplemented with a sensory analysis. This entailed a consumer test with untrained evaluators who, although not formally trained, were accustomed to regularly consuming this type of beverage. The samples were evaluated in the laboratory of sensory analysis at the University of Chemistry and Technology, Prague, Department of Food Analysis and Nutrition with 12 boxes, which is equipped according to the relevant international standard ISO 8589 [35]. The procedures of all sensory analyses were in accordance with the international ISO standards. All samples were tested by sensory profile assessment, according to ISO 13299 [36], by a total of 60 people. A 100-point unstructured scale with 12 descriptors was used for the quantitative description of aroma: harmony, delicacy, overall pleasantness (0, unpleasant–100, very pleasant), intensity of the fruity, herbal, spicy, resinous, bitter almond, pungent (sour), technical (chemical) and negative fragrance (0, not noticeable–100, very strong) and overall impression (0, very poor–100, exceptional), as well as by the following 11 descriptors used for the evaluation of flavor: harmony, delicacy, aftertaste, overall pleasantness (0, unpleasant–100, very pleasant), intensity of the fruity, bitter, sour, alcoholic and negative aftertaste (0, not noticeable–100, very strong) and overall impression (0, very poor–100, exceptional). RedJade software (RedJade Sensory Solutions, LLC, Martinez, CA, USA) was used for the collection of sensory analysis data and their processing.

2.5. Statistical Analysis

Statistical analysis was performed using MetaboAnalyst 6.0 software, where a one-way analysis of variance (ANOVA) with Tukey's post hoc test was used for the comparison of the brandy composition. Dixon's Q test was used for the detection of outliers in data obtained (the analysis of each sample was performed in three parallels, the deviation of the five three analyses was less than 5%).

3. Results

3.1. Analytical Profile

Numerous volatile substances detected in fruit distillates are regulated by EU legislation [7,37]. As per this decree, fruit spirits are classified as spirits with a minimum alcohol content of 37.5% (vol.), and their distillation is conducted to achieve an alcohol content of less than 86% vol. No pure ethanol or even distillate of other agricultural origin may be added to fruit distillates. Fruit distillates contain at least 200 g of volatile substances per hectoliter of 100% vol. ethanol. The addition of flavoring substances is not permitted. Legislation also defines the limits on the content of some substances toxic to the human organism, which could potentially occur in undesirable concentrations in the distillate. Distillates produced from pitted fruit must not exceed a hydrocyanic acid content of 7 g/hL of 100% vol. ethanol. The methanol content for distillate made from plums is regulated to a maximum of 12 g/La (methanol per liter of 100% vol. ethanol), according to this regulation. Fruit distillates may contain ethyl carbamate, which is potentially mutagenic for the human body. Its content in fruit distillates is regulated by Commission Recommendation (EU) 22/2016 [38], which states the recommended maximum concentration of ethyl carbamate as 1 mg/L. The contents of methanol and total volatile substances were significantly lower in all varietal plum brandies in this study.

To observe the influence of the plum variety on the analytical profile of fruit distillates, distillates from plum varieties Presenta, Valjevka, Čačanská lepotica, and Čačanská rodná were selected. Depending on the plum variety, between 71 and 85 analytes were detected in the distillates. The highest number of substances was detected in the sample of Presenta plum distillate. Table 1 lists the most important substances affecting the resulting aroma. An ANOVA was used to assess the significance of difference between plum varieties, which is presented in Table 1 by p-value. As can be seen from Table 1, a statistically significant difference in analyte concentration between plum varieties ($p < 0.05$) was detected for 11 analytes. A table with the quantification of the listed compounds is in the Supplementary Materials (Table S1). The sum of higher alcohol concentrations were, for individual varieties, as follows: Presenta 538.16 mg/La, Valjevka 1188 mg/La, Čačanská lepotica 570 mg/La, and Čačanská rodná 4098 mg/La.

Table 1. Differences in the analyte concentration among the different plum varieties determined as *p*-value and ANOVA test. Significant difference: $p < 0.05$. * Statistically significant differences in analyte concentration.

Major Compounds	RT (min)	*p*	Minor Esters	RT (min)	*p*
Methyl acetate	4.813	0.678	Methyl acetate	4.813	0.732
Ethyl acetate	6.019	0.725	Propyl acetate	9.29	0.047 *
Methanol	6.305	0.341	Butyl acetate	13.509	0.824
2-Butanol	11.47	0.379	Isoamyl acetate	15.257	0.052
1-Propanol	12.058	0.013 *	Ethyl caprylate	25.167	0.240
Isobutanol	14.305	0.041 *	Ethyl caprate	30.258	0.259
Isoamyl alcohol	18.404	0.237	Ethyl benzoate	31.092	0.026 *
Ethyl (-)-L-lactate	22.692	0.240	Diethyl succinate	31.224	0.021 *
Acetic acid	25.590	0.374	Ethyl (*E*)-2-decenoate	33.136	0.065
Minor alcohols			Ethyl salicylate	34.352	0.034 *
1-Butanol	16.216	0.283	2-Phenylethyl acetate	34.407	0.053
1-Pentanol	18.404	0.233	Ethyl laurate	34.82	0.238
1-Hexanol	22.88	0.605	Ethyl myristate	38.975	0.082
Linalool	28.034	0.620	Ethyl cinnamate	40.744	0.214
1-Octanol	28.282	0.045 *	Ethyl palmitate	42.794	0.058
1-Nonanol	30.723	0.0004 *	Ethyl stearate	46.322	0.027 *
1-Decanol	33.034	0.001 *	Ethyl oleate	46.717	0.336
Nerol	33.949	0.298	Ethyl linoleate	47.478	0.086
Geraniol	34.923	0.098	Carbonyl compounds		
Benzyl alcohol	35.661	0.118	Acetaldehyde	3.538	0.019 *
2-Phenylethanol	36.392	0.181	Acetone	4.615	0.144
Eugenol	41.381	0.107	2,3-Butandione	9.591	0.365
			Furfural	26.135	0.298
			Benzaldehyde	27.682	0.283

Figures 1–4 show the mean relative substance context (%) of analytical profiles for the different groups of substances in the four varietal plum brandies. The concentrations of many compounds varied significantly among the samples. The plum brandy made from Presenta plums, was characterized by the highest concentrations of methyl acetate, ethyl acetate, methanol (Figure 1), 1-pentanol, 1-nonanol, geraniol, and benzyl alcohol among the samples (Figure 4). The sample of plum brandy made from the variety Valjevka had a moderately high content of 1-propanol (Figure 1) and, on the contrary, it had the lowest contents of acetaldehyde, 2,3-butanedione, and also other carbonyl compounds (Figure 2). The plum brandy produced from the Čačanská lepotica plum variety contained the highest concentrations of isobutyl alcohol, isoamyl alcohol, 1-decanol, acetone, 2,3-butanedione, benzaldehyde, ethyl decanoate and ethyl laurate compared to the other samples, and, on the other hand, the lowest concentrations of methyl and ethyl acetate, 1-propanol, acetic acid, linalool, benzoic acid, isoamyl acetate, ethyl benzene, ethyl palmitate, and ethyl oleate. The brandy produced from plums of the Čačanská rodná variety contained very high concentrations of 1-propanol, acetaldehyde, propyl acetate, isoamyl acetate and linalool. Compared to the other samples, it contained the lowest concentrations of methanol, 1-butanol, geraniol, ethyl octanoate, ethyl decanoate, ethyl laurate, ethyl myristate and ethyl linoleate.

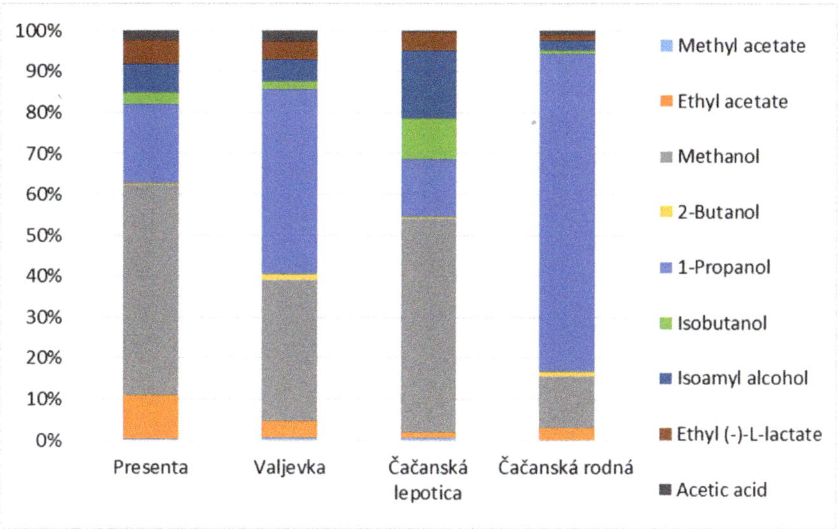

Figure 1. Mean relative contribution (%) of major compounds in plum brandies produced from Presenta, Valjevka, Čačanská lepotica, and Čačanská rodná plum varieties.

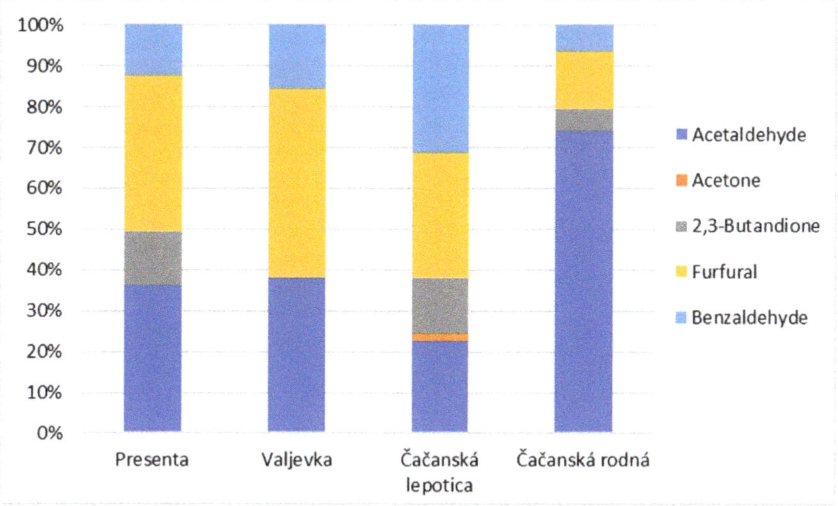

Figure 2. Mean relative contribution (%) of carbonyl compounds in plum brandies produced from Presenta, Valjevka, Čačanská lepotica, and Čačanská rodná plum varieties.

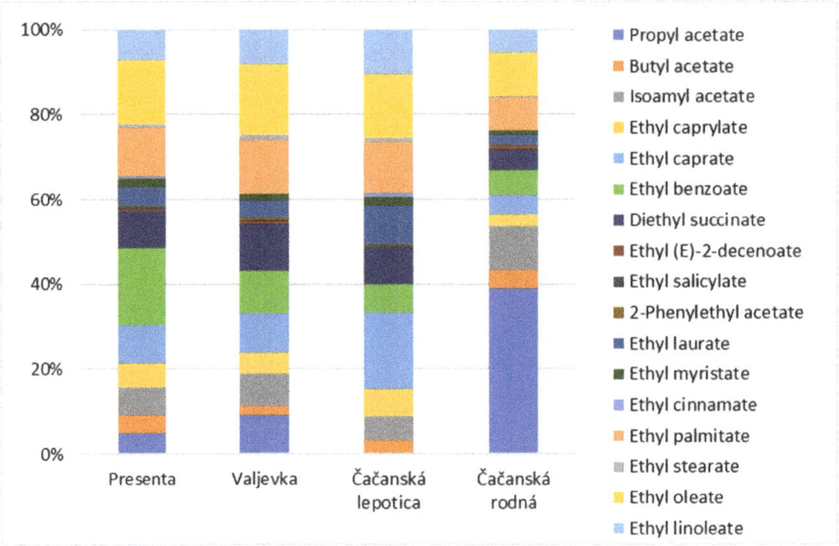

Figure 3. Mean relative contribution (%) of minor esters in plum brandies produced from Presenta, Valjevka, Čačanská lepotica, and Čačanská rodná plum varieties.

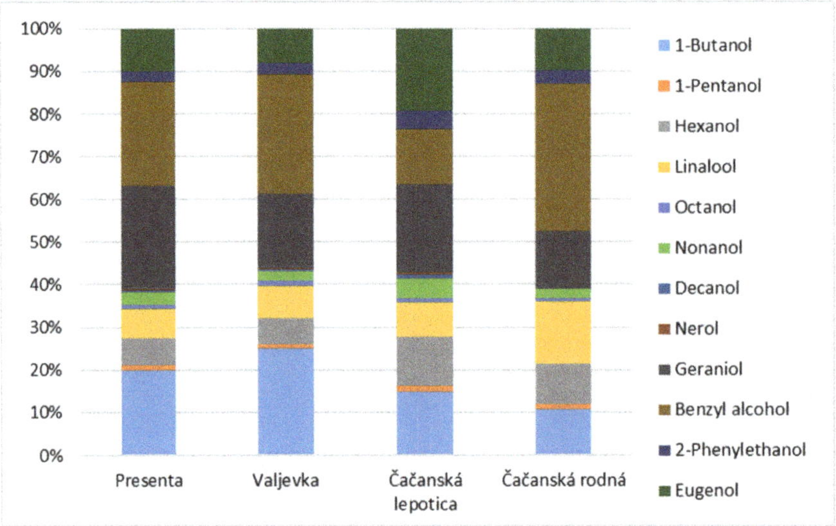

Figure 4. Mean relative contribution (%) of minor alcohols in plum brandies produced from Presenta, Valjevka, Čačanská lepotica, and Čačanská rodná plum varieties.

3.2. Sensory Profiles

During the sensory evaluation, involving 60 assessors, the sensory profiles of the four studied plum brandies were determined. The samples were ranked according to the personal preferences of the assessors, from best to worst. According to the ranking test, the best-ranked sample was plum brandy made from the Presenta variety (Figure 5). This result was determined at a probability value of $p = 98.5\%$, so it can be assessed as conclusive. A statistically significant difference was found in the intensity of the bitter taste between samples Čačanská rodná and Čačanská lepotica and samples Presenta and Valjevka, which can be regarded as moderate evidence of a difference, at the established probability level of $p = 97.2\%$. There was established a statistically significant difference in the intensity of the fruit flavor for the Presenta variety (probability level of $p = 96.3\%$). The taste and flavor profiles of these samples are shown in Figures 5 and 6.

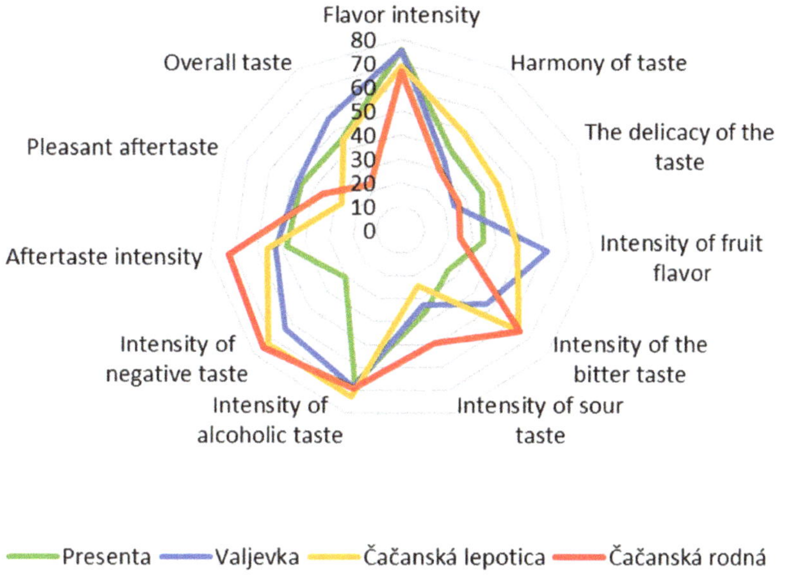

Figure 5. Flavor profiles of plum brandies produced from Presenta, Valjevka, Čačanská lepotica, and Čačanská rodná varieties.

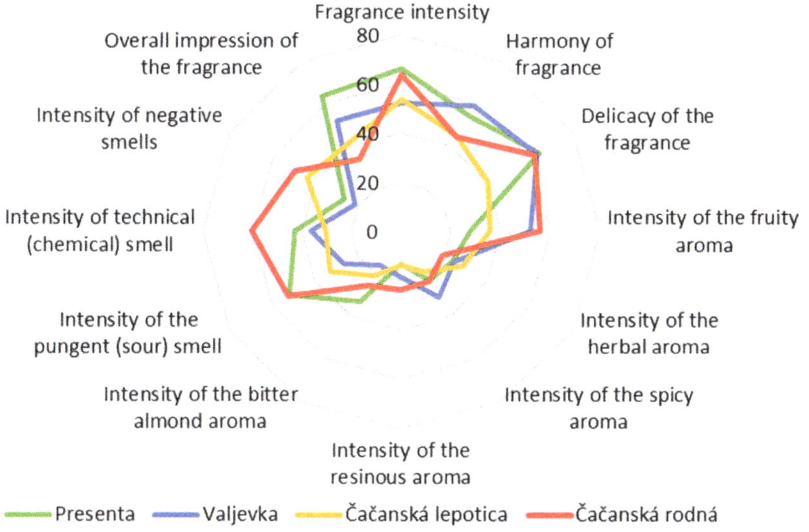

Figure 6. Aroma profiles of plum brandies produced from Presenta, Valjevka, Čačanská lepotica, and Čačanská rodná plum varieties.

Based on statistically significant differences in analyte concentration between plum varieties, Čačanská rodná, the worst-rated variety, exhibited differences in the concentrations of five compounds. Here, the highest concentrations were observed for 1-propanol, acetaldehyde, and propyl acetate. Conversely, 1-decanol and ethyl salicylate were not detected. Acetaldehyde and propyl acetate are known to have pleasant, fruity odors. 1-propanol is characterized by a sharp smell. The highest-rated variety Presenta had the highest concentrations for 1-nonanol (fruity odor), diethyl succinate (fruit flavors), and ethyl benzoate (fruit flavors), as well as high concentration of geraniol (rose-like odor) and 1-octanol (waxy type odor). All these substances, which were present in high and statistically significant concentrations in the Presenta variety, contribute to its fruity aroma and taste. Plum brandy of the Valjevka variety was rated as the second best. It had high concentrations of 1-octanol, diethyl succinate, 1-butanol, ethyl stearate, and ethyl benzoate. These are also substances responsible for pleasant, fruity aromas and the smell of wax.

Correlation analysis was used to investigate the correlation between sensory evaluation scores and aroma compounds detected in plum spirits. The results showed that the main indicators of positive sensory evaluation were more significantly related to the esters, terpenes, aldehydes, and ketones than to the other compounds (see Table 2). As can be seen from the table, the perception of flavor and fragrance intensity correlates well with terpene and ester content. On the other hand, a higher content of higher alcohols has a strong influence on the perception of resinous and technical smells, as well as on bitter and negative tastes. A low content of terpenes, aldehydes, and ketones contributed positively to the sensory quality of plum spirits. Generally, many aroma compounds in plum spirits will have synergistic effects on its flavor; an unbalanced aroma compound level would have adverse effects on the sensory quality of the plum spirit.

Table 2. Correlation analysis between sensory evaluation scores and aroma compounds in plum spirits.

Compounds Categories	Fragrance Intensity	Harmony of Fragrance	Delicacy of Fragrance	Intensity of Fruity Aroma	Intensity of Herbal Aroma	Intensity of Spicy Aroma	Intensity of Resinous Aroma	Intensity of Bitter Almond Aroma	Intensity of Pungent Smell	Intensity of Technical Smell	Intensity of Negative smells	Overall Impression of Fragrance
Higher alcohols	0.355	−0.474	0.342	0.776	−0.864	0.077	0.945	0.011	0.466	0.899	0.612	−0.723
Total esters	0.753	0.455	0.927	−0.029	−0.677	0.399	0.176	0.625	0.643	0.577	−0.418	0.517
Acetic acid	0.471	−0.927	−0.459	−0.222	−0.088	−0.914	0.027	0.503	0.603	0.353	0.867	−0.488
Terpenes	0.610	0.420	0.383	−0.768	0.049	−0.086	−0.611	0.791	0.472	−0.057	−0.525	0.873
Aldehydes and ketones	0.177	−0.846	−0.753	−0.448	0.332	−0.991	−0.306	0.342	0.294	−0.066	0.737	−0.360
Compounds Categories	**Flavor Intensity**	**Harmony of Taste**	**Delicacy of the Taste**	**Intensity of Fruit Flavor**	**Intensity of Bitter Taste**	**Intensity of Sour Taste**		**Intensity of Alcoholic Taste**	**Intensity of Negative Taste**	**Aftertaste Intensity**	**Pleasant Aftertaste**	**Overall Taste**
Higher alcohols	0.100	−0.274	−0.057	−0.872	0.862	0.682		0.396	0.947	0.858	0.257	−0.534
Total esters	0.908	0.219	0.076	−0.082	−0.090	0.639		0.261	0.130	0.484	0.991	0.662
Acetic acid	0.124	0.648	0.845	0.109	0.743	0.547		0.883	0.613	0.558	−0.090	−0.667
Terpenes	0.818	0.711	0.482	0.696	−0.481	0.266		0.294	−0.402	−0.021	0.673	0.775
Aldehydes and ketones	−0.112	0.660	0.814	0.393	0.468	0.174		0.669	0.268	0.165	−0.385	−0.623

4. Discussion

The selection of suitable raw materials is the most important technological step influencing the quality of fruit brandy. Rotten or damaged fruit can introduce unwanted microorganisms, contaminating the mash [4,39,40]. Ethyl acetate, propyl acetate, fatty acid esters, isoamyl alcohol, 2-methyl-1-butanol, and 2-phenylethanol (Figures 1–4), which are present in fresh plum distillate, are generated during fermentation by the activity of yeasts [1,41–43].

Butanol can also be produced during fermentation by certain microorganisms, such as bacteria of the genus *Clostridium* [44]. In yeast, only endogenous pathways for the production of 1-butanol (Figure 4) have been described. Butanol can therefore, appear in the plum brandy as a result of cell lysis during fermentation and further processing [44].

The presence of benzaldehyde (Figure 2) in plum distillates is due to the enzymatic degradation of its precursors found in the stones of fruit occurring during fermentation [41]. Benzaldehyde is typically found in small amounts in plum pulp; however, it can be converted into benzyl alcohol by yeasts during fermentation [45]. Delfini reports that at low glucose concentrations, wine yeasts metabolize benzyl alcohol to benzoic acid [46]. Thus, the concentration of benzaldehyde and its derivatives can increase during fermentation.

Geraniol (Figure 4), which according to many authors is present in plum brandy [12,47,48], is not found in plums [18,49]. Some fragrances, such as terpene alcohols geraniol, linalool, nerol, and α-terpineol, are typically found bound in fruit in the form of glycosides [50]. El Hadi suggests that these substances are predominantly present in fruit in bound form rather than free form and are released spontaneously during ripening or by the action of low pH, heat, or enzymes [3]. Yeasts have also been shown to produce terpenoids and/or transform them [51]. Thus, geraniol concentration can serve as an indicator of the content of aromatic terpene alcohol precursors in plums. The highest concentration of geraniol was found in the variety Presenta, followed by the variety Valjevka (Table S1), which were also the best-rated.

Compounds detected in the plum brandies can, to some extent, serve as markers of the quality of the plums in relation to their fragrance content. With this aim in mind, the most important compounds selected for analysis were as follows: 1-pentanol, 1-hexanol, 1-octanol, 1-nonanol, linalool, geraniol, nerol, eugenol, and ethyl cinnamate [23,25,52–54]. All of these substances, except 1-pentanol, were present in the samples of varietal plum brandies at concentrations higher than their threshold of olfactory perception in water, and thus probably contribute to the organoleptic properties of these plum brandies and their characteristic fruity aroma [18,55]. 1-Pentanol is characterized by balsamic, almond, and alcoholic odors [56]. Pino and Quijano report its olfactory perception threshold in water as 4 mg/L [18]. This compound could also influence the organoleptic properties of the distillate through a synergistic effect with other fragrant volatiles [15]. Concentration of 1-octanol and 1-nonanol were in statistically significant differences concentration in all four plum varieties (Table 1). When the concentrations of aromatic compounds are expressed relative to the amount of alcohol, plum brandy made from the Presenta variety plums exhibits the highest abundance of all these compounds, compared to the other studied brandies. The Presenta variety also had the statistically highest concentration of substances responsible for the fruity aroma, as described by Popovic [26] and Ivanovic [24]. The plum brandy made from plums of the Čačanská lepotica variety also had a high content of 1-hexanol, 1-nonanol, nerol, and eugenol. Plum brandy made from plums of the Čačanská rodná variety had a significantly high linalool content and a relatively high 1-hexanol content, but at the same time contained low amounts of other fruit volatiles.

The fragrance content is not the only factor associated with the raw material that influences the analytical profile of fruit spirits. Amino acids in plums are precursors for the formation of higher alcohols during fermentation, and their composition in fruits of different varieties can thus influence the analytical profiles of higher alcohols in various distillates [10]. The concentration of higher alcohols was highly variable in the four varietal distillates (Table S1). Variability in higher alcohol contents has also been presented by other distillates [6,13]. Spaho suggests that the ratios between the contents of certain higher alcohols in plum brandies are characteristic of plums from different varieties, enabling their identification based on these ratios [11]. Higher contents of isobutyl alcohol, isoamyl alcohol, and 2-phenylethanol were found in plum brandy made from plums of the Čačanská lepotica variety compared to the others, which may indicate a high content of nitrogenous precursors of these alcohols in fruit of this variety. The total content of higher alcohols was the highest in the Čačanská rodná variety, where it reached a value of 410 g/hL of 100% vol. ethanol, above the value of 350 g/hL of 100% vol. ethanol, which is stated as the amount required for a negative sensorial assessment [31]. 1-Propanol contributed significantly to the negative sensory evaluation of plum brandy from the Čačanská rodná variety. The concentrations of other sensory active substances with a positive perception are in higher concentrations than the threshold values and often in higher concentrations than Valjevka, which was rated as the second best. The excellent analytical and sensory qualities of the Čačanská rodná variety are also mentioned by Popović et al., 2019 [23]. The intensity of production of higher alcohols also depends on the composition of fermenting microorganisms, and, therefore, the composition of amino acids in the fruit may not be the only determinant for the analytical profile of higher alcohols in the distillates [13,52].

A sample of plum brandy made from plums of the Čačanská rodná variety had very high concentrations of acetaldehyde, 1-propanol, 2-butanol, and propyl acetate, compared to the other varieties. All four varieties were fermented on the same production scale (4 t), so the difference in the composition of the microflora of this variety may be the determining factor [32,57] in the resulting analytical and sensory profile of the product from this variety [1,58]. A high concentration of 1-propanol is generally a sign of contamination, according to Apostolopolou et al., 2005, as this compound may originate from the development of bacteria during the storage phase of the raw material. From the results, the variety Čačanská rodná is apparently overripe, and the data show how it is difficult to ensure optimal maturity and a short storage time on a large scale of production to avoid the development of epiphytic bacterial microflora [59]. Weather conditions in each season have a much greater effect on the yeast community in the fruit. Due to the high content of yeasts other than *Saccharomyces*, the fermentation process of plums must be monitored [32]. As confirmed by correlation analysis (Table 2) and in accordance with the literature [31], high levels of higher alcohols are considered to be indicators of sensory defects in plum brandies. However, apart from the negative effect of 1-propanol on the sensory profile of the distillate from the variety Čačanská rodná, it can be noted that all distillates were characterized by small differences in sensory evaluation. Satora et al., 2017, reached similar conclusions by comparing different varieties of plums [13].

For the further use of operational control using GC-FID, it would be good to identify unknown substances occurring in higher concentrations, which may contribute to the resulting sensory character of the distillate [31]. It is also possible to connect Electronic nose (E-NOSE) [57–59] and convert the detected data again for GC-FID, usable by manufacturers for in-time evaluation of distillation.

5. Conclusions

Fruit spirits are a sought-after consumer product. The commercial success of plum brandy hinges on sensory evaluation, which is influenced by the analytical profile of volatile substances. Therefore, there is a need for reliable, independent, and quick analytical methods that can be utilized by major brandy producers, such as GC-FID, during the production process. We found that high values of 1-propanol fundamentally negatively affected the sensory evaluation of the plum brandy. In large operations, it is often not possible to control the optimal maturity of the raw material as well as the immediate fermentation of large volumes without storing the raw material and thus enabling the possible development of bacterial microflora. This further increases the need for the rapid and affordable quantification of problematic compounds. Statistically significant concentrations of ethyl benzoate, 1-octanol, 1-nonanol, diethyl succinate, and ethyl stearate have a significantly positive effect on the sensory perception of the final product. Hence, when implementing laboratory control on a production scale with GC-FID, the emphasis should be on determining and regulating the concentrations of compounds perceived negatively, rather than focusing solely on the concentrations of fruit markers.

Supplementary Materials: The following supporting information can be downloaded at: https://www.mdpi.com/article/10.3390/fermentation10050235/s1, Table S1: Quantification of the listed volatile compounds.

Author Contributions: I.J.K.: conceptualization, investigation, supervision, writing—original draft, and writing—review and editing. J.B.: methodology, formal analysis, and writing—original draft preparation. L.D.: software, validation, and data analysis. O.M.: data curation, writing—original draft preparation, and writing—review and editing. M.D.: data analysis. D.M.: writing—original draft preparation and writing—review and editing. All authors have read and agreed to the published version of the manuscript.

Funding: This research was funded by specific university research FPBT 2024 067.

Institutional Review Board Statement: Not applicable.

Informed Consent Statement: Not applicable.

Data Availability Statement: Data are available upon reasonable request from the corresponding author.

Conflicts of Interest: The authors declare no conflicts of interest.

References

1. Christoph, N.; Bauer-Christoph, C. Flavour of Spirit Drinks: Raw Materials, Fermentation, Distillation, and Ageing. In *Flavours and Fragrances: Chemistry, Bioprocessing and Sustainability*; Berger, R.G., Ed.; Springer: Berlin/Heidelberg, Germany, 2007; pp. 219–239.
2. Querol, A.; Fleet, G.H. *Yeasts in Food and Beverages*; Springer: Berlin/Heidelberg, Germany, 2006.
3. El Hadi, M.A.; Zhang, F.J.; Wu, F.F.; Zhou, C.H.; Tao, J. Advances in fruit aroma volatile research. *Molecules* **2013**, *18*, 8200–8229. [CrossRef] [PubMed]
4. Walker, G.M.; Stewart, G.G. *Saccharomyces cerevisiae* in the production of fermented beverages. *Beverages* **2016**, *2*, 30. [CrossRef]
5. Ismail, H.M.; Williams, A.A.; Tucknott, O.G. The flavour of plums (*Prunus domestica* L.). An examination of the aroma components of plum juice from the cultivar victoria. *J. Sci. Food Agric.* **1981**, *32*, 613–619. [CrossRef]
6. Ismail, H.M.; Williams, A.A.; Tucknott, O.G. The flavour components of plums: An examination of the aroma components present in the headspace above four cultivars of intact plums, Marjorie's seedling, Merton Gem, NA 10 and Victoria. *J. Sci. Food Agric.* **1981**, *32*, 498–502. [CrossRef]
7. Blumenthal, P.; Steger, M.C.; Einfalt, D.; Rieke-Zapp, J.; Quintanilla Bellucci, A.; Sommerfeld, K.; Schwarz, S.; Lachenmeier, D.W. Methanol mitigation during manufacturing of fruit spirits with special consideration of novel coffee cherry spirits. *Molecules* **2021**, *26*, 2585. [CrossRef] [PubMed]
8. Food and Agriculture Organization of the United Nations. *Regulation (EU) 2019/787 of the European Parliament and of the Council of 17 April 2019 on the Definition, Description, Presentation and Labelling of Spirit Drinks, the Use of the Names of Spirit Drinks in the Presentation and Labelling of Other Foodstuffs, the Protection of Geographical Indications for Spirit Drinks, the Use of Ethyl Alcohol and Distillates of Agricultural Origin in Alcoholic Beverages, and Repealing Regulation (EC) No 110/2008*; FAO: Rome, Italy, 2019.
9. Eden, A.; Van Nedervelde, L.; Drukker, M.; Benvenisty, N.; Debourg, A. Involvement of branched-chain amino acid aminotransferases in the production of fusel alcohols during fermentation in yeast. *Appl. Microbiol. Biotechnol.* **2001**, *55*, 296–300. [CrossRef] [PubMed]
10. Hazelwood, L.A.; Daran, J.-M.; Van Maris, A.J.; Pronk, J.T.; Dickinson, J.R. The Ehrlich pathway for fusel alcohol production: A century of research on *Saccharomyces cerevisiae* metabolism. *Appl. Environ. Microbiol.* **2008**, *74*, 2259–2266. [CrossRef]
11. Spaho, N.; Dürr, P.; Grba, S.; Velagić-Habul, E.; Blesić, M. Effects of distillation cut on the distribution of higher alcohols and esters in brandy produced from three plum varieties. *J. Inst. Brew.* **2013**, *119*, 48–56. [CrossRef]
12. Nikićević, N. Effects of some production factors on chemical composition and sensory qualities of Williams pear brandy. *J. Agric. Sci.* **2005**, *50*, 193–206.
13. Satora, P.; Kostrz, M.; Sroka, P.; Tarko, T. Chemical profile of spirits obtained by spontaneous fermentation of different varieties of plum fruits. *Eur. Food Res. Technol.* **2017**, *243*, 489–499. [CrossRef]
14. Czerny, M.; Christlbauer, M.; Christlbauer, M.; Fischer, A.; Granvogl, M.; Hammer, M.; Hartl, C.; Hernandez, N.M.; Schieberle, P. Re-investigation on odour thresholds of key food aroma compounds and development of an aroma language based on odour qualities of defined aqueous odorant solutions. *Eur. Food Res. Technol.* **2008**, *228*, 265–273. [CrossRef]

15. Singh, V.; Murphy, N.R.; Balasubramanian, V.; Mainland, J.D. Competitive binding predicts nonlinear responses of olfactory receptors to complex mixtures. *Proc. Natl. Acad. Sci. USA* **2019**, *116*, 9598–9603. [CrossRef] [PubMed]
16. Yu, H.; Xie, T.; Xie, J.; Chen, C.; Ai, L.; Tian, H. Aroma perceptual interactions of benzaldehyde, furfural, and vanillin and their effects on the descriptor intensities of Huangjiu. *Food Res. Int.* **2020**, *129*, 108808. [CrossRef] [PubMed]
17. Solís-Solís, H.; Calderón-Santoyo, M.; Schorr-Galindo, S.; Luna-Solano, G.; Ragazzo-Sánchez, J. Characterization of aroma potential of apricot varieties using different extraction techniques. *Food Chem.* **2007**, *105*, 829–837. [CrossRef]
18. Pino, J.A.; Quijano, C.E. Study of the volatile compounds from plum (*Prunus domestica* L. cv. Horvin) and estimation of their contribution to the fruit aroma. *Food Sci. Technol.* **2012**, *32*, 76–83. [CrossRef]
19. Espino-Díaz, M.; Sepúlveda, D.R.; González-Aguilar, G.; Olivas, G.I. Biochemistry of apple aroma: A review. *Food Technol. Biotechnol.* **2016**, *54*, 375. [CrossRef] [PubMed]
20. Nikićević, N.; Velickovic, M.; Jadranin, M.; Vučković, I.; Novaković, M.; Vujisić, L.V.; Stanković, M.; Urosevic, I.; Tešević, V. The effects of the cherry variety on the chemical and sensorial characteristics of cherry brandy. *J. Serbian Chem. Soc.* **2011**, *76*, 1219–1228. [CrossRef]
21. Zierer, B.; Schieberle, P.; Granvogl, M. Aroma-active compounds in Bartlett pears and their changes during the manufacturing process of Bartlett pear brandy. *J. Agric. Food Chem.* **2016**, *64*, 9515–9522. [CrossRef] [PubMed]
22. Butac, M.; Bozhkova, V.; Zhivondov, A.; Milosevic, N.; Bellini, E.; Nencetti, V.; Blazek, J.; Balsemin, E.; Lafarque, B.; Kaufmane, E. Overview of plum breeding in Europe. In Proceedings of the II Balkan Symposium on Fruit Growing 981, Pitesti, Romania, 5–7 September 2011; pp. 91–98.
23. Popović, B.T.; Mitrović, O.V.; Leposavić, A.P.; Paunović, S.A.; Jevremović, D.R.; Nikićević, N.J.; Tesević, V.V. Chemical and sensory characterization of plum spirits obtained from cultivar Čačanska Rodna and its parent cultivars. *J. Serbian Chem. Soc.* **2019**, *84*, 1381–1390. [CrossRef]
24. Logistica Fruit. *European Statistics Handbook 2024*; Fruit Logistica: Berlin, Germany, 2024.
25. Vyviurska, O.; Matura, F.; Furdíková, K.; Špánik, I. Volatile fingerprinting of the plum brandies produced from different fruit varieties. *J. Food Sci. Technol.* **2017**, *54*, 4284–4301. [CrossRef]
26. Heller, D.; Einfalt, D. Reproducibility of Fruit Spirit Distillation Processes. *Beverages* **2022**, *8*, 20. [CrossRef]
27. Nunes, C.; Coimbra, M.A.; Saraiva, J.; Rocha, S.M. Study of the volatile components of a candied plum and estimation of their contribution to the aroma. *Food Chem.* **2008**, *111*, 897–905. [CrossRef]
28. Niimi, J.; Guixer, B.; Splivallo, R. Odour active compounds determined in the headspace of yellow and black plum wines (*Prunus domestica* L.). *LWT* **2020**, *130*, 109702. [CrossRef]
29. Ivanović, S.; Simić, K.; Tešević, V.; Vujisić, L.; Ljekočević, M.; Gođevac, D. GC-FID-MS based metabolomics to access plum brandy quality. *Molecules* **2021**, *26*, 1391. [CrossRef]
30. Popović, B.; Mitrović, O.; Nikićević, N.; Tešević, V.; Urošević, I.; Miletić, N.; Milojević, S. Influence of Different Pre-Distillation Steps on Aromatic Profile of Plum Spirits Produced by Traditional and Modified Methods. *Processes* **2023**, *11*, 863. [CrossRef]
31. Balcerek, M.; Pielech-Przybylska, K.; Patelski, P.; Dziekońska-Kubczak, U.; Strąk, E. The effect of distillation conditions and alcohol content in 'heart'fractions on the concentration of aroma volatiles and undesirable compounds in plum brandies. *J. Inst. Brew.* **2017**, *123*, 452–463. [CrossRef]
32. Skotniczny, M.; Satora, P.; Pańczyszyn, K.; Cioch-Skoneczny, M. Growth dynamics and diversity of yeasts during spontaneous plum mash fermentation of different varieties. *Foods* **2020**, *9*, 1054. [CrossRef] [PubMed]
33. Liebminger, A.; Philipp, C.; Sari, S.; Holstein, M.; Dietrich, V.; Goessinger, M. In-line conductivity measurement to select the best distillation technique for improving the quality of apricot brandies. *Eur. Food Res. Technol.* **2021**, *247*, 1987–1997. [CrossRef]
34. Filatova, M.; Bechynska, K.; Hajslova, J.; Stupak, M. A comprehensive characterization of volatile profiles of plum brandies using gas chromatography coupled to high resolution mass spectrometry. *LWT* **2022**, *167*, 113864. [CrossRef]
35. Botelho, G.; Anjos, O.; Estevinho, L.M.; Caldeira, I. Methanol in Grape Derived, Fruit and Honey Spirits: A Critical Review on Source, Quality Control, and Legal Limits. *Processes* **2020**, *8*, 1609. [CrossRef]
36. Fleet, G.H. Wine yeasts for the future. *FEMS Yeast Res.* **2008**, *8*, 979–995. [CrossRef] [PubMed]
37. Di Cagno, R.; Coda, R.; De Angelis, M.; Gobbetti, M. Exploitation of vegetables and fruits through lactic acid fermentation. *Food Microbiol.* **2013**, *33*, 1–10. [CrossRef] [PubMed]
38. Spaho, N. Distillation techniques in the fruit spirits production. In *Distillation: Innovative Applications and Modeling*; IntechOpen: London, UK, 2017; pp. 129–152.
39. Suomalainen, H. Yeast esterases and aroma esters in alcoholic beverages. *J. Inst. Brew.* **1981**, *87*, 296–300. [CrossRef]
40. Jones, D.T.; Woods, D.R. Acetone-butanol fermentation revisited. *Microbiol. Rev.* **1986**, *50*, 484–524. [CrossRef] [PubMed]
41. Si, T.; Luo, Y.; Xiao, H.; Zhao, H. Utilizing an endogenous pathway for 1-butanol production in *Saccharomyces cerevisiae*. *Metab. Eng.* **2014**, *22*, 60–68. [CrossRef] [PubMed]
42. Bečvářová, H.; Hanč, O.; Macek, K. Course of transformation of benzaldehyde by *Saccharomyces cerevisiae*. *Folia Microbiol.* **1963**, *8*, 165–169. [CrossRef]
43. Delfini, C.; Gaia, P.; Bardi, L.; Mariscalco, G.; Contiero, M.; Pagliara, A. Production of benzaldehyde, benzyl alcohol and benzoic acid by yeasts and *Botrytis cinerea* isolated from grape musts and wines. *Vitis* **1991**, *30*, 253–263.
44. Velíšek, J.; Pudil, F.; Davídek, J.; Kubelka, V. The neutral volatile components of Czechoslovak plum brandy. *Z. Lebensm.-Unters. Forsch.* **1982**, *174*, 463–466. [CrossRef]

45. Miličević, B.; Lukić, I.; Babić, J.; Šubarić, D.; Miličević, R.; Ačkar, Đ.; Miličević, D. Aroma and sensory characteristics of Slavonian plum brandy. *Technol. Acta* **2012**, *5*, 1–7.
46. Gomez, E.; Ledbetter, C.A.; Hartsell, P.L. Volatile compounds in apricot, plum, and their interspecific hybrids. *J. Agric. Food Chem.* **1993**, *41*, 1669–1676. [CrossRef]
47. Krammer, G.; Winterhalter, P.; Schwab, M.; Schreier, P. Glycosidically bound aroma compounds in the fruits of Prunus species: Apricot (*P. armeniaca*, L.), peach (*P. persica*, L.), yellow plum (*P. domestica*, L. ssp. *Syriaca*). *J. Agric. Food Chem.* **1991**, *39*, 778–781. [CrossRef]
48. Carrau, F.M.; Medina, K.; Boido, E.; Farina, L.; Gaggero, C.; Dellacassa, E.; Versini, G.; Henschke, P.A. De novo synthesis of monoterpenes by *Saccharomyces cerevisiae* wine yeasts. *FEMS Microbiol. Lett.* **2005**, *243*, 107–115. [CrossRef] [PubMed]
49. Pielech-Przybylska, K.; Balcerek, M.; Nowak, A.; Patelski, P.; Dziekońska-Kubczak, U. Influence of yeast on the yield of fermentation and volatile profile of 'Węgierka Zwykła'plum distillates. *J. Inst. Brew.* **2016**, *122*, 612–623. [CrossRef]
50. Śliwińska, M.; Wiśniewska, P.; Dymerski, T.; Wardencki, W.; Namieśnik, J. The flavour of fruit spirits and fruit liqueurs: A review. *Flavour Fragr. J.* **2015**, *30*, 197–207. [CrossRef]
51. Bajer, T.; Hill, M.; Ventura, K.; Bajerová, P. Authentification of fruit spirits using HS-SPME/GC-FID and OPLS methods. *Sci. Rep.* **2020**, *10*, 18965. [CrossRef]
52. Schnabel, K.-O.; Belitz, H.-D.; von Ranson, C. Untersuchungen zur Struktur-Aktivität-Beziehung bei Geruchsstoffen. 1. Mitteilung: Wahrnehmungsschwellenwerte und Geruchsqualitäten von gesättigten aliphatischen und alicyclischen Verbindungen mit Sauerstoff-Funktion. *Z. Lebensm.-Unters. Forsch.* **1988**, *187*, 215–223. [CrossRef]
53. Sánchez-Palomo, E.; Gómez García-Carpintero, E.; Alonso-Villegas, R.; González-Viñas, M.A. Characterization of aroma compounds of Verdejo white wines from the La Mancha region by odour activity values. *Flavour Fragr. J.* **2010**, *25*, 456–462. [CrossRef]
54. Nguyen, N.T.H.; Wang, W.-Y.; Huang, W.-L.; Huang, C.-L.; Chiang, T.-Y. Metagenomics analyses of microbial dynamics associated with putative flavor development in mash fermentation of sake. *LWT* **2022**, *163*, 113570. [CrossRef]
55. Gschaedler, A. Contribution of non-conventional yeasts in alcoholic beverages. *Curr. Opin. Food Sci.* **2017**, *13*, 73–77. [CrossRef]
56. Baleiras-Couto, M.M.; Caldeira, I.; Gomes, F.; Botelho, G.; Duarte, F.L. *Saccharomyces cerevisiae* Diversity in *Arbutus unedo* L. Fermentations in Association with the Volatile and Sensory Similarities of the Distillates. *Foods* **2022**, *11*, 1916. [CrossRef] [PubMed]
57. Stanzer, D.; Hanousek Čiča, K.; Blesić, M.; Smajić Murtić, M.; Mrvčić, J.; Spaho, N. Alcoholic Fermentation as a Source of Congeners in Fruit Spirits. *Foods* **2023**, *12*, 1951. [CrossRef] [PubMed]
58. Russmayer, H.; Marx, H.; Sauer, M. Microbial 2-butanol production with Lactobacillus diolivorans. *Biotechnol. Biofuels* **2019**, *12*, 262. [CrossRef] [PubMed]
59. Apostolopoulou, A.A.; Flouros, A.I.; Demertzis, P.G.; Akrida-Demertzi, K. Differences in concentration of principal volatile constituents in traditional Greek distillates. *Food Control* **2005**, *16*, 157–164. [CrossRef]

Disclaimer/Publisher's Note: The statements, opinions and data contained in all publications are solely those of the individual author(s) and contributor(s) and not of MDPI and/or the editor(s). MDPI and/or the editor(s) disclaim responsibility for any injury to people or property resulting from any ideas, methods, instructions or products referred to in the content.

beverages

MDPI

Article

Chemical Characterization of Cider Produced in Hardanger—From Juice to Finished Cider

Ingunn Øvsthus [1,*], Mitja Martelanc [2], Alen Albreht [3], Tatjana Radovanović Vukajlović [2], Urban Česnik [2] and Branka Mozetič Vodopivec [2]

1 Norwegian Institute of Bioeconomy Research, P.O. Box 115, NO-1431 Aas, Norway
2 Wine Research Centre, University of Nova Gorica, Glavni Trg 8, SI-5271 Vipava, Slovenia; mitja.martelanc@ung.si (M.M.); tatjana.radovanovic@ung.si (T.R.V.); urban.cesnik@ung.si (U.Č.); branka.mozetic@ung.si (B.M.V.)
3 Laboratory for Food Chemistry, Department of Analytical Chemistry, National Institute of Chemistry, Hajdrihova 19, SI-1000 Ljubljana, Slovenia; alen.albreht@ki.si
* Correspondence: ingunn.ovsthus@nibio.no; Tel.: +47-48-20-72-50

Abstract: Our investigation delves into the previously uncharted territory of cider composition from Norway. This study aimed to obtain an overview of the qualitative and quantitative compositions of general chemical parameters, polyphenols (individual and total expressed as gallic acids equivalents), selected esters, and selected C6-alcohols in ciders with the PDO label Cider from Hardanger. In total, 45 juice and cider samples from the fermentation process were collected from 10 cider producers in Hardanger in 2019, 2020, and 2021. Individual sugars, acids, ethanol, and 13 individual phenols were quantified using HPLC-UV/RI. Seven ethyl esters of fatty acids, four ethyl esters of branched fatty acids, ten acetate esters, two ethyl esters of hydroxycinnamic acids, and four C6-alcohols were quantified using HS-SPME-GC-MS. For samples of single cultivars ('Aroma', 'Discovery', 'Gravenstein', and 'Summerred'), the sum of the measured individual polyphenols in the samples ranges, on average, from 79 to 289 mg L^{-1} (the lowest for 'Summerred' and highest for 'Discovery' and 'Gravenstein'). Chlorogenic acid was the most abundant polyphenol in all samples. Ethyl butyrate, ethyl hexanoate, ethyl octanoate, ethyl decanoate, ethyl isobutyrate, ethyl 2-methylbutyrate, isoamyl acetate, and hexanol were present at concentrations above the odour threshold and contributed to the fruity flavour of the Cider from Hardanger.

Keywords: cider; volatile aroma; esters; higher alcohols; polyphenols

1. Introduction

Over the last decade, the consumption of ciders has increased worldwide. Norwegian ciders, especially ciders produced in the Hardanger region, have become popular and have received attention in international cider competitions. The Hardanger region is the leading fruit-growing area in Norway, contributing to approximately 50% of Norwegian fruit production. The typography, with orchards planted in steep hills and a mild coastal climate, makes the Hardanger fjord area suitable for growing fruit at a latitude of approximately 60° N. There are three milestones that are important for the increasing interest in cider from the Hardanger region: (1) in 2003, the Hardanger Cider Association was established; (2) in 2009, Cider from Hardanger became a geographically protected designation of origin; and (3) a change in Norwegian alcohol regulations in 2016, allowed cider producers to sell bottled ciders from the farm. In the period from 2012 to 2022, the sale of Norwegian cider in the state wine monopoly (Vinmonopolet) had a 20-fold increase. In 2023, the total sale of Norwegian cider (including sales at Vinmonopolet, the HORECA market, and farms) corresponded to 550,000 L. The total amount of cider produced in Hardanger sold corresponds to 80% of the total Norwegian sales. In 2023, the production of cider in the Hardanger region was approximately 700,000 L. The goal is to increase cider production to 1 million litres in the Hardanger region by 2025 (personal communication, Anna Gursli Langesæter, Hardanger Cider Association).

The main sensory drivers in cider are phenolics such as flavan-3-ols (catechin and procyanidins); hydroxycinnamic acids (5-caffeoylquinic acid or chlorogenic acid); 4-*p*-coumarylquinic acid; dihydrochalcones (phloretin); dihydroflavanols, and anthocyanins [1]; as well as organic acids and aroma compounds like esters, higher alcohols, fatty acids, aldehydes, and ketons [2]. In recent years, norisoprenoids have also been confirmed in cider [2]. The sensory profiles and chemical quality of ciders are influenced by various factors, including (1) the apple cultivar; (2) orchard-related agronomic factors; (3) processing technologies; (4) fermentation conditions (alcoholic and malolactic), such as nutrition and temperature; and (5) ageing [2–4]. To achieve the cider production target, differentiating the cider and ensuring uniform sensory quality is of importance to increase sales [2]. A comprehensive study [5] analysed sensory profiles of 14 traditional and commercial ciders from Sweden, Denmark, and the UK. The researchers identified unique characteristics in industrially produced and traditionally made ciders, grouping them based on their production technology, including wood ageing, spontaneous fermentations, controlled malolactic fermentation, and industrial production with commercial yeast and addition of other fruit and herb aromas.

Due to the numerous processing steps affecting cider quality, significant variation exists among products from different producers and years. Establishing product differentiation and consistent quality requires identifying the essential chemical composition contributing to the sensory quality in ciders from Hardanger. Unfortunately, limited information is available about the chemical compositions of ciders from both Hardanger and Norway [3,6,7]. To gain a better insight and distinguish Ciders from Hardanger, further research efforts and consistent monitoring are essential. In a groundbreaking 2024 study, researchers compared the general chemical composition of Norwegian ciders, including those from Hardanger, to selected French ciders for the first time. The samples were purchased from wine monopoly stores (Vinmonopolet) in Norway [3]. Notably, the Norwegian ciders were found to be fresher and lighter than their French counterparts, with observed differences in total phenol content and aroma profiles [3].

The Cider from Hardanger is mainly made of apple cultivars produced for fresh consumption in Norway, including 'Aroma', 'Discovery', 'Gravenstein', and 'Summerred'. Some producers also cultivate specific cider apples and old cultivars. The resulting Norwegian consumer apples are typically low in phenols and high in acids, contributing to a fresh cider style [3,6,8]. A key question is how cultivars and season influence the chemical parameters and diversity of ciders among producers with varying production technologies and accessibility to equipment. In order to gain a deeper insight into the chemical properties of Cider from Hardanger, it is essential to obtain samples from real market producers.

Our goal was to enhance the understanding of the chemical properties of ciders from Hardanger by identifying and quantifying taste-specific compounds present in the juice and final cider samples collected from producers in Hardanger. Juice and cider samples were from various apple cultivars commonly used in Hardanger cider production. Moreover, the collected samples from different producers and seasons show variations of chemical composition in apple juice and ciders produced in Hardanger.

2. Materials and Methods

2.1. Samples

Juice from pressed apple fruits and end-fermented cider were sampled from 10 cider producers in Hardanger (Norway) during the 2019, 2020, and 2021 seasons. The samples were mainly from single apple cultivars 'Aroma', 'Gravenstein', 'Discovery', and 'Summerred'. However, some samples were from blends of the main cultivars or other cultivars ('Gravenstein'/'Elstar', 'Gravenstein'/Cider apple, 'Gravenstein'/'Signe Tillisch', 'Gul Gravenstein', and 'Rubinstep'). Samples were collected from cider-makers' fermentation tanks during the fermentation process. A total of 45 samples were collected and subjected to different chemical analyses. Results in the current manuscript's tables and figures present chemical properties of juices and end-fermentation ciders of the four main cultivars 'Aroma', 'Discovery', 'Gravenstein', and 'Summerred', which were represented by a higher number of samples among the producers, while in Supplementary Materials, we present the whole sampling and chemical database collected. The juice and fermented apple juice were produced according to cider-makers' 'in-house' method. They used different pressing methods (from belt to basket press, with or without enzyme addition); the maturity stage of the apple fruits was different and was not assessed in this experiment (it was determined based on cider producer experience). Some used spontaneous fermentation, and some added commercial yeast. Fermentation temperatures were also different (in the range from 12 to 18 °C). Information about producers and samples is presented in Table 1. Samples (apple juice and cider) were analysed for general chemical compounds at Norwegian Institute of Bioeconomy Research (NIBIO), location Ullensvang (Norway) (total phenols, yeast-assimilable nitrogen (YAN), titratable acidity). Each sample was analysed twice. NIBIO Ullensvang prepared the samples for measuring ethanol, individual sugars, acids, polyphenols, and aroma compounds using high-performance liquid chromatography (HPLC) and gas chromatography (GC) coupled to different detectors (described in later subchapters). These samples (juice and cider) were stored in small dark glass containers at −18 °C at NIBIO Ullensvang before shipment under cold conditions to the University of Nova Gorica, Slovenia. All samples were analysed two months after being sampled from producers.

Table 1. Overview of the 10 producers from Hardanger (Norway) and samples.

Producer	Variety	Year	Sample Type (Juice-J, Cider-C)	Production System and Production Volume (Litre of Cider)
Producer 1	Summerred	2019	J and C	Hydropress, steel tank, production volume per year 10,000 L
Producer 2	Discovery Gravenstein Aroma	2019 2019 2021	J and C J and C C	Steel tank, added yeast, production volume per year 5000 L
Producer 3	Summerred Gravenstein Aroma Aroma	2019 2019 2019 2021	J and C J J and C C	Belt press, steel tank, added yeast, production volume per year 7000 L
Producer 4	Discovery Aroma Gravenstein/ciderapple Gravenstein/Signe Tillisch Gul Gravenstein	2019 2021 2019 2019 2021	J and C J and C J and C J and C J and C	Belt press, steel tank, spontaneous fermentation, production volume per year 12,000 L
Producer 5	Gravenstein/Elstar	2021	J and C	Belt press, steel tank, spontaneous fermentation, centrifugation, production volume per year 150,000 L
Producer 6	Aroma Summerred Discovery Gravenstein Rubinstep	2020 2021 2021 2021 2021	J and C J and C J and C J and C J and C	Belt press, steel tank, added yeast and yeast nutrition, production volume per year 150,000 L
Producer 7	Aroma Gravenstein	2020 2020	J and C J and C	Pack cider press, steel tank with floating lid, spontaneous fermentation, production volume per year 4000 L
Producer 8	Summerred	2020	J and C	Belt press, steel tank, added yeast and nutrition, production volume per year 100,000 L
Producer 9	Aroma Rubinstep	2020 2021	J and C J	Belt press, steel tank, added yeast and nutrition, production volume per year 130,000 L
Producer 10	Gravenstein	2020	J	Belt press, steel tank, added yeast, production volume per year 60,000 L

2.2. HPLC-UV/RI Analysis of Sugars, Acids, Ethanol, as Well as Titratable Acidity

Chemicals for high-performance liquid chromatography (HPLC) coupled to ultraviolet (UV)/refractive index (RI) analysis: glucose (99%) (Acros Organics, Fair Lawn, NJ, USA), fructose (99%) (Acros Organics, Fair Lawn, NJ, USA), sucrose (99.9%) (Acros Organics, NJ, USA), tartaric acid (Alfa Aesar, Karlsruhe, Germany), lactic acid (30%) (Sigma, Steinheim, Germany), D-L malic acid (99%) (Sigma-Aldrich, Steinheim, Germany), and citric acid (99.9%) (Sigma, Steinheim, Germany) as standards. Concentrated sulfuric acid (VI) was purchased from VWR Chemicals (Leuven, Belgium).

HPLC–UV/RI analysis was performed using an Agilent 1100 series HPLC system (Agilent Technologies©, Palo Alto, CA, USA) equipped with Agilent OpenLab CDS ChemStation 2.3.54 software, a UV detector (G1314A VWD) for the analysis of organic acids (detection at 210 nm) and a refractive index detector (model G7162A) for the analysis of sugars (fructose, glucose, and sucrose) and ethanol. Samples were filtered using polytetrafluoroethylene (PTFE) 0.45 µm syringe filters (VWR® International, Radnor, PA, USA). The organic acids were determined using a reversed-phase HPLC-UV method [4]. Glucose, fructose, and sucrose determination was performed using the HPLC method based on hydrophilic interaction liquid chromatography [4], while the assessment of ethanol was performed on a Phenomenex ROA Organic Acid H^+ (8%) column with a size of 300 × 7.8 mm with detailed method description published elsewhere [9]. All HPLC quantifications were performed with external calibration in the range of the expected values for the juice and cider samples [4].

Titratable acidity was determined by mixing 1 mL of the samples with 49 mL of distilled water and titrating the sample to pH 8.2 using 0.1 M NaOH (T5 titrator, Mettler-Toledo, Zurich, Switzerland), and the result was calculated as malic acid equivalents (g L^{-1}) [10].

2.3. *Enzymatic Determination of YAN*

Megazyme L-Arginine/Urea/Ammonia (Rapid) determined via the reaction of amino nitrogen in juice spectrophotometric assay kits (Megazyme, Bray, Ireland) and the Megazyme primary amino nitrogen (PAN) were used to measure yeast-assimilable nitrogen (YAN) in the juice and cider samples according to the manufacturer's specifications and read on a VWR-UV 3100PC spectrophotometer at λ 340 nm (VWR, Leuven, Belgium).

2.4. *GC-MS Analyses of Volatile Compounds*

In juice and ciders, we determined selected esters and C6-alcohols using an automatic robotic system for head space (HS) solid-phase microextraction (SPME) combined with gas chromatography coupled to mass spectrometry (GC-MS). The compounds were quantified and identified using validated MS-SPME_GC-MS methods described elsewhere [11].

The samples were extracted using headspace solid-phase micro extraction (HS-SPME) using SPME Fiber Assembly (50/30 µm DVB/CAR/PDMS, Stableflex, 24 Ga, Autosampler, Gray (Supelco, St. Louis, MO, USA)). Into a 20 mL SPME vial, 1 mL of juice and cider samples was added to 2 g of NaCl and 9 mL of deionized water and spiked with 20 µL of internal standard solution. The vials were tightly sealed with a PTFE-lined cap. The solution was then homogenized using a vortex mixer, and the samples were loaded onto a Gerstel MPS Robotic autosampler (CTC Analytics AG, Zwingen, Switzerland). The fibre was inserted into the SPME Arrow Conditioning Module for 2 min at 270 °C. Thereafter, the fibre was inserted into the headspace of the sample vial for 30 min at 40 °C with simultaneous sample swirling at 250 rpm using the Agigator module. Afterwards, the fibre was transferred to the injector for desorption at 250 °C for 15 min. The sample injection time into the GC column was set to 30 s, followed by fibre cleaning in the SPME Arrow Conditioning Module for 10 min at 270 °C.

HS-SPME-GC-MS analyses were carried out on an HP 8890 GC System coupled to a 5977 B GC/MSD quadrupole mass spectrometer (Agilent, Santa Clara, CA, USA) equipped with a Gerstel MPS Robotic autosampler. For quantification, a Mass Hunter Workstation Software (Version 10.1) (Agilent Technologies) was used. Identification of all analysed compounds was performed by comparing the retention times and mass spectra with the pure analysed standards. Library from the NIST Mass Spectral Search Program

(Version 2.4, 2020) was also used for additional confirmation of the analysed compounds. For the purpose of quantification of analysed compounds, internal standards were spiked accordingly. Each internal standard was dissolved in ethanol solution separately to obtain a stock solution.

Injections were performed in the splitless mode for 0.5 min, using ultra-inert inlet single taper liner. A DB-WAX UI capillary column (60 m × 0.25 mm × 0.25 μm film thickness, Agilent Technologies, USA) was used, and the carrier gas was helium (H 6.0) with a flow of 3 mL min^{-1}. The oven temperature was programmed at 40 °C for 5 min, then raised to 200 °C at 3 °C min^{-1}, further increased to 240 °C at a rate of 10 °C min^{-1}, and held for 10 min. The mass spectrometer was operated in electron ionization mode at 70 eV with selected-ion-monitoring (SIM) mode.

2.5. HPLC-UV and HPLC-MS Analysis of Phenolic Compounds and Total Phenol Using Folin–Ciocâlteu Method

Total phenols were measured spectrophotometrically using the Folin–Ciocâlteu method following the protocol of Singleton [6]. The juice and cider samples were filtered using syringe filters (MACHAREY-NAGEL GmbH & Co. KG, Düren, Germany), Folin–Ciocalteu reagent (Sigma-Aldrich, Steinheim, Germany) was used in accordance with Singleton et al. [12], and a gallic acid standard (Merck KGaA, Darmstadt, Germany) was used for calibration. All results are expressed as gallic acid equivalents (GAE mg L^{-1}). Total phenol content was determined using a VWR-UV 3100PC spectrophotometer at 765 nm (VWR, Leuven, Belgium).

Quantitative and qualitative analyses of individual phenolic compounds in juice and cider samples were carried out using the HPLC-UV method, employing an Agilent 1100 series HPLC system (described in Section 2.2) at a wavelength of 285 nm for identification (comparisons to the standard compound) and quantification with external curves, prepared with HPLC grade standards from Extransynthese (Genay, France), as listed in Table S1.

Briefly, the studied compounds were separated on a Phenomenex Luna C18(2) HPLC column (150 mm × 2 mm i.d., 3 μm) protected by a guard column. The mobile phase consisted of solvent A, 0.5% acetic acid, and solvent B, HPLC-grade methanol, and the following solvent gradient was applied: 100% A, 0–6 min; 90% A, 6–7 min; 85% A, 7–35 min; 70% A, for 35–53 min; 55% A, 53–68 min; 100% B for, 68.1–77 min; 100% A, and 78–90 min. The flow rate was set to 0.3 mL min^{-1}, the tray and column oven temperatures were maintained at 13 °C and 30 °C, respectively, and 10 μL injections were made for each analysis. Samples were thawed and filtered through a 0.45 μm polyvinylidene difluoride (PVDF) syringe membrane filter prior to analysis. The data from the UV detector were acquired at 285 nm.

Qualitative analysis of phenolic compounds in cider samples was performed using HPLC-UV-MS^2. The analytical system consisted of an Accela 1250 UHPLC system (Thermo Scientific, Waltham, MA, USA) coupled with an LTQ Velos MS system (Thermo Scientific, Waltham, MA, USA). This method was adopted according to Ivanović et al. [13] with some modifications. The studied compounds were separated on the same column as for HPLC-UV analysis, the Luna C18(2) HPLC column from Phenomenex (150 × 2 mm i.d., 3 μm), which was protected by a guard column. The mobile phase consisted of solvent A, 0.5% acetic acid, and solvent B, methanol, and the following solvent gradient was applied: 0–6 min (0% B), 6–7 min (0–10% B), 7–35 min (10–35% B), 35–53 min (35–45% B), 53–68 min (45–70% B), 68–68.1 min (70–0% B), and 68.1–88 min (0% B). The flow rate was set to 0.3 mL/min, tray and column oven temperatures were maintained at 13 °C and 38 °C, respectively, and 5 μL injections were used for each analysis. Samples were thawed and filtered through a 0.45 μm PVDF syringe membrane filter prior to analysis. Data were acquired with the PDA detector at 280 nm, and UV-Vis spectra were recorded in the range of 230–600 nm. Ionization of the compounds was achieved with ESI operating in negative mode. The settings for the MS system were as follows: source voltage, 3 kV; source temperature, 325 °C; capillary temperature, 350 °C; sheath gas, 60 arbitrary units (a.u.); auxiliary gas, 10 a.u. The presence of the analytes in the samples was confirmed by comparison of the retention time, targeted MS^2 spectra, and UV-Visible (Vis) spectra (where possible) with those of the reference standard materials (Table S2).

2.6. Statistical Analysis

For each cultivar and collection time (juice or finished cider), results are expressed as mean and minimum and maximum values (mean (min–max)) in the tables. The values were calculated using Minitab 20 (version 20, State College, PA, USA) and Microsoft Excel.

Principal component analysis (PCA) was performed using Minitab 20 to evaluate the overall impacts of varieties on the contents of selected volatile aromas.

3. Results and Discussion

3.1. Chemical Composition of Apple Juices and End Fermented Cider of Different Apple Fruit Cultivars

We present the multi-year composition of ciders produced from individual apple cultivars to understand the differences between them.

3.1.1. Acids, Sugars, Ethanol, and YAN

The concentration and composition of the organic acids in the cider affect the pH, taste, and microbial stability. The acids in the finished cider were present in the raw materials; however, their concentrations can change during fermentation by yeasts or malolactic bacteria [14–16]. Malic acid content decreased during fermentation for all cultivars (Tables 2 and S3). Such a decrease is common because of yeast metabolism of malic acid and conversion of malic acid to lactic acid [16]. The latter is also supported by the data in Table 2, showing some extent of malolactic fermentation conversions. In juices, the average titratable acidity (expressed as malic acid equivalent) ranged on average per cultivar from 6 ('Discovery') to 8.9 ('Aroma') g L^{-1}. The level of titratable acidity decreased during fermentation, likely because of possible chemical transformations and precipitation reactions [17]. After fermentation, the average titratable acidity ranged from 5.2 ('Discovery') to 7.2 ('Summerred') g L^{-1}.

Malic acid is the dominant acid in apples [18]. The malic acid contents in apple juices varied from 2.5 g L^{-1} ('Discovery') and up to 22.1 g L^{-1} ('Aroma'). The average concentrations of malic acid in the juice of the four cultivars ranged from low to high in the following order: 'Discovery', 'Summerred', 'Gravenstein', and 'Aroma'. During fermentation, the content of different acids generally decreases [18]. Malic acid contents were reduced by 53%, 32%, 70%, and 9% on average during the fermentation of 'Aroma', 'Discovery', 'Gravenstein', and 'Summerred', respectively. The reduction in malic acid in cider from Hardanger was higher than that found by Whiting et al. [14], who found a reduction of malic acid during fermentation of up to 40%, while Maslov Bandić [16] reported a slight decrease in malic acid in two ciders from Croatia during fermentation. This reduction, as previously mentioned, is the result of many biochemical reactions and yeast assimilation of malic acid. Malic acid can also be reduced by lactic acid bacteria, and the amount of lactic acid produced at the end of fermentation indicates malolactic fermentation, as mentioned previously. The highest measured level of lactic acid in all ciders in the study was over 4.8 g L^{-1} ('Aroma'). The reduction in malic acid content reduces the sour taste of the cider [19]. Lactic acid, on average, increased during the fermentation of 'Aroma', 'Discovery' and 'Gravenstein', but not for 'Summerred' ciders. The citric acid content was low in the monitored samples, which agrees with Maslov Bandić et al. [16]. We also detected small amounts of tartaric acid in all the samples, and no acetic acid.

Table 2. Mean values for titratable acidity, acids (malic acid, lactic acid, citric acid, and tartaric acid), sugars (fructose, glucose, and sucrose), ethanol content, and yeast-assimilable nitrogen (YAN). The results are presented as mean and minimal to maximal values in brackets.

Parameter	Variety	N	Juice Mean (Min–Max)	N	Cider Mean (Min–Max)
Titratable acidity (g L^{-1} g of malic acid)	Aroma	5	8.92 (6.54–12.8)	7	7.07 (5.19–9.33)
	Discovery	3	5.95 (5.20–6.54)	3	5.22 (3.89–6.73)
	Gravenstein	5	7.43 (6.14–9.25)	3	5.73 (5.47–5.90)
	Summerred	4	7.99 (5.98–9.91)	4	7.20 (7.79–9.05)
Malic acid (g L^{-1})	Aroma	5	13.0 (5.90–22.1)	7	5.61 (0.60–10.7)
	Discovery	3	4.38 (2.50–7.15)	3	2.96 (0.70–6.70)
	Gravenstein	5	12.5 (6.46–21.6)	3	3.52 (0.10–6.30)
	Summerred	4	7.23 (3.72–14.6)	4	6.57 (2.80–11.2)
Lactic acid (g L^{-1})	Aroma	5	0.16 (0.02–0.32)	7	1.11 (0.10–4.81)
	Discovery	3	0.17 (0.06–0.36)	3	1.06 (0.06–3.05)
	Gravenstein	5	0.16 (0.09–0.32)	3	0.49 (0.04–1.19)
	Summerred	4	0.84 (0.04–3.03)	4	0.77 (0.02–2.80)
Citric acid (g L^{-1})	Aroma	5	0.09 (ND–0.14)	7	0.17 (0.05–0.60)
	Discovery	3	0.02 (ND–0.07)	3	0.05 (ND–0.10)
	Gravenstein	5	0.14 (0.09–0.20)	3	0.20 (ND–0.50)
	Summerred	4	0.04 (ND–0.10)	4	0.07 (ND–0.10)
Tartaric acid (g L^{-1})	Aroma	5	0.17 (0.05–0.40)	7	0.12 (0.07–0.21)
	Discovery	3	0.05 (0.03–0.09)	3	0.10 (0.08–0.14)
	Gravenstein	5	0.20 (0.10–0.38)	3	0.16 (0.13–0.21)
	Summerred	4	0.13 (0.05–0.27)	4	0.13 (0.03–0.22)
Acetic acid (g L^{-1})	Aroma	5	0.02 (ND–0.10)	7	0.01 (ND–0.10)
	Discovery	3	0.07 (ND–0.20)	3	ND
	Gravenstein	5	0.08 (ND–0.30)	3	ND
	Summerred	4	0.05 (ND–0.10)	4	0.05 (ND–0.20)
Fructose (g L^{-1})	Aroma	5	79.0 (31.9–141)	7	16.6 (1.30–38.9)
	Discovery	3	38.8 (22.1–67.9)	3	7.90 (ND–18.2)
	Gravenstein	5	90.2 (55.1–142)	3	13.0 (ND–37.5)
	Summerred	4	41.0 (22.8–63.4)	4	6.38 (ND–18.1)
Glucose (g L^{-1})	Aroma	5	20.3 (8.20–31.3)	7	9.24 (0.30–21.0)
	Discovery	3	12.6 (7.80–21.0)	3	1.67 (0.70–3.10)
	Gravenstein	5	26.3 (18.1–38.3)	3	4.47 (1.4–9.1)
	Summerred	4	9.82 (4.60–16.8)	4	1.55 (0.40–2.80)
Sucrose (g L^{-1})	Aroma	5	38.7 (0.60–94.1)	7	2.46 (ND–14.6)
	Discovery	3	9.90 (4.80–17.9)	3	ND
	Gravenstein	5	27.6 (10.2–67)	3	ND
	Summerred	4	33.5 (19.8–57.3)	4	0.28 (ND–1.10)
Fructose + Glucose (g L^{-1})	Aroma	5	99.2 (40.0–172)	7	25.8 (3.70–53.8)
	Discovery	3	51.5 (30.0–88.9)	3	9.60 (1.20–21.4)
	Gravenstein	5	117 (75.2–180)	3	17.5 (2.90–46.5)
	Summerred	4	50.8 (27.4–80.2)	4	7.92 (1.10–20.9)
Ethanol (vol %)	Aroma	5	ND	7	8.45 (6.33–11.3)
	Discovery	3	ND	3	7.11 (5.33–8.12)
	Gravenstein	5	ND	3	8.34 (6.33–10.3)
	Summerred	4	ND	4	7.07 (2.77–11.4)
YAN Total (mg L^{-1})	Aroma	5	38.1 (17.7–62.3)	7	9.37 (ND–22.5)
	Discovery	3	23.1 (0.80–35.5)	3	12.9 (9.74–18.6)
	Gravenstein	5	28.4 (3.81–51.7)	3	17.0 (6.88–36.6)
	Summerred	4	69.7 (49.9–101)	4	12.5 (108–14.6)

The main sugars in the juices were fructose, glucose, and sucrose, as expected for apple juice and fruit [9]. In juices of four different apple cultivars ('Aroma', 'Discovery', 'Gravenstein', and 'Summerred'), the content of glucose, fructose, and sucrose ranged on average, respectively, as follows: 9.8–26.3, 38.8–90.2, and 9.9–38.7 g L^{-1}. Juice of 'Gravenstein' and 'Aroma' cultivars had the highest level of total sugar, and that of 'Discovery' was the lowest (Table 2). The sugar content decreases during fermentation, mainly due to yeast assimilation and translation into ethanol, but also to other metabolites [17]. The yeasts are mainly glucophilic [17], so it is no surprise that end products have more fructose than glucose, regardless of the cultivar or the amount of total residual sugar. Residual sugars (sum of glucose and fructose) ranged on average from 7.9 ('Summerred' ciders) to 25.8 g L^{-1} ('Aroma' ciders) in finished cider. Technologies and decisions regarding the target of residual sugars in end-fermented ciders differ among producers.

Chaptalization with sucrose is a traditional technique used in cider production in the Hardanger area to increase the alcohol content to reach 6–7% vol. [6]. Some producers add sucrose before the onset of fermentation, and some during fermentation. This explains the variation of sucrose in juices. In all the final products, the sucrose content was zero (or near zero).

Ethanol ranged in final ciders on average from 7 to 8.5% vol. Naturally present sugars in apple juices from Hardanger that undergo fermentation enable up to 5% natural ethanol, which is in agreement with the theoretical yield of alcoholic fermentation [17]. This is why most of the cider producers, as already mentioned, add sucrose to increase the start density of apple juice and final ethanol contents from 6 to 7% vol. [3,6]. Our study is a case study, focusing on the chemical composition of apple juices and ciders like they are produced in different seasons by real producers. We are, of course, aware that the level of chaptalization was not monitored particularly, and that could have influenced the chemical properties of ciders. However, the study's findings on the natural compositional variations of Cider from Hardanger remain valid and significant part of the established database.

3.1.2. Total Phenols and Individual Polyphenols

Polyphenols are important for cider colour, stability, taste (bitterness and astringency), and aroma. Processing apple fruits to juice impacts the phenolic content due to the presence of enzymes. This is explained by differences in the extraction of phenolic compounds [20] and the degradation or transformation of phenols through oxidation [21–24]. Table apples have lower levels of phenols than cider apples [25,26]. The apple fruits produced in Hardanger are mostly table apples, and we already know that ciders from Hardanger have lower amounts of total phenols than French ciders [2]. As evident from Table 3, the single-cultivar juices had total phenol content ranges on average from 198 mg L^{-1} ('Summerred') to 789 mg L^{-1} ('Discovery'), determined in gallic acid equivalents using the *Folin–Circulate* method. These results agree with earlier reports of other Norwegian groups of scientists [27], who found the concentration of total phenols in the juice of cv. 'Summerred' to be 272 mg L^{-1}. Table apples have lower levels of phenols than cider apples [25,26]. In the juice from mixtures of Gravenstein and cider apples (Table S4), the total phenol reached 1772 mg of gallic acid L^{-1}. The contents of phenolic compounds analysed using the *Folin–Ciocâlteu* method showed quantitative changes during fermentation but not the same pattern for all cultivars. The juice with the highest total phenols was from 'Discovery' and 'Gravenstein', where we observed a similar decrease during fermentation, reaching 496 and 306 mg of gallic acid L^{-1} in the final ciders, while in 'Aroma' and 'Summerred', the total phenols did not change. The literature presents various findings regarding changes in phenolic compounds during fermentation. Some studies indicate a decrease in these compounds during fermentation [1,28], while others indicate an increase [16,29]. Our data are in the range of previous findings about Norwegian ciders [6,27] and from the other ciders made from table apples, containing total phenols from 110 mg L^{-1} up to 706 mg L^{-1} (expressed as gallic acid equivalents [16,28,30].

Table 3. Total phenol content in gallic acid equivalent (GAE mg L^{-1}) and polyphenol profile from HPLC (mg L^{-1}) in apple juice and finished cider. Values are the mean concentrations, with the minimal and maximal values shown in parentheses.

Parameter	Variety	Juice (mg L^{-1}) N	Juice (mg L^{-1}) Mean (Min–Max)	Cider (mg L^{-1}) N	Cider (mg L^{-1}) Mean (Min–Max)
Total Phenol	Aroma	5	483 (306–680)	7	473 (334–685)
	Discovery	3	789 (746–836)	3	496 (146–761)
	Gravenstein	5	737 (572–933)	3	306 (153–418)
	Summerred	4	198 (160–217)	4	215 (159–279)
Gallic acid	Aroma	5	0.25 (ND–1.00)	7	1.00 (ND–3.00)
	Discovery	3	0.67 (ND–2.00)	3	0.67 (ND–2.00)
	Gravenstein	5	ND	3	ND
	Summerred	4	0.67 (ND–2.00)	4	1.25 (ND–5.00)
Protocatechuic acid	Aroma	5	ND	7	ND
	Discovery	3	ND	3	ND
	Gravenstein	5	ND	3	ND
	Summerred	4	ND	4	ND
Procyanidin B1	Aroma	5	8.48 (ND–15.5)	7	12.3 (ND–31.0)
	Discovery	3	12.0 (ND–18.0)	3	21.7 (12.0–39.0)
	Gravenstein	5	10.3 (ND–21.0)	3	18.7 (15.2–22.0)
	Summerred	4	5.47 (ND–16.4)	4	2.8.0 (ND–10.0)
Catechin	Aroma	5	8.87 (ND–18.3)	7	8.20 (ND–12.7)
	Discovery	3	12.7 (8.00–15.0)	3	12.0 (10.0–14.0)
	Gravenstein	5	7.90 (ND–14.0)	3	15.0 (12.0–16.9)
	Summerred	4	1.17 (ND–3.5)	4	0.18 (ND–0.70)
Caffeic acid	Aroma	5	3.30 (0.50–6.70)	7	7.63 (0.90–16.0)
	Discovery	3	3.67 (ND–7.00)	3	22.3 (4.00–32.0)
	Gravenstein	5	4.55 (2.00–10.0)	3	6.40 (3.00–12.0)
	Summerred	4	1.93 (ND–5.00)	4	3.15 (0.60–10.0)
Chlorogenic acid	Aroma	5	85.6 (56.0–135)	7	52.6 (25.0–101)
	Discovery	3	177 (151–214)	3	146 (75.0–231)
	Gravenstein	5	100 (46.6–142)	3	139 (122–171)
	Summerred	4	47.4 (22.3–97.0)	4	62.5 (10.0–150)
Procyanidin B2	Aroma	5	17.9 (ND–32.0)	7	15.0 (ND–38.0)
	Discovery	3	31.3 (9.00–43.0)	3	20.7 (12.0–31.0)
	Gravenstein	5	15.5 (ND–33.0)	3	24.9 (21.0–30.6)
	Summerred	4	4.53 (ND–9.00)	4	2.20 (ND–7.00)
Epicatechin	Aroma	5	26.9 (ND–42.1)	7	34.2 (ND–47.8)
	Discovery	3	23.3 (ND–36.0)	3	17.7 (ND–30.0)
	Gravenstein	5	11.9 (ND–26.6)	3	23.4 (23.0–24.2)
	Summerred	4	4.50 (ND–13.5)	4	0.58 (ND–2.30)
Procyanidin C1	Aroma	5	0.75 (ND–3.00)	7	2.00 (ND–5.00)
	Discovery	3	ND	3	1.00 (ND–3.00)
	Gravenstein	5	1.25 (ND–4.00)	3	1.67 (ND–5.00)
	Summerred	4	ND	4	ND
p-Coumaric acid	Aroma	5	0.50 (ND–1.60)	7	1.07 (ND–3.00)
	Discovery	3	2.67 (ND–4.00)	3	2.00 (ND–4.00)
	Gravenstein	5	1.13 (ND–3.00)	3	0.83 (ND–2.50)
	Summerred	4	ND	4	0.08 (ND–0.30)
Procyanidin A2	Aroma	5	0.33 (ND–1.30)	7	0.64 (ND–2.30)
	Discovery	3	ND	3	ND
	Gravenstein	5	ND	3	0.97 (ND–2.90)
	Summerred	4	ND	4	ND

Table 3. *Cont.*

Parameter	Variety	N	Juice (mg L^{-1}) Mean (Min–Max)	N	Cider (mg L^{-1}) Mean (Min–Max)
Ferulic acid	Aroma	5	0.75 (ND–3.00)	7	0.43 (ND–3.00)
	Discovery	3	ND	3	ND
	Gravenstein	5	ND	3	ND
	Summerred	4	ND	4	0.23 (ND–0.90)
Phloridzin	Aroma	5	4.25 (ND–11.0)	7	4.91 (ND–7.00)
	Discovery	3	3.67 (ND–11.0)	3	4.00 (ND–12.0)
	Gravenstein	5	4.10 (ND–12.0)	3	8.03 (ND–14.0)
	Summerred	4	3.33 (ND–10.0)	4	4.00 (ND–11.0)
Total Polyphenols	Aroma	5	236 (68.0–401)	7	186 (49.8–414)
	Discovery	3	269 (238–287)	3	248 (179–360)
	Gravenstein	5	174 (106–238)	3	289 (209–382)
	Summerred	4	77.0 (23.0–124)	4	79.2 (11.0–193)

In our study, we monitored the contents of 13 individual polyphenols in the collected samples using the HPLC-UV method. The sum of the measured individual polyphenols in the juice ranged from 77 to 269 mg L^{-1} (the lowest for 'Summerred' and highest for 'Discovery' and 'Gravenstein') (Table 3). The total polyphenols measured using the *Folin–Ciocâlteu* method are at a higher level than the total polyphenols measured by HPLC. An overestimation of the polyphenol concentration when using the *Folin–Ciocâlteu* method is well known, as reagents react with other reducing compounds, such as reducing sugars, sulfur dioxides, and proteins [31]. In our samples, there were sugars, which are even more abundant in the juice than in the finished cider.

The main individual polyphenols in the juices and ciders were chlorogenic acid (ranging from 22.3 to 214 mg L^{-1} in juices and from 10 to 231 mg L^{-1} in the ciders), followed by catechin, epicatechin, and dimeric procyanidins B1 and B2 (Table 3). This is in accordance with the findings of Maslov Bandic et al., Nogueira et al., and Tsao et al. [1,16,32] who found flavanols (catechin, epicatechin, and procyanidins), hydroxycinnamic acids (chlorogenic acids, 5-caffeoylquinic acid and 4-*p*-coumaroylquinic acid), dihydrochalcones (phloridzin, phloretin glucoside, and xyloglucoside), flavonols (quercetin glucosides), and anthocyanin (cyanidin galactoside) to be the main polyphenols in apples. The concentrations of chlorogenic acid in our ciders were in similar concentration ranges as those in a Croatian study for Idared and Cripps Pink apple ciders [16]. The average level of chlorogenic acid in 'Aroma' and 'Discovery' shows a tendency to decrease from juice to cider, most likely due to oxidation or because of hydrolysis into caffeic acid and quinic acid. An increasing trend (on average) was observed for the other two cultivars, 'Gravenstein' and 'Summerred'. This variation appears to be cultivar- and yeast-dependent [1,16]. Changes in caffeic acid have been observed in all ciders (compared to juices) and are in agreement with the literature reporting on increasing trends of caffeic acids during the fermentation of apple juice [1,27], assumably due to the hydrolysis process of chlorogenic acid, as already mentioned before. With respect to juices from the same cultivar, the most obvious increase was observed in the 'Discovery' ciders whereas for the other samples, it was not as obvious.

Catechin and epicatechin and their dimeric (B and A types) and trimeric derivatives (C type) contribute to the bitter taste of alcoholic beverages like wine [17,23]. In wines, procyanidins with longer chains of epicatechin, catechin, and their derivatives (up to four subunits) are known to be responsible for a bitter taste, while high molecular tannins (procyanidins with >7 subunits of flavan-3-ols) are responsible for astringency. The main proanthocyanidin was B2, ranging in average from 2.2 mg L^{-1} in 'Summerred' ciders up to 24.9 mg L^{-1} in 'Gravenstein' ciders; similar concentrations ranges (slightly lower) were observed for B1 proanthocyanin (2.8 mg L^{-1} in 'Summerred' and 21.7 mg L^{-1} in 'Discovery' ciders.). The level of catechin seems to be quite stable in all cultivars during fermentation, except for 'Gravenstein', where an average increase was observed from the initial 7.9 to the final 15.0 mg L^{-1}. A slightly higher concentration of epicatechin in comparison to catechin was observed in all ciders, but only for 'Aroma' can we see the trends of increase during fermentation from the initial 25.7 to 34.2 mg L^{-1} [1,4].

3.2. Volatile Compounds in Apple Juice and in Cider

In this study, seven ethyl esters of fatty acids, four ethyl esters of branched fatty acids, ten acetate esters, four C6-Alcohols, and two ethyl esters of hydroxycinnamic acids were analysed in juice and finished ciders. The aromatic compounds analysed are listed in Table S5, which include descriptions of odour and odour thresholds from the literature, and the results are presented in Tables 4 and S6.

Table 4. Contents of esters and C6-alcohol (mg L^{-1}). Values are mean, and minimal and maximal values are shown in brackets. Bold values indicate contents over the threshold limit according to Table S5.

		Juice (mg L^{-1})		Cider (mg L^{-1})	
Parameter	**Variety**	**N**	**Mean (Min–Max)**	**N**	**Mean (Min–Max)**
		Ethyl esters of aliphatic acids			
Ethyl propanoate	Aroma	5	105 (39.6–253)	7	270 (84.8–589)
	Discovery	3	141 (38.5–316)	3	543 (338–763)
	Gravenstein	5	65.7 (38.7–122.7)	3	543 (291–897)
	Summerred	4	45.1 (36.0–63.7)	4	219 (121–357)
Ethyl butyrate	Aroma	5	**709 (163–1346)**	7	**648 (240–1149)**
	Discovery	3	**712 (42.0–1887)**	3	**860 (304–1926)**
	Gravenstein	5	**474 (63.0–859)**	3	**796 (610–1149)**
	Summerred	4	**195 (56.1–336)**	4	**882 (409–1805)**
Ethyl hexanoate	Aroma	5	**47.4 (11.8–59.0)**	7	**839 (311–2697)**
	Discovery	3	**53.5 (4.90–94.8)**	3	**964 (798–1114)**
	Gravenstein	5	**49.8 (9.90–66.4)**	3	**1109 (687–1626)**
	Summerred	4	**45.7 (12.1–68.5)**	4	**2533 (630–6055)**
Ethyl octanoate	Aroma	5	1.70 (ND–8.50)	7	**816 (179–2202)**
	Discovery	3	6.17 (ND–18.5)	3	**841 (344–1639)**
	Gravenstein	5	2.32 (ND–11.6)	3	**1258 (514–1885)**
	Summerred	4	5.50 (ND–22.0)	4	**2171 (535–4523)**
Ethyl decanoate	Aroma	5	30.7 (10.0–40.2)	7	**198 (8.60–608)**
	Discovery	3	48.3 (40.8–62.2)	3	**400 (64.0–1005)**
	Gravenstein	5	40.6 (17.4–75.8)	3	**440 (175–910)**
	Summerred	4	38.4 (29.2–53.1)	4	**550 (148–866)**
Ethyl valerate	Aroma	5	0.72 (0.40–1.5)	7	5.69 (2.10–12.5)
	Discovery	3	1.7 (0.40–3.70)	3	5.80 (5.20–6.50)
	Gravenstein	5	1.36 (0.30–2.90)	3	16.0 (7.20–23.3)
	Summerred	4	1.00 (ND–2.5)	4	6.62 (4.80–10.1)
Ethyl dodecanoate	Aroma	5	10.2 (4.10–16.7)	7	61.0 (6.20–239)
	Discovery	3	16.9 (12.0–22.9)	3	190 (23.1–302)
	Gravenstein	5	19.1 (5.80–49.2)	3	96.1 (13.6–248)
	Summerred	4	9.63 (7.90–13.0)	4	105 (23.3–264)
		Ethyl esters of branched acids			
Ethyl isobutyrate	Aroma	5	1.88 (0.50–3.40)	7	**29.4 (5.30–58.4)**
	Discovery	3	4.37 (ND–12.6)	3	**38.7 (34.3–42.0)**
	Gravenstein	5	2.24 (0.80–4.00)	3	**57.3 (41.5–68.0)**
	Summerred	4	0.63 (0.30–1.00)	4	**51.3 (31.1–64.1)**

Table 4. *Cont.*

		Juice (mg L^{-1})		Cider (mg L^{-1})	
Parameter	**Variety**	**N**	**Mean (Min–Max)**	**N**	**Mean (Min–Max)**
Ethyl isovalerate	Aroma	5	**3.06 (1.50–8.60)**	7	**9.34 (4.10–16.6)**
	Discovery	3	**3.87 (1.30–8.60)**	3	**7.43 (4.90–9.30)**
	Gravenstein	5	**3.66 (1.50–11.4)**	3	**12.9 (9.80–16.0)**
	Summerred	4	**3.45 (1.50–8.70)**	4	**7.08 (4.80–9.50)**
Ethyl 2-methylbutyrate	Aroma	5	**75.9 (17.4–165)**	7	**68.5 (24.9–125)**
	Discovery	3	**347 (20.0–986)**	3	**153 (65.5–312)**
	Gravenstein	5	**54.0 (17.4–164)**	3	**56.1 (47.3–61.5)**
	Summerred	4	**40.3 (16.8–101)**	4	**52.9 (29.5–108)**
Ethyl leucate	Aroma	5	4.70 (0.40–17.6)	7	97.0 (25.6–340)
	Discovery	3	482 (2.00–1442)	3	21.1 (11.4–26.4)
	Gravenstein	5	2.56 (1.60–4.10)	3	141 (28.1–320)
	Summerred	4	1.58 (0.70–2.30)	4	30.0 (16.9–40.1)
		Acetate esters			
propyl acetate	Aroma	5	98.5 (23.3–170)	7	86.5 (16.3–114)
	Discovery	3	635 (167–1150)	3	152 (54.9–279)
	Gravenstein	5	10.1 (6.90–18.2)	3	49.0 (19.1–85.7)
	Summerred	4	8.95 (5.10–16.3)	4	74.0 (20.7–130)
Isobutyl acetate	Aroma	5	47.0 (16.6–98.4)	7	124 (30.8–352)
	Discovery	3	64.0 (14.0–131)	3	76.7 (55.3–88.1)
	Gravenstein	5	5.80 (3.50–12.0)	3	70.7 (30.8–107)
	Summerred	4	7.18 (4.10–16.1)	4	173 (47.3–297)
Phenylethyl acetate	Aroma	5	29.8 (9.2–69.2)	7	401 (100–1402)
	Discovery	3	45 (18.2–97.7)	3	282 (228–330)
	Gravenstein	5	18.56 (9.1–32.6)	3	419 (246–570)
	Summerred	4	13.25 (8.1–26.1)	4	532 (221–1120)
Butyl acetate	Aroma	5	**2249 (261–6480)**	7	**1573 (144–2821)**
	Discovery	3	**3286 (1366–4996)**	3	924 (308–1948)
	Gravenstein	5	82.3 (28.9–128.7)	3	231 (108–342)
	Summerred	4	46.02 (26.5–55.1)	4	315 (121–623)
Isoamyl acetate	Aroma	5	**150 (47.6–386)**	7	**1250 (314–2980)**
	Discovery	3	**537 (324–938)**	3	**1128 (769–1525)**
	Gravenstein	5	6.56 (2.90–12.9)	3	**1962 (347–3381)**
	Summerred	4	5.2 (1.70–14.3)	4	**4138 (2590–6243)**
Hexyl acetate	Aroma	5	**796 (ND–2772)**	7	**517 (ND–1171)**
	Discovery	3	**1147 (ND–2650)**	3	**290 (164–382)**
	Gravenstein	5	53.7 (ND–89.2)	3	**307 (78.0–615)**
	Summerred	4	27.5 (ND–40.3)	4	**1241 (715–1817)**
z-3-hexenyl acetate	Aroma	5	3.40 (1.20–8.80)	7	3.73 (0.90–8.30)
	Discovery	3	2.97 (0.90–5.40)	3	1.67 (0.90–2.20)
	Gravenstein	5	0.92 (0.40–1.30)	3	1.80 (1.00–3.00)
	Summerred	4	0.65 (0.30–0.80)	4	2.63 (1.60–4.30)
E-2-hexenyl acetate	Aroma	5	3.96 (1.10–9.30)	7	0.97 (0.50–1.70)
	Discovery	3	6.50 (3.00–11.5)	3	0.83 (0.60–1.10)
	Gravenstein	5	1.36 (0.50–2.30)	3	0.87 (0.60–1.00)
	Summerred	4	0.68 (0.50–1.00)	4	1.05 (0.60–1.70)
Ethyl phenyl acetate	Aroma	5	0.44 (0.30–0.60)	7	2.56 (1.00–4.60)
	Discovery	3	0.40 (0.20–0.70)	3	4.30 (2.80–6.20)
	Gravenstein	5	0.42 (0.20–0.90)	3	6.53 (2.20–13.5)
	Summerred	4	0.30 (0.20–0.40)	4	2.15 (1.70–2.60)

Table 4. *Cont.*

Parameter	Variety	Juice (mg L^{-1}) N	Juice (mg L^{-1}) Mean (Min–Max)	Cider (mg L^{-1}) N	Cider (mg L^{-1}) Mean (Min–Max)
Octyl acetate	Aroma	5	5.06 (0.30–12.2)	7	1.04 (0.30–2.20)
	Discovery	3	6.80 (2.60–15.2)	3	1.40 (0.80–2.30)
	Gravenstein	5	0.64 (0.10–1.10)	3	1.27 (0.90–1.90)
	Summerred	4	0.48 (0.20–0.60)	4	1.25 (1.00–1.70)
Ethyl Esters of hydroxycinnamic acid					
Ethyl cinnamate	Aroma	5	0.76 (0.40–1.50)	7	0.84 (0.40–1.60)
	Discovery	3	0.73 (0.40–1.30)	3	0.63 (0.40–0.80)
	Gravenstein	5	0.60 (0.10–1.10)	3	1.03 (0.30–2.10)
	Summerred	4	0.50 (0.40–0.60)	4	4.47 (0.30–11.6)
Ethyl dihydrocinnamate	Aroma	5	0.22 (0.20–0.30)	7	0.37 (ND–1.10)
	Discovery	3	0.20 (ND–0.40)	3	0.10 (0.10–0.10)
	Gravenstein	5	0.16 (ND–0.40)	3	0.43 (0.10–0.90)
	Summerred	4	0.08 (ND–0.30)	4	0.60 (0.10–1.80)
C6-Alcohols					
Hexanol	Aroma	5	**15,778 (2895–33,297)**	7	7799 (4602–12,850)
	Discovery	3	**13,441 (2556–27,675)**	3	**10,022 (6958–11,849)**
	Gravenstein	5	**15,380 (6166–28,187)**	3	**11,655 (6824–20,464)**
	Summerred	4	**9101 (1752–24,950)**	4	**8183 (6963–9490)**
Z-3-hexenol	Aroma	5	142 (7.60–272)	7	80.8 (10.2–187)
	Discovery	3	29.5 (11.6–42.2)	3	31.9 (17.8–48.3)
	Gravenstein	5	78.4 (11.7–164)	3	58.6 (15.4–106)
	Summerred	4	32.9 (5.60–97.1)	4	34.9 (26.9–39.1)
E-2 hexenol	Aroma	5	648 (84.0–1839)	7	6.57 (ND–14.5)
	Discovery	3	1267 (67.0–2059)	3	13.4 (5.60–21.6)
	Gravenstein	5	634 (75.0–1717)	3	13.4 (8.70–17.8)
	Summerred	4	552 (188–1207)	4	10.4 (7.70–14.7)
E-3 hexenol	Aroma	5	19.8 (ND–37.6)	7	23.1 (6.80–56.8)
	Discovery	3	25.8 (4.80–62.5)	3	35.6 (21.8–51.8)
	Gravenstein	5	18.1 (4.80–32.1)	3	36.0 (11.5–50.4)
	Summerred	4	11.5 (1.90–26.6)	4	37.4 (28.4–57.8)

To evaluate the overall correlation between the measured volatile aroma compounds and apple cultivars, principal component analysis (PCA) was conducted for juice (Supplementary Materials Figure S1) and finished ciders (Figure 1). The PCA plot shows variation between different samples, indicating that production practice impacts the contents of volatile aromas. However, there is a grouping in the PCA plot. The score plot in Figure 1 shows that 'Summerred' has a wide grouping at the top of the score plot, with higher levels of acetate esters (hexyl acetate and isoamyl acetate) and ethyl esters of aliphatic acid (ethyl hexanoate, ethyl octanoate, ethyl decanoate, and ethyl butyrate). Also, some 'Aroma' samples have high levels of these esters. In contrast, 'Gravenstein' and 'Discovery' showed sample clusters in the middle/lower part of the PCA plot with higher levels of hexanol and ethyl isovalerate. One explanation for these wide groupings could be the nitrogen availability from amino acids for the yeasts. Only a few of the cider producers added nitrogen supplements to support fermentation. Nitrogen content and the amino acid profile, which is mainly composed of asparagine, aspartic acid, glutamic acid, proline, and serine in apples [33], are important for ester formations during fermentation [34]. Esters and nitrogen will be a topic for future cider research.

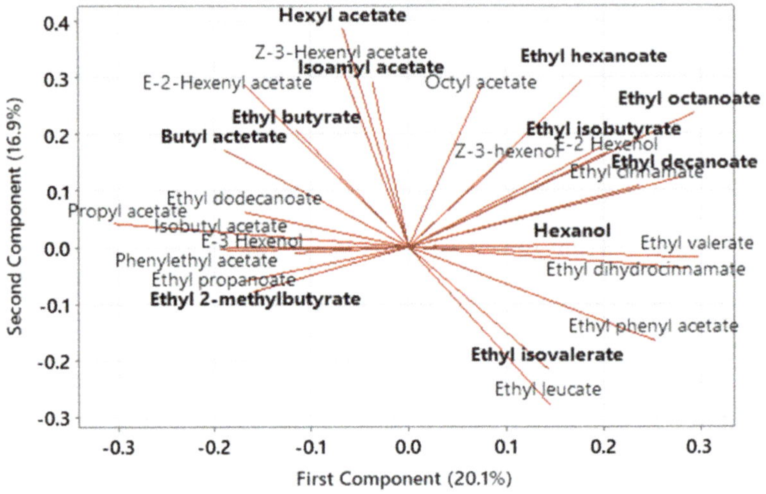

(**A**) Loading plot

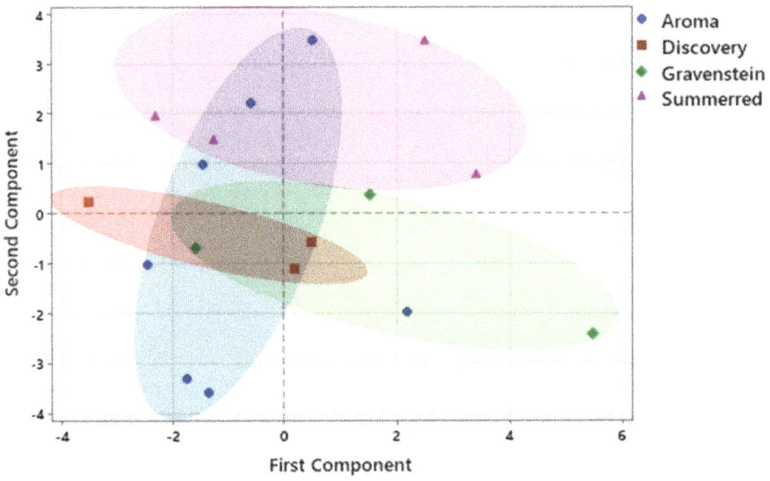

(**B**) Score plot-cultivars

Figure 1. Principal component analysis loading plot (**A**) and score plot for cultivars (**B**) for measured esters and C6-alcohols in the collected samples in 2019, 2020, and 2021 for ciders. Aroma compounds in bold letters are in concentrations over the threshold limit.

The quantitative headspace results for the esters and C6-alcohols are presented in Table 4. In the juice and ciders of the cultivar 'Aroma', 'Discovery', 'Gravenstein', and 'Summerred', the ethyl esters of aliphatic acids ethyl butyrate, ethyl hexanoate, and ethyl octanoate, and the ethyl esters of branched fatty acids isovalerate and ethyl 2-methylbutyrate, were in concentrations over the odour threshold limit (Tables 4 and S5). 'Aroma' and 'Discovery' juice and ciders had butyl acetate concentrations above the threshold limit. In addition, isoamyl acetate, ethyl decanoate, ethyl isobutyrate, and hexyl acetate were present at concentrations above the threshold limit in ciders but not in juice for all cultivars. The volatile aroma in the juices is in accordance with other research findings that contribute to the flavour of apples, which is mainly ethyl butyrate, ethyl hexanoate, butyl acetate, hexyl acetate, 2-methylbutyl acetate, and ethyl 2-methylbutyrate [35]. In general, the total ester content increased (e.g., ethyl isobutyrate, ethyl isovalerate, and ethyl cinnamate) or decreased (e.g., E-2-hexenyl acetate and octyl acetate) during the fermentation process. However, some first increased and then decreased (e.g., ethyl butyrate, ethyl hexanoate, ethyl octanoate, ethyl decanoate, and ethyl dodecanoate).

Fatty acid ethyl esters are esters found in the highest concentrations (ethyl butyrate, ethyl octanoate, ethyl hexanoate, and ethyl decanoate). They give a fruity, apple, and sweet flavour (Tables 4 and S5), and are the result of the esterification process of sugar alcohols with fatty acids during fermentation [36,37]. Among the ethyl esters, ethyl acetate is important for the pleasant and fruity cider flavour at appropriate concentrations [36]; however, this ester was not included in the study.

In this case study, statistical differences in individual esters between the cultivars were not produced because of too many varietal factors in the dataset. However, the trend is that ethyl hexanoate, ethyl octanoate, and ethyl decanoate are highest in 'Summerred' cider compared to the other cultivars (as illustrated in the PCA plot, Figure 1). Similar trends between apple cultivars for ciders have also been identified by Hinkley et al. [38]. The increases in ethyl esters during fermentation are many-folded compared to juice and are affected by the apple variety, pH, sugar, nitrogen, and amino acid composition [38]. Different apple cultivars have different chemical compositions and thus different ester precursors. The precursors for different aromatic compounds are amino acids, carbohydrates, and fatty acids [39]. The YAN concentration in 'Summerred' was, on average, higher in the juices (70 mg L^{-1}) than that of the other cultivars (23–38 mg L^{-1}) (Table 2). However, further investigation of the amino acid composition is needed to determine the relationship between the YAN/amino acid profile and ester production. The same is true for yeast strains and other technological factors.

The aroma profile of apples is dependent on the degree of maturation. Aldehydes are the most abundant aromatic compounds before maturation, and esters are the main group of aromatic compounds in ripe apples. In our study, the ripeness of apple fruit was not identified as the juice samples were delivered by producers. According to Rosend [40], unripe apple fruits contain isoamyl acetate, hexyl acetate, butyl acetate, and ethyl hexanoate.

C6-alcohols result in a green flavour of cider, and at high concentrations, this can be negative. C6-alcohols are produced during the processing of the fruit via the bioremediation of unsaturated fatty acids, while during fermentation, the C6-alcohol content normally decreases [41]. The hexanol concentrations identified in this study varied a lot between the producers, years, and cultivars. Some increase, some decrease, and some decrease and increase. In apple juice, hexanol is an indicator of ripeness [42].

4. Conclusions

The primary objective of this case study was to investigate the chemical makeup of ciders with the PDO designation Cider from Hardanger by analysing selected chemical parameters in juice and cider samples collected from 10 producers in Hardanger, Norway. We measured high acidity in all ciders, with malic acid being the major one. Cultivars responded differently to fermentation changes in malic acid. Thus, on average, a 70% and 9% decrease in the initial malic acid was observed in 'Gravenstein' and 'Summered', respectively. The concentration of individual polyphenols varied among single cultivar samples, with chlorogenic acid being the most prevalent. Our data about total phenols in juices and ciders agree with previous findings on differences in the total phenol contents among apple juices and ciders from Hardanger. However, our study showed different changes (an increase or decrease) in total phenols during fermentation among cultivars monitored.

The field sampling involved many sources of variation on selected secondary metabolites, enabling conclusions only as trends. However, to understand the big set of data, we performed a PCA on aroma compounds that are behind the aromatic profile of ciders.

The score plot for apple cultivars showed that the cultivars have a wide range of results, indicating that technological factors (e.g., the chosen yeast strain and temperature) and the seasonal variation of the fruit (e.g., maturity degree, climate, and agronomical factors) impact the composition of esters in ciders. However, 'Summerred' showed tight clusters at the top of the PCA plot, indicating that the aroma profiles of ciders from 'Summerred' apple have more of the esters hexyl acetate, isoamyl acetate, ethyl hexanoate, ethyl octanoate, ethyl decanoate, and ethyl butyrate than in other ciders. The reason for this difference could be the naturally present sources of nitrogen in the cultivars; however, more research efforts employing controlled experiments are needed to understand and confirm the observed differences. This study found the presence of several ethyl esters and acetate esters at concentrations above the odour threshold, which contribute to the fruity flavour of the ciders.

This research emphasizes the significance of identifying key chemical components that influence the sensory quality of Cider from Hardanger in order to establish product differentiation and consistent quality. Variations in ciders are expected to arise due to factors such as cultivars, production methods, fermentation conditions, and ageing.

Supplementary Materials: The following supporting information can be downloaded at: https://www.mdpi.com/article/10.3390/beverages10030073/s1. Table S1: HPLC-UV calibration curves and concentration ranges of individual polyphenols; Table S2: Targeted MS^2 spectra and UV-Vis spectra of standard phenolic compounds; Table S3: Measured values for titratable acidity, acids (malic acid, lactic acid, citric acid, tartaric acid), and sugars (fructose, glucose, and sucrose), ethanol content, and yeast assimilable nitrogen (YAN) for Gul Gravenstein, Rubinstep, and blends of Gravenstein/cider apples, Gravenstein/Elstar, Gravenstein/Signe Tillisch; Table S4: Total phenol content in gallic acid equivalent (GAE mg L^{-1}) and polyphenol profile from HPLC (mg L^{-1}) in apple juice, and finished cider of Gul Gravenstein, Rubinstep, and blends of Gravenstein/cider apples, Gravenstein/Elstar, Gravenstein/Signe Tillisch; Table S5. Odour description and odour threshold from the literature; Table S6: Contents of esters and C6 alcohols in juice and cider of Gul Gravenstein, Rubinstep, and blends of Gravenstein/cider apples, Gravenstein/Elstar, Gravenstein/Signe Tillisch collected i Hardanger in 2019, 2020 and 2021. Figure S1: Principal component analysis loading plot (A) and score plot for cultivars (B) for measured esters and C6-alcohols in the collected samples in 2019, 2020, and 2021.

Author Contributions: Conceptualization. I.Ø. and B.M.V.; methodology. M.M., A.A., T.R.V., I.Ø. and B.M.V.; software. I.Ø.; formal analysis. T.R.V., M.M., U.Č. and A.A.; investigation. I.Ø. and B.M.V.; resources. I.Ø.; data curation. I.Ø.; writing—original draft preparation U.Č., M.M., A.A., I.Ø. and B.M.V.; writing—review and editing. I.Ø. and B.M.V.; visualization. I.Ø.; supervision. B.M.V.; project administration. I.Ø.; funding acquisition. I.Ø. All authors have read and agreed to the published version of the manuscript.

Funding: This research was funded by Vestland Regional Research Fund (RFF Vestland) grant number 299284 and Slovenian Research and Innovation Agency—research programme P1-0005 (Functional food and food supplements).

Data Availability Statement: The data presented in this study are available on request from the corresponding author.

Acknowledgments: The authors would like to thank Marianne Hotle for taking care of the samples at NIBIO Ullensvang.

Conflicts of Interest: The authors declare no conflicts of interest. The funders had no role in the design of the study; in the collection. analyses. or interpretation of data; in the writing of the manuscript; or in the decision to publish the results.

References

1. Nogueira, A.; Guyot, S.; Marnet, N.; Lequéré, J.M. Effect of Alcoholic Fermentation in the Content of Phenolic Compounds in Cider Processing. *Braz. Arch. Biol. Technol.* **2008**, *51*, 1025–1032. [CrossRef]
2. Ferremi Leali, N.; Salvetti, E.; Luzzini, G.; Salini, A.; Slaghenaufi, D.; Fusco, S.; Ugliano, M.; Torriani, S.; Binati, R.L. Differences in the Volatile Profile of Apple Cider Fermented with Schizosaccharomyces pombe and Schizosaccharomyces japonicus. *Fermentation* **2024**, *10*, 128. [CrossRef]

3. Øvsthus, I.; Martelanc, M.; Radovanović Vukajlović, T.; Lesica, M.; Butinar, L.; Mozetič Vodopivec, B.; Antalick, G. Chemical composition of apple cider: A comparative study of Norwegian and French ciders. *Acta Hortic.* **2024**, *1387*, 303–310. [CrossRef]
4. Česnik, U.; Martelanc, M.; Øvsthus, I.; Radovanović Vukajlović, T.; Hosseini, A.; Mozetič Vodopivec, B.; Butinar, L. Functional Characterization of Saccharomyces Yeasts from Cider Produced in Hardanger. *Fermentation* **2023**, *9*, 824. [CrossRef]
5. Qin, Z.; Petersen, M.A.; Bredie, W.L.P. Flavor profiling of apple ciders from the UK and Scandinavian region. *Food Res. Int.* **2018**, *105*, 713–723. [CrossRef] [PubMed]
6. Vangdal, E.; Kvamm-Lichtenfeld, K. Ciders produced from Norwegian fresh consumption apple cultivars. *Acta Hortic.* **2018**, *1205*, 527–532. [CrossRef]
7. Wicklund, T.; Skottheim, E.R.; Remberg, S.F. Various Factors Affect Product Properties in Apple Cider Production. *Int. J. Food Stud.* **2020**, *9*, a719. [CrossRef]
8. Fotirić Akšić, M.; Zagorac Dabić, D.; Gašić, U.; Tosti, T.; Natić, M.; Meland, M. Analysis of Apple Fruit (Malus × domestica Borkh.) Quality Attributes Obtained from Organic and Integrated Production Systems. *Sustainability* **2022**, *14*, 5300. [CrossRef]
9. McGinley, M.; Nguyen, N.; Mott, J. Real-Time Response to Bacteria Infection of Bioethanol Fermentation Using a Short Rezex™ ROA Column. *Appl. Noteb.* **2008**.
10. Sadler, G.; Murphy, P. PH and Titratable Acidity. In *Food Analysis*; Springer: Boston, MA, USA, 2010; pp. 219–238. [CrossRef]
11. Martelanc, M.; Antalick, A.; Radovanović Vukajlović, T.; Mozetič Vodopivec, B.; Sternad Lemut, M.; Hosseini, A.; Obradović, V.; Mesić, J.; Butinar, L. Aromatic Characterization of Graševina Wines from Slavonia and Podunavlje Sub-Regions. *Beverages* **2024**, *10*, 24. [CrossRef]
12. Singleton, V.L.; Orthofer, R.; Lamuela-Raventós, R.M. Analysis of total phenols and other oxidation substrates and antioxidants by means of folin-ciocalteu reagent. *Methods Enzymol.* **1999**, *299*, 152–178. [CrossRef]
13. Ivanović, M.; Albreht, A.; Krajnc, P.; Vovk, I.; Islamčević Razboršek, M. Sustainable ultrasound-assisted extraction of valuable phenolics from inflorescences of *Helichrysum arenarium* L. using natural deep eutectic solvents. *Ind. Crops Prod.* **2021**, *160*, 113102. [CrossRef]
14. Whiting, G.C. Organic Acid Metabolism of Yeasts During Fermentation of Alcoholic Beverages—A Review. *J. Inst. Brew.* **1976**, *82*, 84–92. [CrossRef]
15. Satora, P.; Sroka, P.; Duda-Chodak, A.; Tarko, T.; Tuszyński, T. The profile of volatile compounds and polyphenols in wines produced from dessert varieties of apples. *Food Chem.* **2008**, *111*, 513–519. [CrossRef] [PubMed]
16. Maslov Bandić, L.; Žulj Mihaljević, M.; Fruk, G.; Babojelić Skendrović, M.; Jemrić, T.; Jeromel, A. The profile of organic acids and polyphenols in apple wines fermented with different yeast strains. *J. Food Sci. Technol.* **2019**, *56*, 599–606. [CrossRef] [PubMed]
17. Moreno, J.; Peinado, R. *Enological Chemistry*, 1st ed.; Academic Press: London, UK; Waltham, MA, USA, 2012; ISBN 978-0-12-388438-1.
18. Zhang, H.; Zhou, F.; Ji, B.; Nout, M.J.R.; Fang, Q.; Yang, Z. Determination of organic acids evolution during apple cider fermentation using an improved HPLC analysis method. *Eur. Food Res. Technol.* **2008**, *227*, 1183–1190. [CrossRef]
19. Whiting, G.C.; Coggins, R.A. Organic acid metabolism in cider and perry fermentations. III.—Keto-acids in cider-apple juices and ciders. *J. Sci. Food Agric.* **1960**, *11*, 705–709. [CrossRef]
20. Guyot, S.; Marnet, N.; Sanoner, P.; Drilleau, J.-F. Variability of the Polyphenolic Composition of Cider Apple (*Malus domestica*) Fruits and Juices. *J. Agric. Food Chem.* **2003**, *51*, 6240–6247. [CrossRef] [PubMed]
21. Boulton, R.B.; Singleton, V.L.; Bisson, L.F.; Kunkee, R.E. *Principles and Practices of Winemaking*; Chapman and Hall: Boca Raton, FL, USA, 1999.
22. Reynolds, A.G. *Managing Wine Quality*, 2nd ed.; Woodhead Publishing: Duxford, UK, 2022; Volume 2: Oenology and Wine Quality, pp. 527–531.
23. Waterhouse, A.; Sacks, G.L.; Jeffery, D.W. *Flavan-3-ols and Condensed Tannin. Understanding Wine Chemistry*; Wiley: Hoboken, NJ, USA, 2016; pp. 117–126. [CrossRef]
24. Oszmianski, J.; Lee, C.Y. Inhibition of polyphenol oxidase activity and browning by honey. *J. Agric. Food Chem.* **1990**, *38*, 101892–101895. [CrossRef]
25. Alberti, A.; Machado Dos Santos, T.P.; Ferreira Zielinski, A.A.; Eleutério Dos Santos, C.M.; Braga, C.M.; Demiate, I.M.; Nogueira, A. Impact on Chemical Profile in Apple Juice and Cider Made from Unripe, Ripe and Senescent Dessert Varieties. *LWT Food Sci. Technol.* **2016**, *65*, 436–443. [CrossRef]
26. Wicklund, T.; Guyot, S.; Le Quéré, J.-M. Chemical Composition of Apples Cultivated in Norway. *Crops* **2021**, *1*, 8–19. [CrossRef]
27. Kunicka-Styczyńska, A.; Pogorzelski, E. L-Malic Acid Effect on Organic Acid Profiles and Fermentation By-Products in Apple Wines. *Czech J. Food Sci.* **2009**, *27*, S228–S231. [CrossRef]
28. Everette, J.D.; Bryant, Q.M.; Green, A.M.; Abbey, Y.A.; Wangila, G.W.; Walker, R.B. Thorough Study of Reactivity of Various Compound Classes toward the Folin-Ciocalteu Reagent. *J. Agric. Food Chem.* **2010**, *58*, 8139–8144. [CrossRef] [PubMed]
29. Tsao, R.; Yang, R.; Xie, S.; Sockovie, E.; Khanizadeh, S. Which polyphenolic compounds contribute to the total antioxidant activities of apple? *J. Agric. Food Chem.* **2005**, *53*, 4989–4995. [CrossRef] [PubMed]
30. Guth, H. Quantitation and Sensory Studies of Character Impact Odorants of Different White Wine Varieties. *J. Agric. Food Chem.* **1997**, *45*, 3027–3032. [CrossRef]
31. Ferreira, V.; López, R.; Cacho, J.F. Quantitative Determination of the Odorants of Young Red Wines from Different Grape Varieties. *J. Sci. Food Agric.* **2000**, *80*, 1659–1667. [CrossRef]

32. Bingman, M.; Stellick, C.; Pelkey, J.; Scott, J.; Cole, C. Monitoring Cider Aroma Development throughout the Fermentation Process by Headspace Solid Phase Microextraction (HS-SPME) Gas Chromatography–Mass Spectrometry (GC-MS) Analysis. *Beverages* **2020**, *6*, 40. [CrossRef]
33. Santos, C.; Pietrowski, G.; Braga, C.; Rossi, M.; Ninow, J.; Santos, T.; Wosiacki, G.; Jorge, R.; Nogueira, A. Apple Aminoacid Profile and Yeast Strains in the Formation of Fusel Alcohols and Esters in Cider Production. *J. Food Sci.* **2015**, *80*, C1170–C1177. [CrossRef]
34. Belda, I.; Ruiz, J.; Esteban-Fernández, A.; Navascués, E.; Marquina, D.; Santos, A.; Moreno-Arribas, M.V. Microbial Contribution to Wine Aroma and Its Intended Use for Wine Quality Improvement. *Molecules* **2017**, *22*, 189. [CrossRef] [PubMed]
35. Espino-Díaz, M.; Sepulveda, D.R.; Aguilar, G.; Olivas, G. Biochemistry of Apple Aroma: A Review. *Food Technol. Biotechnol.* **2016**, *54*, 375–397. [CrossRef] [PubMed]
36. Peng, B.; Li, F.; Cui, L.; Guo, Y. Effects of Fermentation Temperature on Key Aroma Compounds and Sensory Properties of Apple Wine. *J. Food Sci.* **2015**, *80*, S2937–S2943. [CrossRef] [PubMed]
37. Antón, M.J.; Suárez Valles, B.; García Hevia, A.; Picinelli Lobo, A. Aromatic Profile of Ciders by Chemical Quantitative, Gas Chromatography-Olfactometry, and Sensory Analysis. *J. Food Sci.* **2014**, *79*, S92–S99. [CrossRef] [PubMed]
38. Hinkley, J.L.; Bingman, M.T.; Lee, J.S.; Bradley, C.P.; Cole, C.A. Volatile Profile Survey of Five Apple Varieties Grown in Southwest Colorado from Juice to Finished, Dry-Hopped Cider. *J. Am. Soc. Brew. Chem.* **2023**, *81*, 131–140. [CrossRef]
39. Sanz, C.; Olias, J.M.; Perez, A.G. Aroma biochemistry of fruits and vegetables. In *Phytochemistry of Fruit and Vegetables: Proceedings of the Phytochemical Society of Europe*; Tomás-Barberán, F.A., Robins, R.J., Eds.; Oxford University Press: Oxford, UK, 1997; pp. 125–156.
40. Rosend, J.; Kuldjärv, R.; Rosenvald, S.; Paalme, T. The Effects of Apple Variety, Ripening Stage, and Yeast Strain on the Volatile Composition of Apple Cider. *Heliyon* **2019**, *5*, e01953. [CrossRef] [PubMed]
41. Aung, M.T.; Lee, P.-R.; Yu, B.; Liu, S.-Q. Cider Fermentation with Three Williopsis Saturnus Yeast Strains and Volatile Changes. *Ann. Microbiol.* **2015**, *65*, 921–928. [CrossRef]
42. Sapers, G.M.; Abbott, J.; Massie, O.; Watada, A.; Finney, E.E. Volatile Composition of Mcintosh Apple Juice as A Function of Maturity and Ripeness Indices. *J. Food Sci.* **1977**, *42*, 44–47. [CrossRef]

Disclaimer/Publisher's Note: The statements, opinions and data contained in all publications are solely those of the individual author(s) and contributor(s) and not of MDPI and/or the editor(s). MDPI and/or the editor(s) disclaim responsibility for any injury to people or property resulting from any ideas, methods, instructions or products referred to in the content.

fermentation

Article

Effects of Aging in Wood Casks on Anthocyanins Compositions, Volatile Compounds, Colorimetric Properties, and Sensory Profile of *Jerez* Vinegars

Reyhan Selin Uysal

Department of Genetics and Bioengineering, Istanbul Bilgi University, Istanbul 34060, Turkey; selin.uysal@bilgi.edu.tr

Abstract: The *Jerez (Sherry)* vinegars, including *Vinagre de Jerez, Reserva,* and *Gran Reserva*, are crafted from *Sherry* wines and are protected under the Denomination of Origin in Spain. The aim of this study was to (i) characterize the physicochemical properties and composition; (ii) investigate the impact of the aging process on color properties, phenolics, volatile compounds, and sensorial profiles; and (iii) find a marker for tracing the authenticity of non-aged (*Apto*) and aged *Jerez* vinegars. The phenolic components were identified through LC-MS/MS, whereas the volatile compounds were examined using the GC-MS/MS technique. As the aging progressed, a decrease was observed in the levels of flavonol and phenolic acids, with anthocyanin components being undetectable in non-aged and aged samples. In the *Gran Reserva* variety, 2-methylbutyl acetate, acetic acid, and ethanol emerged as the predominant volatile substances. The presence of oaklactone and ethyl butanoate components served as marker substances to authenticate the *Gran Reserva*. Additionally, alterations in color properties were noted, marked by a decrease in yellow content and an increase in the red component depending on aging. Furthermore, novel sensory descriptors, such as vanilla, clove, woody, and nutty notes, and winy character emerged in the samples with prolonged aging.

Keywords: *Sherry* vinegar; LC-MS/MS; GC-MS/MS; phenolics; acetification; oxidation; copolymerization; *Maillard* reaction; lexicon; authentication

1. Introduction

Wine vinegar is an enological product that is highly appreciated by the consumer, and it has different varieties depending on the raw materials used and the preparation process. *Jerez* (also known as *Sherry*) wine vinegar is one of the most well-known products of this type in the world [1]. *Jerez* vinegar (JV) is a very valuable and high-quality product produced in the Vinagre de Jerez Denomination of Origin (DO) (geographical indication of origin "*Vinagre de Jerez*") region in the southwest of Spain, protected by Spain and Europe, and of primary importance in the region of origin [2]. Besides JV, the other wine vinegars produced and marketed under a specific Protected Designation of Origin (PDO) are "*Vinagre de Condado de Huelva*" and "*Vinagre de Montilla-Moriles*" [3]. JV, produced under geographically indicated regions, is obtained from *Sherry* wines produced from Palomino, Pedro Ximenez, and/or Muscatel grapes produced by following the traditional aging method called "*Soleras and Criaderas*" [4].

Jerez vinegars have an extraordinary organoleptic property (bright, highly aromatic, and rich) because of the traditional production technique. This special production technique includes a dynamic aging system. In this method, *Jerez vinegars* are aged in wooden barrels with an oxidative process simultaneously with a slow acetification process. As aging and acetification occur simultaneously in oak barrels, major changes in composition and sensory properties are observed throughout the process. Meanwhile, depending on the aging period in barrels, *Jerez vinegars* are called "*Vinagre de Jerez (VJ)*" for at least 6 months to 2 years, "*Reserva*" for at least 2 years to 10 years, and "*Gran Reserva*" for at least 10 years and more [5]. Analyzing the chemical changes that occur in vinegars with such a long aging process and evaluating their effects on sensory properties are also particularly important in terms of studying and revealing the effects of acetification and oxidation processes in vinegar.

In the analysis of aged vinegars, volatile compounds serve as significant indicators of their distinctive properties [6]. In this context, to characterize the olfactory impact of odorants, gas chromatography–mass spectrometry–olfactometry (GC-MS-O) has been studied for just the "*Reserva*" category (aged up to two years) [7], for Shanxi aged vinegar [8], and for Italian balsamic vinegar [9]. On the other hand, the volatile compounds and aroma composition have been analyzed from the "*VJ*" type (aged up to six months) using the techniques of headspace sorptive extraction (HSSE)-GC-MS, liquid–liquid extraction GC-MS (LLE-GC-MS), and GC-FID by several pieces of research [10,11].

Moreover, volatile and aroma compositions, anthocyanin profiles of vinegar during fermentation, and the aging process in particular remain unknown [12]. Compared to volatile component analysis, anthocyanin analysis studies in vinegar are less common. Anthocyanin composition was analyzed in Cabernet Sauvignon wine vinegar [12], Condado de Huelva, and Montilla Moriles [13] by means of liquid chromatography coupled to a diode array detector.

On the other hand, colorimetric characteristics and sensorial profiles of aged vinegars are another component for identifying aged vinegars. Especially the sensorial profile of aged vinegars is affected highly by the fermentation process, oxidation process, and the structure of storage containers, such as different kinds of wood casks [12,14–19]. In addition, colorimetric features from the physicochemical qualities of aged vinegars show a significant change in terms of visual color because of the transformation of bioactive compounds [1,15,20–22]. As a result, during the microbiological and aging processes of wine vinegars, they undergo various influences and reactions, resulting in a new product with a transformed character. As mentioned earlier, some of these changes have been examined in a few *Jerez* samples, specifically looking at volatile compounds in vinegars aged for 6 months and 2 years. However, essential bioactive components such as anthocyanins, additional physicochemical properties (colorimetric), and the sensory profile of both non-aged and aged *Jerez* vinegars remain unexplored in the literature despite their importance for thorough characterization.

Consequently, the objectives of the study were (i) to determine the content and composition of coloring agents (red, yellow, and blue), anthocyanins, flavonol, volatile compounds, and sensory profiles in non-aged (*Apto*) and all types of *Jerez* vinegars (*VJ*, *Reserva*, and *Gran Reserva*) aged for different years in oak barrels; (ii) to show the effect of acetification and the aging process on other colorimetric characteristics (color intensity, color density, and tonality), phenolics, volatile compounds, and sensory profiles; and (iii) to propose a new approach to the authentication of *Jerez* vinegars by following new distinctive markers of aged vinegars.

2. Materials and Method

2.1. Reagents and Chemicals

Methanol, acetonitrile (all LC-MS grade, ≥99.9%), sodium chloride and 1-hexanol (all reagent grade, 98%) for chromatographic analysis were purchased from Panreac Applichem (Darmstadt, Germany). Formic acid (LC-MS grade, ≥99.9%) was supplied from VWR Chemicals BDH (Leuven, Belgium). Anthocyanins: Cyanidin-3-*O*-glucoside, delphinidin-3-*O*-glucoside, malvidin-3-*O*-glucoside, pelargonidin-3-*O*-glucoside, peonidin-3-*O*-glucoside, petunidin-3-*O*-glucoside, cyanidin-3,5-di-*O*-glucoside, delphinidin-3,5-di-*O*-glucoside, malvidin-3,5-di-*O*-glucoside, pelargonidin-3,5-di-*O*-glucoside, peonidin-3,5-di-*O*-glucoside; malvidin-3-*O*-galactoside, cyanidin-3-*O*-rutinoside, cyanidin-3-*O*-arabinoside, and pelargonidin-3-*O*-rutinoside; *flavonol*: myricetin-3-*O*-glucoside, quercetin-3-*O*-glucoside, quercetin-3-*O*-galactoside, quercetin-3-*O*-rutinoside, quercetin-3-*O*-glucuronide, and quercetin-3-*O*-glucopyranoside; *flavone:* luteolin-7-*O*-glucoside; *phenolic acid:* caffeic acid, neochlorogenic acid, chlorogenic acid, and gallic acid; *flavanone*: eriotricin, hesperidin, cynarine, and rutin; and *stilbene*: (E)-resveratrol were purchased from Cymit Química (Barcelona, Spain).

2.2. Vinegar Samples

Four traditional *Jerez* wine vinegar samples non-aged (young) and aged at different periods (VJ (*Crianza*), *Reserva*, and *Gran Reserva*) were kindly provided by a *Jerez* vinegar producer in the Jerez de la Frontera district of Cádiz, Spain. The *Jerez* vinegar was obtained from *Sherry* wine produced with Palomino grapes. All of the samples were provided in duplicate (from two batches of the same kind of vinegar).

All *Jerez* vinegar samples can be produced only using the origin of *Sherry wines* under the Protected Designation of Origin (PDO) and European system with the application of different aging periods defined by the PDO. *Jerez* vinegar is aged in American oak wooden casks according to a process known as the "*Criadera y Solera*" system, and the analyzed samples were aged using this technique. According to the PDO, the aging period of *Sherry* wine vinegars must be processed as follows: (i) VJ (*Crianza*) vinegars must have an aging period of at least 6 months to 2 years with residual alcohol <3% (v/v) and total acidity >70 g/L acetic acid; (ii) the other sample, *Reserva* vinegar, can have a range from 2 to 10 years for the aging period with residual alcohol <3% (v/v) and total acidity >70 g/L acetic acid; and for the last Jerez vinegar, (iii) *Gran Reserva* must be aged for more than 10 years with less than 3% (v/v) residual alcohol and total acidity >80 g/L acetic acid.

2.3. Anthocyanins and the Other Polyphenol Compositions (LC-MS/MS)

Vinegar samples were diluted (1:1, w/v) by the addition of an extractant composed of methanol/water/formic acid (80:19.9:0.1, v/v). After that, diluted vinegar samples were filtered through a 0.45 μm pore size membrane filter before injection in the LC-MS/MS system. Anthocyanins and polyphenol composition analyses were performed on a Shimadzu LC-MS/MS 8050 triple quadrupole mass spectrometer (Shimadzu Corporation, Kyoto, Japan) equipped with an electrospray ionization (ESI) as the source operating in negative and positive modes. The method of LC-MS/MS analysis was carried out according to the procedure performed in an article by Uysal et al. [23]. The mobile phase for anthocyanins and polyphenols consisted of two solvents: (i) Solvent A, water/formic acid (99.9:0.1, v/v), and (ii) Solvent B, acetonitrile/formic acid (99.9:0.1, v/v). Anthocyanin compounds were eluted under the following conditions: 0.4 mL min^{-1} flow rate and 50 °C, isocratic conditions for 2 min with 95% A, from 2 to 10 min linear gradient of 5% to 95% acetonitrile with 0.1% formic acid (B); isocratic conditions with solvent B continued between 10 and 11 min and then returned to initial conditions of 95% A for 1 min and isocratic conditions with 95% of 1% aqueous formic acid for 4 min followed by washing and reconditioning the column. On the other hand, the elution conditions of polyphenol compounds were as follows: 0.4 mL min^{-1} flow rate and 40 °C, isocratic conditions for 2 min with 90% A, from 2 to 17 min linear gradient of 10% to 100% acetonitrile with 0.1% formic acid (B), and then returned to initial conditions of 90% A for 4 min, followed by washing and reconditioning the column. The sample volume injected was 10 μL both anthocyanin and polyphenol compounds. Two MS experiments were implemented for anthocyanins in positive mode and for polyphenols in negative mode before and after fragmentation. The MS conditions were as follows: Analyses were performed with full scan mode, and data-dependent MS scanning from *m/z* 100 to 1800, via collision-induced fragmentation experiments carried out using argon as the collision gas. The capillary voltage was 4.0 kV. The capillary temperature was set to 300 °C, and the source heater temperature was 250 °C, while the desolvation gas flow rate was 160 L/h. Characterization of the single components was performed considering retention time and the accurate molecular masses. In addition, concentrations of the individual compounds were determined by external calibration with corresponding standard compounds. The stock solution of all individual standards was prepared with the extractant in the range of concentrations of 0.1, 0.3, 0.5, 0.8, and 1 mg/L. The limit of detection (LOD) and limit of quantification (LOQ) values for the present chromatographic method were determined at a signal-to-noise ratio (S/N) of 3 and 10, respectively. The analysis for anthocyanin and polyphenols was performed in duplicate for each sample.

2.4. Volatile Analysis by Gas Chromatography–Mass Spectrometry (GC-MS)

The volatile substances of vinegar samples were determined by using a gas chromatography (GC-2030)–mass spectrometer (GCMS-TQ8040 NX, Shimadzu Corporation, Kyoto, Japan) coupled with the head space solid-phase microextraction (HS-SPME) technique. Volatile analysis was performed with a procedure described in one study [24] with a small modification. Before analyzing, samples were prepared briefly as follows: A 10 mL aliquot of the vinegar sample (10 mL) was transferred into a 20 mL vial, followed by the addition of 100 μL of 1-hexanol as internal standard. The sample was saturated by adding 1 g of sodium chloride). After that, the vial was tightly capped with a polypropylene cap PTFE/silicone septum. The sample was extracted at a constant temperature of 40 °C for 40 min with agitation (250 rpm) in an AOC-6000 Plus autosampler (Shimadzu Corporation, Kyoto, Japan) with SPME capability. Separation and identification of analytes were performed with both InertCap Pure-WAX (30 m × 0.25 mm × 0.25 μm) and Sapiens X5MS (30 m × 0.25 mm × 0.25 μm) capillary columns (Teknokroma, Barcelona, Spain). Following extraction, volatile analytes were desorbed from DVB/CAR/PDMS fiber coating in the injection port of the GC at 210 °C (Pure-WAX) and at 230 °C (X5MS) for 1 min in 1:10 split mode. The condition of column temperature for both columns was adjusted as follows: initial temperature of 50 °C, held for 1 min, followed by increases to 100 °C at a rate of 2 °C/min, increasing by 3 °C/min to 180 °C, and increases from 180 to 230 °C at a rate of 20 °C/min, and then this temperature was held constant for 5 min. Helium was used as the carrier gas at 0.6 mL/min. The MS conditions of the Pure-WAX column were as follows: capillary direct interface temperature, 240 °C; ion source temperature, 210 °C; ionization mode was positive; and *m*/*z* scan range, 40–400 amu. The differences in the injection port and MS conditions of the other column were as follows: capillary direct interface temperature, 280 °C, and ion source temperature, 230 °C.

Identification of the compounds was performed based on the retention index, comparison of EI mass spectra, and C6–C40 alkane (Sigma-Aldrich, Steinheim, Germany) mixing. Each volatile compound was determined by using the relative area to the internal standard (1-hexanol, 1000 mg/L) of each compound. The relative concentration of an analyte = peak area of compound × concentration of internal standard/peak area of the internal standard. The analysis of each sample was performed in triplicate.

2.5. Chromatic Characteristics of Sherry Vinegars

The chromatic characteristics of (i) tonality, (ii) color intensity, and (iii) color density in the *Jerez* vinegar samples were examined according to the procedure described in [25]. Absorbance measurements were taken using a UV–visible spectrophotometer (UV-1280, Shimadzu Corporation, Kyoto, Japan). Colorimetric properties of vinegar samples were measured at wavelengths of 420 nm (yellow components), 520 nm (red components), and 620 nm (blue components). From these measurements, the color intensity (IC), tonality (T), and color density (D) were calculated with the following formulas:

$$IC = A420 + A520 + A620$$

$$T = A420/A520$$

$$D = A420 + A520$$

The Glories color index percentages of yellow (Y %), red (R %), and blue (B %) of the vinegars were calculated using the data obtained at 420, 520, and 620 nm [26]. The measurements of colorimetric properties were performed in triplicate.

2.6. Descriptive Sensory Analysis with Trained Personnel

Nine panelists consisting of five females and four males, aged between 28 and 60 years, assessed *Jerez* vinegar samples at the facilities of the Food Quality and Safety (CSA) research group of Universidad Miguel Hernández de Elche (UMH). The descriptors (attributes) and questionnaire have been created according to the Foundation of Organization of Evaluation Conformity and Food Certification (OECCA, Cádiz, Spain) [27] and a study described in an article by Tesfaye et al. [2]. While samples of 20 mL of vinegar diluted three times with water were served in a non-transparent standard wine-tasting cup for odor analysis (detection of volatile compounds with vinegar outside the mouth), samples of 20 mL of vinegar diluted five times with water were also served for flavor analysis [combination of odor, aroma (detection of volatile compounds with vinegar inside the mouth), basic tastes, and chemical feeling factors] and global attributes. On the other hand, for appearance analysis, samples were served in transparent cups. Sensorial analyses were performed in room conditions (18–20 °C) and under white light. The samples were coded with 3-digit numbers and served one by one with the appropriate questionnaire and randomly, waiting for 5 min between samples. For palate cleansing, water and unsalted crackers were offered to the panelists between the samples. The attributes of sensorial evaluation were as follows: odor (odor intensity, vinegar ID, winy character, raisin, ethyl acetate (chemical), alcohol/liquor, woody, fruity, spicy, vanilla, clove, toasted, nuts, and leather/old); flavor (odor intensity, vinegar ID, winy character, raisin, ethyl acetate (chemical), alcohol/liquor, woody, fruity, spicy, vanilla, clove, toasted, nuts, and leather/old); basic tastes (sweetness, sourness, and bitterness); chemical sensations (astringent and pungent); global attributes (aftertaste); appearance (color and untuoso (texture)); and defects (dirty (sucio), bacteria, cheese, and sawdust (wood shavings)). Panelists used a 0 to 10 scale for assessment, where 10 is extremely high intensity and 0 is extremely low intensity or unnoticeable.

2.7. Statistical Analysis

All experiments were performed in triplicate, and the results are presented as mean ± standard deviation (SD). Statistical analysis was performed using XLSTAT Premium version 2016 (Addingsoft, Barcelona, Spain). The one-way analysis of variance (ANOVA) test was applied to examine the significant differences in chromatic characteristics, anthocyanins, polyphenols, and volatile compounds of the vinegar samples. Then, Tukey's multiple range tests were applied for the post hoc test, where the effect is commonly considered significant if the resulting p-value is below 0.05.

3. Results and Discussion

3.1. Phenolic Compounds

The anthocyanin, *t*-resveratrol (*trans*-stilbene), flavanones, flavonols, and phenolic acid compounds were determined using the positive mode of ESI, which has a mode of flavylium cations under acidic conditions. Table 1 shows the phenolic compounds identified and quantified in the vinegar samples. In anthocyanin analysis, fifteen different derivatives of major mono-anthocyanin compounds were analyzed by LC-MS/MS, but only six forms of these compounds, which are peonidin-3,5-di-*O*-glucoside, cyanidin-3-*O*-rutinoside, pelargonidin-3-*O*-glucoside, peonidin-3-*O*-glucoside, malvidin-3-*O*-glucoside, and malvidin-3-*O*-galactoside, could be detected with a trace amount in some of the vinegar samples. However, as can be seen from the table, none of the anthocyanin compounds were found even in trace amounts in *Gran Reserva* samples.

Table 1. Retention time, mass spectral characteristics, and concentration of anthocyanins and other polyphenols (μg/L) present in *Jerez* vinegars.

Code	Compound	Retention Time (min)	$[M-H]^-$ (m/z)	MS/MS (m/z)	ANOVA ǂ	Apto	*Jerez*	*Reserva*	*Gran Reserva*
Stilbenes									
1	(E)-resveratrol	8.16	229.00	135.10/**107.10** †/91.10	***	ND ‡ b	25.04 a	ND b	ND b
Anthocyanins									
2	Peonidin-3,5-di-*O*-glucoside	4.89	625.15	462.95/**300.95**/285.95	NS	ND a	ND a	tr ¶ a	ND a
3	Cyanidin-3-*O*-rutinoside	4.94	595.10	449.00/**287.10**	NS	tr a	ND a	ND a	ND a
4	Pelargonidin-3-*O*-glucoside	5.04	433.00	**271.10**/121.10	***	tr b	15.70 a	ND b	ND b
5	Peonidin-3-*O*-glucoside	5.14	463.00	**301.10**/286.00/201.05	NS	ND a	ND a	tr a	ND a
6	Malvidin-3-*O*-glucoside	5.16	493.10	**331.10**/315.10/287.10	NS	tr a	ND a	tr a	ND a
7	Malvidin-3-*O*-galactoside	5.21	493.20	**331.10**/315.10/287.10	NS	tr a	ND a	tr a	ND a
Flavones									
8	Luteolin-7-*O*-glucoside	6.68	448.90	**287.10**/153.10/135.15	*	7.28 a	4.19 ab	2.39 b	1.77 b
Flavanones									
9	Eriotricin	5.95	595.20	**287.05**/150.09/135.05	***	65.82 b	ND c	84.67 a	ND c
10	Hesperidin	7.02	609.20	**301.00**/163.90/150.95	***	tr b	ND b	65.05 a	ND b
Flavonols									
11	Myricetin-3-*O*-glucoside	5.69	481.10	**319.10**/273.10/153.00	***	31.72 b	28.59 c	45.50 a	28.69 c
12	Quercetin-3-*O*-glucoside	5.95	463.25	**300.15**/271.15/255.20	*	153.89 a	36.44 b	21.80 b	ND c
13	Quercetin-3-*O*-glucuronide	5.97	478.95	**303.05**/229.00/153.00/112.80	***	1806.09 a	1008.81 b	142.32 c	61.44 c
14	Quercetin-3-*O*-galactoside	5.98	465.00	**303.10**/229.10/164.00/153.00	***	239.92 a	61.19 b	39.11 c	ND d
15	Quercetin-3-*O*-rutinoside	6.20	608.90	301.00/**299.95**/270.95	NS	ND a	ND a	tr a	ND a
16	Quercetin-3-*O*-glucopyranoside	6.41	463.10	300.95/**300.00**/270.95	***	79.71 a	34.85 b	30.71 b	nd c
Phenolic acids and derivatives									
17	Gallic acid	1.18	169.10	**124.95**/124.30/78.95	NS	421.84 a	539.01 a	196.17 a	177.05 a
18	Chlorogenic acid	3.22	353.30	**191.05**/92.95/85.05	NS	tr a	nd a	73.25 a	nd a
19	Caffeic acid	4.12	179.10	**135.00**/134.00/106.95	***	907.27 a	992.92 a	202.6 b	92.73 b
	Total of polyphenolic compounds				***	3713.56 a	2731.06 b	903.59 c	361.69 c

† The m/z values of the dominant ions are highlighted using bold font. ¶ tr: trace. ‡ ND: not detected. ǂ NS: not significant at $p < 0.05$; *, ***, significant at $p < 0.05$ and $p < 0.001$, respectively. Mean values were obtained from the duplicate analysis of all vinegar samples. Values followed by the same letter, within the same row, were not significantly different ($p > 0.05$) according to Tukey's least significant difference test.

Grapes are the unique source of these compounds. On the other hand, anthocyanins are not stable substances and can be degraded under many kinds of processes. First, the vinification process of wine production promotes several reactions between anthocyanins and other molecules, such as pyruvic acid, vinyl phenol, acetone, and glyoxylic acid, leading to a decrease in free anthocyanin contents and copolymerization reactions between tannins and anthocyanins [23,28]. Second, the acetification process of *Sherry* wine to a vinegar product [12] was observed with a decrease (~50%) in the content of monomeric anthocyanins during the acetification of red wine. In many studies, the acetification process was related to a decrease in the total phenolic content of 13%. Surprisingly, most of the individual phenolic compounds, the anthocyanins, showed a much higher decrease of ~50% due to low oxidizable tannins contributing to the total phenols. Moreover, it was stated that the concentration of the monomeric phenols might be reduced by oligo- and polymerization [29–31], yielding products that still contribute to the total phenol content. Therefore, these mechanisms might explain why vinegar samples include a lower content of anthocyanins than wine products. Obviously, while these processes are taking place, there are many conditions, such as temperature, pH, ethanol content, and the presence of oxygen, that can affect the degradation of anthocyanins. Although four anthocyanin compounds were found in trace amounts in the *Apto* sample as seen in the table, this result contributed to the reduction in anthocyanin compounds during the acetification process. This can be explained by the fact that *Apto* also contains lower amounts of anthocyanins. Another raw substance obtained from grapes is t-resveratrol. This component was detected only in the *VJ* sample with a content of 25.04 μg L^{-1}. This result can be explained by the acetification process of red wine vinegar and the loss of t-resveratrol in wood. Additionally, this may be due to polymerization and precipitation [12,29].

A statistically significant decrease and difference was found for flavonol and flavone compounds of the *Apto* to *Gran Reserva* samples depending aging period, for instance, quercetin-3-*O*-glucoside (from 153.89 μg/L to ND), quercetin-3-*O*-glucuronide (1806.09 to 61.44 μg/L), quercetin-3-*O*-galactoside (239.92 μg/L to ND), quercetin-3-*O*-glucopyranoside (from 79.71 μg/L to ND), and luteolin-7-*O*-glucoside (from 7.28 to 1.77 μg/L), which have been identified for the first time in *Jerez* samples. It can be stated that there is a significant effect of aging on flavonol compounds even in wood stored for at least six months (*VJ*). It can be concluded that some reactions between phenolic molecules reduce the flavonoid content of the samples depending on aging time, which can be explained by some reactions taking place between phenolic molecules. One of these reactions may occur by the polycondensation mechanism between flavonoid and tannin compounds and lead to the loss of monomeric flavonoid compounds [21]. Moreover, gallic acid and caffeic acid, among the phenolic acid derivatives, showed a significant decrease in the content of the *VJ* and *Gran Reserva* samples from 539.01 to 177.05 μg/L and 992.92 to 92.73 μg/L, respectively. The decrease in these compounds might be explained by the second mechanism, which is the oxidation process during the aging period [5]. Moreover, the total phenolic compound content shows a sharp decrease depending on the aging period in the sample from *Apto* (3713.56 μg/L) to *Gran Reserva* (361.69 μg/L). A similar result was also obtained for aged balsamic vinegar samples with reduced phenolic compound content from 1500.6 to 1321.4 mg/kg [21]. On the other hand, an increase, although not significant, in the gallic content of *Apto* and *Vinagre de Jerez* samples was shown (from 421.84 to 539.01 μg/L) during the six-month aging period in a wood cask. This phenomenon is explained by the hydrolysis of gallic tannins to free gallic during the aging period. In a study, some phenolic compounds were analyzed in *VJ* and *Reserva* samples, and a similar increasing trend was obtained in the content of gallic acid [5]. Besides gallic acid, a similar trend was also detected for the caffeic acid component (907.27 to 992.92 μg/L) between Apto and *VJ* samples. In another study, the concentration of gallic acid was found to increase after 360 days of aging, and the increase in caffeic acid concentration in the first year was explained by the hydrolysis of their corresponding tartaric esters [2]. Additionally, eriotricin and hesperidin, from flavanone compounds, were detected for the first time in some *Jerez* samples. While eriotricin was determined as 65.82 and 84.67 μg/L in *Apto* and *Reserva* samples, hesperidin was detected as 65.05 μg/L in only *Reserva*.

Although the effect of aging on anthocyanins is not seen much, it is observed that it has a significant effect on other phenolic components. Major flavonol and phenolic acid compounds show statistical differences according to the aging process. In conclusion, as a notable study, many phenolic compounds from *Jerez* samples were identified and determined in this study. As a result of this analysis, the impact of aging on these compounds through a series of chemical interactions revealed that some chemical transformations occur between free and polymerized forms of anthocyanins, flavonol, and phenolic acids.

3.2. Volatile Compounds

Volatile compounds of vinegar samples were identified and quantified using GC-MS/MS by applying the HS-SPME technique. A total of 62 volatile compounds could be quantified by using DB-Wax and DB-5 columns, and the results are shown in Table 2. As can be seen from the table, 24 volatile compounds could be found for both columns, while the remaining 27 and 11 volatile compounds were identified only in the DB-Wax and DB-5 columns, respectively. Thus, it can be said that most of the compounds can be determined with the DB-Wax column. These volatile compounds are classified into seven chemical families: esters, acids, alcohols, aldehydes, ketones, lactones, phenolic compounds, and others (not in any specific family). The total contents of these chemical groups according to the samples are given in Figure 1.

Table 2. Volatile compounds identified in *Jerez* vinegars (mg/L).

Chemical Family	Compound	Code	RT ‡		RI †		ANOVA ǂ	*Apto*	*Vinagre de Jerez*	*Reserva*	*Gran Reserva*
			DB-Wax	DB-5	DB-Wax	DB-5					
Esters	Methyl acetate	V3	2.80	N ¶	857	N	***	3.232 a	1.904 b	1.318 c	0.811 d
	Ethyl acetate	V4	3.16	2.72	888	624	***	37.87 a	35.91 a	36.14 a	24.54 b
	Ethyl propanoate	V7	3.88	N	947	N	***	0.082 b	0.235 b	0.045 b	1.179 a
	Ethyl isobutyrate	V8	4.0	4.62	957	752	***	0.282 b	0.211 bc	0.101 c	1.420 a
	sec-Butyl acetate	V10	4.29	N	980	N	***	ND ᵡ c	0.346 a	ND c	0.111 b
	Isobutyl acetate	V11	4.7	4.95	1007	766	***	10.56 a	5.575 b	4.803 b	5.484 b
	Methyl isovalerate	V12	4.81	N	1012	N	***	0.202 a	0.125 b	ND c	ND c
	Ethyl butanoate	V13	5.13	N	1027	N	***	ND b	ND b	ND b	0.623 a
	Ethyl 2-methylbutanoate	V14	5.45	N	1041	N	***	0.166 b	0.118 b	0.087 b	1.848 a
	Ethyl isovalerate	V16	5.77	7.396	1056	841	***	4.568 b	2.477 b	1.967 b	22.15 a
	Butyl acetate	V17	5.84	N	1059	N	***	0.009 b	ND b	ND b	0.162 a
	2-Methylbutyl acetate	V20	7.21	8.39	1113	868	***	107.65 a	51.07 c	60.45 c	83.55 b
	Ethyl hexanoate	V22	11.42	14.64	1225	992	***	1.580 b	0.447 c	0.220 d	1.942 a
	Hexyl acetate	V23	13.19	15.48	1262	1006	**	0.705 b	0.737 b	1.598 a	0.909 b
	2-Methylbutyl isovalerate	V25	14.39	N	1288	N	***	1.147 a	0.270 c	0.152 c	0.464 b
	(Z)-3-Hexenol acetate	V26	15.38	N	1308	N	***	ND b	ND b	0.090 a	ND b
	Ethyl heptanoate	V27	16.29	N	1325	N	***	ND b	ND b	ND b	0.190 a
	Ethyl lactate	V28	16.60	N	1330	N	***	ND c	0.104 b	ND c	0.186 a
	Acetoin Acetate	V29	18.76	8.76	1370	877	**	0.446 a	0.385 ab	0.328 b	0.208 c
	Ethyl decanoate	V35	32.75	38.68	1630	1386	***	4.367 a	0.408 c	0.096 c	1.232 b
	1,3-Propylene diacetate	V36	33.29	21.41	1642	1096	***	0.406 b	0.754 a	ND c	0.871 a
	Diethyl succinate	V38	34.28	26.59	1664	1172	***	0.345 c	1.101 a	0.367 c	0.722 b
	Benzyl acetate	V40	36.18	25.34	1706	1154	***	0.570 a	0.500 a	0.191 b	0.481 a
	Methyl salicylate	V41	37.74	N	1745	N	***	0.174 b	0.340 a	ND d	0.103 c
	Ethyl benzeneacetate	V42	38.57	30.45	1765	1234	***	0.232 b	0.177 b	0.139 b	1.809 a
	Phenethyl acetate	V43	39.68	31.14	1792	1245	***	23.64 a	10.98 c	8.435 d	17.82 b
	Isopropyl myristate	V49	48.22	N	2010	N	***	0.349 a	0.308 a	0.335 a	0.079 b
	Ethyl butyrate	V56	N	5.64	N	793	***	ND b	ND b	ND b	0.114 a
	1-Methoxy-2-propanol acetate	V58	N	10.54	N	917	**	0.234 ab	0.286 a	0.161 b	0.289 a
	Isopentyl isovalerate	V60	N	21.65	N	1100	***	0.074 a	0.037 b	0.032 b	0.069 a
	Ethyl octanoate	V62	N	27.81	N	1190	***	15.20 a	0.252 b	0.110 b	1.066 b
Acids	Acetic acid	V31	21.93	2.99	1427	661	***	634.95 a	438.53 b	437.57 b	352.84 c
	Isobutyric acid	V34	28.98	4.55	1554	750	***	0.812 b	1.285 a	0.300 d	0.540 c
	Isovaleric acid	V37	33.85	7.39	1654	841	***	23.12 a	25.35 a	9.331 b	9.567 b
	Hexanoic acid	V44	41.03	N	1828	N	***	0.620 a	0.459 b	0.213 d	0.328 c
	Octanoic acid	V50	48.61	26.32	2013	1168	**	0.943 a	0.577 ab	0.350 b	0.875 a
	Isobutyric acid	V54	N	4.55	N	750	**	0.210 a	0.200 a	ND b	0.085 ab
	α-Methylbutyric acid	V57	N	7.69	N	849	***	0.363 a	0.426 a	0.110 b	ND c
Alcohols	Ethanol	V6	3.57	N	922	N	***	3.573 b	4.063 b	4.960 b	41.39 a
	Isobutyl alcohol	V18	6.43	N	1085	N	***	1.278 a	0.993 b	0.778 c	1.020 b
	Active amyl alcohol	V21	10.24	4.25	1200	738	**	13.35 bc	14.19 ab	11.244 c	16.06 a
	l-α-Terpineol	V39	34.95	N	1679	N	***	ND b	ND b	0.225 a	ND b
	Benzyl alcohol	V45	41.95	N	1852	N	***	0.357 b	0.530 a	0.144 c	0.475 a
	Phenylethyl alcohol	V46	43.19	21.83	1885	1102	***	10.93 a	11.13 a	4.791 c	8.761 b
	Dodecyl alcohol	V48	45.72	N	1957	N	*	0.165 b	0.266 a	0.241 ab	0.233 ab
	Isopentyl alcohol	V53	N	4.18	N	735	***	1.016 d	1.379 c	1.675 b	3.302 a
	2,3-Butanediol	V55	N	5.17	N	774	**	0.089 b	0.060 bc	0.028 c	0.144 a
	Linalool	V59	N	21.19	N	1093	NS	0.060 a	ND a	ND a	ND a
	α-Terpineol	V61	N	27.56	N	1186	***	0.097 a	0.022 ab	0.075 ab	ND b
Aldehydes	Acetaldehyde	V1	2.36	N	820	N	***	0.098 b	0.040 c	ND c	0.288 a
	Isovaleraldehyde	V5	3.44	N	911	N	***	0.741 a	0.051 c	ND c	0.330 b
	Furfural	V32	22.95	6.61	1445	821	***	0.560 a	0.357 b	0.164 c	0.304 b
	Benzaldehyde	V33	25.92	12.43	1496	951	***	2.168 a	0.327 b	0.437 b	0.495 b
Ketones	Acetone	V2	2.73	N	851	N	***	0.169 b	0.182 b	0.249 b	1.813 a
	2-Acetylpropane	V9	4.10	N	965	N	***	1.239 b	0.775 c	0.693 c	3.365 a
	Acetoin	V24	13.43	3.73	1268	718	**	3.036 a	3.573 a	2.876 a	1.686 b
	2-Nonanone	V30	19.05	N	1375	N	***	0.398 b	0.161 c	0.196 c	0.548 a
Lactones	(Z)-Oaklactone	V47	44.45	N	1920	N	***	ND b	ND b	ND b	0.286 a
Phenolic compounds	2-Ethylphenol	V51	52.11	N	2047	N	***	0.275 b	0.373 a	0.088 d	0.183 c
Others	2-Methyl-1,3-dioxane	V15	5.69	N	1052	N	***	0.037 b	0.027 b	ND b	0.253 a
	Linalool 3,7-oxide	V19	6.81	13.01	1102	962	***	0.775 a	0.233 b	0.116 c	0.074 c
	2,4,5-Trimethyl-1,3-dioxolane	V52	N	4.04	N	730	***	0.201 b	ND c	ND c	0.924 a
Total							***	915.72 a	620.62 b	594.02 b	616.60 b

‡ RT: Retention times on DB-Wax and DB-5 columns. † RI: Retention index on DB-Wax and DB-5 columns. ¶ N: Not detected by DB-Wax or DB-5 columns. ǂ NS: not significant at $p < 0.05$; *, **, ***, significant at $p < 0.05$, $p < 0.01$, and $p < 0.001$, respectively. Mean values obtained from the triplicate analysis of all vinegar samples. Values followed by the same letter, within the same row, were not significantly different ($p > 0.05$), according to Tukey's least significant difference test. ᵡ ND: Not detected.

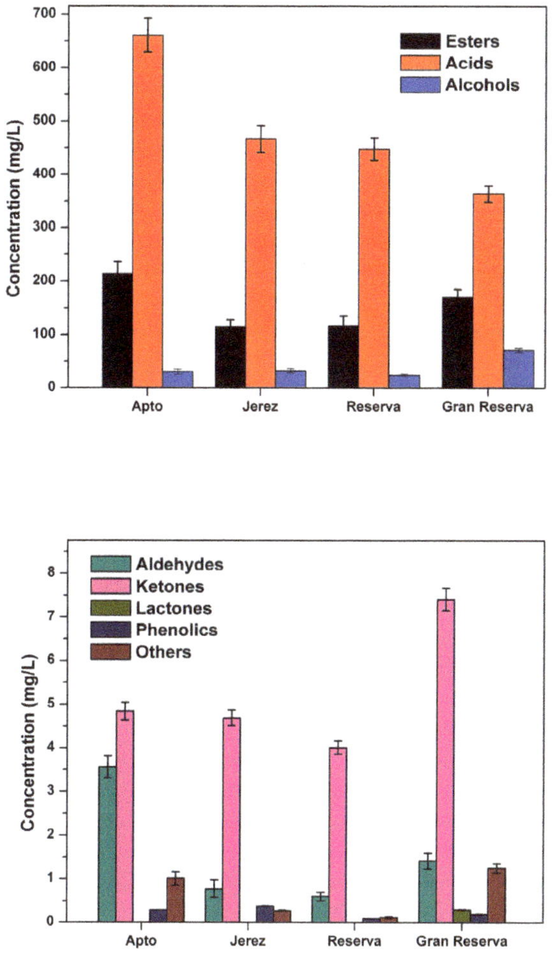

Figure 1. Total concentration of chemical families of esters, acids, alcohols, aldehydes, ketones, lactones, phenolics, and others found in *Jerez* vinegar samples.

Acids and esters constitute the main volatile substances in the samples and show a notable change during the aging process. A decrease in the acid content by approximately half was observed due to aging (from *Apto*, 661.02, to *Gran Reserva*, 364.24 mg/L). A slight decrease in the amount of ester was detected from *Apto*, 214.09, to *VJ*, 115.06 mg/L, and then an increase again (from 115.06 to 170.43 mg/L, *Gran Reserva*) was found during the long aging process. Esters are the most diverse compounds with 31 substances, among the volatile compound families, and they give off flower, fruity, and rose-like odors. Ethyl acetate, phenethyl acetate, and ethyl octanoate were found to be the most abundant components in the *Apto* sample with the amount of 37.87, 23.64, and 15.20 mg/L, respectively. Esters are mostly produced during the fermentation process of vinegars by turning off ethanol and acid production [32,33]. Ethyl acetate, the main ester component, may be formed by the condensation of ethanol and acetic acid with the loss of water and was found in the range of 60–1451 mg/L in aged *Jerez* samples [34]. Ethyl acetate was also found in other types of aged vinegars, namely, Zhenjiang and Shanxi vinegar in increasing concentrations with aging [8,35]. However, in this study, ethyl acetate exhibited a declining profile with a content of 37.87 to 24.54 mg/L in the *Gran Reserva* sample, and this result can be explained because of volatilization or due to the change in esterification balance [36]. Some specific components have been identified for the first time in *Jerez* samples and can be used as a marker to authenticate the *Gran Reserva* sample, which are ethyl butanoate (with the aroma of banana and apple), 0.623 mg/L, and ethyl heptanoate, 0.190 mg/L. During post-processing of vinegar (aging), these compounds may be produced from the esterification process between ethanol-butanoic acid and ethanol-heptanoic acid. On the other side, (Z)-3-hexenol acetate was identified for the first time with an amount of 0.090 mg/L. Additionally, it can be accepted as an authenticator compound for the *Reserva* sample. A similar condensation reaction between acetic acid and hexanol can be stated in the formation of this compound during the aging period.

Secondly, as can be seen from the table, the most abundant chemical family in terms of quantity are acids with seven components. The most dominant component was obtained in acids and all chemical substances for acetic acid with a varying concentration of 634.95–352.84 mg/L from *Apto* to *Gran Reserva* samples, respectively. For acetic acid, a main flavor product of the acetification process of vinegar, a sharp decline in its content can be explained due to evaporation and condensation reactions with ethanol [37]. Moreover, a similar decrease in the concentration of acetic acid was observed during the aging of traditional balsamic vinegar in oak barrels [21]. The second main component is isovaleric acid with a content of 23.12 mg/L and 25.35 mg/L in the samples of *Apto* and *Jerez*, respectively. A quantity of 9.567 mg/L was quantified in *Gran Reserva*. Due to this observed decrease, it is believed that isovaleric acid is involved in the esterification reactions of short-chain carboxylic esters during the aging process [38]. As can be seen from Figure 1, during the aging time, it can be implied that all of these factors, including vaporization, esterification, and condensation reactions, play a significant role in the decrease in the content of acids.

In the case of alcohols, it has been observed that certain components detected in very small amounts decrease over time, such as α-terpineol and linalool (not found in *Gran Reserva*), while some main components found in higher amounts significantly increase, namely, ethanol (from 3.573 to 41.39 mg/L) and amyl alcohol (from 13.35 to 16.06 mg/L). Additionally, as can be seen from Figure 1, there is a significant sharp increase in the number of alcohols in the *Gran Reserva* sample. The aging period can have an impact on the ethanol content in certain types of vinegar. In one study, ethanol content was not detected in Sherry vinegar aged for 6 months [34]. In another study, the concentration of ethanol was found to show a declining profile during the aging period of traditional balsamic vinegar [21]. However, in this study, the statistically significant increase in the content of ethanol may have resulted from the hydrolysis of ethyl acetate into ethanol, resulting in the amount of ethyl acetate.

Four components, namely, acetaldehyde, valeraldehyde, furfural, and benzaldehyde, can be identified and quantified in low amounts in aldehydes. During the aging process, aldehydes exhibited a decreasing trend from *Apto* to *Reserva* (from 3.567 to 0.601 mg/L), followed by a modest increase from *Reserva* (0.601 mg/L) to *Gran Reserva* (1.417 mg L^{-1}), as depicted in Figure 1. A statistically significant difference was observed for benzaldehyde. Initially, a decrease from 2.168 to 0.327 mg/L (between *Apto* and *Jerez*) was noted, followed by an increase to 0.495 mg/L in the *Gran Reserva* sample observed throughout the aging period. An increase in benzaldehyde between *VJ* and *Gran Reserva* samples may be explained by oxidation reactions and the decarboxylation of phenylalanine and methionine [39]. Benzaldehyde is also stated as one of the main flavor aldehyde components in many vinegars, such as Chinese traditional bran vinegar (Cupei), Italian balsamic vinegar, and Shanxi aged vinegar [8,9,40]. A change in the concentration of acetaldehyde was found from 0.098 to 0.288 mg/L, which is an intermediary and is produced by ethanol oxidation, resulting in flavor differences [41]. One of the aromatic wood compounds, furfural, was found to have a content of 0.164 and 0.304 mg/L for *Reserva* and *Gran Reserva*, respectively. It may be extracted from the oak barrels during the aging process, and, additionally, furfural comes from the degradation of monosaccharides resulting from the partial hydrolysis of hemicellulose [34]. Like aldehydes, four ketone substances could be detected in all samples. While acetoin was the most abundant compound (3.036 mg/L) in the *Apto* sample, 2-acetylpropane was found to be the predominant compound in *Gran Reserva*, with its level increasing from 1.239 to 3.365 mg/L during the aging process. As in aldehydes, ketone substances exhibited a low decrease between *Apto* and *Reserva* samples but then a high increase in *Gran Reserva* samples, as shown in Figure 1. Over the aging time, the reason for the increase in the number of ketones can be explained by the *Maillard* reaction, that is, Strecker degradation [33].

Oaklactone, identified for the first time in the *Jerez* vinegars, was exclusively detected in the *Gran Reserva* sample (0.286 mg/L). Its presence serves as an indicator of long aging in traditional wooden barrels and shares similarities with compounds extracted, such as whisky lactones [34,42]. Additionally, this component can be used as a marker for *Gran Reserva* samples. 2-Ethylphenol, a volatile phenol, was found at a content of 0.275 for the *Apto* sample and 0.183 mg/L for the *Gran Reserva* sample. Additionally, other compounds identified in different categories included 2-methyl-1,3-dioxane (0.253 mg/L), linalool 3,7-oxide (0.074 mg/L), and 2,4,5-trimethyl-1,3-dioxolane (0.924 mg/L) in the *Gran Reserva* samples.

As a result, the total content of volatile compounds exhibited a decreasing trend in the range of 915.72 to 616.60 mg/L from the *Apto* to *Gran Reserva* samples. This decline is indicative of various processes such as oxidation, esterification, wood extraction, hydrolysis, and fermentation playing a significant role during this period. In a previous study by Morales et al. [10], 18 volatile compounds were identified in *Sherry* wine vinegars aged up to 24 months using gas chromatography coupled with a flame ionization detector (GC-FID). In the current study, the GC-HSPME fiber method was utilized, enabling the capture and identification of even minor volatile components in all samples.

This study identified an additional forty-four volatile compounds, totaling sixty-two volatile components in all samples, including the detection of these substances in *Gran Reserva*. While prior research extensively studied aroma components in aged vinegars to understand their complexity, certain unidentified volatile components remained undiscovered until this examination. The methodology utilized in this study facilitated the quantification of sixty-two volatile substances using two distinct types of columns, providing new insights into the comprehensive volatile profile of all *Jerez* vinegars.

3.3. Chromatic Characteristics

Absorbance measurements (A420, A520, and A620), color intensity (CI), tonality (T), color density (D), and yellow (%), red (%), and blue (%) percentages of vinegar samples were taken using UV-VIS spectroscopy. The values of all chromatic parameters of vinegar samples are shown in Table 3. The table illustrates a notable difference in color composition between non-aged and aged samples. The *Apto* sample exhibited the highest yellow pigment content (73%) but the lowest levels of red (21%) and blue pigments (6%) compared to other samples. Conversely, the *Reserva* sample displayed the lowest yellow pigment content (60%) and the highest levels of red (29%) and blue pigments (11%). These changes can be attributed to the increase in absorbance values at 420 nm (from 0.879 to 2.049) and at 520 nm (from 0.257 to 0.784), indicating an overall increase in color concentrations. However, the rise in red components may have proportionally exceeded that of yellow components during aging, resulting in this observed difference. The change in the color for these absorbance measurements might be related to *Maillard* and caramelization reactions and finalized with browning agents [43,44]. In a study on *Sherry* vinegar, absorbance measurements were taken at 470 nm to assess the impact of aging time. A simultaneous increase in absorbance values was observed, which was attributed to the oxidation of polyphenolic compounds during the aging process [1]. Furthermore, an increase in the blue pigment was observed in both *Apto* and aged samples, with percentages of 6% and 11% for *Apto* and *Gran Reserva*, respectively. The change in absorbance values at 420 nm, 520 nm, and 620 nm from *Apto* to *Reserva* indicated an overall increase in all CI values (from 1.210 to 4.676) and D values (from 1.136 to 4.164). In contrast, there was a decline in the T values of non-aged and aged samples, decreasing from 3.419 to 2.616 between the *Apto* and *VJ* samples. The increase in the T value may be attributed to the rise in red pigment and a decrease in yellow pigment values. The color characteristics of wine vinegar originate from grapes and the fermented product of wine. Grapes are rich in anthocyanins, tannins, and resveratrol compounds, which contribute to their unique color. However, as detailed in the phenolic section of the results, mono-anthocyanins were not detected even in the non-aged vinegar sample due to various reactions that occur during the acetification process, leading to the formation of anthocyanin derivatives. Furthermore, the aging process facilitates copolymerization reactions between tannins and anthocyanin derivatives, resulting in the loss of these monomeric compounds [23].

Table 3. The chromatic characteristics of *Jerez* vinegars are determined by Glories method.

Vinegar Type	A_{420} ¶	A_{520} ¶	A_{620} ¶	Color Intensity	Tonality	Color Density	Y ¶ (%)	R ¶ (%)	B ¶ (%)
ANOVA †	***	***	***	***	***	***	***	***	***
Apto	0.879 d ‡	0.257 d	0.074 d	1.210 d	3.419 a	1.136 d	73 a	21 d	6 d
Vinagre de Jerez	2.049 c	0.784 c	0.294 c	3.127 c	2.616 b	2.833 c	66 b	25 c	9 c
Reserva	2.825 a	1.339 a	0.512 a	4.676 a	2.109 d	4.164 a	60 d	29 a	11 a
Gran Reserva	2.786 b	1.190 b	0.474 b	4.450 b	2.341 c	3.976 b	62 c	27 b	11 b

† *** Significant at $p < 0.001$. ‡ Values (mean of three replications) followed by the same letter, within the same column, were not significantly different ($p > 0.05$) according to Tukey's least significant difference test. ¶ Y: yellow color, R: red color, and B: blue color. A_{420}: absorbance at 420 nm, A_{520}: absorbance at 520 nm, and A_{620}: absorbance at 620 nm.

Notably, even though there is a statistical difference in the yellow (60% to 62%) and red color pigments (29% to 27%) during long-term aging and storage, the change in absorbance values at 420 nm (from 2.825 to 2.786) and 520 nm (from 1.339 to 1.190) is less pronounced for *Reserva* and *Gran Reserva* samples, respectively.

3.4. Sensory Analysis

Jerez vinegar is officially renowned for its amber hue with mahogany undertones, offering subtle notes of nuts, wood, and a distinct acetic aroma. With prolonged aging, it develops intricate hints of vanilla, nuts, and matured wood. The *Gran Reserva* variant stands out for its intense aroma rich in nutty and spicy notes [45]. While certain characteristics have been acknowledged by regulators, this study delves into previously unexplored sensory descriptors, leading to the development of precise lexicons, some of which were refined in prior works [46]. These terminologies were further tailored to suit each vinegar variety, aiding in effectively characterizing the sensory profile of all *Jerez* vinegars using descriptive analysis. The sensory profile of *Jerez* vinegars, encompassing appearance (color and texture), aroma, flavor (sweetness, sourness, bitterness, pungency, and other specific descriptors), overall impression (aftertaste), and qualification (liking), including a total of 38 attributes, is presented in Table 4. In addition, consumer preferences were assessed under the qualification category. Table 4 reveals that in the appearance category, the darkest color, scoring 9.55 points, was observed in the *Reserva* samples. This outcome aligns with the color characteristics, indicating a high intensity of red pigments. Additionally, the non-aged sample secured the highest vinegar ID scores for odor (8.00) and flavor (8.05), likely due to its potentially higher acetic acid content in volatile compounds. During aging, many attributes showed a meaningful change in the odor feature. These differences mainly with an increasing quantity were observed in the attributes of winy character (from 1.27 to 7.16), raisin (0.27 to 4.27), liquor (0.11 to 2.38), woody (from 0 to 2.88), toasted (from 0 to 2.50), and nuts (from 0 to 0.72) but not fruity (from 6.61 to 4.05). An increase in the attributes of woody, toasted, and nut corresponded to the aging of the samples taking place in wood casks [47,48]. Although no statistically significant difference was found in the attributes of clove (0 to 0.33), spicy, and leather, distinctions were noted between the *Reserva* and *Gran Reserva* samples. Furthermore, no defects were identified in either the odor or taste categories.

Table 4. Descriptive sensory analysis of "*Jerez*" vinegars.

Attribute	ANOVA †	Sample			
		Apto	*Vinagre de Jerez*	*Reserva*	*Gran Reserva*
Appearance					
Color	***	4.16 d	6.05 c	9.55 a	8.66 b
Untuoso (texture)	***	2.00 d	4.55 b	7.00 a	3.22 c
Odor					
Odor intensity	***	2.38 b	1.55 b	1.72 b	4.50 a
Vinegar ID	***	8.00 a	4.61 b	4.50 b	3.72 b
Winy character	***	1.27 c	1.27 c	2.44 b	7.16 a
Raisin	***	0.27 c	1.22 bc	1.33 b	4.27 a
Ethyl acetate	***	0.22 c	0.05 c	0.77 b	1.38 a
Alcohol/liquor	***	0.11 b	0.55 b	0.27 b	2.38 a
Woody	***	0.00 b	0.00 b	0.27 b	2.88 a
Fruity	***	6.61 a	2.61 c	2.00 c	4.05 b
Spicy	NS	0.00 a	0.00 a	0.22 a	0.77 a
Vanilla	***	0.00 b	0.00 b	0.11 b	1.50 a
Clove	NS	0.00 a	0.16 a	0.22 a	0.33 a
Toasted	***	0.00 b	0.38 b	0.38 b	2.50 a
Nuts	*	0.00 b	0.00 b	0.11 ab	0.72 a
Leather/old	NS	0.00 a	0.00 a	0.00 a	0.16 a
Defects	NS	0.00 a	0.00 a	0.00 a	0.00 a
Flavor					
Flavor intensity	***	8.27 a	5.55 b	6.44 b	5.75 b
Vinegar ID	***	8.05 a	6.00 c	6.83 b	5.00 d
Winy character	***	2.55 b	2.33 b	2.16 b	5.06 a
Raisin	***	0.00 b	0.00 b	0.38 b	2.43 a
Ethyl acetate	***	0.00 b	0.00 b	0.00 b	0.43 a
Alcohol/liquor	***	0.00 b	0.00 b	0.00 b	1.81 a
Woody	***	0.00 b	0.00 b	0.00 b	1.43 a
Fruity	**	5.16 a	3.22 b	3.27 b	4.00 ab
Spicy	NS	0.00 a	0.00 a	0.00 a	0.18 a
Vanilla	NS	0.00 a	0.00 a	0.00 a	0.31 a
Clove	NS	0.00 a	0.00 a	0.11 a	0.12 a
Toasted	**	0.00 b	0.16 b	0.22 b	0.81 a
Nuts	*	0.77 a	0.16 ab	0.05 b	0.50 ab
Leather/old	NS	0.00 a	0.00 a	0.00 a	0.00 a
Sweetness	***	1.11 b	2.50 a	2.22 a	2.33 a
Sourness	***	8.11 a	6.11 bc	6.61 b	5.38 c
Bitterness	NS	0.33 a	0.38 a	0.27 a	0.33 a
Astringent	NS	0.00 a	0.00 a	0.00 a	0.38 a
Pungent	***	3.77 a	1.88 b	2.83 ab	2.22 b
Defects	NS	0.00 a	0.00 a	0.00 a	0.00 a
Global					
Aftertaste	***	6.44 a	3.88 b	4.16 b	4.83 b
Qualification					
Liking ‡		14	23	21	32
Ranking ‡	***	b	a	b	a

† NS: not significant at $p < 0.05$; *, **, ***, significant at $p < 0.05$, $p < 0.01$, and $p < 0.001$, respectively. Values (mean of 9 trained panelists) followed by the same letter, within the same row, were not significantly different ($p > 0.05$) according to Tukey's least significant difference test. ‡ Satisfaction degree of 9 panelists is denoted by liking, and the statistical results of Friedman's test are denoted by ranking.

Following odor identification, the flavor attributes exhibited a statistically significant change, particularly noted between the *Apto* and *Gran Reserva* samples. The highest winy character, scoring 5.06 points, was detected in the *Gran Reserva* sample. Additionally, woody attributes were tasted in the *Gran Reserva* sample, scoring 1.43 points. The woody character of the sample may stem from the (*Z*)-oaklactone compounds derived from the wood casks [49,50]. Through the aging process, a decrease in fruity taste (from 5.16 to 4.0), sourness (from 8.11 to 5.38), and pungent (from 3.77 to 2.22) profiles was observed between *Apto* and *Gran Reserva* samples. The decrease in the sourness profile may be attributed to the mostly acidic contributors, mainly acetic acid substances, in vinegar [51–53]. The fruity profile in the *Apto* sample is attributed to the presence of fresh fruit notes from grapes in the volatile substances of ethyl esters and acetates [54]. As can be reviewed in volatile compounds, a significant decrease in many ester components, namely, 2-methylbutyl acetate, isobutyl acetate, and ethyl acetate, affects the change in the concentration of fruity notes in the aged samples. By the point of 0.38, *Gran Reserva* was identified as the most astringent sample akin to its winy character. An increase in the sweetness attribute was observed in both non-aged and *Gran Reserva* samples, with points ranging from 1.11 to 2.33. Moreover, the rise in alcohol notes in the Gran Reserva samples (1.81) correlates with the increase in alcohol content of volatile compounds like ethanol. In a separate study, certain attributes of the sensorial profile—such as aroma intensity, woody flavor, ethyl acetate, wine character, and pungent sensation—were examined in *VJ* samples aged up to six months. Consistent results were obtained with heightened winy character, woody notes, and aroma intensity in the *Reserva* sample aged up to 24 months [34]. To the best of my knowledge, the sensory profile of *Gran Reserva*, along with many new lexicons developed specifically for odor and flavor profiles, has been identified and examined in the analysis of *Jerez* vinegar in this study.

4. Conclusions

Jerez vinegars of varying maturation durations, including both aged (*VJ*, *Reserva*, and *Gran Reserva*) and non-aged varieties, have been subjected to a thorough analysis encompassing physicochemical properties, composition, and sensorial attributes. A unique facet of this investigation involves the discernment of chromatic features in the *Jerez* vinegar samples, with a focus on the proportional representation of color components, specifically CI and D. Comparative analysis revealed a notable shift in the hue profile, characterized by a diminishing yellow hue and an amplifying red hue across the aging continuum, attributed to the progressive accumulation of *Maillard* reaction byproducts. Over 40 volatile components present in *Apto*, *VJ*, and *Reserva* vinegar samples have been meticulously identified and quantified. Noteworthy trends emerged within the volatile compounds, with primary compounds such as esters (including ethyl acetate, 2-methylbutyl acetate, and isobutyl acetate), acids (i.e., acetic acid and isovaleric acid), and aldehydes (such as benzaldehyde) exhibiting a decrease in concentration over the aging process. Conversely, alcohols (specifically ethanol), ketones (including acetone and 2-acetylpropane), and lactone compounds displayed an augmentation in levels with prolonged aging, particularly evident in samples subjected to extended maturation periods.

However, the analysis of anthocyanin composition has presented challenges in quantification due to potential reactions occurring during acetification and oxidation processes that result in the loss of anthocyanin. Notably, phenolics have emerged as prominent components, particularly in the chemical family of flavonol substances within the samples. Furthermore, a comprehensive sensorial evaluation has been conducted across all samples, revealing a substantial impact of aging on various sensory descriptors, evidenced by numerous changes in lexicon terms. Noteworthy intensities were observed in woody, nutty, and wine-like characteristics, whereas fruity and vinegar notes displayed a notable decline. In summary, this investigation underscores the noteworthy influence of chemical reactions throughout fermentation and aging stages, shaping the physical and chemical attributes of aged vinegars and contributing significantly to defining the distinctive characteristics of the final product.

Funding: This work was supported partially by The Scientific & Technological Research Council of Turkey (TUBITAK), project number: 124O687 and the APC was funded partially by Istanbul Bilgi University.

Institutional Review Board Statement: Not applicable.

Informed Consent Statement: Not applicable.

Data Availability Statement: The original contributions presented in the study are included in the article, further inquiries can be directed to the corresponding author.

Acknowledgments: Reyhan Selin Uysal thanks Esther Sendra, Ángel A Carbonell-Barrachina, and CIAGRO-UMH for their support during this study.

Conflicts of Interest: The author declares that there are no conflicts of interest.

References

1. Palacios, V.; Valcárcel, M.; Caro, I.; Pérez, L. Chemical and biochemical transformations during the industrial process of sherry vinegar aging. *J. Agric. Food. Chem.* **2002**, *50*, 4221–4225. [CrossRef] [PubMed]
2. Tesfaye, W.; García-Parrilla, M.; Troncoso, A. Sensory evaluation of Sherry wine vinegar. *J. Sens. Stud.* **2002**, *17*, 133–144. [CrossRef]
3. Council Regulation (EC) No 510/2006 of 20 March 2006 on the Protection of Geographical Indications and Designations of Origin for Agricultural Products and Foodstuffs. 2006. Available online: https://eur-lex.europa.eu/LexUriServ/LexUriServ.do?uri=OJ:L:2006:093:0012:0025:en:PDF (accessed on 1 September 2024).
4. Consejo Regulador del Vino de Jerez (2023, November 7). *Sherry Wine.* Available online: https://www.sherry.wine/sherry-wine (accessed on 1 September 2024).
5. Parrilla, M.G.; Heredia, F.J.; Troncoso, A.M. Sherry wine vinegars: Phenolic composition changes during aging. *Food Res. Int.* **1999**, *32*, 433–440. [CrossRef]
6. Al-Dalali, S.; Zheng, F.; Li, H.; Huang, M.; Chen, F. Characterization of volatile compounds in three commercial Chinese vinegars by SPME-GC-MS and GC-O. *LWT* **2019**, *112*, 108264. [CrossRef]
7. Ríos-Reina, R.; Segura-Borrego, M.P.; Morales, M.L.; Callejón, R.M. Characterization of the aroma profile and key odorants of the Spanish PDO wine vinegars. *Food Chem.* **2020**, *311*, 126012. [CrossRef] [PubMed]
8. Liang, J.; Xie, J.; Hou, L.; Zhao, M.; Zhao, J.; Cheng, J.; Wang, S.; Sun, B.-G. Aroma constituents in Shanxi aged vinegar before and after aging. *J. Agric. Food. Chem.* **2016**, *64*, 7597–7605. [CrossRef]
9. Corsini, L.; Castro, R.; Barroso, C.G.; Durán-Guerrero, E. Characterization by gas chromatography-olfactometry of the most odour-active compounds in Italian balsamic vinegars with geographical indication. *Food Chem.* **2019**, *272*, 702–708. [CrossRef]
10. Morales, M.L.; Tesfaye, W.; García-Parrilla, M.C.; Casas, J.A.; Troncoso, A.M. Evolution of the aroma profile of sherry wine vinegars during an experimental aging in wood. *J. Agric. Food. Chem.* **2002**, *50*, 3173–3178. [CrossRef]
11. Callejon, R.M.; Morales, M.L.; Ferreira, A.C.S.; Troncoso, A.M. Defining the typical aroma of sherry vinegar: Sensory and chemical approach. *J. Agric. Food. Chem.* **2008**, *56*, 8086–8095. [CrossRef]
12. Cerezo, A.B.; Cuevas, E.; Winterhalter, P.; Garcia-Parrilla, M.C.; Troncoso, A.M. Anthocyanin composition in *Cabernet Sauvignon* red wine vinegar obtained by submerged acetification. *Food Res. Int.* **2010**, *43*, 1577–1584. [CrossRef]
13. García-Parrilla, M.C.; Cerezo, A.B.; Tesfaye, W.; Troncoso, A.M. Phenolic compounds as markers for the authentication of *Sherry* vinegars: A foresight for high quality vinegars characterization. In *Progress in Authentication of Food and Wine*; ACS Publications: Washington, DC, USA, 2011; pp. 201–213.
14. Gao, Y.; Jo, Y.; Chung, N.; Gu, S.-Y.; Jeong, Y.-J.; Kwon, J.-H. Physicochemical qualities and flavor patterns of traditional Chinese vinegars manufactured by different fermentation methods and aging periods. *Prev. Nutr. Food Sci.* **2017**, *22*, 30. [CrossRef] [PubMed]
15. Giudici, P.; Gullo, M.; Solieri, L.; Falcone, P.M. Technological and microbiological aspects of traditional balsamic vinegar and their influence on quality and sensorial properties. *Adv. Food Nutr. Res.* **2009**, *58*, 137–182.
16. Góamez, M.L.M.; Bellido, B.B.; Tesfaye, W.; Fernandez, R.M.C.; Valencia, D.; Fernandez-Pachón, M.S.; García-Parrilla, M.C.; González, A.M.T. Sensory evaluation of sherry vinegar: Traditional compared to accelerated aging with oak chips. *J. Food Sci.* **2006**, *71*, S238–S242. [CrossRef]
17. Durán-Guerrero, E.; Castro, R.; García-Moreno, M.d.V.; Rodríguez-Dodero, M.d.C.; Schwarz, M.; Guillén-Sánchez, D. Aroma of sherry products: A review. *Foods* **2021**, *10*, 753. [CrossRef]
18. Cejudo-Bastante, M.J.; Durán, E.; Castro, R.; Rodríguez-Dodero, M.C.; Natera, R.; García-Barroso, C. Study of the volatile composition and sensory characteristics of new *Sherry* vinegar-derived products by maceration with fruits. *LWT-Food Sci. Technol.* **2013**, *50*, 469–479. [CrossRef]
19. González-García, J.A.; Cereceda, M.; Durán-Guerrero, E.; Rodríguez-Dodero, M.C.; Castro, R. Comparative study on the use of seasoned or unseasoned casks made of wood from different origins for the ageing of *Sherry* vinegar. *J. Sci. Food Agric.* **2024**, *104*, 1271–1281. [CrossRef] [PubMed]
20. Xia, T.; Yao, J.; Zhang, J.; Duan, W.; Zhang, B.; Xie, X.; Xia, M.; Song, J.; Zheng, Y.; Wang, M. Evaluation of nutritional compositions, bioactive compounds, and antioxidant activities of Shanxi aged vinegars during the aging process. *J. Food Sci.* **2018**, *83*, 2638–2644. [CrossRef]

21. Verzelloni, E.; Tagliazucchi, D.; Conte, A. Changes in major antioxidant compounds during aging of traditional balsamic vinegar. *J. Food Biochem.* **2010**, *34*, 152–171. [CrossRef]
22. Durán-Guerrero, E.; Schwarz, M.; Fernández-Recamales, M.Á.; Barroso, C.G.; Castro, R. Characterization and differentiation of Spanish vinegars from jerez and condado de huelva protected designations of origin. *Foods* **2019**, *8*, 341. [CrossRef]
23. Uysal, R.S.; Issa-Issa, H.; Sendra, E.; Carbonell-Barrachina, Á.A. Changes in anthocyanin pigments, trans-resveratrol, and colorimetric characteristics of Fondillón wine and other "Monastrell" wines during the aging period. *Eur. Food Res. Technol.* **2023**, *249*, 1821–1831. [CrossRef]
24. Carbonell-Barrachina, Á.A.; Szychowski, P.J.; Vásquez, M.V.; Hernández, F.; Wojdyło, A. Technological aspects as the main impact on quality of quince liquors. *Food Chem.* **2015**, *167*, 387–395. [CrossRef] [PubMed]
25. OIV. Compendium of international methods of wine and must analysis. In Proceedings of the International Organisation of Vine and Wine, Paris, France, 6 July 2021; p. 196.
26. Glories, Y. The colour of red wines. *Conn. Vigne Vin* **1984**, *18*, 195–217.
27. OECCA. Ficha de Cata de Vinagres. 2022. Available online: https://www.oecca.es/ (accessed on 1 September 2024).
28. Ferreira, I.M.; Pérez-Palacios, M.T. Anthocyanic Compounds and Antioxidant Capacity in Fortified Wines. In *Processing and Impact on Antioxidants in Beverages*; Elsevier: Amsterdam, The Netherlands, 2014; pp. 3–14.
29. Escribano-Bailón, T.; Dangles, O.; Brouillard, R. Coupling reactions between flavylium ions and catechin. *Phytochemistry* **1996**, *41*, 1583–1592. [CrossRef]
30. Saucier, C.; Bourgeois, G.; Vitry, C.; Roux, D.; Glories, Y. Characterization of (+)-catechin– acetaldehyde polymers: A model for colloidal state of wine polyphenols. *J. Agric. Food. Chem.* **1997**, *45*, 1045–1049.
31. Andlauer, W.; Stumpf, C.; Fürst, P. Influence of the acetification process on phenolic compounds. *J. Agric. Food. Chem.* **2000**, *48*, 3533–3536. [CrossRef]
32. Shi, H.; Zhou, X.; Yao, Y.; Qu, A.; Ding, K.; Zhao, G.; Liu, S.Q. Insights into the microbiota and driving forces to control the quality of vinegar. *LWT* **2022**, *157*, 113085. [CrossRef]
33. Jiang, Y.; Lv, X.; Zhang, C.; Zheng, Y.; Zheng, B.; Duan, X.; Tian, Y. Microbial dynamics and flavor formation during the traditional brewing of Monascus vinegar. *Food Res. Int.* **2019**, *125*, 108531. [CrossRef]
34. Tesfaye, W.; Morales, M.L.; García-Parrilla, M.C.; Troncoso, A.M. Jerez vinegar. In *Vinegars of the World*; Springer: Berlin/Heidelberg, Germany, 2009; pp. 179–195.
35. Al-Dalali, S.; Zheng, F.; Sun, B.; Chen, F. Comparison of aroma profiles of traditional and modern Zhenjiang aromatic vinegars and their changes during the vinegar aging by SPME-GC-MS and GC-O. *Food Anal. Methods* **2019**, *12*, 544–557. [CrossRef]
36. Durán Guerrero, E.; Chinnici, F.; Natali, N.; Marín, R.N.; Riponi, C. Solid-phase extraction method for determination of volatile compounds in traditional balsamic vinegar. *J. Sep. Sci.* **2008**, *31*, 3030–3036. [CrossRef]
37. Chen, T.; Gui, Q.; Shi, J.J.; Zhang, X.Y.; Chen, F.S. Analysis of variation of main components during aging process of Shanxi Aged Vinegar. *Acetic Acid Bact.* **2013**, *2*, 31–38. [CrossRef]
38. Larios, A.; García, H.S.; Oliart, R.M.; Valerio-Alfaro, G. Synthesis of flavor and fragrance esters using *Candida antarctica* lipase. *Appl. Microbiol. Biotechnol.* **2004**, *65*, 373–376. [CrossRef] [PubMed]
39. Tie, Y.; Zhu, W.; Zhang, C.; Yin, L.; Li, L.; Liu, J. Effect of temperature on chemical compounds of Cupei (precursor of bran vinegar) during in-situ aging and revelation of functional microorganisms in the process. *LWT* **2023**, *182*, 114912. [CrossRef]
40. Hongwen, L.; Xuping, W.; Xiaolan, Y. Influence of different fumigation processes on aroma compounds of Shanxi aged vinegar. *Food Sci* **2015**, *36*, 90–94.
41. Al-Dalali, S.; Zheng, F.; Sun, B.; Rahman, T.; Chen, F. Tracking volatile flavor changes during two years of aging of Chinese vinegar by HS-SPME-GC-MS and GC-O. *J. Food Compos. Anal.* **2022**, *106*, 104295. [CrossRef]
42. Callejón, R.; Torija, M.; Mas, A.; Morales, M.; Troncoso, A. Changes of volatile compounds in wine vinegars during their elaboration in barrels made from different woods. *Food Chem.* **2010**, *120*, 561–571. [CrossRef]
43. Zhu, H.; Falcone, P.M.; Qiu, J.; Ren, C.-Z.; Li, Z.-G. Effect of ageing on rheological properties and quality of Shanxi aged vinegar. *IOP Conf. Ser. Earth Environ. Sci.* **2020**, *615*, 012096. [CrossRef]
44. Bozkurt, H.; Göğüş, F.; Eren, S. Nonenzymic browning reactions in boiled grape juice and its models during storage. *Food Chem.* **1999**, *64*, 89–93. [CrossRef]
45. España. Denominación de Origen de España. *Dossier de Informacion.Vinagre de Jerez.* Available online: https://www.vinagredejerez.org/noticias/Dossier_de_prensa_Vinagre_de_Jerez.pdf (accessed on 1 September 2024).
46. Issa-Issa, H.; Noguera-Artiaga, L.; Sendra, E.; Pérez-López, A.J.; Burló, F.; Carbonell-Barrachina, Á.A.; López-Lluch, D. Volatile composition, sensory profile, and consumers' acceptance of Fondillón. *J. Food Qual.* **2019**, *2019*, 5981762. [CrossRef]
47. Tarko, T.; Krankowski, F.; Duda-Chodak, A. The impact of compounds extracted from wood on the quality of alcoholic beverages. *Molecules* **2023**, *28*, 620. [CrossRef]
48. García-Moreno, M.V.; Sánchez-Guillén, M.M.; Ruiz de Mier, M.; Delgado-González, M.J.; Rodríguez-Dodero, M.C.; García-Barroso, C.; Guillén-Sánchez, D.A. Use of alternative wood for the ageing of brandy de Jerez. *Foods* **2020**, *9*, 250. [CrossRef] [PubMed]
49. Masson, G.; Guichard, E.; Fournier, N.; Puech, J.-L. Stereoisomers of ß-methyl-γ-octalactone. II. Contents in the wood of French (Quercus robur and Quercus petraea) and American (Quercus alba) oaks. *Am. J. Enol. Vitic.* **1995**, *46*, 424–428. [CrossRef]
50. Piggott, J.; Conner, J.; Melvin, J. The contribution of oak lactone to the aroma of wood-aged wine. In *Developments in Food Science*; Elsevier: Amsterdam, The Netherlands, 1995; Volume 37, pp. 1695–1702.

51. Vilela, A. Microbial dynamics in sour–sweet wine vinegar: Impacts on chemical and sensory composition. *Appl. Sci.* **2023**, *13*, 7366. [CrossRef]
52. Salles, C. Acids in foods and perception of sourness. In *Handbook of Molecular Gastronomy*; CRC Press: Boca Raton, FL, USA, 2021; pp. 7–12.
53. Junge, J.Y.; Bertelsen, A.S.; Mielby, L.A.; Zeng, Y.; Sun, Y.-X.; Byrne, D.V.; Kidmose, U. Taste interactions between sweetness of sucrose and sourness of citric and tartaric acid among Chinese and Danish consumers. *Foods* **2020**, *9*, 1425. [CrossRef] [PubMed]
54. Lytra, G.; Tempere, S.; Marchand, S.; de Revel, G.; Barbe, J.-C. How do esters and dimethyl sulphide concentrations affect fruity aroma perception of red wine? Demonstration by dynamic sensory profile evaluation. *Food Chem.* **2016**, *194*, 196–200. [CrossRef]

Disclaimer/Publisher's Note: The statements, opinions and data contained in all publications are solely those of the individual author(s) and contributor(s) and not of MDPI and/or the editor(s). MDPI and/or the editor(s) disclaim responsibility for any injury to people or property resulting from any ideas, methods, instructions or products referred to in the content.

molecules

MDPI

Article

Comparison of Flavor Differences between the Juices and Wines of Four Strawberry Cultivars Using Two-Dimensional Gas Chromatography-Time-of-Flight Mass Spectrometry and Sensory Evaluation

Wei Lan [1,2,3], Wei Cheng [1,2], Ruilong Li [1,2], Mei Zhang [1,2], Mengmeng Li [1,2], Yuan Zhang [1,2] and Yibin Zhou [3,*]

1 School of Biology and Food Engineering, Fuyang Normal University, Fuyang 236037, China; lanwei@fynu.edu.cn (W.L.); 13805585071@163.com (W.C.); pennylrl@163.com (R.L.); 22211310@stu.fynu.edu.cn (M.Z.); limengmeng@fynu.edu.cn (M.L.); zy7704@foxmail.com (Y.Z.)

2 Anhui Engineering Research Center for Functional Fruit Drink and Ecological Fermentation, Fuyang 236037, China

3 College of Food and Nutrition, Anhui Agricultural University, Hefei 230036, China

* Correspondence: zhouyibin@ahau.edu.cn

Abstract: Fruit wine production is a practical approach for extending the shelf life and enhancing the value of strawberries (*Fragaria* × *ananassa*). Fruit cultivars and juices are important sources of volatile organic compounds (VOCs) that determine fruit wine sensory quality. In this study, VOCs in the juices and wines of four strawberry cultivars were identified using two-dimensional gas chromatography-time-of-flight mass spectrometry, and a sensory analysis of the wines was performed. A total of 1028 VOCs were detected. PCA and OPLS-DA distinguished the four cultivars from which the juices and wines were made. Six VOCs with variable importance in projection values greater than one were the main aroma and flavor components of strawberry wines. ZJ wine had the highest sensory scores for coordination (9.0) and overall evaluation (8.9) among the 18 descriptors of strawberry wine evaluated. Overall, the ZJ wine had the highest alcohol content (13.25 ± 0.59%, v/v) and sensory evaluation score, indicating that the ZJ cultivar is more suitable for fermentation. This study reflects the differences between wines made from four strawberry cultivars and provides a reference for brewing fruit wines.

Keywords: strawberry juice; strawberry wine; volatile organic compounds; two-dimensional gas chromatography–time-of-flight mass spectrometry; sensory evaluation

1. Introduction

Fruit wine, one of the oldest fermented beverages made from non-grape fruits, is generally produced by first extracting juice from the fruit and then brewing it through processes such as fermentation, soaking, distillation, or storage [1,2]. The quality of fruit wine is affected by many factors, with the quality of the raw materials being the main factor that directly affects the taste, flavor, and nutritional content of the product. The nutritional composition of fruits (sugars, organic acids, etc.) differs according to the fruit variety, growing region, climate, and harvest season [2–4]. Therefore, the fermentation process varies for particular fruits selected for wine production, resulting in content differences in volatile organic compounds (VOCs), which ultimately determine the sensory quality of the wine [3]. During the brewing process, the main factors affecting the quality of fruit wine include the ratio, initial sugar content, and inoculation amount of the raw material, as well as fermentation temperature, time, and pH [2–5]. Moreover, the optimal fermentation conditions for different raw fruit materials differ slightly [6–8]. Therefore, the development of fermentation techniques to enhance the VOCs and sensory attributes of fruit wines would help increase the market share of such wines, especially products with Chinese-favored characteristics.

Strawberry (*Fragaria* × *ananassa*) wine production is a practical approach to extend the shelf life and enhance the value of the fruit, with the fruit nutrients preserved through the fermentation process [9,10]. However, although the concentrated harvesting period and short shelf life of strawberries can be resolved by turning the fruits into wine, the appropriate cultivars to use for alcoholic fermentation remain unknown [6,11–13]. Although cultivar selection is a prerequisite for quality control of raw materials for fruit winemaking, knowledge regarding the fermentation quality of different strawberry cultivars remains limited. Moreover, aside from the basic characteristics that provide fruit wine with nutritional and health benefits, flavor components and their composition are also important for determining the quality and typical flavor of wine [14,15]. More than 360 volatile flavor compounds have been identified in strawberries and strawberry wine, and the aromas of different fruit cultivars are characterized by distinct combinations of compounds such as alcohols, esters, organic acids, aldehydes, ketones, terpenes, and sulfur compounds [15]. However, the differences in VOCs between strawberry juice and wines of different cultivars have not been fully elucidated, and the influence of juice on the VOCs of wine remains unknown.

VOC profiles of different fruit wines are typically studied using various pretreatment methods and detection equipment. Headspace solid-phase microextraction (HS-SPME), which has been widely used to study VOCs in fruit wine and other alcoholic beverages, is a simple and fully automatic operation with no complex pretreatment steps [15–17]. Moreover, two-dimensional gas chromatography (GC × GC) has a higher peak capacity, higher sensitivity, and better substance-separating power than those of its one-dimensional counterpart, and coupled with time-of-flight mass spectrometry (TOF-MS), GC × GC–TOF-MS has been successfully applied to detect thousands of compounds in distilled liquors or wines [17–19]. For example, HS-SPME coupled with GC × GC-TOF-MS was used to identify 196 biomarkers among 1695 compounds detected in Italian sparkling wines, thereby facilitating the determination of the influence of grape cultivars, pedoclimatic conditions, and metabolomic space on these wines [19].

In this study, we aimed to determine the VOC profiles of the juices and wines of four typical strawberry cultivars widely cultivated in China: Snow White (BX), Sweet Charlie (TCL), Tongzhougongzhu (TZ), and Zhangji (Akihime, ZJ). GC × GC–TOF-MS combined with HS-SPME was used to detect and identify VOCs. Sensory analysis of the four wines produced (designated as BX, TCL, TZ, and ZJ) was performed to distinguish the source and variety of strawberry raw materials for wine fermentation. Our results provide valuable information for the in-depth brewing of strawberry wine.

2. Results and Discussion

2.1. Sensory Evaluation of Strawberry Wines

In sensory evaluations, both volatile and non-volatile components are assessed simultaneously, and strawberry wine is mainly characterized by its appearance, aroma, taste, and overall body [11,15]. In this study, the sensory characteristics of four strawberry wines were quantitatively and descriptively analyzed, and 18 descriptors and their corresponding reference samples were identified (Figure 1). The sensory profiles of the four wines differed significantly based on fruit and flower notes. In terms of aroma attributes, TCL had the highest scores for fruity and floral notes, whereas ZJ had the highest scores for coordination (9.0) and overall evaluation (8.9). In contrast, BX had the lowest coordination and overall evaluation scores.

Floral aromas are associated with the presence of phenylethanol, geraniol, citronellol, and nonanal compounds, as well as the aromatic activities of ethyl acetate, ethyl propanoate, 3-methyl 1-butanol, methyl butanoate, and ethyl hexanoate [20]. In summary, the sensory profiles of strawberry wines are primarily influenced by their VOC composition.

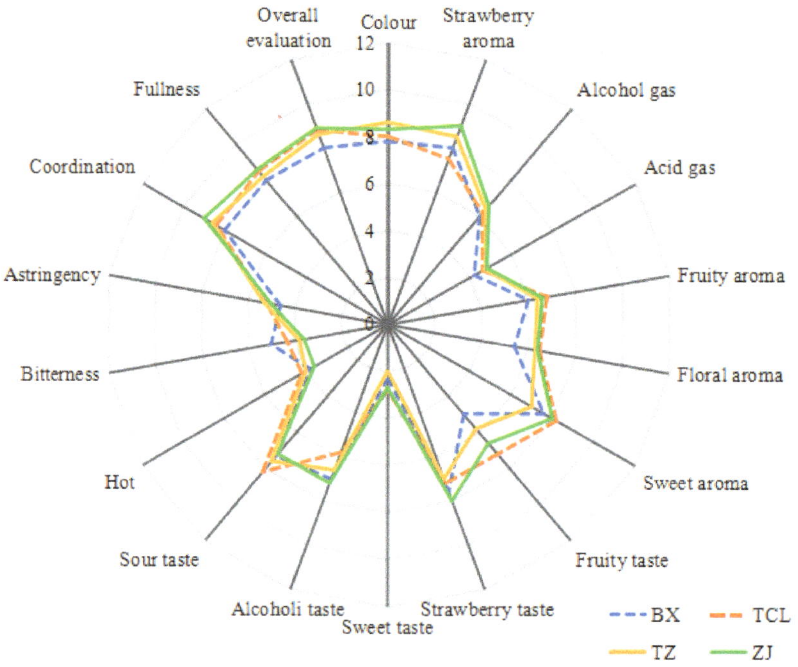

Figure 1. Sensory evaluation of four strawberry wines designated as BX, TCL, TZ, and ZJ. The 18 descriptors and their corresponding reference samples are shown. BX, Snow White; TCL, Sweet Charlie; TZ, Tongzhougongzhu; ZJ, Akihime.

In terms of taste, ZJ had the highest score for alcohol thickness, TCL had the highest score for sour and sweet palatability, and BX had low sour and sweet palatability scores and lacked alcohol thickness and persistence. In addition, ZJ had the highest scores for harmony, fullness, and overall evaluation. In general, the sensory analysis highlighted the different aromas and taste descriptors of the four strawberry wines, and six of the 18 sensory attributes (strawberry aroma, fruity aroma, sweet aroma, strawberry taste, sour taste, and bitterness) were emphasized in two or three samples. Therefore, these sensory attributes may be characteristic of strawberry varieties.

2.2. Physicochemical Properties of Juices and Wines

For fruit wines, ethanol and residual sugar contents are the main quality indicators of fermentation [2,11]; those of the fermented strawberry juices of the four cultivars are shown in Table 1. The tastes of fruit juices and wines were also affected by the pH and total acid content; those of the four samples are shown in Table 2. After fermentation, the ethanol content of the four strawberry cultivars was >13.00% (v/v) (Table 1), and the residual sugar content was <4.00 g/L (Table 1), which is the standard for dry fruit wines. The pH values of the non-fermented strawberry juices varied among the different cultivars, with BX juice (BX_J) having the highest value (4.06) and TCL_J having the lowest (3.36) (Table 2). With regard to wines, BX had the highest pH value (4.03), whereas TCL had the lowest pH (3.53), indicating that the fermentation process had little effect on pH. In alcoholic fermentation, pH is a critical parameter, as it affects the growth of yeasts and the characteristics (e.g., color, aroma, and taste) of the final alcohol product. For most types of fermentation, a pH range of 2.80–4.00 is considered ideal [21,22] and is suitable for the growth of yeasts and other brewing microbes.

Table 1. Ethanol and residual sugar contents in fermented juices of four different strawberry cultivars.

Cultivar/Sample	Ethanol (%, *v/v*)	Residual Sugar (g/L)
BX	13.25 ± 0.59	3.47 ± 0.65 [a]
TCL	13.02 ± 0.6	3.44 ± 0.17 [a]
TZ	13.12 ± 0.08	1.86 ± 0.02 [b]
ZJ	13.69 ± 0.1	3.22 ± 0.04 [a]

Note: Means ± standard deviations (n = 3) followed by different letters within each row indicate significant differences (Duncan's test, $p < 0.05$). BX, Snow White; TCL, Sweet Charlie; TZ, Tongzhougongzhu; ZJ, Akihime.

Table 2. pH values and total acid contents of wines and juices produced from four different strawberry cultivars.

Cultivar/ Sample	pH		Total Acid (g/L)	
	Juice	Wine	Juice	Wine
BX	4.06 ± 0.05 [a]	4.03 ± 0.05 [a]	4.03 ± 0.05 [d]	5.46 ± 0.37 [b]
TCL	3.36 ± 0.05 [d]	3.53 ± 0.05 [d]	6.63 ± 0.15 [a]	7.7 ± 0.1 [a]
TZ	3.7 ± 0 [b]	3.83 ± 0.05 [b]	6.26 ± 0.05 [b]	7.53 ± 0.23 [a]
ZJ	3.6 ± 0 [c]	3.73 ± 0.05 [c]	5.46 ± 0.32 [c]	6.56 ± 0.11 [b]

Note: Means ± standard deviations (n = 3) followed by different letters within each row indicate significant differences (Duncan's test, $p < 0.05$). BX, Snow White; TCL, Sweet Charlie; TZ, Tongzhougongzhu; ZJ, Akihime.

The ZJ wine had the highest alcohol content (13.25 ± 0.59%, v/v) among the four wines, whereas its residual sugar content, pH value, and total acid content were at comparatively moderate levels among the different samples. These indicators may affect the harmony, fullness, and overall evaluation of wines. Additionally, differences in the pH, residual sugar content, and total acidity of strawberry juice resulted in significant differences in the physicochemical properties of the wines after fermentation. These indicators may affect substance metabolism during brewing and further influence the quality characteristics (e.g., aroma and taste) of wine [23].

The physicochemical properties of strawberry wines are closely related to the raw materials (cultivars, origins, and states), microbial strains, and fermentation technologies used, whereas other factors such as the color, antioxidative properties, and VOCs of strawberry wines fermented with juice have been found to differ from those of wines fermented with pulp [11,14,24]. Furthermore, aside from having a lighter color, strawberry wines fermented from the juice of pulp macerated at 50 °C have more desirable qualities and pleasant floral and fruity notes than those of wines fermented from the juice of pulp macerated at normal temperatures [24].

2.3. *Identification and Statistical Analysis of Volatile Organic Compounds*

2.3.1. Identification of Volatile Organic Compounds

VOCs determine the sensory characteristics of food and are of great utility in the food industry as markers of authenticity. Aroma plays a vital role in shaping flavor profiles and is a key attribute that affects consumer satisfaction [11]. The combined qualitative and quantitative analysis of VOCs using chromatography and MS techniques overcomes the shortcomings of separate traditional chromatography, MS, and qualitative data and provides the added benefits of simplified operation and shortened analysis time [25,26]. Compared to GC-MS, GC × GC-TOF-MS uses two columns with different properties to separate compounds that have not been completely separated in the first dimension, thereby facilitating the detection of a more comprehensive range and a higher quantity of volatile flavor compounds [17,25]. As shown in Figure 2, a good classification of VOCs in strawberry wines and juices was obtained (Figure 2A,B), and the number of VOCs identified was higher than that identified by GC-MS [25].

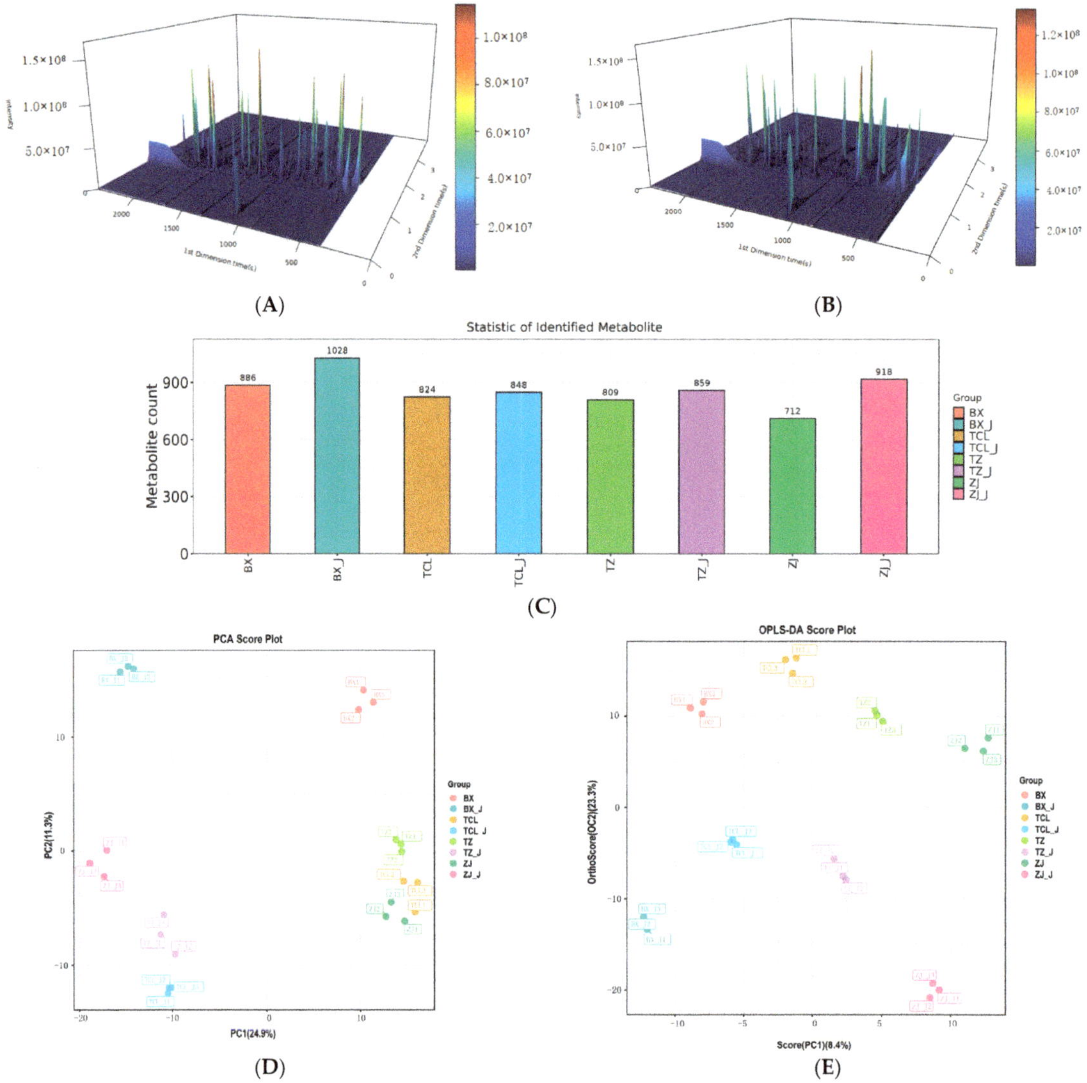

Figure 2. VOCs identified in four strawberry juices (_J) and wines. (**A**) Three-dimensional total ion chromatogram (TIC) of BX_J. (**B**) Three-dimensional total TIC of BX wine. (**C**) Quantitative comparison of VOCs. (**D**) PCA of VOCs. (**E**) OPLS-DA scores of VOCs.

The quantities of VOCs detected in strawberry wines and juices by GC $\times$ GC-TOF-MS are shown in Table 3. In total, 886 metabolites were identified in the BX wine (Figure 2C), including 48 ketones, 88 hydrocarbons, 16 heterocyclic compounds, 9 aldehydes, 130 esters, 82 alcohols, 35 carboxylic acids, and 478 other compounds (Table 3). In contrast, 1028 substances were identified in BX_J (Figure 2C), including 63 ketones, 109 hydrocarbons, 26 heterocyclic compounds, 20 aldehydes, 123 esters, 96 alcohols, 39 carboxylic acids, and 552 other compounds (Table 3). Strawberry aroma is characterized by a combination of different compounds, each of which contributes to the characteristics of different cultivars [15]. In the present study, the VOC detection results indicated that the aroma of strawberry wine was characterized by a combination of ketones, hydrocarbons, heterocyclic compounds, aldehydes, esters, alcohols, and carboxylic acids.

Table 3. Quantification of VOCs detected in strawberry wines and juices using GC × GC–TOF-MS.

Group		Ketones	Hydrocarbons	Heterocyclic Compounds	Aldehydes	Esters	Alcohols	Carboxylic Acids	Others	Total
Wine	BX	48	88	16	9	130	82	35	478	886
	TCL	47	74	15	6	125	86	30	441	824
	TZ	43	73	20	9	121	79	30	434	809
	ZJ	40	58	23	9	100	73	27	382	712
Juice	BX_J	63	109	26	20	123	96	39	552	1028
	TCL_J	61	90	22	10	103	90	16	456	848
	TZ_J	53	88	16	10	132	84	20	456	859
	ZJ_J	63	90	11	17	112	102	24	499	918

Esters, which impart a characteristic aroma and flavor to strawberries, were the main VOCs detected in the four wines and juices, with hexanoic acid ethyl ester, heptanoic acid methyl ester, ethyl acetate, and methyl hexanoate being the most abundant (Table S3). Additionally, the primary furanone detected was dibenzofuran, the primary aldehydes were lilac aldehyde A and 2-hexenal, and the main sulfur compounds and terpenes were dimethyl disulfide, beta-myrcene, gamma-terpinene, DL-panlactone, and D-limonene. Ketones and long-chain fatty acids such as 2-heptanone, 2-pentanone, and *n*-decanoic acid were also detected. Because these volatiles have low sensory thresholds, other low-abundance volatiles may also contribute significantly to their distinctive aroma [11]. These compounds are the main contributors to the strawberry-like aroma, imparting both fruity and caramel-like notes [11,27]. In summary, these compounds defined the dominant and characteristic sensory impressions of strawberry wine and juice.

Esters are a group of volatile constituents present in fruits and vegetables, and those with oral and fruity aromas are the main constituents of strawberry wine [11]. A comparison of the relative content of VOCs in strawberry wines and juices is shown in Table 4, where the total ester content differed significantly among the different samples. Among the wines, TCL had the highest relative amount of esters (37.77 μg/L), whereas ZJ had the lowest (19.27 μg/L), which matched the sensory evaluation results shown in Figure 2. Among the juices, BX_J had the highest relative amount of esters (31.36 μg/L), and TZ_J had the lowest (21.57 μg/L). The most abundant esters in strawberry wine are ethyl octanoate, ethyl decanoate, isoamyl acetate, and ethyl laurate, which are characteristic VOCs [11,15]. In general, strawberry wines contain the largest variety and the highest concentrations of ethanol esters, whereas Merlot wines contain higher than other types of esters [28].

Table 4. Comparison of the relative amounts of VOCs between strawberry wines and juices (μg/L).

Group		Alcohols	Aldehydes	Carboxylic Acids	Esters	Heterocyclic Compounds	Hydrocarbons	Ketones	Others
Wine	BX	29.77	0.09	5.82	27.92	5.78	1.14	0.81	28.62
	TCL	21.80	0.14	5.51	37.77	0.25	1.58	1.40	31.50
	TZ	29.75	0.10	3.78	28.38	1.56	2.30	1.01	33.09
	ZJ	38.81	0.16	5.09	19.27	6.30	0.51	0.84	29.00
Juice	BX_J	18.82	0.69	7.16	31.36	0.53	1.18	3.12	37.11
	TCL_J	22.19	0.20	8.98	18.13	3.37	1.47	2.23	43.40
	TZ_J	26.37	0.08	1.79	21.57	0.28	0.63	4.26	44.98
	ZJ_J	17.44	0.64	3.72	21.99	0.07	2.58	6.45	47.07

Additionally, the ZJ wine had the highest relative amount of alcohol (38.81 μg/L), whereas TCL had the lowest (21.80 μg/L). Alcohols are important aromatic constituents of fermented beverages because they contribute substantially to odor intensity, although they generally tend to impart unpleasant odors [29]. Furans, ketones, aldehydes, and aromatic species were also detected in all four wines (Table S1), albeit in low amounts. These VOCs may play complementary, coordinated, and modifying roles in the aroma and sensory characteristics of strawberry wines [24].

2.3.2. Principal Component Analysis and Orthogonal Partial Least Square-Discriminant Analysis of the Volatile Organic Compounds

Unconstrained principal component analysis (PCA) and constrained orthogonal partial least squares discriminant analysis (OPLS-DA) were performed to characterize the differences and similarities in VOCs among the juices and wines made from the four cultivars. PCA transforms component data into comprehensive indicators via dimensionality reduction and determines the weight of key components according to their correlations and degrees of variation [30]. As shown in Figure 2D, the combined interpretation rate of PC1 and PC2 was 36.2%, indicating that the differences in VOCs among the samples could be effectively described. The four types of strawberry juice and wine were separated without significant overlap, indicating that the samples from the different cultivars could be well distinguished. The VOCs differed significantly between the juice and wine samples, with the sample points located in different quadrants and regions. Sample points BX and BX_J, which were in the first and second quadrants, respectively, were relatively far from the other sample points, indicating that their differences from the other cultivars were greater.

As shown in Figure 2E, the OPLS-DA results were generally consistent with the PCA results (Figure 2D), with strawberry juice and wine samples located in the upper and lower quadrants, respectively. The BX wine samples were located in the first quadrant, whereas TZ, TCL, and BX were in the fourth quadrant. The distances between them were relatively close, indicating that their VOC components were relatively similar. Overall, the differences in the VOCs in strawberry juices and wines were obvious, and PCA and OPLS-DA could clearly distinguish the juices and wines of the four strawberry cultivars based on their VOCs.

2.4. Comparison of Key Differential Volatile Organic Compounds and Their Relative Odor Activity Values in the Strawberry Juices and Wines

2.4.1. Comparison of the Quantities and Relative Amounts of Key Differential Volatile Organic Compounds in Strawberry Juices

Figure 3 shows the relative amounts of VOCs in strawberry juice and wine, with differences indicated by different colors, where a darker red color indicates a higher relative amount and a darker blue color indicates a lower relative amount. The inner columns represent the samples, the rows represent the metabolites, and the cluster tree on the left represents the clustering of the differential VOC species. In BX_J, the levels of 1-hexanol, 1-butanol-3-methyl-acetate, 5-methylhexanoic acid, 2-pentanone, and 2,3-butanedione were relatively high. In contrast, butanedioic acid diethyl ester, 2-pentanone, ethyl-5-methylhexanoate, 2-ethyl-1-propanol, 2-methyl-butanoic acid, and linalool levels were relatively high in BX wine. In the ZJ wine, the levels of linalool, 2-ethyl-1-propanol, ethyl-2-hydroxy-3-phenylpropanoate, butanedioic acid diethyl ester, and 2,3-butanedione were relatively high, imparting cheesy, floral, and flowery aromas to the wine [11,23].

Among strawberry wines, ZJ had a significantly higher alcohol content, TCL had higher amounts of esters and ketones, and BX had a significantly higher carboxylic acid content (Table 3, Figure 3B). Among the four strawberry juices, BX_J was rich in aldehydes and carboxylic acids (Figure 3A), making it the most suitable for fresh consumption. However, the ZJ wine had a relatively high content of different VOCs (Figure 3B), indicating that the ZJ cultivar may be more suitable for strawberry wine fermentation.

Additionally, among the differential VOCs, 157 were common in strawberry juice and 190 were common in strawberry wine, and the number of differential VOCs increased after fermentation (Tables S1 and S2). The differential VOCs may be an important reason for the differences in sensory flavors between strawberry wines and juices. However, microbial strains, inoculation methods, and fermentation processes may also influence VOCs during the fruit winemaking process [11,14,23,31]. Therefore, studies on the optimal fermentation processes are needed after suitable cultivars for strawberry wine production have been identified [2,5,23,32].

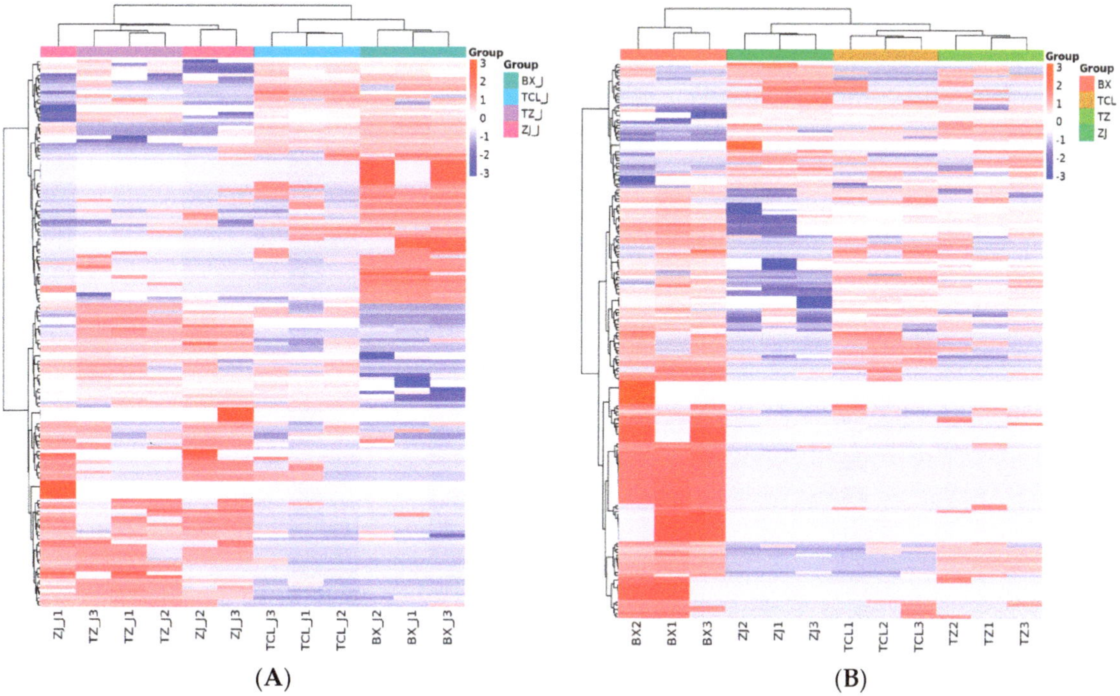

Figure 3. Cluster heat map of key differential VOCs in strawberry juices (**A**) and wines (**B**). Adjusted $p < 0.05$ (Tukey's test).

2.4.2. Comparison of Relative Odor Activity Values of Key Differential Volatile Organic Compounds in Strawberry Wines

In the PCA, the combined interpretation rates of PC1 and PC2 reached 47.7% and 25.5%, respectively (Figure 4A), indicating differences in VOCs among the samples. The OPLS-DA results were consistent with the PCA results (Figure 4B) with strawberry wine samples located in different quadrants.

In the selection of key differential VOCs, those with a variable importance in projection (VIP) value exceeding one are usually considered the key volatiles of the group. The VIP value represents the weight ascribed to the variables within the PLS-DA model, where a higher value correlates with a more substantial contribution magnitude. Generally, VIP values exceeding 1 indicate pivotal differentiating components among samples, whereas values below 1 indicate negligible influence on differentiation [33]. As shown in Figure 4C, the VIP values of six volatile aroma components (butanoic acid ethyl ester, hexanoic acid, 3-methyl-1-butanol, benzenepropanoic acid ethyl ester, butanoic acid 3-methyl-ethyl ester, and butanedioic acid diethyl ester) exceeded 1 ($p < 0.05$). The VIP values were calculated using OPLS-DA based on the relative odor activity values (ROAVs). The differences in VOCs among strawberry juices and wines were statistically significant and were the key indicators of the differences in volatile aroma components in the wines of the different cultivars. Moreover, together with the aroma characteristics of these six aroma compounds, we found that fruit aroma was the main flavor component in strawberry wine (Table S3), indicating that these six compounds could influence the differences between strawberry wine samples, which is similar to the results of the sensory evaluation.

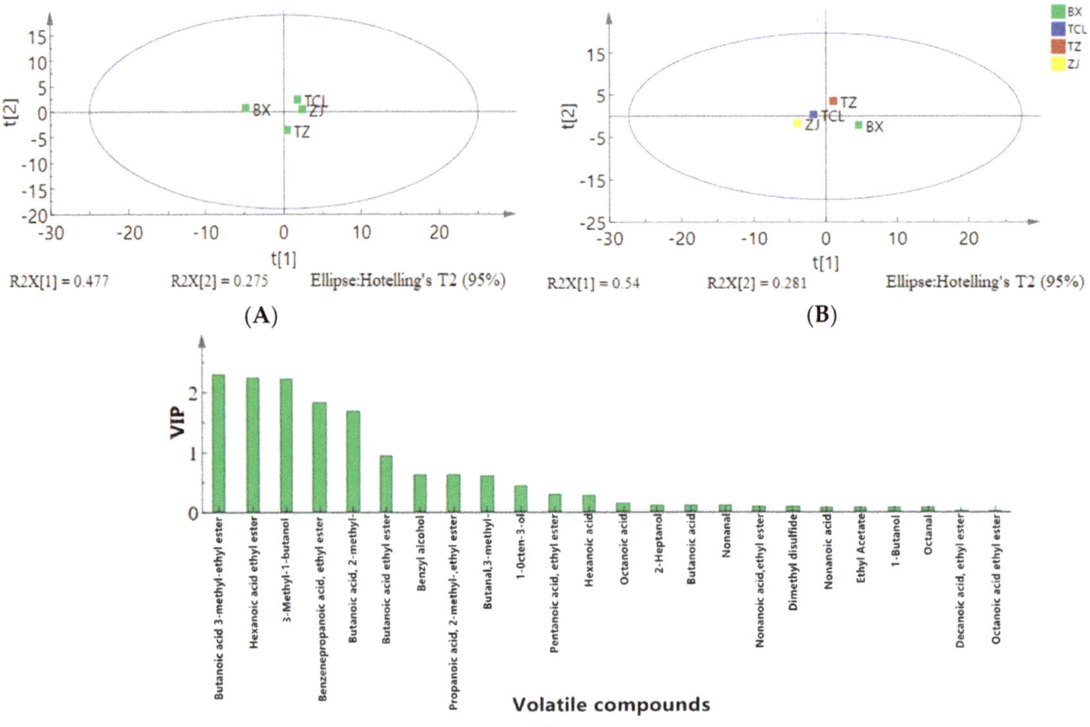

Figure 4. Coordinate and variable importance in projection (VIP) analyses based on the relative odor activity values of key differential VOCs in strawberry wines of different cultivars. (**A**) PCA. (**B**) OPLS-DA. (**C**) VIP value.

Five key differential VOCs with ROAVs exceeding 1 (butanoic acid 3-methyl-ethyl ester, 3-methyl-1-butanol, hexanoic acid ethyl ester, butanoic acid ethyl ester, and octanoic acid ethyl ester) played important roles in the sensory features of the strawberry juice and wine samples (Table S3). Moreover, six key differential VOCs with ROAVs exceeding 1 (the five key differential VOCs listed above and 2-methyl-propanoic acid ethyl ester) played important roles in the sensory features of the strawberry wine samples (Table S3).

2.4.3. Network Diagram of the Relationships between Various Volatile Organic Compounds for Imparting Unique Sensory Flavor Characteristics in Strawberry Wines

With the FlavorDB2 database [34], the iGraph tool was used to build a network diagram of the relationships between flavor substances to impart unique sensory flavor characteristics. In Figure 5, the green circles represent sensory features, and the red circles represent flavor compounds. A larger green circle indicates that more types of flavor compounds are associated with that sensory feature, rendering that feature more important. By contrast, a larger red circle indicates that more sensory features are associated with the flavor compound, rendering it an important flavor substance. By default, the top 10 sensory features were used to create a network diagram.

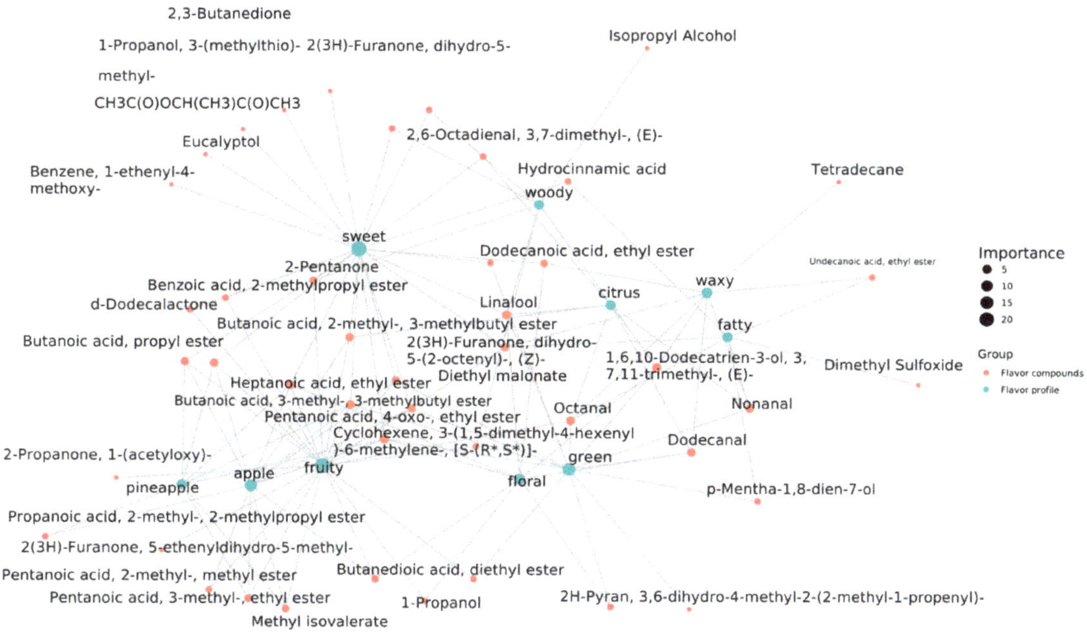

Figure 5. Network showing the relationships between the sensory flavor characteristics and flavor substances of strawberry wines.

As shown in Figure 5, fruity, sweet, apple, green, and floral sensory features were associated with more than nine types of flavor compounds. For example, the fruity sensory feature was associated with butanoic acid propyl ester, 3-methyl butyl ester, 1-propanol, and 19 other flavor compounds. Floral sensory features were associated with dodecanoic acid ethyl ester, linalool, 1-propanol, and six other flavor substances. Overall, substances related to fruit aroma were the main flavor components of strawberry wine. At the same time, we also noticed that the flavor compound linalool was associated with floral, sweet, green, woody, and citrus sensory features. Linalools exhibit antimicrobial, anti-inflammatory, anticancer, and antioxidative properties [35]. However, linalool was not detected in strawberry juice, indicating that it might be produced by yeast metabolism during fermentation [36].

Few studies have examined the correlation between cultivar origin, sensory quality, fermentation process, and characteristics of different strawberry cultivars [4]. Sensory evaluations based on consumer preferences are indispensable for the production and marketability of strawberry wine. Given that VOCs significantly influence sensory attributes, a comprehensive analysis using GC-MS combined with electronic tongue, electronic nose, and sensory evaluations is vital for ensuring the quality of strawberry wines.

3. Materials and Methods

3.1. Samples and Reagents

3.1.1. Strawberry Cultivars and Yeast Strain for Brewing

Four strawberry cultivars, named Snow White (BX), Sweet Charlie (TCL), Tongzhougongzhu (ZJ), and Zhangji (Akihime, ZJ), were collected in Wenji Town (N: 33°03′22.50″, E: 115°38′43.60″; Fuyang City, Anhui Province, China) between March and April 2022. These four typical strawberry cultivars are widely cultivated in China, and the sampled strawberry fruits had a normalized shape, 90% maturity, no damage, and approximately the same size, and were stored at −20 °C after screening until processing for juice press and fermentation. The yeast strain Lalvin Rhône 2323 (Lallemand Inc., Montreal, Canada) was used to brew the strawberry wine.

3.1.2. Strawberry Juice Preparation and Wine Fermentation

Frozen strawberry fruit (5 kg) was thawed at 25 °C for 2 h and then crushed using a juice press. Samples (5 mL) of the juice were collected for GC × GC-TOF-MS analysis of VOCs. The juices were designated as ZJ_J, TCL_J, BX_J, and TZ_J, and each sample was tested three times.

Before the fermentation process, 30 ppm of SO_2 and 300 mg of the commercial pectinase Lallzyme EX-V (30 mg/L) were added to the raw juice, and the mixture was incubated for 2 h at 40 °C [15]. The enzyme-treated juice was centrifuged at 8000× *g* for 10 min at 4 °C and the Brix was then adjusted to 23° with sucrose [31]. The mixture was pasteurized for 30 s in a hot water bath at 97 °C and then immediately cooled to ambient temperature in an ice bath. For fermentation, 300 mL of pasteurized strawberry juice was immediately transferred to a 500 mL sterile jar [4].

The Lalvin Rhône 2323 yeast strain was activated for 12 h at 28 °C in yeast extract-peptone-dextrose liquid medium and diluted strawberry juice (with purified water), and 1.0% (*v*/*v*) strawberry juice preculture was inoculated into the pasteurized juice to start fermentation. The fermentation temperature was controlled at 25 °C (± 2 °C) [37]. During brewing, Brix was measured every 2 days until there was no change in three consecutive measurements, which signaled the completion of alcoholic fermentation [4]. Then, 5 mL samples of fermented strawberry wine were used for VOC analysis. The wines were designated ZJ, TCL, BX, and TZ, and each sample was tested three times.

3.1.3. Chemicals and Reagents

Typical C7 to C30 alkanes (≥99.8%) were purchased from Sigma (St. Louis, MO, USA). Analytical-grade sodium chloride was purchased from Sinopharm (Shanghai, China), and chromatography-grade *n*-hexyl-d13 alcohol (≥98.5%) was purchased from C/D/N Isotopes (Quebec, QC, Canada). Chromatography-grade anhydrous ethanol (≥99.8%) was purchased from Aladdin (Shanghai, China). Ultrapure water was obtained using a Milli-Q system (Millipore, Billerica, MA, USA).

3.2. Descriptive Sensory Analysis

Sensory analysis was performed using the methods described by Zhao et al. [20], and wine samples were assessed by a panel of eight experts (four females and four males, between 25 and 40 years of age, from Anhui Jinzhongzi Distillery, Anhui, China), all of whom had more than 5 years of experience in the sensory evaluation of various drinks (including Baijiu and wines). Before carrying out the sensory evaluation of the strawberry wines, the eight experts underwent training for 4 h per week over 2 months using the "Le Nez du Vin" aroma kit (54 aromas; Yixiangle, Hong Kong, China) [15]. Sensory evaluation was performed in a well-ventilated, odor-free, quiet room, and a mouthwash was provided to the panelists during the evaluation.

The sensory qualities of the four strawberry wine varieties were analyzed quantitatively and descriptively. The evaluators discussed and identified the 18 descriptors and their corresponding reference samples. Descriptive terms for color, six aroma attributes, eight taste attributes, and three overall evaluations of the wine body were scored on a scale of 0–9 points (0–2, very weak; 3–5, medium; and 6–9, very strong) [38]. Each panelist individually scored the samples in the order mentioned above, and a mean score was obtained [20,38].

3.3. Physicochemical Parameter Detection

The °Brix and pH values were measured using a refractometer (ATGO, Tokyo, Japan) and pH meter (FE28, Mettler Toledo, Greifensee, Switzerland), respectively. A fully automatic analyzer (Y15, BioSystem, Barcelona, Spain) and relevant chemical kits were used to analyze the total acid content. The glucose and ethanol contents were determined using high-performance liquid chromatography on an LC-16 system (Shimadzu, Kyoto, Japan) equipped with an RID-20A refractive index detector and a Bio-Rad Aminex HPX-87H resin-based column (300 × 7.8 mm) at 55 °C, with 5 mM H_2SO_4 used for elution at a flow rate of 0.5 mL/min [39].

3.4. GC × GC–TOF-MS Analysis of Volatile Organic Compounds

3.4.1. Preparation of Internal Standard Solution

An appropriate amount of *n*-hexyl-d 13 alcohol was transferred to a volumetric flask, dissolved in 50% (*v*/*v*) ethanol solution to a final concentration of 10 mg/L, and stored at 4 °C. A stock solution with 10 mg/L of *n*-alkanes was prepared in *n*-hexane and stored at 4 °C.

3.4.2. HS-SPME Method

An appropriate amount of sample was transferred to a 15 mL glass test tube and diluted to a 10% ethanol concentration (*v*/*v*) with an aqueous solution of saturated NaCl [40]. Then, 5 mL of each diluted sample was transferred to a 20 mL headspace vial. Subsequently, 10 μL of the internal standard solution was added to each sample, and the mixtures were incubated at 50 °C for 10 min. The samples were extracted using a headspace solid-phase microextracter coated with a DVB/CAR/PDMS fiber head (50/30 μm × 1 cm; Supelco, Bellefonte, PA, USA) and incubated at 50 °C for 20 min.

The extracted samples were desorbed in the GC injection port at 250 °C for 5 min before GC × GC–TOF-MS analysis using set parameters [41–43]. After the injection step, SPME fiber was placed in the chamber at 270 °C for 10 min. Then, 10 μL of the *n*-alkanes was transferred to a 20 mL headspace vial and extracted and injected into the GC port as described above. Each sample was subjected to three parallel tests to ensure the reproducibility of the experimental results.

3.4.3. GC × GC–TOF-MS Method

GC × GC-TOF-MS analyses were performed using a LECO Pegasus 4D instrument (LECO, St. Joseph, MI, USA) consisting of an Agilent 8890A GC system (Agilent Technologies, Palo Alto, CA, USA) equipped with a split/splitless injector and dual-stage cryogenic modulator (LECO) coupled to a TOF-MS detector (LECO). A DB-Heavy Wax column (30 m × 250 μm I.D., 0.5 μm; Agilent) was used for the first dimension, and a Rxi-5Sil MS column (2.0 m × 150 μm I.D., 0.15 μm; Restek, Bellefonte, PA, USA) for the second dimension.

For GC × GC, high-purity helium (>99.999%) was used as a carrier gas at a constant flow rate of 1.0 mL/min. The temperature program of the oven was as follows: the oven temperature was first held at 40 °C for 3 min, then raised to 200 °C at the rate of 6 °C/min, then raised to 250 °C at the rate of 10 °C/min, and finally held for 5 min. The second oven temperature was set to 5 °C higher than that of the first oven. The temperature of the modulator was always 15 °C higher than that of the second column. The modulator was operated with a 4.0 s modulation period. The GC injector temperature was 250 °C.

For the TOF-MS, the flavor substances were analyzed on a LECO Pegasus BT 4D system, with the transfer line and TOF-MS ion source temperatures both set at 250 °C, respectively. The acquisition frequency was 200 spectra/s. The mass spectrometer was operated in electron impact ionization mode at 70 eV using an *m*/*z* range of 35–550 and a detector voltage of 1960 V [41–43].

3.5. Statistical Analysis

All measurements were performed in triplicate and presented as mean values and standard deviations. The statistical significance of the data was evaluated via variance analysis by comparative averages (ANOVA) using the Duncan test and SPSS 22.0, program for Windows (IBM, New York, NY, USA).

3.5.1. Data Processing

Data were collected using a Pegasus 4D workstation (LECO) and analyzed and processed using the Chroma TOF software program built into the instrument. This program automatically identified peaks with a signal-to-noise ratio of over 50 and compared them with the NIST 14 and Wiley 9 MS libraries to automatically generate a peak table. Compounds with halogen and silicon elements were removed, and chromatographic peaks with more than 700 forward and reverse similarities were screened using MS [18].

The retention index (RI) of each VOC was calculated for C7–C30 *n*-alkanes and compared with the reference RI values in the NIST online database (https://webbook.nist.gov/; accessed on 21 September 2022). RI differences of 50 or less were selected, and compounds with greater than 50% occurrence rates were considered reliable results [44]. The internal standard method was used to calculate the content of each flavor compound, as shown in Equation (1).

$$\text{Volatile organic compound content } (\mu g/L) = \frac{\text{Peak area of compound} \times \text{Internal standard amount } (\mu g/L)}{\text{Peak area of internal standard}} \quad (1)$$

The ROAVs were used to estimate the contribution of aromatic compounds to the overall flavor of the wine. VOCs with ROAVs exceeding 1 were generally the key flavor compounds, whereas those with ROAVs between 0.1 and 1 were not considered important for the overall flavor of the sample [45]. ROAVs were calculated by dividing the relative concentration of an aromatic compound by its odor threshold value. The largest defined group was divided into a ROAV of 100, and the ROAVs of the other substances were corrected according to Equation (2):

$$\text{ROAV} = (\text{Peak B}/\text{TB})/(\text{Peak A}/\text{TA}) \quad (2)$$

where Peak A is the maximum peak area of the component, Peak B is the peak area of the material to be measured, TA is the odor value of the largest component, and TB is the odor value of the compound to be measured.

3.5.2. Chemometric Analysis of the GC × GC–TOF-MS Data

GC × GC-TOF-MS data were mined using a chemometric method, and irrelevant and redundant variables were filtered through preprocessing. The variables were analyzed semi-quantitatively based on internal standards. A small portion of the missing values was interpolated using the K-nearest neighbor algorithm based on machine learning to simplify the univariate and multivariate analyses. Variables with no statistical significance were removed by univariate analysis, and potentially important compounds were preliminarily screened. Single-factor analysis of variance (ANOVA) was performed using SPSS software (version 22.0; IBM, New York, NY, USA).

The error detection rate was used to correct the *p*-values to reduce false-positive results, and the variables were preliminarily screened on the basis of a *p*-value of less than 0.05 and a Pearson correlation coefficient $|r|$ greater than 0.6. Data selected by one-way ANOVA were used for multivariate statistical analyses after scaling by unit variance. PCA and OPLS-DA were performed using SIMCA-P 14.1 software. Cluster heat and radar maps were plotted using the BioDeep online analysis platform (https://www.biodeep.cn/, accessed on 16 August 2024.). The correlative network of data pairs satisfying $|\rho| > 0.6$ and $p < 0.05$ was visualized with Gephi (Version 0.9.1).

4. Conclusions

This study was conducted to reveal the VOC characteristics of the juices and wines of four typical strawberry cultivars and to distinguish the sources and varieties of strawberry raw materials for wine fermentation. The findings revealed that the samples had obvious differences in VOC content, and PCA and OPLS-DA of the VOCs could distinguish strawberry juices and wines of the four different cultivars. In strawberry wines, six VOCs had VIP values exceeding 1: butanoic acid ethyl ester, hexanoic acid, 3-methyl-1-butanol, benzenepropanoic acid ethyl ester, butanoic acid 3-methyl-ethyl ester, and butanedioic acid diethyl ester. Overall, the ZJ wine had a relatively high alcohol content (13.25 ± 0.59%, v/v) and sensory evaluation score, indicating that the ZJ cultivar is more suitable for fermentation. The main factors affecting the differences in flavor-related VOCs in wine include cultivars and cultivation techniques, fermentation strains and processes, storage containers and conditions, and aging times and processes. Further studies will be conducted to determine the effects of these factors on the quality of strawberry wine with the aim of providing reference data for the future production of such fruit wines.

Supplementary Materials: The following supporting information can be downloaded at: https://www.mdpi.com/article/10.3390/molecules29194691/s1, Table S1. Classification and relative content of different VOCs in strawberry wine samples. Table S2. Classification and relative content of different VOCs in strawberry juice samples. Table S3. ROAV values of different VOCs in samples.

Author Contributions: Y.Z. (Yibin Zhou): conceptualization, methodology, supervision, review, and editing; W.L.: data curation, investigation, methodology, and writing—original draft; R.L. and W.C.: investigation and data curation; M.Z. and M.L.: investigation and review; Y.Z. (Yuan Zhang): review and editing. All authors have read and agreed to the published version of the manuscript.

Funding: This research was supported by the open research project of the Anhui Ecological Fermentation Engineering Research Center for Functional Fruit Beverage (Grant number: FSKFKT032D), the "Key Technology and Industrialization Demonstration of Fruit Wine Quality Improvement" cooperation project between Fuyang Normal University and the Government of Fuyang City (Grant number: SXHZ202004), and the key natural science research projects in Anhui universities (Grant number: 2022AH051326).

Institutional Review Board Statement: Not applicable.

Informed Consent Statement: No applicable.

Data Availability Statement: The original contributions presented in the study are included in the article/Supplementary Materials, further inquiries can be directed to the corresponding author.

Acknowledgments: This work was supported by members of the Anhui Engineering Research Center for Functional Fruit Drinks and Ecological Fermentation.

Conflicts of Interest: The authors declare no conflicts of interest.

References

1. Yang, H.; Cai, G.; Lu, J.; Gómez Plaza, E. The production and application of enzymes related to the quality of fruit wine. *Crit. Rev. Food Sci.* **2021**, *61*, 1605–1615. [CrossRef] [PubMed]
2. Saranraj, P.; Sivasakthivelan, P.; Naveen, M. Fermentation of fruit wine and its quality analysis: A review. *Aust. J. Sci. Technol.* **2017**, *1*, 85–97. Available online: https://aujst.com/vol-1-2/07_AJST.pdf (accessed on 10 August 2024).
3. Tan, J.; Ji, M.; Gong, J.; Chitrakar, B. The formation of volatiles in fruit wine process and its impact on wine quality. *Appl. Microbiol. Biotechnol.* **2024**, *108*, 420. [CrossRef] [PubMed]
4. Lan, W.; Zhang, M.; Xie, X.; Li, R.; Cheng, W.; Ma, T.; Zhou, Y. Effects of cultivar factors on fermentation characteristics and volatile organic components of strawberry wine. *Foods* **2024**, *13*, 2874. [CrossRef] [PubMed]
5. Cordenunsi, B.R.; Genovese, M.I.; do Nascimento, J.R.O.; Hassimotto, N.M.A.; dos Santos, R.J.; Lajolo, F.M. Effects of temperature on the chemical composition and antioxidant activity of three strawberry cultivars. *Food Chem.* **2005**, *91*, 113–121. [CrossRef]
6. Yang, B.; Liu, S.; Zang, H.; Dai, Y.; Zhang, S.; Lin, X.; Liang, H.; Chen, Y. Flavor profile and quality of strawberry wine are improved through sequential fermentation with indigenous non-*Saccharomyces* yeasts and *Saccharomyces cerevisiae*. *Food Biosci.* **2024**, *59*, 104021. [CrossRef]
7. Sturm, K.; Koron, D.; Stampar, F. The composition of fruit of different strawberry varieties depending on maturity stage. *Food Chem.* **2003**, *83*, 417–422. [CrossRef]
8. Wang, X.; Wu, L.; Qiu, J.; Qian, Y.; Wang, M. Comparative metabolomic analysis of the nutritional aspects from ten cultivars of the strawberry fruit. *Foods* **2023**, *12*, 1153. [CrossRef]
9. Giampieri, F.; Tulipani, S.; Alvarez-Suarez, J.M.; Quiles, J.L.; Mezzetti, B.; Battino, M. The strawberry: Composition, nutritional quality, and impact on human health. *Nutrition* **2012**, *28*, 9–19. [CrossRef]
10. Zhao, Z.; Wu, X.; Chen, H.; Liu, Y.; Xao, Y.; Chen, H.; Tang, Z.; Li, Q.; Yao, H. Evaluation of a strawberry fermented beverage with potential health benefits. *Peer J.* **2021**, *9*, e11974. [CrossRef]

11. Bezerra, M.; Ribeiro, M.; Cosme, F.; Nunes, F.M. Overview of the distinctive characteristics of strawberry, raspberry, and blueberry in berries, berry wines, and berry spirits. *Compr. Rev. Food Sci. Food Saf.* **2024**, *23*, e13354. [CrossRef] [PubMed]
12. Yang, W.; Liu, S.; Marsol-Vall, A.; Tähti, R.; Laaksonen, O.; Karhu, S.; Yang, B.; Ma, X. Chemical composition, sensory profile and antioxidant capacity of low-alcohol strawberry beverages fermented with *Saccharomyces cerevisiae* and *Torulaspora delbrueckii*. *LWT* **2021**, *149*, 111910. [CrossRef]
13. Zhang, H.; Li, J.; Xu, X.; Zhang, X.; Lan, W.; Wang, Y.; Gao, X. Volatile compositions and sensorial properties of strawberry fruit wines fermented with *Torulaspora delbrueckii* and *Saccharomyces cerevisiae* in sequential and simultaneous inoculations. *Eur. Food Res. Technol.* **2024**, 1–13. [CrossRef]
14. Wang, F.; Zhao, P.; Du, G.; Zhai, J.; Guo, Y.; Wang, X. Advancements and challenges for brewing aroma-enhancement fruit wines: Microbial metabolizing and brewing techniques. *Food Chem.* **2024**, *456*, 139981. [CrossRef]
15. Song, Y.; Zhang, Y.J.; Liu, N.; Ye, D.Q.; Gong, X.; Qin, Y.; Liu, Y.L. Volatile compounds in wild strawberry and their odorants of wild strawberry wines: Effects of different stages of fermentation. *Int. J. Food Prop.* **2017**, *20*, S399–S415. [CrossRef]
16. Ulrich, D.; Kecke, S.; Olbricht, K. What do we know about the chemistry of strawberry aroma? *J. Agr. Food Chem.* **2018**, *66*, 3291–3301. [CrossRef]
17. Liu, Z.; Yang, K.; He, Z.; Zhao, D.; Zheng, J.; Qian, M.C. Comparison of two data processing approaches for aroma marker identification in different distilled liquors using comprehensive two-dimensional gas chromatography-time-of-flight mass spectrometry dataset. *J. Food Sci.* **2023**, *88*, 2870–2881. [CrossRef] [PubMed]
18. Mu, X.; Lu, J.; Gao, M.; Li, C.; Chen, S. Optimization and validation of a headspace solid-phase microextraction with comprehensive two-dimensional gas chromatography time-of-flight mass spectrometric detection for quantification of trace aroma compounds in Chinese liquor (Baijiu). *Molecules* **2021**, *26*, 6910. [CrossRef]
19. Carlin, S.; Vrhovsek, U.; Franceschi, P.; Lotti, C.; Bontempo, L.; Camin, F.; Toubiana, D.; Zottele, F.; Toller, G.; Fait, A.; et al. Regional features of northern Italian sparkling wines, identified using solid-phase micro extraction and comprehensive two-dimensional gas chromatography coupled with time-of-flight mass spectrometry. *Food Chem.* **2016**, *208*, 68–80. [CrossRef]
20. Zhao, Y.; Yu, X.; Zhu, F.; Liu, G.; Chu, L.; Yan, X.; Ma, Y.; He, F.; Li, G.; Zhang, Y.; et al. Effects of different winemaking yeasts on the composition of aroma-active compounds and flavor of the fermented jujube wine. *Processes* **2021**, *9*, 970. [CrossRef]
21. Sepúlveda, E.; Sáenz, C.; Peña, Á.; Robert, P.; Bartolomé, B.; Gómez-Cordovés, C. Influence of the genotype on the anthocyanin composition, antioxidant capacity and color of Chilean pomegranate (*Punica granatum* L.) juices. *Chil. J. Agric. Res.* **2010**, *70*, 50–57. Available online: https://digital.csic.es/bitstream/10261/44992/1/Sepulveda_Elena.pdf (accessed on 10 August 2024). [CrossRef]
22. Zarei, M.; Azizi, M.; Bashir-Sadr, Z. Evaluation of physicochemical characteristics of pomegranate (*Punica granatum* L.) fruit during ripening. *Fruits* **2011**, *66*, 121–129. [CrossRef]
23. Scutarașu, E.C.; Teliban, I.V.; Zamfir, C.I.; Luchian, C.E.; Colibaba, L.C.; Niculaua, M.; Cotea, V.V. Effect of different winemaking conditions on organic acids compounds of white wines. *Foods* **2021**, *10*, 2569. [CrossRef] [PubMed]
24. Lv, Z.; Liu, H.; Yang, W.; Zhang, Q.; Chen, D.; Jiao, Z.; Liu, J. Comprehensive analysis of physicochemical properties and volatile compounds in different strawberry wines under various pre-treatments. *Molecules* **2024**, *29*, 2045. [CrossRef] [PubMed]
25. Wang, J.; Chen, H.; Wu, Y.; Zhao, D. Uncover the flavor code of strong-aroma Baijiu: Research progress on the revelation of aroma compounds in strong-aroma Baijiu by means of modern separation technology and molecular sensory evaluation. *J. Food Compos. Anal.* **2022**, *109*, 104499. [CrossRef]
26. Jia, W.; Fan, Z.; Du, A.; Li, Y.; Zhang, R.; Shi, Q.; Shi, L.; Chu, X. Recent advances in Baijiu analysis by chromatography based technology–A review. *Food Chem.* **2020**, *324*, 126899. [CrossRef]
27. Song, C.; Hong, X.; Zhao, S.; Liu, J.; Schulenburg, K.; Huang, F.C.; Franz-Oberdorf, K.; Schwab, W. Glucosylation of 4-hydroxy-2,5-dimethyl-3(2H)-furanone, the key strawberry flavor compound in strawberry fruit. *Plant Physiol.* **2016**, *171*, 139–151. [CrossRef]
28. Massera, A.; Assof, M.; Sari, S.; Ciklic, I.; Mercado, L.; Jofré, V.; Combina, M. Effect of low temperature fermentation on the yeast-derived volatile aroma composition and sensory profile in Merlot wines. *LWT* **2021**, *142*, 111069. [CrossRef]
29. Wu, J.; Liu, Y.; Zhao, H.; Huang, M.; Sun, Y.; Zhang, J.; Sun, B. Recent advances in the understanding of off-flavors in alcoholic beverages: Generation, regulation, and challenges. *J. Food Compos. Anal.* **2021**, *103*, 104117. [CrossRef]
30. Cheng, W.; Chen, X.; Guo, Y.; Zhou, D.; Zeng, H.; Fu, H. The microbial diversity and flavour metabolism of Chinese strong flavour Baijiu: A review. *J. Inst. Brew.* **2023**, *129*, 15–38. [CrossRef]
31. Huang, J.; Wang, Y.; Ren, Y.; Wang, X.; Li, H.; Liu, Z.; Yue, T.; Gao, Z. Effect of inoculation method on the quality and nutritional characteristics of low-alcohol kiwi wine. *LWT* **2022**, *156*, 113049. [CrossRef]
32. Hwang, H.; Kim, Y.-J.; Shin, Y. Influence of ripening stage and cultivar on physicochemical properties, sugar and organic acid profiles, and antioxidant compositions of strawberries. *Food Sci. Biotechnol.* **2019**, *28*, 1659–1667. [CrossRef] [PubMed]
33. Cheng, W.; Chen, X.; Xue, X.; Lan, W.; Zeng, H.; Li, R.; Pan, T.; Li, N.; Gong, Z.; Yang, H. Comparison of the correlations of microbial community and volatile compounds between pit-mud and fermented grains of compound-flavor Baijiu. *Foods* **2024**, *13*, 203. [CrossRef] [PubMed]
34. Grover, N.; Goel, M.; Batra, D.; Garg, N.; Tuwani, R.; Sethupathy, A.; Bagler, G. Flavor DB2: An updated database of flavor molecules. *arXiv* **2022**, arXiv:2205.05451.
35. Kamatou, G.P.; Viljoen, A.M. Linalool–A review of a biologically active compound of commercial importance. *Nat. Prod. Commun.* **2008**, *3*, 1183–1192. [CrossRef]

36. Amiri, P.; Shahpiri, A.; Asadollahi, M.A.; Momenbeik, F.; Partow, S. Metabolic engineering of *Saccharomyces cerevisiae* for linalool production. *Biotechnol. Lett.* **2016**, *38*, 503–508. [CrossRef]
37. Chen, D.; Chia, J.Y.; Liu, S.-Q. Impact of addition of aromatic amino acids on non-volatile and volatile compounds in lychee wine fermented with *Saccharomyces cerevisiae* MERIT.ferm. *Int. J. Food Microbiol.* **2014**, *170*, 12–20. [CrossRef]
38. Cao, W.; Shu, N.; Wen, J.; Yang, Y.; Jin, Y.; Lu, W. Characterization of the key aroma volatile compounds in nine different grape varieties wine by headspace gas chromatography–ion mobility spectrometry (HS-GC-IMS), odor activity values (OAV) and sensory analysis. *Foods* **2022**, *11*, 2767. [CrossRef]
39. Sun, X.; Li, L.; Ma, T.; Liu, X.; Huang, W.; Zhan, J. Profiles of phenolic acids and flavan-3-ols for select Chinese red wines: A comparison and differentiation according to geographic origin and grape variety. *J. Food Sci.* **2015**, *80*, C2170–C2179. [CrossRef]
40. Robinson, A.L.; Boss, P.K.; Heymann, H.; Solomon, P.S.; Trengove, R.D. Development of a sensitive non-targeted method for characterizing the wine volatile profile using headspace solid-phase microextraction comprehensive two-dimensional gas chromatography time-of-flight mass spectrometry. *J. Chromatogr. A* **2011**, *1218*, 504–517. [CrossRef]
41. He, X.; Jeleń, H.H. Comprehensive two-dimensional gas chromatography time-of-flight mass spectrometry (GC × GC-TOFMS) in conventional and reversed column configuration for the investigation of Baijiu aroma types and regional origin. *J. Chromatogr. A* **2021**, *1636*, 461774. [CrossRef] [PubMed]
42. Perestrelo, R.; Petronilho, S.; Camara, J.S.; Rocha, S.M. Comprehensive two-dimensional gas chromatography with time-of-flight mass spectrometry combined with solid phase microextraction as a powerful tool for quantification of ethyl carbamate in fortified wines. The case study of Madeira wine. *J. Chromatogr. A* **2010**, *1217*, 3441–3445. [CrossRef] [PubMed]
43. Yan, Y.; Chen, S.; Nie, Y.; Xu, Y. Characterization of volatile sulfur compounds in soy sauce aroma type Baijiu and changes during fermentation by GC×GC-TOFMS, organoleptic impact evaluation, and multivariate data analysis. *Food Res. Int.* **2020**, *131*, 109043. [CrossRef] [PubMed]
44. Liu, S.P.; Zhou, Z.L.; Yu, H.B.; Mao, L.G.; Pan, X.X.; Ji, Z.W.; Mao, J. Characterization of volatile aging markers in traditional semi-dry Huangjiu by comprehensive two-dimensional gas chromatography-time of flight mass spectrometry and chemometrics. *Food Sci.* **2023**, *44*, 249–258. [CrossRef]
45. Liang, H.; Gao, D.; Wang, C.; Gao, H.; Guo, Y.; Zhao, Z.; Shi, H. Effect of fermentation strategy on the quality and aroma characteristics of yellow peach wines. *Fermentation* **2022**, *8*, 604. [CrossRef]

Disclaimer/Publisher's Note: The statements, opinions and data contained in all publications are solely those of the individual author(s) and contributor(s) and not of MDPI and/or the editor(s). MDPI and/or the editor(s) disclaim responsibility for any injury to people or property resulting from any ideas, methods, instructions or products referred to in the content.

beverages

MDPI

Article

Influence of Terroir on the Grain Composition, and Volatile Profile of Irish Grain (Wheat) New Make Spirit

Anukriti Vashishtha [1], Kieran N. Kilcawley [2,3,4], Iwona Skibinska [2], Stephen Whelan [5], John L. Byrne [1], Guiomar Garcia-Cabellos [1] and Sinead Morris [1,*]

[1] Department of Applied Science, South East Technological University, R93 V960 Carlow, Ireland; c00290664@setu.ie (A.V.); john.byrne@setu.ie (J.L.B.); guiomar.garcia-cabellos@setu.ie (G.G.-C.)

[2] Food Quality & Sensory Science Department, Teagasc Food Research Centre, Moorepark, Fermoy, P61 C996 Cork, Ireland; kieran.kilcawley@teagasc.ie (K.N.K.); iwona.skibinska@teagasc.ie (I.S.)

[3] The School of Food Science and Environmental Health, Technical University of Dublin, D07 EWV4 Dublin, Ireland

[4] School of Food and Nutritional Sciences, University College Cork, T12 CY82 Cork, Ireland

[5] South East Technological University, Y35 KA07 Wexford, Ireland; stephen.whelan@setu.ie

* Correspondence: sinead.morris@setu.ie

Abstract: Terroir refers to the combination of environmental factors, such as climate, soil, and agricultural practices, that shape the characteristics of a crop, contributing to the unique qualities of the final product. The concept has been traditionally linked to wine, but some recent findings suggest that it also holds importance for distilled spirits. The expanding Irish distilling sector is shifting towards local raw materials such as wheat and rye, driven by regulatory changes, economic benefits, and consumer demand for sustainable local products. This research examines the effects of wheat variety, geographical location, and harvest year on grain composition and volatile composition of the new make spirit. For this study, twenty lab-scale wheat whiskey samples were produced from five different wheat varieties grown at two different locations in Ireland over two consecutive years. The wheat samples were analysed for grain composition and the volatile profiling of new make spirit samples by headspace solid-phase microextraction (HS-SPME) followed by gas chromatography–mass spectrometry (GC-MS). A total of fifty-one volatile compounds were detected, with ethanol, ethyl acetate, phenyl ethyl alcohol, and 3-methyl-1-butanol being predominant. Principal component analysis revealed that both the harvest year and geographical location moderately influenced the volatile compound distribution of the new make spirit, which is explained by a 43.25% variance. ANOVA analysis revealed that grain composition was significantly influenced by harvest year, location, and wheat variety. The 2020 samples showed higher protein and β-glucan content, whereas samples from the location Tipperary had higher starch content. This study indicates that terroir—specifically seasons (year) and geography (location)—affects the characteristics of wheat-based Irish whiskey, highlighting opportunities for distillers to differentiate their products by leveraging local environmental factors.

Keywords: Irish whiskey; volatile organic compound; wheat variety; terroir; GC-MS

1. Introduction

The Irish whiskey sector is experiencing a renaissance characterised by rapid growth and innovation, with an increasing emphasis on local raw material and grain diversity. For decades, imported maize dominated the Irish distilling business, particularly in the production of blended whiskey, which is by far the largest volume of whiskey style produced and exported from Ireland. Notably, grain whiskey constitutes approximately 80% of all blends and is traditionally made using maize. However, the Irish technical file, which sets out the geographical area covered by the GI and the systems and materials used in the production of Irish whiskey, is currently under review because future regulations may require Irish grain whiskey to be produced solely from the locally produced grains. To accommodate these anticipated changes and the expansion in the distilling sector, there is a growing need to explore alternative grains such as wheat, especially due to the high production of Irish wheat [1]. Ireland produces 0.68 million tonnes of wheat per year, a production volume high enough to make it a suitable candidate as an alternative grain [2]. The promotion of local grains can also significantly enhance rural economic growth and provide stable employment opportunities [3]. This shift to alternative crops reflects not only the dynamic nature of the industry but also the evolving economic landscape and the pressing need for distilleries to distinguish their service in a crowded market. Amidst this context, the concept of terroir emerges as a potential unique selling point (USP), offering a novel dimension to the narrative and marketing of the new whiskey product [4]. These evolving factors create the framework for further investigation of the elements that make Irish whiskey unique, with an emphasis on understanding the role of terroir in shaping its flavour.

Terroir, a French term traditionally associated only with wine, consists of the influence of a locale's environmental characteristics, such as climate, topography, and soil composition, on the phenotype of crops, therefore affecting the flavour profile of the end product derived from them [5,6]. The concept has been extensively applied to wine, with research suggesting a molecular correlation between terroir and the unique characteristics of wines. Just as in wine, in the beer industry, the influence of barley genotype on sensory descriptors has demonstrated terroir's significance in affecting how the flavour ranges from cereal and floral to malty and fruity [7,8]. Based on these foundations and existing research, recent studies have expanded the application of terroir to distilled beverages, including whiskey [9,10]. Given the growing interest in grain diversity and local sourcing within the industry, the investigation of terroir's impact on whiskey production is very pertinent. The interest is not only driven by the quest for quality and authenticity but also by the strategic and marketing need to differentiate products in a highly competitive environment.

The intricate process of whiskey production, from grain selection to maturation, can significantly influence the sensory attributes of the final product. Factors such as the selection of raw materials and production processes such as malting, mashing, distilling, and, most importantly, maturation shape the whiskey's flavour [11]. Esters primarily produced during alcoholic fermentation by yeast infuse the whiskey with fruity and floral notes, such as those of ethyl hexanoate and isoamyl acetate, while alcohols derived both directly from the raw materials and produced during the malting process, add to the whiskey's complexity [12]. The malting process, through a sequence of enzymatic reactions known as the lipoxygenase pathway, transforms lipids in the presence of oxygen to produce distinctive alcohols such as hexanol. Additional alcohols, including 2-methyl-1-butanol and 3-methyl-1-butanol, emerge during fermentation, further diversifying the spirit's aromatic profile [13]. The maturation process in casks is another critical phase where the new make spirit acquires depth, character, and a harmonious blend of flavours; however, these aspects during maturation are beyond the scope of this study. Gas chromatography–mass spectrometry (GCMS) is widely used to identify volatile compounds in distilled spirits, providing a greater understanding of aromatic components generated from the grains or the production processes [14,15].

Critics of terroir may argue that the distillation process can mask the subtle influences of the original raw material. However, some previous research on barley and maize whiskey have shown that if the production parameters are kept constant, there is a measurable variance in the whiskey flavours, attributable to the different grain types, and grain production environment/geography, reinforcing the idea that terroir does indeed play a critical role in shaping the sensory profiles of whiskey. By investigating the effects of the wheat varieties harvested over two consecutive years in distinct Irish locations, such as Carlow and Tipperary, this paper seeks to contribute to the ongoing debate on terroir in whiskey, aiming to elucidate its potential as a distinctive selling point in the evolving landscape of the Irish whiskey sector.

The overarching aim of this research is to study the effect of terroir factors such as season (year), geography (location), and crop variety on the grain composition and volatile profile of Irish Whiskey. In the subsequent sections, we will be discussing the methodology, results, and discussions centered around the findings of the above objective.

2. Materials and Methods

2.1. Grain Samples

Wheat samples were sourced from field-based research trials by Teagasc, Oak park Rd, Oakpark Or Painestown, Carlow. These included the following varieties: Revelation, Viscount, Elation, Torp, and LG Astronomer, collected from two different locations (Carlow and Tipperary, Ireland) and across two harvest years (2020 and 2021).

In this study, the two different locations from where the crops were harvested (Carlow and Tipperary) differ in the environment, making the location a suitable factor for terroir analysis. Carlow, located in the southeast of Ireland, enjoys a temperate maritime climate characterised by mild winters and cool summers due to its position in the "Sunny Southeast". This climate, combined with a mix of fertile alluvial soils along the Barrow Valley and patches of stonier land, makes it suitable for a diverse range of agricultural practices. Alternatively, Tipperary, situated in the midwest and part of the Golden Vale, experiences slightly cooler temperatures and more precipitation, influenced by its varied topography, including several mountain ranges. The soils here are predominantly deep, fertile, and well-drained, enriched by limestone, which is ideal for intensive agricultural operations.

2.2. Grain Composition Analysis

The grain composition analysis was conducted to quantify the levels of β-Glucan, arabinoxylans, protein, and starch within the wheat samples. Protein measurements were taken using a whole grain analyser (Infratec 1241 grain analyser; Foss Tecator AB, Hoganas, Sweden).

- β-Glucan, Arabinoxylans, and starch analysis:

The quantification of β-glucan K-BGLU, SKU: 700004269, Arabinoxylans, and total starch using a Total starch assay kit (K-TSTA-100A; SKU: 700004351) was performed using specific assay kits supplied by Megaenzyme (Bray, Ireland). All the samples were analysed in triplicate, and their average mean and standard deviation values were recorded, as referred to in Table A2, Appendix A. The methodology for each component was executed according to the manufacturer's protocols.

- Moisture content determination:

The samples were first grounded using a Buhler Miag disc mill (Buhler Group Dublin, Ireland) to achieve the mean particle size of 0.2 mm. The moisture content method of the wheat was adapted from the European Brewing Convention (EBC: 6.2.2) [16] method for measuring the moisture content of maize. Approximately 5 g of the ground grain sample was spread evenly in a thin layer in pre-weighed aluminium moisture dishes about 50 mm in diameter and not more than 20 mm in depth. The samples were then dried in an oven set at 130–135 °C for 1 h. After drying, the moisture was calculated based on the weight difference before and after drying, using the following Equation (1).

$$Moisture\ content\ (\%) = \frac{Initial\ weight - Dry\ weight}{Initial} \times 100 \tag{1}$$

2.3. Whiskey Sample Preparation

Whiskey samples were prepared using the method adopted by Morris et al., 2022, [1], which simulated the "typical" lab-scale production of Irish whiskey. Wheat flour (30 g) was obtained by milling the grains in a Buhler Miag Disc mill (setting 0.2 mm) (Buhler Group, Dublin, Ireland). The flour was then transferred to the mash utensils and slurried with water (86 mL preheated to 40 °C), with the addition of α-amylase (39.6 μL, Kerry

BiofermTM LC Alpha-amylase, sourced from BSG CraftBrewing) and 141 mg/L ca ions in the form of calcium chloride dihydrate ($CaCl2{\cdot}2H2O$; ranged from 0–733.66 mg/L; PanReac Appliedchem ITW reagents, Dublin, Ireland). The contents were gradually heated to 78 °C (temperature rises to 2 °C/min) in a water bath and cooked for 123 min. The cooked slurry was then cooled to 66 °C and given a second treatment of α-amylase (14.4 μL, Kerry BiofermTM LC Alpha-amylase, sourced from BSG CraftBrewing), and amyloglucosidase (26.5 μL, AmyloTM 300, sourced from BSG CraftBrewing). This was mashed with an inclusion rate of 5% using high diastatic power-distilling malted barley (cv. Laureate, Miag setting 0.2 mm). After this step, the mash was cooled to 40 °C, followed by additional β-Glucanase (45 μL BioglucanaseR GB sourced from BSG CraftBrewing), and the mash was allowed to rest for 60 min. The mash was then cooled to 22 °C and was made up to 250 mL with water. After this, all the samples were fermented with distillers' yeast (Pinnacle 'M' type) at a pitching rate of 0.4% for 72 h. The fermented wort samples were distilled using the EBC Method 9.2.1 [17] in a still steam distillation apparatus. The final ABV% for the final samples was read on an Anton Paar 5000 density meter (Anton Paar, Dublin, Ireland), which was not more than 10% for any sample, preventing further dilution at the stage of volatile analysis using headspace solid phase microextraction (HS-SPME) GC-MS analysis. The distillates/fresh-made spirits were stored in sterile falcon tubes. The samples were stored under optimal conditions in a 0–5 °C refrigerator to preserve the chemical integrity for subsequent analysis.

2.4. Volatile Profile Analysis

- HS-SPME GCMS analysis

Twenty whiskey samples were analysed in triplicates for volatile components using headspace solid-phase microextraction gas chromatography–mass spectroscopy; sample IDs are shown in Table A1, Appendix A. The ABV% of these samples post-distillation was measured using an Anton Paar 5000 density meter (Anton Paar, Dublin, Ireland), which is also recorded in Table A1.

- SPME: Sample preparation

A 5 mL sample and 100 μL standard (4-methyl, 2-pentanol, and 2-methyl, 3-heptanone at 10 ppm) sample were added to a 200 mL screw-capped amber SPME vial with a magnetic cap and silicone/polytetrafluoroethylene septa (Element, Maynooth, Ireland) and equilibrated to 40 °C for 10 min with pulsed agitation of 5 s at 500 rpm. The sample was introduced using a Gerstel MPS autosampler.

- GCMS Method

A single 50/30 μm CarboxenTM/divinylbenzene/polydimethylsiloxane (DVB/CAR/PDMS) fibre was used (Agilent Technologies Ltd., Cork, Ireland). The SPME fibre was exposed to the headspace above the samples for 40 min at a depth of 1 cm at 40 °C. The fibre was retracted and injected into the GC inlet and desorbed for 2 min at 250 °C. Injections were made on a Shimadzu 2010 Plus GC (Mason Technology, Dublin, Ireland) with a DB-624 UI (60 m × 0.32 mm × 1.8 μm) (Agilent Technologies Ltd.) column using a split/split less injector with a 1/10 split. A Merlin micro seal was used as the septum (Agilent Technologies Ltd.). The temperature of the column oven was set at 40 °C, held for 5 min, increased at 5 °C/min to 230 °C, then increased at 15 °C/min to 260 °C, and held for 15 min, yielding a total GC run time of 60 min. The carrier gas was helium, held at a constant flow of 1.2 mL/min. The detector was a Shimadzu TQ8030 mass spectrometer detector (Mason Technology), run in single quad mode. The ion source temperature was 220 °C, and the interface temperature was set at 260 °C. The MS mode was electronic ionisation (70 v) with the mass range scanned between 35 and 250 amu. Compounds were identified using mass spectra comparisons to the NIST 2014 mass spectral library, a commercial flavour and fragrance library (FFNSC 2, Shimadzu Corporation, Japan), and an in-house library created using authentic compounds with target and qualifier ions and linear retention indices (LRI) for each compound using Kovats index [18]. Spectral deconvolution was also performed to confirm the identification of compounds using AMDIS [19]. Batch processing of samples was carried out using MetaMS [20]. MetaMS is an open-source pipeline for GC-MS-based untargeted metabolomics. An auto-tune of the GCMS was carried out before the analysis to ensure optimal GCMS performance. A set of external standards was run at the start and end of the sample set, and abundances were compared to known amounts to ensure that both the SPME extraction and MS detection were performed within specifications.

2.5. Statistical Analysis

Statistical analysis was carried out using R studio, R version 4.2.2 (2022-10-31 ucrt) [21] for principal component analysis (PCA), normality test, constant variance test, and statistical analysis of variability (ANOVA). Three-way ANOVA was conducted to determine if there were statistically significant differences in the grain composition and volatile profile based on factors such as wheat variety, geographical location, and harvest year and due to the interaction of these factors. Turkey's post hoc HSD was conducted on the most significant factors that affected the grain composition. PCA was used for volatile profile analysis because of its ability to reduce the dimensionality of large data sets, help identify clusters, and visualise how different terroir factors influence the samples' characteristics.

3. Results and Discussion

The following section includes the results of the effect of terroir factors, such as season (year), geography (location), and wheat variety, on grain composition and volatile profile of whiskey.

3.1. Grain Composition Analysis

The effect of variety, location, and year on the response variables—protein, β-glucan, and arabinoxylans—was evaluated using a three-way ANOVA, followed by post hoc analysis using Turkey's HSD. Before conducting the ANOVA, the assumptions of normality and homogeneity of variance were assessed for all response variables. Normality was evaluated using the Shapiro–Wilk test. The results indicated that protein and β-glucan required log transformations to meet normality assumptions, while starch and arabinoxylans were sufficiently normal. Homogeneity of variance was assessed using Levene's test, which confirmed that the variances were homogeneous across groups for each variable.

Statistical Analysis of Variability

ANOVA was conducted to assess the significance of the influence of site (Tipperary and Carlow), variety (LG Astronomer, Torp, Viscount, Elation, and Revelation), and harvest year (2020 and 2021) on the grain composition, as summarised in Table 1 below.

From the above table, it can be concluded that the protein content is primarily affected by the harvest year and wheat variety, which suggests that both the genetic makeup of the crop and climatic conditions during each harvest season play an important role in determining protein levels. There was a significant interaction between varieties and location, suggesting that the impact of variety on protein content depended on the location where it was grown. It aligns with the literature, which suggests that geography is known to affect the protein components of wheat, majorly gliadin and glutenin [22]. Additionally, interactions between variety and year, as well as location and year, were also significant. Post hoc analysis revealed that Torp and revelation had a higher protein content than Elation, and protein content was significantly lower in samples from 2021 than in 2020.

Table 1. ANOVA results for the effects of year, variety, and site on the composition of wheat grain.

Response Variable	Source	Df	Sum Sq	Mean Sq	F value	Pr (>F)
Protein	Variety	4	0.05	0.01	15.10	1.78×10^{-7} ***
	Location	1	0.0	1×10^{-5}	0.01	0.91
	Year	4	0.05	0.12	153.55	$<2 \times 10^{-16}$ ***
	Variety: Location	3	0.02	0.01	6.76	0.00091 ***
	Variety: Year	3	0.01	0.01	3.91	0.01 *
	Location: Year	1	0.01	0.01	21.81	3.70×10^{-5} ***
	Variety: Location: Year	3	0.01	0.00	2.31	0.09
	Residuals	38	0.03	0.00		
Beta Glucan	Variety	4	0.56	0.14	5.7	0.001076 **
	Location	1	2.85	2.85	114.74	4.88×10^{-13} ***
	Year	4	0.76	0.19	7.72	0.000118 ***
	Variety: Location	3	0.33	0.11	4.47	0.008769 **
	Variety: Year	3	0.77	0.25	10.36	4.04×10^{-5} ***
	Location: Year	1	0.32	0.32	12.96	0.000905 ***
	Variety: Location: Year	3	0.36	0.12	4.95	0.005355 **
	Residuals	38	0.94	0.02		
Starch	Variety	4	1288.9	322.2	8.77	4.06×10^{-5} ***
	Location	1	2213.0	2213.0	60.23	2.36×10^{-9} ***
	Year	4	727.9	182.0	4.95	0.00259 **
	Variety: Location	3	372.2	124.1	3.38	0.02806 *
	Variety: Year	3	161.1	53.7	1.46	0.24
	Location: Year	1	0.4	0.4	0.01	0.91
	Variety: Location: Year	3	296.3	98.8	2.69	0.06
	Residuals	38	1396.2	36.7		
Arabinoxylans	Variety	4	363.0	90.74	17.48	3.24×10^{-8} ***
	Location	1	31.3	31.26	6.02	0.01883 *
	Year	4	123.2	30.81	5.93	0.000821 ***
	Variety: Location	3	128.9	42.97	8.28	0.000231 ***
	Variety: Year	3	51.9	17.31	3.335	0.029365 *
	Location: Year	1	199.9	199.9	38.509	2.97×10^{-7} ***
	Variety: Location: Year	3	215.5	71.83	13.839	2.99×10^{-6} ***
	Residuals	38	197.3	5.19		

Significance level: ***: $p < 0.001$, highly significant; **: $0.001 \leq p \leq 0.01$, moderately significant, *: $0.01 \leq p \leq 0.05$, significant.

Starch content, on the other hand, was significantly influenced by all the factors (variety, location, and year). Factors such as environmental conditions, including temperature, rainfall, and soil characteristics, play a critical role in determining the synthesis and quality of starch in wheat grains. Studies have shown that even environmental stress factors, such as heat and drought, can directly impact starch metabolism by altering the enzyme activity during crucial periods like anthesis and grain filling, which in turn affects starch composition and quality across different years [23]. Additionally, factors such as variation in the timing of sowing and the specific environmental conditions during different seasons have also been found to significantly alter starch properties, such as viscosity and thermal characteristics, indicating the importance of site-specific agricultural practices and seasonal timing [24]. Post hoc analysis revealed that the starch content in Tipperary was significantly higher than that in Carlow.

For the non-starch polysaccharides (arabinoxylans and β-Glucan), all factors, such as wheat variety, location of harvest, and harvest year, had a significant effect, and even the interaction effect between all three factors was significant for the non-starch polysaccharides. Tipperary showed a significantly lower β-glucan content compared to Carlow, and 2021 samples had lower β-glucan content compared to 2020.

While the above findings suggest an influence of environmental factors as an integral part of terroir, additional factors such as soil nutrient composition might also play a role, highlighting the complex interplay of environmental influence.

3.2. Volatile Profile Analysis

In this section, the distribution of the volatile profiles was studied to determine the effect of terroir factors such as season (year), geography (location), and wheat variety.

In total, 51 volatile compounds were identified in the twenty newly made spirit samples (Tables A3 and A4—Appendix A). The aroma profile consisted of acetals (2), alcohols (12), aldehydes (9), benzenes (2), esters (19), ethers (1), furan (1), ketones (3), lactone (1), phenol (1).

The volatile compounds in Tables A3 and A4 in Appendix A offer a detailed volatile structure of the new-made samples, which were used for multivariate analysis. Ethanol, ethyl acetate, phenylethyl alcohol, 3-methyl-1-butanol, and 2-methyl-1-butanol were identified as the most abundant volatiles across the samples. Ethanol is the main alcohol in all distillates, so it was expected to be a dominant compound (the abundance of ethanol is not a reflection of the true ethanol content as the sample was introduced as a 1:20 split onto the GCMS so as not to saturate the column and detector with ethanol to prevent ethanol overlapping with other early eluting volatile components). Ethyl acetate, with its sweet and fruity attributes, was prominent, adding to the complex sensory profile of the whiskey. The concentration of ethyl acetate is indicative of a balance between yeast metabolism and the conditions during fermentation and distillation [25].

Phenylethyl alcohol, known for its rose-like honeyed odour profile, was also detected in significant quantities. This aromatic alcohol is associated with higher perceived aromatic complexity and quality in distilled spirits [26]. The higher alcohols, or fusel alcohols, such as 3-methyl-1-butanol and 2-methyl-1-butanol, are important components due to their impact on the overall character of the spirit. These compounds are often linked to malty and roasted sensory notes, contributing to the complexity of the aroma. Their concentration can serve as an indicator of yeast metabolism during fermentation, as well as the quality and characteristics of the raw material employed [27].

Principal Component Analysis

The analysis helped in examining clusters formed by whiskey in regards to the distribution of volatile compounds based on the location, year, and variety, as shown in Figure 1.

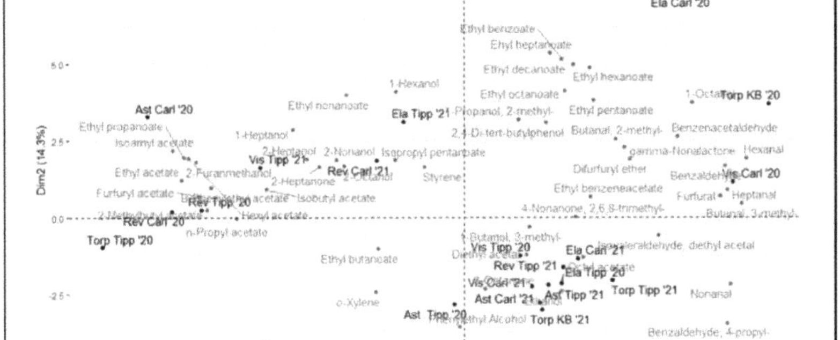

Figure 1. PCA Bi plot of separation based on the volatile profile (samples + volatiles); sample IDs present in Table A1.

Figure 1 presents a PCA plot, where each point represents an individual whiskey sample plotted against the first two principal components (PC1 and PC2), which together capture 43.3% of the total variance in the volatile compound data. The samples are distributed across all four quadrants, indicating a wide range of volatile compositions. However, the lack of distinct clustering suggests that some samples share more similarities than differences in their volatile profiles.

- Volatiles distribution

(a) Aldehyde and Ester

Elation and Viscount from the location Carlow and Torp from Tipperary, from the year 2020, had a higher concentration of aldehydes (straight chain, branched chain, and aromatic) and esters. These compounds are known to contribute to the fruity, floral, and sometimes nutty flavour of whiskey. In comparison, a group of samples from 2021-LG Astronomer, Torp, Viscount from Carlow, LG Astronomer, and Torp from Tipperary had a higher abundance of mainly aldehydes and a very low abundance of esters. Elation Tipperary 2021, Viscount Tipperary 2021, Revelation Carlow 2021, and LG Astronomer Carlow 2020, Revelation Carlow 2020, Revelation Tipperary 2020, and Torp Tipperary 2020 located on the negative side of the PC-1 plot were characterised by a low abundance of aldehydes and have a high abundance of specific esters.

(b) Acetals and Alcohol Ester

There is one more cluster of samples mainly associated with acetals and alcohols, especially ethanol and phenyl ethyl alcohol at the bottom right corner of the PCA (Elation Tipperary 2020, Elation Carlow 2021, LG Astronomer Tipperary 2021, LG Astronomer Carlow 2021, Viscount Tipperary 2020, Viscount Carlow 2021, Torp Carlow 2021, Torp Tipperary 2021, and Revelation Tipperary 2021). The 2021 samples had a higher abundance of these volatiles compared to the 2020 samples.

Total peak Area Variability: The peak areas also varied considerably across sample sets. Torp Tipperary 2020 and Viscount Tipperary had the largest peak area, whereas Viscount Carlow 2020 and LG Astronomer Tipperary 2021 had the lowest peak area. Having larger peak areas suggests richer or more intensive flavour profiles, possibly due to the region's specific conditions in that year.

Further analysis, which separated the samples based on a single factor such as year and location, as shown in Figures 2 and 3, highlighted that year and location had more impact on the volatiles compared to the variety. Figure 2 highlights the impact of the harvest year on the volatile profiles. The red (2020) and blue (2021) ellipses show a distinct separation, indicating that the harvest year significantly influenced the volatile composition of the newly made spirit. These results can be attributed to the differences in growing conditions between 2020 and 2021, such as variations in temperature and rainfall, which influenced the grain composition and, ultimately, the volatile profile of whiskey. Figure 3 shows the influence of location (Carlow vs. Tipperary) on the volatile profile. The samples from Carlow (red) are more dispersed compared to those from Tipperary (blue), suggesting greater variability in the environmental conditions in Carlow, potentially due to its heterogeneous soil types and microclimate. In contrast, the more uniform conditions in Tipperary may have contributed to a narrower range of volatile profiles.

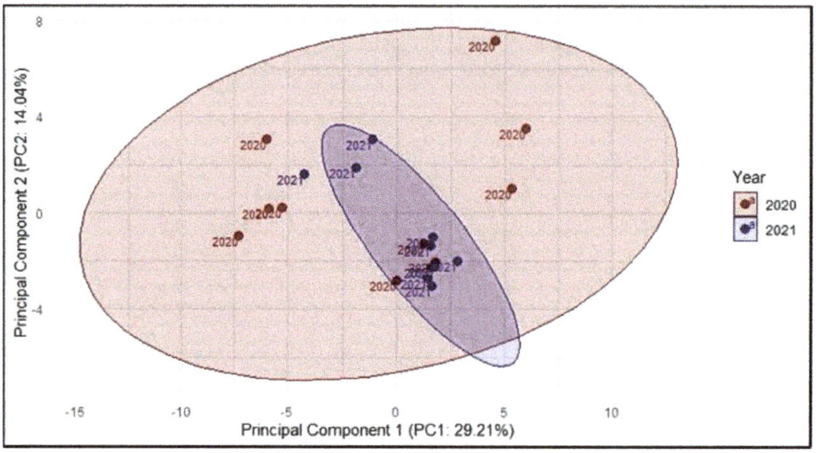

Figure 2. PCA plot and groups of spirit according to harvest year (2020 and 2021).

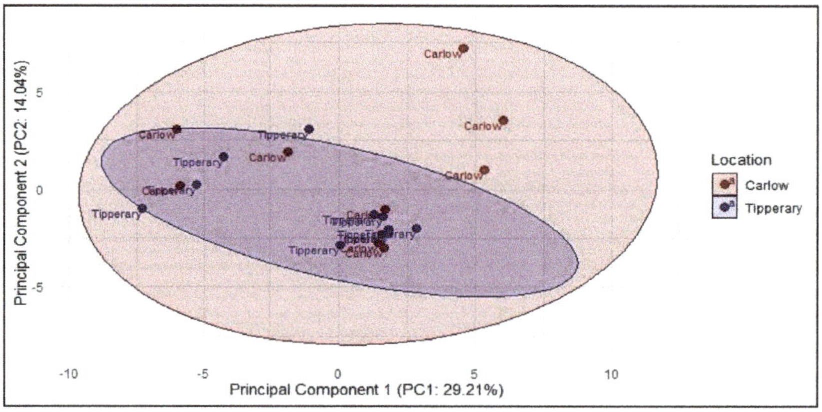

Figure 3. PCA plot and groups of spirits according to location (Carlow and Tipperary).

4. Conclusions

This study provides valuable insights into the influence of terroir factors on the grain composition and volatile profile of the wheat-based newly made spirit, highlighting the relevance of environmental factors such as season, geography, and variety in spirit production.

- Influence of Terroir on grain composition:

The grain composition was significantly affected by the variety, year, and location where the grains were grown. Understanding the effect of these factors is crucial, as grain composition not only affects the processing characteristics (such as mashing and fermentation efficiency) but also ultimately influences the flavour perception of the final whiskey. Key findings include the following:

1. The harvest year (vintage effect) significantly impacted the protein, starch, β-glucan, and arabinoxylan contents. The 2020 samples exhibited higher protein and β-glucan content compared to 2021, likely due to favourable climatic conditions such as rainfall and temperature. However, caution needs to be taken as a high β-glucan content in grain can also result in processability issues due to increased viscosity.

2. Location (Tipperary vs. Carlow): The location of wheat cultivation significantly influenced its starch, β-glucan, and arabinoxylan content. Tipperary showed consistently higher starch content compared to Carlow, likely due to its nutrient-rich, loamy soils. The higher starch content supports greater sugar availability during mashing, favouring the production of esters, which impart fruity and floral notes. In contrast, arabinoxylans and β-glucan can influence viscosity and affect fermentation kinetics, ultimately contributing to flavour complexity.
3. Variety: The variety of wheat plays an essential role in defining the starch, protein, and non-starch polysaccharide content. Torp and Revelation varieties had higher protein content, which enhances yeast activity and results in increased fusel alcohol production. Viscount, with its higher starch content, contributes more to a light and fruity character, and increased alcohol content. Understanding these varietal effects allows distillers to control grain characteristics that ultimately affect the flavour development and mouthfeel of the whiskey.
4. The interaction effect of variety x location x year significantly affected the non-starch polysaccharide content.

- Influence of Terroir on Volatile Distribution:
    1. The result illustrates a clear separation between the 2020 and 2021 samples, with the year contributing to the distinct volatile profiles. This difference aligns with the variation in environmental conditions such as rainfall and temperature; the 2020 season was warmer compared to 2021, affecting the grain composition and volatile formation.
    2. The PCA plot by location shows greater dispersion in Carlow samples compared to Tipperary, suggesting more variable environmental conditions, potentially due to its heterogeneous soil types and microclimate. In contrast, Tipperary's more uniform soil and climate conditions contribute to a narrower range of volatiles.
    3. The wheat spirit was characterised by a wide range of volatile compounds, but the overall variability in the volatile profile was relatively low compared to the malt whiskey profile, as during the malting process, barley undergoes enzymatic changes that lead to the development of numerous flavour compounds [9].

The above findings suggest that wheat may be a less complex contributor to the aroma and flavour of the whiskey, aligning with its role as a "diluent" in blended whiskeys. This simplicity could be advantageous in blends where the desired flavour profile comes primarily from malt or pot still whiskey, allowing for the wheat to contribute to a smooth, neutral base without overpowering the palate with strong flavours [28].

Understanding this can be of importance to tillage farmers and distillers regarding the selection of wheat varieties and cultivation sites to optimise the production of high-quality Irish whiskey. The study also reinforces the applicability of the terroir concept beyond viticulture, extending it to the production of distilled spirits, particularly grain-based Irish whiskey. It also aligns with the growing consumer interest in authenticity and locality in food and beverage products.

Future Research Directions: To better understand the specific contributions of individual terroir components, future studies should aim to isolate variables such as soil type, temperature, or nutrient availability and conduct controlled experiments. This would help to more precisely quantify the influence of each factor on the whiskey's aromatic and sensory profile. Moreover, expanding the analysis to include sensory evaluation alongside chemical profiling could provide further insight into how terroir affects the consumer perception of whiskey, thereby linking chemical composition to consumer preference and marketability.

In summary, while terroir influences wheat-based whiskey production, its impact is quite subtle. Nevertheless, these findings highlight an important opportunity for the Irish whiskey sector to differentiate its products by leveraging local environmental factors, particularly as consumer preferences continue to shift towards sustainability and locality.

Author Contributions: Conceptualisation, A.V. and S.M.; methodology, A.V. and S.M.; validation, A.V., K.N.K., I.S., J.L.B. and S.M.; formal analysis, S.M. and A.V.; data curation, A.V., S.M. and K.N.K.; writing—original draft preparation, A.V.; writing—review and editing, A.V., S.M., S.W., J.L.B., G.G.-C. and K.N.K.; supervision, S.M., S.W., J.L.B. and G.G.-C. All authors have read and agreed to the published version of the manuscript.

Funding: The research is funded by President's fellowship awards, South East Technological University, Carlow, under account PES3821.

Data Availability Statement: The authors confirm that the data supporting the findings of this study are available within the article.

Conflicts of Interest: The authors declare no conflicts of interest.

Appendix A

Table A1. IDs and plot codes of the samples analysed.

S.No.	Table and Plot Codes	Sample ID	ABV%
1	Rev Carl '21	Revelation Carlow 2021	5
2	Rev Tipp '21	Revelation Tipperary 2021	8
3	Rev Tipp '20	Revelation Tipperary 2020	4
4	Rev Carl '20	Revelation Carlow 2020	6
5	Vis Tipp '20	Viscount Tipperary 2020	5
6	Vis Tipp '21	Viscount Tipperary 2021	5
7	Vis Carl '20	Viscount Carlow 2020	3
8	Vis Carl '21	Viscount Carlow 2021	4
9	Ela Carl '20	Elation Carlow 2020	6
10	Ela Carl '21	Elation Carlow 2021	8
11	Ela Tipp '20	Elation Tipperary 2020	8
12	Ela Tipp '21	Elation Tipperary 2021	4
13	Ast Carl '20	LG Astronomer Carlow 2020	5
14	Ast Carl '21	LG Astronomer Carlow 2021	7
15	Ast Tipp '20	LG Astronomer Tipperary 2020	3
16	Ast Tipp '21	LG Astronomer Tipperary 2021	5
17	Torp Tipp '20	Torp Tipperary 2020	5
18	Torp Tipp '21	Torp Tipperary 2021	5
19	Torp KB '20	Torp Carlow 2020	5
20	Torp KB '21	Torp Carlow 2021	5

Table A2. Grain composition (protein, starch, arabinoxylans, and β-glucans for 4 different wheat varieties grown at two different sites and harvested in two different years, all readings were recorded in triplicate and presented as their average ± standard deviation.

Variety	Year	Location	Protein ± SD	Starch ± SD	Arabinoxylans ± SD	β-glucan ± SD
Elation	2020	Tipperary	9.82 ± 0.44	77.07 ± 3.76	6.172 ± 1.42	3.16 ± 0.2
Elation	2020	Carlow	10.06 ± 0.11	62.44 ± 0.12	3.86 ± 0.45	7.12 ± 1.45
Elation	2021	Tipperary	8.38 ± 0.22	74.04 ± 3.72	0.23 ± 0.05	4.97 ± 0.92
Elation	2021	Carlow	7.64 ± 0.22	73.76 ± 2.03	16.42 ± 2.46	7.69 ± 1.04
LG Astronomer	2020	Tipperary	9.83 ± 0.21	85.90 ± 1.87	11.097 ± 2.56	4.97 ± 2.48
LG Astronomer	2020	Carlow	10.36 ± 0.28	70.58 ± 1.20	15.94 ± 1.88	5.22 ± 0.65
LG Astronomer	2021	Tipperary	8.76 ± 0.4	73.82 ± 5.07	15.85 ± 3.71	3.15 ± 0.57
LG Astronomer	2021	Carlow	8.31 ± 0.22	54.25 ± 0.69	9.25 ± 1.3	7.85 ± 1.04
Revelation	2020	Tipperary	10.23 ± 0.39	81.53 ± 7.67	11.166 ± 2.56	2.97 ± 0.15
Revelation	2020	Carlow	10.07 ± 0.24	69.17 ± 3.05	15.7 ± 1.83	5.62 ± 1.03
Revelation	2021	Tipperary	8.69 ± 0.18	89.18 ± 1.96	10.91 ± 2.56	2.2 ± 0.08
Revelation	2021	Carlow	8.5 ± 0.34	63.08 ± 10.54	12.08 ± 2.84	4.84 ± 0.62
Torp	2020	Tipperary	10.31 ± 0.36	77.76 ± 1.6	15.811 ± 3.63	3.4 ± 0.33
Torp	2020	Carlow	11.78 ± 0.09	73.65 ± 8.29	10.11 ± 0.77	10.05 ± 0.83
Torp	2021	Tipperary	8.8 ± 0.45	82.55 ± 21.75	10.15 ± 2.39	3.9 ± 0.05
Torp	2021	Carlow	8.64 ± 0.21	68.61 ± 3.46	16.76 ± 2.11	3.7 ± 0.15
Viscount	2020	Tipperary	10.024 ± 0.48	88.279 ± 2.14	15.32 ± 3.51	5.98 ± 1.1
Viscount	2021	Carlow	10.35 ± 0.32	75.86 ± 2.044	11.34 ± 0.72	7.51 ± 2.3
Viscount	2021	Tipperary	8.44 ± 0.26	91.89 ± 1.97	12.01 ± 2.81	3.79 ± 0.28
Viscount	2021	Carlow	7.79 ± 0.13	63.89 ± 41.09	14.39 ± 3.61	3.85 ± 0.77

Table A3. Compounds identified by HS-SPME GCMS in the New- Make spiritsamples.

Name	CAS	RI	Ref RI	Rev Carl '21	Rev Tipp '21	Rev Tipp '20	Rev Carl '20	Vis Tipp '20	Vis Tipp '21	Vis Carl '20	Vis Carl '21	Ela Carl '20	Ela Carl '21
Acetal													
Diethyl acetal	105-57-7	742	747	0	30,871	0	0	0	0	0	0	0	28,584
3-Methylbutanal, diethyl acetal	03-03-3842	960	*	0	4581	0	0	0	0	2507	0	2453	5855
Alcohol													
Ethanol	64-17-5	489	506	3,658,448	5,926,241	2,089,463	2,570,413	1,870,733	3,232,613	3,2867,58	2,136,321	1,996,144	7,608,948
1-Propanol, 2-methyl-	78-83-1	673	678	74,491	139,037	103,936	103,228	85,015	77,912	124,427	53,882	112,148	126,817
1-Butanol, 3-methyl-	123-51-3	779	784	1,579,432	2,367,820	1,063,291	1,169,229	1,147,384	1,275,580	1,119,081	1,219,898	1,510,003	1,965,940
1-Butanol, 2-methyl-	137-32-6	782	789	887,119	1,288,811	868,408	937,931	929,593	757,457	751,201	683,873	973,218	1,193,111
1-Hexanol	111-27-3	911	915	77,158	102,252	88,872	96,213	93,247	105,157	90,652	64,701	122,381	63,404
2-Heptanol	543-49-7	938	947	0	0	20,480	20,626	0	0	0	0	0	0
1-Heptanol	111-70-6	1011	1016	7241	6211	6772	7187	4754	7153	5771	2940	6842	0
2-Octanol, (S)	08-06-6169	1032	*	0	0	0	0	3042	0	0	0	0	0
1-Octanol	111-87-5	1112	1118	6704	7362	6242	5738	7183	7256	14,204	7239	14,528	7484
2-Nonanol	628-99-9	1136	1143	0	0	0	0	0	0	0	0	0	0
Phenylethyl Alcohol	60-12-8	1194	1201	1,320,950	1,588,485	1,603,174	1,430,394	1,842,483	1,354,686	941,109	1,080,744	1,226,542	2,278,321
2-Furanmethanol	98-00-0	924	*	4180	0	8398	9722	0	3163	0	0	0	0
Aldehyde													
Butanal, 3-methyl-	590-86-3	687	692	5285	14,163	6824	11,337	36,088	5969	29,434	18,860	39,224	16,673
Butanal, 2-methyl-	96-17-3	695	700	23,714	4951	5761	7115	18,954	4706	14,973	10,322	16,807	10,211
Hexanal	66-25-1	834	839	0	8433	0	0	19,255	0	32,237	16,804	38,371	9770
Furfural	98-01-1	894	899	0	12,769	0	0	73,454	0	366,244	81,629	261,486	31,537
Heptanal	111-71-7	936	943	0	0	0	0	2805	0	4694	2637	3916	0
Benzaldehyde	100-52-7	1024	1031	0	2173	0	0	6426	0	43,866	12,277	28,300	3051
Benzenacetaldehyde	122-78-1	1114	*	0	4305	0	0	25,416	0	279,059	2585	108,461	18,543
Nonanal	124-19-6	1143	1150	3448	9734	2385	1370	6796	772	13,537	14,016	6980	8872
Benzaldehyde, 4-propyl-	28785-06-0	1357	*	0	18,739	4338	6002	23,560	3689	17,438	17,509	10,527	14,337
Benzene													
Styrene	100-42-5	923	929	0	0	0	0	0	0	0	0	0	0
o-Xylene	108-38-3	922	929	5084	3676	4291	5052	7449	4622	5330	3605	0	0
Ester													

Table A3. *Cont.*

Name	CAS	RI	Ref RI	Rev Carl '21	Rev Tipp '21	Rev Tipp '20	Rev Carl '20	Vis Tipp '20	Vis Tipp '21	Vis Carl '20	Vis Carl '21	Ela Carl '20	Ela Carl '21
Ethyl acetate	141-78-6	636	642	3,285,572	265,629	6,266,471	6,384,798	235,939	7,142,368	103,024	147,009	252,082	544,533
Ethyl propanoate	105-37-3	732	737	14,319	0	31,339	40,564	10,547	28,353	0	4230	12,874	9020
n-Propyl acetate	109-60-4	738		0	0	7563	20,808	0	17,335	0	0	0	0
Isobutyl acetate	110-19-0	795	800	0	0	0	0	0	0	0	0	0	0
Ethyl butanoate	105-54-4	820	826	5402	4974	4399	4945	7823	4521	0	3166	6464	7613
Isoamyl acetate	123-92-2	898	902	320,930	75,612	411,020	464,540	121,647	518,464	75,832	82,041	205,476	115,743
2-Methylbutyl acetate	624-41-9	901	906	108,183	0	182,515	201,899	54,666	173,550	18,491	0	77,946	115,595
Ethyl pentanoate	539-82-2	920	924	2414	3737	2684	0	5392	8063	3790	2651	10,624	4030
Isopropyl pentanoate	18362-97-5	957	*	4603	3246	5769	5439	8164	4947	4323	4423	7377	2101
Ethyl hexanoate	123-66-0	1017	1024	193,226	140,281	82,842	62,344	142,815	285,002	141,487	116,807	611,323	173,458
Hexyl acetate	142-92-7	1034	*	24,461	0	40,009	38,010	0	62,198	0	0	0	0
Ethyl heptanoate	106-30-9	1115	*	5185	2417	2440	9399	1928	6243	2427	1804	20,061	4262
Ethyl octanoate	106-32-1	1216	1222	115,602	4251	2955	0	3102	20,604	5364	8243	102,467	60,572
Ethyl benzoate	93-89-0	1225	1232	1636	1648	0	0	0	0	1938	0	3867	2176
Octyl acetate	112-14-1	1215	*	0	0	0	0	0	0	10,159	4470	0	0
B-Phenylethyl acetate	103-45-7	1313	1322	40,527	12,721	73,283	69,985	8259	77,165	20,195	21,820	16,808	22,582
Ethyl nonanoate	123-29-5	1314	*	4470	0	1651	0	0	2519	0	0	1710	0
Ethyl decanoate	110-38-3	1414	1422	5072	0	0	0	0	0	0	2012	10,031	3325
Ethyl benzeneacetate	101-97-3	1298	*	10,189	14,785	0	0	1767	2412	6214	0	3793	18,290
						Ether							
Difurfuryl ether	4437-22-3	926	*	0	0	0	0	0	0	4023	0	0	0
						Furan							
Furfuryl acetate	623-17-6	1030	*	0	0	3377	3303	0	1358	0	0	0	0
						Ketone							
2-Heptanone	110-43-0	930	936	24,503	0	133,423	127,576	12,130	9350	12,933	7078	10,681	0
3-Octanone	106-68-3	1022	*	0	0	0	0	0	0	0	0	0	0
4-Nonanone, 2,6,8-trimethyl-	123-18-2	1247	*	3942	4874	5234	4072	4654	5058	4151	3839	5206	4817
						Lactone							
γ-Nonalactone	104-61-0	1484	*	31,873	52,879	22,125	18,433	28,241	64,719	51,897	25,520	61,895	46,106
						Phenol							
2,4-Di-tert-butylphenol	96-76-4	1592	1644	33,250	36,216	33,091	30,931	35,051	38,713	29,773	20,547	61,121	34,506

Compound identification, chemical class, and average abundance values measured (n = 3); CAS: chemical CAS (chemical abstract service) (blanks relate to isomers where we could not be 100% sure of identification and therefore could not provide full identification. LRI: linear retention indices as determined using the method by Van Den Dool and Kratz (1963); REF LRI: These values were obtained from published papers or NIST 2014. *: No published reference available to date (not many published as yet on a DB624 column), tentative identification, might be isomer of this chemical compound.

Table A4. Compounds identified by HS-SPME GCMS in the New-Make spiritsamples.

Name	CAS	RI	Ref RI	Ela Tipp '20	Ela Tipp '21	Ast Carl '20	Ast Carl '21	Ast Tipp '20	Ast Tipp '21	Torp Tipp '20	Torp Tipp '21	Torp KB '20	Torp KB '21
Acetal													
Diethyl acetal	105-57-7	742	747	0	0	0	0	0	0	0	0	0	0
3-Methylbutanal, diethyl acetal	03-03-3842	960	*	4985	0	0	2719	0	0	0	2437	2441	0
Alcohol													
Ethanol	64-17-5	489	506	5,257,459	2,715,097	1,970,918	4,789,790	3,056,767	2,330,556	2,532,878	3,514,650	2,738,950	3,216,485
1-Propanol, 2-methyl-	78-83-1	673	678	69,527	113,733	107,524	80,899	81,212	59,137	101,917	89,568	142,289	76,400
1-Butanol, 3-methyl-	123-51-3	779	784	1,262,577	1,032,851	1,622,604	1,514,584	952,109	1,239,367	1,112,751	1,792,022	1,134,854	1,221,304
1-Butanol, 2-methyl-	137-32-6	782	789	669,247	57,3149	889,173	908,836	779,533	693,110	946,563	872,999	740,315	642,925
1-Hexanol	111-27-3	911	915	78,012	78,875	161,625	74,317	76,276	75,196	62,859	91,487	86,675	54,189
2-Heptanol	543-49-7	938	947	0	0	152,703	0	0	0	0	0	0	0
1-Heptanol	111-70-6	1011	1016	5164	4483	19,280	7970	5433	0	6165	3730	3766	0
2-Octanol, (S)	08-06-6169	1032	*	0	0	12,432	0	0	0	0	0	0	0
1-Octanol	111-87-5	1112	1118	5764	4588	6748	6375	4019	8275	4588	8849	15,465	6091
2-Nonanol	628-99-9	1136	1143	0	0	249,369	0	0	0	0	0	0	0
Phenylethyl Alcohol	60-12-8	1194	1201	1,543,105	868,736	969,069	1,597,147	1,325,589	1,336,868	1,560,983	2,055,401	847,927	1,446,128
2-Furanmethanol	98-00-0	924	*	0	0	11,720	0	0	0	3496	0	0	0
Aldehyde													
Butanal, 3-methyl-	590-86-3	687	692	14,080	2615	2705	22,557	21,961	21,800	0	23,375	33,697	20,258
Butanal, 2-methyl-	96-17-3	695	700	5962	6438	5845	9985	7644	7890	0	0	18,058	9378
Hexanal	66-25-1	834	839	9945	0	0	7481	0	18,822	0	20,502	40,568	17,604
Furfural	98-01-1	894	899	78,683	29,958	0	253,996	86,332	91,317	0	367,941	345,793	183,414
Heptanal	111-71-7	936	943	2142	0	0	2707	0	3592	0	3977	7005	2489
Benzaldehyde	100-52-7	1024	1031	4457	7481	0	21,626	7158	14,468	0	28,580	48,370	13,763
Benzenacetaldehyde	122-78-1	1114	*	28,305	1730	0	22,275	2557	0	0	30,919	214,466	26,191
Nonanal	124-19-6	1143	1150	8963	1569	0	6690	7273	13,796	992	22,463	16,959	15,245
Benzaldehyde, 4-propyl-	28785-06-0	1357	*	17,233	5843	0	22,308	16,027	19,601	0	25,444	18,691	20,997
Benzene													
Styrene	100-42-5	923	929	0	10,184	0	0	0	0	2299	0	1097	1222
o-Xylene	108-38-3	922	929	2443	0	7250	6023	6233	4349	5714	5984	4769	6240
Ester													

Table A4. *Cont.*

Name	CAS	RI	Ref RI	Ela Tipp '20	Ela Tipp '21	Ast Carl '20	Ast Carl '21	Ast Tipp '20	Ast Tipp '21	Torp Tipp '20	Torp Tipp '21	Torp KB '20	Torp KB '21
Ethyl acetate	141-78-6	636	642	196,102	2,593,055	2,150,842	225,220	175,361	165,889	7,520,894	425,306	141,353	221,336
Ethyl propanoate	105-37-3	732	737	0	39,820	22,481	5393	13,938	6314	45,558	0	0	0
n-Propyl acetate	109-60-4	738		0	0	0	0	0	0	38,102	0	0	0
Isobutyl acetate	110-19-0	795	800	0	0	35,914	0	0	0	37,543	0	0	0
Ethyl butanoate	105-54-4	820	826	2978	1117	5188	4538	5253	4665	5082	4245	885	4442
Isoamyl acetate	123-92-2	898	902	40,886	237,443	433,820	92,222	92,689	104,991	644,723	134,687	105,384	117,554
2-Methylbutyl acetate	624-41-9	901	906	27,674	64,857	151,657	91,890	68,706	100,866	305,136	132,971	45,609	116,627
Ethyl pentanoate	539-82-2	920	924	4076	7055	2462	1397	4025	3380	0	3630	3975	2148
Isopropyl pentanoate	18362-97-5	957	*	2171	4833	7043	6145	5798	5562	6000	5459	5364	5906
Ethyl hexanoate	123-66-0	1017	1024	121,759	302,117	60,166	79,233	90,102	159,323	22,728	114,735	196,110	64,722
Hexyl acetate	142-92-7	1034	*	0	24,781	24,106	0	0	0	35,690	0	0	0
Ethyl heptanoate	106-30-9	1115	*	2851	9359	0	0	0	2935	0	1921	11,153	0
Ethyl octanoate	106-32-1	1216	1222	11,834	28,235	1463	5590	2409	13,317	0	5475	61,218	3130
Ethyl benzoate	93-89-0	1225	1232	520	1270	3160	0	0	0	0	0	2254	0
Octyl acetate	112-14-1	1215	*	0	0	0	0	0	9045	0	0	0	4501
B-Phenylethyl acetate	103-45-7	1313	1322	14,485	33,770	21,010	16,997	6091	24,876	91,152	15,647	19,622	17,318
Ethyl nonanoate	123-29-5	1314	*	0	4403	2340	0	0	0	0	0	0	0
Ethyl decanoate	110-38-3	1414	1422	0	7139	0	0	0	0	0	0	5250	0
Ethyl benzeneacetate	101-97-3	1298	*	13,088	6780	0	0	0	0	0	0	8167	0
Ether													
Difurfuryl ether	4437-22-3	926	*	0	0	0	0	0	0	0	0	3294	0
Furan													
Furfuryl acetate	623-17-6	1030	*	0	0	1317	0	0	0	2157	0	0	0
Ketone													
2-Heptanone	110-43-0	930	936	0	21,446	35,054	9131	12,311	7900	22,187	8363	78,087	8480
3-Octanone	106-68-3	1022	*	0	0	0	10,084	8369	0	0	0	0	0
4-Nonanone, 2,6,8-trimethyl-	123-18-2	1247	*	4114	4229	4253	5490	4449	4719	4049	5298	5124	5212
Lactone													
γ-Nonalactone	104-61-0	1484	*	80,355	94,137	22,099	32,877	23,637	34,166	16,892	65,773	64,032	30,489
Phenol													
2,4-Di-tert-butylphenol	96-76-4	1592	1644	23,969	19,170	28,656	39,295	28,938	25,377	34,329	40,321	38,226	29,813

Compound identification, chemical class, and average abundance values measured (n = 3); CAS: chemical CAS (chemical abstract service) (blanks relate to isomers where we could not be 100% sure of identification and therefore could not provide full identification. LRI: linear retention indices as determined using the method by Van Den Dool and Kratz (1963). REF LRI: These values were obtained from published papers or NIST 2014. *: No published reference available to date (not many published as yet on a DB624 column), tentative identification, might be isomer of this chemical compound.

References

1. Morris, S.; Byrne, J.; Murphy, B.; Whelan, S.; Carroll, J.; Ryan, D. Optimization of Mashing Parameters Used during the Conversion of Irish Wheat Grain to Spirit Alcohol. *J. Am. Soc. Brew. Chem.* **2022**, *81*, 383–395. [CrossRef]
2. Morris, S.; Byrne, J.L.; Murphy, B.; Whelan, S.J.; Carroll, J.P.; Ryan, D. The optimisation of cooking parameters for spirt whiskey production from native Irish wheat: A response surface method approach. *Foods* **2022**, *11*, 1199. [CrossRef] [PubMed]
3. Woodhill, J.; Kishore, A.; Njuki, J.; Jones, K.; Hasnain, S. Food systems and rural wellbeing: Challenges and opportunities. *Food Secur.* **2022**, *14*, 1099–1121. [CrossRef] [PubMed]
4. Murphy, B.; Keaney, R. The Rise of Whiskey Tourism in Ireland: Developing a Terroir Engagement Template. *J. Gastron. Tour.* **2018**, *3*, 107–121. [CrossRef]
5. Van Leeuwen, C. Terroir: The effect of the physical environment on vine growth, grape ripening, and wine sensory attributes. In *Managing Wine Quality*; Elsevier: Amsterdam, The Netherlands, 2022; pp. 341–393. [CrossRef]
6. Johnston-Monje, D.; Vergara, L.I.; Lopez-Mejia, J.; White, J.F. Plant microbiomes as contributors to agricultural terroir. *Front. Agron.* **2023**, *5*, 1216520. [CrossRef]
7. Herb, D.; Filichkin, T.; Fisk, S.; Helgerson, L.; Hayes, P.; Meints, B.; Jennings, R.; Monsour, R.; Tynan, S.; Vinkemeier, K.; et al. Effects of Barley (Hordeum vulgare L.) Variety and Growing Environment on Beer Flavor. *J. Am. Soc. Brew. Chem.* **2017**, *75*, 345–353. [CrossRef]
8. Van Leeuwen, C.; Barbe, J.-C.; Darriet, P.; Geffroy, O.; Gomès, E.; Guillaumie, S.; Helwi, P.; Laboyrie, J.; Lytra, G.; Le Menn, N.; et al. Recent advancements in understanding the terroir effect on aromas in grapes and wines. *OENO One* **2020**, *54*, 985–1006. [CrossRef]
9. Kyraleou, M.; Herb, D.; O'reilly, G.; Conway, N.; Bryan, T.; Kilcawley, K.N. The impact of terroir on the flavour of single malt whisk(e)y new make spirit. *Foods* **2021**, *10*, 443. [CrossRef]
10. Arnold, R.J.; Ochoa, A.; Kerth, C.R.; Miller, R.K.; Murray, S.C. Assessing the impact of corn variety and Texas terroir on flavor and alcohol yield in new-make bourbon whiskey. *PLoS ONE* **2019**, *14*, e0220787. [CrossRef]
11. Wanikawa, A. Flavors in malt whisky: A review. *J. Am. Soc. Brew. Chem.* **2020**, *78*, 260–278. [CrossRef]
12. Holt, S.; Miks, M.H.; de Carvalho, B.T.; Foulquié-Moreno, M.R.; Thevelein, J.M. The molecular biology of fruity and floral aromas in beer and other alcoholic beverages. *FEMS Microbiol. Rev.* **2019**, *43*, 193–222. [CrossRef] [PubMed]
13. Feussner, I.; Wasternack, C. The lipoxygenase pathway. *Annu Rev Plant Biol.* **2002**, *53*, 275–297. [CrossRef] [PubMed]
14. Fitzgerald, G.; James, K.J.; MacNamara, K.; Stack, M.A. Characterisation of whiskeys using solid-phase microextraction with gas chromatography–mass spectrometry. *J. Chromatogr. A* **2000**, *896*, 351–359. [CrossRef] [PubMed]
15. Gerretzen, J.; Buydens, L.M.; Beukel, A.O.T.v.D.; Koussissi, E.; Brouwer, E.R.; Jansen, J.J.; Szymańska, E. A novel approach for analyzing gas chromatography-mass spectrometry/olfactometry data. *Chemom. Intell. Lab. Syst.* **2015**, *146*, 290–296. [CrossRef]
16. European Brewery Convention. *Analytica EBC: Cereal Adjuncts, Sugars, Syrups and Caramel, 6.2.2-Moisture Content of Maize.* Available online: https://brewup.eu/ebc-analytica/cereal-adjuncts-sugars-syrups-and-caramel/moisture-content-of-maize/6.2.2 (accessed on 12 June 2024).
17. European Brewery Convention. *Analytica EBC. 9.2.1-Alcohol in Beer by Distillation.* 2018. Available online: https://brewup.eu/ebc-analytica/beer/alcohol-in-beer-by-distillation/9.2.1 (accessed on 28 February 2024).
18. van Den Dool, H.; Kratz, P.D. A generalization of the retention index system including linear temperature programmed gas—Liquid partition chromatography. *J. Chromatogr. A* **1963**, *11*, 463–471. [CrossRef]
19. AMDIS Is (No Date) AMDIS Is a New (Easy to Use) Sophisticated Software for GC-MS Data Interpretation from NIST, Amdis.net. Available online: http://www.amdis.net/About/AMDIS-net--by-Tobias-Kind.pdf (accessed on 6 March 2024).
20. Wehrens, R.; Weingart, G.; Mattivi, F. metaMS: An open-source pipeline for GC–MS-based untargeted metabolomics. *J. Chromatogr. B* **2014**, *966*, 109–116. [CrossRef]
21. RStudio Team. *RStudio: Integrated Development for R. RStudio, PBC*; RStudio Team: Boston, MA, USA, 2020; Available online: http://www.rstudio.com/ (accessed on 5 June 2024).
22. Haiying, T. Distribution of protein components of wheat from different regions. *Afr. J. Biotechnol.* **2012**, *11*, 10568–10574. [CrossRef]
23. Kim, K.-H.; Kim, J.-Y. Understanding wheat starch metabolism in properties, environmental stress condition, and molecular approaches for value-added utilization. *Plants* **2021**, *10*, 2282. [CrossRef]
24. Huang, X.; Zhou, X.; Liu, X.; Zhong, W.; Wang, X.; Ju, Z.; Yin, Y.; Xin, Q.; Liu, N.; Liu, X.; et al. Structural and physicochemical effects on the starch quality of the high-quality wheat genotype caused by delayed sowing. *Front. Nutr.* **2024**, *11*, 1389745. [CrossRef]
25. Bujake, J.E. Beverage spirits, distilled. In *Kirk-Othmer Encyclopedia of Chemical Technology*, 1st ed.; Kirk-Othmer, Ed.; Wiley: Hoboken, NJ, USA, 2007. [CrossRef]
26. Stewart, G.G. The production of secondary metabolites with flavour potential during brewing and distilling wort fermentations. *Fermentation* **2017**, *3*, 63. [CrossRef]

27. Lee, K.-Y.M.; Paterson, A.; Piggott, J.R.; Richardson, G.D. Perception of whisky flavour reference compounds by Scottish distillers†. *J. Inst. Brew.* **2000**, *106*, 203–208. [CrossRef]
28. Biernacka, P.; Wardencki, W. Volatile composition of raw spirits of different botanical origin: Volatile composition of raw spirits. *J. Inst. Brew.* **2012**, *118*, 393–400. [CrossRef]

Disclaimer/Publisher's Note: The statements, opinions and data contained in all publications are solely those of the individual author(s) and contributor(s) and not of MDPI and/or the editor(s). MDPI and/or the editor(s) disclaim responsibility for any injury to people or property resulting from any ideas, methods, instructions or products referred to in the content.

foods

MDPI

Article

Antioxidant-Enhanced Alginate Beads for Stabilizing Rapeseed Oil: Utilizing Extracts from Post-Distillation Waste Residues of Rosemary

Petroula Tsitlakidou [1], Despina Kamplioni [1], Anastasia Kyriakoudi [1], Maria Irakli [2], Costas G. Biliaderis [1] and Ioannis Mourtzinos [1,*]

[1] Laboratory of Food Chemistry and Biochemistry, Department of Food Science and Technology, School of Agriculture, Faculty of Agriculture, Forestry and Natural Environment, Aristotle University of Thessaloniki (AUTH), 54124 Thessaloniki, Greece; ptsitlak@agro.auth.gr (P.T.); despwkamb@gmail.com (D.K.); ankyria@agro.auth.gr (A.K.); biliader@agro.auth.gr (C.G.B.)

[2] Institute of Plant Breeding and Genetic Resources, Hellenic Agricultural Organization—Demeter, P.O. Box 60458, Thermi, 57001 Thessaloniki, Greece; mirakli@elgo.gr

* Correspondence: mourtzinos@agro.auth.gr; Tel.: +30-23-1099-1637

Abstract: An eco-friendly extraction process of polyphenols from conventional dried rosemary tissues and post-distillation waste residues was applied using β-cyclodextrin as a co-solvent. The aqueous extracts were characterized by measuring the total phenolic content, and their phenolic compounds were identified and quantified by LC-MS. Sodium alginate solutions (2% *w/w*) with/without incorporation of rosemary aqueous extracts were prepared and used for the preparation of O/W emulsions containing 20% rapeseed oil and an 80% water phase. Hydrogel beads were then stored at 20 °C for 28 days. The quality of encapsulated oil during storage was evaluated by measurements of the peroxide value, p-anisidine value, free fatty acids, total oxidation value, and fatty acid composition, whilst the aqueous phase of the beads was analyzed for its total extractable phenolic content (TEPC). The experimental findings indicate that the incorporation of aqueous extracts from post-distillation rosemary residues in emulsion-filled hydrogel beads resulted in the lowest level of oxidation products in the encapsulated rapeseed oil (PV = 10.61 ± 0.02 meq/Kg oil, p-AnV = 4.41 ± 0.09, and FFA = 0.14 ± 0.00, expressed as % oleic acid content), indicating an acceptable oil quality until the end of the storage period.

Keywords: alginate; emulsion gels; rosemary distillate residues; rapeseed oil; encapsulation; oxidation stability

1. Introduction

Consumer awareness towards healthier diets and the strong link between nutrition and health have triggered an increase in the production and commercialization of foods with increasing amounts of polyunsaturated fats [1] as well as the substitution of "chemical additives" with "natural" alternatives in food formulations. As a result, aromatic and medicinal plants (AMPs) have been gaining a growing interest as a source of natural bioactive compounds with potential uses in the food industry; moreover, aromatic plants and mainly their essential oils and extracts have been used for centuries as flavouring ingredients and for food preservation purposes because of their well-documented antimicrobial and antioxidant properties [2].

Rosemary officinalis L. (rosemary) is a medicinal and aromatic plant, and several studies have revealed its distinguishing polyphenolic diversity and antioxidative potency; the main bioactive ingredients of this plant are carnosic acid, carnosol, and rosmarinic acid [3,4]. Moreover, rosemary is heavily exploited in the production of essential oil (e.g., via hydrodistillation), generating high amounts of solid residues due to the low yield of essential oil [5]. The production of essential oil rarely yields more than 0.3–2.5% *w*/*w* of dry plant in the case of rosemary [6], whilst the essential oil yield of AMPs generally varies between 0.5 and 8% *w*/*w* of dry biomass [7].

EOs are natural, complex mixtures of volatile compounds, which are formed in the plant material as secondary plant metabolites. Terpenoids are the most important chemical group of EOs and are compounds formed by the conjunction of isoprene units. As a rule, isoprene units are joined in one direction, coupling head-to-tail, i.e., terpenoids contain in their structure several five-carbon atom entities. They may be hydrocarbons, or oxygenated derivatives (alcohols, aldehydes, ketones, or acids) or reaction products (esters and ethers). Additionally, non-terpenic compounds generated via the phenyl propanoid pathway are EO constituents, i.e., eugenol, cinnamaldehyde, safrole, etc. [8]. Oxygenated compounds are mostly responsible for flavouring properties, demonstrating a high odour intensity score [9]. The chemical composition of EOs is an important criterion in their utilization for several uses and a valuable quality marker for marketing, thus contributing to their potential valorization. Several factors (extraction techniques, plant raw material varying in composition due to physiological, biological, genetic, and environmental conditions) influence EO composition, and the major constituents of rosemary's essential oil are 1,8-cineole (43.77–88.9%), camphor (2.4–12.53%), α-pinene (2.7–11.51%), and β-pinene (0–8.16%) [10].

The current tendency for sustainable development, involving cyclic economy principles, largely focuses on the valorization of agro-industrial by-products and the recovery of valuable compounds with bioactivities, mostly related to their antioxidant and other human health-promoting effects. Previous studies suggested the potential exploitation of the solid residues from the hydrodistillation of rosemary as a valuable source of polyphenols with antioxidant activities [11–13]. Whilst the main focus of those studies was the recovery of phenolic compounds and the techniques applied for their extraction, there have been no studies on the use of phenolic extracts obtained from rosemary solid waste streams of the EOs industry in food applications.

The potential exploitation of bioactive compounds present in distillation residuals by extraction of phytochemicals and further conversion into value-added products could address a number of Sustainable Development Goals (SDGs) laid out in the 2030 Agenda for Sustainable Development [14]. The 17 Sustainable Development Goals (SDGs) are intended to address unfulfilled issues such as extreme poverty, inequality, social injustice, and protection of the environment by 2030 [15]. The development of eco-friendly extraction techniques and valorization of solid waste streams of the EOs industry could promote the achievement of SDG 9 and SDG 12, i.e., to build resilient infrastructure, promote inclusive and sustainable industrialization, and foster innovation (SDG 9), as well as to ensure sustainable consumption and production patterns (SDG 12).

Sodium alginate is a renewable, biodegradable, biocompatible, and non-toxic polysaccharide extracted from brown seaweed. Alginates, being ionic polysaccharides, are capable of forming hydrogels (Ca^{2+}-mediated) in various morphologies (e.g., beads, fibres) ideal for functioning as carriers of bioactive compounds with the additional benefit of controlled release under the typical pH conditions encountered in the human GI tract [16]. They are linear polymers composed of 1,4-linked β-D-mannuronic acid (M) and 1,4 α-L-guluronic acid (G) anhydro residues interlinked in homogeneous (poly-G, poly-M) or heterogeneous (mixed MG) block-like patterns on the chains [17]. The relative amounts of α-L-guluronic acid (G) and β-D-mannuronic acid (M) groups in the alginate chains (i.e., the G/M ratio) determine the permeability and physicochemical properties of the ionotropic hydrogels formed with these polysaccharides [18]. The G/M ratio varies depending on their natural source (seaweed), location in the plant, geographical location, and the extraction process adopted [19], i.e., different types of alginates are commercially available with varying structural features depending on the seaweed species.

Besides biocompatibility, biodegradability, and environmental friendliness, alginate hydrogels are cost-effective, simple to produce [20], and have relatively low capital and operating costs [21]. Alginates demonstrate great capability to be cross-linked with other natural or non-natural materials and build a wide range of alginate-based systems such as hydrogels, emulsions, emulsion-filled hydrogels, liposomes, edible films, and aerogel [22]. In terms of food applications, alginates can be used as viscosifiers or gelling agents as well as an edible coating material to extend the shelf-life of foods by maintaining the moisture, flavour, and reducing fat oxidation [23]. Another emerging application of alginate in the food industry involves creating emulsion-filled hydrogels as carriers of health-promoting ingredients [24] or fat replacers due to their thermal stability [22], enabling the development of "clean label" products and the use of nutritional claims. Moreover, recent studies on double emulsions fortified with olive leaf extract [25] and emulsion gels containing a high content of unsaturated fatty acids and tannins [26]—as functional fat replacers in meat systems—suggested that the development of those food systems is a promising strategy for enhancing antioxidant properties and, thus, preventing lipid oxidation in those reformulated systems. In this context, encapsulation of olive oil [27,28], ginger oil [29], linseed oil [30], and sunflower oil [31] by means of alginate gelation provided enhanced stability of the respective oil against oxidation. Yet, little information is available on the encapsulation of flavours in sodium alginate beads; in this respect, previously published works demonstrated that encapsulation of flavour components [32] and complex mixtures [33] in alginate hydrogels improved the stability and retention, and permitted the controlled-release of flavours during thermal processing, which are important properties affecting the flavour loss in final products.

In view of the strong antioxidant potency of rosemary extracts and the ability of alginate to form emulsion-filled hydrogels, the incorporation of antioxidant compounds in the aqueous phase of an emulsion and the use of rapeseed oil with high content of oleic and linoleic acids in the oil phase may serve as a dual strategic approach in formulating emulsion systems: first, to prevent the oxidative deterioration of the oil phase and, thus, enhancing its nutritional value, and second, to optimize the presence of bioactive compounds in the dispersion in which polyphenols extracted from rosemary will be included as an ingredient. Therefore, the objective of the present study was to encapsulate in alginate beads rapeseed oil (in an O/W emulsion) loaded with polyphenols extracted from rosemary using two different streams of plant materials (tissues from the conventional raw plant material of rosemary and post-distillation waste residues) and evaluate the effect of the recovered antioxidants on storage stability of the oil and the retention of the extracted polyphenols during storage. The information obtained in this study might be also useful for the development of novel encapsulation approaches by valorizing post-distillation waste residues of aromatic plants while working towards the "zero-waste" biorefinery concept.

2. Materials and Methods

2.1. Plant Material

Dry-rubbed rosemary leaves (*Rosmarinus officinalis* L.) were obtained from Physis Ingredients; the plants were organically cultivated in the region of Serres, in northern Greece, and shade-dried after harvesting. Cold-pressed rapeseed oil (*Brassica campestris* seed oil) was purchased from AOT Organic products (Wiggensbach, Germany).

2.2. Chemicals, Reagents, and Solvents

Alginic acid in the form of sodium salt with low viscosity (viscosity: 66 mPas at 1% *w*/*w* in aq. solution) was obtained from Thermo Fisher Scientific (Athens, Greece) and β-cyclodextrin (CAS No: 7585-39-9, MW: 1134.99 g/mol) was sourced from TCI-Tokyo Chemical Industry Co., Ltd., Tokyo, Japan. Tween 80- Polyoxyethylene sorbitan monooleate (100%, CAS No: 9005-66-6) was purchased from Duchefa Biochemie B.V. (Haarlem, Netherlands). Calcium chloride (>98%, CAS No: 10043-52-4), anhydrous citric acid (>99.5%, CAS No: 5949-29-1), methanol (>99%), hexane (>95%), acetic acid (99–100%), chloroform (>99%), and sodium thiosulfate, $Na_2S_2O_3$ (>98%, CAS No: 10102-17-7), were obtained from Chem-Lab NV (Zedelgem, Belgium). Corn starch (CAS No: 9005-25-8) and potassium iodide (≥99%, CAS No: 7681-11-0, MW: 166 g/mol) were purchased from Sigma Aldrich (Taufkirchen, Germany).

2.3. Isolation of Rosemary Essential Oil

Hydrodistillations of rosemary leaves were conducted at a laboratory scale according to the 10th edition of the European Pharmacopoeia [34]. Specifically, 25 g of dried rosemary leaves was added in a 500 mL round bottom flask, 300 mL of distilled water was added, and then it was submitted to water distillation for 2 h using a Clevenger apparatus. The distilled essential oil was collected in a vial, which was hermetically sealed and stored at −18 °C until further use.

2.4. Preparation of Post-Distillation Rosemary Residues

For the preparation of post-distillation rosemary residues, the hot water/rosemary leaves mixture was removed from the round bottom flask immediately after the end of hydrodistillation and the residues of rosemary leaves were collected. The residual leaves were dried in an air drier for 24 h at 40 °C until the weight of rosemary biomass was stable (~10% moisture content). The dried material was subsequently milled to pass through a 2.0 mm sieve and stored in heat-sealed bags at ambient temperature until further analysis. A flow diagram describing the experimental procedure is displayed in Figure 1.

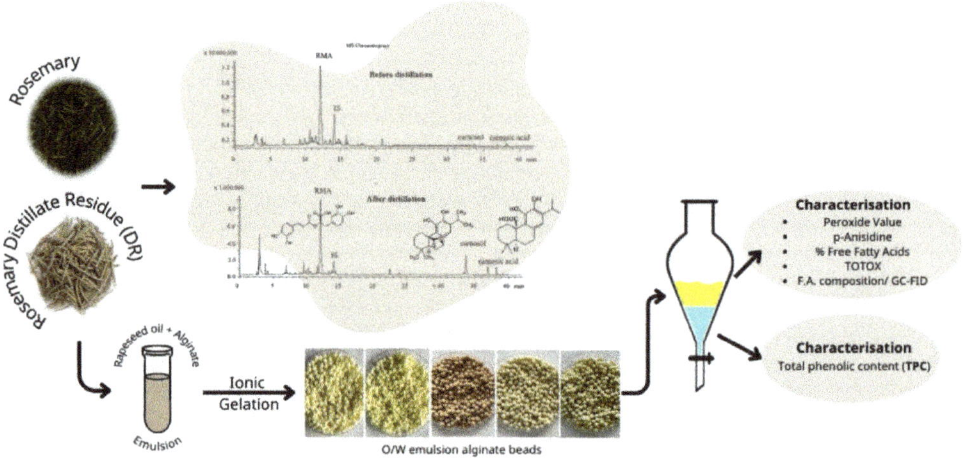

Figure 1. Flow diagram describing the experimental procedure.

2.5. Green Extraction using Aqueous Solutions of β-Cyclodextrin

Samples of dried and ground post-distillation rosemary solid residues were extracted with an aqueous solution of β-cyclodextrin (0.925% *w*/*v*) at a liquid-to-solid ratio: L/S = 8 mL/g. The extraction was held at 50 °C for 124 min and under constant stirring at 600 rpm. The experimental conditions for extraction were chosen based on previous research with slight modifications [13,35]. Following extraction, the extracts were obtained by filtration through filter paper (Whatman No. 1), removing the main solid biomass, and subsequently centrifuged at 6000× *g* rpm for 10 min using a bench centrifuge (Hettich Universal 32, DJB Labcare Ltd., Newport Pagnell, UK). The clarified supernatants were placed in a stoppered glass bottle and stored at −18 °C until analysis.

2.6. Determination of Total Phenolic Content (TPC) in Aqueous Extracts

The total phenolic content of the samples was determined according to a published protocol [36] using the Folin–Ciocalteu methodology and gallic acid as the standard; the results were expressed as milligrams of gallic acid equivalents per g sample on a dry weight basis (mg GAE/g dw) using a calibration curve. Measurements were performed in triplicate.

2.7. LC-MS Analysis of Aqueous Extracts

LC-MS analysis was performed using a Shimadzu Nexera HPLC system (Shimadzu, Kyoto, Japan) equipped with a diode array detector (DAD) and a single quadrupole mass spectrometer, combined with an electrospray ionization source [12]. The separation of phenolic compounds was achieved on a Poroshell 120 EC-C18 column (4.6 × 150 mm, 4-micrometre particle size), thermostatted at 35 °C, with mobile phase A (0.1% aqueous formic acid) and B (acetonitrile) at a flow rate of 0.5 mL/min, following a 45-min linear gradient elution as follows: 15–25% B, 0–5 min; 25–35% B, 5–10 min; 35–60% B, 10–28 min; 60–100% B, 28–35 min; 100–15% B, 35.01–40 min; followed by an isocratic elution (15% B) until 45 min. The mass spectrometry was carried out in negative ion mode and the quantitative determination of target compounds in the phenolic extracts was performed in the selected ion monitoring (SIM) mode, using calibration curves of the respective standard solutions at five concentration levels within the linear range of 0.01 to 10 μg/mL for rosmarinic acid, and 0.05 to 200 μg/mL for carnosol and carnosic acid. Data acquisition and processing were performed using the Lab Solutions LC-MS software version 5.97.SP1 (Shimadzu, Kyoto, Japan). Analyses were performed in triplicate and the results were expressed as mg per g of extract.

2.8. Preparation and Storage of O/W Emulsion Alginate Beads

2.8.1. Preparation of O/W Emulsions with Alginate

For preparation of the aqueous continuous phase, sodium alginate (2% wt) was dissolved in the two types of aqueous rosemary extracts (aqueous extract of dry rosemary and aqueous extract of post-distillation rosemary waste residues) as well as in deionized water in the case of control samples. Both aqueous extracts and deionized water were previously adjusted at pH 5.00 by means of a dilute citric acid solution (pH = 2.00). The solubilization of sodium alginate was carried out under magnetic stirring at 500 rpm at room temperature for 2–3 h, and once dissolution of the alginate powder particles was achieved, the alginate solutions were stored overnight at below 8 °C prior to their use for the preparation of the emulsions. For preparation of O/W emulsions, rapeseed oil was firstly mixed with a specified amount of Tween 80, and then the water phase (alginate alone or alginate dissolved in aqueous rosemary extracts) was slowly added and sheared at 1000 rpm for 20 min. The final emulsion system consisted of 80% aqueous phase (containing 2% sodium alginate and 0.5% Tween 80) and 20% rapeseed oil (in the presence or absence of rosemary essential oil incorporated in the rapeseed oil at a level of 0.1% *w/w*). Table 1 illustrates all the formulations used for preparing the O/W emulsions. The emulsions were freshly made each time and no phase separation was visually observed after placing them into cylinders of 100 mL when left for at least 1 h.

Table 1. Formulation of experimental emulsions used for encapsulation of rapeseed oil in alginate beads.

Sample Identification *	AERDR	AERDR/EO	AER	EO	Control	RPSO
Rosemary raw plant aqueous extract			+			
Post-distillation rosemary residue aqueous extract	+	+				
Water				+	+	
Alginate conc. (% wt)	2.0	2.0	2.0	2.0	2.0	
Rosemary essential oil at 0.1% *w/w*		+		+		
Rapeseed oil	+	+	+	+	+	+
Tween 80	+	+	+	+	+	

* AERDR: aqueous extract of post-distillation rosemary waste residues; AERDR/EO: aqueous extract of post-distillation rosemary waste residues with rosemary essential oil; AER: aqueous extract of rosemary (raw plant material); EO: deionized water with rosemary essential oil; RPSO: free, non-encapsulated rapeseed oil; (+) dictates ingredients used for each formulation.

2.8.2. Preparation and Storage of Alginate Beads

For producing the emulsion alginate beads, the formulated O/W emulsions were placed into a square mould (9 cm × 9 cm × 6 cm) fitted with a frame consisting of 100 pores of 0.3-centimetre diameter (as illustrated in Figure A2). Ionic gelation (cross-linking) was elaborated by dropping the emulsion droplets into a 2.0% (*w*/*w*) $CaCl_2 \cdot 2H_2O$ solution. The distance between the mould (generating the emulsion droplets) and the surface of the $CaCl_2$ solution was 6 cm. The immersed emulsion droplets in the $CaCl_2$ solution were allowed to gellify ("hardened" via ionic exchange for the alginate chain cross-linking) for 30 min without any stirring. After gelation, the resultant beads were rinsed with deionized water, dried by an absorbent paper towel, placed in plastic Petri dishes, and stored in a cupboard at ambient temperature (20 ± 2.0 °C).

2.9. *Oxidative Stability Study*

To study the ability of the different emulsion formulations to protect rapeseed oil from oxidation, each emulsion alginate bead preparation was placed in a Petri dish, whereas a small amount of free, non-encapsulated rapeseed oil was put in another dish as the control preparation (RPSO). All dishes were stored at 20 °C and away from sunlight for 28 days to carry out the oxidative stability study. All chemical assays were performed at the beginning of storage (Day 0), and periodically after 7 (Day 7), 15 (Day 15), and 28 (Day 28) days of storage. Analysis of the samples was performed in quadruplicate at every sampling time interval in the oxidative stability study ($n = 4$).

2.9.1. Extraction of Rapeseed Oil from the Alginate Beads

The rupture of emulsion alginate beads was managed according to Flaminii et al.'s method (2020) [37] with slight modifications. At every sampling time interval, a Petri dish from all the different sample preparations was used for the analysis. Briefly, a 10% (*w*/*v*) sodium citrate solution was prepared and mixed with the alginate beads at a weight ratio of 5:1. The dispersions were left for 3 h at ambient temperature until the beads were greatly deformed. Extraction of rapeseed oil was then performed following the method reported by Sun-Waterhouse et al. (2011) [28] and Atencio et al. (2020) [29] with slight modifications. The dispersion of the deformed alginate beads was mixed with 100 mL of methanol and slightly shaken until aggregation of the polysaccharide matrix was noted, and the mixture was transferred into a separatory funnel. Hexane (100 mL) was then added, and the mixture was vigorously agitated to enable the transfer of oil into the hexane top layer, which was collected. The extraction steps were repeated three to four times, with the obtained hexane layers being combined, and the hexane was finally removed using a rotary evaporator (Rotavapor R114, Waterbath B480, Buchi, Flawil, Switzerland). Small amounts of the remaining hexane evaporated from the oil under the flow of a nitrogen gas stream.

2.9.2. Peroxide Value (PV) Determination

The PVs of all extracted oil samples were determined according to the AOCS Official Method [38], with all reagents at 10% of the amount recommended for the standard protocol [39] and expressed as peroxide milliequivalents per kg of oil. Oil (0.5 g) was mixed in a stoppered flask with acetic acid-chloroform solution (3 mL, 3:2 *v*/*v*). Then, saturated potassium iodide KI solution (50 μL) was added, and the mixture was shaken vigorously and left to stand for 1 min. An aliquot (3 mL) of distilled water was subsequently added, and the mixture was titrated with 0.001 N standardized sodium thiosulphate solution (for liberated iodine) until the light yellow colour almost disappeared; a small portion (0.2 mL) of a starch solution (1% *w*/*v*) was used as the indicator, with the titration being continued until the blue colour derived from the iodine disappeared. A blank sample as a control reagent was set up and analyzed similarly according to all the previous steps. The PV was obtained from the formula:

$$PV = \frac{1000(V - Vo)\,C}{m} \quad (1)$$

where V = volume of $Na_2S_2O_3$ titrant of sample (mL); Vo = volume of $Na_2S_2O_3$ titrant of blank (mL); C = normality of $Na_2S_2O_3$ (moles/L); m = sample mass (g).

2.9.3. p-Anisidine Value (p-AnV) Determination

The determination of p-AnVs (a measure of aldehyde levels in oils) was performed according to the AOCS Official Method [40]. The procedure was carried out in such a way to minimize the exposure of all preparations to sunlight. Briefly, p-anisidine solution was prepared by mixing 0.25 g of the reagent with 100 mL glacial acetic acid and the solution was kept in the dark. Approximately 0.5 g of the extracted oil samples was first dissolved in iso-octane (25 mL). The solution was allowed to stand for 10 min and the absorbance was measured (UV-1800, UV/Visible scanning spectrophotometer, SHIMADZU, Kyoto, Japan) at 350 nm and denoted as A0, using isooctane as the blank solution (A2). Then, an aliquot (5 mL) of the oil solution was mixed with 1 mL of the 0.25% p-AnV reagent. After 10 min, the absorbance of the reacted mixture (yellowish reaction products) was read at 350 nm and denoted as A1. The p-AnV of the sample was then calculated using the following equation:

$$\text{p-AnV} = \frac{25 \times [1.2(A_1 - A_2) - A_0]}{m} \quad (2)$$

where m = sample mass (g).

2.9.4. TOTOX (Total Oxidation) Value (TV) Determination

TVs were calculated from the sum of twice the peroxide value (PV) and the anisidine value (p-AnV) as follows [28,41]:

$$\textit{Totox value} = 2PV + (\text{p-AnV}) \quad (3)$$

2.9.5. Free Fatty Acids (FFA) determination

FFAs of the extracted oil samples were determined using the direct titration method of AOCS [42] with slight modifications. Neutralized alcohol was prepared with absolute ethanol (0.625 mL), oil (0.1 mL), and 1% (*w*/*v*) phenolphthalein indicator (50 µL) in a conical flask. The flask was then placed in a water bath (60 °C) until warmed. Sufficient NaOH (0.005 M) was added to produce a faint permanent pink colour. Extracted oil samples were weighed (0.7 g), dissolved into the neutralized alcohol, and titrated with the NaOH solution until a faint permanent pink colour appeared (phenolphthalein-determined endpoint). The FFA content was expressed as oleic acid (percentage):

$$\%\text{FFA}(\textit{as oleic acid}) = \left(\frac{V_{NaOH} \times C_{NaOH} \times 282.46}{W}\right) \times 100 \quad (4)$$

where V = volume of NaOH titrant (ml); C = molarity of NaOH titrant (mol/1000 mL); 282.46 = MW of oleic acid (g/mol); W = sample mass (g).

2.9.6. Determination of the Total Extractable Phenolic Content (TEPC)

The total phenolic content of the aqueous phase, obtained after rupture of the emulsion alginate beads and removal of oil, was determined according to the method described in Section 2.5. The results were expressed as milligrams of gallic acid equivalents per L of sample (mg GAE/L) using a calibration curve. Measurements were performed in triplicate.

2.9.7. Fatty Acid Profile/Composition Analyses

For the determination of the fatty acid profile, transesterification was carried out for the extracted oil samples to proceed with gas chromatographic analysis. In particular, 0.1 g of extracted lipids was transferred in a test tube with a screw cap and 2 mL of *n*-hexane was added, followed by addition of 0.2 mL 2.0 M methanolic solution of potassium hydroxide for preparation of the fatty acid methyl esters (FAMEs). The mixture was vortexed for 1 min and was left to settle until the upper phase that contains the FAMEs became transparent. The phase that contained the methyl esters was collected, filtered (0.45 µm PTFE hydrophobic filters), and analyzed by a gas chromatograph (TRACE GC 2000 Series, Thermo Quest CE Instruments, Rodano (MI), Italy) with a flame ionization detector (FID) equipped with an autosampler (TRIPLUS AS Thermo Quest CE Instruments). FAMEs were analyzed on a BPX70 GC capillary column (30 m length, 0.32 mm, i.d., 0.25 µm film thickness, SGE Analytical Science, Stockbridge, GA, USA). Helium was the carrier gas at a flow rate of 2.0 mL/min. The injector port and detector temperature were maintained at 250 °C. The split ratio was 1:20. The column oven was initially set at 46 °C for 2 min; then, the temperature was raised to 130 °C at a rate of 50 °C/min and maintained there for 10 min; then increased to 175 °C at 2 °C/min and maintained at that temperature for 2 min; then increased to 200 °C at 3 °C/min and maintained at that temperature for 3.5 min; before raising finally the temperature to a plateau of 240 °C at a rate of 10 °C/min and maintaining it at this level for 5 min, i.e., the total run time was 60 min. Identification of FAMEs was carried out by comparing the retention times (RT) with those of a standard mixture (AccuStandard, New Haven, CT, USA) containing 37 fatty acids and analyzed under the same chromatographic conditions [43]. Chromatograms were acquired and processed with the aid of the Chrom Quest 5.0 software (ver. 3.2.1, Thermo Separation Products, Thermo Scientific, West Palm Beach, FL, USA).

2.10. Statistical Analysis

Basic statistical parameters (average and standard deviations) were calculated by Microsoft Excel 2016. At every sampling time interval in the oxidative stability study, the obtained results were analyzed by one-way ANOVA with Tukey's ad hoc test by using minitab 19 to determine significant differences ($p < 0.05$) between the mean values obtained for the measured parameters of beads with different compositions.

3. Results

3.1. Total Phenolic Content (TPC) in Aqueous Extracts

The concentration of phenolic compounds in the aqueous extract of undistilled rosemary (R) was two times higher than that of the rosemary post-distillate residues (DR) (Figure A1). These levels are in agreement with the findings of Christaki et al. (2022) [13], who reported a higher phenolic concentration (expressed in mg GAE/g extract) in aqueous methanolic (70%) extracts obtained by ultrasound treatment of the post-distillation (EO hydrodistillation) plant materials of rosemary, sage, and spearmint compared with the extracts of the respective undistilled aromatic medicinal plants. Moreover, they reported that hydrodistillation showed the greatest negative impact on the recovery of TPC from solid residues, compared with two other methods studied, i.e., steam distillation and microwave-assisted hydrodistillation. The considerable losses of bioactive compounds from the solid residues following hydrodistillation have been previously attributed to the extended distillation time, leading to the destruction of thermolabile compounds, as well as the nature of some compounds that are water-extractable solubles and, thus, partition in the residual wastewater phase [44,45].

Besides the distillation method, the extraction conditions and the method applied to treat post-distillation rosemary waste residues can also affect the recovery yield and the composition of the extracts. Among the extraction methods, accelerated solvent extraction at elevated temperature and pressure [46], ultrasound-assisted extraction [12,13], and solid–liquid extraction with organic solvents [47] have been previously examined to exploit the rosemary solid residues as a source of phenolics after removal of the essential oil; the reported TPC values in the present study are in agreement with the data available in the scientific literature. Regardless of the applied extraction method and conditions, rosemary solid residues could be generally considered as a valuable post-distillation waste by-product that could be valorized for the recovery of important bioactive compounds.

3.2. LC-MS Analysis

LC-MS analysis was performed to identify the phenolic compounds present in two typical rosemary aqueous extracts (raw plant tissues vs. post-distillation plant residues, Figure 2), as the Folin–Ciocalteu assay only gives an estimate of the total phenolic functionalities and cannot be directly related to compositional analysis data. Table 2 presents the content of the major phenolics identified in the aqueous extracts. In total, eleven phenolic compounds were identified and quantified, including five phenolic acids, four flavonoids, and two phenolic diterpenes in both of the aqueous extracts of rosemary tissues. The major phenolic acids identified were rosmarinic acid, salvianolic acid isomer B, and quinic acid, whilst minor peaks of caffeic acid and neochlorogenic acid were also found in both aqueous extracts. The rosmarinic acid was the most dominant compound, constituting more than 82% of the total quantified phenolic acids in both samples. Rosmarinic acid was previously described as the most abundant compound in rosemary infusions [48] and, in fact, is considered a "family marker" for the *Lamiaceae* family [49]. The aqueous extract of undistilled (raw) rosemary (R) contained a higher total content of phenolic acids (65.2 ± 0.38 mg/g) than the aqueous extract from the post-distillation rosemary waste (DR) residues (42.73 ± 0.28 mg/g). This observation can be largely attributed to differences in rosmarinic acid content between the two extracts as the initial (R extract) rosmarinic acid content of 56.6 mg/g decreased significantly to 35.3 mg/g in the rosemary distillate extract (DR). Previous studies on rosmarinic acid content in extracts obtained prior to and after the hydro- and steam-distillation of *Rosmarinus officinalis* L. have reported similar findings [13,50]. The decrease in the recovery of rosmarinic acid can be explained by its thermal degradation during distillation and its water solubility as it has been previously shown that substantial amounts of rosmarinic acid are detected in the water phase during the hydrodistillation of *Rosmarinus officinalis* L. residues [50]. Nevertheless, considerable amounts of rosmarinic acid were present in both extracts; in fact, it was the most abundant compound among the identified polyphenols, thus, it can be considered as a relatively stable compound [44,51].

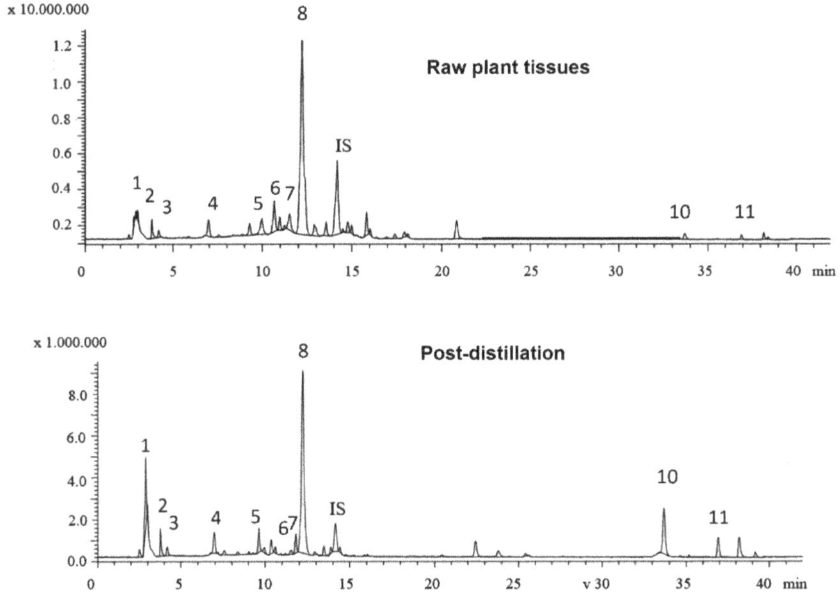

Figure 2. Total ion chromatograms of the phenolic compounds in aqueous extracts from raw plant tissues and post-distillation residues of rosemary. (1: quinic acid; 2: neochlorogenic acid; 3: vicenin-2; 4: caffeic acid; 5: luteolin-7-O-rutinoside; 6: apigenin-7-O-glucoside; 7: hesperidin; 8: rosmarinic acid; 9: salvianolic acid B; 10: carnosol; and 11: carnosic acid).

Table 2. Phenolic compounds identified and quantified by LC-MS in aqueous extract of dried ground rosemary (R) and aqueous extract of dried ground post-distillation rosemary residues (DR).

Chemical Group	Compound Name	Concentration (mg/g DW)	
		R [1]	DR [1]
Phenolic acids	Quinic acid	2.80 ± 0.07 [a]	2.29 ± 0.06 [b]
	Neochlorogenic acid	0.47 ± 0.02 [a]	0.17 ± 0.01 [b]
	Caffeic acid	0.20 ± 0.01 [b]	0.36 ± 0.01 [a]
	Rosmarinic acid	56.6 ± 0.6 [a]	35.3 ± 0.5 [b]
	Salvianolic acid B	5.13 ± 0.13 [a]	4.61 ± 0.15 [b]
Total Phenolic acids		65.2 ± 0.38	42.73 ± 0.28
Flavonoids	Vicenin-2	0.16 ± 0.01 [a]	0.12 ± 0.01 [a]
	Luteolin-7-O-rutinoside	1.73 [a] ± 0.09 [a]	0.14 ± 0.00 [b]
	Apigenin-7-O-glucoside	0.03 ± 0.002 [a]	0.01 ± 0.002 [b]
	Hesperidin	1.29 ± 0.075 [a]	0.19 ± 0.01 [b]
Total Flavonoids		3.21 ± 0.01 [a]	0.47 ± 0.01 [b]
Phenolic diterpenes	Carnosol	0.58 ± 0.02 [b]	5.67 ± 0.22 [a]
	Carnosic acid	2.51 ± 0.10 [b]	8.22 ± 0.08 [a]
Total Phenolic diterpenes		3.09 ± 0.08 [b]	13.89 ± 0.15 [a]

[1] Mean values for each compound with different superscript letters are statistically different ($p < 0.05$); analyses were performed in duplicate.

On the other hand, the total content of phenolic diterpenes in the aqueous extract of post-distillation rosemary residues (DR) was four times higher than that of the undistilled rosemary (R). There is a large body of literature stating that the diterpenoids, carnosol, and carnosic acid are the most important active components of rosemary extracts [3,4,51–53]. Oreopoulou et al. (2018) [54] reported that after alkaline extraction of hydrodistillation residues, *Rosmarinus officinalis* demonstrated the highest antioxidant activity in sunflower oil, and this was attributed to the presence of carnosic acid. In this context, the concentration of carnosic acid was shown to be highly correlated with the oxidative stability of the sunflower oil [55]. Both carnosic acid and carnosol have been proposed as compounds comprising over 90% of the antioxidant potential of rosemary extracts [54]. The aqueous extract of raw rosemary contained 0.58 mg/g of carnosol and 2.51 mg/g of carnosic acid, whilst the amounts found in the extract of post-distillate rosemary residues were 5.67 mg/g and 8.22 mg/g, respectively. The amount of carnosol present in both aqueous extracts is dependent on the applied extraction process; carnosol, as well as rosmanol, are generated by oxidative degradation of carnosic acid and thus as native constituents of unprocessed (raw) rosemary leaves exist in trace or very small amounts [50]. Moreover, carnosol content may increase at the expense of carnosic acid when the plant biomass is subjected to drying, heating, storage, extraction, and distillation [4]. Non-polar solvents were shown to extract considerably higher amounts of phenolic diterpenes [6,12,13] in comparison to the levels obtained by aqueous media. Apparently, only small amounts of carnosol have been found in the hydrodistillation water following the distillation of rosemary leaves, presumably due to its high hydrophobicity [50]. On the other hand, organic solvents suitable for the extraction of less-polar compounds are not consumer/environment-friendly, and as a result, more "green solvents"—often combined with innovative extraction methods—are preferable nowadays for isolation purposes. In the current study, the use of β-cyclodextrin in water as a "green" extraction medium could have enhanced the extraction efficiency of the aforementioned polyphenols due to the formation of inclusion complexes with the cyclic dextrin, as this has been previously reported [56–58].

3.3. Protection of Oil against Oxidation

The evolution of oxidative deterioration of rapeseed oil with time was determined in either free, non-encapsulated rapeseed oil (RPSO) or in encapsulated forms of various emulsified preparations (O/W oil/aqueous phase, 20/80 *w/w*), i.e., rapeseed oil alone (control), rapeseed oil containing rosemary essential oil (EO), rapeseed oil with aqueous phenolic extract from raw rosemary plant material (AER), rapeseed oil with aqueous phenolic extract from the post-distillation of rosemary waste residue (AERDR), and rapeseed oil with a combination of rosemary essential oil and the aqueous phenolic extract from the post-distillation of rosemary waste residue (AERDR/EO). The composition of the various formulations of encapsulated rapeseed oil emulsions in the alginate beads is presented in Table 1.

3.3.1. Peroxide Values

The PV is the most often used indicator to evaluate the state of oxidation of oils and fats. It is measured by a relatively simple assay and is especially utilized to assess oil quality after processing [59] and storage [28] or to examine the antioxidant potential of additives incorporated in oils [60,61]. The PV is determined by titration of liberated iodine, produced from the reaction of hydroperoxides with potassium iodide; hydroperoxides are formed at the early stages of lipid oxidation—the so-called "induction period/primary oxidation"—and thus, determination of PV can give an early indication of rancidity in fats and oils. In this study, the determination of the gradual progression of PVs for each type of rapeseed oil formulation (unencapsulated and encapsulated forms) with storage time (Figure 3A) implies that the rate of hydroperoxide formation is greater than their decomposition (propagation reactions) [62]. Significant differences ($p < 0.05$) in PVs of the oil samples were noted throughout storage. Over the 28-day storage, the PV value of the unencapsulated oil (RPSO) was significantly higher than the PV of all the encapsulated rapeseed oil preparations in alginate beads, indicating that the encapsulation of emulsified oil droplets in the polysaccharide matrix was effective to slow down the formation of primary oxidation products. Similarly, the encapsulation of olive oil during storage at 37 °C [28] and ginger oil during 15-day storage at 4 °C [29] using alginate and blends with other polymers were effective in slowing down the formation of primary oxidation products.

Among the encapsulated oil samples, the PVs decreased in the order of AERDR/EO <AERDR <control $\approx$ AER < EO and most of the PVs were below 18 meq/Kg over the storage time—indicative of acceptable oil quality. The initial PV of the encapsulated control sample was 11.02 meq/Kg and increased to 15.94 meq/Kg on Day 28, whilst the encapsulated oil with the addition of rosemary essential oil only (EO) increased from 12.8 (Day 0) to 20.7 meq/Kg by the end of storage time, suggesting that the incorporation of rosemary essential oil in the encapsulated oil did not have a positive impact against oxidative deterioration. Likewise, the encapsulated emulsified oil with the aqueous extract of phenolic compounds of rosemary raw tissues (AER) in the aqueous phase showed similar values of PVs to the control encapsulated oil sample during 28-day storage, indicating negligible ability to inhibit the generation of primary oil oxidation products. The lowest PVs were observed in the oil samples encapsulated with phenolic compounds of rosemary distillate aqueous extract (AERDR) as well as when a combination of AERDR with rosemary essential oil was used, on all days of measurements, indicating that the latter combined treatment (AERDR/EO) was the most effective to slow down the formation of primary oxidation products among all formulations tested. Additive effects between the phenols of the aqueous extract of the rosemary distillate by-products and the antioxidant compounds present in the rosemary essential oil result in greater ability to interact with free radicals, and thereby improve the protection of rapeseed oil against the undesirable oxidative deterioration. In general, the encapsulated oil samples in emulsified form (O/W) with aqueous extracts from the post-distillation rosemary residues as the continuous phase (AERDR) or in combination with EO (AERDR/EO) exhibited the lowest PVs; these observations of the oxidative stability of rapeseed oil could be attributed to the relatively high content of carnosic acid in the rosemary distillate aqueous extract, as the LC-MS analysis revealed, and is further evidenced by findings from previous studies [54,55]. Earlier works also showed that carnosic acid has a very high reactivity toward reactive oxygen species (ROS), and upon scavenging ROS, carnosic acid is converted into a variety of secondary antioxidants (such as carnosol), amplifying the antioxidant power of the precursor compound [63].

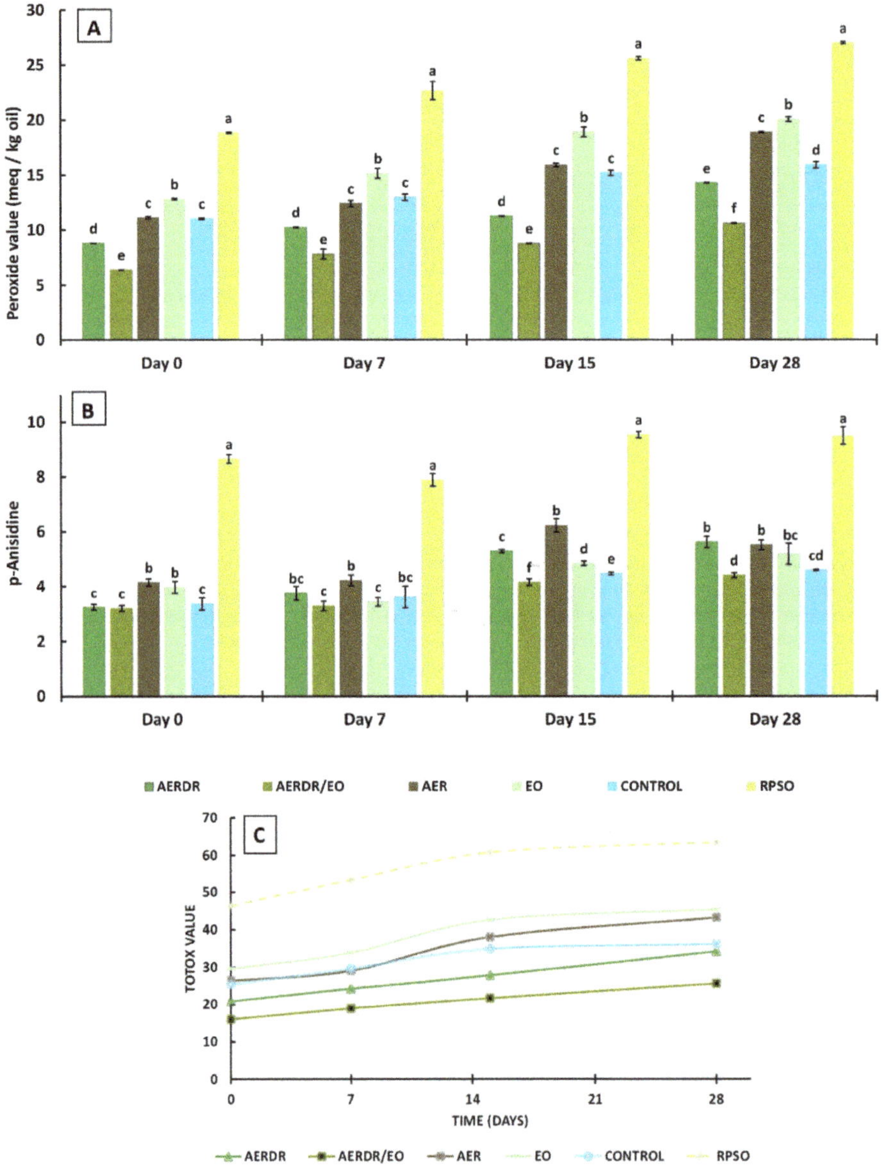

Figure 3. Peroxide values (**A**), p-Anisidine values (**B**), and TOTOX values (**C**) of encapsulated rapeseed oils (emulsified preparations in alginate beads) and free rapeseed oil (RPSO) when stored at 20 °C for 28 days. Data are expressed as means ± standard deviation; different letters above bars for the data sets of each specified storage time indicate significant differences ($p < 0.05$).

3.3.2. p-Anisidine Values

The p-AnV index measures the secondary oxidation products, including aldehydes, ketones, and various other products. It is a useful measurement of the oxidation status of "abused" fats and serves as a good predictor of storage stability when accelerated storage tests are used [60]. Over the 28-day storage, the p-AnV value of free unencapsulated oil (RPSO) was found higher by two-fold than the encapsulated oil samples ($p < 0.05$) (Figure 4), irrespective of the day of measurement, indicating that encapsulation led to significantly lowered p-AnV values, as previously reported [28,64]. The results also revealed a gradual increase in the secondary oxidation products with the storage time for all formulations; the p-AnV values of the encapsulated oil samples ranged between 3.2 and 4.15 on Day 0 without pronounced changes on Day 7, whereas the p-AnV values were found to vary between 4.41 and 5.62 on Day 28 for these preparations. At any day of storage, small differences were noted in p-AnV values among the encapsulated oil samples, with the preparation consisting of the aqueous extract of phenolic compounds from the post-distillation rosemary residues and the rosemary essential oil (AERDR/EO) exhibiting the lowest p-AnV values (Figure 3B), in accordance with the trends of the PV values (Figure 3A). Generally, the p-AnV should be below 10 for the acceptable quality of edible oils [41]. Overall, the p-AnV values of all oil samples did not exceed this maximum threshold value for any given storage time, indicating that secondary oxidation was still in its early stages.

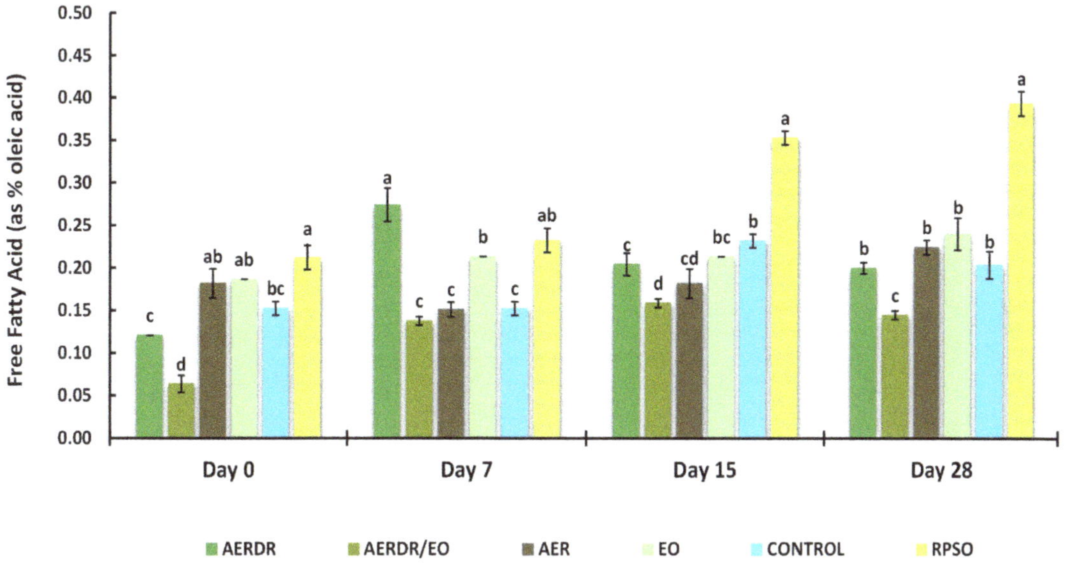

Figure 4. Free fatty acid (FFA) of encapsulated rapeseed oils (emulsified preparations in alginate beads) and free rapeseed oil (RPSO) when stored at 20 °C for 28 days. Data are expressed as means (wt. % of the oil) ± standard deviation; different letters above bars of each cluster indicate significant differences ($p < 0.05$).

3.3.3. TOTOX Values

The TOTOX is calculated from the obtained PV and p-AnV values of the oil for a given storage time, thus providing an overall indication of the oil oxidation status. Specifically, this value provides insight regarding the oxidation history of the oil (PV) and the potential for further deterioration (p-AnV). As a general guidance, an oil is considered of acceptable quality when the TOTOX value is less than 30 [41]. It was noticed that the TOTOX values of all samples increased as the storage proceeded (Figure 3C). In general, the encapsulated forms of rapeseed oil were found to be less susceptible to oxidation than RPSO.

After 28 days of storage at 20 °C, the TOTOX values decreased in the order of free unencapsulated oil RPSO (63.50) > EO (45.39) > AER (43.29) > control (36.12) > AERDR (34.14) > AERDR/EO (25.63). Thus, only the rapeseed oil encapsulated with phenolic compounds of the rosemary distillate aqueous extract as well as the inclusion of rosemary essential oil in the oil phase (AERDR/EO preparation) had an acceptable TOTOX value (<30) until the end of storage time. In the absence of any rosemary-derived matter, the control formulation (encapsulated oil in emulsified form with water and Tween) achieved a shelf-life of 7 days. Similarly, the encapsulated oil sample prepared by emulsification with the aqueous phenolic extract from the raw rosemary tissues (AER) also displayed a maximum shelf-life of 7 days. These observations suggest that the alginate encapsulation of emulsified rapeseed oil alone cannot prevent the oxidative deterioration of the oil and neither can the use of the aqueous extract of rosemary in the aqueous phase; instead, the combination of an aqueous extract (phenolic compounds) of post-distillation rosemary residues (water phase) along with the rosemary essential oil dissolved in the oil phase of the O/W emulsions, followed by alginate encapsulation, substantially improved rapeseed's oil oxidative stability. These findings concur with observations made by Sun-Waterhouse and co-workers who reported greater stability of olive [28] and avocado [65] oil degradation when using alginate as an encapsulating agent in the presence of an antioxidant.

3.3.4. Free Fatty Acids Content

FFAs can act as prooxidants in both bulk and emulsified oils by accelerating the decomposition rate of hydroperoxides [66]. The presence of FFAs is a result of hydrolytic rancidity caused by the reaction between water and oil. Thus, high FFA content in oils can cause further oxidation contributing to the perception of "rancidity" by the development of unpleasant taste and odour [60]. The Codex Alimentarius International Food Standards (2109) [67] set a standard of 0.3% (the relationship FFA = Acid value/1.989 was applied) for good-quality refined oils. The effect of encapsulation and the studied rosemary derivatives on FFA formation is shown in Figure 4. Over the 28-day storage time, a gradual increase in the FFA content of the RPSO sample was observed; moreover, the rate of increase in the unencapsulated oil was significantly higher ($p < 0.05$) on days 15 and 28 of measurement and found at levels above 0.3%, suggesting unacceptable quality. In contrast, the rate of FFA increase for the encapsulated oil samples was found to be much lower than RPSO, showing an FFA content of < 0.3% after 28 days of storage. Among the studied rosemary adjuncts, the preparation with AERDR/EO resulted in the lowest FFA levels across all days of measurement. The FFA content of the encapsulated oil sample with the AERDR/EO varied significantly ($p < 0.05$) from all the remaining encapsulated samples at the end of the storage period. An earlier study reported that encapsulation of avocado oil with alginate and BHT or phloridzin had a subtle additional effect on the FFA content compared with the encapsulated oil in the absence of an antioxidant [28]; however, a greater decrease in FFA content was observed when the oil was encapsulated with a combined matrix of alginate and hydroxypropyl methylcellulose (HPMC). On the other hand, encapsulation with alginate and fortification with caffeic acid of olive oil did not result in delaying the formation of FFA after 30-day storage at 20 or 37 °C [65]. That was also consistent with the study of Wang and co-workers (2013) [68], who did not find any distinct differences in suppressing hydrolytic rancidity of rapeseed oil encapsulated with alginate when the oil preparations were enriched with three different antioxidants, i.e., BHT, quercetin, and vitamin E.

3.3.5. Total Extractable Phenolic Content after Storage

TEPC from the alginate beads were determined by the Folin–Ciocalteu assay and are presented in Figure 5. Significant differences ($p < 0.05$) in the TEPC content of the aqueous phase were observed between the beads made with AER and AERDR (i.e., the preparations with or without inclusion of EO in the oil phase of the O/W emulsions) over the storage time; the TEPC content of the beads treated with AER was found to be two-fold higher than AERDR, which is consistent with the original TPC results obtained for the aqueous rosemary extracts "R" and "DR" (as shown in Section 3.1). In addition, the "DR" aqueous rosemary extract, utilized in the AERDR formulation, was earlier revealed to contain a relatively high content of the hydrophobic compounds carnosol and carnosic acid; therefore, the observed difference in TEPC could be attributed to the oil extraction methods [69,70] and/or the potential partitioning of phenolic compounds in the lipid phase [71].

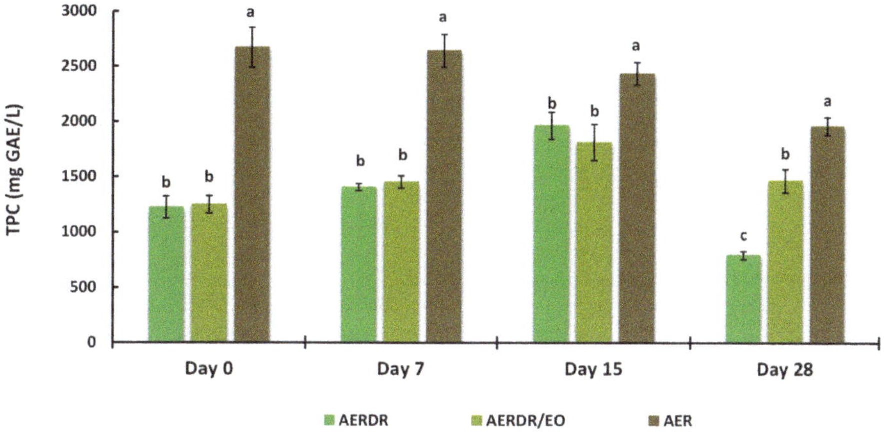

Figure 5. Total phenolic content (TPC) of the aqueous phase extracted from the alginate beads when stored at 20 °C for 28 days. Data are expressed as means ± standard deviation; different letters above bars of each cluster indicate significant differences ($p < 0.05$).

Regarding the beads formulated with AERDR and AERDR/EO, the TEPC of the two treatments were found at similar levels (1220–1250 mg GAE/L), a gradual increase in TEPC was observed over the first 15 days, and a noticeable decline in TEPC was finally noted on Day 28 of storage for both preparations. These results are in agreement with the study of Sun-Waterhouse et al. [28], in which olive oil (fortified with caffeic acid) extracted from sodium alginate beads showed an increasing trend of TEPC for the first 14 days and then a decline after 30 days of storage at 20 °C. In the case of the aqueous phase from the beads formulated with the AER extract, a high and constant TEPC content was revealed over the first 7 days and then TEPC also followed a downward trend until the end of the storage period. The "R" aqueous rosemary extract, utilized in the AER formulation, was earlier revealed to contain high TPC content, with rosmarinic acid as the main antioxidant in abundancy; it is hypothesized that rosmarinic acid, due to its hydrophilicity, might be predominantly bound to the alginate network (mannuronic and guluronic units of alginate) resulting in increased exposure to environmental conditions (oxygen), which subsequently might lead to accelerated antioxidant depletion and poor oxidative stability of the oil. As mentioned previously, the ability of free radical scavenging antioxidants to inhibit lipid degradation in food systems not only depends on their chemical reactivity but also on factors such as polarity, physical location, interactions with other components, and environmental conditions [72]. Thus, the antioxidant compounds present in the AER formulation might have reacted with oxygen and scavenged free radicals in the process of inhibiting lipid oxidation, resulting in levelled-off TEPC over storage time. Overall, influencing factors for the observed differences in the oxidative stability of the rapeseed oil in different alginate bead preparations could be the different total phenolic content in the aqueous extracts, exhibiting varying free radical scavenging activity [73], the polarity of the antioxidant compounds present in the extracts [74], and the physical location of the antioxidant compounds as different types of interfaces in oil-in-water emulsions can exist depending on phenolic composition [3,75]. Overall, in the present study, high amounts of phenolics were still present (extractable fraction) until the end of the stability test in all samples, implying a good retention capacity of the phenolic compounds following the encapsulation process and during storage; therefore, free phenolic compounds would be available in beads until the end of the storage period.

3.3.6. Fatty Acid Composition

Fatty acid composition analysis was performed using gas chromatography to determine whether the levels of the unsaturated fatty acids are affected by the encapsulation process and the inclusion of rosemary antioxidants in the alginate beads of the emulsified rapeseed oil preparations. In total, 11 fatty acids (i.e., palmitic acid, palmitoleic acid, stearic acid, oleic acid, linoleic acid, γ-linolenic acid, cis-11-eicosenoic acid, linolenic acid, cis-11,14,17-eicosatrienoic acid, cis-5,8,11,14,17-eicosapentaenoic acid, and nervonic acid) were identified by the GC-FID method and their respective peak area percentages were calculated (Table A1). Rapeseed oil samples exhibited the typical FA profile of canola oil [76], which is mostly composed of oleic acid (C18:1) followed by linoleic acid (C18:2), corresponding to about 60% and 20% of the total FA composition, respectively.

As a general trend, oil samples, irrespective of the treatment, had shown a gradual decrease in the peak area percentage of the identified fatty acids with storage time, which is in line with previous oxidative stability studies [28,61,77]. This observation is more evident by the calculated parameters of the sum of saturated fatty acids (SFA), sum of monounsaturated fatty acids (MUFA), sum of polyunsaturated fatty acids (PUFA), and their respective ratios in the oil samples encapsulated with rosemary antioxidants (Table 3). On Day 28, the MUFA content of the encapsulated oil samples was found to be higher than in the RPSO, i.e., significantly higher MUFA content of the oil samples encapsulated with AERDR and AERDR/EO was observed when compared to the RPSO sample ($p < 0.05$). This observation was mainly attributed to the content of oleic acid (C18:1) (detailed results in Table A1). However, decreases occurred in the PUFA content in the case of encapsulated oils, whilst the PUFA content of the RPSO oil sample showed a minor increase after the 28-day storage period. The detected changes are partially in agreement with the findings of Sun-Waterhouse and co-workers (2011) [28] who reported that both encapsulation in alginate microspheres and the incorporation of caffeic acid (300 ppm) in olive oil helped with preserving the MUFA and PUFA content. Overall, the observed differences in the fatty acid profile between the encapsulated rapeseed oil samples were relatively small in the current study; elevated storage temperatures and longer storage times might reveal greater differences for losses in the mono- and polyunsaturated fatty acids in the absence of rosemary antioxidants, as the relevant literature reports also suggest [61,77–79]. Moreover, it has been previously described that the oxidative resistance of lipids is higher with increasing extract concentrations [77,80]. There have been several studies aiming to elucidate the potential effects of natural antioxidants on lipid oxidation and fatty acid profiles determined by gas chromatography. The study of Sun-Waterhouse et al. (2011) [28] has shown that unsaturated fatty acids of olive oil, including C18:1 (ω-9), C18:2 (ω-6), and C18:3 (ω-3), were protected by encapsulation in alginate beads and the addition of caffeic acid when stored at 20 and 37 °C for 30 days. Tohma and Turan [81] reported that hazelnut oil treated with various rosemary additives (plant, ethanolic extract, and essential oil), showed higher oxidative stability during the frying of French fries. Fhaner et al. (2016) [77] also showed that the presence of 0.84 mM phenolic diterpenes in rosemary extract preserved the eicosapentaenoic acid (EPA) and docosahexaenoic acid (DHA) levels during the storage of fish oil for 14 days at 30 and 50 °C.

Table 3. Saturated (SFA), monounsaturated (MUFA), and polyunsaturated (PUFA) proportions of fatty acids (% of total peak area in GC-FID chromatographs) as means ± standard deviations; values in the same column with different letters (a,b,c) for each storage time are significantly different ($p < 0.05$).

Storage Day	Oil Sample	SFA (%) [1]	MUFA (%) [1]	PUFA (%) [1]	UFA/SFA	MUFA/SFA
Day 0	AERDR	6.14 ± 0.08 ab	61.79 ± 0.09 a	32.07 ± 0.01 b	15.44 ± 0.04 ab	10.07 ± 0.15 ab
	AERDR/EO	6.05 ± 0.02 b	61.91 ± 0.02 a	32.03 ± 0.02 b	15.52 ± 0.04 a	10.23 ± 0.03 a
	AER	6.06 ± 0.01 b	61.87 ± 0.02 a	32.07 ± 0.01 b	15.50 ± 0.03 a	10.21 ± 0.02 a
	EO	6.24 ± 0.06 a	61.18 ± 0.43 b	32.58 ± 0.42 a	15.04 ± 0.15 c	9.81 ± 0.13 c
	Control	6.14 ± 0.01 ab	61.79 ± 0.01 a	32.07 ± 0.01 b	15.28 ± 0.03 b	10.06 ± 0.02 ab
	RPSO	6.24 ± 0.02 a	61.90 ± 0.01 a	31.87 ± 0.02 b	15.03 ± 0.05 c	9.92 ± 0.03 bc
Day 7	AERDR	6.08 ± 0.01 ab	61.91 ± 0.01 a	32.01 ± 0.02 a	15.44 ± 0.04 ab	10.18 ± 0.02 a
	AERDR/EO	6.10 ± 0.02 ab	61.91 ± 0.02 a	31.99 ± 0.01 a	15.39 ± 0.04 ab	10.15 ± 0.03 a
	AER	6.10 ± 0.04 ab	61.88 ± 0.03 a	32.02 ± 0.02 a	15.40 ± 0.10 ab	10.15 ± 0.07 a
	EO	6.12 ± 0.01 a	61.92 ± 0.00 a	31.97 ± 0.01 a	15.35 ± 0.03 b	10.12 ± 0.02 a
	Control	6.12 ± 0.01 a	61.90 ± 0.01 a	31.98 ± 0.00 a	15.35 ± 0.02 b	10.12 ± 0.01 a
	RPSO	6.02 ± 0.07 b	61.48 ± 0.61 a	32.50 ± 0.68 a	15.61 ± 0.19 a	10.21 ± 0.02 a
Day 15	AERDR	6.10 ± 0.04 bc	61.84 ± 0.13 a	32.06 ± 0.09 a	15.39 ± 0.10 ab	10.14 ± 0.08 a
	AERDR/EO	6.14 ± 0.00 ab	61.80 ± 0.02 a	32.06 ± 0.02 a	15.30 ± 0.01 bc	10.07 ± 0.01 ab
	AER	6.12 ± 0.00 b	61.78 ± 0.00 a	32.09 ± 0.01 a	15.34 ± 0.01 b	10.09 ± 0.01 a
	EO	6.24 ± 0.06 a	61.18 ± 0.43 a	32.58 ± 0.42 a	15.04 ± 0.15 c	9.81 ± 0.13 c
	Control	6.23 ± 0.06 a	61.28 ± 0.42 a	32.49 ± 0.36 a	15.05 ± 0.15 c	9.83 ± 0.16 bc
	RPSO	6.01 ± 0.01 c	61.80 ± 0.01 a	32.19 ± 0.01 a	15.64 ± 0.03 a	10.29 ± 0.02 a
Day 28	AERDR	6.15 ± 0.03 ab	62.27 ± 0.01 a	31.58 ± 0.03 c	15.25 ± 0.09 bc	10.12 ± 0.05 ab
	AERDR/EO	6.07 ± 0.06 bc	62.17 ± 0.35 a	31.75 ± 0.29 bc	15.47 ± 0.15 ab	10.24 ± 0.15 a
	AER	6.11 ± 0.01 abc	61.91 ± 0.01 ab	31.97 ± 0.02 ab	15.36 ± 0.03 abc	10.13 ± 0.02 ab
	EO	6.20 ± 0.01 a	61.90 ± 0.01 ab	31.84 ± 0.02 bc	15.13 ± 0.03 c	9.99 ± 0.02 b
	Control	6.06 ± 0.03 c	62.02 ± 0.03 ab	31.92 ± 0.03 ab	15.50 ± 0.08 a	10.23 ± 0.06 a
	RPSO	6.05 ± 0.02 c	61.74 ± 0.00 b	32.21 ± 0.02 a	15.54 ± 0.05 a	10.21 ± 0.03 a

[1] SFA represents the sum of palmitic acid (C16:0) and stearic acid (C18:0); MUFA corresponds to the sum of palmitoleic acid (C16:1), oleic acid (C18:1), eicosenoic acid (C20:1) and nervonic acid (C24:1); and PUFA represents the sum of linoleic acid (C18:2), g-linolenic acid (C18:3α), linolenic acid (C18:3γ), eicosatrienoic acid (C20:3), and eicosapentaenoic acid (C20:5).

4. Conclusions

Hydrodistillation has a strong influence on the antioxidant potency of rosemary leaves; during distillation, antioxidants are lost by the wastewater stream or thermal degradation. A solid–liquid extraction using water as a solvent and β-cyclodextrin as a co-solvent resulted in considerable amounts of phenolic compounds in the extract of undistilled rosemary leaves and the extract of post-distillation rosemary waste residues as well. The LC-MS analysis of the extracts revealed that both aqueous extracts primarily contained rosmarinic acid, salvianolic acid B, quinic acid, carnosol, and carnosic acid. The aqueous extract of raw rosemary tissues was mainly abundant in rosmarinic acid (56.60 ± 0.6 mg/g dw), and the extract of rosemary distillate residues contained both high amounts of rosmarinic acid (42.73 ± 0.28 mg/g dw) and the phenolic diterpenes of carnosol and carnosic acid (13.89 ± 0.15 mg/g dw). The differences in the polyphenols profile of the extracts may have contributed to the observed variation in oxidative stabilities among the diverse preparations of the encapsulated rapeseed oil samples in alginate beads. The encapsulation of rapeseed oil emulsions in an alginate-based gel matrix with the use of aqueous rosemary extracts, rich in phenolic compounds, has enhanced the oxidative stability of the rapeseed oil during a 28-day storage period, as monitored by the evolution of the primary and secondary products of oxidation. The alginate beads prepared with the aqueous extract of the post-distillation rosemary residues—also comprising the aqueous phase of the rapeseed oil emulsions—when combined with the rosemary essential oil dissolved in the rapeseed oil phase (AERDR/EO) exhibited the lowest lipid deterioration (PV = 10.61 ± 0.02 meq/Kg oil, p-AnV = 4.41 ± 0.09, and FFA = 0.14 ± 0.00, expressed as % oleic acid content), indicating an acceptable oil quality until the end of the storage period. The observed high protective ability of the aqueous extract of post-distillation rosemary residues in rapeseed's oil oxidation is probably attributed to the proportionally higher amounts of carnosic acid detected in the rosemary extracts, a compound with strong antioxidant activity and known for its effectiveness in protecting vegetable oils rich in UFA from oxidation [81,82]. Moreover, high amounts of phenolics were also detected until the end of the stability tests for all rapeseed oil preparations encapsulated in alginate beads, implying a constant antioxidant environment throughout storage. Overall, the phenolic extracts obtained by applying green extraction techniques and valorizing waste residues of the essential oil industry could find applications in the food and drug industries due to their antioxidant properties and the enhanced protection they can offer for the preservation of oil-based nutraceutical formulations.

Author Contributions: Conceptualization, P.T. and I.M.; methodology, P.T., C.G.B. and I.M.; software, P.T., A.K. and M.I; validation, P.T., I.M. and C.G.B.; formal analysis, P.T., D.K., A.K. and M.I.; investigation, P.T. and D.K.; resources, I.M.; data curation, P.T., D.K., A.K. and M.I.; writing—original draft preparation, P.T. and D.K.; writing—review and editing, C.G.B. and I.M.; visualization, P.T.; supervision, C.G.B. and I.M; project administration, I.M. All authors have read and agreed to the published version of the manuscript.

Funding: This research received no external funding.

Institutional Review Board Statement: Not applicable.

Informed Consent Statement: Not applicable.

Data Availability Statement: The original contributions presented in the study are included in the article, further inquiries can be directed to the corresponding author.

Acknowledgments: The authors would like to thank Georgia Koletsou who substantially contributed during the development of the protocols for the evaluation of lipid oxidation. We also thank Konstantinos Klimantakis for his graphic design skills and Huw Griffiths (Besmoke Ltd.) for his sincere support during this project.

Conflicts of Interest: The authors declare no conflicts of interest.

Appendix A

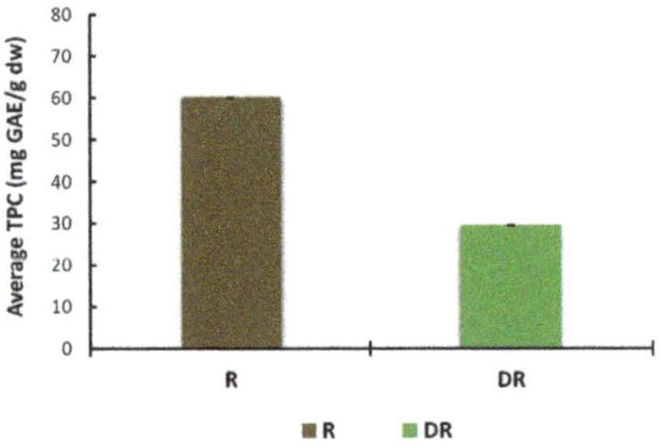

Figure A1. Total phenolic content (TPC) in aqueous extracts of dried ground raw rosemary (R) and dried ground post-distillation residue (DR). Data are expressed as means ± standard deviation; measurements were performed in triplicate.

Appendix B

Table A1. Levels of identified fatty acids (% of total fatty acid area) in encapsulated (emulsified preparations in alginate beads) and unencapsulated rapeseed oil (RPSO) samples stored for 28 days at 20 °C; comparisons of the reported mean values are made for each fatty acid separately; values in the same column with different letters (a,b,c) for each storage time are significantly different ($p < 0.05$).

Storage Day	Oil Sample	C16:0 (%)	C16:1 (%)	C18:0 (%)	C18:1 (%)	C18:2 (%)	C18:3 α (%)	C18:3 γ (%)	C20:1 (%)	C20:3 (%)	C20:5 (%)	C24:1 (%)	Total (%)
0	AERDR	4.56 ± 0.05 b	0.18 ± 0.03 ab	1.58 ± 0.03 abc	60.92 ± 0.16 ab	20.55 ± 0.03 ab	9.94 ± 0.01 ab	1.18 ± 0.01 b	0.56 ± 0.03 a	0.31 ± 0.03 a	0.09 ± 0.02 a	0.13 ± 0.02 a	100.00
	AERDR/EO	4.50 ± 0.01 b	0.16 ± 0.01 b	1.55 ± 0.00 bc	61.06 ± 0.02 a	20.57 ± 0.01 a	9.92 ± 0.01 c	1.18 ± 0.01 b	0.54 ± 0.01 a	0.29 ± 0.01 a	0.08 ± 0.00 a	0.15 ± 0.03 a	100.00
	AER	4.52 ± 0.01 b	0.17 ± 0.01 b	1.55 ± 0.01 c	61.04 ± 0.02 ab	20.58 ± 0.01 a	9.96 ± 0.01 a	1.17 ± 0.01 b	0.54 ± 0.01 a	0.28 ± 0.00 a	0.08 ± 0.01 a	0.12 ± 0.01 a	100.00
	H_2O/EO	4.53 ± 0.02 b	0.17 ± 0.00 b	1.56 ± 0.00 bc	61.06 ± 0.02 a	20.56 ± 0.01 ab	9.92 ± 0.01 bc	1.17 ± 0.01 b	0.55 ± 0.01 a	0.28 ± 0.01 a	0.08 ± 0.00 a	0.12 ± 0.00 a	100.00
	Control	4.53 ± 0.01 b	0.21 ± 0.00 a	1.61 ± 0.01 a	60.86 ± 0.04 b	20.51 ± 0.03 b	9.92 ± 0.01 c	1.20 ± 0.01 a	0.57 ± 0.01 a	0.25 ± 0.16 a	0.19 ± 0.12 a	0.14 ± 0.05 a	100.00
	Rapeseed	4.64 ± 0.03 a	0.16 ± 0.01 b	1.60 ± 0.01 ab	61.04 ± 0.02 ab	20.43 ± 0.01 c	9.87 ± 0.00 d	1.18 ± 0.01 b	0.55 ± 0.01 a	0.30 ± 0.01 a	0.09 ± 0.01 a	0.13 ± 0.01 a	100.00
7	AERDR	4.52 ± 0.02 a	0.16 ± 0.00 b	1.56 ± 0.00 ab	61.09 ± 0.02 a	20.56 ± 0.02 a	9.94 ± 0.01 a	1.17 ± 0.01 a	0.53 ± 0.01 a	0.27 ± 0.00 c	0.08 ± 0.01 b	0.12 ± 0.00 a	100.00
	AERDR/EO	4.54 ± 0.01 a	0.16 ± 0.01 ab	1.56 ± 0.01 ab	61.09 ± 0.02 a	20.56 ± 0.00 a	9.91 ± 0.01 a	1.17 ± 0.01 a	0.53 ± 0.01 a	0.27 ± 0.01 bc	0.08 ± 0.01 ab	0.12 ± 0.01 a	100.00
	AER	4.52 ± 0.03 a	0.17 ± 0.01 ab	1.57 ± 0.01 ab	61.05 ± 0.05 a	20.57 ± 0.01 a	9.91 ± 0.00 a	1.18 ± 0.01 a	0.54 ± 0.01 a	0.28 ± 0.00 b	0.08 ± 0.01 ab	0.13 ± 0.01 a	100.00
	H_2O/EO	4.54 ± 0.01 a	0.17 ± 0.00 ab	1.57 ± 0.01 a	61.07 ± 0.01 a	20.56 ± 0.01 a	9.88 ± 0.01 a	1.17 ± 0.01 a	0.54 ± 0.01 a	0.27 ± 0.01 c	0.09 ± 0.01 ab	0.13 ± 0.01 a	100.00
	Control	4.55 ± 0.00 a	0.17 ± 0.01 a	1.57 ± 0.01 ab	61.06 ± 0.01 a	20.57 ± 0.01 a	9.89 ± 0.01 a	1.17 ± 0.00 a	0.54 ± 0.01 a	0.28 ± 0.00 bc	0.09 ± 0.01 ab	0.13 ± 0.00 a	100.00
	Rapeseed	4.48 ± 0.06 a	0.17 ± 0.01 ab	1.54 ± 0.01 b	60.64 ± 0.61 a	20.48 ± 0.21 a	10.47 ± 0.89 a	1.17 ± 0.01 a	0.54 ± 0.01 a	0.30 ± 0.01 a	0.09 ± 0.01 a	0.13 ± 0.00 a	99.99
15	AERDR	4.54 ± 0.03 c	0.19 ± 0.01 ab	1.55 ± 0.00 c	60.47 ± 0.78 a	20.53 ± 0.08 b	9.91 ± 0.04 a	1.19 ± 0.01 ab	0.55 ± 0.01 a	0.32 ± 0.02 a	0.10 ± 0.01 a	0.63 ± 0.83 a	100.00
	AERDR/EO	4.58 ± 0.01 abc	0.18 ± 0.00 bc	1.56 ± 0.00 abc	60.93 ± 0.03 a	20.59 ± 0.01 ab	9.91 ± 0.01 a	1.18 ± 0.01 bc	0.55 ± 0.01 a	0.29 ± 0.00 b	0.10 ± 0.01 a	0.15 ± 0.01 a	100.00
	AER	4.56 ± 0.01 bc	0.18 ± 0.00 bc	1.56 ± 0.00 bc	60.90 ± 0.01 a	20.60 ± 0.01 ab	9.91 ± 0.01 a	1.18 ± 0.01 abc	0.55 ± 0.01 a	0.30 ± 0.01 ab	0.10 ± 0.00 a	0.15 ± 0.01 a	100.00
	H_2O/EO	4.61 ± 0.03 ab	0.19 ± 0.01 a	1.57 ± 0.01 ab	60.29 ± 0.43 a	20.68 ± 0.12 ab	9.91 ± 0.06 a	1.21 ± 0.01 a	0.55 ± 0.01 a	0.31 ± 0.01 a	0.42 ± 0.55 a	0.15 ± 0.01 a	99.90
	Control	4.64 ± 0.04 a	0.20 ± 0.01 a	1.58 ± 0.00 a	60.38 ± 0.44 a	20.86 ± 0.22 a	9.96 ± 0.11 a	1.20 ± 0.02 a	0.55 ± 0.01 a	0.31 ± 0.02 ab	0.09 ± 0.02 a	0.16 ± 0.02 a	99.93
	Rapeseed	4.48 ± 0.01 d	0.17 ± 0.00 c	1.54 ± 0.01 d	60.98 ± 0.01 a	20.69 ± 0.01 ab	9.97 ± 0.00 a	1.17 ± 0.01 c	0.53 ± 0.00 b	0.29 ± 0.00 b	0.08 ± 0.01 a	0.13 ± 0.01 a	100.00
28	AERDR	4.59 ± 0.04 ab	0.18 ± 0.01 a	1.56 ± 0.00 ab	61.42 ± 0.02 a	20.38 ± 0.02 c	9.67 ± 0.01 d	1.17 ± 0.01 b	0.54 ± 0.01 bc	0.28 ± 0.01 ab	0.08 ± 0.01 b	0.13 ± 0.01 b	100.00
	AERDR/EO	4.51 ± 0.04 bc	0.66 ± 0.85 a	1.56 ± 0.01 bc	60.84 ± 0.49 b	20.43 ± 0.18 bc	9.80 ± 0.09 bc	1.17 ± 0.02 b	0.55 ± 0.01 ab	0.27 ± 0.01 b	0.09 ± 0.01 b	0.13 ± 0.01 b	100.00
	AER	4.55 ± 0.01 abc	0.18 ± 0.00 a	1.56 ± 0.00 bc	61.06 ± 0.02 ab	20.59 ± 0.01 ab	9.85 ± 0.01 b	1.16 ± 0.01 b	0.54 ± 0.01 abc	0.28 ± 0.00 ab	0.09 ± 0.01 ab	0.13 ± 0.00 b	100.00
	H_2O/EO	4.61 ± 0.00 a	0.18 ± 0.01 a	1.59 ± 0.01 a	61.01 ± 0.02 ab	20.51 ± 0.01 bc	9.75 ± 0.01 cd	1.19 ± 0.00 a	0.56 ± 0.01 a	0.29 ± 0.01 ab	0.10 ± 0.00 a	0.15 ± 0.00 a	99.94
	Control	4.51 ± 0.03 bc	0.17 ± 0.01 a	1.55 ± 0.00 bc	61.19 ± 0.03 ab	20.58 ± 0.02 abc	9.80 ± 0.01 bc	1.17 ± 0.01 b	0.54 ± 0.01 bc	0.29 ± 0.01 a	0.08 ± 0.01 b	0.12 ± 0.01 b	100.00
	Rapeseed	4.51 ± 0.02 c	0.17 ± 0.01 a	1.54 ± 0.00 c	60.92 ± 0.01 ab	20.72 ± 0.01 a	9.96 ± 0.00 a	1.16 ± 0.00 b	0.52 ± 0.01 c	0.28 ± 0.01 ab	0.08 ± 0.01 b	0.12 ± 0.01 b	100.00

Appendix C

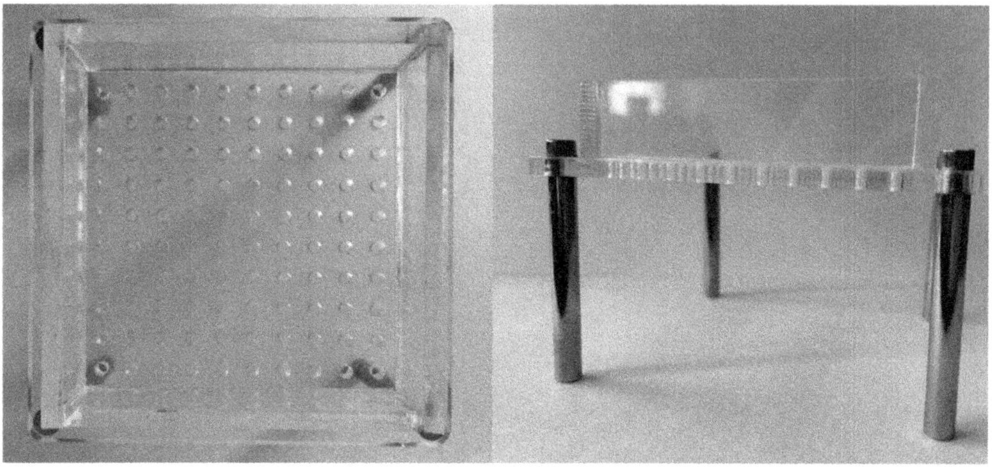

Figure A2. Illustration of the mould used for the preparation of the O/W emulsion alginate beads.

References

1. Ganesan, B.; Brothersen, C.; McMahon, D.J. Fortification of foods with omega-3 polyunsaturated fatty acids. *Crit. Rev. Food Sci. Nutr.* **2014**, *54*, 98–114. [CrossRef] [PubMed]
2. Tsitlakidou, P.; Tasopoulos, N.; Chatzopoulou, P.; Mourtzinos, I. Current status, technology, regulation and future perspectives of essential oils' usage in food and drink industry. *J. Sci. Food Agric.* **2023**, *103*, 6727–6751. [CrossRef]
3. Frankel, E.N.; Huang, S.W.; Aeschbach, R.; Prior, E. Antioxidant activity of a rosemary extract and its constituents, carnosic acid, carnosol, and rosmarinic acid, in bulk oil and oil-in-water emulsion. *J. Agric. Food Chem.* **1996**, *44*, 131–135. [CrossRef]
4. Richheimer, S.L.; Bernart, M.W.; King, G.A.; Kent, M.C.; Beiley, D.T. Antioxidant activity of lipid-soluble phenolic diterpenes from rosemary. *J. Am. Oil Chem. Soc.* **1996**, *73*, 507–514. [CrossRef]
5. Skendi, A.; Irakli, M.; Chatzopoulou, P.; Bouloumpasi, E.; Biliaderis, C.G. Phenolic extracts from solid wastes of the aromatic plant essential oil industry: Potential uses in food applications. *Food Chem. Adv.* **2022**, *1*, 100065. [CrossRef]
6. Psarrou, I.; Oreopoulou, A.; Tsimogiannis, D.; Oreopoulou, V. Extraction kinetics of phenolic antioxidants from the hydro distillation residues of rosemary and effect of pretreatment and extraction parameters. *Molecules* **2020**, *25*, 4520. [CrossRef] [PubMed]
7. Saha, A.; Basak, B.B. Scope of value addition and utilization of residual biomass from medicinal and aromatic plants. *Ind. Crops Prod.* **2020**, *145*, 111979. [CrossRef]
8. Dhifi, W.; Bellili, S.; Jazi, S.; Bahloul, N.; Mnif, W. Essential oils' chemical characterization and investigation of some biological activities: A critical review. *Medicines* **2016**, *3*, 25. [CrossRef] [PubMed]
9. Silva, A.C.; Bizzo, H.R.; Vieira, R.F.; Bringel, J.B., Jr.; Azevedo, D.A.; Uekane, T.M.; Rezende, C.M. Characterization of volatile and odor-active compounds of the essential oil from *Bidens graveolens* Mart. (Asteraceae). *Flavour Fragr. J.* **2020**, *35*, 79–87. [CrossRef]
10. Lahlou, M.; Berrada, R. Composition and niticidal activity of essential oils of three chemotypes of *Rosmarinus officinalis* L. acclimatized in Morocco. *Flavour Fragr. J.* **2003**, *18*, 124–127. [CrossRef]
11. Santana-Méridas, O.; Polissiou, M.; Izquierdo-Melero, M.E.; Astraka, K.; Tarantilis, P.A.; Herraiz-Peñalver, D.; Sánchez-Vioque, R. Polyphenol composition, antioxidant and bioplaguicide activities of the solid residue from hydrodistillation of *Rosmarinus officinalis* L. *Ind. Crops Prod.* **2014**, *59*, 125–134. [CrossRef]
12. Irakli, M.; Skendi, A.; Bouloumpasi, E.; Chatzopoulou, P.; Biliaderis, C.G. LC-MS Identification and Quantification of Phenolic Compounds in Solid Residues from the Essential Oil Industry. *Antioxidants* **2021**, *10*, 2016. [CrossRef] [PubMed]
13. Christaki, S.; Bouloumpasi, E.; Lalidou, E.; Chatzopoulou, P.; Irakli, M. Bioactive Profile of Distilled Solid By-Products of Rosemary, Greek Sage and Spearmint as Affected by Distillation Methods. *Molecules* **2022**, *27*, 9058. [CrossRef] [PubMed]
14. UN General Assembly. Transforming Our World: The 2030 Agenda for Sustainable Development. A/RES/70/1. 2015. Available online: https://www.refworld.org/docid/57b6e3e44.html (accessed on 19 January 2024).
15. Carrillo, M. Measuring Progress towards Sustainability in the European Union within the 2030 Agenda Framework. *Mathematics* **2022**, *10*, 2095. [CrossRef]
16. Calvo, T.R.A.; Perullini, M.; Santagapita, P.R. Encapsulation of betacyanins and polyphenols extracted from leaves and stems of beetroot in Ca (II)-alginate beads: A structural study. *J. Food Eng.* **2018**, *235*, 32–40. [CrossRef]

17. Zhang, R.; Belwal, T.; Li, L.; Lin, X.; Xu, Y.; Luo, Z. Recent advances in polysaccharides stabilized emulsions for encapsulation and delivery of bioactive food ingredients: A review. *Carbohydr. Polym.* **2020**, *242*, 116388. [CrossRef] [PubMed]
18. Amsden, B. Solute diffusion within hydrogels. Mechanisms and models. *Macromolecules* **1998**, *31*, 8382–8395. [CrossRef]
19. Fernando, I.P.S.; Lee, W.; Han, E.J.; Ahn, G. Alginate-based nanomaterials: Fabrication techniques, properties, and applications. *J. Chem. Eng.* **2020**, *391*, 123823. [CrossRef]
20. Ching, S.H.; Bansal, N.; Bhandari, B. Alginate gel particles—A review of production techniques and physical properties. *Crit. Rev. Food Sci. Nutr.* **2017**, *57*, 1133–1152. [CrossRef]
21. Strobel, S.A.; Knowles, L.; Nitin, N.; Scher, H.B.; Jeoh, T. Comparative technoeconomic process analysis of industrial-scale microencapsulation of bioactives in cross-linked alginate. *J. Food Eng.* **2020**, *266*, 109695. [CrossRef]
22. Li, D.; Wei, Z.; Xue, C. Alginate-based delivery systems for food bioactive ingredients: An overview of recent advances and future trends. *Compr. Rev. Food Sci. Food Saf.* **2021**, *20*, 5345–5369. [CrossRef]
23. Saji, S.; Hebden, A.; Goswami, P.; Du, C. A brief review on the development of alginate extraction process and its sustainability. *Sustainability* **2022**, *14*, 5181. [CrossRef]
24. Paglarini, C.D.S.; Vidal, V.A.S.; Martini, S.; Cunha, R.L.; Pollonio, M.A.R. Protein-based hydrogelled emulsions and their application as fat replacers in meat products: A review. *Crit. Rev. Food Sci. Nutr.* **2022**, *62*, 640–655. [CrossRef]
25. Robert, P.; Zamorano, M.; González, E.; Silva-Weiss, A.; Cofrades, S.; Giménez, B. Double emulsions with olive leaves extract as fat replacers in meat systems with high oxidative stability. *Food Res. Int.* **2019**, *120*, 904–912. [CrossRef] [PubMed]
26. Freire, M.; Cofrades, S.; Pérez-Jiménez, J.; Gómez-Estaca, J.; Jiménez-Colmenero, F.; Bou, R. Emulsion gels containing n-3 fatty acids and condensed tannins designed as functional fat replacers. *Food Res. Int.* **2018**, *113*, 465–473. [CrossRef]
27. Carpenter, J.; George, S.; Saharan, V.K. A comparative study of batch and recirculating flow ultrasonication system for preparation of multilayer olive oil in water emulsion stabilized with whey protein isolate and sodium alginate. *Chem. Eng. Process Process Intensif.* **2018**, *125*, 139–149. [CrossRef]
28. Sun-Waterhouse, D.; Zhou, J.; Miskelly, G.M.; Wibisono, R.; Wadhwa, S.S. Stability of encapsulated olive oil in the presence of caffeic acid. *Food Chem.* **2011**, *126*, 1049–1056. [CrossRef]
29. Atencio, S.; Maestro, A.; Santamaria, E.; Gutiérrez, J.M.; Gonzalez, C. Encapsulation of ginger oil in alginate-based shell materials. *Food Biosci.* **2020**, *37*, 100714. [CrossRef]
30. Mania, S.; Tylingo, R.; Michałowska, A. The drop-in-drop encapsulation in chitosan and sodium alginate as a method of prolonging the quality of linseed oil. *Polymers* **2018**, *10*, 1355. [CrossRef]
31. Martins, E.; Renard, D.; Davy, J.; Marquis, M.; Poncelet, D. Oil core microcapsules by inverse gelation technique. *J. Microencapsul.* **2015**, *32*, 86–95. [CrossRef]
32. Wang, M.; Doi, T.; McClements, D.J. Encapsulation and controlled release of hydrophobic flavors using biopolymer-based microgel delivery systems: Sustained release of garlic flavor during simulated cooking. *Food Res. Int.* **2019**, *119*, 6–14. [CrossRef]
33. Petzold, G.; Gianelli, M.P.; Bugueño, G.; Celan, R.; Pavez, C.; Orellana, P. Encapsulation of liquid smoke flavoring in ca-alginate and ca-alginate-chitosan beads. *J. Food Sci. Technol.* **2014**, *51*, 183–190. [CrossRef]
34. European Directorate for the Quality of Medicines & Healthcare of the Council of Europe. *European Pharmacopoeia*, 10th ed.; Council of Europe: Strasbourg, France, 2019; Volume I, ISBN 978-92-871-8912-7.
35. Mourtzinos, I.; Makris, D.P.; Yannakopoulou, K.; Kalogeropoulos, N.; Michali, I.; Karathanos, V.T. Thermal stability of anthocyanin extract of *Hibiscus sabdariffa* L. in the presence of β-cyclodextrin. *J. Agric. Food Chem.* **2008**, *56*, 10303–10310. [CrossRef]
36. Arnous, A.; Makris, D.P.; Kefalas, P. Correlation of pigment and flavanol content with antioxidant properties in selected aged regional wines from Greece. *J. Food Compos. Anal.* **2002**, *15*, 655–665. [CrossRef]
37. Flamminii, F.; Di Mattia, C.D.; Nardella, M.; Chiarini, M.; Valbonetti, L.; Neri, L.; Difonzo, G.; Pittia, P. Structuring alginate beads with different biopolymers for the development of functional ingredients loaded with olive leaves phenolic extract. *Food Hydrocoll.* **2020**, *108*, 105849. [CrossRef]
38. AOCS (American Oil Chemists' Society). AOCS Official method Cd 8-53. Peroxide Value (acetic acid-chloroform method). In *Official Methods and Recommended Practices of the American Oil Chemists' Society*, 5th ed.; Firestone, D., Ed.; AOCS: Champaign, IL, USA, 1998.
39. Crowe, T.D.; White, P.J. Adaptation of the AOCS official method for measuring hydroperoxides from small-scale oil samples. *J. Am. Oil Chem. Soc.* **2001**, *78*, 1267–1269. [CrossRef]
40. AOCS (American Oil Chemists' Society). AOCS Official method Cd 18-90. p-Anisidine value method Cd 18-90. In *Official Methods and Recommended Practices of the American Oil Chemists' Society*, 5th ed.; Firestone, D., Ed.; AOCS: Champaign, IL, USA, 1998.
41. O'Connor, C.J.; Lal, S.N.D.; Eyres, L. *Handbook of Australasian Edible Oils. Oils and Fats*, 1st ed.; Specialist Group of NZIC: Auckland, New Zealand, 2007; pp. 5–6.
42. Firestone, D. *Official Methods and Recommended Practices of the American Oil Chemists' Society*, 5th ed.; AOCS: Champaign, IL, USA, 1998.
43. Alvanou, M.V.; Kyriakoudi, A.; Makri, V.; Lattos, A.; Feidantsis, K.; Papadopoulos, D.K.; Georgoulis, I.; Apostolidis, A.P.; Michaelidis, B.; Mourtzinos, I.; et al. Effects of dietary substitution of fishmeal by black soldier fly (*Hermetia illucens*) meal on growth performance, whole-body chemical composition, and fatty acid profile of Pontastacus leptodactylus juveniles. *Front. Physiol.* **2023**, *14*, 1156394. [CrossRef]

44. Celano, R.; Piccinelli, A.L.; Pagano, I.; Roscigno, G.; Campone, L.; De Falco, E.; Russo, M.; Rastrelli, L. Oil distillation wastewaters from aromatic herbs as new natural source of antioxidant compounds. *Food Res. Int.* **2017**, *99*, 298–307. [CrossRef] [PubMed]
45. Alice, G.; Corina, B.; Lucia, P.; Sultana, N.; Bazdoaca, C.; Nicoleta, D. Polyphenol content dynamics in hydrodistillation water residues of lamiaceae species. *J. Essent. Oil-Bear. Plants* **2019**, *22*, 858–864. [CrossRef]
46. Hossain, M.B.; Barry-Ryan, C.; Martin-Diana, A.B.; Brunton, N.P. Optimisation of accelerated solvent extraction of antioxidant compounds from rosemary (*Rosmarinus officinalis* L.), marjoram (*Origanum majorana* L.) and oregano (*Origanum vulgare* L.) using response surface methodology. *Food Chem.* **2011**, *126*, 339–346. [CrossRef]
47. Nieto, G.; Huvaere, K.; Skibsted, L.H. Antioxidant activity of rosemary and thyme by-products and synergism with added antioxidant in a liposome system. *Eur. Food Res. Technol.* **2011**, *233*, 11–18. [CrossRef]
48. Peixoto, J.A.B.; Álvarez-Rivera, G.; Alves, R.C.; Costa, A.S.G.; Machado, S.; Cifuentes, A.; Ibáñez, E.; Oliveira, M.B.P.P. Comprehensive Phenolic and Free Amino Acid Analysis of Rosemary Infusions: Influence on the Antioxidant Potential. *Antioxidants* **2021**, *10*, 500. [CrossRef]
49. Napoli, E.; Siracusa, L.; Ruberto, G. New tricks for old guys: Recent developments in the chemistry, biochemistry, applications and exploitation of selected species from the Lamiaceae Family. *Chem. Biodivers.* **2020**, *17*, e1900677. [CrossRef] [PubMed]
50. Wollinger, A.; Perrin, É.; Chahboun, J.; Jeannot, V.; Touraud, D.; Kunz, W. Antioxidant activity of hydro distillation water residues from *Rosmarinus officinalis* L. leaves determined by DPPH assays. *CR Chim.* **2016**, *19*, 754–765. [CrossRef]
51. Zhang, Y.; Smuts, J.P.; Dodbiba, E.; Rangarajan, R.; Lang, J.C.; Armstrong, D.W. Degradation study of carnosic acid, carnosol, rosmarinic acid, and rosemary extract (*Rosmarinus officinalis* L.) assessed using HPLC. *J. Agric. Food Chem.* **2012**, *60*, 9305–9314. [CrossRef] [PubMed]
52. Wei, G.J.; Wang, M.; Rosen, R.T.; Ho, C.T. Identification of thermal decomposition products of carnosol, an antioxidant in rosemary and sage. *J. Food Lipds* **1999**, *6*, 173–179. [CrossRef]
53. Xie, J.; VanAlstyne, P.; Uhlir, A.; Yang, X. A review on rosemary as a natural antioxidation solution. *Eur. J. Lipid Sci. Technol.* **2017**, *119*, 1600439. [CrossRef]
54. Oreopoulou, A.; Papavassilopoulou, E.; Bardouki, H.; Vamvakias, M.; Bimpilas, A.; Oreopoulou, V. Antioxidant recovery from hydrodistillation residues of selected Lamiaceae species by alkaline extraction. *J. Appl. Res. Med. Aromat. Plants* **2018**, *8*, 83–89. [CrossRef]
55. Zhang, Y.; Yang, L.; Zu, Y.; Chen, X.; Wang, F.; Liu, F. Oxidative stability of sunflower oil supplemented with carnosic acid compared with synthetic antioxidants during accelerated storage. *Food Chem.* **2010**, *118*, 656–662. [CrossRef]
56. Medronho, B.; JMValente, A.; Costa, P.; Romano, A. Inclusion complexes of rosmarinic acid and cyclodextrins: Stoichiometry, association constants, and antioxidant potential. *Colloid Polym. Sci.* **2014**, *292*, 885–894. [CrossRef]
57. Korompokis, K.; Igoumenidis, P.E.; Mourtzinos, I.; Karathanos, V.T. Green extraction and simultaneous inclusion complex formation of *Sideritis scardica* polyphenols. *Int. Food Res. J.* **2017**, *24*, 1233–1238. Available online: https://food-science.agro.auth.gr/wp-content/uploads/sites/81/2018/11/2017-18-Green-extraction-and-simultaneous-inclusion-complex-formation-of-Sideritis-scardica-polyphenols.pdf (accessed on 19 August 2023).
58. Fateminasab, F.; Bordbar, A.K.; Asadi, B.; Shityakov, S.; Karizak, A.Z.; Mohammadpoor-Baltork, I.; Saboury, A.A. Modified β-cyclodextrins: Rosmarinic acid inclusion complexes as functional food ingredients show improved operations (solubility, stability and antioxidant activity). *Food Hydrocoll.* **2022**, *131*, 107731. [CrossRef]
59. Roman, O.; Heyd, B.; Broyart, B.; Castillo, R.; Maillard, M.N. Oxidative reactivity of unsaturated fatty acids from sunflower, high oleic sunflower and rapeseed oils subjected to heat treatment, under controlled conditions. *LWT-Food Sci. Technol.* **2013**, *52*, 49–59. [CrossRef]
60. Irwin, J.W.; Hedges, N. Measuring lipid oxidation. In *Understanding and Measuring the Shelf-Life of Food*; Woodhead Publishing: Sawston, UK, 2004; pp. 289–316. [CrossRef]
61. Kowalski, R.; Kowalska, G.; Pankiewicz, U.; Mazurek, A.; Włodarczyk-Stasiak, M.; Sujka, M.; Wyrostek, J. The effect of an addition of marjoram oil on stabilization fatty acids profile of rapeseed oil. *LWT-Food Sci. Technol.* **2019**, *109*, 225–232. [CrossRef]
62. Choe, E. Effects and mechanism of minor compounds in oil on lipid oxidation. In *Food lipids: Chemistry, Nutrition and Biotechnology*, 3rd ed.; Akoh, C.C., min, D.B., Eds.; CRC Press: Boca Raton, FL, USA, 2008; pp. 449–467.
63. Loussouarn, M.; Krieger-Liszkay, A.; Svilar, L.; Bily, A.; Birtić, S.; Havaux, M. Carnosic acid and carnosol, two major antioxidants of rosemary, act through different mechanisms. *Plant Physiol.* **2017**, *175*, 1381–1394. [CrossRef] [PubMed]
64. Leong, M.H.; Tan, C.P.; Nyam, K.L. Effects of accelerated storage on the quality of kenaf seed oil in chitosan-coated high methoxyl pectin-alginate microcapsules. *J. Food Sci.* **2016**, *81*, C2367–C2372. [CrossRef]
65. Sun-Waterhouse, D.; Penin-Peyta, L.; Wadhwa, S.S.; Waterhouse, G.I. Storage stability of phenolic-fortified avocado oil encapsulated using different polymer formulations and co-extrusion technology. *Food Bioprocess Technol.* **2012**, *5*, 3090–3102. [CrossRef]
66. Waraho, T.; McClements, D.J.; Decker, E.A. Impact of free fatty acid concentration and structure on lipid oxidation in oil-in-water emulsions. *Food Chem.* **2011**, *129*, 854–859. [CrossRef]
67. *CXS 210-1999*; Standard for Named Vegetable Oils. Codex Alimentarius Commission: Rome, Italy, 2019.
68. Wang, W.; Waterhouse, G.I.; Sun-Waterhouse, D. Co-extrusion encapsulation of canola oil with alginate: Effect of quercetin addition to oil core and pectin addition to alginate shell on oil stability. *Food Res. Int.* **2013**, *54*, 837–851. [CrossRef]
69. Naczk, M.; Shahidi, F. Extraction and analysis of phenolics in food. *J. Chromatogr. A* **2004**, *1054*, 95–111. [CrossRef]

70. Pérez-Jiménez, J.; Saura-Calixto, F. Effect of solvent and certain food constituents on different antioxidant capacity assays. *Food Res. Int.* **2006**, *39*, 791–800. [CrossRef]
71. Richards, M.P.; Chaiyasit, W.; McClements, D.J.; Decker, E.A. Ability of surfactant micelles to alter the partitioning of phenolic antioxidants in oil-in-water emulsions. *J. Agric. Food Chem.* **2002**, *50*, 1254–1259. [CrossRef] [PubMed]
72. Sasaki, K.; Alamed, J.; Weiss, J.; Villeneuve, P.; Giraldo, L.J.L.; Lecomte, J.; Figueroa-Espinoza, M.-C.; Decker, E.A. Relationship between the physical properties of chlorogenic acid esters and their ability to inhibit lipid oxidation in oil-in-water emulsions. *Food Chem.* **2010**, *118*, 830–835. [CrossRef]
73. Decker, E.A.; McClements, D.J.; Bourlieu-Lacanal, C.; Durand, E.; Figueroa-Espinoza, M.C.; Lecomte, J.; Villeneuve, P. Hurdles in predicting antioxidant efficacy in oil-in-water emulsions. *Trends Food Sci. Technol.* **2017**, *67*, 183–194. [CrossRef]
74. Laguerre, M.; Bayrasy, C.; Panya, A.; Weiss, J.; McClements, D.J.; Lecomte, J.; Decker, E.A.; Villeneuve, P. What makes good antioxidants in lipid-based systems? The next theories beyond the polar paradox. *Crit. Rev. Food Sci. Nutr.* **2015**, *55*, 183–201. [CrossRef] [PubMed]
75. Martínez-Aranda, N.; Losada-Barreiro, S.; Bravo-Díaz, C.; Romsted, L.S. Influence of Temperature on the Distribution of Catechin in Corn Oil-in-Water Emulsions and Some Relevant Thermodynamic Parameters. *Food Biophys.* **2014**, *9*, 380–388. [CrossRef]
76. Biliaderis, C.G.; Eskin, N.A.M. Canola. In *Encyclopedia of Food Science and Technology*; Hui, Y.H., Ed.; John Wiley & Sons, Inc.: Hoboken, NJ, USA, 1991; Volume 1, pp. 264–277.
77. Fhaner, M.; Hwang, H.S.; Winkler-Moser, J.K.; Bakota, E.L.; Liu, S.X. Protection of fish oil from oxidation with sesamol. *Eur. J. Lipid Sci. Technol.* **2016**, *118*, 885–897. [CrossRef]
78. Przybylski, R.; Malcolmson, L.J.; Eskin, N.A.M.; Durance-Tod, S.; Mickle, J.; Carr, R. Stability of low linolenic acid canola oil to accelerated storage at 60 C. *LWT Food Sci. Technol.* **1993**, *26*, 205–209. [CrossRef]
79. Keene, K.A.; Ruddy, R.M.; Fhaner, M.J. Investigating the relationship between antioxidants and fatty acid degradation using a combination approach of GC-FID and square-wave voltammetry. *ACS Omega* **2019**, *4*, 983–991. [CrossRef] [PubMed]
80. Gramza-Michalowska, A.N.N.A.; Sidor, A.; Hes, M. Herb extract influence on the oxidative stability of selected lipids. *J. Food Biochem.* **2011**, *35*, 1723–1736. [CrossRef]
81. Tohma, S.; Turan, S. Rosemary plant (*Rosmarinus officinalis* L.), solvent extract and essential oil can be used to extend the usage life of hazelnut oil during deep frying. *Eur. J. Lipid Sci. Technol.* **2015**, *117*, 1978–1990. [CrossRef]
82. Yang, Y.; Song, X.; Sui, X.; Qi, B.; Wang, Z.; Li, Y.; Jiang, L. Rosemary extract can be used as a synthetic antioxidant to improve vegetable oil oxidative stability. *Ind. Crops Prod* **2016**, *80*, 141–147. [CrossRef]

Disclaimer/Publisher's Note: The statements, opinions and data contained in all publications are solely those of the individual author(s) and contributor(s) and not of MDPI and/or the editor(s). MDPI and/or the editor(s) disclaim responsibility for any injury to people or property resulting from any ideas, methods, instructions or products referred to in the content.

foods

MDPI

Article

The Characteristic Aroma Compounds of GABA Sun-Dried Green Tea and Raw Pu-Erh Tea Determined by Headspace Solid-Phase Microextraction Gas Chromatography–Mass Spectrometry and Relative Odor Activity Value

Chenyang Ma [1], Chang Gao [1], Yuanda Li [2], Xiaohui Zhou [1], Guofu Fan [1], Di Tian [3], Yuan Huang [4], Yali Li [1,*] and Hongjie Zhou [3,*]

1 College of Tea Science, Yunnan Agricultural University, Kunming 650500, China; machenyang0305@icloud.com (C.M.); 15825226653@163.com (C.G.); 18887449386@163.com (X.Z.); f1179643966@163.com (G.F.)
2 Laboratory of Tea Science of Ministry of Education, Hunan Agricultural University, Changsha 410128, China; 13398802207@163.com
3 College of Food Science and Technology, Yunnan Agricultural University, Kunming 650500, China; m17853435722@163.com
4 College of Pu-Erh Tea, West Yunnan University of Applied Sciences, Puer 671000, China; 17787107713@139.com
* Correspondence: 2008019@ynau.edu.cn (Y.L.); 1984020@ynau.edu.cn (H.Z.); Tel.: +86-13759169886 (Y.L.); +86-13708478322 (H.Z.)

Abstract: We aim to improve the product quality of GABA raw Pu-erh tea during development and processing. In this study, headspace solid-phase microextraction gas chromatography–mass spectrometry technology combined with relative odor activity evaluations was used to compare the volatile compounds of GABA sun-dried green tea and GABA raw Pu-erh tea. Sensory evaluation showed a higher aroma score of GABA raw Pu-erh tea than that of GABA sun-dried green tea, with significant differences in aroma type and purity. A total of 147 volatile compounds of 13 categories were detected, which differed in composition and quantity between the two teas. 2-Buten-1-one,1-(2,6,6-trimethyl-1,3-cyclohexadien-1-yl)-,(E)- and beta.-myrcene largely contributed to the aroma formation of both teas. Five volatile compounds were screened as potential markers for tea aroma. Metabolic pathway analysis showed that monoterpenoid biosynthesis may be beneficial to the formation of flowery and fruity aromas in the teas. We suggest that the findings of this study may provide important guidance for the processing and optimization of GABA tea.

Keywords: GABA sun-dried green tea; GABA raw Pu-erh tea; characteristic aroma compounds; metabolic pathways; relative odor activity

1. Introduction

Yunnan, China, is the origin center of the tea tree, contributing to a culture characterized by tea drinking for more than 1000 years [1]. Pu-erh tea is a traditional and popular type of tea with a long history in Yunnan. According to the processing type and quality characteristics, Pu-erh tea can be divided into raw and ripe forms [2]. In recent years, Pu-erh tea has become increasingly popular with consumers because of its unique health benefits and quality characteristics [3]. Raw Pu-erh tea is a tightly pressed tea that is prepared by steaming and drying Yunnan big-leaf sun-dried green tea. Raw Pu-erh tea has the quality characteristics of a pure and lasting aroma, green and yellow clear soup color, and strong and sweet taste [4]. The aroma type of raw Pu-erh tea is mainly clean and refreshing, accompanied by nutty, grassy, and fatty aromas. The composition of volatile substances is similar to that of sun-dried green tea, albeit with some distinct differences [5,6]. After dry storage, the aroma of raw Pu-erh tea gradually changes to a woody, sweet, and unique aging aroma [7]. With the goal of developing raw Pu-erh tea products with unique flavor, many recent studies have explored the influence of traditional processing technology on the flavor of the tea. Fan et al. collected samples at different processing stages for determination of volatile compounds (VCs), showing that deactivation and autoclaving were the key processing steps to form the flavor substances of raw Pu-erh tea. Deactivation significantly increased the relative content of aldehydes and significantly reduced the relative content of olefins. Autoclaving further reduced the contents of aldehydes and esters and significantly increased the contents of nitrogen-containing compounds and ketones. Therefore, exploring appropriate processing parameters may offer new methods to regulate the flavor of raw Pu-erh tea [8]. Feng et al. found that the VCs in raw Pu-erh tea were mainly alcohols, terpenes, aldehydes, and ketones. Through aroma recombination experiments, it was determined that light, fat, flower, and fruit aromas were the characteristic aroma attributes of raw Pu-erh tea, providing a theoretical basis for processing different flavors of raw Pu-erh tea [9].

In 1987, Tsushida and colleagues discovered the method of gamma-aminobutyric acid (GABA) enrichment in tea, leading to the successful development of GABA teas (with a GABA content above 1.5 mg/g tea) [10,11]. Since the content of GABA in ordinary tea is generally low, nitrogen-filled anaerobic methods are usually used to enrich GABA in tea. Under anaerobic conditions, GABA in tea is mainly enriched by GABA shunt and polyamine degradation [12]. GABA tea has also been favored by the public because of its unique flavor and pharmacological effects. Zhen found that the flavor of GABA green tea processed by an anaerobic method changed significantly compared with that of ordinary green tea, resulting in an obvious sweet and sour fruit aroma, accompanied by increased contents of methylphenol, methyl myristate, methyl laurate, and methyl palmitate [13]. Li et al. found that GABA white tea processed by an anaerobic method exhibited a strong floral and fruity aroma, which was mainly due to 2-heptanol [14]. Our previous study demonstrated that the GABA content in the fresh leaves of Yunnan big-leaf tea trees can be effectively accumulated after 6 h of anaerobic treatment. The processed GABA sun-dried green tea and GABA raw Pu-erh tea products exhibited unique flavor qualities with certain promotion value [15].

To date, more than 700 VCs have been identified in tea, but only a small number of VCs have been confirmed to contribute to the formation of the characteristic tea flavor [16]. The formation of the characteristic flavor of tea is not only affected by the concentration and taste but is also the result of the interaction between the constituent compounds and is closely related to the olfactory threshold (OT) [17]. Therefore, determination of the odor activity value (OAV) and relative odor activity value (ROAV) is considered to be an important method for screening and forming characteristic aroma compounds [18,19]. In addition, headspace solid-phase microextraction (HS-SPME) technology, gas chromatography–mass spectrometry (GC-MS) technology, gas chromatography–olfactometry technology, electronic nose technology, sensory evaluation, Kyoto Encyclopedia of Genes and Genomes (KEGG) pathway analysis, aroma recombination and omission experiments, and other methods [20,21] are commonly used for the screening, identification, and testing of characteristic aroma compounds in food research. Some of these techniques have also been applied to the analysis of tea. For example, Hong et al. used HS-SPME-GC-MS technology combined with sensory evaluation results and then applied OAV and multivariate statistical analysis to explore the characteristic aroma compounds of Chinese yellow tea with different aroma types. They identified 8, 14, 7, and 18 VCs in yellow tea with flowery, high-fired, fresh, and corny aromas, respectively, which could be used as the characteristic aroma compounds of yellow tea with different aroma types [22].

In this study, sensory evaluation, HS-SPME-GC-MS combined with ROAV, and multivariate analysis were used to compare and analyze the VCs of GABA sun-dried green tea and GABA raw Pu-erh tea. The key VCs of aroma formation and the main metabolic pathways of biosynthesis in GABA sun-dried green tea and GABA raw Pu-erh tea were clarified. These results are expected to provide a theoretical basis for optimizing the processing technology of GABA sun-dried green tea and GABA raw Pu-erh tea.

2. Materials and Methods

2.1. Tea Samples

Four groups of samples, including fresh tea leaves as a control (CK), anaerobic-processed tea leaves (CA), GABA sun-dried green tea leaves (SGT), and GABA raw Pu-erh tea leaves (PRT), were prepared and compared. Three replicates were established for each group. In July 2021, all samples (one bud, two or three leaves of the same tea variety *Camellia assamica* vs. Yun kang 10) were picked and processed in the tea garden of Dalishu Tea Factory in Yunlong County, Dali, China.

In the CK group, the fresh leaves of tea plants were deactivated by steam at 100 °C for 5 min and then dried at 45 °C for 2 h. In the CA group, the fresh leaves of tea plants were placed in a special machine filled with N_2 for anaerobic treatment for 6 h, vaporized at 100 °C for 5 min, and then dried at 45 °C for 2 h. In the SGT group, the fresh leaves of the tea plant were air-dried for 2 h and then placed in a special machine filled with N_2 for anaerobic treatment for 6 h, followed by steam deactivation at 100 °C for 3 min, rolling for 45 min, and finally dried by sunshine for 8 h. In the PRT group, the fresh leaves of the tea plant were aired for 2 h, placed in a special machine filled with N_2 for anaerobic treatment for 6 h, and then vaporized at 100 °C for 3 min. After rolling for 45 min, the tea leaves were dried by sunshine for 8 h, autoclaved into a cake type, cooled to room temperature, dried for 3 h, and finally dried at 60 °C for 1 h. The manufacturing procedures are specified in the Chinese national standard procedure and are shown in Figure 1.

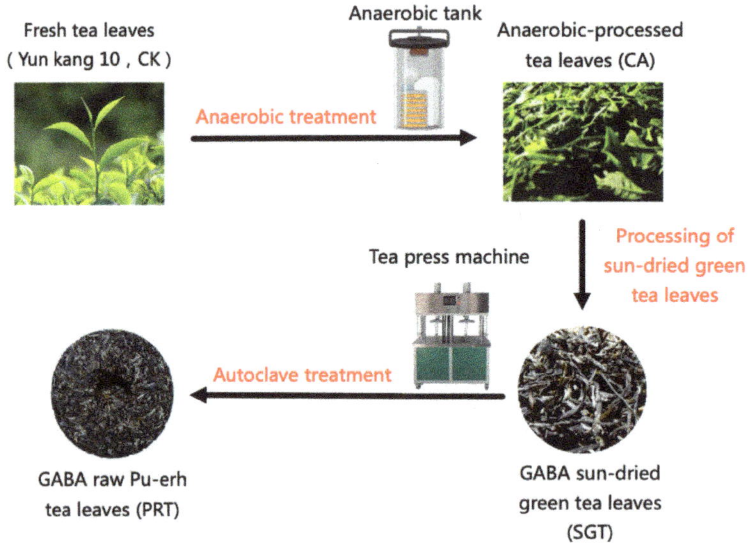

Figure 1. The process of GABA sun-dried green tea and GABA raw Pu-erh tea manufacturing.

2.2. Chemicals

N-hexane (chromatographical purity) was purchased from Merk (Darmstadt, Germany). NaCl (analytical purity) was obtained from China National Pharmaceutical Group Co., Ltd. (Kunming, China). C7-C40 (chromatographical purity) saturated alkanes and standards were purchased from Sigma-Aldrich (Darmstadt, Germany), which were used to determine linear retention indices.

2.3. Sensory Panel Evaluation of Teas

Sensory evaluation of tea samples was carried out according to Chinese national standard procedure [23]. The members of the tea review team were all from the Tea College of Yunnan Agricultural University. The team included 10 trained reviewers (five females and five males with an average age of 30 years and professional experience of 5–35 years) who evaluated the appearance and quality of tea samples according to the Chinese national standard procedure. The score was calculated by the method of percentage weighting. First, the tea evaluation terms were discussed, and then the sensory evaluation was carried out in the special tea review room at room temperature (25 °C). The tea samples were delivered in random order to ensure that the team members did not know the sample number [24]. Then, 100 g samples were randomly selected to be evaluated and have their appearance scored (A). Next, 3 g of each tea sample was taken and brewed with 150 mL of boiling water. After 5 min, the beverage was used to evaluate and score the internal quality, including aroma (B), beverage color (C), taste (D), and leaf base (E). Finally, the opinions of the group members were summarized, and the quality of tea samples was quantified according to the following formulas:

$$\text{SGT review score} = A \times 25\% + B \times 25\% + C \times 10\% + D \times 30\% + E \times 10\%$$

$$\text{PRT review score} = A \times 20\% + B \times 30\% + C \times 10\% + D \times 35\% + E \times 5\%$$

2.4. Sample Preparation

All sample preparation and detection methods were improved by referring to the methods in the existing papers [25]. Materials were harvested, weighed, immediately frozen in liquid nitrogen, and stored at −80 °C until needed. Samples were ground to a powder in liquid nitrogen. One gram of the powder was immediately transferred to a 20 mL headspace vial (Agilent, Palo Alto, CA, USA) containing NaCl saturated solution to inhibit any enzyme reaction. The vials were sealed using crimp-top caps with TFE-silicone headspace septa (Agilent). At the time of SPME analysis, each vial was placed in a 100 °C water bath for 5 min, and then a 120 µM divinylbenzene/carboxen/polydimethylsilioxan fiber (Agilent) was exposed to the headspace of the sample for 15 min at 100 °C.

2.5. GC-MS Conditions

After sampling, desorption of the VCs from the fiber coating was carried out in the injection port of the GC apparatus (Model 8890; Agilent) at 250 °C for 5 min in splitless mode. The identification and quantification of VCs were carried out using an Agilent Model 8890 GC and a 5977B mass spectrometer (Agilent), equipped with a 30 m × 0.25 mm × 0.25 µM DB-5MS (5% phenyl-polymethylsiloxane) capillary column. Helium was used as the carrier gas at a linear velocity of 1.2 mL/min. The injector temperature was kept at 250 °C, and the detector temperature was maintained at 280 °C. The oven temperature was programmed from 40 °C (3.5 min), increasing at 10 °C/min to 100 °C, at 7 °C/min to 180 °C, at 25 °C/min to 280 °C, and held for 5 min. Mass spectra were recorded in electron impact ionization mode at 70 eV. The quadrupole mass detector, ion source, and transfer line temperatures were, respectively, set at 150 °C, 230 °C, and 280 °C. Mass spectra were scanned in the m/z range of 50–500 amu at 1 s intervals. Identification of VCs was achieved by comparing the mass spectra with the data system library (MWGC or NIST) and linear retention index.

2.6. ROAV Calculation

Based on the relative content of aroma compounds and the threshold value of each aroma component in water, the ROAV calculation formula is determined as follows:

$$\text{ROAV} = (C_n/C_{max}) \times (T_{max}/T_n) \times 100$$

where C_n represents the relative content of any VC (%), C_{max} represents the relative content of the largest VC (%), T_n represents the OT of any VC, and T_{max} represents the maximum OT of VCs; the greater the ROAV value, the greater the contribution value of the compounds to the aroma. VCs with ROAV > 1 can be considered to have an important contribution to the composition of the aroma, whereas VCs with 0.1 < ROAV < 1 can be considered to have a modification effect on the composition of the aroma [26].

2.7. KEGG Pathway Analysis

To explore the biosynthesis pathways of aroma compounds during the processing of tea samples in the SGT and PRT groups, the detected VCs were compared with the known compounds in the KEGG database (https://www.kegg.jp/kegg/pathway.html (accessed on 15 August 2023)), and metabolic pathway analysis was performed using MBRole 2.0 (http://csbg.cnb.csic.es/mbrole2/ (accessed on 15 August 2023)) [27].

2.8. Statistical Analysis

All data were measured three times in parallel, and the results are expressed as mean ± standard deviation. SPSS v.26 was used for analysis of variance, and Tukey's post-hoc test. $p < 0.05$ was used as the standard for judging a significant difference between groups. Metware Cloud was used for principal component analysis (PCA) and orthogonal partial least-squares discriminant analysis (OPLS-DA). Draw Venn and Origin 2022 were used for drawing the clustering heat map and other data analysis and mapping.

3. Results and Discussion

3.1. Sensory Evaluation of Aroma Characteristics of GABA Sun-Dried Green Tea and GABA Raw Pu-Erh Tea

The sensory evaluation method, as the most basic tea quality evaluation method, is based on people's intuitive feelings. Nevertheless, this method remains irreplaceable and is widely used for the evaluation of all tea types [28,29]. The evaluation of aroma factors in tea sensory evaluation usually includes four aspects: aroma type, concentration, purity, and persistence [30].

As shown in Table 1, the aroma characteristics of the SGT and PRT were significantly different. The aroma of SGT was fragrant, high, pure, and lasting, and the flowery aroma was rich, while the fruity aroma was weak. The aroma of the PRT was purer and longer-lasting and had flowery, fruity, and sweet characteristics. The aroma factor score and total scores of PRT were higher than those of SGT. Since a previous study showed that the processing type (autoclaved) can enhance the flavor of raw Pu-erh tea [31], and the autoclaved process is a unique part of the PRT, the autoclaving process may be the cause of the difference in aroma characteristics between the two tea samples.

Table 1. Sensory evaluation results of GABA sun-dried green tea and GABA raw Pu-erh tea.

Sample	Appearance	s	Aroma	s	Beverage Color	s	Taste	s	Infused Leaves	s	Total
SGT	Approach tight, yellowish green	86	Flowery aroma, fruity aroma (inferior), fragrant, high, pure, lasting	89	Lightly apricot, bright	90	Mellow, thick	86	Soft, bright	91	87.65
PRT	Normal cake, yellowish green	88	Flowery aroma, fruity aroma and sweet aroma, fragrant, high, more pure and lasting	90	Greenish yellow, bright	91	Sweet, fresh, thick	90	More soft and bright	90	89.7

3.2. Quantitative and Qualitative Analysis of VCs in GABA Sun-Dried Green Tea and GABA Raw Pu-Erh Tea

Using HS-SPME-GC-MS technology to detect GABA sun-dried green tea and GABA raw Pu-erh tea, a total of 148 VCs were identified (Table 2). Among them, 145, 145, and 143 VCs were detected in SGT, PRT, and CK samples, respectively, and all 148 VCs could be detected in CA samples. After comparing the types of VCs between tea samples, it was found that after anaerobic treatment of CK, a total of five VCs were newly generated, including 9,12-octadecadienoic acid, (Z,Z)-, methyl ester; 9,12,15-octadecatrienoic acid, (Z,Z,Z)-; (Z)-9,17-octadecadienal; linoleic acid ethyl ester; and (E)-2-hexenoic acid, butyl ester. The first four VCs were also detected in both SGT and PRT, and their contents increased. Therefore, further study is needed to determine whether these four compounds affect the formation of SGT and PRT aroma characteristics. However, the last VC was only detected in CA, indicating that anaerobic treatment can increase the type of VCs formed. After SGT was autoclaved into PRT, two new VCs were generated, 2-(1,1-dimethylethyl)-6-(1-methylethyl) phenol and 4H-pyran-4-one,2,3-dihydro-3,5-dihydroxy-6-methyl-. Further analysis showed that tea samples with different treatments could be clustered, indicating that the data were representative [32], and there were significant differences in the contents of VCs among different tea samples (Figure S1). The results of PCA showed that the four groups of tea samples could be clustered separately with obvious separation between the groups (Figure 2a). In addition, there were also some differences in the types of VCs among different tea samples (Figure 2b).

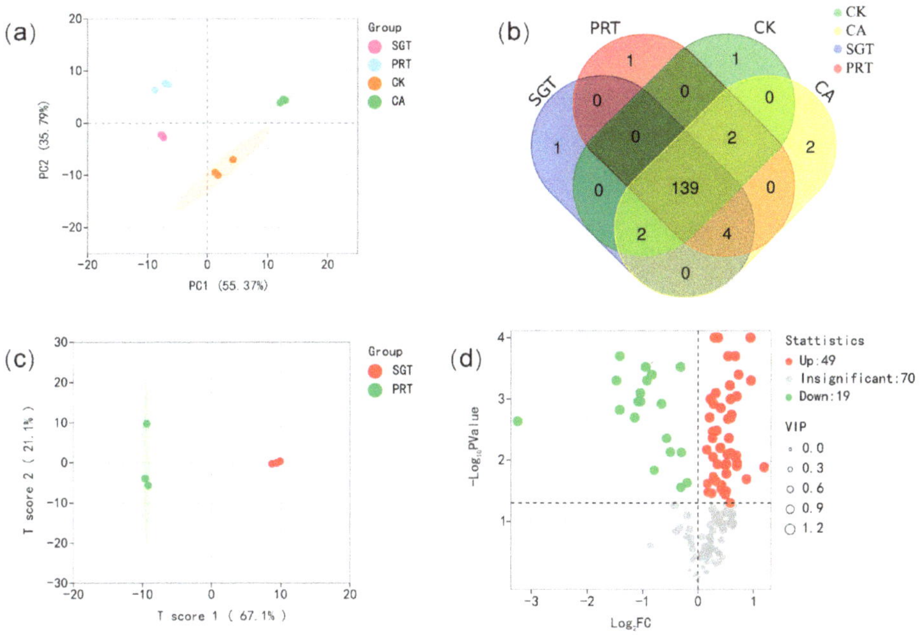

Figure 2. Multivariate statistical analysis of volatile organic compounds in the processing of GABA sun-dried green tea and GABA raw Pu-erh tea. (**a**) PCA model score scatter plot (total). (**b**) Venn diagram of volatile organic compounds. (**c**) Scatter plot of scores of OPLS-DA model (SGT vs. PRT). (**d**) Volcano plot of differential compounds (SGT vs. PRT). CK refers to fresh tea leaves, CA refers to anaerobic tea leaves, SGT refers to GABA sun-dried green tea, and PRT refers to GABA raw Pu-erh tea.

Table 2. Volatile compound composition and absolute contents.

No.	Aroma Compounds	Retention Time	Retention Index	Identification Methods	Absolute Content (μg/g)			
					SGT	PRT	CK	CA
1	[2H3]-beta.-Ionone	16.66	1477.90	MS, RI	7.41 ± 0.18 [b]	8.27 ± 0.14 [a]	7.96 ± 0.48 [a]	6.79 ± 0.15 [c]
2	Phytol, acetate	22.32	1837.31	MS, RI	0.16 ± 0.00 [b]	0.20 ± 0.01 [a]	0.06 ± 0.01 [c]	0.16 ± 0.00 [c]
3	Phthalic acid, butyl hept-4-yl ester	23.09	1958.70	MS, RI	0.30 ± 0.00 [c]	0.24 ± 0.02 [c]	0.39 ± 0.09 [b]	1.05 ± 0.01 [a]
4	Linoleic acid ethyl ester	24.50	2160.32	MS, RI	0.12 ± 0.01 [b]	0.18 ± 0.03 [a]	—	0.03 ± 0.00 [c]
5	Isophytol	23.08	1946.90	MS, RI	0.07 ± 0.00 [b]	0.08 ± 0.01 [a]	0.04 ± 0.01 [c]	0.07 ± 0.00 [ab]
6	Phytol	22.80	2113.79	MS, RI	0.12 ± 0.01 [a]	0.15 ± 0.05 [a]	0.04 ± 0.01 [b]	0.11 ± 0.01 [a]
7	3,7,11,15-Tetramethyl-2-hexadecen-1-ol	23.45	1879.20	MS, RI	0.09 ± 0.00 [b]	0.11 ± 0.01 [a]	0.03 ± 0.01 [c]	0.08 ± 0.00 [b]
8	9,12-Octadecadienoic acid (Z,Z)-, methyl ester	23.67	2092.49	MS, RI	0.15 ± 0.01 [b]	0.20 ± 0.03 [a]	—	0.01 ± 0.00 [c]
9	Hexadecanoic acid, ethyl ester	24.40	1990.70	MS, RI	0.73 ± 0.02 [b]	1.05 ± 0.08 [a]	0.01 ± 0.00 [d]	0.14 ± 0.00 [c]
10	Hexadecane, 2,6,10,14-tetramethyl-	22.09	1805.46	MS, RI	0.15 ± 0.00 [c]	0.02 ± 0.00 [d]	0.16 ± 0.01 [b]	0.20 ± 0.00 [a]
11	Hexadecane, 2,6,11,15-tetramethyl-	15.29	1745.74	MS, RI	0.16 ± 0.00 [c]	0.06 ± 0.00 [d]	0.24 ± 0.02 [b]	0.26 ± 0.01 [a]
12	9,12,15-Octadecatrienoic Acid, (Z,Z,Z)-	24.56	2167.24	MS, RI	0.06 ± 0.01 [b]	0.09 ± 0.02 [a]	—	0.01 ± 0.00 [c]
13	Sulfurous acid, 2-ethylhexyl hexyl ester	24.45	1998.17	MS, RI	0.01 ± 0.00 [b]	0.02 ± 0.00 [a]	0.02 ± 0.01 [a]	0.02 ± 0.00 [ab]
14	7,9-Di-tert-butyl-1-oxaspiro(4,5)deca-6,9-diene-2,8-dione	23.59	1907.29	MS, RI	0.05 ± 0.00 [b]	0.05 ± 0.00 [b]	0.07 ± 0.01 [a]	0.05 ± 0.01 [b]
15	Hexadecanoic acid, methyl ester	23.67	1923.63	MS, RI	1.23 ± 0.03 [b]	1.62 ± 0.08 [a]	0.12 ± 0.02 [d]	0.30 ± 0.01 [c]
16	Pentadecane, 2,6,10,14-tetramethyl-	22.33	1701.98	MS, RI	0.29 ± 0.02 [c]	0.03 ± 0.00 [d]	0.38 ± 0.04 [b]	0.50 ± 0.05 [a]
17	2-Pentadecanone, 6,10,14-trimethyl-	21.33	1842.46	MS, RI	0.38 ± 0.00 [b]	0.56 ± 0.03 [a]	0.17 ± 0.04 [c]	0.11 ± 0.00 [d]
18	(Z)-9,17-Octadecadienal	24.21	2098.35	MS, RI	0.07 ± 0.00 [b]	0.09 ± 0.01 [a]	—	0.01 ± 0.00 [c]
19	Farnesol, acetate	18.28	1560.48	MS, RI	0.16 ± 0.00 [b]	0.19 ± 0.00 [a]	0.17 ± 0.02 [ab]	0.17 ± 0.00 [ab]
20	Heptadecane, 2-methyl-	20.42	1674.10	MS, RI	0.13 ± 0.01 [b]	0.07 ± 0.01 [c]	0.18 ± 0.05 [a]	0.15 ± 0.01 [ab]
21	Heptadecane, 3-methyl-	21.11	1772.65	MS, RI	0.15 ± 0.00 [b]	0.04 ± 0.00 [c]	0.27 ± 0.04 [a]	0.26 ± 0.00 [a]
22	Pentadecane, 2,6,10-trimethyl-	21.17	1798.71	MS, RI	0.13 ± 0.00 [b]	0.15 ± 0.02 [a]	0.18 ± 0.03 [c]	0.17 ± 0.01 [c]
23	2-Oxobicyclo(3.2.2)nona-3,6-dien-1-yl benzoate	20.48	1588.30	MS, RI	0.09 ± 0.00 [b]	0.09 ± 0.01 [b]	0.10 ± 0.01 [b]	0.21 ± 0.01 [a]
24	Hexadecane, 4-methyl-	19.85	1660.99	MS, RI	0.08 ± 0.01 [c]	0.04 ± 0.00 [d]	0.12 ± 0.01 [b]	0.14 ± 0.00 [a]
25	Hexadecane, 2-methyl-	19.89	1666.84	MS, RI	0.10 ± 0.00 [c]	0.05 ± 0.01 [d]	0.14 ± 0.01 [b]	0.16 ± 0.01 [a]
26	Heptadecane	20.65	1698.50	MS, RI	0.40 ± 0.02 [b]	0.36 ± 0.03 [b]	0.36 ± 0.05 [b]	0.47 ± 0.02 [a]
27	4-ethyl-Tetradecane	18.58	1542.70	MS, RI	0.05 ± 0.00 [c]	0.05 ± 0.01 [c]	0.10 ± 0.01 [b]	0.13 ± 0.00 [a]
28	Hexadecane	20.04	1598.46	MS, RI	0.83 ± 0.01 [c]	0.53 ± 0.03 [d]	0.98 ± 0.06 [b]	1.17 ± 0.02 [a]
29	Pentadecane, 4-methyl-	20.26	1556.14	MS, RI	0.02 ± 0.00 [c]	0.02 ± 0.00 [d]	0.03 ± 0.00 [b]	0.04 ± 0.00 [a]
30	Nonane, 2,2,4,4,6,8,8-heptamethyl-	17.33	1409.57	MS, RI	0.07 ± 0.00 [c]	0.09 ± 0.01 [c]	0.11 ± 0.02 [b]	0.16 ± 0.01 [a]
31	3-methyl-Pentadecane	18.99	1569.34	MS, RI	0.34 ± 0.01 [c]	0.27 ± 0.01 [d]	0.57 ± 0.04 [b]	0.72 ± 0.01 [a]

Table 2. *Cont.*

No.	Aroma Compounds	Retention Time	Retention Index	Identification Methods	Absolute Content (μg/g)			
					SGT	PRT	CK	CA
32	2,6,10-Trimethyltridecane	17.96	1459.04	MS, RI	0.85 ± 0.01 [b]	1.56 ± 0.18 [a]	0.57 ± 0.04 [c]	0.37 ± 0.00 [c]
33	diethyl(decyloxy)-Borane	9.55	999.10	MS, RI	0.01 ± 0.00 [b]	0.01 ± 0.01 [b]	0.01 ± 0.00 [b]	0.03 ± 0.00 [a]
34	Pentadecanal	12.93	1716.20	MS, RI	0.01 ± 0.00 [a]	0.01 ± 0.00 [a]	0.01 ± 0.00 [a]	0.01 ± 0.00 [a]
35	Nerolidol 1	13.43	1911.51	MS, RI	0.06 ± 0.00 [b]	0.07 ± 0.01 [a]	0.05 ± 0.01 [b]	0.03 ± 0.00 [c]
36	(3R,3aS,6S,7R)-3,6,8,8-Tetramethyloctahydro-1H-3a,7-methanoazulen-6-ol	19.10	1618.06	MS, RI	0.03 ± 0.00 [b]	0.01 ± 0.00 [c]	0.04 ± 0.00 [a]	0.04 ± 0.00 [a]
37	alpha.-Cadinol	19.88	1664.61	MS, RI	0.05 ± 0.00 [b]	0.06 ± 0.00 [a]	0.03 ± 0.00 [c]	0.06 ± 0.00 [a]
38	Diethyl Phthalate	12.87	1586.64	MS, RI	0.17 ± 0.00 [c]	0.02 ± 0.00 [d]	1.09 ± 0.15 [b]	1.59 ± 0.02 [a]
39	(3E,7E)-4,8,12-Trimethyltrideca-1,3,7,11-tetraene	19.58	1572.66	MS, RI	0.09 ± 0.00 [c]	0.15 ± 0.01 [a]	0.06 ± 0.01 [d]	0.11 ± 0.00 [b]
40	(1R,4S,9aS)-1-Methyl-4-((Z)-pent-2-en-4-yn-1-yl)octahydro-1H-quinolizine	19.75	1565.64	MS, RI	0.29 ± 0.01 [c]	0.44 ± 0.01 [a]	0.31 ± 0.03 [c]	0.36 ± 0.01 [b]
41	3-methyl-Tetradecane	19.71	1654.55	MS, RI	0.05 ± 0.00 [c]	0.03 ± 0.00 [d]	0.08 ± 0.00 [b]	0.10 ± 0.00 [a]
42	Tetradecane, 4-methyl-	17.56	1419.36	MS, RI	0.03 ± 0.00 [b]	0.03 ± 0.00 [a]	0.05 ± 0.01 [c]	0.07 ± 0.00 [c]
43	Pentadecane	18.65	1498.34	MS, RI	0.25 ± 0.00 [b]	0.34 ± 0.02 [a]	0.30 ± 0.02 [a]	0.34 ± 0.02 [a]
44	2,2′,5,5′-tetramethyl-1,1′-Biphenyl	19.78	1660.37	MS, RI	0.03 ± 0.00 [c]	0.05 ± 0.00 [b]	0.05 ± 0.00 [b]	0.08 ± 0.00 [a]
45	6-Methyl-6-(5-methylfuran-2-yl)heptan-2-one	17.45	1416.72	MS, RI	0.07 ± 0.00 [b]	0.14 ± 0.00 [a]	0.07 ± 0.00 [b]	0.05 ± 0.00 [c]
46	4′,6′-Dimethoxy-2′,3′-dimethylacetophenone	19.24	1628.78	MS, RI	0.08 ± 0.00 [b]	0.12 ± 0.01 [a]	0.05 ± 0.01 [c]	0.01 ± 0.00 [d]
47	2,4-Di-tert-butylphenol	18.79	1504.35	MS, RI	0.74 ± 0.16 [b]	0.67 ± 0.15 [b]	1.41 ± 0.21 [a]	0.75 ± 0.15 [b]
48	2-(2-butoxyethoxy)-Ethanol,acetate	12.45	1167.75	MS, RI	1.27 ± 0.03 [b]	1.49 ± 0.03 [a]	1.39 ± 0.11 [ab]	1.30 ± 0.04 [b]
49	3-Hexen-1-ol benzoate	19.09	1573.04	MS, RI	0.15 ± 0.00 [c]	0.19 ± 0.01 [b]	0.11 ± 0.01 [d]	0.25 ± 0.00 [a]
50	Hexanoic acid, hexyl ester	17.32	1382.70	MS, RI	0.04 ± 0.00 [b]	0.04 ± 0.00 [b]	0.02 ± 0.00 [c]	0.27 ± 0.00 [a]
51	alpha.-Calacorene	18.89	1545.24	MS, RI	0.09 ± 0.00 [b]	0.13 ± 0.00 [a]	0.07 ± 0.01 [c]	0.10 ± 0.00 [b]
52	Tetradecane	17.38	1398.41	MS, RI	0.40 ± 0.00 [b]	0.44 ± 0.06 [a]	0.59 ± 0.06 [c]	0.81 ± 0.01 [c]
53	Dodecane, 4,6-dimethyl-	16.76	1319.70	MS, RI	0.10 ± 0.00 [c]	0.14 ± 0.03 [bc]	0.16 ± 0.03 [b]	0.23 ± 0.00 [a]
54	Tridecane, 2-methyl-	18.43	1493.53	MS, RI	0.23 ± 0.00 [b]	0.25 ± 0.02 [b]	0.10 ± 0.01 [c]	0.27 ± 0.00 [a]
55	3,5-Dimethyldodecane	17.08	1368.99	MS, RI	0.13 ± 0.00 [c]	0.20 ± 0.03 [b]	0.23 ± 0.04 [b]	0.36 ± 0.00 [a]
56	Hexanoic acid, 3-hexenyl ester, (Z)-	17.30	1377.24	MS, RI	0.33 ± 0.00 [c]	0.34 ± 0.02 [c]	0.48 ± 0.05 [b]	1.23 ± 0.02 [a]
57	(E)-Hexanoic acid, 2-hexenyl ester	17.37	1385.40	MS, RI	0.11 ± 0.00 [b]	0.12 ± 0.01 [b]	0.05 ± 0.01 [c]	0.28 ± 0.01 [a]
58	5,9-Undecadien-2-one, 6,10-dimethyl-, (E)-	17.88	1445.93	MS, RI	1.38 ± 0.01 [b]	1.66 ± 0.04 [a]	1.13 ± 0.04 [c]	0.79 ± 0.02 [d]
59	1-Oxaspiro [4.5]dec-6-ene, 2,6,10,10-tetramethyl-	16.51	1302.26	MS, RI	0.17 ± 0.00 [b]	0.26 ± 0.05 [a]	0.16 ± 0.03 [b]	0.24 ± 0.00 [a]
60	Caffeine	23.19	1853.02	MS, RI	11.83 ± 0.50 [b]	13.43 ± 0.62 [ab]	15.52 ± 2.04 [a]	12.74 ± 0.60 [b]
61	alpha.-Ionone	17.46	1423.20	MS, RI	0.25 ± 0.00 [b]	0.30 ± 0.01 [a]	0.16 ± 0.00 [c]	0.12 ± 0.01 [d]
62	5-Methyl-2,4-diisopropylphenol	16.77	1337.69	MS, RI	0.09 ± 0.00 [b]	0.16 ± 0.00 [a]	0.08 ± 0.01 [c]	0.04 ± 0.00 [c]

Table 2. *Cont.*

No.	Aroma Compounds	Retention Time	Retention Index	Identification Methods	Absolute Content (μg/g)			
					SGT	PRT	CK	CA
63	2-(1,1-Dimethylethyl)-6-(1-methylethyl)phenol	15.47	1258.48	MS, RI	——	0.01 ± 0.00 [a]	0.01 ± 0.00 [b]	0.01 ± 0.00 [b]
64	2-Buten-1-one, 1-(2,6,6-trimethyl-1,3-cyclohexadien-1-yl)-, (E)-	17.32	1380.04	MS, RI	0.06 ± 0.00 [b]	0.08 ± 0.00 [a]	0.09 ± 0.00 [a]	0.05 ± 0.01 [b]
65	4-(2,6,6-Trimethylcyclohexa-1,3-dienyl)but-3-en-2-one	17.65	1407.67	MS, RI	0.10 ± 0.00 [d]	0.12 ± 0.00 [c]	0.14 ± 0.02 [b]	0.17 ± 0.01 [a]
66	1-Dodecanol	19.94	1591.17	MS, RI	0.04 ± 0.00 [c]	0.02 ± 0.00 [d]	0.08 ± 0.01 [b]	0.10 ± 0.00 [a]
67	Undecane, 2,9-dimethyl-	17.01	1362.43	MS, RI	0.09 ± 0.01 [c]	0.11 ± 0.01 [b]	0.08 ± 0.01 [c]	0.13 ± 0.00 [a]
68	Nonane, 5-(2-methylpropyl)-	12.82	1162.73	MS, RI	0.03 ± 0.00 [b]	0.03 ± 0.01 [b]	0.03 ± 0.01 [b]	0.06 ± 0.00 [a]
69	Decane, 3-ethyl-3-methyl-	15.59	1260.38	MS, RI	0.08 ± 0.00 [b]	0.09 ± 0.03 [b]	0.10 ± 0.03 [b]	0.18 ± 0.00 [a]
70	Undecane, 4,6-dimethyl-	13.51	1211.40	MS, RI	0.08 ± 0.00 [b]	0.11 ± 0.03 [b]	0.10 ± 0.03 [b]	0.18 ± 0.00 [a]
71	Undecane, 5,7-dimethyl-	16.81	1329.38	MS, RI	0.03 ± 0.00 [b]	0.04 ± 0.01 [b]	0.05 ± 0.01 [b]	0.07 ± 0.00 [a]
72	Decane, 2,3,5-trimethyl-	16.53	1304.88	MS, RI	0.05 ± 0.00 [c]	0.06 ± 0.01 [c]	0.07 ± 0.01 [b]	0.08 ± 0.00 [a]
73	n-Valeric acid cis-3-hexenyl ester	13.76	1228.24	MS, RI	0.01 ± 0.00 [c]	0.02 ± 0.00 [bc]	0.02 ± 0.00 [b]	0.06 ± 0.00 [a]
74	2(4H)-Benzofuranone, 5,6,7,7a-tetrahydro-4,4,7a-trimethyl-, (R)-	19.02	1533.13	MS, RI	0.96 ± 0.03 [b]	1.08 ± 0.04 [a]	0.60 ± 0.07 [c]	0.20 ± 0.00 [d]
75	1,3-Benzenediol, 5-pentyl-	18.96	1518.37	MS, RI	1.88 ± 0.01 [b]	1.88 ± 0.13 [b]	3.04 ± 0.24 [a]	2.04 ± 0.06 [b]
76	Aspirin	22.67	1807.58	MS, RI	0.01 ± 0.00 [a]	——	0.01 ± 0.00 [b]	0.01 ± 0.00 [a]
77	Butyl benzoate	19.87	1580.75	MS, RI	0.02 ± 0.00 [bc]	0.02 ± 0.00 [b]	0.02 ± 0.00 [c]	0.04 ± 0.00 [a]
78	1, 1, 5-Trimethyl-1, 2-dihydronaphthalene	16.99	1357.23	MS, RI	0.03 ± 0.00 [c]	0.05 ± 0.00 [b]	0.04 ± 0.00 [c]	0.06 ± 0.00 [a]
79	Ethyl 4-(ethyloxy)-2-oxobut-3-enoate	15.87	1284.27	MS, RI	0.03 ± 0.00 [b]	0.03 ± 0.00 [b]	0.03 ± 0.00 [c]	0.04 ± 0.00 [a]
80	Undecane, 5-methyl-	10.73	1053.58	MS, RI	0.07 ± 0.01 [b]	0.09 ± 0.04 [b]	0.08 ± 0.03 [b]	0.16 ± 0.01 [a]
81	Decane, 2,4-dimethyl-	16.09	1298.50	MS, RI	0.06 ± 0.00 [c]	0.08 ± 0.02 [bc]	0.10 ± 0.02 [b]	0.13 ± 0.01 [a]
82	trans-Linalool oxide (furanoid)	11.08	1070.91	MS, RI	0.97 ± 0.01 [b]	1.29 ± 0.03 [a]	1.17 ± 0.07 [c]	1.36 ± 0.03 [c]
83	(E)-2-Hexenoic acid, butyl ester	17.68	1437.17	MS, RI	——	——	——	0.01 ± 0.00 [a]
84	(3R,6S)-2,2,6-Trimethyl-6-vinyltetrahydro-2H-pyran-3-ol	13.12	1175.18	MS, RI	0.98 ± 0.02 [b]	0.80 ± 0.04 [c]	0.43 ± 0.02 [d]	1.22 ± 0.04 [a]
85	2H-Pyran-2-one, tetrahydro-6-(2-pentenyl)-, (Z)-	18.30	1491.40	MS, RI	0.13 ± 0.01 [a]	0.14 ± 0.02 [a]	0.07 ± 0.01 [b]	0.13 ± 0.01 [a]
86	2,6-Octadienoic Acid, 3,7-dimethyl-, (E)-	16.86	1353.99	MS, RI	1.24 ± 0.01 [a]	1.18 ± 0.30 [a]	0.29 ± 0.05 [b]	1.04 ± 0.06 [a]
87	2,6,6-trimethyl-1-Cyclohexene-1-acetAldehyde	15.45	1257.39	MS, RI	0.04 ± 0.00 [a]	0.03 ± 0.00 [ab]	0.03 ± 0.00 [b]	0.02 ± 0.00 [c]
88	BenzAldehyde, 2,4-dihydroxy-3,6-dimethyl-	19.97	1593.22	MS, RI	0.02 ± 0.00 [b]	0.02 ± 0.00 [a]	0.01 ± 0.00 [d]	0.01 ± 0.00 [c]

Table 2. *Cont.*

No.	Aroma Compounds	Retention Time	Retention Index	Identification Methods	Absolute Content (μg/g)			
					SGT	PRT	CK	CA
89	2-Cyclopenten-1-one, 3-methyl-2-(2-pentenyl)-, (Z)-	17.44	1394.17	MS, RI	0.37 ± 0.01 [a]	0.37 ± 0.04 [a]	0.26 ± 0.01 [c]	0.32 ± 0.01 [b]
90	2-(formyloxy)-1-phenyl-Ethanone	19.65	1573.33	MS, RI	0.10 ± 0.00 [c]	0.12 ± 0.00 [b]	0.05 ± 0.00 [d]	0.16 ± 0.00 [a]
91	Nonanoic Acid	15.73	1268.14	MS, RI	0.52 ± 0.04 [a]	0.42 ± 0.08 [a]	0.41 ± 0.08 [a]	0.27 ± 0.01 [b]
92	Decane, 5-methyl-	9.86	1011.55	MS, RI	0.02 ± 0.00 [b]	0.02 ± 0.00 [a]	0.02 ± 0.01 [c]	0.04 ± 0.00 [c]
93	Decanal	13.46	1205.23	MS, RI	0.03 ± 0.00 [a]	0.03 ± 0.00 [a]	0.01 ± 0.00 [b]	0.01 ± 0.00 [b]
94	Geraniol	15.37	1249.83	MS, RI	4.11 ± 0.09 [b]	5.50 ± 0.46 [a]	5.27 ± 0.30 [a]	5.76 ± 0.11 [a]
95	L-alpha.-Terpineol	13.34	1197.22	MS, RI	4.79 ± 0.08 [c]	6.97 ± 0.17 [a]	6.40 ± 0.31 [b]	6.52 ± 0.15 [b]
96	2,6-Octadien-1-ol, 3,7-dimethyl-, (Z)-	13.70	1225.24	MS, RI	0.85 ± 0.03 [b]	1.25 ± 0.08 [a]	1.18 ± 0.08 [a]	1.26 ± 0.02 [a]
97	3-Cyclohexen-1-ol, 4-methyl-1-(1-methylethyl)-, (R)-	13.06	1182.77	MS, RI	0.20 ± 0.00 [d]	0.25 ± 0.00 [c]	0.30 ± 0.03 [b]	0.35 ± 0.01 [a]
98	Linalool	11.60	1101.31	MS, RI	6.06 ± 0.14 [c]	9.11 ± 0.35 [a]	7.61 ± 0.80 [b]	9.10 ± 0.07 [a]
99	1,5,7-Octatrien-3-ol, 3,7-dimethyl-	11.78	1103.50	MS, RI	1.13 ± 0.02 [c]	1.71 ± 0.07 [a]	1.36 ± 0.11 [b]	1.38 ± 0.03 [b]
100	2,6,6-trimethyl-1-Cyclohexene-1-carboxaldehyde	13.66	1220.95	MS, RI	0.35 ± 0.00 [a]	0.37 ± 0.02 [a]	0.21 ± 0.02 [b]	0.12 ± 0.00 [c]
101	2,6-Octadienal, 3,7-dimethyl-, (E)-	15.69	1267.57	MS, RI	0.14 ± 0.00 [a]	0.12 ± 0.01 [ab]	0.12 ± 0.01 [ab]	0.11 ± 0.00 [b]
102	Methyl salicylate	13.26	1194.38	MS, RI	0.57 ± 0.00 [b]	0.53 ± 0.04 [a]	0.56 ± 0.05 [c]	0.87 ± 0.02 [c]
103	Methyl anthranilate	16.75	1345.34	MS, RI	0.09 ± 0.00 [b]	0.10 ± 0.02 [b]	0.06 ± 0.00 [c]	0.13 ± 0.00 [a]
104	Ethanone, 1-(2-hydroxy-5-methylphenyl)-	16.50	1313.12	MS, RI	0.51 ± 0.01 [b]	0.56 ± 0.06 [b]	0.13 ± 0.01 [c]	0.63 ± 0.02 [a]
105	2-Methyl-7-exo-vinylbicyclo [4.2.0]oct-1(2)-ene	12.33	1144.57	MS, RI	0.02 ± 0.00 [b]	0.03 ± 0.00 [a]	0.03 ± 0.00 [a]	0.03 ± 0.00 [a]
106	Benzene, pentamethyl-	15.83	1280.53	MS, RI	0.05 ± 0.00 [c]	0.07 ± 0.01 [b]	0.07 ± 0.01 [b]	0.14 ± 0.01 [a]
107	Anethole	15.90	1287.43	MS, RI	11.39 ± 0.11 [b]	6.37 ± 0.34 [d]	14.32 ± 0.35 [a]	7.89 ± 0.14 [c]
108	4-Formyl-3,5-dimethyl-1H-pyrrole-2-carbonitrile	15.81	1277.95	MS, RI	0.06 ± 0.00 [b]	0.09 ± 0.02 [ab]	0.11 ± 0.02 [a]	0.07 ± 0.01 [b]
109	4H-Pyran-4-one, 2,3-dihydro-3,5-dihydroxy-6-methyl-	12.25	1148.70	MS, RI	——	0.01 ± 0.00 [b]	0.01 ± 0.00 [ab]	0.01 ± 0.00 [a]
110	Naphthalene, 2-methyl-	16.33	1300.17	MS, RI	0.16 ± 0.00 [d]	0.20 ± 0.00 [c]	0.21 ± 0.01 [b]	0.30 ± 0.01 [a]
111	(+)-4-Carene	10.03	1016.79	MS, RI	0.09 ± 0.00 [c]	0.14 ± 0.02 [b]	0.13 ± 0.02 [b]	0.22 ± 0.00 [a]
112	D-Limonene	10.20	1029.13	MS, RI	0.73 ± 0.02 [d]	1.09 ± 0.15 [b]	0.93 ± 0.17 [c]	1.52 ± 0.01 [a]
113	beta.-Myrcene	9.38	989.67	MS, RI	0.78 ± 0.01 [c]	1.27 ± 0.09 [a]	1.01 ± 0.13 [b]	1.39 ± 0.04 [a]
114	(S)-(+)-alpha-Phellandrene	9.77	1006.64	MS, RI	0.09 ± 0.00 [c]	0.14 ± 0.01 [b]	0.13 ± 0.02 [b]	0.19 ± 0.00 [a]
115	2,4,6-Octatriene, 3,4-dimethyl-	11.84	1127.36	MS, RI	0.04 ± 0.00 [d]	0.07 ± 0.01 [b]	0.06 ± 0.01 [c]	0.09 ± 0.00 [a]
116	.gamma.-Terpinene	10.65	1058.03	MS, RI	0.15 ± 0.00 [b]	0.18 ± 0.04 [b]	0.17 ± 0.04 [b]	0.34 ± 0.00 [a]
117	beta.-Ocimene	10.49	1045.17	MS, RI	0.42 ± 0.02 [b]	0.53 ± 0.06 [a]	0.32 ± 0.04 [c]	0.49 ± 0.01 [ab]
118	Benzene, 1,2,3,5-tetramethyl-	12.07	1120.46	MS, RI	0.13 ± 0.00 [c]	0.21 ± 0.05 [b]	0.17 ± 0.04 [bc]	0.32 ± 0.01 [a]
119	1,3,8-p-Menthatriene	12.97	1130.49	MS, RI	0.02 ± 0.00 [c]	0.04 ± 0.00 [a]	0.03 ± 0.00 [c]	0.03 ± 0.00 [b]
120	Benzene, 1-methyl-3-(1-methylethyl)-	10.16	1024.38	MS, RI	0.14 ± 0.00 [c]	0.21 ± 0.04 [b]	0.19 ± 0.05 [bc]	0.41 ± 0.01 [a]
121	Benzene, 1-methyl-3-(1-methylethenyl)-	11.28	1091.00	MS, RI	0.08 ± 0.00 [c]	0.13 ± 0.01 [a]	0.10 ± 0.01 [b]	0.14 ± 0.00 [a]
122	Benzene, 4-ethenyl-1,2-dimethyl-	12.66	1149.59	MS, RI	0.05 ± 0.00 [bc]	0.08 ± 0.01 [b]	0.07 ± 0.01 [b]	0.12 ± 0.00 [a]

Table 2. *Cont.*

No.	Aroma Compounds	Retention Time	Retention Index	Identification Methods	Absolute Content (μg/g)			
					SGT	PRT	CK	CA
123	Cyclohexanol, 2,6-dimethyl-	11.99	1111.77	MS, RI	0.21 ± 0.00 [a]	0.19 ± 0.01 [b]	0.15 ± 0.01 [c]	0.10 ± 0.00 [d]
124	1-Octen-3-ol	9.11	980.83	MS, RI	7.80 ± 0.14 [a]	2.58 ± 0.16 [b]	1.51 ± 0.24 [c]	0.28 ± 0.02 [d]
125	Naphthalene	13.43	1188.68	MS, RI	1.06 ± 0.01 [c]	1.49 ± 0.08 [b]	1.46 ± 0.14 [b]	2.10 ± 0.06 [a]
126	3,5,5-trimethyl-2-Hexene	8.89	975.03	MS, RI	2.74 ± 0.02 [a]	0.60 ± 0.05 [b]	0.56 ± 0.07 [b]	0.10 ± 0.01 [c]
127	Benzyl chloride	9.90	1012.98	MS, RI	0.06 ± 0.00 [b]	0.09 ± 0.01 [a]	0.07 ± 0.01 [b]	0.06 ± 0.00 [b]
128	3-Formyl-4,5-dimethyl-pyrrole	10.52	1047.82	MS, RI	0.16 ± 0.00 [c]	0.05 ± 0.00 [d]	0.27 ± 0.03 [b]	0.60 ± 0.02 [a]
129	Propane, 2-chloro-2-nitro-	6.03	709.00	MS, RI	6.38 ± 0.20 [a]	5.84 ± 0.49 [a]	5.28 ± 0.53 [a]	5.55 ± 0.91 [a]
130	1,3-Cyclopentadiene, 5,5-dimethyl-1-ethyl-	7.54	841.47	MS, RI	0.04 ± 0.00 [b]	0.07 ± 0.02 [a]	0.03 ± 0.01 [c]	0.03 ± 0.00 [c]
131	Phenylethyl Alcohol	11.80	1114.93	MS, RI	0.43 ± 0.02 [b]	0.29 ± 0.03 [c]	0.20 ± 0.02 [d]	0.63 ± 0.00 [a]
132	Phenol, 3,5-dimethyl-	8.40	934.00	MS, RI	0.02 ± 0.00 [b]	0.03 ± 0.00 [a]	0.02 ± 0.00 [c]	0.01 ± 0.00 [c]
133	2,3-dihydro-Benzofuran	13.64	1219.63	MS, RI	1.91 ± 0.03 [b]	2.72 ± 0.27 [a]	1.20 ± 0.17 [d]	1.57 ± 0.06 [c]
134	Indole	16.01	1295.10	MS, RI	2.28 ± 0.03 [a]	1.11 ± 0.11 [c]	1.46 ± 0.07 [b]	1.42 ± 0.04 [b]
135	2-Heptanol	8.22	902.75	MS, RI	0.02 ± 0.00 [a]	0.02 ± 0.01 [a]	0.02 ± 0.00 [b]	0.03 ± 0.00 [a]
136	Heptanal	8.20	901.74	MS, RI	0.08 ± 0.01 [a]	0.06 ± 0.01 [b]	0.03 ± 0.00 [c]	0.03 ± 0.01 [c]
137	Cyclobutanone, 2,2,3-trimethyl-	6.44	751.40	MS, RI	0.02 ± 0.00 [b]	0.03 ± 0.00 [b]	0.04 ± 0.02 [b]	0.11 ± 0.01 [a]
138	5-methyl-1,2,5,6-Tetrahydropyridin-2-one	10.06	1020.73	MS, RI	0.19 ± 0.02 [a]	0.11 ± 0.02 [c]	0.16 ± 0.04 [ab]	0.11 ± 0.02 [bc]
139	Benzyl Alcohol	10.24	1036.04	MS, RI	0.36 ± 0.00 [b]	0.44 ± 0.03 [a]	0.33 ± 0.03 [b]	0.43 ± 0.00 [a]
140	BenzAldehyde	8.86	963.89	MS, RI	0.31 ± 0.01 [a]	0.33 ± 0.03 [a]	0.23 ± 0.02 [b]	0.23 ± 0.01 [b]
141	Dihydro-3-(2H)-thiophenone	8.67	955.93	MS, RI	0.02 ± 0.00 [b]	0.02 ± 0.00 [a]	0.01 ± 0.00 [c]	0.02 ± 0.00 [c]
142	3-Hexen-1-ol, (E)-	8.70	856.29	MS, RI	0.14 ± 0.02 [c]	0.16 ± 0.01 [b]	0.09 ± 0.01 [d]	0.39 ± 0.01 [a]
143	Hexanal	6.98	797.72	MS, RI	0.07 ± 0.00 [ab]	0.08 ± 0.03 [a]	0.05 ± 0.01 [b]	0.04 ± 0.01 [b]
144	3-Hexen-2-one	6.66	797.57	MS, RI	0.15 ± 0.02 [a]	0.08 ± 0.07 [ab]	0.07 ± 0.01 [b]	0.13 ± 0.01 [ab]
145	2-methoxy-Furan	6.64	797.09	MS, RI	0.12 ± 0.00 [c]	0.08 ± 0.01 [d]	0.19 ± 0.03 [b]	0.27 ± 0.01 [a]
146	Dimethylphosphinic fluoride	6.01	701.52	MS, RI	0.03 ± 0.00 [a]	0.02 ± 0.00 [ab]	0.02 ± 0.00 [c]	0.02 ± 0.00 [bc]
147	3-ethyl-1H-Pyrrole	7.11	809.67	MS, RI	0.05 ± 0.00 [b]	0.02 ± 0.00 [b]	0.16 ± 0.04 [a]	0.19 ± 0.01 [a]
148	1-Pentanol	6.56	769.54	MS, RI	0.51 ± 0.02 [a]	—	0.28 ± 0.03 [c]	0.34 ± 0.02 [b]

Represents that the post-test has been carried out, different lowercase letters in the same line indicate significant difference ($p < 0.05$), and '—' indicates that it is not detected. Each tea sample was measured in parallel 3 times, and all data were expressed as mean value ± SD.

The functional groups of VCs in GABA sun-dried green tea and GABA raw Pu-erh tea could be divided into 13 categories, including 35 hydrocarbons, 26 terpenes, 22 esters, 12 alcohols, 12 heterocyclics, 11 ketones, 10 aromatics, 8 aldehydes, 4 phenols, 3 halogenated hydrocarbons, 3 acids, 1 nitrogenous compound, and 1 miscellaneous category. These 13 VCs were detected in SGT, PRT, CK, and CA samples, with a total of 145 VCs detected in 13 categories in SGT and PRT, 143 VCs detected in 13 categories in CK, and 148 VCs detected in 13 categories in CA. The 10 VCs with the highest contents in the PRT samples were caffeine (13.43 μg/g), linalool (9.11 μg/g), [2H3]-beta.-ionone (8.27 μg/g), L-alpha.-terpineol (6.97 μg/g), geraniol (5.50 μg/g), 2,3-dihydro-benzofuran (2.72 μg/g), anethole (6.37 μg/g), propane,2-chloro-2-nitro- (5.84 μg/g), 1-octen-3-ol (2.58 μg/g), and 1,3-benzenediol,5-pentyl- (1.88 μg/g). Compared with the SGT, the contents of the first six VCs increased significantly, and the others increased in the SGT. This study represents the first use of HS-SPME-GC-MS technology to detect and determine the VCs of GABA raw Pu-erh tea, demonstrating that the contents of terpenes such as linalool, [2H3]-beta.-ionone, L-alpha.-terpineol, and geraniol could be effectively increased by autoclave treatment after anaerobic treatment. At present, it is speculated that these four terpenes may be the key compounds contributing to the distinct aroma of GABA raw Pu-erh tea, which required further validation by OAV or ROAV.

According to the isoprene rules in organisms, the carbon units are condensed to form precursors for the synthesis of terpenes, which have basic functions in the growth and development of organisms and continue to form more terpenes through metabolic pathways [33]. According to existing research, terpenes are important contributors to the VCs of green tea, black tea, Pu-erh tea, and other teas, but they are rarely reported as the first-class compound category [34,35]. Therefore, the molecular configuration and metabolic pathway of terpenes can be used as an important direction for the study of characteristic aroma compounds in tea in the future. In this study, 26 terpenes were detected in SGT, PRT, CK, and CA samples. Compared with the SGT, the contents of 18 terpenes in the PRT increased significantly, and the total content of terpenes increased by 1.4 times, indicating that the autoclave technology contributed to the formation of terpenes in PRT. A variety of terpenes have been found and confirmed to be the characteristic aroma compounds in tea, such as linalool with a sweet, woody, flowery, and fruity aroma [36]; [2H3]-beta.-ionone with a woody and violet flowery aroma, accompanied by a fruity aroma [37]; L-alpha.-terpineol with a flowery and light grassy aroma [38]; and geraniol with a sweet and light rose aroma [39]. The aroma characteristics of these terpenes can help to better explain the aroma type of PRT. In addition, most of the increased terpenes have a relatively low OT [40], which can be used as a basis for further verification of whether these terpenes contribute to the formation of aroma in PRT.

OPLS-DA is generally used to distinguish the differences of variables between two groups. In this study, OPLS-DA was used to distinguish and analyze the different VCs between SGT and PRT samples (Figure 2c). A total of 68 differential VCs between the SGT and the PRT were screened (variable importance in projection (VIP) > 1, $p < 0.05$, fold change > 0). By comparing the SGT with the PRT (Figure 2d), 49 VCs were found to be up-regulated, including linalool, geraniol, alpha.-ionone, [2H3]-beta.-ionone, 6-methyl-6-(5-methylfuran-2-yl)heptan-2-one, 2-buten-1-one,1-(2,6,6-trimethyl-1,3-cyclohexadien-1-yl)-,(E)-, 1,3,8-p-menthatriene, undecane, 2,9-dimethyl-, hexadecanoic acid, methyl ester, alpha.-cadinol, and others; 19 VCs were down-regulated, including anethole, 1-octen-3-ol, 1-dodecanol, 3-methyl-tetradecane, hexadecane, 2-methyl-, phenylethyl alcohol, 2-methoxy-furan, cyclohexanol, 2,6-dimethyl-, (3R,6S)-2,2,6-trimethyl-6-vinyltetrahydro-2H-pyran-3-ol, diethyl phthalate, and others.

3.3. *Key Active Compounds of GABA Sun-Dried Green Tea and GABA Raw Pu-Erh Tea Aroma*

The types of VCs in SGT and PRT are very complex, and many of the VCs identified are odorless or have low odor activity with no contribution to or minimal effect on the overall aroma formation of tea samples. Therefore, the key active compounds of SGT and PRT aroma formation were screened to reveal the key role of VCs in the formation of tea aroma. A total of 14 VCs were screened based on ROAV and aroma types (Table 3). In the SGT, there were six VCs with ROAV > 1 and eight compounds with 0.1 < ROAV < 1. In the PRT, there were five VCs with ROAV > 1 and eight compounds with 0.1 < ROAV < 1. The Pearson correlation circle showed that there was a weak correlation between 2-buten-1-one,1-(2,6,6-trimethyl-1,3-cyclohexadien-1-yl)-, (E)- and other compounds in PRT, which was more prominent (Figure 3a). In SGT, 1-octen-3-ol and D-limonene showed a weak correlation with other compounds (Figure 3b). Further analysis of ROAV by construction of the cluster heat map showed that the key active compounds in SGT and PRT were significantly different (Figure 3c).

Table 3. ROAV and aroma types of key aroma compounds.

No.	Index	ROAV		VIP	Fold Change	Aroma Type
		SGT	PRT			
1	[2H3]-beta.-Ionone	0.39 ± 0.02	0.29 ± 0.02	1.13	1.12	Woody, flowery
2	5,9-Undecadien-2-one, 6,10-dimethyl-, (E)-	0.68 ± 0.01	0.55 ± 0.03	1.19	1.20	Fresh, flowery, sweet (weak)
3	alpha.-Ionone	1.09 ± 0.02	0.88 ± 0.05	1.16	1.21	Flowery, sweet, weak
4	2-Buten-1-one, 1-(2,6,6-trimethyl-1,3-cyclohexadien-1-yl)-, (E)-	10.51 ± 0.69	10.68 ± 0.66	1.18	1.52	Flowery, fruity, lasting
5	Geraniol	13.54 ± 0.15	12.11 ± 1.50	1.13	1.34	Flowery, sweet, weak, and normal
6	L-alpha.-Terpineol	0.39 ± 0.00	0.38 ± 0.02	1.21	1.46	Woody, flowery, weak
7	Linalool	100.00 ± 0.00	100.00 ± 0.00	1.20	1.50	Woody, flowery, fruity (weak), sweet (weak)
8	Anethole	0.28 ± 0.01	0.11 ± 0.01	1.21	0.56	Licorice
9	D-Limonene	0.53 ± 0.01	0.53 ± 0.05	1.09	1.50	Flowery, lemony, weak
10	beta.-Myrcene	1.16 ± 0.03	1.26 ± 0.07	1.19	1.63	Fatty (weak)
11	Naphthalene	0.52 ± 0.01	0.49 ± 0.04	1.18	1.41	Aromatic, normal
12	Phenylethyl Alcohol	0.24 ± 0.02	0.11 ± 0.02	1.16	0.67	Sweet (weak), flowery (weak)
13	Indole	0.11 ± 0.00	0.04 ± 0.00	1.21	0.49	Flowery, fresh
14	1-Octen-3-ol	96.51 ± 4.17	21.21 ± 0.57	1.05	0.33	Clean, fatty, mushroomy

Each tea sample was measured in parallel for 3 times, and all data were expressed as mean value ± SD.

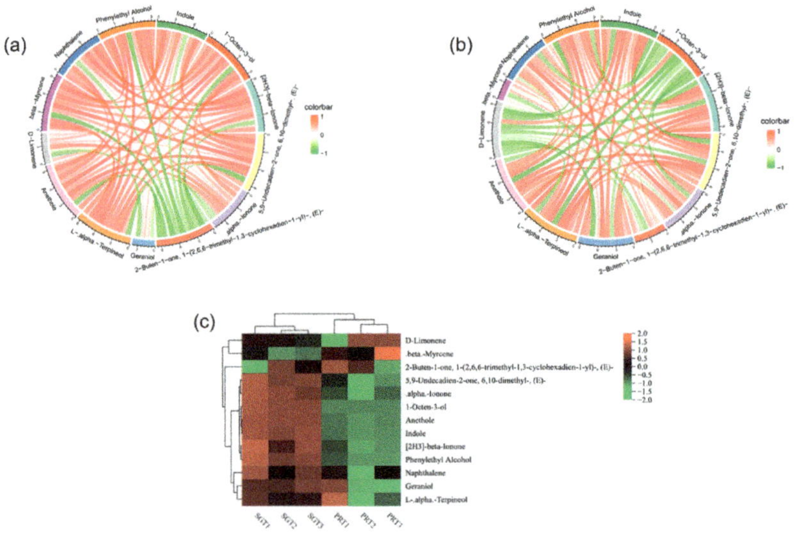

Figure 3. Based on the ROAV value, the correlation statistics of the characteristic aroma compounds of tea samples were carried out. SGT refers to GABA sun-dried green tea, and PRT refers to GABA raw Pu-erh tea. (**a**) Pearson correlation circle of characteristic aroma compounds of PRT. (**b**) Pearson correlation circle of characteristic aroma compounds of SGT. (**c**) Heat map of hierarchical clustering of relative odor activity value.

In the SGT, there were two types of VCs with ROAV > 90, which were linalool (woody, flowery, fruity (weak), and sweet (weak)) and 1-octen-3-ol (clean, fatty, and mushroomy).

The VCs of 1 < ROAV < 90 included alpha.-ionone (flowery, sweet, and weak), 2-buten-1-one,1-(2,6,6-trimethyl-1,3-cyclohexadien-1-yl)-, (E)- (flowery, fruity, and lasting), geraniol (flowery, sweet, weak, and normal), and beta.-myrcene (fatty and weak). The compounds with 0.1 < ROAV < 1 included [2H3]-beta.-ionone (woody and flowery), 5,9-undecadien-2-one,6,10-dimethyl-, (E)- (fresh, flowery, and sweet (weak)), L-alpha.-terpineol (woody, flowery, and weak), anethole (licorice), D-limonene (flowery, lemony, and weak), naphthalene (aromatic and normal), phenylethyl alcohol (sweet (weak) and flowery (weak)), and indole (flowery and fresh). The aroma types of SGT mainly included flowery, clean, fatty, fruity (weak), and sweet (weak). Because there were two main VCs with ROAV > 90, and the two compounds show completely different aroma types, SGT forms the aroma characteristics of flowery and pure aroma.

In the PRT group, linalool (woody, flowery, fruity (weak), and sweet (weak)) was the only VC with ROAV > 90. The VCs of 1 < ROAV < 90 included 2-buten-1-one,1-(2,6,6-trimethyl-1,3-cyclohexadien-1-yl)-, (E)- (flowery, fruity, and lasting), geraniol (flowery, sweet, weak, and normal), beta.-myrcene (fatty (weak)), and 1-octen-3-ol (clean, fatty, and mushroomy). The compounds with 0.1 < ROAV < 1 included [2H3]-beta.-ionone (woody and flowery), 5,9-undecadien-2-one,6,10-dimethyl-, (E)- (fresh, flowery, and sweet (weak)), alpha.-ionone (flowery, sweet, and weak), L-alpha.-terpineol (woody, flowery, and weak), anethole (licorice), D-limonene (flowery, lemony, and weak), naphthalene (aromatic and normal), and phenylethyl alcohol (sweet (weak) and flowery (weak)). Compared with SGT, the ROAV of linalool and 1-octen-3-ol in PRT showed a fourfold difference, and the aroma of 1-octen-3-ol was relatively weakened so that the aroma of PRT showed pure characteristics. Because the main VCs in PRT have better aroma persistence, such as 2-buten-1-one,1-(2,6,6-trimethyl-1,3-cyclohexadien-1-yl)-,(E)-, PRT shows persistent aroma characteristics. These results indicate that autoclave technology can improve the purity and persistence of tea aroma and effectively improve the quality of tea aroma. From the perspective of aroma types, the main VCs in PRT were dominated by linalool, which relatively weakened the VCs exhibiting aroma characteristics such as clean, refreshing, and fatty aroma, and finally conferred PRT with a flowery and fruity aroma characterized as light and sweet.

Linalool (ROAV = 100) in SGT had the greatest contribution to the aroma, followed by 1-octen-3-ol (ROAV = 96.51), geraniol (ROAV = 13.54), 2-buten-1-one,1-(2,6,6-trimethyl-1,3-cyclohexadien-1-yl)-,(E)-(ROAV = 10.51), beta.-myrcene (ROAV = 1.16), and alpha.-ionone (ROAV = 1.09). According to previous screening results of GC-MS and differential VCs, linalool and alpha.-ionone greatly contributed to the aroma of sun-dried green tea, which was consistent with the results of this study [41]. However, this study also found other compounds that contribute more to the aroma, such as 2-buten-1-one,1-(2,6,6-trimethyl-1,3-cyclohexadien-1-yl)-,(E)- and beta.-myrcene; these compounds may be the reason why GABA sun-dried green tea has no obvious sun flavor. Linalool (ROAV = 100) had the greatest contribution to the aroma in PRT, followed by 1-octen-3-ol (ROAV = 21.21), geraniol (ROAV = 12.11), 2-buten-1-one,1-(2,6,6-trimethyl-1,3-cyclohexadien-1-yl)-,(E)-(ROAV = 10.68), and beta.-myrcene (ROAV = 1.26). In addition to linalool, methoxy compounds and ketones also contributed to the aroma of raw Pu-erh tea [42]. This study found that the compounds with the main contribution to the aroma of GABA raw Pu-erh tea were terpenes, which indicated that anaerobic technology could significantly change the flavor of raw Pu-erh tea. In previous studies on the aroma compounds of sun-dried green tea, most of the compounds were identified to contribute to the aroma, such as methoxy, aldehydes, and ketones [43]. However, terpenes such as 2-buten-1-one,1-(2,6,6-trimethyl-1,3-cyclohexadien-1-yl)-,(E)- and beta.-myrcene were identified to have a great contribution to the aroma of the tea in this study for the first time, with an ROAV of more than one in both PRT and SGT, although the ROAV of the two compounds in PRT was higher than that in SGT. This indicates that these compounds play an important role in aroma formation, which is not consistent with previous studies demonstrating that the terpenes linalool, geraniol, and alpha.-ionone are important contributors to the aroma formation of sun-dried green tea and raw Pu-erh tea [44]. This difference may be due to the fact that previous studies tested a lower content of GABA sun-dried green tea and GABA raw Pu-erh tea. Another reason may be that there are differences between tea varieties. In this study, the contribution of 2-buten-1-one,1-(2,6,6-trimethyl-1,3-cyclohexadien-1-yl)-,(E)- and beta.-myrcene to the aroma formation of GABA sun-dried green tea and GABA raw Pu-erh tea was clarified by ROAV; however, determining the effect of dynamic changes of VCs on the quality of the tea requires further study.

3.4. Potential Markers of Differential Aroma

Finding and identifying potential marker aroma compounds is an important quantitative research method in the study of flavor differences [45]. There were significant differences in aroma characteristics between SGT and PRT. OPLS-DA and ROAV analysis (VIP > 1, ROAV > 1, $p < 0.05$, fold change > 0) identified differential VCs of SGT and PRT, including linalool, 1-octen-3-ol, geraniol, 2-buten-1-one,1-(2,6,6-trimethyl-1,3-cyclohexadien-1-yl)-,(E)-, and beta.-myrcene. These compounds may therefore be used as potential markers of SGT and PRT aroma. Among them, linalool, geraniol, and 1-octen-3-ol were confirmed as potential markers for the aroma of sun-dried green tea and raw Pu-erh tea in a previous study [9].

The production of potential markers is usually related to the variety, processing technology, or unique active ingredients. The identification of potential markers of tea can help tea quality control or test repeatability [46]. According to previous studies, potential marker aroma compounds may be formed during tea storage or processing, and even biological pathways such as oxidation and degradation dominated by environmental factors may also promote the formation of potential marker aroma compounds [47,48]. According to the five potential marker aroma compounds found in this study, we believe that it is likely to be due to anaerobic treatment. Potential marker aroma compounds have practical significance for explaining the aroma results of this study and will provide a reference for future research on aroma compounds of GABA tea.

3.5. Metabolic Evolution Pathway of Main Aroma Compounds

The main VCs in SGT and PRT were terpenes and alcohols, which was similar to the results of previous studies [49]. KEGG pathway analysis was used to further explore the metabolic pathways for the formation of key aroma compounds in SGT and PRT. The results showed that the formation of major VCs mainly involves monoterpenoid biosynthesis, biosynthesis of phytochemical compounds, and amino acid metabolism (Figure 4).

Combined with the characteristic aroma compounds map and metabolic pathway map of PRT (Figure 5), it can be found that terpenes in SGT and PRT are mainly formed through monoterpenoid biosynthesis. Most of the terpenes in tea have woody, flowery, and fruity aromas, which can better coordinate the aroma of SGT and PRT. According to a previous study, the product of glycoside hydrolysis, geranyl diphosphate, is an important precursor of terpenoids [50]. Hydrolases hydrolyze glycoside to form geranyl diphosphate in the process of water loss and anaerobic digestion of SGT and PRT, which leads to the formation of various terpenes in the tea through the enzymatic action of corresponding synthases. In addition, linalool forms geraniol under the action of isomerase. Indole is mainly formed through the amino acid metabolism pathway, and anethole is mainly formed through the biosynthesis of phytochemical compounds, which has a certain contribution to SGT aroma; on the one hand, it enriches the SGT aroma type, and on the other hand, it affects the purity of the SGT aroma. However, the contribution of indole to the PRT aroma was not significant. A previous study found that high-temperature drying will lead to the degradation of a large number of amino acids in black tea [51]. In the process of SGT deactivation and autoclaving, tryptophan, as a precursor for the synthesis of indole, is greatly degraded, thus reducing the contribution of indole to the PRT aroma.

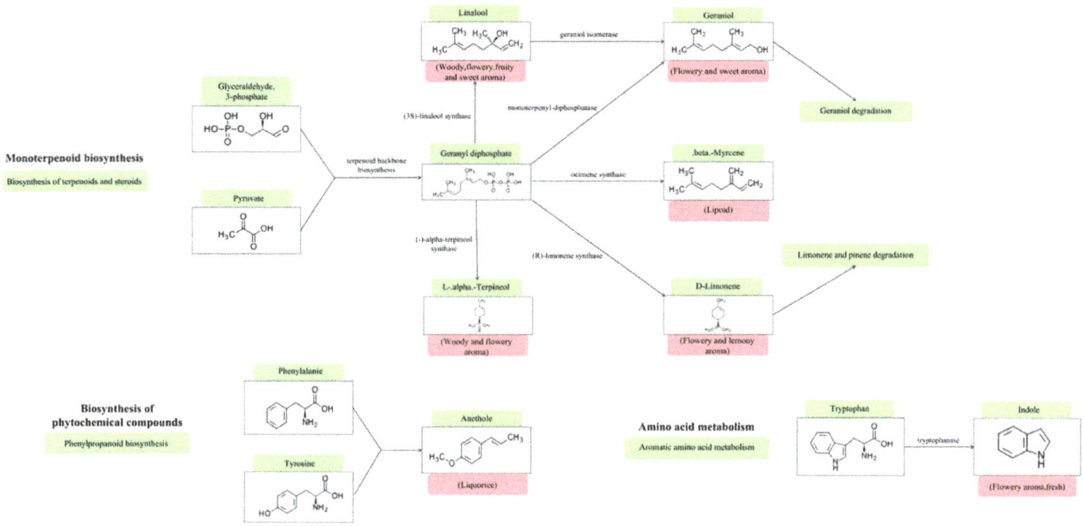

Figure 4. Metabolic evolution pathway of main flavor compounds in GABA sun-dried green tea and GABA raw Pu-erh tea.

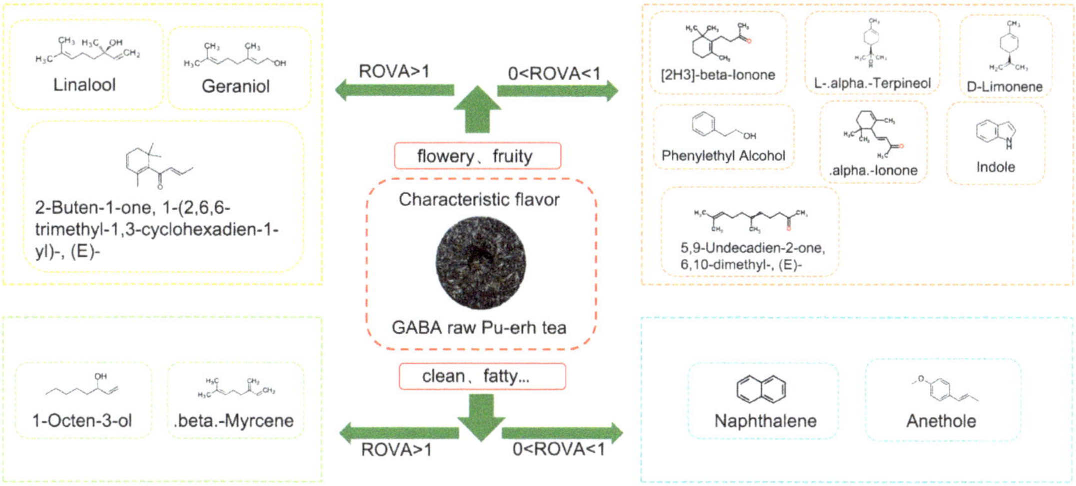

Figure 5. Characteristic aroma compounds and characteristic aroma in GABA raw Pu-erh tea.

In addition, some VCs are produced by non-enzymatic reactions. For example, 2-buten-1-one,1-(2,6,6-trimethyl-1,3-cyclohexadien-1-yl)-,(E)- is mainly formed by the non-enzymatic reaction of neoxanthin under acidic conditions such as thermal degradation or oxidation [52]. Therefore, after autoclaving, the content of 2-buten-1-one,1-(2,6,6-trimethyl-1,3-cyclohexadien-1-yl)-,(E)- increased significantly, which was beneficial to the formation of the floral and fruity aroma of PRT. Linoleic acid is oxidized by lipoxygenase and then reduced by alcohol dehydrogenase to form 1-octen-3-ol; however, the key enzyme is still unclear [53]. A previous study showed that the content of 1-octen-3-ol in wheat food significantly reduced after steam heating or baking heating [54]. Although the specific degradation mechanism remains unclear, this may explain the difference in 1-octen-3-ol between SGT and PRT.

4. Conclusions

In summary, the main VCs in SGT and PRT were formed by monoterpenoid biosynthesis, biosynthesis of phytochemical compounds, and amino acid metabolism during processing. We believe that terpenes play an important role in promoting the aroma of SGT and PRT, mainly including linalool, geraniol, 2-buten-1-one,1-(2,6,6-trimethyl-1,3-cyclohexadien-1-yl)-,(E)-, and beta.-myrcene, resulting in flowery and fruity aroma characteristics. Compared with PRT, 1-octen-3-ol, indole, and anethole may contribute to the more impure aroma of SGT, which may be related to the autoclave processing technology. In addition, the qualitative and quantitative results of VCs were quite different between CK and CA samples, indicating that anaerobic technology could effectively change the aroma compounds and contents in fresh tea leaves. Overall, the findings of this study provide a basis for the formation of flowery and fruity aromas, help to clarify the characteristic aroma compounds, and serve as a guide to further improve the quality of SGT and PRT.

In the future, aroma recombination experiments can be explored, and response surface experiments can be designed to provide more clear evidence for determining the actual aroma contribution compounds and to determine improved quality optimization methods. Finally, the dynamic change mechanism of terpenoids and other compounds that have significant contributions to aroma in tea processing remains to be elucidated.

Supplementary Materials: The following supporting information can be downloaded at https://www.mdpi.com/article/10.3390/foods12244512/s1, Figure S1: Heat map of volatile compounds.

Author Contributions: Conceptualization, Y.L. (Yali Li) and H.Z.; methodology, X.Z.; software, Y.L. (Yuanda Li); validation, C.M., C.G. and G.F.; formal analysis, C.M.; investigation, C.G.; resources, G.F.; data curation, C.M.; writing—original draft preparation, C.M.; writing—review and editing, Y.L.(Yali Li); visualization, H.Z.; supervision, D.T. and Y.H.; project administration, D.T.; funding acquisition, C.M. All authors have read and agreed to the published version of the manuscript.

Funding: This research was funded by national natural science fund project (NSFC-funded project), Funding number: 31460215.

Institutional Review Board Statement: Sensory evaluation of tea samples was carried out according to Chinese national standard procedure. The experimental scheme involving sensory evaluation is in line with Chinese national law. In the course of the implementation of the study, no human body, animal and violation of law, morality or the "Declaration of Helsinki" were involved. And all participants have written consent.

Informed Consent Statement: Informed consent was obtained from all subjects involved in the study.

Data Availability Statement: Data is contained within the article and Supplementary Materials.

Conflicts of Interest: The authors declare no potential conflict of interest with respect to the research, authorship, and/or publication of this article.

References

1. Lewis, R. *Tea: History, Terroirs, Varieties*; Library Journal; No. 127; Firefly Books: Richmond Hill, ON, Canada, 2012; Volume 137. Available online: https://lccn.loc.gov/2014415257 (accessed on 27 August 2023).
2. Wu, X.; Liu, Y.; Guo, J.; Wang, J.; Li, M.; Tan, Y.; Zheng, Q.; Feng, Y. Differentiating Pu-erh raw tea from different geographical origins by 1 H-NMR and U-HPLC/Q-TOF-MS combined with chemometrics. *J. Food Sci.* **2021**, *86*, 779–791. [CrossRef] [PubMed]
3. Lv, H.; Zhang, Y.; Lin, Z.; Liang, Y. Processing and chemical constituents of Pu-erh tea: A review. *Food Res. Int.* **2013**, *53*, 608–618. [CrossRef]
4. Ahmed, S.; Stepp, J. Pu-erh Tea: Botany, Production, and Chemistry. In *Tea in Health and Disease Prevention*; Elsevier: Amsterdam, The Netherlands, 2013; pp. 59–72.
5. Zheng, X.; Li, Q.; Xiang, L.; Liang, Y. Recent Advances in Volatiles of Teas. *Molecules* **2016**, *21*, 338. [CrossRef] [PubMed]
6. Yang, X.; Liu, Y.; Mu, L.; Wang, W.; Zhan, Q.; Luo, M.; Tian, H.; Lv, C.; Li, J. Discriminant research for identifying aromas of non-fermented Pu-erh tea from different storage years using an electronic nose. *J. Food Process. Preserv.* **2018**, *42*, e13721. [CrossRef]
7. Xu, S.; Zeng, X.; Wu, H.; Shen, S.; Yang, X.; Deng, W.; Ning, J. Characterizing volatile metabolites in raw Pu'er tea stored in wet-hot or dry-cold environments by performing metabolomic analysis and using the molecular sensory science approach. *Food Chem.* **2021**, *350*, 129186. [CrossRef]

8. Fan, X.; Chen, N.; Cai, F.; Ren, F.; Zhong, J.; Wang, D.; Shi, L.; Ren, D.; Yi, L. Effects of manufacturing on the volatile composition of raw Pu-erh tea with a focus on de-enzyming and autoclaving–compressing treatments. *LWT* **2020**, *137*, 110461. [CrossRef]
9. Feng, T.; Sun, J.; Wang, K.; Song, S.; Chen, D.; Zhuang, H.; Lu, J.; Li, D.; Meng, X.; Shi, M.; et al. Variation in Volatile Compounds of Raw Pu-erh Tea upon Steeping Process by Gas Chromatography-Ion Mobility Spectrometry and Characterization of the Aroma-Active Compounds in Tea Infusion Using Gas Chromatography-Olfactometry-Mass Spectrometry. *J. Agric. Food Chem.* **2022**, *70*, 13741–13753. [CrossRef]
10. Tsushida, T.; Murai, T. Conversion of Glutamic Acid to γ-Aminobutyric Acid in Tea Leaves under Anaerobic Conditions. *Agric. Biol. Chem.* **1987**, *51*, 2865–2871. [CrossRef]
11. Norio, N.; Katsuhiro, H.; Toshihiro, M.; Hirokazu, F. An Improvement of the Qualities of Anaerobically Treated Tea (Gabaron Tea) by Heating. *Chagyo Kenkyu Hokoku Tea Res. J.* **1988**, *28*, 40–42. [CrossRef]
12. Ren, T.; Zheng, P.; Zhang, K.; Liao, J.; Xiong, F.; Shen, Q.; Ma, Y.; Fang, W.; Zhu, X. Effects of GABA on the polyphenol accumulation and antioxidant activities in tea plants (*Camellia sinensis* L.) under heat-stress conditions. *Plant Physiol. Biochem.* **2021**, *159*, 363–371. [CrossRef]
13. Zhen, Z. Study on flavour volatiles of γ-aminobutyric acid (GABA) green tea. *Afr. J. Biotechnol.* **2012**, *11*, 11333–11341. [CrossRef]
14. Li, Y.; Wu, T.; Deng, X.; Tian, D.; Ma, C.; Wang, X.; Li, Y.; Zhou, H. Characteristic aroma compounds in naturally withered and combined withered γ-aminobutyric acid white tea revealed by HS-SPME-GC-MS and relative odor activity value. *LWT* **2023**, *176*, 114467. [CrossRef]
15. Yang, S. Study on GABA *Pu'er* Tea. Master's Thesis, Yunnan Agricultural University, Kunming, China, 2012. Available online: https://kns.cnki.net/KCMS/detail/detail.aspx?dbname=CMFD2012&filename=1012030358.nh (accessed on 27 August 2023).
16. Yang, P.; Yu, M.; Song, H.; Xu, Y.; Lin, Y.; Granvogl, M. Characterization of Key Aroma-Active Compounds in Rough and Moderate Fire Wuyi Rock Tea by Sensory-Directed Flavor Analysis and Elucidation of the Influences of Roasting on Aroma. *J. Agric. Food Chem.* **2022**, *70*, 267–278. [CrossRef] [PubMed]
17. Zhu, J.; Chen, F.; Wang, L.; Niu, Y.; Xiao, Z. Evaluation of the synergism among volatile compounds in Oolong tea infusion by odour threshold with sensory analysis and E-nose. *Food Chem.* **2017**, *221*, 1484–1490. [CrossRef] [PubMed]
18. Guo, X.; Ho, C.T.; Schwab, W.; Wan, X. Aroma profiles of green tea made with fresh tea leaves plucked in summer. *Food Chem.* **2021**, *363*, 130328. [CrossRef] [PubMed]
19. Fang, Z.Y.; Li, G.Z.; Gu, Y.; Wen, C.; Ye, H.; Ma, J.L.; Liang, Z.Y.; Yang, L.; Wu, J.W.; Chen, H.Y. Flavour analysis of different varieties of camellia seed oil and the effect of the refining process on flavour substances. *LWT* **2022**, *170*, 114040. [CrossRef]
20. Xie, J.; Wang, L.; Deng, Y.; Yuan, H.; Zhu, J.; Jiang, Y.; Yang, Y. Characterization of the key odorants in floral aroma green tea based on GC-E-Nose, GC-IMS, GC-MS and aroma recombination and investigation of the dynamic changes and aroma formation during processing. *Food Chem.* **2023**, *427*, 136641. [CrossRef]
21. Wang, Y.; Li, C.; Zhao, Y.; Li, L.; Yang, X.; Wu, Y.; Chen, S.; Cen, J.; Yang, S.; Yang, D. Novel insight into the formation mechanism of volatile flavor in Chinese fish sauce (Yu-lu) based on molecular sensory and metagenomics analyses. *Food Chem.* **2020**, *323*, 126839. [CrossRef]
22. Hong, X.; Wang, C.; Jiang, R.; Hu, T.; Zheng, X.; Huang, J.; Liu, Z.; Li, Q. Characterization of the Key Aroma Compounds in Different Aroma Types of Chinese Yellow Tea. *Foods* **2022**, *12*, 27. [CrossRef]
23. *GB/T 23776*; Methodology for Sensory Evaluation of Tea. China Standards Press: Beijing, China, 2018.
24. Li, J.; Wang, J.; Yao, Y.; Hua, J.; Zhou, Q.; Jiang, Y.; Deng, Y.; Yang, Y.; Wang, J.; Yuan, H.; et al. Phytochemical comparison of different tea (*Camellia sinensis*) cultivars and its association with sensory quality of finished tea. *LWT* **2020**, *117*, 108595. [CrossRef]
25. Deng, X.; Huang, G.; Tu, Q.; Zhou, H.; Li, Y.; Shi, H.; Wu, X.; Ren, H.; Huang, K.; He, X. Evolution analysis of flavor-active compounds during artificial fermentation of Pu-erh tea. *Food Chem.* **2021**, *357*, 129783. [CrossRef] [PubMed]
26. Bi, J.; Li, Y.; Yang, Z.; Lin, Z.; Chen, F.; Liu, S.; Li, C. Effect of different cooking times on the fat flavor compounds of pork belly. *J. Food Biochem.* **2022**, *46*, 14184. [CrossRef]
27. Jin, Q.; Wang, Z.; Chen, Y.; Luo, Y.; Tian, N.; Liu, Z.; Huang, J.; Liu, S. Transcriptomics analysis reveals the signal transduction mechanism of brassinolides in tea leaves and its regulation on the growth and development of Camellia sinensis. *BMC Genom.* **2022**, *23*, 29. [CrossRef] [PubMed]
28. Ou, Y.; Zhang, Y.; Qin, L.; Miao, Y.; Xiao, L. Research progress on evaluation methods of tea color, aroma and quality. *Sci. Technol. Food Ind.* **2019**, *40*, 59. (In Chinese)
29. Qin, Z.; Pang, X.; Chen, D.; Cheng, H.; Hu, X.; Wu, J. Evaluation of Chinese tea by the electronic nose and gas chromatography–mass spectrometry: Correlation with sensory properties and classification according to grade level. *Food Res. Int.* **2013**, *53*, 864–874. [CrossRef]
30. Li, Y.; Zhang, J.; Jia, H.; Pan, Y.; Xu, Y.; Wang, Y.; Deng, W. Metabolite analysis and sensory evaluation reveal the effect of roasting on the characteristic flavor of large-leaf yellow tea. *Food Chem.* **2023**, *427*, 136711. [CrossRef] [PubMed]
31. Su, D.; Xu, T.; Li, Y.; Zhou, H. Flavor evolution in raw Pu-erh tea during manufacturing using different processing types. *LWT* **2022**, *154*, 112905. [CrossRef]
32. Biasetton, N.; Disegna, M.; Barzizza, E.; Salmaso, L. A new adaptive membership function with CUB uncertainty with application to cluster analysis of Likert-type data. *Expert Syst. Appl.* **2023**, *213*, 118893. [CrossRef]
33. Tholl, D. Biosynthesis and biological functions of terpenoids in plants. *Adv. Biochem. Eng. Biotechnol.* **2015**, *148*, 63–106. [CrossRef]

34. Wen, S.; Sun, L.; Zhang, S.; Chen, Z.; Chen, R.; Li, Z.; Lai, X.; Zhang, Z.; Cao, J.; Li, Q.; et al. The formation mechanism of aroma quality of green and yellow teas based on GC-MS/MS metabolomics. *Food Res. Int.* **2023**, *172*, 113137. [CrossRef]
35. Lv, S.; Wu, Y.; Li, C.; Xu, Y.; Liu, L.; Meng, Q. Comparative analysis of Pu-erh and Fuzhuan teas by fully automatic headspace solid-phase microextraction coupled with gas chromatography-mass spectrometry and chemometric methods. *J. Agric. Food Chem.* **2014**, *62*, 1810–1818. [CrossRef] [PubMed]
36. Matsubara, E.; Morikawa, T.; Kusumoto, N.; Hashida, K.; Matsui, N.; Ohira, T. Subjective Effects of Inhaling Kuromoji Tea Aroma. *Molecules* **2021**, *26*, 575. [CrossRef] [PubMed]
37. Xu, J.; Zhang, Y.; Yan, F.; Tang, Y.; Yu, B.; Chen, B.; Lu, L.; Yuan, L.; Wu, Z.; Chen, H. Monitoring Changes in the Volatile Compounds of Tea Made from Summer Tea Leaves by GC-IMS and HS-SPME-GC-MS. *Foods* **2022**, *12*, 146. [CrossRef] [PubMed]
38. Shao, C.; Zhang, Y.; Lv, H.; Zhang, Z.; Zeng, J.; Peng, Q.; Zhu, Y.; Lin, Z. Aromatic profiles and enantiomeric distributions of chiral odorants in baked green teas with different picking tenderness. *Food Chem.* **2022**, *388*, 132969. [CrossRef] [PubMed]
39. Wu, H.; Chen, Y.; Feng, W.; Shen, S.; Wei, Y.; Jia, H.; Wang, Y.; Deng, W.; Ning, J. Effects of Three Different Withering Treatments on the Aroma of White Tea. *Foods* **2022**, *11*, 2502. [CrossRef] [PubMed]
40. Abbas, F.; Ke, Y.; Yu, R.; Yue, Y.; Amanullah, S.; Jahangir, M.M.; Fan, Y. Volatile terpenoids: Multiple functions, biosynthesis, modulation and manipulation by genetic engineering. *Planta* **2017**, *246*, 803–816. [CrossRef]
41. Zhang, W.; Liu, C.; Yang, R.; Zheng, T.; Zhao, M.; Ma, L.; Yan, L. Comparison of volatile profiles and bioactive components of sun-dried Pu-erh tea leaves from ancient tea plants on Bulang Mountain measured by GC-MS and HPLC. *J. Zhejiang Univ. Sci. B* **2019**, *20*, 563–575. [CrossRef]
42. Pang, X.; Yu, W.; Cao, C.; Yuan, X.; Qiu, J.; Kong, F.; Wu, J. Comparison of Potent Odorants in Raw and Ripened Pu-erh Tea Infusions Based on Odor Activity Value Calculation and Multivariate Analysis: Understanding the Role of Pile Fermentation. *J. Agric. Food Chem.* **2019**, *67*, 13139–13149. [CrossRef]
43. Du, L.; Li, J.; Li, W.; Li, Y.; Li, T.; Xiao, D. Characterization of volatile compounds of Pu-erh tea using solid-phase microextraction and simultaneous distillation–extraction coupled with gas chromatography–mass spectrometry. *Food Res. Int.* **2014**, *57*, 61–70. [CrossRef]
44. Shen, S.; Wu, H.; Li, T.; Sun, H.; Wang, Y.; Ning, J. Formation of aroma characteristics driven by volatile components during long-term storage of An tea. *Food Chem.* **2023**, *411*, 135487. [CrossRef]
45. Toci, A.; Farah, A. Volatile fingerprint of Brazilian defective coffee seeds: Corroboration of potential marker compounds and identification of new low quality indicators. *Food Chem.* **2014**, *153*, 298–314. [CrossRef] [PubMed]
46. Hu, C.; Li, D.; Ma, Y.; Zhang, W.; Lin, C.; Zheng, X.; Liang, Y.; Lu, J. Formation mechanism of the oolong tea characteristic aroma during bruising and withering treatment. *Food Chem.* **2018**, *269*, 202–211. [CrossRef] [PubMed]
47. Guo, X.; Ho, C.; Schwab, W.; Song, C.; Wan, X. Aroma compositions of large-leaf yellow tea and potential effect of theanine on volatile formation in tea. *Food Chem.* **2019**, *280*, 73–82. [CrossRef] [PubMed]
48. Su, D.; He, J.; Zhou, Y.; Li, Y.; Zhou, H. Aroma effects of key volatile compounds in Keemun black tea at different grades: HS-SPME-GC-MS, sensory evaluation, and chemometrics. *Food Chem.* **2022**, *373*, 131587. [CrossRef] [PubMed]
49. Liu, M.; Hu, C.; Lu, Y.; Zhang, H.; Zhang, Z. Research progress on aroma components of *Pu'er* tea. *Chin. Tea Process.* **2013**, *2*, 38–41. (In Chinese)
50. Wang, D.; Yoshimura, T.; Kubota, K.; Kobayashi, A. Analysis of glycosidically bound aroma precursors in tea leaves. 1. Qualitative and quantitative analyses of glycosides with aglycons as aroma compounds. *J. Agric. Food Chem.* **2000**, *48*, 5411–5418. [CrossRef]
51. Zhou, J.; Fang, T.; Li, W.; Jiang, Z.; Zhou, T.; Zhang, L.; Yu, Y. Widely targeted metabolomics using UPLC-QTRAP-MS/MS reveals chemical changes during the processing of black tea from the cultivar *Camellia sinensis* (L.) O. Kuntze cv. Huangjinya. *Food Res. Int.* **2022**, *162 Pt B*, 112169. [CrossRef]
52. Ho, C.; Zheng, X.; Li, S. Tea aroma formation. *Food Sci. Hum. Wellness* **2015**, *4*, 9–27. [CrossRef]
53. Wang, K.; Liu, F.; Liu, Z.; Huang, J.; Xu, Z.; Li, Y.; Yang, X. Comparison of catechins and volatile compounds among different types of tea using high performance liquid chromatograph and gas chromatograph mass spectrometer. *Int. J. Food Sci. Technol.* **2011**, *46*, 1406–1412. [CrossRef]
54. Klensporf, D.; Jeleń, H.H. Effect of heat treatment on the flavor of oat flakes. *J. Cereal Sci.* **2008**, *48*, 656–661. [CrossRef]

Disclaimer/Publisher's Note: The statements, opinions and data contained in all publications are solely those of the individual author(s) and contributor(s) and not of MDPI and/or the editor(s). MDPI and/or the editor(s) disclaim responsibility for any injury to people or property resulting from any ideas, methods, instructions or products referred to in the content.

foods

Article

GC-MS Combined with Proteomic Analysis of Volatile Compounds and Formation Mechanisms in Green Teas with Different Aroma Types

Xiaojun Niu †, Cun Ao †, Jizhong Yu, Yun Zhao and Haitao Huang *

Tea Research Institute, Hangzhou Academy of Agricultural Science, Hangzhou 310024, China; xiaojunwords@126.com (X.N.); aocun123@163.com (C.A.); jz310024@126.com (J.Y.); yunz2431@163.com (Y.Z.)
* Correspondence: hthuang309@hotmail.com
† These authors contributed equally to this study.

Abstract: Aroma is one of the key factors for evaluating the quality of green tea. A tender aroma (NX) and floral-like aroma (HX) are two types of high-quality aroma of green tea. In this work, the different aroma types of baked green tea were classified by sensory evaluation. Then, seven tea samples with a typical tender or floral-like aroma were selected for further volatile component analysis by GC-MS. A total of 43 aroma compounds were identified in two different aroma types of baked green tea samples. The PCA showed that linalool, geraniol, 3-hexenyl butyrate, and 3-hexenyl hexanoate were the major volatiles contributing to the HX. On the other hand, most of the alcohol volatiles, such as 1-octanol, 1-octen-3-ol, 1-dodecanol, 1-hexadecanol, phenylethyl alcohol, benzyl alcohol, aldehydes and some hydrocarbons contributed more to the NX. In addition, the chemical composition analysis showed that the content of free amino acids was higher in NX green tea samples, while the content of catechins was relatively higher in HX tea samples. A proteomic analysis revealed that most of the enzymes involved in VPBs pathways, such as phenylalanine ammonialyase, peroxidase, and shikimate-O-hydroxycinnamoyl transferase, were more abundant in NX than in HX tea samples. These results laid a foundation for the aroma formation mechanism of different aroma types of baked green tea and provided some theoretical guidance for the breeding of specific aroma varieties.

Keywords: baked green tea; tender aroma; floral-like aroma; volatile component; proteomic

1. Introduction

Tea is one of the world's most popular beverages. Based on the degree of fermentation by endogenous enzymes in the tea leaves, tea can be classified into three main categories, including green tea (non-fermented), oolong tea (semi-fermented), and black tea (fully fermented) [1]. Of these, green tea is also the most widely consumed tea in most East Asian countries (especially China) due to its delightful flavor, many health benefits, and a long cultural history. In 2022, the tea plantation area in China was approximately 3.2641 million ha, with a yield of 3.1810 million tonnes, of which green tea (1.8538 tonnes) accounted for 58.3% of the total output. Meanwhile, its production value accounts for 64.71% of the total tea production in China [2]. According to different manufacturing processes, green tea can generally be divided into four categories: baked green tea, roasted green tea, steamed green tea, and sun-dried green tea [3]. There are significant differences in the flavor and other quality characteristics of different categories of green tea due to the tea plant variety, growing conditions, and processing technology.

Aroma is one of the key factors in judging the quality and character of tea, and it also influences consumer choice. The type of aroma characteristics of green tea can be classified into various aroma types, such as clean and refreshing (Chinese: Qingxiang, QX), chestnut-like (Chinese: Lixiang, LX), tender (Chinese: Nenxiang, NX), and floral-like

(Chinese: Huaxiang, HX) aromas, among others, categories based on sensory characteristics [4–6]. It is well known that the aroma type of tea is a comprehensive effect of numerous aroma components, which is determined by its chemical composition. The formation of different aroma types is mainly due to the comprehensive effects of different types and concentrations of aroma compounds on the human olfactory nerves.

Recently, over 700 volatile components have been identified from various types of tea, including alcohols, aldehydes, ketones, esters, lactones, acids, phenols, heterocyclic compounds, sulfur compounds, nitrogen compounds, and hydrocarbons, among others [7]. Although over 300 aroma components have been identified in green tea, not all of them are involved in the formation of tea aroma quality and tea aroma type. Only a few key aroma components play an important role, such as alcohols (linalool and its oxides, geraniol, nerolidol, etc.), aldehydes (decanal, hexanal, nonanal, etc.), ketones (1-octen-3-one, β-ionone, etc.), esters (capro-cis-3-hexelate, methyl salicylate, methyl jasmonate, etc.) and so on [3,8,9]. Therefore, the identification of the key volatile compounds in different flavors of green tea is particularly relevant.

During the past decade, gas chromatography–mass spectrometry (GC-MS) has become the most widely used analytical technique for the detection of volatile components in teas and other foods, due to its rapid qualitative and accurate quantitative capabilities [10–12]. Moreover, the methodology of headspace solid-phase microextraction combined with gas chromatography–mass spectrometry (HS-SPME/GC-MS) has also been widely applied for the analysis of volatile and semi-volatile compounds in environmental, food, and biological samples, due to its ease of operation and good extraction performance [13]. Numerous aroma compounds have been identified using this technology, to elucidate the formation of different types of tea aroma in green tea [7,8,14,15]. In recent years, omics technology has attracted researchers in the field of food and nutrition, because it can reveal complex biochemical processes and regulatory mechanisms in living organisms. It has been widely used in the field of food flavor analysis [16–18]. Moreover, the use of multi-omics combined analysis technology to reveal the flavor of tea has also become an effective method [19]. The aim of this work was to elucidate the major volatile compounds contributing to the specific aromas of baked green tea, including NX and HX. Firstly, the aroma type classification of tea samples made from different varieties was carried out through sensory evaluation. Then, the volatile components, chemical compositions, and the differences of related metabolic pathway enzymes of tea samples with different aroma types were identified using GC-MS, PCA, and label-free proteomic techniques.

2. Materials and Methods

2.1. Tea Samples

A total of 26 tea samples were used in this study. Fresh tea leaves were picked based on the picking standards of a bud with one leaf in the spring of 2017 (10 samples) and 2019 (16 samples) at the experimental base of Daqing valley, the Tea Institute, Hangzhou Academy of Agricultural Sciences. Then, the manufacturing process of baked green tea samples was performed according to the following process: firstly, the fresh tea leaves were withered for 4–6 h, and then they were fixed by a flat-shape tea-fixing machine to terminate the activities of the endogenous enzymes at 200–230 °C for about 2.5 min. They were allowed to cool and rolled manually for 10 min. Finally, they were dried in a dryer at 100 °C for 12 min and then at 80 °C until the moisture content was about 5%. Some of the fresh tea leaves were freeze-dried in liquid nitrogen for proteomic analysis, and the rest of the fresh tea leaves were processed into the corresponding baked green tea samples using a consistent processing method. The fresh leaves and dried tea samples were stored in a refrigerator at −80 °C and −20 °C, respectively, until further analysis.

2.2. Sensory Evaluation

The sensory evaluation was carried out at the Sensory Analysis Unit of the Tea Institute, Hangzhou Academy of Agricultural Sciences, China. The aroma characteristics of the tea samples were described based on the Chinese national standard in "Methodology of sensory evaluation of tea" (GB/T 23776—2018) by three well-trained tea experts. The specific steps of the sensory evaluation are mainly based on the description of Zhu et al. [20]. The results of the sensory evaluation were based on the principle of minority rule.

2.3. Chemicals and Standards

Folin phenol, anhydrous sodium carbonate, potassium dihydrogen phosphate, and disodium hydrogen phosphate were purchased from Sinopharm (Shanghai, China). Acetic acid, acetonitrile, ethylenediaminetetraacetic acid disodium salt, and L-ascorbic acid were purchased from Merck KGaA, Darmstadt, Germany. Catechin monomers and theanine standards were purchased from Sigma-Aldrich (Shanghai, China). Distilled water was purchased from the Wahaha Group Company (Hangzhou, China). Solid-phase microextraction (SPME) fiber (CAR-DVB-PDMS), 50/30 μm, with a length of 2 cm, was purchased from Supelco (Bellefonte, PA, USA). Standard samples of catechins, including catechin (C), epicatechin (EC), gallocatechin (GC), epigallocatechin (EGC), catechin gallate (CG), epicatechin gallate (ECG), epigallocatechin gallate (GCG), epigallocatechin gallate (EGCG), and caffeine (CAF), were purchased from Sigma-Aldrich (Shanghai, China).

2.4. Chemical Composition Analyses

An estimation of the total free amino acid content in the tea leaves was determined using the ninhydrin method described by Chen et al. [21]. The total polyphenol content was determined by the Folin colorimetric method described by Teng et al. [22]. Catechin components and caffeine were estimated using a high-performance liquid chromatography (HPLC, Agilent 1100, Santa Clara, CA, USA) system equipped with a Hypsial ODS C18 column (4.6 mm × 250 mm, 5 μm). The detection wavelength and column temperature were set at 280 nm and 35 °C, respectively. Mobile phase A (acetic acid–deionized water, 1:49) and B (acetonitrile 100%) were run in linear gradients, with A decreasing from 93.5% to 85% in the first 25 min, A decreasing from 85% to 75% in the next 5 min, and then A increasing from 75% to 93.5% in 3 min and maintained for 2 min. Each injection volume was 10 μL, and the flow rate was 1.0 mL/min. The content of the different components was quantified based on the peak area using the external standard method. The external standard samples included C, EC, GC, EGC, CG, ECG, GCG, EGCG, and CAF. The experimental data were calculated using three independent replicates.

2.5. Determination of Volatile Compounds

2.5.1. Extraction of Aroma Compounds

Accurately, 3 g of tea sample was weighed and placed in a 250 mL conical flask, and 150 mL boiling water was added. Subsequently, the conical flask was sealed with tin foil and immersed in a 70 °C water bath for 10 min, after which the tea broth was filtered with 4 layers of gauze and transferred to another conical flask. After cooling to room temperature, 10 mL of tea broth was removed from the conical flask and transferred to a 20 mL headspace flask. Then, 5 μL of the internal standard (ethyl decanoate 25 mg/L) and 2 g of sodium chloride was added into the headspace flask and subsequently shaken. The headspace bottle cap was immediately tightened and placed in a thermostatized water bath to equilibrate for 10 min at 60 °C, and then the SPME fiber was exposed to the sample vial headspace for 40 min.

2.5.2. GC-MS Analysis

A TRACE™ 2000 gas chromatograph equipped with a Trace DSQ mass spectrometer (Thermo Finnigan, San Jose, CA, USA) was used. Desorption was performed on a GC analyzer in splitless mode at 250 °C for 5 min. Volatile separation was performed using an HP-INNOWAX column (30 m × 0.32 mm × 0.50 μm, Agilent, USA). The oven temperature programmer was initially set at 50 °C (held for 5 min), then ramped at 3 °C/min to 210 °C (held for 5 min), then increased at a rate of 15 °C/min to 230 °C (held for 5 min). The carrier gas was helium (>99.999%) at a constant flow rate of 1.0 mL/min. Mass spectrometer conditions: Ion source temperature, 250 °C. Interface temperature, 230 °C. The mass spectrometer was run in a scanning range from 35 to 540 m/z at 70 eV.

2.5.3. Identification and Quantification of Aroma Compounds

Volatile compounds were tentatively identified by comparing their mass spectra with the NIST mass spectrometry library, retention index (RI), and by comparing the peak type and peak emergence time with data available in the literature [23]. The relative content of volatile compounds was calculated using the formula of A_i/A total, where A_i represents the peak area of each volatile compound and A represents the total peak area of all volatile compounds in a single sample.

2.6. *Proteomic Analysis*

2.6.1. Protein Extraction

Total protein was extracted as described by Wisniewski et al. with slight modifications [24]. About 0.1 g of frozen fresh leaves was ground to powder using liquid nitrogen. The tissue was transferred to 2 mL Eppendorf tubes and kept frozen, then 1.5 mL protein extraction solution I (40 g trichloroacetic acid dissolved in 400 mL protein extraction solution II) was added to the tube, extracted at −20 °C for 1 h, shaken 5~6 times uniformly, and spun at 13,000 rpm at 4 °C for 20 min. The supernatant was removed and 1.5 mL protein extraction solution II (500 mL acetone and 350 μL β-mercaptoethanol) was added, extracted at −20 °C for 1 h, and spun at 13,000 rpm at 4 °C for 20 min. The above steps were repeated twice. The precipitates were washed twice with 90% ice-cold acetone and dried under vacuum to obtain dry protein powder (the drying time is about half an hour). Approximately 20 mg of protein powder was solubilized in protein lysate solution (8 M urea and 4% sodium dodecyl sulphate (SDS)) at room temperature for about 1 h, and shaken 5~6 times evenly during this time. After centrifugation at 13,000 rpm for 30 min, the protein supernatants were filtered through a 0.22 μm membrane. The final protein filtrates were quantified using a bovine serum albumin protein assay kit (Thermo Fisher Scientific, Waltham, MA, USA) and determined by SDS–polyacrylamide gel electrophoresis (SDS-PAGE). Proteins were stored at −80 °C until further analysis.

2.6.2. Protein Digestion and Deionization

A total of 150 μg of protein samples was placed in a 1.5 mL EP tube, and 100 mM NH_4HCO_3 was added to make it up to 100 μL. Then, 11 μL of 100 mM dithiothreitol (DTT) was added to a final concentration of 10 mM, gently vortexed, and incubated in a metal bath at 37 °C for 1 h. Then, 12 μL of 500 mM iodoacetamide (IAA) was added to the tube. Samples were incubated at room temperature for 30 min in the dark and transferred to an ultrafiltration filter (10 kD, Sartorius; Göttingen, Germany), centrifuged at 12,000× g for 15 min at 4 °C, and the filtrates were discarded. A total of 150 μL of 100 mM NH_4HCO_3 was added to the ultrafiltration and centrifuged twice at 12,000× g for 15 min at 4 °C. Finally, proteins were digested with 6 μg trypsin (protein–trypsin = 25:1) in 100 μL 100 mM NH_4HCO_3 at 37 °C for 12–16 h. After trypsin digestion, the peptide segments were dried using a vacuum pump.

2.6.3. Peptide Desalination and Quantification

Peptides were eluted with 5% trifluoroacetic acid (TFA) solution, until the final TFA concentration was 0.1–1%. The peptides were then desalted on C18 filter cartridges (EmporeTM SPE Cartridges C18, 7 mm internal bed diameter, 3 mL volume, Sigma, St. Louis, MO, USA) and washed three times with 0.1% formic acid (FA). The wash components were collected by vacuum centrifugation and a step gradient elution was started. The lyophilized sample was redissolved in 20 μL of 0.1% FA, and the peptide concentration was measured by OD_{280}.

2.6.4. Liquid Chromatography–Mass Spectrometry Analysis

QExactive HF-X mass spectrometry, combined with Easy nLC (Thermo Fisher, USA), was used for the LC-MS/MS analysis. The detailed steps of the LC-MS/MS analysis are based on the report by Chen et al. [25].

Mass spectrometry data analysis: Raw file data were analyzed using Maxquant software (version 1.6.0.1) for label-free qualitative and quantitative analysis, and protein databases of species such as Prunus, Rosaceae, and green plants were searched. All differentially expressed proteins (DEPs) were mapped to the Gene Ontology (GO) database (http://www.geneontology.org/, accessed on 4 April 2024) for functional annotation. A Kyoto Encyclopedia of Genes and Genomes (KEGG) pathway enrichment analysis (http://www.genome.jp/kegg/, accessed on 4 April 2024) was performed using the KEGG Automatic Annotation Server.

2.7. Statistical Analysis

Statistical analysis was performed by SPSS software (version 19.0) and Microsoft Excel 2019. The measured content of the aroma compounds was expressed as mean ± standard deviation (SD). Student's *t*-test was used to determine the differences between the different tea samples at a 5% and 1% significance level, respectively. Principal component analysis was performed using Origin 8.0 (MicroCal, Northampton, MA, USA). Proteomic data analysis was performed using the Meiji Biological Cloud platform (https://cloud.majorbio.com/, accessed on 4 April 2024).

3. Results and Discussion

3.1. Sensory Evaluation Results for the Baked Green Tea Samples

The tea plant variety is one of the key factors determining the aroma of tea. As shown in Table 1, the different tea samples exhibited different aroma characteristics (such as high brisk, tender aroma, clean aroma, and floral-like, etc.) when using the same manufacturing process. Four of the tea samples showed a typical tender aroma, namely, Fuding, Lifeng, 741103, and 741006. Three of the tea samples, 730206, 730120, and 741209, had a typical floral-like aroma. These seven tea samples were selected for further aroma component analysis.

Table 1. Aroma sensory evaluation results of baked green teas in 2017.

NO.	Cultival	Aroma Characteristics
1	Fuding	High brisk, with tender aroma
2	Lifeng	Clean and refreshing, tender aroma
3	741103	High brisk, obvious tender aroma
4	741006	Obvious tend aroma and tip aroma
5	J1130	Clean and refreshing, fresh and brisk
6	730206	Fragrant and lasting
7	730120	With flowery aroma
8	741209	Obvious flowery aroma
9	XGL16-14	High brisk
10	740209	High brisk, with tip aroma

3.2. Detected Volatile Compounds in Baked Green Teas by GC-MS

The volatile compounds in each tea sample were analyzed by HS-SPME/GC-MS, and the major volatile components and content are presented in Figure 1 and Table S1. A total of 43 volatile compounds were detected and tentatively identified, including 9 alcohols, 5 aldehydes, 4 ketones, 14 alkanes, 10 esters, and 1 heterocyclic compound. As shown in Figure 1A,B, the composition of volatile compounds in two aroma types of baked green tea was similar, but the aroma content differed slightly. Alcohols were the richest volatile compounds in NX green tea samples (average of 36%), followed by esters (average of 30%). In HX green tea samples, esters were the most abundant volatile compounds (average of

40%), followed by alcohols (average of 36%). The total content of alcohols and esters was more than 60% in the two aroma types of baked green tea, but the aldehydes, ketones, and heterocyclic compounds were relatively low. In addition, there was a significant difference in the relative content of hydrocarbons between the two different aroma types of baked green tea (Figure 1C).

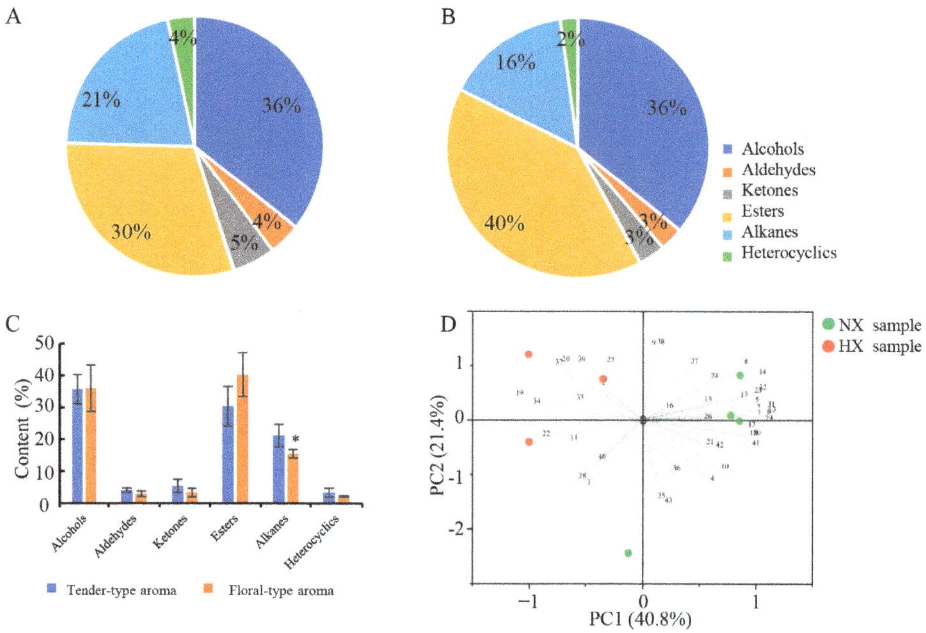

Figure 1. Analysis of aroma components in different aroma-type tea samples. (**A**) Proportion of aroma components in NX tea samples. (**B**) Proportion of aroma components in HX tea samples. (**C**) The relative content of aroma components in two different aroma-types of tea samples. * Indicates significant difference ($p < 0.05$) in t-test. (**D**) Principal component analysis of aroma components in two different aroma-types of tea samples. The number represents the code of aroma components.

As shown in Table S1, there were significant differences in the relative content of different volatile compounds in the baked green tea samples. Among them, linalool had the highest content of tea aroma compounds in these tea samples, while there was no significant difference between the two types of tea samples. The content of linalool in HX tea samples was slightly higher than that in NX tea samples. Linalool is a colorless liquid with a strong grassy, sweet woody, floral, and citrus-like aroma. Several previous studies have revealed linalool as a key volatile of specific aroma types in green tea [20,26–28]. 3-hexenyl hexanoate had the second highest content of tea aroma components in HX tea samples; the average content ranged from 8.59 to 16.07%, significantly higher than in NX tea samples (average of 5.68 to 9.11%). Xue et al. reported that 3-hexenyl hexanoate contributes to the formation of the floral flavor in black tea [29]. In addition, 1-dodecanol, nonanal, and many esters also promoted the formation of the unique floral aromas in green teas [30–32].

In recent years, there have been few studies on the mechanism of aroma formation in NX green tea. Zhou et al. found that the tender aroma characteristics of Longjing tea were significantly positively correlated with the content of compounds such as styrene, hexadecane, and 3-methoxy-1,2-propanediol in the aroma components [33]. Zheng et al. found that linalool oxide, 3-hexenyl butyrate, and 3-hexenyl isovalerate can be used as characteristic volatile components of tender-aroma green tea [34]. In this study, 3-hexenyl butyrate and 3-hexenyl hexanoate showed significant differences in the two different aroma types of green tea. However, the levels of these two aroma compounds were higher in the HX tea samples than in the NX tea samples. Furthermore, there was no significant difference in the content of hexadecane between the two different-flavored tea samples, but its content was higher in the NX tea samples than in the HX tea samples.

In addition, some low-content aroma compounds such as α-cubebene also showed significant differences in the two different aroma types of baked green tea (Table S1). These results indicate that these low levels of aroma compounds may play a key role in distinguishing different aroma types of tea samples. Previous studies have also reported that these low-content volatile compounds also play a crucial role in the formation of green tea aroma [28,35–37]. Therefore, the formation of tea aroma is not only related to the content of aroma compounds and aroma components, but also the odor threshold values of each determined compound, as well as their odor activity value (OAV), were also very important factors.

3.3. PCA

To explore the differences in aroma compounds in different aroma types of baked green tea, a principal component analysis (PCA) was performed on the content of aroma components. As shown in Figure 1D, there was a clear separation in the different aroma types of the tea samples, indicating that the volatile profile of the NX tea sample was quite different from that of the HX tea sample. The first principal component (PC1) accounted for 40.8%, and the second principal component (PC2) accounted for 21.4%. As shown in Figure 1D and Table S1, 1-octanol, 1-octen-3-ol, 1-dodecanol, 1-hexadecanol, phenylethyl alcohol, benzyl alcohol, nonanal, decanal, β-cyclocitral, β-ionone, hexadecane, 3-hexenyl isovalerate, etc., mainly contributed to PC1. Meanwhile, linalool, geraniol, benzaldehyde, 3-hexenyl butyrate, 3-hexenyl hexanoate, 3-hexen-1-ol acetate, 3-hexenyl-3-hexenoate, hex-3-enyl-2-methylbut-2-enoate, 3-carene, d-cadinene, α-cubebene, and α-muurolene contributed strongly to PC2. PC2 contains mainly HX-type samples. Liu et al. suggested that 1-hexanol, linalool oxide, linalool, geraniol, β-ionone, isoamyl acetate, and 2-methylpropanal were significantly enriched in flowery-type tea and contributed to the floral-like aroma [14]. β-ionone, linalool, geraniol, methyl salicylate, and other components may jointly contribute to the floral aroma of baked green tea [28]. Wang et al. reported that volatile components such as linalool, furfural, β-ionone, hexanoic acid-3-hexene ester, ethylbenzene, naphthalene, and 2-n-pentylfuran made important contributions to the aroma quality of fresh-scented green tea [38]. Our results also showed that linalool, geraniol, 3-hexenyl butyrate, and 3-hexenyl hexanoate were important volatiles contributing to a floral-like aroma. In addition, according to PC1, most of the alcohol aroma components, such as 1-octanol, 1-octen-3-ol, 1-dodecanol, 1-hexadecanol, phenylethyl alcohol, benzyl alcohol, aldehydes, and some hydrocarbons, contributed more to the tender aroma.

3.4. *Chemical Compositions Detected in Different Flavors of Baked Green Tea*

To investigate whether there were differences in the chemical composition of different flavors of baked green tea, the chemical composition content was analyzed. As shown in Figure 2 and Table S2, the content of total amino acids in NX tea samples (average of 4.42 to 5.15%) was significantly ($p < 0.05$) higher than that in HX tea samples (average of 3.15 to 4.30%). Meanwhile, the content of tea polyphenols in NX tea samples (average of 17.72 to 20.46%) was lower in HX tea samples (average of 23.17 to 23.59%). In addition, there were significant differences in the levels of total catechins, simple catechins, and ester-type catechins between the two different aroma types of baked green tea. The levels of catechins, simple catechins, and ester-type catechins in the HX samples were higher than those in the NX samples. Amino acids are thought to play a major role in making green tea taste refreshing. During tea processing, amino acids may also undergo decarboxylation or deamination reactions, or Maillard reactions with sugars, to generate new flavors and aromas [39]. In addition, the content of catechins, especially epigallocatechin gallate and epigallocatechin, play a key role in the taste and aroma quality of green tea [40]. Therefore, our results indicated that tea varieties with a high content of free amino acids are suitable for the production of NX green tea, while tea varieties with a high content of catechins are suitable for the production of HX green tea.

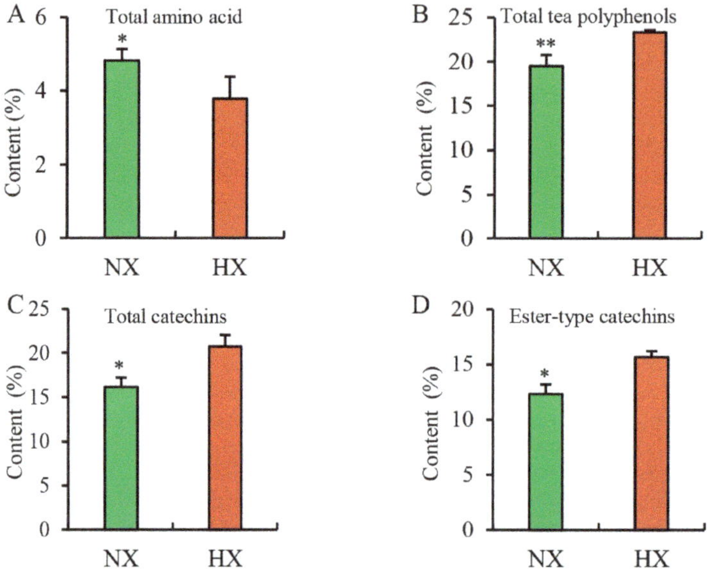

Figure 2. The relative content of representative chemical compositions in two different aroma types of baked green tea samples. * and ** indicate significant difference ($p < 0.05$) and extremely significant difference ($p < 0.01$) in *t*-test, respectively. (**A**) Total amino acid; (**B**) Total tea polyphenols; (**C**) Total catechins; (**D**) Ester-type catechins.

3.5. Proteomic Analysis

Sixteen tea samples exhibiting a typical tender aroma and floral-like aroma were selected for proteomic analysis in 2019 (Table S3). A total of 4911 related proteins were identified in this study, of which 4652 proteins were detected in two different aroma types of baked green tea samples (Figure 3A). Fold modification values larger than 2.0 times and a p value < 0.05 (from Student's *t*-tests) were used to identify differentially expressed proteins (DEPs). As shown in Figure 3B and Table S4, 365 proteins showed significant differences between NX tea samples and HX tea samples, including 117 up-regulated and 248 down-regulated proteins.

A bar graph of the top 10 GO enrichment results is shown in Figure 4A and Table S5. The results showed that the defense response, protein phosphorylation, reactive oxygen species metabolic process, cell wall macromolecule metabolic process, and carbohydrate-derivative catabolic process were significantly enriched terms of biological processes. Lyase activity, UDP-glycosyl transferase activity, carbon-oxygen lyase activity, and chitin binding were mainly involved under the molecular functions category. In addition, transferase complexes, transferring phosphorus-containing groups, were significantly enriched proteins under the cellular component category. Among these, the most dominant categories were lyase activity, UDP-glycosyltransferase activity, and the defense response. Moreover, as shown in Figure 4B and Table S5, the top 10 enriched KEGG pathways of these DEPs were mainly involved in sesquiterpenoid and triterpenoid biosynthesis; phenylpropanoid biosynthesis; glycerophospholipid metabolism; alpha-linolenic acid metabolism; ether lipid metabolism; snare interactions in vesicular transport; amino sugar and nucleotide sugar metabolism; terpenoid backbone biosynthesis; arachidonic acid metabolism and the anthocyanin biosynthesis signaling pathway. Among these pathways, sesquiterpenoid and triterpenoid biosynthesis, phenylpropanoid biosynthesis, and glycerophospholipid metabolism-related pathways were most enriched. Sesquiterpenes, terpenes, and their derivatives are important volatile secondary metabolites in plants; these compounds often play a decisive role in determining the type of tea aroma [41,42]. A recent study also showed a positive correlation between tea tree varieties and linalool content [43]. As shown in Table S5, three DEPs were directly involved in the metabolic pathways of ko00909 (sesquiterpenoid and triterpenoid biosynthesis), ko00100 (steroid biosynthesis) and ko01110 (biosynthesis of secondary metabolites), including GERD, FDFT1, and LUP4.

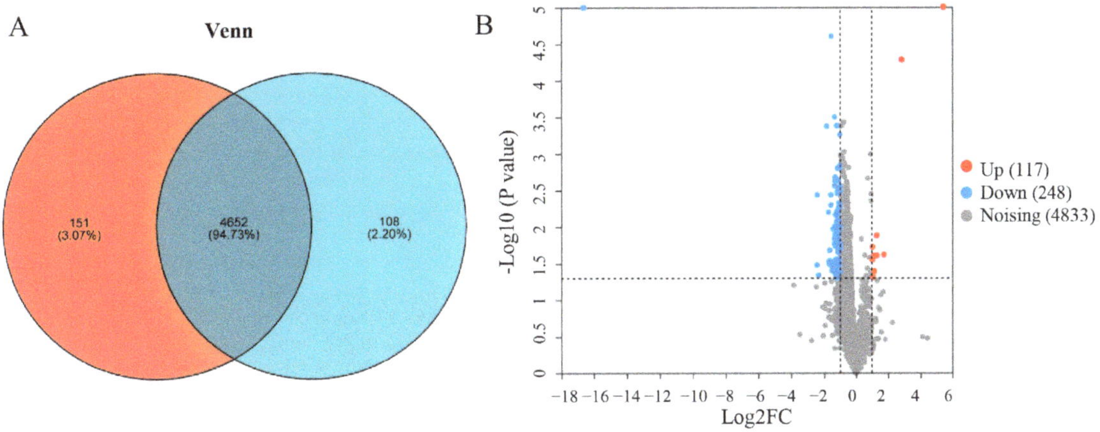

Figure 3. The number of proteins identified in the study. (**A**) The number of proteins identified in different aroma types of tea samples. (**B**) DEPs as a volcano plot in different aroma types of tea samples.

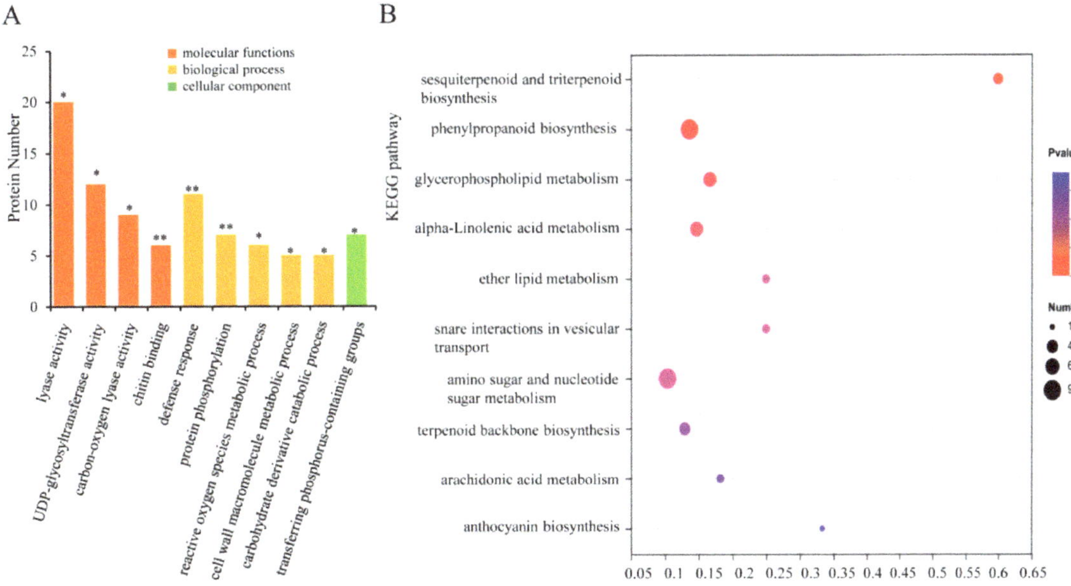

Figure 4. GO (**A**) and KEGG (**B**) analysis of differentially expressed proteins. * and ** indicate significant difference ($p < 0.05$) and extremely significant difference ($p < 0.01$), respectively.

Moreover, volatile phenylpropanoids/benzenoids (VPBs) are also important contributors to the formation of tea aroma [44]. Research has found that most VPBs belong to the shikimic acid metabolism pathway. The first step, which plays a crucial role in this metabolic pathway, is the formation of cinnamic acid from L-phenylalanine, catalyzed by phenylalanine lyase [45]. As shown in Table S5, nine DEPs were involved in phenylpropanoid biosynthesis, including ko00360 (phenylalanine metabolism), ko00940 (phenylpropanoid biosynthesis), ko00941 (flavonoid biosynthesis) and ko00945 (stilbenoid, diarylheptanoid, and gingerol biosynthesis). Most of the enzymes involved in these pathways, such as phenylalanine ammonia-lyase, peroxidase, shikimate-O-hydroxycinnamoyl transferase, and coniferyl aldehyde dehydrogenase, were more abundant in NX tea samples than in floral HX tea samples. This suggests that the high expression of enzymes involved in phenylpropanoid biosynthesis may contribute to the release of high-intensity aroma compounds. The major phenylpropanoids in tea are phenylethanol and phenylacetaldehyde [46]. As shown in Table S1, the relative content of phenylethanol was higher in NX tea samples (average of 2.59%) than in HX tea samples (average of 0.54%). Therefore, the expression levels of the above enzymes in fresh tea leaves can be used as indicators in future work for breeding specific aroma varieties.

4. Conclusions

In this study, a total of 26 tea varieties/samples were processed into baked green teas using a consistent process, and the samples were subjected to sensory evaluation. From these, seven tea samples with typical aromas were selected for further aroma component analysis. The result of the PCA showed that linalool, geraniol, 3-hexenyl butyrate, and 3-hexenyl hexanoate were the major volatiles contributing to the floral-like aroma. Meanwhile, the major alcohol aroma components such as 1-octanol, 1-octen-3-ol, 1-dodecanol, 1-hexadecanol, phenylethyl alcohol, benzyl alcohol, aldehydes, and partly hydrocarbons contributed more to the tender aroma. Furthermore, it was found that tea varieties with a high content of free amino acids may be suitable for the production of NX baked green tea, while tea varieties with high levels of catechins are suitable for the production of HX baked green tea. The proteomic analysis showed that most of the enzymes involved in VPBs pathways, such as phenylalanine ammonia-lyase, peroxidase, and shikimate O-hydroxycinnamoyl transferase, were higher in NX tea samples than in HX tea samples. Therefore, the expression levels of these enzymes in fresh tea leaves can be used as indicators in future work to breed specific aroma varieties.

Supplementary Materials: The following supporting information can be downloaded at: https://www.mdpi.com/article/10.3390/foods13121848/s1, Table S1: Aroma components and content in different aroma-type baked green tea samples; Table S2: Analysis of chemical compositions in different aroma-type baked green tea samples; Table S3: Aroma sensory evaluation results of different baked green teas in 2019; Table S4: Differential proteins in two groups of baked green tea with different aroma types. Table S5: GO and KEGG pathway analysis of differentially expressed proteins.

Author Contributions: Conceptualization, J.Y. and H.H.; methodology, X.N. and C.A.; software, X.N.; datal analysis, C.A. and X.N.; investigation, J.Y. and Y.Z.; resources, J.Y. and H.H.; writing—original draft preparation, X.N. and C.A.; writing—review and editing, C.A. and X.N. All authors have read and agreed to the published version of the manuscript.

Funding: This research was funded by the Zhejiang Science and Technology Major Program on Agricultural New Variety Breeding—Tea Plant (2021C02067-4), the National Tea Industry Technology System (CARS-19), and the Science and Technology Innovation, Demonstration and Promotion Fund of Hangzhou Academy of Agricultural Sciences (2022HNCT-04).

Data Availability Statement: The original contributions presented in the study are included in the article, further inquiries can be directed to the corresponding authors.

Conflicts of Interest: The authors declare no conflicts of interest.

References

1. Baba, R.; Kumazawa, K. Characterization of the potent odorants contributing to the characteristic aroma of Chinese green tea infusions by aroma extract dilution analysis. *J. Agric. Food Chem.* **2014**, *62*, 8308–8313. [CrossRef] [PubMed]
2. Mei, Y.; Zhang, S. Analysis of China's tea production and domestic sales in 2022. *China Tea* **2023**, *45*, 25–30.
3. You, Q.; Li, Q.; Zhu, Y.; Shi, Y.; Ma, W.; Lin, Z.; Lu, H. Research progress of aroma components in green tea with different aroma types and their influencing factors. *China Tea* **2022**, *44*, 7–18.
4. Ye, G.; Yuan, H.; Jiang, Y.; Yin, J.; Wang, F.; Chen, J. Application of Bayes Stepwise Discrimination Analysis on Chemical Recognition of Green Tea with Chestnut-like Aroma. *J. Tea Sci.* **2009**, *29*, 27–33.
5. Yin, H.; Yang, Y.; Yao, Y.; Zhang, M.; Wang, J.; Jiang, Y.; Yuan, H. Discrimination of Different Characteristics of Chestnut-like Green Tea Based on Gas Chromatography-Mass Spectrometry and Multivariate Statistical Data Analysis. *Food Sci.* **2019**, *40*, 192–198.
6. Wang, H.; Hua, J.; Jiang, Y.; Yang, Y.; Wang, J.; Yuan, H. Influence of fixation methods on the chestnut-like aroma of green tea and dynamics of key aroma substances. *Food Res. Int.* **2020**, *136*, 109479. [CrossRef] [PubMed]
7. Zeng, L.; Watanabe, N.; Yang, Z. Understanding the biosyntheses and stress response mechanisms of aroma compounds in tea (*Camellia sinensis*) to safely and effectively improve tea aroma. *Crit. Rev. Food Sci. Nutr.* **2019**, *59*, 2321–2334. [CrossRef] [PubMed]
8. Wang, M.; Ma, W.; Shi, J.; Zhu, Y.; Lin, Z.; Lv, H. Characterization of the key aroma compounds in Longjing tea using stir bar sorptive extraction (SBSE) combined with gas chromatography-mass spectrometry (GC-MS), gas chromatography-olfactometry (GC-O), odor activity value (OAV), and aroma recombination. *Food Res. Int.* **2020**, *130*, 108908. [CrossRef] [PubMed]
9. Yang, P.; Yu, M.; Song, H.; Xu, Y.; Lin, Y.; Granvogl, M. Characterization of key aroma-active compounds in rough and moderate fire Rougui Wuyi rock tea (Camellia sinensis) by sensory-directed flavor analysis and elucidation of the influences of roasting on aroma. *J. Agric. Food Chem.* **2022**, *70*, 267–278. [CrossRef]
10. Xiao, Y.; Huang, Y.; Chen, Y.; Xiao, L.; Zhang, X.; Yang, C.; Li, Z.; Zhu, M.; Liu, Z.; Wang, Y. Discrimination and characterization of the volatile profiles of five Fu brick teas from different manufacturing regions by using HS–SPME/GC–MS and HS–GC–IMS. *Curr. Res. Food Sci.* **2022**, *5*, 1788–1807. [CrossRef]
11. Yin, X.; Huang, J.A.; Huang, J.; Wu, W.; Tong, T.; Liu, S.; Zhou, L.; Liu, Z.; Zhang, S. Identification of volatile and odor-active compounds in Hunan black tea by SPME/GC-MS and multivariate analysis. *LWT-Food Sci. Technol.* **2022**, *164*, 113656. [CrossRef]
12. Ma, L.; Sun, Y.; Wang, X.; Zhang, H.; Zhang, L.; Yin, Y.; Wu, Y.; Du, L.; Du, Z. The characteristic of the key aroma-active components in white tea using GC-TOF-MS and GC-olfactometry combined with sensory-directed flavor analysis. *J. Sci. Food Agric.* **2023**, *103*, 7136–7152. [CrossRef] [PubMed]
13. Sílvia, P.; Marcelo, M.; Ivonne, D.; Manuel, A.; Sílvia, M. Sesquiterpenic composition of the inflorescences of Brazilian chamomile (*Matricaria recutita* L.): Impact of the agricultural practices. *Ind. Crops Prod.* **2011**, *34*, 1482–1490.
14. Liu, N.; Shen, S.; Huang, L.; Deng, G.; Wei, Y.; Ning, J.; Wang, Y. Revelation of volatile contributions in green teas with different aroma types by GC-MS and GC-IMS. *Food Res. Int.* **2023**, *169*, 112845. [CrossRef] [PubMed]
15. Xie, J.; Wang, L.; Deng, Y.; Yuan, H.; Zhu, J.; Jiang, Y.; Yang, Y. Characterization of the key odorants in floral aroma green tea based on GC-E-Nose, GC-IMS, GC-MS and aroma recombination and investigation of the dynamic changes and aroma formation during processing. *Food Chem.* **2023**, *427*, 136641. [CrossRef] [PubMed]
16. Yao, L.; Yu, Q.; Huang, M.; Hung, W.; Grosser, J.; Chen, S.; Wang, Y.; Gmitter, F. Proteomic and metabolomic analyses provide insight into the off-flavour of fruits from citrus trees infected with 'Candidatus Liberibacter asiaticus'. *Hortic. Res.* **2019**, *6*, 31. [CrossRef] [PubMed]
17. Chen, D.; Sun, Z.; Gao, J.; Peng, J.; Wang, Z.; Zhao, Y.; Lin, Z.; Dai, W. Metabolomics combined with proteomics provides a novel interpretation of the compound differences among Chinese tea cultivars (*Camellia sinensis* var. *sinensis*) with different manufacturing suitabilities. *Food Chem.* **2022**, *377*, 131976. [CrossRef] [PubMed]
18. Li, Z.; Zhou, B.; Zheng, T.; Zhao, C.; Shen, X.; Wang, X.; Qiu, M.; Fan, J. Integrating metabolomics and proteomics technologies provides insights into the flavor precursor changes at different maturity stages of arabica coffee cherries. *Foods* **2023**, *12*, 1432. [CrossRef] [PubMed]
19. Wu, L.; Wang, Y.; Liu, S.; Sun, Y.; Li, C.; Lin, J.; Wei, S. The stress-induced metabolites changes in the flavor formation of oolong tea during enzymatic-catalyzed process: A case study of Zhangping Shuixian tea. *Food Chem.* **2022**, *391*, 133192. [CrossRef]
20. Zhu, Y.; Lv, H.; Shao, C.; Kang, S.; Zhang, Y.; Guo, L.; Dei, W.; Tan, J.; Peng, Q.; Lin, Z. Identification of key odorants responsible for chestnut-like aroma quality of green teas. *Food Res. Int.* **2018**, *108*, 74–82. [CrossRef]
21. Chen, Y.; Fu, X.; Mei, X.; Zhou, Y.; Cheng, S.; Zeng, L.; Dong, F.; Yang, Z. Proteolysis of chloroplast proteins is responsible for accumulation of free amino acids in dark-treated tea (*Camellia sinensis*) leaves. *J. Proteom.* **2017**, *157*, 10–17. [CrossRef]
22. Teng, R.; Ao, C.; Huang, H.; Shi, D.; Mao, Y.; Zheng, X.; Zhao, Y. Research of Processing Technology of Longjing Tea with 'Baiye 1' Based on Non-Targeted Aroma Metabolomics. *Foods* **2024**, *13*, 1338. [CrossRef] [PubMed]
23. An, H.; Ou, X.; Zhang, Y.; Li, S.; Xiong, Y.; Li, Q.; Huang, J.; Liu, Z. Study on the key volatile compounds and aroma quality of jasmine tea with different scenting technology. *Food Chem.* **2022**, *385*, 132718. [CrossRef] [PubMed]
24. Wisniewski, J.R.; Zougman, A.; Nagaraj, N.; Mann, M. Universal sample preparation method for proteome analysis. *Nat. Methods* **2009**, *6*, 359–362. [CrossRef] [PubMed]
25. Chen, Q.; Shi, J.; Mu, B.; Chen, Z.; Dai, W.; Lin, Z. Metabolomics combined with proteomics provides a novel interpretation of the changes in nonvolatile compounds during white tea processing. *Food Chem.* **2020**, *332*, 127412. [CrossRef] [PubMed]

26. Zhu, J.; Niu, Y.; Xiao, Z. Characterization of the key aroma compounds in Laoshan green teas by application of odour activity value (OAV), gas chromatography-mass spectrometry-olfactometry (GC-MS-O) and comprehensive two-dimensional gas chromatography mass spectrometry (GC×GC-qMS). *Food Chem.* **2021**, *339*, 128136. [CrossRef] [PubMed]
27. Yin, P.; Kong, Y.; Liu, P.; Wang, J.; Zhu, Y.; Wang, G.; Sun, M.; Chen, Y.; Guo, G.; Liu, Z. A critical review of key odorants in green tea: Identification and biochemical formation pathway. *Trends Food Sci. Technol.* **2022**, *129*, 221–232. [CrossRef]
28. You, Q.; Shi, Y.; Zhu, Y.; Yang, G.; Yan, H.; Liu, Z.; Lv, H. Effect of different processing technologies on the key aroma-active compounds of green tea. *Food Sci.* **2023**, *44*, 194–200.
29. Xue, J.; Liu, P.; Yin, J.; Wang, W.; Zhang, J.; Wang, W.; Le, T.; Ni, D.; Jiang, H. Dynamic Changes in Volatile Compounds of Shaken Black Tea during Its Manufacture by GC × GC-TOFMS and Multivariate Data Analysis. *Foods* **2022**, *11*, 1228. [CrossRef]
30. Wang, C.; Lv, S.; Wu, Y.; Lian, M.; Gao, X.; Meng, Q. Study of aroma formation and transformation during the manufacturing process of Biluochun green tea in Yunnan Province by HS-SPME and GC-MS. *J. Sci. Food Agric.* **2016**, *96*, 4492–4498. [CrossRef]
31. He, Y.; Liu, Z.; Qian, M.; Yu, X.; Xu, Y.; Chen, S. Unraveling the chemosensory characteristics of strong-aroma type Baijiu from different regions using comprehensive two-dimensional gas chromatography-time-of-flight mass spectrometry and descriptive sensory analysis. *Food Chem.* **2020**, *331*, 127335. [CrossRef]
32. Zhang, M.; Yang, Y.; Yuan, H.; Hua, J.; Deng, Y.; Jiang, Y.; Wang, J. Contribution of addition theanine/sucrose on the formation of chestnut-like aroma of green tea. *LWT-Food Sci. Technol.* **2020**, *129*, 109512. [CrossRef]
33. Zhou, S.; Huang, C.; Li, C.; Gong, S.; Guo, H.; Tong, Y.; Fan, F. Aroma compositions and comparison of their related components of Longjing tea with different aroma types. *J. Zhejiang Univ. (Agric. Life Sci.)* **2021**, *47*, 203–211.
34. Zheng, L.; Liu, P.; Gong, Z.; Wang, X.; Cheng, Y.; Yang, W.; Tang, Q.; Zheng, P. Analysis on the difference of main chemical components of different aromatic green tea from Xiangyang. *Sci. Technol. Food Ind.* **2021**, *42*, 252–258.
35. Nose, M.; Nakatani, Y.; Yamanishi, T. Studies on the flavor of green tea Part IX. Identification and composition of intermediate and high boiling constituents in green tea flavor. *J. Agric. Chem. Soc. Jpn.* **1971**, *35*, 261–271.
36. Tan, H.; Lau, H.; Liu, S.; Tan, L.; Sakumota, S.; Lassabliere, B.; Leong, K.; Sun, J.; Yu, B. Characterisation of key odourants in Japanese green tea using gas chromatography-olfactometry and gas chromatography-mass spectrometry. *LWT-Food Sci. Technol.* **2019**, *108*, 221–232. [CrossRef]
37. Hu, D.; Chen, W.; Ma, C.; Fang, Z.; Huang, H.; Miao, A. Effect of high-temperature and long-time final roasting on aroma of roasted green tea. *Food Sci.* **2022**, *43*, 255–261.
38. Wang, M.; Zhu, Y.; Zhang, Y.; Shi, J.; Lin, Z.; Lv, H. Analysis of volatile composition and key aroma compounds of green teas with fresh scent flavor. *Food Sci.* **2019**, *40*, 219–228.
39. Ho, C.; Zheng, X.; Li, S. Tea aroma formation. *Food Sci. Hum. Wellness* **2015**, *4*, 9–27. [CrossRef]
40. Xu, Y.; Zhang, Y.; Chen, J.; Wang, F.; Du, Q.; Yin, J. Quantitative analyses of the bitterness and astringency of catechins from green tea. *Food Chem.* **2018**, *258*, 16–24. [CrossRef]
41. Mei, X.; Wan, S.; Lin, C.; Zhou, C.; Hu, L.; Deng, C.; Zhang, L. Integration of metabolome and transcriptome reveals the relationship of benzenoid-phenylpropanoid pigment and aroma in Purple tea flowers. *Front. Plant Sci.* **2021**, *12*, 762330. [CrossRef] [PubMed]
42. Chen, S.; Li, X.; Liu, Y.; Chen, J.; Ma, J.; Chen, L. Identification of QTL controlling volatile terpene contents in tea plant (Camellia sinensis) using a high-aroma 'Huangdan' x 'Jinxuan' F1 population. *Front. Plant Sci.* **2023**, *14*, 1130582. [CrossRef] [PubMed]
43. Zeng, L.; Xiao, Y.; Zhou, X.; Yu, J.; Jian, G.; Li, J.; Chen, J.; Tang, J.; Yang, Z. Uncovering reasons for differential accumulation of linalool in tea cultivars with different leaf area. *Food Chem.* **2021**, *345*, 128752. [CrossRef] [PubMed]
44. Yang, Z.; Baldermann, S.; Watanabe, N. Recent studies of the volatile compounds in tea. *Food Res. Int.* **2013**, *53*, 585–599. [CrossRef]
45. Pichersky, E.; Noel, J.; Dudareva, N. Biosynthesis of plant volatiles: Nature's diversity and ingenuity. *Science* **2006**, *311*, 808–811. [CrossRef]
46. Yang, Z.; Sakaia, M.; Sayama, H.; Shimeno, T.; Yamaguchi, K.; Watanabe, N. Elucidation of the biochemical pathway of 2-phenylethanol from shikimic acid using isolated protoplasts of rose flowers. *J. Plant Physiol.* **2009**, *166*, 887–891. [CrossRef]

Disclaimer/Publisher's Note: The statements, opinions and data contained in all publications are solely those of the individual author(s) and contributor(s) and not of MDPI and/or the editor(s). MDPI and/or the editor(s) disclaim responsibility for any injury to people or property resulting from any ideas, methods, instructions or products referred to in the content.

chemosensors

Article

Evaluating the Intensity of Bitter–Astringent Interactive Perception in Green Tea Based on the Weber–Fechner Law

Siying Li [1,2], Dongdong Du [1,2], Shaoming Cheng [1] and Zhenbo Wei [1,2,*]

[1] Department of Biosystems Engineering, Zhejiang University, 866 Yuhangtang Road, Hangzhou 310058, China; lisiying@zju.edu.cn (S.L.); dudd@zju.edu.cn (D.D.); chengshaoming@zju.edu.cn (S.C.)
[2] Zhejiang Key Laboratory of Intelligent Sensing and Robotics for Agriculture, Department of Biosystems Engineering and Food Science, Zhejiang University, Hangzhou 310058, China
* Correspondence: weizhb@zju.edu.cn

Abstract: This study aimed to quantify the bitter–astringent interactive perceived intensity (BAIPI) of green tea based on the Weber–Fechner law by two-alternative forced choice test. Caffeine, (-)-epigallocatechin-3-gallate (EGCG), and Quercetin-3-o-rutinoside (Que-rut) were identified as the key bitter–astringent compounds. Then, BAIPI values were assessed based on the concentration ratios and ranges of caffeine, EGCG, and Que-rut in green tea according to the Weber–Fechner law. The BAIPI values were further fitted to a logarithmic curve: $y_{\text{BAIPIvalue}} = a \times \ln(x_{concentration\ of\ caffeine} - b) - c + IV_{\text{initialBAIPIvalue}} - 1$. The constant terms a, b, and c were derived from the ratio of caffeine to EGCG, as the slope and X-intercept of the curves correlated exponentially with this ratio. The initial BAIPI value (IV) was calculated as: $IV_{\text{initialBAIPIvalue}} = 18.7960 - 0.3956q_{(the\ ratio\ of\ caffeine\ to\ Que-rut)}$. Validation showed a standard error of 1.5865 between estimated and actual BAIPI values. This method enables the estimation of bitter–astringent intensity in green tea by analyzing the concentrations of caffeine, EGCG, and Que-rut.

Keywords: astringency; bitterness; green tea; interactive perceived intensity; quantification curve; Weber–Fechner law

1. Introduction

Tea, as the most widely consumed beverage globally following water, has unique flavor profiles [1]. Tea's distinct flavor arises from secondary metabolites—flavonoids, phenolic acids, and alkaloids—in tender leaves, which also contribute to bitterness and astringency [2,3]. Green tea, as a non-fermented tea, retains the highest concentration of these secondary metabolites due to manufacturing process, thereby preserving more significant bitterness and astringency, which are the key to affect the choice of consumers [2,4,5].

Bitterness is a basic taste produced by the activation of taste receptors encoded by the TAS2Rs gene when the bitter tastants in tea entering the oral cavity [2,6]. The compounds responsible for bitterness in green tea are mainly caffeine and catechins [2], which can enhance the umami taste of green tea caused by theanine [7] and relieve the sweetness of desserts when drinking green tea [4,8]. Astringency refers to a characteristic drying, rough, or puckery sensation in the oral cavity that arises from two distinct mechanisms: (1) the physicochemical interaction between polyphenols and salivary proteins/mucosal pellicle proteins, leading to precipitation and reduced oral lubrication [2,5,9]; or (2) a trigeminal nerve-mediated sensation triggered when astringents activate G protein-coupled receptor signaling pathways in trigeminal ganglion cells [10]. This complex sensory phenomenon involves both physical protein–tannin interactions and neural sensory transduction pathways.

https://doi.org/10.3390/chemosensors13040137

Astringency enhances the sweet aftertaste (Huigan) of tea [5] and influences retronasal aroma perception [11]. Additionally, the bitter and astringent perception of consumers varied due to genetic differences in taste receptors and receptor expression [6]. Therefore, it is urgent to develop a standardized quantitative scale for the intensity of bitterness and astringency in green tea to help consumers choose products that suit their taste preferences.

Previous studies have employed two primary approaches to measure bitterness and astringency in green tea: conventional and intelligent sensory evaluation. Conventional sensory evaluation is to assess the level of bitterness and astringency through subjects rating samples using standardized taste solutions as references [8,12–14]. However, variability in reference solution concentrations across studies prevents unified intensity standards. Intelligent sensory evaluation is mainly to quantify bitter and astringent intensity through electrochemical [15], optical [16], mass [5], and biological [17] sensing techniques based on the reaction of taste compounds with the corresponding receptors or saliva in the oral cavity. However, it fails to directly link instrumental signals to human perceptual experience [18,19].

The Weber–Fechner law is central to the field of psychophysics in quantifying the internal perceptual intensity of external stimuli, specifically that the psychological perceptual intensity of a stimulus is proportional to the logarithm of the physical intensity by experimental verification [20]. At present, there were many studies to assess the internal psychological perception of taste stimuli based on the Weber–Fechner law. The sweetness index [14] and sourness index [21] of five sugars and acids were determined by converting absolute thresholds and difference thresholds obtained through triangle tests and paired comparisons based on the Weber–Fechner law. The sweetness of sucrose under different citric acid backgrounds and the sourness of citric acid under different sucrose backgrounds were calculated by fitting absolute threshold and weber fraction equations, respectively [22]. A quantitative equation for overall umami strength in foods and condiments was also defined by sensory evaluation based on the Weber–Fechner law [23]. Therefore, the perceived intensity of taste stimuli could be deduced according to the content of taste compounds. The content range and ratio of bitter and astringent compounds in green tea is defined. If the relationship between the content of bitter and astringent compounds in green tea and the perceived intensity of taste is derived based on the Weber–Fechner law, the perceived intensity of bitterness and astringency in green tea can be quantified under uniform criteria.

Additionally, the bitter and astringent perception of green tea is often inseparable for consumers [17]. Catechins have both bitterness and astringency, and caffeine and (-)-epigallocatechin-3-gallate (EGCG; the most abundant catechin in green tea) can strengthen each other's bitterness and astringency [7]. Flavonol glycosides contributed astringency to green tea and can also increase the bitterness of caffeine [24]. Moreover, a subregion within the insular cortex of the cerebral cortex exhibited responsiveness to both astringent and bitter stimuli, indicating that these inputs underwent neural information integration in the human brain to promote people to produce bitter–astringent interactive perception [25]. Therefore, the overall perceived intensity of bitterness and astringency produced by people after drinking green tea was defined as bitter–astringent interactive perceived intensity (BAIPI) in this study. To establish a universal quantification scale for BAIPI, this study is (1) to determine the concentration of bitter and astringent compounds in green tea; and then, (2) to quantify the BAIPI values of green tea and fit curves based on the Weber–Fechner law according to the concentration; and next, (3) to verify the applicability of the BAIPI quantification curve by green tea samples in the market.

2. Materials and Methods

2.1. Chemicals

The standards (alkaloids, gallic acid (GA), catechins, flavonol glycosides, and organic acids) for bitter and astringent compounds analysis, other taste indicators analysis, and sensory panel test of green tea were purchased from GlpBio Co., Ltd. (Montclair, NJ, USA). The mobile phase components for high performance liquid chromatography (HPLC) analysis consisted of formic acid (Macklin, Shanghai, China) and acetonitrile (Aladdin, Shanghai, China). Ultrapure water used in the experiments was produced by an AWL-6000-M purification system (Aquapro, FL, USA). For analysis of other taste indicators in green tea, we used a Total Carbohydrate Assay Kit (Solarbio, Beijing, China) and Amino Acid Assay Kit (Ninhydrin colorimetry method, Yuanye, Shanghai, China).

2.2. Green Tea Collection and Sample Preparation

Green tea samples were purchased from different tea factories, and the specific information is shown in Table 1. From the complete collection, we selected 11 representative green tea samples as model-samples (encompassing all fixation methods; indicated in bold italic font), while the remaining 32 samples served as the validation set. The tea leaves were crushed into particles (0.83 mm < the particle size of tea leaves $\leq$ 1 mm). Precisely 1 g of processed tea particles was combined with 100 mL of ultrapure water (maintained at 95 $\pm$ 1 °C) in a standardized 440 mL tea bowl. The mixture was homogenized using a DF-101S magnetic stirring system (Lichen Instrument, Shanghai, China) at a constant 300 rpm for exactly 3 min to produce the tea infusion.

Table 1. Data on the green tea samples used.

Name	Code	Cultivated Variety	Picking Time	Fixation Ways	Manufacturer
West Lake Long Jing Spring	WLS	Longjing 43	March 2023	Stir-frying	Hangzhou Zhijiang Tea Co. Ltd. (Zhejiang, China)
Long Jing Spring	***LJS***	***Longjing 43***	***March 2023***	***Stir-frying***	***Hangzhou LvYi Tea Industry Co. Ltd.*** (Zhejiang, China)
Long Jing Summer	***LJM***	***Longjing 43***	***June 2023***	***Stir-frying***	***Hangzhou LvYi Tea Industry Co. Ltd.*** (Zhejiang, China)
Yunwu Spring	FWS	Fuyun No.6	March 2023	Stir-frying	Fujian Runmutang Tea Manor Co. Ltd. (Fujian, China)
Yuhua before Tomb Sweeping	YBTS	Longjing 43	March 2023	Stir-frying	Nanjing Chunrun Tea Industry Co. Ltd. (Jiangsu, China)
Yuhua before Grain Rain	YBGR	Longjing 43	April 2023	Stir-frying	Nanjing Chunrun Tea Industry Co. Ltd. (Jiangsu, China)
Laoshan tea Before Grain Rain	LTB	Local population	April 2023	Stir-frying	Qingdao Lao-Tea Industry Co. Ltd. (Shandong, China)
Laoshan tea After Grain Rain	LTA	Local population	May 2023	Stir-frying	Qingdao Lao-Tea Industry Co. Ltd. (Shandong, China)
Anji White tea High grade	AWH	White Leaf No.1	March 2023	Stir-frying	Anji Bai Tea Factory (Anhui, China)
Anji White tea Low grade	AWL	White Leaf No.1	March 2023	Stir-frying	Anji Bai Tea Factory (Anhui, China)
Biluochun Spring	***BCS***	***Dongshan population***	***March 2023***	***Stir-frying***	***Shandong YIBEIXIANG Tea Industry Co. Ltd.*** (Shandong, China)
Biluochun High grade	BCH	Dongshan population	March 2023	Stir-frying	Zhejiang Yifutang Tea Industry Co. Ltd. (Zhejiang, China)
Biluochun Low grade	BCL	Dongshan population	March 2023	Stir-frying	Zhejiang Yifutang Tea Industry Co. Ltd. (Zhejiang, China)
Xinyang Maojian before Tomb Sweeping	XTS	Local population	March 2023	Stir-frying	Farmhouse Tea (Henan, China)
Xinyang Maojian before Grain Rain	XGR	Local population	April 2023	Stir-frying	Farmhouse Tea (Henan, China)

Table 1. *Cont.*

Name	Code	Cultivated Variety	Picking Time	Fixation Ways	Manufacturer
Guizhou Maojian Spring	GNS	Local Small Leaf	March 2023	Stir-frying	Guizhou Lenton Agricultural Development Co. Ltd. (Guizhou, China)
Guizhou Maofeng Spring	GMS	Fuding big white tea	March 2023	Stir-frying	Guizhou Meitan Daye Xin Tea Co. Ltd. (Guizhou, China)
Yunnan tea Spring	YTS	Yunnan Large Leaf	February 2023	Baking	Yunnan Wanhongyang Tea Co. Ltd. (Yunnan, China)
Huangshan Maofeng High grade	HMH	Huangshan Large Leaf	March 2023	Baking	Xie Yuda Tea Co. Ltd. (Anhui, China)
Huangshan Maofeng Low grade	HML	Huangshan Large Leaf	April 2023	Baking	Xie Yuda Tea Co. Ltd. (Anhui, China)
Taiping Houkui	THK	Huangshan Large Leaf	April 2023	Baking	Huangshan Mingpingxing Tea Co. Ltd. (Anhui, China)
Taiping Bujian	TBJ	Mountain Native	April 2023	Baking	Huangshan Hukeng Tea Co., Ltd. (Anhui, China)
Jingting Lvxue Tea	JLT	Southern Anhui Small Leaf	March 2023	Baking	Anhui Jingting Lvxue Tea Co. Ltd. (Anhui, China)
Liu 'an Guapian Spring	***LGS***	***Longjing 43***	***April 2023***	***Baking***	***Jinzhai County Red Xin Tea Professional Cooperative*** (Anhui, China)
Liu 'an Guapian Low grade	LGL	Dushan Small Leaf	April 2023	Baking	Anhui Lu 'an Guapian Tea Industry Co. Ltd. (Anhui, China)
Sichuan Maofeng	SMF	Huangshan Large Leaf	March 2023	Baking	Sichuan Sanhua Tea Industry Co. Ltd. (Sichuan, China)
Huangshan Maofeng Spring	***HMS***	***Huangshan Large Leaf***	***March 2023***	***Baking***	***Huangshan XIQINGFENG Agricultural Development Co. Ltd.*** (Anhui, China)
Huangshan Maofeng Summer in Last year	***HMML***	***Huangshan Large Leaf***	***June 2022***	***Baking***	***Huangshan MUFAN Trading Co. Ltd.*** (Anhui, China)
Hunan Yunwu	HYW	Small Leaf Population	April 2023	Baking	Hunan Gulou Xuefeng Yunwu Tea Co. Ltd. (Hunan, China)
Liangshan Mahu	LMH	Local population	March 2023	Sunning	Leibo county Huang Lang Tea Factory (Sichuan, China)
Sun-dried raw tea Spring	***SRS***	***Yunnan Large Leaf***	***March 2023***	***Sunning***	***Biyun Spring Tea Factory*** (Yunnan, China)
Sun-dried raw tea Spring in Last year	***SRSL***	***Yunnan Large Leaf***	***March 2022***	***Sunning***	***Yunnan Cangyuan Wa Camellia Factory Co. Ltd.*** (Yunnan, China)
Tuo Tea Spring	TTS	Yunnan Large Leaf	March 2023	Sunning	Yunnan Xiaguantuo Tea (Group) Co. Ltd. (Yunnan, China)
Dian Green Spring	DGS	Yunnan Large Leaf	March 2023	Sunning	Lancang Jing Tea Factory (Yunnan, China)
Yunnan Qingzhen Spring	YQS	Yunnan Large Leaf	March 2023	Steaming	Yunnan Wanhongyang Tea Co. Ltd. (Yunnan, China)
Enshi Gyokuro	EGK	Gyokuro	March 2023	Steaming	Enshi Huazhishan Ecological Agriculture Co. Ltd. (Hubei, China)
Enshi Gyokuro Spring	***EGS***	***Gyokuro***	***March 2023***	***Steaming***	***Enshi Chunzhihuo Tea Co. Ltd.*** (Hubei, China)
Enshi Gyokuro Spring in Last year	***EGSL***	***Gyokuro***	***April 2022***	***Steaming***	***Jinhua Jade Tea Co. Ltd.*** (Zhejiang, China)
Fried Green Tea	FGT	Uji	May 2023	Steaming	ITO Hisako Shimen Co., Ltd. (Japan)
Fukuoka Yamecha	FKY	Fukuoka	March 2023	Steaming	Kamuyuan Co., Ltd. (Japan)
Shizuoka Sencha Spring	***SSS***	***Shizuoka***	***March 2023***	***Steaming***	***Harada Tea Industry Ltd. (Japan)***
Shizuoka Kakegawa	SKK	Shizuoka	May 2023	Steaming	Ryowa Garden Co., Ltd. (Japan)
Kagoshima Chiran	KSC	Kagoshima	March 2023	Steaming	Yamashiro Co., Ltd. (Japan)

The font of model-samples is bold and italic.

2.3. Bitter and Astringent Compounds Analysis of Green Tea

The bitter and astringent compounds of green tea were determined by HPLC (Waters 1525; Milford, MA, USA) equipped with a 250 × 4.6 mm, 5 μm, Agilent 5 TC-C18 (2) column (CA, USA). All samples were filtered through 0.45 μm filters (mixed cellulose, Navigator, Tianjin, China) prior to analysis by HPLC. The determination method of alkaloids, GA, and catechins was adapted from Huang et al. [26] with modifications. A gradient elution with 0.1% formic acid as mobile phase A and acetonitrile as mobile phase B at a flow rate of 1 mL/min was performed. The gradient was carried out as follows: initial 5% B for 2 min, linear change to 13% B for 6 min, linear change to 28% B for 14 min, linear change to 95% B for 3 min, and finally linear change to 5% B for 5 min. Detection was performed at 278 nm using a Waters 2489 UV-Visible detector (Milford, MA, USA).

Flavonol glycosides were determined according to a modified method from Liu et al. [27], utilizing the following mobile phases: 0.15% formic acid in 3% acetonitrile/water (mobile phase A) and 0.15% formic acid in 30% acetonitrile/water (mobile phase B). The gradient elution (1 mL/min) was carried out as follows: 0% B, 0–20 min; 0–50% B, 20–28 min; 50–100% B, 28–47 min; 100% B, 47–55 min; 0–100% B, 55–58 min. Detection wavelength was set at 360 nm. The specific detection parameters for all target compounds are shown in Table 2.

Table 2. The detection parameters of bitter and astringent compounds.

Compounds	Perceived Types and Thresholds (mg/L)	CAS	Purity	MW	Calibration Curves $^{\alpha}$	$R^{2\ \beta}$	Linear Range (mg/L)
Caffeine	Bitter, 97.1	58-08-02	HPLC ≥ 98	194.19	y = 23,234x + 1223.2	0.9995	13.38–462
Theobromine	Bitter, 144.1	83-67-0	HPLC ≥ 98	180.16	y = 57,560x − 11,190	0.9995	1.07–86.25
Theophylline	Bitter, not reported	58-55-9	HPLC ≥ 98	180.16	y = 48,170x − 6702.7	0.9961	0.13–13.88
Gallic acid (GA)	Astringent, 49.3	149-91-7	HPLC ≥ 98	170.12	y = 28,540x + 25,118	0.9985	1.31–31.36
(−)-Gallocatechin (GC)	Astringent, 165.39	3371-27-5	HPLC ≥ 98	306.27	y = 177.04x + 881.17	0.9996	3.25–96.84
(−)-Epigallocatechin (EGC)	Astringent, 159.26	970-74-1	HPLC ≥ 98	306.27	y = 1435x − 77,330	0.9997	115.01–684.6
(+)-Catechin (C)	Astringent, 119.01; Bitter, 249.63	154-23-4	HPLC ≥ 98	290.27	y = 6346.7x − 20,348	0.9999	10.46–311.5
(−)-Epicatechin (EC)	Astringent, 284.83; Bitter, 284.83	490-46-0	HPLC ≥ 98	306.27	y = 6247.5x − 7629.6	0.9995	31.65–188.4
(−)-Epigallocatechin-3-gallate (EGCG)	Astringent, 87.09; Bitter, 174.18	989-51-5	HPLC ≥ 98	458.37	y = 12,043x − 679,287	0.9980	131.12–780.5
(−)-Gallocatechin gallate (GCG)	Astringent, 178.76; Bitter, 178.76	4233-96-9	HPLC ≥ 98	458.37	y = 11,012x − 81,486	0.9979	10.46–62.3
(−)-Epicatechin-3-gallate (ECG)	Astringent, 115.02; Bitter, 79.63	1257-08-5	HPLC ≥ 98	442.37	y = 15,400x − 117,766	0.9993	8.24–245.25
(−)-Catechin gallate (CG)	Astringent, 110.59; Bitter, 75.20	130405-40-2	HPLC ≥ 98	442.37	y = 16,715x − 26,469	0.9978	1.59–47.6
Myricetin-3-o-rhamnoside (Myr-rha)	Astringent, 4.6	17912-87-7	HPLC ≥ 98	464.38	y = 8549.6x − 4314.5	0.9992	0.57–2.74
Quercetin-3-o-galactoside (Que-gal)	Astringent, 0.2	482-36-0	HPLC ≥ 98	464.38	y = 16,461x − 5756.4	0.9992	0.89–26.67
Quercetin-3-o-glucoside (Que-glu)	Astringent, 0.3	21637-25-2	HPLC ≥ 98	464.38	y = 18,239x − 1582.5	0.9976	0.06–10.39
Quercetin-3-o-rutinoside (Que-rut)	Astringent, 0.0007	153-18-4	HPLC ≥ 98	610.52	y = 13,286x − 5294.3	0.9995	0.91–27.25
Kaempferol-3-o-rutinoside (Kae-rut)	Astringent, 0.15	17650-84-9	HPLC ≥ 98	594.52	y = 7351.2x − 1655.8	0.9990	0.44–13.17
Kaempferol-3-o-β-D-glucoside (Kae-glu)	Astringent, 0.3	480-10-4	HPLC ≥ 98	448.38	y = 15,641x − 884.87	0.9990	0.13–3.21

α: x refers to the concentration of compounds (mg/L), and y refers to the peak area of chromatogram; β: is the goodness of fit for the calibration curves.

2.4. Matrix Composition Analysis of Green Tea

In addition to bitterness and astringency, the basic taste profile of green tea comprises sweetness, umami, and sourness. Accordingly, the matrix composition of the bitter–astringent simulation solution was designed to reflect the concentrations of sweet, umami, and sour compounds typically found in green tea. Total carbohydrate (TC) and total free amino acid (TFAA) were measured using commercial assay kits (UV-Vis spectrophotometry, WFJ 7200; Unico, Shanghai, China) at absorbance wavelengths of 540 nm and 570 nm, respectively. These metrics represent the collective sweetness (TC) and umami (TFAA) contributions in green tea. Organic acids, representing sourness, were quantified via HPLC following the method of Yu et al. [28]. Sample pretreatment and HPLC conditions aligned with Section 2.3. Detailed analytical parameters are provided in Table 3. Sucrose, theanine, and citric acid were selected as the primary matrix components for the simulation solution, as they are the most abundant sugar, amino acid, and organic acid in green tea, respectively [26,29]. Their concentrations corresponding to the levels of TC, TFAA, and total organic acids in green tea (shown in Table 4).

Table 3. The detection parameters of other taste indicators.

Taste Indicators	Calibration Curves	*x*	y	$R^{2\,\alpha}$	Linear Range (mg/L)
Total carbohydrate (TC)	y = 0.0006x − 0.0128	The concentration of glucose (mg/L)	The absorbance difference of spectrum	0.9973	10–500
Total free amino acid (TFAA)	y = 0.0593x + 0.0014	The concentration of amino acid standard solution (mg/L)	The absorbance difference of spectrum	0.9990	2–10
Oxalic acid	y = 13,017x − 104,627	The concentration of oxalic acid (mg/L)	The peak area of chromatogram	0.9953	14.09–208.7
Tartaric acid	y = 1383.3x − 16,178	The concentration of tartaric acid (mg/L)	The peak area of chromatogram	0.9948	10.96–121.77
Malic acid	y = 2370.2x − 4067.5	The concentration of malic acid (mg/L)	The peak area of chromatogram	0.9934	25.01–250.1
Citric acid	y = 1001x − 5925.8	The concentration of citric acid (mg/L)	The peak area of chromatogram	0.9999	46.31–385.95
Succinic acid	y = 564.49x + 9557.8	The concentration of succinic acid (mg/L)	The peak area of chromatogram	0.9946	11.52–214.25

α: The goodness of fit for the calibration curves.

Table 4. The concentration of the other taste indicators in all green tea samples.

Samples	Total Free Amino Acid (mg/L)	Total Carbohydrate (mg/L)	Oxalic Acid (mg/L)	Tartaric Acid (mg/L)	Malic Acid (mg/L)	Citric Acid (mg/L)	Succinic Acid (mg/L)	Total Organic Acids (mg/L)
LJS	36.09 ± 0.78	74.01 ± 1.61	79.71 ± 7	51.24 ± 2.26	144.85 ± 4.03	141.96 ± 8.8	109.23 ± 6.14	526.99 ± 7.52
LJM	36.32 ± 0.67	112.56 ± 3.21	91.55 ± 4.33	30.95 ± 2.52	90.89 ± 6.19	173.62 ± 7.9	112.87 ± 6.44	499.87 ± 11.4
BCS	29.13 ± 0.78	102.93 ± 1.61	81.85 ± 1.5	21.56 ± 1.01	138.65 ± 10.51	183.98 ± 6.15	111.76 ± 4.44	537.8 ± 13.05
HMS	37.22 ± 1.03	100.52 ± 2.41	99.01 ± 1.48	46.28 ± 4.81	152.32 ± 8.05	231 ± 3.7	132.53 ± 8.21	661.13 ± 9.07
HMML	23.07 ± 1.4	62.77 ± 1.61	73.64 ± 2.24	22.17 ± 1.63	33.02 ± 6.94	128.97 ± 8.89	95.85 ± 3.25	353.64 ± 5.97
LGS	23.07 ± 1.7	63.57 ± 2.41	105.6 ± 3.42	37.14 ± 2.62	76.53 ± 4.51	280.05 ± 3.6	82.33 ± 10.08	581.65 ± 12.1
SRS	29.35 ± 1.7	85.26 ± 1.61	99.84 ± 6.73	29.3 ± 2.3	63.09 ± 0.38	142.77 ± 7.87	188.67 ± 6.92	523.68 ± 16.69
SRSL	21.49 ± 0.67	49.92 ± 1.61	83.73 ± 1.24	50.4 ± 3.34	136.59 ± 2.84	123.53 ± 0.01	133.92 ± 8.01	528.17 ± 7.23
EGS	26.43 ± 1.4	55.54 ± 4.02	78.46 ± 0.75	41.7 ± 3.37	122.07 ± 1.55	131.53 ± 0.25	89.97 ± 2.79	463.74 ± 1.13
EGSL	8.74 ± 0.08	115.78 ± 3.21	158.02 ± 2.54	52.55 ± 7.53	81.14 ± 6.82	139.46 ± 3.42	92.15 ± 8.96	523.32 ± 24.08
SSS	18.57 ± 1.4	90.88 ± 2.41	111.83 ± 5.08	37.21 ± 3.75	139.26 ± 6.52	126.8 ± 1.5	78.15 ± 9.18	493.25 ± 23.55
Mean	21.77 ± 5.36	96.65 ± 10.41	86.16 ± 28.16	39.58 ± 12.85	93.02 ± 40.41	154.41 ± 66.19	103.02 ± 34.92	475.5 ± 117.73
FWS	26.14 ± 0.81	55.54 ± 0.8	113.46 ± 1.1	29.35 ± 2.05	117.39 ± 5.4	128.72 ± 9.02	72.98 ± 7.06	461.91 ± 2.96
WLS	27.56 ± 0.34	89.27 ± 0.8	107.25 ± 11.83	28.16 ± 1.16	111.59 ± 6.53	129.41 ± 8.79	110.3 ± 5.41	486.71 ± 22.34
GMS	15.93 ± 0.37	79.63 ± 2.41	73.15 ± 2.97	29.02 ± 3.75	86.86 ± 7.16	76.05 ± 7.06	76.04 ± 9.84	341.12 ± 16.92
GNS	19.81 ± 0.94	70 ± 2.41	72.54 ± 4.85	35.55 ± 1.8	144.42 ± 5.72	61.87 ± 0.63	78.36 ± 7.11	392.75 ± 6.86
XTS	15.26 ± 0.98	62.77 ± 3.21	84.31 ± 3.29	23.67 ± 1.2	116.49 ± 2.78	62.16 ± 1.42	82.69 ± 7.17	369.31 ± 8.93
XGR	18.7 ± 0.64	67.59 ± 1.61	84.04 ± 1.04	28.32 ± 6.34	118.96 ± 3.05	139.71 ± 12.42	112.05 ± 1.52	483.08 ± 4.64
YBTS	19.44 ± 0.37	78.03 ± 2.41	90.5 ± 6.36	36.04 ± 7.89	141.96 ± 9.12	202.4 ± 19.2	133.45 ± 3.06	604.36 ± 30.55
YBGR	20.38 ± 0.57	147.9 ± 4.82	71.97 ± 10.34	41.93 ± 1.7	99.38 ± 1.86	145.52 ± 7.43	72.89 ± 1.77	431.69 ± 16.71
LTB	15.6 ± 1.18	83.65 ± 1.61	81.96 ± 1.72	33.5 ± 5.22	46.58 ± 3.86	182.67 ± 7.06	17.04 ± 2.2	361.74 ± 2.41
LTA	16 ± 0.51	169.59 ± 2.41	75.14 ± 1.96	42.49 ± 1.97	41.08 ± 0.6	130.24 ± 12.03	39.71 ± 2.55	328.67 ± 12.29
AWH	37.56 ± 0.84	65.18 ± 2.41	86.54 ± 8.24	37.51 ± 5.05	147.54 ± 3.92	157.05 ± 6.37	162.08 ± 2.71	590.73 ± 8.08
AWL	23.65 ± 0.27	53.93 ± 2.41	64.47 ± 2.88	31.39 ± 4.98	116.48 ± 6.12	147.86 ± 4.21	76.65 ± 2.98	436.85 ± 13.54
BCH	25 ± 0.34	106.14 ± 3.21	34.41 ± 5.68	36.56 ± 1.48	118.05 ± 2.81	139.2 ± 7.98	77.16 ± 2.25	405.38 ± 9.48
BCL	9.47 ± 1.04	71.6 ± 0.8	75.8 ± 3.13	39.44 ± 3.12	47.48 ± 2.69	78.74 ± 2.24	158.41 ± 6.41	399.87 ± 6.32
YTS	15.23 ± 0.54	71.6 ± 2.41	53.51 ± 8.44	33.4 ± 5.13	37.64 ± 0.51	71.43 ± 2.28	109.98 ± 6.44	305.96 ± 7.68
HMH	23.21 ± 0.57	57.95 ± 1.61	54.56 ± 1.28	30.11 ± 2.06	121.91 ± 4.54	122.76 ± 2.64	114.49 ± 9.43	443.83 ± 13.36
HML	20.01 ± 0.34	70.8 ± 3.21	78.01 ± 7.06	33.21 ± 1.21	147.27 ± 8.35	89.4 ± 4.75	80.36 ± 0.81	428.25 ± 18.06
THK	36.39 ± 0.74	110.15 ± 4.02	84.61 ± 2.34	49.74 ± 4.13	115.49 ± 3.34	64.49 ± 2.78	114.33 ± 6.05	428.66 ± 14.12
TBJ	23.38 ± 0.4	159.95 ± 2.41	56.03 ± 1.76	30.33 ± 0.37	101.99 ± 4.93	86.66 ± 8.37	85.72 ± 5.03	360.72 ± 16
JLT	11.28 ± 0.51	188.06 ± 1.61	62.7 ± 3.89	31.14 ± 3.63	140.66 ± 12.23	81.41 ± 3.14	97.15 ± 3.92	413.07 ± 15.26
LGL	10 ± 0.37	155.13 ± 0.8	83.49 ± 4.7	64.21 ± 1.85	53.47 ± 4.25	170.47 ± 1.42	86.45 ± 6.96	458.1 ± 13.67
SMF	22.07 ± 0.17	120.55 ± 2.63	66.57 ± 2.33	59.36 ± 3.39	69.81 ± 4.47	220.43 ± 2.05	124.03 ± 7.47	540.21 ± 5.59
HYW	25.77 ± 0.3	147.89 ± 2.63	76 ± 6.4	63.5 ± 3.3	106.64 ± 4.61	279.3 ± 18.34	121.55 ± 8.87	646.99 ± 6
LMH	24.8 ± 0.54	143.84 ± 3.82	81.39 ± 1.68	64.16 ± 3.5	128.92 ± 4.41	273.43 ± 18.26	169.46 ± 5.59	717.37 ± 31.93
TTS	13.95 ± 0.4	133.21 ± 4.39	90.42 ± 3.31	44.26 ± 2.17	48.8 ± 6.56	224.12 ± 6.95	127.75 ± 7	535.35 ± 16.6
DGS	9.47 ± 0.24	140.8 ± 3.16	75.81 ± 4.41	27.95 ± 1.56	31.91 ± 0.17	123.42 ± 0.26	21.75 ± 0.15	280.85 ± 2.79
YQS	17.45 ± 0.47	167.14 ± 4.39	51.18 ± 0.69	24.49 ± 1.48	55.83 ± 5.72	83.4 ± 7.4	52.61 ± 5.44	267.5 ± 17.35
EGK	27.09 ± 0.74	107.89 ± 3.16	106.09 ± 6.12	51.36 ± 1.67	128.91 ± 5.75	263.86 ± 6.92	135.26 ± 5	685.48 ± 14.07
FGT	22.5 ± 0.2	107.38 ± 0.88	128.14 ± 2.24	37.7 ± 2.7	96 ± 4.82	207.76 ± 6	56.94 ± 4.09	526.54 ± 6.36
FKY	15.36 ± 0.54	107.38 ± 3.51	116.14 ± 2.68	29.84 ± 7.57	109.88 ± 10.75	71.64 ± 10.02	78.52 ± 7.27	406.03 ± 7.23
SKK	8.66 ± 0.1	113.46 ± 2.32	69.2 ± 3.32	32.21 ± 4.72	34.17 ± 1.41	155.17 ± 7.41	67.46 ± 1.79	358.2 ± 10.92
KSC	10.61 ± 0.57	83.58 ± 2.32	167.04 ± 6.07	32.74 ± 1.85	36.14 ± 3.16	207.52 ± 13.85	104.08 ± 9.28	547.52 ± 11.01
Mean	19.62 ± 7.24	91.51 ± 36.86	81.76 ± 25.2	37.9 ± 11.78	94.37 ± 38.65	143.07 ± 64.51	94.3 ± 37.01	451.4 ± 112.31

The results are reported as mean ± standard deviation of triplicate analysis.

2.5. Sensory Panel

A total of 20 experienced subjects (9 males and 11 females, from 18 to 28 years old) passed the screening and training about the test rules following an explanation of the study's purpose. Subjects were initially recruited based on surveys of health (i.e., food allergies), interest, expressiveness, availability, attitudes toward green tea, and frequency of green tea consumption [30]. All subjects exhibited favorable attitudes toward green tea consumption and had a minimum of three years' experience in regular tea tasting, with specific expertise in green tea sensory evaluation protocols. Subsequently, screening tests to assess their sensory acuity were carried out [31]. The triangle test and ranking test of caffeine and EGCG solutions were employed to assess subjects' discrimination ability for bitter and astringent sensations, with a required accuracy rate >80%. The above test was conducted in two sessions, each lasting about two hours. Following the screening process, a two-week training program was subsequently conducted. Subjects underwent daily training consisting of three sets of 2-AFC tests with caffeine and EGCG solutions at concentration ratios of 0.5, 0.75, and 1 (targeting identification of the more astringent and bitter solution). The caffeine concentration range was 150–310 mg/L with a gradient interval of 120%, 115%, 110%, 105%, and 100%. All subjects ultimately achieved >80% correct identification rates. All subjects were instructed not to have cold symptoms, smoke, or eat for at least a half hour prior to the test.

2.6. Composition Determination of Simulation Solution

The composition of simulation solutions 1–7 is described in Section 3.1.3 in detail, and the concentration of compounds in simulation solution were consistent with those in the corresponding model-samples. A total of 7 groups of simulation solutions containing different bitter and astringent compounds were tested with 11 types of green tea based on the two-alternative forced choice (2-AFC) test. The entire testing procedure spanned 11 days, with one sample analyzed per day, and each evaluation session was completed within a 3-h time frame. The specific steps to determine the composition of simulation solution were as follows:

Step 1. The selected green tea sample was provided to the subjects in pairs with the corresponding simulation solutions 1–7 in a random order with covers at room temperature (25 ± 1 °C). Samples were divided into 20 mL in 50 mL cups marked with three random numbers.

Step 2. Subjects with a nose clip were instructed to take a small sip of the sample with their eyes closed after removing the covers, hold it in the mouth for 7–8 s and spit it out. Completing the evaluation of one sample, subjects were required to gargle with ultrapure water and have some unsalted biscuits to relieve taste bud stress for 15 min.

Step 3. The next sample was evaluated after rinsing the mouth with ultrapure water. Subjects were asked to write down the number of the sample with stronger BAIPI following the two samples tasting.

Step 4. When 11 green tea samples had been tested, it was necessary to record the number, denoted as m, at which the selection rates of the simulation solution are below 75% and above 25% [32]. The composition of the simulation solution with the largest m value was optimal. When multiple m maxima were the same, the composition of the simulation solution with the fewest compounds was optimal.

2.7. BAIPI Quantification Curves Establishment

The BAIPI quantification curves were grouped according to the compound ratios selected in Section 2.6 (Table 5). Each recombinant solution group maintained the corresponding compound concentration ratios and ranges specified in Table 6. The recombinant solution matrix contained sucrose, theanine, and citric acid at concentrations matching the average levels of total carbohydrates (TC), total free amino acids (TFAA), and total organic acids in the model samples (Table 4). Within each group, caffeine served as the independent variable (x) for establishing the BAIPI quantification curves. And the specific steps were as follows:

Table 5. The ratio and the close ratio of caffeine, EGCG, and Que-rut in model-samples used for grouping.

Samples	The Ratio of Caffeine, EGCG, and Que-Rut	The Close Ratio of Caffeine, EGCG, and Que-Rut
SSS	9:18:1	10:20:1
EGS	45:83:1	45:85:1 45:80:1
LGS	18:33:1	20:35:1
BCS	44:78:1	45:80:1 45:75:1
HMML	44:75:1	45:75:1
EGSL	16:23:1	15:25:1 15:20:1
LJM	38:53:1	40:55:1
LJS	25:34:1	25:35:1
HMS	34:42:1	30:40:1 35:45:1 35:40:1
SRSL	12:13:1	10:10:1 15:15:1
SRS	13:13:1	10:10:1 15:15:1

Table 6. The grouping basis and limitation of the BAIPI quantification curve of green tea.

Groups	Ratio of Caffeine, EGCG, and Que-Rut	Concentration Range of Caffeine (mg/L)	Concentration Range of EGCG (mg/L)	Concentration Range of Que-Rut (mg/L)
1	10:20:1	150–240	300–480	15–24
2	10:10:1	230–240	230–240	15–24
3	15:25:1	150–310	250–517	15–21
4	15:20:1	172.5–310	230–414	11.5–21
5	15:15:1	230–310	230–310	15–21
6	20:35:1	150–310	262.5–542.5	7.5–15.5
7	25:35:1	164.25–310	230–434	6.57–12.4
8	30:40:1	172.5–310	230–413.33	5.75–10.33
9	35:45:1	178.89–310	230–398.57	5.11–8.86
10	35:40:1	201.25–310	230–354.29	5.75–8.86
11	40:55:1	167.27–310	230–426.25	4.18–7.75
12	45:85:1	153–296.47	289–560	3.4–6.82
13	45:80:1	153–310	272–551.11	3.4–6.89
14	45:75:1	153–310	255–516.67	3.4–6.89

Step 1. The lowest caffeine concentration in the selected group was used as the initial concentration, and the corresponding BAIPI value was defined as 1.

Step 2. Based on the initial concentration of caffeine, the first recombinant solution was prepared according to the proportion of compounds in this group. Subsequently, a series of recombinant solutions with varying compound concentrations within this group were established based on the lg11 increment.

Step 3. The first recombinant solution was provided to the subjects in pairs with other recombinant solutions. The tasting sequence between the two reconstituted solutions was randomized. With the first reconstituted solution held constant, the other recombinant solutions were paired with it for comparative tasting, with their presentation order following an ascending gradient of compound concentration.

Step 4. Tasting, rinsing, and recording followed Section 2.6 protocols. Until the selection rates, the first recombinant solution was less than or equal to 25% comparing with the nth recombinant solution, and the caffeine content in the nth recombinant solution was recorded as C_n.

Step 5. The BAIPI value was increased by one unit, and the caffeine concentration, denoted as C, corresponding to the updated BAIPI value was calculated according to this formula:

$$C = \sqrt{C_n \times C_{n-1}} \quad (1)$$

where C_{n-1} was the concentration of caffeine in the n − 1th recombinant solution.

Step 6. The caffeine concentration C calculated in Step 5 was used as the initial concentration of the next round. Sequentially, subjects were instructed to advance to Steps 2 to 6 if all compounds were within the concentration range or to Step 7 otherwise.

Step 7. The initial caffeine concentration and the caffeine concentration C calculated in each round were recorded as the x variable, and the BAIPI value was recorded as the y variable. Subsequently, the curve was fitted for this group.

After establishing the BAIPI quantification curve for a given group, Steps 1 to 7 were repeated for the next group. In the initial group, participants performed duplicate tests—both with and without nose clips. Results indicated that nasal occlusion had no significant effect on BAIPI curve establishment ($p > 0.05$). Consequently, we omitted nose clip usage in following trials.

2.8. Initial BAIPI Values Calibration

The group of BAIPI quantification curves corresponding to the initial compound with the lowest dose-over-threshold (DoT) values was designated as 'Group r'. The DoT value represents the ratio of a compound's concentration to its corresponding perception threshold [24]. The compound concentrations associated with BAIPI values within this group was used as references to calibrate the initial BAIPI values of other groups by 2-AFC tests. The specific steps were as follows:

Step 1. A series of recombinant solutions was prepared at the compound concentrations corresponding to the BAIPI values of 'Group r'.

Step 2. The recombinant solution described in Step 1 was provided to the subjects one by one in ascending order of BAIPI values. Concurrently, the first recombinant solution from each group (as described in Section 2.7) was also provided to subjects as a comparison sample. These solutions were presented in ascending order according to their initial compound DoT values.

Step 3. Tasting, rinsing, and recording followed Section 2.6 protocols. The BAIPI value of the recombinant solution in 'Group r' with the selection rates exceeding 25% but below 75% was the initial BAIPI value of the other group in the pairing comparison.

Step 4. When the selection rate remained at 25% or lower following the tests of all recombinant solutions in 'Group r', the group with the highest initial BAIPI value was chosen as a new 'Group r'. Steps 1 to 4 were repeated until the initial BAIPI values standardization was completed across all groups.

2.9. Direct Assessment for BAIPI Values

The concentration of bitter and astringent compounds corresponding to BAPI values at certain intervals was used as the scale (the specific scale was described in Section 3.4). Subjects underwent a three-week training period to become acquainted with the varying levels of bitter and astringent intensity. Subsequently, a 9-point intensity scale (1 = lowest, 9 = highest) was used for direct evaluation. The specific steps of providing samples, tasting, and gargling were carried out according to Section 2.6.

3. Results

3.1. Bitter–Astringent Simulation Solution of Green Tea

3.1.1. Bitter Compounds Analysis of Green Tea

Figure 1 shows the concentration of bitter and astringent compounds in green tea, as detected by HPLC, along with their corresponding bitter (red dashed line) and astringent (blue dotted line) perceived thresholds. Bitterness in green tea is mainly caused by alkaloids and catechins [2]. The primary alkaloids in green tea are caffeine, theobromine, and theophylline. Caffeine accounts for over 90% of the total alkaloid content and was the only alkaloid exceeding the bitter threshold (Figure 1A). In Figure 1B,C, the levels of theobromine and theophylline in green tea samples were very low. Theophylline's threshold was not found, but quantitative structure-activity relationship analysis suggests its content was below the threshold [33,34].

Figure 1. The concentration and perceived thresholds of bitter and astringent compounds (**A**) Caffeine; (**B**) Theobromine; (**C**) Theophylline; (**D**) EGCG; (**E**) ECG; (**F**) C; (**G**) EC; (**H**) GCG; (**I**) CG; (**J**) EGC; (**K**) GC; (**L**) GA; (**M**) Myr-rha; (**N**) Que-gal; (**O**) Que-glu; (**P**) Que-rut; (**Q**) Kae-rut; (**R**) Kae-glu in model samples. (Error lines were based on standard deviation; 'a–k' referred to the significant differences

between the groups were determined by one-way ANOVA analysis based on Duncan's multiple range test at $p \leq 0.05$).

Most of catechins contribute to both bitterness and astringency [2], including EGCG, (−)-epicatechin-3-gallate (ECG), (+)-catechin (C), (−)-epicatechin (EC), (−)-gallocatechin gallate (GCG), and (−)-catechin gallate (CG) (Figure 1). EGCG (65% of total catechins [35]) exceeded bitter and astringent thresholds in all samples (Figure 1D). ECG surpassed the bitter threshold in half of the samples but only exceeded the astringency threshold in three (Figure 1E), likely due to fewer hydroxyl groups on its B-ring compared to EGCG [2,36]. Non-galloylated catechins (C and EC) have higher thresholds than galloylated ones [2,36]. Only two green tea samples exceeded the astringency threshold (119.01 mg/L) for C (Figure 1F). EC showed a similar distribution to C, suggesting minimal impact on the bitterness and astringency (Figure 1G). GCG and CG were present in low amounts (Figure 1H,I) due to their conversion to EGCG and ECG at high temperatures [37,38].

3.1.2. Astringent Compounds Analysis of Green Tea

Catechins, phenolic acids, and flavonol glycosides mainly contribute to the astringency of green tea [2,29]. Apart from the above catechins, (-)-Epigallocatechin (EGC) and (-)-gallocatechin (GC) can only provide astringency to green tea, likely because they weakly interact with bitter receptors [36]. EGC (second most abundant after EGCG) exceeded its astringent threshold (159.26 mg/L) (Figure 1J). GC (a minor catechin) fell below the threshold due to processing losses [38] (Figure 1K). GA (linked to astringency, sourness, and sweet aftertaste [2]) had concentrations < 10 mg/L, making its contribution negligible (Figure 1L). Flavonol glycosides with extremely low astringent thresholds can impart a silky astringency to green tea [2]. All flavonol glycosides except for myricetin-3-o-rhamnoside (Myr-rha) (Figure 1M) exceeded thresholds (Figure 1N–R). In summary, 9 compounds contributed significantly to the bitterness and astringency of green tea, which were caffeine, EGCG, ECG, EGC, Quercetin-3-o-galactoside (Que-gal), Quercetin-3-o-glucoside (Que-glu), Quercetin-3-o-rutinoside (Que-rut), Kaempferol-3-o-rutinoside (Kae-rut), and Kaempferol-3-o-β-D-glucoside (Kae-glu).

3.1.3. Composition of Bitter–Astringent Simulation Solution

Additionally, the contribution of the 9 compounds to the overall bitter and astringent profiles of green tea was determined by 2-AFC test. Simulation solution 1 contained all 9 compounds. Simulation solutions 2 to 7 were created by progressively removing the compound with the lowest DoT value from the previous solution, while maintaining at least one representative from each perceived type. Finally, simulation solution 7 was composed of caffeine, EGCG, and Que-rut. According to the 2-AFC test results between green tea samples and simulation solutions in Table 7, the selection rates of simulation solutions 1, 2, and 3 were nearly all greater than or equal to 75%. This indicates that nearly all subjects perceived the simulation solution as exhibiting significantly greater bitterness and astringency ($\alpha \leq 0.05$) [32]. This may be attributed to GA, non-galloylated catechins, and galloylated catechin hydrolysis products in green tea, which likely reduce BAIPI through sweet aftertaste effects [2,29]. Five green tea samples showed no significant difference ($\alpha > 0.05$) in bitter and astringent intensity compared to simulation solution 4 and 5, with selectivity ranging from 25% to 75%. Simulation solutions 6–7 matched over 80% of samples after removing Kae-rut and Que-gal and 9 out of the 11 green tea samples showed no significant difference in bitter and astringent intensity from simulated solutions 6 and 7. Therefore, the minimal 3-compound combination (caffeine, EGCG, Que-rut) in solution 7 was identified as optimal. This result was similar to the taste reconstitution and omission experiment of black tea in the study of [24] and our study extends prior work by specifically examining individual flavonol glycosides' effect. Meanwhile, caffeine, EGCG, and Que-rut were used to develop recombinant solutions for BAIPI quantification curves.

Table 7. The percentage of responses considering the simulation solution more bitter and astringent compared to samples (n = 20).

Samples	Simulation Solution 1	Simulation Solution 2	Simulation Solution 3	Simulation Solution 4	Simulation Solution 5	Simulation Solution 6	Simulation Solution 7
LJS	85	80	75	35	15	55	35
LJM	80	75	85	25	20	50	45
BCS	80	70	75	20	20	35	25
HMS	75	80	65	40	35	65	40
HMML	90	75	85	75	60	75	65
LGS	85	70	65	25	25	70	50
SRS	95	80	75	35	65	70	55
SRSL	75	80	70	70	55	75	45
EGS	85	75	80	20	10	65	25
EGSL	80	90	85	25	35	60	55
SSS	75	65	75	45	25	55	60

Numbers highlighted in red (25 < x < 75) in the table denote no statistically significant difference between the two samples (α > 0.05).

3.2. BAIPI Quantification Curves in Green Tea

According the actual ratio of caffeine, EGCG, and Que-rut in model-samples (Table 5), 14 groups with similar ratios were selected to construct BAIPI quantification curves (Table 6). Table 8 shows the BAIPI quantification curves for different groups divided based on the ratio of caffeine, EGCG, and Que-rut of model-samples. When the BAIPI value of Group 2 reached 2, the concentration of caffeine, EGCG, and Que-rut exceeded the content range of model-samples specified in Table 6. As a result, it was not feasible to establish the curve in Group 2. Logarithmic curve fitting for several BAIPI plots across different groups showed that they all followed the same equation form, except for Group 2:

$$y = a \times \ln(x - b) - c \quad (2)$$

where a, b, and c were constants, x was the concentration of caffeine in this group, and y was the value of BAIPI. It can be observed that all parameter values of a are greater than zero ($a > 0$), indicating a positive correlation between BAIPI values and caffeine concentration, where larger magnitudes of a correspond to accelerated rates of BAIPI change. The parameter b defines the effective concentration threshold of caffeine, as the logarithmic term $\ln(x - b)$ requires $x > b$ to maintain mathematical validity. The parameter c modulates the vertical offset of the baseline, effectively shifting the entire dose-response curve to calibrate BAIPI reference values. As caffeine concentration approaches b ($x \to b^+$), BAIPI values plummet, reflecting heightened sensitivity to minimal concentration variations. When caffeine concentration substantially exceeds b ($x \gg b$), the growth rate of BAIPI diminishes progressively due to the asymptotic nature of logarithmic functions, ultimately approaching saturation.

The goodness of fit (R^2) and adjusted R^2 of the other groups were greater than 0.99, and the standard error of estimate (SEE) and residual sum of squares (RSS) were less than 0.3, indicating that a high level of curve fitting and predictive accuracy. However, in this study, the Weber fraction increased with BAIPI values (Figure 2A–D), indicating reduced perceptual sensitivity at higher BAIPI levels. This phenomenon also appeared in the study to determine the human sweetness and sourness interactive response [22]. This could be attributed to the interactive effect of the two sensations, resulting in alterations to the

Weber fraction when the intensity of stimulation surpassed a certain threshold [7,22,24,39]. Equation (2) conformed to the complete form of Fechner law [39], which indicated that the interactive perceived intensity of the two sensations (e.g., bitterness and astringency) conformed to the Weber–Fechner law.

Table 8. The specific information of BAIPI quantification curves of green tea.

Group	Code of Curve	m^{α}	n^{β}	Equation	$R^{2\ \gamma}$	Adjust R^2	SEE $^{\delta}$	RSS $^{\varepsilon}$
Group 1	Fit curve 1	10:20:1	0.50	$y = 3.21 \times \ln(x - 142.4) - 5.47$	0.9969	0.9960	0.1920	0.2580
Group 2	Fit curve 2	10:10:1	1.00	-	-	-	-	-
Group 3	Fit curve 3	15:25:1	0.60	$y = 4.20 \times \ln(x - 120.17) - 13.3$	0.9993	0.9991	0.0833	0.0416
Group 4	Fit curve 4	15:20:1	0.75	$y = 2.90 \times \ln(x - 155.98) - 6.95$	0.9970	0.9958	0.1592	0.1268
Group 5	Fit curve 5	15:15:1	1.00	$y = 4.20 \times \ln(x - 183.86) - 15.13$	0.9981	0.9968	0.1062	0.0338
Group 6	Fit curve 6	20:35:1	0.57	$y = 3.56 \times \ln(x - 121.79) - 10.87$	0.9991	0.9988	0.0852	0.0363
Group 7	Fit curve 7	25:35:1	0.71	$y = 3.69 \times \ln(x - 137.43) - 11.22$	0.9988	0.9983	0.0999	0.0499
Group 8	Fit curve 8	30:40:1	0.75	$y = 4.24 \times \ln(x - 137.59) - 14.11$	0.9990	0.9986	0.0905	0.0410
Group 9	Fit curve 9	35:45:1	0.78	$y = 2.90 \times \ln(x - 161.77) - 7.05$	0.9970	0.9958	0.1593	0.1268
Group 10	Fit curve 10	35:40:1	0.88	$y = 4.42 \times \ln(x - 157.55) - 15.74$	0.9987	0.9980	0.0960	0.0369
Group 11	Fit curve 11	40:55:1	0.73	$y = 3.64 \times \ln(x - 140.55) - 11.04$	0.9988	0.9983	0.1019	0.0519
Group 12	Fit curve 12	45:85:1	0.53	$y = 3.79 \times \ln(x - 126.73) - 11.44$	0.9991	0.9988	0.0960	0.0552
Group 13	Fit curve 13	45:80:1	0.56	$y = 3.56 \times \ln(x - 124.22) - 10.94$	0.9991	0.9988	0.0852	0.0363
Group 14	Fit curve 14	45:75:1	0.60	$y = 4.64 \times \ln(x - 111.1) - 16.28$	0.9991	0.9988	0.0940	0.0530

α—the concentration ratio of caffeine, EGCG, and Que-rut; β—the concentration ratio of caffeine and EGCG; γ—goodness of fit; δ—standard error of estimate; ε—residual sum of squares.

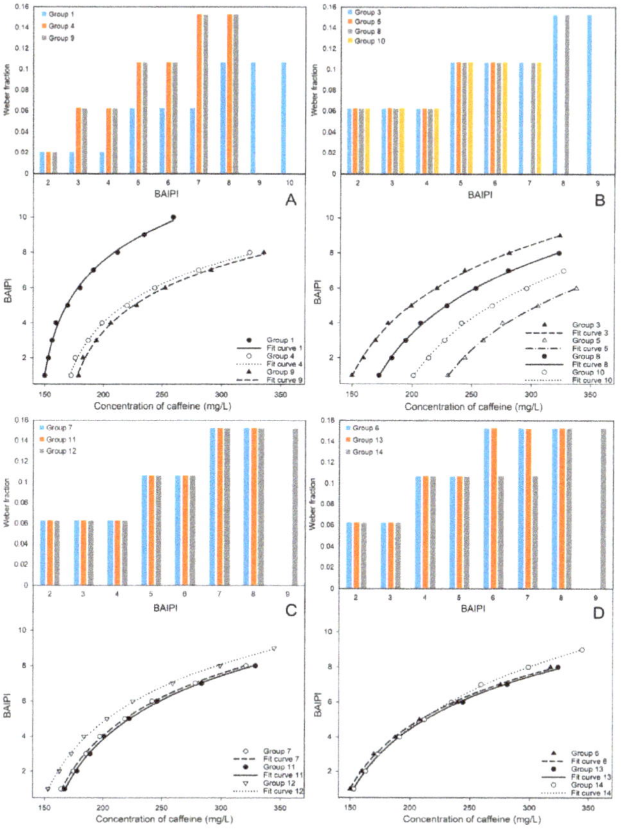

Figure 2. The Weber fraction and fit curves for 13 groups of BAIPI quantification curves (**A**) Group 1, 4 and 9; (**B**) Group 3, 5, 8 and 10; (**C**) Group 7, 11 and 12; (**D**) Group 6, 13 and 14.

3.2.1. X-Intercept of Curves

In Figure 2A, Groups 1, 4, and 9 showed an initial Weber fraction of 0.02. Groups 4 and 9 exhibited identical trends in Weber fraction changes. Additionally, it was also observed that fit curves 4 and 9 had similar degree of curvature and inclination, while fit curve 1 was quite different from them. In Figure 2B, a similar phenomenon was also observed in Groups 3, 5, 8, and 10, where the change of Weber fraction and the curvature and inclination of fit curves were similar. However, a distinct observation was the gradual rightward shift along the X-axis of fit curves 3, 8, 10, and 5 as the ratio of caffeine and EGCG (n) increased. In Figure 2C, fit curves 12, 7, and 11 were also arranged in sequence along the X-axis, with the corresponding n being 0.53, 0.71, and 0.73, respectively. Among them, the position of fit curves 7 and 11 in the figure was closer due to the n of Groups 7 and 11 was relatively close. The n of Groups 13, 6, and 14 were very close, about 0.58, resulting in their nearly overlapping positions in Figure 2D. Therefore, the curvature and inclination of the fit curve was correlated with the change trend of Weber fraction, and the position of the fit curve on the X-axis was closely associated with n.

Figure 3A shows the BAIPI value points and fit curves of all groups, and it was found that the fit curve gradually shifted to the right with the increase of n in general. This phenomenon appeared due to the gradual increase of n from 0.5 to 1.0, where the molecular quantities of caffeine and EGCG approached stoichiometric equilibrium. Under these conditions, caffeine and EGCG preferentially formed 2:2 heterotypic complexes [40]. These complexes competitively inhibited EGCG's binding affinity for salivary proteins, thereby attenuating astringency perception [41]. To maintain equivalent BAIPI values, compensatory elevation of total caffeine concentration (x) became requisite to achieve parity in overall sensory stimulation. The X-intercept, as an approximate indicator of the position of the fit curve on the X-axis, was used to conduct a correlation analysis with the n of all groups as shown in Figure 3B. It was found that the correlation fitted into exponential curves, and the equation form was as follows:

$$IC = 123.0742 + 3.3603 \times \exp 3.3852n \tag{3}$$

where n was the ratio of caffeine and EGCG in different groups, and IC was the X-intercept of the fit curve. The IC value, representing the caffeine concentration required to achieve a null BAIPI (relatively complete bitter–astringent suppression), exhibited an exponential amplification relationship with n. Specifically, at n = 0.5, $IC \approx$ 141.2 mg/L; at n = 1.0, $IC \approx$ 222.5 mg/L. This 57% increase in IC despite caffeine proportion doubling revealed a nonlinear caffeine-mediated bitter–astringent suppression efficacy reduction. The R^2 and SEE of the exponential curve were 0.9786 and 3.6627, respectively, indicating a strong correlation between n and X-intercept of the fit curve. The X-intercept of Equation (2) was as follows:

$$IC = \exp(c/a) + b \tag{4}$$

Substitute Equation (3) into Equation (4), the constant term c of Equation (2) could be calculated according to the following equation:

$$\exp(c/a) + b = 123.0742 + 3.3603 \times \exp 3.3852n \tag{5}$$

$$c = a \times \ln(123.0742 + 3.3603 \times \exp 3.3852n - b) \tag{6}$$

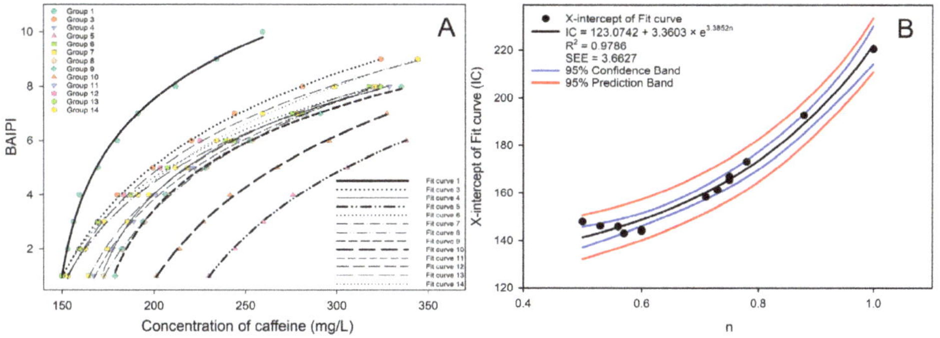

Figure 3. All the fit curves (**A**) and the correlation analysis between the concentration ratio of caffeine and EGCG and the X-intercept (**B**) for 13 groups of BAIPI quantification curves.

3.2.2. Slope of Curves

It was difficult to observe the regularity for the curvature and inclination of fit curves in Figure 3A, so an attempt was made to establish a correlation between the n and the slopes at different points of all the fit curves. Since the b of fit curve 5 was 183.86 (requiring N of ln N to be greater than 0), the slopes of the fitted curves at x = 190, 200, and 210 were calculated for correlation analysis with the n, which conformed the exponentially correlation (Figure 4A–C):

$$k_{190} = 0.0639 + 3.8717 \times 10^{-8} \times \exp 16.5879n \quad (7)$$

$$k_{200} = 0.0495 + 5.5375 \times 10^{-6} \times \exp 10.544n \quad (8)$$

$$k_{210} = 0.0420 + 2.0690 \times 10^{-5} \times \exp 8.6555n \quad (9)$$

where k was the slope of the fit curve at x = 190, 200, and 210. As x increased (from 190 to 210), the exponential coefficient decreased (from 16.5879 to 8.6555), indicating that in the high-concentration range, the sensitivity of k to n is reduced.

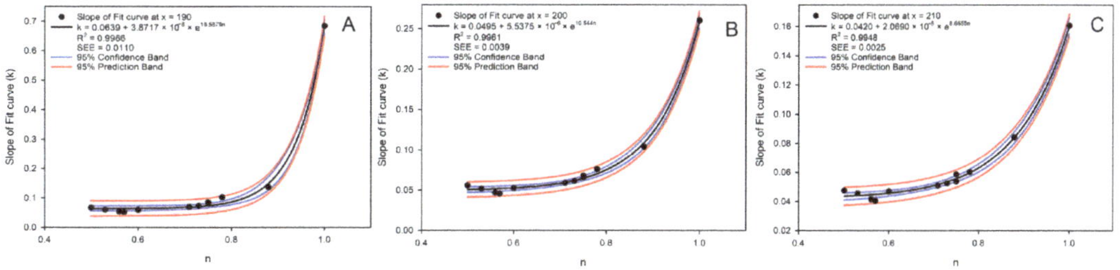

Figure 4. The correlation analysis between the concentration ratio of caffeine and EGCG and the slope at x = 190 (**A**), 200 (**B**), and 210 (**C**) for 13 groups of BAIPI quantification curves.

The R^2 of the above exponential curves were 0.9966, 0.9961, and 0.9948, respectively, and the SEEs were 0.0110, 0.0039, and 0.0025, respectively. The R^2 for the slope of the fit curve at the point following x = 220 was lower than 0.99, so the slopes of the three points were used to calculate a and b of Equation (2). Additionally, the slope of the fit curve at $x = x_0$ was as follows:

$$k_{x_0} = a/(x_0 - b) \quad (10)$$

Substitute Equations (7)–(9) into Equation (10), the constant term a and b of Equation (2) could be calculated according to the following equation:

$$a/(190-b) = 0.0639 + 3.8717 \times 10^{-8} \times \exp 16.5879n \quad (11)$$

$$a/(200-b) = 0.0495 + 5.5375 \times 10^{-6} \times \exp 10.544n \quad (12)$$

$$a/(210-b) = 0.0420 + 2.0690 \times 10^{-5} \times \exp 8.6555n \quad (13)$$

Equations (10)–(12) were combined in pairs, and the n values of all groups were substituted into to calculate *a* and *b*. These values then substituted into Equation (6) to calculate the corresponding *c*, as shown in Table 9. It was found that only the *a* and *b* values calculated by the combination of Equations (12) and (13) could substituted into Equation (6), so *a* and *b* were:

$$a = \frac{20790 + 2.3258 \times \exp 10.544n + 10.2416 \times \exp 8.6555n + 0.0011 \times \exp 19.1995n}{7500 + 5.5375 \times \exp 10.544n - 20.69 \times \exp 8.6555n} \quad (14)$$

$$b = \frac{1080000 - 4344.9 \times \exp 8.6555n + 1107.5 \times \exp 10.544n}{7500 + 5.5375 \times \exp 10.544n - 20.69 \times \exp 8.655n} \quad (15)$$

Table 9. Different sets of equations solving *a*, *b* and *c* of Equation (2).

n^{α}	Equations (6), (11) and (12)			Equations (6), (11) and (13)			Equations (6), (12) and (13)		
	a	*b*	*c*	*a*	*b*	*c*	*a*	*b*	*c*
0.50	2.40	152.47	-	2.72	147.47	-	3.14	137.86	3.91
0.53	2.48	151.30	-	2.81	146.23	-	3.23	136.62	6.13
0.56	2.59	149.70	-	2.92	144.65	−0.66	3.34	135.27	7.74
0.57	2.64	149.05	-	2.96	144.03	2.32	3.37	134.81	8.21
0.60	2.81	146.59	2.08	3.12	141.84	6.00	3.50	133.45	9.53
0.71	4.28	127.95	14.87	4.12	130.25	14.01	3.97	133.11	13.09
0.73	4.74	123.19	17.44	4.35	128.71	15.35	4.01	134.91	13.35
0.75	5.21	119.35	19.97	4.54	128.33	16.45	4.02	137.54	13.42
0.78	5.70	118.77	22.45	4.70	131.23	17.22	3.99	142.96	13.18
0.88	4.07	162.61	13.34	3.87	163.93	12.50	3.65	166.04	11.48
1.00	4.19	183.87	15.28	4.21	183.84	15.35	4.12	183.73	15.06

α: The concentration ratio of caffeine and EGCG.

3.3. Initial BAIPI Values Calibration of Quantification Curves

The initial BAIPI values of all quantification curves were calibrated using 2-AFC tests (Table 10). While previous studies on taste intensity quantification typically used absolute thresholds as starting points [14,42,43], this approach may not adequately represent the actual concentration ranges of compounds in specific foods, potentially introducing bias in practical applications [44]. Therefore, the minimum concentration of compounds found in green tea was employed as the quantification baseline in this research. Based on the perceived thresholds and concentration of caffeine, EGCG, and Que-rut in Tables 2 and 6, Group 14 (with the lowest initial compound DoT values) was designated as the first reference group ('Group r'). Groups 6–13 were calibrated against 'Group r' via 2-AFC tests (Table 10). Groups 6 and 7 showed the highest initial BAIPI value (9), qualifying them as potential new reference groups and Group 6 was selected to be the new 'Group r' because its DoT values were closer to the remaining groups. Thus, the initial BAIPI values of Groups 1–5 were then calibrated against the new 'Group r'. A linear correlation was observed between the initial BAIPI values of all groups and the ratio of caffeine to Que-rut (q) (Figure 5):

$$IV = 18.7960 - 0.3956q \tag{16}$$

where q was the ratio of caffeine and Que-rut of different groups, and IV was initial BAIPI values.

Table 10. Standardized BAPI values.

q^{α}		Group	BAPI Values																							
45	1	14	1	2	3	4	5	6	7	8	9															
45	1	13	1	2	3	4	5	6	7	8																
45	1	12	1	2	3	4	5	6	7	8	9															
40	1	11			3	4	5	6	7	8	9	10														
35	1	10						6	7	8	9	10	11	12												
35	1	9					5	6	7	8	9	10	11	12												
30	1	8						6	7	8	9	10	11	12	13											
25	1	7									9	10	11	12	13	14	15	16								
20	1	6									9	10	11	12	13	14	15	16								
15	1	5															15	16	17	18	19	20				
15	1	4													13	14	15	16	17	18	19	20				
15	1	3												12	13	14	15	16	17	18	19	20				
10	1	1															15	16	17	18	19	20	21	22	23	24

α: The concentration ratio of caffeine and Que-rut.

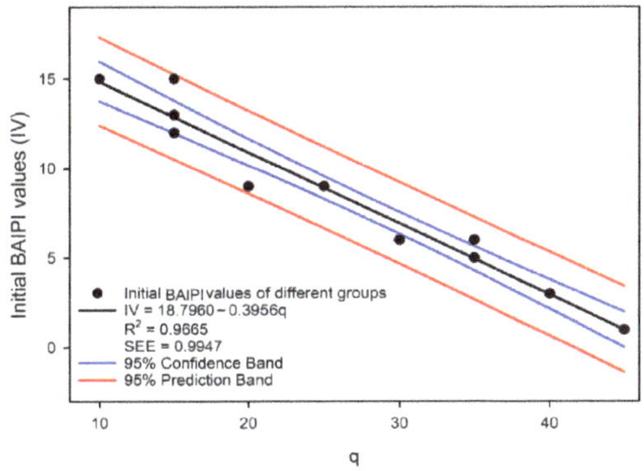

Figure 5. The correlation analysis between the concentration ratio of caffeine and Que-rut and the initial BAIPI values for 13 groups of BAIPI quantification curves.

The R^2 of the linear correlation model was 0.9665, and the SEE was 0.9947, indicating that the model was acceptable. Since the initial BAIPI value of all the fit curves in Table 8 was set to 1, the Equation (2) should be modified as follow:

$$y = a \times \ln(x - b) - c + IV - 1 \tag{17}$$

3.4. Verification of BAIPI Quantification Curves in Green Tea

The applicability of the BAIPI quantification curve was assessed through correlation analysis between calculated and directly assessed BAIPI values (Figure 6). The concentration of caffeine, EGCG, and Que-rut of green tea samples were input into Equations (6) and (14)–(16) calculated the corresponding a, b, c and IV. These parameters were then substituted into Equation (17) to calculate final BAIPI values (Table 11). Direct assessment of BAIPI values was based on a 9-point intensity scale (1 = lowest, 9 = highest). Scale reference points corresponded to the concentration of caffeine, EGCG, and Que-rut corresponding to the BAIPI values of 1, 3, 6 (in Group 14), 9, 12, 15 (in Group 6), 18, 21 and 24 (in Group 1) in Table 10. In Figure 6, the R^2 of the correlation model based on the model, validation and all samples were 0.9475, 0.9418 and 0.9439, respectively, and SEEs were 1.4537, 1.5865 and 1.5099, respectively. These results demonstrate the model's ability to reliably estimate the comprehensive bitter–astringent perception in green tea.

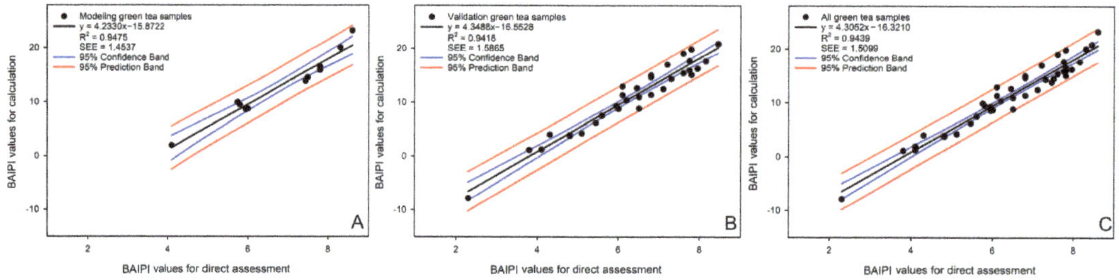

Figure 6. The correlation analysis between the BAIPI values calculated by BAIPI quantification curves and those obtained by direct assessment based on model (**A**), validation (**B**), and all samples (**C**).

Table 11. BAIPI values calculated by BAIPI quantification curves and those obtained by direct assessment.

Samples	Concentration of Caffeine (x, mg/L)	m^{α}	n^{β}	q^{γ}	$y = a \times \ln(x - b) - c + IV - 1$				BAIPI Values for Calculation (y)	BAIPI Values for Direct Assessment
					IV	a	b	c		
Model										
LJS	282.31	25:34:1	0.74	25	8.91	4.02	136.12	13.41	14.54	7.5
LJM	258.74	38:53:1	0.73	38	3.76	4.01	134.91	13.35	8.73	6
BCS	303.93	44:78:1	0.57	44	1.39	3.37	134.81	8.21	9.49	5.8
HMS	227.70	34:42:1	0.82	34	5.35	3.86	152.08	12.41	8.63	5.95
HMML	157.18	44:75:1	0.60	44	1.39	3.50	133.45	9.53	1.94	4.1
LGS	306.55	18:33:1	0.56	18	11.68	3.34	135.27	7.74	20.09	8.3
SRS	235.81	13:13:1	1.00	13	13.65	4.12	183.73	15.06	13.89	7.45
SRSL	257.26	12:13:1	0.96	12	14.05	3.78	179.44	12.88	16.61	7.8
EGS	300.35	45:83:1	0.54	45	0.99	3.26	136.18	6.70	9.94	5.75
EGSL	215.27	16:23:1	0.70	16	12.47	3.94	132.50	12.90	15.95	7.8
SSS	215.57	9:18:1	0.51	9	15.24	3.17	137.46	4.77	23.28	8.6
Validation										
FWS	283.56	20:22:1	0.91	20	10.88	3.63	171.33	11.54	15.44	7.6
WJS	306.13	26:37:1	0.70	26	9.30	3.92	132.32	12.82	15.05	6.8
GMS	213.78	33:56:1	0.60	33	5.74	3.50	133.47	9.52	10.37	6.2
GNS	192.33	34:58:1	0.59	34	5.74	3.45	134.00	9.01	9.34	5.95
XTS	269.28	60:96:1	0.62	60	−4.94	3.61	132.51	10.49	1.25	4.1
XGR	272.68	40:76:1	0.53	40	3.37	3.25	136.42	6.40	11.35	6.1
YTS	283.85	23:26:1	0.90	23	10.09	3.63	170.29	11.49	14.37	7.3
YGR	195.97	17:26:1	0.65	17	12.47	3.75	131.75	11.63	15.13	7.8
LTB	209.87	17:27:1	0.63	17	12.47	3.63	132.37	10.65	16.30	7.95
LTA	201.61	17:31:1	0.55	17	12.47	3.29	135.82	7.13	17.70	7.75
AWH	255.99	40:47:1	0.84	40	3.37	3.79	156.17	12.06	7.54	5.6
AWL	204.33	51:70:1	0.73	51	−1.38	4.01	134.99	13.36	1.13	3.8
BCH	239.47	50:72:1	0.69	50	−0.59	3.90	132.08	12.67	3.77	4.8
BCL	249.82	55:97:1	0.56	55	−2.57	3.34	135.24	7.77	4.25	5.1
YTS	306.85	55:87:1	0.64	55	−2.57	3.66	132.15	10.93	3.96	4.8
HMH	224.07	39:48:1	0.83	39	3.37	3.84	153.43	12.29	6.22	5.45
HML	232.93	35:45:1	0.76	35	6.53	4.01	139.95	13.35	8.89	6.5
THK	204.17	22:31:1	0.70	22	9.70	3.94	132.62	12.95	12.95	6.1
TBJ	277.26	15:16:1	0.92	15	13.26	3.64	174.17	11.77	17.03	7.2
JLT	228.59	26:37:1	0.71	26	8.51	3.97	133.23	13.12	12.47	7.1
LGL	206.88	22:36:1	0.60	22	10.49	3.52	133.28	9.70	14.71	6.8
SMF	216.72	45:57:1	0.78	45	0.99	3.98	143.44	13.14	4.01	4.3
HYW	250.67	37:47:1	0.78	37	4.55	3.99	142.86	13.18	8.81	6
LMH	260.01	35:56:1	0.63	35	5.35	3.62	132.40	10.62	10.97	6.5
TTS	238.36	20:20:1	1.00	20	11.28	4.11	183.67	15.01	11.40	6.8
DGS	246.54	22:23:1	0.94	22	10.49	3.69	177.12	12.25	12.67	6.45
YQS	261.78	79:97:1	0.82	79	−14.43	3.86	152.11	12.41	−7.84	2.3
EGK	294.86	17:18:1	0.96	17	12.07	3.77	179.33	12.84	15.95	7.75
FGT	299.18	14:16:1	0.85	14	13.65	3.75	159.02	11.84	19.07	7.6
FKY	271.52	13:21:1	0.62	13	14.05	3.60	132.55	10.46	19.88	7.8
SKK	172.98	12:23:1	0.50	12	14.44	3.13	138.02	3.48	20.88	8.45
KSC	270.69	20:32:1	0.61	20	11.28	3.52	133.21	9.77	17.67	8.15

α–the concentration ratio of caffeine, EGCG, and Que-rut; β–the concentration ratio of caffeine and EGCG; γ–the concentration ratio of caffeine and Que-rut.

4. Conclusions

This study developed a quantification curve to evaluate BAIPI in green tea, based on its unique flavor profile. Caffeine, EGCG, and Que-rut were identified as primary contributors to BAIPI, which were also subsequently employed to develop the quantification curve. The utilization of 2-AFC test enhanced the assessment reliability of subjects in this research, with all the fit curves conforming to the Weber–Fechner law. The slope and X-intercept of the fit curve were exponentially related to the concentration ratio of caffeine and EGCG (n). The initial BAIPI values of the fit curve were linearly associated with the concentration ratio of caffeine and Que-rut (q). Thus, BAIPI values were calculated by inputting concentrations of the three key compounds into the BAIPI quantification curve. Validation showed strong correlation with direct assessment. The R^2 and SEE of the correlation analysis were about 0.94 and 1.51, respectively. This work advances the objective assessment of tea sensory properties while offering practical tools for industry applications. However, further research integrating biochemical, genetic, and larger-scale sensory data will strengthen the robustness of this approach. These findings pave the way for more precise flavor profiling, quality control, and consumer-driven product design in the tea and broader beverage sector.

Author Contributions: Conceptualization, S.L. and Z.W.; methodology, S.L.; software, D.D.; validation, S.C.; formal analysis, D.D.; investigation, S.C.; resources, D.D.; data curation, S.L.; writing—original draft preparation, S.L.; writing—review and editing, S.L.; visualization, Z.W.; supervision, Z.W.; project administration, Z.W.; funding acquisition, Z.W. All authors have read and agreed to the published version of the manuscript.

Funding: The work was funded by the National Natural Science Foundation of China [grant number 31972200].

Institutional Review Board Statement: Ethical approval for the involvement of human subjects in this study was granted by the Medical Ethics Committee in School of Medicine of Zhejiang University, Reference number 2019-046, dtd 3 April 2019. All subjects signed a consent form to participate in the sensory and consumer tests.

Informed Consent Statement: Informed consent was obtained from all subjects involved in the study. Written informed consent has been obtained from the patients to publish this paper.

Data Availability Statement: Data will be made available on request.

Acknowledgments: The authors are grateful to the financial support from the National Natural Science Foundation of China (NSFC, 31972200) and technical support from Zhejiang University.

Conflicts of Interest: All authors declared that there was no conflicts of interest.

References

1. Wang, J.Q.; Fu, Y.Q.; Chen, J.X.; Wang, F.; Feng, Z.H.; Yin, J.F.; Zeng, L.; Xu, Y.Q. Effects of baking treatment on the sensory quality and physicochemical properties of green tea with different processing methods. *Food Chem.* **2022**, *380*, 132217. [CrossRef] [PubMed]
2. Ye, J.-H.; Ye, Y.; Yin, J.-F.; Jin, J.; Liang, Y.-R.; Liu, R.-Y.; Tang, P.; Xu, Y.-Q. Bitterness and astringency of tea leaves and products: Formation mechanism and reducing strategies. *Trends Food Sci. Technol.* **2022**, *123*, 130–143. [CrossRef]
3. Lu, L.; Wang, L.; Liu, R.Y.; Zhang, Y.B.; Zheng, X.Q.; Lu, J.L.; Wang, X.C.; Ye, J.H. An efficient artificial intelligence algorithm for predicting the sensory quality of green and black teas based on the key chemical indices. *Food Chem.* **2024**, *441*, 138341. [CrossRef]
4. Romeo-Arroyo, E.; Mora, M.; Noguera-Artiaga, L.; Vazquez-Araujo, L. Tea pairings: Impact of aromatic congruence on acceptance and sweetness perception. *Curr. Res. Food Sci.* **2023**, *6*, 100432. [CrossRef]
5. Chong, P.H.; Chen, J.; Yin, D.; Qin, L. Tea compound-saliva interactions and their correlations with sweet aftertaste. *Npj Sci. Food* **2022**, *6*, 13. [CrossRef]
6. Nolden, A.A.; Feeney, E.L. Genetic Differences in Taste Receptors: Implications for the Food Industry. *Annu. Rev. Food Sci. Technol.* **2020**, *11*, 183–204. [CrossRef]

7. Yin, J.F.; Zhang, Y.N.; Du, Q.Z.; Chen, J.X.; Yuan, H.B.; Xu, Y.Q. Effect of Ca^{2+} concentration on the tastes from the main chemicals in green tea infusions. *Food Res. Int.* **2014**, *62*, 941–946. [CrossRef]
8. Zhang, Y.N.; Yin, J.F.; Chen, J.X.; Wang, F.; Du, Q.Z.; Jiang, Y.W.; Xu, Y.Q. Improving the sweet aftertaste of green tea infusion with tannase. *Food Chem.* **2016**, *192*, 470–476. [CrossRef]
9. Chong, P.H.; Chen, J.; Yin, D.; Upadhyay, R.; Mo, L.; Han, L. "Oral" tribology study on saliva-tea compound mixtures: Correlation between sweet aftertaste (Huigan) perception and friction coefficient. *Food Res. Int.* **2019**, *125*, 108642. [CrossRef]
10. Jiang, Y.; Gong, N.N.; Matsunami, H. Astringency: A More Stringent Definition. *Chem. Senses* **2014**, *39*, 467–469. [CrossRef]
11. Lyu, J.; Fu, J.; Chen, S.; Xu, Y.; Nie, Y.; Tang, K. Impact of tannins on intraoral aroma release and retronasal perception, including detection thresholds and temporal perception by taste, in model wines. *Food Chem.* **2022**, *375*, 131890. [CrossRef] [PubMed]
12. Zeng, Y.; Liang, J.H.; Zhang, J.Y.; Chen, W.; Hu, D.; Xia, H.L.; Ma, C.Y.; Qiao, X.Y. Improvement of the sensory characteristics of Dancong tea using a dry-heating post-treatment approach. *LWT—Food Sci. Technol.* **2023**, *185*, 115118. [CrossRef]
13. De Toffoli, A.; Spinelli, S.; Monteleone, E.; Arena, E.; Di Monaco, R.; Endrizzi, I.; Toschi, T.G.; Laureati, M.; Napolitano, F.; Torri, L.; et al. Influences of Psychological Traits and PROP Taster Status on Familiarity with and Choice of Phenol-Rich Foods and Beverages. *Nutrients* **2019**, *11*, 1329. [CrossRef] [PubMed]
14. Mao, Y.Z.; Tian, S.Y.; Qin, Y.M.; Han, J.Z. A new sensory sweetness definition and sweetness conversion method of five natural sugars, based on the Weber-Fechner Law. *Food Chem.* **2019**, *281*, 78–84. [CrossRef]
15. Hayashi, N.; Ujihara, T.; Chen, R.; Irie, K.; Ikezaki, H. Objective evaluation methods for the bitter and astringent taste intensities of black and oolong teas by a taste sensor. *Food Res. Int.* **2013**, *53*, 816–821. [CrossRef]
16. Ye, Q.Q.; Chen, G.S.; Pan, W.; Cao, Q.Q.; Zeng, L.; Yin, J.F.; Xu, Y.Q. A predictive model for astringency based on in vitro interactions between salivary proteins and (−)-Epigallocatechin gallate. *Food Chem.* **2021**, *340*, 127845. [CrossRef]
17. Hayashi, N.; Chen, R.; Hiraoka, M.; Ujihara, T.; Ikezaki, H. β-Cyclodextrin/Surface Plasmon Resonance Detection System for Sensing Bitter-Astringent Taste Intensity of Green Tea Catechins. *J. Agric. Food Chem.* **2010**, *58*, 8351–8356. [CrossRef]
18. Li, S.; Du, D.; Wang, J.; Wei, Z. Application progress of intelligent flavor sensing system in the production process of fermented foods based on the flavor properties. *Crit. Rev. Food Sci. Nutr.* **2022**, *64*, 3764–3793. [CrossRef]
19. Zhang, J.; Zhao, X.; Li, L.; Chen, W.; Zhao, Q.; Su, G.; Zhao, M. Application of electronic tongue in umami detection and soy sauce refining process. *Food Chem. X* **2023**, *18*, 100652. [CrossRef]
20. Portugal, R.D.; Svaiter, B.F. Weber-Fechner Law and the Optimality of the Logarithmic Scale. *Minds Mach.* **2011**, *21*, 73–81. [CrossRef]
21. Mao, Y.; Tian, S.; Qin, Y.; Cheng, S. An optimized organic acid human sensory sourness analysis method. *J. Sci. Food Agric.* **2021**, *101*, 5880–5887. [CrossRef] [PubMed]
22. Mao, Y.Z.; Tian, S.Y.; Qin, Y.M.; Chen, S.W. Sensory sweetness and sourness interactive response of sucrose-citric acid mixture based on synergy and antagonism. *Npj Sci. Food* **2022**, *6*, 33. [CrossRef] [PubMed]
23. Wang, Y.; Wan, S.; Pan, S.; Luan, P.; Cen, X.; Sun, G.; Feng, Y.; Zhu, M. A quantitative equation for umami evaluation of food and condiments based on the Weber-Fechner law. *Food Qual. Prefer.* **2025**, *123*, 105347. [CrossRef]
24. Scharbert, S.; Hofmann, T. Molecular definition of black tea taste by means of quantitative studies, taste reconstitution, and omission experiments. *J. Agric. Food Chem.* **2005**, *53*, 5377–5384. [CrossRef]
25. Kishi, M.; Sadachi, H.; Nakamura, J.; Tonoike, M. Functional magnetic resonance imaging investigation of brain regions associated with astringency. *Neurosci. Res.* **2017**, *122*, 9–16. [CrossRef]
26. Huang, Y.L.; Vivian, R.M.; Goh, A.; Quan, S.; Sakumoto, S.; Yun, H.; Huey, K.; Sun, J.C.; Lassabliere, B.; Yu, B. Effect of three milling processes (cyclone-, bead- and stone-millings) on the quality of matcha: Physical properties, taste and aroma. *Food Chem.* **2022**, *372*, 131202. [CrossRef]
27. Liu, Y.; Luo, L.H.; Liao, C.X.; Chen, L.; Wang, J.; Zeng, L. Effects of brewing conditions on the phytochemical composition, sensory qualities and antioxidant activity of green tea infusion: A study using response surface methodology. *Food Chem.* **2018**, *269*, 24–34. [CrossRef]
28. Yu, J.Y.; Liu, Y.; Zhang, S.R.; Luo, L.Y.; Zeng, L. Effect of brewing conditions on phytochemicals and sensory profiles of black tea infusions: A primary study on the effects of geraniol and β-ionone on taste perception of black tea infusions. *Food Chem.* **2021**, *354*, 129504. [CrossRef]
29. Zhang, L.; Cao, Q.Q.; Granato, D.; Xu, Y.Q.; Ho, C.T. Association between chemistry and taste of tea: A review. *Trends Food Sci. Technol.* **2020**, *101*, 139–149. [CrossRef]
30. Daniel Estevez-Lopez, R.; Garcia-Gomez, B.; Lourdes Vazquez-Oderiz, M.; Munoz Ferreiro, N.; Angeles Romero-Rodriguez, M. Influence of bread shape on the sensory characteristics of Galician breads: Development of lexicon, efficacy control of the trained panel and establishment of a sensory profile. *LWT—Food Sci. Technol.* **2021**, *135*, 110024.
31. *ISO 8586*; Sensory Analysis. General Guidelines for the Selection, Training and Monitoring of Selected Assessors and Expert Sensory Assessors. International Organization for Standardization: Geneva, Switzerland, 2012.

32. *ISO 5495*; Sensory Analysis. Methodology. Paired Comparison Test. International Organization for Standardization: Geneva, Switzerland, 2005.
33. Kim, H.O.; Li-Chan, E.C.Y. Quantitative structure-activity relationship study of bitter peptides. *J. Agric. Food Chem.* **2006**, *54*, 10102–10111. [CrossRef] [PubMed]
34. Feng, B.L.; Ren, H.B.; Duan, J.H.; Zhang, H.S.; Wen, C.H.; Bai, X.S.; Gao, F.; Wang, Y.T. Molecular recognition and threshold prediction model of bitterness in natural compounds. *Sci. Technol. Food Ind.* **2022**, *43*, 24–32.
35. Shitandi, A.A.; Muigai Ngure, F.; Mahungu, S.M. Chapter 16—Tea Processing and its Impact on Catechins, Theaflavin and Thearubigin Formation. In *Tea in Health and Disease Prevention*; Preedy, V.R., Ed.; Academic Press: Cambridge, MA, USA, 2013; pp. 193–205.
36. Narukawa, M.; Kimata, H.; Noga, C.; Watanabe, T. Taste characterisation of green tea catechins. *Int. J. Food Sci. Technol.* **2010**, *45*, 1579–1585. [CrossRef]
37. Rong, W.; Zhou, W.; Wen, R.A.H. Kinetic study of the thermal stability of tea catechins in aqueous systems using a microwave reactor. *J. Agric. Food Chem.* **2006**, *54*, 5924–5932.
38. Ananingsih, V.K.; Sharma, A.; Zhou, W. Green tea catechins during food processing and storage: A review on stability and detection. *Food Res. Int.* **2013**, *50*, 469–479. [CrossRef]
39. Norwich, K.H.; Wong, W. Unification of psychophysical phenomena: The complete form of Fechner's law. *Percept. Psychophys.* **1997**, *59*, 929–940. [CrossRef]
40. Ishizu, T.; Tsutsumi, H.; Sato, T. Mechanism of Creaming Down Based on Chemical Characterization of a Complex of Caffeine and Tea Catechins. *Chem. Pharm. Bull.* **2016**, *64*, 676–686. [CrossRef]
41. Zhou, Z.; Ou, M.; Shen, W.; Jin, W.; Yang, G.; Huang, W.; Guo, C. Caffeine weakens the astringency of epigallocatechin gallate by inhibiting its interaction with salivary proteins. *Food Chem.* **2024**, *460*, 140753. [CrossRef]
42. Li, C.X.; Ren, Y.N.; Yao, J.; Wang, Y.L.; Gao, X.L.; Li, H.; Guo, X.F.; Liu, R.X.; Li, X.L. Study on quantitative method of specific bitterness of bitter compounds based on traditional human taste panel method. *Chin. Tradit. Herb. Drugs* **2023**, *54*, 2758–2764.
43. Li, X.L.; Gui, X.J.; Liu, R.X.; Gao, X.J.; Meng, X.L.; Chen, P.J.; Kang, B.Y.; Zhang, L.; Shi, J.H. Bitterness intensity prediction of bitter compounds of traditional Chinese medicine based on an electronic tongue. *Chin. J. New Drugs* **2016**, *25*, 1307–1314.
44. Zhu, Y.W.; Zhou, X.R.; Chen, Y.P.; Liu, Z.Y.; Jiang, S.; Chen, G.L.; Liu, Y. Exploring the relationships between perceived umami intensity, umami components and electronic tongue responses in food matrices. *Food Chem.* **2022**, *368*, 130849. [CrossRef] [PubMed]

Disclaimer/Publisher's Note: The statements, opinions and data contained in all publications are solely those of the individual author(s) and contributor(s) and not of MDPI and/or the editor(s). MDPI and/or the editor(s) disclaim responsibility for any injury to people or property resulting from any ideas, methods, instructions or products referred to in the content.

foods

MDPI

Article

Electronic Nose and Gas Chromatograph Devices for the Evaluation of the Sensory Quality of Green Coffee Beans

Gema Cascos [1], Jesús Lozano [2], Ismael Montero-Fernández [3], Jhunior Abrahan Marcía-Fuentes [4], Ricardo S. Aleman [5], Antonio Ruiz-Canales [6] and Daniel Martín-Vertedor [1,*]

1 Technological Institute of Food and Agriculture CICYTEX-INTAEX, Junta of Extremadura, Avda Adolfo Suárez s/n, 06007 Badajoz, Spain; gcascosc01@educarex.es
2 Industrial Engineering School, University of Extremadura, 06006 Badajoz, Spain; jesuslozano@unex.es
3 Department of Chemical Engineering and Physical Chemistry, University of Extremadura, Avda de Elvas s/n, 06006 Badajoz, Spain; ismonterof@unex.es
4 Faculty of Technological Sciences, Universidad Nacional de Agricultura, Catacamas 16201, Honduras; jmarcia@unag.edu.hn
5 School of Nutrition and Food Sciences, Louisiana State University Agricultural Center, Baton Rouge, LA 70803, USA; rsantosaleman@lsu.edu
6 Engineering Department, Politechnic High School of Orihuela, Miguel Hernández University of Elche, 03312 Elche, Spain; acanales@umh.es
* Correspondence: daniel.martin@juntaex.es

Abstract: The aim of this work is to discriminate between the volatile org9anic compound (VOC) characteristics of different qualities of green coffee beans (*Coffea arabica*) using two analysis approaches to classify the fresh product. High-quality coffee presented the highest values for positive attributes, the highest of which being fruity, herbal, and sweet. Low-quality samples showed negative attributes related to roasted, smoky, and abnormal fermentation. Alcohols and aromatic compounds were most abundant in the high-quality samples, while carboxylic acids, pyrazines, and pyridines were most abundant in the samples of low quality. The VOCs with positive attributes were phenylethyl alcohol, nonanal and 2-methyl-propanoic acid, and octyl ester, while those with negative attributes were pyridine, octanoic acid, and dimethyl sulfide. The aroma quality of fresh coffee beans was also discriminated using E-nose instruments. The PLS-DA model obtained from the E-nose data was able to classify the different qualities of green coffee beans and explained 96.9% of the total variance. A PLS chemometric approach was evaluated for quantifying the fruity aroma of the green coffee beans, obtaining an R_P^2 of 0.88. Thus, it can be concluded that the E-nose represents an accurate, inexpensive, and non-destructive device for discriminating between different coffee qualities during processing.

Keywords: green coffee; coffee quality; tasting panel; coffee aroma quality; electronic nose

1. Introduction

Coffee is one of the most consumed beverages in the world due to its organoleptic characteristics and stimulating effects [1,2].

The botanical family of coffee consists of around 500 genera and more than 6000 species. The most important genus from an economic viewpoint is Coffea, with more than 25 species. About 98% of the nine million tons of green coffee harvested annually are of the Arabica and Robusta varieties.

The two main coffee species have important sensory differences. Arabica is more aromatic, with a moderate body, average bitterness, high acidity, and less caffeine, while Robusta coffee is less aromatic, fuller-bodied, and more bitter [3]. The physical and/or organoleptic qualities of coffee are impacted by climate conditions, soil type, crop management, post-harvest treatment, and grain damage due to insect attacks, such as those of the coffee berry borer [4].

Its originality, strength, and character have inspired much research over the years in search of the principle chemical responsible for its aroma [5]. The particular flavor and aroma of coffee are the result of around a thousand chemical compounds already present in the product [6]. We have to take into account that there are many more volatile compounds that are generated during the roasting process that differ from those of green coffee, such as amino acids and proteins, among others [7]. Furthermore, the chemical composition of the grain changes with the temperature. However, only a small number of these factors contribute to the perceived aroma. The volatile compounds allow us to determine the variety of coffee, the processing techniques used by farmers, its geographical origin and, of course, its quality [8].

The most characteristic volatile compounds belong to the family of furans, pyrroles, pyrazines, ketones, aldehydes, alcohols, acids, pyridines, thiazoles, esters, hydrocarbons, and/or aromatics in both green and roasted coffee [9]. These compounds are determined using laboratory techniques, such as gas chromatography. This technique is widely used for the separation of volatiles, semi-volatiles, and thermally stable components at temperatures of up to 350–400 °C. Gas chromatography establishes the number of individual components present in a sample by using calibration curves of the corresponding standards. Some of these techniques are present in certain laboratories, while other techniques are selected when detecting compounds in different matrixes [10].

A sensory evaluation is another method of analysis that aims to collect, measure, and analyze food properties that are captured by the sense organs [11]. This analysis requires specialized personnel trained in the product to be tasted. A sensory evaluation involves the preparation and tasting of foods and beverages under controlled conditions in such a way that factors that could cause bias are minimized. Attributes such as aroma, flavor, acidity, and body are the sensory characteristics that describe coffee quality [8].

Electronic devices such as electronic noses (E-nose) can be used in the coffee industry to classify green beans of different qualities prior to processing. They are a combination of technology that detects and analyzes chemical compounds in the air and enables the discrimination between agroalimentary food with different aroma profiles, such as wine, citrus [12], fungi [13], or olives, including virgin olive oil [14] and table olives [15–17], with the use of matrices. The main technology involved in this work is the E-nose sensors, which are designed to react to specific molecules and generate electrical signals proportional to the detected aroma. When these sensors come into contact with coffee beans, they can discriminate between the type of production based on its quality. The electrical signals generated by the sensors are converted into digital data that can be processed by the system. An important part of E-nose technology is the use of algorithms and machine learning techniques to analyze and process the data obtained from the sensors. These algorithms are essential for the identification of the different odors and chemical compounds present in the coffee bean samples. Finally, the data collected are compared against previously established databases that contain information on the different odors and chemical compounds to enable the identification of the odors and the issuance of results.

An E-nose has a fast response time and is a low-cost device that can be used as an alternative to gas chromatography and sensory evaluation that enables the classification of food according to its odor profile. In an attempt to improve quality and profitability, the coffee industry seeks to implement such engineering advances in all processes, a large part of which has already been documented with regard to establishing quality standards associated with each stage. In this sense, the electronic device could detect the odor characteristics of green coffee, classifying them into different qualities. Furthermore, this equipment could help industries standardize the harvested green coffee before it enters the industry or the roasting process [18], thereby helping control the quality [19,20]. Thus, two different E-noses were used to increase the discriminatory power of the studied fresh coffee samples of different qualities. Therefore, the aim of this work is to demonstrate that this technology is capable of obtaining a data matrix that allows us to establish a correlation between the acquired signals of the electronic sensors of the sensory analysis and volatile organic compounds' characteristics from coffee samples of different qualities, making it a useful measuring instrument for a first classification.

2. Materials and Methods

2.1. Samples

Samples of *Coffea arabica* (arabica coffee) were harvested at the red stage of ripeness by a company in the Choluteca region (Honduras). Once collected, they were transported to the plant to be washed and dried using natural processes. Many of the volatile compounds in coffee are generated during the roasting process. This is very different and less numerous in fresh coffee [7]. In addition, it prevents the sensors from being affected by humidity. Green coffee beans were vacuum-packed in portions and stored at room temperature until the analyses were carried out.

2.2. Sensory Analysis

The sensory characteristics of the samples were evaluated by a tasting panel made up of eight experts of the CICYTEX research center who were trained in the descriptive analysis of products derived from coffee provided by industry. The tasters had given their informed consent to carry out the olfactory evaluation of the samples studied. This evaluation was performed in different cabins in a tasting room [16]. Green coffee beans (10 g) from different categories were placed into a standard glass jar, covering the bottom of the glass, arranged in a single layer, and kept at 28 °C. Samples were sensory evaluated by the panel to be classified into three different sensory categories: high, medium, and low quality. The positive attributes evaluated by the tasting panel were fruity, herbal, sweet, nutty, and spiced, while the negative ones were roasted, smoky, fermented, and earthy.

The high-quality green coffee beans presented a positive aroma higher than 6 points and a defect attribute less than 1. The medium-quality beans presented a positive aroma between 4 and 6 points and a defect between 2 and 3. The low-quality beans presented a positive aroma of less than 4 and a defect higher than 3 points.

First, a triangular evaluation test was carried out on green coffee beans of different qualities. Two sessions per day and four in one week were conducted. Tasters had to identify which of the three tasting glasses was different. This test allowed us to determine if two samples were significantly different. The minimum number of correct answers was indicated to establish significant differences between the two samples ($p < 0.05$).

Secondly, a descriptive analysis was carried out. The different intensity of positive and negative attributes perceived by the tasters was evaluated on a structured scale from 0–10. The result of each attribute was expressed as average values when the coefficient of variation was less than 20. A total of 33 samples were studied in triplicates.

2.3. Analysis of Volatile Compounds

The profiles of volatile compounds of green coffee beans were analyzed using the method described by Sánchez et al. [15] using static headspace and gas chromatography–mass spectrometry. A sample weighing 2 g was placed into a vial, to which 7 mL of NaCl solution (30% w/v) was added. A polydimethylsiloxane/divinylbenzene (PDMS/DVB) StableFlex fiber (65 µm, Supelco) was used to sample the volatile compounds. The vials were closed and kept at 40 °C for 30 min according to the methodology described by Sánchez et al. [15].

Determinations were performed using gas chromatograph with a triple quadrupole mass spectrometry detector model 456-GC using a capillary column Agilent DB WAXetr (60 m × 0.25 mm; DI: 0.25 mm). Peaks were tentatively identified with the help of the NIST 2.0 MS reference spectral library.

2.4. E-Nose Measurements

An E-nose is a piece of miniaturized electronic equipment composed of 11 sensors [15]. It has low power consumption and is the size of a hockey puck, thereby making it easily portable and convenient. The device contains novel metal oxide semiconductor (MOX) chip gas sensors which exhibit global selectivity to odor patterns for each sample: (i) BME680 (Bosch Sensortech GmbH, Reutlingen, Germany): temperature (°C), pressure (hPa), humidity (%RH), and gas measurement (Ω); (ii) SGP30 (Sensirion AG, Stäfa, Switzerland): eCO_2 (ppm), TVOC (ppb), H_2(2), and ethanol; (iii) CCS811 (ScioSense B.V., Eindhoven, The Netherlands): eCO_2 (ppm), TVOC (ppb), and sensor resistance (Ω); and (iv) iAQ-Core (ScioSense B.V., Eindhoven, The Netherlands): eCO_2 (ppm), TVOC (ppb), and sensor resistance (Ω).

The electronic analysis required 5 g of green coffee beans to be placed into standard tasting glasses, covered with a watch glass, and placed on a block kept at 25 °C. Eleven measurements were taken per sample. The E-nose was placed on top of the cup and the adsorption phase was carried out to adsorb the headspace of the sample for 60 s. Once this time had elapsed, the E-nose was moved to another empty cup to carry out the desorption phase for 30 s to obtain the baseline. The E-nose received data in one second intervals and eight measurements were carried out for each roasted bean sample. After the analyses were conducted, the microprocessor sent the values obtained to a specific application in an intelligent device (mobile phone) via Bluetooth, and a multivariate analysis was completed with a computer. The algorithm used to characterize the sensor response curves to obtain the raw data for the E-nose was maximum signal value minus minimum 1, plus 100, minus 1. As a result, a vector of data with 11 rows (sensors) for each sample was obtained and a multivariate analysis was performed with the data obtained. The E-nose also had temperature and humidity sensors that enabled the monitorization and correction of the data with compensation algorithms. The results of both analyses were analyzed together to obtain more comprehensive information.

2.5. Multivariate Data Analysis

Principal component analysis (PCA), partial least squares discriminant analysis (PLS-DA), and partial least squares (PLS) methods were carried out with the data obtained by the E-nose. The chemometric method was that described by Sánchez et al. [17]. The program used for the multivariate analysis was MATLAB version R2016b, version 9.1 (The Mathworks Inc. Natick, MA, USA) with PLS_Toolbox 8.2.1 (Eigenvector Research Inc., Wenatchee, WA, USA).

2.6. Statistical Analysis

An analysis of variance (ANOVA) was carried out to establish statistical differences between the different treatments studied. Next, the Tukey method (univariate analysis) was performed to ascertain the differences between each of the classes studied. SPSS 18.0 software was used for the statistical analysis (SPSS Inc., Chicago, IL, USA).

3. Results and Discussion

3.1. Sensory Analysis of Green Coffee Beans

First, a sensory analysis was conducted with a triangular test to verify the capacity of the panelist to differentiate samples of different qualities (Table 1). The results showed that tasters were able to discriminate between green coffee beans of different qualities. As can be seen in Table 1, the tasters had to identify which of the three tasting glasses were different. The table shows that the different qualities of the coffee samples were tentatively identified in the different assays carried out. Thus, the tasters were able to classify samples according to their smell. It should be noted that in the triangle test, for samples of maximum quality, the percentage of success was 100%, while for the medium and low quality, this percentage decreased to 95–98%. This is a very interesting result since tasters can discriminate samples based on their olfactory quality.

Table 1. Results of the triangular sensory analysis test for green coffee beans (p-value < 0.05 for each triangle).

	High	Medium	Low
High	---	$p < 0.05$ $n = 5$ 95%	$p < 0.05$ $n = 5$ 98%
Medium	$p < 0.05$ $n = 5$ 100%	---	$p < 0.05$ $n = 5$ 95%
Low	$p < 0.05$ $n = 5$ 100%	$p < 0.05$ $n = 5$ 96%	---

Subsequently, the coffee samples were selected individually to perform a descriptive evaluation of certain positive and negative attributes of the beans. It is well known that arabica coffee has a higher sensory characteristic that gives it the greatest commercial value. For this reason, green coffee beans were sensorially evaluated descriptively by the tasting panel (Table 2). The panel aimed to define whether the green coffee beans had pleasant or unpleasant attributes that could discriminate samples based on their sensory quality.

Table 2. Sensory olfactory evaluation of green coffee beans (mean ± standard deviation). Different small letters indicate significant statistical differences according to the coffee qualities (Tukey's Test, $p < 0.05$).

Quality	Positive Attributes				
	Fruity	Herbal	Sweet	Nutty	Spiced
High	7.8 ± 0.5^{a}	4.5 ± 0.4^{a}	3.2 ± 0.3^{a}	2.3 ± 0.2^{a}	2.2 ± 0.2^{a}
Medium	5.6 ± 0.7^{b}	3.4 ± 0.2^{b}	2.5 ± 0.1^{b}	1.6 ± 0.3^{b}	1.4 ± 0.2^{b}
Low	3.6 ± 0.4^{c}	2.6 ± 0.3^{c}	2.0 ± 0.1^{b}	1.5 ± 0.2^{b}	1.1 ± 0.2^{b}

Quality	Negative Attributes			
	Roasted	Smoky	Fermented	Earthy
High	0.6 ± 0.2^{c}	0.5 ± 0.1^{c}	n.d.	n.d.
Medium	2.5 ± 0.2^{b}	1.5 ± 0.3^{b}	1.0 ± 0.1^{b}	1.5 ± 0.3^{b}
Low	3.6 ± 0.2^{a}	2.6 ± 0.1^{a}	2.6 ± 0.3^{a}	2.3 ± 0.2^{a}

n.d.: not detected.

The tasting panel can be seen to be very accurate. The main positive attributes were related to fruity, herbal, sweet, nutty, and spiced, while the negative attributes were roasted, smoky, fermented, and earthy. The high-quality coffee presented the highest values of the positive attributes with the highest scores in the fruity attribute, followed by the herbal one. The sweet smell was also an important attribute in these coffee samples with differences between the different qualities. Nutty and spiced odors were only significantly different in the highest-quality samples.

Regarding the negative attributes, roasted was the most significant, followed by smoky. In this case, the low-quality samples showed higher values of this type of attribute. We have to take into account that the roasted odor is a descriptor related to burnt food, while smoky describes a smell associated with smoke from burning wood [3]. Negative attributes were also found in the low-quality samples related to abnormal fermentation products and those with an earthy or musty odor. The coffee industry elaboration process has possibly brought about the development of certain microorganisms that have spoiled these coffee beans. These results concur with those obtained by Illy and Viani [21] who affirmed that the contact of the fruit with soil favors the development of undesirable microbial fermentation.

The results show that the sensory analysis carried out was able to identify the characteristics of the samples tested. Researchers [9] have evaluated sensory attributes in green coffee from different varieties submitted to different industrial conditions. However, the results obtained in this research can be used as a development strategy for classifying this product in the industry. Therefore, these results highlight the market potential for high-quality green coffee. Although there is wide variation of the coffee quality harvested in Colombia, it is evident that the aroma characteristics of high-quality coffee are far more acceptable than the others. Other researchers [9] have indicated that stinker alteration is produced by microbial fermentation of samples during the elaboration process that produces green coffee seeds with fermented, earthy, or putrid aromas, which makes the quality look diminished. The study findings demonstrate that the sensory analysis successfully identified the specific characteristics of the tested samples. Previous research [9] also evaluated sensory attributes in green coffee from different varieties and under various industrial conditions. These results can serve as a valuable development strategy for product classification in the industry, highlighting the potential for high-quality coffee in the market.

3.2. VOCs of Green Coffee Beans

As can be seen in Figure 1, significant differences were observed between the VOCs of the different green coffee quality analyzed by the GC-MS technique. The families of VOCs that were found in greater proportion were alcohols, aromatics, and aldehydes, while the minor groups were pyrroles, acids derivate, and sulfur compounds.

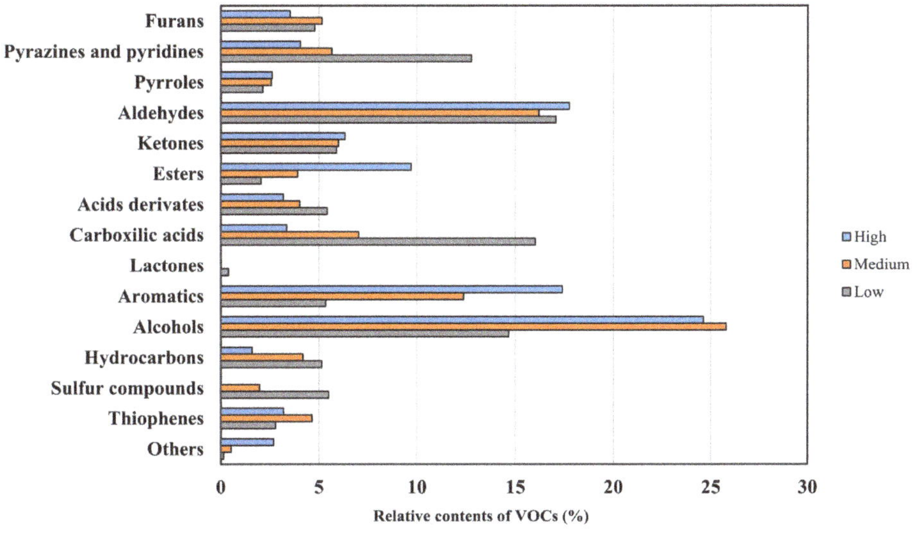

Figure 1. Chemical distribution of volatile compounds (%) in green coffee beans of different qualities.

There are volatiles whose presence gains as the quality of the coffee increases, which could be attributed to the fact that these groups are linked to a positive aroma in coffee beans [9]. That is the case with the alcohol groups, which represented the most abundant compounds in the samples studied. The content of alcohols was greater in the high- and medium-quality beans than in the low-quality beans, which was linked to floral aromas [22]. The aromatic compounds were the second most abundant VOCs in the high-quality samples, with 3.2 times more content compared with the rest of the green coffee beans [23]. Ketones clearly increased to 79.6% and 48.2% in high- and medium-quality coffees, respectively. Similarly, esters increased their content 2.1 times more in the high-quality coffee, although not in the medium-quality group, whose values were slightly lower than in that of the low-quality group. The other group of compounds represented a low percentage in all the qualities and was only moderately higher in the high-quality groups. Moreira et al. [24] obtained similar results.

On the contrary, there were other VOC families that appeared in greater proportion in the low-quality samples. This was the case with carboxylic acids; although they are present in all the samples, their proportion was much higher in the low-quality ones. Their content represented 4.5 times more in the high-quality samples than in the other qualities. This family represented the second most abundant VOCs in low-quality coffees, after the aldehydes group. Pyrazines and pyridines were the third most abundant group in the low-quality coffee. Their content rose 2.0 times more with regard to the high-quality coffee, which had similar values to the medium-quality coffees. These compounds emit negative earthy notes [9,25]. The group of hydrocarbons was the fourth most abundant in the low-quality coffee, representing 7% of the VOCs. They are usually associated with the formation of tar and soot in a cooking or roasting process. Lactones only appeared in low-quality samples in very low proportion. Despite their low concentrations, they were significant in determining the quality of coffee [26]. The odor of this family is associated with a defective odor in seeds [9]. Furthermore, the sulfur compound only appeared in the medium- and low-quality samples, with the greater content in the latter (4.5%). Sulfur compounds are extremely important due to their low odor thresholds. Finally, furans, pyrroles and aldehydes appeared in all the samples and presented no significant differences between the different qualities analyzed. These families appeared in the volatile compound's literature with roasted or smoked characteristics in different coffee varieties [9,27]. Thiophenes appeared in all the samples but slightly more in those of medium and low quality. These compounds emit negative smoky and roasted notes [9].

The VOCs for the green coffee beans of different quality are shown in Table 3. In total, 70 VOCs were tentatively identified, which were grouped according to different families of VOCs. The main VOCs were phenylethyl alcohol and 2-methoxy-4-vinylphenol, and the minor ones were (E)-7-methyl-1,6-dioxaspiro [4.5], decane, acetic acid, and 2-methyl-butanal.

Table 3. Content of volatile compounds (mean %, $n = 3$) obtained from green coffee beans of different qualities. R.T. = retention time. Different lowercase letters indicate significant differences (one-way ANOVA followed by Tukey's test. $p < 0.05$) between coffee bean samples qualities.

R.T. (Min)	CAS Number	Volatile Compounds	Quantities High	Medium	Low
		Furans			
9.0	108-29-2	Dihydro-5-methyl-2(3H)-furanone	0.70 ^{a}	0.76 ^{a}	0.28 ^{b}
12.4	3777-69-3	2-Methyl-furan	1.04 ^{b}	2.07 ^{a}	2.24 ^{a}
22.0	51080-20-7	Furan	0.81 ^{b}	2.33 ^{a}	2.27 ^{a}
38.1	104-61-0	Dihydro-2-methyl-2(3H)-furanone	1.00 ^{ns}	0.00	0.00
		Pyrazines and pyridines			
13.7	110-86-1	Pyridine	1.52 ^{c}	1.94 ^{b}	6.97 ^{a}
15.9	75354-36-8	(E)-7-Methyl-1,6-dioxaspiro [4.5]decane	0.08 ^{c}	0.26 ^{b}	0.90 ^{a}
24.1	24683-00-9	2,5-Dimethyl-pyridine	2.46 ^{c}	3.44 ^{b}	4.88 ^{a}
		Pyrroles			
24.9	3146-84-7	3,4-Dihydro-2,2,3-trimethyl-2H-pyrrole 1-oxide	0.00 ^{ns}	0.00	1.28
28.0	20189-42-8	1H-Pyrrole-2,5-dione, 3-ethyl-4-methyl-	1.31 ^{a}	1.17 ^{a}	0.54 ^{b}
30.5		1H-Pyrrole	1.30 ^{ns}	1.40	0.00
32.4	120-72-9	1H-Indole	0.00 ^{ns}	0.00	0.32
		Aldehydes			
2.0	123-72-8	Butanal	1.49 ^{ns}	0.00	0.00
2.8	96-17-3	2-Methyl-butanal	2.27 ^{a}	1.27 ^{c}	1.99 ^{b}
5.3	66-25-1	Hexanal	2.03 ^{c}	2.99 ^{b}	4.29 ^{a}

Table 3. *Cont.*

R.T. (Min)	CAS Number	Volatile Compounds	Quantities		
			High	Medium	Low
10.9	100-52-7	Benzaldehyde	0.00 [ns]	0.68	0.80
16.0	2548-87-0	(E)-2-Octenal	0.00 [ns]	0.00	0.84
18.9	124-19-6	Nonanal	7.46 [a]	7.07 [a]	4.17 [b]
22.9	56114-69-3	Benzaldehyde, 2,5-bis[(trimethylsilyl)oxy]-	2.00 [a]	1.40 [b]	2.02 [a]
22.6	18829-56-6	(E)-2-Nonenal	0.00 [ns]	0.00	1.62
26.6	5910-87-2	(E,E)-2,4-Nonadienal	0.00 [ns]	0.00	0.66
40.7	121-33-5	4-Hydroxy-3-methoxybenzaldehyde (Vanilene)	2.49 [a]	2.78 [a]	0.65 [b]
		Ketones			
20.2	1000302-96-9	Propenone, 1-(4-nitrophenyl)-3-phenylamino-	3.17 [ns]	0.00	0.00
27.6	2758-18-1	3-Methyl-2-cyclopenten-1-one	1.19 [ns]	0.00	0.00
45.4	105-86-2	Geraniol	1.23 [b]	1.71 [a]	0.81 [c]
71.3	502-69-2	2-Pentadecanone, 6,10,14-trimethyl-	0.73 [c]	4.26 [b]	5.07 [a]
		Esters			
8.4	141-32-2	2-Propenoic acid, butyl ester	0.44 [ns]	0.38	0.26
22.2	103-09-3	Acetic acid, 2-ethylhexyl ester	2.02 [a]	1.59 [b]	0.32 [c]
22.2	72218-58-7	3-Methylheptyl acetate	0.84 [ns]	0.00	0.00
26.1	109-15-9	2-Methyl-propanoic acid, octyl ester	4.33 [ns]	0.00	0.00
54.3	1000298-25-6	1,3-Dimethylbutyl isopropylphosphonofluoridate	0.66 [ns]	0.00	0.00
74.2	628-97-7	Hexadecanoic acid, ethyl ester	1.41 [a]	1.43 [a]	0.88 [b]
80.7	544-35-4	Linoleic acid ethyl ester	0.00 [ns]	0.53	0.59
		Acids derivates			
6.9	503-74-2	3-Methyl-butanoic acid	0.34 [c]	2.32 [b]	2.84 [a]
6.7	4536-23-6	2-Methyl-hexanoic acid	1.51 [a]	1.02 [b]	0.39 [c]
7.2	116-53-0	2-Methyl-butanoic acid	0.00 [ns]	0.00	1.80
8.2	541-47-9	3-Methyl-2-butenoic acid	1.34 [a]	0.69 [b]	0.38 [c]
		Carboxilic acids			
2.3	64-19-7	Acetic acid	0.12 [c]	1.02 [b]	1.32 [a]
6.5	109-52-4	Pentanoic acid	0.00	0.59 [b]	1.80 [a]
12.1	142-62-1	Hexanoic acid	0.00 [ns]	0.00	1.06
12.1	107-92-6	Butanoic acid	0.00 [ns]	0.00	1.29
24.9	124-07-2	Octanoic acid	0.83 [c]	1.99 [b]	5.72 [a]
31.6	112-05-0	Nonanoic acid	2.43 [c]	3.42 [b]	4.80 [a]
		Lactones			
9.0	96-48-0	γ-Butyrolactone	0.00 [ns]	0.00	0.39
		Aromatics			
12.0	108-95-2	Phenol	2.34 [a]	1.54 [b]	0.00
14.2	5989-27-5	(D-Limonene)	1.98 [a]	2.01 [a]	1.36 [b]
17.8	90-05-1	2-Methoxy-phenol (Guaiacol)	0.66 [ns]	0.00	0.00
34.2	7786-61-0	2-Methoxy-4-vinylphenol	9.92 [a]	7.01 [b]	2.27 [c]
37.2	584-84-9	Benzene, 2,4-diisocyanato-1-methyl-	2.48 [a]	1.79 [b]	0.00
75.1	33777-97-8	3-Phenyl-4-azafluorene	0.00 [ns]	0.00	0.33
80.2	15089-22-2	N-Benzyl-N-ethyl-p-isopropylbenzamide	0.00 [ns]	0.00	1.38
		Alcohols			
3.8	123-51-3	3-Methyl-1-butanol,	1.51 [b]	2.33 [a]	2.29 [a]
4.5	71-41-0	1-Pentanol	0.56 [ns]	0.00	0.00
4.8	513-85-9	2,3-Butanediol	0.00 [ns]	0.00	0.92
7.4	111-27-3	1-Hexanol	2.66 [a]	2.38 [b]	0.65 [c]
11.9	3391-86-4	1-Octen-3-ol	0.96 [b]	1.29 [a]	1.24 [a]
14.4	104-76-7	2-Ethyl-1-hexanol	1.81 [a]	1.97 [a]	1.25 [b]
14.5	100-51-6	Benzyl alcohol	2.02 [a]	2.06 [a]	1.26 [b]
19.3	60-12-8	Phenylethyl Alcohol	11.60 [a]	11.82 [a]	4.85 [b]
22.0	768-95-6	1-Adamantanol	0.98 [c]	2.14 [a]	1.53 [b]
23.6	143-08-8	1-Nonanol	1.07 [ns]	0.00	0.00
26.9	122-99-6	2-Phenoxy-ethanol	1.45 [b]	1.80 [a]	0.67 [c]

Table 3. *Cont.*

R.T. (Min)	CAS Number	Volatile Compounds	Quantities		
			High	Medium	Low
		Hydrocarbons			
18.6	78-70-6	3,7-Dimethyl-1,6-octadien-3-ol (Linalool)	0.73 [a]	0.65 [a]	0.36 [b]
23.6	124-11-8	1-Nonene	0.00 [ns]	0.00	0.90
25.7	112-40-3	Dodecane	0.00 [ns]	0.00	0.53
41.4	629-59-4	Tetradecane	0.85 [c]	3.55 [a]	3.35 [b]
		Sulfur compounds			
1.9	75-18-3	Dimethyl sulfide	0.00	1.98 [b]	5.48 [a]
		Thiophenes			
16.8	2557-78-0	o-Fluorothiophenol	2.88 [ns]	2.99	0.00
25.2	1708-32-3	2,5-Dihydro-thiophene	0.33 [c]	1.67 [b]	2.79 [a]
		Others			
8.1	100-42-5	Styrene	1.01 [a]	0.52 [n]	0.14 [c]
11.6	127-91-3	Beta-Pinene	1.67 [ns]	0.00	0.00

The furan compounds significantly decreased in low-quality coffee, with 2-methyl-furan and furan representing the most abundant compounds, increasing 2.5 times more than in high-quality coffee. In the literature, these compounds provide negative aromas (smoky or roasted) [28] and are linked to causing mutagenesis in mice [29]; therefore, the European Commission published a recommendation regarding monitoring the presence of this compound in foods [30]. Thus, in green coffee, this family is shown to decrease coffee quality. Pyridine (5.73%) and 2,5-di methyl-pyridine (4.01%) were found in much higher proportion in low-quality coffee in pyrazines and pyrimidines families, with significant results. These VOCs have roasted and burnt notes [21]. As for the group of pyrroles, the main volatile composition varied in the different samples. In medium- and high-quality coffees, 1H-pyrrole (1.12–1.13%) was the main VOC found, and it smells like dried fruit [31]. However, in low-quality coffee, 3,4-dihydro-2,2,3-trimethyl-2H-pyrrole 1-oxide (1.05%) and 1H-indole (0.26%) appeared in low proportion. These compounds are linked to a roasted odor [9]. High contents of pyrazines, pyridines, and pyrroles indicate a decrease in the quality of green coffee. In the group of aldehydes, butanal was the only VOC detected in high-quality coffee (1.28%), while nonanal and 4-hydroxy-3-methoxybenzaldehyde appeared in higher proportions in the medium- and high-quality samples. There were also several compounds that are associated with low-quality coffee, such as 2-octenal, (E)-2-nonenal, and (E,E)-2,4-nonadienal, that appeared exclusively in these samples, and they are associated with buttery fragrances. Hexanal (3.52%) and benzaldehyde (0.65%) were aldehydes found in a higher proportion, and these are markers of poor quality, a rancid odor, and/or oxidative processes [5,9]. The VOCs of the ketone family were found exclusively in the high-quality coffee; specifically, propenone, 1-(4-nitrophenyl)-3-phenylamino (2.73%), and 3-methyl-2-cyclopenten-1-one (1.02%). Both VOCs are associated with sweet smells [4,9]. This family is usually found in coffee of the highest quality. The esters family is linked to positive fruity aromas, mainly due to 3-methylheptyl acetate, propanoic acid, 2-methyl- and octyl ester that appeared in high-quality samples but not in medium- and low-quality ones [4]. The acid derivatives such as 2-methyl-butanoic acid and 3-methyl-2-butenoic acid emit unpleasant odors [32] and only appeared in a greater proportion in the low-quality samples. Regarding the carboxylic acids, nonanoic acid and octanoic acid increased significantly in the low-quality samples. The literature links this family to pungent and rancid odors [32]. γ-Butyrolactone is the only compound within the lactone family that was detected in green coffee beans. Furthermore, it only appeared in the low-quality samples. The literature associates lactones with coffee quality [26] because its odor is linked to a defective odor in seeds [9]. 2-Methoxy-4-vinylphenol and benzene are the main VOCs within the aromatics family, which have a spicy clove odor. The main VOC in the alcohols groups was phenylethyl alcohol, which was the most abundant in green coffee and has a floral smell [33]. Tetradecane, 1-nonene and dodecane, which smell of solvent [4], were the VOCs that appeared in the majority of the hydrocarbons group. These negative attributes only appeared in the low-quality group. Dimethyl sulfide was the only VOC detected in the sulfur compounds family in the low-quality samples. This compound has a putrid or decomposing odor [34]. Finally, in other groups, we must highlight that styrene and beta-pinene presented aromas linked to fresh fruit, citrus, herbs, and green onions [17].

3.3. E-Nose Application to Discriminate Quality from Green Coffee Beans

The volatile compounds of green coffee beans were also evaluated with E-nose devices to discriminate samples based on their quality. The results showed that the PCA was able to identify green coffee beans with different quality categories (Figure 2). The 62.9% of the total variance of the data was explained by PC1, while PC2 explained 21.7%. This is an interesting result since the combined sensors of the electronic devices used were able to discriminate Arabica samples based on their sensory quality. This result could prove useful for companies given that these devices enable a quick and efficient preliminary classification of the quality of the coffee beans before they are subjected to a roasting process. This would allow a quality classification of green coffee to be conducted at the first phases of coffee processing.

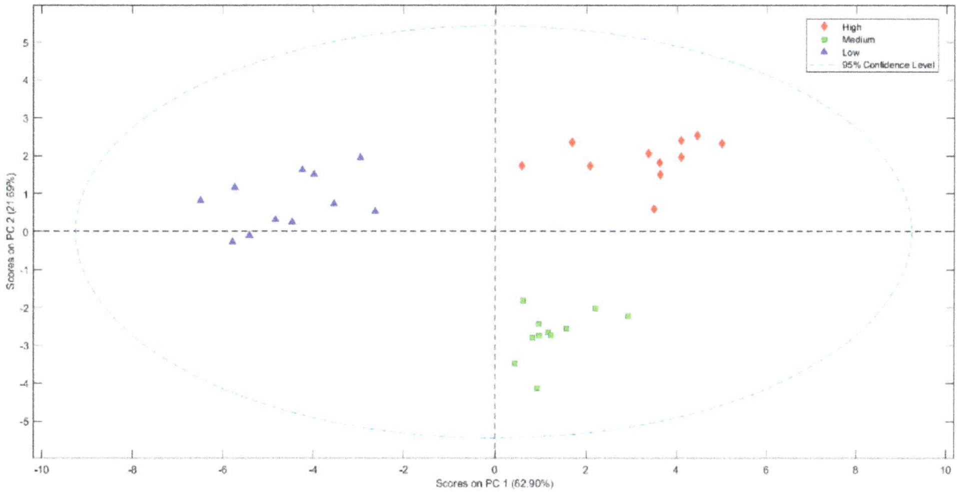

Figure 2. Score plot of the PCA analysis in fresh coffee beans of different qualities.

Next, a PLS-DA was performed using a leave-one-out cross-validation. The confusion matrix of the model indicates that the sum of the diagonal elements gave a hit rate of 100%. Eleven samples from each class were used to construct the model allowing green coffee beans with different sensorial quality to be discriminated. The confusion matrix of the PLS-DA model for green coffee beans shows that the sum of the diagonal elements gave a hit rate of 100% for high and low quality, while giving 93.8% for medium quality. The medium-quality group presented aromas from the other samples whose classification was not 100%. The rest of the samples were successfully predicted with an accuracy of 100%. Therefore, the model established that it could predict a correct discrimination of the green coffee beans of different quality.

3.4. Relation between E-Nose Data and Coffee Bean Aroma

A multivariate analysis (PLS chemometric model) was performed between all the results obtained with the electronic nose sensors and the fruity aroma obtained by the sensory panel. These results are shown in Figure 3. This regression was used to predict the fruity aroma perceived by panelist and data from E-nose signals. R^2_{CV} values for the models established for the fruity aroma were 0.88. Low RMSECV values of 0.64 were also estimated. The PLS calibration model was validated using samples that were not included in the calibration test. The validation parameters obtained were also acceptable. R^2_P values were 0.95 for the fruity aroma perceived whilst RMSEP values were 0.53. Thus, this model led to the quantification of the fruity aroma of green coffee bean samples and its classification of different qualities. The literature points to some results related to the discrimination of roasted coffee with electronic devices. Rodríguez et al. [35] were also able to classify roasted coffee samples of different qualities using an E-nose aimed at testing the quality of the Colombian coffee. These researchers discriminated coffee with different seed defects. Gardner et al. [36], using oxide sensors, discriminated and classified three commercial coffees, as well as one coffee which had been subjected to a range of six roasting times. Brudzewski et al. [37] applied an electronic nose for the recognition of coffee with two different quality coffee brands: a mediocre product and the high-quality coffee type. Gonzalez Viejo et al. [38] estimated the intensity and aromas of roasted coffee using a low-cost and portable electronic nose and machine learning modelling.

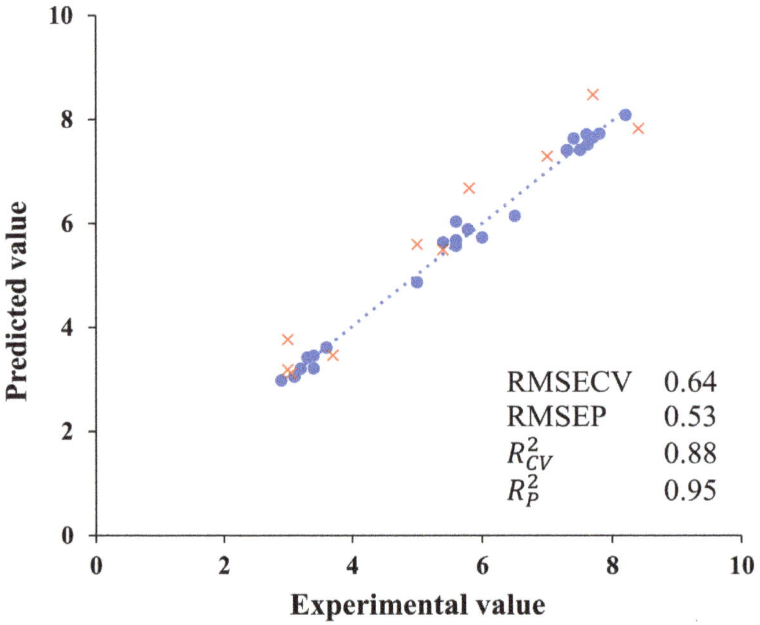

Figure 3. Experimental values against PLS cross-validation predictions (•) and validation set predictions (×) for fruity aroma perceived.

4. Conclusions

Post-harvest, fresh coffee beans have different sensory characteristics that determine the quality of the final product. The positive attributes of the highest-quality coffee are linked to fruity, herbal, and sweet odors, while the negative odors are associated with roasted, smoky, earthy, and fermented odors. All these odors are represented by different VOCs. Furthermore, results obtained by the PLS models suggest that these electronic devices may be used to estimate the fruity aroma perceived by the tasting panel. The E-nose has the advantage of being an objective model with which to detect the olfactory patterns of coffee samples of different qualities. It is considered a useful discrimination tool that can be applied to green coffee beans; therefore, it can be useful for classifying coffee qualities in a production factory. This electronic device can be used in combination with a tasting panel to produce qualitative analysis, and it can be regarded as rapid, inexpensive, non-destructive, and environmentally friendly.

Author Contributions: Conceptualization, A.R.-C., J.L. and D.M.-V.; Data curation, G.C., I.M.-F., J.A.M.-F. and D.M.-V.; Formal analysis, R.S.A., A.R.-C., I.M.-F. and D.M.-V.; Funding acquisition, A.R.-C., J.L. and D.M.-V.; Investigation, A.R.-C., G.C., J.L. and D.M.-V.; Methodology, R.S.A., G.C. and D.M.-V.; Project administration, D.M.-V.; Resources, R.S.A., A.R.-C., J.A.M.-F. and D.M.-V.; Supervision, J.L. and D.M.-V.; Validation, J.L., A.R.-C. and D.M.-V.; Visualization, A.R.-C. and D.M.-V.; Writing—original draft, G.C. and D.M.-V.; Writing—review and editing, G.C. and D.M.-V. All authors have read and agreed to the published version of the manuscript.

Funding: This research was funded by the National University of Agriculture (Honduras), the local government of Extremadura (ref. GR21121—AGA008) and the European Regional Development Fund (FEDER).

Data Availability Statement: The authors confirm that the data supporting the findings of this study are available within the article and the raw data that support the findings are available from the corresponding author, upon reasonable request.

Conflicts of Interest: The authors declare no conflict of interest.

References

1. Alves, R.C.; Casal, S.; Oliveira, M.B.P.P. Tocopherols in coffee brews: Influence of coffee species roast degree and brewing procedure. *J. Food Compos. Anal.* **2010**, *23*, 802–808. [CrossRef]
2. Di Bella, G.; Potortì, A.G.; Turco, V.L.; Saitta, M.; Dugo, G. Plasticizer residues by HRGC–MS in espresso coffees from capsules, pods and moka pots. *Food Control* **2014**, *41*, 185–192. [CrossRef]
3. Stokes, C.N.; O'Sullivan, M.G.; Kerry, J.P. Hedonic and descriptive sensory evaluation of instant and fresh coffee products European. *Food Res. Technol.* **2017**, *243*, 331–340. [CrossRef]
4. Puerta, G.I. Composición química de una taza de café Centro Nacional de Investigaciones de Café (Cenicafé). *Prog. Investig. Cient.* **2013**, *414*, 10.
5. Holscher, W.; Vitzthum, O.G.; Steinhart, H. Identification and sensorial evaluation of aroma-impact-compounds in roasted Colombian coffee. *AGRIS* **1990**, *34*, 205–212.
6. Gotteland, M.; Saturnino de Pablo, V. Some trues concerning coffee. *Rev. Chil. Nutr.* **2007**, *34*, 105.
7. Wei, F.; Tanokura, M. Chemical changes in the components of coffee beans during roasting. In *Coffee in Health and Disease Prevention*; Academic Press: Cambridge, MA, USA, 2015; pp. 83–91. [CrossRef]
8. Gutierrez, N.; Barrera, O. Selección y entrenamiento de un panel en análisis sensorial de café *Coffea arabica* L. *Rev. Cienc. Agrar.* **2015**, *32*, 77–87. [CrossRef]
9. Toledo, P.R.; Pezza, L.; Pezza, H.R.; Toci, A.T. Relationship between the different aspects related to coffee quality and their volatile compounds. *Compr. Rev. Food Sci. Food Saf.* **2016**, *15*, 705–719. [CrossRef]
10. Sánchez, R.; Martín-Tornero, E.; Lozano, J.; Arroyo, P.; Meléndez, F.; Martín-Vertedor, D. Evaluation of the olfactory pattern of black olives stuffed with flavored hydrocolloids. *LWT-Food Sci. Technol.* **2022**, *163*, 113556. [CrossRef]
11. Stone, H.; Bleibaum, R.; Thomas, H.A. *Sensory Evaluation Practices*; Academic Press: Cambridge, MA, USA, 2020.
12. Martínez Gila, D.M.; Gámez García, J.; Bellincontro, A.; Mencarelli, F.; Gómez Ortega, J. Fast tool based on electronic nose to predict olive fruit quality after harvest. *Postharvest Biol. Technol.* **2020**, *160*, 111058. [CrossRef]
13. Portalo-Calero, F.; Arroyo, P.; Suárez, J.I.; Lozano, J. Triangular test of amanita mushrooms by using electronic nose and sensory panel. *Foods* **2019**, *8*, 414. [CrossRef] [PubMed]
14. Escuderos, M.E.; García, M.; Jiménez, A.; Horrillo, M.C. Edible and non-edible olive oils discrimination by the application of a sensory olfactory system based on tin dioxide sensors. *Food Chem.* **2013**, *136*, 1154–1159. [CrossRef] [PubMed]
15. Sánchez, R.; Martín-Tornero, E.; Lozano, J.; Boselli, E.; Arroyo, P.; Meléndez, F.; Martín-Vertedor, D. E-Nose discrimination of abnormal fermentations in Spanish-Style Green Olives. *Molecules* **2021**, *26*, 5353. [CrossRef] [PubMed]
16. Sánchez, R.; Boselli, E.; Fernández, A.; Arroyo, P.; Lozano, J.; Martín-Vertedor, D. Determination of the masking effect of the 'zapateria' defect in flavoured stuffed olives using E-nose. *Molecules* **2022**, *27*, 4300. [CrossRef] [PubMed]
17. Sánchez, R.; Martín-Tornero, E.; Lozano, J.; Fernández, A.; Arroyo, P.; Meléndez, F.; Martín-Vertedor, D. Electronic nose application for the discrimination of sterilization treatments applied to Californian-style black olive varieties. *J. Sci. Food Agric.* **2022**, *102*, 2232–2241. [CrossRef]
18. Barea-Ramos, J.D.; Cascos, G.; Mesías, M.; Lozano, J.; Martín-Vertedor, D. Evaluation of the Olfactory Quality of Roasted Coffee Beans Using a Digital Nose. *Sensors* **2022**, *22*, 8654. [CrossRef]

19. Ceballos, D.A.C.; Meneses, J.A.M.; López, C.A.G.; Narváez, J.A.G.; Luna, D.A.R.; García, J.H. Estudio de fragancia y aroma del café tostado con la nariz electrónica Coffee-NOSE. In Proceedings of the 2020 IX International Congress of Mechatronics Engineering and Automation (CIIMA), Cartagena, Colombia, 4–6 November 2020; pp. 1–6.
20. Oates, M.J.; Abu-Khalaf, N.; Molina-Cabrera, C.; Ruiz-Canales, A.; Ramos, J.; Bahder, B.W. Detection of Lethal Bronzing Disease in Cabbage Palms (Sabal palmetto) Using a Low-Cost Electronic Nose. Comput. Electron. *Agriculture* **2020**, *155*, 348–358.
21. Illy, A.; Viani, R. *Espresso Coffee: The Science of Quality*; Academic Press: Cambridge, MA, USA, 2005.
22. Martínez-Avila, O.; Sánchez, A.; Font, X.; Barrena, R. 2-phenylethanol (rose aroma) production potential of an isolated pichia kudriavzevii through solid-state fermentation. *Process Biochem.* **2020**, *93*, 94–103. [CrossRef]
23. Rostagno, M.A.; Celeghini, R.M.; Debien, I.C.; Nogueira, G.C.; Meireles, M.A.A. Phenolic compounds in coffee compared to other beverages. In *Coffee in Health and Disease Prevention*; Academic Press: Cambridge, MA, USA, 2015; pp. 137–142.
24. Moreira, R.F.A.; Trugo, L.C.; De Maria, C.A.B. Volatile components in roasted coffee Part II Aliphatic alicyclic and aromatic compounds. *Quim. Nova* **2000**, *23*, 195–203. [CrossRef]
25. Toci, A.T.; Farah, A.; Deliza, R. Investigação da composição volátil dos defeitos intriínsecos do café em relação aos grãos de boa qualidade. *Quim. Nova* **2007**, *32*, 1–5.
26. Perrone, D.; Farah, A.; Donangelo, C.M.; de Paulis, T.; Martin, P.R. Comprehensive analysis of major and minor chlorogenic acids and lactones in economically relevant Brazilian coffee cultivars. *Food Chem.* **2008**, *106*, 859–867. [CrossRef]
27. Wang, X.; Wang, Y.; Hu, G.; Hong, D.; Guo, T.; Li, J.; Li, Z.; Qiu, M. Review on factors affecting coffee volatiles: From seed to cup. *J. Sci. Food Agric.* **2022**, *102*, 1341–1352. [CrossRef]
28. Clarke, R.J. The Flavour of Coffee Dev. *Food Sci.* **1986**, *3B*, 1–47.
29. Bejarano Roncancio, J.J.; Suárez Latorre, L.M. Algunos peligros químicos y nutricionales del consumo de los alimentos de venta en espacios públicos. *Rev. Univ. Ind. Santander Salud* **2015**, *47*, 349–360.
30. European Commission. *Commission Recommendation (EC) No 2007/196 of 28 March 2007 Concerning on the Monitoring of the Presence of Furan in Foodstuffs, L 88/56*; European Commission: Brussels, Belgium, 2007.
31. Yeretzian, C.; Jordan, A.; Badoud, R.; Lindinger, W. From the green bean to the cup of coffee: Investigating coffee roasting by on-line monitoring of volatiles. *Eur. Food Res. Technol.* **2002**, *214*, 92–104. [CrossRef]
32. Caporaso, N.; Whitworth, M.B.; Cui, C.; Fisk, I.D. Variability of single bean coffee volatile compounds of Arabica and robusta roasted coffees analysed by SPME-GC-MS. *Food Res. Int.* **2018**, *108*, 628–640. [CrossRef] [PubMed]
33. Fabre, C.E.; Blanc, R.J.; Gorna, G. Phenylethyl Alcohol: An Aroma Profile Perfumer. *Flavorist* **1998**, *23*, 43–46.
34. Gutiérrez, M.C.; Droguet, M. La cromatografía de gases y la espectrometría de masas: Identificación de compuestos causantes de mal olor. *BoletÍN Intexter* **2022**, *122*, 35–41.
35. Rodríguez, J.; Durán, C.; Reyes, A. Electronic nose for quality control of Colombian coffee through the detection of defects in "Cup Tests". *Sensors* **2009**, *10*, 36–46. [CrossRef]
36. Gardner, J.W.; Shurmer, H.V.; Tan, T.T. Application of an electronic nose to the discrimination of coffees. *Sens. Actuators B Chem.* **1992**, *6*, 71–75. [CrossRef]
37. Brudzewski, K.; Osowski, S.; Dwulit, A. Recognition of coffee using differential electronic nose. *IEEE Trans. Instrum. Meas.* **2012**, *61*, 1803–1810. [CrossRef]
38. Gonzalez Viejo, C.; Tongson, E.; Fuentes, S. Integrating a low-cost electronic nose and machine learning modelling to assess coffee aroma profile and intensity. *Sensors* **2021**, *21*, 2016. [CrossRef] [PubMed]

Disclaimer/Publisher's Note: The statements, opinions and data contained in all publications are solely those of the individual author(s) and contributor(s) and not of MDPI and/or the editor(s). MDPI and/or the editor(s) disclaim responsibility for any injury to people or property resulting from any ideas, methods, instructions or products referred to in the content.

beverages

MDPI

Article

Synergetic Effects of *Coffea liberica* and *Curcuma zanthorrhiza*: Study of Sensory Profile, Proximate, and Chemical Compound

Madyawati Latief [1,2,3], Retno Widya Muntasir [1], Dhian Eka Wijaya [4], Indra Lasmana Tarigan [1,2,3] and Sutrisno Sutrisno [1,3,*]

[1] Department of Chemistry, Faculty of Science and Technology, Universitas Jambi, Muaro Jambi 36361, Jambi, Indonesia; madyawatilatief@unja.ac.id (M.L.); retnowidyamuntasir24@gmail.com (R.W.M.); indratarigan@unja.ac.id (I.L.T.)
[2] Natural Product and Bioactive Compound Laboratory, Faculty of Science and Technology, Universitas Jambi, Muaro Jambi 36361, Jambi, Indonesia
[3] The University Centre of Excellences, E2-KOLIM, Universitas Jambi, Muaro Jambi 36361, Jambi, Indonesia
[4] Department of Chemical Analyst, Faculty of Science and Technology, Universitas Jambi, Muaro Jambi 36361, Jambi, Indonesia; dhianekawijaya@unja.ac.id
* Correspondence: herasutrisno@unja.ac.id

Abstract: Liberica coffee (*Coffea liberica*) thrives on peat soil, unlike Arabica and Robusta, making it an essential commodity in Jambi, Indonesia, where it is known as Liberika Tungkal Komposit (Libtukom) and serves as a primary livelihood. This low-caffeine coffee has a higher economic value than high-caffeine varieties. Adding spices such as temulawak (*Curcuma zanthorrhiza*) to Liberica coffee can create a unique, functional beverage with new aromas and health benefits, thanks to the curcuminoid antioxidants in *Curcuma zanthorrhiza* (*C. zanthorrhiza*). This research aims to develop a spice-infused Liberica coffee with enhanced sensory qualities and health benefits. Methods include spice coffee formulation; sensory evaluation; phytochemical screening; and analysis of phenolics, flavonoids, antioxidants, caffeine, and chlorogenic acid levels. Statistical analysis (ANOVA and Duncan's post hoc test) reveals that *C. zanthorrhiza*-enhanced Liberica coffee is preferred by panelists, with spice coffee outperforming plain Liberica in sensory tests. This study showed that Liberica coffee mixed with *C. zanthorrhiza* affected the sensory performance of Liberica coffee, where the coffee most favored by panelists was spice coffee. Based on the cupping test results, spice coffee had excellent results, while Liberica coffee had very good results. The phytochemical analysis revealed that coffee–*C. zanthorrhiza* exhibited higher concentrations of total phenolics, flavonoids, and antioxidant activity than the original Liberica coffee. However, the caffeine and chlorogenic acid levels in all spiced coffee were lower compared to the original Liberica coffee.

Keywords: *Curcuma zanthorrhiza*; Liberica coffee; sensory profile; synergetic

1. Introduction

Coffee not only plays an important role as a source of foreign exchange but also as a source of income for coffee farmers in Indonesia [1]. There are various types of coffee grown in Indonesia, including Arabica, Robusta, and Liberica. Liberica coffee contains less caffeine than Arabica and Robusta [2]. Generally, Liberica coffee is known to have a caffeine content of 1.32%, lower than Arabica coffee, at 1.77%, and Robusta, at 2.15% [3]. Liberica coffee is a plantation crop well-suited for cultivation on peat soils, unlike other coffee varieties such as Arabica and Robusta, which are unable to thrive in such conditions [4]. Liberica coffee has been cultivated for generations across various provinces in Indonesia, including Jambi, Riau, and South Sumatera. It is grown in nutrient-rich peat soil in harmony with native forests; Liberica thrives within a unique agroforestry system. Unlike the more widely known Arabica and Robusta varieties, Liberica offers a truly unique taste experience with its unmistakable hint of jackfruits, which has earned it the nickname "jack coffee" in some regions. This rare flavor profile, combined with its deep cultural roots, makes Liberica coffee a treasured gem in Indonesia coffee [5]. Coffee has high caffeine levels that can affect health. Therefore, decaffeinated coffee (lower caffeine) is a good and safe alternative drink for consumers who are sensitive to caffeine [6]. A more sensible alternative approach to producing decaffeinated coffee involves the addition of spices to coffee beverages. This method is promising to reduce the caffeine content effectively. Spices commonly used in coffee making are cardamom, cinnamon [7], ginger [8], swallow's nest, black cumin, and cloves [9].

Indonesia is also rich in spices, one of which is temulawak. Temulawak (*Curcuma zanthorrhiza*) is one of the medicinal plants of the Zingiberaceae family that is widely grown and used as raw material for traditional medicine in Indonesia [10]. *C. zanthorrhiza* plants are empirically widely used as single or mixed drugs. There are more than 50 traditional medicine recipes using temulawak [11]. The existence of *C. zanthorrhiza* as a medicinal plant has long been recognized, especially among the community. The rhizome part of temulawak is the main ingredient in this research.

In vitro testing of *C. zanthorrhiza* rhizomes has shown its many benefits, one of which is its potential as an antioxidant [11]. The bioactive components in temulawak rhizomes that are responsible for antioxidants are curcumin, demethoxycurcumin, and bisdemethoxycurcumin [12]. In addition, a number of compounds were obtained in temulawak rhizomes, including water, 19.98%; starch, 41.45%; fiber, 12.62%; ash, 4.62%; acid insoluble ash, 0.56%; water essence, 10.96%; alcohol essence, 9.48%; and curcumin, 2.29%. From the test results, it was also found that the content of alkaloids, flavonoids, phenolics, triterpenoids, tannin glycosides, saponins, and steroids was thought to be the cause of the efficacy of temulawak [13].

Coffee with added spices can reduce caffeine levels and enhance radical scavenging activity. The combination of species in coffee plays a role in providing a synergetic effect so that it can produce new aromas and flavors. Moreover, adding spices to coffee can enhance its pharmaceutical properties, transforming it into a functional beverage that is beneficial for health. The coffee–spice combination plays a key role in improving the overall coffee quality by changing its chemical content and flavor profile [14]. Adding spices can impart an acceptable aroma; offer refreshing and warming effects; aid in detoxifying the body; and help eliminate toxins through urine, feces, and sweat [15]. Research on the effects of adding spices to Liberica coffee still needs to be completed. Notwithstanding, our previous study demonstrated that the addition of cinnamon exhibits significantly enhanced total flavonoid and phenolic content, as well as antioxidant activity, in Liberica coffee [16]. Several previous studies have only focused on the analysis of spice combinations with Arabica or Robusta coffee. This study examines how the addition of *C. zanthorrhiza* to Liberica coffee affects the sensory profile of coffee, chemical compounds, antioxidant activity, and proximate analysis. Combining spices and ingredients with coffee, particularly Liberica coffee, can enhance its quality, positioning it as a functional and lifestyle beverage. This combination improves the sensory attributes and increases the health benefits, resulting in antioxidant-rich herbal coffee products. Antioxidants play a crucial role in inhibiting oxidative processes in the body, thereby reducing the risk of diseases associated with oxidative stress, such as cardiovascular disorders, neurodegenerative diseases, and certain cancers. Liberica coffee can offer a flavorful experience and significant health-promoting properties by incorporating herbal elements.

2. Materials and Methods

2.1. Materials and Equipment

The materials used in this study were coffee Liberica beans (*Coffea liberica*) purchased from CV. Haji Bangun Kopi Liberika from Tanjung Jabung Timur Regeny and *C. zanthorrhiza* from CV. Kawa Kerinci from Kerinci Regency, Jambi Province, Indonesia. Other materials used consisted of distilled water (ddH_2O), MeOH (p.a), acetone, HCl, gallic acid standard, ethanol P.A., Folin–Ciocâlteu reagent, Na_2CO_3, quercetin standard, $AlCl_3$, CH_3COONa, formic acid, acetonitrile, caffeine standard, chlorogenic acid standard, trigonelline standard, K_2SO_4, $CuSO_4$, H_2SO_4 concentrated, NaOH, PP indicator, n-hexane, Luff-Schoorl solution, CaCl, Na_2SO_3, amylum indicator, 2,2-diphenyl-1-picrylhydrazy (DPPH), dichloromethane (DCM), and ascorbic acid standard (Sigma-Aldrich, Singapore). Equipment used during the study included a Rotary Evaporator (BUCHI R-300 System, Büchi Labortechnik AG, Flawil, Switzerland), Coffee Roaster Machine Pratter 1.5 (Pratter, Banjarsari Surakarta, Central Java, Indonesia), Coffee Grinder (ICRRI, Jember, Indonesia), Analytical balance, centrifuge tubes, glassware, vortex mixer tools, centrifugation tools, Whatman 54, 41 or 541 filter paper, 0.45 μm membrane filter, crucibles and lids, oven, desiccator, furnace, Kjeldahl flask, a set of distillation apparatus, a set of titration apparatus, a set of Soxhlet apparatus, Buchner funnel, reflux apparatus, glassware, drop pipette, and stirring rod. Analysis in this study using Ultraviolet–Visible spectroscopy (UV-Vis) and Gas Chromatography–Mass Spectrometry (GC-MS) (Thermo-Fischer Orion Scientific, Waltham, MA, USA). Some limitations in this study: This study did not use raw material variability, roasting temperature, or component interactions. Phytochemical analysis is only on total phenolics, flavonoids, caffeine, chlorogenic acid, and antioxidants.

2.2. Sample Preparation

Roasting coffee: The samples studied were the coffee beans of Liberica (*Coffea liberica*) and *C. zanthorrhiza*. The dried coffee beans were roasted at 203 °C for 12 min. Furthermore, the cooled coffee beans were pulverized using a grinder until a fine coffee powder was obtained [17]. *C. zanthorrhiza* drying: *C. zanthorrhiza* was washed, cut, and then dried using an oven-drying technique, at 60 °C for 25 min to 30 min [16]. The combination coffee–*C. zanthorrhiza* was formulated with different ratios (Table 1).

Table 1. Sample codes.

No.	Sample Codes	Information Ratio (%)
1	CB	*C. zanthorrhiza* (100%)
2	CO	Coffee Liberica (100%)
3	CH1	Coffee: *C. zanthorrhiza* (99:1)
4	CH2	Coffee: *C. zanthorrhiza* (97:3)
5	CH3	Coffee: *C. zanthorrhiza* (95:5)

2.3. Cupping Test

The roasted coffee and ground coffee, processed to a fine powder (20-mesh size), were brewed using 150 mL of hot water at a temperature between 94 and 96 °C to prepare samples for sensory profile evaluation. The sensory analysis was conducted by a panel of three experts from the Jambi Cupper Team, Q-Graders from Lymphocyte Coffee Workshop Shah Alam Jaya Ltd, Jambi, Indonesia. The sensory profile was carried out using the SCAA (Specialty Coffee Association of America) scoring system. The scoring ranged from 1 to 10, depending on the coffee's quality. This evaluation method assesses the coffee's aroma, flavor, body, acidity, and aftertaste. It provides a comprehensive profile of its sensory qualities that is crucial for determining the overall quality and consumer acceptability. Panelists rated each sensory attribute with a score from 6.00 to 6.75 as good, from 7.00 to 7.75 as very good, from 8.00 to 8.75 as excellent, and from 9.00 to 10.00 as outstanding. The final score was obtained by summing the scores of each attribute [18].

2.4. Sample Extraction

The sample (1 g) was extracted into 40 mL of a MeOH-ddH_2O (with ratio 1:1 v/v) solution. This solvent mixture is commonly used to efficiently extract bioactive compounds, including phenolics and flavonoids, due to its ability to dissolve polar and non-polar substances. The resulting extract is often subjected to further analysis to determine the concentration of crucial phytochemicals and assess antioxidant properties, critical for evaluating the sample's functional and health-related potential. Hydrochloric acid (HCl) was added until the mixture reached a pH of 2, followed by vortexing for 3 min. The solution was centrifuged at 4500× g rpm for 10 min, and the resulting supernatant was collected as "Extract 1". The remaining pellet was treated with 40 mL of acetone (70%; acetone/water), vortexed for 3 min, and then centrifuged again. The supernatant from the second extraction was combined with Extract 1 to obtain the final liquid extract [19].

2.5. Total of Phenolics Content (TPC) and Total of Flavonoids (TFC)

Total phenolics content (TPC): A 1000 ppm liquid extract sample and a series of gallic acid (GA) standard solutions at 10, 20, 30, 40, and 50 ppm were each measured at 1 mL. To each, 0.4 mL Folin–Ciocâlteu was added, followed by gentle shaking and incubation at RT for 4–8 min. Next, 4 mL of a 7% Na_2CO_3 solution was added, and the volume was adjusted to 10 mL with distilled water. The mixture was then kept in a dark environment at room temperature for 2 h. Finally, the absorbance was measured at a wavelength of 745 nm [20].

Total Flavonoids Content (TFC): A 1 mL aliquot of 1000 ppm liquid extract and a 1 mL aliquot from a series of quercetin standard solutions (concentrations of 10, 20, 30, 40, and 50 ppm) were each combined with MeOH (3 mL), $AlCl_3$ 10% (0.2 mL), and CH_3COONa 1M (0.2 mL). The mixture was then diluted to a final volume of 10 mL with distilled water. Samples were incubated in a dark environment at room temperature for 30 min. Absorbance readings were obtained using a UV-Vis spectrophotometer at a wavelength of 510 nm [21]. The absorbance data from the gallic acid standard solutions were used to generate a calibration curve correlating concentration and absorbance, yielding a linear equation for calculating the total phenolic content in the samples. The calibration curve for the gallic acid standards is shown in Figure 1.

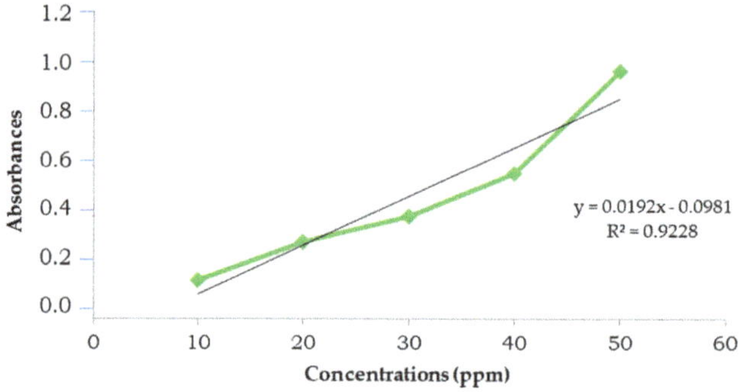

Figure 1. Calibration curve of gallic acid standard solution. The green line represents a linear regression, it means a straight line has been calculated to best fit the data points. Linear regression minimizes the sum of squared differences (errors) between observed data points and the predicted values on the line.

From the calibration curve of gallic acid standard solution in Figure 1 which uses 5 series of standard solutions with a concentration range of 10–50 ppm, the regression line equation y = 0.0192x − 0.0981 is obtained, with an R^2 value of 0.9228. In measuring the absorbance of the sample, two replications were made for data accuracy purposes.

Quercetin was used as a standard solution, which was prepared in five different concentrations: 10, 20, 30, 40, and 50 ppm. Quercetin was measured at a peak of 431 nm. Then, we obtained the absorbance value of quercetin standard at each concentration. The results of the absorbance of the standard solution were used to make a calibration curve that connects the concentration to the absorbance so as to obtain a linear line equation used to calculate the total flavonoid content in the sample. The calibration curve of quercetin standard solution is presented in Figure 2.

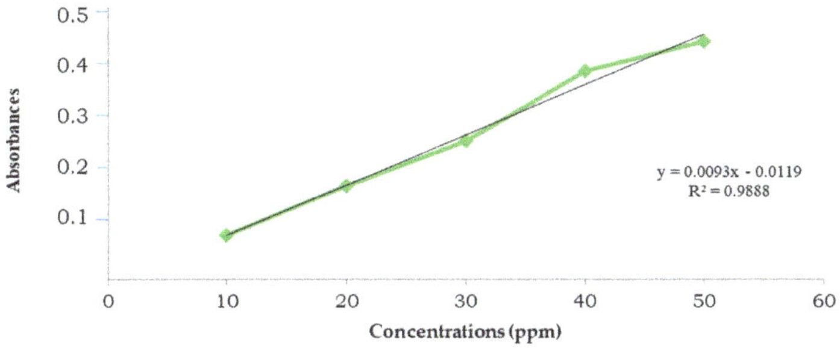

Figure 2. Calibration curve of quercetin standard solution. The green line represents a linear regression, it means a straight line has been calculated to best fit the data points. Linear regression minimizes the sum of squared differences (errors) between observed data points and the predicted values on the line.

2.6. Total Caffeine and Chlorogenic Acids

Caffeine content: A total of 1 g of the samples was dissolved in 150 mL of boiled ddH_2O (approximately 93 °C), with continuous stirring. The samples were then filtered to obtain a filtrate and then transferred into a separatory funnel. Next, 1.5 g of $CaCO_{3(S)}$ was added to the filtrate and then extracted with $CHCl_3$ for 15 min. It was evaporated in 60 °C water bath system. The extract was transferred to a volumetric flask, diluted with ddH_2O up to 100 mL, and thoroughly mixed. Then, 50 mL of the extract was diluted with 100 mL ddH_2O. The caffeine content was measured by a UV-Vis spectrophotometer at 280 nm, the maximum wavelength measured before. Determination of the maximum wavelength was carried out in the wavelength range of 200–400 nm, and the measurement results obtained the maximum wavelength value of the caffeine standard solution, i.e., 286 nm. Absorbance measurements of both the test solution and a series of caffeine standard solutions (at concentrations of 5, 10, 15, 20, and 25 ppm) were taken at the determined maximum wavelength using the UV-Vis spectrophotometer [22]. Then, the absorbance value of caffeine was set at each concentration. The results of the absorbance of the standard solution made a calibration curve that connects the concentration to the absorbance to obtain a linear equation used to calculate the caffeine content in the sample. The calibration curve of the caffeine standard solution is presented in Figure 3.

From the calibration curve of the caffeine standard solution in Figure 3, which uses 5 series of standard solutions with a concentration range of 5–25 ppm, the regression line equation y = 0.0151x − 0.0768 is obtained, and R^2 is 0.9959. Two replications were made to measure the absorbance of the sample for data accuracy purposes.

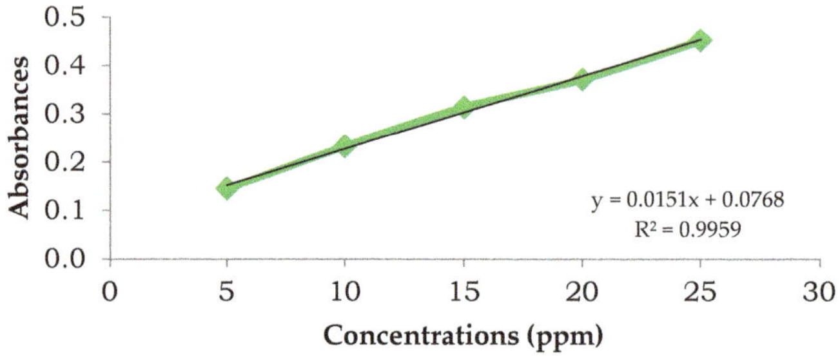

Figure 3. Calibration curve of caffeine standard solution.

Chlorogenic acid (CGA) content: A total of 4.8 mL of the sample coffee extract was added to 25 mL of distilled water. The mixture was stirred continuously for 1 h, using a magnetic stirrer, while heating to 100 °C to ensure thorough extraction. The extraction process was conducted by adding 25 mL of DCM to the sample to separate caffeine and chlorogenic acid. The solution was stirred for an additional 10 min to enhance phase separation. Subsequently, the aqueous and DCM phases were separated using a separatory funnel. The aqueous phase, containing chlorogenic acid, was carefully collected for further analysis. The maximum absorbance wavelength of the CGA solution was determined by scanning over a range of 200–400 nm using a UV-Vis spectrophotometer. For quantification, the absorbance of the test solution and a series of CGA standard solutions (10, 20, 30, 40, and 50 ppm) was measured at the identified maximum wavelength. The calibration curve constructed from the standards was used to calculate the chlorogenic acid concentration in the sample, ensuring accurate and reliable quantification. This method is critical for assessing the bioactive compound profile of coffee and related products, particularly given chlorogenic acid's antioxidant properties and potential health benefits [23].

2.7. Antioxidant Activities

A total of 1.97 mg of DPPH was dissolved in 100 mL of MeOH to obtain 50 µM of DPPH. The standard of ascorbic acid solution was prepared at four concentration variations of 0, 10, 30, and 50 ppm. The test-sample solution was also prepared at four concentration variations. Aliquots containing 0.2 mL of test sample and standard were each added to 3.8 mL of 50 µM DPPH solution. The mixture was homogenized and incubated in the dark for 30 min (RT). Furthermore, the absorbance value was measured using UV-Vis spectroscopy at a wavelength of 517 nm, using a UV-Vis spectrophotometer [24]. The scavenging activity was calculated as the percentage of inhibition using Equation (1).

$$\text{DPPH scavenging activity (\% Inhibition)} = (A_0 - A_1)/A_0 \times 100\% \quad (1)$$

where A_0 is the absorbance of the control (DPPH solution without sample), and A_1 is the absorbance of the sample (DPPH solution with antioxidant samples).

The inhibition curve was created by plotting % inhibition against the logarithm of the sample concentrations. Linear regression is observed using Equation (2a), and IC_{50} is calculated using Equation (2b).

$$y = ax + b \quad (2a)$$

$$IC_{50} = (50 - b)/a \quad (2b)$$

where y is % inhibition, x is log of concentration, m is slope, and c is intercept.

2.8. Proximate Analysis

Determination of moisture content: A petri dish was initially weighed and dried in an oven at 105 °C for 24 h. After cooling for 1 h, 2 g of sample was added to the petri dish, which was then heated again at 1050 °C for 48 h. The final mass was measured, and the water content percentage was calculated using Equation (3).

$$\%\text{Moisture content} = \frac{\text{IM(g)} - \text{F(g)}}{\text{IM(g)} - \text{MEC(g)}} \times 100\% \quad (3)$$

where FM = final mass, MEC = mass of the empty cup, and IM = initial mass.

Determination of ash content: The petri dish was first weighed and then dried in an oven at 105 °C for 24 h. After allowing it to cool for 1 h, 2 g of sample was placed in the dish, which was subsequently heated at 550 °C in a furnace for an additional 3 h. Furthermore, the sample was allowed to cool in a desiccator for 30 min [25]. The final weight was recorded, and the percentage of ash content was calculated using Equation (4).

$$\%\text{Ash content} = \frac{\text{FM} - \text{MEC}}{\text{IMs} - \text{MEC}} \times 100\% \quad (4)$$

where FM = final mass, MEC = mass of the empty cup, and IM = initial mass.

Determination of protein content: A 1 g sample, accompanied by 7 g of K_2SO_4 and 0.8 g of $CuSO_4$, was placed into a 100 mL Kjeldahl flask. Subsequently, 15 mL of concentrated H_2SO_4 was carefully added. The flask was gradually heated, beginning at room temperature, and the temperature was steadily increased until the mixture reached a boiling point. The digestion process continued until the solution became clear and greenish, signifying completion. The digested solution was diluted with 100 mL of distilled water for the distillation step. Following this, 10 mL of NaOH solution (30%) was introduced along the inner wall of the distillation flask. Steam generated from the boiling mixture was collected and added to an Erlenmeyer flask containing 10 mL of HCl (0.1N) and phenolphthalein (pp) indicator. Furthermore, the distillate collected in the Erlenmeyer flask was titrated with 0.1 N NaOH. The titration endpoint was identified by a color change from colorless to pink. A blank sample, prepared without the test substance, was titrated under identical conditions. Finally, the protein content was calculated using Equation (5).

$$\%\text{Protein content} = \frac{(\text{V} - \text{V}_0)}{\text{IN}} \times 0.014 \times \text{CF} \times \text{DF} \times 100\% \quad (5)$$

where IN is the mass of the sample (g), V is the volume of NaOH used for titration of the sample (mL), V_0 is the volume of NaOH used for titration of the blank (mL), N is the normality of NaOH (mol/L), CF is the conversion factor, and DF is the dilution factor.

Determination of carbohydrate content: A 1 g sample was combined with 50 mL of distilled water and 10 mL of 25% HCl. The mixture was refluxed at 100 °C (3 h) and cooled to room temperature. The phenolphthalein (PP) indicator was added, and the solution was neutralized with 20% NaOH while monitoring the pH using a universal indicator. The neutralized solution was transferred to a 250 mL volumetric flask and diluted to the calibration mark with distilled water. A 10 mL aliquot of the prepared solution was mixed with 10 mL of Luff-Schoorl reagent and refluxed (10 min), starting from the onset of boiling. After refluxing, the solution was cooled in an ice bath.

Into the sample, about 5 mL of 20% KI and 5 mL of 25% H_2SO_4 were carefully added. The mixture was immediately titrated with 0.1 N $Na_2S_2O_3$ until it developed a straw-yellow color. At this point, three drops of the starch indicator were introduced, and titration continued until the blue color completely disappeared. A blank titration, prepared using distilled water in place of the sample, was conducted under identical conditions. The carbohydrate content was determined using Equations (6) and (7).

$$\text{VNt} = (\text{VNb} - \text{VNs}) \times [\text{N}_t] \times 10 \quad (6)$$

$$\%\ \text{Carbohydrate} = \frac{\text{WN} \times \text{DF}}{\text{WS}} \times 0.9 \times 100\% \quad (7)$$

where VNt = $Na_2S_2O_3$ used (mL), VNb = blank titration of $Na_2S_2O_3$ (mL), VNs = sample titration of $Na_2S_2O_3$ (mL), $[N_t]$ = normality of $Na_2S_2O_3$ (N), WS = weight of sample (mg), WN = glucose contained per mL of $Na_2S_2O_3$ (mg), and DF = dilution factor.

Determination of fat content: About 5 g of sample was extracted using n-hexane, in 69 °C. The extraction process ran for approximately 12 cycles within a 6 h extraction time. Additionally, the solvent was distilled to obtain the fat, and the fat was heated at 105 °C for 1 h. Once heated, the fat was cooled in a desiccator for 15 min before weighing. The remaining mass represented the fat weight, which was calculated as the fat content using Equation (8).

$$\%\ \text{Fat} = \frac{\text{IFE(g)}}{\text{IMS(g)}} \times 100\% \quad (8)$$

where IFE = initial fat extracted (g), and IMS = initial mass of sample (g)

2.9. Data Analysis

The data obtained from the phenol content, caffeine content, and antioxidant activity tests were analyzed by ANOVA (analysis of variance), and if the treatment was affected by the observed variables, it was continued with the Duncan's Multiple Range Test (DMRT) test. Sensory with organoleptic tests was carried out using the Chi-Square Test to determine the treatment ranking [16].

3. Results

3.1. Sensory Profile

The sensory profile is a test performed by Q-Grader experts to identify and assess the sensory attributes of coffee. This study measured the sensory profile to assess the quality and characteristics of coffee mixed with *C. zanthorrhiza*. There are several aspects to consider, including aroma, flavor, texture, acidity, sweetness, balance, body, uniformity, aftertaste, clean cup, and an overall score of coffee. The average score was calculated, and the results were used to determine coffee quality standards. The sensory profile present in Table 2 shows that, of the two samples with very good results, the CO (original Liberica coffee) and CH1 (coffee/*C. zanthorrhiza* with a ratio of 99:1). The other two samples, CH2 (coffee/*C. zanthorrhiza* with a ratio of 97:3) and CH3 (coffee: *C. zanthorrhiza* with a ratio of 95:5), are in the excellent category. The data results (Table 2) state that CH3 has the highest total score value compared to CO, CH1, and CH2. Compared to the existing samples, CH3 outperformed the aroma, flavor, and balance parameters.

Table 2. Sensory profile of coffee/*C. zanthorrhiza*.

Parameters	Samples			
	CO ± SD	CH1 ± SD	CH2 ± SD	CH3 ± SD
Aroma	7.58 ± 0.02	7.13 ± 0.12	7.17 ± 0.12	7.17 ± 0.11
Flavor	7.67 ± 0.04	6.92 ± 0.01	7.08 ± 0.01	7.42 ± 0.01
Aftertaste	7.67 ± 0.15	6.92 ± 0.08	7.17 ± 0.03	7.25 ± 0.02
Acidity	7.77 ± 0.09	6.83 ± 0.08	7.08 ± 0.03	7.42 ± 0.05
Body	7.67 ± 0.09	6.58 ± 0.08	7.17 ± 0.04	7.08 ± 0.05

Table 2. *Cont.*

Parameters	Samples			
	CO ± SD	**CH1 ± SD**	**CH2 ± SD**	**CH3 ± SD**
Uniformity	8.00 ± 0.00	10.00 ± 0.00	10.00 ± 0.03	10.00 ± 0.00
Balance	7.67 ± 0.08	6.58 ± 0.15	6.92 ± 0.05	7.33 ± 0.05
Clean cup	8.00 ± 0.00	10.00 ± 0.00	10.00 ± 0.00	10.00 ± 0.00
Sweetness	8.00 ± 0.00	10.00 ± 0.00	10.00 ± 0.00	10.00 ± 0.00
Overall	7.67 ± 0.08	6.25 ± 0.07	6.75 ± 0.07	7.08 ± 0.07
Total score	7.77 ± 0.073	7.72 ± 0.078	7.93 ± 0.051	8.08 ± 0.051
Category	**	**	***	***
Final score	77.7 ± 0.073 ^{a}	77.2 ± 0.078 ^{a}	79.3 ± 0.051 ^{b}	80.8 ± 0.051 ^{c}
Group	Premium	Premium	Specialty	Specialty

CO = coffee Liberica; CH1 = coffee + *C. zanthorrhiza* (99:1); CH2 = Cofffe + *C. zanthorrhiza* (97:3); CH3 = coffee + *C. zanthorrhiza* (95:5). ^{a,b,c} superscripts with distinct letters within the same column significant differences ($p < 0.05$). SD represents the standard deviation; ** very good; *** excellent.

3.2. Total Phenolic and Flavonoid

Phenolic compounds are one of the compounds present in plants and have various benefits, one of which is as that they have an antioxidant effect. Phenols help reduce free radicals by giving up one electron from their -OH group, which helps stabilize the free radical. There is a positive correlation between the number of phenolic compounds and antioxidant activity; the more phenolic compounds present, the greater the antioxidant capacity. One example of a phenolic compound is gallic acid, which has three phenolic hydroxyl groups. Gallic acid can be oxidized by Folin–Ciocâlteu reagent in an alkaline atmosphere. This reagent will oxidize the phenolic hydroxyl group in gallic acid and form a molybdenum–tungsten complex that produces a yellow color, which is an indicator of the presence of phenolic compounds. This study uses additional Na_2CO_3 in the sample solution in the phenolic test. This aims to create an alkaline atmosphere, as the Folin–Ciocâlteu reagent reacts to the sample in alkaline conditions caused by phenolic hydroxyl groups in the sample [26].

The results revealed that there was an increase observed in the samples which was directly proportional to the increase in the levels of ginger added to Liberica coffee. This is in accordance with the previous research [27]: The addition of *C. zanthorrhiza* substances to food will increase the pH value and total phenolics but has the potential to reduce the value of total titrated acid and total soluble solids. Further explained, the relationship of phenolic content in a solution mixed with *C. zanthorrhiza* will increase the antioxidant activity that captures free radicals DPPH and IC_{50}. The phenolic content of CH3, measured at 49.146 mgGAE/g, was higher than that of CO. This increase is attributed to the presence of *C. zanthorrhiza*, which has the highest phenolic content among the combinations.

Total flavonoids: Flavonoids are the most widely found group of phenol compounds in nature. The total flavonoid content in all extract samples was determined by using a UV-Vis spectrophotometer. In this test, quinones are used as the standard because quinine is a type of flavonol flavonoid with a ketone group at the C-4 atom and a hydroxyl group at the neighboring C-3 and C-5 atoms.

The calibration curve of quercetin standard solution in Figure 2 uses five series of standard solutions with a concentration range of 10–50 ppm, and the regression line equation $y = 0.0093x - 0.0119$ was obtained. The R^2 value of 0.9888 is nearly equal to 1, indicating that the regression equation is linear, and the absorbance and concentration have a robust correlation. Two replications were made to measure the absorbance of the sample for data accuracy. The results of the total flavonoids are presented in Table 3.

Table 3. Result of total phenolics and flavonoids content.

Samples	TPC (mgGAE/g ± SD)	TFC (mgQE/g ± SD)
CB	103.313 ^{a} ± 0.037	34.452 ^{a} ± 0.076
CO	42.271 ^{b} ± 0.037	8.43 ^{b} ± 0.001
CH1	43.313 ^{c} ± 0.037	18.323 ^{d} ± 0.001
CH2	46.542 ^{d} ± 0.037	20.903 ^{e} ± 0.009
CH3	49.146 ^{e} ± 0.010	24.667 ^{c} ± 0.076

CO = Coffee Liberica; CH1 = coffee + *C. zanthorrhiza* (99:1); CH2 = coffee + *C. zanthorrhiza* (97:3); CH3 = coffee + *C. zanthorrhiza* (95:5); CB = *C. zanthorrhiza*. ^{a,b,c,d,e} Superscripts with distinct letters within the same column significant differences ($p < 0.05$). SD represents the standard deviation.

The results indicate that the highest average flavonoid content was 24.667 QE/g, whereas CO had the lowest flavonoid content among all samples, with an average of 8.43 QE/g. Research by [28] shows that coffee is a rich source of flavonoids, especially in organic coffee. The flavonoid concentration in coffee varies depending on the roasting conditions, with levels ranging from 8.6 mg per 100 mL under light roast conditions, 6.3 mg per 100 mL for medium roasts, and 5.1 mg per 100 mL in dark roasts. Based on these data, it can be concluded that temulawak has high flavonoid compounds compared to original coffee. The increase in flavonoids is due to the addition of temulawak to the original Liberica coffee.

3.3. *Caffeine and Chlorogenic Acid Content*

Caffeine content: Caffeine is a type of alkaloid that belongs to the group of xanthine derivatives and is found in almost all types of coffee. Caffeine, 1,3,7-trimethyl xanthine, is usually found in tea, cocoa, and coffee. Determination of the maximum wavelength aims to obtain a wavelength that gives maximum absorption for the analyte to be analyzed. The maximum wavelength is used to determine the calibration curve and the caffeine content in the sample.

Based on Table 4, it is known that the order of highest caffeine to lowest caffeine is CO, CH1, CH2, and CH3, with a value of 0.858, 0.761, 0.657, and 0.559, respectively. The high caffeine content in CO is because CO specimens consist entirely of coffee. The same was reported by previous studies [29], that the ratio between coffee and its ingredients, such as water, milk, or mixed ingredients, affects the caffeine content contained in coffee. The highest percentage of caffeine content will always be obtained by coffee with more coffee powder than mixed coffee.

Table 4. Results of caffeine and chlorogenic acid content (CGA).

Samples	Caffeine (%) ± SD	CGA (%) ± SD
CH3	0.559 ^{a} ± 0.041	2.611 ^{a} ± 0.002
CH2	0.657 ^{b} ± 0.031	2.788 ^{b} ± 0.001
CH1	0.761 ^{c} ± 0.044	2.893 ^{c} ± 0.001
CO	0.858 ^{d} ± 0.001	2.984 ^{d} ± 0.005
CB	0.000 ^{d} ± 0.000	0.000 ^{d} ± 0.000

CO = coffee Liberica; CH1 = coffee + *C. zanthorrhiza* (99:1); CH2 = coffee + *C. zanthorrhiza* (97:3); CH3 = coffee + *C. zanthorrhiza* (95:5); CB = *C. zanthorrhiza*. ^{a,b,c,d} Superscripts with distinct letters within the same column significant differences ($p < 0.05$). SD represents the standard deviation.

Chlorogenic acid (CGA): Chlorogenic acid is a polyphenolic compound with polar properties that allow it to dissolve in water. Chlorogenic acid affects the flavor and aroma characteristics of processed coffee [30]. The presence of chlorogenic acid is instrumental in determining coffee beans' quality, flavor, and aroma. In this study, UV-Vis spectrophotometric was used to determine the CGA at a 323.4 nm wavelength. Selecting the maximum wavelength was essential to identify the wavelength that produces peak absorption for the analyte. This maximum wavelength was then applied to create a calibration curve and to quantify chlorogenic acid in the sample. The optimal wavelength was determined within the range of 200–400 nm, resulting in a maximum absorption for chlorogenic acid at 323.4 nm. Chlorogenic acid standard solutions with concentrations of 10, 20, 30, 40, and 50 ppm were then measured at this wavelength, and the absorbance for each concentration was recorded. These absorbance values were used to construct a calibration curve, which plotted concentration against absorbance, producing a linear equation. This equation was subsequently used to calculate chlorogenic acid levels in the sample. The calibration curve for the chlorogenic acid standard solution is presented in Table 4.

It was found that the highest chlorogenic acid content in the specimen was obtained by CO (2.984%), followed by CH1 (2.808%), CH2 (2.788%), and CH3 (2.611). The high chlorogenic acid in original coffee is due to the presence of compounds that form chlorogenic acid in coffee. There are at least 30 types of chlorogenic acids contained in coffee, such as *caffeoylquinic acids* (CQAs), *dicaffeoylquinic acids* (DCQAs), tricaffeoylquinic acids (TCQAs), *feruloylquinic acids* (FQAs), and *p-coumaroylquinic acids* (p-CoQAs) [31]. On the other hand, some studies suggest that 5-caffeoylquinic acid (5-CQA) and 3,5-dicaffeoylquinic acid (3,5-DCQA) are the main chlorogenic acids found in coffee by-products [32].

The difference in chlorogenic acid content in the composition of coffee blends may be due to the form of solvent and the roasting effect on coffee. Different blends or solvents in coffee can reduce or increase chlorogenic acid levels in coffee blends. Meanwhile, the roasting effect decreases chlorogenic acid substances in coffee. This is due to the effect of temperature on chlorogenic acid. Most of the chlorogenic acid becomes caffeine and quinic acid during roasting [33].

3.4. Antioxidant Activity

Antioxidants are defined as chemical substances that are able to neutralize free radical agents. These substances work by providing electrons to achieve stability, thus inhibiting oxidative processes that can cause degenerative diseases [34]. Antioxidants can include various types of compounds, both of natural origin and artificially synthesized (synthetic). Antioxidant activity produced from food ingredients can be measured using the DPPH method. IC_{50} (inhibition concentration) is the concentration of sample solution needed to inhibit 50% of DPPH free radicals. The smaller the IC_{50} states, the stronger the antioxidant is in warding off free-radical DPPH, i.e., in terms of having stronger antioxidant activity.

It seems that CH3 has a higher antioxidant activity compared to *C. zanthorrhiza*, IC_{50} 4.984 ppm and 6.1408 ppm, as coffee contains various antioxidant compounds, such as chlorogenic acid, caffeine, and trigonelline (Table 5). Chlorogenic acid, in particular, is an abundant antioxidant compound in coffee and has strong antioxidant activity [28,34]. On the other hand, *C. zanthorrhiza* contains curcumin, also known as an antioxidant compound [13]. Phenolics and flavonoids play a role as antioxidant agents since they can donate hydrogen atoms from hydroxy to radical compounds so that they can be more stable [35]. The coffee with the most spices added has the highest total phenol yield. This is in addition to its large total phenol content. Phenolic and flavonoid compounds play a major role in antioxidant activity, where the greater the phenolic and flavonoid levels, the better the antioxidant activity [36].

Table 5. Antioxidant activity.

Samples	IC_{50} (ppm) ± SD	Antioxidant Activity
CH3	4.984 ^{a} ± 0.000	Very Strong
CH2	8.39 ^{c} ± 0.188	Very Strong
CH1	12.221 ^{d} ± 0.021	Very Strong
Ascorbic Acid	45.393 ^{e} ± 0.000	Very Strong
CO	72.122 ^{b} ± 0.004	Strong
CB	6.1408 ^{f} ± 0.051	Very Strong

CO = Coffee Liberica; CH1 = coffee + *C. zanthorrhiza* (99:1); CH2 = coffee + *C. zanthorrhiza* (97:3); CH3 = coffee + *C. zanthorrhiza* (95:5); CB = *C. zanthorrhiza*. ^{a,b,c,d,e,f} Superscripts with distinct letters within the same column have significant differences ($p < 0.05$). SD represents the standard deviation.

3.5. Physical–Chemical Analysis

Water content: Water content is the amount of water in a food product or ingredient. Water content is an important parameter in determining the quality of a food ingredient. Water content itself also determines the quality of the shelf life of food ingredients, including food. The lower the water content, the higher the resistance to damage caused by microorganisms [37]. The water content of CO is 4.302%, and that of CB is 8.205%. Meanwhile, the water content of CH1 is 4.302%, that of CH2 is 4.412%, and that of CH3 is 4.522%.

Ash content: Ash content is part of the proximate analysis used to determine the nutritional value of a food ingredient or product. Ash content is one parameter used to evaluate the nutritional value of a food ingredient and an indicator of total food minerals. The higher the ash content, the worse the quality of the food product. Table 6 presents the water content of the test samples. The measurement results show that CO is 4.175% and CB is 3.209%. Meanwhile, the water content for CH1 is 4.232%, that fo CH2 is 4.244%, and that for CH3 is 4.254%.

Table 6. Results of physical–chemical analysis.

Samples	% Waters	% Ash	% Proteins	% Fats	% Carbohydrates
CO	4.302	4.175	17.61	10.17	5.940
CB	8.205	3.209	1.52	1.35	5.185
CH1	4.302	4.232	17.09	10.13	5.300
CH2	4.412	4.244	16.80	9.75	4.742
CH3	4.522	4.254	15.78	9.14	4.318

Total protein: Protein content analysis is the result of protein surface activity, which is related to its hydrophobic and hydrophilic properties. The protein content for Liberica coffee is 17.61%; for curcuma, it is 1.52%; for CH1, it is 17.09%; for CH2, it is 16.80%; and for CH3, it is 15.78%.

Total fat: Fat is the fat contained in food samples' fat content. The principle is to dissolve a material with an organic solvent to separate the crude fat from the other components and then evaporate the organic solvent again to obtain crude fat in the material. The total fat for CO is 10.17%; for CB, it is 1.35%; for CH1, it is 10.13%; for CH2, it is 9.75%; and for CH3, it is 9.14%.

Total carbohydrate content: This aims to classify food ingredients that contain carbohydrates and are used as a source of carbohydrates by observing color changes in the tested ingredients. Table 6 shows that the % carbohydrate for CO is 5.940%; for CB, it is 5.185%; for CH1, it is 5.300%; for CH2, it is 4.742%; and for CH3, it is 4.318%

Adding *C. zanthorrhiza* to Liberica coffee in various formulations led to increased water content, which is attributed to the higher moisture content of curcuma compared to Liberica coffee. The measured water content in the spiced coffee formulations complied with the quality standards set by SNI 01-3542-2004 (Indonesian National Standard), which specifies a maximum moisture content of 7% *w/w* for ground coffee [25]. Similarly, the ash content increased with each formulation, reflecting the higher ash content of *C. zanthorrhiza* relative to coffee.

3.6. Volatile Compound Analysis Results

The spectra data from the GC-MS analysis of Liberica coffee revealed multiple peaks with varying area percentages, indicating differences in the concentration of detected compounds (Table 7). The peak area (%) in the GC-MS spectra correlates directly with the relative abundance of each compound. One of the identified compounds is dodecanoic acid, 1,2,3-propanetriyl ester ($C_{39}H_{74}O_6$), commonly referred to as glyceryl tridodecanoate, glyceryl trilaurate, or lauric acid triglyceride. This compound displayed the highest peak, indicating its predominant presence in the sample. For the chromatogram of *Curcuma zanthorrhiza* (Table 8), 25 peaks were observed, with dodecanoic acid, 1,2,3-propanetriyl ester, again registering as the most abundant compound. Notably, the spiced coffee sample, a blend of Liberica coffee and *C. zanthorrhiza*, exhibited 29 peaks (Table 9), indicating the presence of additional compounds introduced through the mixture. Despite the increase in compound diversity, dodecanoic acid, 1,2,3-propanetriyl ester, remained the predominant compound in the spiced coffee formulation. This suggests that the blending process enriches the chemical profile while maintaining the dominance of key compounds.

Table 7. Results of volatile compound analysis of Liberica coffee.

Peak Compounds	R. Time (Minutes)	Area (%)	Molecular Formula	Compound Names
1	22.307	7.98	$C_{39}H_{74}O_6$	Dodecanoic acid, 1,2,3-propanetriyl ester
			$C_{27}H_{52}O_5$	Dodecanoic acid, 1-(hydroxymethyl)-1,2-ethanediyl ester
			$C_{57}H_{10}O_6$	Glyceryl tridodecanoate
2	22.447	30.97	$C_{39}H_{74}O_6$	Dodecanoic acid, 1,2,3-propanetriyl ester
			$C_{27}H_{52}O_5$	Dodecanoic acid, 1-(hydroxymethyl)-1,2-ethanediyl ester
			$C_{57}H_{10}O_6$	Glyceryl tridodecanoate
3	22.622	10.30	$C_{39}H_{74}O_6$	Dodecanoic acid, 1,2,3-propanetriyl ester
			$C_{57}H_{10}O_6$	Glyceryl tridodecanoate
			$C_{49}H_{94}O_6$	Octadecanoic acid, 2,3-bis[(1-oxotetradecyl)oxy]propyl ester
4	22.800	7.16	$C_{39}H_{74}O_6$	Dodecanoic acid, 1,2,3-propanetriyl ester
			$C_{57}H_{10}O_6$	Glyceryl tridodecanoate
			$C_{35}H_{66}O_6$	2-Lauro-1,3-didecoin
5	22.875	5.79	$C_{39}H_{74}O_6$	Dodecanoic acid, 1,2,3-propanetriyl ester
			$C_{57}H_{10}O_6$	Glyceryl tridodecanoate
			$C_{27}H_{52}O_5$	Dodecanoic acid, 1-(hydroxymethyl)-1,2-ethanediyl ester
6	22.945	9.84	$C_{39}H_{74}O_6$	Dodecanoic acid, 1,2,3-propanetriyl ester
			$C_{57}H_{10}O_6$	Glyceryl tridodecanoate
			$C_{35}H_{66}O_6$	2-Lauro-1,3-didecoin
7	23.091	15.82	$C_{39}H_{74}O_6$	Dodecanoic acid, 1,2,3-propanetriyl ester
			$C_{57}H_{10}O_6$	Glyceryl tridodecanoate
			$C_{26}H_{48}F_3NO_2$	Dodecanamide, N-Dodecyl-N-(Trifluoroacetyl)
8	23.152	8.41	$C_{39}H_{74}O_6$	Dodecanoic acid, 1,2,3-propanetriyl ester
			$C_{57}H_{10}O_6$	Glyceryl tridodecanoate
			$C_{26}H_{48}F_3NO_2$	Dodecanamide, N-Dodecyl-N-(Trifluoroacetyl)
9	23.285	2.35	$C_{39}H_{74}O_6$	Dodecanoic acid, 1,2,3-propanetriyl ester
			$C_{47}H_{90}O_6$	Hexadecanoic acid, 2-[(1-oxododecyl)oxy]-1,3-propanediyl ester
			$C_{55}H_{106}O_6$	Eicosanoic acid, 2-[(1-oxohexadecyl)oxy]-1-[[(1-oxohexadecyl)oxy]methyl]ethyl ester
10	23.444	1.03	$C_{35}H_{66}O_6$	2-Lauro-1,3-didecoin
			$C_{14}H_{26}O_2$	Dodecanoic acid, ethenyl ester
			$C_{17}H_{36}$	2,6,10-Trimethyltetradecane

Table 7. *Cont.*

Peak Compounds	R. Time (Minutes)	Area (%)	Molecular Formula	Compound Names
11	23.545	0.35	$C_{39}H_{74}O_6$	Dodecanoic acid, 1,2,3-propanetriyl ester
			$C_{14}H_{26}O_2$	Dodecanoic acid, ethenyl ester
			$C_{27}H_{52}O_5$	Dodecanoic acid, 1-(hydroxymethyl)-1,2-ethanediyl ester

Table 8. Analysis results of Curcuma volatile compounds.

Peak Compounds	R. Time (Minute)	Area (%)	Molecular Formula	Compound Name
1	15.563	0.40	$C_{39}H_{74}O_6$	Dodecanoic acid, 1,2,3-propanetriyl ester
			$C_{57}H_{10}O_6$	Glyceryl tridodecanoate
			$C_{51}H_{98}O_6$	Octadecanoic acid, 3-[(1-oxododecyl)oxy]-1,2-propanediyl ester
2	16.533	0.64	$C_{39}H_{74}O_6$	Dodecanoic acid, 1,2,3-propanetriyl ester
			$C_{57}H_{10}O_6$	Glyceryl tridodecanoate
			$C_{51}H_{98}O_6$	Octadecanoic acid, 3-[(1-oxododecyl)oxy]-1,2-propanediyl ester
3	17.123	0.98	$C_{39}H_{74}O_6$	Dodecanoic acid, 1,2,3-propanetriyl ester
			$C_{57}H_{10}O_6$	Glyceryl tridodecanoate
			$C_{49}H_{94}O_6$	Octadecanoic acid, 2,3-bis[(1-oxotetradecyl)oxy]propyl ester
4	17.758	1.91	$C_{39}H_{74}O_6$	Dodecanoic acid, 1,2,3-propanetriyl ester
			$C_{57}H_{10}O_6$	Glyceryl tridodecanoate
			$C_{14}H_{26}O_2$	Dodecanoic acid, ethenyl ester
5	17.865	3.05	$C_{39}H_{74}O_6$	Dodecanoic acid, 1,2,3-propanetriyl ester
			$C_{57}H_{10}O_6$	Glyceryl tridodecanoate
			$C_{51}H_{98}O_6$	Octadecanoic acid, 3-[(1-oxododecyl)oxy]-1,2-propanediyl ester
6	18.602	0.93	$C_{39}H_{74}O_6$	Dodecanoic acid, 1,2,3-propanetriyl ester
			$C_{57}H_{10}O_6$	Glyceryl tridodecanoate
			$C_{51}H_{98}O_6$	Octadecanoic acid, 3-[(1-oxododecyl)oxy]-1,2-propanediyl ester
7	18.732	0.65	$C_{39}H_{74}O_6$	Dodecanoic acid, 1,2,3-propanetriyl ester
			$C_{57}H_{10}O_6$	Glyceryl tridodecanoate
			$C_{51}H_{98}O_6$	Octadecanoic acid, 3-[(1-oxododecyl)oxy]-1,2-propanediyl ester
8	19.249	3.15	$C_{39}H_{74}O_6$	Dodecanoic acid, 1,2,3-propanetriyl ester
			$C_{57}H_{10}O_6$	Glyceryl tridodecanoate
			$C_{51}H_{98}O_6$	Octadecanoic acid, 3-[(1-oxododecyl)oxy]-1,2-propanediyl ester
9	19.367	2.21	$C_{39}H_{74}O_6$	Dodecanoic acid, 1,2,3-propanetriyl ester
			$C_{27}H_{52}O_5$	Dodecanoic acid, 1-(hydroxymethyl)-1,2-ethanediyl ester
			$C_{49}H_{98}O_4$	Lauric acid, 2-(hexadecyloxy)-3-(octadecyloxy)propyl ester
10	19.414	2.87	$C_{39}H_{74}O_6$	Dodecanoic acid, 1,2,3-propanetriyl ester
			$C_{57}H_{10}O_6$	Glyceryl tridodecanoate
			$C_{27}H_{52}O_5$	Dodecanoic acid, 1-(Hydroxymethyl)-1,2-Ethanediyl Ester
11	19.696	15.20	$C_{39}H_{74}O_6$	Dodecanoic acid, 1,2,3-propanetriyl ester
			$C_{57}H_{10}O_6$	Glyceryl tridodecanoate
			$C_{51}H_{98}O_6$	Octadecanoic acid, 3-[(1-oxododecyl)oxy]-1,2-propanediyl ester
12	19.833	1.51	$C_{39}H_{74}O_6$	Dodecanoic acid, 1,2,3-propanetriyl ester
			$C_{57}H_{10}O_6$	Glyceryl tridodecanoate
			$C_{27}H_{52}O_5$	Dodecanoic acid, 1-(hydroxymethyl)-1,2-ethanediyl ester
13	19.925	1.41	$C_{39}H_{74}O_6$	Dodecanoic acid, 1,2,3-propanetriyl ester
			$C_{49}H_{94}O$	Octadecanoic acid, 2,3-bis[(1-oxotetradecyl)oxy]propyl ester
			$C_{21}H_{36}$	14-β-H-Pregna
			$C_{51}H_{98}O_6$	Octadecanoic acid, 3-[(1-oxohexadecyl)oxy]-2-[(1-oxotetradecyl)oxy]propyl ester
14	20.092	12.48	$C_{39}H_{74}O_6$	Dodecanoic acid, 1,2,3-propanetriyl ester
			$C_{57}H_{10}O_6$	Glyceryl tridodecanoate
			$C_{51}H_{98}O_6$	Octadecanoic acid, 3-[(1-oxododecyl)oxy]-1,2-propanediyl ester
15	20.281	4.05	$C_{39}H_{74}O_6$	Dodecanoic acid, 1,2,3-propanetriyl ester
			$C_{57}H_{10}O_6$	Glyceryl tridodecanoate
			$C_{51}H_{98}O_6$	Octadecanoic acid, 3-[(1-oxododecyl)oxy]-1,2-propanediyl ester

Table 8. *Cont.*

Peak Compounds	R. Time (Minute)	Area (%)	Molecular Formula	Compound Name
16	20.333	0.86	$C_{45}H_{86}O_6$	Tetradecanoic acid, 1,2,3-propanetriyl ester
			$C_{36}H_{72}NO_8P$	L-Dimyristoyl lecithin
			$C_{49}H_{94}O_6$	Octadecanoic acid, 2,3-bis[(1-oxotetradecyl)oxy]propyl ester
			$C_{45}H_{86}O_6$	Tetradecanoic acid, 1,2,3-propanetriyl ester
			$C_{51}H_{98}O_6$	Octadecanoic acid, 3-[(1-oxododecyl)oxy]-1,2-propanediyl ester
17	20.425	4.72	$C_{39}H_{74}O_6$	Dodecanoic acid, 1,2,3-propanetriyl ester
			$C_{57}H_{10}O_6$	Glyceryl tridodecanoate
			$C_{51}H_{98}O_6$	Octadecanoic acid, 3-[(1-oxododecyl)oxy]-1,2-propanediyl ester
18	20.463	4.03	$C_{39}H_{74}O_6$	Dodecanoic acid, 1,2,3-propanetriyl ester
			$C_{57}H_{10}O_6$	Glyceryl tridodecanoate
			$C_{51}H_{98}O_6$	Octadecanoic acid, 3-[(1-oxohexadecyl)oxy]-2-[(1-oxotetradecyl)oxy]propyl ester
19	20.747	8.35	$C_{39}H_{74}O_6$	Dodecanoic acid, 1,2,3-propanetriyl ester
			$C_{47}H_{90}O_6$	Hexadecanoic acid, 2-[(1-oxododecyl)oxy]-1,3-propanediyl ester
			$C_{39}H_{74}O_6$	Dodecanoic acid, 1,2,3-propanetriyl ester
			$C_{57}H_{10}O_6$	Glyceryl tridodecanoate
20	20.828	9.46	$C_{47}H_{90}O_6$	Hexadecanoic acid, 2-[(1-oxododecyl)oxy]-1,3-propanediyl ester
			$C_{39}H_{74}O_6$	Dodecanoic acid, 1,2,3-propanetriyl ester
			$C_{35}H_{66}O_6$	2-Lauro-1,3-didecoin
			$C_{47}H_{90}O_6$	Hexadecanoic acid, 2-[(1-oxododecyl)oxy]-1,3-propanediyl ester
21	21.197	6.07	$C_{39}H_{74}O_6$	Dodecanoic acid, 1,2,3-propanetriyl ester
			$C_{51}H_{98}O_6$	Octadecanoic acid, 3-[(1-oxododecyl)oxy]-1,2-propanediyl ester
22	21.312	2.45	$C_{39}H_{74}O_6$	Dodecanoic acid, 1,2,3-propanetriyl ester
			$C_{47}H_{90}O_6$	Hexadecanoic acid, 2-[(1-oxododecyl)oxy]-1,3-propanediyl ester
			$C_{57}H_{10}O_6$	Glyceryl tridodecanoate
23	21.434	7.81	$C_{51}H_{98}O_6$	Octadecanoic acid, 3-[(1-oxododecyl)oxy]-1,2-propanediyl ester
			$C_{39}H_{74}O_6$	Dodecanoic acid, 1,2,3-propanetriyl ester
24	21.491	2.57	$C_{51}H_{98}O_6$	Octadecanoic acid, 3-[(1-oxododecyl)oxy]-1,2-propanediyl ester
			$C_{39}H_{74}O_6$	Dodecanoic acid, 1,2,3-propanetriyl ester
25	21.571	2.24	$C_{51}H_{98}O_6$	Octadecanoic acid, 3-[(1-oxododecyl)oxy]-1,2-propanediyl ester
			$C_{39}H_{74}O_6$	Dodecanoic acid, 1,2,3-propanetriyl ester
			$C_{14}H_{26}O_2$	Dodecanoic acid, ethenyl ester

Table 9. Volatile compound of coffee–*C. zanthorrhiza* CH3.

Peak Compounds	R. Time (Minute)	Area (%)	Molecular Formula	Compound Name
1	15.535	0.16	$C_{39}H_{74}O_6$	Dodecanoic acid, 1,2,3-propanetriyl ester
			$C_{57}H_{10}O_6$	Glyceryl tridodecanoate
			$C_{49}H_{94}O_6$	Octadecanoic acid, 2,3-bis[(1-oxotetradecyl)oxy]propyl ester
2	17.849	0.72	$C_{39}H_{74}O_6$	Dodecanoic acid, 1,2,3-propanetriyl ester
			$C_{57}H_{10}O_6$	Glyceryl tridodecanoate
			$C_{51}H_{98}O_6$	Octadecanoic acid, 3-[(1-oxododecyl)oxy]-1,2-propanediyl ester
3	18.508	0.65	$C_{39}H_{74}O_6$	Dodecanoic acid, 1,2,3-propanetriyl ester
			$C_{57}H_{10}O_6$	Glyceryl tridodecanoate
			$C_{27}H_{52}O_5$	Dodecanoic acid,1-(hydroxymethyl)
4	18.603	1.13	$C_{39}H_{74}O_6$	Dodecanoic acid, 1,2,3-propanetriyl ester
			$C_{57}H_{10}O_6$	Glyceryl tridodecanoate
			$C_{51}H_{98}O_6$	Octadecanoic acid, 3-[(1-oxododecyl)oxy]-1,2-propanediyl ester
5	18.724	0.44	$C_{39}H_{74}O_6$	Dodecanoic acid, 1,2,3-propanetriyl ester
			$C_{57}H_{10}O_6$	Glyceryl tridodecanoate
			$C_{27}H_{52}O_5$	Dodecanoic acid, 1-(hydroxymethyl)-1,2-ethanediyl ester
6	19.049	1.03	$C_{39}H_{74}O_6$	Dodecanoic acid, 1,2,3-propanetriyl ester
			$C_{57}H_{10}O_6$	Glyceryl tridodecanoate
			$C_{51}H_{98}O_6$	Octadecanoic acid, 3-[(1-oxododecyl)oxy]-1,2-propanediyl ester

Table 9. *Cont.*

Peak Compounds	R. Time (Minute)	Area (%)	Molecular Formula	Compound Name
7	19.247	2.10	$C_{39}H_{74}O_6$	Dodecanoic acid, 1,2,3-propanetriyl ester
			$C_{57}H_{10}O_6$	Glyceryl tridodecanoate
			$C_{51}H_{98}O_6$	Octadecanoic acid, 3-[(1-oxododecyl)oxy]-1,2-propanediyl ester
8	19.367	2.92	$C_{39}H_{74}O_6$	Dodecanoic acid, 1,2,3-propanetriyl ester
			$C_{57}H_{10}O_6$	Glyceryl tridodecanoate
			$C_{27}H_{52}O_5$	Dodecanoic acid, 1-(hydroxymethyl)-1,2-ethanediyl ester
9	19.411	3.84	$C_{39}H_{74}O_6$	Dodecanoic acid, 1,2,3-propanetriyl ester
			$C_{57}H_{10}O_6$	Glyceryl tridodecanoate
			$C_{51}H_{98}O_6$	Octadecanoic acid, 3-[(1-oxododecyl)oxy]-1,2-propanediyl ester
10	19.592	2.76	$C_{39}H_{74}O_6$	Dodecanoic acid, 1,2,3-propanetriyl ester
			$C_{57}H_{10}O_6$	Glyceryl tridodecanoate
			$C_{27}H_{52}O_5$	Dodecanoic acid, 1-(hydroxymethyl)-1,2-ethanediyl ester
11	19.647	3.96	$C_{39}H_{74}O_6$	Dodecanoic acid, 1,2,3-propanetriyl ester
			$C_{57}H_{10}O_6$	Glyceryl tridodecanoate
			$C_{51}H_{98}O_6$	Octadecanoic acid, 3-[(1-oxododecyl)oxy]-1,2-propanediyl ester
12	20.070	7.74	$C_{39}H_{74}O_6$	Dodecanoic acid, 1,2,3-propanetriyl ester
			$C_{57}H_{10}O_6$	Glyceryl tridodecanoate
			$C_{51}H_{98}O_6$	Octadecanoic acid, 3-[(1-oxododecyl)oxy]-1,2-propanediyl ester
13	20.276	3.54	$C_{39}H_{74}O_6$	Dodecanoic acid, 1,2,3-propanetriyl ester
			$C_{57}H_{10}O_6$	Glyceryl tridodecanoate
			$C_{51}H_{98}O_6$	Octadecanoic acid, 3-[(1-oxohexadecyl)oxy]-2-[(1-oxotetradecyl)oxy]propyl ester
14	20.333	1.10	$C_{39}H_{74}O_6$	Dodecanoic acid, 1,2,3-propanetriyl ester
			$C_{57}H_{10}O_6$	Glyceryl tridodecanoate
			$C_{49}H_{94}O_6$	Octadecanoic acid, 2,3-bis[(1-oxotetradecyl)oxy]propyl ester
15	20.455	9.43	$C_{39}H_{74}O_6$	Dodecanoic acid, 1,2,3-propanetriyl ester
			$C_{57}H_{10}O_6$	Glyceryl tridodecanoate
			$C_{51}H_{98}O_6$	Octadecanoic acid, 3-[(1-oxododecyl)oxy]-1,2-propanediyl ester.
16	20.721	12.90	$C_{39}H_{74}O_6$	Dodecanoic acid, 1,2,3-propanetriyl ester
			$C_{57}H_{10}O_6$	Glyceryl tridodecanoate
			$C_{51}H_{98}O_6$	Octadecanoic acid, 3-[(1-oxododecyl)oxy]-1,2-propanediyl ester
17	20.815	8.81	$C_{39}H_{74}O_6$	Dodecanoic acid, 1,2,3-propanetriyl ester
			$C_{57}H_{10}O_6$	Glyceryl tridodecanoate
			$C_{49}H_{94}O_6$	Octadecanoic acid, 2,3-bis[(1-oxotetradecyl)oxy]propyl ester
18	20.950	2.38	$C_{39}H_{74}O_6$	Dodecanoic acid, 1,2,3-propanetriyl ester
			$C_{47}H_{90}O_6$	Hexadecanoic acid, 2-[(1-oxododecyl)oxy]-1,3-propanediyl ester
			$C_{51}H_{98}O_6$	Octadecanoic acid, 3-[(1-oxododecyl)oxy]-1,2-propanediyl ester
19	21.087	8.62	$C_{39}H_{74}O_6$	Dodecanoic acid, 1,2,3-propanetriyl ester
			$C_{47}H_{90}O_6$	Hexadecanoic acid, 2-[(1-oxododecyl)oxy]-1,3-propanediyl ester
			$C_{57}H_{10}O_6$	Glyceryl tridodecanoate
20	21.315	1.13	$C_{47}H_{90}O_6$	Hexadecanoic acid, 2-[(1-oxododecyl)oxy]-1,3-propanediyl ester
			$C_{39}H_{74}O_6$	Dodecanoic acid, 1,2,3-propanetriyl ester
			$C_{45}H_{86}O_6$	Tetradecanoic acid, 1,2,3-propanetriyl ester
21	21.408	2.18	$C_{39}H_{74}O_6$	Dodecanoic acid, 1,2,3-propanetriyl ester
			$C_{51}H_{98}O_6$	Octadecanoic acid, 3-[(1-oxohexadecyl)oxy]-2-[(1-oxotetradecyl)oxy]propyl ester
			$C_{20}H_{38}O_2$	Octadecanoic acid, ethenyl ester
			$C_{19}H_{40}$	n-Nonadecane
			$C_{32}H_{66}$	Dotriacontane
22	21.489	1.46	$C_{39}H_{74}O_6$	Dodecanoic acid, 1,2,3-propanetriyl ester
			$C_{51}H_{98}O_6$	Octadecanoic acid, 3-[(1-oxododecyl)oxy]-1,2-propanediyl ester
23	21.582	2.65	$C_{51}H_{98}O_6$	Octadecanoic acid, 3-[(1-oxododecyl)oxy]-1,2-propanediyl ester
			$C_{39}H_{74}O_6$	Dodecanoic acid, 1,2,3-propanetriyl ester
24	21.695	5.40	$C_{51}H_{98}O_6$	Octadecanoic acid, 3-[(1-oxododecyl)oxy]-1,2-propanediyl ester
			$C_{39}H_{74}O_6$	Dodecanoic acid, 1,2,3-propanetriyl ester
25	21.784	3.43	$C_{51}H_{98}O_6$	Octadecanoic acid, 3-[(1-oxododecyl)oxy]-1,2-propanediyl ester
			$C_{39}H_{74}O_6$	Dodecanoic acid, 1,2,3-propanetriyl ester
26	21.918	3.27	$C_{51}H_{98}O_6$	Octadecanoic acid, 3-[(1-oxododecyl)oxy]-1,2-propanediyl ester

Table 9. *Cont.*

Peak Compounds	R. Time (Minute)	Area (%)	Molecular Formula	Compound Name
27	22.014	3.98	$C_{39}H_{74}O_6$	Dodecanoic acid, 1,2,3-propanetriyl ester
			$C_{47}H_{90}O_6$	Hexadecanoic acid, 2-[(1-oxododecyl)oxy]-1,3-propanediyl ester
			$C_{57}H_{10}O_6$	Glyceryl tridodecanoate
28	22.216	1.77	$C_{39}H_{74}O_6$	Dodecanoic acid, 1,2,3-propanetriyl ester
			$C_{47}H_{90}O_6$	Hexadecanoic acid, 2-[(1-oxododecyl)oxy]-1,3-propanediyl ester
			$C_{57}H_{10}O_6$	Glyceryl tridodecanoate
29	22.514	0.50	$C_{51}H_{98}O_6$	Octadecanoic acid, 3-[(1-oxododecyl)oxy]-1,2-propanediyl ester
			$C_{39}H_{74}O_6$	Dodecanoic acid, 1,2,3-propanetriyl ester
			$C_{47}H_{90}O_6$	Hexadecanoic acid, 2-[(1-oxododecyl)oxy]-1,3-propanediyl ester

4. Discussion

Plant foods and beverages are the main sources of antioxidants in our diet. Nowadays, coffee is not only a lifestyle drink but also acts as a health drink. Studies on bioactive compounds and antioxidant activity of coffee have been widely reported. New properties of coffee beverages that make it beneficial as nutraceutical and pharmaceutical products continue to be discovered [38]. Brewed coffee has many critical functional ingredients that have been known, such as flavonoids (catechins and anthocyanins), ferulic acid, and caffeic acid. In addition, other biological bioactive compounds are also found in coffee, such as trigonelline, caffeine, chlorogenic acid, nicotinic acid, quinolinate, ester groups, hydroxycinnamic acids, and melanoidins, which are known to have good antioxidants.

In the present work, the antioxidant potential of Liberica coffee enriched with curcuma extract was also determined, as well as the ability to inhibit DPPH radicals. The recent literature shows a lack of information concerning DPPH inhibition in Liberica coffee enriched with curcuma extract. The interaction of coffee with herbs has been reported in several previous studies using Arabica and Robusta coffee species [39]. Curcuma contains various polyphenolic compounds and has high antioxidant activity [40]. In this work, we found that the total phenolics and total flavonoids of Curcuma were higher compared to Liberica coffee. Phenolic compounds might interact with many components in the food matrix. The combination might induce synergistic and antagonistic effects in varying circumstances [41]. Investigated the antioxidant capacity of the interaction between three primary phenolic compounds found in coffee—chlorogenic acid, caffeic acid, and ferulic acid—compared to α-tocopherol, within L-α-phosphatidylcholine liposome systems rather than within a food matrix [42].

Along with the increase in the proportion of curcuma in the coffee–Curcuma combination, there was an increase in the total phenolics and flavonoids of Liberica coffee (Table 2). This result is in line with the increase in antioxidant activity. The decrease in the IC_{50} value when adding curcuma shows a synergetic interaction between coffee and curcuma. The same results were found in various studies of herbal coffee interactions [39,43,44]. Moreover, another study reported that parallel to the cocoa proportion in the mixture, antioxidant capacity increases, indicating a more prominent antioxidant activity [45]. The addition of herbs to coffee also has an impact on reducing caffeine and chlorogenic acid. Currently, the market need for low-caffeine coffee is very high because caffeine has negative effects; it will cause adverse effects, such as insomnia, worry, increased blood pressure, and heart rate that is too fast [46]. Similar research was conducted by Kuswardhani et al. [47] adding ginger to reduce the caffeine content of Robusta coffee.

On the other hand, previous studies on combining coffee with spices looked at their interactions with polyphenols and their antioxidant activity. However, there is a lack of research looking at other sensory effects [39,43,47]. In this study, a sensory test was conducted to see the effect of Curcuma on coffee taste and its effect on antioxidant activity. The addition of spices to coffee has an impact on increasing the taste profile of coffee [7]. The addition of Curcuma to Liberica coffee has an impact on sensory enhancement. Other research also shows that adding ginger and lemongrass improves the organoleptic profile of coffee [48]. Curcuma belongs to herb that is generally added in multiherbal combinations. Adding spices to coffee can enhance its sensory profile since the complex interaction between volatile compounds in coffee and the aromatic compounds in spices creates a more intricate flavor and aroma profile. Spices contribute unique volatile compounds—such as terpenes, phenol, and aldehydes—that introduce new layers of aroma and taste, making the coffee experience more complex and appealing. Previous studies have shown that spices like cardamom, cinnamon, and ginger contain eucalyptol, cinnamaldehyde, and zingiberene, respectively. Depending on the spice added, these compounds blend with coffee's existing flavors and aromas, enhancing sweetness, warmth, or spiciness. Adding spices to coffee significantly impacts its perceived sensory qualities, including flavor complexity and aroma intensity. This synergy between coffee and spices can elevate the sensory experience by creating a more layered and enjoyable beverage.

The GC-MC data show the volatile compounds found in Liberica coffee, *C. zanthorrhiza*, and the combination, with dodecanoic and octadecanoic acid derivatives being the primary constituents. These results align with previous research that underscores the prominence of medium- and long-chain fatty acids in contributing to coffee's aroma and flavor profile. Studies on Liberica coffee show that these fatty acids, along with triglycerides like glyceryl tri-dodecanoate, are crucial in providing the distinct taste and scent of Liberica, differentiating it from other coffee varieties that contain lower levels of these compounds. Similarly, research on Curcuma species emphasizes the role of fatty acid esters like octadecanoic acid esters in enhancing the therapeutic qualities of these plants. Curcuma is known for its anti-inflammatory and antioxidant properties, often attributed to these volatile compounds, which may be beneficial in medicinal and cosmetic applications. The analyzed *C. zanthorrhiza* CH3 sample exhibits a profile close to Curcuma species, suggesting its potential use in similar therapeutic contexts, especially for health-related products leveraging both aroma and bioactive properties. These findings illustrate the chemical richness and functional versatility of these compounds across different applications in food, health, and fragrance industries.

Furthermore, studies on polyphenols and other antioxidant-rich compounds in spices have shown they can enhance coffee's sensory and health profiles. The presence of spices like Curcuma, rich in curcuminoids or cinnamon, with its cinnamic acid derivatives, not only adds to the flavor complexity but also amplifies the beverage's antioxidant activity. This enhancement has been observed in studies where coffee mixed with spices showed improved aroma and taste complexity scores due to these bioactive compounds [42]. Additionally, the Maillard reaction, which contributes to the characteristic coffee aroma during roasting, may interact with certain spice-derived compounds, resulting in more prosperous and desirable sensory characteristics [47]. Thus, adding spices to coffee achieves a sensory boost by adding new aromatic and flavor-enhancing compounds, resulting in an enjoyable beverage that aligns with consumer preferences for functional foods with added health benefits.

Coffee as a beverage also has proximate standards, including water content, ash content, total protein, total fat, and carbohydrates. The addition of Curcuma to Liberica coffee has SNI standard proximate values. Evaluating water and ash content in Curcuma-flavored coffee indicates compliance with the quality standards outlined in SNI 01-3542-2004, which stipulates that ground coffee's maximum permissible ash content is 5% w/w [25]. In the fat content test, the fat content decreased due to the fact that Curcuma has a lower fat content than coffee, so the fat content in spiced coffee in each formulation decreased. The incorporation of curcuma into coffee at varying ratios resulted in a reduction in fat content, attributed to the lower fat composition of curcuma compared to coffee. This decrease in fat content highlights the impact of ingredient ratios on the overall lipid profile of the blend. Furthermore, carbohydrate compounds in coffee are crucial contributors to the development of its characteristic aroma and sweetness during brewing. The interplay between fatty acids and sugars plays a significant role in creating the distinct sensory attributes of coffee. Fatty acids contribute to the richness and complexity of flavor, while sugars enhance the sweetness and balance the overall taste profile, reinforcing the unique aromatic qualities associated with coffee. Furthermore, it seems that sugars undergo a caramelization process during roasting. On the other hand, some reducing sugars form other aromas and sweetness through interaction with amino acids in the Maillard reaction process [49].

5. Conclusions

Based on the results of the analysis, it was found that among the coffee and curcuma blends, the CH3 sample was the best sample, with a balanced assessment in the aspects of aroma, flavor, aftertaste, acidity, and body. The increase in total phenols was from 42.271 mgGAE/g (CO) to 49.146 mgGAE/g (CH3), and the increase in total flavonoids was from 8.43 (CO) to 24.667c $\pm$ 0.076 mgQE/g (CH3). The alteration was accompanied by a significant improvement in antioxidant activity, as evidenced by the reduction in the IC50 value from 72.122 ppm (CO) to 4.984 ppm (CH3). Liberica coffee added with *Curcuma zanthorrhiz* affected the sensory of Liberica coffee, where the coffee most favored by panelists was spice coffee. Based on the cupping test results, CH3 spice–coffee had excellent results, while Liberica coffee had very good results. The results of phytochemical showed that coffee–*C. zanthorrhiza* contains higher levels of total phenolics, flavonoids, and antioxidant activity than original Liberica coffee. Meanwhile, spice coffee's caffeine and chlorogenic acid levels were lower than those of the original Liberica coffee

Author Contributions: Conceptualization, M.L., D.E.W., I.L.T. and S.S.; methodology, R.W.M.; software, D.E.W.; validation, I.L.T., S.S. and M.L.; formal analysis, R.W.M.; investigation, R.W.M. and M.L.; resources, S.S.; data curation, R.W.M. and D.E.W.; writing—original draft preparation, R.W.M. and I.L.T.; writing—review and editing, S.S.; visualization, R.W.M.; supervision, M.L.; project administration, I.L.T.; funding acquisition, S.S. All authors have read and agreed to the published version of the manuscript.

Funding: This research was funded by LPPM Universitas Jambi, Grant number 272/UN21.11/PT.01.05/SPK/2022, and a research Grant from Universitas Jambi funded the APC.

Data Availability Statement: All data are contained within the article.

Acknowledgments: Acknowledgments are extended to the Faculty of Science and Technology, Universitas Jambi, for their laboratory support. Appreciation is also given to the University Center of Excellence, E2-KOLIM, for their contributions.

Conflicts of Interest: The authors declare no conflicts of interest.

References

1. Darma, R.; Salman, D. Problems identification of Arabica coffee commodities on traditional farming in Indonesia: A review. *IOP Conf. Ser. Earth Environ. Sci.* **2021**, *886*, 012069. [CrossRef]
2. Umiyati, E.; Zulfanetti, Z.; Marselina, E. Sustainability Strategy of Small and Medium Micro Business During the Covid19 Pandemic in Jambi Province. In Proceedings of the 3rd Progress in Social Science, Humanities and Education Research Symposium (PSSHERS 2021), West Sumatra, Indonesia, 3 November 2021; Atlantis Press SARL: Paris, France, 2023; pp. 256–266. [CrossRef]

3. Anshori, N.I.; Zainol, M.K.; Zin, Z.M. Antioxidant Activities of Different Varieties of Spent Coffee Ground (Scg) Extracted Using Ultrasonic-Ethanol Assisted Extraction Method. *Univ. Malays. Teren. J. Undergrad. Res.* **2021**, *3*, 33–42. [CrossRef]
4. Saidi, B.B.; Suryani, E. Evaluasi Kesesuaian Lahan Untuk Pengembangan Kopi Liberika Di Kabupaten Tanjung Jabung Timur Jambi. *J. Ilm. Ilmu. Terap. Univ. Jambi* **2021**, *5*, 1–15.
5. Mawardhi, A.D.; Setiadi, D. Strategi Pemanfaatan Lahan Gambut melalui Pengembangan Agroforestri Kopi Liberika (*Coffea liberica*) Strategy on Peatland Utilisaton through Development of Coffea Liberica Agroforestry. In Proceedings of the Seminar Nasional Lahan Suboptimal, Palembang, Indonesia, 18–19 October 2018; pp. 43–51.
6. Mubarok, F.; Suwasono, S.; Palupi, N.W. Perubahan Kadar Kafein Biji Kopi Arabika Hasil Pengolahan Semi Basah Dengan Perlakuan Variasi Jenis Wadah Dan Lama Fermentasi. *Berk. Imliah Pertan.* **2014**, 1–7.
7. Artha, B.A.P.; Wulandari, Y.W.; Suhartatik, N. Aktivitas Antioksidan Kopi Rempah dengna Penambahan Kapulaga (*Amomum compactum*) dan Kayu Manis (*Cinnamomum verum*). *JITIPARI J. Ilm. Teknol. Dan. Ind. Pangan. UNISRI* **2020**, *5*, 48–58. [CrossRef]
8. Mardhatilah, D. Pengaruh Penambahan Konsentrasi Jahe dan Rempah Pada Pembuatan Sirup Kopi. *Agroteknose* **2015**, *6*, 55–61.
9. Sripoo, A.R.; Maheswari, U.T.N.; Rajeshkumar, S. Preparation of oregano, coffee and black cumin aqueous formulation and its anti inflammatory activity. *Int. J. Health Sci. Qassim* **2022**, *6*, 288–295. [CrossRef]
10. Syamsudin, R.A.M.R.; Farid, P.; Farly, S.M.; Rina, A.P.A.; Cahyani, N.D.; Aprilya, S.; Yanti, R.; Khendri, F. Temulawak Plant (*Curcuma xanthorrhiza* Roxb) as a Traditional Medicine. *J. Ilm. Farm. Bahari* **2019**, *10*, 51–65. [CrossRef]
11. Fathir, A.; Haikal, M.; Wahyudi, D. Ethnobotanical study of medicinal plants used for maintaining stamina in madura ethnic, East Java, Indonesia. *Biodiversitas* **2021**, *22*, 386–392. [CrossRef]
12. Rosidi, A.; Khomsan, A.; Setiawan, B.; Riyadi, H.; Briawan, D. Antioxidant potential of temulawak (*Curcuma xanthorrhiza* roxb). *Pak. J. Nutr.* **2016**, *15*, 556–560. [CrossRef]
13. Widyastuti, I.; Luthfah, H.Z.; Hartono, Y.I.; Islamadina, R.; Can, A.T.; Rohman, A. Antioxidant Activity of Temulawak (*Curcuma xanthorrhiza* Roxb.) and its Classification with Chemometrics. *Indones. J. Chemom. Pharm. Anal.* **2021**, *1*, 28–41. [CrossRef]
14. Sumardi, S.; Rasdiansyah, R. Kualitas Fisik dan Kimia Kopi Celup Arabika Rasa Kayu Manis pada Tingkat Penyangraian yang Berbeda. *J. Ilm. Mhs. Pertan.* **2022**, *7*, 323–329. [CrossRef]
15. F Fatmawati, F.; Muhammad, M.; Fokaya, R. Feasibility Analysis of Spice Coffee Processing Business in Home Industries in Tabahawa Village, Ternate City. *Agrikan J. Agribisnis Perikan.* **2020**, *13*, 344–351. [CrossRef]
16. Tarigan, I.L.; Munawaroh, S.; Sutrisno; Yusnaidar Latief, M. Liberica coffee enriched with Cinnamon (*Cinnamomum verum*): Synergetic study of sensory, antioxidant activity, and chemical components. *Coffee Sci.* **2023**, *18*, e182149. [CrossRef]
17. Perdana, B.M.; Manihuruk, R.; Ashyar, R.; Heriyanti; Sutrisno. Evaluation of the effect of roasting process on the energy transition and the crystalline structures of Arabica, Robusta, and Liberica coffee from Jambi Indonesia Evaluation of the effect of roasting process on the energy transition and the crystalline. *IOP Conf. Ser. Mater. Sci. Eng.* **2018**, *345*, 012021. [CrossRef]
18. Isnidayu, A.V.; Sukartiko, A.C.; Ainuri, M. Indicator of Sensory Attributes of Speciality Coffee Originated from West Java Based on Biochemical Component. *J. Tanam. Ind. Dan. Penyegar.* **2020**, *7*, 1. [CrossRef]
19. Somporn, C.; Kamtuo, A.; Theerakulpisut, P.; Siriamornpun, S. Effects of roasting degree on radical scavenging activity, phenolics and volatile compounds of Arabica coffee beans (*Coffea arabica* L. cv. Catimor). *Int. J. Food Sci. Technol.* **2011**, *46*, 2287–2296. [CrossRef]
20. Ahmad, A.R.; Juwita, J.; Ratulangi, S.A.D.; Malik, A. Penetapan Kadar Fenolik dan Flavonoid Total Ekstrak Metanol Buah dan Daun Patikala (Etlingera elatior (Jack) R.M.SM). *Pharm. Sci. Res.* **2015**, *2*, 1–10. [CrossRef]
21. Haile, M.; Kang, W.H. Antioxidant activity, total polyphenol, flavonoid and tannin contents of fermented green coffee beans with selected yeasts. *Fermentation* **2019**, *5*, 29. [CrossRef]
22. Arwangga, A.F.; Asih, I.A.R.A.; Sudiarta, I.W. Analisis Kandungan Kafein Pada Kopi di Desa Sesaot Narmada Menggunakan Spektrofotometri UV-Vis. *J. Kim.* **2016**, *10*, 110–114. [CrossRef]
23. Furqan, M.; Nurman, S. Ekstrak Polar Kopi Hijau Arabika (*Coffea arabica* L.) sebagai Antihiperglikemi pada Mencit (*Mus musculus*). *J. Healthc. Technol. Med.* **2020**, *6*, 1323–1331.
24. Selvi, A.T.; Joseph, G.S.; Jayaprakasha, G.K. Inhibition of growth and aflatoxin production in Aspergillus flavus by Garcinia indica extract and its antioxidant activity. *Food Microbiol.* **2003**, *20*, 455–460. [CrossRef]
25. Suseno, H.; Haryanto, Galih, N.R.P.; Hidayati, N.; Alonto, C.; Irfan, M. *Panduan Penerapan Dan Sertifikasi SNI Produk Kopi Bubuk*; Badan Standarisasi Nasional: Jakarta, Indonesia, 2020.
26. Maizura, M.; Aminah, A.; Aida, W.M.W. Total phenolic content and antioxidant activity of kesum (*Polygonum minus*), ginger (*Zingiber officinale*) and turmeric (*Curcuma longa*) extract. *Int. Food Res. J.* **2011**, *18*, 529–534.
27. Cepeda, G.N.; Lisangan, M.M.; Silamba, I. Antibacterial Activity of Akway Bark Essential Oil (*Drimys piperita* Hook. f.) at Several Levels of Concentration, Acidity (pH) and Salt Content. *J. Apl. Teknol. Pangan.* **2019**, *8*, 149. [CrossRef]
28. Górecki, M.; Hallmann, E. The antioxidant content of coffee and its in vitro activity as an effect of its production method and roasting and brewing time. *Antioxidants* **2020**, *9*, 308. [CrossRef]

29. Rashidinejad, A.; Tarhan, O.; Rezaei, A.; Capanoglu, E.; Boostani, S.; Khoshnoudi-Nia, S.; Samborska, K.; Garavand, F.; Shaddel, R.; Akbari-Alavijeh, S.; et al. Addition of milk to coffee beverages; the effect on functional, nutritional, and sensorial properties. *Crit. Rev. Food Sci. Nutr.* **2022**, *62*, 6132–6152. [CrossRef]
30. Husniati, H.; Oktiani, D. Chlorogenic Acid Isolation from Coffee as Affected by the Homogeneity of Cherry Maturity. *Pelita Perkeb Coffee Cocoa Res J.* **2019**, *35*, 119–124. [CrossRef]
31. Rojas-González, A.; Figueroa-Hernández, C.Y.; González-Rios, O.; Suarez-Quiroz, M.L.; Gonzalez-Amaro, R.M.; Hernandez-Estrada, Z.J.; Rayas-Duarte, P. Coffee Chlorogenic Acids Incorporation for Bioactivity Enhancement of Foods: A Review. *Molecules* **2022**, *27*, 3400. [CrossRef]
32. Yang, X.W.; Wang, N.; Li, W.; Xu, W.; Wu, S. Biotransformation of 4,5-O-dicaffeoylquinic acid methyl ester by human intestinal flora and evaluation on their inhibition of NO production and antioxidant activity of the products. *Food Chem. Toxicol.* **2013**, *55*, 297–303. [CrossRef]
33. Mills, C.E.; Oruna-Concha, M.J.; Mottram, D.S.; Gibson, G.R.; Spencer, J.P.E. The effect of processing on chlorogenic acid content of commercially available coffee. *Food Chem.* **2013**, *141*, 3335–3340. [CrossRef]
34. Jomova, K.; Raptova, R.; Alomar, S.Y.; Alwasel, S.H.; Nepovimova, E.; Kuca, K.; Valko, M. Reactive Oxygen Species, Toxicity, Oxidative Stress, and Antioxidants: Chronic Diseases and Aging. *Arch. Toxicol.* **2023**, *97*, 2499–2574. [CrossRef] [PubMed]
35. Hassanpour, S.H.; Doroudi, A. Review of the antioxidant potential of flavonoids as a subgroup of polyphenols and partial substitute for synthetic antioxidants. *Avicenna J. Phytomed.* **2023**, *13*, 354–376. [CrossRef] [PubMed]
36. Ahwan, A.; Suwarni, A.; Ariastuti, R.; Hafidz, R.; Mei Enjelina, S. Effect of Total Phenolic and Total Flavonoid Levels on the Antioxidant Power od Water Extract, Ethanol, and Chloroform of Green Tea Leaves (*Camellia sinensis* L). *Med. Sains J. Ilm. Kefarmasian.* **2024**, *9*, 17–28. [CrossRef]
37. Seninde Chambers, E. Coffee flavor: A review. *Beverages* **2020**, *6*, 44. [CrossRef]
38. Esquivel, P.; Jiménez, V.M. Functional properties of coffee and coffee by-products. *Food Res. Int.* **2012**, *46*, 488–495. [CrossRef]
39. Durak, A.; Gawlik-Dziki, U.; Pecio, Ł. Coffee with cinnamon—Impact of phytochemicals interactions on antioxidant and anti-inflammatory in vitro activity. *Food Chem.* **2014**, *162*, 81–88. [CrossRef]
40. Yurasbe, N.Q.; Din, N.A.; Palaniveloo, K.; Manikam, S.; Nagappan, T. Phytochemical diversity and biological activities of Curcuma species from the East Coast of Peninsular Malaysia. *Biodiversitas* **2023**, *24*, 4243–4252. [CrossRef]
41. Pinarli, B.; Simge Karliga, E.; Ozkan, G.; Capanoglu, E. Interaction of phenolics with food matrix: In vitro and in vivo approaches. *Med. J. Nutr. Metab.* **2020**, *13*, 63–74. [CrossRef]
42. Neunert, G.; Górnaś, P.; Dwiecki, K.; Siger, A.; Polewski, K. Synergistic and antagonistic effects between alpha-tocopherol and phenolic acids in liposome system: Spectroscopic study. *Eur. Food Res. Technol.* **2015**, *241*, 749–757. [CrossRef]
43. Durak, A.; Gawlik-Dziki, U.; Sugier, D. Coffee enriched with willow (*Salix purpurea* and *Salix myrsinifolia*) bark preparation—Interactions of antioxidative phytochemicals in a model system. *J. Funct. Foods.* **2015**, *18*, 1106–1116. [CrossRef]
44. Erskine, E.; Gültekin Subaşl, B.; Vahapoglu, B.; Capanoglu, E. Coffee Phenolics and Their Interaction with Other Food Phenolics: Antagonistic and Synergistic Effects. *ACS Omega* **2022**, *7*, 1595–1601. [CrossRef] [PubMed]
45. Acosta-Otálvaro, E.; Domínguez-Perles, R.; Mazo-Rivas, J.C.; García-Viguera, C. Bioavailability and radical scavenging power of phenolic compounds of cocoa and coffee mixtures. *Food Sci. Technol. Int.* **2022**, *28*, 514–523. [CrossRef] [PubMed]
46. Rodak, K.; Kokot, I.; Kratz, E.M. Caffeine as a factor influencing the functioning of the human body—Friend or foe? *Nutrients* **2021**, *13*, 3088. [CrossRef] [PubMed]
47. Kuswardhani, N.; Mukti, N.P.; Sari, P. Antioxidant and sensory properties of ready to drink coffee-ginger made from decaffeinated and non-decaffeinated robusta coffee beans. *IOP Conf. Ser. Earth Environ. Sci.* **2021**, *653*, 012050. [CrossRef]
48. Saifulloh, S.; Pamungkas, R.; Sari, D.A. Sensory Analysis of the Panelist's Favorite Spiced Coffee Drinks. *Gontor Agrotech. Sci. J.* **2023**, *8*, 118–124. [CrossRef]
49. Santosa, K.M.; Supriyadi; Anggrahini, S.; Rahmadian, Y. Sensory Analysis, Caffeine, Chlorogenic Acid and Non-Volatile Taste Compounds of Arabica Coffee (*Coffea arabica*) Fermented with Sugar Addition for Brew Taste. *Indones. Food Nutr. Prog.* **2021**, *17*, 37–44. [CrossRef]

Disclaimer/Publisher's Note: The statements, opinions and data contained in all publications are solely those of the individual author(s) and contributor(s) and not of MDPI and/or the editor(s). MDPI and/or the editor(s) disclaim responsibility for any injury to people or property resulting from any ideas, methods, instructions or products referred to in the content.

foods

MDPI

Article

Exploring the Characteristic Aroma Components of Traditional Fermented Koumiss of Kazakh Ethnicity in Different Regions of Xinjiang by Combining Modern Instrumental Detection Technology with Multivariate Statistical Analysis Methods for Odor Activity Value and Sensory Analysis

Yongzhen Gou, Xinmiao Ma, Xiyue Niu, Xiaopu Ren , Geminguli Muhatai and Qian Xu *

College of Food Science and Engineering, Tarim University, Alar 843301, China; 19190316975@163.com (Y.G.)
* Correspondence: xuqiantaru@126.com

Abstract: To investigate the characteristic aromatic compounds, present in the traditional fermented koumiss of the Kazakh ethnic group in different regions of Xinjiang, GC-IMS, and GC-MS were used to analyze the volatile compounds in koumiss from four regions. A total of 87 volatile substances were detected, and esters, acids, and alcohols were found to be the main aroma compounds in koumiss. While the types of aroma compounds in koumiss were similar across different regions, the differences in their concentrations were significant and displayed clear regional characteristics. The fingerprint spectrum of GC-IMS, combined with PLS-DA analysis, indicates that eight distinctive volatile compounds, including ethyl butyrate, can be utilized to distinguish between different origins. Additionally, we analyzed the OVA value and sensory quantification of koumiss in different regions. We found that aroma components such as ethyl caprylate and ethyl caprate, which exhibit buttery and milky characteristics, were prominent in the YL and TC regions. In contrast, aroma components such as phenylethanol, which feature a floral fragrance, were more prominent in the ALTe region. The aroma profiles of koumiss from the four regions were defined. These studies provide theoretical guidance for the industrial production of Kazakh koumiss products.

Keywords: koumiss; GC-IMS; GC-MS; PLS-DA; OVA; volatile components; QDA; origin discrimination; aroma-presenting substances

1. Introduction

The vast grasslands in northern Xinjiang, where the developed animal husbandry industry and ancient nomadic Kazakh people reside, have nurtured a long history of making and consuming koumiss [1]. Some traditional fermented dairy products are still an indispensable part of people's lives today. Fresh mare's milk is fermented under specific temperatures and time through the joint action of lactic acid bacteria, yeast, and other microorganisms to form koumiss [2]. Koumiss has high nutritional value [3], can effectively improve intestinal flora [4], and has certain auxiliary therapeutic effects on cardiovascular disease, gastrointestinal infections, diabetes, and tuberculosis [5]. Xinjiang Kazakh koumiss differs from the Western industrialized koumiss [6] and relies on traditional natural fermentation, which has a unique fermentation process. The flavor and sensory attributes of Kazakh koumiss are influenced by factors, such as fermentation agents, regional conditions, and fermentation process [7]. Koumiss in different regions has its own characteristics, forming unique sensory properties and flavors.

In recent years, with the burgeoning development of the equine industry in Xinjiang, koumiss, as a high-value-added food, has gradually transformed from a mere pastoral self-sustenance beverage to a constantly selling commodity [8]. With the improvement of people's economic level, the demand for food with medical and health benefits is increasing [9], and the market prospects for koumiss are also becoming broader. However, the uncertainty of the flavor of koumiss in different regions has become a bottleneck that restricts the development of the koumiss industry. Due to the differences in the composition and content of substances in koumiss in different regions, the differences in flavor and quality of koumiss are determined [10]. Therefore, in-depth research on these substances can guide the targeted processing and flavor quality control of koumiss products, thereby producing corresponding economic benefits. This will also help promote local horse industry economic growth, assist in rural revitalization strategies, and increase the income of herdsmen, which has important economic and social significance.

Gas chromatography–ion mobility spectrometry (GC-IMS) is an emerging detection technology in food aroma detection, which has the advantages of high sensitivity and strong convenience and is widely used in food quality control and traceability [11]. Zhang et al. [12] analyzed the volatile compounds of different years of brewed liquor using GC-IMS and chemometrics. By comparing the differences in the content of volatile compounds, fingerprint spectra were quickly generated, and rapid differentiation of different years of brewed liquor was achieved through principal component analysis (PCA). Liu et al. [13] analyzed the fermented milk of *Lactobacillus delbrueckii subsp. bulgaricus* of cheese type and fermentation type by GC-IMS and flavoromics, finally screening out six volatile compounds, such as 2,3-butanedione, as key aroma active substances. Literature studies have shown that the application of GC-IMS in fermented dairy products is in its preliminary stage, and there are almost no literature reports on the application of GC-IMS in koumiss. Therefore, this study innovatively uses GC-IMS to analyze the changes in volatile substances in koumiss in different regions.

In recent years, research on the aroma of koumiss has been gradually developing. Meng and colleagues [14], through their study of the microbial composition and aroma changes during the fermentation process of koumiss in Xilingol League, Inner Mongolia, found that the core microbial groups responsible for producing volatile flavor compounds in koumiss are lactobacillus, Brettanomyces, and Torulaspora. Additionally, the formation of koumiss aroma depends on the type and amount of acid, ester, and alcohol present. Xia and colleagues [15] used high-throughput sequencing and liquid-liquid extraction-gas chromatography-mass spectrometry to analyze the time-dependent changes in microbial groups and volatile compounds in koumiss from Xilingol League, Inner Mongolia. They identified lactobacillus and Dekkera as the core microorganisms in koumiss. During the fermentation process of koumiss, 52 volatile compounds were identified, including 15 different metabolic products of volatile compounds. These showed a significant positive correlation with the synthetic metabolism of lactobacillus and Dekkera.

To our knowledge, the aroma compounds of traditional fermented Kazakh-style koumiss from different regions in Xinjiang have not been systematically studied. This study focuses on the traditional fermented koumiss of the Kazakh ethnic group in four different regions of Xinjiang. By utilizing the convenient and rapid GC-IMS technique and employing multivariate statistical analysis, we have established fingerprint profiles to differentiate the volatile compounds of koumiss from different regions. In tandem with gas chromatograph-mass spectrometer (GC-MS) analysis, utilizing volatile compounds as the foundation, we have integrated odor activity value (OAV) analysis and quantitative descriptive analysis (QDA) based on a sensory evaluation to characterize the sensory aromatic attributes of koumiss from various regions. This study provides a theoretical basis for understanding the volatile substance and sensory characteristics of traditional fermented Kazakh-style koumiss in Xinjiang and improving its quality.

2. Materials and Methods

2.1. Sample Collection and Material Reagents

The samples were collected in early July 2022 from the households of Kazakh farmers and herdsmen in four regions (Table 1) surrounding the Altai and Tianshan Mountains in northern Xinjiang, China. To ensure the representative significance of samples from different regions, the linear distance between sampling locations is guaranteed to exceed 350 km. The collected samples were stored in sealed containers and transferred to the laboratory, where they were kept at −80 °C in liquid nitrogen for subsequent experiments. The fermentation methods of the koumiss samples from the four regions varied, as shown in Figure 1. Chemicals used in the study included 3-octanol (Aladdin Biochemical Technology Co., Ltd., Shanghai, China), normal ketones C4~C9 (China National Pharmaceutical Chemical Reagent Co, Ltd., Beijing, China), and normal alkanes (C7~C30) Sigma-Aldrich (St. Louis, MO, USA). All chemicals used were of chromatographic grade purity.

Table 1. Koumiss sample information.

Abbreviation of Koumiss	Locality	Fermentation Method	Fermentation Temperature °C
YL	Ili	machinery	20~26
TC	Tacheng	machinery	22~26
ALTe	Altai east	manually	22~27
ALTw	Altai west	manually	21~27

Figure 1. Depicts the fermentation methods of traditional koumiss from different regions. Panel (**a**) represents the YL region, panel (**b**) the TC region, panel (**c**) the ALTe region, and panel (**d**) the ALTw region.

2.2. Analysis by Gas Chromatography-Ion Mobility Spectrometry

2.0 mL of koumiss sample was introduced into a 20 mL headspace vial (Agilent Technologies, Santa Clara, CA, USA). A 10 μL aliquot of 3-octanol solution (300 mg/L) was added to the sample, followed by incubation at 50 °C with stirring at 500 rpm for 15 min. After the incubation, 500 μL of headspace gas was automatically extracted without splitting into the GC-IMS instrument (FlavorSpec, GAS, Dortmund, Germany). The sample was separated using a WAT-Wax capillary column (15 m × 0.53 mm) with a temperature of 60 °C. Nitrogen gas (99.99% purity) was used as the carrier gas. The flow rate was controlled as follows: initially kept at 2 mL/min for 2 min, increased to 10 mL/min over 8 min, increased to 100 mL/min over 5 min, and finally increased to 150 mL/min over 5 min and kept constant for 10 min. The drift tube was 98.0 mm long, with a working temperature of 45 °C and a constant voltage of 5 kV. The entire analysis took 45 min, and all analyses were performed in triplicate.

The GC-IMS data obtained were analyzed using the laboratory analytical viewer (LAV). The Reporter plugin was used to directly compare the two-dimensional top view and difference spectra, while the gallery plot was used for fingerprint spectrum comparison. The retention index (RI) of the volatile compounds was calculated using the internal standard of normal ketones C4–C9. The identified volatile compounds were compared with the IMS database (G.A.S, Dortmund, Germany) using the retention index (RI), normalized drift time (RIP), and other parameters.

2.3. Gas Chromatography-Mass Spectrometry Analysis

8 g of koumiss sample was placed in a 25 mL headspace vial (Agilent Technologies) with a silicon septum, and 200 μL of 300 mg/L 3-octanol solution was added. The sample was equilibrated at 65 °C for 30 min. A solid-phase microextraction fiber coated with 50/30 μm divinylbenzene/carboxen/polydimethylsiloxane (DVB/CAR/PDMS) (Shanghai ANPEL Laboratory Technologies Co., Ltd., Shanghai, China) was inserted into the headspace of the glass vial for 30 min at 65 °C. After extraction, the fiber was inserted into the inlet of the GC-MS and thermally desorbed for 3 min at 25 °C. The GC-MS analysis was performed using an Agilent 7890B–7000D gas chromatography-mass spectrometry instrument, and volatile compounds was separated on a DB-WAX column (60 mm × 0.25 mm, 0.25 μm, Agilent). The temperature program started with a 5 min hold at 40 °C, followed by an increase to 200 °C at a rate of 3 °C/min, and a final increase to 250 °C at a rate of 6 °C/min. The total analysis time for each sample was 122 min. The retention indices (RI) were determined using normal alkane hydrocarbons (C7~C30), and volatile compounds were identified by mass spectrometry and RI values.

2.4. Quantitative Descriptive Analysis (QDA)

Following the method outlined in ISO 8586:2012 (ISO, 2012) [16], 20 students and faculty members from the School of Food Science and Engineering at Tarim University in China were selected as sensory panelists. These individuals received relevant training and had previous experience in sensory analysis. Over one month, with an average of one hour per day, the 20 panelists underwent training to familiarize themselves with and identify the common odors in fermented dairy products. Based on their sensory perceptual abilities, 10 panelists (4 male and 6 female; average age of 24 years) were selected to undergo additional sensory training, where they were required to detect standard reference materials and grade the intensity of different aroma attributes. Further, the training and selection of the panelists, as well as the subsequent sensory descriptive analysis, were conducted in the sensory laboratory at the School of Food Science and Engineering at Tarim University, with room temperature maintained at 20 °C.

According to ISO 5492:2008 (ISO, 2008) [17], six sensory descriptive terms including fruitiness, creaminess, floral, off-flavor, dairy flavor, and sourness were selected to measure the odor intensity using a continuous scale ranging from 0 (no attribute) to 10 (extremely intense). Ten grams of koumiss were weighed into a disposable cup for sensory evaluation. To prevent olfactory fatigue, the panel evaluated six samples during a period of 60 min, taking a break after assessing three samples. Three replicates were performed for each sample, and six experiments were conducted, with sensory data collected by trained assessors. To ensure the accuracy and reliability of sensory evaluation, we engaged professional personnel from Alar Xin Nong Dairy Industry Co., Ltd., Alar, China, who specialize in dairy product research and development, to provide guidance and supervision on our sensory analysis work.

2.5. Odor Activity Value Calculation

Utilize the methodology of OAV to assess the olfactory contribution of various regional koumiss. OAV refers to the ratio between the mass concentration of an aroma compound and its sensory threshold in an odor system.

2.6. Statistical Analysis

The differences in key components among the koumiss products from different regions were explored using partial least squares discriminant analysis (PLS-DA) with SIMCA-P 13.0 software (Umetrics, Umea, Sweden). The differences in volatile compounds and threshold values among different regions were evaluated using one-way analysis of variance (ANOVA) with SPSS Statistics 20.0 software (SPSS Inc., Chicago, IL, USA), with a significance level of $p < 0.05$. Bar graphs, Venn diagrams, and radar charts were produced using Origin 2021 software (OriginLab Corporation, Northampton, MA, USA).

3. Results and Analysis

3.1. GC-IMS Was Used to Establish the Fingerprint of Volatile Substances in Koumiss from Different Regions

Currently, GC-IMS is a new emerging instrument analysis technique that can effectively separate ions based on their migration speed at atmospheric pressure, eliminating the need for a vacuum environment and increasing convenience. It has the advantages of high sensitivity and data visualization and is widely used for origin tracing and quality control.

We utilized GC-IMS to investigate the volatile fingerprint of traditional fermented koumiss of the Kazakh ethnic group in four regions of Xinjiang. Data was generated using the FlavourSpec® instrument, as shown in Figure 2. Figure 2a presents a top-down view of the 3D spectrum of the volatile components of koumiss. The blue background of the entire figure represents the ion migration time along the X-axis and the retention time of the ions in the gas chromatography on the Y-axis. The red vertical line at 1.0 on the X-axis represents the normalized response ion peak of the volatile components because subtle changes in capillary column temperature and carrier gas flow rate can lead to changes in retention time. The point to the right of the cleavage represents a volatile substance, with the color indicating the signal strength of individual compounds. The higher the concentration of the compound, the redder the color, and conversely, the lower the concentration, the bluer the color. As shown in Figure 2a, most signals are distributed within the retention time range of 100–1200 s, and many peaks have similar signal distributions in the samples. Still, the peak intensities differ significantly between koumiss from different regions, indicating differences in aroma content.

To compare the differences between samples, Figure 2b shows that a differential contrast method can be used. One spectrum from a sample is chosen as the reference sample, and other spectra are subtracted for comparison. If the volatile substances in the two samples are the same, the background after subtraction is white, with red representing higher concentrations than the reference sample, and blue representing lower concentrations. As shown in Figure 2b, when comparing sample YL-2 with reference sample YL-1 and other samples, the blue and red areas are significantly reduced, indicating that the volatile substances of YL-1 and YL-2 are similar in concentration, while there are significant differences in concentration between YL-1 and other samples. By establishing differential contrast maps for reference samples from different regions using GC-IMS, we can preliminarily discriminate and trace the koumiss from different regions.

Additionally, Figure 2c presents the GC-IMS volatile organic compound fingerprint patterns of different samples, which were created by piecing together screenshots of each volatile organic compound from Figure 2a. Each row represents a sample, and each column represents a compound. By drawing the fingerprint patterns, we can clearly compare the volatile organic compound contents and changes in each region's samples. In total, 35 volatile organic compounds were qualitatively determined. Based on Figure 3 and previous research results [18], the results of volatile organic compounds identified by GC-IMS were considered insufficient. Although the results cannot fully demonstrate the flavor characteristics of koumiss from different regions, they can establish fingerprint patterns for grading and discrimination. By selecting the region outlined by the red box in Figure 2c, it can be observed that the concentrations of dimethyl sulfide, ethyl butyrate, ethyl hexanoate, ethyl octanoate, and 2-methylbutanal in the samples from the TC region decreased significantly with the change of the region. In the region outlined by the yellow box, the concentrations of isobutyl acetate, ethyl propanoate, 2-methylpropanoic acid ethyl ester, propyl acetate, and 1-propanol in the samples from the ALTe region were significantly higher than those in other regions. In the region outlined by the green box, the concentrations of trans-2-pentenal, methyl 2-methylbutyrate, hexanal, (E)-2-hexenal, 1-penten-3-ol, 1-penten-3-one, and acetone in the samples from the ALTw region were significantly higher than those in other regions.

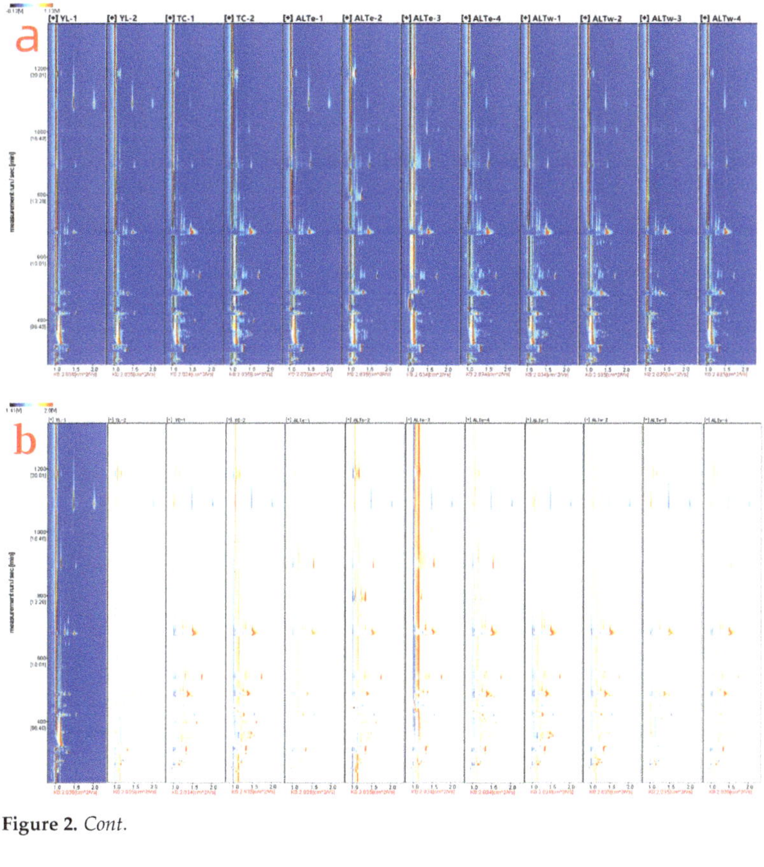

Figure 2. *Cont.*

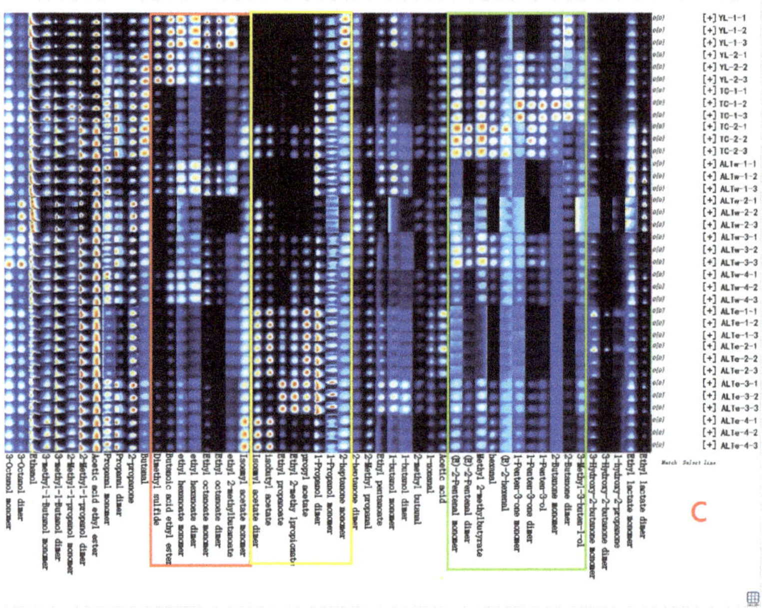

Figure 2. GC-IMS analysis of koumiss from different regions. (**a**) Top view of the 3D spectrum of the volatile components of koumiss from different regions. (**b**) 2D comparison chart of the volatile components of koumiss from different regions. (**c**) The fingerprint spectra of the volatile constituents in koumiss from distinct regions. The delineated color zones represent the variances in volatile substances among different regions.

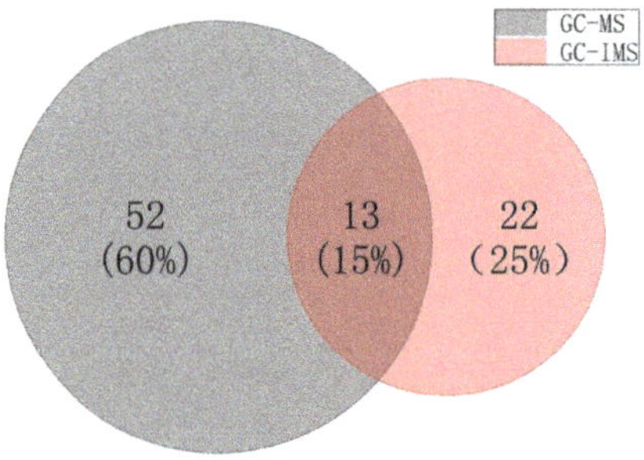

Figure 3. The Wayne diagram of the number of aroma substances determined by GC-MS and GC-IMS.

3.2. Analysis of Differences in Volatile Compounds of Koumiss from Different Regions Based on PLS-DA Technology

PLS-DA is a supervised discriminant analysis statistical technique. It establishes a relationship model between volatile compounds and sample categories to predict sample categories [19]. Therefore, in this study, a PLS-DA model was adopted to identify specific marker compounds for the volatile substance differences of koumiss samples from four regions based on the detected volatile substance content data matrix.

As shown in Figure 4a, the four groups of samples were successfully distinguished in the quadrant plot. Specifically, the samples from the YL region were in the fourth quadrant, those from the TC region were in the first quadrant, samples from the ALTw region were in the second quadrant, and samples from the ALTe region were in the third quadrant. Additionally, as shown in Figure 4b, 200 permutation tests were performed, and the model predictive index (Q2) was 0.69 for the prediction of different origins of koumiss, indicating a prediction accuracy of 69%. A Q2 value exceeding 0.5 indicates an acceptable fit of the model. The intersection of the Q2 regression line and the vertical axis is less than zero, indicating the absence of overfitting, thus validating the model's effectiveness [20]. The variable importance in projection (VIP) method was employed to select the most statistically significant variables. It is generally recognized that volatile substances with VIP values >1.0 play a crucial role in distinguishing geographical origins [21]. As shown in Figure 4c, 18 substances were selected based on their VIP values, including 2-heptanone, ethyl butyrate, dimethyl sulfide, 2-methylbutanal, ethyl acetate, isobutyraldehyde, methyl ethyl ketone, isobutyl acetate, isopentyl acetate, acetone, ethyl propionate, nonanal, ethyl caproate, isobutanol, propyl acetate, amyl alcohol, ethyl 2-methylpropanoate, and n-propanol.

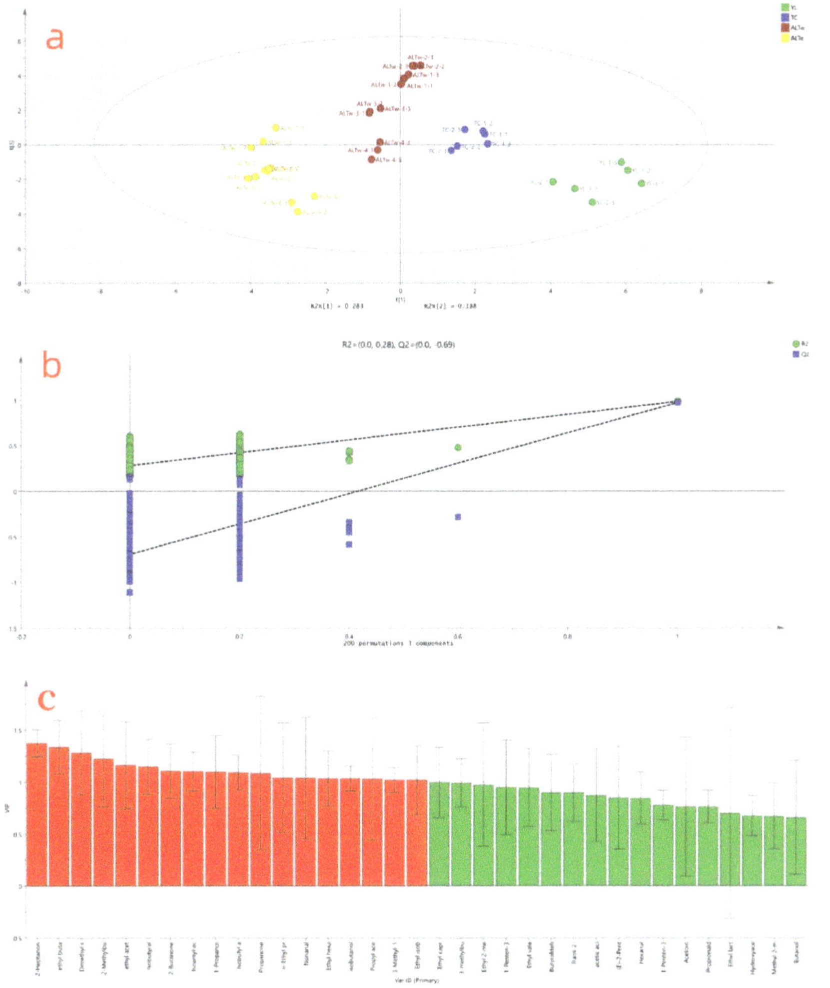

Figure 4. PLS-DA analysis of koumiss from different regions. (**a**) Score plot of PLS-DA; (**b**) Cross-validation results based on 200 permutation tests; (**c**) Ranking of volatile substances based on VIP scores. The red part is the substance with VIP value greater than 1. The green area is for substances with VIP finger less than 1.

To ensure the accuracy of identifying the selected key volatile compounds' traceability for koumiss from different regions, a standard was adopted that included volatile compounds with VIP values greater than 1 and shared by GC-IMS fingerprinting discrimination of volatile compounds. Based on this standard, the following eight volatile compounds were selected as the final markers for distinguishing koumiss from different regions: dimethyl sulfide, ethyl butyrate, 2-methylbutanal, ethyl propionate, ethyl 2-methylpropanoate, isobutyl acetate, propan-1-ol, and methyl ethyl ketone.

3.3. Comparison of GC-MS and GC-IMS Results

Both GC-MS and GC-IMS are excellent instruments for determining volatile compounds, and the research results can better understand the characteristics of the two instruments. The results in Table 2 show that 65 volatile compounds were identified by GC-MS in koumiss samples from four regions. 35 volatile compounds were identified by GC-IMS, of which IMS identified 76 peaks, but 26 peaks were not characterized. Although IMS has a strong sensitivity to the ion drift of volatile compounds without fragmentation and can identify 76 peaks, 26 volatile compounds cannot be confirmed due to incomplete databases. Although GC-IMS is much smaller than GC-MS in the number of substances determined, GC-IMS can conveniently sample and test and quickly obtain results. Among the 50 qualitatively determined volatile compounds, 15 were diploids. When the charged volatile substance is given water and a hydrogen ion, there is a chance that a monomer and two monomers will combine to form a haploid and a diploid [22], making the quantification of volatile compounds by IMS more accurate. Analyzing the types of identified compounds, among the 35 volatile compounds identified by GC-IMS, there are 14 ester compounds, six alcohol compounds, six ketone compounds, and eight aldehyde compounds. Compared with the pattern of identified compounds by GC-MS, GC-IMS is more sensitive to aldehyde and ketone compounds and is more suitable for samples with more aldehyde and ketone compounds. This is consistent with the research results of Zhang [23]. In addition, as shown in Figure 3, 13 common volatile compounds were identified by these two types of analysis methods, and a total of 87 volatile compounds were identified, which is much more abundant than the substances determined by a single detection method. Therefore, the combination of GC-IMS and GC-MS can more comprehensively identify and quantify volatile compounds.

Table 2. Volatile substance content of koumiss in different regions.

No.	Cas	VOCs	Identification	Contents (μg/kg)			
				YL	TC	ALTw	ALTe
C1	60-12-8	Phenethyl alcohol	MS	941.79 ± 640.57 c	2483.21 ± 442.29 ab	1373.66 ± 1395.45 bc	3568.09 ± 1500.03 a
C2	78-83-1	isobutanol	MS, IMS e	47.98 ± 31.67 c	158.28 ± 55.17 b	122.32 ± 131.17 bc	403.57 ± 103.51 a
C3	24347-58-8	(2R,3R)-(−)-2,3-Butanediol	MS	43.97 ± 29.01 a	61.04 ± 2.46 a	47.33 ± 38.37 a	52.97 ± 12.09 a
C4	123-51-3	3-Methyl-1-butanol	MS, IMS e	552.43 ± 314.88 c	1761.30 ± 372.16 b	1110.76 ± 944.97 bc	3464.17 ± 717.10 a
C5	1565-80-6	(S)-(−)-2-Methylbutanol	MS	183.61 ± 201.13 a	0	178.36 ± 341.27 a	295.30 ± 534.22 a
C6	67-63-0	Isopropanol	MS	44.14 ± 74.77 a	18.82 ± 6.38 a	112.60 ± 163.46 a	23.51 ± 34.63 a
C7	50551-88-7	5-METHYL-5-HEXEN-2-OL	MS	7.27 ± 8.60 a	18.63 ± 12.22 a	19.76 ± 41.02 a	10.66 ± 23.26 a
C8	137-32-6	2-Methylbutan-1-ol	MS	177.12 ± 194.24 b	876.01 ± 158.21 a	237.95 ± 322.88 b	1017.48 ± 282.73 a
C9	2313-61-3	4-methyl-2-hexanol	MS	30.83 ± 33.99 a	26.65 ± 1.61 a	61.71 ± 112.77 a	0
C10	19132-06-0	(2S,3S)-(+)-2,3-BUTANEDIOL	MS	0	80.78 ± 25.91 a	21.46 ± 39.04 a	124.81 ± 250.74 a
C11	108-82-7	2,6-Dimethyl-4-heptanol	MS	0	0	82.38 ± 84.49 a	117.19 ± 137.57 a
C12	71-36-3	Butanol	MS	5.04 ± 0.52 a	2.87 ± 0.21 a	5.22 ± 3.58 a	4.39 ± 0.62 a
C13	616-25-1	1-Penten-3-ol	MS	1.23 ± 0.26a	3.79 ± 1.26 a	2.93 ± 3.29 a	3.61 ± 3.28 a
C14	71-23-8	1-Propanol	MS	46.37 ± 10.19 b	42.89 ± 14.84 b	43.83 ± 15.33 b	52.61 ± 23.54 a
C15	763-32-6	3-methylbut-3-en-1-ol	MS	0.66 ± 0.18 b	0.83 ± 0.16 ab	0.76 ± 0.35 ab	1.01 ± 0.32 a
C16	112-60-7	2,2′-((Oxybis(ethane-2,1-diyl))bis(oxy))diethanol	MS	647.70 ± 764.98 a	858.08 ± 220.96 a	762.76 ± 940.57 a	0
Z1	110-38-3	Ethyl caprate	MS	28,937.46 ± 2643.22 a	14,466.14 ± 1822.56 b	4222.47 ± 2435.52 c	2445.60 ± 1629.19 c
Z2	106-32-1	Ethyl caprylate	MS, IMS e	24,254.33 ± 2941.10 a	15,167.06 ± 2536.12 b	3865.41 ± 1682.78 c	2890.76 ± 1198.86 c
Z3	106-33-2	Ethyl laurate	MS	7819.93 ± 1002.54 a	5144.83 ± 376.36 b	589.22 ± 323.47 c	274.81 ± 278.27 c
Z4	50816-18-7	9-Decen-1-ol, 1-acetate	MS	8430.05 ± 753.10 a	3831.73 ± 279.66 b	1707.36 ± 1320.09 c	941.79 ± 906.60 c
Z5	687-47-8	(−)-Ethyl L-lactate	MS	1877.55 ± 853.21 a	1009.89 ± 606.72 b	461.19 ± 366.60 c	1129.09 ± 357.94 b
Z6	124-06-1	Ethyl tetradecanoate	MS	626.97 ± 99.88 a	467.33 ± 74.17 b	18.30 ± 14.13 c	14.60 ± 18.24 c
Z7	123-29-5	Ethyl nonanoate	MS	445.54 ± 192.25 a	0	260.64 ± 474.45 a	0
Z8	123-66-0	Ethyl hexanoate	MS, IMS	306.92 ± 22.82 a	263.36 ± 44.09 b	46.65 ± 28.70 c	62.93 ± 14.65 c
Z9	141-78-6	ethyl acetate	MS, IMS	651.07 ± 446.58 b	1265.80 ± 450.34 ab	1177.51 ± 1129.16 b	2071.71 ± 693.74 a
Z10	54546-22-4	9-Hexadecenoic acid,ethyl ester	MS	167.27 ± 29.15 a	94.63 ± 37.19 b	0	0
Z11	2035-99-6	3-methylbutyl octanoate	MS	126.86 ± 39.42 a	70.28 ± 23.46 b	0	0
Z12	628-97-7	Ethyl palmitate	MS	73.81 ± 20.29 a	55.57 ± 5.10 b	0	0
Z13	105-54-4	ethyl butanoate	MS, IMS	59.25 ± 29.14 b	135.77 ± 22.49 a	2.45 ± 4.43 c	7.66 ± 8.12 c

Table 2. *Cont.*

No.	Cas	VOCs	Identification	Contents (μg/kg)			
				YL	TC	ALTw	ALTe
Z14	10348-47-7	ethyl 2-hydroxy-4-methyl valerate	MS	121.85 ± 53.43 b	538.80 ± 119.68 a	113.12 ± 113.58 b	68.03 ± 23.09 b
Z15	624-13-5	propyl octanoate	MS	69.30 ± 17.51 a	60.29 ± 5.17 a	8.21 ± 10.02 b	9.15 ± 6.10 b
Z16	106-30-9	Ethyl heptanoate	MS	46.49 ± 6.88 a	32.76 ± 2.65 a	13.89 ± 25.46 b	0
Z17	2277-23-8	1-Decanoyl-rac-glycerol	MS	16.81 ± 7.78 a	12.29 ± 13.46 ab	0	12.78 ± 20.58 ab
Z18	110-19-0	Isobutyl acetate	MS, IMS	157.60 ± 82.56 a	59.80 ± 9.95 b	0	8.38 ± 9.42 c
Z19	142-91-6	Isopropyl palmitate	MS	9.62 ± 3.12 a	68.20 ± 97.21 a	0	0
Z20	547-64-8	1-(Dibromomethyl)-3-phenoxybenzene	MS	198.66 ± 222.76 a	0	139.81 ± 111.72 ab	28.10 ± 55.04 b
Z21	97-64-3	Ethyl lactate	MS, IMS	49.63 ± 55.55 b	368.21 ± 182.02 a	52.18 ± 61.10 b	96.82 ± 122.33 b
Z22	15399-05-0	ethyl phenyl lactate	MS	30.12 ± 33.53 b	104.55 ± 17.42 a	6.42 ± 11.65 c	0
Z23	55130-16-0	benzyl oleate	MS	27.34 ± 30.21 bc	43.65 ± 13.95 ab	74.60 ± 60.89 a	1.46 ± 2.64 c
Z24	103-45-7	Phenethyl acetate	MS	0	733.98 ± 95.20 ab	320.60 ± 584.85 b	1088.80 ± 583.44 a
Z25	123-92-2	Isoamyl acetate	MS, IMS	0	44.12 ± 49.27 b	68.29 ± 67.81 b	130.11 ± 46.85 a
Z26	19329-89-6	lactic acid isoamyl ester	MS	0	53.88 ± 59.30 ab	36.60 ± 39.34 ab	80.05 ± 70.30 a
Z27	624-41-9	2-Methylbutylacetat	MS	0	49.68 ± 21.70 a	11.91 ± 22.99 b	26.15 ± 36.63 ab
Z28	539-82-2	Ethyl valerate	IMS	129.20 ± 16.61 a	66.45 ± 10.05 b	85.55 ± 42.10 b	99.60 ± 47.47 ab
Z29	7452-79-1	Ethyl 2-methylbutanoate	IMS	58.05 ± 23.25 a	16.78 ± 3.14 b	8.69 ± 3.06 ab	8.00 ± 4.90 b
Z30	97-62-1	Ethyl isobutyrate	IMS	15.72 ± 1.84 ab	97.88 ± 51.61 ab	58.80 ± 72.58 b	118.66 ± 118.51 a
Z31	868-57-5	Methyl 2-methylbutyrate	IMS	11.66 ± 0.56 c	51.97 ± 16.46 a	30.34 ± 17.33 b	47.28 ± 16.41 a
Z32	105-37-3	n-Ethyl propanoate	IMS	10.18 ± 2.16 b	41.22 ± 12.76 b	95.06 ± 89.10 ab	184.14 ± 202.16 a
Z33	109-60-4	Propyl acetate	IMS	4.07 ± 1.26 a	9.09 ± 6.11 a	30.12 ± 31.96 a	24.58 ± 26.11 a
S1	124-07-2	Octanoic acid	MS	11,231.01 ± 3510.55 a	4556.99 ± 720.53 b	5045.42 ± 3687.15 b	747.79 ± 617.20 c
S2	334-48-5	Decanoic acid	MS	5546.54 ± 1855.71 a	1946.03 ± 309.87 b	2062.60 ± 1556.39 b	158.60 ± 302.53 c
S3	14436-32-9	9-Decenoic acid	MS	1459.96 ± 407.24 a	294.40 ± 322.50 b	411.40 ± 302.25 b	0
S4	142-62-1	1-Hexanoic acid	MS	1043.66 ± 53.25 a	409.12 ± 107.38 b	278.01 ± 138.91 c	81.90 ± 87.16 d
S5	64-19-7	acetic acid	MS, IMS	472.64 ± 88.24 bc	726.91 ± 141.36 ab	1045.17 ± 621.60 a	196.39 ± 144.50 c
S6	143-07-7	Lauric acid	MS	464.89 ± 142.63 a	205.86 ± 297.01 b	18.22 ± 32.95 c	0
S7	3004-93-1	S-2-methyl octanoic acid	MS	49.38 ± 38.43 a	114.60 ± 45.29 a	0	0
S8	503-74-2	3-Methylbutanoic acid	MS	71.11 ± 55.90 a	64.17 ± 15.38 a	5.47 ± 9.95 b	51.20 ± 7.75 a
S9	65-85-0	benzoic acid	MS	148.38 ± 162.54 c	481.30 ± 528.87 b	572.45 ± 375.91 a	0
S10	629-56-1	Hexadecenoic acid	MS	647.70 ± 764.98 a	858.08 ± 220.96 a	762.76 ± 940.57 a	0
S11	79-31-2	Isobutyric acid	MS	33.76 ± 37.10 b	91.65 ± 29.33 a	27.73 ± 32.63 b	127.28 ± 82.73 a
S12	116-53-0	2-Methylbutyric acid	MS	28.25 ± 31.72 c	73.02 ± 4.55 b	39.12 ± 25.58 c	120.02 ± 42.69 a
T1	110-43-0	2-Heptanone	MS, IMS	30.92 ± 34.20 a	22.89 ± 25.07 a	42.90 ± 47.54 a	24.16 ± 26.96 a
T2	105-42-0	4-Methyl-2-hexanone	MS	20.09 ± 22.06 b	17.42 ± 19.09 b	0	44.40 ± 8.02 a
T3	19093-20-0	(E)-5-octenone	MS	0	0	40.32 ± 40.37 a	16.37 ± 22.08 a
T4	821-55-6	2-Nonanone	MS	0	0	76.28 ± 72.95 a	11.45 ± 7.02 b
T5	67-64-1	Propanone	IMS	212.78 ± 38.55 b	332.06 ± 64.46 a	265.48 ± 70.82 b	326.51 ± 16.94 a
T6	513-86-0	Acetoin	IMS e	144.37 ± 8.56 b	229.70 ± 60.35 ab	484.79 ± 490.56 a	242.56 ± 102.04 ab
T7	78-93-3	2-Butanone	IMS e	26.54 ± 16.34 a	16.72 ± 10.94 ab	10.52 ± 11.76 b	7.14 ± 2.92 b
T8	1629-58-9	1-Penten-3-one	IMS e	6.28 ± 0.67 b	56.96 ± 35.47 a	33.87 ± 34.63 ab	45.78 ± 41.89 a
T9	116-09-6	Hydroxyacetone	IMS	13.12 ± 1.31 a	5.36 ± 0.51 b	8.25 ± 6.46 ab	8.32 ± 4.51 ab
F1	100-52-7	Benzaldehyde	MS	66.50 ± 5.41 a	29.28 ± 6.05 b	55.76 ± 8.12 a	25.35 ± 12.53 b
F2	4313-03-5	(E,E)-2,4-Heptadienal	MS	86.63 ± 37.43 a	93.38 ± 63.69 a	187.31 ± 161.50 a	61.91 ± 112.06 a
F3	66-25-1	Hexanal	MS, IMS	113.35 ± 88.89 a	87.66 ± 27.18 a	150.26 ± 73.26 a	115.66 ± 56.13 a
F4	4748-78-1	4-Ethylbenzaldehyde	MS	11.98 ± 13.20 b	8.16 ± 8.94 c	39.99 ± 6.21 ab	47.62 ± 36.29 a
F5	142-83-6	2,4-Hexadienal	MS	30.91 ± 33.91 a	30.37 ± 33.27 a	57.15 ± 25.37 a	42.66 ± 49.06 a
F6	74094-61-4	octanal propylene glycol acetal	MS	7.93 ± 8.68 b	14.55 ± 16.08 b	126.23 ± 141.30 a	27.61 ± 10.45 b
F7	124-19-6	Nonanal	MS, IMS	0	58.98 ± 13.51 a	94.43 ± 74.07 a	55.89 ± 42.23 a
F8	2548-87-0	(2E)-2-Octenal	MS	0	0	194.19 ± 149.75 a	64.14 ± 63.23 b
F9	3913-81-3	(2E)-2-Decenal	MS	0	0	56.02 ± 39.01 a	42.09 ± 16.90 a
F10	123-38-6	Propionaldehyde	IMS e	57.03 ± 11.14 b	153.87 ± 11.72 a	89.36 ± 38.27 b	161.55 ± 52.85 a
F11	123-72-8	Butyraldehyde	IMS	78.67 ± 6.57 ab	104.61 ± 53.51 a	48.95 ± 26.14 b	69.10 ± 32.39 b
F12	6728-26-3	Trans-2-Hexenal	IMS	6.96 ± 1.40 ab	8.14 ± 1.98 a	4.34 ± 1.88 c	5.66 ± 2.74b c
F13	96-17-3	2-Methylbutanal	IMS	17.35 ± 11.30 a	15.85 ± 8.66 a	30.62 ± 38.90 a	31.57 ± 20.05 a
F14	1576-87-0	(E)-2-Pentenal	IMS e	22.81 ± 2.74 b	92.04 ± 24.48 ab	84.41 ± 40.15 b	171.79 ± 129.48 a
F15	78-84-2	Isobutyraldehyde	IMS	10.56 ± 0.67 c	38.58 ± 26.62 bc	73.35 ± 66.59 ab	94.48 ± 21.60 a
Q1	106-44-5	p-Cresol	MS	0	12.57 ± 13.80 b	13.36 ± 18.02 b	56.41 ± 57.52 a
Q2	629-59-4	Tetradecane	MS	0	0	26.48 ± 34.26 a	20.26 ± 15.90 a
Q3	75-18-3	Dimethyl sulfide	IMS	46.42 ± 25.96 a	48.77 ± 42.66 a	5.89 ± 3.98 b	9.63 ± 4.45 b

Note: The letters a, b, c, and d represent significant differences ($p < 0.05$) in the volatile compounds of koumiss from different regions. The letter e indicates that the compound was detected as a dimer in GC-IMS analysis.

3.4. The Characteristic Volatile Compounds of Koumiss from Different Regions Were Analyzed by GC-MS Combined with GC-IMS

To investigate the aroma characteristics of traditional fermented Kazakh koumiss from different regions, a combined GC-IMS and GC-MS method was used to analyze and identify the volatile compounds and their relative contents in samples from four regions in Xinjiang. As shown in Table 2, a total of 87 volatile flavor substances were detected, including 16 kinds of alcohol compounds (excluding ethanol), 33 kinds of ester compounds, 15 kinds of aldehyde compounds, nine kinds of ketone compounds, three kinds of acid compounds, and one kind each of alkane, phenol, and ether compounds. Compared with the report by Wang [24], more volatile compounds were detected using the combined detection method. As shown in Figure 5, the aroma component types of koumiss samples from the four regions were similar, but the total concentrations of each type varied. The highest content of aroma components was ester compounds, followed by acid compounds, alcohol compounds, and ketone compounds. Ester compounds, acid compounds, and alcohol compounds accounted for a large proportion of the total and were important volatile compounds in koumiss.

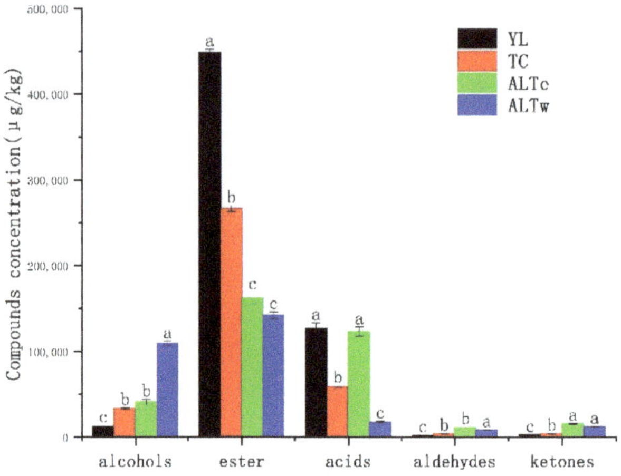

Figure 5. Changes in the total content of alcohols, acids, esters, aldehydes, and ketones in koumiss from different regions. Letters a, b, and c indicate significant differences ($p > 0.05$) in the content of these compounds among different samples.

Sourness is the main component of the flavor of koumiss. The relative content of caprylic acid, capric acid, 9-decenoic acid, caproic acid, and acetic acid is relatively high in koumiss samples from different regions. Among them, medium-chain fatty acids caprylic acid, capric acid, 9-decenoic acid, and caproic acid account for 97% of the total volatile acids, giving koumiss a pungent, sour taste and a faint odor, similar to the results of Du's research [25]. This may be due to the abundant free fatty acids in mare's milk or the breakdown of fats by microorganisms. Compared with cereal wines, which are also fermented by lactic acid bacteria and yeast, the fatty acids in koumiss are mostly oleic acid and linoleic acid [26], different from the metabolism and breakdown of fatty acids in fermented dairy products. The acetic acid in koumiss is sour but not astringent, formed by the metabolism of lactose in mare's milk by some β-galactosidase-producing microorganisms [27]. The total volatile acid content of the YL and ALTe regions is significantly higher than that of the TC and ALTw regions (see Figure 5). Due to the special process of fermenting koumiss in the YL region, the machine stirring method was used in the fermentation process (Table 1 and Figure 1). Compared with hand-stirring fermentation, mechanical stirring fermentation extends the stirring time, so that microorganisms do not settle at the bottom and obtain more oxygen, which can fully ferment and produce volatile acid compounds. However, the TC region also used mechanical stirring to ferment koumiss. Still, the content of volatile acid compounds is significantly lower than that of other regions, which may be related to ester volatile compounds.

The esters of volatile substances are important flavor compounds in koumiss due to their very low threshold. Esters make up an average of 63% of the total content of koumiss, and impart a buttery, fruity, and floral aroma to it, helping to reduce the off-flavors of fatty acids and miscellaneous alcohols in koumiss [28]. Therefore, this class of compounds plays an important role in forming the characteristic aroma of koumiss. Most of the esters in koumiss are produced by two pathways. The first pathway involves the esterification of short-chain and medium-chain fatty acids, free fatty acids, and primary and secondary alcohols produced during lactose fermentation or amino acid metabolism in koumiss, and the second pathway involves the formation of esters by alcohol acyltransferases produced by microorganisms from alcohols and acyl-CoA [29,30]. As shown in Figure 5, the total content of esters of volatile substances in the YL and TC regions is significantly higher than that in the ALT region, which may be due to the diversity of fermentation agents and the special nature of the process used in the fermentation of koumiss in these regions. The activity and enzymatic activity of yeast in the YL and TC regions are enhanced by the climate and fermentation methods used there, which can fully promote the formation of esters. The esters with relatively high relative contents are ethyl caprylate, ethyl octanoate, ethyl laurate, and ethyl 9-decenoate, which are positively correlated with the types and contents of volatile acids. Yu's [31] research also showed that ethyl octanoate and ethyl caprate are important components of esters of volatile substances in koumiss.

Alcohols with volatile substances are one of the main flavor compounds differentiating koumiss from different regions. Most high-level alcohols with high content in koumiss include phenylethanol, isopentanol, and isobutanol. Phenylethanol is a widely used edible fragrance with a rose scent, usually employed as a flavoring agent for various food preparations. Dairy products contain a rich amount of phenylalanine. In the fermentation process of koumiss, the majority of phenylethanol is generated by the transamination and decarboxylation metabolism of phenylalanine via the Ehrlich reaction by yeast [32]. The content of phenylethanol in the ALTe region is as high as 3568.9 μg/kg, significantly higher than the 941 μg/kg in the YL region, possibly due to the different fermentation environments and fermentation strains. Phenylethanol, isopentanol, and isobutanol are common fusel oils in brewing, and the content of fusel oils determines some aroma characteristics and taste of koumiss. Excessively high content can cause severe headaches and pose a hazard to human health [33]. Similar to phenylethanol, most of the isopentanol and isobutanol are produced through the Ehrlich reaction with leucine and valine as precursors.

Table 2 shows that aldehydes and ketones are the main volatile compounds that differentiate the flavors of koumiss in different regions. Among the high-content aldehydes in koumiss, hexanal, trans-2-octenal, and acetaldehyde are the most prominent. Aldehydes in the fermentation process mainly come from the metabolism of fatty acids, the transamination of amino acids, or the Strecker degradation [34]. However, aldehydes are unstable in koumiss and can be easily reduced to the corresponding alcohols and fatty acids under certain conditions, resulting in relatively low relative content [35]. Trans-2-octenal usually has a fatty and meaty aroma. It was found to be highest in samples from the ALTw region and was not detected in samples from the TC and YL regions. The content of ketones in koumiss is relatively low, mainly consisting of acetoin and acetone. Acetoin is a metabolic product of some microorganisms that balance the acid-base level and store energy, produced by the glycolytic pathway of pyruvate [36]. The highest content of acetoin was found in the ALTw region, which was significantly higher than in other regions.

In summary, it can be speculated that geographical environment, fermentation temperature and method, and the type of fermentation agent may be important factors that influence the differences in the aroma of traditional fermented Kazakh koumiss in different regions.

3.5. Analysis of OVA in Koumiss from Different Regions

Different regions of koumiss have unique flavors, mainly determined by the types, amounts, and corresponding thresholds of volatile flavor substances they contain. Studies have shown that the contribution of a single aroma component to the overall aroma of koumiss depends on its concentration and odor threshold [37]. Although the above analysis provides a quantitative characterization of the flavor substances in koumiss, it is only by understanding the threshold of each substance that the key flavor substances can be directly linked to their contribution to the overall aroma of koumiss. The OAV can be used to evaluate the composite contribution of flavor substances to food. Compounds with an OAV $\geq$1 are generally considered to be aroma-contributing components in koumiss, while substances with an OAV $\geq$10 are important aroma-contributing components [38]. The results (shown in Table 3) indicate that 20 volatile flavor substances can be used to calculate OAV, and their OAV values are all greater than 1. Among them, the OAV values of caprylic acid ethyl ester, caprylic acid, 9-decenyl acetate, decanoic acid ethyl ester, decanoic acid, isoamyl alcohol, phenethyl alcohol, and hexanoic acid ethyl ester in koumiss from different regions are all higher than 10, indicating that these volatile substances may be key flavor components in koumiss.

Table 3. OVA content of volatile substances in koumiss from different regions.

VOCs	Cas	Threshold	OVA			
			YL	TC	ALTw	ALTe
Ethyl caprate	110-38-3	0.02 e	1446.87 ± 120.65 a	723.31 ± 83.19 b	221.12 ± 16.59 c	112.28 ± 77.99 c
Ethyl caprylate	106-32-1	0.022 g	1102.47 ± 122.04 a	689.41 ± 105.23 b	175.70 ± 73.23 c	131.40 ± 52.17 c
Ethyl laurate	106-33-2	3.5 f	2.23 ± 0.26 a	1.47 ± 0.09 b	0.17 ± 0.09 c	0.08 ± 0.07 c
Octanoic acid	124-07-2	0.5 e	22.46 ± 6.41 a	9.11 ± 1.32 b	10.09 ± 7.06 b	1.50 ± 1.18 c
9-Decen-1-ol, 1-acetate	50816-18-7	0.1 g	84.30 ± 6.87 a	38.32 ± 2.55 b	17.07 ± 12.64 c	9.42 ± 8.68 c
Decanoic acid	334-48-5	0.5 e	11.09 ± 3.39 a	3.89 ± 0.57 b	4.13 ± 2.98 b	0.32 ± 0.06 c
1-Hexanoic acid	142-62-1	0.2 e	5.22 ± 0.24 a	2.05 ± 0.49 b	1.39 ± 0.66 b	0.41 ± 0.04 c
Ethyl tetradecanoate	124-06-1	0.5 e	1.25 ± 0.18 a	0.93 ± 0.14 b	0.04 ± 0.03 c	0.03 ± 0.03 c
acetic acid	64-19-7	1.2 e	0.39 ± 0.07 bc	0.61 ± 0.11 ab	1.87 ± 0.50 a	0.16 ± 0.12 c
Phenethyl alcohol	60-12-8	0.045 e	20.93 ± 12.99 b	55.18 ± 8.97 ab	30.53 ± 20.69 b	79.29 ± 31.91 a
Ethyl hexanoate	123-66-0	0.005 f	61.38 ± 4.17 a	52.67 ± 8.05 b	9.33 ± 5.50 c	12.59 ± 2.81 c
3-Methyl-1-butanol	123-51-3	0.25 e	2.21 ± 1.15 c	7.05 ± 1.36 b	4.44 ± 3.62 bc	13.86 ± 2.75 a
3-methylbutyl octanoate	2035-99-6	0.125 f	1.01 ± 0.29 a	0.56 ± 0.17 b	0	0
ethyl butanoate	105-54-4	0.025 f	2.37 ± 1.06 b	5.43 ± 0.82 a	0.1 ± 0.17 c	0.31 ± 0.31 c
(E,E)-2,4-Heptadienal	4313-03-5	0.1 f	0.87 ± 0.34 a	0.93 ± 0.58 a	1.87 ± 1.55 a	0.62 ± 1.07 a
Hexanal	66-25-1	0.005 f	2.27 ± 1.62 a	1.75 ± 0.50 a	3.01 ± 1.40 a	2.31 ± 1.07 a
benzoic acid	65-85-0	0.5 e	0.30 ± 0.30 b	0.96 ± 0.27 ab	1.14 ± 0.72 a	0
Acetoin	513-86-0	0.04 e	3.61 ± 0.20 b	5.74 ± 1.38 b	12.12 ± 11.74 a	6.06 ± 2.44 b
Ethyl valerate	539-82-2	0.025 f	5.17 ± 0.61 a	2.66 ± 0.37 b	3.42 ± 1.61 ab	3.98 ± 1.82 ab
Ethyl 2-methylbutanoate	7452-79-1	0.018 f	3.22 ± 1.18 a	0.93 ± 0.16 b	0.48 ± 0.16 b	0.44 ± 0.26 b

Note: The significant differences ($p < 0.05$) in the OVA values of different volatile compounds in koumiss from various regions are represented by letters a, b, c, and d. The letter e refers to the threshold value as per reference [16]. The letter f refers to the threshold value as per reference [39] with milk as the solvent. The letter g refers to the threshold value as per reference [40].

The results are shown in Table 3. The YL region stands out from other regions regarding its OVA values, with notable components including ethyl decanoate, ethyl octanoate, and ethyl 9-decenoate. In particular, ethyl decanoate and ethyl octanoate in the YL region have OVA values more than seven times higher than those in the ALT region and are therefore the most important characteristic aromas of the YL region. Ethyl decanoate and ethyl octanoate impart fruity and creamy aromas and have OVA values exceeding 100 in all regions of koumiss and have repeatedly been confirmed as key aroma components in koumiss from different regions of China [41]. The OVA value of phenethyl alcohol in the ALTe region is significantly higher than in other areas, making it one of the main aroma components that affect koumiss in this region. It is worth noting that previous studies on koumiss in other regions did not find phenethyl alcohol as a leading aroma sample, making it a unique flavor in the traditional fermented koumiss of the Kazakh ethnic group in Xinjiang. Both octanoic acid and decanoic acid, as sour aromas with OVA values greater than 10, contribute positively to the koumiss flavor and the dairy aroma of koumiss, providing a unique and distinctive taste.

3.6. Sensory Analysis of Koumiss in Different Regions

As the aroma components of koumiss samples measured by instruments cannot well explain the overall flavor of koumiss, and the aroma components contained in koumiss cannot explain their interactions, the human olfactory sensory system can provide comprehensive and sensitive feedback on the sensory characteristics of koumiss. Quantitative descriptive analysis (QDA) can obtain a complete sensory description of the sample, thus clearly showing the differences between samples [42]. To quantitatively describe the aroma of koumiss, it is crucial to develop sensory vocabulary with definitions and scales. Our sensory research team eventually determined the sensory evaluation vocabulary of fruit, cream, floral, odor, milk, and sourness. A radar chart was drawn based on the obtained sensory analysis data, as shown in Figure 6. For the attributes of fruit and cream, the scores of YL and TC regions were significantly higher than those of ALTe and ALTw regions, which were the main attributes of YL and TC regions. This may be due to the high content of ethyl caprylate and decanoic acid ethyl ester in YL and TC regions, consistent with the prediction of OVA values. The odor attribute had only a relatively high intensity in the

YL region, which may be a manifestation of excessively high levels of Octanoic acid and n-decanoic acid. For the milky attribute, the ALTw region had the highest score, with an intensity of 4.2. The ALTe region had the highest score for the floral attribute, which was related to the content of phenylethanol. Interestingly, the sensory profile characteristics of koumiss in different regions are quite different. These phenomena indicate that the metabolic capacity of flavor substances produced in the fermentation process of koumiss differs due to different fermentation conditions and other factors, which provides a theoretical basis for establishing and differentiating koumiss fermentation techniques with specific sensory characteristics.

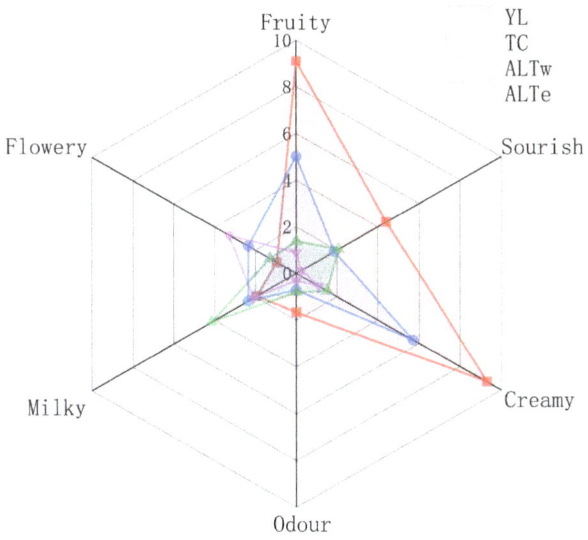

Figure 6. Radar chart of sensory analysis of koumiss from different regions.

Referring to Liu's classification method for fermented dairy products [13], the following rules were applied for aroma classification: if a specific aroma attribute of a fermented dairy sample scored the highest, it was categorized under that aroma type. Therefore, the four different regions of koumiss were classified into three aroma types, with the TC and YL regions falling under the fruit-cream type, ALTw under the milk type, and ALTe under the floral type. Currently, there is no other research on the aroma types of koumiss and the aroma classification results of this study can provide suggestions for establishing aroma classification of koumiss in the future.

4. Conclusions

Using gas chromatography coupled with ion mobility spectrometry, we established fingerprint profiles of koumiss from four different regions. Combination of the application of PLS-DA analysis, we identified dimethyl sulfide, ethyl butyrate, 2-methylbutanal, ethyl propionate, 2-methylpropyl propionate, isobutyl acetate, n-propanol, and methyl ethyl ketone as indicative markers for the rapid detection of traditionally fermented koumiss from these regions. In conclusion, it is worth emphasizing that the use of GC-IMS makes it possible to identify regions of origin, and thus control the authenticity of food origin. This is important for different food groups but for regional and traditional food. By utilizing GC-IMS and GC-MS to analyze the volatile components of the koumiss from these regions, we found that the volatile components were similar in terms of variety. However, due to differences in the fermentation process between regions, the volatile components exhibited significant differences in concentration and displayed clear regional characteristics. A total of 87 volatile components were detected, with esters, acids, and alcohols being the main aromatic substances in the koumiss. Furthermore, comparing these two detection methods regarding volatile compound analysis revealed that GC-IMS exhibited a more sensitive response towards aldehydes and ketones. Subsequently, an analysis of the OAV for koumiss from different regions identified ethyl caprylate, ethyl caprate, octanoic acid, 9-decenoic acid ethyl ester, n-decanoic acid, isopentanol, phenethyl alcohol, and ethyl hexanoate as key aroma components in traditional fermented koumiss of the Kazakh ethnic group. Interestingly, the aroma profile of the ALTw region was primarily dominated by phenethyl alcohol, while the YL and TC regions showcased a prominent presence of ethyl caprylate and ethyl caprate. Additionally, a qualitative descriptive analysis (QDA) was conducted to evaluate and categorize the aroma profiles of koumiss from the four regions, resulting in three distinct aroma types: fruit-cream type, milk type, and floral type.

Author Contributions: Conceptualization, X.R. and Y.G.; methodology, G.M. and Q.X.; software, X.N.; validation, X.N. and X.R.; formal analysis, Q.X.; investigation, X.M.; resources, G.M. and X.M.; data curation, Y.G.; writing—original draft preparation, Y.G.; writing—review and editing, Y.G.; visualization, X.M.; supervision, Q.X.; project administration, Q.X. All authors have read and agreed to the published version of the manuscript.

Funding: This work was supported by the National Natural Science Foundation of China (31760442) and Xinjiang First Division High-Tech Project (2022GJJ03).

Data Availability Statement: All related data and methods are presented in this paper. Additional inquiries should be addressed to the corresponding author.

Conflicts of Interest: This article encompasses human olfactory sensory experiments, conducted with the necessary consent obtained from all participants, along with appropriate training provided to the participants. The article does not involve experimental research on animal subjects.

References

1. Li, X.; Li, K.; Zou, S. Koumiss. *China Dairy* **2006**, 58–60. [CrossRef]
2. Sun, T.; Menhebilige; Wang, J.; Chen, Y.; Zhao, R.W.; Hepin, Z. Analysis of chemical composition and microorganism flora of traditionally home-made Koumiss in Xinjiang. *China Dairy Ind.* **2005**, *33*, 10–14.
3. He, Q.; Hou, Q.; Wang, Y.; Shen, L.; Sun, Z.; Zhang, H.; Liong, M.-T.; Kwok, L.-Y. Long-term administration of *Lactobacillus casei* Zhang stabilized gut microbiota of adults and reduced gut microbiota age index of older adults. *J. Funct. Foods* **2020**, *64*, 103682. [CrossRef]
4. Yizaitiguli, W.; Alimijiang, A.; MA, J.; Kalibinuer, A.; Gulibahaer, K. Development of Compound Probiotic Sour Mare Milk and Its Effect on Intestinal Flora. *China Dairy Ind.* **2021**, *49*, 13–20. [CrossRef]
5. Cao, K.; Wu, J.; Li, Y.; Han, Q.; Yu, B.; Qi, S.; Wu, R.; Mang, l. Research Progress on the Mechanism of Koumiss in Treating Tuberculosis. *China Dairy Ind.* **2022**, *50*, 34–40. [CrossRef]
6. Bogoyavlenskiy, A.; Alexyuk, M.; Alexyuk, P.; Amanbayeva, M.; Anarkulova, E.; Imangazy, A.; Bektuganova, A.; Berezin, V.J.M.R.A. Metagenomic Exploration of Koumiss from Kazakhstan. *Microbiol. Resour. Announc.* **2022**, *11*, e01082-21. [CrossRef] [PubMed]
7. Simujide; Aorigele; Wang, C.; Aricha; Zang, J.; Chen, H.; Liu, B.; He, L. Isolation, identification and diversity analysis of lactic acid bacteria in fresh koumiss. *Feed Res.* **2021**, *44*, 79–82. [CrossRef]
8. Luo, P. Xinjiang horse industry development present situation analysis. *Xinjiang Anim. Husb.* **2022**, *37*, 8–12. [CrossRef]
9. Liu, S.; Pan, L. Research on the sales market of koumiss—Based on the consumption market of koumiss in Hohhot. *China Dairy* **2022**, 8–12.
10. Tang, H.; Ma, H.; Hou, Q.; Li, W.; Xu, H.; Liu, W.; Sun, Z.; Haobisi, H.; Menghe, B. Profiling of koumiss microbiota and organic acids and their effects on koumiss taste. *BMC Microbiol.* **2020**, *20*, 85. [CrossRef]
11. Wang, S.; Chen, H.; Sun, B. Recent progress in food flavor analysis using gas chromatography–ion mobility spectrometry (GC–IMS). *Food Chem.* **2020**, *315*, 126158. [CrossRef] [PubMed]
12. Zhang, M.; Lu, Y.; Zhao, Z.; Choi, L.; Yan, H.; Wang, X.; Zhao, H. Rapid Discrimination of Different Years of Brewing Liquor by Gas Chromatography-Ion Mobility Spectroscopy Combined with Chemometrics Method. *Sci. Technol. Food Ind.* **2021**, *42*, 226–232. [CrossRef]
13. Liu, A.; Liu, Q.; Bu, Y.; Hao, H.; Liu, T.; Gong, P.; Zhang, L.; Chen, C.; Tian, H.; Yi, H. Aroma classification and characterization of *Lactobacillus delbrueckii* subsp. bulgaricus fermented milk. *Food Chem. X* **2022**, *15*, 100385. [CrossRef]
14. Meng, Y.; Chen, X.; Sun, Z.; Li, Y.; Chen, D.; Fang, S.; Chen, J. Exploring core microbiota responsible for the production of volatile flavor compounds during the traditional fermentation of Koumiss. *LWT* **2021**, *135*, 110049. [CrossRef]

15. Xia, Y.; Yu, J.; Liu, H.; Feng, C.; Shuang, Q. Novel insight into physicochemical and flavor formation in koumiss based on microbial metabolic network. *Food Res. Int.* **2021**, *149*, 110659. [CrossRef] [PubMed]
16. *ISO 8586: 2012*; Sensory Analysis—General Guidelines for the Selection, Training and Monitoring of Selected Assessors and Expert Sensory Assessors. International Organization for Standardization: Geneva, Switzerland, 2012.
17. *ISO 5492: 2008*; Sensory Analysis—Vocabulary. International Organization for Standardization: Geneva, Switzerland, 2008.
18. Qi, H.; Ding, S.; Pan, Z.; Li, X.; Fu, F. Characteristic volatile fingerprints and odor activity values in different citrus-tea by HS-GC-IMS and HS-SPME-GC-MS. *Molecules* **2020**, *25*, 6027. [CrossRef]
19. Jiménez-Carvelo, A.M.; Martín-Torres, S.; Ortega-Gavilán, F.; Camacho, J. PLS-DA vs. sparse PLS-DA in food traceability. A case study: Authentication of avocado samples. *Talanta* **2021**, *224*, 121904. [CrossRef]
20. Shao, S.; Xu, M.; Lin, Y.; Chen, X.; Fang, D.; Cai, J.; Wang, J.; Jin, S.; Ye, N. Differential Analysis of Aroma Components of Huangguanyin Oolong Tea from Different Geographical Origins Using Electronic Nose and Headspace Solid-Phase Microextraction-Gas Chromatography-Mass Spectrometry. *Food Sci.* **2023**, *44*, 232–239.
21. Yang, Y.; Zhu, H.; Chen, J.; Xie, J.; Shen, S.; Deng, Y.; Zhu, J.; Yuan, H.; Jiang, Y. Characterization of the key aroma compounds in black teas with different aroma types by using gas chromatography electronic nose, gas chromatography-ion mobility spectrometry, and odor activity value analysis. *LWT* **2022**, *163*, 113492. [CrossRef]
22. Yin, J.; Wu, M.; Lin, R.; Li, X.; Ding, H.; Han, L.; Yang, W.; Song, X.; Li, W.; Qu, H. Application and development trends of gas chromatography–ion mobility spectrometry for traditional Chinese medicine, clinical, food and environmental analysis. *Microchem. J.* **2021**, *168*, 106527. [CrossRef]
23. Zhang, K.; Zhang, C.; Gao, L.; Zhuang, H.; Feng, T.; Xu, G. Analysis of volatile flavor compounds of green wheat under different treatments by GC-MS and GC-IMS. *J. Food Biochem.* **2022**, *46*, e13875. [CrossRef] [PubMed]
24. Wang, Z.; Tian, H.; Wang, P.; Zhang, F.; Wei, Z. Comparative Analysis of Volatile Compounds of Traditional Koumiss from Different Areas in Xinjiang. *Food Res. Dev.* **2018**, *39*, 131–138.
25. Du, X.; Liu, W.; Zhang, H. Study on the dynamic change of volatile flavor compounds in traditional fermented koumiss. *China Dairy Ind.* **2017**, *45*, 4–9.
26. Xue, L.; Wu, H.; Hu, Z. Effect of Lipase Contained in Glutinous Rice Wine on Lipolysis in Traditional Chinese-style Cheese. *Food Sci.* **2010**, *31*, 71–73.
27. Delgado, F.J.; González-Crespo, J.; Cava, R.; Ramírez, R. Formation of the aroma of a raw goat milk cheese during maturation analysed by SPME–GC–MS. *Food Chem.* **2011**, *129*, 1156–1163. [CrossRef]
28. Li, N.; Sun, B.; Liang, M. Analysis of Volatile Compounds in a Dairy Flavoring Using SPME-GC/MS. *Food Sci.* **2011**, *32*, 221–225.
29. Zhang, G.; Xie, M.; Kang, X.; Wei, M.; Zhang, Y.; Li, Q.; Wu, X.; Chen, Y. Optimization of ethyl hexanoate production in *Saccharomyces cerevisiae* by metabolic engineering. *LWT* **2022**, *170*, 114061. [CrossRef]
30. Zhang, X.; Ai, N.; Wang, J.; Tong, L.; Zheng, F.; Sun, B. Lipase-catalyzed modification of the flavor profiles in recombined skim milk products by enriching the volatile components. *J. Dairy Sci.* **2016**, *99*, 8665–8679. [CrossRef]
31. Yu, J. Analysis of the Microbial Community Structure and the Correlation with Flavor Substances during Koumiss Fermentation. MD Thesis, Inner Mongolia Agricultural University, Hohhot, China, 2021. [CrossRef]
32. Fox, P.F.; Guinee, T.P.; Cogan, T.M.; McSweeney, P.L.H. *Fundamentals of Cheese Science*; Springer: Boston, MA, USA, 2000.
33. Zeng, C.; Zhang, Y.; Kang, S.; Zhang, J.; Zhang, F.; Zhang, H. Research progress of higher alcohols in fermented wine. *China Brew.* **2015**, *34*, 11–15.
34. Ma, Y.; Cao, Y.; Yang, Z.; Wang, B. Comparative Analysis of Volatile Compounds in Different Cheese Samples by SPME-GC-MS. *Food Sci.* **2013**, *34*, 103–107.
35. Fernández-García, E.; Carbonell, M.A.; Gaya, P.; Nuñez, M. Evolution of the volatile components of ewes raw milk Zamorano cheese. Seasonal variation. *Int. Dairy J.* **2004**, *14*, 701–711. [CrossRef]
36. Chen, Y.; Wu, Y.; Liu, X. The regulation and application of acetoin biosynthesis. *J. Biol.* **2014**, *31*, 76–80+84.
37. Plotto, A.; Margaría, C.A.; Goodner, K.L.; Baldwin, E.A. Odour and flavour thresholds for key aroma components in an orange juice matrix: Esters and miscellaneous compounds. *Flavour Fragr. J.* **2008**, *23*, 398–406. [CrossRef]
38. Zhong, M.; Han, Y.; Huang, Y.; Shi, Y.; Li, D.; Li, X. Aroma Composition and Synergistic Effect among Key Aroma Substances of Pine Wine. *Food Res. Dev.* **2022**, *43*, 179–188.
39. Van Gemert, L.J. *Odour Thresholds: Compilations of Odour Threshold Values in Air, Water and Other Media*; Oliemans Punter & Partners BV: Utrecht, The Netherlands, 2011.
40. Xia, Y.; Zhao, Y.; Wang, H.; Yu, J.; Shuang, Q. Analysis on the key volatile flavor compounds of fermented koumiss by SPME-GC/MS combined with OAV. *Food Sci. Technol.* **2019**, *44*, 318–325. [CrossRef]
41. Wu, R.; Bao, L.; Bao, X.; Li, S.; Wang, X.; Jia, S. Dynamic changes of volatile flavor compounds during the fermentation of koumiss for food therapy in horqin area. *China Dairy Ind.* **2019**, *47*, 10–16.
42. Guld, Z.; Nyitrainé Sárdy, D.; Gere, A.; Rácz, A. Comparison of sensory evaluation techniques for Hungarian wines. *J. Chemom.* **2020**, *34*, e3219. [CrossRef]

Disclaimer/Publisher's Note: The statements, opinions and data contained in all publications are solely those of the individual author(s) and contributor(s) and not of MDPI and/or the editor(s). MDPI and/or the editor(s) disclaim responsibility for any injury to people or property resulting from any ideas, methods, instructions or products referred to in the content.

molecules

Article

New Insight into the Substrate Selectivity of Bovine Milk γ-glutamyl Transferase via Structural and Molecular Dynamics Predictions

Lichuang Cao [1], Cameron J. Hunt [2], Anne S. Meyer [2] and René Lametsch [1,*]

1 Department of Food Science, Faculty of Science, University of Copenhagen, 1958 Frederiksberg C, Denmark; lichuang.cao@food.ku.dk
2 Department of Biotechnology and Biomedicine, Technical University of Denmark, 2800 Kgs. Lyngby, Denmark; cjahu@dtu.dk (C.J.H.); asme@dtu.dk (A.S.M.)
* Correspondence: rla@food.ku.dk

Abstract: Bovine milk γ-glutamyltransferase (BoGGT) can produce γ-glutamyl peptides using L-glutamine as a donor substrate, and the transpeptidase activity is highly dependent on both γ-glutamyl donors and acceptors. To explore the molecular mechanism behind the donor and acceptor substrate preferences for BoGGT, molecular docking and molecular dynamic simulations were performed with L-glutamine and L-γ-glutamyl-*p*-nitroanilide (γ-G*p*NA) as donors. Ser450 is a crucial residue for the interactions between BoGGT and donors. BoGGT forms more hydrogen bonds with L-glutamine than γ-G*p*NA, promoting the binding affinity between BoGGT and L-glutamine. Gly379, Ile399, and Asn400 are crucial residues for the interactions between the BoGGT intermediate and acceptors. The BoGGT intermediate forms more hydrogen bonds with Val-Gly than L-methionine and L-leucine, which can promote the transfer of the γ-glutamyl group from the intermediate to Val-Gly. This study reveals the critical residues responsible for the interactions of donors and acceptors with the BoGGT and provides a new understanding of the substrate selectivity and catalytic mechanism of GGT.

Keywords: γ-glutamyl donor; γ-glutamyl acceptor; molecular docking; γ-glutamyl-enzyme intermediate; molecular dynamic simulation

1. Introduction

Because of its ability to catalyze the transpeptidase reaction to produce γ-glutamyl peptides, γ-glutamyl transferase from bovine milk (BoGGT) has recently received increasing attention [1–3]. γ-Glutamyl peptides can be applied as taste enhancers to enhance the mouthfeel and aftertaste of food [4,5]. Furthermore, it has been demonstrated that γ-glutamyl peptides exhibit a variety of biological activities, comprising anti-inflammatory, hypoglycemic, and appetite-suppressing effects [6]. In 2009, γ-glutamyl peptides were identified from Swiss Gruyere and Gouda cheeses [7,8]. Toelstede and Hofmann [7] found an association between the transpeptidase activity and concentration of γ-glutamyl peptides in various cheese types. They demonstrated that BoGGT might produce γ-glutamyl peptides in raw milk cheeses. It has been established that γ-glutamyl peptides in Parmesan cheese are mainly produced by the transpeptidase reaction of BoGGT [3]. Moreover, Yang et al. [9] identified γ-glutamyl peptides from protein hydrolysates of enzyme-modified butter and revealed that BoGGT could produce γ-glutamyl peptides. Our previous study identified BoGGT (G3N2D8) using LC-MS/MS and validated its ability to generate γ-glutamyl peptides from whey protein and casein hydrolysates with L-glutamine as a donor [1,2].

The types of γ-glutamyl donors and acceptors affect the transpeptidase activity of GGT. A recent study documented that *Escherichia coli* GGT has stereoselectivity for γ-glutamyl acceptors, suggesting that the transpeptidase activity of *E. coli* GGT depends on the acceptor substrate [10]. Hillmann, Behr, Ehrmann, Vogel and Hofmann [3] proposed that BoGGT has a higher substrate specificity for L-methionine and L-phenylalanine than L-lysine, L-histidine, L-leucine, L-aspartic acid, L-threonine, and L-glutamic acid. The GGT activity is typically measured using a colorimetric method with γ-glutamyl-*p*-nitroanilide (γ-G*p*NA) as a donor [11–14]. Our previous study showed that BoGGT has a lower transpeptidase activity with γ-G*p*NA as the γ-glutamyl donor than that of a L-glutamine donor; further, the transpeptidase activity of BoGGT is highly dependent on the acceptor with both L-glutamine and γ-G*p*NA donors [2].

However, no previous study has been conducted into the interaction mechanism between BoGGT and its γ-glutamyl donor, as well as the interaction between the BoGGT intermediate and γ-glutamyl acceptor. It is well-established that human γ-glutamyl transferase (GGT) works by sequential mechanisms to catalyze a bi-substrate reaction [10,15]. Firstly, hydroxyl oxygen of threonine in the active site of GGT will initiate a nucleophilic attack for the amide carbon on the donor, forming a γ-glutamyl enzyme intermediate with the γ-glutamyl moiety. Then the γ-glutamyl moiety will transfer from the intermediate to the γ-glutamyl acceptor. γ-Glutamyl peptides will be produced in case the acceptor is a peptide or an amino acid, whereas L-glutamic acid will be generated in case the acceptor is a water molecule [10,15]. It has been established that the catalytic nucleophile of human GGT is Thr381 [16]. Analogously, Thr380 was confirmed to be the catalytic nucleophile in the active site of BoGGT by multiple sequence alignment [1].

To explore the binding mechanism between BoGGT and substrates, molecular docking and molecular dynamics (MD) simulations have been utilized [17,18]. The computational work requires a three-dimensional (3D) structure of BoGGT; however, the 3D structure of BoGGT is unresolved. Recently, it was proposed that AlphaFold could reliably construct 3D structures from protein sequences [19], which could predict the 3D structure of BoGGT. This study aims to (1) explain the donor and acceptor substrate preferences for BoGGT via modeling of the interaction between key active site residues of BoGGT and different donors and acceptors; (2) provide new insight into the catalytic reaction trajectory by molecular dynamics simulations of the BoGGT substrate binding and the bond distance between the key amino acids and substrate reactant in the active site.

2. Results and Discussion

2.1. Mechanism behind the Donor Selectivity of BoGGT

2.1.1. Prediction of the 3D Structure of BoGGT

Our previous study identified BoGGT as the bovine enzyme (G3N2D8) [1]. In the present study, AlphaFold v2.0 was applied to construct the 3D structure of BoGGT. As illustrated in Figure 1A, the majority of the BoGGT residues were predicted with very high confidence, supporting that the 3D structure was accurately predicted.

2.1.2. Interactions between BoGGT and Donor

The catalytic nucleophile of human GGT has already been reported to be Thr381 [15,16]; multiple sequence alignment confirmed that the catalytic nucleophile of BoGGT is Thr380 [1]. It is generally accepted that the hydroxyl oxygen (-OH) on Thr380 from BoGGT will attack the amide carbon of the donor, resulting in the formation of the γ-glutamyl–BoGGT intermediate [20]. Our previous study demonstrated that BoGGT has lower transpeptidase activity with γ-G*p*NA as a γ-glutamyl donor than the L-glutamine donor [2]. However, the molecular mechanism underlying the difference in donor affinity for BoGGT remains unclear.

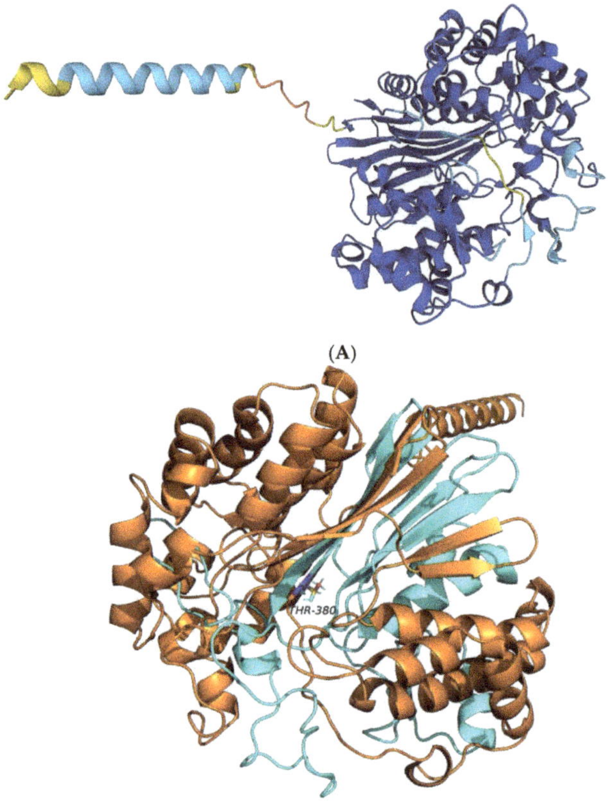

(A)

(B)

Figure 1. AlphaFold prediction on the three-dimensional structure of bovine milk γ-glutamyl transferase. (**A**) Three-dimensional structure of bovine milk γ-glutamyl transferase displayed in various colors based on model confidence. Note: The per-residue confidence score (pLDDT) is utilized to assign a color to each residue in the model. Residues with extremely high confidence (pLDDT > 90) are shown in blue, while those with high confidence (70 < pLDDT < 90) are shown in cyan. On the contrary, residues with low confidence (50 < pLDDT < 70) are displayed in yellow, while very low confidence residues (pLDDT < 50) are represented in orange. (**B**) Three-dimensional structure of bovine milk γ-glutamyl transferase with light and heavy chains shown in different colors.

Molecular docking results showed eight hydrogen bonds between BoGGT and L-glutamine, whereas there were four hydrogen bonds between BoGGT and γ-G*p*NA. These hydrogen bonds are responsible for glutamine binding at the active site. This might explain why BoGGT has a higher transferase activity with L-glutamine as a γ-glutamyl donor than γ-G*p*NA. Figure 2A displays a PyMol visualization of the interaction between BoGGT and L-glutamine. The oxygen atom on Gly379 and Thr380 formed a hydrogen bond (H-bond) with the L-glutamine's amide hydrogen atom, respectively. H-bonds were formed between the hydrogen atoms on Ser450, Ser451, and Met452 and the carboxy oxygen atom on L-glutamine. The transpeptidase activity of human GGT was reduced to 1% of the wild-type GGT after site mutagenesis of Ser451 or Ser452 [21], and Ser451 or Ser452 also formed H-bonds with the substrate; this indicates that Ser451 and Ser452 promote the interaction between the active site and substrate. The corresponding residues of Ser451 and Ser452 in human GGT are Ser450 and Ser451 in BoGGT, respectively. Therefore, H-bonds between Ser450, Ser451, and the carboxy oxygen atom on L-glutamine seem to promote the donor substrate binding in the active site, resulting in the formation of the γ-glutamyl–BoGGT intermediate. In addition, the amide hydrogen and carbonyl oxygen atoms on Gly473 formed H-bonds with the L-glutamine's carbonyl oxygen and amide hydrogen atoms, respectively. It has been proposed that the H-bonds between Gly473 and the donor substrate promote the formation of the γ-glutamyl-GGT intermediate [15].

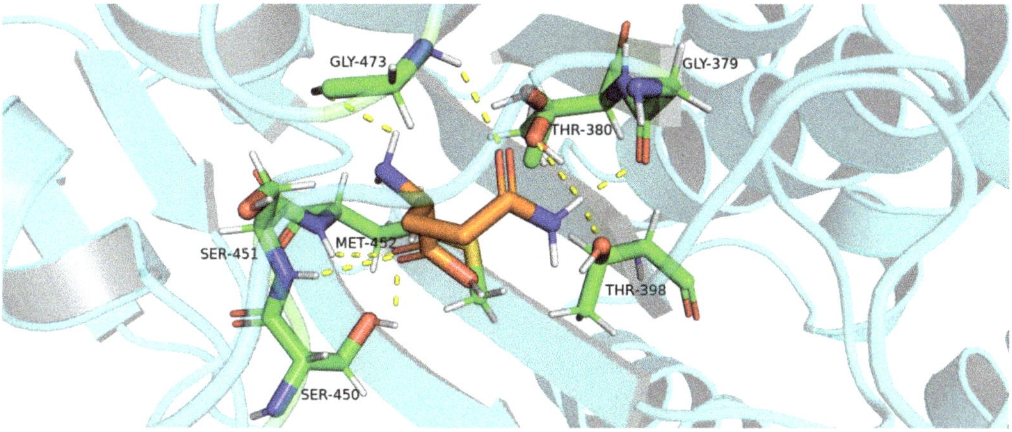

(**A**)

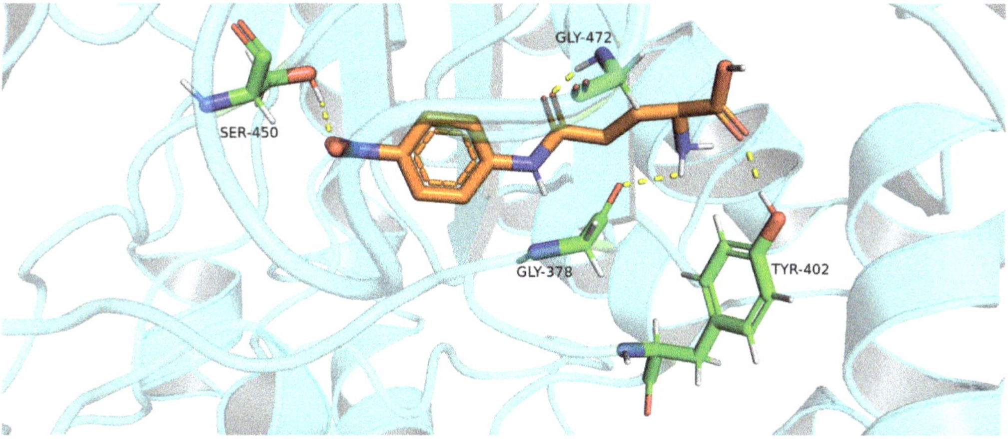

(**B**)

Figure 2. Interactions between γ-glutamyl transferase from bovine milk and γ-glutamyl donor. (**A**) Interactions between bovine γ-glutamyl transferase and L-glutamine. (**B**) Interactions between bovine γ-glutamyl transferase and L-γ-glutamyl-*p*-nitroanilide. Note: The atoms in silver represent hydrogen; the atoms in blue and red represent nitrogen and oxygen, respectively.

An H-bond occurs between the oxygen atom on Thr398 and the amide hydrogen on L-glutamine. It is universally accepted that the formation of H-bonds between enzyme and substrates greatly affects the specificity of the enzyme and thus facilitates the enzymatic reaction. PyMol visualization of the interaction between BoGGT and γ-G*p*NA is presented in Figure 2B. The oxygen atom on Gly378 formed an H-bond with the amide hydrogen atom of γ-G*p*NA. The hydrogen atom on Tyr402, Ser450, and Gly472 formed H-bonds with the oxygen atom on γ-G*p*NA. Based on the interactions between BoGGT and the γ-glutamyl donor, it could be concluded that Ser450 is a crucial residue for donor substrate recognition.

2.1.3. MD Simulations of BoGGT–Donor Complex

After the molecular docking, the BoGGT–donor complex was employed to conduct 80 ns MD simulations to explore the interaction between BoGGT and the donor substrate. The root-mean-square deviation (RMSD) represents how structures change over time compared to the starting point. The RMSD of the complexes (C_α of BoGGT and the donor substrate) was used to assess the system stability of BoGGT–donor complexes during an 80 ns simulation. As demonstrated in Figure 3A, both complexes had steady RMSDs after 40 ns simulations, indicating that the L-glutamine and the γ-G*p*NA donor were tightly bound to BoGGT. The RMSDs of the donor substrates and backbone of BoGGT were calculated to evaluate the stability of the donor substrate and backbone of BoGGT further (Figure 3B). After a 60 ns simulation, the RMSD of the BoGGT backbone and ligand (L-glutamine and γ-G*p*NA) from complexes stabilized, indicating that the protein had equilibrated. Nevertheless, the RMSDs of the complexes were notably greater than that of the BoGGT backbone, which might be due to the opening and closing motion of a lid from BoGGT observed during the simulations. The above results indicate that the BoGGT–glutamine and BoGGT–G*p*NA complexes reached a steady state after 60 ns of simulations. For each BoGGT–donor substrate complex, the root-mean-square fluctuations (RMSFs) of its constituent amino acids were determined, as well. Amino acids 121–124 and 431–434 of the BoGGT–Glutamine complexes were found to have higher fluctuation values than those of BoGGT–G*p*NA, suggesting these residues might be involved in the conformational alteration of the BoGGT–Glutamine complexes (Figure 3C).

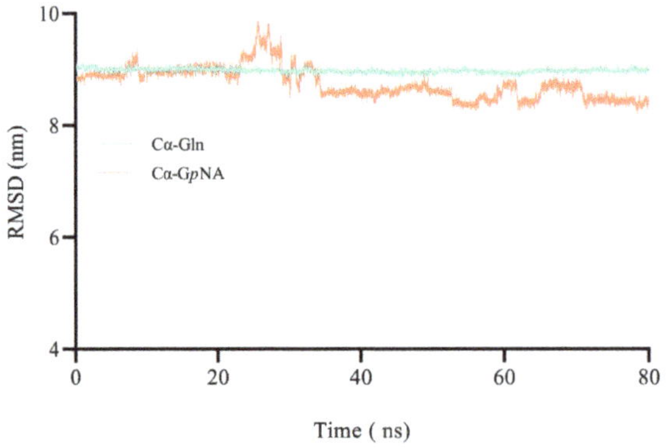

(A)

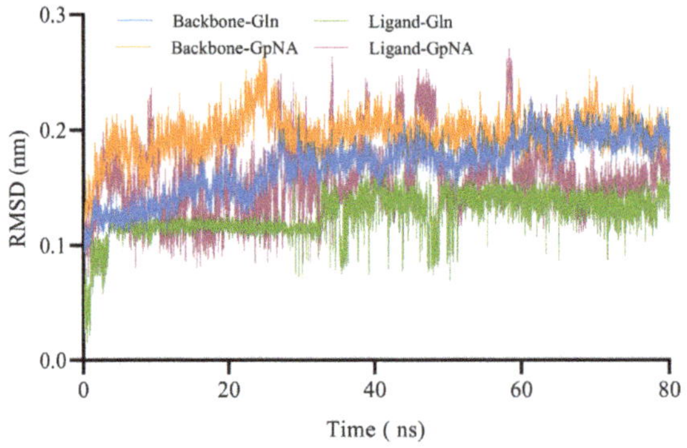

(B)

Figure 3. *Cont.*

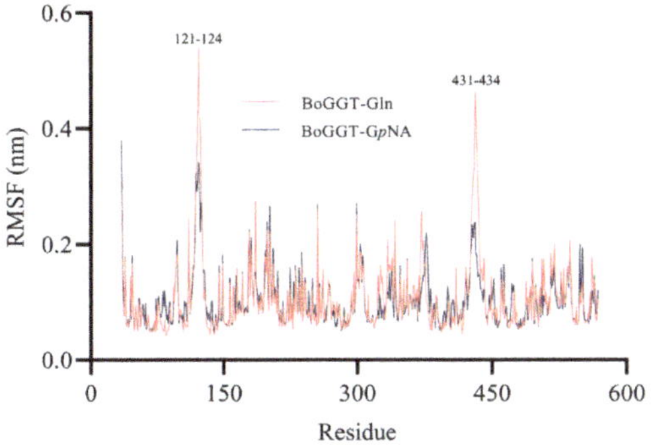

(**C**)

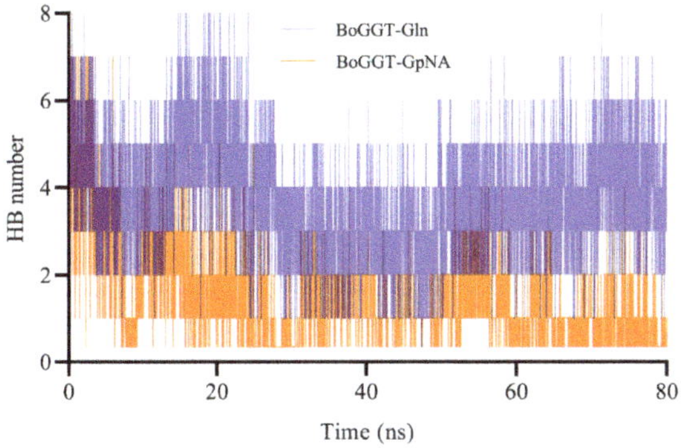

(**D**)

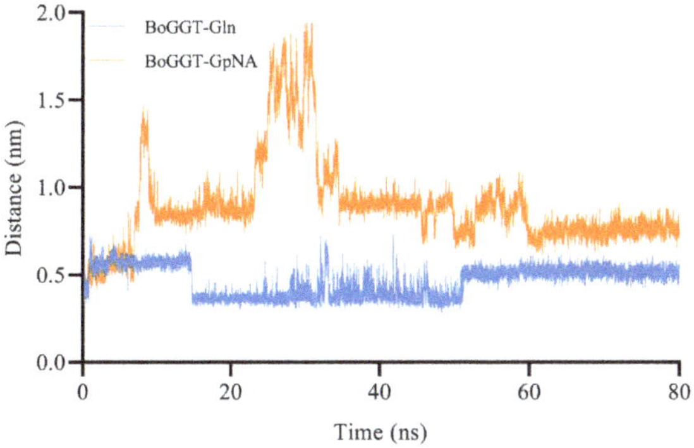

(**E**)

Figure 3. Molecular dynamic simulations of BoGGT–donor complexes. (**A**) Root-mean-square deviation of C_{α} of BoGGT–donor. (**B**) Root-mean-square deviation of BoGGT backbone and donor substrate. (**C**) Root-mean-square fluctuations (RMSFs) of BoGGT from BoGGT–Gln and BoGGT–G*p*NA complexes. (**D**) Hydrogen bond numbers between BoGGT and the donor substrate. (**E**) Distance between the hydroxyl oxygen atom on Thr380 of BoGGT and the amide carbon on the donor substrate.

H-bonds between the enzyme and substrate are widely considered to be the primary factor that facilitate the enzymatic reaction [22]. Consequently, H-bond numbers were also utilized to evaluate the binding between the donor substrate and BoGGT. As displayed in Figure 3D, the H-bond numbers between BoGGT and L-glutamine were significantly higher than those between BoGGT and γ-G*p*NA during the 60–80 ns simulation ($p < 0.05$), which was consistent with the docking results. This suggests that BoGGT has a higher interaction with L-glutamine than γ-G*p*NA.

GGT first forms γ-glutamyl–GGT intermediate during the transpeptidase reaction when the hydroxyl oxygen on the active site attacks the carbonyl carbon on the donor substrate [23]. Consequently, during the MD simulations, the distance between the hydroxyl oxygen atom (OG_1) on Thr380 and the amide carbon (C) on the γ-glutamyl donor was inspected to evaluate the possibility of the interaction between BoGGT and the donor. As displayed in Figure 3E, the distance between OG_1 on Thr380 and the amide carbon on L-glutamine (0.514 ± 0.031 nm) was significantly ($p < 0.05$) lower than that for BoGGT and γ-G*p*NA (0.749 ± 0.042 nm) during the 60–80 ns simulations.

Previous research has reported that Pro427-Ser438 of human GGT form a lid-loop, which may affect the substrate binding in the active site [24–26]. Sequence alignment showed that the lid-loop of BoGGT is Pro426-Ser437, and the opening and closing motion was observed during the MD simulation of BoGGT–donor complexes, which could recognize the donor substrate and regulate the access of the donor to the active site. γ-G*p*NA is a bigger molecule than L-glutamine, allowing L-glutamine to access the active site easily. This could also explain why BoGGT has higher transpeptidase activity with L-glutamine as a donor than γ-G*p*NA. The H-bond numbers between L-glutamine and BoGGT were significantly ($p < 0.05$) greater than those of γ-G*p*NA, which can decrease the OG_1-C distance and thus promote the binding affinity between BoGGT and L-glutamine.

2.2. Mechanism behind the Acceptor Selectivity of BoGGT

2.2.1. Construction and Optimization of the BoGGT Intermediate

Our previous study revealed that the transpeptidase activity of BoGGT is highly dependent on the acceptor with both L-glutamine and γ-G*p*NA donors [2]. In the second step of the transpeptidation reaction, it is well established that the γ-glutamyl–GGT intermediate reacts with acceptors to produce γ-glutamyl peptides. To explore the interaction mechanism between the γ-glutamyl–BoGGT intermediate and acceptors, the γ-glutamyl–BoGGT intermediate was constructed and optimized using GROMACS. After constructing the γ-glutamyl–BoGGT intermediate, the intermediate was optimized using 100 ns MD simulations. Figure 4B presents the γ-glutamyl–BoGGT intermediate, which is the combination of the hydroxyl oxygen on Thr380 of BoGGT with the γ-glutamyl group from the donor. As demonstrated in Figure 4C, the RMSDs of the intermediate's backbone and alpha-carbon atom stabilized after 60 ns of simulation, indicating that the 3D structure of BoGGT reached a steady state. The intermediate's 3D structure at 100 ns was used to perform molecular docking.

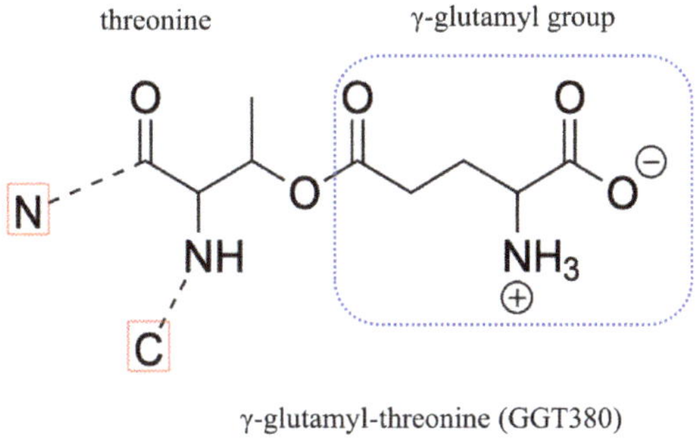

(**A**)

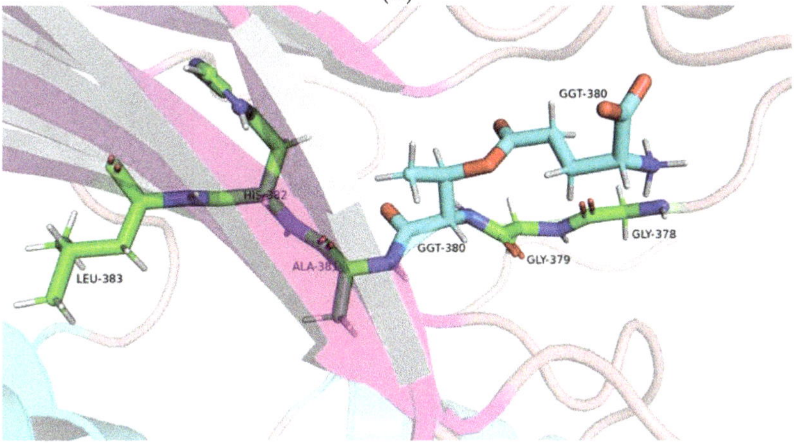

(**B**)

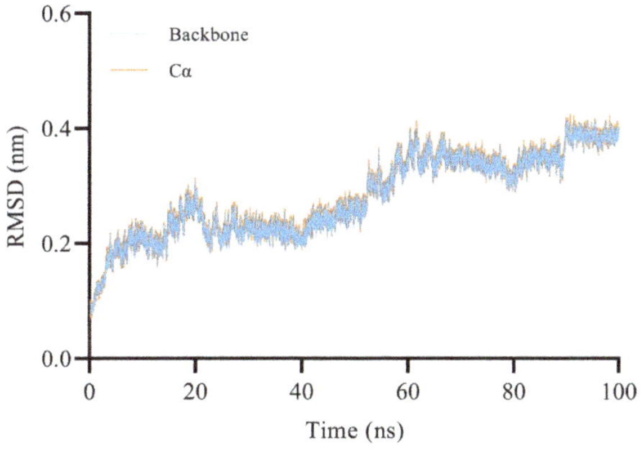

(**C**)

Figure 4. Molecular dynamic simulations of the γ-glutamyl–BoGGT intermediate. (**A**) Diagram of the γ-glutamyl-threonine on the γ-glutamyl–BoGGT intermediate. (**B**) Three-dimensional structure of the γ-glutamyl–BoGGT intermediate. Note: The atoms in silver represent hydrogen; the atoms in blue and red represent nitrogen and oxygen, respectively. (**C**) Root-mean-square deviation of backbone and alpha-carbon from the γ-glutamyl–BoGGT intermediate.

2.2.2. Interactions between the BoGGT Intermediate and Acceptors

Our previous study demonstrated that the transpeptidase activity of BoGGT with L-methionine as an acceptor substrate was significantly lower than that of Val-Gly, whereas it was significantly higher than that of L-leucine [2]. Hillmann, Behr, Ehrmann, Vogel and Hofmann [3] also proposed that L-methionine was a better acceptor substrate than L-leucine for BoGGT. However, the molecular mechanism behind the difference in acceptor affinity for the BoGGT intermediate remains unclear. In the current study, molecular docking was utilized to explore the interactions between the BoGGT intermediate and γ-glutamyl acceptors.

As shown in Figure 5A, molecular docking of the BoGGT intermediate and Val-Gly showed five hydrogen bonds; three residues (Gly379, Ile399, and Asn400) were involved in the formation of H-bonds. The oxygen atom on Gly379 formed an H-bond with the hydroxyl hydrogen atom on Val-Gly. Two H-bonds were formed between the hydrogen atoms on Ile399 and Asn400, and the carboxyl oxygen atom on Val-Gly, and one H-bond was formed between the hydrogen atom on Asn400 and the hydroxyl oxygen atom on Val-Gly. Four H-bonds were formed between four residues (Ser81, Gly379, Ile399, and Asn400) from the intermediate and L-methionine (Figure 5B), whereas there were three H-bonds between Gly379, Ile399, and Asn400 on the intermediate and L-leucine (Figure 5C).

The hydrogen atom on Ile399 and Asn400 formed one H-bond with the same hydroxyl oxygen atom on L-methionine, respectively. The oxygen atom on Ser81 formed an H-bond with the amide hydrogen atom on L-methionine, and carboxyl oxygen on Gly379 formed an H-bond with the hydroxyl hydrogen atom on L-methionine. Regarding the interaction between the intermediate and L-leucine, the hydrogen atom on Ile399 and Asn400 formed one H-bond with the same hydroxyl oxygen atom on L-leucine, respectively. The carboxyl oxygen on Gly379 formed an H-bond with the amide hydrogen atom on L-leucine. It could be concluded that Gly379, Ile399, and Asn400 are key residues responsible for the interactions between the intermediate and acceptors, thereby regulating the transfer of the γ-glutamyl moiety from the γ-glutamyl–BoGGT intermediate to the acceptor.

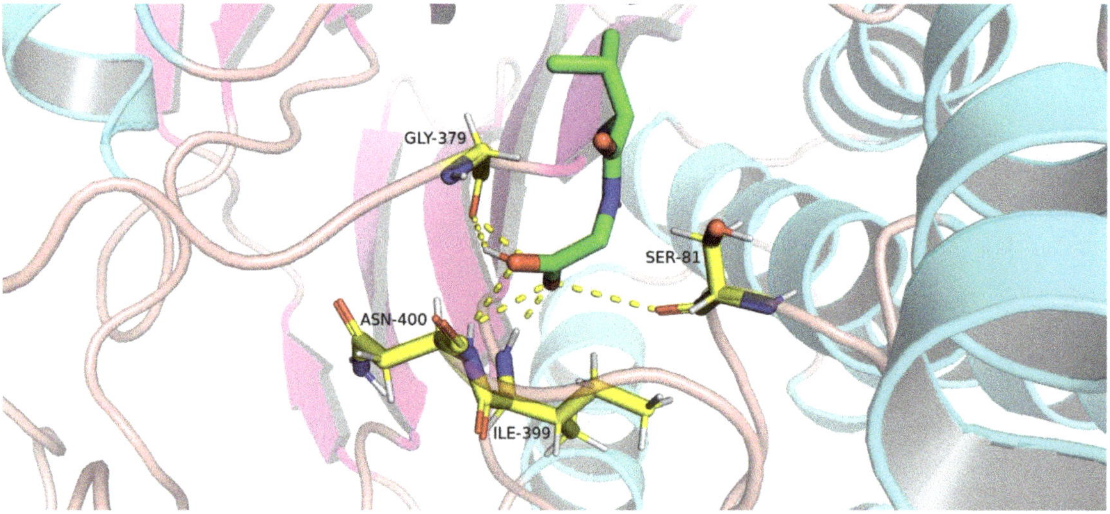

(A)

Figure 5. *Cont.*

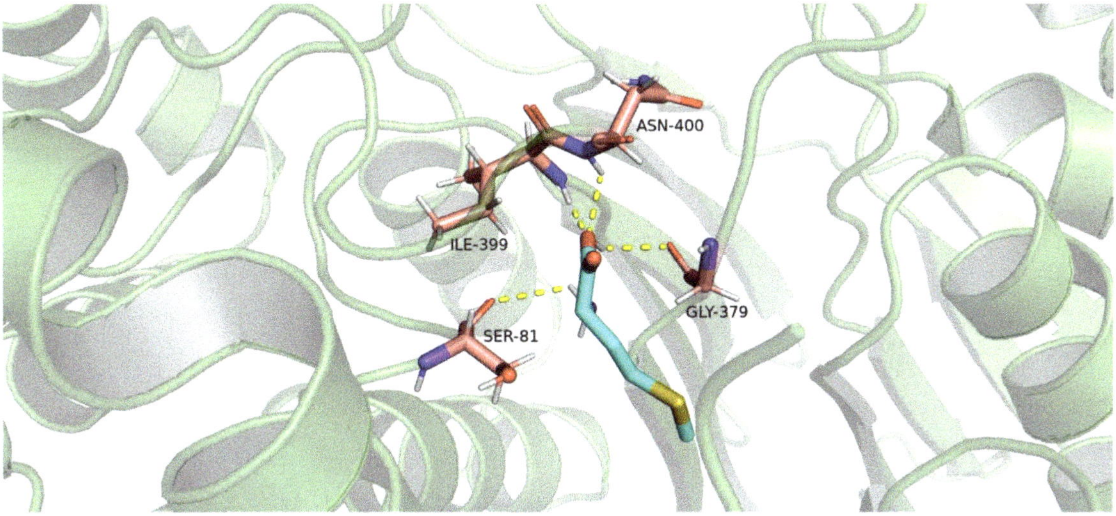

(**B**)

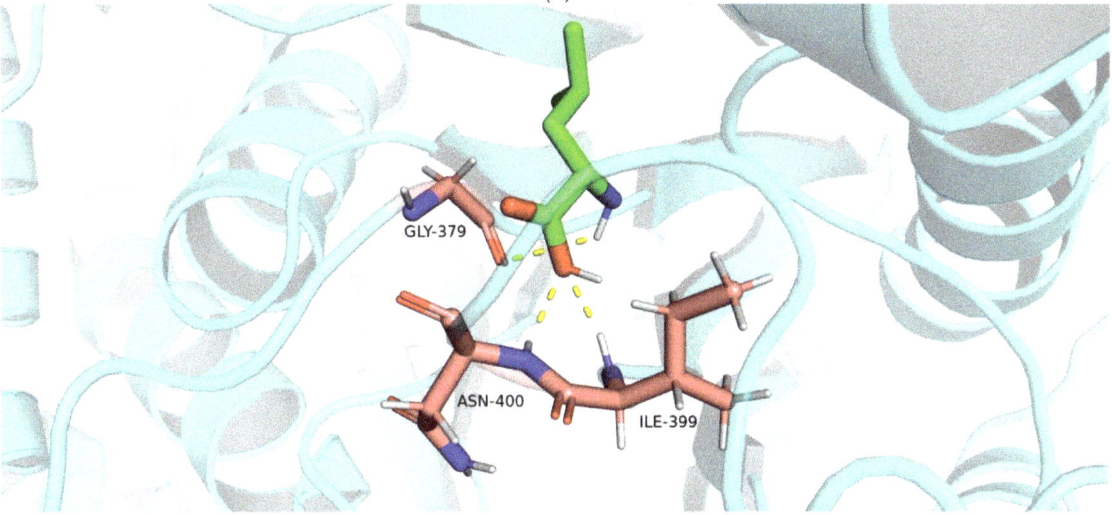

(**C**)

Figure 5. Interactions between the γ-glutamyl–BoGGT intermediate and γ-glutamyl acceptors. (**A**) Interactions between the γ-glutamyl–BoGGT intermediate and Val-Gly. (**B**) Interactions between the γ-glutamyl–BoGGT intermediate and L-methionine. (**C**) Interactions between the γ-glutamyl–BoGGT intermediate and L-leucine. Note: The atoms in silver and yellow represent hydrogen and sulfur, respectively. The atoms in blue and red represent nitrogen and oxygen, respectively.

2.2.3. MD Simulations of BoGGT Intermediate–Acceptor Complex

After the molecular docking, the BoGGT intermediate–acceptor complexes were used to perform 40 ns MD simulations to explore the interaction between the intermediate and acceptors. As shown in Figure 6A, the RMSD of the BoGGT intermediate–acceptor complexes (C_α of the BoGGT intermediate–acceptor) stabilized after 20 ns of simulations, indicating that the BoGGT intermediate and the acceptor reached a steady state.

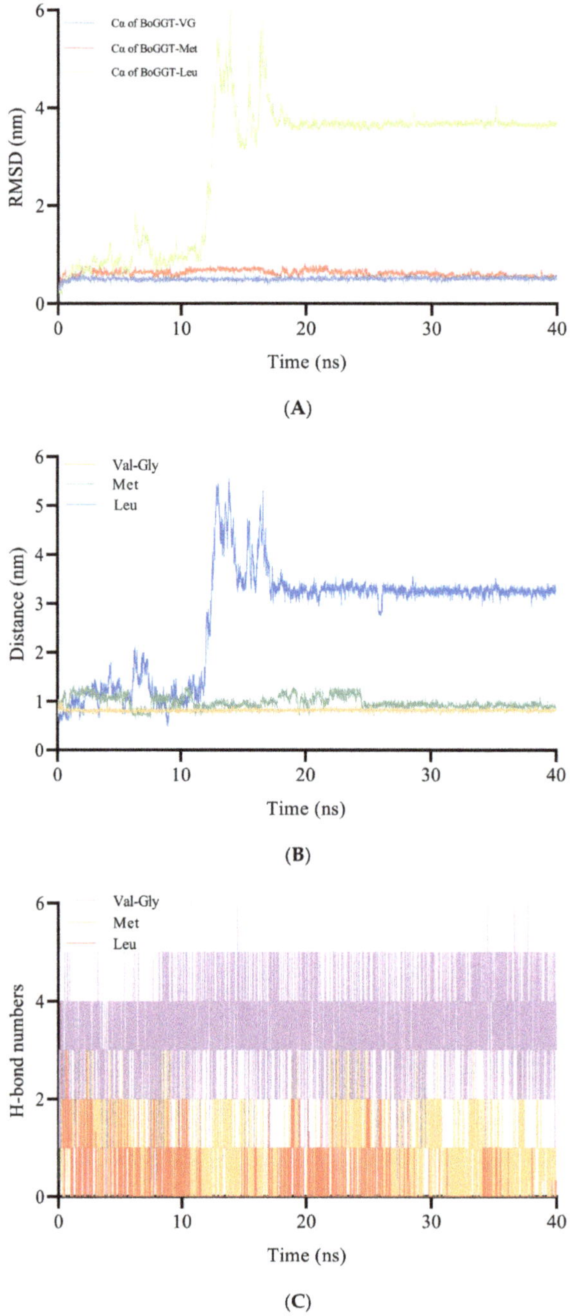

Figure 6. Molecular dynamic simulations of the γ-glutamyl–BoGGT intermediate–acceptor complexes. (**A**) Root-mean-square deviation of Cα of the γ-glutamyl–BoGGT intermediate–acceptor. (**B**) Distance between the hydroxyl oxygen atom on residue380 of the γ-glutamyl–BoGGT intermediate and the nitrogen on the acceptor substrate. (**C**) Hydrogen bond numbers between the γ-glutamyl–BoGGT intermediate and the acceptor substrate.

In the second step, the nitrogen atom on the acceptor will attack the oxygen atom on GGT380 (γ-glutamyl-threonine) to produce γ-glutamyl peptides. Therefore, the distance between the oxygen atom on residue 380 and the nitrogen atom on the acceptor is very important to produce γ-glutamyl peptides. As presented in Figure 6B, during the 30–40 ns simulations, the distance between the oxygen atom (O) on GGT380 and the nitrogen atom on Val-Gly (N) (0.814 ± 0.022 nm) was significantly ($p < 0.05$) lower than that of L-methionine (0.917 ± 0.039 nm) and L-leucine (3.244 ± 0.094 nm), indicating that the intermediate has a higher interaction possibility with Val-Gly. It is generally accepted that H-bonds between enzyme and substrate are the main driving force for the enzymatic reaction. Consequently, the number of H-bonds was also employed to evaluate the interactions between the intermediate and acceptors. As shown in Figure 6C, during the 30–40 ns simulations, the number of H-bonds between the BoGGT intermediate and Val-Gly were significantly higher than that of L-methionine and L-leucine ($p < 0.05$), which could decrease the O-N distance and thus promote the transfer of the γ-glutamyl moiety from the BoGGT intermediate to Val-Gly. To further validate the binding affinity between the acceptor substrate and the BoGGT intermediate, MMPBSA analysis was employed to estimate the binding affinity between the complexes during MD simulations. As shown in Table 1, the total binding energy of the BoGGT intermediate–Val-Gly complexes (−141.062 ± 14.829 kJ/mol) was significantly lower than those of the BoGGT intermediate–Met (−110.784 ± 11.275 kJ/mol) and BoGGT intermediate–Leu complexes (−75.481 ± 19.973 kJ/mol), indicating that BoGGT has a higher affinity for the Val-Gly acceptor. Moreover, the total binding energy of the BoGGT intermediate–Met complexes was significantly lower than that of the BoGGT intermediate–Leu complexes, denoting that BoGGT has a lower affinity for the L-leucine acceptor.

Table 1. Total binding energies for acceptor substrates against the BoGGT intermediate.

Complex	Total Binding Energy (kJ/mol)	Van der Waals Energy (kJ/mol)	Electrostatic Energy (kJ/mol)	Polar Solvation Energy (kJ/mol)	SASA Energy (kJ/mol)
BoGGT intermediate–Val-Gly	−141.062 ± 14.829	−176.361 ± 25.742	−102.037 ± 36.825	165.193 ± 41.980	−27.857 ± 5.184
BoGGT intermediate–Met	−110.784 ± 11.275	−157.906 ± 32.609	−76.074 ± 23.259	134.209 ± 33.527	−11.013 ± 4.513
BoGGT intermediate–Leu	−75.481 ± 19.973	−123.524 ± 27.128	−62.598 ± 19.937	124.372 ± 32.252	−13.731 ± 4.235

3. Materials and Methods

3.1. Structure Construction of BoGGT and Molecular Docking between γ-glutamyl Donor and BoGGT

In the current study, the three-dimensional (3D) structure of bovine GGT (UniProt no. G3N2D8) was produced by AlphaFold (v2.0) on an in-house system equipped with an A40 GPU running Docker [19,27]. Multiple sequence alignment was performed against the reduced database (db_preset = reduced_dbs), max_template_date set to the enzymes and model_preset as monomer. Each of the five relaxed structures was manually inspected for defects, and the structure with the highest score was selected for further analysis.

The 2D structures of the donor substrates (L-glutamine or L-γ-glutamyl-*p*-nitroanilide) were obtained from PubChem, and the 3D structures were converted using ChemSketch (ACD Labs, Toronto, ON, Canada). This allowed us to generate models for donor substrate–enzyme complexes by molecular docking using Autodock Vina with default parameters [28,29]. The docking results of ligand–enzyme complexes were ranked based on the binding energy, and the best match was chosen to analyze the interactions between BoGGT and the donors in PyMOL (Schrödinger Inc., New York, NY, USA).

3.2. MD Simulation of BoGGT–Donor Complex

GROMACS (version 2022.2) was utilized to conduct MD simulations for BoGGT–Gln and BoGGT–G*p*NA complexes, based on our previous study [30]. Force field parameters of BoGGT were written in Charmm36 all-atom force field format (July 2021), and force field parameters of L-glutamine and γ-G*p*NA were generated by the CHARMM General

Force Field (CGenFF, https://cgenff.umaryland.edu, accessed on 6 September 2022). In the first stage, a steep descent approach of 50,000 steps was adopted to minimize energy and achieve a stable conformation. After minimizing the system's energy, we performed an isochoric–isothermal equilibration (300 K) for 100 ps, followed by an isothermal–isobaric equilibration (300 K, 1.0 bar) for 100 ps. The complexes were subsequently subjected to 80 ns MD simulations, meaning that the simulations were run in blocks of 80 ns until RMSD simulations were stable within this time course.

The RMSD, RMSF, and H-bond number between BoGGT and donors (L-glutamine and γ-G*p*NA) were calculated after the MD simulations. To evaluate their interactions, the distance between the amide carbon (C) on the donor and hydroxyl oxygen (OG_1) on the active site was also calculated.

3.3. Construction of the BoGGT Intermediate and Optimization of the Intermediate

GGT first interacts with a γ-glutamyl group to form γ-glutamyl–GGT intermediate, which then combines with acceptors to synthesize γ-glutamyl peptides [15]. To reveal the mechanism between the γ-glutamyl–BoGGT intermediate and acceptors, the γ-glutamyl–BoGGT intermediate was constructed and optimized using GROMACS. The γ-glutamyl–BoGGT intermediate was formed when hydroxyl oxygen on the active site of BoGGT (Thr380) attacked the carbonyl carbon of the γ-glutamyl group from the donor. In other words, the γ-glutamyl–BoGGT intermediate is the combination of the hydroxyl oxygen of Thr380 on BoGGT with the γ-glutamyl group from the donor. The γ-glutamyl-threonine is not a standard residue in GROMACS; we need to define it in the existing force field for GROMACS to recognize this residue as an amino acid. Briefly, the atoms and linkage information for the new residue was added to the RTP file (aminoacids.RTP), and then the information about how to add hydrogens to the new residue was supplemented to the HDB file (aminoacids.HDB). Afterward, the information about modifying the residue if it is a terminal residue was added to the r2b file. The new residue was supplemented to residuetypes.dat with the name of GGT into the working directory. The γ-glutamyl-threonine residue was built and aligned with the Thr380 of BoGGT, and then the original Thr380 was deleted to obtain the γ-glutamyl–BoGGT intermediate. Subsequently, a 100 ns MD simulation (300 K, 1.0 bar) was applied to optimize the γ-glutamyl–BoGGT intermediate.

3.4. Molecular Docking between the BoGGT Intermediate and γ-glutamyl Acceptors

Molecular docking between the BoGGT intermediate and acceptors was performed using Autodock Vina [28,29]. Briefly, the 2D structures of acceptor substrates (Val-Gly, L-methionine, and L-leucine) were obtained from PubChem, and the 3D structures were converted using ChemSketch (ACD Labs, Toronto, ON, Canada). This allowed us to generate models for acceptor substrate–enzyme complexes by molecular docking using Autodock Vina with default parameters [28,29]. The docking results of ligand–enzyme complexes were ranked based on the binding energy, and the best match was chosen to analyze the interactions between the BoGGT intermediate and acceptors in PyMOL (Schrödinger Inc., New York, NY, USA).

3.5. MD Simulation of BoGGT Intermediate–Acceptor Complexes

GROMACS was utilized to conduct MD simulations for BoGGT intermediate–acceptor complexes, based on our previous study [30]. Force field parameters of the BoGGT intermediate were written in Charmm36 all-atom force field format (July 2021), and force field parameters of acceptor substrates were generated by the CHARMM General Force Field (CGenFF, https://cgenff.umaryland.edu, accessed on 15 November 2022). In the first stage, a steep descent approach of 50,000 steps was adopted to minimize energy and achieve a stable conformation. After minimizing the system's energy, we performed an isochoric–isothermal equilibration (300 K) for 100 ps, followed by an isothermal–isobaric equilibration (300 K, 1.0 bar) for 100 ps. The complexes were subsequently subjected to

MD simulations that were run in 40 ns blocks until a time course of relatively stable RMSD was observed.

RMSDs and H-bond numbers between the intermediate and acceptors were calculated after the MD simulations. To evaluate their interactions, the distance between the amide nitrogen (N) on the acceptor and hydroxyl oxygen on the active site was also calculated. The molecular mechanics Poisson Boltzmann surface area (MMPBSA) was applied to determine the binding free energy using the g_mmpbsa package [31]. Binding energy was calculated using the final 10 ns from the 40 ns MD simulations. Bonded and non-bonded interactions in the vacuum were utilized to estimate the binding affinity of the BoGGT intermediate with acceptors in the solvent stage. The Poisson Boltzmann equation and Solvent Accessible Surface Area (SASA) were utilized to determine the polar and non-polar solvation energy.

3.6. Statistical Analysis

The MD simulations were performed in triplicate. The obtained data are presented as mean $\pm$ standard deviation. To test differences among means of multiple groups, one-way analysis of variance (ANOVAs) with Duncan's analysis was utilized. To determine whether the difference in means between the two groups was statistically significant, a t-test was used. A p-value less than 0.05 ($p < 0.05$) indicates that there are significant differences between the samples. The figures and ANOVAs were created using GraphPad Prism 9.3.1 and SAS 9.4, respectively.

4. Conclusions

BoGGT is an enzyme that could produce γ-glutamyl peptides with L-glutamine as a donor substrate and has the potential as a catalyst to produce taste enhancers, whereas the transpeptidase activity is highly dependent on the γ-glutamyl donors and acceptors. In summary, to explore the molecular mechanism behind the donor and acceptor substrate preferences for BoGGT, structure prediction of BoGGT, molecular docking, and molecular dynamic simulations were performed. The light subunit of BoGGT (Thr380, Thr398, Ser450, Ser451, Met452, and Gly473) is responsible for forming the majority of the H-bonds with the L-glutamine donor. Ser450 is a crucial residue for both L-glutamine–BoGGT and G*p*NA–BoGGT interactions. H-bond numbers between L-glutamine and BoGGT were significantly higher than that of γ-G*p*NA, which can decrease the distance between the hydroxyl oxygen atom on the active site of BoGGT and the amide carbon on the donor substrate (OG_1-C), thereby promoting the binding affinity between BoGGT and L-glutamine. Gly379, Ile399, and Asn400 are crucial residues for the interactions between the BoGGT intermediate and acceptors. H-bond numbers between the BoGGT intermediate and Val-Gly were significantly higher than that of L-methionine and L-leucine, which can decrease the O_1-N distance, thereby promoting the transfer of the γ-glutamyl group from the intermediate to Val-Gly. This study reveals the interactions between BoGGT and the donor, as well as the interactions between the intermediate and the acceptor substrate, and provides a meaningful method for selecting donors and acceptors with higher affinity for GGT, which can extend the understanding of the catalytic mechanism and promote the yields of γ-glutamyl peptides.

Author Contributions: Conceptualization, L.C.; methodology, L.C. and C.J.H.; software, L.C. and C.J.H.; formal analysis, L.C.; investigation, L.C.; data curation, L.C.; writing—original draft preparation, L.C.; writing—review and editing, A.S.M. and R.L.; supervision, R.L.; project administration, R.L.; funding acquisition, A.S.M. and R.L. All authors have read and agreed to the published version of the manuscript.

Funding: This research was funded by the China Scholarship Council, grant number 201903250091, and the Novo Nordisk Foundation to the Proferment project, grant application no.—NNF21OC0066330. Data was generated though accessing research infrastructure at University of Copenhagen, including FOODHAY (Food and Health Open Innovation Laboratory, Danish Roadmap for Research Infrastructure).

Institutional Review Board Statement: Not applicable.

Informed Consent Statement: Not applicable.

Data Availability Statement: Not applicable.

Acknowledgments: The authors would like to thank the China Scholarship Council for financial support.

Conflicts of Interest: The authors declare no conflict of interest.

Sample Availability: Not applicable.

References

1. Cao, L.; Li, Q.; Lametsch, R. Identification and Activity Characterization of gamma-Glutamyltransferase from Bovine Milk. *J. Agric. Food Chem.* **2021**, *69*, 15325–15333. [CrossRef]
2. Cao, L.; Li, Q.; Lametsch, R. Comparative analysis of substrate affinity and catalytic efficiency of gamma-glutamyltransferase from bovine milk and *Bacillus amyloliquefaciens*. *Food Chem.* **2023**, *405 Pt B*, 134930. [CrossRef]
3. Hillmann, H.; Behr, J.; Ehrmann, M.A.; Vogel, R.F.; Hofmann, T. Formation of Kokumi-Enhancing gamma-Glutamyl Dipeptides in Parmesan Cheese by Means of gamma-Glutamyltransferase Activity and Stable Isotope Double-Labeling Studies. *J. Agric. Food Chem.* **2016**, *64*, 1784–1793. [CrossRef]
4. Yang, J.; Bai, W.; Zeng, X.; Cui, C. Gamma glutamyl peptides: The food source, enzymatic synthesis, kokumi-active and the potential functional properties—A review. *Trends Food Sci. Technol.* **2019**, *91*, 339–346. [CrossRef]
5. Zhao, C.J.; Schieber, A.; Ganzle, M.G. Formation of taste-active amino acids, amino acid derivatives and peptides in food fermentations—A review. *Food Res. Int.* **2016**, *89 Pt 1*, 39–47. [CrossRef]
6. Lu, Y.; Wang, J.; Soladoye, O.P.; Aluko, R.E.; Fu, Y.; Zhang, Y. Preparation, receptors, bioactivity and bioavailability of γ-glutamyl peptides: A comprehensive review. *Trends Food Sci. Technol.* **2021**, *113*, 301–314. [CrossRef]
7. Toelstede, S.; Hofmann, T. Kokumi-active glutamyl peptides in cheeses and their biogeneration by *Penicillium roquefortii*. *J. Agric. Food Chem.* **2009**, *57*, 3738–3748. [CrossRef] [PubMed]
8. Toelstede, S.; Dunkel, A.; Hofmann, T. A series of kokumi peptides impart the long-lasting mouthfulness of matured Gouda cheese. *J. Agric. Food Chem.* **2009**, *57*, 1440–1448. [CrossRef]
9. Yang, Z.; Wang, J.; Han, Z.; Blank, I.; Meng, F.; Wang, B.; Cao, Y.; Tian, H.; Chen, C. Isolation, identification and sensory evaluation of kokumi peptides from by-products of enzyme-modified butter. *J. Sci. Food Agric.* **2022**, *102*, 6668–6675. [CrossRef]
10. Somma, V.; Calvio, C.; Rabuffetti, M.; Rama, E.; Speranza, G.; Morelli, C.F. An overall framework for the *E. coli* gamma-glutamyltransferase-catalyzed transpeptidation reactions. *Bioorg. Chem.* **2021**, *115*, 105217. [CrossRef]
11. Shuai, Y.; Zhang, T.; Jiang, B.; Mu, W. Development of efficient enzymatic production of theanine by gamma-glutamyltranspeptidase from a newly isolated strain of *Bacillus subtilis*, SK11.004. *J. Sci. Food Agric.* **2010**, *90*, 2563–2567. [CrossRef]
12. Blel, M.; Guingamp, M.-F.; Gaillard, J.-L.; Humbert, G. Studies on the thermal sensitivity of γ-glutamyl transpeptidase measured with a modified test procedure and compared with that of alkaline phosphatase and lactoperoxidase in milk. *Le Lait* **2002**, *82*, 555–566. [CrossRef]
13. Baumrucker, C.R. gamma-Glutamyl transpeptidase of bovine milk membranes: Distribution and characterization. *J. Dairy Sci.* **1979**, *62*, 253–258. [CrossRef] [PubMed]
14. Tate, S.S.; Meister, A. Interaction of gamma-glutamyl transpeptidase with amino acids, dipeptides, and derivatives and analogs of glutathione. *J. Biol. Chem.* **1974**, *249*, 7593–7602. [CrossRef]
15. West, M.B.; Chen, Y.; Wickham, S.; Heroux, A.; Cahill, K.; Hanigan, M.H.; Mooers, B.H. Novel insights into eukaryotic gamma-glutamyltranspeptidase 1 from the crystal structure of the glutamate-bound human enzyme. *J. Biol. Chem.* **2013**, *288*, 31902–31913. [CrossRef]
16. Terzyan, S.S.; Burgett, A.W.; Heroux, A.; Smith, C.A.; Mooers, B.H.; Hanigan, M.H. Human gamma-Glutamyl Transpeptidase 1: Structures of the free enzyme, inhibitor-bound tetrahedral transition states, and glutamate-bound enzyme reveal novel movement within the active site during catalysis. *J. Biol. Chem.* **2015**, *290*, 17576–17586. [CrossRef] [PubMed]
17. Taylor, R.D.; Jewsbury, P.J.; Essex, J.W. A review of protein-small molecule docking methods. *J. Comput. Aided Mol. Des.* **2002**, *16*, 151–166. [CrossRef] [PubMed]
18. Sledz, P.; Caflisch, A. Protein structure-based drug design: From docking to molecular dynamics. *Curr. Opin. Struct. Biol.* **2018**, *48*, 93–102. [CrossRef]
19. Jumper, J.; Evans, R.; Pritzel, A.; Green, T.; Figurnov, M.; Ronneberger, O.; Tunyasuvunakool, K.; Bates, R.; Zidek, A.; Potapenko, A.; et al. Highly accurate protein structure prediction with AlphaFold. *Nature* **2021**, *596*, 583–589. [CrossRef] [PubMed]
20. Castonguay, R.; Halim, D.; Morin, M.; Furtos, A.; Lherbet, C.; Bonneil, E.; Thibault, P.; Keillor, J.W. Kinetic characterization and identification of the acylation and glycosylation sites of recombinant human gamma-glutamyltranspeptidase. *Biochemistry* **2007**, *46*, 12253–12262. [CrossRef]

21. Ikeda, Y.; Fujii, J.; Anderson, M.E.; Taniguchi, N.; Meister, A. Involvement of Ser-451 and Ser-452 in the catalysis of human gamma-glutamyl transpeptidase. *J. Biol. Chem.* **1995**, *270*, 22223–22228. [CrossRef]
22. Chen, D.; Oezguen, N.; Urvil, P.; Ferguson, C.; Dann, S.M.; Savidge, T.C. Regulation of protein-ligand binding affinity by hydrogen bond pairing. *Sci. Adv.* **2016**, *2*, e1501240. [CrossRef]
23. Hibi, T.; Nii, H.; Nakatsu, T.; Kimura, A.; Kato, H.; Hiratake, J.; Oda, J. Crystal structure of gamma-glutamylcysteine synthetase: Insights into the mechanism of catalysis by a key enzyme for glutathione homeostasis. *Proc. Natl. Acad. Sci. USA* **2004**, *101*, 15052–15057. [CrossRef]
24. Okada, T.; Suzuki, H.; Wada, K.; Kumagai, H.; Fukuyama, K. Crystal structures of gamma-glutamyltranspeptidase from *Escherichia coli*, a key enzyme in glutathione metabolism, and its reaction intermediate. *Proc. Natl. Acad. Sci. USA* **2006**, *103*, 6471–6476. [CrossRef]
25. Morrow, A.L.; Williams, K.; Sand, A.; Boanca, G.; Barycki, J.J. Characterization of *Helicobacter pylori* gamma-glutamyltranspeptidase reveals the molecular basis for substrate specificity and a critical role for the tyrosine 433-containing loop in catalysis. *Biochemistry* **2007**, *46*, 13407–13414. [CrossRef]
26. Wada, K.; Irie, M.; Suzuki, H.; Fukuyama, K. Crystal structure of the halotolerant gamma-glutamyltranspeptidase from *Bacillus subtilis* in complex with glutamate reveals a unique architecture of the solvent-exposed catalytic pocket. *FEBS J.* **2010**, *277*, 1000–1009. [CrossRef]
27. Varadi, M.; Anyango, S.; Deshpande, M.; Nair, S.; Natassia, C.; Yordanova, G.; Yuan, D.; Stroe, O.; Wood, G.; Laydon, A.; et al. AlphaFold Protein Structure Database: Massively expanding the structural coverage of protein-sequence space with high-accuracy models. *Nucleic Acids Res.* **2022**, *50*, D439–D444. [CrossRef] [PubMed]
28. Trott, O.; Olson, A.J. AutoDock Vina: Improving the speed and accuracy of docking with a new scoring function, efficient optimization, and multithreading. *J. Comput. Chem.* **2010**, *31*, 455–461. [CrossRef]
29. Eberhardt, J.; Santos-Martins, D.; Tillack, A.F.; Forli, S. AutoDock Vina 1.2.0: New Docking Methods, Expanded Force Field, and Python Bindings. *J. Chem. Inf. Model.* **2021**, *61*, 3891–3898. [CrossRef] [PubMed]
30. Cao, L.; Hunt, C.J.; Lin, S.; Meyer, A.S.; Li, Q.; Lametsch, R. Elucidation of the Molecular Mechanism of Bovine Milk gamma-Glutamyltransferase Catalyzed Formation of gamma-Glutamyl-Valyl-Glycine. *J. Agric. Food Chem.* **2023**, *71*, 2455–2463. [CrossRef] [PubMed]
31. Abdulrashid, N.I.; Aminu, S.; Adamu, R.M.; Tajuddeen, N.; Isah, M.B.; Jatau, I.D.; Aliyu, A.B.; Simelane, M.B.C.; Onyike, E.; Ibrahim, M.A. Phloroglucinol as a Potential Candidate against *Trypanosoma congolense* Infection: Insights from In Vivo, In Vitro, Molecular Docking and Molecular Dynamic Simulation Analyses. *Molecules* **2022**, *27*, 469. [CrossRef] [PubMed]

Disclaimer/Publisher's Note: The statements, opinions and data contained in all publications are solely those of the individual author(s) and contributor(s) and not of MDPI and/or the editor(s). MDPI and/or the editor(s) disclaim responsibility for any injury to people or property resulting from any ideas, methods, instructions or products referred to in the content.

fermentation

MDPI

Article

A Study of Key Aroma Compounds in Hurood Cheese and Their Potential Correlations with Lactic Acid Bacteria

Yadong Wang, Hong Zeng, Yanping Cao, Shaojia Wang * and Bei Wang *

School of Food and Health, Beijing Technology and Business University, No. 11, Fucheng Road, Beijing 100048, China; wyd2020bgs@163.com (Y.W.); zenghong@btbu.edu.cn (H.Z.); caoyp@th.btbu.edu.cn (Y.C.)
* Correspondence: wangshaojia@btbu.edu.cn (S.W.); wangbei@th.btbu.edu.cn (B.W.); Tel.: +86-10-68984547 (S.W. & B.W.)

Abstract: Hurood cheese (namely Hurood) is a traditional acid-coagulated cheese in China. This work investigated key aroma compounds and their potential correlations with dominant species of Hurood sampled from three distinct geographical origins. Key aroma compounds were determined according to Gas chromatography–mass spectrometry (GC–MS), gas chromatography–olfactometry (GC–O), and relative odor active values (ROAVs) analyses. In addition, 16S rDNA sequencing was used to identify the dominant species. Furthermore, Pearson correlation analysis was used to determine the potential relationships between key aroma compounds and dominant species. A total of 31 key aroma compounds were identified in the Hurood samples from three regions. *Lactobacillus paracasei*, *Lactobacillus crispatus*, and *Leuconostoc citreum* were found to be significantly correlated with the key aroma compounds ($p < 0.05$) and were identified as the core species. This study shows the link between the presence of presumptive functional core microbes and the unique aroma profiles of this traditional dairy product.

Keywords: key aroma compounds; lactic acid bacteria (LAB); gas chromatography–mass spectrometry (GC–MS); Hurood

1. Introduction

Hurood cheese (namely Hurood) is a traditional acid-coagulated cheese with a unique aroma, mainly manufactured and consumed in North China—especially in the Inner Mongolia region [1]. Hurood is usually made with fresh bovine milk. The processing steps of making Hurood are to first naturally ferment and acidify the curd without the heat treatment of fresh milk (Figure 1). Subsequently followed by whey draining, the raw Hurood is pressed in a specialized mold for 24 h, which forms the final product cheese. Hurood is not aged; therefore, it belongs to the fresh cheese family.

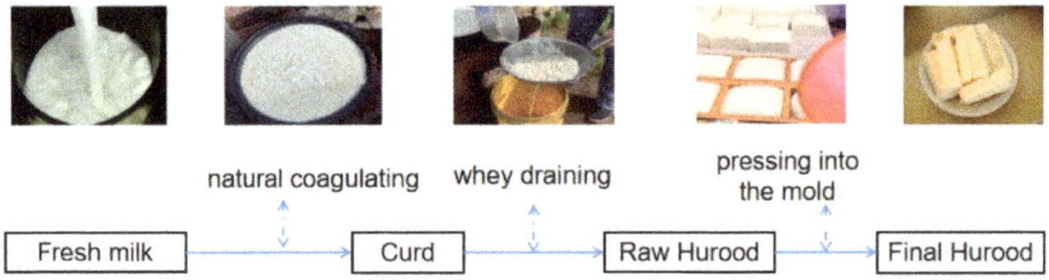

Figure 1. The process of making Hurood in the Inner Mongolia region, China.

Aroma is one of the key indexes that determines the cheese's overall quality. Aroma compounds in cheese are mainly derived from microbial metabolism [2]. Dicrepancies in fermenting cultures can lead to a deviation in the aroma of the final product [3]. Through the analysis of the relationship between the microbial diversity and the aroma of milk fan samples from six different geographical origins, Chen et al. found that *Lactobacillus* spp. was greatly associated with short even-chain fatty acids such as butyric acid, caprylic acid, and caproic acid [4,5]. Jiang et al. reported that *Lactobacillus* spp. displayed a significant correlation with aroma compounds such as benzaldehyde, 2,3-pentanedione, ethanol, and ethyl acetate in traditional yak yogurt [6]. However, only a few efforts have been made to investigate how microorganisms can affect the aroma profiling of Hurood [7]. Functional microbes that contribute to the formation of key aroma compounds of Hurood with different production locations remains unclear.

Lactic acid bacteria (LAB) are widely used starter cultures for the production of fermented dairy products [8]. Traditional Chinese fermented dairy products, such as Hurood, are rich in LAB resources [1,9]. Several studies have explored the microbial diversity of Hurood-grown LAB [1,10]. *Lactococcus* and *Lactobacillus* were found to be the main bacterial genera in Hurood samples collected in the Inner Mongolian region [11], while microbial communities of Hurood sampled from the Xinjiang Uygur Autonomous Region (Northwest China) were mainly composed of *Lactobacillus*, *Acetobacter*, and *Enterobacter* [12]. The microbial compositions of Hurood can vary significantly across different regions [13].

This work aims to explore the microbial diversities, key aroma compounds, and potential relationships between dominant species and key aroma compounds of Hurood sampled from three regions of Inner Mongolia, i.e., the Ulanqab League, Bayannur League, and Xilingol League. To do this, 16S rDNA sequencing was used to identify the dominant species in each region. SAFE–GC–MS was used to detect their aroma compounds. Finally, Pearson correlation analysis was adopted to assess the possible links between core microbial species and key aroma compounds.

2. Materials and Methods

2.1. Sample Collection

Hurood samples (n = 10) were collected from the Bayannur league (sample B1, B2, B3), Ulanqab league (sample U1, U2, U3, U4), and Xilingol league (sample X1, X2, X3) of Inner Mongolia in China—freshly made by local herdsmen, with basically the same production process (Figure 1). Hurood samples were vacuum-packed and transported by plane to the laboratory within 24 h at 4 °C. Hurood samples were cut into cubes (2 × 2 × 2 cm^3) and stored at −20 °C and 48% relative humidity.

2.2. Sample Preparation

Hurood samples were prepared according to the method described by Majcher [14], with some modifications. Each Hurood sample was cut into 2 cm^3 cubes, frozen in liquid nitrogen, and ground to obtain a homogenized sample. Then, 100 g of ground Hurood was transferred into an Erlenmeyer flask and extracted with diethyl ether (200 mL) at 350 rpm in an incubator shaker at room temperature (24 ± 2 °C) for 7 h. The residual substances were separated by centrifuging at 3000× *g* for 15 min at 4 °C. To further enrich the aroma compounds, solvent extraction was performed using a solvent-assisted flavor evaporation (SAFE) method.

2.3. Solvent-assisted Flavor Evaporation (SAFE)

Volatile compounds from Hurood solvent extract with a 10 μL internal standard solution containing 2-methyl-3-heptanone (154.3 μg/mL) and 3-methyl-pentanoic acid (1266.8 μg/mL) dissolved in diethyl ether were distilled using SAFE, as described by Majcher [15]. The SAFE apparatus was connected to a receiving tube and a waste tube. The glassware was then connected to an Edwards nxds6i rotary vane pump as the vacuum source. The SAFE apparatus was kept thermostated at 40 °C with a circulating water bath. Liquid nitrogen was poured into the cooling trap and the receiver flasks. Distillation was carried out for 2 h under vacuum (~10^{-4} Pa). After distillation, the distillate was dried over anhydrous sodium sulphate overnight and then concentrated to 1 mL under a gentle stream of high-purity nitrogen. The concentrated distillate was washed three times with 1 mL 0.5 M sodium bicarbonate and mixed thoroughly. After each wash, the bottom layer—the water phase of the distillate—was removed and collected in a separate test tube.

2.4. Gas Chromatography–Mass Spectrometry (GC–MS) Analysis

GC–MS analysis was carried out using an Agilent 7890/5977 instrument. Samples (1 μL) for MS analysis were injected using a CombiPal autosampler with a 10 μL direct injection syringe attached. Compounds were separated on DB-WAX (30 m × 0.25 mm i.d. × 0.25 μm film, Agilent Technologies, Santa Clara, CA, USA). The GC oven temperature program used was as follows: 35 °C for 3 min, raised at 4 °C/min to 100 °C held for 2 min, then 3 °C/min to 150 °C, and finally raised at 10 °C/min to 230 °C held for 3 min to confirm the cleanness of the column. The GC–MS transfer line temperature was maintained at 250 °C. The mass spectrometer was operated in the electron impact (EI) mode and the source temperature was set at 220 °C. The mass spectrometer operated in full scan mode from 33 to 450 amu. All the analyses were performed using 70 eV.

2.5. Gas Chromatography Olfactometry (GC–O) Analysis

An Agilent 7890B series GC coupled with a sniffing system (ODP 3 Gerstel, Mülheim, Germany) was used to locate odor-active components. The temperatures of the olfactory port and transfer line were kept at 230 °C and 250 °C, respectively. During the GC–O analysis, four experienced panelists recorded the aroma descriptor and the time when the aroma compound occurred. If 2 or more panelists detected the aroma, an odor-active location was identified.

2.6. Qualitative and Quantitative Methods

Identification of the constituents was performed by comparing mass spectra, linear retention indices (RI), and GC–O. Tentative identification was based on comparisons of their mass spectra with those of the known compounds from the standard NIST 14 library and the RI sourced from the NIST Standard Reference Database (the reference column type is DB-WAX). RI was calculated using the C_7–C_{33} n-alkane series under the same chromatographic conditions. The internal standard method was conducted to quantify the compounds.

2.7. Extraction of DNA from Cheese Samples

The genomic DNA of all bacteria present in the cheese samples was extracted using the OMEGA DNA isolation kit (Omega Bio-tek, Inc., Norcross, Georgia, USA), following the manufacturer's instructions, and triple DNA extraction was performed. The quality of the extracted DNA was checked by 1% agarose gel electrophoresis and spectrophotometry (optical density at 260 nm/280 nm ratio). The final DNA concentration was above 100 ng/μL and a 260 nm/280 nm ratio at 1.8–2.0. All extracted DNA samples were stored at −20 °C for further analysis.

2.8. Microbial Analysis

The V3 domains of the 16S rDNA were amplified by PCR (95 °C for 5 min; followed by 25 cycles of 95 °C for 30 s, 55 °C for 30 s, and 72 °C for 40 s with a final extension of 72 °C for 10 min) using primer pairs 515F (5′-GTGYCAGCMGCCGCGGTAA-3′) and 806R (5′-GGACTACHVGGGTWTCTAAT-3′) with a specific barcode. The purified amplicons were pooled at equimolar concentrations, and further paired-end sequencing was performed using an Illumina MiSeq instrument (Illumina Inc., San Diego, CA, USA)

2.9. Statistical Analysis

All statistical analyses were carried out with three biological replicates. The principal component analysis and Pearson correlation of the aroma compounds were analyzed using OriginPro 2021 64-bit (OriginPro Lab Corp., Northampton, MA, USA) and IBM SPSS Statistics 23.

3. Results and Discussion

3.1. Quantitation of Odor-Active Compounds and Associated ROAVs

The olfactory intensity of an aroma compound is related to both the compound concentration and odor threshold. The relative odor activity value (ROAV) [16] is the most commonly used quantification index of olfactory intensity. When the ROAV of an aroma compound is bigger than one, this compound will likely be the key contributor to the overall aroma [17,18]. In addition, ROAVs are also suitable indicators for the determination of aroma differences in foods [19].

To better study the key aroma contributors in Hurood, the original 75 aroma compounds detected by SAFE–GC–MS were refined by their ROAVs. A total number of 31 compounds with ROAV > 1 were identified as the key aroma compounds, including six carboxylic acids, five ketones, six aldehydes, one aromatic compound, four alcohols, six esters, and three sulfur compounds (Table 1).

Carboxylic acids are important, sometimes predominant, contributors to the overall aroma of cheeses [20–22]. They are also the precursors of other aroma substances such as esters, aldehydes, and methyl ketones [23]. Among all the carboxylic acids detected, acetic acid, hexanoic acid, and octanoic acid were the top three ROAVs. These volatile carboxylic acids are typical aroma compounds of cheese [2]. The average ROAVs of these three carboxylic acids showed significant differences between the three regions ($p < 0.05$). The ROAVs of octanoic acid were the highest (between 8.94 and 43.34) compared to the other acids. The ROAV of hexanoic acid in samples U1–U4 was higher than that of the others. In addition, the ROAV of acetic acid in the samples from the Xilingol League (X1–X3) were all less than one, indicating that acetic acid does not contribute much to the overall aroma of Hurood produced from this region. Besides this, decenoic acid—which has an unpleasant rancid odor—was detected only in samples from the Xilingol League (X1–X3). The means of the ROAVs of the acetic acid, hexanoic acid, and decanoic acid were significantly different between the Hurood samples of the three different regions ($p < 0.05$). These three acid compounds may be one of the characteristic aroma compounds that distinguish Hurood samples from the three different regions in terms of their aroma characteristics.

Esters, often with a low perception threshold, are another major contributor to cheese's aroma. They generally have sweet, fruity, and floral notes [23]. Esterification reactions can occur between short- to medium-chain fatty acids and primary and secondary alcohols derived from lactose fermentation and amino acid catabolism [24]. In cheese production, esters are generally produced during the re-maturation process. Matured cheeses, such as cheddar and blue cheeses, usually have high ester concentrations. As Hurood is a fresh cheese with no aging process, the ester content was relatively lower—ranging from 18.89 μg/kg to 1096.89 μg/kg for 10 samples (Table S1). Previous research on the aroma of traditional Chinese dairy products also found that ester contents in the milk fan are minor; only ethyl butyrate, isobutyl butyrate, ethyl octanoate, and ethyl hexanoate were reported to have an ROAV bigger than one [25].

Ketones are another group of important flavor compounds with a unique aroma and low sensory thresholds [26]. Five ketones with an ROAV greater than one were detected in 10 Hurood samples, including 2,3-butanedione, 3-hydroxy-2-butanone, 2-butanone, 2-nonanone, and 2-undecanone. Of these, 2,3-Butanedione (i.e., diacetyl) is one of the most important diketones for dairy products [20,21]—mainly produced by the fermentation of citrate. Diacetyl has a buttery and nutty note and has been identified as the key odorant of Camembert, Cheddar, and Emmental cheese [27,28]. Diacetyl was the only ketone compound with significant differences ($p < 0.05$) in the ROAV analyses among the Hurood samples from the three different regions. This may indicate that there is little difference in the aroma quality of Hurood samples from different regions in terms of ketone compounds.

Table 1. The relative odor activity values (ROAVs) of 31 key aroma compounds detected in the Hurood samples (* $p < 0.05$).

RT	Compound *	Sample										Threshold (μg/kg)	Aroma Description	Identification
		U1	U2	U3	U4	B1	B2	B3	X1	X2	X3			
	Carboxylic Acids													
21.71	Acetic acid *	1.85 ± 0.97	25.71 ± 21.89	23.49 ± 10.32	21.16 ± 15.62	1.23 ± 0.58	0.25 ± 0.17	2.83 ± 1.35	0.25 ± 0.15	1.35 ± 0.86	1.48 ± 1.06	700.00	sour, vinegar	MS, RI, O
28.83	Butanoic acid	0.97 ± 0.24	0.87 ± 0.06	2.66 ± 1.22	3.67 ± 2.54	1.34 ± 0.24	1.10 ± 1.00	0.90 ± 0.54	1.37 ± 1.23	0.22 ± 0.08	0.62 ± 0.34	410.00	pungent	MS, RI, O
36.82	Hexanoic acid *	11.41 ± 5.34	13.85 ± 2.54	21.66 ± 10.32	30.88 ± 14.63	5.76 ± 2.34	4.73 ± 1.32	9.79 ± 4.23	17.80 ± 6.35	4.58 ± 1.23	7.30 ± 2.64	225.00	sour, cheesy	MS, RI, O
42.00	Octanoic acid	21.53 ± 19.63	14.44 ± 9.45	33.12 ± 16.87	43.34 ± 24.78	9.55 ± 6.49	10.59 ± 3.28	10.21 ± 2.64	43.07 ± 26.59	16.23 ± 6.37	8.94 ± 6.33	250.00	rancid	MS, RI, O
44.84	n-Decanoic acid	3.21 ± 2.64	1.45 ± 0.64	1.12 ± 0.24	2.46 ± 1.23	0.59 ± 0.32	5.59 ± 1.36	4.09 ± 1.24	2.55 ± 1.33	1.97 ± 0.16	1.11 ± 0.54	2200.00	sour, cheesy	MS, RI, O
44.81	Decenoic acid *	-	-	-	-	-	-	-	149.21 ± 32.54	36.71 ± 11.34	54.30 ± 23.45	4.30	sour, cheesy, rancid	MS, RI, O
	Alcohols													
5.33	Ethanol	5.54 ± 1.26	8.06 ± 2.12	10.46 ± 3.26	12.08 ± 2.15	25.24 ± 10.26	4.29 ± 1.33	4.89 ± 1.24	7.53 ± 2.95	13.57 ± 6.49	7.90 ± 5.29	8.00	alcohol	MS, RI, O
38.89	Phenyl ethanol	1.18 ± 0.58	0.58 ± 0.42	1.46±	1.18 ± 1.09	6.36 ± 4.27	1.61 ± 1.26	0.26 ± 0.08	0.24 ± 0.19	0.39 ± 0.27	0.29 ± 0.03	15.00	fruity, fermented	MS, RI, O
18.79	Pentanol	16.76 ± 4.29	39.74 ± 10.48	190.92 ± 62.84	195.62 ± 98.49	1337.25 ± 526.34	25.03 ± 11.85	54.44 ± 12.65	-	-	-	0.15		MS, RI
23.57	2-ethyl-n-hexanol *	-	-	-	-	60.28±	51.15±	26.33±	-	-	-	0.80		MS, RI
	Ketones													
9.72	2-Butanone	2.52 ± 1.03	-	-	-	-	-	-	-	4.89 ± 2.49	-	1.30		MS, RI
10.49	2,3-Butanedione *	1.04 ± 0.26	-	-	-	2.98 ± 1.62	2.08 ± 1.05	2.22 ± 1.06	-	-	-	15.00	creamy, buttery	MS, RI, O
18.53	3-Hydroxy-2-butanone	44.60 ± 10.26	0.91 ± 0.26	2.70 ± 1.62	-	-	15.33 ± 3.45	13.38 ± 2.48	4.91 ± 1.26	-	0.02 ± 0.00	14.00	creamy	MS, RI, O
19.55	2-Nonanone	-	0.46 ± 0.32	-	0.38 ± 0.24	1.36 ± 1.49	-	0.38 ± 0.29	-	0.11 ± 0.00	-	420.00	fruity	MS, RI, O
27.72	2-Undecanone	0.04 ± 0.01	0.11 ± 0.02	0.24 ± 0.06	0.23 ± 0.08	2.16 ± 1.09	0.53 ± 0.37	0.05 ± 0.00	0.11 ± 0.04	0.25 ± 0.03	0.09 ± 0.01	82.00	fruity	MS, RI, O
	Aldehydes													
9.27	Hexanal	1.02 ± 0.23	0.07 ± 0.01	-	-	-	1.86 ± 1.03	0.02 ± 0.00	0.01 ± 0.00	0.93 ± 0.25	2.20 ± 1.06	50.00	grass	MS, RI, O
16.19	Octanal	1.51 ± 0.52	-	9.84 ± 2.45	-	-	1.26 ± 0.65	-	-	-	-		fruity	MS, RI, O

Table 1. *Cont.*

RT	Compound *	Sample										Threshold (µg/kg)	Aroma Description	Identification
		U1	U2	U3	U4	B1	B2	B3	X1	X2	X3			
19.66	Nonanal	-	0.87 ± 0.25	1.06 ± 0.26	1.08 ± 0.32	9.70 ± 2.65	0.22 ± 0.05	0.21 ± 0.03	0.06 ± 0.01	-	3.06 ± 1.62	42.00		MS, RI
22.13	Furfural	2.71 ± 0.89	1.28 ± 0.41	-	-	-	3.14 ± 1.29	0.85 ± 0.45	0.85 ± 0.32	3.54 ± 1.62	2.37 ± 0.85	2.80		MS, RI
24.46	Benzaldehyde	16.98 ± 4.16	14.66 ± 5.96	13.38 ± 2.85	22.75 ± 6.84	20.23 ± 4.85	11.35 ± 2.98	5.84 ± 1.54	1.73 ± 0.82	10.69 ± 6.12	19.00 ± 5.26	3.00	almond	MS, RI, O
24.99	2-Nonenal	75.90 ± 15.26	-	-	-	-	119.30 ± 20.84	-	-	-	-	0.10	Roast	MS, RI, O
	Esters													
6.22	Ethyl acetate	1.51 ± 0.85	16.27 ± 1.65	12.99 ± 3.48	19.80 ± 5.92	819.31 ± 105.46	24.93 ± 2.16	37.47 ± 11.85	8.39 ± 2.06	2.16 ± 1.25	22.08 ± 0.58	38.00		MS, RI
11.23	Methyl hexanoate	-	-	-	-	-	-	96.71 ± 36.15	22.57 ± 12.48	25.14 ± 10.65	50.43 ± 8.45	0.07		MS, RI
13.44	Ethyl butyrate	16.42 ± 0.58	-	-	-	-	-	316.60 ± 107.40	268.30 ± 98.45	41.32 ± 12.45	-	0.05		MS, RI
14.39	Ethyl hexanoate	0.06 ± 0.01	0.29 ± 0.12	5.33 ± 1.52	5.28 ± 2.16	6.45 ± 2.12	0.49 ± 0.16	0.67 ± 0.05	2.94 ± 1.20	0.06 ± 0.02	0.37 ± 0.01	0.88	fruity, fermented	MS, RI, O
21.31	Ethyl octanoate	0.04 ± 0.01	0.42 ± 0.12	2.67 ± 1.26	3.64 ± 1.41	4.02 ± 1.26	-	-	6.99 ± 1.24	0.11 ± 0.06	-	220.00	flower, fruity	MS, RI, O
41.72	Ethyl decanoate	0.03 ± 0.02	0.20 ± 0.14	1.23 ± 0.85		2.16 ± 1.26	0.03 ± 0.02	0.08 ± 0.01	2.77 ± 1.24	0.03 ± 0.01	-	180.00	apple	MS, RI, O
	Sulfur compounds													
14.32	Dimethyl disulfide	4.76 ± 1.26	-	-	-	-	5.95 ± 2.58	4.76 ± 3.49	-	2.38 ± 1.32	-	0.0084	rotten eggs	MS, RI, O
35.26	Dimethyl trisulfide	-	-	-	-	-	12.50 ± 3.26	-	-	-	-	0.0080		MS, RI
4.23	Methyl mercaptan	-	-	-	-	-	-	-	-	28.32 ± 5.64	-	0.28		MS, RI
	Aromatic hydrocarbon													
15.01	Styrene	-	-	0.16 ± 0.05	-	2.23 ± 0.51	0.12 ± 0.03	-	0.07 ± 0.05	-	1.02 ± 0.32	26.40	aromatics, reagents	MS, RI, O

Method of identification: MS = mass spectrum comparison using NIST 14; RI = retention index in agreement with literature value (the reference column type is DB-WAX); O = odor.

Aldehydes can contribute to freshness and floral aromas. Linear aldehydes, such as hexanal, heptanal, octanal, and nonanal, are commonly found in cheeses. Branched-chain aldehydes can originate from the degradation of amino acids via enzymatic processes [29] or non-enzymatic reactions (Strecker degradation). The chemical properties of aldehydes are relatively active; they can easily be reduced into alcohols or be oxidized to their corresponding acids. Such an unstable feature coincides with the observed low content of aldehydes (25.34–2394.56 μg/kg) in the 10 Hurood samples. Nevertheless, the low odor thresholds of aldehydes—such as furfural (threshold = 2.8 μg/kg), benzaldehyde (threshold = 3 μg/kg), and 2-nonenal (threshold = 0.1 μg/kg)—make them still important contributors to the overall aroma.

Other groups of aroma active compounds—e.g., sulfur compounds, alcohols, and aromatic heterocyclic compounds—were also detected. The content of sulfur compounds (0.04–7.1 μg/kg) in the fresh Hurood samples was low. The ROAVs of sulfur compounds in samples B1–B3 were the highest. Dimethyl trisulfide was only detected in sample B2. Pentanol and 2-ethyl-n-hexanol were not detected in samples X1–X3. There was a significant difference in the ROAVs of 2-ethyl-n-hexanol between the Hurood samples from the three different regions ($p < 0.05$). In summary, the volatility and ROAV values of these compounds were relatively low, and they could not be or were only slightly recognized by the GC–O experiment, which may indicate that their impact on overall aroma characteristics is relatively small.

3.2. Analysis of Bacterial Communities

We carried out a microbiological analysis of the samples from the different regions by high-throughput sequencing (16S rDNA), where samples U1–U4 (U), B1–B3 (B), and X1–X3 (X) were collected from the Ulanqab league, Bayannur league, and Xilingol league, respectively (Figure 2a).

A total of 79 species were detected at the level of bacterial community species in the Hurood. As shown in Figure 2b, the top 10 species in the species level were *Lactococcus lactis, Lactobacillus delbrueckii, Lactobacillus psittaci, Lactobacillus helveticus, Lactobacillus crispatus, Lactobacillus paracasei, Lactobacillus amylovorus, Leuconostoc mesenteroides,* and *Leuconostoc pseudomesenteroides.* Among them, the dominant species in the samples collected from the Ulanqab league (i.e., U) were *Lactobacillus delbrueckii, Lactococcus lactis,* and *Leuconostoc mesenteroides.* The dominant species in the samples collected from the Bayannur league (i.e., B) were *Lactobacillus delbrueckii, Lactobacillus amylovorus. Lactobacillus psittaci,* and *Leuconostoc mesenteroides.* The dominant species in the samples collected from the Xilingol league (i.e., X) were *Lactobacillus amylovorus, Leuconostoc mesenteroides, Lactobacillus helveticus, Lactobacillus psittaci,* and *Lactobacillus delbrueckii. Lactobacillus delbrueckii* is a dominant bacterium commonly found in three regions. *Lactobacillus delbrueckii* is an important microorganism for cheese or yogurt production and is widely used as a starter for yogurt and cheese worldwide, playing important roles in acid production and flavor formation in fermented dairy products [30]. In addition, there were certain differences in the dominant bacteria among the three regions, which may be one of the direct reasons for the differences in the flavor of the Hurood samples in the different regions.

The α-diversity index shows the differences in microorganisms within a group, while the β-diversity index can display microbial differences between groups. To further investigate the microbial diversities in the Hurood samples, an analysis of the microbiota α-diversity (indicated by the Shannon index) indices suggests that the microbial diversity of the Bayannur League (B) was higher than that of the samples from the other two regions—suggesting a higher microbial richness and evenness. In addition, based on the weighted uniFrac distance metric, the PCA of the microorganisms' β-diversity in the Hurood samples from the different regions is presented in Figure 2d; it shows the distribution map for the first two principal components determined by PCA, which describe 48.4% and 35.1% of the accumulative variance contribution rate. A visualization of the data was also obtained; it showed that the microbial diversity of the Xilingol League (X) was significantly different to the other two regions ($p < 0.05$).

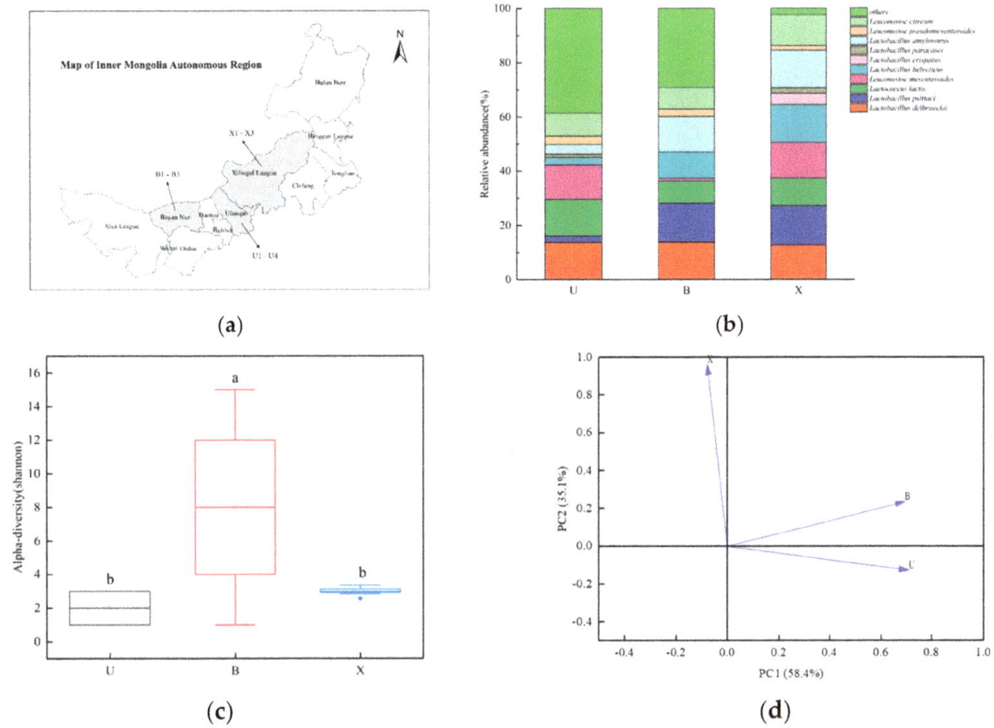

Figure 2. (**a**) An overview of the sampling location of the Hurood studied in this work. Samples U1–U4 (U) were obtained in the Ulanqab league; samples B1–B3 (B) were obtained in the Bayannur; samples X1–X3 (X) were obtained in the Xilingol league; (**b**). The relative abundance of bacterial communities at the species level of the Hurood; (**c**). Changes in bacteria community diversity of the Hurood (within the community). Based on the Shannon index of the Tukey test of α-Diversity (Different letters corresponding to statistically significant differences ($p < 0.05$)); (**d**). Changes in the bacteria community diversity of the Hurood, based on weighted uniFrac distance metric β-Diversity.

3.3. Correlation Analysis of Dominant Species and Key Aroma Compounds

As illustrated in Figure 3, the Pearson correlation coefficients were calculated to better understand the potential relationships between key aroma compounds and the dominant microbial species found in the Hurood samples. From this, 13 significant pair-wise correlations ($|r| > 0.6$, $p < 0.05$) were captured. Nine bacterial species were noted to be significantly correlated with the aroma properties of the Hurood ($p < 0.05$). Among these, *Lactobacillus paracasei*, *Lactobacillus crispatus*, and *Leuconostoc citreum* were significantly correlated with two or more key aroma compounds ($p < 0.05$), such as methyl hexanoate or ethyl butyrate.

Lactobacillus paracasei was correlated with the highest number (three) of key aroma compounds in the Hurood. Specifically, *Lactobacillus paracasei* was positively correlated with methyl mercaptan ($r = 1.0$, $p < 0.001$) and 2-butanone ($r = 0.87$, $p < 0.05$), while it was negatively correlated with butanoic acid ($r = -0.79$, $p < 0.05$). *Lactobacillus paracasei* is a probiotic that is widely present in the human gut and in fermented foods such as cheeses and pickles. It can enhance secondary proteolysis in soft and semi-hard cheese as well as the production of aroma compounds in mini-soft cheese [31]. Barouei et al. found that

Lactobacillus paracasei could improve the sensory quality of cheeses; in this study, after 8 weeks of ripening, cheese that was supplemented with *Lactobacillus paracasei* received significantly higher scores for taste, aroma, texture, and overall preference than the control cheese [32,33].

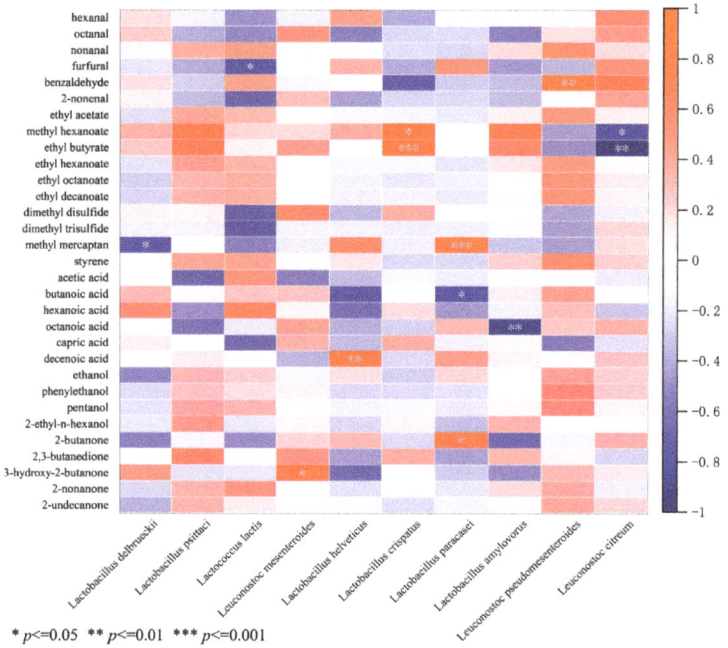

Figure 3. The correlation between core microbes and key aroma compounds in the Hurood samples. Blue represents negative correlations between aroma compounds and microorganisms, red represents positive correlations between aroma compounds and microorganisms.

Lactobacillus crispatus was positively correlated with methyl hexanoate (r = 0.86, $p < 0.05$) and ethyl butyrate (r = 0.99, $p < 0.001$). *Lactobacillus crispatus* is considered a probiotic [34]; it may help the body break down food, absorb nutrients, and fight off harmful organisms that might cause diseases [33]. These bacteria are sometimes added to fermented foods such as cheese and are also found in dietary supplements [35]. *Leuconostoc citreum* possesses a negative correlation with methyl hexanoate (r = −0.79, $p < 0.05$) and ethyl butyrate (r = −0.93, $p < 0.01$), which indicates that the growth of *Leuconostoc citreum* is in opposition to the formation and accumulation of ester compounds in Hurood. However, *Leuconostoc citreum* is closely related to food fermentation. Previous studies have shown that *Leuconostoc citreum* is one of the most prevalent LAB found in the manufacturing process of kimchi [36]. Genes in *Leuconostoc citreum* that are likely involved in fermentation and its probiotic effects have been revealed by a complete genome sequencing study [37].

In summary, *Lactobacillus paracasei*, *Lactobacillus crispatus*, and *Leuconostoc citreum* were identified as the core functional microorganisms in Hurood due to their significant correlation with key aroma compounds. In the future, more in-depth studies are needed to validate this preliminary inference. For instance, mono-culture fermentation tests should be conducted to verify the role of each strain on aroma formation.

4. Conclusions

This study characterized the aroma profiles, microbial diversities, and presumptive correlations between the key aroma compounds and dominant microorganisms of Hurood cheeses sampled from three distinct geographical origins. There were significant differences in LAB between different regions ($p < 0.05$) These dominant microbes were found to be significantly correlated with key aroma compounds such as methyl hexanoate and ethyl butyrate. *Lactobacillus paracasei*, *Lactobacillus crispatus*, and *Leuconostoc citreum* were identified as the core functional microorganisms of Hurood based on their P values and Pearson correlation coefficients. This work provides a theoretical basis of how microbial diversity can affect the final aroma profiles of Hurood cheeses, with special attention paid to their different production locations, paving the way to improving the overall aromas of traditional Chinese cheese products via the modification of the composition of functional LAB strains.

Supplementary Materials: The following supporting information can be downloaded at: https://www.mdpi.com/article/10.3390/fermentation9070670/s1, Table S1: GC–MS analysis results of aroma compounds in different Hurood samples. Table S2: Correlation coefficient r.

Author Contributions: Y.W.: Formal analysis, Investigation, Methodology, Writing—original draft and editing. H.Z.: Methodology, Writing—review and editing, Supervision. Y.C.: Supervision. S.W.: Supervision, Validation, Methodology, Writing—review and editing. B.W.: Supervision, Validation, Methodology, Writing—review and editing, Funding acquisition, Project administration. All authors have read and agreed to the published version of the manuscript.

Funding: This research was funded by the National Natural Science Foundation of China (Bei Wang, project No. 32072345; Beijing) and the Cultivation Project of Double First-Class Disciplines of Food Science and Engineering, Beijing Technology & Business University (Bei Wang, project No. BTBUYXTD202205).

Institutional Review Board Statement: Not applicable.

Informed Consent Statement: Not applicable.

Data Availability Statement: Data is contained within the article.

Conflicts of Interest: The authors declare no conflict of interest.

References

1. Zhang, J.; Zhang, X.; Zhang, L.; Zhao, Y.; Niu, C.; Yang, Z.; Li, S. Potential probiotic characterization of Lactobacillus plantarum strains isolated from Inner Mongolia "Hurood" cheese. *J. Microbiol. Biotech.* **2014**, *24*, 225–235. [CrossRef] [PubMed]
2. Fox, P.F.; Guinee, T.P.; Cogan, T.M.; McSweeney, P. *Biochemistry of Cheese Ripening*; Springer: New York, NY, USA, 2017. [CrossRef]
3. Hailu, Y.; Hansen, E.B.; Seifu, E.; Eshetu, M.; Petersen, M.A.; Lametsch, R.; Rattray, F.; Ipsen, R. Rheological and sensory properties and aroma compounds formed during ripening of soft brined cheese made from camel milk. *Int. Dairy J.* **2018**, *81*, S0958694618300311. [CrossRef]
4. Chen, Y.; Ning, H.E.; Yi-Jun, M.U.; Liu, Q.Y.; Qin, X.J.; Yun, M.A.; Xing, C.; Bai, L.J. Effect of Lactobacillus helveticus on the quality of Cheddar cheese. *Food Sci. Tech.-Braz.* **2018**, *39*, 122–125+131. [CrossRef]
5. Wei, Y.; Xiang, W.U.; Zhou, H.; Baocai, X.U.; Xinfu, L.I. Changes of Microbial Flora during the Fermentation of Salami and Their Effect on Flavor. *Meat Res.* **2018**, *32*, 48–54.
6. Chen, C.; Huang, K.; Yu, H.; Tian, H. The diversity of microbial communities in Chinese milk fan and their effects on volatile organic compound profiles. *J. Dairy Sci.* **2020**, *104*, 2581–2593. [CrossRef]
7. Jiang, Y.; Li, N.; Wang, Q.; Liu, Z.; Chen, W. Microbial diversity and volatile profile of traditional fermented yak milk. *J. Dairy Sci.* **2019**, *103*, 87–97. [CrossRef]
8. Gao, X.; Zhang, L.; Bo, L.I. Effect of fermentation starter addition on the flavor of Mongolian hard cheese. *China Brew.* **2015**, *34*, 77–81. [CrossRef]
9. Esen, Y.; Cetin, B. Bacterial and yeast microbial diversity of the ripened traditional middle east surk cheese. *Int. Dairy J.* **2021**, *117*, 105004. [CrossRef]
10. Hasiqimuge; Zhang, H.M.; Fang, X.; Zhu, J.J.; Chen, Y.F.; Mei, Y.A. Study of the production process and the nutrition analysis of the traditional cheese from different regions of Inner Mongolia. *China Dairy Ind.* **2016**, *44*, 35–37.
11. Zhu, L.; Hou, Z.; Hu, X.; Liu, X.; Dai, T.; Wang, X.; Zeng, C.; Wang, Y.; Wang, C.; Yang, S.; et al. Genomic and Metabolic Features of an Unexpectedly Predominant, Thermophilic, Assistant Starter Microorganism, Thermus thermophilus, in Chinese Inner Mongolian Cheese. *Foods* **2021**, *10*, 2962. [CrossRef]
12. Lei, Y.; Zhou, Y.; Shan, C.; Cai, W.; Song, K.; Guo, Z. Study on the core bacterial microbiota of dairy toufu based on high-throughput sequencing technology. *China Dairy Ind.* **2020**, *48*, 9–13. [CrossRef]

13. Gao, M.L.; Hou, H.M.; Teng, X.X.; Zhu, Y.L.; Hao, H.S.; Zhang, G.L. Microbial diversity in raw milk and traditional fermented dairy products (Hurood cheese and Jueke) from Inner Mongolia, China. *Genet. Mol. Res.* **2017**, *16*, gmr16019451. [CrossRef] [PubMed]
14. Zz, A.; Chao, N.; Long, C.D.; Yz, A.; Ge, Y.A.; Chao, W.A.; Nc, D.; Zz, D.; Sl, A. Impacts of manufacture processes and geographical regions on the microbial profile of traditional Chinese cheeses. *Food Res. Int.* **2021**, *148*, 110600. [CrossRef]
15. Majcher, M.A.; Jeleń, H. Key Odorants of Oscypek, a Traditional Polish Ewe's Milk Cheese. *J. Agric. Food Chem.* **2011**, *59*, 4932–4937. [CrossRef]
16. Zhao, D.; Hu, J.; Chen, W. Analysis of the relationship between microorganisms and flavour development in dry-cured grass carp by high-throughput sequencing, volatile flavour analysis and metabolomics. *Food Chem.* **2022**, *368*, 130889. [CrossRef]
17. Engel, W.; Bahr, W.; Schieberle, P. Solvent assisted flavour evaporation—A new and versatile technique for the careful and direct isolation of aroma compounds from complex food matrices. *Eur. Food Res. Technol.* **1999**, *209*, 237–241. [CrossRef]
18. Trabue, S.L.; Anhalt, J.C.; Zahn, J.A. Bias of Tedlar Bags in the Measurement of Agricultural Odorants. *J. Environ. Qual.* **2006**, *35*, 1668–1677. [CrossRef]
19. Friedrich, J.E.; Acree, T.E. Gas Chromatography Olfactometry (GC/O) of Dairy Products. *Int. Dairy J.* **1998**, *8*, 235–241. [CrossRef]
20. Wang, J.; Yang, Z.J.; Xu, L.Y.; Wang, B.; Tan, L. Key aroma compounds identified in Cheddar cheese with different ripening times by aroma extract dilution analysis, odor activity value, aroma recombination, and omission. *J. Dairy Sci.* **2020**, *104*, 1576–1590. [CrossRef]
21. Wang, B.; Wang, J.; Xu, L.Y.; Zhang, J.H.; Ai, N.S.; Cao, Y.P. Characterization of the key odorants in kurut with aroma recombination and omission studies. *J. Dairy Sci.* **2020**, *103*, 4164–4173. [CrossRef]
22. Wang, J.; Yang, Z.J.; Wang, Y.D.; Cao, Y.P.; Wang, B.; Liu, Y. The key aroma compounds and sensory characteristics of commercial Cheddar cheeses. *J. Dairy Sci.* **2021**, *104*, 7555–7571. [CrossRef] [PubMed]
23. Delgado, F.J.; Gonzalez-Crespo, J.; Cava, R.; Ramirez, R. Formation of the aroma of a raw goat milk cheese during maturation analysed by SPME–GC–MS. *Food Chem.* **2011**, *129*, 1156–1163. [CrossRef] [PubMed]
24. Liu, S.Q.; Holland, R.; Crow, V.L. Esters and their biosynthesis in fermented dairy products: A review. *Int. Dairy J.* **2004**, *14*, 923–945. [CrossRef]
25. Kruis, A.J.; Bohnenkamp, A.C.; Patinios, C.; Nuland, Y.; Weusthuis, R.A. Microbial production of short and medium chain esters: Enzymes, pathways, and applications. *Biotechnol. Adv.* **2019**, *37*, 107407. [CrossRef] [PubMed]
26. Tian, H.; Xu, X.; Ch En, C.; Yu, H. Flavoromics approach to identifying the key aroma compounds in traditional Chinese milk fan—ScienceDirect. *J. Dairy Sci.* **2019**, *102*, 9639–9650. [CrossRef]
27. Rincon-Delgadillo, M.I.; Lopez-Hernandez, A.; Wijaya, I.; Rankin, S.A. Diacetyl levels and volatile profiles of commercial starter distillates and selected dairy foods. *J. Dairy Sci.* **2012**, *95*, 1128–1139. [CrossRef]
28. Midje, D.L.; Bastian, E.D.; Morris, H.A.; Martin, F.B.; Bridgeman, T.; Vickers, Z.M. Flavor enhancement of reduced fat cheddar cheese using an integrated culturing system. *J. Agric. Food Chem.* **2000**, *48*, 1630–1636. [CrossRef]
29. Papaioannou, G.; Kosma, I.; Badeka, A.V.; Kontominas, M.G. Profile of Volatile Compounds in Dessert Yogurts Prepared from Cow and Goat Milk, Using Different Starter Cultures and Probiotics. *Foods* **2021**, *10*, 3153. [CrossRef]
30. Tsuchihashi, H.; Ichikawa, A.; Takeda, M.; Koizumi, A.; Mizoguchi, C.; Ishida, T.; Kimura, K. Genetic diversity of Lactobacillus delbrueckii isolated from raw milk in Hokkaido, Japan. *J. Dairy Sci.* **2021**, *105*, 2082–2093. [CrossRef]
31. Baptista, I.; Santos, M.; Rudnitskaya, A.; Saraiva, J.A.; Almeida, A.; Rocha, S.M. A comprehensive look into the volatile exometabolome of enteroxic and non-enterotoxic Staphylococcus aureus strains. *Int. J. Biochem. Cell Biol.* **2019**, *108*, 40–50. [CrossRef]
32. Poveda, J.M.; Chicón, R.; Cabezas, L. Biogenic amine content and proteolysis in Manchego cheese manufactured with Lactobacillus paracasei subsp. paracasei as adjunct and other autochthonous strains as starters. *Int. Dairy J.* **2015**, *47*, 94–101. [CrossRef]
33. Barouei, J.; Karbassi, A.; Ghoddusi, H.B.; Mortazavi, A.; Moussavi, M. Impact of native Lactobacillus paracasei subsp. paracasei and Pediococcus spp. as adjunct cultures on sensory quality of Iranian white brined cheese. *Int. Dairy J.* **2011**, *64*, 526–535. [CrossRef]
34. Peralta, G.H.; Wolf, I.V.; Bergamini, C.V.; Perotti, M.A.C.; Hynes, E.R. Evaluation of volatile compounds produced byLactobacillus paracaseiI90 in a hard-cooked cheese model using solid-phase microextraction. *Dairy Sci. Technol.* **2014**, *94*, 73–81. [CrossRef]
35. Li, T.; Liu, Z.; Zhang, X.; Chen, X.; Wang, S. Local Probiotic Lactobacillus crispatus and Lactobacillus delbrueckii Exhibit Strong Antifungal Effects Against Vulvovaginal Candidiasis in a Rat Model. *Front. Microbiol.* **2019**, *10*, 1033. [CrossRef] [PubMed]
36. Muhammad, I.; Ahmad, A.A.; Shah, T. Health Promoting and Disease Preventing Properties of Probiotics with Special Reference to Lactobacillus: A Review. *J. Probiotics Health* **2018**, *6*, 1. [CrossRef]
37. Patrignani, F.; Siroli, L.; Parolin, C.; Serrazanetti, D.I.; Vitali, B.; Lanciotti, R.; Shukla, P. Use of Lactobacillus crispatus to produce a probiotic cheese as potential gender food for preventing gynaecological infections. *PLoS ONE* **2019**, *14*, e0208906. [CrossRef]

Disclaimer/Publisher's Note: The statements, opinions and data contained in all publications are solely those of the individual author(s) and contributor(s) and not of MDPI and/or the editor(s). MDPI and/or the editor(s) disclaim responsibility for any injury to people or property resulting from any ideas, methods, instructions or products referred to in the content.

foods

MDPI

Article

Discriminant Analysis of Aroma Differences between Cow Milk Powder and Special Milk Powder (Donkey, Camel, and Horse Milk Powder) in Xinjiang Based on GC-IMS and Multivariate Statistical Methods

Yongzhen Gou [1,2], Yaping Han [1,2], Jie Li [1,2], Xiyue Niu [1,2], Guocai Ma [3] and Qian Xu [1,2,*]

[1] College of Food Science and Engineering, Tarim University, Alar 843300, China
[2] Corps Key Laboratory of Deep Processing of Specialty Agricultural Products in Southern Xinjiang, Alar 843300, China
[3] Instrumental Analysis Center, Tarim University, Alar 843300, China
* Correspondence: xuqiantaru@126.com

Abstract: In order to explore the aromatic differences between Xinjiang cow milk powder and specialty milk powder (donkey, camel, and horse milk powder), Gas Chromatography-Ion Mobility Spectrometry (GC-IMS) analysis was employed to investigate the volatile compounds in these four types of milk powders. A total of 61 volatile substances were detected, with ketones, aldehydes, and alcohols being the primary flavor components in the milk powders. While the aromatic components of the different milk powders showed similarities in terms of types, there were significant differences in their concentrations, exhibiting distinct characteristics for each type. The Partial Least Squares Discriminant Analysis (PLS-DA) showed that there were 15, 14, and 23 volatile compounds that could be used for discrimination of cow milk powder against specialty milk powders, respectively. And it was validated by Receiver Operating Characteristic (ROC) analysis, and finally, 8, 6, and 19 volatile compounds were identified as valid differential marker substances. To facilitate visual discrimination between the different milk powders, we established GC-IMS fingerprint spectra based on the final discriminant markers. These studies provide theoretical guidance for the application of volatile compounds to discriminate adulteration of milk powder marketed in Xinjiang.

Keywords: milk powder; finger printing; volatile compounds; GC-IMS; PLS-DA; ROC analysis

1. Introduction

Special milk refers to the milk obtained from milch animals other than cows, such as camel milk, horse milk, donkey milk, etc., which possess unique flavors and nutritional value [1]. Xinjiang, as a prominent province in animal husbandry in China, holds a significant advantage in livestock numbers and milk production [2]. In recent years, with the rapid development of the dairy industry and continuous advancements in dairy science and technology, scholars from both domestic and international domains have delved deeper into the nutritional value and physiological functions of special milk, contributing to the increasing recognition of the immense potential of specialty milks like donkey milk, horse milk, and camel milk. These specialty milks have found applications in disease adjunctive therapy, nutritional health, and food processing [3–5], boasting an economic value far superior to that of cow milk. However, due to the relatively low yield of specialty milks compared to cow milk and the significant impact of seasonal fluctuations on milk supply [6], along with the challenges in controlling milk source quality [7], these factors have made specialty milks and their dairy products susceptible targets for intentional adulteration driven by economic interests. As a result, incidents of adulteration motivated by financial gains occur with some frequency [8].

Milk powder serves as the most common means for preserving and transporting dairy products, carries high added value, and is widely produced and sold. Among the various fraudulent practices, the most prevalent and easily achievable one is adulterating high-value specialty milk powder with lower-value cow milk powder. This adulteration not only leads to economic losses for consumers but also poses potential health risks, especially for individuals allergic to cow milk [9]. Unfortunately, there are currently no clear standards for detecting adulteration in specialty milk powders [10], and market supervision lacks reliable criteria and testing methods. Furthermore, there is a paucity of research concerning the detection of adulteration in specialty milk powders. Thus, there is an urgent need to address the issue of adulteration in Xinjiang's specialty milk powders by establishing screening, analytical testing methods, and corresponding standards to safeguard the interests of general consumers.

Currently, the most common technique used to discriminate between different milk powder sources is based on DNA sequencing. Species-specific DNA sequences have been widely applied in polymerase chain reaction (PCR) detection, enabling the identification of animal milk sources and thereby detecting adulteration in milk [11,12]. However, PCR methods involve intricate DNA separation procedures and typically do not yield quantitative information. Specialty milk powders and cow milk powders share similarities in appearance and major chemical components, but their physiological metabolic lactation activities vary among different animal species [13]. Additionally, their milk powder processing methods also differ, wherein volatile compounds can serve as a basis for distinguishing specialty milk powders from cow milk powder. Compared to DNA sequencing-based methods, studying the differences in volatile compounds between cow milk powder and specialty milk powder proves more practical, convenient, and rapid. Commonly employed techniques include Gas Chromatography-Mass Spectrometry (GC-MS) and Gas Chromatography-Ion Mobility Spectrometry (GC-IMS).

GC-IMS is an emerging detection technique in food aroma analysis, known for its high sensitivity, strong convenience, short analysis time, and data visualization advantages, making it widely applied in food quality control and traceability [14]. Gomez et al. [15] non-destructively analyzed volatile compounds in spoiled and unspoiled ham using GC-IMS. They established a volatile compound model to classify the freshness of ham through PLS-DA. He et al. [16] used a combination of Gas Chromatography-Mass Spectrometry (GC-MS) and GC-IMS to analyze volatile compounds during different distillation stages of white spirit. By employing PLS-DA, they identified four aroma-active markers and validated them using ROC analysis. Ethyl butyrate and ethyl hexanoate were ultimately determined as the most effective discriminant markers for distinguishing different distillation stages. Feng et al. [17] compared the influence of two drying methods (spray drying and freeze drying) on the volatile organic compounds of yak milk powder using GC-IMS and Principal Component Analysis (PCA). They established characteristic fingerprint spectra for yak milk powder under different drying methods.

The literature suggests that the application of GC-IMS in milk powder analysis is feasible and is still in its preliminary stage. There are scarce reports on the use of GC-IMS for discriminating between different types of milk powders. Therefore, this study innovatively employs GC-IMS to analyze and investigate the variations in volatile compounds between cow milk powder and specialty milk powder. The research aims to utilize PLS-DA to identify volatile compounds that can effectively discriminate between cow milk powder and specialty milk powder. The selected volatile compounds will undergo validation and screening through ROC analysis. Subsequently, the final chosen volatile compounds will be used to establish a rapid identification fingerprint spectrum and provide a theoretical basis for the application of volatile compounds to discriminate against the adulteration of commercially available milk powder in Xinjiang.

2. Materials and Methods

2.1. Preparation and Collection of Milk Powder Samples

Samples of cow milk, camel milk, horse milk, and donkey milk were collected by 2022 Xinnong Enterprises (Aral, China) in February, June, and October on behalf of the three batches of collection, and three samples were collected in each batch, and the collected milk samples were then transported to the laboratory at −80 °C in a liquid nitrogen tank to be preserved and stored at −80 °C, and experiments were carried out after all the samples were collected.

In the laboratory of Xinnong Enterprises (Aral, China), experiments were carried out to produce milk powder. This aimed to simulate the market products of milk powder production companies in Xinjiang. Cow and camel milk powder were produced using spray drying, while horse and donkey milk powders were produced through freeze-drying on the market. Drying experiments were conducted on different milk types, as detailed in Table 1.

Table 1. Information on different milk powder samples.

Description of Sample	Abbreviation	Sampling Site	Processing Methods
Cow milk	M	Aksu Xinjiang	Spray drying
Donkey milk	D	Hetian Xinjiang	Freeze drying
Camel milk	C	Hetian Xinjiang	Spray drying
Horse milk	H	Altay Xinjiang	Freeze drying

Spray drying: The spray drying equipment (Oumeng Laboratory specialized small spray dryer OM-1500A, Shanghai, China) operated with the following parameters: inlet air temperature of 120 °C, an outlet air temperature of 65 °C, and a flow rate of 5 mL/min. The milk powder samples were spray-dried until their moisture content reached 5–6%.

Freeze drying: Firstly, the milk samples were pre-frozen and then placed in a −20 °C freezer for 6 h until they were completely frozen. After the initial freezing, the samples were transferred to a freeze dryer (Jinan Junde FD-503 box-type vacuum freeze dryer, Jinan, China) with the following settings: vacuum at 20 Pa, sublimation temperature at 40 °C, and freeze trap temperature at −40 °C. The milk powder's moisture content was reduced to 5–6%. Before analysis, each sample was stored in dry containers.

2.2. GC-IMS Analysis

2.0 g of sample was placed into a 20 mL headspace vial (Agilent, Santa Clara, CA, USA) and incubated at 80 °C for 20 min while stirring at 500 rpm. After the incubation was completed, the temperature of the injection needle was 85 °C, and 500 μL of gas from the headspace vial was automatically aspirated into the GC-IMS instrument (FlavorSpec, GAS, Dortmund, Germany) without shunting. The samples were subjected to compound separation using a WAT—Wax capillary column (30 m × 0.53 mm, 1 μm) at a constant temperature of 60 °C, thereby standardizing the entire experimental procedure. The flow rate was controlled as follows: initially held at a constant 2 mL/min for 2 min, increased to 10 mL/min over 8 min, and then increased to 100 mL/min over 10 min, held for 10 min. The drift tube was 98.0 mm long, the operating temperature was 45 °C, and the constant voltage was 5 kV. The entire analysis took 45 min. All analyses were repeated three times. The obtained GC-IMS data were observed using Laboratory Analytical Viewer (G.A.S, Dortmund, Germany). The Reporter plug-in was used to directly compare the 2D top view and difference plots. The retention indices (RI) of volatile compounds were calculated using ortho-ketones C4~C9 (Sinopharm Chemical Reagent Co., Ltd., Beijing, China) as external standards. The volatile compounds were identified by comparing the retention index (RI) with the normalized drift time (RIP) in the IMS database (G.A.S., Dortmund, Germany). Fingerprints were made and compared using the Gallery Plot plugin.

2.3. Statistical Analysis

The key differential components between different milk powders were explored using SIMCA-P 13.0 software (Umetrics, Stockholm, Sweden) with Partial Least Squares Discriminant Analysis (PLS-DA). To evaluate the differences in volatile compounds among the various milk powders, One-Way Analysis of Variance (ANOVA) was conducted using SPSS Statistics 20.0 software (SPSS Inc., Chicago, IL, USA). The significance level was set at $p < 0.05$. The identification and validation of discriminant volatile compounds were carried out through ROC analysis.

3. Results and Analysis

3.1. GC-IMS Profiling of Different Milk Powders

Currently, GC-MS is one of the most commonly used techniques for analyzing volatile compounds. However, the requirement for a vacuum environment and its bulky size limit its portability [18]. On the other hand, GC-IMS, as an emerging instrument analysis technology, effectively separates ions based on their migration velocity at atmospheric pressure. It offers advantages such as high sensitivity, low environmental requirements, portability, and data visualization. Consequently, GC-IMS finds widespread application in food classification and quality control [19].

In this study, gas chromatography-ion mobility spectrometry (GC-IMS) analysis was conducted using the FlavorSpec instrument to generate data, as shown in Figure 1a. The figure presents a 2D topographical map obtained from a 3D spectrum scan, comparing the volatile compounds in different milk powders. The X-axis represents ion migration time, and the Y-axis represents the retention time of ions in gas chromatography. To eliminate the influence of capillary column temperature and carrier gas flow rate, the response ion peak at X = 1.0 was normalized, as indicated by the red vertical line in the plot. The color represents the signal intensity of individual compounds, with red indicating a higher concentration and blue indicating a lower concentration. Most of the signals are distributed within the retention time range of 100–1600 s, and many peaks show similar distributions within the samples, but with varying intensities among the same type of samples. Additionally, there are evident differences in peak signals among different samples, indicating variations in the content of volatile compounds in different milk powders. Particularly, there are distinct differences in the drying methods, with samples D and H (freeze-dried) having more peaks compared to samples M and C (spray-dried). This discrepancy may be attributed to the production of some volatile substances during the vacuum freezing process [20], or it could be due to the pre-freezing of water before drying, potentially reducing the diffusion of surface volatile compounds. Spray drying may also cause changes and the loss of some volatile compounds due to the high temperature conditions [21].

To clearly compare the differences between samples from different milk powders, this study utilized a difference comparison mode, as depicted in Figure 1b. The ion mobility chromatogram of sample M-1 was chosen as the reference, and the ion mobility chromatograms of other samples were subtracted from the reference. If two volatile compounds are the same, the subtracted background would be white. Red dots indicate that the concentration of a compound is higher than the reference concentration, while blue dots indicate that the concentration is lower. Comparing samples M-2 and M-3 to other different milk powder samples, the areas of red and blue are significantly reduced. This suggests that the volatile compound content of the reference sample M-1 is similar to samples M-2 and M-3 but significantly differs from other samples. By establishing difference comparison plots for different milk powders using gas chromatography-ion mobility spectrometry, preliminary and simple discrimination between different milk powders can be achieved.

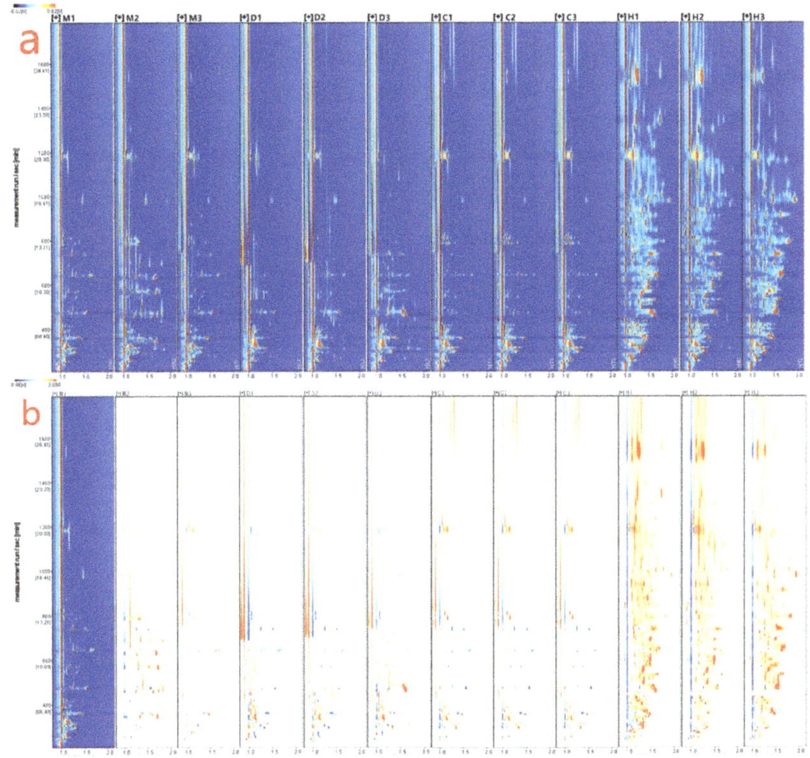

Figure 1. GC−IMS analysis of different milk powders. (**a**) 3D Spectrogram of volatile components in different milk powders−top view; (**b**) 2D Comparative chart of volatile components in different milk powders. Wherein, M, D, C, and H denote bovine milk powder, donkey milk powder, camel milk powder, and equine milk powder, respectively.

3.2. Analysis of Volatile Matter Content Based on GC-IMS

In order to study the characteristics of volatile compounds in different milk powders, GC-IMS was employed to analyze and identify the volatile compound components and their relative concentrations in four different samples. As depicted in Figure 1, a total of 108 peak signals were detected. Despite GC-IMS's high sensitivity, due to the limitations of the database, only 61 compounds could be qualitatively identified. Among these identified compounds, there were 22 dimers, where water and hydrogen ions have the possibility of combining with one monomer and two monomers to form a monomer and a dimer when charged volatile compounds are given [22]. This enhances the accuracy of GC-IMS for the quantification of volatile compounds. Analyzing the types of identified compounds, Table 2 shows the 61 volatile flavor compounds identified by GC-IMS, including 13 alcohols, 7 esters, 12 aldehydes, 7 ketones, 1 ether, 1 acid, and 2 furan compounds. Compared to previous reports [23], GC-IMS demonstrated higher sensitivity to aldehydes, ketones, and alcohols than to alkanes, which is consistent with the results reported by Gou [24]. The aroma components of the four milk powder samples were similar in type, but their total concentrations varied. Aldehydes and ketones had the highest content, followed by alcohols and esters. Among them, aldehydes and ketones accounted for a significant proportion of the overall composition, making them important volatile compounds in milk powders. This aligns with the findings of Bai et al. [25] in their analysis of donkey milk powder using GC-MS, as well as with the results of Chi et al. [26] in their examination of bovine milk powder. It is worth noting that GC-IMS did not detect a significant proportion of hydrocarbon compounds, owing to the fundamental principles of GC-IMS analysis [27], which exhibit limited sensitivity towards hydrocarbons.

Ketones are common flavor compounds in milk powder, often present in the form of methyl ketones [28]. They have a low threshold value and play an extremely important role in the overall flavor of milk powder [29]. During the processing and storage of milk powder, free fatty acids in the sample are first oxidized to 8-ketonic acids, which are then decarboxylated into corresponding methyl ketones [30]. As shown in Table 2, the detected ketones in the four milk powders accounted for 23.41%, 10.83%, 15.46%, and 9.69% of the total volatile compounds, respectively. Among them, acetone, 3-hydroxy-2-butanone, 2-butanone, and 2-heptanone were relatively abundant. 3-Hydroxy-2-butanone imparts a milky aroma and is a representative volatile flavor compound in dairy products [20]. 2-heptanone and 2-butanone are associated with fruity, sweet, and slightly milky aromas [31]. In terms of differences between different milk powders, the total ketone content of sample M and sample C was significantly higher than that of sample D and sample H. This could be due to the conditions of spray drying, which may disrupt the coalescence of milk droplets, releasing free fatty acids that support oxidative reactions and increasing the total ketone content [32]. As for the content of 1-penten-3-one, sample M and sample C had significantly lower levels than sample D and sample H. This discrepancy may be attributed to the differences in the content and types of free fatty acids in different types of milk [33].

Aldehydes are one of the main flavor compounds in milk powder. The generation of aliphatic and enal aldehydes with more than 5 carbon atoms in milk powder mainly occurs through the oxidation of higher homologous fatty acids [34]. For example, in this study, aldehydes such as 1-hexanal, 1-nonanal, heptanal, and 1-octanal were detected. According to Li et al. [35], these substances can be used as indicators to evaluate the oxidative flavor of milk powder. Aldehydes with less than six carbon atoms are produced by PUFA oxidation [36]. Aldehydes in milk powder have a lower overall content than ketones but have a lower threshold and a significant impact on the overall flavor of the milk powder. Low concentrations of aliphatic aldehydes typically exhibit herbal aromas, giving the milk powder a fresh taste, while high concentrations may lead to undesirable odors [37]. Among the four milk powders, the aliphatic aldehydes accounted for 16.65%, 22.21%, 21.00%, and 23.81% of the total volatile compounds, respectively. The total aldehyde content of sample M was significantly lower than that of sample D, sample C, and sample H. This difference could be attributed to variations in the content of free fatty acids and free amino acids in different liquid milk types.

Alcohol compounds have a higher threshold and contribute to the flavor of milk only when they are present in an unsaturated state or at high concentrations [38], but they are still important components of milk powder flavor. Alcohol compounds in milk powder may be the result of the reduction reaction of corresponding aldehyde compounds [39], and their content is positively correlated with the content of aldehyde compounds. 4-terpenol is likely to be synthesized by plants and some microorganisms and transferred from feed to milk through the rumen [40], so its content is unrelated to aliphatic aldehydes in milk powder but rather related to the forage fed to the livestock.

Ester compounds, due to their low threshold value, are essential flavor compounds in milk powder, imparting buttery, fruity, and floral aromas and helping to reduce the off-flavors of fatty acids and miscellaneous alcohols in milk powder [41]. Ester compounds are mainly formed by the esterification reaction between free fatty acids and alcohols present in milk fat. Some ester compounds are produced by bacterial esterases, and their detection can indicate the degree of bacterial contamination during the raw material processing and storage of milk powder. Dimethyl sulfide is the only sulfur compound detected by GC-IMS. It may be through the Strecker reaction involving riboflavin and methionine [42]. Low concentrations of dimethyl sulfide contribute to the ideal flavor of fresh milk and milk powder, but at high concentrations, it can cause undesirable off-flavors. The detected furan derivatives include 2-pentylfuran and 2-ethylfuran, which result from the Maillard reaction and oxidation of lactose and amino acids [43]. They have nutty and vanilla-like flavors [44].

Table 2. Volatile matter content of different milk powders.

Compound	Formula	MW	RI	Dt	Contents (%)			
					M	D	C	H
Ketones					23.41 ± 4.63^{a}	10.83 ± 2.46^{c}	15.46 ± 2.27^{b}	9.69 ± 1.24^{c}
Acetone	C_3H_6O	58.1	809.6	1.11185	9.28 ± 0.85^{a}	6.80 ± 1.24^{c}	8.06 ± 0.50^{b}	2.86 ± 0.07^{d}
3-Hydroxy-2-butanone *	$C_4H_8O_2$	88.1	1285.1	1.0663	5.60 ± 1.58^{a}	0.55 ± 0.14^{c}	3.16 ± 0.31^{b}	1.41 ± 0.31^{c}
2-Butanone *	C_4H_8O	72.1	898.1	1.06083	3.10 ± 0.72^{a}	1.30 ± 0.59^{b}	2.62 ± 0.44^{a}	1.21 ± 0.41^{b}
2-Heptanone *	$C_7H_{14}O$	114.2	1172.8	1.26016	3.51 ± 0.52^{a}	1.20 ± 0.23^{c}	0.78 ± 0.20^{d}	1.51 ± 0.06^{b}
2-Pentanone	$C_5H_{10}O$	86.1	973.4	1.36654	1.61 ± 0.93^{a}	0.29 ± 0.13^{b}	0.42 ± 0.15^{b}	1.38 ± 0.21^{a}
1-Penten-3-one *	C_5H_8O	84.1	1018.2	1.0789	0.24 ± 0.02^{d}	0.61 ± 0.12^{b}	0.34 ± 0.01^{c}	1.39 ± 0.07^{a}
2-Nonanone	$C_9H_{18}O$	142.2	1394.8	1.87621	0.06 ± 0.009^{b}	0.08 ± 0.003^{b}	0.06 ± 0.01^{b}	0.20 ± 0.11^{a}
Aldehydes					16.65 ± 6.96^{b}	22.21 ± 7.96^{a}	21.00 ± 6.24^{a}	23.81 ± 2.88^{a}
1-Hexanal *	$C_6H_{12}O$	100.2	1095.9	1.26878	4.38 ± 2.78^{b}	9.02 ± 2.47^{a}	6.19 ± 2.51^{b}	4.69 ± 0.07^{b}
Propanal *	C_3H_6O	58.1	773.8	1.05825	3.996 ± 0.94^{b}	3.86 ± 1.16^{b}	4.27 ± 1.00^{a}	2.18 ± 0.24^{b}
1-Nonanal *	$C_9H_{18}O$	142.2	1408.2	1.47784	1.93 ± 0.63^{a}	1.43 ± 0.64^{ab}	1.26 ± 0.14^{b}	1.53 ± 0.45^{ab}
Heptanal *	$C_7H_{14}O$	114.2	1173.6	1.33622	1.77 ± 0.70^{bc}	2.10 ± 1.19^{b}	3.35 ± 1.02^{a}	1.19 ± 0.17^{c}
n-Pentanal	$C_5H_{10}O$	86.1	975.7	1.42324	0.95 ± 0.39^{b}	0.86 ± 0.07^{b}	1.53 ± 0.64^{a}	0.98 ± 0.18^{b}
(E)-2-Hexenal *	$C_6H_{10}O$	98.1	1207.7	1.17953	0.44 ± 0.23^{b}	0.52 ± 0.09^{b}	0.46 ± 0.07^{b}	1.31 ± 0.31^{a}
(E)-2-Pentenal *	C_5H_8O	84.1	1129.4	1.1047	0.75 ± 0.15^{c}	1.30 ± 0.39^{b}	0.95 ± 0.08^{c}	3.19 ± 0.16^{a}
Butanal	C_4H_8O	72.1	898.5	1.28609	1.12 ± 0.69^{a}	0.87 ± 0.25^{a}	1.29 ± 0.45^{a}	1.09 ± 0.27^{a}
1-Octanal *	$C_8H_{16}O$	128.2	1283.7	1.40792	0.50 ± 0.16^{bc}	0.44 ± 0.06^{b}	0.59 ± 0.09^{b}	0.86 ± 0.19^{a}
(E,E)-2,4-Heptadienal *	$C_7H_{10}O$	110.2	1541	1.19486	0.54 ± 0.19^{b}	1.00 ± 0.34^{a}	0.59 ± 0.03^{b}	1.30 ± 0.55^{a}
(Z)-2-Methylpent-2-enal	$C_6H_{10}O$	98.1	1153.5	1.49675	0.22 ± 0.008^{b}	0.24 ± 0.03^{b}	0.19 ± 0.01^{b}	5.34 ± 0.25^{a}
2-Methyl butanal	$C_5H_{10}O$	86.1	916.1	1.40205	0.10 ± 0.02^{b}	0.55 ± 0.06^{a}	0.36 ± 0.01^{ab}	0.14 ± 0.01^{b}
Alcohols					9.84 ± 2.87^{a}	8.83 ± 1.79^{a}	9.85 ± 2.82^{a}	13.10 ± 1.95^{b}
1-Pentanol *	$C_5H_{12}O$	88.1	1247.7	1.25373	2.95 ± 0.18^{ab}	1.92 ± 1.60^{a}	2.92 ± 1.13^{ab}	3.48 ± 0.20^{b}
1-Butanol *	$C_4H_{10}O$	74.1	1136.8	1.18259	2.41 ± 0.13^{a}	2.27 ± 0.73^{a}	1.61 ± 0.59^{ab}	1.15 ± 0.02^{b}
1-Propanol *	C_3H_8O	60.1	1026.8	1.11126	1.11 ± 0.63^{b}	0.67 ± 0.16^{d}	1.58 ± 0.28^{a}	1.05 ± 0.05^{c}
1-Penten-3-ol	$C_5H_{10}O$	86.1	1154.6	0.94065	1.03 ± 0.28^{a}	1.28 ± 0.49^{a}	1.32 ± 0.32^{a}	0.65 ± 0.12^{b}
1-Hexanol *	$C_6H_{14}O$	102.2	1366.7	1.32911	0.48 ± 0.22^{c}	0.41 ± 0.07^{c}	0.70 ± 0.24^{b}	1.07 ± 0.07^{a}
4-Terpinenol *	$C_{10}H_{18}O$	154.3	1590.2	1.22368	0.58 ± 0.08^{b}	0.95 ± 0.13^{b}	0.68 ± 0.06^{b}	3.63 ± 1.12^{a}

Table 2. *Cont.*

Compound	Formula	MW	RI	Dt	Contents (%)			
					M	D	C	H
2-Methyl-1-propanol	$C_4H_{10}O$	74.1	1089.8	1.17393	0.26 ± 0.15^{a}	0.32 ± 0.01^{a}	0.28 ± 0.06^{a}	0.03 ± 0.004^{b}
(Z)-2-Pentenol	$C_5H_{10}O$	86.1	1315.9	0.94353	0.42 ± 0.08^{b}	0.33 ± 0.03^{c}	0.32 ± 0.04^{c}	0.82 ± 0.01^{a}
(E)-2-Heptenal	$C_7H_{12}O$	112.2	1321.8	1.2579	0.29 ± 0.10^{b}	0.25 ± 0.05^{bc}	0.18 ± 0.02^{c}	0.39 ± 0.09^{a}
3-Methyl-3-buten-1-ol	$C_5H_{10}O$	86.1	1254.8	1.17277	0.12 ± 0.005^{b}	0.12 ± 0.001^{b}	0.10 ± 0.02^{b}	0.21 ± 0.08^{a}
3-Methyl-1-butanol	$C_5H_{12}O$	88.1	1196.5	1.24773	0.13 ± 0.06^{ab}	0.23 ± 0.02^{a}	0.057 ± 0.01^{b}	0.03 ± 0.003^{b}
(Z)-4-heptenal *	$C_7H_{12}O$	112.2	1232.8	1.14759	0.064 ± 0.001^{b}	0.10 ± 0.003^{b}	0.10 ± 0.03^{b}	0.6 ± 0.14^{a}
Esters					9.45 ± 2.88^{a}	8.17 ± 2.04^{b}	9.03 ± 2.25^{a}	6.94 ± 2.37^{c}
Ethyl acetate	$C_4H_8O_2$	88.1	869.4	1.33873	4.3 ± 2.44^{ab}	2.66 ± 3.31^{b}	5.05 ± 1.60^{a}	3.15 ± 0.27^{ab}
Acetic acid butyl ester *	$C_6H_{12}O_2$	116.2	1065.2	1.23989	2.3 ± 0.24^{ab}	4.69 ± 2.8^{ab}	0.78 ± 0.009^{b}	2.19 ± 1.24^{a}
Propyl acetate	$C_5H_{10}O_2$	102.1	969.8	1.47649	0.57 ± 0.11^{c}	0.34 ± 0.03^{c}	2.93 ± 0.50^{a}	1.03 ± 0.64^{b}
Butyl propanoate	$C_7H_{14}O_2$	130.2	1134.9	1.71778	1.67 ± 2.28^{a}	0.20 ± 0.03^{b}	0.14 ± 0.02^{b}	0.07 ± 0.004^{b}
Butanoic acid, butyl ester	$C_8H_{16}O_2$	144.2	1203.1	1.8173	0.44 ± 0.05^{a}	0.09 ± 0.02^{b}	0.06 ± 0.008^{b}	0.04 ± 0.002^{b}
Ethyl propanoate	$C_5H_{10}O_2$	102.1	952.7	1.45226	0.089 ± 0.05^{b}	0.082 ± 0.03^{b}	0.046 ± 0.006^{b}	0.37 ± 0.15^{a}
Butanoic acid ethyl ester	$C_6H_{12}O_2$	116.2	1032.3	1.55874	0.048 ± 0.002^{b}	0.10 ± 0.04^{a}	0.032 ± 0.004^{b}	0.09 ± 0.006^{a}
Other					6.07 ± 1.87^{a}	4.92 ± 1.54^{b}	2.71 ± 0.76^{c}	1.79 ± 0.48^{d}
Dimethyl sulfide	C_2H_6S	62.1	760.5	0.95479	1.83 ± 0.08^{a}	0.37 ± 0.08^{c}	1.33 ± 0.51^{b}	0.17 ± 0.02^{c}
2-Methylpropanoic acid	$C_4H_8O_2$	88.1	1503.7	1.38303	4.05 ± 0.69^{a}	4.42 ± 0.83^{a}	1.30 ± 0.24^{a}	1.41 ± 0.40^{a}
2-Pentylfuran	$C_9H_{14}O$	138.2	1219.4	1.25334	0.17 ± 0.14^{a}	0.10 ± 0.07^{ab}	0.05 ± 0.004^{b}	0.10 ± 0.01^{ab}
2-Ethylfuran	C_6H_8O	96.1	775.9	1.30114	0.03 ± 0.004^{b}	0.03 ± 0.007^{b}	0.03 ± 0.005^{b}	0.10 ± 0.05^{a}

Note: In the context of the provided information, the abbreviations represent the following; MW: Molecular Weight; RI: Retention Index; Dt: Relative Migration Time Compounds with dimer formation is denoted by the symbol "*"; The letters "a", "b", "c", and "d" represent significant differences ($p < 0.05$) in volatile compounds among different regions of milk powder. Wherein, M, D, C, and H denote bovine milk powder, donkey milk powder, camel milk powder, and equine milk powder, respectively.

However, merely observing the content of volatile compounds in various milk powders makes it challenging to classify them effectively. Therefore, the volatile compound content of different milk powders is utilized as feature data for subsequent PLS-DA statistical multivariate analysis.

3.3. Analysis of the Differences between Cow Milk Powder and Other Kinds of Milk Powder Based on PLS-DA

PLS-DA is a supervised discriminant analysis statistical method used to establish a relationship model between volatile compounds and sample categories, enabling the prediction of sample categories [18]. In this study, a PLS-DA model was employed to identify specific marker compounds in cow milk powder samples that differentiate them from the other three samples based on the detected volatile compound content data matrix. The results are shown in Figure 2, where the contributions of Partial Least Squares Discriminant Analysis (PLS-DA) for the three score plots are 64%, 56.8%, and 60.4%, respectively. Contributions exceeding 50% indicate successful differentiation between these three groups of analyses. The samples are stably distributed in the first and fourth quadrants and the second and third quadrants. Additionally, a permutation test was conducted on these three models, and the model prediction index (Q2) for each was found to be 0.67, 0.70, and 0.60, respectively, with prediction rates of 67%, 70%, and 60% for distinguishing different milk powders. A Q2 value exceeding 0.5 indicates an acceptable model fit, and the Q2 regression line intersecting the y-axis below zero suggests no overfitting, making the model validation reliable [22]. Therefore, these results can be used for milk powder identification analysis.

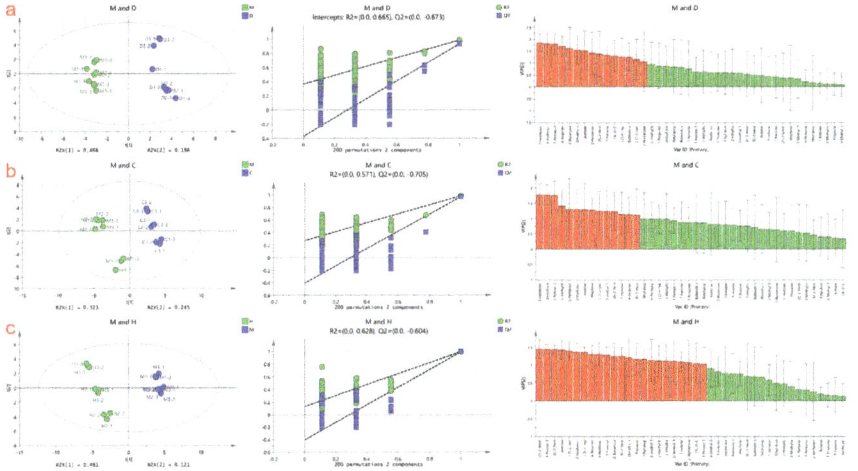

Figure 2. Presents the PLS−DA analysis based on cow milk powder samples compared to the other three types of milk powder samples. (**a**) Displays the score plot, cross-validation results obtained from 200 permutation tests, and the plot of volatile compounds arranged according to VIP scores for discriminating between cow milk powder and donkey milk powder; (**b**) Represents the PLS−DA analysis plot for discriminating between cow milk powder and camel milk powder; (**c**) Illustrates the PLS−DA analysis plot for discriminating between cow milk powder and horse milk powder.Where the red part is VIP > 1 and the green part is VIP < 1.

To determine the most statistically significant variables, the Variable Importance in the Projection (VIP) value variable selection method was applied. Generally, volatile compounds with VIP values > 1.0 are considered to play an important role in discrimination [45]. In Figure 2a, the red markers indicate a total of 15 substances, namely 2-Nonanone, (Z)-2-pentenol, ethyl butyrate, (Z)-4-heptenal, (E, E)-2,4-heptadienal, 1-hexanal, (E)-2-Pentenal, 2-Pentanone, acetone, dimethyl sulfide, 2-Butanone, 4-Terpinenol, 1-Penten-

3-ol, 3-Hydroxy-2-butanone, and 2-heptanone, as the marker volatile compounds distinguishing cow milk powder from donkey milk powder. In Figure 2b, the red markers indicate a total of 14 substances, namely (E)-2-Heptenal, 4-Terpinenol, 1-nonanal, (E)-2-Pentenal, 3-methyl-1-Butanol, (Z)-2-pentenol, Heptanal, acetone, 3-Hydroxy-2-butanone, 2-Pentanone, 2-methyl butanal, 1-Penten-3-one, propyl acetate, and 2-heptanone, as the marker volatile compounds distinguishing cow milk powder from camel milk powder. In Figure 2c, the red markers indicate a total of 23 substances, namely 1-Penten-3-ol, (E,E)-2,4-heptadienal, 2-Nonanone, 1-octanal, 2-Methyl-1-propanol, 2-ethylfuran, 2-methyl butanal, 3-methyl-1-Butanol, Ethyl propanoate, Propanal, dimethyl sulfide, (E)-2-hexenal, 2-Butanone, 1-hexanol, 3-Hydroxy-2-butanone, 4-Terpinenol, (Z)-4-heptenal, 2-heptanone, (Z)-2-pentenol, acetone, (E)-2-Pentenal, and (Z)-2-Methylpent-2-enal, as the marker volatile compounds distinguishing cow milk powder from horse milk powder.

However, the models derived from the PLS-DA analysis have not undergone precise validation. Therefore, it is imperative to employ ROC analysis for more accurate model calibration.

3.4. Validation of PLS-DA Screening Substances Based on Roc Analysis and Yoden Index Discrimination

The ROC curve is a convenient and efficient tool used in the study of binary classification problems. It plots the true positive rate against the false positive rate at different cutoff points or thresholds, constructing a monotonically increasing curve. The area under the ROC curve (AUC) serves as a measure of discriminatory performance, where a larger AUC value indicates a more effective classification method [46]. In order to assess the discriminative ability of the key volatile compounds identified by PLSD-DA between cow milk powder and the other three types of milk powders, ROC curve analysis was conducted using SPSS software. The Jordon index of these volatile compounds is also calculated, and the Jordon index maximum value selection is performed to determine the maximum threshold for the discrimination of the substance.

In the application of the ROC curve analysis, by considering both sensitivity and specificity results at various cutoff points [47], we selected the volatile compounds with AUC values between 0.95 and 1. These compounds have good discriminatory ability for different types of milk powder and are more representative of distinguishing milk powders. For cow milk powder versus donkey milk powder discrimination, the final discriminatory markers were determined as (E)-2-Pentenal, 2-Pentanone, acetone, dimethyl sulfide, 2-Butanone, 4-Terpinenol, 3-Hydroxy-2-butanone, and 2-heptanone, as shown in Figure 3a. For cow milk powder versus camel milk powder discrimination, the final discriminatory markers were determined as 3-methyl-1-butanol, 3-hydroxy-2-butanone, 2-methyl butanal, 1-penten-3-one, propyl acetate, and 2-heptanone, as shown in Figure 3b. For cow milk powder versus horse milk powder discrimination, the final discriminatory markers were determined as 2-nonanone, 2-methyl-1-propanol, 2-methyl butanal, 3-methyl-1-Butanol, ethyl propanoate, Propanal, dimethyl sulfide, (E)-2-hexenal, 2-Butanone, 1-hexanol, 3-hydroxy-2-butanone, 4-Terpinenol, (Z)-4-heptenal, 2-heptanone, (Z)-2-pentenol, acetone, (E)-2-pentenal, 1-penten-3-ol, and (Z)-2-methylpent-2-enal, as shown in Figure 3c. The thresholds determined by the Youden index in Figure 3 also indicate the maximum values for each compound during the corresponding discrimination process, making the classification more accurate.

Among these, (E)-2-pentenal and 4-terpinenol are downregulated volatile compounds in both cow milk powder and donkey milk powder, and they are also downregulated in the comparison between cow and horse milk powder. 3-hydroxy-2-butanone, 2-heptanone, and ketone-like substances are upregulated in the comparison between cow and specialty milk powder. This phenomenon is attributed to the intestinal microbiota, which breaks down plant materials and transfers them into milk [48]. It may be due to differences in the food breakdown process between cows and donkeys. Cows and camels, being ruminants, primarily digest and absorb nutrients in their four stomach compartments, whereas for horses and donkeys, which lack a complex multi-chambered stomach, the primary site for digesting and absorbing nutrients is in the intestines [49]. These distinct species-specific processes result in variations in the composition of the produced milk, with the most significant impact on volatile compounds being the differences in homologous fatty acids and unsaturated fatty acids. Lu et al. [33] found significant differences in the types of fatty acids in different milk powders, especially between cow and horse milk powder. Dimethyl sulfide, a substance produced in large quantities through a high-temperature oxidation [50], is upregulated in the comparison between cow and horse milk powder but is not shown in the comparison with camel milk powder.

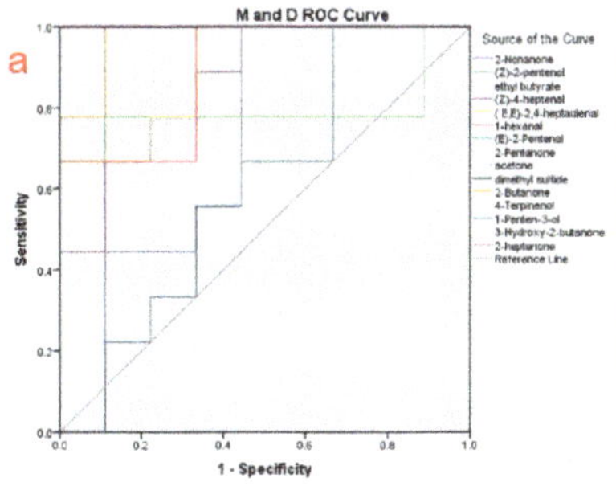

M and D

Test Result Variable(s)	Area	Agreement index max	Asymptotic Sig. a	threshold value(%)
2-Nonanone	0.716	0.556	0.122	0.07919
(Z)-2-pentenol	0.790	0.667	0.038	0.35619
ethyl butyrate	0.889	0.667	0.005	0.03668
(Z)-4-heptenal	0.864	0.667	0.009	0.08316
(E, E)-2,4-heptadienal	0.901	0.667	0.004	0.97976
1-hexanal	0.889	0.667	0.005	8.26642
(E)-2-Pentenal	0.975	0.889	0.001	1.06555
2-Pentanone	1.000	1	0.000	0.47780
acetone	1.000	1	0.000	8.05595
dimethyl sulfide	1.000	1	0.000	0.78928
2-Butanone	0.975	0.889	0.001	2.17147
4-Terpinenol	1.000	1	0.033	0.72472
1-Penten-3-ol	0.605	0.556	0.453	0.01558
3-Hydroxy-2-butanone	1.000	1	0.000	2.48239
2-heptanone	1.000	1	0.000	2.26961

a: Null hypothesis: true area = 0.5

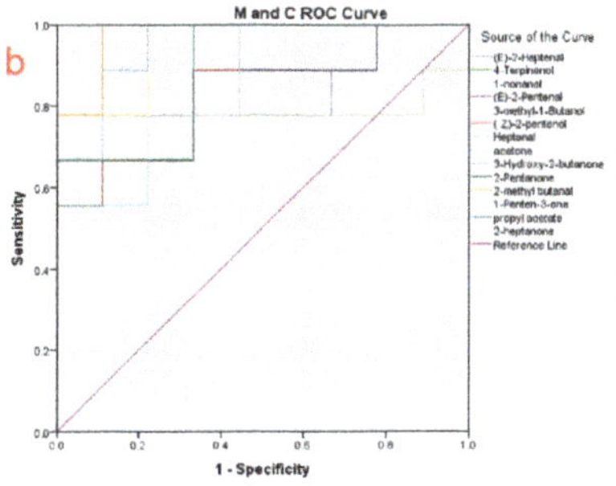

M and C

Test Result Variable(s)	Area	Agreement index max	Asymptotic Sig.ᵃ	threshold value(%)
(E)-2-Heptenal	0.840	0.778	0.015	0.22143
4-Terpinenol	0.827	0.556	0.019	0.66491
1-nonanal	0.753	0.667	0.070	1.78499
(E)-2-Pentenal	0.877	0.778	0.007	0.80086
3-methyl-1-Butanol	0.951	0.778	0.001	0.06588
(Z)-2-pentenol	0.864	0.556	0.009	0.31566
Heptanal	0.864	0.667	0.009	2.85025
acetone	0.877	0.667	0.007	8.87858
3-Hydroxy-2-butanone	0.951	0.778	0.001	4.28623
2-Pentanone	0.889	0.667	0.005	1.13717
2-methyl butanal	0.975	0.889	0.001	0.14342
1-Penten-3-one	1.000	1	0.000	0.30277
propyl acetate	1.000	1	0.000	1.16484
2-heptanone	1.000	1	0.000	2.01987

a: Null hypothesis: true area = 0.5

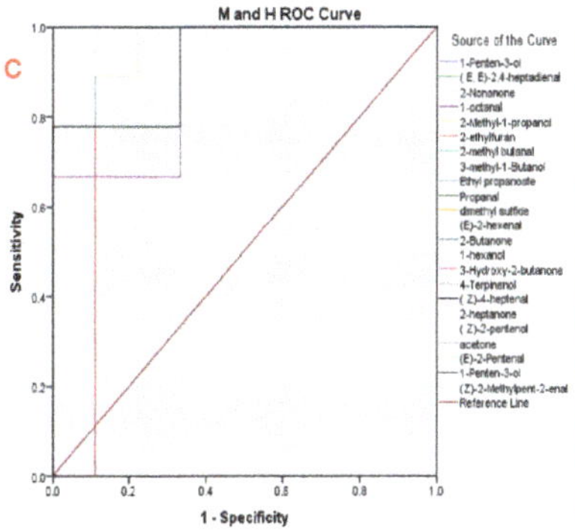

M and H

Test Result Variable(s)	Area	Agreement index max	Asymptotic Sig.ᵃ	threshold value(%)
1-Penten-3-ol	0.926	0.778	0.002	0.84933
(E, E)-2,4-heptadienal	0.926	0.778	0.002	0.61366
2-Nonanone	0.963	0.778	0.001	0.06648
1-octanal	0.889	0.667	0.005	0.54845
2-Methyl-1-propanol	1.000	1	0.000	0.07564
2-ethylfuran	0.889	0.889	0.005	0.05034
2-methyl butanal	0.975	0.889	0.001	0.12759
3-methyl-1-Butanol	1.000	1	0.000	0.04920
Ethyl propanoate	1.000	1	0.000	0.16353
Propanal	1.000	1	0.000	2.75773
dimethyl sulfide	1.000	1	0.000	0.35963
(E)-2-hexenal	0.988	0.889	0.000	0.72955
2-Butanone	1.000	1	0.000	1.95817
1-hexanol	1.000	1	0.000	0.85350
3-Hydroxy-2-butanone	1.000	1	0.000	2.24522
4-Terpinenol	1.000	1	0.000	1.42648
(Z)-4-heptenal	1.000	1	0.000	0.23431
2-heptanone	1.000	1	0.000	2.10207
(Z)-2-pentenol	1.000	1	0.000	0.66168
acetone	1.000	1	0.000	4.37596
(E)-2-Pentenal	1.000	1	0.000	1.99598
1-Penten-3-ol	0.926	0.778	0.002	0.84933
(Z)-2-Methylpent-2-enal	1.000	1	0.000	2.66229

a: Null hypothesis: true area = 0.5

Figure 3. The ROC analysis graphs and the maximum Youden index threshold discrimination table are based on the initial screening of discriminative compounds using PLS-DA analysis for cow milk powder samples compared to the other three types of milk powder samples. (**a**) The ROC analysis validation graph for discriminating between cow milk powder and donkey milk powder; (**b**) The ROC analysis validation graph for discriminating between cow milk powder and camel milk powder; (**c**) The ROC analysis validation graph for discriminating between cow milk powder and horse milk powder. Note: The lines with colors not appearing in the ROC analysis graphs coincide with the upper-left boundary.

However, the discriminative substances obtained are not represented specifically with numerical values. Therefore, the decision was made to employ GC-IMS fingerprint spectra to showcase the discriminative substances.

3.5. Validation of the Production of GC-IMS Discriminant Fingerprints of Cow Milk Powder against Other Milk Powders Based on ROC Analysis

For the expeditious and convenient discrimination of cow milk powder from other milk powders, the GC-IMS fingerprint spectra of volatile compounds validated through ROC analysis were established, as illustrated in Figure 4. The fingerprint spectra were created by assembling cutouts of volatile compounds from Figure 1a. Each row represents a sample, and each column represents a specific compound, with brighter and redder shades indicating higher concentrations, while darker and bluer shades suggest lower concentrations. By comparing the textual and quantitative attributes of volatile compounds, the GC-IMS fingerprint spectra enable to clearly assess the varying levels of volatile compounds among different milk powders and discern distinctive characteristics in their volatile profiles. In Figure 4a, the distinctive features of volatile compounds such as 2-butanone facilitate rapid and precise discrimination between cow and donkey milk powder. In Figure 4b, the distinctive features of volatile compounds like 3-hydroxy-2-butanone allow for clear and swift discrimination between cow and camel milk powder. Similarly, in Figure 4c, the distinctive features of volatile compounds including dimethyl sulfide enable a clear and rapid discrimination between cow and camel milk powder.

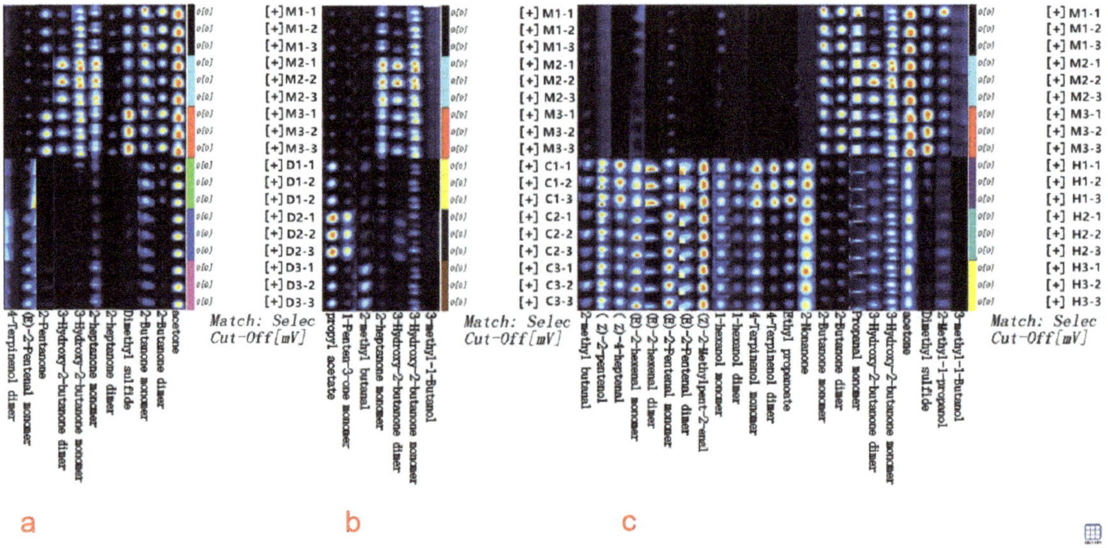

Figure 4. The GC-IMS fingerprint spectra are based on ROC validation for discriminating volatile compounds. (**a**) The fingerprint spectra for discriminating between cow and donkey milk powder; (**b**) The fingerprint spectra for discriminating between cow and camel milk powder; (**c**) The fingerprint spectra for discriminating between cow and horse milk powder.

4. Conclusions

Using GC-IMS, the volatile compounds of cow milk powder, donkey milk powder, horse milk powder, and camel milk powder were detected, resulting in a total of 61 volatile compounds. Ketones, aldehydes, and alcohols were found to be the primary flavor components in the milk powders. While the aroma compositions of different milk powders were similar in type, significant differences were observed in their concentrations, exhibiting distinct characteristics for each type. GC-IMS demonstrated a higher sensitivity to aldehydes and ketones compared to alkanes, based on the preference in detecting various substances. Through multivariate statistical analysis (PLS-DA and ROC validation), specific volatile compounds were identified that can discriminate between different types of milk powders and the maximum threshold for discrimination was established. For cow and donkey milk powder, (E)-2-pentenal and other eight volatile compounds were identified as the ultimate discriminatory markers. For cow and camel milk powder, 3-methyl-1-butanol and other six volatile compounds served as the final discriminatory markers. For cow and horse milk powder, 2-nonanone and other 19 volatile compounds were determined as the ultimate discriminatory markers. These findings provide a theoretical basis for detecting adulterated cow milk powder in special milk powders. To facilitate the swift and convenient analysis of research results, a fingerprint spectrum was established for distinguishing cow milk powder from different types of milk powders. It is essential to emphasize that the application of GC-IMS enables the identification of different food products, ensuring the authenticity of food and detecting potential adulteration, especially for food items with complex testing methods.

Author Contributions: Y.G. and Y.H. have made substantial contributions to the conception of design of the work and prepared the original draft. Y.G., Y.H., X.N. and J.L. were responsible for conducting the research and investigation process. G.M., J.L. and Q.X. revised this manuscript critically for important intellectual content and the interpretation of data for the work. X.N. and G.M. were responsible for the supervision and coordination of this project. Q.X. was responsible for all aspects of the work by ensuring that questions related to the accuracy or integrity of any part of the work were appropriately investigated and resolved. All authors have read and agreed to the published version of the manuscript.

Funding: This study was supported by the National Natural Science Foundation of Beijing, China (31760442), the High and New Technology Project of the First Division of Xinjiang Production and Construction Corps (2022GJJ03), and the Financial Support for the Construction of Key Disciplines in Food Science and Engineering (130000304). The authors acknowledge the assistance of the Instrumental Analysis Center of Tarim University. Thanks Xinnong Enterprises (Aral, China) for their support.

Data Availability Statement: The data used to support the findings of this study can be made available by the corresponding author upon request.

Conflicts of Interest: The authors declare that they have no known competing financial interests or personal relationships that could have appeared to influence the work reported in this paper.

References

1. Li, Y.; Zhang, X.; Li, Y.; Feng, F. Research progress of special milk processing technology. *Sci. Technol. Food Ind.* **2018**, *39*, 347–352. [CrossRef]
2. Liu, X.; Chen, F. Development status and countermeasures of dairy products processing industry in Xinjiang. *Food Ind.* **2020**, *41*, 262–265.
3. Li, M.; Liu, A.; Zhao, S.; Zhao, H.; Zhang, J.; Song, W.; Yue, X.; Zheng, Y. Research progress on nutritional components of donkey milk. *J. Dairy Sci. Technol.* **2022**, *45*, 42–49.
4. Lu, D.; Liu, P.; Wang, S.; Feng, W.; Zhang, J. Nutritional value and development and utilization of horse milk. *Xinjiang Anim. Husb.* **2012**, *170*, 4–9. [CrossRef]
5. He, S.; Chen, X.; Guo, W.; Wang, S. Research progress on nutritional value of camel milk. *Dairy Humankind* **2023**, *129*, 32–37. [CrossRef]
6. Wang, L.; Xiu, B. Development Status and Countermeasures of Horse Milk Industry in Inner Mongolia. *Inn. Mong. Sci. Technol. Econ.* **2020**, *464*, 3–4+14.
7. Lu, D.; Xu, M.; Li, J.; He, X.; Ye, D. Development status, problems and countermeasures of special dairy industry in Xinjiang. *Xinjiang Anim. Husb.* **2017**, *240*, 4–7. [CrossRef]
8. Souhassou; Bassbasi, S.; Hirri, M.; Kzaiber, A.; Oussama, F. Detection of camel milk adulteration using Fourier transformed infrared spectroscopy FT-IR coupled with chemometrics methods. *Int. Food Res. J.* **2018**, *25*, 1213–1218.
9. Kumar, D.; Verma, A.K.; Chatli, M.K.; Singh, R.; Kumar, P.; Mehta, N.; Malav, O.P. Camel milk: Alternative milk for human consumption and its health benefits. *Nutr. Food Sci.* **2016**, *46*, 217–227. [CrossRef]
10. Lu, D.; Xu, M.; Li, J.; He, X.; Li, X.; Wang, X.; Ye, D. The development and utilization status and development prospect of special milk in Xinjiang. *China Dairy* **2017**, *185*, 72–77. [CrossRef]

11. Domenico, M.D.; Giuseppe, M.D.; Rodriguez, J.D.W.; Camma, C. Validation of a fast real-time PCR method to detect fraud and mislabeling in milk and dairy products. *J. Dairy Sci.* **2017**, *100*, 106. [CrossRef] [PubMed]
12. Guo, L.; Ya, M.; Hai, X.; Guo, Y.S.; Li, C.D.; Xu, W.L.; Liao, C.S.; Feng, W.; Cai, Q. A simultaneous triplex TaqMan real-time PCR approach for authentication of caprine and bovine meat, milk and cheese. *Int. Dairy J.* **2019**, *95*, 58–64. [CrossRef]
13. Ge, L.; Sun, X.; Wang, Y. Comparison of volatile components of instant soybean powder, milk powder and goat milk powder. *Sci. Technol. Food Ind.* **2019**, *40*, 248–254. [CrossRef]
14. Wang, S.; Chen, H.; Sun, B. Recent progress in food flavor analysis using gas chromatography–ion mobility spectrometry (GC–IMS). *Food Chem.* **2020**, *315*, 126158. [CrossRef] [PubMed]
15. Martín-Gómez, A.; Segura-Borrego, M.P.; Ríos-Reina, R.; Cardador, M.J.; Callejón, R.M.; Morales, M.L.; Rodríguez-Estévez, V.; Arce, L. Discrimination of defective dry-cured Iberian ham determining volatile compounds by non-destructive sampling and gas chromatography. *LWT* **2021**, *154*, 112785. [CrossRef]
16. He, F.; Shiqi, Y.; Guihu, Z.; Ling, X.; Hehe, L.; Jinyuan, S.; Mingquan, H.; Fuping, Z.; Baoguo, S. Exploration of key aroma active compounds in strong flavor Baijiu during the distillation by modern instrument detection technology combined with multivariate statistical analysis methods. *J. Food Compos. Anal.* **2022**, *110*, 104577. [CrossRef]
17. Feng, D.; Wang, J.; Ji, X.; Min, W.-X.; Yan, W.-J. HS-GC-IMS detection of volatile organic compounds in yak milk powder processed by different drying methods. *LWT* **2021**, *141*, 110855. [CrossRef]
18. Yang, Y.; Zhu, H.; Chen, J.; Xie, J.; Shen, S.; Deng, Y.; Zhu, J.; Yuan, H.; Jiang, Y. Characterization of the key aroma compounds in black teas with different aroma types by using gas chromatography electronic nose, gas chromatography-ion mobility spectrometry, and odor activity value analysis. *LWT* **2022**, *163*, 113492. [CrossRef]
19. Yang, Y.; Chen, J.; Jiang, Y.; Qian, M.C.; Deng, Y.; Xie, J.; Li, J.; Wang, J.; Dong, C.; Yuan, H. Aroma dynamic characteristics during the drying process of green tea by gas phase electronic nose and gas chromatography-ion mobility spectrometry. *LWT* **2022**, *154*, 112691. [CrossRef]
20. Newton, A.E.; Fairbanks, A.J.; Golding, M.; Andrewes, P.; Gerrard, J.A. The role of the Maillard reaction in the formation of flavour compounds in dairy products–not only a deleterious reaction but also a rich source of flavour compounds. *Food Funct.* **2012**, *3*, 1231–1241. [CrossRef]
21. Fyfe, K.; Kravchuk, O.; Nguyen, A.V.; Deeth, H.; Bhandari, B. Influence of dryer type on surface characteristics of milk powders. *Dry. Technol.* **2011**, *29*, 758–769. [CrossRef]
22. Yin, J.; Wu, M.; Lin, R.; Li, X.; Ding, H.; Han, L.; Yang, W.; Song, X.; Li, W.; Qu, H. Application and development trends of gas chromatography–ion mobility spectrometry for traditional Chinese medicine, clinical, food and environmental analysis. *Microchem. J.* **2021**, *168*, 106527. [CrossRef]
23. Ye, M.; Li, R.; Jiang, Z.; Wang, Y.; Tan, J.; Tang, S. Analysis Classification and Prediction of Volatile Flavor Components in Milk Powders for Different Age Groups. *Food Sci.* **2022**, *43*, 242–252.
24. Gou, Y.; Ma, X.; Niu, X.; Ren, X.; Muhatai, G.; Xu, Q. Exploring the Characteristic Aroma Components of Traditional Fermented Koumiss of Kazakh Ethnicity in Different Regions of Xinjiang by Combining Modern Instrumental Detection Technology with Multivariate Statistical Analysis Methods for Odor Activity Value and Sensory Analysis. *Foods* **2023**, *12*, 2223. [CrossRef] [PubMed]
25. Bai, X.; Chen, Y.; Liu, C.; Du, W. Volatile flavor compounds of donkey milk powder were extracted by simultaneous distillation extraction and solid phase microextraction. *J. Xinjiang Norm. Univ. Nat. Sci. Ed.* **2016**, *35*, 40–45. [CrossRef]
26. Chi, X.; Guo, H.; Zhang, Y.; Zheng, N.; Liu, H.; Wang, J. E-nose, E-tongue Combined with GC-IMS to Analyze the Influence of Key Additives during Processing on the Flavor of Infant Formula. *Foods* **2022**, *11*, 3708. [CrossRef]
27. Zhu, H.; Zhu, D.; Sun, J. Application of GC-IMS coupled with chemometric analysis for the classification and authentication of geographical indication agricultural products and food. *Front. Nutr.* **2023**, *10*, 1247695. [CrossRef]
28. Vagenas, G.; Roussis, I.G. Fat-derived volatiles of various products of cows', ewes', and goats' milk. *Int. J. Food Prop.* **2012**, *15*, 665–682. [CrossRef]
29. Karagül-Yüceer, Y.; Cadwallader, K.R.; Drake. Volatile flavor components of stored nonfat dry milk. *J. Agric. Food Chem.* **2002**, *50*, 305–312. [CrossRef]
30. Li, Y.; Zhang, L.; Wang, W. Heat-induced changes in volatiles of milk and effects of thermal processing on microbial metabolism of yogurt. *J. Food Biochem.* **2013**, *37*, 409–417. [CrossRef]
31. Licón, C.C.; de Mendoza, J.H.; Maggi, L.; Berruga, M.I.; Aranda, R.M.M.; Carmona, M. Optimization of headspace sorptive extraction for the analysis of volatiles in pressed ewes' milk cheese. *Int. Dairy J.* **2012**, *23*, 53–61. [CrossRef]
32. Li, Y.; Zhang, L.; Wang, W.; Han, X. Differences in particle characteristics and oxidized flavor as affected by heat-related processes of milk powder. *J. Dairy Sci.* **2013**, *96*, 4784–4793. [CrossRef]
33. Lu, D.; Ye, D.; Li, J.; Xu, M. Comparative Analysis of Fatty Acid Content and Composition in Four Kinds of Livestock Milk Powder in Xinjiang. *Grass-Feed. Livest.* **2017**, *185*, 7–14. [CrossRef]
34. Contador, R.; Delgado, F.; García-Parra, J.; Garrido, M.; Ramírez, R. Volatile profile of breast milk subjected to high-pressure processing or thermal treatment. *Food Chem.* **2015**, *180*, 17–24. [CrossRef] [PubMed]
35. Li, Y.; Zhang, L.; Wang, W. Formation of aldehyde and ketone compounds during production and storage of milk powder. *Molecules* **2012**, *17*, 9900–9911. [CrossRef] [PubMed]
36. Mottram, D.S. Volatile compounds in food—Qualitative and quantitative data. *Food Chem.* **1991**, *39*, 120–122. [CrossRef]

37. Gioacchini, A.M.; De Santi, M.; Guescini, M.; Brandi, G.; Stocchi, V. Characterization of the volatile organic compounds of Italian 'Fossa'cheese by solid-phase microextraction gas chromatography/mass spectrometry. *Rapid Commun. Mass Spectrom.* **2010**, *24*, 3405–3412. [CrossRef]
38. Wang, Y.; Yang, R.; Zhao, W.; Liang, Q. Comparative Study of Effects of Pulsed Electric Field (PEF) and Ultra-high Temperature (UHT) Processing on Flavor Compounds in Milk. *Food Sci.* **2009**, *30*, 43–46.
39. Dong, Y.; Liu, Y.; Yu, N.; Sun, Y.; Yan, X.R.; Deng, T.; Zhang, J.; Pan, M.; Chen, Y. Analysis of defatting treatment on volatile flavor compounds of milk based on headspace-gas chromatography-ion mobility spectrometry. *J. Food Saf. Qual.* **2022**, *13*, 6155–6162. [CrossRef]
40. Tornambé, G.; Cornu, A.; Pradel, P.; Kondjoyan, N.; Carnat, A.; Petit, M.; Martin, B. Changes in terpene content in milk from pasture-fed cows. *J. Dairy Sci.* **2006**, *89*, 2309–2319. [CrossRef] [PubMed]
41. Zhang, X.; Ai, N.; Wang, J.; Tong, L.; Zheng, F.; Sun, B. Lipase-catalyzed modification of the flavor profiles in recombined skim milk products by enriching the volatile components. *J. Dairy Sci.* **2016**, *99*, 8665–8679. [CrossRef] [PubMed]
42. Badings, H.; de Jong, C. Headspace analysis for the study of aroma compounds in milk and dairy products. *Anal. Volatiles Methods-Appl.* **1984**, 401–419. [CrossRef]
43. Shiratsuchi, H.; Shimoda, M.; Imayoshi, K.; Noda, K.; Osajima, Y. Volatile flavor compounds in spray-dried skim milk powder. *J. Agric. Food Chem.* **1994**, *42*, 984–988. [CrossRef]
44. Ye, D.; Zheng, X.; Li, Y.; Duan, C.; Liu, Y. Influence of Aging in Different Oak Barrels on Volatile Sulfur Compounds in Wines. *Food Sci.* **2020**, *41*, 285–291.
45. Zhang, K.; Zhang, C.; Gao, L.; Zhuang, H.; Feng, T.; Xu, G. Analysis of volatile flavor compounds of green wheat under different treatments by GC-MS and GC-IMS. *J. Food Biochem.* **2022**, *46*, e13875. [CrossRef] [PubMed]
46. Deng, F.; Zhang, R.; Yu, J.; He, J.; Hong, S. Cutting Value Optimization Method of Logistic Regression Based on ROC Curve. *J. Kunming Univ. Sci. Technol.* **2023**, *48*, 53–57. [CrossRef]
47. Chen, W.; Ni, Z.; Pan, X.; Liu, Y.; Xia, Y. The ROC curve was used to determine the optimal critical point and the range of suspicious values. *Mod. Prev. Med.* **2005**, *7*, 729–731.
48. Zhang, X.; Tong, L.; Chi, X.; Ai, N.; Wang, J.; Sun, B. Analysis of volatile flavor substances in cow milk, buffalo milk and yak milk. *Food Res. Dev.* **2017**, *38*, 126–131.
49. Barłowska, J.; Szwajkowska, M.; Litwińczuk, Z.; Król, J. Nutritional value and technological suitability of milk from various animal species used for dairy production. *Compr. Rev. Food Sci. Food Saf.* **2011**, *10*, 291–302. [CrossRef]
50. Sue, C.D.; Yang, M. Odor components of milk and its products. *China Dairy Ind.* **1987**, *15*, 212–225+197.

Disclaimer/Publisher's Note: The statements, opinions and data contained in all publications are solely those of the individual author(s) and contributor(s) and not of MDPI and/or the editor(s). MDPI and/or the editor(s) disclaim responsibility for any injury to people or property resulting from any ideas, methods, instructions or products referred to in the content.

foods

MDPI

Article

Effects of Reducing Sugars on Colour, Amino Acids, and Volatile Flavour Compounds in Thermally Treated Minced Chicken Carcass Hydrolysate

Xing Zhang [1] and Shao-Quan Liu [2,*]

[1] Department of Food Science and Technology, National University of Singapore, Science Drive 3, Singapore 117543, Singapore; e0669543@u.nus.edu
[2] National University of Singapore (Suzhou) Research Institute, 377 Lin Quan Street, Suzhou Industrial Park, Suzhou 215213, China
* Correspondence: fstlsq@nus.edu.sg;

Abstract: This study investigated the changes in colour, amino acids, and volatile flavour compounds in the enzymatic hydrolysates of chicken carcasses containing different types and amounts of reducing sugars (xylose, arabinose, glucose, and fructose), so as to develop a chicken-based flavouring agent. Before heat treatment at 100 °C for 60 min, the chosen reducing sugars were separately added to the chicken carcass hydrolysate at its natural pH. Pentoses decreased pH more significantly than hexoses in the chicken carcass hydrolysate. The browning degree followed the pattern of pH decline, as pentoses caused more intense browning than hexoses, with xylose dosage having the greatest effect on the colour changes (ΔE). Fructose addition notably reduced free amino acids (FAAs) and cystine contents. Furthermore, phenylalanine decreased with increasing dosages of arabinose, xylose, and fructose. Glutamic acid content decreased significantly with fructose addition but showed insignificant changes with xylose. At the same dosage, the addition of pentoses resulted in the production of more sulphur-containing volatile compounds like methional, 2-[(methylthio) methyl] furan, and dimethyl disulphide than hexoses. Methional and furfural, which provide a roasted, savoury flavour, were produced by adding more xylose. Heat treatment with xylose also removed hexanal, the main off-odourant.

Keywords: chicken carcass; enzymatic hydrolysis; Maillard reaction; flavour compounds

1. Introduction

A chicken carcass or frame refers to the remains of a chicken after it has been processed for its meat. It includes the skeletal structure and other parts of the chicken that are typically not consumed as meat. These carcasses can be used in various ways, such as making chicken broth or stock, pet food, or even in some industrial applications to extract valuable components [1]. Kampong chicken is the main chicken in Southeast Asia, which is used in dishes like Hainanese chicken rice. Free-range Kampong chicken is known for its toughness and unique flavour as well as its resistance to several diseases and high selling price [2,3]. Moreover, due to the lower growth rates, aged Kampong chickens have lower nutritional value and meat quality, making them difficult to market [4,5]. Thus, our study was focused on fresh Kampong chicken carcasses. Some studies have been conducted to assess the suitability of chicken by-products for the manufacture of value-added products such as protein hydrolysates and polyunsaturated fatty acids [6–8]. The efficient utilisation of by-products represents a comprehensive strategy for minimizing losses. Presently, various methods exist for maximizing the utility of chicken by-products, including their incorporation into animal feed, pet food, and soil enhancement; however, these avenues may entail considerable expenses. Moreover, the majority of these by-products are unsuitable for direct human consumption, underscoring the imperative to explore alternative means of enhancing their value. Additionally, inadequate reaction conditions in previous studies have led to the persistence of undesirable bitterness. Furthermore, a systematic comparison is lacking regarding the influence of pentoses and hexoses on the flavour of chicken carcass enzymatic hydrolysates during the Maillard reaction (MR).

Some peptides and amino acids from the enzymatically hydrolysed animal protein are recognised as taste-active substances and/or important flavour precursors [9]. For instance, some researchers developed a flavouring substance that could be added in instant noodle seasonings by the microencapsulation of mussel protein hydrolysates [9]. Previous research has indicated that most meaty flavour compounds have a similar structural unit to sulphur or oxygen heteroatoms on the adjacent two carbon atoms (one of the heteroatoms should be a sulphur atom) [10,11]. These distinctive structural units are further classified into six subclasses based on different bonding types, including 2-methyl-3-furanthiol (MFT) and 2-furfurylthiol (FFT), 1-alkylthio-2-butanethiols, 1-alkylthio-2-butanols, 2-methoxybutylalkyl sulphides, and 1-alkylthio-2-butanones [10].

Maillard reaction is one of the most important non-enzymic reactions in the food industry, which often occurs during the thermal processing of foodstuffs such as cooking meat, baking bread, and roasting coffee. Sun et al. (2014) suggested that the heat treatment of chicken bone hydrolysates was a promising way to prepare natural meat flavour enhancers [12]. The nature of the sugars and amino acids, as well as the pH, and the reaction temperature, time, and the water activity of the reaction system greatly affect the formation of volatiles and therefore influence the flavour profile of the final product [13]. The Maillard reaction products (MRPs) produced from chicken bone extracts may be exploited as possible natural flavourings in food applications [14]. Further, chicken bones were transformed into hydrolysates that could be nutritional and flavouring substances [15]. Recently, researchers demonstrated the effect of degree of thermal treatment on flavour generation from the Maillard reaction of xylose and chicken peptides; high temperature ($\geq$100 °C) could remarkably increase the formation of meaty aroma, while lower temperature and longer heating duration tended to generate a broth-like taste (i.e., umami and kokumi) [16]. Furthermore, the taste enhancers that could lead to bitterness would be produced at a higher temperature (>120 °C); hence, relatively lower temperatures and longer heating time were needed [16]. This study aimed to further explore the valorisation of chicken carcass enzymatic hydrolysates, building upon our previous optimisation research [17]. To be specific, the objective was to promote the thermal generation of both volatile and non-volatile flavour compounds by incorporating four reducing sugars as flavour precursors (xylose, arabinose, glucose, and fructose). This approach was driven by several reasons: (1) in view of the lack of a reducing sugar in the chicken carcass hydrolysate, the supplementation of reducing sugars would be of great importance for both the colour and the volatile compound formation [18]; (2) previous research showed that pentoses tended to be more reactive than hexoses in the Maillard reaction [19], contributing to different volatile flavour compounds; (3) C-2 and C-3 OH groups are the key differences between xylose and arabinose, which could bring about the different stabilities in the hemiacetal ring that may cause different cleavages during heat treatment, resulting in diverse reactivity ratios and browning degrees [19]; and (4) the order of reactivity according to which aldoses are more reactive than ketoses; however, fructose (ketose) and glucose (aldose) are the counter examples; in other words, more reactive and more numerous reaction products can be generated from ketose (i.e., fructose) compared to aldose (i.e., glucose) [19]. The findings from the present study would be expected to help the poultry industry towards by-product valorisation by developing chicken-based flavouring substances from chicken carcasses.

2. Materials and Methods

2.1. Preparation of Chicken Carcass Hydrolysates

Fresh Kampong chicken carcasses were purchased from a local supermarket (Aw's Market Fresh Anxin Kampong Chicken Bones, NTUC Fairprice Co Ltd., Singapore). The protease used was MEAP (mixed enzymes PB02 for animal proteolysis, Nanning Pangbo

Biological Engineering Co., Ltd., Nanning, China), which is a protease/peptidase composed of fungal protease and peptidase derived from *A. oryzae*, endoprotease, and mixed exopeptidases, 200,000 units g^{-1}, and the optimised hydrolysis conditions were described by Zhang et al. [17]. In general, prior to the experiment, 40 g of defrosted minced chicken carcasses was suspended in distilled water in a Duran bottle (Merck, Darmstadt, Germany) at a ratio of 1:1 (*w/v*). The content was manually mixed and then heated to 50 °C for 10 min in a water bath (SW22, Julabo, Seelbach, Germany). Then, a 4 h hydrolysis was immediately conducted, and the optimal parameters were determined to be a protease/substrate ratio of 3:100 (*w/w*), 51.20 °C, natural pH of 6.62 ± 0.05, and substrate/water ratio of 1:1 (*w/v*). The resulting slurry was filtered to eliminate the insoluble components using a coconut cloth filter to obtain a clear, transparent, light yellow-brown liquid. The highest DH for this hydrolysate was 45.44%. On a dry weight basis, the protein recovery was 50.45 ± 2.05%, including 7.29 ± 0.01% ash, 67.67 ± 1.55% crude protein, and 1.25 ± 0.04% fat. The majority of the peptides in this hydrolysate had smaller molecular weights (<5 kDa) [17].

2.2. Thermal Treatment of Chicken Carcass Hydrolysates

A 40 mL volume of fresh chicken carcass hydrolysates was transferred into a 100 mL blue-capped Duran® laboratory bottle (Merck, Darmstadt, Germany) with a specific amount of different reducing sugars (L-(+)-xylose, L-(+)-arabinose, D-(+)-glucose, and D-(-)-fructose, Merck, Darmstadt, Germany) as shown in Table 1 without any pH adjustments. The thermal treatment conditions were determined according to the previous research and preliminary test [16]. To be specific, all samples except the unheated control were heated at 100 °C for 1 h in a vertical autoclave (Hirayama MFG Corp., Kasukabe-ShiSaitama, Japan), to allow the formation of MRPs. At the end of autoclaving, the samples were cooled immediately in ice water. Unheated samples were prepared at room temperature (25 °C). At the end of heating, all samples were immediately cooled in an ice bath to prevent further reactions.

Table 1. Heat treatment conditions for chicken carcass hydrolysates.

Sample Name [1]	Sugar Added (g/100 g)				Temperature (°C)	Time (min)
	Xylose	Arabinose	Glucose	Fructose		
Unheated control [2,3]	0	0	0	0	-	-
Heated Control	0	0	0	0	100	60
Xyl-05	0.5	0	0	0	100	60
Xyl-15	1.5	0	0	0	100	60
Xyl-25	2.5	0	0	0	100	60
Xyl-35	3.5	0	0	0	100	60
Ara-05	0	0.5	0	0	100	60
Ara-15	0	1.5	0	0	100	60
Ara-25	0	2.5	0	0	100	60
Ara-35	0	3.5	0	0	100	60
Glu-05	0	0	0.5	0	100	60
Glu-15	0	0	1.5	0	100	60
Glu-25	0	0	2.5	0	100	60
Glu-35	0	0	3.5	0	100	60
Fru-05	0	0	0	0.5	100	60
Fru-15	0	0	0	1.5	100	60
Fru-25	0	0	0	2.5	100	60
Fru-35	0	0	0	3.5	100	60

"-": not applied. [1] there were four replications of each treatment for data acquisitions (n = 4). [2] unheated control was prepared and incubated at room temperature (25 °C) for 1 h. [3] the unadjusted pH was 6.62 ± 0.05.

2.3. Colour Analysis and pH Measurement

A reflectance spectrometer (CM-3500 d; Konica Minolta Sensing Inc., Osaka, Japan) connected to a D65 illumination with 10 observations was used to measure the colour. a* (red > 0, green < 0), b* (yellow > 0, blue < 0), and L* (lightness, black = 0, white = 100) were the three colour distribution parameters used in the measurement. The following equation was used to calculate the colour difference (ΔE) using the L*, a*, and b* values obtained from unheated control samples [20].

$$\Delta E = \sqrt{(L_t - L_0)^2 + (a_t - a_0)^2 + (b_t - b_0)^2} \quad (1)$$

The pH values of each sample were measured with a SevenCompact™ pH meter (Mettler-Toledo 135 International Inc., Columbus, OH, USA).

2.4. Determination of Sugars

The reducing sugar (xylose, arabinose, glucose, and fructose) contents were measured before and after heat treatment using a 1260 Infinity II preparative ultra-high performance liquid chromatography and reflective index detector (UHPLC-RID) system (Agilent, Santa Clara, CA, USA) with a modified protocol described by Chua et al. [21]. Before injection, 1.0 mL of each sample was filtered through a 0.2 μm Minisart RC 15 syringe filter (Sartorius, Goettingen, Germany), and then, 200 μL of each filtered sample was mixed with 200 μL of deionised (DI) water (arium®, Sartorius Stedim Biotech, Goettingen, Germany) and 600 μL of pure acetonitrile (Tedia, Fairfield, OH, USA) and kept at 4 °C for more than 2 h to allow protein precipitation. Each sample was centrifuged at 10,000× *g* at 4 °C for 15 min to remove the precipitate, and the collected supernatant was injected to the UHPLC system for sugar analysis. The sugars were separated using a Zorbax Eclipse Plus C18 column (150 × 4.6 mm, Agilent, Santa Clara, CA, USA) using a mobile phase containing 80% acetonitrile and 20% DI water at a flow rate of 1.4 mL/min at 40 °C. A series of external xylose, arabinose, glucose, and fructose standards were used for identification and quantification.

2.5. Analysis of Amino Acids

Amino acids were determined using the ARACUS Amino Acid Analyzer's pre-set physiological separation algorithm (MembraPure, Berlin, Germany) following the procedure of Zhou et al. [22]. A lithium-cation exchange column was used along with ninhydrin (a reagent for post-column derivatisation) and eluents (Eluent A to Eluent F) provided by the manufacturer. Except for proline and hydroxyproline, which were identified at 440 nm, all other amino acids were detected at 570 nm. The amino acids were quantified using a calibration factor generated by MembraPure.

2.6. Analysis of Volatile Compounds

The analysis of volatile compounds in thermally treated chicken carcass hydrolysates was conducted using a modified protocol described by Vong and Liu [23]. The volatile compounds were extracted by headspace solid-phase microextraction (HS-SPME), coupled with gas chromatography–mass spectrometry and semi-quantified by flame ionisation detector (GC-MS/FID, Agilent 5975C, Santa Clara, CA, USA). An aliquot of 5 mL of samples was added to a 20 mL glass vial with a PTFE septum. An 85 mL carboxen/polydimethylsiloxane SPME fibre (CAR/PDMS, Supelco, Bellefonte, PA, USA) was used to extract the headspace volatile compounds at 60 °C for 30 min under 250 rpm agitation using a Combi PAL autosampler (CTC Analytics, Zwingen, Switzerland). The extracted compounds were separated on a DB-FFAP capillary column (60 m length, 0.25 mm i.d., 0.25 μm film thickness, Agilent, Santa Clara, CA, USA) with helium as the carrier gas at a flow rate of 1.2 mL/min. The conditions of separation were adopted from the study of Gao et al. [24]. Oven temperature was set at 40 °C for 3 min, then increased to 90 °C at a rate 5 °C per min, without holding, and increased to 230 °C at a rate 10 °C min per min with a holding time of 7 min. The identification of compounds was based on the comparison of their mass spectra with a database in NIST 17.0 and Wiley 275 MS libraries and their linear retention index (LRI) values. The LRI value of each compound was calculated using its retention time and a series of alkane standards determined in our laboratory. GC-FID peak areas were used to semi-quantify the volatiles, and the relative peak area (RPA) expressed in percentage was calculated in each major group of compounds.

2.7. Statistical Analysis

All results were obtained from three independent experiments (conducted at different times). Two-way analysis of variance (ANOVA) and Duncan's multiple range tests were performed for each data set by using SPSS® 20.0 (SPSS Inc. Chicago, IL, USA). The bar graphs were plotted by GraphPad Prism 8.0. The FID peak areas were subjected to missing value estimation and normalisation by data scaling, that is, auto-scaling by mean-centring and dividing by the standard deviation of each variable prior to analysis.

3. Results and Discussion

3.1. pH Changes

A general decreasing trend of pH was observed in all heated samples (from 6.14 to 5.01, 5.19, 5.63, and 5.97, respectively). Figure 1 shows an inverse correlation between the pH and the sugar dosage; pentoses showed a more significant pH reduction than hexoses. To be specific, the lowest pH of 5.01 ± 0.01 was found in the sample of xylose (Xyl-35), followed by arabinose (Ara-35) (5.19 ± 0.03), glucose (Glu-35) (5.67 ± 0.03), and fructose (Fru-35) (5.97 ± 0.01). The consumption of the amino groups and the accumulation of various organic acidic molecules such as formic acid and acetic acid during the Maillard reaction might be the major reason of pH decline [25]. Pentoses typically have greater impacts on decreasing pH compared to hexoses during the Maillard reaction for several reasons: some researchers revealed pentoses have less steric hindrance, so that the reaction with amino groups is promoted more than that with hexoses, and there is a greater probability for pentoses to be in their open-chain form because of the lower stability of the furanose ring (higher tension within the molecule, which forces the ring to open) compared to the pyranose ring. Pentoses may release more protons (H^+ ions), causing the pH to drop more noticeably. Additionally, the disposition of OH groups affect the stability of the molecule (including the proportion of acyclic form), the initial steps of the Maillard reaction, and intermediate conversion, ultimately influencing global sugar reactivity [19,26].

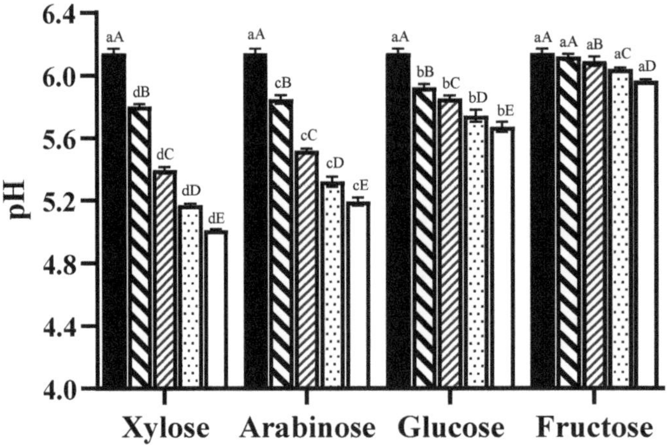

Figure 1. The changes in pH in chicken carcass hydrolysates with different added sugars (■ 0%, ▨ 0.5%, ▨ 1.5%, ▭ 2.5%, ▭ 3.5%) during heat treatment at 100 °C for 1 h. ^{a,b,c,d} values within the same dosage on different types of sugars followed by the same letters are not significantly different ($p > 0.05$). ^{A,B,C,D,E} values between the different sugar dosages on the same sugar followed by the same letters are not significantly different ($p > 0.05$).

The pH value affects the Maillard reaction significantly. It is known that when the pH is below seven, the Amadori rearrangement products (ARPs) undergo 1,2-enolisation to create 3-deoxyosone, which may then cyclise and dehydrate to produce 5-hydroxymethylfurfural (HMF) from hexoses and furfural from pentoses [27]. When the pH is greater than seven, ARPs undergo 2,3-enolisation to create 1-deoxyosone, which can subsequently break down into α-dicarbonyls (i.e., α-DCs) and reducing ketones [27]. In contrast, solutions with higher pH values exhibit a quicker reaction rate and a more intense colour development [28].

3.2. Colour Changes

The colour changes in controls and heated samples are shown in Figure S1 and Table 2. In general, ΔE values rose as each sugar dose increased. The relationship between the colour changes and the dosage of reducing sugar was evident. The browning degree shared the same trend as pH decline in that the pentoses had more intense browning than the hexoses. At 0.5%, fructose addition had deeper browning than glucose; however, all the fructose-added samples with higher dosages (1.5%, 2.5%, 3.5%) had overall colour changes (ΔE) opposite to those of the glucose samples, with the corresponding dosages being consistent with the previous research [29]. To be specific, the heated control showed the highest value of lightness (L*) and the lowest value of redness (a*), with a moderate yellowness (b*) level. After heat treatment with different types and dosages of reducing sugars, pentoses could result in lower L* and b* than hexoses. Pentoses-added samples, on the other hand, could also result in a relatively higher a*; among which, Xyl-15 sample had the lower L* (28.64 ± 0.52) and the highest a* (45.81 ± 0.18), with moderate b* (49.13 ± 0.95), indicating a dark brownish red and yellow colour. L* was mainly affected by the amount of browning pigments (i.e., melanoidins), while the amount of these pigments was influenced by the reactive degree of the Maillard reaction and caramelisation. Fructose has less accessible carbonyl function and a much lower energy open-chain form, which is characterised by a much lower reactive degree of Maillard reaction. Moreover, at the temperature used in the present study (100 °C, 1 h), caramelisation, which could have led to an overestimation of browning measurements in a fructose system, may have not occurred.

Table 2. Colour changes in control and heat-treated chicken carcass hydrolysates.

Sample Name	Colour			
	L*	a*	b*	ΔE [a]
Unheated control	83.77 ± 1.33	1.51 ± 0.32	42.23 ± 0.92	-
Heated control	80.55 ± 0.99	2.96 ± 0.44	49.22 ± 0.46	10.57 ± 0.82
Xyl-05	62.11 ± 0. 20 Ab	30.40 ± 0. 16 Cc	84.22 ± 0. 13 Aa	55.38 ± 0.21 Da
Xyl-15	28.64 ± 0. 52 Bc	45.81 ± 0. 18 Ab	49.13 ± 0. 95 Ba	71.06 ± 0.21 Ca
Xyl-25	13.96 ± 0. 19 Cc	38.49 ± 0. 21 Ba	23.70 ± 0. 40 Ca	81.15 ± 0.17 Ba
Xyl-35	6.29 ± 0. 09 Dc	30.18 ± 0. 18 Ca	10.22 ± 0. 14 Db	88.59 ± 0.07 Aa
Ara-05	59.38 ± 0. 32 Ab	22.01 ± 0. 31 Cc	71.65 ± 0. 34 Aa	43.37 ± 0.49 Db
Ara-15	34.81 ± 0. 43 Bc	41.19 ± 0. 37 Ab	59.27 ± 0. 66 Ba	65.29 ± 0.26 Cb
Ara-25	21.74 ± 0. 77 Cc	41.43 ± 0. 40 Aa	36.96 ± 1. 40 Cb	73.96 ± 0.55 Bb
Ara-35	14.13 ± 0. 42 Dc	37.59 ± 0. 40 Ba	23.76 ± 0. 74 Db	80.58 ± 0.35 Ab
Glu-05	76.92 ± 0. 13 Aa	5.10 ± 0. 05 Dc	53.33 ± 0. 18 Db	13.53 ± 0.12 Dc
Glu-15	75.16 ± 0. 14 Ba	8.01 ± 0. 14 Cc	59.64 ± 0. 34 Cb	20.35 ± 0.28 Cc
Glu-25	72.32 ± 0. 30 Ca	12.08 ± 0. 14 Bc	66.90 ± 0. 09 Bb	29.18 ± 0.17 Bc
Glu-35	69.93 ± 0. 22 Db	15.83 ± 0. 22 Ac	73.15 ± 0. 78 Aa	36.78 ± 0.74 Ac
Fru-05	78.80 ± 0. 49 Aa	5.49 ± 0. 10 Bc	56.62 ± 0. 09 Cb	15.74 ± 0.12 Cd
Fru-15	79.51 ± 0. 13 Aa	5.90 ± 0. 87 Bc	58.26 ± 2. 31 Cb	17.16 ± 2.41 Cd
Fru-25	78.85 ± 0. 21 Aa	7.75 ± 0. 04 Ac	62.43 ± 0. 13 Bb	21.70 ± 0.11 Bd
Fru-35	78.00 ± 0. 53 Aa	9.27 ± 0. 25 Ac	65.28 ± 0. 09 Ab	25.00 ± 0.27 Ad
	Test of significance between effects, P			
	Main effect			
Sugar type	***	***	***	***

Table 2. *Cont.*

Sample Name	Colour			
	L*	a*	b*	ΔE [a]
Sugar dosage	***	***	***	***
Sugar type × dosage	***	***	***	***

[a] Different capital letters of ΔE indicate significant differences among the groups with the same sugar at different dosages, while lowercase letters indicate significant differences among the groups with different sugars at the same dosage. All values are the mean ± standard deviation with three replications of each sample treatment (n = 3). [a,b,c,d] values within the same sugar dosage on different types of sugars followed by the same letters are not significantly different ($p > 0.05$). [A,B,C,D] values between the different sugar dosages on the same sugar followed by the same letters are not significantly different ($p > 0.05$). ***: Significant differences ($p < 0.001$).

Besides the caramelisation of sugars, the Maillard reaction between reducing sugars and amino acids results in the production of melanoidins. Amadori rearrangement products (ARPs)/Heyns rearrangement products (Heyns) are crucial intermediates for the ensuing chain of events that generate volatile compounds and brown pigments (i.e., melanoidins), which directly affect the organoleptic qualities of foods [30]. Interestingly, previous research showed that cysteine was effective in preventing browning, but this phenomenon was only observed when cysteine was added to the system during the first mild reaction step [31,32], but the final reactions already happened in our study after 1 h heat treatment, which might be the reason why browning inhibition did not happen in the current study. The very low content of cysteine in the original chicken carcass hydrolysate may be another reason for the lack of browning inhibition [17].

3.3. *Sugar Changes after Heat Treatment*

The four sugars were barely detectable in both the original chicken carcass hydrolysate (unheated control) and the heated control (xylose: 0 mg/mL; arabinose: 0.8 mg/mL; glucose: 0.72 mg/mL; fructose: 0 mg/mL). Thus, reducing sugars were added to enable the Maillard reaction. All four reducing sugars were consumed in variable amounts, as shown in Figure 2, and this led to varying levels of sugar reduction. Overall, the more sugar was added, the larger the sugar decreased; the largest sugar reduction was 33.63 mg/mL in the fructose-added sample (Fru-35), followed by 33.59 mg/mL in the arabinose-added sample (Ara-35). Fragmentation of reducing sugars in the Maillard reaction could produce numerous substances including volatile flavour compounds (e.g., furans), melanoidins, and Maillard reaction intermediates (MRIs) [33].

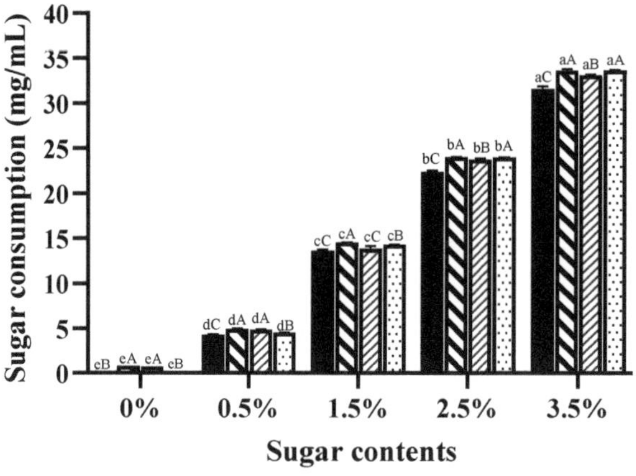

Figure 2. Reducing sugar consumption (mg/mL) in chicken carcass hydrolysates with different added sugars (▬ Xylose, ▬ Arabinose, ▬ Glucose, ▬ Fructose) after heat treatment at 100 °C

for 1 h. [a,b,c,d,e] values between the different dosages on the same sugar followed by the same letters are not significantly different ($p > 0.05$). [A,B,C] values within the same dosages on different types of sugars followed by the same letters are not significantly different ($p > 0.05$).

3.4. Free Amino Acids

The total amount of free amino acids (FAAs) was 446.84 mg/10 mL in the heated control. After heat treatment, there were varying degrees of decline in the total FAAs as the dosage of reducing sugars increased (Figure S2). In general, fructose addition caused the most reduction in FAAs (189.07 mg/10 mL in sample Fru-35), followed by 180.30 mg/10 mL in sample Ara-35, likely due to the higher reactivity of these sugars as mentioned earlier. During the early stage of the Maillard reaction, the terminal α-amino groups of peptides and ε-NH_2 groups of amino acids interact with the carbonyl group of reducing sugars in the reaction mixture. To compare the sugar reactivity in the MR, another measure that is employed is the loss of accessible primary amino groups. Similar to hexose, fructose should have a lower reduction in amino acids than pentoses; on the other hand, as mentioned in the introduction, fructose tended to be more reactive than glucose, which was in line with other studies [19]. Past research has shown conflicting results on the reactivity of fructose and glucose, with fructose sometimes exhibiting a higher reactivity based on the reaction conditions. This finding may also have been caused by the thermal treatment conditions; it has been demonstrated that conditions such as pH, heating duration, temperature, and moisture/a_w factors have a significant impact on the decrease in amino acids [19,28,34]. This study may be notable for its attempt to preserve the highest concentration of taste-active amino acids, while simultaneously producing a high level of aroma flavour components.

To create a meat-based liquid seasoning with desirable aromas such as savoury and meaty notes, it would be necessary to monitor the changes in sulphur-containing amino acids such as cysteine (Cyst), cystine ($(Cys)_2$), and methionine (Met) [33]. As shown in Figure 3a, there exists a dose-dependent relationship between cystine reduction and the dosage of arabinose and fructose; the lowest residual $(Cys)_2$ amount was 2.73 mg/10 mL in sample Fru-35. Nevertheless, the cystine content in sample Xyl-05 decreased from 8.44 mg/10 mL to 6.67 mg/10 mL, which was statistically significant ($p < 0.05$), but more significant reduction was not achieved if more xylose was added ($p > 0.05$). Interestingly, there was no clear relationship between glucose dosage and cystine reduction. As shown in Figure 3b, in the hexose-added group, the lowest amount of methionine (Met) was found in sample Fru-35 at 5.28 mg/10 mL with an evident dose-dependent relationship between fructose and Met decrease, which was not the case for glucose. Similar to fructose, Met consumption was proportional to the dosage of pentoses.

In addition, the change in phenylalanine (Phe) was investigated (Figure 3c). There was a general trend of decreasing Phe with increasing dosage of Ara, Xyl, and especially Fru, but not in the case of glucose in which Phe fluctuated. A previous investigation demonstrated that Xyl-Phe had the capability to form an intermediate of the Maillard reaction with a molecular weight of 297 [35]. Through subsequent processes such as dehydration, decarboxylation, and the removal of the furan ring, Xyl-Phe can undergo degradation, ultimately resulting in the formation of benzaldehyde and benzeneacetaldehyde, which could frequently be encountered as flavour components that enhance the overall flavour profile while simultaneously diminishing the perception of bitterness [12,35]. Interestingly, very little recent Ara-Phe related research has been reported. As for the Fru-Phe system, previous research in a model system showed that furfural was the only furan derivative detected, and the formation of phenylacetaldehyde can be due to the Strecker breakdown of phenylalanine [36].

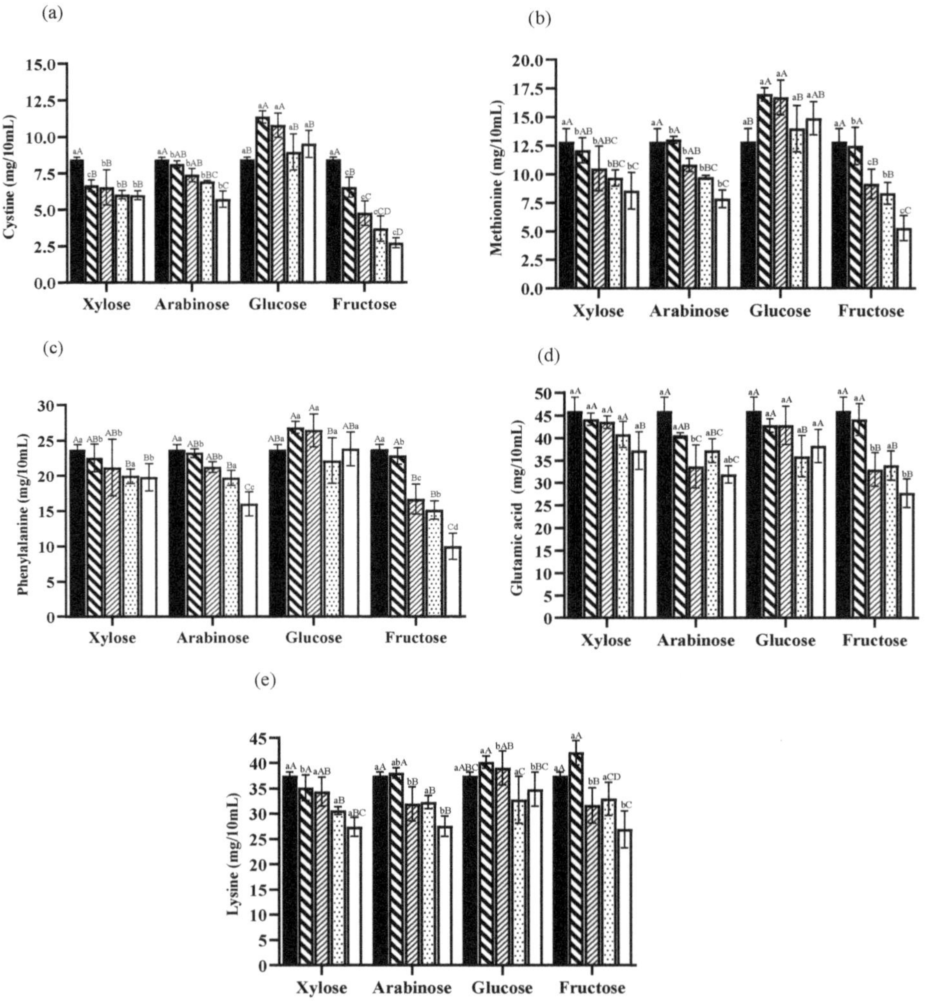

Figure 3. Changes in selected amino acids: cystine (**a**), methionine (**b**), phenylalanine (**c**), glutamic acid (**d**), and lysine (**e**) in heat-treated samples added with different sugars at different dosages (0%, 0.5%, 1.5%, 2.5%, 3.5%) after heat treatment at 100 °C for 1 h. ^{a,b,c,d} values within the same dosage on different types of sugars followed by the same letters are not significantly different ($p > 0.05$). ^{A,B,C,D} values between the different sugar dosages on the same sugar followed by the same letters are not significantly different ($p > 0.05$).

The content of glutamic acid (Glu) decreased consistently but insignificantly in the xylose group ($p > 0.05$) except at 3.5% ($p < 0.05$) (Figure 3d). On the contrary, Glu in sample Fru-35 reduced markedly at 27.72 mg/10 mL (Figure 3d). Significant reductions in Glu mainly occurred at higher dosages of reducing sugars. Lysine (Lys) also plays an important role in MR, including the antioxidation activity and browning degree [37]. Figure 3e shows the changes in Lys. In general, there was no linear relationship between the two hexoses dosage and the Lys content. Furthermore, among the xylose-added samples, the Lys content decreased steadily from 37.47 mg/10 mL to 27.37 mg/10 mL. Furthermore, Fru-35 sample had the lowest Lys content of 26.89 mg/10 mL.

3.5. Volatile Compounds

The volatile compounds are grouped into six groups: acids, alcohols, aldehydes, ketones, furans, and sulphur-containing volatiles. Figure 4 presents the relative peak areas (RPAs, %, Table S1) of each group in the total volatile compounds among the heat-treated samples added with reducing sugars at 2.5%. In general, furans were the dominant compounds in arabinose, glucose, and fructose-added samples (58.39%, 66.90%, 74.48% RPA), while aldehydes had the highest number (nine types) in the xylose-added sample, accounting for 57.31%. Interestingly, the highest number of sulphur-containing volatiles (four types) was found in the xylose-added sample, accounting for 5.88%, followed by arabinose, whereas only one sulphur-containing volatile (2-pentyl-thiophene, accounting for 0.62%) was found in the fructose-added sample, despite it consuming a considerable amount of sulphur-containing amino acids. Fructose addition at high concentration (3.5%) consumed the highest number of amino acids but produced the lowest level of sulphur-containing volatile flavour compounds.

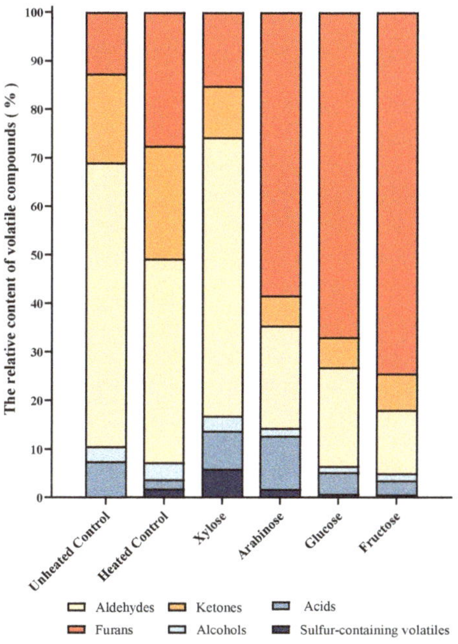

Figure 4. The relative peak areas (RPAs, %) of six groups of volatiles in the total volatile compounds with different sugars added at 2.5% after heat treatment at 100 °C for 1 h.

Farmer et al. (1994) found that Cyst breakdown during heat treatment would be most desirable since it was anticipated that Cyst would combine with xylose to form meat-like flavour compounds, such as 2-methyl-3-furanthiol (MFT) and 2-furfurylthiol (FFT) [38]. However, the chicken carcass hydrolysate used in our study contained little amount of Cyst, which might explain the lack of MFT and FFT after heat treatment. The highest methional amount was found in sample Xyl-35 (peak area of 89.45×10^5), accounting for 4.09%, which was nearly 2.7 times of the methional content in the control (peak area of 33.22×10^5). Only xylose could boost the generation of methional, and there was no direct relationship between other sugars added and the methional formation (Figure 5a). Previous study showed that Phe and xylose could generate benzeneacetaldehyde after heating at 80 °C for 70 min [35], which was consistent with the result of this study (Table S1).

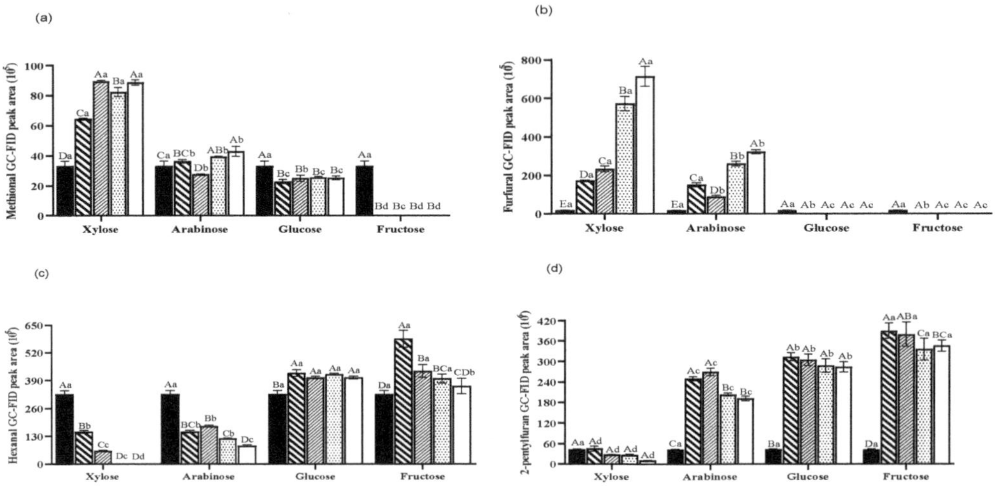

Figure 5. Changes in GC-MS/FID peak areas of methional (**a**) furfural (**b**), hexanal (**c**), and 2-pentylfuran (**d**) in heat-treated samples added with sugars (xylose, arabinose, glucose, and fructose) (0%, 0.5%, 1.5%, 2.5%, 3.5%). ^{a,b,c,d} values within the same dosage on different types of sugars followed by the same letters are not significantly different ($p > 0.05$). ^{A,B,C,D} values between the different dosages on the same sugar followed by the same letters are not significantly different ($p > 0.05$).

The change in furfural was obvious (Figure 5b) in pentose-added samples, and the amount of furfural increased significantly with rising xylose and arabinose dosage, except that the furfural content reduced suddenly from a peak area of 169.63×10^5 in sample Ara-05 to 92.32×10^5 in sample Ara-15. On the other hand and as expected, the furfural contents of hexoses-added samples were below the detectable level because pentoses are supposed to be dehydrated into furfural, whereas hexoses to hydroxymethylfurfural (HMF) [39,40]. HMF imparts sensory properties (colour, taste, and aroma); however, HMF was not detected in both glucose- and fructose-added samples, which might be caused by the lower heating temperature (100 °C) or higher initial pH value (6.62) and/or the detection method. Furfural is an intermediate formed in the MR in the presence of a pentose; it imparts a robust, spicy, and burned flavour [40,41]. During the cooking of meat, furfural often interacts with hydrogen sulphide, which is produced from the breakdown of Cyst and/other sources, to generate 2-furanmethanethiol, which imparts a strong and distant "roasted meat" aroma to the product [42]. However, 2-furanmethanethiol was not found in this investigation, which may be the result of the low concentration of Cyst in the chicken carcass hydrolysate [17].

There were two opposite trends of hexanal changes between pentoses and hexoses, i.e., with different amounts of added pentoses (Figure 5c), the hexanal level decreased to varying levels, with peak areas of 62.47×10^6 and 177.61×10^6, accounting for 2.91% and 3.81% RPA, respectively. When the amount of xylose exceeded 2.5%, the hexanal content in the heat treatment sample was below the detected level. Contrary to the pentoses, the hexanal levels in hexose-added samples increased first and then were relatively stable in the glucose-added samples, but decreased in the fructose-added samples, with the highest hexanal peak area of 585.78×10^6 in sample Fru-05, accounting for 8.93% RPA. Hexanal is an aliphatic aldehyde that has a grassy odour when present in untreated meat hydrolysate [43]. The high level of hexanal found in different meat products such as cooked poultry meat can be a marker to distinguish the type of meat [44]. These oxidised lipids played a crucial part in the Maillard process and the formation of volatile compounds such as pyrazines, thiazoles, and thiols [42,45]. According to a previous investigation [42], the reduction in hexanal could also indicate its interference in the interaction between cysteine and reducing sugars (Figures 3a and 5c).

Overall, there were dramatic increases in 2-pentylfuran in sugar-added samples, except xylose (Figure 5d). In contrast, the xylose-added sample had the highest level of furfural (Figure 5b), which was predominant in the compounds generated by the heat treatment. Furfural could easily participate in further Strecker degradation and condensations with amino acids, leading to the formation of melanoidins that may have strong antioxidative properties, whereas 2-pentylfuran is often produced by the oxidation of linoleic acid and imparts buttery, green, and beany characteristics to foodstuffs when present at 1–10 ppm [46,47]. Xu et al. (2011) also revealed that MRPs affected the production of lipid oxidation products, which means 2-pentylfuran could be a marker for the lipid oxidation degree [42]. Although 2-pentylfuran generated after heat treatment does not directly provide a desired "meaty" aroma, it has been identified as a contributor to the overall broiled or roasted-meat aroma in certain sauce products [48].

4. Conclusions

The addition of two pentoses (xylose and arabinose) and two hexoses (glucose and fructose) led to different degrees of pH decrease and browning in the chicken carcass hydrolysate after heat treatment. Variable quantities of added sugars were used. There was a dose-dependent relationship between each sugar dosage and the colour change (ΔE values). The volatile flavour compound profiles were affected by the sugar type significantly. Pentoses addition at the same dosage could generate more sulphur-containing volatile compounds than hexoses, such as methional, 2-[(methylthio) methyl] furan and dimethyl disulphide. However, typical meat-like volatile flavour compounds such as 2-methyl-3-furanthiol (MFT) were not found due to the deficiency of cysteine.

Supplementary Materials: The following supporting information can be downloaded at: https://www.mdpi.com/article/10.3390/foods13070991/s1, Figure S1: Colour changes in unheated control and heat-treated chicken carcass hydrolysates; Figure S2: Change in total free amino acids in heat-treated samples added with different sugars at different dosages (0%, 0.5%, 1.5%, 2.5%, 3.5%) after heat treatment at 100 °C for 1 hr. ^{A,B,C,D} values within the same sugar dosage on different sugars followed by the same letters are not significantly different ($p > 0.05$). ^{a,b,c,d} values between the different sugar dosages on the same sugar added followed by the same letters are not significantly different ($p > 0.05$). Table S1: Average FID-GC peak areas (×106) and relative peak areas (RPAs, %) of volatiles in the blank, control, and heated chicken carcass hydrolysates with different added sugars (2.5%).

Author Contributions: Conceptualisation, X.Z. and S.-Q.L.; methodology, X.Z. and S.-Q.L.; software, X.Z. and S.-Q.L.; validation, X.Z., and S.-Q.L.; formal analysis, X.Z.; investigation, X.Z. and S.-Q.L.; resources, X.Z.; data curation, X.Z. and S.-Q.L.; writing—original draft preparation, X.Z.; writing—review and editing, X.Z. and S.-Q.L.; visualisation, X.Z. and S.-Q.L.; supervision, S.-Q.L.; project administration, X.Z. and S.-Q.L. All authors have read and agreed to the published version of the manuscript.

Funding: This research received no external funding.

Institutional Review Board Statement: Not applicable.

Informed Consent Statement: Not applicable.

Data Availability Statement: The original contributions presented in the study are included in the article/Supplementary Materials, further inquiries can be directed to the corresponding author.

Conflicts of Interest: The authors declare no conflicts of interest.

References

1. Livestock and Poultry: World Markets and Trade. Available online: https://www.fas.usda.gov/data/livestock-and-poultry-world-markets-and-trade (accessed on 15 June 2022).
2. Saragih, H.; Salsabila, N.; Deliaputri, R.; Firdaus, A.B.I.; Kurnianto, H. Growth Morphology of the Gastrointestinal Tract, Pectoralis Thoracicus Muscle, Lymphoid Organ and Visceral Index of Kampong Chicken. *J. Saudi Soc. Agric. Sci.* **2024**, *23*, 34–41. [CrossRef]
3. Wardiny, T.M.; Sumiati, S.; Retnani, Y.; Setiyono, A. Production of Functional Kampong Chicken Meat with Low Cholesterol, High Antioxidant, and Unsaturated Fatty Acids Fed Noni (Morinda Citrifolia) Leaf Shoot Meal. *Trop. Anim. Sci. J.* **2020**, *43*, 35–42. [CrossRef]
4. Suyatno, S.; Sujono, S.; Winaya, A.; Zalizar, L.; Pangestu, M. Characterization of Qualitative and Quantitative Traits of Four Types of Indonesian Native Chickens as Ancestor of New Strains of Local Super Laying Hens. *Jordan J. Biol. Sci.* **2023**, *16*, 171–179. [CrossRef]
5. Sarjana, T.A.; Suprijatna, E.; Mahfudz, L.D.; Sunarti, D.; Kismiati, S.; Ma'rifah, B.; Muryani, R.; Susanti, S. Effect of Transportation Distance on the Physiological Condition and Carcass Traits of Kampong Chickens. *S. Afr. J. Anim. Sci.* **2022**, *52*, 530–538. [CrossRef]
6. Bravo, F.I.; Calvo, E.; López-Villalba, R.A.; Torres-Fuentes, C.; Muguerza, B.; García-Ruiz, A.; Morales, D. Valorization of Chicken Slaughterhouse Byproducts to Obtain Antihypertensive Peptides. *Nutrients* **2023**, *15*, 457. [CrossRef] [PubMed]
7. Ibarz-Blanch, N.; Alcaide-Hidalgo, J.M.; Cortés-Espinar, A.J.; Albi-Puig, J.; Suárez, M.; Mulero, M.; Morales, D.; Bravo, F.I. Chicken Slaughterhouse By-Products: A Source of Protein Hydrolysates to Manage Non-Communicable Diseases. *Trends Food Sci. Technol.* **2023**, *139*, 104125. [CrossRef]
8. Zou, Y.; Shahidi, F.; Shi, H.; Wang, J.; Huang, Y.; Xu, W.; Wang, D. Values-Added Utilization of Protein and Hydrolysates from Animal Processing by-Product Livers: A Review. *Trends Food Sci. Technol.* **2021**, *110*, 432–442. [CrossRef]
9. Breternitz, N.R.; Bolini, H.M.A.; Hubinger, M.D. Sensory Acceptance Evaluation of a New Food Flavoring Produced by Microencapsulation of a Mussel (Perna Perna) Protein Hydrolysate. *LWT-Food Sci. Technol.* **2017**, *83*, 141–149. [CrossRef]
10. Sun, B.; Tian, H.; Zheng, F.; Liu, Y.; Xie, J. Meaty Aromas: Characteristic Structural Unit of Sulfur-Containing Compounds with a Basic Meat Flavor. *Meaty Aromas Charact. Struct. Unit Sulfur-Contain. Compd. Basic Meat Flavor* **2005**, *30*, 36–45.
11. Brehm, L.; Frank, O.; Jünger, M.; Wimmer, M.; Ranner, J.; Hofmann, T. Novel Taste-Enhancing 4-Amino-2-Methyl-5-Heteroalkypyrimidines Formed from Thiamine by Maillard-Type Reactions. *J. Agric. Food Chem.* **2019**, *67*, 13986–13997. [CrossRef]
12. Sun, H.-M.; Wang, J.-Z.; Zhang, C.-H.; Li, X.; Xu, X.; Dong, X.-B.; Hu, L.; Li, C.-H. Changes of Flavor Compounds of Hydrolyzed Chicken Bone Extracts during Maillard Reaction: Flavor Compounds of Chicken Bone Extrac. *J. Food Sci.* **2014**, *79*, C2415–C2426. [CrossRef]
13. Jousse, F.; Jongen, T.; Agterof, W.; Russell, S.; Braat, P. Simplified Kinetic Scheme of Flavor Formation by the Maillard Reaction. *J. Food Sci.* **2002**, *67*, 2534–2542. [CrossRef]
14. Chiang, J.H.; Yeo, M.T.Y.; Ong, D.S.M.; Henry, C.J. Comparison of the Molecular Properties and Volatile Compounds of Maillard Reaction Products Derived from Animal- and Cereal-Based Protein Hydrolysates. *Food Chem.* **2022**, *383*, 132609. [CrossRef]
15. Dong, X.B.; Li, X.; Zhang, C.H.; Wang, J.Z.; Tang, C.H.; Sun, H.M.; Jia, W.; Li, Y.; Chen, L.L. Development of a Novel Method for Hot-Pressure Extraction of Protein from Chicken Bone and the Effect of Enzymatic Hydrolysis on the Extracts. *Food Chem.* **2014**, *157*, 339–346. [CrossRef] [PubMed]
16. Liu, J.; Liu, M.; He, C.; Song, H.; Chen, F. Effect of Thermal Treatment on the Flavor Generation from Maillard Reaction of Xylose and Chicken Peptide. *LWT-Food Sci. Technol.* **2015**, *64*, 316–325. [CrossRef]
17. Zhang, X.; Li, X.; Liu, S.-Q. Enzymatic Hydrolysis of Minced Chicken Carcasses for Protein Hydrolysate Production. *Poult. Sci.* **2023**, *102*, 102791. [CrossRef] [PubMed]
18. Murata, M. Browning and Pigmentation in Food through the Maillard Reaction. *Glycoconj. J.* **2021**, *38*, 283–292. [CrossRef] [PubMed]
19. Laroque, D.; Inisan, C.; Berger, C.; Vouland, É.; Dufossé, L.; Guérard, F. Kinetic Study on the Maillard Reaction. Consideration of Sugar Reactivity. *Food Chem.* **2008**, *111*, 1032–1042. [CrossRef]
20. Li, X.; Liu, S.-Q. Effect of pH, Xylose Content and Heating Temperature on Colour and Flavour Compound Formation of Enzymatically Hydrolysed Pork Trimmings. *LWT* **2021**, *150*, 112017. [CrossRef]
21. Chua, J.-Y.; Tan, S.J.; Liu, S.-Q. The Impact of Mixed Amino Acids Supplementation on *Torulaspora Delbrueckii* Growth and Volatile Compound Modulation in Soy Whey Alcohol Fermentation. *Food Res. Int.* **2021**, *140*, 109901. [CrossRef]
22. Zhou, R.Y.; Chua, J.-Y.; Liu, S.-Q. Exploring the Feasibility of Biotransforming Salted Soy Whey into a Soy Sauce-like Condiment Using Wine Yeast Torulaspora Delbrueckii and Soy Sauce Yeasts Zygosaccharomyces Rouxii and Candida Versatilis as Single Starter Cultures. *Food Res. Int.* **2022**, *156*, 111350. [CrossRef]
23. Vong, W.C.; Liu, S.-Q. Changes in Volatile Profile of Soybean Residue (Okara) upon Solid-State Fermentation by Yeasts. *J. Sci. Food Agric.* **2017**, *97*, 135–143. [CrossRef]
24. Gao, P.; Wang, W.; Jiang, Q.; Xu, Y.; Xia, W. Effect of Autochthonous Starter Cultures on the Volatile Flavour Compounds of Chinese Traditional Fermented Fish (Suan Yu). *Int. J. Food Sci. Technol.* **2016**, *51*, 1630–1637. [CrossRef]
25. Liu, S.-C.; Yang, D.-J.; Jin, S.-Y.; Hsu, C.-H.; Chen, S.-L. Kinetics of Color Development, pH Decreasing, and Anti-Oxidative Activity Reduction of Maillard Reaction in Galactose/Glycine Model Systems. *Food Chem.* **2008**, *108*, 533–541. [CrossRef]

26. Esteban, J.; Yustos, P.; Ladero, M. Catalytic Processes from Biomass-Derived Hexoses and Pentoses: A Recent Literature Overview. *Catalysts* **2018**, *8*, 637. [CrossRef]
27. Martins, S.I.F.S.; Jongen, W.M.F.; van Boekel, M.A.J.S. A Review of Maillard Reaction in Food and Implications to Kinetic Modelling. *Trends Food Sci. Technol.* **2000**, *11*, 364–373. [CrossRef]
28. Kaewtathip, T.; Wattana-Amorn, P.; Boonsupthip, W.; Lorjaroenphon, Y.; Klinkesorn, U. Maillard Reaction Products-Based Encapsulant System Formed between Chitosan and Corn Syrup Solids: Influence of Solution pH on Formation Kinetic and Antioxidant Activity. *Food Chem.* **2022**, *393*, 133329. [CrossRef] [PubMed]
29. Reyes, F.G.R.; Poocharoen, B.; Wrolstad, R.E. Maillard Browning Reaction of Sugar-Glycine Model Systems: Changes in Sugar Concentration, Color and Appearance. *J. Food Sci.* **1982**, *47*, 1376–1377. [CrossRef]
30. Sun, F.; Cui, H.; Zhan, H.; Xu, M.; Hayat, K.; Tahir, M.U.; Hussain, S.; Zhang, X.; Ho, C. Aqueous Preparation of Maillard Reaction Intermediate from Glutathione and Xylose and Its Volatile Formation During Thermal Treatment. *J. Food Sci.* **2019**, *84*, 3584–3593. [CrossRef] [PubMed]
31. Huang, M.; Zhang, X.; Karangwa, E. Comparation Sensory Characteristic, Non-Volatile Compounds, Volatile Compounds and Antioxidant Activity of MRPs by Novel Gradient Temperature-Elevating and Traditional Isothermal Methods. *J. Food Sci. Technol.* **2015**, *52*, 858–866. [CrossRef]
32. Huang, M.G.; Zhang, X.M.; Eric, K.; Abbas, S.; Hayat, K.; Liu, P.; Xia, S.Q.; Jia, C.S. Inhibiting the Color Formation by Gradient Temperature-Elevating Maillard Reaction of Soybean Peptide-Xylose System Based on Interaction of l-Cysteine and Amadori Compounds. *J. Pept. Sci.* **2012**, *18*, 342–349. [CrossRef]
33. Nooshkam, M.; Varidi, M.; Verma, D.K. Functional and Biological Properties of Maillard Conjugates and Their Potential Application in Medical and Food: A Review. *Food Res. Int.* **2020**, *131*, 109003. [CrossRef]
34. Chang, Y.-W.; Zeng, X.Y.; Sung, W.-C. Effect of Chitooligosaccharide and Different Low Molecular Weight Chitosans on the Formation of Acrylamide and 5-Hydroxymethylfurfural and Maillard Reaction Products in Glucose/Fructose-Asparagine Model Systems. *LWT* **2020**, *119*, 108879. [CrossRef]
35. Cui, H.; Jia, C.; Hayat, K.; Yu, J.; Deng, S.; Karangwa, E.; Duhoranimana, E.; Xia, S.; Zhang, X. Controlled Formation of Flavor Compounds by Preparation and Application of Maillard Reaction Intermediate (MRI) Derived from Xylose and Phenylalanine. *RSC Adv.* **2017**, *7*, 45442–45451. [CrossRef]
36. Seck, S.; Crouzet, J. Formation of Volatile Compounds in Sugar-Phenylalanine and Ascorbic Acid-Phenylalanine Model Systems during Heat Treatment. *J. Food Sci.* **1981**, *46*, 790–793. [CrossRef]
37. Cao, J.; Yan, H.; Liu, L. Optimized Preparation and Antioxidant Activity of Glucose-Lysine Maillard Reaction Products. *LWT* **2022**, *161*, 113343. [CrossRef]
38. Farmer, L.J. The Role of Nutrients in Meat Flavour Formation. *Proc. Nutr. Soc.* **1994**, *53*, 327–333. [CrossRef] [PubMed]
39. Nielsen, S.S. Phenol-Sulfuric Acid Method for Total Carbohydrates. In *Food Analysis Laboratory Manual*; Nielsen, S.S., Ed.; Food Science Texts Series; Springer: Boston, MA, USA, 2010; pp. 47–53. ISBN 978-1-4419-1463-7.
40. Agcam, E. A Kinetic Approach to Explain Hydroxymethylfurfural and Furfural Formations Induced by Maillard, Caramelization, and Ascorbic Acid Degradation Reactions in Fruit Juice-Based Mediums. *Food Anal. Methods* **2022**, *15*, 1286–1299. [CrossRef]
41. Hua, D.-R.; Wu, Y.-L.; Liu, Y.-F.; Chen, Y.; Yang, M.-D.; Lu, X.-N.; Li, J. Preparation of Furfural and Reaction Kinetics of Xylose Dehydration to Furfural in High-Temperature Water. *Pet. Sci.* **2016**, *13*, 167–172. [CrossRef]
42. Xu, Y.; Chen, Q.; Lei, S.; Wu, P.; Fan, G.; Xu, X.; Pan, S. Effects of Lard on the Formation of Volatiles from the Maillard Reaction of Cysteine with Xylose. *J. Sci. Food Agric.* **2011**, *91*, 2241–2246. [CrossRef] [PubMed]
43. Azarbad, M.H.; Jeleń, H. Determination of Hexanal—An Indicator of Lipid Oxidation by Static Headspace Gas Chromatography (SHS-GC) in Fat-Rich Food Matrices. *Food Anal. Methods* **2015**, *8*, 1727–1733. [CrossRef]
44. Kosowska, M.; Majcher, M.A.; Fortuna, T. Volatile Compounds in Meat and Meat Products. *Food Sci. Technol.* **2017**, *37*, 1–7. [CrossRef]
45. Yang, Z.; Xie, J.; Zhang, L.; Du, R.; Cao, C.; Wang, M.; Acree, T.; Sun, B. Aromatic Effect of Fat and Oxidized Fat on a Meat-like Model Reaction System of Cysteine and Glucose. *Flavour Fragr. J.* **2015**, *30*, 320–329. [CrossRef]
46. Shahidi, F.; Oh, W.Y. Lipid-Derived Flavor and off-Flavor of Traditional and Functional Foods: An Overview. *J. Food Bioact.* **2020**, *10*. [CrossRef]
47. Su, G.; Zheng, L.; Cui, C.; Yang, B.; Ren, J.; Zhao, M. Characterization of Antioxidant Activity and Volatile Compounds of Maillard Reaction Products Derived from Different Peptide Fractions of Peanut Hydrolysate. *Food Res. Int.* **2011**, *44*, 3250–3258. [CrossRef]
48. Moon, S.-Y.; Cliff, M.A.; Li-Chan, E.C.Y. Odour-Active Components of Simulated Beef Flavour Analysed by Solid Phase Microextraction and Gas Chromatography–Mass Spectrometry and –Olfactometry. *Food Res. Int.* **2006**, *39*, 294–308. [CrossRef]

Disclaimer/Publisher's Note: The statements, opinions and data contained in all publications are solely those of the individual author(s) and contributor(s) and not of MDPI and/or the editor(s). MDPI and/or the editor(s) disclaim responsibility for any injury to people or property resulting from any ideas, methods, instructions or products referred to in the content.

foods

MDPI

Review

The Impact of Artificial and Natural Additives in Meat Products on Neurocognitive Food Perception: A Narrative Review

Marius-Mihai Ciobanu [1], Elena-Iuliana Flocea [2,*] and Paul-Corneliu Boișteanu [2]

[1] Department of Food Technology, 'Ion Ionescu de la Brad' Iasi University of Life Sciences, 700490 Iasi, Romania; marius.ciobanu@iuls.ro

[2] Faculty of Food and Animal Sciences, "Ion Ionescu de la Brad" University of Life Sciences, 8 Mihail Sadoveanu Alley, 700489 Iasi, Romania; paul.boisteanu@iuls.ro

* Correspondence: floceaiuliana5@gmail.com

Abstract: The holistic sensory experience creates a unified perception that influences consumer memory. Consumer interest in clean label products underlines an accelerated trend towards products without artificial additives. From a sensory point of view, food appeal is significantly influenced by how additives actively participate in the organoleptic properties of the final product. This research aims to shed light on the impact of artificial and natural additives in meat products on neurocognitive food perception, which is essential for understanding how they influence the consumer's final decision and direct food trends. Different neural mechanisms involved in multimodal sensory integration confirm differences in perception of meat products containing artificial and natural additives. Analysis confirms that neurocognitive perception integrates organoleptic sensations to form a complete sensory experience. The encephalon simultaneously processes multimodally integrated stimuli from organoleptic properties, reaching the orbitofrontal cortex and other regions involved in the neuroprocessing of the final product. The reformulation and development of meat products need a detailed analysis of the impact of additives on sensory properties contributing to the shaping of consumption trends.

Keywords: meat products; artificial additives; natural additives; organoleptic properties; food perception

1. Introduction

Fresh and processed meat offer numerous nutritional health benefits, they are high in protein, balanced in amino acids, and rich in several minerals and vitamins, which play an essential role in metabolism and can be more easily assimilated from meat than from other foods [1]. Meat and meat products offer a unique type of food satisfaction in modern society. Certain products require additives to ensure efficient industrial processing. The additives actively participate in preserving meat products and improving food safety, but they also enrich sensory properties. Utilization of additives such as butylated butylated hydroxyanisole (BHA), butylated hydroxytoluene (BHT), propyl gallate (PG), and tertiary butylated hydroquinone (TBHQ) have long been used to inhibit harmful oxidation-induced changes in meat that cause sensory degradation, but are under increasing scrutiny due to potential genotoxic effects on human health [2]. Consumers associate artificial additives with harmful health effects; however, they are monitored constantly by The European Food Safety Authority (EFSA) in the EU and the FDA (Food and Drug Administration) in the USA. In the EU, for meat products, the list is long, including antimicrobials, antioxidants, and texturizers as the most commonly used, but also other additives (such as colourants, stabilizers, and acidity regulators). All additives used in meat processing are considered safe within the limits set by food safety authorities [3,4]. The current direction is towards natural antioxidants derived from various plant materials rich in polyphenols. The authors in [5] emphasized the antioxidant properties of natural extracts that are proposed as potential replacements for synthetic additives such as nitrites in meat products, further highlighting the impact of additives on sensory properties. Those in [6] supported that the application of natural additives in meat products can maintain similar sensory characteristics as those obtained by artificial additives, but more studies are needed to maximize their sensory productivity in the final product. Food preference and perceptual differences for meat products containing natural or synthetic additives may be distributed over different neural mechanisms responsible for multimodal sensory integration [7]. Fluctuations in neurophysiological states in the sensory neocortex play an important role in perceptual variability, influencing the processing of sensory stimuli and contributing to perceptual differences [8]. Figure 1 illustrates the process of multimodal sensory integration of sensory experience of the consumer. The Figure 1 summarizes how additives alter the sensory experience of the consumer. It demonstrates how chemical interventions influence the physical and organoleptic properties of the final product, which in turn activate complex sensory processes. Figure 1 starts with a symbolic representation of the chemical structure of food additive. Additives are added to meat products to modify or enhance the sensory characteristics of meat products intended for consumption. This schematic representation between the molecules released by meat and human sensory receptors illustrates how chemical signals are transmitted and interpreted by the sensory system. The brain image on the right suggests the integration of sensory signals. The brain processes and combines information from different receptors to generate an overall perception of the product.

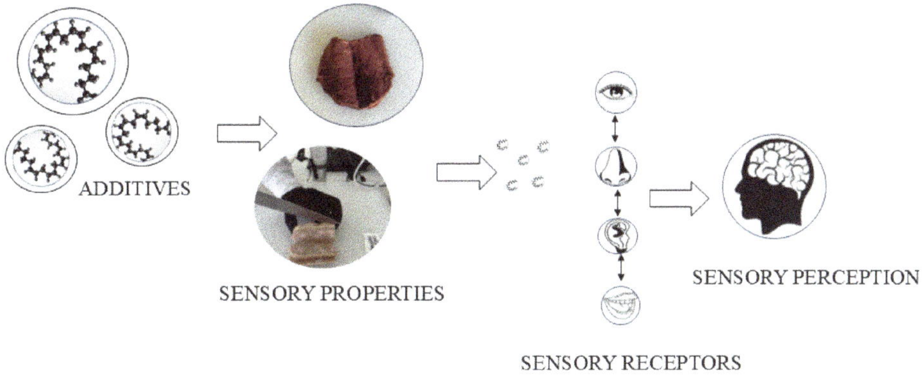

Figure 1. Multimodal sensory integration of sensory properties resulting from the use of additives in meat products. Processing content according to Chen J. [9].

Human ascending sensory systems are involved in food perception. In multimodal sensory integration, the visual system participates through distal recognition and food selection. Striate and extrastriate areas of the ventral system provide information on external appearance. As one progresses from the striate primary visual cortex to the visual inferotemporal areas of the temporooccipital cortex and temporal-anterior cortex, the size of the visual receptive fields and the optimal stimuli become increasingly complex. The visual projections of the inferotemporal cortex move to the amygdala, orbitofrontal cortex, and adjacent perirhinal cortex [10]. Studies analysing food perception generate distinct neuronal activations in the medial prefrontal cortex (MPFC). The medial prefrontal cortex is associated with complex cognitive processes, including preference formation and decision-making. By associating food with health benefits and a positive impact on sustainability, the brain activates neural pathways related to moral evaluation and reward. This phenomenon indicates that organic food perceptions are nuanced and highlights the complexity of brain processes involved in food perception and decision-making, going beyond simply recognizing taste or food appearance [11]. Perception of meat foods with natural and artificial additives involves distinct neural pathways and mechanisms. The brain simultaneously processes stimuli from the organoleptic properties, reaching the orbitofrontal cortex and other regions involved in the neuroprocessing of the flavour of the final product. Research suggests that brain activation during food consumption is characterized by the convergence of olfactory and gustatory sensations, leading to a holistic flavour experience. Depending on the type of additives, different activations in the brain may reflect their influence on how the meat product is perceived. Natural additives are associated with foods perceived as healthy, they may activate reward circuits in the brain more intensely than artificial additives that are associated with health problems. Artificial additives may induce lower or potentially ambiguous activations [12].

When the consumer does not consult the product label, artificial additives show a higher sensory efficacy, favouring acceptance and increased interest in the product compared to natural additives, which have relatively lower sensory efficacy. The sensitivity of the brain to visual cues of spoiled or rotten foods indicates an automatic mechanism for detecting inedibles, triggering enhanced attention capture and early alarm responses [13]. The brain processes and differentiates meat foods with natural and synthetic additives through early visual discrimination and neural responses. The neural responses differ for foods with natural additives compared to those processed with artificial additives, reflecting instantaneous assessments of the perceived value and suitability of the food essential for consumption decisions. Research shows that the brain can distinguish between food and non-food products within milliseconds of stimulus onset, encoding information about the naturalness of the food, the level of processing, and the caloric content. Regardless of the attention or cognitive task imposed, this rapid processing occurs as a function of an automatized mechanism, whereby the encephalon rapidly evaluates possible food sources to determine food safety and consumption decisions [14].

Under the influence of insulin, which plays a modulatory role by amplifying or attenuating, cortical activity responds differently to food stimuli. Images of food with sensory-significant properties generate more intense neural activity compared to control images. Insulin contributes to sensory discrimination of food by adapting neural responses to food type and quality [15]. The brain's reaction to meat products with artificial additives can be multi-sided. Synthetic food colourings, commonly used to enhance the attractiveness of processed foods, have been linked to an increase in behavioural problems in children, such as attention deficit disorder (ADD) and attention-deficit/hyperactivity disorder (ADHD) [16]. Additives, such as sodium nitrite and monosodium glutamate (MSG), used in meat processing, have been implicated in various pathological changes by damaging brain cells and may also affect male fertility [17]. These additives can challenge the immune system, leading to inflammatory responses and may cause neurobehavioral disorders. Therefore, the brain response to artificial meat additives may involve immune reactions, inflammatory cascades and neurological implications, highlighting the importance of understanding and monitoring the impact of these additives on brain perception and health [18]. The inclusion of meat products with artificial additives in the diets of laboratory animals affected by hemorrhagic strokes indicates a potential negative impact on brain tissue. The study shows that these additives may help to stabilize certain destructive changes in brain tissue, particularly in neurons through the proliferation of glial cells in response to injury. Gliosis indicates an attempt by the tissue to adapt and respond to harmful changes caused by artificial additives. This mechanism provides a possible detrimental impact of artificial additives on neuronal function and integrity under conditions of brain vulnerability [19]. On the other hand, studies suggest that the incorporation of natural extracts into meat products may help to reduce carcinogenic compounds by decreasing the formation of N-nitroso compounds [20]. Understanding the neural mechanisms involved in perception and preference may help to elucidate how individuals make food choices based on sensory cues and modification of nutritional value, as demonstrated in studies of food choices and neural activity during decision-making tasks [21].

This review explains consumer responses by exploring the impact of artificial and natural additives in meat products on neurocognitive perception. The integration of food neurocognitive perception provides a complex and interdisciplinary framework for the development and market success of reformulated meat products. The way natural and artificial additives influence neurocognitive mechanisms, and the perception of organoleptic properties of products explains consumption decisions. Analysing the potential of natural additives to replace artificial compounds is a key focus for food industry specialists without compromising the acceptability of reformulated products. Studying the link between food neuroperception and additives in meat products focuses on how added compounds guide consumption trends. This interdisciplinary field fosters innovation in food science.

2. Neurocognitive Mechanisms Involved in Food Perception

Food flavour molecules have the potential to influence people's emotions via olfactory pathways. Current food flavour research mainly focuses on the physicochemical properties and formation mechanisms of flavour components, neglecting the effects of flavour molecules in emotional regulation. The assessment of liking, an essential dimension of emotion, lacks objective assessment methods. In one study [22], sensory evaluations of sensory acceptabilities for 12 aroma compounds were collected from 45 subjects, and their correlation with brain activity responses in the left frontal-temporal lobe (LFT) and right frontal-temporal lobe (RFT) were analysed. The results revealed a close relationship between brain activity in the left frontal-temporal lobe (LFT) and the perception of aroma liking. The authors confirm that they observed a substantial correlation between the α, β and γ frequency bands in the left frontal-temporal (LFT) and subjective pleasure scores. These findings demonstrate that the left frontal-temporal (LFT) plays a critical role in the evaluation of the pleasantness of flavour molecules and that changes in the strength of the α, β, and γ bands serve as important indicators of assessment [22]. Taste processing in primates involves several brain structures. The nucleus of the solitary tract and the primary gustatory cortex provide the direct perception of the characteristics of food by identifying and intensifying taste inputs (sweet, salty, sour, bitter, umami, etc). The orbitofrontal cortex is involved in taste evaluation, influenced by physiological states of the organism, such as hunger. The neuronal structures in this area respond to taste at the sensation of hunger and manifest sensory-specific satiety. In the orbitofrontal cortex, there is a convergence of olfactory and gustatory input, forming a complex perception of food. This region can integrate information about the texture and appearance of food. The orbitofrontal cortex allows rapid associations between taste and other sensory stimuli. This facilitates food recognition based on previous experience, having an important role in determining food learning preferences. The orbitofrontal cortex has a complex role in integrating and processing information about the organoleptic properties of the product, influencing consumer preferences and behaviour [23]. Neurophysiologic recordings and functional neuroimaging show the primary gustatory cortex in the rostral insula and adjacent frontal operculum provide separate and combined representations of taste, temperature, and texture of food. Neurons respond differently to combinations, providing a rich representation of the sensory properties of food. The representation of taste and other food-related stimuli is found mainly in the medial orbitofrontal cortex. Consistently, the activation of parts of the human orbitofrontal cortex correlates with subjective ratings of the pleasantness of food taste and odour. Food intake is mediated by medial orbitofrontal cortex activation through multimodal integration of the sensory properties of products. A neuronal representation of taste is also found in the pregenual cingulate cortex, which receives information from the orbitofrontal cortex, and, in humans, many pleasant stimuli activate the pregenual cingulate cortex, indicating that this is an area important in motivation and emotion [24]. Optogenetic activation of hypothalamic agouti-related hypothalamic peptide provides information on how internal states such as hunger influence odour attractiveness. Neuronal projections to the paraventricular thalamus contribute to sensory integration and regulation of olfactory stimuli. Neuropeptide Y plays an essential role in the selective activation of olfactory subcircuits and influences behaviour depending on the physiological state. The release of neuropeptide Y in the thalamus highlights how internal states can modulate behavioural preferences related to the olfactory sense [25].

According to [26], differentiated perception of food odours activates medial prefrontal and lateral orbitofrontal areas. Medial prefrontal area activation to food odours and scores on the externalness subscale of the Dutch Eating Behaviour Questionnaire subscale under satiety conditions are positively correlated. In individuals who are sensitive to external food cues (organoleptic properties), the medial prefrontal area plays a role in the inappropriate evaluation of food stimuli even when hunger is reduced. Consumers influenced by external stimuli may continue to respond to sensory properties in a way that predisposes them to food consumption even in the absence of a physiological need [24–26]. Figure 2 illustrates a general approach to multimodal sensory integration and consumer response. The food product is the initial stimulus, it interacts with several sensory organs (eyes, nose, mouth) of the consumer. The information collected by the senses is integrated at the level of the brain. The brain combines the signals to form a unified perception of the food product. Sensory processing determines consumer responses by adapting the information. The consumer adjusts their perception over time, based on previous experiences. The response may be positive, with the consumer accepting the product for consumption, or it may be perceived negatively, rejecting consumption of the final product.

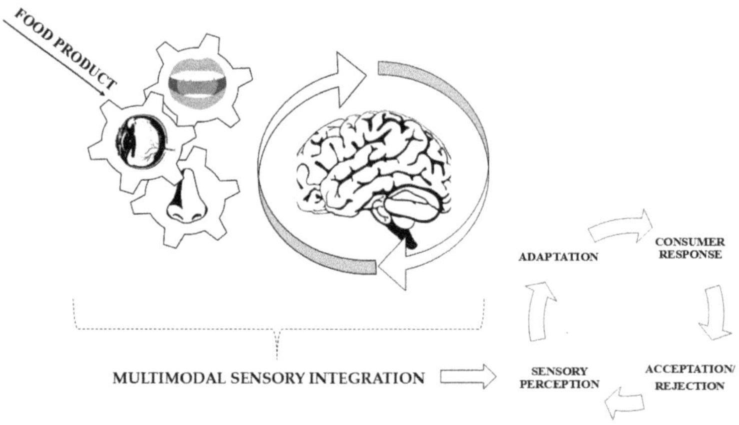

Figure 2. The general approach to multimodal sensory integration and consumer response. Processing content according to [24–26].

3. The Impact of Artificial Additives

Meat and processed meat products are valuable foods for consumers due to their sensory properties and the presence of essential nutritional components such as proteins with high biological value, minerals (Fe, Zn, Se), and some vitamins (mainly B6 and B12) [27]. Additives are intended to enrich the sensory properties of meat products, making them more attractive to consumers. Artificial additives in meat products play an essential role in preservation, colour, and flavour and inhibit bacterial spoilage [28]. Food additives can bring great sensory pleasure and marketing comfort to humans, but they may also cause potential risks to human health [29] through the production of methemoglobin and the potential carcinogenic effects of N-nitrosocompounds [30]. Phosphate additives are widely applied in food processing companies to improve product quality in a time-efficient manner [31]. Common additives in processed meat include nitrites, phosphates, and monosodium glutamate, which improve sensory qualities and shelf life and ensure production in an optimal time [32]. Calcium propionate, sodium nitrate, sodium nitrate, sodium benzoate, sodium nitrite, sulphites (sulphur dioxide, sodium bisulphite, potassium hydrogen sulphate, etc.) and disodium are examples of common chemical preservatives with a high teratogenic effect [33]. Ref. [34] confirmed the adverse effects of sodium nitrate by a dose-dependent decrease in sperm motility, serum testosterone concentration, body weight and organ weights and an increase in abnormal sperm morphology in $NaNO_3$-*treated* groups. In addition, histologic analysis confirmed that NO_3 induced toxicity. Decreased seminiferous tubules and loss of spermatids in the testes, shrinkage of pancreatic acinar cells, sinusoidal congestion and necrosis of the liver, atrophy of glomeruli and congestion of the renal tubules of the kidney were the histologic changes observed.

One of the compounds known as a preservative with a high safety profile is sodium benzoate. While some studies show that it can be used to treat conditions such as depression, pain, schizophrenia, autism spectrum disorders, and neurodegenerative diseases, others report its harmfulness. For example, it has been found to cause mutagenic effects, generate oxidative stress, disrupt hormones, and reduce fertility [35]. Ref. [36] examined the effects of sodium benzoate, a common preservative, on sweet and sour taste perception in healthy adults. They used an alternative forced-choice and time-intensity method to measure the intensity and duration of sweet and sour sensations elicited by sucrose and citric acid solutions with or without sodium benzoate. They found that sodium benzoate significantly reduced the intensity and duration of both sweet and sour tastes, suggesting that it interferes with the activation of taste receptors and/or signal transduction pathways. They recommended consumers should be aware of the potential sensory effects of sodium benzoate and that food manufacturers should consider alternative preservatives or lower concentrations of sodium benzoate [36–38]. Due to the large number of reports on the harmfulness of food additives, more and more consumers are following the so-called "clean label" trend, i.e., preferring and choosing the least processed foods. Monosodium glutamate can be used to reduce sodium intake without compromising taste [37]. For sodium reduction, the use of salt substitutes, such as potassium chloride (*KCl*), is a classic alternative. However, *KCl* imparts bitter and metallic tastes to the product, which can be reduced by the use of flavour enhancers that mask these undesired tastes [37].

Table 1 presents a selection of natural substitutes that can replace most of the synthetic additives used in the meat industry, offering healthier and consumer-acceptable alternatives.

Table 1. Natural replacements for most synthetic additives used in the meat industry.

Synthetic Additive	Natural Substitute	Effect on Meat Product	Ref.
Sodium erythorbate	Acerola Extract	Phenolic compounds; powerful antioxidant and an effective method in slowing down colour and lipid oxidation in meat products.	[39]
Butylated hydroxyanisole (BHA)	Rosemary Extract	Phenolic diterpenes, including carnosol, carnosic acid, epirosmanol, isorosmanol, rosmaridiphenol, and rosmariquinone contribute to the overall antioxidant activity of meat products and provide significant sensory properties.	[40–42]
Butylated hydroxytoluene (BHT)	Grape Seed Extract	Lipid stabilization of turkey meat; the addition of grape seed extract combined with vacuum packaging is considered a good method to improve lipid stability in cooked poultry meat.	[43]
Sodium nitrite	Beetroot and Celery Extracts	Contribute to the overall antibacterial activity of meat products.	[44]
Monosodium glutamate	Tomato by-products	Contain lycopene, which has been shown to have antioxidant activity; potentially improve sensory properties;	[45,46]
Phosphoric acid	Citrus lemon oil	Natural antioxidant and antimicrobial agent; beef has susceptibility to spoilage microorganisms during refrigeration, citrus lemon oil is an essential point for the use of natural antioxidants.	[47]
Di- and triphosphates	Papain; Ficin; Bromelain	Had significant results for meat tenderness and sensory properties.	[48,49]
Polyphosphates	Bamboo leaf Extract	The water-holding capacity and tenderness of poultry have been improved.	[50]

4. The Impact of Natural Additives

A major obstacle to the use of natural plant extracts in food has been the imposition of undesirable flavours and odours. However, technological advances have enabled food ingredient manufacturers to produce extracts with sensory characteristics that do not interfere while maintaining antioxidant properties. These technologies include solvent extraction, hydrodistillation, spray drying, and supercritical fluid extraction [41–43]. Antioxidants are used to minimize lipid oxidation. Antioxidants can act as metal chelators and free radical scavengers or oxygen scavengers, which can slow down the rate of lipid oxidation. Lipid oxidation can have negative effects on the quality of meat and meat products, causing changes in sensory attributes such as colour, texture, odour, and flavour as well as nutritional quality. Several synthetic antioxidants have been successfully used to prevent lipid oxidation in the meat industry, but consumers are concerned about the health risks of consuming certain synthetic antioxidants. Therefore, there is increased interest in natural antioxidants. Currently, compounds obtained from natural sources such as cereals, oilseeds, spices, fruits, and vegetables have been studied to reduce lipid oxidation [51,52]. Plants are rich in bioactive compounds (BACs), mainly polyphenols, which are valuable choices to replace synthetic antioxidants in meat products. These natural plant antioxidants, in the form of extracts and essential oils (EOs), have been obtained from different sources such as fruits (dragon fruit, guarana, and pomegranate), vegetables, (cabbage and onion), herbs and spices (epazot, ginger, rosemary, sage, thyme, turmeric and winter cider) by several extraction processes. However, in the context of current directives, there is a noticeable incentive for 'green' solvents to replace organic and conventional techniques to avoid harm to the environment, operators, and the health of consumers [44]. Nitrate and nitrite-rich natural extracts from sources such as red beet and celery can enhance the sensory properties of meat products while addressing safety and health concerns. The use of both natural sources of nitrate and nitrite to replace commercial nitrate and nitrite salts can develop the expected effects in terms of colour, lipid oxidation, microbial spoilage and safety, and sensory properties, but contrasting effects, in particular for redness and lipid oxidation, are still the main concerns. Advances in the utilization of these extracts support their progress towards higher-scale experiments and their extension towards the development of healthier and more functional meat products [53].

Consumers are becoming increasingly concerned about synthetic preservatives such as nitrites in meat, prompting the meat industry to explore alternatives to reduce nitrite levels. In these syntheses, the effects of hemp meal incorporation on the chemical and preservative characteristics of minced meat products with reduced nitrite content were investigated. Results of the study demonstrate that hemp flour can be utilized effectively as a natural ingredient with antioxidant properties in minced meat products, though some differences in sensory characteristics may occur [53]. Another study reported the effect of grape seed extract on colour stability, inhibition of lipid oxidation, and best overall acceptability after 6 days of storage in meat food products [54]. M. oleifera seed flour improved the physicochemical properties, organoleptic characteristics, consumer preference, and aerobic stability of cold storage of beef meat patties [55]. Onion (*Allium cepa* L.) is a widely cultivated and consumed vegetable due to its rich content of bioactive compounds. Red onion peel powder, which is a by-product derived from the onion industry, has attracted significant interest as a potential functional ingredient to improve the overall quality of food [56]. The bioactive compounds of pumpkin by-products, which are rich in phytochemicals, can prevent the oxidation process in different foods [57]. Avoidance of the use of certain additives may create situations where food safety is jeopardized. Some alternatives are based on the origin of the additive because naturalness is perceived as an essential characteristic for most consumers [3].

5. Influence of Natural and Artificial Additives on Food Perception

Modern consumers are tending towards products with natural additives and are increasingly focusing on the potential risks of consuming synthetic substances, reflecting current trends towards clean label foods. Certain demographic segments, however, are not influenced by these modern trends. Baby boomer consumers reject innovations in taste and tend towards the traditional, explained by a strong link with past food values. This highlights scepticism towards changes perceived as inauthentic [58]. While artificial flavours may be acceptable, artificial colours often evoke negative reactions, affecting the overall perception of the final product. Products with natural organoleptic properties are perceived as healthier, influencing consumer choices [59]. Consumers favour natural additives, their preference emphasizing the link between the type of additive and the quality of the product concerning the well-being of the human body [60]. Natural additives must meet strict food safety specifications similar to synthetic counterparts, complicating direct substitutions. Concerns about synthetic additives have led to increased scrutiny and demand for natural alternatives, although natural additives do not guarantee the safety of the final product [29,61].

Consumers prefer natural rather than artificial food additives, associating the latter with health risks. This perception influences purchasing decisions, as artificially named additives can lead to a perceived lack of naturalness, reducing acceptance of both the additive and the final product [62]. While consumers may prefer natural additives, studies show that artificial flavours can sometimes be more palatable [63]. Perceived quality differences between natural and artificial additives are shaped by cognitive integrations of the organoleptic properties of products, consumers associate natural additives with healthier products [64]. Educating consumers about the benefits of novel additives can stimulate positive emotional connections and acceptance of the final product [3]. Positive perceptions of the health benefits of additives can enhance emotional attachment [65]. The perceived health benefits of food additives can increase consumer acceptance by overcoming perceptions of risk. Consumers value convenience and time-saving aspects, stimulating emotional attachment to products, especially when additives are associated with positive attributes such as naturalness and safety, influencing purchasing decisions [66]. In contrast, scepticism about food additives persists, particularly among older demographics that favour traditional flavours. This scepticism may hinder acceptance, suggesting that consumer education and transparent labelling are key to promoting trust and acceptance of novel additives [58,61].

Consumers increasingly prefer natural additives to synthetic ones because of perceived health benefits. However, the presence of other additives in natural formulations may complicate this view, as consumers may not always recognize that natural does not inherently mean inherently safer or healthier. Direct replacement of synthetic colours requires technology development [67]. Consumers expressed concerns about food additives, with 63.7% linking these concerns to human health. This perception influences their purchasing decisions as they prefer products labelled as natural and free of artificial ingredients, indicating a high demand for clean label products [68]. Human health risk perceptions significantly influence their evaluation, attitude, and willingness to purchase, ultimately affecting their purchasing decisions on food products containing artificial additives [69].

6. Conclusions

Meat products are an essential component of the human diet, and the use of additives remains a widespread practice due to their impact on the overall quality of the final products. Some meat products require careful processing using various synthetic additives. The key aspect is to raise consumer awareness of the health impact of artificial additive and to encourage reformulation, more natural and healthier options for meat products. Consumers turn their attention to a clean label but are nonetheless guided by the sensory properties that the final product exhibits. Certain natural additives do not perform as well as synthetic additives in terms of the sensory properties of meat products. Synthetic additives are effective on the sensory properties of some meat products, but many studies confirm their harmful effects. The term artificial activates areas of the brain associated with rejection, even if the amount used under legislative standards states it is safe for consumption. The brain interprets multimodally involved information and provides an integrated, unified, and complex sensory experience about the product. The orbitofrontal cortex allows rapid associations between taste and other sensory stimuli. This helps to recognize foods based on previous experience, playing an important role in preference determination and food learning. Exploring the potential of natural additives to substitute artificial additives is an important step in accepting reformulated meat products.

Consumers associate additives with health effects, and the emotional response is the defining factor for product acceptability. Depending on demographic segments, consumers respond differently to products reformulated with new additives. How natural and artificial additives influence neurocognitive mechanisms and the perception of organoleptic properties of products explain consumption decisions. In general, consumers turn to the most natural products because they associate them with health benefits. The growing trend towards clean label products is committing the food industry to reformulate products to replace artificial additives with natural ingredients that have similar functions and productivity. Understanding food neurocognitive perception provides a complex, transparent framework. Optimizing the sensory experience for the development and market success of reformulated products helps to overcome consumer scepticism and contributes to the acceptance of additives, especially when they have significant results in the safety, quality, and nutritional value of the final food products that reach the consumer.

Future studies will focus on the impact on food quality, acceptability, and efficiency and will concentrate on overcoming the technical and economic hurdles in implementing natural alternatives. Challenges will include balancing the performance requirements of the food industry with consumer preferences for clean label products. This research will contribute to the efficient and sustainable development of safe and healthy meat products.

Author Contributions: Conceptualization, M.-M.C. and P.-C.B.; methodology, M.-M.C.; software, E.-I.F.; validation, M.-M.C., E.-I.F. and P.-C.B.; formal analysis, P.-C.B.; investigation, E.-I.F.; resources, P.-C.B.; data curation, M.-M.C.; writing—original draft preparation, E.-I.F.; writing—review and editing, E.-I.F.; visualization, M.-M.C.; supervision, P.-C.B.; project administration, M.-M.C.; funding acquisition, M.-M.C. All authors have read and agreed to the published version of the manuscript.

Funding: This research received no external funding.

Data Availability Statement: No new data were created or analyzed in this study. Data sharing is not applicable to this article.

Conflicts of Interest: The authors declare no conflicts of interest.

References

1. Wood, J. Meat composition and nutritional value. In *Lawrie's Meat Science*, 8th ed.; Toldra, F., Ed.; Woodhead Publishing Limited: Duxford, UK, 2017; pp. 635–659.
2. Jiang, J.; Xiong, Y.L. Natural antioxidants as food and feed additives to promote health benefits and quality of meat products: A review. *Meat Sci.* **2016**, *120*, 107–117. [CrossRef] [PubMed]
3. Delgado-Pando, G.; Ekonomou, S.I.; Stratakos, A.C.; Pintado, T. Clean Label Alternatives in Meat Products. *Foods* **2021**, *10*, 1615. [CrossRef] [PubMed]
4. Commission Regulation (EU) No 1129/2011 of 11 November 2011 Amending Annex II to Regulation (EC) No 1333/2008 of the European Parliament of the Council by Establishing a Union List of Food Additives Text with EEARelevance Off. *J. Eur. Union* **2011**, *L295*, 1–177.
5. Bolognesi, V.J.; Garcia, C.E. Annatto carotenoids as additives replacers in meat products. In *Alternative and Replacement Foods*; Academic Press: Cambridge, MA, USA, 2018; pp. 355–384.
6. Pollini, L.; Blasi, F.; Ianni, F.; Grispoldi, L.; Moretti, S.; Di Veroli, A.; Cossignani, L.; Cenci-Goga, B.T. Ultrasound-Assisted Extraction and Characterization of Polyphenols from Apple Pomace, Functional Ingredients for Beef Burger Fortification. *Molecules* **2022**, *27*, 1933. [CrossRef]
7. Zhou, L.; Tang, Z.; Zuo, Z.; Zhou, K. Neural mechanism underlying the sleep deprivation-induced abnormal bistable perception. *Cereb. Cortex.* **2022**, *32*, 583–592. [CrossRef]
8. Morton, M.P.; Denagamage, S.; Blume, I.J.; Reynolds, J.H.; Jadi, M.P.; Nandy, A.S. Brain state and cortical layer-specific mechanisms underlying perception at threshold. *eLife* **2024**, *12*, RP91722. [CrossRef]

9. Chen, J. Food oral processing: Some important underpinning principles of eating and sensory perception. *Food Struct.* **2014**, *1*, 91–105. [CrossRef]
10. Verhagen, J.V.; Engelen, L. The neurocognitive bases of human multimodal food perception: Sensory integration. *Neurosci. Biobehav. Rev.* **2006**, *30*, 613–650. [CrossRef]
11. Huerta, C.I.; Sarkar, P.R.; Duong, T.Q.; Laird, A.R.; Fox, P.T. Neural bases of food perception: Coordinate-based meta-analyses of neuroimaging studies in multiple modalities. *Obesity* **2014**, *22*, 1439–1446. [CrossRef]
12. Fehse, K.; Simmank, F.; Gutyrchik, E.; Sztrókay-Gaul, A. Organic or popular brands—Food perception engages distinct functional pathways. *Fmri Study. Cogent Psychol.* **2017**, *4*, 1284392. [CrossRef]
13. Key, A.P.; Charboneau, E.J.; Cowan, R.L. Food perception in adults: Neuroimaging findings. In *Handbook of Behavior, Food and Nutrition*; Springer: New York, NY, USA, 2011; pp. 515–530.
14. Coricelli, C.; Toepel, U.; Bielser, M.L.; Murray, M.M.; Rumiati, R.I. Distinct Brain Representations of Natural and Manufactured Foods: A Spatio-Temporal Brain Dynamics Investigation. *TSPC*. 2016. Available online: https://www.openstarts.units.it/server/api/core/bitstreams/b8712168-be63-418a-b584-5c56392a83a2/content (accessed on 15 October 2024).
15. Coricelli, C.; Toepel, U.; Notter, M.L.; Murray, M.M.; Rumiati, R.I. Distinct brain representations of processed and unprocessed foods. *Eur. J. Neurosci.* **2019**, *50*, 3389–3401. [CrossRef] [PubMed]
16. Doguc, D.K.; Deniz, F.; İlhan, İ.; Ergonul, E.; Gultekin, F. Prenatal exposure to artificial food colorings alters NMDA receptor subunit concentrations in rat hippocampus. *Nutr. Neurosci.* **2021**, *24*, 784–794. [CrossRef] [PubMed]
17. Kyakma, S.S.; Tella, A.K.; Sanwo, K.A. Some meat quality parameters of broiler chickens fed diets containing different additives. *Niger. J. Anim. Prod.* **2022**, *49*, 33–45. [CrossRef]
18. Voirol, E.; Daget, N. Direct nasal and oronasal profiling of a meat flavouring: Influence of temperature, concentration and additives. *Lebensm.-Wiss. Technol.* **1989**, *22*, 399–405.
19. Rasha, A.E.; Reham, A.A.; Amin, A. Health Risks of Some Meat Additives on Male Rats. *World J. Dairy Food Sci.* **2014**, *9*, 285–298.
20. van Breda, S.G.; Mathijs, K.; Pieters, H.J.; Sági-Kiss, V.; Kuhnle, G.G.; Georgiadis, P.; Saccani, G.; Parolari, G.; Virgili, R.; Sinha, R.; et al. Replacement of Nitrite in Meat Products by Natural Bioactive Compounds Results in Reduced Exposure to N-Nitroso Compounds: The PHYTOME Project. *Mol. Nutr. Food Res.* **2021**, *65*, 2001214. [CrossRef]
21. Gawda, B.; Bernacka, R.; Gawda, A. The Neural Mechanisms Underlying Personality Disorders. *Neuroquantology* **2016**, *14*, 348–356. [CrossRef]
22. Ke, Q.; Zhang, J.; Huang, T.; Huang, X.; Li, Q.; Lu, Z.; Kou, X. Unveiling the Neurocognitive Impact of Food Aroma Molecules on Pleasantness Perception: Insights From EEG and Key Brain LFT Activation. *Flavour Fragr. J.* **2024**. [CrossRef]
23. Rolls, E.T. Taste and olfactory processing in the brain and its relation to the control of eating. *Crit. Rev. Neurobiol.* **1997**, *11*, 263–287. [CrossRef]
24. Rolls, E. Functions of the orbitofrontal and pregenual cingulate cortex in taste, olfaction, appetite and emotion. *Acta Physiol. Hung.* **2008**, *95*, 131–164. [CrossRef]
25. Horio, N.; Liberles, S.D. Hunger enhances food-odour attraction through a neuropeptide Y spotlight. *Nature* **2021**, *592*, 262–266. [CrossRef] [PubMed]
26. Eiler, W.J.; Dzemidzic, M.; Case, K.R.; Considine, R.V.; Kareken, D.A. Correlation between ventromedial prefrontal cortex activation to food aromas and cue-driven eating: An fMRI study. *Chemosens. Percept.* **2012**, *5*, 27–36. [CrossRef] [PubMed]
27. Pérez-Palacios, T.; Ruiz-Carrascal, J.; Solomando, J.C.; Antequera, T. Strategies for enrichment in ω-3 fatty acids aiming for healthier meat products. *Food Rev. Int.* **2019**, *35*, 485–503. [CrossRef]
28. Sorour, M.A.; Mehanni AH, E.; Mahmoud ES, A.; El-Hawashy, R.M. Nitrate, nitrite and N-nitrosamine in meat products. *J. Sohag Agriscience* **2023**, *8*, 121–135. [CrossRef]
29. Wu, L. Analysis of food additives. In *Innovative Food Analysis*; Academic Press: Cambridge, MA, USA, 2021; pp. 157–180.
30. Dissanayake, K.; Rifky, M.; Hunupolagama, D.; Harris, J.M.; Zokirov, K.; Azizova, K.; Ermat, S.; Samadiy, M. Inorganic additives in meat production and processing. In Proceedings of the International Conference on Ensuring Sustainable Development in the Context of Agriculture, Energy, Ecology and Earth Science, Moscow, Russia, 23–25 December 2024; Volume 510, p. 01028.
31. Dissanayake, K.; Rifky, M.; Farmonov, J.; Boynazarova, Y.; Meyliyev, A.; Zokirov, K.; Narziyev, F.; Samadiy, M. Impact of incorporating inorganic additives in processed meat production. In Proceedings of the International Conference on Ensuring Sustainable Development in the Context of Agriculture, Energy, Ecology and Earth Science, Moscow, Russia, 23–25 December 2024; Volume 510, p. 8.
32. Sambu, S.; Hemaram, U.; Murugan, R.; Alsofi, A.A. Toxicological and Teratogenic Effect of Various Food Additives: An Updated Review. *BioMed Res. Int.* **2022**, *2022*, 6829409. [CrossRef]
33. Sv, R.; Basavaiah, S. Assessment of the effects of sodium nitrate ($NaNO_3$) on the reproductive system, liver, pancreas, and kidney of male rats. *Toxicol. Ind. Health* **2022**, *38*, 702–711. [CrossRef]
34. Walczak-Nowicka, Ł.J.; Herbet, M. Sodium Benzoate—Harmfulness and Potential Use in Therapies for Disorders Related to the Nervous System: A Review. *Nutrients* **2022**, *14*, 1497. [CrossRef]
35. Zhang, Y.; Wang, Q.; Liu, Y.; Chen, Q. Effects of sodium benzoate on sweet and sour taste perception in humans. *Food Chem.* **2019**, *275*, 715–722.
36. Augusto, M. Analysis of Food Additives and Their Effects on Flavor Perception in Brazil. *Int. J. Food Sci.* **2023**, *6*, 39–49. [CrossRef]

37. Lee, S.; Kim, J.; Lim, J. Effects of monosodium glutamate on saltiness perception and preference in young and elderly adults. *Food Qual. Prefer.* **2020**, *80*, 103828.
38. Saldaña, E.; Merlo, T.C.; Patinho, I.; Rios-Mera, J.D.; Contreras-Castillo, C.J.; Selani, M.M. Use of sensory science for the development of healthier processed meat products: A critical opinion. *Curr. Opin. Food Sci.* **2021**, *40*, 13–19. [CrossRef]
39. Realini, C.E.; Guàrdia, M.D.; Díaz, I.; García-Regueiro, J.A.; Arnau, J. Effects of acerola fruit extract on sensory and shelf-life of salted beef patties from grinds differing in fatty acid composition. *Meat Sci.* **2015**, *99*, 18–24. [CrossRef] [PubMed]
40. Dorman HJ, D.; Peltoketo, A.; Hiltunen, R.; Tikkanen, M.J. Characterisation of the antioxidant properties of de-odourised aqueous extracts from selected *Lamiaceae herbs*. *Food Chem.* **2003**, *83*, 255–262. [CrossRef]
41. McBride, N.T.M.; Hogan, S.A.; Kerry, J.P. Comparative addition of rosemary extract and additives on sensory and antioxidant properties of retail packaged beef. *Int. J. Food Sci. Technol.* **2007**, *42*, 1201–1207. [CrossRef]
42. Hernández-Hernández, E.; Ponce-Alquicira, E.; Jaramillo-Flores, M.E.; Legarreta, I.G. Antioxidant effect rosemary (*Rosmarinus officinalis* L.) and oregano (*Origanum vulgare* L.) extracts on TBARS and colour of model raw pork batters. *Meat Sci.* **2009**, *81*, 410–417. [CrossRef]
43. Mielnik, M.B.; Olsen, E.; Vogt, G.; Adeline, D.; Skrede, G. Grape seed extract as antioxidant in cooked, cold stored turkey meat. *LWT-Food Sci. Technol.* **2006**, *39*, 191–198. [CrossRef]
44. Munekata PE, S.; Cittadini, A.; Domínguez, R.; Teixeira, A.; Fernández-López, J.; Campagnol PC, B.; Pateiro, M. Natural alternatives and use of nitrate and nitrate salts. In *Strategies to Improve the Quality of Foods*; Academic Press: Cambridge, MA, USA, 2024; pp. 145–169.
45. dos Santos, B.A.; Campagnol, P.C.B.; Morgano, M.A.; Pollonio, M.A.R. Monosodium glutamate, disodium inosinate, disodium guanylate, lysine and taurine improve the sensory quality of fermented cooked sausages with 50% and 75% replacement of NaCl with KCl. *Meat Sci.* **2014**, *96*, 509–513. [CrossRef]
46. Viuda-Martos, M.; Sanchez-Zapata, E.; Sayas-Barberá, E.; Sendra, E.; Pérez-Álvarez, J.A.; Fernández-López, J. Tomato and tomato byproducts. Human health benefits of lycopene and its application to meat products: A review. *Crit. Rev. Food Sci. Nutr.* **2014**, *54*, 1032–1049. [CrossRef]
47. Ben Hsouna, A.; Ben Halima, N.; Smaoui, S.; Hamdi, N. Citrus lemon essential oil: Chemical composition, antioxidant and antimicrobial activities with its preservative effect against Listeria monocytogenes inoculated in minced beef meat. *Lipids Health Dis.* **2017**, *16*, 1–11. [CrossRef]
48. Sullivan, G.A.; Calkins, C.R. Application of exogenous enzymes to beef muscle of high and low-connective tissue. *Meat Sci.* **2010**, *85*, 730–734. [CrossRef]
49. Marques, A.Y.C.; Maróstica, M.R., Jr.; Pastore, G.M. Some nutritional, technological and environmental advances in the use of enzymes in meat products. *Enzym. Res.* **2010**, *2010*, 480923. [CrossRef] [PubMed]
50. Shen, M.M.; Zhang, L.L.; Chen, Y.N.; Zhang, Y.Y.; Han, H.L.; Niu, Y.; He, J.T.; Zhang, Y.L.; Cheng, Y.F.; Wang, T. Effects of bamboo leaf extract on growth performance, meat quality, and meat oxidative stability in broiler chickens. *Poult. Sci.* **2019**, *98*, 6787–6796. [CrossRef] [PubMed]
51. Karakaya, M.; Bayrak, E.; Ulusoy, K. Use of natural antioxidants in meat and meat products. *J. Food Sci. Eng.* **2011**, *1*, 1.
52. Pateiro, M.; Gómez-Salazar, J.A.; Jaime-Patlán, M.; Sosa-Morales, M.E.; Lorenzo, J.M. Plant extracts obtained with green solvents as natural antioxidants in fresh meat products. *Antioxidants* **2021**, *10*, 181. [CrossRef]
53. Papatzimos, G.; Mitlianga, P.; Basdagianni, Z.; Kasapidou, E. Hemp Flour as a Functional Ingredient for the Partial Replacement of Nitrites in a Minced Meat Model: Effect on Nutrient Composition, Antioxidant Profile and Sensory Characteristics. *Appl. Sci.* **2024**, *14*, 3925. [CrossRef]
54. Ahmad, S.R.; Gokulakrishnan, P.; Giriprasad, R.; Yatoo, M.A. Fruit-based Natural Antioxidants in Meat and Meat Products: A Review. *Crit. Rev. Food Sci. Nutr.* **2015**, *55*, 1503–1513. [CrossRef]
55. Al-Juhaimi, F.; Ghafoor, K.; Hawashin, M.D.; Alsawmahi, O.N.; Babiker, E.E. Effects of different levels of Moringa (*Moringa oleifera*) seed flour on quality attributes of beef burgers. *CyTA—J. Food.* **2015**, *14*, 1–9. [CrossRef]
56. Lipșa, F.D.; Stoica, F.; Rațu, R.N.; Veleșcu, I.D.; Cârlescu, P.M.; Motrescu, I.; Usturoi, M.G.; Râpeanu, G. Red Onion Peel Powder as a Functional Ingredient for Manufacturing Ricotta Cheese. *Foods* **2024**, *13*, 182. [CrossRef]
57. Aziz, A.; Noreen, S.; Khalid, W.; Ejaz, A.; Faiz ul Rasool, I.; Maham Munir, A.; Farwa Javed, M.; Ercisli, S.; Okcu, Z.; Marc, R.A.; et al. Pumpkin and Pumpkin Byproducts: Phytochemical Constitutes, Food Application and Health Benefits. *ACS Omega* **2023**, *8*, 23346–23357. [CrossRef]
58. Petit, S. Evaluation of Novel Food Additives on Product Acceptability in France. *Int. J. Food Sci.* **2024**, *7*, 18–28. [CrossRef]
59. Murley, T.; Chambers, I.V.E. The influence of colorants, flavorants and product identity on perceptions of naturalness. *Foods* **2019**, *8*, 317. [CrossRef] [PubMed]
60. Szűcs, V.; Szabo, E.; Tarcea, M.; Banati, D. Romanian consumers' willingness to buy foodstuff containing food additives: Results of Conjoint Study. *Bull. USAMV Food Sci. Technol.* **2015**, *72*, 153–161. [CrossRef] [PubMed]
61. Scotter, M.J. Emerging and persistent issues with artificial food colours: Natural colour additives as alternatives to synthetic colours in food and drink. *Qual. Assur. Saf. Crops Foods* **2011**, *3*, 28–39. [CrossRef]
62. Bearth, A.; Hartmann, C. Consumers' Perception and Acceptance of Food Additives. In *Reference Module in Food Science*; Elsevier: Amsterdam, The Netherlands, 2017.

63. Khantyanissa, H.S.; Ervina, E. Consumer preferences of artificial and natural flavours: A case in soft ice cream. In *IOP Conference Series: Earth and Environmental Science*; IOP Publishing: Bristol, UK, 2024; Volume 1352.
64. Perkovic, S.; Otterbring, T.; Schärli, C.; Pachur, T. The perception of food products in adolescents, lay adults, and experts: A psychometric approach. *J. Exp. Psychol. Appl.* **2022**, *28*, 555. [CrossRef]
65. Bearth, A.; Cousin, M.E.; Siegrist, M. The consumer's perception of artificial food additives: Influences on acceptance, risk and benefit perceptions. *Food Qual. Prefer.* **2014**, *38*, 14–23. [CrossRef]
66. Wang, Y. Research on consumers' understanding and attitudes to food additives-Taking Shanghai as an example. *Theor. Nat. Sci.* **2023**, *15*, 131–136. [CrossRef]
67. Park, M.K.; Lee, H.; Kim, B.K.; Kang, M.C.; Kim, T.K.; Sung, J.M.; Jeon, Y.E.; Choi, Y.S. A survey on domestic consumer's awareness of food additives and Clean Label concept. *Food Life* **2023**, *2023*, 95–103. [CrossRef]
68. Wiedmann, K.P.; Walten, L. Consumers' Perception of Product Information and Its Effect on Product Evaluation and Behavioral Intention: An Abstract. In *From Micro to Macro: Dealing with Uncertainties in the Global Marketplace. AMSAC 2020. Developments in Marketing Science: Proceedings of the Academy of Marketing Science*; Pantoja, F., Wu, S., Eds.; Springer: Cham, Switzerland, 2022.
69. Jaiswal, S.; Tripathi, A.P.; Tripathi, A. The Impact of Product Display on Consumer Attention and Buying Intention. In *Impact of Sensory Marketing on Buying Behavior*; IGI Global: Hershey, PA, USA, 2025; pp. 219–236.

Disclaimer/Publisher's Note: The statements, opinions and data contained in all publications are solely those of the individual author(s) and contributor(s) and not of MDPI and/or the editor(s). MDPI and/or the editor(s) disclaim responsibility for any injury to people or property resulting from any ideas, methods, instructions or products referred to in the content.

foods

MDPI

Article

Analysis and Evaluation of Muscle Quality in Different Parts of the Bighead Carp (*Aristichthys nobilis*)

Jiaqi Peng [1,2], Xiaorong Lu [3], Ruiqi Fan [2], Yuanyuan Ren [2], Huiwu Sun [2], Yali Yu [4,*] and Bo Cheng [2,*]

1 College of Fisheries and Life Science, Shanghai Ocean University, 999 Hucheng Ring Road, Nanhui New Town, Pudong New Area, Shanghai 201306, China
2 Chinese Academy of Fishery Sciences, No.150, Qingta West Road, Fengtai District, Beijing 100141, China
3 College of Fisheries, Huazhong Agricultural University, Wuhan 430070, China
4 Yangtze River Fisheries Research Institute, Chinese Academy of Fishery Sciences, Wuhan 420322, China
* Correspondence: ylyu8811@yfi.ac.cn (Y.Y.); chengb@cafs.ac.cn (B.C.)

Abstract: In this study, the bighead carp (*Aristichthys nobilis*) was the object of research to compare and analyze the contents of conventional nutrients, amino acids, fatty acids, inosinic acid, and earthy-smelling compounds (geosmin and 2-methylisoborneol) in muscles of its dorsal, belly, tail, opercula, eye socket, and mandible in order to evaluate their quality. The findings could inform recommendations for the consumption and processing of different muscle parts of the bighead carp. The results showed that the water content in the abdominal muscle was significantly lower than that in other parts, and the crude fat content was significantly higher than that in other parts ($p < 0.05$, the same below). Seventeen kinds of amino acids were detected in the muscles of the six parts of the fish, and the dorsal muscles had the highest umami amino acids, essential amino acids and total amino acids, which were 6.45 g/100 g, 6.82 g/100 g and 17.26 g/100 g, respectively. The total amount of essential amino acids in the muscle was higher than that in the FAO/WHO standard model. According to the AAS standard, the first limiting amino acid in the muscle of the six parts was valine (Val). There were 26 kinds of fatty acids in the abdomen, under the gill cover and in the eye socket muscles, and the content of polyunsaturated fatty acids in the mandibular muscles was the highest (45.41%). The content of inosine in the dorsal muscle was significantly higher than that in other parts. Geosmin was the main substance in the muscle. There was no correlation between the distribution of earthy-smelling compounds and fat content, but the content of earthy-smelling compounds in the muscle of the belly and eye socket was the highest. Therefore, the muscle quality of different parts of the bighead carp has its own characteristics, and targeted development and utilization can make more efficient use of bighead carp resources.

Keywords: *Aristichthys nobilis*; different parts; nutritional composition; volatile component

1. Introduction

Aristichthys nobilis is commonly known as chubby fish, bighead fish and baotou fish, as one of "the four major Chinese carps", and is a unique freshwater economic fish in China. The species is extensively distributed across a range of freshwater habitats nationwide, spanning northern and southern regions. The bighead carp exhibits rapid growth, strong disease resistance, and high nutritional value, making it highly sought after in the market. In 2022, bighead carp production reached 3.269 million metric tons, securing its position as the third-most-produced species, following the grass carp (*Ctenopharyngodon idellus*) and the silver carp (*Hypophthalmichthys molitrix*). This volume far surpassed the production of other common freshwater species, such as the black carp (*Mylopharyngodon piceus*), common carp (*Cyprinus carpio*), and crucian carp (*Carassius auratus*) [1].

The bighead carp is primarily sold fresh, with the edible portion constituting 67.3% of its total weight. It is renowned for its richness in docosahexaenoic acid (DHA), eicosapentaenoic acid (EPA), and essential amino acids [2], which makes it a relatively good source of such nutrients. In China, it serves as a fundamental ingredient in various traditional culinary preparations and enjoys considerable consumer preference [3].

Although the head of the bighead carp is widely favored by consumers, the other parts of its body are typically sold at lower prices. The prevalence of intermuscular bones and pronounced earthy odors contributes to this diminished marketability. This presents significant challenges, including a low processing yield, market volatility, and marginal product value addition, which cumulatively impede the growth of the bighead carp's aquaculture and processing sectors [4].

The shift in consumer preferences towards convenient, ready-to-cook freshwater fish products, such as fillets, tails, and heads, has been catalyzed by increasing disposable incomes and a faster pace of life [5]. Segmenting products not only facilitates transportation but also allows for the strategic processing of less desirable sections, based on their unique nutritional profiles, to optimize their commercial worth. Regarding the nutritional components of the bighead carp, the head and dorsal of the bighead carp were analyzed. It was found that the content of essential amino acids in the muscle of the bighead carp was above 40%, and that the content of calcium in dorsal meat was the highest, reaching 1195.36 mg/kg [6]. The nutritional composition differences among *Aristichthys nobilis* muscles, bones, skin, scales, viscera, and fin rays have been examined, with fish skin displaying the highest crude protein content of 24.28%, while the viscera and bones showed the lowest crude protein content [7]. The composition of volatile components in Nanwan bighead carp was analyzed, and 41 flavor substances were detected in the muscle, with the highest number being aldehydes (49.15%), which results in the unique flavor of bighead carp [8]. In addition to volatile substances, nucleotides also have a significant impact on the flavor of bighead carp [9]. Inosine monophosphate (IMP), also known as inosinic acid, is a nucleotide that is widely recognized for its role in the flavor of foods, particularly its contribution to the umami taste [10]. However, apart from its unique flavor, bighead carp also has a relatively noticeable earthy and musty odor, which can be attributed to geosmin (GSM) and 2-methylisoborneol (2-MIB) [11]. GSM and 2-MIB, known for imparting earthy-musty scents, are common in freshwater fish, and have a significant impact on the earthy and musty odor [12]. A previous study determined the content of earthy-smelling compounds in the entire bighead carp. It was discovered that the average level of these compounds in bighead carp is 5.4 μg/kg, a concentration significantly exceeding the smell threshold for earthy odors in fish [13].

Although prior research has shed light on the nutritive and volatile components and earthy odor compounds of the bighead carp, comprehensive analysis pertaining to the distribution of these compounds, particularly the geosmin and 2-MIB, across different fish parts is absent. Thus, the present study endeavors to conduct a thorough examination of the nutritional constituents, IMP, GSM, and 2-MIB levels in different sections of six bighead carp, coupled with an assessment of their quality. The objective is to generate pivotal nutritional insights for bighead carp consumption and to provide empirical support for the judicious utilization of its various segments.

2. Materials and Methods

2.1. Sample Collection and Preparation

Six live bighead carp (2.0 ± 0.1 kg) were purchased from Xinzhu Road Vegetable Market in Wuhan City, and they were cultured using semi-intensive aquaculture practices. The muscle tissues from different parts were collected from the dorsal, belly, tail, below the opercula, the eye socket, and the mandible and homogenized (n = 6) for each part. The dorsal, belly, and tail muscles were considered parts of the trunk, while the opercula, eye socket, and mandible were categorized as parts of the head. This categorization is used to analyze whether differences still exist between the head and trunk regions for various parameters. The sampling sites are indicated in Figure 1.

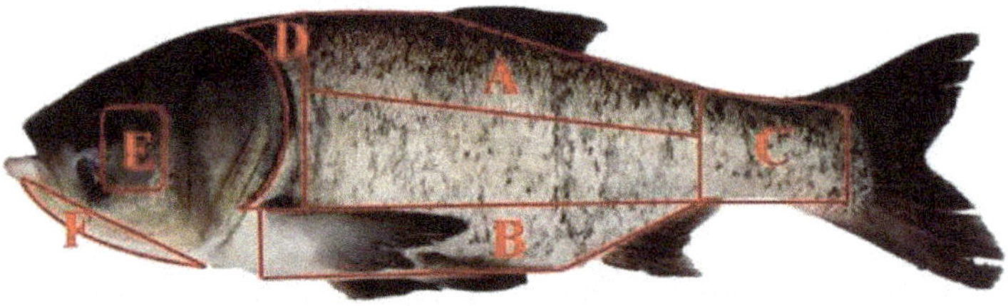

Figure 1. The sampling sites of muscle of a bighead carp. A: dorsal muscle; B: belly muscle; C: tail muscle; D: opercula muscle; E: eye socket muscle; F: mandible muscle.

2.2. Determination of Muscle Composition

The determination of moisture content adhered to the protocol specified in the China National Standard GB/T 5009.3-2016 [14], involving a thermal procedure at 105 °C. For ascertaining ash content, the methodology conformed to the standard GB/T 5009.4-2016 [15]. This entailed the complete carbonization of the sample, followed by transfer to a muffle furnace maintained at 550 °C for high-temperature ashing. Crude protein content was quantified via the Kjeldahl method, a reference method for nitrogen analysis. Finally, the estimation of crude fat content was conducted in accordance with GB 5009.6-2016 [16], utilizing a Soxhlet extraction method with petroleum ether as the solvent.

2.3. Amino Acid Content Determination and Quality Evaluation

The determination of amino acid content was based on the reference method GB 5009.124-2016 by the L-8900 automatic amino acid analyzer (Hitachi Ltd., Tokyo, Japan). The nutritional evaluation of amino acids was conducted according to the FAO/WHO (1973) pattern and the amino acid pattern of whole egg protein [17,18]. The amino acid score (AAS), chemical score (CS), and essential amino acid index (EAAI) were calculated using the following formulas [19]:

$$\text{AAS} = \frac{\text{Content of a specific amino acid in the sample (mg/gN))}}{\text{(Content of the corresponding amino acid in the FAO/WHO pattern (mg/gN))}} \quad (1)$$

$$\text{CS} = \frac{\text{Content of a specific amino acid in the sample (mg/gN)}}{\text{Content of the same amino acid in whole egg protein (mg/gN)}} \quad (2)$$

$$\text{EAAI} = \sqrt[n]{\frac{100A}{AE} \times \frac{100B}{BE} \times \frac{100C}{CE} \times \cdots \times \frac{100H}{HE}} \quad (3)$$

Note: n represents the number of essential amino acids being compared, A, B, C, ...; H represent the content of essential amino acids in the sample protein (%); AE, BE, CE, ..., HE represent the content of essential amino acids in whole egg protein (%).

2.4. Determination of Fatty Acids

The determination of fatty acid content was based on the reference method GB 5009.168-2016 "National Food Safety Standard—Determination of Fatty Acids in Foods" [20]. The sample preparation process can be briefly summarized as follows: A 2 g sample was weighed and hydrolyzed with 10 mL of 8.3 mol/L hydrochloric acid in a water bath at 70 °C to 80 °C for 40 min. Post-hydrolysis, the sample was extracted with 50 mL of an ether–petroleum ether mixture (1:1, *v*/*v*). The extract was then rotary evaporated to dryness, and the residue was dissolved in 8 mL of 2% sodium hydroxide in methanol solution. Subsequently, 7 mL of 15% boron trifluoride in methanol solution was added, and the mixture was refluxed in a water bath at 80 °C to facilitate the methylation of fatty acids. Finally, the methylated fatty acids were extracted with 5 mL of n-heptane. An amount of 1 mL of the upper layer of the resultant solution was then collected and placed in a sample vial for analysis.

The sample was determined by an Agilent 7890A gas chromatograph equipped with flame ionization detector (FID) (Agilent Technologies, Santa Clara, CA, USA). The instrumental conditions were set as follows: The temperature of the injector was maintained at 270 °C, and the FID temperature was set at 280 °C. The temperature program was established with an initial temperature of 100 °C, held for 13 min; this was followed by a ramp up from 100 °C to 180 °C at a rate of 10 °C/min, with a hold time of 6 min. The temperature was then increased from 180 °C to 200 °C at a rate of 1 °C/min, holding for 20 min, and finally from 200 °C to 230 °C at a rate of 4 °C/min, with a hold time of 10.5 min. The fatty acid content was determined using the external standard method, and the fatty acid methyl ester standards were purchased from Shanghai ANPEL Scientific Instrument Co., Ltd. (Shanghai, China).

2.5. Determination of Inosine Monophosphate (IMP)

To determine IMP, a 0.4 g sample was taken and transferred to a 50 mL centrifuge tube. Then, 20 mL of 10% perchloric acid was added, followed by vortex shaking for 1 min. The tube was centrifuged at 8000 rpm for 10 min at 4 °C. The extraction was repeated twice, and the supernatants were combined. The pH was adjusted to 6.0 with 10 mol/L NaOH and then further adjusted to 6.0–6.4 using 1 mol/L NaOH. After pH adjustment, the solution was transferred to a 50 mL volumetric flask and diluted with deionized water at 4 °C. The mixture was centrifuged at 8000 rpm for 10 min at 4 °C, and the supernatant was filtered through a 0.22 μm filter membrane and then analyzed using a Thermo TSQ QUANTUM LC/MS analyzer (Thermo Fisher Scientific, Waltham, MA, USA)

2.6. Determination of Geosmin and 2-Methylisoborneol

Samples weighing 0.1 g were put into headspace vials. Then, 1 mL of saturated NaCl solution and 0.15 g NaCl were added, and the bottle was sealed with a C10 rotor. The headspace vial was placed on a magnetic stirrer and incubated at 70 °C for 30 min, and an extraction needle was inserted into the headspace vial. The fiber was then inserted into the injection port of the gas chromatograph-mass spectrometer (GC-MS) for 5 min of sample desorption.

The gas chromatograph-mass spectrometer was equipped with an HP-5MS capillary column (30 mm × 0.25 mm × 0.25 μm). The injection port temperature was set at 250 °C. The temperature program was as follows: initial temperature of 60 °C, held for 2 min, increased at a rate of 10 °C/min to 120 °C, then increased at a rate of 3 °C/min to 150 °C, held for 1 min, and finally increased at a rate of 20 °C/min to 260 °C, held for 1 min (total time 25.5 min). Helium was used as the carrier gas at a flow rate of 1 mL/min without splitting. The mass spectrometer conditions were as follows: ion source temperature of 230 °C, transfer line temperature of 280 °C. The detection was performed in electron impact (EI) mode using single ion monitoring (SIM), with characteristic m/z values of 112.0 for geosmin and 95.0 for 2-MIB.

2.7. Statistical Analysis

Statistical analysis of the data was performed using SPSS 26.0 and was presented as mean ± standard deviation (X ± SD). To verify the normality and homogeneity of variances of the data, the Shapiro–Wilk test and Levene's test were used. Depending on whether these conditions were met, one-way ANOVA with a post-hoc LSD (least significant difference) test was applied for normally distributed data, or the Kruskal–Wallis non-parametric test followed by a post-hoc Bonferroni test was used otherwise. Different letters indicate significant differences ($p < 0.05$).

3. Results

3.1. Muscle Composition Variation

The contents of moisture, ash, crude protein, and crude fat in the muscles of different parts of bighead carp are detailed in Table 1. The belly muscle exhibited a significantly lower moisture content compared to the other parts ($p < 0.05$). The lowest values of ash content and crude protein were found in the belly muscle. In contrast, the belly muscle contained a crude fat content of 28.65%, markedly higher than the other parts, with a factor of 24.49 times greater than that of the dorsal muscle, which had a crude protein content of 17.51% and a crude fat content slightly lower than that of the head muscles.

Table 1. Essential nutrient components of muscle in different parts of *Aristichthys nobilis* (% wet weight).

Index	Trunk			Head		
	Dorsal	Belly	Tail	Opercula	Eye Socket	Mandible
Moisture	79.51 ± 0.71^{a}	60.81 ± 5.13^{b}	76.51 ± 4.75^{a}	78.96 ± 0.43^{a}	80.98 ± 3.32^{a}	83.09 ± 0.41^{a}
Ash	1.26 ± 0.02^{ab}	0.80 ± 0.01^{c}	1.34 ± 0.25^{a}	1.03 ± 0.13^{bc}	1.19 ± 0.03^{ab}	1.05 ± 0.06^{bc}
Crude protein	17.51 ± 0.46^{a}	12.30 ± 0.22^{d}	16.60 ± 0.65^{ab}	15.77 ± 0.52^{bc}	12.36 ± 1.48^{d}	14.76 ± 0.18^{c}
Crude lipid	1.17 ± 0.14^{c}	28.65 ± 0.08^{a}	3.07 ± 1.20^{b}	3.47 ± 0.65^{b}	4.05 ± 0.98^{b}	3.66 ± 0.07^{b}

Note: distinct letters denote statistically significant differences ($p < 0.05$).

3.2. Amino Acids Distribution

Seventeen amino acids were detected in all muscle samples, which are categorized into essential, non-essential, and semi-essential amino acids as shown in Table 2. Dorsal and tail muscles had higher levels of umami taste, essential, and non-essential amino acids than belly muscles. The total umami amino acid content (aspartic acid, glutamic acid, glycine, and alanine) was higher in these parts. The muscles under the opercula had a higher amino acid content compared to the eye socket and mandible muscles but a slightly lower content than in the dorsal and tail muscles.

Table 2. The amino acid content in muscle of different parts of *Aristichthys nobilis* (g/100 g wet weight).

Amino Acid	Trunk			Head		
	Dorsal	Belly	Tail	Opercula	Eye Socket	Mandible
Aspartic acid Asp △	1.82 ± 0.11^{a}	1.04 ± 0.15^{d}	1.65 ± 0.01^{ab}	1.37 ± 0.06^{bc}	1.13 ± 0.20^{cd}	1.34 ± 0.03^{c}
Glutamic acid Glu △	2.69 ± 0.13^{a}	1.54 ± 0.18^{e}	2.50 ± 0.00^{ab}	2.20 ± 0.08^{bc}	1.77 ± 0.28^{de}	2.08 ± 0.01^{cd}
Glycine Gly △	0.96 ± 0.04^{b}	0.68 ± 0.00^{c}	1.15 ± 0.01^{a}	0.87 ± 0.06^{b}	0.95 ± 0.09^{b}	0.96 ± 0.11^{b}
Alanine Ala △	0.98 ± 0.06^{a}	0.60 ± 0.03^{d}	1.00 ± 0.01^{a}	0.88 ± 0.01^{b}	0.78 ± 0.04^{c}	0.82 ± 0.05^{bc}
Threonine Thr *	0.80 ± 0.04^{a}	0.51 ± 0.01^{d}	0.76 ± 0.00^{ab}	0.74 ± 0.01^{b}	0.57 ± 0.02^{c}	0.61 ± 0.00^{c}
Valine Val *	0.83 ± 0.05^{a}	0.52 ± 0.08^{c}	0.84 ± 0.03^{a}	0.70 ± 0.03^{ab}	0.60 ± 0.12^{bc}	0.68 ± 0.04^{abc}
Methionine Met *	0.29 ± 0.01^{b}	0.08 ± 0.04^{d}	0.28 ± 0.04^{b}	0.40 ± 0.00^{a}	0.19 ± 0.03^{c}	0.32 ± 0.01^{b}
Isoleucine Ile *	0.78 ± 0.04^{a}	0.45 ± 0.04^{c}	0.79 ± 0.01^{a}	0.81 ± 0.00^{a}	0.61 ± 0.04^{b}	0.63 ± 0.06^{b}
Leucine Leu *	1.47 ± 0.08^{a}	0.91 ± 0.06^{c}	1.40 ± 0.02^{a}	1.38 ± 0.04^{a}	1.07 ± 0.10^{b}	1.13 ± 0.04^{b}
Phenylalanine Phe *	0.84 ± 0.05^{a}	0.70 ± 0.06^{bc}	0.80 ± 0.02^{ab}	0.82 ± 0.04^{a}	0.63 ± 0.08^{c}	0.61 ± 0.01^{c}
Lysine Lys *	1.65 ± 0.08^{a}	0.92 ± 0.13^{d}	1.52 ± 0.04^{ab}	1.23 ± 0.04^{bc}	1.03 ± 0.23^{cd}	1.25 ± 0.06^{bc}
Histidine His #	0.54 ± 0.04^{a}	0.27 ± 0.01^{e}	0.47 ± 0.00^{ab}	0.43 ± 0.01^{bc}	0.32 ± 0.05^{de}	0.37 ± 0.01^{cd}
Arginine Arg #	1.04 ± 0.04^{a}	0.71 ± 0.02^{e}	1.05 ± 0.01^{a}	0.97 ± 0.01^{b}	0.84 ± 0.01^{d}	0.92 ± 0.03^{b}
Cysteine Cys	0.45 ± 0.02^{b}	0.45 ± 0.11^{b}	0.48 ± 0.02^{b}	0.92 ± 0.09^{a}	0.58 ± 0.27^{ab}	0.37 ± 0.07^{b}
Serine Ser	0.76 ± 0.04^{a}	0.47 ± 0.04^{d}	0.73 ± 0.01^{ab}	0.63 ± 0.01^{bc}	0.56 ± 0.08^{cd}	0.60 ± 0.01^{c}
Tyrosine Tyr	0.70 ± 0.04^{a}	0.47 ± 0.09^{c}	0.67 ± 0.01^{a}	0.65 ± 0.02^{ab}	0.50 ± 0.07^{c}	0.54 ± 0.02^{bc}
Proline Pro	0.65 ± 0.03^{b}	0.59 ± 0.07^{b}	0.76 ± 0.04^{b}	1.00 ± 0.06^{a}	0.79 ± 0.15^{b}	0.64 ± 0.06^{b}
UAA	6.45 ± 0.31^{a}	3.87 ± 0.36^{d}	6.29 ± 0.02^{a}	5.31 ± 0.07^{b}	4.61 ± 0.41^{c}	5.21 ± 0.11^{bc}
EAA	6.82 ± 0.37^{a}	4.17 ± 0.21^{c}	6.55 ± 0.06^{a}	6.26 ± 0.08^{a}	4.87 ± 0.52^{bc}	5.37 ± 0.11^{b}
NEAA	9.01 ± 0.44^{a}	5.86 ± 0.30^{c}	8.94 ± 0.09^{a}	8.51 ± 0.42^{a}	7.04 ± 0.16^{b}	7.36 ± 0.26^{b}
TAA	17.26 ± 0.88^{a}	10.94 ± 0.60^{c}	16.85 ± 0.18^{a}	15.99 ± 0.06^{a}	12.91 ± 0.81^{b}	13.86 ± 0.38^{b}
EAA/TAA(%)	39.51%	45.62%	38.87%	39.15%	37.72%	38.74%
UAA/TAA(%)	37.37%	35.37%	37.33%	33.21%	35.71%	37.60%

Note: * indicates essential amino acids; △ denotes umami taste amino acid; # stands for semi-essential amino acid. UAA: umami taste amino acids. EAA: essential amino acids. NEAA: non-essential amino acids. TAA: total amino acids. Distinct letters denote statistically significant differences ($p < 0.05$).

3.3. Essential Amino Acids and Protein Quality

As Table 3 indicates, the total amount of essential amino acids in the muscle tissues of different parts of the bighead carp ranged from 2546 to 3032 mg/g N, and all were higher than the amino acid standards recommended by FAO/WHO (2190 mg/g N). The total essential amino acids in the muscles around the eye socket and gill cover were the closest to the standard of whole egg protein (2960 mg/g N). The amino acid score (AAS), chemical score (CS), and essential amino acid index (EAAI) for the muscles of different parts of the trunk are presented in Table 4. There are certain differences in the limiting amino acids of the muscles from different parts of the trunk under various scoring standards. When evaluated using the AAS, the first limiting amino acid in the muscles of all six regions was valine (Val), with the second limiting amino acid being threonine (Thr) for the eye socket, gill cover, lower jaw, and tail muscles, while for the dorsal and abdominal muscles it was isoleucine (Ile). When utilizing the chemical score (CS) method, the amino acid with the lowest CS is identified as the first limiting amino acid, and the one with the second lowest CS is identified as the second limiting amino acid. In this context, methionine (Met) + cysteine (Cys) were the first limiting amino acids for the dorsal and tail muscles, with Val being the second. For the eye socket, gill cover, lower jaw, and abdominal muscles, Val was the first limiting amino acid. With the exception of the lower jaw muscle, where Met + Cys was the second limiting amino acid, the second limiting amino acid for the remaining three parts was Ile.

Table 3. Comparison of muscle essential amino acids of *Aristichthys nobilis* in different parts (mg/g N).

Essential Amino Acid	Trunk			Head			FAO/WHO	Egg Protein
	Dorsal	Belly	Tail	Opercula	Eye Socket	Mandible		
Thr	286	259	286	293	288	258	250	292
Val	296	264	316	277	303	288	310	411
Ile	278	229	297	321	308	267	250	331
Leu	525	462	527	547	541	478	440	534
Lys	589	467	572	487	521	529	340	441
Met + Cys	264	269	286	523	389	292	220	386
Phe + Try	550	595	553	583	571	487	380	565
Total volume	2788	2546	2839	3032	2923	2600	2190	2960

Table 4. Comparison of AAS and CS in different parts of *Aristichthys nobilis*.

	Trunk						Head					
	Back		Belly		Tail		Opercula		Eye Socket		Mandible	
	AAS	CS	AAS	CS	AAS	CS	AAS	CS	AAS	CS	AAS	CS
Thr	1.14	0.98	1.04	0.89	1.14 ▲▲	0.98	1.17 ▲▲	1.00	1.15 ▲▲	0.99	1.03 ▲▲	0.88
Val	0.96 ▲	0.72 ▲▲	0.85 ▲	0.64 ▲	1.02 ▲	0.77 ▲▲	0.89 ▲	0.68 ▲	0.98 ▲	0.74 ▲	0.93 ▲	0.70 ▲
Ile	1.11 ▲▲	0.84	0.91 ▲▲	0.69 ▲▲	1.19	0.90	1.28	0.97 ▲▲	1.23	0.93 ▲▲	1.07	0.81
Leu	1.19	0.98	1.05	0.87	1.20	0.99	1.24	1.02	1.23	1.01	1.09	0.90
Lys	1.73	1.34	1.37	1.06	1.68	1.30	1.43	1.11	1.53	1.18	1.56	1.20
Met + Cys	1.20	0.68 ▲	1.22	0.70	1.30	0.74 ▲	2.38	1.36	1.77	1.01	1.33	0.76 ▲▲
Phe + Try	1.45	0.97	1.56	1.05	1.46	0.98	1.53	1.03	1.50	1.01	1.28	0.86
EAAI	91.03		82.75		93.62		100.57		97.36		86.09	

Note: ▲ denotes the first limiting amino acid; ▲▲ represents the second limiting amino acid.

3.4. Fatty Acid Profile

The fatty acid analysis presented in Table 5 showed 26 types of fatty acids in the belly, under opercula, and eye socket muscles, while the tail muscle had 24, and the dorsal and mandible muscles contained 20 types. As for the content and significant differences between sections of the bighead carp, it is indicated that the belly muscle had high levels of SFAs (saturated fatty acids, 33.12%) and MUFAs (monounsaturated fatty acids, 31.93%), with the PUFAs (polyunsaturated fatty acids) relatively high in the mandibular muscle (45.41%) and dorsal muscle (44.99%). The distribution of fatty acids in different sections of the bighead carp showed variation. SFAs exhibit a consistent presence across both trunk and head regions, but with some variations. In the trunk, the mean SFA content is slightly higher in the belly section (33.12%) and tail sections (32.33%) compared to the dorsal (30.89%). This suggests a marginally richer SFA concentration in the belly area. In the head, there were no significant differences between each section.

Table 5. Fatty acid composition and content in different parts of *Aristichthys nobilis* (%).

Fatty Acid	Trunk			Head		
	Dorsal	Belly	Tail	Opercula	Eye Socket	Mandible
C11:0	-	0.09 ± 0.03 ^{a}	0.04 ± 0.06 ^{ab}	0.07 ± 0.01 ^{ab}	0.07 ± 0.02 ^{ab}	-
C12:0	0.24 ± 0.01 ^{c}	0.43 ± 0.04 ^{a}	0.29 ± 0.00 ^{bc}	0.35 ± 0.03 ^{b}	0.33 ± 0.03 ^{b}	0.17 ± 0.03 ^{d}
C13:0	0.17 ± 0.01 ^{b}	0.30 ± 0.05 ^{a}	0.22 ± 0.03 ^{ab}	0.24 ± 0.04 ^{ab}	0.26 ± 0.04 ^{ab}	0.17 ± 0.04 ^{b}
C14:0	4.12 ± 0.13 ^{cd}	6.61 ± 0.50 ^{a}	4.95 ± 0.21 ^{bc}	5.83 ± 0.54 ^{ab}	5.57 ± 0.33 ^{ab}	3.84 ± 0.54 ^{d}
C15:0	1.21 ± 0.05 ^{bc}	1.66 ± 0.16 ^{a}	1.41 ± 0.01 ^{abc}	1.48 ± 0.11 ^{ab}	1.54 ± 0.14 ^{a}	1.19 ± 0.09 ^{c}
C16:0	17.84 ± 0.35	18.29 ± 0.19	18.19 ± 0.68	17.93 ± 0.41	17.80 ± 0.40	18.14 ± 0.18
C17:0	0.98 ± 0.05	1.07 ± 0.17	1.10 ± 0.04	0.99 ± 0.08	1.17 ± 0.03	1.06 ± 0.04
C18:0	6.33 ± 0.18 ^{b}	3.96 ± 0.16 ^{e}	5.75 ± 0.18 ^{c}	4.81 ± 0.20 ^{d}	4.80 ± 0.25 ^{d}	7.34 ± 0.23 ^{a}
C20:0	-	0.36 ± 0.01	0.36 ± 0.01	0.33 ± 0.02	0.33 ± 0.02	-
C22:0	-	0.38 ± 0.01	-	0.38 ± 0.07	0.37 ± 0.07	-
∑SFAs	30.89 ± 0.07 ^{c}	33.12 ± 0.59 ^{a}	32.33 ± 0.58 ^{ab}	32.43 ± 0.04 ^{ab}	32.25 ± 0.22 ^{ab}	31.91 ± 0.24 ^{b}
C14:1n5	0.15 ± 0.01	0.28 ± 0.02	0.18 ± 0.05	0.23 ± 0.03	0.25 ± 0.01	0.29 ± 0.17
C16:1n7	7.29 ± 0.16 ^{c}	10.12 ± 0.00 ^{a}	8.64 ± 0.59 ^{b}	9.05 ± 0.20 ^{b}	9.13 ± 0.01 ^{b}	6.91 ± 0.37 ^{c}
C17:1n7	-	-	-	1.00 ± 0.01	1.06 ± 0.01	-
C18:1n9t	0.39 ± 0.03 ^{b}	0.52 ± 0.06 ^{a}	0.45 ± 0.06 ^{ab}	0.47 ± 0.05 ^{ab}	0.48 ± 0.04 ^{ab}	0.40 ± 0.02 ^{ab}
C18:1n9c	15.20 ± 2.35	19.18 ± 3.16	17.23 ± 1.77	17.71 ± 2.88	17.62 ± 1.97	13.54 ± 1.44
C22:1n9	-	0.20 ± 0.04	-	-	-	-
C24:1n9	-	-	-	-	-	0.57 ± 0.04
C20:1	1.10 ± 0.31 ^{ab}	1.65 ± 0.27 ^{a}	1.25 ± 0.22 ^{ab}	1.53 ± 0.27 ^{ab}	1.43 ± 0.23 ^{ab}	0.98 ± 0.12 ^{b}
∑ MUFAs	24.13 ± 2.85 ^{bc}	31.93 ± 3.51 ^{a}	27.77 ± 2.69 ^{abc}	29.99 ± 3.00 ^{ab}	29.96 ± 2.20 ^{ab}	22.69 ± 1.42 ^{c}
C18:2n6c	4.78 ± 0.12 ^{ab}	5.42 ± 0.40 ^{a}	4.90 ± 0.15 ^{ab}	5.01 ± 0.31 ^{a}	5.01 ± 0.29 ^{a}	4.20 ± 0.26 ^{b}
C18:3n6	-	0.45 ± 0.01a	0.32 ± 0.01	0.36 ± 0.04	0.37 ± 0.03	-
C18:3n3	6.83 ± 0.21 ^{cd}	9.36 ± 0.54 ^{a}	7.56 ± 0.11 ^{bc}	8.54 ± 0.77 ^{ab}	8.33 ± 0.51 ^{ab}	6.02 ± 0.62 ^{d}
C20:2	0.65 ± 0.06	0.64 ± 0.06	0.75 ± 0.10	0.64 ± 0.04	0.69 ± 0.03	0.72 ± 0.05
C20:3n6	-	0.72 ± 0.03	0.72 ± 0.04	0.71 ± 0.01	0.71 ± 0.01	-
C20:3n3	1.18 ± 0.45	0.96 ± 0.13	0.99 ± 0.09	0.99 ± 0.11	1.02 ± 0.09	1.08 ± 0.12
C20:4n6	6.51 ± 0.27 ^{b}	3.29 ± 0.21 ^{e}	5.40 ± 0.63 ^{c}	4.22 ± 0.11 ^{d}	4.41 ± 0.06 ^{d}	7.75 ± 0.48 ^{a}
C22:2n6	0.35 ± 0.09	-	-	-	-	-
C20:5n3	9.52 ± 0.64	7.55 ± 0.86	8.07 ± 0.77	7.98 ± 1.02	7.88 ± 0.78	9.43 ± 0.50
C22:6n3	15.18 ± 0.94 ^{a}	6.55 ± 0.69 ^{c}	11.17 ± 1.58 ^{b}	9.10 ± 0.64 ^{b}	9.36 ± 0.61 ^{b}	16.20 ± 0.11 ^{a}
∑PUFAs	44.99 ± 2.78 ^{a}	34.96 ± 2.92 ^{b}	39.90 ± 3.26 ^{ab}	37.57 ± 3.04 ^{b}	37.79 ± 2.43 ^{b}	45.41 ± 1.18 ^{a}
∑n-3 PUFA	32.70 ± 2.23 ^{a}	24.42 ± 2.22 ^{b}	27.79 ± 2.35 ^{ab}	26.62 ± 2.53 ^{b}	26.60 ± 2.00 ^{b}	32.74 ± 1.34 ^{a}
∑n-6 PUFA	11.64 ± 0.49 ^{ab}	9.89 ± 0.65 ^{c}	11.36 ± 0.81 ^{ab}	10.32 ± 0.47 ^{bc}	10.51 ± 0.40 ^{bc}	11.95 ± 0.21 ^{a}

Note: - indicates that the data were below the limit of quantification. Distinct letters denote statistically significant differences ($p < 0.05$).

Regarding MUFAs, the highest content of MUFAs was found in the belly muscle (31.93%), with slightly lower levels in the tail, opercula, and eye socket muscles, which were relatively close to each other (27.77%, 29.99%, and 29.96%, respectively). The lowest content was in the mandible (22.69%), followed by the dorsal muscle (24.13%). PUFAs exhibit a trend opposite to that of MUFAs, with relatively higher concentrations in the dorsal (45.41%) and mandible (44.99%) muscles, and the lowest in the belly muscle (34.96%). This trend is consistent across n-3 PUFA, n-6 PUFA, and DHA (docosahexaenoic acid, C22:6n3), with the highest concentrations found in the mandible muscle and the lowest in the belly muscle. As for EPA (eicosapentaenoic acid, C20:5n3), its distribution across various parts showed no significant differences, remaining within the range of 7.55% to 9.52%.

3.5. Flavor Compound Concentrations

As shown in Table 6, the IMP (inosine monophosphate) content in bighead carp muscles was highest in the dorsal muscle at 7.76 mg/g. The mandibular muscle followed with 5.65 mg/g, and the belly muscle had the lowest content at 1.86 mg/g. The distribution of IMP in the head and trunk regions does not show a specific trend.

Table 6. IMP content in different parts of *Aristichthys nobilis* (mg/g wet weight).

Index	Trunk			Head		
	Dorsal	Belly	Tail	Opercula	Eye Socket	Mandible
IMP	7.76 ± 0.14 ^{a}	1.86 ± 0.21 ^{de}	3.24 ± 1.54 ^{cd}	3.42 ± 0.17 ^{cd}	3.35 ± 0.17 ^{cd}	5.65 ± 0.99 ^{b}

Note: distinct letters denote statistically significant differences ($p < 0.05$).

3.6. Earthy Odor Determinants

GSM and 2-MIB were detected in all six muscle parts (Table 7). The concentrations of these compounds were notably higher in the belly and eye socket muscles, with 2-MIB at 0.19 μg/kg and GSM at 1.44 μg/kg in the belly muscles and 1.85 μg/kg in the eye socket muscles. These levels exceed the sensory thresholds of 0.085 μg/kg for 2-MIB and 0.59 μg/kg for GSM in fish [21]. In the tail and mandible muscles, the levels of GSM (0.28 μg/kg in the tail, 0.35 μg/kg in the mandible) and 2-MIB (both 0.07 μg/kg) were relatively low, with neither exceeding the sensory thresholds.

Table 7. GSM and 2-MIB content in different parts of *Aristichthys nobilis* (μg/kg wet weight).

Index	Trunk			Head		
	Dorsal	Belly	Tail	Opercula	Eye Socket	Mandible
2-MIB	0.10 ± 0.01 ^{bc}	0.19 ± 0.01 ^{a}	0.07 ± 0.00 ^{c}	0.16 ± 0.05 ^{ab}	0.19 ± 0.02 ^{a}	0.07 ± 0.02 ^{c}
GSM	0.15 ± 0.02 ^{c}	1.44 ± 0.26 ^{a}	0.28 ± 0.00 ^{c}	0.88 ± 0.20 ^{b}	1.85 ± 0.40 ^{a}	0.35 ± 0.14 ^{bc}

Note: distinct letters denote statistically significant differences ($p < 0.05$).

4. Discussion

4.1. Muscle Composition Variation

The findings regarding the muscle composition of the bighead carp have several implications. The high crude fat content of the belly muscle, coupled with its simple component profile, few impurities, and richness in polyunsaturated fatty acids [22], underscores its potential as a raw material for fish oil processing. This could significantly contribute to the advancement of the deep processing industry of freshwater fish, thereby enhancing economic benefits. Moreover, given that the processing of freshwater fish predominantly targets surimi and its products, the utilization of by-products such as the belly, head, and viscera—which constitute 40–55% [23] of the raw fish—is of great interest.

In contrast, the crude protein content in other parts of the bighead carp, ranging from 12.30% to 17.51%, and the crude lipid content between 1.17% and 4.05%, categorizes these areas as low-fat (oil content <5%) according to Stansby [24]. This distinction is vital for nutritional exploitation, as the dorsal (17.51%) and tail (16.60%) sections, with their high protein content, are ideal for protein or amino acid extraction. The moisture content across different parts of the carp varies from 60.81% to 83.09%, and the ash content from 0.80% to 1.34%. These moisture and ash levels are consistent with findings for other species, such as the common carp (77.20% moisture, 1.30% ash) reported by Mohanty et al. [25] and the African catfish (75.92% moisture, 0.83% ash) by Teame et al. [26]. For mineral extraction, the tail and dorsal sections, with the highest ash content, emerge as promising areas.

However, the reasons for the higher fat content in the belly of the bighead carp and strategies for its regulation require further research, especially due to limited studies segmenting fish into such detailed parts for nutritional comparison. Overall, the data present a clear stratification of nutritional components across different sections of the bighead carp, offering valuable insights for targeted nutritional exploitation.

4.2. Amino Acids Distribution

The nutritional quality of the muscle proteins is gauged by the type, content, and ratio of amino acids present. The delicacy of muscle is largely influenced by umami amino acids, such as aspartic acid, glutamic acid, glycine, and alanine, with the content of glutamic acid being particularly pivotal for the umami taste [27]. Analysis of the amino acids in bighead carp muscles revealed that the dorsal and tail muscles not only had a higher glutamic acid content (2.69 g/100 g and 2.50 g/100 g, respectively) but also exhibited significantly greater total, essential, and umami amino acid levels compared to other parts. This suggests that the dorsal and tail portions of the bighead carp could offer superior taste and nutritional benefits.

4.3. Essential Amino Acids and Protein Quality

The protein quality of bighead carp muscles is considerably influenced by their anatomical location. When compared with different types of bighead carp, the EAAI for trunk muscles in this study was found to be superior to that of red and black bighead carp (80.47 and 80.13) as reported by Liu et al. [28], but inferior to that for bighead carp from the Miyun Reservoir (98.76) as documented by Jia et al. [29]. These muscles were, however, on par with American bighead carp (92.69) as per the findings of Ma et al. [30]. Age-related variations in EAAI were also evident in the study by Zheng et al. [31]. Furthermore, the disparity in limiting amino acids in the muscles, especially Val, Thr, Ile, and Met+Cys, as opposed to other regions and studies, suggests that the nutritional content is not only species-specific but also significantly affected by environmental conditions and cultivation practices, as seen in comparisons with bighead carp from the Miyun Reservoir (Trp, Val) [29], Shanxi Reservoir (Trp, Val, Met, and Cys) [31], and in cultivated bighead carp (Val, Met, and Cys) [32]. Additionally, the limiting amino acids (Val, Thr) in the three head muscle parts studied here show differences compared to the limiting amino acids (Met, Cys, and Val) in the head of bighead carp as measured by Jiang et al. [6], which indicates that the head muscle limiting amino acids vary. This suggests that further research is needed to compare the effects of various factors such as temperature, water quality, and nutritional intake on the nutritional composition of bighead carp muscle.

4.4. Fatty Acid Profile

The muscle tissue of bighead carp presents a beneficial fatty acid composition, with an adequate balance of saturated (SFAs), monounsaturated (MUFAs), and polyunsaturated fatty acids (PUFAs). With unsaturated fatty acids accounting for 66.88–69.11% of the total fatty acids across the six muscle parts analyzed, the muscles of the mandible and dorsal part exhibit the highest content of these beneficial fats. This finding indicates that the nutritional quality of the dorsal muscle is comparable, if not superior, to that of the head muscle, which has been traditionally preferred by consumers for its rich protein and unsaturated fat content [33]. The mandibular muscle, with the highest percentage of EPA and DHA at 25.63%, surpasses the content found in other fish species like the grass carp (9.43%), carp (13.91%), and tilapia (*Oreochromis mossambicus*) (11.49%) [34]. The significant presence of PUFAs in the mandibular and dorsal muscles, representing 45.41% and 44.99%, respectively, suggests that these parts have high fatty acid nutritional value and contribute to the flavor profile of the fish, which correlates with findings on flavor enhancement by PUFA in muscle [35].

4.5. IMP Content and Flavor Enhancement

The importance of IMP as a key umami substance has been well documented, especially its role in enhancing the flavor of fish when combined with adenosine diphosphate [36,37]. In the present study, IMP was highest in the dorsal muscle, followed by the mandibular muscle, and the belly muscle had the lowest concentration. These findings are consistent with the IMP contents in different fish species and tissues, as reported in previous studies by Chen et al. [38]. Except for belly muscles, all other examined muscle parts of bighead carp had an IMP content exceeding 3.24 mg/g. This is notably higher than in other fish species, such as Siniperca chuatsi (*Hypophthalmichthys molitrix*) with 2.46 mg/g [39], the grass carp with 2.18 mg/g [40], and the silver carp with 0.91 mg/g [41], highlighting the superior flavor profile of bighead carp. These findings indicate that not only does the IMP content vary significantly between different species and within the tissues of the same species, but also that the bighead carp, particularly its dorsal and mandibular muscles, can be a superior choice for consumers seeking the umami taste in their seafood.

4.6. Earthy Odor Compounds

GSM and 2-MIB were notably higher in the belly and eye socket muscles in our research. These levels exceed the sensory thresholds of 0.085 μg/kg for 2-MIB and 0.59 μg/kg for GSM in fish [21], emphasizing the significant presence of GSM, which ranges from 0.15 to 1.85 μg/kg in bighead carp muscles, contributing to the earthy odor. However, the levels of GSM and 2-MIB are relatively low, with neither exceeding the sensory thresholds. This suggests that these two parts may have a relatively lighter earthy and musty odor. Comparative analysis of GSM and 2-MIB content with other studies indicates that bighead carp might have higher concentrations of these earthy-smelling compounds than other fish species in different conditions reported in literature. For instance, Atlantic salmon in recirculating aquaculture systems (RAS) showed 2-MIB concentrations of 0.55 to 0.99 μg/kg and GSM concentrations of 0.26 to 0.51 μg/kg [42]. Furthermore, tilapia cultured in Hainan ponds displayed GSM levels of 0.11 to 0.28 μg/kg and 2-MIB levels of 0.08 to 0.72 μg/kg [43], which are lower than those found in bighead carp. The findings from this study underscore a pronounced earthy odor issue in bighead carp that could affect its consumer acceptability.

The effect of fish fat content on the concentration of these compounds has been explored in other research [21,44]. While some studies, such as that by Tucker [45], have found a correlation between higher fat content and stronger earthy odors, this study did not find a significant correlation between the crude fat content in various parts of bighead carp muscle and the concentration of earthy-smelling compounds ($p > 0.05$). This lack of correlation echoes the conclusions of Petersen et al. [46], who found no correlation between fat content and the concentration of off-flavor compounds. Such findings suggest that the mechanisms governing the accumulation of GSM and 2-MIB in bighead carp may be different from those in other species or may be influenced by factors other than fat content. Given the economic importance of bighead carp, understanding and mitigating the factors contributing to off-flavors is crucial for improving its marketability and ensuring its palatability for consumers.

5. Conclusions

The dorsal and tail muscles of bighead carp showcase a desirable profile of high protein (17.51% and 16.60%, respectively) and low fat (1.17% and 3.07%, respectively). Essential amino acids in these muscles meet and surpass FAO/WHO standards. Unsaturated fatty acids are prevalent, accounting for 66.88–69.11% of total fatty acids, with the highest concentrations found in mandible and dorsal muscles, affirming their nutritional worth. The dorsal muscle stands out for its high IMP content (7.76 mg/g), contributing to flavor. While the belly and eye socket muscles exhibit elevated levels of GSM and 2-MIB, compounds responsible for the fish's earthy scent, this issue still requires further investigation in subsequent research. Different parts of bighead carp muscle possess distinct qualities, suggesting specialized approaches to their use could enhance resource utilization. Future research could explore targeted applications, such as harnessing protein-rich areas for nutritional enhancements and tapping into lipid-dense sections for oil extraction. Advancements in aquaculture practices to optimize the nutritional quality of bighead carp will be a key area of focus. These approaches hold the potential to maximize the value derived from this species, aligning with sustainable and efficient resource utilization.

Author Contributions: J.P. and X.L.: writing—original draft, conceptualization, investigation, data curation, formal analysis, methodology. R.F. and Y.R.: investigation, data curation. Y.Y.: conceptualization, writing—review and editing, formal analysis, methodology. H.S.: conceptualization. B.C.: conceptualization, funding acquisition, writing—review and editing, project administration, methodology. All authors have read and agreed to the published version of the manuscript.

Funding: This work was supported by the China Agriculture Research System of MOF and MARA (CARS-46) and the Central Public-interest Scientific Institution Basal Research Fund, CAFS (No. 2023XT0704).

Institutional Review Board Statement: Use of laboratory animals was approved by the Animal Welfare and Animal Experimental Ethical Committee at the Shanghai Ocean University (Shanghai, China) (certification no: SHOU-DW-2023-077, certification date: 13 May 2023).

Data Availability Statement: Data is contained within the article.

Acknowledgments: This work is supported by the Chinese Academy of Fishery Sciences—Quality and Standards Research Center and the Beijing Fisheries Research Institute, as well as the collaboration of the Yangtze River Fisheries Research Institute, Chinese Academy of Fishery Sciences, Shanghai Ocean University, and Huazhong Agricultural University.

Conflicts of Interest: The authors declare no conflict of interest.

Abbreviations

AAS	Amino acid score
CS	Chemical score
GSM	Geosmin
2-MIB	2-methylisoborneol
IMP	Inosine monophosphate

References

1. Bureau of Fisheries, Ministry of Agriculture and Rural Affairs of China. *China Fisheries Statistical Yearbook 2022*; China Agriculture Press: Beijing, China, 2022.
2. Li, X.; You, J.; Yin, T.; Liu, R.; Du, H.; Zhao, S.; Xiong, S. Analyzing nutrient composition in edible parts of bighead carp (*Aristichthys nobilis*) head. *J. Huazhong Agric. Univ.* **2023**, *42*, 163–169.
3. Fu, B.; Wang, X.; Feng, X.; Yu, X.; Tong, J. Comparative transcriptomic analyses of two bighead carp (*Hypophthalmichthys nobilis*) groups with different growth rates. *Comp. Biochem. Physiol. Part D Genom. Proteom.* **2016**, *20*, 111–117. [CrossRef]
4. Hong, H. Fatty Acid Profile, Quality Changes and Controlling Techniques of Bighead Carp (*Aristichthys nobilis*) during Storage. Ph.D. Thesis, China Agricultural University, Beijing, China, 2015.
5. Mohamad, R.; Majid, N.A.; Abidin, Z.Z.; Rusli, R. Consumer preference towards fresh water fish product developed by MARDI. *Econ. Technol. Manag. Rev.* **2010**, *5*, 71–77.
6. Jiang, Q.; Wu, J.; Xu, Y.; Tang, F.; Wang, L.; Zhang, H.; Xia, W. Composition Analysis and Nutritional Evaluation of Different Parts of Bighead Carp (*Aristichthys nobilis*). *Food Sci.* **2014**, *35*, 183–187.
7. Liu, Y.; Hong, H.; Li, D.; Luo, Y. Composition and nutritional function evaluation of bighead carp. *Sci. Fish Farming* **2020**, *11*, 73–74. [CrossRef]
8. Zhao, L.; Ma, L. Investigation of chemical volatile compounds of south bay bighead carp by GC/MS. *Food Mach.* **2011**, *27*, 80–82.
9. Gao, Q.; An, Y.; Chen, Z.; Xiong, S. The effect of short-term micro-flow water treatment on the muscle taste quality of bighead carp cultured in ponds. *Acta Hydrobiol. Sin.* **2021**, *45*, 1057–1066. [CrossRef]
10. Fletcher, G.; Bremner, H.; Olley, J.; Statham, J. Umami revisited: The relationship between inosine monophosphate, hypoxanthine and smiley scales for fish flavor. *Food Rev. Int.* **1990**, *6*, 489–503. [CrossRef]
11. Zhang, T.; Xue, Y.; Li, Z.; Wang, Y.; Yang, W.; Xue, C. Effects of ozone on the removal of geosmin and the physicochemical properties of fish meat from bighead carp (*Hypophthalmichthys nobilis*). *Innov. Food Sci. Emerg. Technol.* **2016**, *34*, 16–23. [CrossRef]
12. Schrader, K.K.; Rubio, S.A.; Piedrahita, R.H.; Rimando, A.M. Geosmin and 2-Methylisoborneol Cause Off-Flavors in Cultured Largemouth Bass and White Sturgeon Reared in Recirculating-Water Systems. *N. Am. J. Aquac.* **2005**, *67*, 177–180. [CrossRef]
13. Xue, Y. Study on Antifreeze Denaturants for Bighead Carp Surimi and Methods for Removing Geosmin and Earthy Odors. Ph.D. Thesis, China Ocean University, Qingdao, China, 2007.
14. *GB 5009.3-2016*; National Food Safety Standard—Determination of Moisture in Food. National Health and Family Planning Commission of the People's Republic of China: Beijing, China, 2016.
15. *GB 5009.4-2016*; National Food Safety Standard—Determination of Ash Content in Food. National Health and Family Planning Commission of the People's Republic of China: Beijing, China, 2016.

16. *GB 5009.6-2016*; National Food Safety Standard—Determination of Fat in Food. National Health and Family Planning Commission of the People's Republic of China: Beijing, China, 2016.
17. Millward, D.J. Amino acid scoring patterns for protein quality assessment. *Br. J. Nutr.* **2012**, *108*, S31–S43. [CrossRef] [PubMed]
18. Usydus, Z.; Szlinder-Richert, J.; Adamczyk, M. Protein quality and amino acid profiles of fish products available in Poland. *Food Chem.* **2009**, *112*, 139–145. [CrossRef]
19. Hu, J.; Li, X.; Li, L.; Gao, P.; Xu, J.; Liu, J. Comparative study on growth and muscle composition of *Ctenopharyngodon idella* under two aquaculture patterns. *Feed Res.* **2021**, *44*, 62–67. [CrossRef]
20. *GB 5009.186-2016*; National Food Safety Standard—Determination of Fatty Acid in Food. National Health and Family Planning Commission of the People's Republic of China: Beijing, China, 2016.
21. Persson, P.-E. Sensory properties and analysis of two muddy odour compounds, geosmin and 2-methylisoborneol, in water and fish. *Water Res.* **1980**, *14*, 1113–1118. [CrossRef]
22. Luo, T. Study on the Extraction Technology and Quality Analysis of Fish Oil from the Belly of Silver Carp. Master's Thesis, Jiangnan University, Wuxi, China, 2017.
23. Liu, L. Comprehensive Development and Utilization of Silver Carp Scraps. Master's Thesis, Wuhan University of Light Industry, Wuhan, China, 2015.
24. Stansby, M.E. Composition of certain species of fresh-water fish. *J. Food Sci.* **1954**, *19*, 231–234. [CrossRef]
25. Mohanty, B.P.; Ganguly, S.; Mahanty, A.; Sankar, T.V.; Anandan, R.; Chakraborty, K.; Paul, B.N.; Sarma, D.; Syama Dayal, J.; Venkateshwarlu, G.; et al. DHA and EPA Content and Fatty Acid Profile of 39 Food Fishes from India. *BioMed Res. Int.* **2016**, *2016*, 4027437. [CrossRef]
26. Teame, T.; Natarajan, P.; Tesfay, Z. Proximate and mineral composition of some commercially important fish species of tekeze reservoir and lake Hashenge, Ethiopia. *Int. J. Fish. Aquat. Stud.* **2016**, *4*, 160–164.
27. Wang, X.; Tu, X.; Wu, J.; Dai, Z. Nutritional Evaluation and Analysis of the Volatile Flavor Component of Grass Carp. *J. Chin. Inst. Food Sci. Technol.* **2014**, *14*, 182–189.
28. Liu, F.; Men, Y.; Han, Z.; Xia, H.; Yang, P.; Zhou, G.; Chen, K.; Li, M.; Yang, F. Analysis and evaluation of nutrient components in muscle of red bighead carp and *Aristichthys nobilis*. *Freshw. Fish.* **2017**, *47*, 101–106.
29. Jia, C.; Qu, J.; Li, Y.; Hu, Q.; Liu, P.; Yang, M.; Zhang, Q. Nutritional compositions in muscles of silver carp *Hypophthalmichthys molitrix* and bighead carp *Aristichthys nobilis* in Miyun Reservoir. *Fish. Sci. (Dalian)* **2019**, *38*, 40–47.
30. Ma, X.; Qian, P.; Dai, Z. Analysis and Quality Evaluation of Nutritional Components in the Muscle of Hypophthalmichthys nobilis and Hypophthalmichthys molitri. *J. Chin. Inst. Food Sci. Technol.* **2016**, *16*, 273–280.
31. Zheng, Y.; Zhou, C.; Du, S.; Jiang, Q.; Lu, R.; Ke, A.; Zhu, J.; Hu, Y. Nutritional Composition Analysis and Evaluation of Bighead Carp in Shanxi Reservoir, Wenzhou City. *Chin. Fish. Qual. Stand.* **2021**, *11*, 46–54.
32. Wang, J.; Tang, L.; Liu, K.; Li, H.; Yao, J.; An, M. Comparative analysis of muscle nutritional components between cultured and wild bighead carp. *Hebei Fish.* **2013**, *2*, 8–14+16.
33. Hong, H.; Luo, Y.K.; Zhou, Z.Y.; Bao, Y.L.; Lu, H.; Shen, H.X. Effects of different freezing treatments on the biogenic amine and quality changes of bighead carp (*Aristichthys nobilis*) heads during ice storage. *Food Chem.* **2013**, *138*, 1476–1482. [CrossRef] [PubMed]
34. Han, Y. Analysis of Lipid Characteristics of Main Freshwater Fish Species in China and Study on Gel Properties of Fish Meat. Master's Thesis, Shanghai Ocean University, Shanghai, China, 2019.
35. Zhou, M.; Zhao, L.; Chen, L.; Bai, C.; Hou, Y.; Dai, Y.; Chen, S. Analysis of Nutritional Composition and Quality of Bighead Carp Muscle from Shanggao. *Sci. Fish Farming* **2015**, *10*, 75–78. [CrossRef]
36. Kuda, T.; Fujita, M.; Goto, H.; Yano, T. Effects of freshness on ATP-related compounds in retorted chub mackerel *Scomber japonicus*. *LWT—Food Sci. Technol.* **2007**, *40*, 1186–1190. [CrossRef]
37. Lv, D.; Lu, G.; Shi, Z.; Zhu, R. Discussion on the formation and degradation mechanism of inosinic acid. *China Condiment* **2009**, *34*, 36–40.
38. Chen, S.; Shi, W.; Wang, Z. Comparison of flavor of dorsal meat, belly meat and tail meat of *Mylopharyngodon piceus*. *Fish. Mod.* **2021**, *48*, 57–66.
39. Liu, X. Integrated Study on the Characters of Fish Meat. Master's Thesis, Xiamen University, Xiamen, China, 2009.
40. Liu, Y.; Fu, Z.; Zhang, H.; Gao, S.; Shu, R.; Luo, Y.; Hong, H. Quality comparison of grass carp cultured in containers and ponds. *J. China Agric. Univ.* **2022**, *27*, 193–202.
41. Zhang, L.; Lv, J.; Li, Q.; Luo, Y. Changes of quality, ATP-related compounds and enzyme activities in silver carp (*Hypophthalmichthys molitrix*) fillets during partial freezing. *J. China Agric. Univ.* **2016**, *21*, 77–83.
42. Davidson, J.; Schrader, K.; Ruan, E.; Swift, B.; Aalhus, J.; Juarez, M.; Wolters, W.; Burr, G.; Good, C.; Summerfelt, S.T. Evaluation of depuration procedures to mitigate the off-flavor compounds geosmin and 2-methylisoborneol from Atlantic salmon Salmo salar raised to market-size in recirculating aquaculture systems. *Aquac. Eng.* **2014**, *61*, 27–34. [CrossRef]
43. Zhang, K. Impact of Different Cultivation Systems on the Accumulation of Off-Flavor Compounds in Tilapia Flesh. Master's Thesis, Shanghai Ocean University, Shanghai, China, 2018.
44. Liu, L.; Li, K.; Yan, L. Sources, determination, prevention and elimination of off-flavour compounds in aquatic animals. *J. Fish. China* **2021**, *45*, 813–829.

45. Tucker, C.S. Off-Flavor Problems in Aquaculture. *Rev. Fish. Sci.* **2000**, *8*, 45–88. [CrossRef]
46. Petersen, M.A.; Hyldig, G.; Strobel, B.W.; Henriksen, N.H.; Jørgensen, N.O. Chemical and sensory quantification of geosmin and 2-methylisoborneol in rainbow trout (*Oncorhynchus mykiss*) from recirculated aquacultures in relation to concentrations in basin water. *J. Agric. Food Chem.* **2011**, *59*, 12561–12568. [CrossRef] [PubMed]

Disclaimer/Publisher's Note: The statements, opinions and data contained in all publications are solely those of the individual author(s) and contributor(s) and not of MDPI and/or the editor(s). MDPI and/or the editor(s) disclaim responsibility for any injury to people or property resulting from any ideas, methods, instructions or products referred to in the content.

foods

Article

The Biochemical Composition and Quality of Adult Chinese Mitten Crab *Eriocheir sinensis* Reared in Carbonate-Alkalinity Water

Shihui Wang [1,2], Liang Luo [1,2], Rui Zhang [1,2], Kun Guo [1,2] and Zhigang Zhao [1,2,*]

1 Key Open Laboratory of Cold Water Fish Germplasm Resources and Breeding of Heilongjiang Province, Heilongjiang River Fisheries Research Institute, Chinese Academy of Fishery Sciences, Harbin 150070, China; hrfriwsh@yeah.net (S.W.); luoliang@hrfri.ac.cn (L.L.); zhangrui@hrfri.ac.cn (R.Z.); guokun@hrfri.ac.cn (K.G.)
2 Engineering Technology Research Center of Saline-alkaline Water Fisheries (Harbin), Chinese Academy of Fishery Sciences, Harbin 150070, China
* Correspondence: zhaozhigang@hrfri.ac.cn; Tel.: +86-451-8793-0948

Abstract: Saline–alkaline aquaculture has become an important breakthrough in expanding the space available for aquaculture in China. However, the biochemical composition and quality of *Eriocheir sinensis* reared in carbonate alkalinity water are still unclear. Therefore, this study investigated the edible yield, coloration, and nutritional and flavor quality of *Eriocheir sinensis*. A significantly lower gonadosomatic index (GSI), meat yield (MY), and total edible yield (TEY) were detected in intensive pond (IP) samples than those in semi-intensive reed wetland (SIWR) ($p < 0.05$). Six color parameters in the hepatopancreas ($p < 0.05$) differed between IP and SIRW. The contents of crude protein and fat in the female hepatopancreas of IP were significantly higher than those in SIRW ($p < 0.05$). The concentrations of total monounsaturated fatty acids (∑MUFA), total essential fatty acids (∑EFA), and hypocholesterolaemic/hypercholesterolaemic ratio (h/H) in the female edible tissues checked were higher in IP than those in SIRW, with significant differences including ∑MUFA in the hepatopancreas and ovary, ∑EFA in the muscle, and h/H in the ovary ($p < 0.05$). Higher total free amino acid (∑FAA) contents of muscle were detected in SIRW than that in IP samples. Significantly higher K, Ca, Mg, Fe, and Zn contents in the ovary were detected in SIRW samples compared to IP ($p < 0.05$). Overall, *Eriocheir sinensis* reared in carbonate-alkalinity water is an important source of nutrients.

Keywords: *Eriocheir sinensis*; saline–alkaline; aquaculture; edible yield; mineral element

1. Introduction

Saline–alkaline land, accounting for 0.95 billion ha, covers approximately 7.26% of the total land area worldwide [1], of which 99.13 million ha saline–alkaline land are located in China, including approximately 45.87 million ha of low-lying saline–alkaline water mainly distributed in northeast, northwest, and coastal areas [2]. Salinity and carbonate alkalinity, serving as the most significant stressors in saline–alkaline water, have a substantial impact on the growth, survival, reproduction, and quality of aquatic animals [3–5]. The physiological metabolism, osmoregulation, and intestinal microbiota of aquatic animals can be significantly influenced by high saline–alkaline concentrations [6–11]. Therefore, not all aquatic animals can be reared in the saline–alkaline water. Only a few species, such as Nile tilapia *Oreochromis niloticus* [9], crucian carp *Carassius auratus* [10], Chinese mitten crab *Eriocheir sinensis* [5], naked carp *Gymnocypris przewalskii* [11], Bulatmai barbel *Luciobarbus capito* [12], and White shrimp *Litopenaeus vannamei* [13] are domesticated for saline–alkaline water culture. Due to shrinking aquaculture space, expanding new aquaculture space for aquaculture development is thus essential, especially in saline–alkaline water.

The saline–alkaline water distributed in China mainly comprises three types, including chloride, carbonate, and sulfate, where the main ions are Na^+, K^+, Ca^{2+}, Mg^{2+}, Cl^-, $CO_3{}^{2-}$,

HCO_3^-, OH^- and SO_4^{2-}, separately [14]. Meanwhile, different types of saline–alkaline water also have different ionic compositions. Cl^- is the main anion of chloride alkalinity, while CO_3^{2-} and HCO_3^- are in carbonate alkalinity, and SO_4^{2-} is observed in sulfate alkalinity [15]. Among these, Daqing area in Heilongjiang province is a typical saline–alkaline wetland with abundant carbonate alkalinity ($NaHCO_3$) located in the western Songnen plain of China, where high carbonate alkalinity and pH in the saline–alkaline water are characteristics of the environment. High carbonate alkalinity can reduce the concentration of H^+ in saline–alkaline water due to the high pH, thereby leading ammonia (NH_3) toward the direction of an equilibrium state of $NH_4^+ + OH^- \rightleftharpoons NH_3{\cdot}H_2O$ and causing NH_3 poisoning in aquatic animals [16–18]. Consequently, the growth and quality of aquatic animals may be influenced significantly.

The Chinese mitten crab, *Eriocheir sinensis*, is an important aquatic animal with high economic and nutritional values. In addition, its aquaculture yield reached 815,318 t in 2022 [19]. *E. sinensis* is a migratory aquatic animal that grows in freshwater until it reaches sexual maturity and reproduction occurs in brackish water, thereby leading to high salinity tolerance [20]. Recent studies have also illustrated that *E. sinensis* has high carbonate-alkalinity tolerance characteristics [5]. However, the edible yield and quality of *E. sinensis* reared in carbonate-alkalinity water are still unclear. Therefore, the aim of this study was to investigate the edible yield, coloration, and nutritional and flavor quality of *E. sinensis* reared in carbonate-alkalinity water.

2. Materials and Methods

2.1. Experimental Set Up and Culture Management

The *E. sinensis* was reared in carbonate-alkalinity water from a local aquaculture demonstration farm (124.62° E, 45.70° N) in Zhaoyuan City of Heilongjiang Province, China. The culture experiment began on 1st May 2021 and was completed on 30th September 2021. The megalopa Guanghe No. 1 originated from Panjin Guanghe Crab Industry Co., Ltd., Panjin, China, with an juvenile average body weight of 5.32 ± 0.26 g/ind. These juveniles were reared in carbonate-alkalinity water at a stocking density of 15,000/ha, and the water quality parameters were as follows: intensive pond (IP), salinity 0.68 ± 0.05 ppt, carbonate alkalinity 8.48 ± 0.32 mmol/L, pH 8.72 ± 0.04; semi-intensive reed wetland (SIWR), salinity 0.56 ± 0.02 ppt, carbonate alkalinity 8.88 ± 0.04 mmol/L, and pH 8.65 ± 0.04. The IP was transplanted with Canadian pondweed, *Elodea canadensis*, while the SIWR was planted with natural *Phragmites australis*. During the culture stage, the juveniles were fed twice a day at 8:00 a.m. and 5:00 p.m. with a commercial formulated diet (crude protein $\geq$ 39.0%, crude fat $\geq$ 5.0%, moisture $\leq$ 12.0%, ash $\leq$ 18.0%. Nantong Charoen Pokphand Co., Ltd., Nantong, China). The feeding amount accounted for approximately 2% of the total body weights.

2.2. Sample Collection and Dissection

All *E. sinensis* procedures in this study were conducted according to the Guidelines for the Care and Use of Laboratory Animals of Heilongjiang River Fisheries Research Institute (HRFRI), Chinese Academy of Fishery Sciences (CAFS), Harbin, China. The *E. sinensis* used in the present study were reviewed and approved by the Committee for the Welfare and Ethics of Laboratory Animals of HRFRI, CAFS (Approval code: 20210910-002; approval date: 10 September 2021). On 4 and 29 September 2021, a total of one hundred mature *E. sinensis* (♀:♂= 1:1) reared in IP and SIWR were collected, respectively. Subsequently, these alive *E. sinensis* were transported to HRFRI, CAFS, accurately weighed with an electronic balance (JA2002, Shanghai Puchun measuring instrument Co., Ltd., Shanghai, China), and then the carapace length and carapace width were also measured with a Vernier caliper (605, Harbin measuring tools and cutting tools Co., Ltd., Harbin, China). The anatomical procedures were followed according to the previous study [5]. The hepatosomatic index (HSI, %), gonadosomatic index (GSI, %), meat yield (MY, %), total edible yield (TEY, %), and condition factor (CF, g/cm^3) were also referenced in the Wang et al.'s study [5] and calculated according to the following Equations (1)–(5):

$$\text{GSI}\ (\%) = 100 \times \text{Gonad weight}/\text{Body weight} \tag{1}$$

$$\text{HSI}\ (\%) = 100 \times \text{Hepatopancreas weight}/\text{Body weight} \tag{2}$$

$$\text{MY}\ (\%) = 100 \times \text{Muscle weight}/\text{Body weight} \tag{3}$$

$$\text{TEY}\ (\%) = \text{GSI}\ (\%) + \text{HSI}\ (\%) + \text{MY}\ (\%) \tag{4}$$

$$\text{CF}\ (\text{g/cm}^3) = \text{Body weight}/\text{Carapace length}^3 \tag{5}$$

2.3. Measurements of Color and Nutritional Parameters

The color parameters L^* (brightness), a^* (redness), and b^* (yellowness) of carapace, hepatopancreas, and female gonad (ovary) between IP and SIRW samples were measured by a colorimeter (CR-400, Konica Minolta, Marunouchi, Tokyo, Japan). The measured method was carried out according to Long et al. 's study [21]. The overall collected samples of each edible tissue were randomly selected and combined into three duplicate samples. The proximate composition, fatty acids, free amino acids, and mineral elements of *E. sinensis* reared in IP and SIRW were measured as reported in a previous study [22]. The moisture was checked by using a vacuum freeze-dryer (FD-1A-50, Biocoll, Beijing, China) at −50 °C vacuum freezing to a constant weight. The crude protein (Kjeldahl method) and ash at 550 °C were extracted by muffle furnace burning by using the AOAC method [23]. The crude fat was extracted by using the Soxhlet extraction method. An Agilent 7890B-5977A gas chromatograph (GC–MS, Agilent Technologies Co., Ltd., Santa Clara, CA, USA) was used for the determinations of fatty acids. An automatic amino acid instrument (L-8800, Hitachi Co., Ltd., Tokyo, Japan) was used for the free amino acid analysis. The taste activity value (TAV) was evaluated as the ratio between the determined FAA content and its threshold [24]. The mineral elements were performed by using an inductively coupled plasma-mass spectrometer (7500, ICP–MS, Agilent Technologies Co., Ltd., PA, USA). The hypocholesterolaemic/hypercholesterolaemic ratio (h/H), index of atherogenicity (AI), and index of thrombogenicity (TI) [25] were calculated using the following Equations (6)–(8):

$$\text{h/H} = \sum(18{:}1\text{n}9, 18{:}1\text{n}7, 18{:}2\text{n}6, 18{:}3\text{n}6, 18{:}3\text{n}3, 20{:}3\text{n}6, 20{:}4\text{n}6, 20{:}5\text{n}3, 22{:}4\text{n}6, 22{:}5\text{n}3, 22{:}6\text{n}3)/\sum(14{:}0,16{:}0) \tag{6}$$

$$\text{AI} = (12{:}0 + 4 \times 14{:}0 + 16{:}0)/(\sum\text{n}-6\ \text{PUFA} + \sum\text{n}-3\ \text{PUFA} + \sum\text{MUFA}) \tag{7}$$

$$\text{TI} = (14{:}0 + 16{:}0 + 18{:}0)/(0.5 \times \sum\text{MUFA} + 0.5 \times \sum\text{n}-6\text{PUFA} + 3.0 \times \sum\text{n}-3\text{PUFA} + \text{n}-3/\text{n}-6\ \text{PUFA}) \tag{8}$$

2.4. Statistical Analysis

The results are presented as the mean values ± standard error (SE). SPSS 22.0 software (SPSS Inc., Chicago, IL, USA) was used for statistical analysis. Independent samples *t*-test was used to determine the differences between IP and SIWR. In comparison tests, $p < 0.05$ was regarded as statistically significance, and $p < 0.01$ was regarded as extremely statistically significance.

3. Results

3.1. Total Edible Yield

The edible yield and condition factor of adult *E. sinensis* reared in carbonate-alkalinity water are presented in Figure 1. For females, the HSI and CF in IP were significantly higher than those of SIRW; however, significantly lower GSI and TEY in IP were observed

($p < 0.05$, Figure 1A,C). For males, the GSI, MY, and TEY of SIRW were significantly higher than those of IP ($p < 0.05$, Figure 1B).

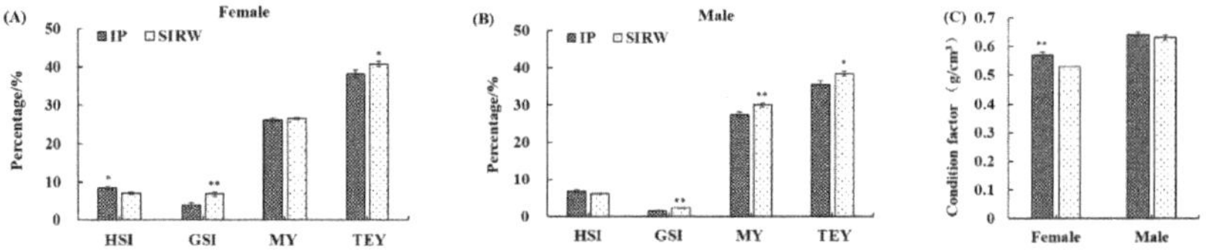

Figure 1. The edible yield (% body weight, (**A**,**B**)) and condition factor (%, (**C**)) of adult *Eriocheir sinensis* reared in carbonate-alkalinity water. Data are presented as means ± standard error (SE) (n = 25). * denotes significant difference ($p < 0.05$), ** denotes extreme significant difference ($p < 0.01$). IP, *Eriocheir sinensis* reared in intensive pond; SIRW, *Eriocheir sinensis* reared in semi-intensive reed wetland; HSI, hepatosomatic index; GSI, gonadosomatic index; MY, muscle yield; TEY, total edible yield.

3.2. Color Parameters

The color parameters of adult *E. sinensis* reared in carbonate-alkalinity water are shown in Table 1. A significantly higher L^* value of male dried carapace was observed in IP than that in SIRW ($p < 0.05$). For hepatopancreas, there were extremely significantly different b^* values of female *E. sinensis* between IP and SIRW ($p < 0.01$), while significant differences were also observed by the a^* value of female wet hepatopancreas, the L^* value of male wet hepatopancreas, and the b^* values of male hepatopancreas ($p < 0.05$). No significant differences were found in ovary color between IP and SIRW ($p > 0.05$).

Table 1. The color comparison of adult *Eriocheir sinensis* reared in carbonate-alkalinity water.

Item		Color	Female		Male	
			IP	SIRW	IP	SIRW
Carapace	Wet sample	L^*	47.18 ± 0.99	48.57 ± 0.95	47.78 ± 0.70	49.51 ± 1.13
		a^*	2.76 ± 0.29	2.78 ± 0.21	3.26 ± 0.24	3.26 ± 0.40
		b^*	15.04 ± 0.39	16.59 ± 0.77	16.08 ± 0.48	17.70 ± 0.62
	Dried sample	L^*	67.09 ± 1.15	64.89 ± 0.95	70.16 ± 1.00 *	65.80 ± 1.11
		a^*	21.30 ± 0.96	21.29 ± 1.02	20.79 ± 1.12	20.95 ± 1.53
		b^*	40.99 ± 1.14	42.19 ± 1.15	45.85 ± 1.10	43.29 ± 0.88
Hepatopancreas	Wet sample	L^*	62.22 ± 1.61	61.52 ± 1.25	54.65 ± 2.18	62.46 ± 2.18 *
		a^*	15.52 ± 1.89	21.65 ± 1.31 *	16.90 ± 1.13	16.59 ± 1.33
		b^*	43.38 ± 2.94	54.23 ± 1.32 **	39.46 ± 2.42	48.41 ± 2.48 *
	Dried sample	L^*	60.60 ± 2.73	63.99 ± 2.12	57.31 ± 4.01	64.41 ± 1.43
		a^*	12.53 ± 3.25	15.43 ± 1.42	14.01 ± 1.48	13.22 ± 0.92
		b^*	37.49 ± 1.49	48.43 ± 1.26 **	35.17 ± 1.31	45.26 ± 2.37 *
Gonad	Wet sample	L^*	28.55 ± 0.48	28.25 ± 0.32	——	——
		a^*	1.78 ± 0.67	1.22 ± 0.27	——	——
		b^*	3.96 ± 0.41	3.88 ± 0.24	——	——
	Dried sample	L^*	74.96 ± 1.77	73.96 ± 0.54	——	——
		a^*	30.93 ± 2.00	32.31 ± 0.86	——	——
		b^*	52.01 ± 0.19	51.34 ± 0.87	——	——

Note: Data are presented as means ± standard error (SE) (n = 25). "——" means no detection. * denotes significant difference ($p < 0.05$), ** denotes extreme significant difference ($p < 0.01$). Abbreviations: IP, *Eriocheir sinensis* reared in intensive pond; SIRW, *Eriocheir sinensis* reared in semi-intensive reed wetland.

3.3. Proximate Composition

The proximate compositions of adult *E. sinensis* reared in carbonate-alkalinity water are presented in Table 2. The contents of crude protein and crude fat in the female hepatopancreas in IP were significantly increased compared with that in SIRW ($p < 0.05$). The male gonad crude protein content of *E. sinensis* from SIRW was significantly higher than that in IP ($p < 0.05$), while the content of crude protein in male muscle in IP was lower than that in SIRW, with an extremely significant difference ($p < 0.01$).

Table 2. The proximate composition of adult *Eriocheir sinensis* reared in carbonate-alkalinity water (%, wet weight).

Item	Female		Male	
	IP	SIRW	IP	SIRW
Hepatopancreas				
Moisture	48.74 ± 5.16	61.72 ± 3.17	61.18 ± 1.82	62.32 ± 1.69
Crude protein	11.17 ± 0.16 *	9.63 ± 0.10	8.79 ± 1.46	8.51 ± 1.03
Crude fa	35.39 ± 0.05 *	23.79 ± 1.21	25.21 ± 1.88	25.11 ± 1.45
Ash	1.31 ± 0.08	1.57 ± 0.27	1.09 ± 0.19	1.17 ± 0.19
Gonad				
Moisture	56.46 ± 2.89	55.25 ± 2.66	76.08 ± 0.62	75.05 ± 0.77
Crude protein	28.90 ± 0.23	28.83 ± 0.10	15.90 ± 0.06	16.62 ± 0.09 *
Crude fat	6.42 ± 0.50	7.23 ± 0.02	0.72 ± 0.07 *	0.37 ± 0.02
Ash	1.89 ± 0.11	1.83 ± 0.04	2.08 ± 0.05	2.20 ± 0.05
Muscle				
Moisture	78.63 ± 0.67	79.73 ± 0.55	83.69 ± 0.56	82.29 ± 0.35
Crude protein	17.76 ± 0.16	17.46 ± 0.03	13.64 ± 0.06	15.07 ± 0.06 **
Crude fat	0.50 ± 0.03	0.42 ± 0.03	0.44 ± 0	0.39 ± 0.02
Ash	1.44 ± 0.03	1.54 ± 0.02	1.38 ± 0.01	1.48 ± 0.08

Notes: Data are presented as means ± standard error (SE) (n = 3). * denotes significant difference ($p < 0.05$), ** denotes extreme significant difference ($p < 0.01$). Abbreviations: IP, *Eriocheir sinensis* reared in intensive pond; SIRW, *Eriocheir sinensis* reared in semi-intensive reed wetland.

3.4. Fatty Acids Profiles

Table 3 illustrates the evaluation and comparison of the main fatty acid composition, concentration, h/H, AI, and TI of adult *E. sinensis* reared in carbonate-alkalinity water. For females, higher concentrations of ∑SFA, ∑n−3 PUFA, ∑LC-PUFA, and ∑DHA + EPA in all the edible tissues, and ∑PUFA in the ovary and muscle, were observed in SIRW samples than those in IP, with no significant difference ($p > 0.05$). Concentrations of ∑MUFA, ∑EFA, and h/H in all the edible tissues were observed to be higher in IP than those in SIRW, with the following significant differences: ∑MUFA in the hepatopancreas and ovary, ∑EFA in the muscle, and h/H in the ovary ($p < 0.05$). For males, the levels of ΣSFA in gonads and muscle, ΣPUFA in gonads, Σn−3 PUFA in muscle, and ΣDHA + EPA in hepatopancreas of SIRW-reared crabs were significantly higher than those in IP-reared crabs ($p < 0.05$). The concentrations of ∑MUFA, h/H in all the edible tissues, and ∑EFA in the hepatopancreas and muscle detected were higher in IP than those in SIRW, with the following significant differences: ∑MUFA, ∑EFA of the hepatopancreas between IP and SIRW ($p < 0.05$), ∑MUFA of the gonad system and muscle, and ∑EFA of muscle between IP and SIRW ($p < 0.01$).

Table 3. The fatty acid composition in the hepatopancreas, gonad, and muscle of adult *Eriocheir sinensis* reared in carbonate-alkalinity water (% of total fatty acids).

Fatty Acid	Hepatopancreas		Gonad		Muscle	
	IP	SIRW	IP	SIRW	IP	SIRW
Female						
C14:0	1.51 ± 0.17	2.17 ± 0.47	0.79 ± 0.03	1.20 ± 0.08 *	0.33 ± 0.05	0.51 ± 0.05
C15:0	0.63 ± 0.03	1.31 ± 0.34	0.47 ± 0.06	0.93 ± 0.10	0.27 ± 0.02	0.47 ± 0.04 *
C16:0	22.26 ± 0.73	23.01 ± 1.24	16.59 ± 0.39	17.73 ± 0.59	12.96 ± 0.40	13.73 ± 0.51
C17:0	0.61 ± 0.01	1.34 ± 0.14 *	0.50 ± 0.06	0.77 ± 0.04	0.73 ± 0.07	1.34 ± 0.14
C18:0	3.14 ± 0.13	3.72 ± 0.34	2.72 ± 0.09	3.05 ± 0.14	8.98 ± 0.01	9.33 ± 0.20
C20:0	0.32 ± 0.02	0.41 ± 0.03	0.09 ± 0.01	0.13 ± 0	0.17 ± 0.02	0.28 ± 0.01 *
∑SFA	29.24 ± 1.11	33.24 ± 2.72	21.25 ± 0.63	24.06 ± 0.47	23.42 ± 0.57	25.65 ± 0.54
C15:1n5	0.23 ± 0.02	0.45 ± 0.10	0.15 ± 0	0.26 ± 0.02 *	0.85 ± 0.09	0.78 ± 0.06
C16:1n7	8.38 ± 0.30	10.95 ± 0.34 *	11.40 ± 0.01	13.65 ± 1.04	3.04 ± 0.21	4.26 ± 0.18 *
C18:1n9	39.15 ± 0.15 **	32.45 ± 0.40	33.69 ± 1.13 *	26.10 ± 1.12	25.92 ± 0.65	22.20 ± 0.63
C20:1n9	1.07 ± 0.03	1.28 ± 0.10	0.45 ± 0.02	0.34 ± 0.03	0.71 ± 0.07	0.81 ± 0.09
∑MUFA	49.42 ± 0.43 *	45.89 ± 0.27	46.07 ± 1.09 *	40.51 ± 0.07	30.51 ± 1.01	28.04 ± 0.48
C18:2n6	15.81 ± 0.20 *	11.56 ± 0.78	17.19 ± 0.80	13.87 ± 0.84	13.46 ± 0.87 *	8.71 ± 0.09
C18:3n3	2.17 ± 0.04	2.93 ± 0.51	3.78 ± 0.58	6.21 ± 1.39	1.89 ± 0.05	2.73 ± 0.14 *
C20:2n6	0.95 ± 0.26	0.88 ± 0.02	0.95 ± 0.06	0.81 ± 0.05	1.50 ± 0.11	1.47 ± 0.06
C20:4n6 (ARA)	0.65 ± 0.10	1.55 ± 0.43	2.33 ± 0.19	3.85 ± 0.18 *	5.40 ± 0.42	8.14 ± 0.47 *
C20:3n3	0.24 ± 0.06	0.48 ± 0.08	0.38 ± 0.03	0.55 ± 0.08	0.43 ± 0.03	0.74 ± 0.05 *
C20:5n3 (EPA)	0.77 ± 0.12	1.61 ± 0.60	4.72 ± 0.53	6.68 ± 0.12	14.05 ± 0.19	16.21 ± 1.16
C22:6n3 (DHA)	0.48 ± 0.04	1.28 ± 0.62	3.20 ± 0.10	2.82 ± 0.01	9.18 ± 0.86	7.92 ± 0
∑PUFA	21.29 ± 0.72	20.89 ± 2.98	32.85 ± 0.51	35.34 ± 0.53	46.08 ± 1.57	46.67 ± 0.65
∑EFA	17.97 ± 0.24	14.48 ± 1.28	20.97 ± 0.22	20.07 ± 0.55	15.35 ± 0.81 *	11.44 ± 0.05
∑LC-PUFA	3.32 ± 0.48	6.21 ± 1.71	11.71 ± 0.81	14.99 ± 0.04	30.55 ± 0.77	34.74 ± 0.63
∑n−3 PUFA	3.65 ± 0.18	6.29 ± 1.27	12.08 ± 0.87	16.26 ± 1.11	25.54 ± 0.72	27.59 ± 0.95
∑n−6 PUFA	17.65 ± 0.27	14.60 ± 0.68	20.78 ± 0.42	19.09 ± 0.60	20.55 ± 0.31	19.08 ± 0.40
n−3/n−6 PUFA	0.21 ± 0.01	0.42 ± 0.09	0.58 ± 0.08	0.86 ± 0.13	1.24 ± 0.02	1.45 ± 0.12
∑DHA + EPA	1.25 ± 0.15	2.89 ± 1.22	7.92 ± 0.62	9.50 ± 0.11	23.22 ± 1.05	24.13 ± 1.16
DHA/EPA	0.63 ± 0.04	0.76 ± 0.11	0.68 ± 0.06 *	0.42 ± 0.01	0.65 ± 0.05	0.49 ± 0.04
h/H	2.48 ± 0.12	2.06 ± 0.27	3.72 ± 0.12 *	3.15 ± 0.05	5.23 ± 0.23	4.65 ± 0.18
AI	0.40 ± 0.03	0.48 ± 0.07	0.25 ± 0.01	0.30 ± 0.01	0.19 ± 0.01	0.21 ± 0.01
TI	0.60 ± 0.04	0.60 ± 0.11	0.29 ± 0	0.28 ± 0.02	0.22 ± 0.01	0.22 ± 0.01
Male						
C14:0	1.35 ± 0.16	1.59 ± 0.22	0.37 ± 0.03	0.48 ± 0.04	0.26 ± 0.01	0.36 ± 0.01 **
C15:0	0.70 ± 0.05	0.89 ± 0.13	0.28 ± 0.07	0.30 ± 0.02	0.25 ± 0.01	0.43 ± 0.03 *
C16:0	22.16 ± 0.12	21.49 ± 1.23	9.26 ± 0.27	10.43 ± 0.33	12.93 ± 0.11	13.18 ± 0.49
C17:0	0.53 ± 0.05	0.94 ± 0.02 *	0.62 ± 0.02	0.92 ± 0.10	0.76 ± 0.03	1.33 ± 0 **
C18:0	3.37 ± 0.01	3.26 ± 0.10	8.06 ± 0.05	7.92 ± 0.26	9.04 ± 0.18	9.90 ± 0.06 *
C20:0	0.37 ± 0	0.34 ± 0.02	0.25 ± 0	0.26 ± 0.01	0.17 ± 0.01	0.26 ± 0 **
∑SFA	29.52 ± 0.33	29.53 ± 0.88	19.32 ± 0.32	20.70 ± 0.02 *	23.39 ± 0.11	25.45 ± 0.46 *
C15:1n5	0.25 ± 0.02	0.24 ± 0.13	0.93 ± 0.05	0.73 ± 0.08	1.08 ± 0.03	1.02 ± 0.02
C16:1n7	7.79 ± 0.99	10.79 ± 1.18	2.03 ± 0.27	2.31 ± 0.10	2.03 ± 0.11	2.64 ± 0.04 *
C18:1n9	36.24 ± 0.77 *	31.27 ± 0.62	23.88 ± 0.24	23.19 ± 0.17	25.31 ± 0.07 **	20.52 ± 0.16
C20:1n9	1.09 ± 0.05	1.01 ± 0.04	1.34 ± 0.04	1.22 ± 0.10	0.79 ± 0.01	0.79 ± 0.07
∑MUFA	46.03 ± 0.21 *	43.86 ± 0.43	35.91 ± 0.02 **	30.20 ± 0.43	29.20 ± 0.06 **	24.96 ± 0.25
C18:2n6	19.19 ± 0.04 *	14.51 ± 0.79	9.84 ± 0.79	10.13 ± 3.01	12.10 ± 0.21 *	8.59 ± 0.35
C18:3n3	1.98 ± 0.07	3.35 ± 0.89	1.00 ± 0.09	1.60 ± 0.17	1.40 ± 0.04	2.10 ± 0.27
C20:2n6	0.76 ± 0.19	0.86 ± 0.04	3.17 ± 0.11 *	2.34 ± 0.08	1.80 ± 0.01	1.61 ± 0.11
C20:4n6 (ARA)	0.65 ± 0.11	2.11 ± 0.28 *	11.78 ± 0.63	15.16 ± 0.52	6.92 ± 0.15	9.96 ± 0.70 *
C20:3n3	0.17 ± 0.03	0.52 ± 0.16	0.61 ± 0.02	0.80 ± 0.09	0.44 ± 0.01	0.70 ± 0.03 **
C20:5n3 (EPA)	0.83 ± 0.08	2.20 ± 0.23 *	11.13 ± 0.10	12.35 ± 1.31	14.74 ± 0.09	16.18 ± 0.02 **
C22:6n3 (DHA)	0.67 ± 0.23	2.38 ± 0.41	6.81 ± 0.37	5.98 ± 0.72	9.82 ± 0.09	9.85 ± 0.58
∑PUFA	24.45 ± 0.54	26.62 ± 1.31	44.78 ± 0.31	49.10 ± 0.45 *	47.41 ± 0.06	49.59 ± 0.72
∑EFA	21.17 ± 0.11 **	17.86 ± 0.11	10.84 ± 0.88	11.73 ± 3.18	13.50 ± 0.17 *	10.67 ± 0.62

Table 3. *Cont.*

Fatty Acid	Hepatopancreas		Gonad		Muscle	
	IP	SIRW	IP	SIRW	IP	SIRW
∑LC-PUFA	3.29 ± 0.64	8.59 ± 1.21	33.70 ± 1.19	36.93 ± 2.73	33.71 ± 0.13	38.56 ± 1.41
∑n−3 PUFA	3.65 ± 0.19	8.44 ± 1.20	19.55 ± 0.28	20.72 ± 1.37	26.39 ± 0.03	28.82 ± 0.30 *
∑n−6 PUFA	20.80 ± 0.16 *	18.18 ± 0.22	25.23 ± 0.05	28.38 ± 1.38	21.02 ± 0.05	20.78 ± 0.43
n−3/n−6 PUFA	0.18 ± 0.01	0.47 ± 0.10	0.78 ± 0.02	0.74 ± 0.13	1.26 ± 0.01	1.39 ± 0.01 *
∑DHA + EPA	1.50 ± 0.31	4.58 ± 0.65 *	17.94 ± 0.47	18.32 ± 2.02	24.56 ± 0	26.03 ± 0.59
DHA/EPA	0.79 ± 0.20	1.08 ± 0.07	0.61 ± 0.03	0.48 ± 0.01	0.67 ± 0.01	0.61 ± 0.03
h/H	2.52 ± 0.08	2.44 ± 0.18	6.65 ± 0.21	6.26 ± 0.13	5.30 ± 0.04	4.97 ± 0.21
AI	0.39 ± 0.01	0.40 ± 0.01	0.13 ± 0.01	0.16 ± 0	0.18 ± 0	0.20 ± 0.01
TI	0.60 ± 0.02	0.47 ± 0.06	0.20 ± 0.01	0.20 ± 0.01	0.21 ± 0	0.21 ± 0.01

Notes: Data are presented as means ± standard error (SE) (n = 3). * denotes significant difference ($p < 0.05$). ** denotes extreme significant difference ($p < 0.01$). Abbreviations: IP, *Eriocheir sinensis* reared in intensive pond; SIRW, *Eriocheir sinensis* reared in semi-intensive reed wetland;∑SFA, total saturated fatty acids; ∑MUFA, total monounsaturated fatty acids; ∑PUFA, total polyunsaturated fatty acids; ∑EFA, total essential fatty acids; ∑LC-PUFA, total long chain polyunsaturated fatty acids; ∑n−3 PUFA, total ω-3 polyunsaturated fatty acids; ∑n−6 PUFA, total ω-6 polyunsaturated fatty acids; DHA, docosahesaenoic acid; EPA, eicosapentaenoic acid, ARA, arachidonic acid; h/H, hypocholesterolaemic/hypercholesterolaemic ratio; AI, index of atherogenicity; TI, index of thrombogenicity.

3.5. Free Amino Acids Composition and Taste Activity Value

The composition and contents of free amino acids (FAAs) in adult *E. sinensis* reared in carbonate-alkalinity water are presented in Table 4. With respect to females, the concentrations of serine (Ser), isoleucine (Ile), lysine (Lys), valine (Val), and ∑EFAA in the hepatopancreas were significantly higher in IP crabs than in SIRW-reared crabs ($p < 0.05$); however, the concentrations of alanine (Ala), histidine (His), Lys, and methionine (Met) observed in the muscle were higher in SIRW compared with those in IP ($p < 0.05$). For males, significant differences were noted by observing Ile in the gonad system and ∑FAA in the muscle ($p < 0.05$), while extremely significant differences were also observed in Ala in the hepatopancreas and muscle, and in glycine (Gly) in the muscle between ($p < 0.01$).

Table 4. The free amino acid composition in the hepatopancreas, gonad, and muscle of adult *Eriocheir sinensis* reared in carbonate-alkalinity water (mg/100 g, wet weight).

Free Amino Acids	Hepatopancreas		Gonad		Muscle	
	IP	SIRW	IP	SIRW	IP	SIRW
Female						
Aspartic acid	61.64 ± 1.28	41.42 ± 6.26	2.80 ± 0.54	3.45 ± 0.50	2.18 ± 0.27	2.94 ± 0.31
Arginine	370.76 ± 18.22	324.80 ± 57.25	222.08 ± 11.32	274.41 ± 6.80	556.48 ± 11.41	548.28 ± 35.14
Alanine	233.89 ± 25.35	285.79 ± 53.52	100.10 ± 10.94	137.58 ± 3.70	338.37 ± 8.75	530.03 ± 25.96 *
Cysteine	17.16 ± 0.92	14.76 ± 2.11	2.68 ± 0.76	2.73 ± 0.27	2.17 ± 0.18	1.68 ± 0
Glutamic acid	135.80 ± 11.69	113.99 ± 0.14	97.72 ± 7.99	86.44 ± 5.72	33.63 ± 3.46	46.17 ± 14.28
Glycine	130.66 ± 21.38	133.69 ± 24.05	60.24 ± 5.88	64.19 ± 10.08	347.70 ± 24.21	453.18 ± 78.36
Histidine	43.34 ± 7.62	39.31 ± 4.41	27.93 ± 4.53	32.22 ± 5.14	22.88 ± 2.42	35.61 ± 1.30 *
Proline	92.75 ± 16.01	107.32 ± 23.94	71.14 ± 5.25	73.86 ± 2.83	135.70 ± 7.48	147.75 ± 23.28
Serine	28.33 ± 0.66 *	21.30 ± 0.68	8.21 ± 0.55	9.60 ± 1.23	9.94 ± 1.17	12.84 ± 1.79
Tyrosine	98.84 ± 8.62	73.88 ± 3.42	18.61 ± 1.70	19.23 ± 1.67	18.08 ± 2.52	20.06 ± 2.60
Isoleucine ▲	67.41 ± 0.04 *	60.24 ± 1.08	9.32 ± 1.38	10.26 ± 2.22	8.61 ± 0.03	13.76 ± 3.10
Leucine ▲	156.88 ± 3.55	130.35 ± 6.04	13.60 ± 1.94	14.58 ± 2.94	15.91 ± 0.94	27.06 ± 6.37
Lysine ▲	155.55 ± 1.00 *	138.35 ± 3.31	38.97 ± 9.97	43.95 ± 11.31	28.99 ± 0.07	39.34 ± 0.96 **
Methionine ▲	50.09 ± 0.52	51.19 ± 1.91	22.30 ± 2.65	22.31 ± 5.50	24.28 ± 2.51	43.56 ± 3.02 *
Phenylalanine ▲	92.75 ± 4.09	74.59 ± 2.04	15.49 ± 2.05	15.18 ± 2.40	11.57 ± 1.02	14.83 ± 1.23
Threonine ▲	94.83 ± 5.28	85.27 ± 7.19	66.20 ± 6.70	94.49 ± 5.91	25.18 ± 0	32.52 ± 10.71
Valine ▲	106.61 ± 2.37 *	92.42 ± 1.26	26.09 ± 3.65	28.94 ± 4.53	21.70 ± 0.40	31.46 ± 5.97
∑EFAA	724.11 ± 14.76 *	632.4 ± 5.94	191.98 ± 28.34	229.71 ± 34.81	136.24 ± 2.23	202.54 ± 31.35
∑FAA	1937.29 ± 85.06	1788.67 ± 146.25	803.48 ± 77.8	933.42 ± 45.92	1603.38 ± 9.64	2001.07 ± 162.46
PETFAA	37.42 ± 0.89	35.62 ± 3.24	23.77 ± 1.22	24.49 ± 2.52	8.50 ± 0.19	10.06 ± 0.75

Table 4. *Cont.*

Free Amino Acids	Hepatopancreas		Gonad		Muscle	
	IP	SIRW	IP	SIRW	IP	SIRW
Male						
Aspartic acid	43.85 ± 3.26	48.34 ± 24.86	24.44 ± 3.22	28.43 ± 5.07	2.27 ± 0.56	2.74 ± 0.04
Arginine	188.15 ± 29.12	266.95 ± 79.80	60.46 ± 5.69	47.63 ± 7.49	408.25 ± 8.95	451.24 ± 24.15
Alanine	139.45 ± 13.65	318.65 ± 0.04 **	97.27 ± 5.28	100.84 ± 10.70	331.50 ± 8.40	471.54 ± 7.67 **
Cysteine	8.73 ± 1.51	16.25 ± 8.61	1.44 ± 0.14	2.31 ± 0.58	3.06 ± 0.07 *	1.83 ± 0.12
Glutamic acid	94.93 ± 3.21	114.46 ± 46.49	68.52 ± 0.56	49.61 ± 5.25	53.15 ± 0.61	53.69 ± 1.81
Glycine	79.84 ± 1.00	135.22 ± 48.32	43.29 ± 6.29	42.69 ± 3.20	321.59 ± 12.02	465.93 ± 6.64 **
Histidine	30.11 ± 0.21	41.39 ± 11.40	8.26 ± 1.31	6.97 ± 0.41	22.76 ± 1.25	25.80 ± 1.42
Proline	66.41 ± 2.52	93.25 ± 8.48	61.24 ± 4.80	41.13 ± 1.57	97.51 ± 6.49	121.40 ± 22.25
Serine	17.96 ± 1.13	22.84 ± 7.64	3.11 ± 0.07 *	2.30 ± 0.14	6.54 ± 0.84	7.81 ± 1.76
Tyrosine	70.26 ± 1.48	83.70 ± 39.02	17.70 ± 0.46	18.37 ± 1.92	26.17 ± 4.63	23.25 ± 1.65
Isoleucine ▲	37.64 ± 2.23	64.30 ± 29.24	7.64 ± 0.37	11.55 ± 0.52 *	14.99 ± 2.55	12.50 ± 0.52
Leucine ▲	99.65 ± 11.15	138.90 ± 64.09	11.01 ± 0.50	13.65 ± 1.10	24.63 ± 3.85	23.31 ± 0.68
Lysine ▲	91.37 ± 13.89	145.85 ± 67.58	13.49 ± 2.08	12.36 ± 1.65	41.75 ± 4.79	30.91 ± 2.93
Methionine ▲	36.67 ± 2.62	53.00 ± 23.41	7.50 ± 1.90	11.32 ± 0.43	31.16 ± 3.14	32.16 ± 3.17
Phenylalanine ▲	60.00 ± 2.35	81.29 ± 37.07	10.27 ± 0.29	13.43 ± 1.35	14.21 ± 1.94	16.85 ± 1.25
Threonine ▲	57.42 ± 2.41	85.36 ± 25.80	15.66 ± 2.54	10.54 ± 0.89	35.47 ± 2.69	27.29 ± 2.66
Valine ▲	60.49 ± 5.91	94.62 ± 36.67	17.26 ± 0.96	17.16 ± 1.50	34.95 ± 1.08	30.06 ± 0.48
∑EFAA	443.23 ± 40.57	663.33 ± 283.85	82.82 ± 8.64	90.01 ± 7.45	197.17 ± 8.38	173.08 ± 5.84
∑FAA	1182.92 ± 95.24	1804.37 ± 558.45	468.56 ± 25.61	430.3 ± 43.77	1469.95 ± 5.52	1798.31 ± 56.21 *
PETFAA	37.44 ± 0.42	35.27 ± 4.81	17.63 ± 0.88	20.96 ± 0.40	13.42 ± 0.62 *	9.62 ± 0.02

Notes: Data are presented as means ± standard error (SE) (n = 3). ▲ essential amino acid. * denotes significant difference ($p < 0.05$). ** denotes extreme significant difference ($p < 0.01$). Abbreviations: IP, *Eriocheir sinensis* reared in intensive pond; SIRW, *Eriocheir sinensis* reared in semi-intensive reed wetland; ∑EFAA, total essential free amino acids; ∑FAA, total free amino acids; PETFAA, percentage of ∑EFAA to ∑FAA.

The flavor characteristics and taste activity values (TAV) of adult *E. sinensis* reared in carbonate-alkalinity water are shown in Table 5. The 17 FAAs were separated into two tastes including pleasant taste (umami and sweetness) and unpleasant taste (bitterness). The concentrations of ∑TUV and ∑TBV in the female hepatopancreas of IP were higher than those in SIRW; however, higher concentrations of ∑TSV were observed in SIRW samples. Higher concentrations of ∑TSV and ∑TBV in the ovary were detected in SIRW samples than in IP. In the muscle and male hepatopancreas, the ∑TUV, ∑TSV, and ∑TBV values of IP were lower than those of SIRW.

Table 5. The threshold and taste activity value of free amino acid composition in the hepatopancreas, gonad, and muscle of adult *Eriocheir sinensis* reared in carbonate-alkalinity water.

Free Amino Acids	Flavor Characteristics	Threshold (mg/100 mL)	Hepatopancreas		Gonad		Muscle	
			IP	SIRW	IP	SIRW	IP	SIRW
Female								
Aspartic acid	umami (+)	100	0.62	0.41	0.03	0.03	0.02	0.03
Glutamic acid	umami (+)	30	4.53	3.80	3.26	2.88	1.12	1.54
∑TUV			5.14	4.21	3.29	2.92	1.14	1.57
Alanine	sweetness (+)	60	3.90	4.76	1.67	2.29	5.64	8.83
Glycine	sweetness (+)	130	1.01	1.03	0.46	0.49	2.67	3.49
Serine	sweetness (+)	150	0.19	0.14	0.05	0.06	0.07	0.09
Threonine	sweetness (+)	260	0.36	0.33	0.25	0.36	0.10	0.13
Proline	sweetness/bitterness (+)	300	0.31	0.36	0.24	0.25	0.45	0.49
∑TSV			5.77	6.62	2.68	3.46	8.93	13.02
Arginine	sweetness/bitterness (−)	50	7.42	6.50	4.44	5.49	11.13	10.97
Lysine	sweetness/bitterness (−)	50	3.11	2.77	0.78	0.88	0.58	0.79
Valine	sweetness/bitterness (−)	40	2.67	2.31	0.65	0.72	0.54	0.79
Methionine	bitterness/sweetness/sulphur (−)	30	1.67	1.71	0.74	0.74	0.81	1.45
Histidine	bitterness (−)	20	2.17	1.97	1.40	1.61	1.14	1.78
Isoleucine	bitterness (−)	90	0.75	0.67	0.10	0.11	0.10	0.15
Leucine	bitterness (−)	190	0.83	0.69	0.07	0.08	0.08	0.14
Phenylalanine	bitterness (−)	90	1.03	0.83	0.17	0.17	0.13	0.16
∑TBV			19.60	17.40	8.40	9.80	14.50	16.20

Table 5. *Cont.*

Free Amino Acids	Flavor Characteristics	Threshold (mg/100 mL)	Hepatopancreas		Gonad		Muscle	
			IP	SIRW	IP	SIRW	IP	SIRW
Male								
Aspartic acid	umami (+)	100	0.44	0.48	0.24	0.28	0.02	0.03
Glutamic acid	umami (+)	30	3.16	3.82	2.28	1.65	1.77	1.79
∑TUV			3.60	4.30	2.53	1.94	1.79	1.82
Alanine	sweetness (+)	60	2.32	5.31	1.62	1.68	5.52	7.86
Glycine	sweetness (+)	130	0.61	1.04	0.33	0.33	2.47	3.58
Serine	sweetness (+)	150	0.12	0.15	0.02	0.02	0.04	0.05
Threonine	sweetness (+)	260	0.22	0.33	0.06	0.04	0.14	0.10
Proline	sweetness/bitterness (+)	300	0.22	0.31	0.20	0.14	0.33	0.40
∑TSV			3.50	7.14	2.24	2.20	8.50	12.00
Arginine	sweetness/bitterness (−)	50	3.76	5.34	1.21	0.95	8.17	9.02
Lysine	sweetness/bitterness (−)	50	1.83	2.92	0.27	0.25	0.84	0.62
Valine	sweetness/bitterness (−)	40	1.51	2.37	0.43	0.43	0.87	0.75
Methionine	bitterness/sweetness/sulphur (−)	30	1.22	1.77	0.25	0.38	1.04	1.07
Histidine	bitterness (−)	20	1.51	2.07	0.41	0.35	1.14	1.29
Isoleucine	bitterness (−)	90	0.42	0.71	0.08	0.13	0.17	0.14
Leucine	bitterness (−)	190	0.52	0.73	0.06	0.07	0.13	0.12
Phenylalanine	bitterness (−)	90	0.67	0.90	0.11	0.15	0.16	0.19
∑TBV			11.40	16.80	2.80	2.70	12.50	13.20

Note: + means pleasant taste; − means unpleasant taste. Abbreviations: IP, *Eriocheir sinensis* reared in intensive pond; SIRW, *Eriocheir sinensis* reared in semi-intensive reed wetland; ∑TUV, total umami values; ∑TSV, total sweetness values; ∑TBV, total bitterness values.

3.6. Mineral Element Composition

The mineral element composition and contents of adult *E. sinensis* reared in carbonate-alkalinity water are presented in Table 6. Regarding the hepatopancreas, there was no significant difference observed between IP and SIRW ($p > 0.05$). The levels of the K, Ca, Mg, Fe, and Zn elements in the ovary of IP-reared crabs were significantly higher than in SIRW crabs ($p < 0.05$); meanwhile, the content of ∑TME in IP was extremely significantly different compared with that in SIRW ($p < 0.01$). The contents of K and Mg in female muscle, as well as Na and ΣTME in male muscle, in IP crabs were significantly higher than in SIRW crabs ($p < 0.05$). Overall, higher contents of hepatopancreas ∑TME were observed in SIRW; however, the contents of gonad and muscle ∑TME in IP were higher than those in SIRW.

Table 6. The mineral element composition in the hepatopancreas, gonad, and muscle of adult *Eriocheir sinensis* reared in carbonate-alkalinity water (mg/kg, wet weight).

Element	Hepatopancreas		Gonad		Muscle	
	IP	SIRW	IP	SIRW	IP	SIRW
Female						
Na	1635.77 ± 148.67	1710.28 ± 312.14	2113.71 ± 440.18	1061.98 ± 78.54	3027.63 ± 495.63	2472.82 ± 56.59
K	2407.05 ± 82.20	2642.24 ± 599.61	2715.20 ± 201.54 *	2090.29 ± 72.20	3947.00 ± 129.33 *	3383.32 ± 118.73
Ca	488.47 ± 97.24	835.46 ± 147.10	487.54 ± 65.58 *	230.54 ± 26.95	1135.93 ± 93.36	894.56 ± 73.72
Mg	349.11 ± 22.89	367.95 ± 77.81	1514.93 ± 100.57 *	705.99 ± 102.58	696.91 ± 50.97 *	491.37 ± 11.33
Fe	79.48 ± 8.06	79.82 ± 9.65	41.61 ± 4.00 *	15.42 ± 4.82	10.28 ± 0.46	9.78 ± 0.24
Zn	14.43 ± 0.88	13.74 ± 0.79	59.64 ± 6.92 *	32.65 ± 1.53	42.05 ± 1.27	39.76 ± 1.36
Cu	5.98 ± 1.41	6.29 ± 1.16	5.97 ± 1.36	5.05 ± 0.40	9.87 ± 0.53	6.63 ± 1.19
Mn	3.69 ± 0.67	1.61 ± 0.34	4.66 ± 1.51	2.48 ± 0.44	0.63 ± 0.02	0.66 ± 0.01
∑TME	4983.97 ± 319.40	5657.39 ± 1101.56	6943.27 ± 592.93 **	4144.40 ± 33.23	8870.30 ± 763.43	7298.90 ± 167.45
Male						
Na	1626.64 ± 411.30	1567.84 ± 306.79	4430.63 ± 240.01	4502.20 ± 49.07	3459.41 ± 235.32 *	1977.00 ± 317.44
K	2414.57 ± 490.21	2510.66 ± 367.45	3397.72 ± 166.22	2954.14 ± 616.72	3617.96 ± 328.32	2828.30 ± 229.34
Ca	467.16 ± 170.51	780.07 ± 114.32	1129.46 ± 76.44	1064.87 ± 160.05	1084.10 ± 88.74	869.42 ± 150.51
Mg	324.33 ± 79.49	256.74 ± 34.56	554.40 ± 45.71	513.87 ± 5.79	607.20 ± 76.02	456.52 ± 34.59
Fe	108.43 ± 11.12	115.75 ± 17.56	8.47 ± 1.03	8.05 ± 1.40	10.10 ± 1.35	14.08 ± 3.65
Zn	12.20 ± 4.04	14.97 ± 2.14	12.11 ± 1.68	10.38 ± 1.39	31.96 ± 8.21	35.04 ± 0.73
Cu	4.54 ± 1.94	4.04 ± 1.28	5.06 ± 1.03	5.71 ± 0.91	5.90 ± 1.83	5.04 ± 0.69
Mn	3.68 ± 0.61	3.14 ± 0.80	3.55 ± 0.42	3.52 ± 0.44	0.80 ± 0.23	1.21 ± 0.63
∑TME	4961.54 ± 1142.91	5253.21 ± 814.49	9541.40 ± 194.67	9062.75 ± 514.79	8817.43 ± 664.45 *	6186.61 ± 612.24

Notes: Data are presented as means ± standard error (SE) (n = 3). * denotes significant difference ($p < 0.05$). ** denotes extreme significant difference ($p < 0.01$). Abbreviations: IP, *Eriocheir sinensis* reared in intensive pond; SIRW, *Eriocheir sinensis* reared in semi-intensive reed wetland; ∑TME, total mineral elements.

4. Discussion

4.1. Total Edible Yield

The hepatopancreas, gonad, and muscle are important edible tissues of *E. sinensis*. However, previous studies have focused on the total edible yield of *E. sinensis* reared in freshwater- [26,27], chloride-, or sulfate-type saline–alkaline water [28], while little attention has been paid to *E. sinensis* reared in carbonate-alkalinity water. The present study indicates that *E. sinensis* can complete gonadal development normally compared with previous studies [5,22]. The GSI value of female *E. sinensis* reared in IP was 3.78 ± 0.67% on 4th September and 6.75 ± 0.55% in SRW on 29th September, while the GSI value of female *E. sinensis* in freshwater pond was 5.30 ± 0.22% on 15th September [22]. It has been demonstrated that the later the sampling period, the usually higher the GSI in the same location, which is consistent with the present study. The specific values between previous and present studies were different, which was mainly caused by the difference in water temperature and sampling period [5,22,26,27]. The MY and TEY values of female *E. sinensis* in the present study were 26.11 ± 0.67%, 26.53 ± 0.28%, 38.19 ± 0.98%, and 40.78 ± 0.68%, respectively, indicating that *E. sinensis* reared in carbonate-alkalinity water have similar MY and TEY values compared with those obtained in previous studies [5,26,27]. Overall, the above results show that rearing of *E. sinensis* in carbonate-alkalinity water will not significantly affect the gonadal development and the total edible yield.

4.2. Color Parameters

Color is one of the important indicators of sensory and quality in evaluation of *E. sinensis*. In the cognition of consumers, the higher reddish values of the dried carapace and ovary, as well as the higher reddish and yellowish values of the wet hepatopancreas, suggest better quality [21]. It has been demonstrated that the reddish and yellowish parameters of *E. sinensis* tissues are significantly related to the deposition of carotenoids [29]. The present study illustrates that the b^* values of IP [female: 40.99 ± 1.14; male: 45.85 ± 1.10] and SIRW [female: 42.19 ± 1.15; male: 43.29 ± 0.88] freeze-dried *E. sinensis* carapace reared in carbonate-alkalinity water were obviously higher than those of wild-caught mitten crabs in the natural Suifenhe and Nanliujinag delta [30], suggesting that *E. sinensis* reared in carbonate-alkalinity water accumulated more carotenoids. Similar b^* values of freeze-dried three-year-old *E. sinensis* carapace were also observed in the pond of Zhaodong city, Heilongjiang Province, China [22]. In addition, the higher b^* values in the ovary of IP [52.01 ± 0.19] and SIRW freeze-dried *E. sinensis* in the present study were also detected in carbonate-alkalinity water compared with previous studies [22,30]. These results suggest that *E. sinensis* reared in carbonate-alkalinity water accumulated more carotenoids and exhibited better quality.

4.3. Biochemical Composition

Biochemical composition, especially crude protein, has become one of the most important indicators for evaluating the nutritional value of aquatic animals, and can be affected by numerous factors, such as the culture environment [22]. The present study demonstrated that although the culture environment of IP and SIRW both belonged to carbonate-alkalinity water, significant differences were still observed between IP and SIRW. The crude protein in the female hepatopancreas from IP [11.17 ± 0.16%] and SIRW [9.63 ± 0.10%] was higher than those cultured in the Shandong, Qinghai, and Shanghai [26,28], as well as wild mitten crabs caught in Suifenhe, Liaohe, and Nanliujiang [30]. At the same time, the crude protein of gonad and muscle from *E. sinensis* reared in carbonate-alkalinity water were similar to those obtained in the previous studies [22,26,28,30]. All these results illustrate that *E. sinensis* reared in carbonate-alkalinity water is a good high-protein seafood source.

4.4. Fatty Acids Composition

Fatty acid composition and contents are also one of the most important nutritional indicators for aquatic animals, particularly essential fatty acids (EFAs) and unsaturated fatty acids (UFAs) [22,30]. The present study showed that the total saturated fatty acids

(∑SFA), total monounsaturated fatty acids (∑MUFA), and total polyunsaturated fatty acids (∑PUFA) of *E. sinensis* reared in carbonate-alkalinity water were similar to those cultured in the other regions [22,27,28,30], implying that water alkalinity did not significantly alter the fatty acid composition of *E. sinensis*. It can be estimated that the differential fatty acid composition and content changes in the edible tissues of *E. sinensis* are mainly a result of heredity, followed by culture environment and diet. Culture environment can slightly regulate fatty acid contents other than that of composition.

Due to the fact that PUFAs are more beneficial to human health, they have received widespread attention, especially DHA, EPA, and ARA [30]. Numerous studies have confirmed that DHA and EPA can play an important role in preventing inflammation and cardiovascular diseases, while ARA can promote the development of the central nervous system [31–33]. The present study demonstrated that water alkalinity affected the fatty acid contents of *E. sinensis* compared with previous studies [26,27]. Furthermore, increasing DHA, EPA, and ARA contents in males, but decreasing DHA and EPA contents in females, were detected, implying the differential effects of alkalinity on each gender of *E. sinensis*. These results are consistent with our previous studies [5]. As *E. sinensis* were reared in similar carbonate-alkalinity water but with a different culture type, the EPA and ARA contents of male hepatopancreas and muscle in SIRW were significantly increased compared with those in IP. This phenomenon can be explained by the different developmental stages because the EPA and ARA contents of the hepatopancreas and muscle increased with the GSI improvement [27]. The h/H, AI, and TI are important indexes used to evaluate the beneficial effects of fatty acids on human health [22,25]. Generally, higher h/H, and the lower AI and TI are illustrated to be of better quality for human health. The three indexes of *E. sinensis* reared in carbonate-alkalinity water were similar to the results of previous studies [22,30].

4.5. FAAs Composition and TAV Analysis

The FAAs composition and contents are important factors affecting the taste and quality of crustaceans [22]. The present study illustrated that the ∑EFAA [IP: (724.11 $\pm$ 14.76) mg/100 g; SIRW: (632.4 $\pm$ 5.94) mg/100 g] and ∑FAA [IP: (1937.29 $\pm$ 85.06) mg/100 g; SIRW: (1788.67 $\pm$ 146.25) mg/100 g] contents of female *E. sinensis* hepatopancreas reared in carbonate-alkalinity water were higher than wild-caught mitten crabs [30] and three-year-old *E. sinensis* [22], indicating better taste quality. These results were consistent with our previous studies, in which prolonged alkalinity stress can improve ∑EFAA and ∑FAA contents [5]. In terms of specific amino acids, differential amino acids present different taste characteristics, while the taste characteristic, TAV, is positively correlated with the ratio between the special amino acid value and its threshold. Table 5 shows that the main umami amino acid Glu and the sweetness amino acid Ala were similar to previous studies [22,28,30]. Although we have classified Arg as a bitter amino acid, the flavor characteristic of Arg is actually significantly related to its concentration, with low concentrations exhibiting bitterness and high concentrations exhibiting umami [34]. The present study illustrates that the Arg content [IP: 7.42; SIRW: 6.50] in the female hepatopancreas was slightly higher than that of previous studies [5,22,30]. In addition, the ∑TUV, ∑TSV, and ∑TBV values of female hepatopancreas were higher than those in the pond-reared and wild-caught mitten crabs [22,30], suggesting stronger flavor characteristics.

4.6. Mineral Element Analysis

Mineral elements are important nutritional substances required to maintain normal growth, development, and metabolism in human beings, ensuring normal life activities [35]. Na, K, Ca, and Mg are macro-elements required by the human body, and they play differential roles in human health. Among these four macro-elements, Na and K play an important role in maintaining the acid–base balance and osmotic pressure of blood and body fluids. Ca is an important component of human bones and teeth. Mg participates in energy metabolism in the human body, catalyzes and activates various enzyme systems, and plays an important role in preventing cardiovascular diseases [36]. Of all the *E. sinensis* edible tissues, Na and K contents were significantly higher than other macro-elements, which was consistent with previous studies [22,37]. In addition, the K content was higher than the Na content except for the testis, implying that the culture environment perhaps has influence on Na and K accumulation. Fe, Zn, Cu, and Mn are micro-elements required by the human body. Among these, Fe plays an important role in the body's hematopoietic, oxygen transport, and fluid balance. Zn is an important coenzyme factor in the human body and is involved in the synthesis of DNA, RNA, and proteins. Cu can promote the production of hemoglobin in humans. Mn plays an important role in the central nervous system of the human brain [37]. A higher Fe content in the *E. sinensis* hepatopancreas was observed compared with that in the gonad and muscle tissues; furthermore, Fe is an important source for humans and has a certain significance in preventing iron deficiency anemia. Higher Zn contents were noted in the *E. sinensis* ovary and muscle compared with those in the testis and hepatopancreas tissues. The above results illustrate that *E. sinensis* reared in carbonate-alkalinity water is a good source of mineral elements.

5. Conclusions

The present study investigated the gonadal development, edible yield, coloration, nutritional, and flavor quality of *E. sinensis* reared in carbonate-alkalinity water (IP and SIRW culture model). Due to the sampling time, differential GSI and TEY values of *E. sinensis* were observed between IP and SIRW. IP samples accumulated more ∑MUFA, ∑EFA, and h/H in the female edible tissues compared with those of SIRW, whereas SIRW had better ∑FAA content in the muscle. IP samples also exhibited higher mineral elements, such as K, Ca, Mg, Fe, and Zn in the ovary. In summary, *E. sinensis* reared in carbonate-alkalinity water can complete gonadal development, accumulate more carotenoids, and are rich in fatty acids, FFAs, and mineral elements.

Author Contributions: Conceptualization, S.W. and Z.Z.; Methodology, S.W. and Z.Z.; Software, S.W. and L.L.; Validation, R.Z.; Formal Analysis, S.W., R.Z., and K.G.; Investigation, S.W. and L.L.; Resources, Z.Z.; Data Curation, S.W. and K.G.; Writing—Original Draft Preparation, S.W.; Writing—Review and Editing, S.W. and Z.Z.; Visualization, R.Z.; Supervision, Z.Z.; Project Administration, Z.Z.; Funding Acquisition, Z.Z. All authors have read and agreed to the published version of the manuscript.

Funding: This work was supported by the National Key Research and Development Project (2023YFD2401003), Central Public-interest Scientific Institution Basal Research Fund (2023TD59). The APC was funded by NO. 2023YFD2401003.

Institutional Review Board Statement: The animal study protocol was approved by the Committee for the Welfare and Ethics of Laboratory Animals of HRFRI CAFS (Approval code: 20210910-002; approval date: 10 September 2021) for studies involving animals.

Informed Consent Statement: Not applicable.

Data Availability Statement: Data is contained within the article.

Conflicts of Interest: The authors declare no conflicts of interest.

References

1. Hu, H.L.; Lai, Q.F.; Yao, Z.L. *Healthy Aquaculture Technology Model for Saline-Alkaline Water*; China Agriculture Press: Beijing, China, 2021; pp. 1–3.
2. Yao, Z.L.; Lai, Q.F.; Zhou, K.; Rizalita, R.E.; Wang, H. Developmental biology of medaka fish (*Oryzias latipes*) exposed to alkalinity stress. *J. Appl. Ichthyol.* **2010**, *26*, 397–402. [CrossRef]
3. Xu, J.; Ji, P.F.; Wang, B.S.; Zhao, L.; Wang, J.; Zhao, Z.X.; Zhang, Y.; Li, J.T.; Xu, P.; Sun, X.W. Transcriptome sequencing and analysis of wild Amur ide (*Leuciscus waleckii*) inhabiting an extreme alkaline-saline lake reveals insights into stress adaptation. *PLoS ONE* **2013**, *8*, e59703. [CrossRef] [PubMed]
4. Yao, Z.L.; Guo, W.F.; Lai, Q.F.; Shi, J.Q.; Zhou, K.; Qi, H.F.; Lin, T.T.; Li, Z.N.; Wang, H. *Gymnocypris przewalskii* decreases cytosolic carbonic anhydrase expression to compensate for respiratory alkalosis and osmoregulation in the saline-alkaline Lake Qinghai. *J. Comp. Physiol. B* **2016**, *186*, 83–95. [CrossRef] [PubMed]

5. Wang, S.H.; Guo, K.; Luo, L.; Zhang, R.; Xu, W.; Song, Y.Y.; Zhao, Z.G. Fattening in saline and alkaline water improves the color, nutritional and taste quality of adult Chinese mitten crab *Eriocheir sinensis*. *Foods* **2022**, *11*, 2573. [CrossRef] [PubMed]
6. Arjona, F.J.; Vargas-Chacoff, L.; Ruiz-Jarabo, I.; del Rio, M.P.M.; Mancera, J.M. Osmoregulatory response of Senegalese sole (*Solea senegalensis*) to changes in environmental salinity. *Comp. Biochem. Phys. A* **2007**, *148*, 413–421. [CrossRef]
7. Long, X.W.; Wu, X.G.; Zhu, S.C.; Ye, H.H.; Cheng, Y.X.; Zeng, C.S. Salinity can change the lipid composition of adult Chinese mitten crab after long-term salinity adaptation. *PLoS ONE* **2019**, *14*, e0219260. [CrossRef] [PubMed]
8. Tian, L.; Tan, P.; Yang, L.; Zhu, W.L.; Xu, D.D. Effects of salinity on the growth, plasma ion concentrations, osmoregulation, non-specific immunity, and intestinal microbiota of the yellow drum (*Nibea albiflora*). *Aquaculture* **2020**, *528*, 735470. [CrossRef]
9. Song, L.Y.; Zhao, Y.; Song, Y.D.; Zhao, L.L.; Ma, C.X.; Zhao, J.L. Effects of saline-alkaline water on growth performance, nutritional processing, and immunity in Nile tilapia (*Oreochromis niloticus*). *Aquaculture* **2021**, *544*, 737036. [CrossRef]
10. Liu, Y.J.; Yao, M.Z.; Li, S.W.; Wei, X.F.; Ding, L.; Han, S.C.; Wang, P.; Lv, B.C.; Chen, Z.X.; Sun, Y.C. Integrated application of multi-omics approach and biochemical assays provides insights into physiological responses to saline-alkaline stress in the gills of crucian carp (*Carassius auratus*). *Sci. Total Environ.* **2022**, *822*, 153622. [CrossRef]
11. Wang, F.; Zhu, L.; Wei, Y.X.; Gao, P.C.; Liu, Y.M.; Zhou, K.; Sun, Z.; Lai, Q.F.; Yao, Z.L. Intestinal ion regulation exhibits a daily rhythm in *Gymnocypris przewalskii* exposed to high saline and alkaline water. *Sci. Rep.* **2022**, *12*, 807. [CrossRef]
12. Shang, X.C.; Geng, L.W.; Yang, J.; Zhang, Y.T.; Xu, W. Transcriptome analysis reveals the mechanism of alkalinity exposure on spleen oxidative stress, inflammation and immune function of *Luciobarbus capito*. *Ecotoxicol. Environ. Saf.* **2021**, *225*, 112748. [CrossRef] [PubMed]
13. Zhu, C.B.; Dong, S.L.; Wang, F.; Huang, G.Q. Effects of Na/K ratio in seawater on growth and energy budget of juvenile *Litopenaeus vannamei*. *Aquaculture* **2004**, *234*, 485–496. [CrossRef]
14. Fang, W.; Liu, L.; Chang, W.; Liu, Z.G.; Yang, J.X.; Liu, X.; Mu, C.K.; Wang, C.L. Comparative analysis of growth and nutritional components of *Scylla paramamosain* cultured in saline-alkali water and marine water. *J. Fish. China* **2022**, *46*, 2143–2157.
15. Liang, L.Q.; Ren, B.; Chang, Y.M.; Tang, R.; Zhang, L.M. Inland brackish (saline-alkaline) water resources and fisheries utilization in China. *Chin. Fish. Econ.* **2013**, *31*, 138–145.
16. Wright, P.A.; Wood, C.M. Seven things fish know about ammonia and we don't. *Respir. Physiol. Neurobiol.* **2012**, *184*, 231–240. [CrossRef]
17. Copatti, C.E.; Bolner, K.C.D.; de Rosso, F.L.; Loro, V.L.; Baldisserotto, B. Tolerance of piava juveniles to different ammonia concentrations. *Semin.-Cienc. Agrar.* **2015**, *36*, 3991–4001. [CrossRef]
18. Wu, K.; Zhong, Z.; Chen, Y.; Weng, S.; He, J. The relationship between climate change, feeding management and ammonia, rdtrite and nitrate nitrogen in the *Litopenaeus vannamei* aquaculture ponds. *Acta Sci. Natural. Univ. Sunyatseni* **2017**, *56*, 102–114.
19. Bureau of Fisheries and Fishery Management, Ministry of Agriculture and Rural affairs of China. *2023 China Fisheries Statistical Yearbook*; China Agriculture Press: Beijing, China, 2023; p. 34.
20. Cheng, Y.X.; Wu, X.G.; Yang, X.Z.; Hines, A.H. Current trends in hatchery techniques and stock enhancement for Chinese mitten crab, *Eriocheir japonica sinensis*. *Rev. Fish. Sci. Aquac.* **2008**, *16*, 377–384. [CrossRef]
21. Long, X.W.; Wu, X.G.; Zhao, L.; Liu, J.G.; Cheng, Y.X. Effects of dietary supplementation with *Haematococcus pluvialis*, cell powder on coloration, ovarian development and antioxidation capacity of adult female Chinese mitten crab, *Eriocheir sinensis*. *Aquaculture* **2017**, *473*, 545–553. [CrossRef]
22. Wang, S.H.; Luo, L.; Zhang, R.; Guo, K.; Bai, S.Y.; Qin, D.L.; Zhao, Z.G. Comparison of edible yield and quality of female Chinese mitten crab between two-year-old and three-year-old. *J. Food Compos. Anal.* **2022**, *112*, 104687. [CrossRef]
23. AOAC. *Official Methods of Analysis of Association of Official Analytical Chemists*, 16th ed.; AOAC Inc.: Arlington, VA, USA, 1995; p. 13.
24. Rotzoll, N.; Dunkel, A.; Hofmann, T. Quantitative studies, taste reconstitution, and omission experiments on the key taste compounds in morel mushrooms (*Morchella deliciosa* Fr.). *J. Agric. Food Chem.* **2006**, *54*, 2705–2711. [CrossRef]
25. Śmietana, N.; Panicz, R.; Sobczak, M.; Śmietana, P.; Nędzarek, A. Spiny-Cheek Crayfish, *Faxonius limosus* (Rafinesque, 1817), as an alternative food source. *Animals* **2021**, *11*, 59. [CrossRef]
26. Long, X.W.; Guo, Q.; Wang, X.C.; Francis, D.S.; Cheng, Y.X.; Wu, X.G. Effects of fattening period on ovarian development and nutritional quality of adult female Chinese mitten crab *Eriocheir sinensis*. *Aquaculture* **2020**, *519*, 734–748. [CrossRef]
27. Wu, X.G.; Zhu, S.C.; Zhang, H.C.; Liu, M.M.; Wu, N.; Pan, J.; Luo, M.; Wang, X.C.; Cheng, Y.X. Fattening culture improves the gonadal development and nutritional quality of male Chinese mitten crab *Eriocheir sinensis*. *Aquaculture* **2019**, *518*, 734865. [CrossRef]
28. Wang, S.H.; Wang, Y.Z.; Wu, X.G.; Zhang, X.B.; Zhao, J.S.; Yang, J.M.; Cheng, Y.X. Gonadal development and biochemical composition of Chinese mitten crabs (*Eriocheir sinensis*) from four sources. *J. Food Sci.* **2021**, *86*, 1066–1080. [CrossRef] [PubMed]
29. Li, Q.Q.; Sun, Q.F.; Liu, Q.; Cheng, Y.X.; Wu, X.G. Estimation of genetic parameters for carotenoid traits in Chinese mitten crab, *Eriocheir sinensis*, females. *Aquaculture* **2021**, *532*, 735990. [CrossRef]
30. Wang, S.H.; Luo, L.; Zhang, R.; Guo, K.; Zhang, X.B.; Kong, L.J.; Zhao, Z.G. Biochemical composition and quality of wild-caught adult mitten crabs from three river basins. *J. Food Compos. Anal.* **2022**, *110*, 104574. [CrossRef]
31. Zhang, Y.L.; Yang, B.; Yang, L.; Jiang, S.T.; Lu, J.F.; Lin, L. Comparison of the nutritional qualities of the pond, rice-field and wild crayfish (*Procambarus clarkii*) meat. *Food Chem. Adv.* **2023**, *2*, 100272. [CrossRef]

32. Muskiet, F.A.J.; Kemperman, R.F.J. Folate and long-chain polyunsaturated fatty acids in psychiatric disease. *J. Nutr. Biochem.* **2006**, *17*, 717–727. [CrossRef]
33. Zhou, S.J.; Yang, R.; Yu, G.; Ma, Z.H.; Meng, X.J. Muscle composition determination and nutrition evaluation of three tuna species near Meiji Reef. *South China Fish. Sci.* **2021**, *17*, 51–59.
34. Dermiki, M.; Phanphensophon, N.; Mottram, D.S.; Methven, L. Contributions of non-volatile and volatile compounds to the umami taste and overall flavour of shiitake mushroom extracts and their application as flavour enhancers in cooked minced meat. *Food Chem.* **2013**, *141*, 77–83. [CrossRef] [PubMed]
35. Jiang, Y.; Liu, X.Z.; Cui, A.J.; Wang, K.J.; Wang, B.; Xu, Y.J. Analysis of muscle components of striped jack (*Pseudocaranx dentex*). *South China Fish. Sci.* **2021**, *17*, 130–135.
36. Cui, L.L.; Lu, S.X.; Gao, H.T.; Wang, B.Q.; Fu, S.W.; Xu, G.F.; Sun, Y.; Li, J.Y.; Mao, L.; Yang, C.L.; et al. Analysis and evaluation of muscle nutritional components of five principal trains of rainbow trout *Oncorhynchus mykiss* in China. *Chin. J. Fish.* **2021**, *34*, 1–10.
37. Zhao, L.; Wu, X.G.; Long, X.W.; Ma, M.J.; Cheng, Y.X. Comparative study on major mineral contents among three mitten crab populations: *Eriocheir sinensis*, *E. japonica* and their hybrids. *Guangdong Agric. Sci.* **2014**, *22*, 109–118.

Disclaimer/Publisher's Note: The statements, opinions and data contained in all publications are solely those of the individual author(s) and contributor(s) and not of MDPI and/or the editor(s). MDPI and/or the editor(s) disclaim responsibility for any injury to people or property resulting from any ideas, methods, instructions or products referred to in the content.

foods

MDPI

Article

Forging Prawn and Salmon Flavours with Non-Animal-Based Ingredients

Jiaqiang Luo [1,*], Damian Frank [2] and Jayashree Arcot [1]

[1] Food and Health Group, School of Chemical Engineering, Faculty of Engineering, UNSW Sydney, Kensington, NSW 2052, Australia; j.arcot@unsw.edu.au
[2] All G, Waterloo, NSW 2017, Australia; damian@allg.com
* Correspondence: xavier.luo@unsw.edu.au

Abstract: The development of plant-based seafood alternatives with authentic flavour profiles remains a significant challenge, limiting their appeal to seafood consumers. This study hypothesised that incorporation of flavour precursors including free amino acids, betaine, and long chain omega-3 fatty acids would enhance the flavour resemblance of plant-based prawn and salmon prototypes to their authentic seafood counterparts. Prototypes were analysed using headspace solid-phase microextraction gas chromatography–mass spectrometry and evaluated by a semi-trained sensory panel. Volatile analysis revealed 64 compounds across prototypes, with significant variations attributed to precursor combinations and thermal treatments. Frying enhanced volatile profiles, particularly in plant-based prawn prototypes fortified with all three flavour precursors, producing key prawn odourants, including pyrazines and trimethylamine. Notably, betaine pyrolysis under moderate cooking conditions was demonstrated as a potential pathway for trimethylamine formation, contributing to fish-like odours. Sensory evaluation showed that the final plant-based prawn prototype exhibited strong cooked crustacean and grilled notes, aligning with the observed volatile profile. While the salmon prototype displayed key salmon odourants, its cooked salmon odour was less pronounced, suggesting a need for a more robust flavouring strategy. This study highlights the potential of targeted flavour precursor formulations to improve the flavour quality of plant-based seafood alternatives, paving the way for their wider acceptance.

Keywords: alternative protein; flavour; formulation; volatile; odour

1. Introduction

In the global plant-based (PB) protein market, the PB seafood alternative (PBSA) sector remains relatively small compared to meat and dairy analogues. However, despite its global market size being valued at only USD 42.1 million in 2021, it is projected to reach USD 1.3 billion by 2031, making it the fastest-growing sector within the PB protein alternative industry [1]. Despite this optimistic forecast, both person- and product-related barriers inhibit the wider acceptance of PBSAs. Factors such as flavour, dietary preferences, product pricing, product availability, and naturalness all play a role [2]. Amongst these barriers, sensory qualities, particularly flavour, are fundamental to product acceptability and largely depend on the use of flavouring agents, whether chemically synthesised or derived from natural sources.

The first step in an effective flavouring process for PBSAs is to gain a thorough understanding of the flavour profile of the target seafood. Numerous studies have been conducted to profile inherent flavour molecules and precursor compounds, both of which

 https://doi.org/10.3390/foods14050820

play a crucial role in shaping the flavour profile of cooked seafood. Due to their high abundance and contribution to umami and sweetness [3], free amino acids (FAAs) together with 5′-nucleotides significantly influence seafood taste profiles [4–6]. In addition to directly contributing to taste, FAAs also serve as precursors to seafood odour formation, especially through thermal processing. Strecker degradation of FAAs generates a range of aldehydes that are important for the seafood flavour profile [7]. For instance, methional, derived from methionine, is a common odourant found in cooked crustaceans, molluscs, and finfish [8]. 2-Acetyl-1-pyrroline, a well-established key odourant of cooked prawns [9,10], can be generated during thermal food processing through the Maillard reaction between proline or ornithine and carbohydrates [11]. Furthermore, trimethylamine oxide (TMAO) and betaine (*N,N,N*-trimethyl glycine) are key osmolyte molecules in seafood species—present at relatively high concentration—that protect cells from osmotic stress, which may contribute to sweet and umami tastes [12]. Both TMAO and betaine can be degraded to trimethylamine (TMA) through microbial activity [13,14], or for betaine, potentially through thermal degradation during cooking. While a high concentration of TMA is considered a biomarker of seafood spoilage, a low concentration of TMA contributes to the fresh fish-like odour, which is essential for shaping fundamental seafood flavour [15]. The thermal degradation of omega-3 polyunsaturated fatty acids (n-3 PUFAs) produces a range of aldehydes and methyl ketones, followed by acids, hydrocarbons, lactones, and alcohols [7]. These compounds contribute to the characteristic volatile profile of cooked oily fish, such as salmon [6,16].

Even with this information available, the formulation of seafood flavour is still far from straightforward, and the flavour quality of commercial PBSAs currently does not live up to expectations of regular seafood consumers and is often disappointing. Recent research has shown that most PBSAs available in the Australian market fail to replicate the flavours of their seafood counterparts, with the exception of PB sashimi products, where flavour plays a relatively minor role [17]. This may be due to the lack of ideal PB seafood flavouring agents and/or flavouring strategies that can accurately introduce the complex and subtle flavour attributes to the PB substrate [17].

Seafood typically has a very mild odour when raw, but its flavour profile changes significantly upon cooking. In contrast, in the production of commercial PBSAs, flavouring agents are generally pre-added during the structuring process and homogenised into the food matrix [18]. However, manufacturers often add excessive amounts of flavouring agents at this stage, one of the important reasons being to mask undesirable flavours like beany odour and bitterness from plant ingredients [19]. This results in overly pronounced odours even before the products are cooked, giving the products an unnatural flavour impression [17]. Upon heating, more complex interactions occur, further shifting the flavour profile away from the intended target.

Numerous studies, including many review articles [20–22], have explored potential solutions for overcoming flavour challenges in PB products. However, much of this research has focused on removing undesirable flavours and exploring new PB ingredients and processing methods. Little emphasis has been placed on how ideal PBSA flavouring agents should be formulated to more closely replicate seafood flavour. Our recent investigation into commercial PBSAs revealed that excessively high levels of particular volatile compounds and a lack of signature seafood odourants in these products are common issues that warrant greater attention in flavour formulation. In this study, we explore a formulation strategy based on using flavour precursors. Prawn and salmon were selected as target species due to their widespread popularity worldwide. Besides, prawn is low in fat (~0.9%) and has a relatively high abundance of flavour precursors such as FAAs and betaine, while salmon is high in fat (~19.8%) and rich in n-3 PUFAs, making them ideal targets for testing this flavouring strategy. Using a gel system consisting of hydrocolloids, starch, faba protein isolate, and canola oil, FAAs, betaine, and encapsulated algal DHA were added at different levels to create PB prawn and salmon prototypes. Additionally, the potential formation of TMA through the thermal degradation of betaine under moderate cooking conditions was also investigated, evaluating its potential for inducing fish-like odour in PBSAs. Headspace solid-phase microextraction coupled with gas chromatography–mass spectrometry (HS-SPME-GC–MS) was used to analyse the volatile compositions. Analyses were performed on both thermoset gels and fried thermoset gels, enabling a comprehensive comparison of volatile profiles of different formulations and changes upon cooking. Olfactory evaluation was conducted to assess the odour qualities of the PB prawn and salmon prototypes perceived by a semi-trained panel.

2. Materials and Methods

2.1. Chemicals

Food-grade FAAs including alanine, arginine, aspartic acid, glutamic acid, glycine, histidine, leucine, lysine, methionine, proline, valine, cystine, cysteine, phenylalanine, and betaine (*N*,*N*,*N*-trimethylglycine) were sourced from various sports nutrition suppliers in Australia. A standard alkane standard mix (C_8–C_{20}) was purchased from Merck (Castile Hill, NSW, Australia). Purified water was obtained from a Milli-Q system (Millipore Australia, Bayswater, VIC, Australia).

2.2. Materials

Australian faba protein powder (82.4% protein, Wellness Road, Coles), sunflower oil, deodorised coconut oil, and tapioca flour were purchased from local supermarkets in Sydney. A hydrocolloid mixture was added to facilitate gelling. Microencapsulated algal oil containing a minimum of 10% docosahexaenoic acid (DHA) was supplied by DSM-Firmenich (life's DHA^{TM} S10-P200).

2.3. Prototype Formulation

Macronutrient targets for the prawn and salmon prototypes were determined using records from the Australian Food Composition Database (available at: https://www.foodstandards.gov.au/science-data/monitoringnutrients/afcd, accessed on 12 December 2024). FAA targets for prawn were adapted from the studies of Vázquez-Ortiz, Caire [23] and Zhang, Chen [24] and then converted into salmon targets based on relative ratios determined in our previous study [8]. Similarly, betaine targets for both prawn and salmon were calculated using literature values from squid [25]. Formulation targets for prawn and salmon prototypes are summarised in Table 1. Prawn flavour mix (PFM) and salmon flavour mix (SFM) were prepared by mixing the listed concentrations of FAAs and betaine. In the formulation, protein, saturated fat, unsaturated fat, and omega-3 oil (algal DHA oil) targets were matched with faba protein isolate, coconut oil, sunflower oil, and encapsulated algal oil, respectively. To develop stable gel structures that could withstand the cooking process, the proportions of protein, fats, and water were slightly adjusted from the targets, with hydrocolloid gelling agents and $CaCl_2$ added. A total of six prototypes were formulated, including (1) prawn matrix (P, control of the prawn prototype group), (2) prawn matrix + encapsulated algal DHA oil (P + DHA), (3) prawn matrix + encapsulated algal DHA oil + prawn flavour mix (P + DHA + PFM), (4) salmon matrix (S, control of the salmon prototype group), (5) salmon matrix + encapsulated DHA algal oil (S + DHA), and (6) salmon matrix + encapsulated DHA algal oil + salmon flavour mix (S + DHA + SFM). Ingredients of each prototype were mixed using a commercial stand mixer (KitchenAid, MI, USA). The recipes for the prototypes are summarised in Table 2. All prototypes were prepared in triplicate.

Table 1. PB prawn and salmon formulation targets.

	Prawn	Salmon
Macronutrients (g/100 g fresh weight) [1]		
Protein	17.5	19.3
Saturated fat	0.2	4.7
Unsaturated fat	0.7	15.1
n-3 PUFAs	0.24	2.8
Sodium	0.25	0.03
FAAs (mg/100 g fresh weight) [2]		
Alanine	83.5	57.1
Arginine	420.3	7.4
Aspartic acid	1.9	0.5
Glutamic acid	31.7	17.7
Glycine	660.3	44.1
Histidine	43.9	31.1
Leucine	33.4	4.5
Lysine	38.7	3.9
Methionine	13.8	1.5
Proline	370.8	2.8
Valine	39.5	10.2
Cystine	1.1	0.1
Cysteine	0.5	0.1
Phenylalanine	16.1	4.7
Other (mg/100 g fresh weight) [3]		
Betaine	357.1	0.6

Notes: [1] Data were obtained from the Australian Food Composition Database (available at https://www.foodstandards.gov.au/science-data/monitoringnutrients/afcd, accessed on 12 December 2024). [2] Values were adapted from the studies of Vázquez-Ortiz, Caire [23] and Zhang, Chen [24] and then converted into salmon targets based on relative ratios determined in our previous study [8]. [3] Betaine targets for both prawn and salmon were calculated using literature values from squid [25].

Table 2. Formulation and odour quality of prototypes.

	Prawn Prototypes			Salmon Prototypes		
	P	*P + D HA*	*P + DHA + PFM*	*S*	*S + DHA*	*S + DHA + SFM*
Ingredients (g/100 g fresh weight)						
Faba protein isolate	10.0	10.0	10.0	10.0	10.0	10.0
Encapsulated algal DHA	0	2.4	2.4	0	10	10
Total fat [1]	0.9	1.1	1.1	19.8	20.8	20.8
- Sunflower oil	0.7	0.7	0.7	15.1	15.1	15.1
- Coconut oil	0.2	0.2	0.2	4.7	4.7	4.7
- DHA (from encapsulated algal DHA)	0	0.2	0.2	0	1	1
Flavour mix	0	0	2.1	0	0	0.2
- FAAs [2]	0	0	1.8	0	0	0.2
- Betaine [3]	0	0	0.3	0	0	<0.1
NaCl	0.2	0.2	0.2	0.3	0.3	0.3
Gelling agent	5.0	5.0	5.0	5.0	5.0	5.0
$CaCl_2$	1.0	1.0	1.0	1.0	1.0	1.0
MQ water	80.4	80.4	78.3	54.0	54.0	53.7
Flavour mix	0	0	2.1	0	0	0.2
Algal DHA	0	2.4	2.4	0	10	10
Tapioca flour	2.4	0	0	10.0	0	0
Odour quality [4]						
Unheated mixture (R)	very mild beany, neutral	very mild beany, neutral	mild fresh seafood, not unpleasant, savoury	very mild fatty, neutral	fresh fish, mild fatty, not unpleasant	neutral, very mild fatty, fresh fish

Table 2. *Cont.*

	Prawn Prototypes			Salmon Prototypes		
	P	*P + D HA*	*P + DHA + PFM*	*S*	*S + DHA*	*S + DHA + SFM*
Thermal stabilised (TS)	mild oxidised fat, very mild beany	very mild fish oil, not unpleasant, neutral	very mild crustacean, not unpleasant	very mild fatty, beany	fatty, fresh fish, mild, not unpleasant	mild oxidised fish oil, slightly fishy, not unpleasant
Fried thermal stabilised (FTS)	mild beany, fatty	mild fried fish, fatty	cooked crustacean, fried fish	mild salmon, fatty, fired food	fried oily fish, fried salmon	fried oily fish, fried salmon

Notes: [1] The total content of sunflower oil, coconut oil, and encapsulated algal DHA (containing no less than 10% DHA) in the prototype. [2] The total content of all listed FAAs in Table 1. [3] Betaine content listed in Table 1. [4] Odour qualities assessed and recorded by two experienced flavour chemists.

2.4. Thermal Treatment

An in-house proprietary blend of hydrocolloids that form heat-set thermo-irreversible gels was used. The ingredient mixtures for the prototypes were shaped into uniform disk shapes using silicon moulds (3 cm in diameter and 1.5 cm in height). Samples at this uncooked status were defined as raw (R). The samples were then vacuum-sealed in plastic bags and then heated in a steam oven at 90 °C for 60 min (Unox Cheftop, Unox, Cadoneghe, Italy). After cooling in an ice-water bath, thermally stabilised (TS) samples were formed. TS samples were then subsequently fried at 200 °C for 3 min on each side in a preheated frying pan lightly greased with canola oil spray to create fried thermally stabilised (FTS) samples. The odour qualities of the R, TS, and FTS samples were assessed and recorded by two experienced flavour chemists, with the descriptors provided in Table 2.

2.5. Thermal Generation of Trimethylamine from Betaine

To investigate the potential of betaine to form TMA under moderate cooking conditions (at 200 °C), six model reaction mixes were prepared as follows: (1) betaine only, (2) betaine with oils (coconut oil and sunflower oil), (3) betaine with oils and encapsulated algal DHA oil, (4) betaine with an FAA mix, (5) betaine with an FAA mix and oils, (6) betaine with an FAA mix, oils, and encapsulated algal DHA oil. The ingredient quantities were based on those used in preparing prawn prototypes (ingredients of each model reaction mix are provided in Supplementary Table S1). Samples were placed in 20 mL amber headspace vials and sealed with aluminium foil. After baking at 200 °C for 10 min, aluminium foil was replaced with headspace vial fitted caps after baking. The pyrolysis samples were prepared in duplicate.

2.6. Volatile Analysis

Volatile compounds were analysed by a GC–MS system (GC–MS-QP-2020 NX, Shimadzu, Kyoto, Japan) coupled to an AOC-6000 autosampler. Volatile extraction was carried out using a 1.10 mm 120 µm DVB/Carbon WR/PDMS SPME arrow fibre. The fibre was exposed to the headspace of 20 mL amber headspace vials of pyrolysis of betaine experiment or containing 3 g of homogenised prototype samples at 40 °C for 40 min, after a 5 min equilibrium period at the same temperature with agitation in the autosampler. The SPME arrow was then desorbed in splitless mode at 240 °C for 5 min. Compound separation was achieved using a Zebron-Wax capillary column (Phenomenex, CA, USA; 30 m × 0.25 mm ID × 0.25 µm) with helium (99.999% purity, Coregas, NSW, Australia) as the carrier gas, maintaining pressure at 100 kPa. The GC oven temperature was programmed as follows: initially set at 35 °C (held for 2 min), followed by an increase to 250 °C at 6 °C/min, with a final holding time of 22 min. The MS operated in positive electron ionisation (EI) mode at

70 eV, scanning a mass range of m/z 35 to 220. The ion source and interface temperatures were set to 200 °C and 240 °C, respectively.

A standard mixture of n-alkanes (C_8–C_{20}) was analysed to calculate the retention indices (RIs) of the compounds. Compound identification was achieved by comparing mass spectra (similarity of at least 90%) and RI values (difference of no more than 50) with those in the National Institute of Standards and Technology (NIST, USA) library (version 11). Semi-quantification was based on the peak area of the target ion. Details of tentatively identified compounds, including quantifiers and qualifiers, are provided in Supplementary Materials.

2.7. Olfactory Evaluation of Samples

Olfactory evaluation of samples followed the procedure described by [17] and utilised the same semi-trained panel (n = 20, balanced by sex). Sensory panel recruitment and the use of collected data for analysis were approved by the UNSW Human Research Ethics Committee (ethics approval number: 9628). All participants provided their written consent prior to training and evaluation sessions. A total of 4 samples, including prawn (P + DHA + PFM) and salmon (S + DHA + SFM) prototypes in their TS and FTS states, were selected for evaluation. Three grams of each sample were minced and placed into 20 mL glass vials. The four samples were labelled with 3-digit random codes and presented to panellists in randomised order. Odour attributes were evaluated on smart phones using Qualtrics XM platform (Qualtrics, Seattle, WA, USA). Panellists were instructed to sniff the samples and evaluate pre-defined odour attributes using a 100 mm line-scale; overall odour intensity, artificial, cooked crustaceans, cooked salmon, fishy, grilled, herbaceous, marine, meaty, muddy, oxidised oil, vegetable, and smoky odours, as well as using continuous hedonic scales. Odour attributes were defined previously [17], through focus groups and input from experienced flavour chemists. The evaluation was conducted in a single session without replication.

2.8. Statistical Analysis

After testing the normality of the data, one-way analysis of variance (ANOVA) followed by Tukey's post hoc test was performed to compare group means of individual volatile compounds. Multivariate analysis using principal component analysis (PCA) was also conducted. Group means of perceived odour intensities were compared using the Mann–Whitney test. These analyses were conducted using XLSTAT software version 2023.3.1 (1416) (Lumivero, Denver, CO, USA). The PCA biplot and radar charts were visualised using XLSTAT, while the results of TMA pyrolysis and the heatmap of volatile evolution in prawn and salmon prototypes were visualised using GraphPad Prism version 9.5.1 (733) (GraphPad Software, Boston, MA, USA).

3. Results and Discussion

3.1. Volatile Composition of PB Prawn and Salmon Prototypes

A total of 64 volatile compounds were identified across PB prawn and salmon prototypes with various treatments. These included eight acids, eight alcohols, nineteen aldehydes, four furans, thirteen ketones, ten N-containing compounds, and two S-containing compounds. Significant ($p < 0.05$) differences across samples were observed. A detailed summary comparison of volatiles in each sample compared by one-way ANOVA is provided in Supplementary Materials S1.

To provide a clearer visualisation of the results, PCA was performed, where about 50% of the total variance was explained by the first two principal components (PCs, Figure 1). The positive side of PC1 was associated with high levels of alcohols, aldehydes, ketones, and furans, while the negative side was dominated by 2,3-pentanedione, acetic acid, and 3-methylbutanoic acid. This distribution placed most raw (R) and thermally stabilised (TS) samples for both prawn and salmon prototypes on the negative side of PC1. After subjecting the samples to frying (FTS), the samples shifted to the positive side of PC1, and this was most apparent for the prawn samples P + DHA + PFM and P + DHA. Within each prototype formulation, thermal treatment resulted in a consistent shift toward the positive side of PC1. Additionally, pyrazines and sulphides drove the distinct separation of the P + DHA + PFM (FTS) sample on the positive end of PC2. This trend was also observed for S + DHA + PFM (FTS), but to a lesser extent.

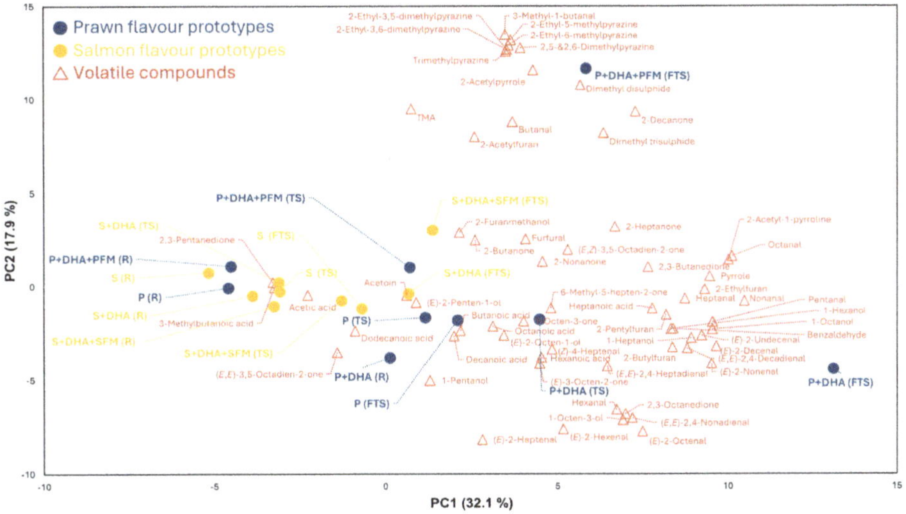

Figure 1. PCA biplot illustrating sample discrimination based on volatile composition. Formulations aiming to mimic prawn and salmon flavours are labelled in blue and yellow, respectively. Thermal treatments are indicated as follows: no thermal treatment applied (R), thermally stabilised (TS), and fried after thermal stabilisation (FTS). Volatile compounds are labelled with red triangles.

The PCA highlighted the impact of formulation and thermal treatment on volatile profiles, being most strongly influenced by the higher FAAs present in the prawn samples. Typical DHA degradation products were higher in the fried prawn samples, even though they had a much lower lipid and DHA content. Raw PB prawn and salmon prototypes exhibited similar volatile profiles, regardless of the inclusion of DHA or FAA-based flavour mixes. Following thermal stabilisation by steaming, relatively small changes in volatile profiles occurred compared to frying. The vacuum-sealing of prototypes prior to steaming likely limited oxygen exposure and oxidative degradation of PUFAs, especially in salmon prototypes such as S (TS), S + DHA (TS), and S + DHA + SFM (TS), which remained on the negative side of PC1. Frying induced more substantial changes, particularly in PB prawn prototypes. The higher FAA content in prawn formulations likely contributed to the formation of reactive intermediates that accelerated PUFA breakdown, even in low-oxygen conditions [26]. This was reflected in the volatile profiles of P (FTS), P + DHA (FTS), and P + DHA + PFM (FTS). Without PFM, P and P + DHA displayed distinct volatile profiles due to differences in PUFA-derived compounds. The inclusion of PFM resulted in significantly higher levels of pyrazines and sulphide compounds, generated through Strecker degradation and Maillard reactions, creating a profile that was markedly distinct from the other prawn prototypes. A similar trend was observed in S + DHA + SFM (FTS).

However, due to the lower FAA content in SFM compared to PFM, the production of pyrazines and sulphide compounds was not as pronounced.

The pan-frying process was designed to replicate a typical cooking method for prawn, salmon, and their PB analogues. Consequently, greater emphasis was placed on the flavour evolutionary patterns of FTS prawn and salmon prototypes, which are visualised in Figure 2. The heatmap reveals a higher abundance of volatiles in the PB prawn prototypes compared to the salmon prototypes, with distinct differences in the evolutionary patterns of volatiles across various chemical classes between the two.

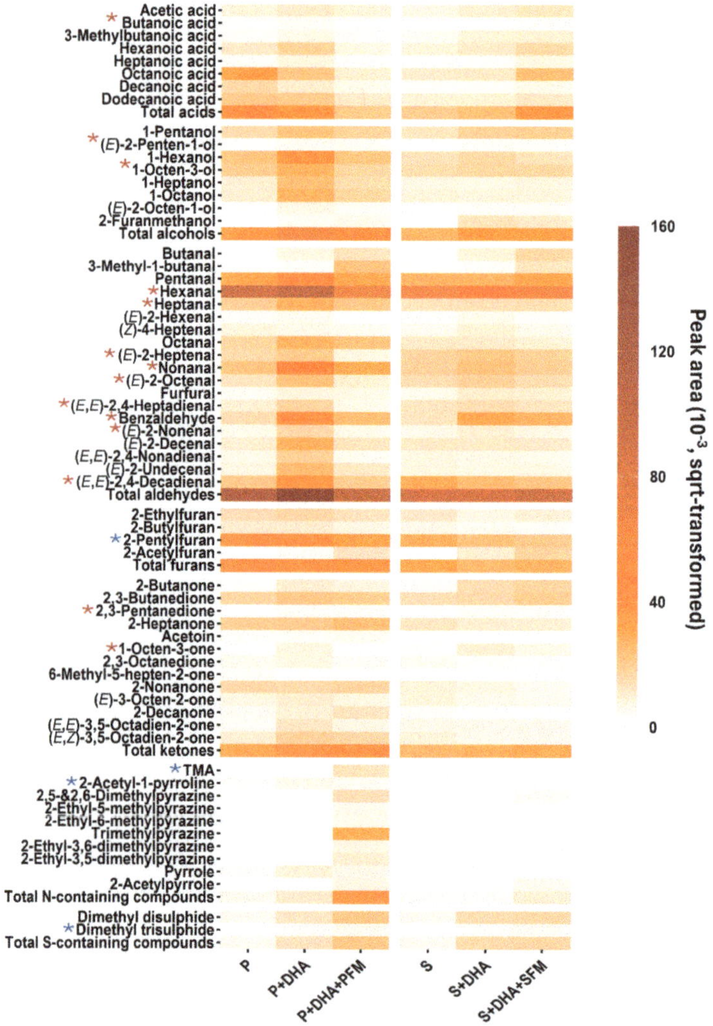

Figure 2. Evolution of volatile compounds in prawn (P, P + DHA, and P + DHA + PFM) and salmon (S, S + DHA, and S + DHA + SFM) prototypes in fried thermally stabilised (FTS) status. Red colours from light to dark indicate concentration gradients from low to high, expressed as 10^{-3} square root-transformed peak area. Odourants identified at particularly high concentrations and classified as key volatiles of prawn and salmon are marked with blue and red "*", respectively.

For prawn prototypes, the fortification with n-3 PUFAs (P + DHA) and the subsequent inclusion of PFM (P + DHA + PFM) gradually reduced the total acid content, primarily driven by decreases in octanoic, decanoic, and dodecanoic acids. In contrast, an opposite pattern was observed for salmon prototypes, where the acid content increased. The reasons for the decreasing trend in prawn prototypes remain unclear. However, for salmon prototypes, these acids are known to be derived from n-3 PUFAs, and their fortification with DHA likely promotes the production of these acids during frying. This reaction is further enhanced in the presence of FAAs [26].

For alcohols, aldehydes, and furans, the highest abundance was consistently observed in P + DHA amongst the prawn prototypes. Notably, 1-hexanol, 1-hexanal, and 2-pentylfuran were the major contributors to their respective chemical classes in this trend. Compared to P, the addition of DHA in P + DHA likely provided more precursors to produce aldehydes, alcohols, and furans. The reduced levels observed in P + DHA + PFM may suggest a protective mechanism of FAAs against DHA oxidative degradation. This hypothesis is supported by previous studies reporting the antioxidant activities of FAAs [27,28]. Similar patterns were observed in the salmon prototypes, although the trends were less pronounced than in the prawn prototypes. For instance, alcohols and aldehydes were generally most abundant in S + DHA. However, no consistent trend was observed for furans across the salmon prototypes. This complexity was further evident in the ketone compounds. While total ketone contents tended to increase from the PB matrix to the treatment group with added DHA and then to the group fortified with a flavour mix, the specific evolution of individual ketones varied. The highly variable profiles of lipid oxidation-derived compounds in PB prawn and salmon prototypes were closely linked to differences in fat composition, particularly DHA content (Table 2), and the distinct FAA compositions, which likely influenced the fat degradation process, as discussed previously. Furthermore, salmon prototypes (S + DHA and S + DHA + SFM) had a much higher concentration of total fat (20.8%) and DHA (1%) compared to the prawn prototype (1.1% and 0.2%). Theoretically, this would result in a higher production of lipid degradation products, which was not observed. This could be attributed to the higher water content in the prawn formulations, creating a more hydrophilic environment that facilitated the release of non-polar volatiles.

Overall, the addition of the flavour mix enhanced the formation of N- and S-containing volatiles, which are typical products of Strecker degradation of amino acids. This was observed in both PB prawn and salmon prototypes. Additionally, the FAAs (1.8%) and betaine (0.3%) content in the final prawn formulation (P + DHA + PFM) was higher than in the salmon prototype (0.2% and <0.1%, respectively). Consequently, the addition of SFM in S + DHA + SFM resulted in a lower production of N- and S-containing compounds compared to P + DHA + PFM. Pyrazines, including 2-ethyl-5-methylpyrazine, 2-ethyl-6-methylpyrazine, 2-ethyl-3,6-dimethylpyrazine, 2-ethyl-3,5-dimethylpyrazine, dimethyl disulphide, and dimethyl trisulphide, were detected in both S + DHA + SFM and P + DHA + PFM. Notably, trimethylpyrazine was exclusively detected in P + DHA + PFM, and at high levels. This result aligns with the high glycine content in PFM, as glycine is a well-established precursor of trimethylpyrazine [29]. TMA was only detected in P + DHA + PFM, likely resulting from betaine degradation.

3.2. Thermal Generation of TMA from Betaine in a Model System

While both TMAO and betaine are precursors of TMA in seafood, only betaine is considered suitable for product formulation due to toxicity concerns associated with TMAO. As shown in Figure 3, TMA was not detected by GC–MS in any samples prior to baking. However, after baking at 200 °C for 10 min, TMA was detected in all six treatment groups. Amongst these, TMA production was most pronounced in the betaine + oil and betaine + FAAs + oil groups, followed by betaine + FAAs. The lowest TMA content was observed in betaine + oils + DHA, which was even lower than that of betaine alone. The inclusion of FAAs (betaine + FAAs + oils + DHA) had higher TMA production compared to betaine + oils + DHA.

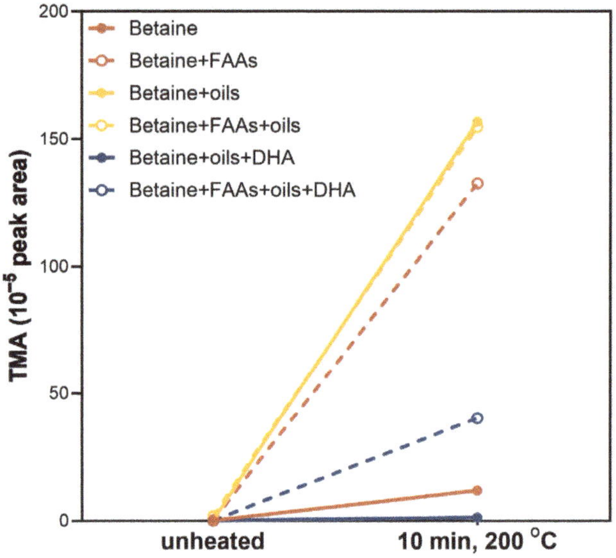

Figure 3. Trimethylamine (TMA) formation through pyrolysis of betaine at 200 °C for 10 min with different ingredients. Treatment groups include betaine alone (betaine, red solid line), betaine with FAA mix (betaine + FAAs, red dashed line), betaine with sunflower and coconut oils (betaine + oils, yellow solid line), betaine with FAA mix and oils (betaine + FAAs + oils, yellow dashed line), betaine with oils and encapsulated algal DHA (betaine + oils + DHA, blue solid line), and betaine with FAA mix, oils, and encapsulated algal DHA (betaine + FAAs + oils + DHA, blue dashed line). TMA content is expressed as 10^{-5} peak area.

Suuronen, Pitkänen [30] reported the pyrolysis of betaine using a filament pyrolyser coupled to GC–MS, noting that TMA production only occurred when the temperature exceeded 242 °C. Our results indicated that thermal degradation of betaine to form TMA can occur at 200 °C after 10 min of heating, which is typical of home baking or frying conditions. This suggests that betaine could be utilised to produce TMA after appropriate thermal processing of PBSAs to create a characteristic fish odour. Furthermore, our results indicated that the presence of FAAs enhances betaine pyrolysis, while DHA appears to inhibit the reaction. However, since TMA is primarily formed through the direct reductive cleavage of betaine, the specific roles of FAAs and DHA in this reaction remain unclear and require further investigation.

3.3. Presence of Key Prawn and Salmon Odourants in the PB Prototypes

As highlighted in Figure 2, a range of key prawn and salmon odourants, previously identified as abundant and essential in shaping typical seafood characteristics, were detected in the PB prototypes [17]. For instance, key salmon odourants such as butanoic acid (*sweaty*), (*E*)-penten-1-ol (*mushroom*), 1-octen-3-ol (*mushroom, green*), hexanal (*green, grass*), heptanal (*green, floral*), (*E*)-2-heptenal (*grass, roasted*), nonanal (*marine, plastic*), (*E*)-2-octenal (*nutty, fatty*), (*E,E*)-2,4-heptadienal (*fatty, mushroom*), benzaldehyde (*nutty, roasted*), (*E*)-2-nonenal (*earthy, green*), (*E,E*)-2,4-decadienal (*fatty, plastic*), 2,3-pentanedione (*butter, caramel*), and 1-octen-3-one (*mushroom, green*) were present. These odourants are known to differentiate cooked salmon from other seafood species [8]. Similarly, typical prawn odourants, including 2-pentylfuran (*green, grass*), TMA (*fishy*), 2-acetyl-1-pyrroline (*roasted, popcorn*), and dimethyl trisulphide (*vegetable, sulphur*), were also detected. The presence of these compounds demonstrates the potential of these flavour formulations to impart typical prawn and salmon flavour attributes to the PB matrix.

A subset of key odourants for prawn and salmon was not detected in the samples. Bromophenols, which are abundant in prawns, particularly wild-caught ones, may play a role in imparting a pleasant marine odour. These compounds are accumulated in prawns through their dietary intake of algae. While the DHA used in our formulation was derived from algae, no other algal ingredients were included in the prototypes, which likely explains the absence of bromophenols. Furthermore, Luo, Frank [8] reported the presence of 2,4-dithiapentane (*truffle*) and indole (*faecal, musty*) in prawns. The occurrence of these compounds in prawns is likely due to microbial or enzymatic activities that lead to the decomposition of amino acids such as cysteine, methionine, and tryptophan [31,32]. However, such reactions are unlikely to have occurred during the formulation of the prawn prototypes. Furthermore, two key salmon odourants, (*E*,*Z*)-2,6-nonadienal (*cucumber, floral*) and (*E*)-3-hexen-1-ol (*fresh, green*), reported in our previous study [8], were not detected. Since these compounds are derived from PUFA degradation, their absence suggests that authentic seafood possesses a more complex PUFA profile, which facilitates the production of these characteristic odourants.

3.4. Odour Qualities of Prototypes

The odour quality of the final prawn (P + DHA + PFM) and salmon (S + DHA + SFM) prototypes was evaluated through olfactory assessment, with detailed results provided in Supplementary Materials S2. Samples in their TS and FTS states, representing typical shelf and pre-consumption statuses, were selected for the assessment. For the prawn prototypes (Figure 4A), a significantly ($p < 0.05$) stronger grilled note was perceived after frying, aligning with the observed increase in N- and S-containing compounds associated with grilled aromas in the volatile analysis. Interestingly, a significant decrease in the strength of marine odour was also noted after frying. Since the formulation did not include typical marine odour-contributing compounds (e.g., bromophenols) apart from algal DHA, this decrease might be attributed to the odour-masking effect of the grilled notes. For the salmon prototypes (Figure 4B), no statistically significant differences in the assessed odour attributes were observed, which could be explained by the relatively similar volatile profiles of S + DHA + SFM before and after frying (Figure 1). Overall, for both prawn and salmon prototypes, the changes in odour quality before and after frying were minimal. This aligns with previous findings indicating that the volatile profile changes in commercial PB seafood alternatives before and after cooking are minor compared to authentic seafood [17].

When comparing the two final formulations after frying (Figure 4C), most odour attributes showed similar perceived intensities. However, the prawn prototype (P + DHA + PFM) exhibited significantly ($p < 0.05$) stronger cooked crustacean and grilled notes. This suggests that fortifying the formulation with FAAs and betaine can effectively enhance crustacean-like flavour attributes in PB products. On the other hand, a significantly stronger cooked salmon odour was not perceived in the DHA-fortified formulation (S + DHA + SFM). This finding may imply that the characteristic salmon odour profile involves a more complex combination of odourants, beyond those derived solely from algal n-3 PUFAs.

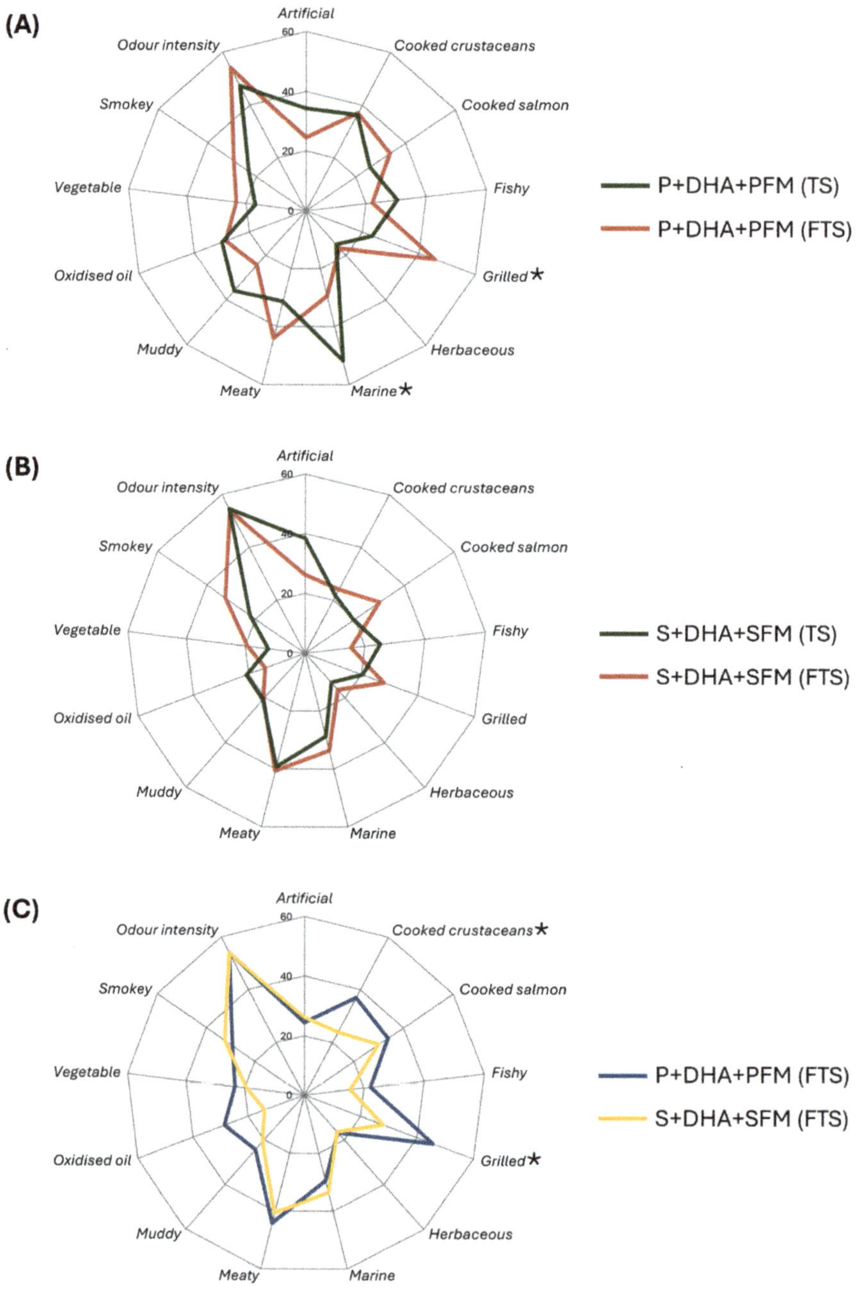

Figure 4. Comparison between odour perception of thermal stablised (TS, green) and fried thermal stablised (FTS, red) P + DHA + PFM prototypes (**A**), TS (green) and FTS (red) S + DHA + SFM prototypes (**B**), and FTS P + DHA + PFM (blue) and FTS S + DHA + SFM prototypes (yellow) (**C**). Significant ($p < 0.05$) differences in perception were tested using the Mann–Whitney test and are marked with "*".

4. Conclusions

Commercial PB seafood flavourings, primarily formulated with plant- and yeast-derived ingredients, struggle to replicate authentic seafood flavours, which significantly contributes to the often-disappointing flavour quality of commercial PBSAs. In this study, we examined the feasibility of a flavouring strategy for PB prawn and salmon products. Using natural levels of n-3 PUFAs, FAAs, and betaine as references, different formulations were designed. Volatile profiles of the prototypes were analysed through HS-SPME-GC–MS, and odour quality was evaluated via olfactory assessment.

Our findings reveal significant differences in the volatile compositions of PB prawn and salmon formulations, depending on the combination of DHA, FAAs, and betaine. A range of key prawn and salmon odourants was detected in prototypes fortified with all three flavour precursors. Notably, this study also indicated the potential production of TMA through the pyrolysis of betaine under moderate cooking conditions. The development of cooked crustacean odours in the final prawn prototype after frying was validated by a semi-trained panel, demonstrating the potential of this flavouring strategy in producing PB prawn products with flavour attributes that more closely resemble authentic prawn. While the final salmon prototype contained key salmon odourants, the cooked salmon odour was not pronounced. Perhaps, fortifying the PB matrix with a more robust n-3 PUFA profile that better mimics authentic salmon, for example, including eicosapentaenoic acid, could better enhance its flavour quality.

This flavouring strategy could also be extended to mimic other popular PBSAs, such as PB scallop and squid, whose key flavourants and flavour precursors have been well-studied. Additionally, future research could focus on incorporating other odourants that, although present in trace amounts, play a critical role in shaping characteristic aromas. For instance, a more pronounced marine odour, attributed to bromophenols, could be achieved by incorporating seaweed or algal oil enriched in bromophenols in the formulation.

Supplementary Materials: The following supporting information can be downloaded at: https://www.mdpi.com/article/10.3390/foods14050820/s1, Supplementary Materials S1: compound list; Supplementary Materials S2: Olfactory evaluation; Supplementary Table S1: Composition of model reaction mixes used in the betaine pyrolysis trial.

Author Contributions: J.L.: conceptualisation, formal analysis, investigation, methodology, visualisation, writing—original draft, writing—review and editing. D.F.: Conceptualisation, funding acquisition, methodology, resources, writing—original draft, writing—review and editing. J.A.: conceptualisation, resources, writing—review and editing. All authors have read and agreed to the published version of the manuscript.

Funding: This work was supported, in whole, by The Good Food Institute, Inc. under GFI grant number 22-PB-AU-FCC-1-361.

Institutional Review Board Statement: Sensory panel recruitment and the use of collected data for analysis were approved by the UNSW Human Research Ethics Committee (ethics approval number: 9628).

Informed Consent Statement: Informed consent was obtained from all subjects involved in the study.

Data Availability Statement: The original contributions presented in the study are included in the article/Supplementary Materials. Further inquiries can be directed to the corresponding author.

Acknowledgments: We acknowledge support from Udayasika Piyasiri, Thomas Hauswald, and Marine Girardot in sample preparation and analysis.

Conflicts of Interest: Damian Frank is employed by All G. The remaining authors declare that the research was conducted in the absence of any commercial or financial relationships that could be construed as a potential conflict of interest.

References

1. Allied Market Research. Plant-Based Seafood Market. Global Opportunity Analysis and Industry Forecast, 2021–2030. 2022. Available online: https://www.alliedmarketresearch.com/plant-based-seafood-market-A17387 (accessed on 9 September 2024).
2. Giacalone, D.; Clausen, M.P.; Jaeger, S.R. Understanding barriers to consumption of plant-based foods and beverages: Insights from sensory and consumer science. *Curr. Opin. Food Sci.* **2022**, *48*, 100919. [CrossRef]
3. Nishimura, T.; Kato, H. Taste of free amino acids and peptides. *Food Rev. Int.* **1988**, *4*, 175–194. [CrossRef]
4. Duppeti, H.; Manjabhatta, S.N.; Martin, A.; Kempaiah, B.B. Effects of different processing methods on the biochemical composition, color and non-volatile taste active compounds of whiteleg shrimp (*Litopenaeus vannamei*). *Food Chem. Adv.* **2022**, *1*, 100118. [CrossRef]
5. Liu, L.; Zhao, Y.; Lu, S.; Liu, Y.; Xu, X.; Zeng, M. Metabolomics investigation on the volatile and non-volatile composition in enzymatic hydrolysates of Pacific oyster (*Crassostrea gigas*). *Food Chem. X* **2023**, *17*, 100569. [CrossRef]
6. Duan, Z.; Dong, S.; Sun, Y.; Dong, Y.; Gao, Q. Response of Atlantic salmon (*Salmo salar*) flavor to environmental salinity while culturing between freshwater and seawater. *Aquaculture* **2021**, *530*, 735953. [CrossRef]
7. Whitfield, F.B.; Mottram, D.S. Volatiles from interactions of Maillard reactions and lipids. *Crit. Rev. Food Sci. Nutr.* **1992**, *31*, 1–58. [CrossRef]
8. Luo, J.; Frank, D.; Peterson, A.L.; Nijagal, B.; Arcot, J. Metabolomic characterisation and flavour profiles of prawn, scallop, squid, barramundi, Salmon, snapper, and tuna. *Food Chem. X* **2025**, *26*, 102284. [CrossRef]
9. Mall, V.; Schieberle, P. Characterization of key aroma compounds in raw and thermally processed prawns and thermally processed lobsters by application of aroma extract dilution analysis. *J. Agric. Food Chem.* **2016**, *64*, 6433–6442. [CrossRef]
10. Mall, V.; Schieberle, P. Evaluation of Key Aroma Compounds in Processed Prawns (Whiteleg Shrimp) by Quantitation and Aroma Recombination Experiments. *J. Agric. Food Chem.* **2017**, *65*, 2776–2783. [CrossRef]
11. Adams, A.; De Kimpe, N. Chemistry of 2-acetyl-1-pyrroline, 6-acetyl-1, 2, 3, 4-tetrahydropyridine, 2-acetyl-2-thiazoline, and 5-acetyl-2, 3-dihydro-4 H-thiazine: Extraordinary Maillard flavor compounds. *Chem. Rev.* **2006**, *106*, 2299–2319. [CrossRef]
12. Sarower, M.G.; Hasanuzzaman, A.F.M.; Biswas, B.; Abe, H. Taste producing components in fish and fisheries products: A review. *Int. J. Food Ferment. Technol.* **2012**, *2*, 113. [CrossRef]
13. Kelly, R.H.; Yancey, P.H. High contents of trimethylamine oxide correlating with depth in deep-sea teleost fishes, skates, and decapod crustaceans. *Biol. Bull.* **1999**, *196*, 18–25. [CrossRef] [PubMed]
14. Heijthuijsen, J.; Hansen, T. Betaine fermentation and oxidation by marine Desulfuromonas strains. *Appl. Environ. Microbiol.* **1989**, *55*, 965–969. [CrossRef]
15. Cadwallader, K.R.; Tan, Q.; Chen, F.; Meyers, S.P. Evaluation of the aroma of cooked spiny lobster tail meat by aroma extract dilution analysis. *J. Agric. Food Chem.* **1995**, *43*, 2432–2437. [CrossRef]
16. Varlet, V.; Knockaert, C.; Prost, C.; Serot, T. Comparison of odor-active volatile compounds of fresh and smoked salmon. *J. Agric. Food Chem.* **2006**, *54*, 3391–3401. [CrossRef]
17. Luo, J.; Frank, D.; Arcot, J. Consumer Perceptions and Volatile Profiles of Plant-Based Seafood Alternatives and Flavouring Agents. *SSRN* **2025**. [CrossRef]
18. Yuan, J.; Qin, F.; He, Z.; Zeng, M.; Wang, Z.; Chen, J. Influences of spices on the flavor of meat analogs and their potential pathways. *Foods* **2023**, *12*, 1650. [CrossRef]
19. Wang, Y.; Tuccillo, F.; Lampi, A.M.; Knaapila, A.; Pulkkinen, M.; Kariluoto, S.; Coda, R.; Edelmann, M.; Jouppila, K.; Sandell, M. Flavor challenges in extruded plant-based meat alternatives: A review. *Compr. Rev. Food Sci. Food Saf.* **2022**, *21*, 2898–2929. [CrossRef]
20. Wang, Y.; Cai, W.; Li, L.; Gao, Y.; Lai, K.-h. Recent advances in the processing and manufacturing of plant-based meat. *J. Agric. Food Chem.* **2023**, *71*, 1276–1290. [CrossRef]
21. Li, X.; Li, J. The flavor of plant-based meat analogues. *Cereal Foods World* **2020**, *65*, 40. [CrossRef]
22. Zioga, E.; Tøstesen, M.; Madsen, S.K.; Shetty, R.; Bang-Berthelsen, C.H. Bringing plant-based Cli-meat closer to original meat experience: Insights in flavor. *Future Foods* **2022**, *5*, 100138. [CrossRef]
23. Vázquez-Ortiz, F.; Caire, G.; Higuera-Ciapara, I.; Hernández, G. High performance liquid chromatographic determination of free amino acids in shrimp. *J. Liq. Chromatogr. Relat. Technol.* **1995**, *18*, 2059–2068. [CrossRef]
24. Zhang, Y.H.; Chen, S.S.; Qiu, W.Q.; Cheng, S.K. Comparative Analysis of Free Amino Acids in Parapenaeopsis hardwickii, Penaeus vannamei and Macrobrachium nipponensis. *Adv. Mater. Res.* **2013**, *781*, 1528–1533. [CrossRef]
25. Kani, Y.; Yoshikawa, N.; Okada, S.; Abe, H. Taste-active components in the mantle muscle of the oval squid Sepioteuthis lessoniana and their effects on squid taste. *Food Res. Int.* **2008**, *41*, 371–379. [CrossRef]
26. Stadtman, E.; Levine, R. Free radical-mediated oxidation of free amino acids and amino acid residues in proteins. *Amino Acids* **2003**, *25*, 207–218. [CrossRef]
27. Matsui, R.; Honda, R.; Kanome, M.; Hagiwara, A.; Matsuda, Y.; Togitani, T.; Ikemoto, N.; Terashima, M. Designing antioxidant peptides based on the antioxidant properties of the amino acid side-chains. *Food Chem.* **2018**, *245*, 750–755. [CrossRef]

28. Meucci, E.; Mele, M. Amino acids and plasma antioxidant capacity. *Amino Acids* **1997**, *12*, 373–377. [CrossRef]
29. Guerra, P.V.; Yaylayan, V.A. Double Schiff base adducts of 2, 3-butanedione with glycine: Formation of pyrazine rings with the participation of amino acid carbon atoms. *J. Agric. Food Chem.* **2012**, *60*, 11440–11445. [CrossRef]
30. Suuronen, J.; Pitkänen, I.; Halttunen, H.; Moilanen, R. Formation of the main gas compounds during thermal analysis and pyrolysis: Betaine and betaine monohydrate. *J. Therm. Anal. Calorim.* **2002**, *69*, 359–369. [CrossRef]
31. Chang, O.; Cheuk, W.L.; Nickelson, R.; Martin, R.; Finne, G. Indole in shrimp: Effect of fresh storage temperature, freezing and boiling. *J. Food Sci.* **1983**, *48*, 813–816. [CrossRef]
32. Mustafa, A.M.; Angeloni, S.; Nzekoue, F.K.; Abouelenein, D.; Sagratini, G.; Caprioli, G.; Torregiani, E. An overview on truffle aroma and main volatile compounds. *Molecules* **2020**, *25*, 5948. [CrossRef] [PubMed]

Disclaimer/Publisher's Note: The statements, opinions and data contained in all publications are solely those of the individual author(s) and contributor(s) and not of MDPI and/or the editor(s). MDPI and/or the editor(s) disclaim responsibility for any injury to people or property resulting from any ideas, methods, instructions or products referred to in the content.

molecules

MDPI

Article

Exploring the Aroma Fingerprint of Various Chinese Pear Cultivars through Qualitative and Quantitative Analysis of Volatile Compounds Using HS-SPME and GC×GC-TOFMS

Wenjun Zhang [1,2], Mengmeng Yan [1,2], Xinxin Zheng [3], Zilei Chen [1,2], Huidong Li [1,2], Jiangsheng Mao [1,2], Hongwei Qin [1,2], Chao Zhu [1,2], Hongxia Du [1,2,*] and A. M. Abd El-Aty [4,5]

1 Institute of Quality Standard and Testing Technology for Agro-Products, Shandong Academy of Agricultural Sciences, Ji'nan 250100, China; zipingguozhang@163.com (W.Z.); ynky202@163.com (M.Y.); czl7274@163.com (Z.C.); lihuidong8066@163.com (H.L.); maojiangsheng@163.com (J.M.); qhw01@163.com (H.Q.); ndytzhuchao@126.com (C.Z.)
2 Shandong Provincial Key Laboratory Test Technology on Food Quality and Safety, Ji'nan 250100, China
3 School of Bioengineering, Qilu University of Technology (Shandong Academy of Sciences), Ji'nan 250353, China; m17861407185@163.com
4 Department of Pharmacology, Faculty of Veterinary Medicine, Cairo University, Giza 12211, Egypt; abdelaty44@hotmail.com
5 Department of Medical Pharmacology, Medical Faculty, Ataturk University, Erzurum 25240, Turkey
* Correspondence: dhx-74@163.com

Abstract: To comprehensively understand the volatile compounds and assess the aroma profiles of different types of *Pyrus ussuriensis* Maxim. Anli, Dongmili, Huagai, Jianbali, Jingbaili, Jinxiangshui, and Nanguoli were detected via headspace solid phase microextraction (HS-SPME) coupled with two-dimensional gas chromatography/time-of-flight mass spectrometry (GC×GC-TOFMS). The aroma composition, total aroma content, proportion and number of different aroma types, and the relative quantities of each compound were analyzed and evaluated. The results showed that 174 volatile aroma compounds were detected in various cultivars, mainly including esters, alcohols, aldehydes, and alkenes: Jinxiangshui had the highest total aroma content at 2825.59 ng/g; and Nanguoli had the highest number of aroma species detected at 108. The aroma composition and content varied among pear varieties, and the pears could be divided into three groups based on principal component analysis. Twenty-four kinds of aroma scents were detected; among them, fruit and aliphatic were the main fragrance types. The proportions of aroma types also varied among different varieties, visually and quantitatively displaying changes of the whole aroma of the different varieties of pears brought by the changes in aroma composition. This study contributes to further research on volatile compound analysis, and provides useful data for the improvement of fruit sensory quality and breeding work.

Keywords: *Pyrus ussuriensis* Maxim; aroma characteristics; aroma profiles; GC×GC-TOFMS; HS-SPME; quality evaluation

1. Introduction

Pyrus ussuriensis Maxim is a pear species native to northeastern China that is known for its cold and drought resistance and the ability to grow in barren areas [1]. It is the primary pear cultivated in cold regions of China, with oblate or nearly round fruits that can be green, yellow, or yellowish-green, and red on the sunny side. Nanguoli, the most commonly grown cultivar, covers approximately 72,000 hectares of land and yields approximately 400,000 tons in China [2]. The fruits of *Pyrus ussuriensis* Maxim typically require postripening to develop a rich and excellent flavor [3]. Varieties such as Nanguoli and Jianbali are also suitable for producing high-quality frozen pears with good sour and sweet flavors [4]. In addition to its distinctive quality [5], *Pyrus ussuriensis* Maxim has various beneficial properties, such as acaricidal, antioxidant, and anti-inflammatory effects [6–8], which can help alleviate cough, gout, and rheumatoid arthritis. Researchers have found that *Pyrus ussuriensis* Maxim contains higher levels of polyphenols and triterpenic acids than other pear species [9], making it an excellent candidate for breeding purposes.

The scent of fruit is a crucial factor that affects the overall aroma and consumer preference [10]. It originates from the secondary metabolites of plants, which is an essential part of fruit quality and reflects the differences between species and varieties [11]. The analysis and evaluation of fruit aroma can aid in variety identification, quality assessment, breeding, and fruit processing [12–15]. Numerous studies have explored the composition of volatile aroma in pears and have shown that the aroma of pears is made up of esters, aldehydes, alkenes, alkanes, and other compounds [13]. Various factors, such as pollination, bagging, harvesting period, storage conditions, and calcium regulation, can impact pear aroma [16–19]. Regarding *Pyrus ussuriensis* Maxim aroma evaluation, some scholars have evaluated the types, quantity, and volatile content of the aroma in *Pyrus ussuriensis* Maxim, including Nanguoli, Huaguai, and Jianbali [13,20,21], mainly using headspace solid-phase microextraction (HS-SPME) in combination with gas chromatography-mass spectrometry (GC-MS) technology. Some researchers have also studied the physiological mechanism of aroma formation and release in "Nanguoli" fruit and identified relevant candidate genes via calcium regulation, providing valuable data for further research into the molecular mechanism of pear fruit aroma regulation [19]. However, the number of aroma components reported in *Pyrus ussuriensis* Maxim is still limited, and the analysis of aroma characteristics has not been adequately reported.

The comprehensive two-dimensional gas chromatography/time-of-flight mass spectrometry (GC×GC-TOFMS) technique was developed in the 1990s to identify volatile substances in a qualitative and quantitative manner. Compared to gas chromatography, GC×GC-TOFMS offers several advantages, including detecting more compounds, having a prominent peak capacity, good sensitivity, high resolution, and fast analysis [22]. Compared to QMS, TOFMS is especially useful in identifying volatile and semi-volatile compounds [23] and providing a comprehensive volatile fingerprint with its higher data acquisition frequency. As a result of these benefits, GC×GC-TOFMS has been widely used in the analysis of wine [24], cloud water [25], and green tea [26]. However, it has not yet been used to analyze *Pyrus ussuriensis* Maxim. Therefore, in order to comprehensively understand and evaluate the aroma quality of *Pyrus ussuriensis* Maxim, this study adopts the HS-SPME combined with GC×GC-TOFMS to demonstrate aroma fingerprint information of Anli, Dongmili, Huagai, Jianbali, Jingbaili, Jinxiangshui, Nanguoli. This study will characterize comprehensively volatile compounds, and provide more information on *Pyrus ussuriensis* Maxim aroma fingerprint. A comprehensive evaluation will be performed for each variety for the aspect of the concentration and number of volatile substances, volatile aroma biosynthetic, and aroma characteristic percentage distributions. Ultimately, we expect to provide valuable data for promoting pear fruit aroma quality, deep processing, and genetic breeding.

2. Results

2.1. Analysis of Aroma Composition in Different Pear Fruits

The GC×GC-TOFMS total ion flow diagram of aroma components of different cultivars pears is shown in Figure 1. After retrieving and analyzing data from the NIST/WILEY spectrum database, the main aroma components of pears were identified and are listed in Table S1. A total of 174 volatile organic compounds were identified in the seven pear cultivars, comprising 102 esters, 14 alcohols, 20 aldehydes, 5 ketones, 18 alkenes, 8 alkanes, and 7 other substances. Hexyl acetate, ethyl acetate, butyl acetate, ethyl caproate, methyl caproate, methyl acetate, ethyl butyrate, ethyl caprylate, ethyl 2-methyl butyrate, octyl acetate, (*Z*)-3-hexenyl acetate, heptyl acetate, pentyl acetate, ethyl phenylacetate, 1-hexanol, 1-octanol, (*Z*,*E*)-α-farnesene, and hexanal were detected in all seven pear cultivars, in relatively high contents. Among the pear cultivars, Jinxiangshui had the highest total content of aroma substances at 2825.59 ng/g, which was 1.5–5 times higher than those of the other six pear cultivars. Anli (1877.71 ng/g), Dongmili (1799.46 ng/g), and Jianbali (1158.07 ng/g) had higher total contents of aroma substances than Nanguoli (657.67 ng/g), Huagai (608.53 ng/g), and Jingbaili (568.84 ng/g). Nanguoli had the largest number of aroma species (108), followed by Jiangubali (93) and Huagai (91), while Dongmili, Jinxiangshui, and Anli had the smallest numbers of aroma species (66, 64, and 64, respectively).

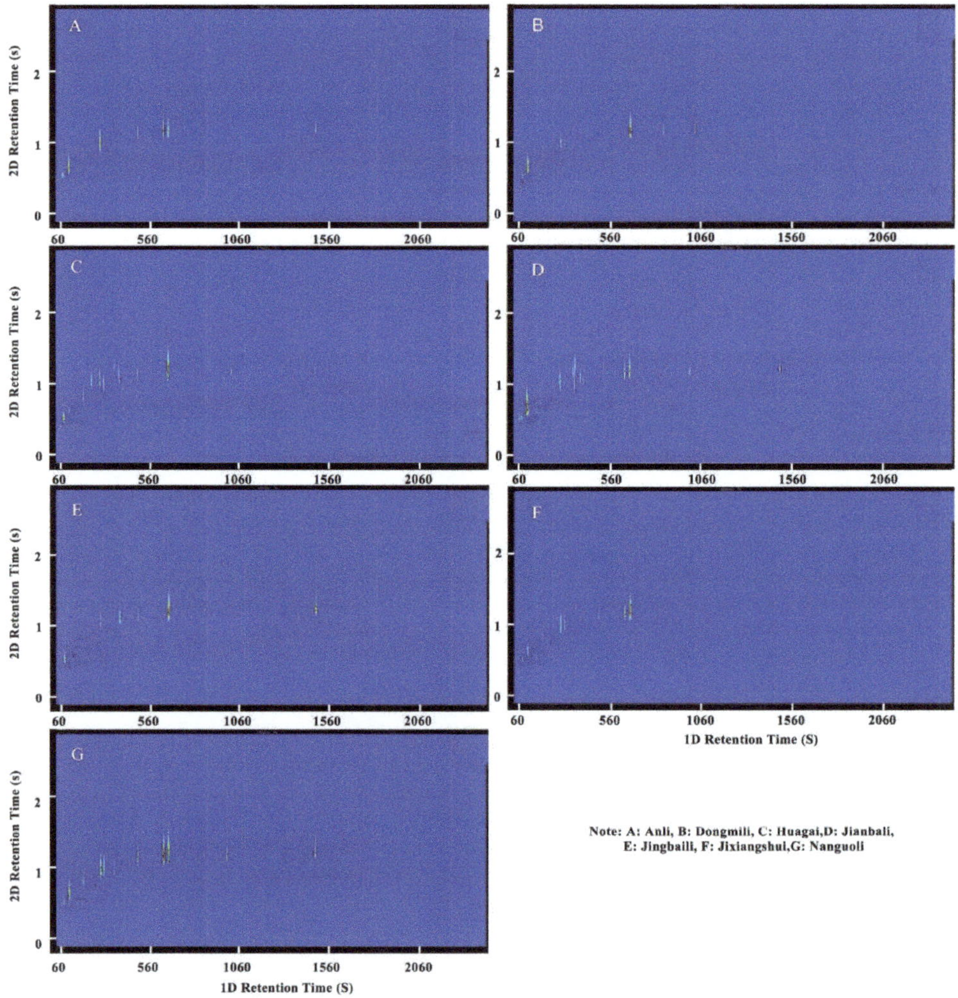

Figure 1. Total ion chromatograms of the aromatic components of different pear cultivars. Note: (**A**): aroma fingerprint of Anli; (**B**): aroma fingerprint of Dongmili; (**C**): aroma fingerprint of Huagai; (**D**): aroma fingerprint of Jianbali; (**E**): aroma fingerprint of Jingbaili; (**F**): aroma fingerprint of Jinxiangshui; (**G**): aroma fingerprint of Nanguoli.

Esters were found to be the main aroma substances in the seven types of pears by comparing the number and relative percentages of different aroma types. The number of esters identified ranged from 40 to 72 (as shown in Figure 2), with the relative content percentage accounting for 56.68–94.92% (as shown in Figure 3). Although only 48 types of esters were identified in Jinxiangshui, its content was the highest, and the total amount of aroma substances was also higher than the other six types of pears (Table S1). This suggests that the relative content of primary esters in Jinxiangshui was generally higher than that in the other six types of pears, which was further supported by the clustering heatmaps (as shown in Figure 4). The first cluster in Figure 4 shows that hexyl acetate, ethyl acetate, ethyl butyrate, ethyl caproate, methyl caproate, methyl acetate, and ethyl butyrate in Jinxiangshui were significantly higher than in the other six types of pears. These aroma components were also the main ester substances in all pear cultivars, as shown in Table S1.

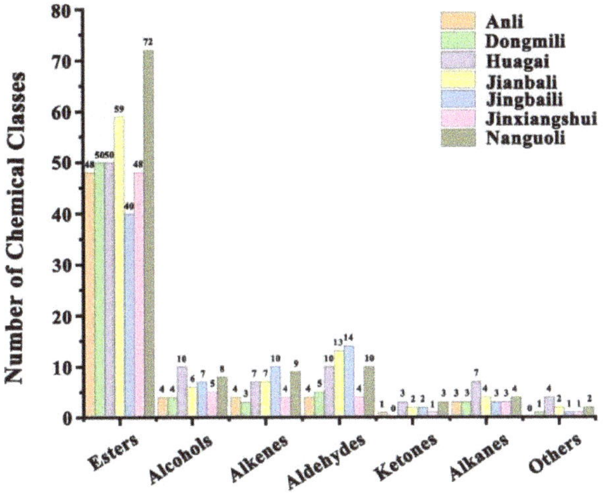

Figure 2. The number of different types of aroma substances in pears.

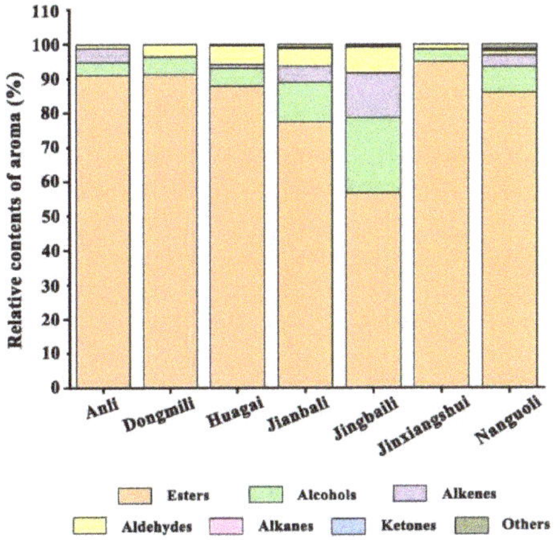

Figure 3. Relative content proportions of different volatile compounds in seven pear cultivars.

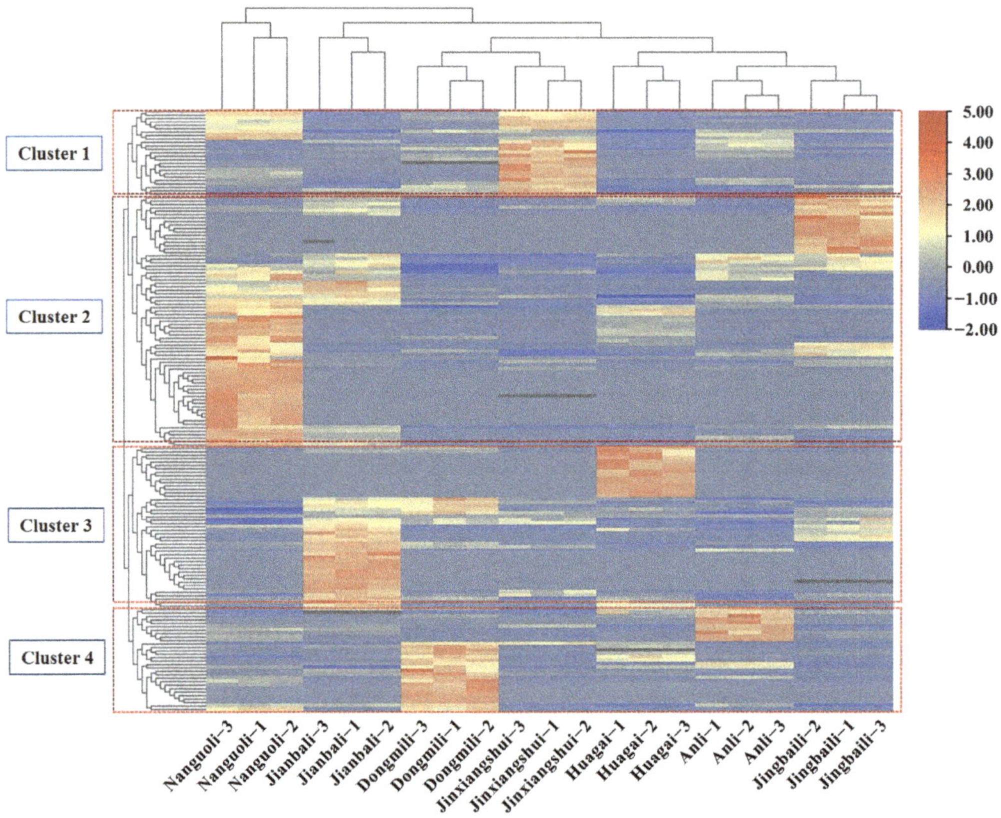

Figure 4. Cluster heatmap analysis of volatile organic compounds in 7 pear cultivars.

The relative percentages of esters in Anli, Dongmili, Huagai, and Nanguoli ranged from 85% to 91%, while in Jianbali and Jingbaili, these were less than 80%. Dongmili had the second-highest percentage of esters (91.17%) after Jinxiangshui, and the number of ester species was 50. Butyl acetate, ethyl acetate, hexyl acetate, ethyl crotonate, methyl acetate, and heptyl acetate were the main ester components of Dongmili. Heptyl acetate and isobutyl acetate had the highest relative content in Dongmili, according to Figure 4 and Table S1. Anli had a percentage of esters of 90.96%, and the number of species was the same as that of Jinxiangshui. Hexyl acetate, ethyl acetate, butyl acetate, ethyl caproate, methyl caproate, methyl crotonate, methyl acetate, and ethyl butyrate were the main ester components of Anli. The relative content of hexyl acetate was significantly higher in Anli than in the other six types of pears (Figure 4 and Table S1). Huagai had a percentage of esters of 87.80%, and the number of species was the same as that of Dongmili. The primary esters were hexyl acetate, butyl acetate, and (*Z*)-hex-2-enyl acetate, with the highest relative content of (*Z*)-hex-2-enyl acetate in Huagai, as shown in Figure 4 and Table S1. Nanguoli had the largest number of ester species at 72, but the relative proportion was 85.87%. The main ester components were ethyl acetate, butyl acetate, methyl acetate, methyl butyrate, ethyl caprylate, ethyl 2-methyl butyrate, hex-2-enyl acetate, and methyl octanoate. The relative contents of ethyl octylate and methyl octylate were the highest in Nanguoli, as shown in Figure 4 and Table S1. Jianbali had a lower proportion of esters (77.42%) and contained 59 esters. The main ester components were hexyl acetate, ethyl acetate, isoamyl acetate, ethyl caproate, methyl acetate, butyl acetate, and propyl acetate. Isoamyl acetate was only detected in Jianbali, as shown in Table S1. Jingbaili had the lowest proportion of esters (56.68%) and the lowest number of species (Figure 3).

The percentage of alcohols varied between 3.43% and 21.94%, and the number of species ranged from 4 to 10 (Figures 2 and 3). Ethanol, 1-hexanol, and 1-octanol were present in all seven types of pears and had higher relative contents than other alcohols (Table S1). Jingbaili had the highest alcohol percentage (21.94%); the relative contents of (*E*)-2-hexanol were higher than those of the other six types of pears, and 4-methyl-1-pentanol (26.08 ng/g) was only detected in Jingbaili (Table S1). The second was Jianbali, accounting for 11.41%, of which (±)-2-methyl-1-butanol was only detected (Table S1). Nanguoli, Huaguai, and Dongmili had proportions of alcohols ranging from 7.54% to 5.06%. Only Nanguoli contained 1-nonanol (Table S1), and only Huagai had 3,7,11-trimethyl-6,10-dodecadiene-1-ol and 3-octanol (Table S1). The proportion of alcohols in Anli and Jinxiangshui were all below 4%. (*Z*)-2-hexene-1-ol was only present in these two pear species, with a higher content in Jinxiangshui (5.52 ng/g) (Table S1 and Figure 4).

In the seven types of pears, aldehydes were also found to be critical volatile compounds, accounting for 1.08%–7.56%, and the number of detected species ranged from 4 to 14 (Figures 2 and 3). The most abundant aldehydes detected were hexaldehyde and (*E*)-2-hexenal. Jingbaili had the highest proportion of aldehydes (7.56%) and the largest number of species (14) (Figures 2 and 3). Croton aldehyde was found only in Jingbaili (Table S1). Huagaili (5.45%) and Jianbali (5.11%) were second, with 10 and 13 aldehyde species, respectively (Figures 2 and 3). (*Z*)-3-hexenal and (*E*)-2-heptenal were only detected in Jianbali (Table S1). Dongmili had a lower proportion of aldehydes (3.35%), with only five species identified (Figures 2 and 3). Anli (1.08%), Nanguoli (1.28%), and Jinxiangshui (1.31%) had low aldehyde percentages with 4, 10, and 4 species, respectively (Figures 2 and 3). Although there were many aldehydes in Nanguoli, the content was low, among which 2-methylbut-2-enal and 2-methyl propenal were unique aldehydes in Nanguoli.

Alkenes accounted for 0.18–12.98% in the different types of pears (Figure 3), and the number of alkenes detected varied from 3 to 10 (Figure 2). Among the alkenes, (*Z,E*)-α-farnesene and α-farnesene were found in all seven pear types with higher contents. Additionally, (*Z,Z*)-α-farnesene and (*E*)-β-farnesene, other isomers of the above two substances, were also identified. (*Z,Z*)-α-farnesene was only detected in Nanguoli (Table S1), and (*E*)-β-farnesene was detected in the five other pear species except for Dongmili and Jinxiangshui and had the highest content in Anli (2.91 ng/g). The highest percentage of alkenes (12.98%) was in Jingbaili, which also had the highest number of alkenes (10) (Figures 2 and 3). Anli (4.09%) and Jianbali (4.74%) had alkenes in proportions above 4% (Figure 3), and their respective numbers of alkenes were 4 and 7. Dongmili (0.23%), Huagai (1.04%), Jinxiangshui (0.18%), and Nanguoli (3.35%) all had alkenes below 4%, with the number of species ranging from 3 to 9 (Figures 2 and 3). Similar to the aldehydes, Nanguoli had a large number of alkenes but a low relative content (Figure 2 and Table S1). Furthermore, some alkenes, such as 3-(4-methylpent-3-enyl)furan, (*Z*)-1,3-pentadiene, (*E*)-1,4-hexadiene, (*E*)-3-hexene, and (*E*)-1-methyl-4-(6-methylhept-5-en-2-ylidene) cyclohex-1-ene, were only detected in Jingbaili, Jianbali, and Huagai (Table S1). Additionally, α-curcumene was only found in Jinxiangshui (Table S1). Finally, this study identified three monoterpenes, limonene, β-myrcene, and (*E*)-β-ocimene, and although their concentration was low, they still played an important role in fruit flavor [27]. Limonene was only present in Huagai (0.04 ng/g) and Nanguoli (0.19 ng/g), while β-myrcene (0.04 ng/g) and (*E*)-β-ocimene (0.03 ng/g) were only detected in Nanguoli (Table S1).

The content and species of ketones in seven types of *Pyrus ussuriensis* Maxim were lower; the percentage of ketones ranged from 0.01% to 0.34%. The higher proportions of ketones were found in Jingbaili (0.34%) and Nanguoli (0.30%), which accounted for over 0.3% (Figure 3). Jingbaili had both methyl heptenone and butenone, while methyl heptenone, a crucial volatile component of Fuzhuan tea [28], was detected in all types except Dongmili. Butenone was only present in Jingbaili and Nanguoli. Moreover, Nanguoli had (*E*)-6,10-dimethyl-5,9-undecadien-2-one, which is commonly found in tobacco and tea and has a distinct floral and woody fragrance [29]. The proportions of ketones in Huagai (0.11%) and Jianbali (0.19%) were all above 0.1% (Figure 3), and the number of ketones in pear species was 3 and 2, respectively (Figure 2). Huagai contained methyl heptenone, (*E*)-6,10-dimethyl-5,9-undecadien-2-one, and 2,3-dioctyl ketone, the latter of which was only detected in Huagai. The ketone substances in Jianbali were methyl heptenone and 2,3-butanedione, with the latter being exclusively found in Jianbali (Table S1). The percentages of ketones in Anli (0.06%) and Jinxiangshui (0.01%) were relatively low, accounting for less than 0.1% (Figure 3), and only methyl heptenone was detected (Table S1).

The content of alkanes in seven types of *Pyrus ussuriensis* Maxim was lower, and the proportion ranged from 0.03% to 0.39% (Figure 3). Except for Anli, different types of alkanes were detected in the six other pears (Table S1). Only *o*-xylene was identified in Dongmili, which has a distinctive aromatic odor [30]. Both Jianbali and Nanguoli contained *p*-xylene, a compound with a sweet and fruity aroma [31]. Denderalasin, which has not been previously reported in pear aroma identification, was the only compound belonging to alkanes detected in Huagai.

2.2. Principal Component Analysis

A total of 174 volatile components present in seven types of *Pyrus ussuriensis* Maxim were analyzed by SIMCA14.1 software. The variance contribution rate, cumulative variance contribution rate and characteristic value of each principal component are shown in Table 1. The results indicate that the six principal components, which had eigenvalues greater than 1 and an accumulative variance contribution rate of 97.0%, reflect all the information of the original variable. Based on the identified aroma substances and their relative content in different types of pears, the seven types were categorized into three groups (Figure 5). The gray spheres in Figure 5 represent the aroma substances, and the closer they are to the hexagon representing the samples, the greater the contribution to sample classification. Table S4 indicates the primary substances contributing to the grouping. Group 1 comprises Jianbali and Jingbaili, which were clustered in the first quadrant, and 2-methylbutyl acetate(*E*)-2-hexenoic acid methyl ester, diethyl carbonate, (*E*)-2-hexen-1-ol, ethyl tiglate, and other substances played a crucial role in the grouping. Nanguoli was the only pear type in Group 2, clustered in the third quadrant with some C_6 and C_9 compounds, such as 2-hexenal, (*E*)-2-nonenal, 1-nonanol, some unsaturated esters, alkenes, sulfur-containing ester compounds, hexyl ester compounds, and ethyl compounds, such as (*Z*)-4-decenoic acid methyl ester, limonene, *β*-myrcene, (*E*)-*β*-ocimene, (*Z*,*Z*)-*α*-farnesene, 3-(Methylthio)propanoic acid ethyl ester, hexyl caprylate, and ethyl heptanoate, which significantly affected the separation between Nanguoli and other pear cultivars. Group 3 comprises Anli, Huagai, Dongmili, and Jinxiangshui, clustered in the fourth quadrant, characterized by high concentrations of hexyl acetate, ethyl (2*Z*)-but-2-enoate, (*Z*)-hex-2-enyl acetate, ethyl (*E*)-hex-3-enoate, methyl tiglate, 1-octanol, *n*-heptanol, and other substances.

Table 1. Principal component variance contribution rate.

Component	Proportion (%)	Cumulative (%)	Eigenvalue
1	27.0	27.0	5.67
2	19.6	46.6	4.11
3	15.5	62.1	3.25
4	14.0	76.1	2.94
5	10.9	87.0	2.3
6	9.9	97.0	2.1

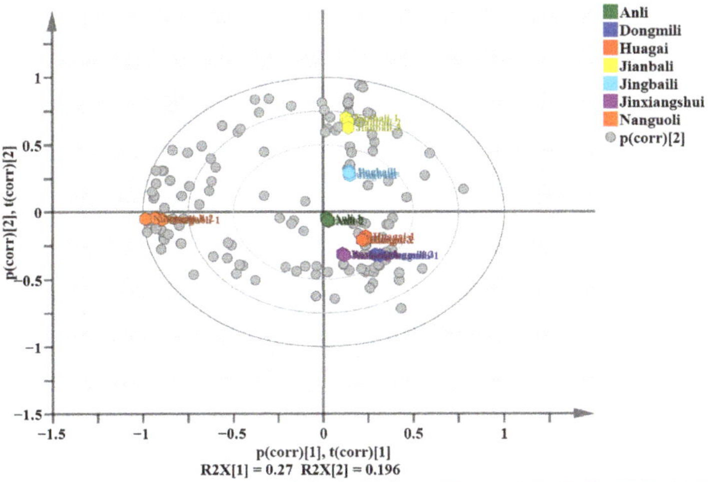

Figure 5. Principal component analysis of volatile organic compounds in 7 pear cultivars.

2.3. Analysis of Aroma Characteristic Percentage Distributions

In seven varieties of pears, a total of 24 types of flavors were identified, with 18 common flavors including aliphatic, ice, citrus, dairy, edible, fruit, green, herb, konifer, it-chem, narcotic, orchid, phenol, rose, animal, wood, earthy, and solvent flavors. Anli, Dongmili, and Jinxiangshui exhibited a higher total distribution values of all aroma characteristics, with Jianbali following closely behind, and the remaining pear varieties had lower values (Table S3). Among all the flavors, fruit aroma, accounting for a significant portion of the total flavor profile (44.12–82.90%), was the most prominent aroma of the seven pear varieties compared with aliphatic aroma (2.79–19.46%), green aroma (5.18–16.76%), dairy aroma (1.81–5.22%), and solvent aroma (1.25–5.67%). Anli had the highest proportion of fruit flavor, followed by Jinxiangshui (76.90%), Huagai (75.94%), Jianbali (71.31%), Dongmili (66.10%), Nanguoli (62.45%), and Jingbaili (44.12%). The main contributors to fruit flavor were short-chain saturated esters, particularly ethyl ester compounds such as hexyl acetate, ethyl acetate, isoamyl acetate, and butyl acetate (Table S1), with the percentage of relative contents of approximately 77.87% and 76.05% in Jinxiangshui and Anli, respectively. The second most abundant aroma types were aliphatic and green. Jingbaili had the highest aliphatic aroma percentage (19.46%), followed by Dongmili (17.38%) and Nanguoli (11.43%), while it was below 10% in the other pear cultivars (Figure 6). Dongmili had a high proportion (75.99%) of saturated esters. However, it also contained a higher proportion of saturated alcohols (5.06%) and aldehydes (3.35%), which significantly contributed to aliphatic and other scents, compared to Jinxinagshui (3.43%, 1.31%) and Anli (3.69%, 1.08%) (Table S1). Therefore, this could reduce the proportion of fruit flavor and the "intensity" of the fruit aroma in Dongmili to some extent. Comparative analysis also revealed that although the total value of aliphatic aroma in Dongmili was significantly larger than that in Jingbaili (Table S3), the proportion of aliphatic aroma in Jingbaili was larger than that in Dongmili. This is because the total aroma value of Dongmili is large, which reduces the proportion of aliphatic aroma. These findings suggest that the combined presence of different aroma types has a significant impact on the overall aroma of pears. The green scent accounted for 5.18–16.76% (Figure 6), contributed by ethyl caproate, methyl caproate, methyl acetate, (*Z*)-3-hexenyl acetate, heptyl acetate, (*E*)-2-hexen-1-ol, hexanal, and some alkene substances. Jingbaili (16.76%) and Dongmili (10.58%) had a higher proportion of green aroma, which was more than 10% of the total aroma characteristics. The percentage of dairy aroma, accounting for 1.81–5.22%, was also significant, mainly contributed by isoamyl acetate, butyl acetate, ethyl caproate, and methyl acetate. Jinxiangshui (5.22%)

and Jianbali (4.46%) had a higher proportion, and the remaining pear species had less than 4% dairy aroma. The percentage of solvent aroma ranged from 1.25% (Dongmili) to 5.67% (Nanguoli) (Figure 6) and was primarily produced by methyl acetate and other short-chain esters, such as *n*-hexanol. The presence of citrus aroma (as in many citrus fruits [32] and green tea [33]), accounting for 0.003% (Jinxiangshui)–0.25% (Jingbaili), was also detected and mainly emitted by (*E*,*Z*)-2,4-decadienoic acid, methyl ester, nonanal, heptanal, and decanal. Similarly, the rose aroma, contributed by phenethyl acetate, diisobutyl phthalate, and *n*-heptanol, accounted for 0.01% (Jinxiangshui)–0.27% (Dongmili). Narcotic flavors, which were reported in *P. communis* L. [34] and were found in seven *Pyrus ussuriensis* Maxim in this study, were present in high proportions in Jinxiangshui, Anli, and Jianbali, accounting for more than 0.5% and contributed by ethyl caproate, ethyl benzoate, methyl benzoate, benzeneacetaldehyde, and benzaldehyde. Conifers and woody aroma types, which are mostly emitted by terpenes [35], were also detected, with Nanguoli (0.55%) having the highest proportion of conifers aroma, and Jingbaili (3.01%), Huagai (2.99%), and Jianbali (1.36%) having the highest proportion of woody aroma.

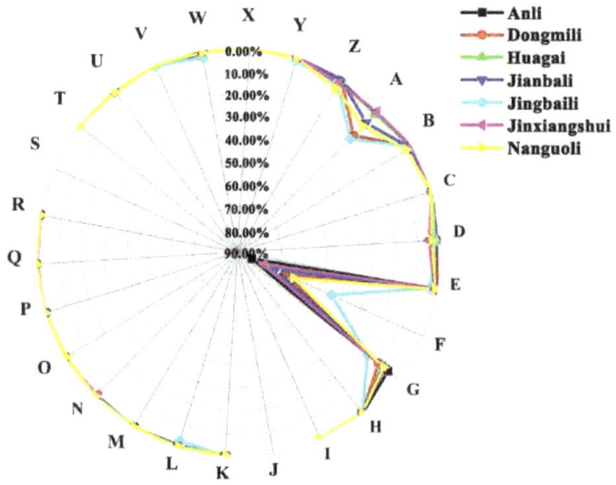

Figure 6. Analysis of aroma characteristic percentage distributions of 7 pear cultivars. Note: A: aliphatic; B: ice; C: citrus; D: dairy; E: edible; F: fruit; G: green; H: herb; I: iris; J: jasmin; K: konifer; L: it-chem; M: muguet; N: narcotic; O: orchid; P: phenol; Q: balsam; R: rose; S: animal; T: smoke; U: animal; W: wood; X: musk; Y: earthy; Z: solvent.

To summarize, Nanguoli were characterized by a dominant fruit and aliphatic aromas, along with some green, dairy, and solvent fragrances. Jinxiangshui, Jiandaili, Huagai, and Anli had a fruit-forward aroma, with some green, aliphatic, dairy, and solvent fragrances. Huagai and Jianbali had traces of it-chen, woody, and solvent aromas. Jingbaili and Dongmili had a complex aroma profile, with a dominant fruit, aliphatic, and green aromas, and with some edible, it-chen, woody, and solvent flavors. The analysis of aroma substance content and aroma type proportions revealed that the aroma of pears was affected not only by the main contributing substances but also by the composition of different aroma types.

Analyzing the number of aroma types is also a crucial aspect of characterizing aroma profiles. This study revealed that Nanguoli had 24 fragrance types, except for jasmine and spice, and had more iris, smoke, and musk than the other six pear species. Iris has been reported in *Pyrus communis* L. [17], and β-myrcene, a monoterpene found in Nanguoli and not presenting in the other pear cultivars (Table S1), is responsible for the iris, burnt tobacco, and musk scents. This explains why Nanguoli has a more diverse flavor profile than other pear species to some extent. Moreover, some fragrances were only detected in a few pear species. For instance, the muguet scent, found in Yuluxiangli [17], was detected in Anli, Dongmili, Huagai, Jianbali, and Nanguoli in this study, primarily contributed by ethyl benzoate and diisobutyl phthalate. Balsam scent, detected in mango [36], was found only in Dongmili (0.16%), Huagai (0.0009%), and Nanguoli (0.004%), primarily contributed by diisobutyl phthalate. Vanilla scent, detected in *Pyrus communis* L. and Yuluxiangli [17], mainly contributed by ethyl phenylacetate and (*E*)-2-hexen-1-ol, was found in all five other pear species, except Dongmili and Jinxiangshui.

2.4. Main Biosynthesis Pathway of Aroma Components

Linear esters, aldehydes, alcohols, and ketones, the main aroma substances of seven pear cultivars, are derived from fatty acids and included in [16]. Short-chain acetate and medium-chain esters (C_6–C_{12}), belonging to linear esters, have relatively high contents in the seven pear varieties, including hexyl acetate (3.05–376.8 ng/g), ethyl acetate (7.14–391.96 ng/g), ethyl caproate (1.72–264.42 ng/g), methyl caproate (9.93–246.00 ng/g), ethyl butyrate (0.22–171.90 ng/g), methyl butyrate (7.88–129.01 ng/g), ethyl caprylate (0.35–42.40 ng/g), and methyl octanoate (4.90–32.96 ng/g) (Table S1). Some of the above compounds give a strong "pear-like" flavor [37]. Volatile substances such as C_6 and C_9 alcohols, C_6 and C_9 aldehydes, have a "green grass" flavor, and are also the intermediate products of the fatty acid oxidation pathway. In this study, the seven pear varieties contained high levels of 1-hexanol (11.40–61.26 ng/g) and hexanal (3.03–65.40 ng/g). 1-nonanol and 2-nonanol were also identified in all seven *Pyrus ussuriensis* Maxim varieties, and nonanal was found in Anli (0.63 ng/g), Huagai (0.12 ng/g), Jianbali (0.52 ng/g), and Jingbaili (0.53 ng/g) (Table S1).

Amino acid-derived compounds include branched-chain compounds, sulfur-containing compounds, and aromatic compounds [17]. In this study, various forms of branched-chain esters, such as 3-methyl-3-buten-1-ol, acetate, propanoic acid, 2-methyl-, 3-hydroxyl-2, 2,4-trimethyl pentyl ester, were detected at concentrations ranging from 0.34 to 1.27 ng/g (Table S1), and ethyl 2-methyl butyrate (36.19 ng/g) had a relatively high content in Jianbali (Table S1). The sulfur-containing ester compounds, such as ethyl 3-(methylthio)-(*E*)-2-propenoate, methyl 3-(methylthio)-(*E*)-2-propenoate-3-(methylthio), and methyl 3-methylthiopropionate, derived from methionine and cysteine [34], were only detected in Huagai and Nanguoli, with contents ranging from 0.21 to 0.68 ng/g. Diisobutyl phthalate, the only one aromatic ester, was detected in Huagai (0.26 ng/g) and Nanguoli (1.03 ng/g).

Carbohydrate-derived volatile substances are also an important group of compounds that can impact the flavor of pears to some extent [38]. Monoterpenoids and sesquiterpenoids were the main alkenes detected in the seven pear types. β-myrcene and (*E*)-β-ocimene, two monoterpene compounds, were identified in Nanguoli, with concentrations ranging from 0.03 ng/g to 0.04 ng/g. (*E,Z*)-2,4-decadienoic acid, methyl ester, and (*E,Z*)-2,4-decadienoic acid ethyl ester were esters composed of monoterpenoids. Among these, (*E,Z*)-2,4-decadienoic acid methyl ester was detected in all pear types except for Jingbaili, with concentrations varying from 0.08 ng/g (Anli) to 1.37 ng/g (Nanguoli). (*E,Z*)-2,4-decadienoic acid ethyl ester, was only identified in Anli, Jianbali, and Nanguoli, with concentrations ranging from 1.13 ng/g to 3.06 ng/g. Five sesquiterpenoids were also identified: α-farnesene, (*Z,E*)-α-farnesene, (*Z,Z*)-α-farnesene, (*E*)-β-farnesene, and limonene. α-farnesene and (*Z,E*)-α-farnesene were present in all seven pear types; α-farnesene had the highest content in Jingbaili (66.11 ng/g) and the lowest content in Dongmili (2.34 ng/g); (*Z,E*)-α-farnesene varied from 0.12 ng/g (Jinxiangshui) to 5.98 ng/g (Nanguoli). (*Z,Z*)-α-farnesene was only found in Nanguoli (0.25 ng/g). In addition to Dongmili and Jinxiangshui, the other five kinds of pear species all contained (*E*)-β-farnesene with contents from 0.14 ng/g to 2.91 ng/g, and limonene was detected in Huagai (0.04 ng/g) and Nanguoli (0.19 ng/g).

Based on the relative contents of the main aroma components in seven pear cultivars, the main aroma synthesis pathway of each pear can be preliminarily deduced. The fatty acid pathway was found to be the main pathway in all seven pear cultivars, with short-chain saturated esters making up over 50% of the aroma compounds in Anli, Dongmili,

Huagai, Jianbali, Jingbaili, Jinshui, and Nanguoli. Other aroma biosynthesis pathways were also detected in the seven *Pyrus ussuriensis* Maxim samples besides the fatty acid synthesis pathway. For instance, the relative contents of amino acid derivatives in Huagai and Jianbali were over 3%, indicating the presence of the amino acid aroma biosynthesis pathway. In Jingbaili and Nanguoli, the relative contents of carbohydrate derivatives were 12.49% and 4.5%, respectively, while it accounted for a relatively lower proportion in the five other pear cultivars.

3. Discussion

A total of 174 volatile organic compounds were identified using HS-SPME in conjunction with GC×GC-TOFMS. Esters, primarily containing short-chain saturated acetates and medium-chain esters, were the dominant aroma components in both species in terms of quantity and proportion. Numerous researchers, including Gaihua Qin [13], Liping Zhang [20], and Jie Li [21], have investigated the aroma of *Pyrus ussuriensis* Maxim. They discovered that esters were the primary aroma compounds, which is consistent with our findings, but the total number of volatile organic compounds detected was only approximately 30 for Nanguoli, Huagai, Jingbaili, and Jianbali. Ethyl 3-methylthiopropionate was discovered by Takeoka in pineapple in 1989 [39] and was also detected in Nanguoli in this study. Furthermore, methyl 3-methylthiopropionate and ethyl 3-methylsulfonamidopropionate were found in Nanguoli. Other specialists found ethyl tiglate, an uncommon plant aroma compound [40], in Nanguoli. In this study, ethyl tiglate was found in all seven pear cultivars, propyl tigelate was found in Dongmili, and methyl tiglate was found in Huagai (Table S1). Compared to other research, this study provides more detailed aroma fingerprint data, offering more extensive theoretical support for analyzing and assessing *Pyrus ussuriensis* Maxim germplasm resources and aroma breeding.

The flavor of pears is strongly influenced by the type and amount of aroma components present, which can be objectively evaluated using sensory tests. Numerous studies have employed sensory evaluation tests to investigate aroma properties in different types of fruits. For example, Hayaloglu et al. analyzed the phenolic compounds and sensory characteristics of 12 sweet cherry varieties [41], while Taiti et al. studied traditional and innovative Japanese plum cultivars grown in Italy [42]. Hernandez et al. explored the nutritional composition and sensory profile of Spanish jujube fruits [43]. Moreover, the critical flavor factors of fruits were investigated by integrating the substance content and sensory evaluation attributes. Sung et al. investigated the relationship between the sensory attributes and chemical components of different mango varieties [44]. Ramirez et al. used a quantitative descriptive-analytical method to investigate the sensory characteristics and factors affecting the freshness of seven watermelon varieties [45], while Siebert et al. studied the volatile compounds in Viognier and Chardonnay wines associated with "stone fruit" aroma properties [46]. Aroma characteristic analysis has been utilized in previous studies to assess fruit flavor attributes, such as Wang's analysis of aroma types in *Pyrus communis* L. [17], and Zhang's study of mango aroma attributes using fragrance characteristic analysis [36]. These studies have provided useful approaches for evaluating fruit flavor quality. This study employs aroma characteristic analysis to illustrate the flavor attributes of *Pyrus ussuriensis* Maxim. The study identifies fruit aroma, aliphatic aroma, and green aroma as the most important aroma types of the seven pears and quantitatively demonstrates the aroma attributes of different *Pyrus ussuriensis* Maxim. Based on the compound information, the main contributors to aroma types were identified. Through comparison, we found that the relative content of contributors to fruit aroma was basically the same, but the flavor characteristics were still different for Jinxiangshui, Anli, and Dongmili. Therefore, it can be concluded that the aroma characteristics were not only related to the content of the main contributors but also closely related to the proportion of aroma types in the total aroma value. Meanwhile, the change in aroma and the reason for aroma difference can be visually and quantitatively displayed according to the aroma characteristics of *Pyrus ussuriensis* Maxim. Thus, the aroma characteristics analysis of *Pyrus ussuriensis* Maxim provides a valuable tool for improving the aroma quality of this species and evaluating germplasm resources. This study also speculates on the major aroma biosynthesis pathways of the seven *Pyrus ussuriensis* Maxim, with the fatty acid oxidation pathway being the most crucial, in addition to the amino acid and carbohydrate biosynthesis pathways. The analysis of the aroma pathway provides some theoretical basis for the subsequent aroma synthesis pathway research on *Pyrus ussuriensis* Maxim.

4. Materials and Methods

4.1. Materials

For this study, seven cultivars of *Pyrus ussuriensis* Maxim were selected for analysis (Table 2 lists the cultivars, their production regions, and harvesting times). The maturity of the pears was determined using standard criteria such as days of growth, appearance, and seed color. All samples were kept refrigerated until treatment. A postripening process was necessary to enhance their flavor and taste. Samples were placed at room temperature before experiments (approximately five days) of this time. In addition, 50/30 μm DVB/CAR/PDMS was supplied by Supelco (Bellefonte, PA, USA). The length of the fiber coating was 1 cm. The internal standard 2-nonanone (>99%) was obtained from Dr. Ehrensorfer (Augsburg, Bavaria, Germany). A mixture of 500 μg/mL C_8–C_{40} *n*-alkanes was purchased from Dr. Ehrenstorfer (Germany), and 2.0 mg/mL C_6–C_{10} *n*-alkanes were obtained from AccuStandard (New Haven, CT, USA).

Table 2. The cultivars, producing areas, and sampling times of 7 pear cultivars.

Cultivars	Producing Region	Sampling Time
Anli	Huludaoshi city of Liaoning Province	27 September 2020
Dongmili	Huludaoshi city of Liaoning Province	27 September 2020
Huagai	Anshanshi city of Liaoning Province	7 October 2020
Jianbali	Huludaoshi city of Liaoning Province	15 September 2020
Jingbaili	Huludaoshi city of Liaoning Province	24 August 2020
Jinxiangshui	Huludaoshi city of Liaoning Province	12 September 2020
Nanguoli	Anshanshi city of Liaoning Province	5 October 2020

4.2. Methods

4.2.1. Preparation of Standard Solution

First, 0.1 g of 2-nonanone (accurate to 0.0001 g) was diluted with 10 mL of methanol (HPLC grade, Thermo Fisher Scientific, Waltham, MA, USA) to prepare a 10,000 mg/L stock solution. From this, a 10 mg/L working solution of 2-nonanone was prepared using methanol (HPLC grade, Thermo Fisher Scientific, Waltham, MA, USA). Then, 500 μg/mL of C_8–C_{40} *n*-alkane mix solutions and 2.0 mg/mL of C_6–C_{10} *n*-alkanes were diluted to 10 mg/L with acetone (Analytical reagent, Sinopharm, China). All standard solutions were stored at −18 °C.

4.2.2. GC×GC-TOFMS Conditions

An Agilent 7890B gas chromatograph (San Jose, California, USA) coupled with a LECO Pegasus 4D-C time-of-flight mass spectrometer (TOFMS) (Saint Joseph, CA, USA) was utilized for the analysis. Two columns were employed for compound separation, a weak polarity Rxi-5MS column (Shimadzu, Kyoto, Japan) (0.25 μm, 30 m × 250 μm) for the first dimension (1D) and a medium polar Rxi-17Sil MS column (Shimadzu, Kyoto, Japan) (0.25 μm, 2 m × 250 μm) for the second dimension (2D). The helium carrier gas (99.999%) was used as a carrier gas at a constant flow rate of 1.4 mL/min, while the split ratio was 10:1. The injection port and transmission line temperatures were 270 °C and 280 °C, respectively. The temperature program started at 40 °C for 2 min, heated to 200 °C at 5 °C/min, then rose at 20 °C/min to 280 °C and kept up for 2 min. During the entire collection, the second column temperature was maintained at 5 °C above the first column temperature, and the modulator temperature exceeded 15 °C compared with that of the second column. The modulation cycle was 3 s, and the thermal pulse time was 0.6 s.

The MS parameters were set as follows: the acquisition delay was 60 s, the acquisition rate was 100 spectra/s, the acquisition voltage was 1450 V, and the electron energy was set to −70 V. The ion source temperature was maintained at 250 °C. The mass spectra were collected over a range of 35–550 amu.

4.2.3. Volatile Extraction

The extraction of isolated volatile organic compounds was performed using HS-SPME. A 50/30 μm DVB/CAR/PDMS SPME fiber was selected based on prior testing and published literature [27] and conditioned according to operational instructions before usage. Each pear sample was quartered following the quartic method, and the flesh without cores was cut into pieces measuring 0.5 cm × 0.5 cm × 0.5 cm. Subsequently, 6.0 g of the mixed sample was placed in a 15 mL headspace bottle, 5 μL of 10 mg/L 2-nonanone was added, and the vial was sealed, allowing it to stand for 10 min. The bottle was then kept at 40 °C for 40 min under thermostatic conditions, after which the SPME fiber was immediately placed into the GC injector for desorption. Each pear variety was tested three times in the experiment.

4.2.4. Analysis of Aroma Characteristic Distributions

The aroma characteristics of seven pear varieties were evaluated using the "Training the ABCs of Perfumery" method [17]. In this method, all aroma compounds were categorized into 26 types and quantified using the "scent ABC". The aroma characteristic percentage (Ac) was calculated using the following formula [36]:

$$A_c = \frac{\sum_i^n (m_i \times I \times P_i)}{T}$$

where A_c represents aroma characteristics (%); m_i represents the relative amount of the aroma component; I represents the intensity of an aroma component compared to linalool with a fixed value of 100; P_i represents the percentage of the odor of the substance in the 26 odors of A–Z (%); n is the number of aroma components with an ABC value; and T is the total distribution value of all aroma characteristics for each pear cultivar.

4.2.5. Data Processing and Statistical Analysis

The NIST/WILEY.11 database was used to search for the mass spectrometry fragments of each chromatographic peak. Volatile aroma substances with matching degrees greater than 700 were selected and quantified using the internal standard method, with 2-nonanone as the standard substance for normalization. The following formula was used for the quantification of aroma substances [27]:

$$C_a = \frac{\frac{PA_a}{PA_{is}} \times C_{is} \times 5\ \mu L}{m}$$

where C_a is the concentration of aroma components (ng/g), PA_a is the peak area of aroma components, PA_{is} is the peak area of the internal standard, C_{is} is the concentration of the internal standard (g/mL), and m is the mass of the sample (g). The concentration of 2-nonanone was 10 mg/L, and the mass of the sample was 6.0 g. The data presented are the means ± standard deviations of three replications.

The retention index was used to further confirm the accuracy of the selected aroma substances with degrees greater than 700. First, five microliters of 10 mg/L n-alkanes were analyzed according to step 2.2.3 to calculate the retention index of each aroma component. Then, the retention index data were confirmed with those searched from the NIST database.

Finally, aroma substances were identified. The formula used to calculate the retention index of each aroma component was followed [27]:

$$RI_a = \frac{RT_a - RT_n}{RT_N - RT_n} 100(N - n) + 100RT_n$$

where RI_a is the retention index of the target compound, a is the target compound, n is the carbon number of the lower normal alkane, N is the carbon number of the higher normal alkane, and RT is the retention time.

SIMCA 14.1 was employed to conduct principal component analysis (PCA) and group pear varieties based on the identified volatile components. Tbtools was utilized to perform heatmap and clustering analysis (HCA) to classify all volatile organic compounds and represent their content through color depth. The data were analyzed using variance analysis in SPSS Statistics 19.0, and the mean values were compared using Tukey's test with a significance level of $p < 0.05$. The results are presented as the means ± SDs with a sample size of 3.

5. Conclusions

The aroma of fruits is a crucial indicator of their quality and can reveal their flavor characteristics and ripeness. This study employed HS-SPME combined with GC×GC-TOFMS to increase the richness of aroma fingerprints in pear. The analysis identified a total of 174 chemical substances, including esters, alcohols, aldehydes, and alkenes, in seven different pear varieties. Based on this information, it was possible to cluster the pear varieties into groups with similar characteristic substances. Through the analysis of aroma characteristics, this study visually demonstrated how changes in the composition of aroma materials could affect the overall aroma quality of the pear. The results suggest that the flavor of pear is influenced not only by the content of major flavor substances but also by the collective effect of all the substances. The conclusion provides a new method and useful data for evaluating pear flavor quality, and the application of the new assessment format can improve the comprehensiveness and rationality of evaluating pear flavor quality, processing quality, and breeding based on flavor characteristics. It is of great significance for research on pear flavor quality and the development of the pear industry.

Supplementary Materials: The following supporting information can be downloaded at: https://www.mdpi.com/article/10.3390/molecules28124794/s1, Table S1: Aroma composition and content of different cultivars of *Pyrus ussuriensis* Maxim; Table S2: "ABC" value of scent of aroma substances in different cultivars of *Pyrus ussuriensis* Maxim; Table S3: Aroma value of different cultivars of *Pyrus ussuriensis* Maxim; Table S4: Primary substances contributing to the grouping.

Author Contributions: W.Z. conceived the experiments, performed the experiments, analyzed the data, and modified the manuscript; X.Z. and H.L. analyzed the data and wrote the manuscript; J.M., H.Q. and C.Z. sampled and preprocessed pears; A.M.A.E.-A. validation, formal analysis, writing—review and editing. Z.C. and M.Y. helped revise the manuscript; H.D. helped with the use of the GC×GC-TOFMS; M.Y. and H.D. supported the manuscript. All authors have read and agreed to the published version of the manuscript.

Funding: This work was supported by the National Natural Science Foundation of China (32102088), the Agricultural Scientific and Technological Innovation Project of Shandong Academy of Agricultural Sciences (CXGC2018E05, CXGC2021B14 and CXGC2022E05), Shandong Key R & D Plan (2020CXGC010804), and National Pear Industry Technology System (CARS-28-23).

Institutional Review Board Statement: This article does not contain any studies with human participants or animals performed by any of the authors.

Informed Consent Statement: Not applicable.

Data Availability Statement: All relevant data are within the manuscript and its supporting files.

Acknowledgments: We thank Wenhui Wang and Yanmin Du from the Research Institute of Pomology of CAAS for contributing materials.

Conflicts of Interest: The authors declare no conflict of interest.

References

1. Lu, M.Y.; Wu, C.H.; Wang, Q.; Yan, X.K.; Liu, M.H.; Zhang, M.J. Analysis of photosynthetic pigments and mineral elements content of leaves of wild *Pyrus ussuriensis* Maxin. with different color grown in saline alkali soil. *Nonwood For. Res.* **2021**, *39*, 10–17.
2. Qiu, D.; Guo, J.; Yu, H.M.; Yan, J.; Yang, S.X.; Li, X.; Zhang, Y.M.; Sun, J.J.; Cong, J.; He, S.L.; et al. Antioxidant phenolic compounds isolated from wild *Pyrus ussuriensis* Maxim. Fruit peels and leaves. *Food Chem.* **2018**, *241*, 182–187. [CrossRef]
3. Wang, Y.; Wang, W.H.; Jia, X.H.; Tong, W.; Wang, Z.H.; Yang, X.L. Evaluation of frozen fruit quality of different pear cultivars. *Sci. Agric. Sin.* **2017**, *50*, 3400–3412.
4. Song, L.; Zhang, Q.; Du, G.D.; Wang, X.Q.; Liu, Z. Present situation and countermeasures of pear industry development in Liaoning Province. *North. Fruits* **2018**, *4*, 36–38.
5. Yang, X.L. Physiological Differences and Regulation Mechanism of Different *Pyrus ussuriensis* Maxim Varieties during Softening. Master's Thesis, Chinese Academy of Agricultural Sciences, Beijing, China, 2017.
6. Jeon, J.H.; Yang, J.Y.; Lee, H.S. Acaricidal activities of materials derived from *Pyrus ussuriensis* fruits against stored food mites. *J. Food Prot.* **2012**, *75*, 1258–1262. [CrossRef] [PubMed]
7. Qu, J.H.; Du, B.; Peng, F.; Wang, T.K.; Yang, Y.D. Optimization of triterpenoids extraction from Anli pears (*Pyrus ussuriensis* Maxim) by pressurized liquid extraction. *Qual. Assur. Ssf. Crop* **2016**, *8*, 105–110. [CrossRef]
8. Li, X.; Wang, T.T.; Zhou, B.; Gao, W.Y.; Cao, J.G.; Huang, L.Q. Chemical composition and antioxidant and anti-inflammatory potential of peels and flesh from 10 different pear varieties (*Pyrus* spp.). *Food Chem.* **2014**, *152*, 531–538. [CrossRef]
9. Sun, L.Q.; Hao, W.J.; Tang, X.Q.; Wang, K.C.; Zhang, S.L. Analysis of Characteristic Polyphenols and Triterpenic Acids in Ripe Pears of 36 Cultivars by UPLC-MS/MS. *Food Sci.* **2020**, *41*, 206–214.
10. Goliáš, J.; Balík, J.; Létal, J. Identification of volatiles formed in Asian pear cultivars subjected to short-term storage using multinomial logistic regression. *J. Food Compos. Anal.* **2021**, *97*, 103793. [CrossRef]
11. Siegmund, B. Biogenesis of aroma compounds: Flavour formation in fruits and vegetables. In *Flavor Development, Analysis and Perception in Food and Beverages*; Parker, J.K., Elmore, J.S., Methven, L., Eds.; Woodhead Press: Sawston Cambridge, UK, 2015; pp. 127–149.
12. Senoussi, A.; Rapisarda, T.; Schadt, I.; Chenchouni, H.; Saoudi, Z.; Senoussi, S.; Zitoun, O.A.; Zidoune, M.N.; Carpino, S. Formation and dynamics of aroma compounds during manufacturing ripening of Bouhezza goat cheese. *Int. Dairy J.* **2022**, *129*, 105349. [CrossRef]
13. Qin, G.H.; Tao, S.T.; Cao, Y.F.; Wu, J.Y.; Zhang, H.P.; Huang, W.J.; Zhang, S.L. Evaluation of the volatile profile of 33 *Pyrus ussuriensis* cultivars by HS-SPME with GC-MS. *Food Chem.* **2012**, *134*, 2367–2382. [CrossRef] [PubMed]
14. Ndikuryayo, C.; Ndayiragije, A.; Kilasi, N.; Kusolwa, P. Breeding for Rice Aroma and Drought Tolerance: A Review. *Agronomy* **2022**, *12*, 1726. [CrossRef]
15. Farag, M.A.; Dokalahy, E.U.; Eissa, T.F.; Kamal, I.M.; Zayed, A. Chemometrics-Based Aroma Discrimination of 14 Egyptian Mango Fruits of Different Cultivars and Origins, and Their Response to Probiotics Analyzed via SPME Coupled to GC-MS. *ACS Omega* **2022**, *7*, 2377–2390. [CrossRef] [PubMed]
16. Liu, W.J.; Zhang, Y.; Zhang, Y.X.; Du, G.Q. Analysis and evaluation of fruit quality and aroma components of 'Yali' pear (*Pyrus bretschneideri* Rehd.) pollinated with eighteen pollinizers. *Food Sci.* **2022**, *43*, 294–302.
17. Chen, Y.Y. Analysis on Volatile Aroma Constituent Characteristics of Pears with Different Cultivars. Master's Thesis, Nanjing Agricultural University, Nanjing, China, 2019.
18. Zlatić, E.; Zadnik, V.; Fellman, J.; Demšar, L.; Hribar, J.; Čejić, Ž.; Vidrih, R. Comparative analysis of aroma compounds in 'Bartlett' pear in relation to harvest date, storage conditions, and shelf-life. *Postharvest Biol. Technol.* **2016**, *117*, 71–80. [CrossRef]
19. Wei, S.W.; Tao, S.T.; Qin, G.H.; Wang, S.M.; Tao, J.H.; Wu, J.; Wu, J.Y.; Zhang, S.L. Transcriptome profiling reveals the candidate genes associated with aroma metabolites and emission of pear (*Pyrus ussuriensis* cv.). *Sci. Hortic.* **2016**, *206*, 33–42. [CrossRef]
20. Zhang, L.P. Effects of Refrigeration and 1-Methylcyclopropene on the Metabolic of Volatile Aroma in Nanguo Pear and Control. Ph.D. Thesis, Shenyang Agricultural University, Shenyang, China, 2013.
21. Li, J.; Han, J.C.; Yang, S.M.; Wang, Y. GC-MS Analysis on aroma components of Huagaili Pear fruits. *J. Hebei Agric. Sci.* **2020**, *24*, 51–55.
22. Farrapeira, R.O.; Andrade, Y.B.; Schena, T.; Schneider, J.K.; Muhlen, C.V.; Bjerk, T.R.; Krause, L.C.; Caramão, E.B. Characterization by Fast–GC×GC-TOFMS of the Acidic/Basic Neutral Fractions of Bio-Oils from Fast Pyrolysis of Green Coconut Fibers. *Am. Chem. Soc.* **2022**, *61*, 9567–9574. [CrossRef]
23. Zhao, M.; Li, T.; Yang, F.; Cui, X.Y.; Zou, T.T.; Song, H.L.; Liu, Y. Characterization of key aroma-active compounds in Hanyuan *Zanthoxylum bungeanum* by GC-O-MS and switchable GC×GC-O-MS. *Food Chem.* **2022**, *385*, 132659. [CrossRef]

24. Welke, J.E.; Zanus, M.; Lazzarotto, M.; Pulgati, F.H.; Zini, C.A. Main differences between volatiles of sparkling and base wines accessed through comprehensive two-dimensional gas chromatography with time-of-flight mass spectrometric detection and chemometric tools. *Food Chem.* **2014**, *164*, 427–437. [CrossRef]
25. Lebedev, A.T.; Polyakova, O.V.; Mazur, D.M.; Artaev, V.B.; Canet, I.; Lallement, A.; Vaitilingom, M.; Deguillaume, L.; Delort, A.M. Detection of semivolatile compounds in cloud waters by GC×GC-TOF-MS. Evidence of phenols and phthalates as priority pollutants. *Environ. Pollut.* **2018**, *241*, 616–625. [CrossRef]
26. Zhu, Y.; Lv, H.P.; Shao, C.Y.; Kang, S.Y.; Zhang, Y.; Guo, L.; Dai, W.D.; Tan, J.F.; Peng, Q.H.; Lin, Z. Identification of key odorants responsible for chestnut-like aroma quality of green teas. *Food Res. Int.* **2018**, *108*, 74–82. [CrossRef]
27. Wang, C.C.; Zhang, W.J.; Li, H.D.; Mao, J.S.; Guo, C.Y.; Ding, R.Y.; Wang, Y.; Fang, L.P.; Chen, Z.L.; Yang, G.S. Analysis of volatile compounds in pears by HS-SPME-GC×GC-TOFMS. *Molecules* **2019**, *24*, 1795. [CrossRef]
28. Li, Y.D. Study on Characteristic Aroma Compounds during the Process of Fu Brick Tea. Master's Thesis, Hunan Agricultural University, Changsha, China, 2019.
29. Kang, S.Y.; Zhu, Y.; Zheng, X.Q.; Liang, Y.R.; Lin, Z. Multivariate statistical analysis of volatiles compounds in green teas from different harvesting seasons. *Food Sci.* **2018**, *39*, 268–275.
30. Zhang, X.N.; Jia, X.H.; Du, Y.M.; Wang, W.H.; Shu, Q. Comparation of fruit quality and aroma components among three red-skinned sand pear varieties. *Food Sci.* **2022**, *39*, 366–375.
31. Liu, P.P.; Zheng, P.C.; Gong, Z.M.; Wang, S.P.; Teng, J.; Gao, S.W.; Wang, X.P.; Ye, F.; Zheng, L. Analysis of aroma components in Qingzhuan Dark Tea. *Food Sci.* **2017**, *38*, 164–170.
32. Zhang, H.; Lu, Z.M.; Wang, J.T.; Guo, X. Comparative aroma components of fruits of four main *Citrus* varieties. *Food Sci.* **2017**, *38*, 192–196.
33. Wang, J.T.; Shi, J.; Zhu, Y.; Ma, W.J.; Yan, H.; Shao, C.Y.; Wang, M.Q.; Zhang, Y.; Peng, Q.H.; Chen, Y.Q.; et al. Insights into crucial odourants dominating the characteristic flavour of citrus-white teas prepared from citrus reticulata Blanco 'Chachiensis' and *Camellia sinensis* 'Fudingdabai'. *Food Chem.* **2022**, *377*, 132048. [CrossRef] [PubMed]
34. Li, M.P.; Li, C.L.; Wang, H.R.; Zhang, S.W.; Yu, L.G.; Guo, C.X. Analysis of aroma compounds in Red Fragrant pear by headspace solid phase microextraction and gas chromatography–olfactometry–mass spectrometry. *Food Res. Dev.* **2020**, *41*, 130–139.
35. Campeloa, P.H.; Alves Filhob, E.G.; Silvac, L.M.A.; Britoc, E.S.; Rodriguesb, S.; Fernandesd, F.A.N. Modulation of aroma and flavor using dielectric barrier discharge plasma technology in a juice rich in terpenes and sesquiterpenes. *LWT-Food Sci. Technol.* **2020**, *130*, 109644. [CrossRef]
36. Zhang, J. Study on Aroma Characterization of Mango Fruit. Master's Thesis, Guangxi University, Nanning, China, 2011.
37. Rizzolo, A.; Cambiaghi, P.; Grassi, M.; Zerbini, P.E. Influence of 1-methylcyclopropene and storage atmosphere on changes in volatile compounds and fruit quality of conference pears. *Agric. Food Chem.* **2005**, *53*, 9781–9789. [CrossRef]
38. Schwab, W.; Davidovich-Rikanati, R.; Lewinsohn, E. Biosynthesis of plant-derived flavor compounds. *Plant J.* **2008**, *54*, 712–732. [CrossRef]
39. Takeoka, G.; Buttery, R.G.; Flath, R.A.; Teranishi, R.; Wheeler, E.L.; Wieczorek, R.L.; Guentertet, M. Volatile constituents of pineapple (*Ananas Comosus* [L.] Merr.). In *Flavor Chemistry Trends and Development*; Teranishi, R., Buttery, R.G., Shahidi, F., Eds.; U.S. ACS Symposium Series Press: Washington, DC, USA, 1989; pp. 223–237.
40. Chen, J.; Zhou, S.; Yan, S.J.; Ma, Y.K.; Hu, X.S. Analysis of aroma components of Fengshui, Dangshan and Nanguo pear by SPME/GC/MS. *Acta Hortic. Sin.* **2005**, *32*, 301–303.
41. Hayaloglu, A.A.; Demir, N. Phenolic compounds, volatiles, and sensory characteristics of twelve sweet cherry (*Prunus avium* L.) cultivars grown in Turkey. *J. Food Sci.* **2016**, *81*, 7–18. [CrossRef] [PubMed]
42. Taiti, C.; Pandolfi, C.; Caparrotta, S.; Dei, M.; Giordani, E.; Mancuso, S.; Nencetti, V. Fruit aroma and sensorial characteristics of traditional and innovative Japanese plum (*Prunus salicina* Lindl.) cultivars grown in Italy. *Eur. Food Res. Technol.* **2019**, *245*, 2655–2668. [CrossRef]
43. Hernández, F.; Noguera-Artiaga, L.; Burló, F.; Wojdyło, A.; Carbonell-Barrachina, Á.A.; Legua, P. Physico-chemical, nutritional, and volatile composition and sensory profile of Spanish jujube (*Ziziphus jujuba* Mill.) fruits. *J. Sci. Food Agric.* **2016**, *96*, 2682–2691. [CrossRef] [PubMed]
44. Sung, J.; Suh, J.H.; Chambers, A.H.; Crane, J.; Wang, Y. Relationship between Sensory Attributes and Chemical Composition of Different Mango Cultivars. *J. Agric. Food Chem.* **2019**, *67*, 5177–5188. [CrossRef] [PubMed]
45. Ramirez, J.L.; Du, X.; Wallace, R.W. Investigating sensory properties of seven watermelon varieties and factors impacting refreshing perception using quantitative descriptive analysis. *Food Res. Int.* **2020**, *138*, 109681. [CrossRef]
46. Siebert, T.E.; Barker, A.; Pearson, W.; Barter, S.R.; de Barros Lopes, M.A.; Darriet, P.; Herderich, M.J.; Francis, I.L. Volatile Compounds Related to 'Stone Fruit' Aroma Attributes in Viognier and Chardonnay Wines. *J. Agric. Food Chem.* **2018**, *66*, 2838–2850. [CrossRef]

Disclaimer/Publisher's Note: The statements, opinions and data contained in all publications are solely those of the individual author(s) and contributor(s) and not of MDPI and/or the editor(s). MDPI and/or the editor(s) disclaim responsibility for any injury to people or property resulting from any ideas, methods, instructions or products referred to in the content.

foods

MDPI

Article

Headspace Solid-Phase Microextraction and Gas Chromatography–Mass Spectrometry Combined with Sensory Evaluation for the Analysis of Volatile Aromatic Compounds in Apricot (*Prunus armeniaca* L.) Germplasm Resources Cultivated in Xinjiang, China

Xueling Zeng [1], Shikui Zhang [2], Wenjuan Geng [1], Jie Jin [1], Kang Liao [1], Zhanghu Tang [2], Shaopeng Wang [1,2] and Weiquan Zhou [1,3,*]

[1] College of Horticulture, Xinjiang Agricultural University, Urumqi 830052, China; xuelingz0118@163.com (X.Z.); gwj0526@xjau.edu.cn (W.G.); 17767622134@163.com (J.J.); liaokang01@163.com (K.L.); 19915127905@163.com (S.W.)
[2] Luntai Fruit Germplasm Resources Garden, Xinjiang Academy of Agricultural Sciences, Luntai 841600, China; zsk2366287@163.com (S.Z.); tangzhanghu@163.com (Z.T.)
[3] Postdoctoral Mobile Station of Horticulture, College of Horticulture, Xinjiang Agricultural University, Urumqi 830052, China
* Correspondence: zhouwq1120@xjau.edu.cn

Abstract: The volatile compounds in the fruits of 24 apricot cultivars were quantitatively and qualitatively determined via headspace solid-phase microextraction and gas chromatography–mass spectrometry (HS-SPME-GC–MS). A total of 429 volatile compounds were detected in these fruits, and the greatest number of detected terpenoids was 77. Significant differences were found among the cultivars in terms of the total volatile compound content of the fruits, with variation from 112.76 ('ZSHYX') to 317.36 μg/g ('JNL'). Using relative odor activity value (rOAV) analysis, 42 key aroma compounds were identified. The rOAVs of (2S,4R)-4-methyl-2-(2-methylprop-1-enyl)oxane, (E)-non-2-enal, (3-methyl-3-sulfanylbutyl) formate, and thiophen-2-ylmethanethiol were above 1000, and most had green, fruity, and woody odors; these results indicated that these substances were important contributors to the overall aroma of the apricot fruits. Our study provides a comprehensive analysis of the volatile compounds from 24 representative apricot cultivars and can aid in the further scientific understanding of the metabolites and aroma in apricots. These findings provide a reference for controlling fruit quality and for future apricot cultivar breeding.

Keywords: apricot; aromatic components; relative odor activity value; volatile metabolomics; gas chromatography–mass spectrometry

1. Introduction

In higher plants, volatile compounds are widely present and important secondary metabolites [1]. Volatile compounds are crucial in the interactions between plants and their surroundings because they attract pollinators and seed dispersers. Plants resist and adapt to their external environment by synthesizing and releasing volatile compounds, thereby protecting themselves from parasites, herbivores, and pathogens [2]. In addition, volatile compounds are a source of olfactory stimulation in humans; they provide aromas to cereals, fruits, and vegetables and have a large impact on flavoring, preservatives, and herbal remedies [3,4]. Aroma is one of the most important organoleptic characteristics of fruits; volatile compounds constitute only 10^{-4} to 10^{-7} of the weight of fresh fruit and are secondary metabolites formed from terpenoids, esters, heterocyclic compounds, and alcohols [5,6]. Aroma is a crucial factor influencing consumer perception and acceptance of fruit products and has a great impact on appetite and the digestive system, which are important for transmitting nutritional signals to humans and can even affect a person's mental state [7,8]. Researchers have increasingly shown great interest in fruit aroma.

The apricot species *Prunus armeniaca* L. is a widely cultivated drupe fruit tree, and Italy, Turkey, the United States, Spain, France, and China are the biggest apricot-producing regions of the world [9]. Apricots are highly appreciated by consumers for their rich nutritional content and unique aroma, and they can be eaten raw or dried and processed [10]. The earliest research on the aroma compounds of apricot fruits was conducted in 1967, and Tang and Jenning were the first to identify and analyze volatile compounds from apricot fruits in a European ecosystem [11]. Many scholars have extensively researched the components of volatile compounds in apricot fruits. Solís-Solís et al. [12] studied the aroma-imparting substances of the fruits of eight apricot varieties via four extraction techniques, and their results revealed that the extraction technique and the variety had large influences on the fruit aroma type and content. Gas chromatography–mass spectrometry (GC–MS) can be used to comprehensively identify volatile compounds in fruits. Zhang et al. [13] analyzed the volatile profiles of the fruits of apricot germplasm resources in different regions of China and reported that the contents and components of volatile profiles in these fruits greatly varied. The volatile compound content was the highest for apricot varieties in Northwest China but was low for apricot varieties in Southwest China. These results revealed that genetic background strongly affected the contents and components of volatile compounds. Using an odor activity value (OAV) analysis, Greger and Schieberle [14] reported that (Z)-1,5-octadien-3-one, β-ionone, γ-decalactone, acetaldehyde, linalool, and (E, Z)-2,6-nonadienal had OAVs greater than 100 and were key aroma components of apricot fruit.

Headspace solid-phase microextraction (HS-SPME) is a simple procedure that has a low monitoring threshold, good detection sensitivity, and good repeatability. This technique can be combined with gas chromatography–mass spectrometry (GC–MS) to obtain more comprehensive and reliable data. It can be used to identify, simultaneously separate, and provide specific flavor information and quantify multiple volatile compounds [15,16]. Although many reports have described the contents and components of volatile compounds in apricot fruits, no systematic studies have been conducted on the composition and content of volatile compounds in Xinjiang apricot fruits, and the use of volatile metabolite analysis to qualitatively and quantitatively evaluate volatile compounds in apricot fruits has not been reported. In this study, sensory analysis combined with HS-SPME-GC–MS was used to qualitatively and quantitatively analyze and compare the volatile compounds in the fruits of 24 apricot cultivars, and key aroma compounds were identified via the relative odor activity value (rOAV) method. The aim of this study was to explore aroma information to understand the quality differences among a wider range of apricot germplasm resources and provide a reference for apricot fruit breeding and quality control in the future. In addition, these results provide valuable data on apricot aroma for future research on genes that regulate aroma components in combination with genetic linkage mapping and genome-wide association study (GWAS) analysis.

2. Materials and Methods

2.1. Sample Preparation and Methods for Determining the Fruit Quality Index

The samples were collected from the Luntai Fruit Germplasm Resources Garden of the Xinjiang Academy of Agricultural Sciences (E 84°13′, N 41°47′). The experimental site was in the Luntai County, Bayingoleng Mongolian Autonomous Prefecture, Xinjiang Uygur Autonomous Region, China, and pictures of the mature fruits of the 24 apricot cultivars are shown in Supplementary Figure S1. Conventional water and fertilizer management practices were employed. From June to July 2023, samples were collected from the fruits of each cultivar during the ripening period, and the criteria for determining the ripening period of the fruits were their hardness, color, and soluble solid content. Three apricot trees of each variety were randomly selected, and 100 apricot fruits were collected from the outer canopy. The cultivar samples were randomly divided into three parts. Each sample was quickly cut into small pieces with a scalpel, frozen, and submerged in liquid nitrogen. After mixing, the apricot samples were immediately placed in a frozen tube and stored at −80 °C to determine the volatile compounds. The other portion was used for quality index determination. The longitudinal diameter, transverse diameter, and lateral diameter of the apricot fruit were measured with a digital display caliper, an electronic balance (MP2001, Shanghai, China) was used to measure the fruit weight (g), and a portable pH meter (Horiba, Kyoto, Japan) was used to measure the pH of the apricot fruit. Fruit hardness was determined with a hardness tester (GY-4, Zhejiang, China). The soluble solids were measured with a portable refractometer (Atago, Tokyo, Japan). The edible rate (pitted fresh fruit quality as a percentage of the total fresh fruit quality) and fruit water content (dried fruit quality as a percentage of fresh fruit quality) were also determined.

2.2. Panel Training and Aroma Characteristic Analysis

The sensory properties of the fruits were evaluated according to the methods of Fan et al. [17]. The sensory assessment team consisted of four female members and four male members who were trained in sensory assessment. Sweet, green, fruity, floral, herbal, woody, waxy, fatty, citrusy, and nutty aromas were evaluated. The sensory evaluation was graded on a 1–60 scale, with 1 representing very low strength and 60 representing the highest strength. Ten fruits were tested for each apricot cultivar, and the average of all scores was calculated.

2.3. Volatile Extraction and Concentration

The extraction steps were as follows: the apricot frozen sample was removed from storage at ultralow temperatures, 500 mg of sample was weighed, −80 °C liquid nitrogen was added, and the sample was ground to powder. Then, 500 mg powder sample was weighed and dissolved in 2 mL saturated NaCl solution (analytical reagent). A 20 μL internal standard solution consisting of 3-hexanone-2, 2, 4, 4-d4 (10 μg/mL) was added, the sample was promptly transferred to an Agilent USA 20 mL headspace vial, and the addition of a saturated NaCl solution inhibited any enzyme reactions. In this study, an SPME extractor with 120 μm DVB/CWR/PDMS (Agilent, Santa Clara, CA, USA) was used with an SPME Arrow (CTC Analytics AG, Basel, Switzerland). The extractor was set to 250 °C for 5 min before use. A rolled cap with a TFE silicone headspace diaphragm (Agilent) was used to seal the vial. First, the headspace vial was shaken at 60 °C for 5 min; afterward, the aging extraction head was placed in the vial for 15 min. A gas chromatograph (Type 7890B, Agilent) injector was used for desorption at 250 °C for 5 min in nonshunt mode, and the sample was then removed [18].

2.4. Gas Chromatography–Mass Spectrometry Analyses

The test instrument used was a 7890B (Agilent) type gas chromatograph and 7000D mass spectrometer (Agilent). The column was a 30 m × 25 mm × 0.25 μm DB5-ms column (Agilent, Santa Clara, CA, USA). The constant flow rate of the high-purity carrier gas was 1.2 mL/min, the injection mode was splitless, and the inlet temperature was maintained at 250 °C. Initially, the temperature of the column was 40 °C, held for 3.5 min, then increased to 100 °C at 10 °C per minute, increased to 180 °C at 7 °C per minute, and finally increased to 280 °C at 25 °C per minute and held for 5 min. The mass spectrum conditions included an electron energy of 70 eV, a quadrupole mass detector temperature of 150 °C, an ion source temperature of 230 °C, and a mass spectrum interface temperature of 280 °C. Selective ion monitoring (SIM) was used as the scanning mode for mass spectrometry, the data were obtained via accurate qualitative and quantitative scanning, and the obtained mass spectra were compared with those in the MWGCSIM 1.0 library to complete the qualitative and quantitative analyses [19].

2.5. Qualitative and Quantitative Analysis

On the basis of the MWDB (MetWare database) 2023 Update of plant volatile substances, qualitative and quantitative analyses of the volatile substances in the cultivar samples were performed via mass spectrometry. Supplementary Figure S2 shows the total ion current (TIC) data of the samples. The standard substances used in this test were prepared by diluting n-hexane (chromatographic purity, CNW) to a certain concentration, and the solution was stored at −20 °C. The volatile substances in the cultivar samples were determined semiquantitatively via the internal standard method, and the relative contents of each volatile substance were determined using 3-hexanone-2, 2, 4, 4-d4 as the internal standard.

Regarding qualitative analysis, based on multiple species, the literature, partial reference standards, and retention indices, an independent database (the updated version of the MWGC database in 2023, provided by Wuhan Maiwei Metabolic Biotechnology Co., Ltd.) (Wuhan, China) was established, and an SIM mode detection protocol was constructed. All ions in each group were detected separately in time intervals according to the elution sequence. If the retention time of the detected chromatographic peak was consistent with that of the standard reference and if all selected ions appeared (2–3 qualitative fragments) in the sample mass spectrum after background subtraction, the sample was determined to be this substance. The qualitative method adopted in this study was supported by the existing methodologies used in the literature.

Quantitative analysis using Mass Hunter was conducted to increase the accuracy of the quantification. Specific quantification ions were carefully selected for integration and calibration. For the relative quantification of volatile compounds, the internal standard (3-hexanone-2,2,4,4-d4) semiquantitative method was employed. The resulting relative content of each volatile compound was as follows:

$$X_i = \frac{V_s \times C_s}{M} \times \frac{I_i}{I_s} \times 10^{-3}$$

where X_i represents the content of compound i in the sample to be measured (μg/g); vs. represents the volume of the added internal standard (μL); C_s represents the concentration of the internal standard (μg/mL); M is the amount of the sample to be measured (g); I_s is the peak area of the internal standard; and I_i is the peak area of compound i in the sample to be measured.

2.6. Volatile Compound Qualification and rOAV Calculation

The relative odor activity value (*rOAV*) is the ratio of the relative content of a volatile compound to its odor threshold (OT). Generally, compounds with an *rOAV* > 1 directly contribute to the aroma characteristics of fruits [20], and the *rOAV* content (μg/g) is determined via $rOAVi = \frac{Ci}{Ti}$, where *rOAVi* is the relative odor activity of compound *i*, *Ci* is the relative content of compound *i* (μg/g), and *Ti* is the defined threshold of the compound (threshold, μg/g).

2.7. Statistical Analysis

Microsoft Excel 2020 and SPSS 23.0 were used for the statistical analysis of all three biological replicates. PCA (principal component analysis), PCC (Pearson correlation coefficient), and HCA (hierarchical cluster analysis) were performed using R language software 3.5.1.

3. Results

3.1. Analysis of the Quality Indices of Apricot Fruits at Maturity

The fruit quality characteristics of the different apricot cultivars are listed in Table 1. The maturity period of the 24 apricot cultivars ranged from 11 June to 11 July, and considerable differences in the single-fruit weight, pH, soluble solids, fruit shape index, water content, hardness, and edible rate were observed. The single-fruit weight varied among the cultivars and ranged from 18.52 to 100.91 g. The soluble solids ranged from 10.96 to 26.04%; here, the fruits of 'MTYLK' had the highest soluble solids, and those of 'KBKYLK' and 'SGJNL' had the lowest soluble solids. The fruit hardness ranged from 1.16 kg/cm^2 to 6.44 kg/cm^2. The fruit pH ranged from 3.63 to 5.10, and great differences were observed among the varieties. The fruit shape index ranged from 0.89 to 1.17, and the fruit shape index greatly varied among the different varieties. The fruit water content ranged from 73.17% to 91.73%. The variety with the highest edible rate was 'ZGYDJX', and the variety with the lowest edible rate was 'DX'.

3.2. Analysis of the Sensory Flavor Characteristics of the Apricot Cultivars

A group of professionals used the aroma sensory evaluation method to evaluate and analyze the aroma types of 24 apricot varieties and conducted a preliminary investigation and evaluation of the aroma intensity of different apricot varieties using sensory evaluation. These professionals agreed that all apricot fruits had recognizable sensory characteristics. The aroma types of the apricot fruits were classified as sweet, green, fruity, floral, herbal, woody, waxy, fatty, citrusy, and nutty (Figure 1a). However, the intensity of each type of aroma differed, and sweet, green, and fruity aromas were more prominent. The sensory results revealed that the scores of the 'LJX', 'MTYLK', 'LTBX', 'AKDLZ', 'KZL', and 'KCBX' cultivars were greater than 50, with scores of 57, 56, 56, 54, 54, and 54, respectively; these results indicated that these cultivars contained more volatile compounds.

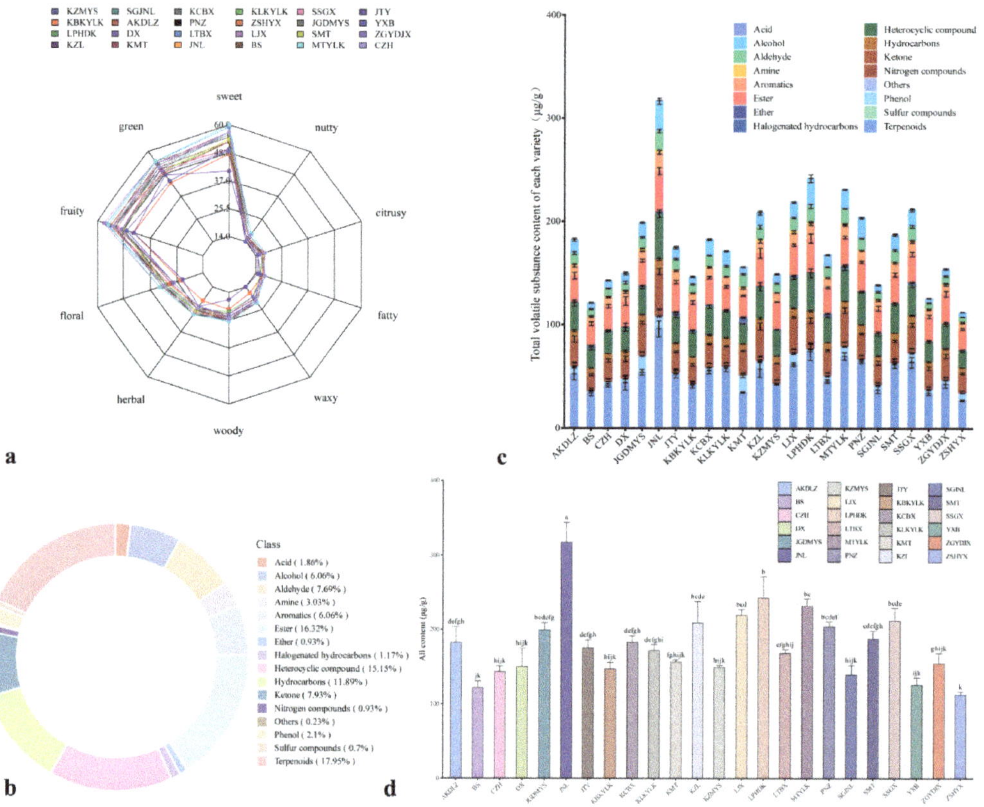

Figure 1. Analysis of the volatile compounds in apricot fruits: (**a**) Radar map of the European rich sensory flavor characteristics of 24 apricot cultivars; (**b**) volatile compound classes of the 24 apricot cultivars from a ring chart, where each color represents a metabolite class and the area of the color block represents the proportion of that class; (**c**) histogram of the metabolic analysis of the different types of volatile compounds in the fruits of 24 apricot cultivars; and (**d**) analysis of the difference in the total volatile compounds in the fruits of 24 apricot cultivars. A statistical analysis was performed via a one-way analysis of variance (ANOVA), and different letters indicate significant differences at the 0.05 level.

Table 1. Fruit quality indices of 24 apricot cultivars [A].

Number	Cultivar	Harvest Date	Single-Fruit Weight (g)	Soluble Solids (%)	pH	Fruit Firmness (kg/cm^2)	Fruit Shape Index (%)	Fruit Water Content (%)	Edible Rate (%)
KZMYS	Kezimayisang	11 June	18.52 ± 0.64^{m}	17.96 ± 0.53^{efg}	4.33 ± 0.17^{cdef}	4.73 ± 0.19^{cd}	0.99 ± 0.00^{efg}	81.76 ± 0.04^{efg}	89.60 ± 0.01^{ijkl}
KBKYLK	Kabakeyulvke	11 June	55.48 ± 2.01^{c}	10.96 ± 0.54^{i}	4.13 ± 0.11^{cdef}	1.16 ± 0.09^{j}	0.96 ± 0.01^{fghi}	91.73 ± 0.01^{a}	93.01 ± 0.01^{bc}
LPHDK	Luopuhongdaike	11 June	23.17 ± 0.55^{jkl}	17.34 ± 1.04^{fg}	4.30 ± 0.12^{cdef}	5.50 ± 0.23^{b}	0.94 ± 0.01^{hij}	85.96 ± 0.01^{bcd}	90.32 ± 0.01^{hijk}
KZL	Kezilang	16 June	48.90 ± 1.40^{ld}	16.46 ± 0.55^{fgh}	4.38 ± 0.13^{bcde}	4.44 ± 0.24^{de}	1.06 ± 0.01^{cd}	83.33 ± 0.01^{def}	92.59 ± 0.01^{bcdef}
SGJNL	Suogejianali	20 June	28.95 ± 1.06^{hi}	12.42 ± 0.57^{i}	5.10 ± 0.04^{a}	5.44 ± 0.20^{b}	0.98 ± 0.01^{efgh}	87.33 ± 0.02^{bc}	92.40 ± 0.01^{bcdefg}
AKDLZ	Akedalazi	21 June	22.63 ± 0.34^{klm}	25.34 ± 0.62^{a}	4.54 ± 0.13^{abc}	5.58 ± 0.29^{b}	1.12 ± 0.01^{b}	73.17 ± 0.03^{j}	90.95 ± 0.02^{fghijk}
DX	Danxing	22 June	30.81 ± 0.74^{h}	21.12 ± 0.55^{bcd}	3.63 ± 0.18^{f}	5.38 ± 0.13^{b}	0.92 ± 0.01^{ijk}	79.23 ± 0.06^{gh}	87.65 ± 0.02^{n}
KMT	Kumaiti	26 June	21.31 ± 0.75^{lm}	24.80 ± 2.33^{a}	5.08 ± 0.22^{ab}	5.58 ± 0.20^{b}	1.07 ± 0.01^{c}	76.41 ± 0.03^{hij}	92.88 ± 0.01^{bcde}
KCBX	Kuchebaixing	26 June	27.22 ± 0.62^{hij}	23.08 ± 1.11^{abc}	4.62 ± 0.18^{abc}	4.68 ± 0.15^{cd}	1.12 ± 0.01^{b}	74.69 ± 0.03^{j}	91.32 ± 0.01^{defgh}
PNZ	Pinaizi	30 June	47.45 ± 2.48^{d}	20.64 ± 0.55^{cde}	4.40 ± 0.21^{abcd}	6.44 ± 0.23^{a}	0.95 ± 0.02^{ghij}	78.69 ± 0.01^{gh}	91.22 ± 0.01^{fghij}
LTBX	Luntaibaixing	1 July	25.24 ± 0.39^{ijkl}	23.66 ± 2.08^{abc}	4.74 ± 0.14^{abc}	5.18 ± 0.14^{bc}	1.02 ± 0.01^{de}	81.81 ± 0.02^{efg}	91.26 ± 0.00^{efghi}
JNL	Jianali	3 July	41.68 ± 1.46^{f}	16.10 ± 0.60^{gh}	4.26 ± 0.08^{abcd}	3.86 ± 0.10^{fg}	1.05 ± 0.01^{cd}	81.80 ± 0.01^{efg}	90.86 ± 0.01^{ghijk}
KLKYLK	Kalakeyulvke	4 July	22.67 ± 0.48^{klm}	25.04 ± 0.75^{a}	3.71 ± 0.13^{def}	4.52 ± 0.06^{de}	1.00 ± 0.03^{ef}	73.89 ± 0.02^{j}	89.38 ± 0.01^{klm}
ZSHYX	Zaoshuheiyexing	5 July	46.30 ± 1.09^{de}	17.68 ± 0.73^{efg}	4.46 ± 0.13^{abc}	4.04 ± 0.33^{efg}	1.07 ± 0.01^{c}	78.40 ± 0.01^{ghi}	89.56 ± 0.00^{jkl}
LJX	Lajiaoxing	6 July	42.69 ± 0.99^{ef}	16.46 ± 1.09^{fgh}	4.57 ± 0.07^{abc}	4.30 ± 0.13^{def}	1.17 ± 0.03^{a}	78.48 ± 0.01^{ghi}	91.89 ± 0.01^{cdefgh}
BS	Beishan	6 July	37.68 ± 1.03^{g}	19.66 ± 0.72^{def}	4.70 ± 0.12^{abc}	4.04 ± 0.33^{efg}	0.99 ± 0.02^{efgh}	84.52 ± 0.01^{bcde}	87.96 ± 0.01^{mn}
SSGX	Shushangganxing	8 July	22.99 ± 0.75^{jkl}	25.54 ± 1.37^{a}	4.59 ± 0.05^{abc}	3.08 ± 0.11^{hi}	0.97 ± 0.01^{fgh}	75.24 ± 0.02^{ij}	90.64 ± 0.01^{hijk}
JGDMYS	Jiagedamayisang	9 July	42.64 ± 1.14^{ef}	24.20 ± 0.63^{ab}	4.56 ± 0.11^{abc}	2.76 ± 0.11^{i}	1.13 ± 0.01^{ab}	78.67 ± 0.05^{gh}	92.95 ± 0.01^{bcd}
SMT	Saimaiti	11 July	26.03 ± 0.89^{ijk}	17.58 ± 1.10^{efg}	4.14 ± 0.11^{cdef}	4.62 ± 0.21^{d}	1.05 ± 0.01^{cd}	79.99 ± 0.01^{fg}	88.08 ± 0.01^{lmn}
MTYLK	Mantouyulvke	9 July	30.90 ± 1.25^{h}	26.04 ± 1.43^{a}	4.54 ± 0.06^{abc}	3.86 ± 0.13^{fg}	0.98 ± 0.01^{efgh}	87.80 ± 0.00^{b}	90.72 ± 0.03^{ghijk}
JTY	Jintaiyang	20 June	100.91 ± 2.56^{a}	16.32 ± 0.87^{fgh}	4.04 ± 0.46^{cdef}	5.58 ± 0.14^{b}	0.99 ± 0.01^{efgh}	84.05 ± 0.01^{cde}	89.61 ± 0.00^{ijkl}
YXB	Yinxiangbai	20 June	58.61 ± 2.22^{c}	16.84 ± 0.72^{fgh}	3.68 ± 0.48^{ef}	5.28 ± 0.16^{b}	0.89 ± 0.01^{k}	84.25 ± 0.01^{cde}	96.02 ± 0.01^{a}
ZGYDJX	Zhanggongyuandajiexing	3 July	70.33 ± 2.55^{b}	13.72 ± 0.40^{hi}	4.11 ± 0.46^{cdef}	3.76 ± 0.18^{fg}	0.91 ± 0.02^{jk}	80.37 ± 0.01^{fg}	91.43 ± 0.01^{cdefgh}
CZH	Chuanzhihong	4 July	42.85 ± 2.32^{ef}	13.72 ± 0.57^{hi}	4.03 ± 0.15^{cdef}	3.58 ± 0.11^{gh}	1.06 ± 0.01^{cd}	85.52 ± 0.01^{bcd}	93.68 ± 0.01^{b}

[A] Data are expressed as the mean ± standard deviation of 3 biological replicates ($p < 0.05$). Different lowercase letters indicate significant differences at the 0.05 level.

3.3. Analysis of the Contents and Components of Volatile Compounds

The contents and components of volatile compounds in the fruits of 24 cultivars of apricot were quantitatively and qualitatively analyzed via HS-SPME-GC–MS. The contents and components of the volatile compounds in the apricots at maturity were compared (Supplementary Table S1). In this study, 429 volatile compounds were detected. The 429 volatile compounds were classified into 16 categories: terpenoids, heterocyclic compounds, esters, hydrocarbons, ketones, aldehydes, alcohols, aromatics, amines, phenols, acids, nitrogen compounds, halogenated hydrocarbons, ethers, sulfur compounds, and other unclassified compounds. Here, terpenoids were predominant, accounting for 77 species and 17.95% of all volatile compounds. In addition, 70 esters, 65 heterocyclic compounds, 51 hydrocarbons, 34 ketones, 33 aldehydes, 26 alcohols, and 26 aromatic compounds accounted for 16.32%, 15.15%, 11.89%, 7.93%, 7.69%, 6.06%, and 6.06%, respectively, of all volatile compounds. The above eight types of substances accounted for 89.05% of all the volatile compounds. The remaining volatile compounds included 13 amines, 9 phenols, 8 acids, 5 halogenated hydrocarbons, 4 nitrogen compounds, 4 ethers, 3 sulfur compounds, and 1 other unclassified compound. These compounds accounted for 3.03%, 2.1%, 1.86%, 1.17%, 0.93%, 0.93%, 0.7%, and 0.23% of all volatile compounds, respectively (Figure 1b). Among the 16 types of volatile compounds, terpenoids were the main volatile compounds in the 24 cultivars of apricot, followed by esters, heterocyclic compounds, hydrocarbons, ketones, and aldehydes; these compounds accounted for more than 70% of the total volatile compounds. The contents of the ether compounds, sulfur compounds, and other compounds in the fruits of the 24 apricot cultivars were less than 1.00 μg/g (Figure 1c). An analysis revealed a significant difference in the total content of volatile compounds in the fruits of the 24 apricot cultivars. The average total content of the aroma volatiles in these fruits was 179.37 ± 49.57 μg/g, and the range of volatile compound contents was 112.76 to 317.36 μg/g. The total content of the aromatic volatiles in the fruit of the 'JNL' cultivar was the highest (317.36 ± 26.36 μg/g), followed by 'LPHDK' and 'MTYLK' fruits (242.37 ± 29.07 μg/g and 231.63 ± 9.68 μg/g, respectively). ZSHYX had the lowest volatile compound content, at 112.76 ± 3.99 μg/g (Figure 1d).

3.4. Analysis of the Total Content of Volatile Compounds in the Fruits of 24 Apricot Cultivars

The aroma characteristics of fruits are not only related to the types of volatile compounds present but also closely related to the content of volatile compounds. We analyzed the metabolism of various volatile compounds in the fruits of different apricot cultivars, and the components and contents widely varied among the cultivars (Supplementary Figure S3). Terpenes were the main volatile compounds in the apricot fruits, with an average total content of 51.93 ± 16.45 μg/g, ranging from 27.23 μg/g ('ZSHYX') to 96.03 μg/g ('JNL'). Esters were the second most dominant aroma compound, with an average total volatile content of 26.73 ± 5.35 μg/g and a range of 39.95 μg/g ('JNL') to 19.99 μg/g (ZSHYX'). The heterocyclic compounds ranked third, with an average concentration of 26.44 ± 6.69 μg/g and a range of 15.939 μg/g ('ZSHYX') to 43.836 μg/g ('JNL'). The ketone content was 21.19 ± 6.40 μg/g on average and ranged from 13.11 μg/g ('BS') to 36.55 μg/g ('JNL'). Alcohols were the fifth most abundant volatile compound and had an average total content of 12.61 ± 6.65 μg/g and a range of 4.20 μg/g ('YXB') to 28.91 μg/g ('JNL') (Figure 2a).

To confirm the contribution of the above volatile compounds to the total volatile compound content, we conducted Pearson's correlation analysis between the total volatile components and all volatile components. Figure 2b shows that the contents of total volatile compounds were significantly positively correlated with those of terpenoids ($r = 0.97$), heterocyclic compounds ($r = 0.97$), aromatic compounds ($r = 0.94$), aldehydes ($r = 0.93$), nitrogen compounds ($r = 0.93$), hydrocarbons ($r = 0.92$), and alcohols ($r = 0.90$). Moreover, the total volatile compound content was weakly positively correlated with the halogenated hydrocarbon content ($r = 0.13$), ether compound content ($r = 0.36$), and phenolic compound content ($r = 0.46$). These results indicated that the terpenoids, heterocyclic compounds, aromatics, nitrogen compounds, aldehydes, alcohols, and hydrocarbons were the main volatile compounds in apricot fruits.

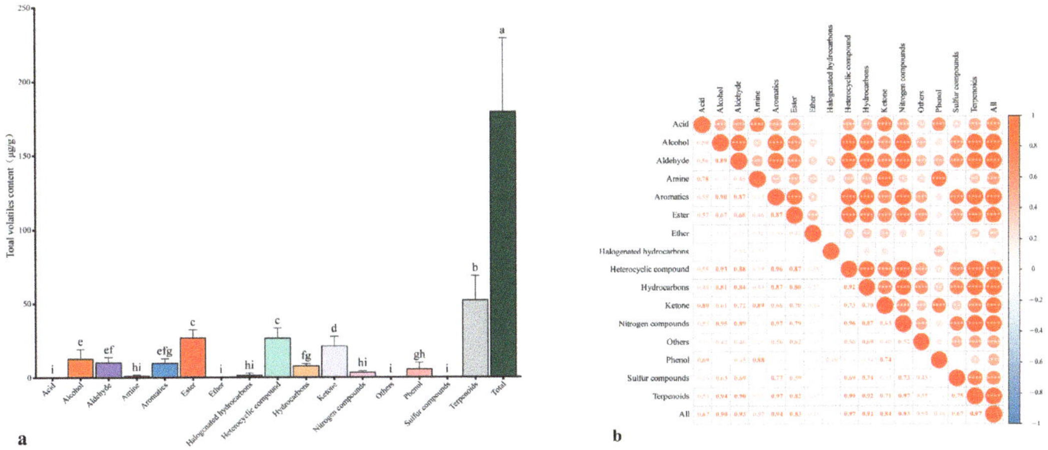

Figure 2. Volatile compound content and correlation analysis of the fruits of 24 apricot cultivars. (**a**) Histogram of the volatile compound contents in the fruits of 24 apricot cultivars. A statistical analysis was performed via a one-way analysis of variance (ANOVA), and different letters indicate significant differences at the 0.05 level. (**b**) Heatmap of the correlation between the content of 16 volatile compounds and the content of total volatile compounds; * represents a significant correlation when $p < 0.05$, ** $p < 0.01$, *** $p < 0.001$, and **** $p < 0.0001$, and color depth represents Pearson's phase relationship value. Each value is the mean ± standard deviation of three biological replicates.

3.5. Identification of the Relative Odor Activity Values in Apricot Cultivars

The rOAV can be used to analyze the contributions of volatile components to the aroma of apricots. The levels of all volatile components detected in the fruits of 24 apricot cultivars were evaluated. A total of 42 key aroma compounds with rOAVs greater than 1 were found, and most of the compounds had outstanding aroma properties, such as green, fruity, and woody odors (Table 2). The 42 key aroma compounds could be separated into 7 categories, including 9 terpenoids, 6 aldehydes, 7 esters, 8 heterocyclic compounds, 6 alcohols, 2 aromatics, 1 ketone, and 3 phenol compounds. To clarify the fruit flavor characteristics of the different apricot germplasms, the aroma characteristics of the different apricot germplasms were determined according to the total rOAV of the aroma components of the same aroma; the total rOAV represents the total intensity of each aroma type (Supplementary Figure S4). The 24 apricot germplasm resources included 10 main types of fragrances: nutty, citrusy, fatty, waxy, woody, herbal, floral, fruity, green, and sweet. Citrusy, fatty, and green aromas contributed the most to the aroma characteristics of apricot fruits.

We analyzed the rOAV contributions of the compounds, and the (2S,4R)-4-methyl-2-(2-methylprop-1-enyl)oxane, (E)-non-2-enal, (3-methyl-3-sulfanylbutyl) formate, and thiophen-2-ylmethanethiol contents were greater than 1000; these results indicated that these substances greatly contributed to the overall aroma of the apricot fruits. The compounds with 100 < rOAV < 1000 included 4-methoxybenzaldehyde, 2-nonenal, 2-ethoxy-3-methylpyrazine, 1-thiophen-2-ylethanone, and 4-(2,6,6-trimethylcyclohexen-1-yl) butan-2-one and 4-methylphenol.

Table 2. Key volatile compounds with rOAVs greater than 1 in 24 apricot cultivars [a,b].

Volatile Compounds	Odor Description	Threshold μg/g	KZMYS	KBKYLK	LPHDK	KZL	SGJNL	AKDLZ	DX	KMT	KCBX	PNZ	LTBX	JNL	KLKYLK	ZSHYX	LJX	BS	SSGX	JGDMYS	SMT	MTYLK	JTY	YXB	ZGYDJX	CZH
Terpenoids																										
(5R)-2-methyl-5-prop-1-en-2-ylcyclohex-2-en-1-one	sweet, spearmint, herbal, minty	0.085	22.41	21.14	20.91	21.45	18.29	20.40	19.83	16.84	19.58	23.03	17.97	25.48	17.91	17.21	20.89	18.66	20.11	17.83	20.50	18.35	24.49	21.77	20.78	20.47
(2S,4R)-4-methyl-2-(2-methylprop-1-enyl)oxane	rose, cortex, green, floral, geranium, powdery, metallic	0.0002	3771.60	3403.88	5905.12	4688.88	2823.62	3901.08	3359.75	2507.70	3927.37	4460.41	3160.84	6420.09	3878.22	2045.42	4027.48	2773.81	4043.41	3211.68	3883.88	3873.36	4537.55	2449.56	2690.79	3003.40
3-methylidene-6-propan-2-ylcyclohexene	terpenic, herbal	0.036	9.62	7.19	53.43	27.03	8.22	37.03	18.86	15.82	41.74	49.30	24.07	76.22	50.00	1.28	40.42	12.43	50.89	31.35	40.02	62.15	20.44	1.58	2.96	16.76
(4R)-1-methyl-4-prop-1-en-2-ylcyclohexene	citrus	0.034	11.32	8.38	49.24	25.27	8.80	34.70	18.86	16.45	42.84	49.20	26.93	79.08	55.79	3.74	43.55	14.63	58.70	35.95	47.45	68.73	21.65	4.53	7.22	20.45
(2E)-3,7-dimethylocta-2,6-dienal	citrus, lemon	0.028	1.54	1.07	1.25	1.16	1.09	1.23	1.22	1.10	1.18	1.24	1.11	1.57	1.26	1.03	1.32	1.15	1.38	1.20	1.22	1.08	1.46	1.34	1.26	1.29
7-methyl-3-methylideneocta-1,6-diene	musty, balsamic, spice	0.015	10.17	9.56	36.02	18.03	9.41	23.72	14.28	14.23	29.03	34.11	19.75	52.33	33.67	5.89	31.35	14.29	40.51	26.18	31.43	48.65	18.74	6.84	8.53	16.31
(3Z)-3,7-dimethylocta-1,3,6-triene	warm, floral, herbal, flowery, sweet	0.034	9.15	6.25	53.63	22.55	7.00	26.45	16.60	12.04	32.41	45.89	18.44	80.70	48.71	1.51	37.53	8.79	49.58	27.66	37.81	64.65	17.21	1.70	2.33	12.10
6,6-dimethyl-2-methylidenebicyclo[3.1.1]heptane	dry, woody, resinous, pine, hay, green	0.14	5.51	4.09	30.18	13.65	4.65	18.25	9.62	8.59	21.15	25.00	13.54	42.92	26.61	1.02	20.84	6.92	29.93	17.69	23.11	36.52	12.88	1.08	2.41	9.41
4-methyl-2-(2-methylprop-1-enyl)-3,6-dihydro-2H-pyran	green, weedy, cortex, herbal, diphenyl, narcissus, celery	0.08	7.95	7.54	6.52	7.46	5.50	4.99	6.00	3.28	5.16	5.83	5.30	6.84	5.22	4.60	6.02	3.17	5.54	4.70	5.19	4.29	5.16	5.38	5.23	5.32
Aldehyde																										
(E)-non-2-enal	fatty, green, cucumber, aldehydic, citrus	0.00008	26,867.54	26,660.46	24,441.08	27,909.35	21,260.59	21,089.75	22,704.10	14,688.58	20,704.07	23,124.84	20,907.21	26,470.78	18,856.92	17,103.23	21,362.19	15,076.52	20,721.52	18,117.62	21,222.03	20,796.90	23,529.57	22,915.28	22,765.57	22,388.08
4-methoxybenzaldehyde	sweet, powdery, mimosa, floral, hawthorn, balsamic	0.0002	687.29	692.46	642.06	738.19	595.01	614.24	630.03	510.19	554.14	579.79	498.00	694.20	515.63	467.00	510.25	514.58	575.76	541.59	651.82	540.45	739.52	686.05	619.79	600.92
2-Nonenal	fatty, green, waxy, cucumber, melon	0.0001	509.62	492.87	462.81	489.54	410.19	420.18	440.02	283.47	410.50	425.58	467.63	480.70	348.99	361.49	499.42	347.51	369.76	368.05	442.95	416.72	602.18	584.35	525.56	391.75
(E)-oct-2-enal	fresh, cucumber, fatty, green, herbal, banana, waxy, leafy	0.003	21.47	5.61	6.69	4.62	4.25	6.22	3.24	3.30	5.04	5.27	16.13	5.54	2.95	2.95	4.42	3.99	5.31	7.66	3.18	5.30	3.27	5.01	2.95	2.92
(Z)-dec-7-enal	citrus, aldehydic, cucumber	0.0022	27.24	20.36	24.17	19.88	19.35	13.92	9.89	6.42	13.69	18.90	19.79	16.57	17.05	8.87	19.63	8.76	19.47	13.78	25.11	12.81	8.86	22.16	12.18	13.07
(E)-dec-4-enal	fresh, aldehydic, citrus, orange, mandarin, tangerine, green, fatty	0.025	4.50	4.46	4.41	4.34	3.14	3.67	3.80	2.81	4.18	5.43	3.42	4.98	5.13	2.82	3.02	3.12	4.17	3.13	5.74	4.40	2.63	3.13	3.26	2.83
Ester																										
2-methylbutyl hexanoate	ethereal	0.032	10.65	10.36	9.44	11.20	9.04	9.18	9.54	7.19	9.21	9.37	7.92	10.93	7.46	7.42	8.86	7.89	7.60	7.45	8.84	7.30	10.91	9.73	10.27	8.64
3-methylbutyl hexanoate	fruity, banana, apple, pineapple, green	0.32	1.97	2.02	1.81	2.07	1.60	1.69	1.74	1.39	1.66	1.83	1.69	2.04	1.47	1.44	1.73	1.52	1.65	1.70	1.71	1.60	2.08	1.87	1.92	1.63
methyl benzoate	phenol, wintergreen, almond, floral, canga	0.00052	529.56	321.85	1861.74	973.59	297.90	937.73	575.36	427.16	1102.52	1201.88	783.10	2468.27	1172.07	77.88	1286.40	396.75	1335.26	1099.57	1257.31	1187.62	962.02	99.98	183.67	497.97
hexyl propanoate	pear, green, fruity, musty, rotten	0.008	13.86	16.24	14.64	8.99	11.26	10.30	8.92	8.38	7.26	11.07	14.23	11.38	8.18	5.94	9.67	9.34	8.98	10.26	7.85	7.25	5.65	11.80	7.94	9.13
2-methylbutyl 2-methylpropanoate	fruity, ethereal, tropical, banana	0.014	5.42	5.27	5.10	4.86	3.80	5.71	4.55	4.08	6.22	4.19	9.59	5.06	4.96	3.64	5.17	3.55	4.84	4.11	4.65	4.32	4.54	4.15	4.24	4.50
cyclohexyl acetate	fruity, sweet, musty, ethereal	0.0016	57.11	60.73	76.80	73.71	55.86	83.92	72.61	66.38	86.54	94.66	70.69	103.19	85.92	54.59	90.51	70.00	92.82	88.92	94.89	127.54	72.12	61.01	74.73	75.50
(3-methyl-3-sulfanylbutyl) formate	sulfur, catty, caramel, onion, roasted coffee, roasted meat, tropical	0.000002	23,748.42	22,618.76	61,174.24	59,536.11	22,496.22	82,847.39	41,518.40	44,328.44	80,447.02	116,085.97	46,860.80	159,809.00	105,693.96	159,809.00	87,642.72	59,317.63	113,129.02	69,779.02	67,231.62	129,673.12	44,795.11	27,531.45	30,570.70	38,475.38

Table 2. *Cont.*

Volatile Compounds	Odor Description	Threshold µg/g	KZMYS	KBKYLK	LPHDK	KZL	SGJNL	AKDLZ	DX	KMT	KCBX	PNZ	LTBX	JNL	KLKYLK	ZSHYX	LJX	BS	SSGX	JGDMYS	SMT	MTYLK	JTY	YXB	ZGYDJX	CZH
Heterocyclic compound																										
2-ethoxy-3-methylpyrazine	hazelnut, roasted, almond, pineapple, earthy	0.0008	377.56	376.63	293.38	369.85	318.33	329.39	318.45	287.05	298.04	338.20	317.55	404.54	285.32	302.12	324.49	326.91	317.54	337.78	323.54	274.42	364.35	365.76	366.42	330.26
2-pentylfuran	fruity, green, earthy, beany, vegetable, metallic	0.006	94.33	37.64	61.23	42.78	39.38	54.02	36.43	39.58	42.79	58.19	78.20	75.74	41.99	29.12	54.94	48.83	64.12	53.31	51.09	52.08	42.36	39.37	35.26	47.38
thiophen-2-ylmethanethiol	roasted, coffee, fishy	0.00004	21,883.49	22,186.77	34,310.19	26,980.58	18,304.39	22,783.75	20,901.61	16,405.07	22,218.44	25,532.74	22,454.56	38,795.11	22,196.90	12,530.77	25,316.13	16,206.21	25,363.86	22,637.74	24,791.75	22,827.29	26,362.86	14,593.03	16,272.72	17,936.69
1-pyrazin-2-ylethanone	popcorn, nutty, corn, chip, bread, crust, chocolate, hazelnut, coffee	0.01	3.71	4.57	10.50	5.15	4.50	6.56	5.18	5.91	7.12	8.52	4.75	14.93	9.51	2.62	7.22	6.01	9.40	5.67	7.39	12.42	4.49	2.91	2.72	7.44
1-thiophen-2-ylethanone	sulfur, nutty, hazelnut, walnut	0.001	245.17	173.41	920.68	392.43	150.61	395.49	243.01	181.06	467.11	522.41	308.86	976.43	474.25	55.61	441.52	142.74	496.00	367.68	426.20	517.22	320.80	64.52	84.82	182.63
6-methyloxan-2-one	creamy, fruity, coconut	0.02683	6.47	5.49	16.03	8.38	4.55	7.34	5.90	4.90	8.73	10.16	6.98	18.31	9.72	2.91	8.73	4.65	10.01	7.80	9.14	10.46	8.60	3.44	3.85	5.44
2-ethyl-2H-furan-5-one	spice	0.0097	53.56	40.72	45.98	38.12	32.87	41.00	31.05	38.06	36.43	39.14	49.19	47.18	31.11	29.76	38.18	35.47	40.50	45.29	33.08	37.60	38.60	35.95	32.77	29.17
(5-methylfuran-2-yl)methanethiol	sulfur, roasted, coffee	0.00005	976.56	1491.94	974.18	1439.70	844.42	1074.18	1447.34	976.59	1004.76	1223.74	1081.20	1198.54	1149.23	850.56	1100.24	1081.61	1047.08	906.11	1016.13	1233.14	882.90	1080.02	1224.78	1187.38
Aromatics																										
naphthalene	pungent, dry, tarry	0.05	15.74	18.40	16.54	17.98	16.82	17.64	18.27	14.30	15.70	19.33	13.98	18.95	15.05	14.61	15.99	17.51	15.51	15.47	18.18	13.17	16.68	16.75	16.86	19.99
1-methyl-4-propan-2-ylbenzene	woody, citrus	0.0114	26.03	21.56	118.31	64.07	22.64	87.59	49.85	42.56	95.98	110.34	55.28	168.43	113.80	4.24	94.34	38.21	118.88	74.00	96.05	150.57	53.31	4.80	10.94	48.40
Alcohol																										
octan-1-ol	intense citrus, rose	0.022	3.47	2.38	2.46	2.58	2.23	1.94	2.10	1.91	2.21	2.33	3.33	3.08	1.98	1.96	2.43	2.01	2.79	2.36	2.35	2.29	2.74	2.64	2.27	2.14
undecan-6-ol	-	0.0086	8.93	6.31	6.16	6.36	5.42	5.83	5.59	5.48	4.82	4.85	5.08	6.95	6.00	4.94	5.54	5.50	5.42	5.23	5.22	4.22	6.05	5.74	5.94	6.05
decan-1-ol	fatty, waxy, floral, orange, sweet, watery	0.023	5.65	4.37	4.26	4.41	3.74	3.72	3.49	3.14	3.46	4.15	4.07	4.98	4.17	3.73	3.98	3.74	3.84	3.50	3.92	2.91	4.34	3.90	4.03	4.13
(5E)-3,7-dimethylocta-1,5,7-trien-3-ol	sweet, tropical, ocimene, fennel, ginger, myrcene	0.11	32.22	23.34	130.65	55.31	20.59	56.71	32.84	25.34	74.59	89.28	48.69	140.93	63.98	4.60	66.12	19.36	69.80	63.65	61.81	74.05	46.54	5.19	7.76	24.20
(3E,6Z)-nona-3,6-dien-1-ol	fatty, green, cucumber, green pepper, fruity, watermelon	0.003	36.73	26.89	26.92	28.91	19.57	19.71	20.80	12.75	19.70	24.66	23.55	29.13	20.63	17.40	22.78	12.12	15.88	17.26	20.39	16.86	18.25	22.74	22.84	17.39
4-phenylbutan-2-ol	floral, peony, foliage, sweet, mimosa, heliotrope	0.0043	7.97	10.61	7.37	8.55	6.21	6.31	7.06	6.87	5.95	6.23	4.94	7.18	5.37	5.14	6.02	5.21	6.50	5.08	8.29	5.16	11.32	6.26	7.24	5.23
Ketone																										
4-(2,6,6-trimethylcyclohexen-1-yl)butan-2-one	earthy, woody, mahogany, orris, dry, amber	0.0036	19.15	74.11	270.01	570.78	257.37	418.26	24.81	246.32	262.53	419.65	91.23	235.73	11.93	296.37	149.82	16.21	40.86	569.51	93.80	1241.93	3.65	240.95	687.99	267.79
Phenol																										
4-methylphenol	phenol, narcissus, animalic, mimosa	0.00024	234.68	170.26	1186.66	532.67	160.03	557.54	316.89	211.83	686.42	898.13	420.03	1574.38	935.60	53.81	783.76	223.10	983.15	545.64	822.98	1162.24	364.63	78.72	122.28	303.18
2-methylphenol	phenol	0.0039	10.86	11.38	19.88	16.08	10.85	17.23	14.70	12.98	15.87	17.71	13.45	24.47	16.71	10.08	15.51	12.41	17.62	13.61	17.64	20.59	14.97	11.64	13.50	14.56
phenol	phenol, medicinal	0.03	4.32	4.22	19.19	9.04	4.04	13.37	6.82	7.76	15.01	17.53	9.00	25.05	19.49	2.81	14.30	7.46	18.24	11.15	14.95	23.62	7.86	3.90	3.40	9.20

[a] All odor thresholds were obtained from Tan et al. [21]. [b] The aroma of the material is described at http://www.thegoodscentscompany.com

3.6. Principal Component Analysis and Heatmap Analysis

To obtain more accurate and intuitive classifications of the volatile components present, PCA was performed for the volatile compounds in the fruits of the 24 apricot cultivars. The PCA diagram clearly reflects the clustering and distribution of different apricot cultivars (Figure 3a), and the principal components explain the changes in the volatile compounds in apricot fruits. Principal component analysis revealed three principal components (PCs), and PC1, PC2, and PC3 accounted for 29.18%, 16.66%, and 12.45% of the total variables, respectively. The cumulative variance contribution rate reached 58.29%, and the distribution of samples in each group was relatively concentrated, indicating good repeatability of samples within groups. Evident separation was found among the 24 tested apricot cultivars. The results indicated that the volatile compounds differed among the cultivars. MTYLK and KMT were distant from those of the other 22 apricot cultivars; these results indicated that the volatile compound contents of the MTYLK and KMT fruits were considerably different from those of the fruits of the other cultivars.

a b

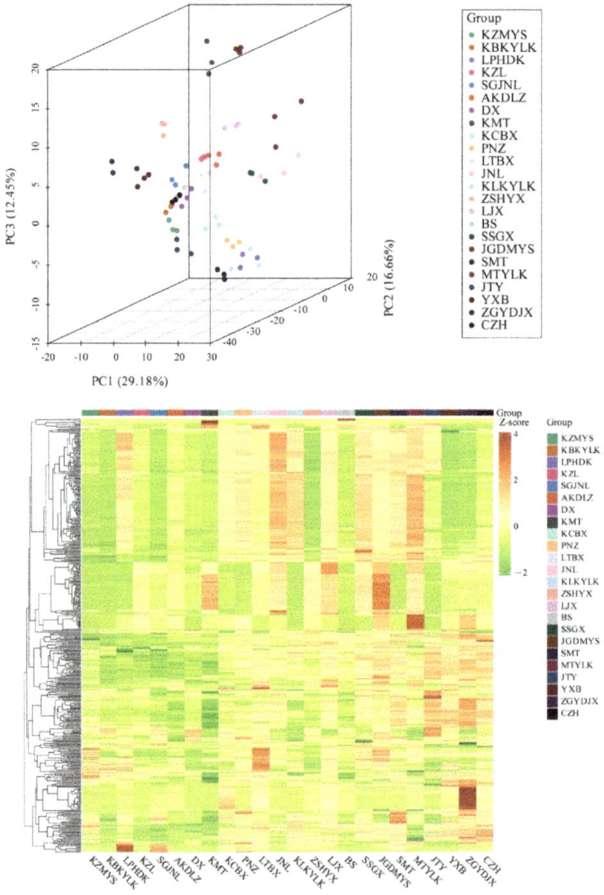

Figure 3. (**a**) Principal component analysis of the volatile compounds in the fruits of 24 apricot cultivars and (**b**) clustering heatmaps of the volatile compounds in the fruits of 24 apricot cultivars. In this figure, the cultivar name is shown in the horizontal direction, the volatile compound information is presented in the vertical direction, the group is under the group column, and the different colors correspond to different values obtained after the standardization treatment of different relative contents (red represents high content, and green represents low content).

To directly show subtle differences in the contents and components of the mature fruits of the 24 apricot cultivars, a cluster heatmap of 429 volatile aroma compounds was generated (Figure 3b). Lower contents of the volatile aroma compounds correlate to a greener color, whereas higher contents correlate to a redder color; these results reflect differences between the different cultivars and the contents of the volatile compounds.

3.7. Correlation Between the Sensory Evaluations and HS–SPME–GC–MS

Due to their high rOAV based on the GC–MS analysis, forty-two key aroma volatile compounds were selected to be associated with the sensory evaluation results via Pearson's rank correlation, as shown in Figure 4. Most terpenoids, such as 6,6-dimethyl-2-methylidenebicyclo [3.1.1]heptane, 7-methyl-3-methylideneocta-1,6-diene, and 3-methylidene-6-propan-2-ylcyclohexene, were positively correlated with nutty, herbal, and green aromas. In addition, esters such as thiophen-2-ylmethanethiol, 1-thiophen-2-ylethanone, and 6-methyloxan-2-one were also positively associated with floral, herbal, and nutty compounds; these results were consistent with the HS-SPME-GC–MS results.

Figure 4. Correlation analysis of the aroma compounds with sensory evaluation results on the basis of Pearson's rank correlation (* represents a significant difference at $p < 0.05$, ** $p < 0.01$, and *** $p < 0.001$, and color depth represents the Pearson phase relationship value).

4. Discussion

Fruit flavor involves aroma and taste; aroma is determined by the numerous volatile substances, and taste is determined by the organic acids and sugars [22]. At present, many reports exist on the use of HS-SPME-GC–MS to analyze volatile components in fruits. For example, Li et al. [23] used HS-SPME-GC–MS to analyze the volatile organic compounds of passion fruit at different maturity stages and identified 148 volatile compounds in passion fruit at the maturity stage; among these compounds, the contribution rate of the ester compound content was the highest. Lan et al. [24] used HS-SPME-GC–MS to detect 172 aroma chemical compositions from 23 kiwifruit varieties, and the identified components could be divided into 9 categories. Sensory assessment is the most reliable technique for converting human perceptions of the aroma, appearance, taste, and aftertaste of food into digital data [25]. In this study, the volatile compounds of 24 apricot cultivars were analyzed via sensory evaluation combined with volatile metabolomic analysis via HS-SPME-GC–MS, and 429 volatile aromatic compounds were detected and could be separated into 16 main categories. Tang et al. [11] detected and identified 15 volatile compounds in apricot fruits. Solís-Solís et al. [12] used simultaneous distillation extraction (SDE), solid-phase extraction (SPE), and liquid–liquid extraction (LLE) methods to detect the volatile components of 8 different varieties of apricot fruits and identified 21, 16, and 8 volatile compounds, respectively. These results were likely caused by the different detection varieties or detection methods and resulted in differences in the amount of compounds detected. For example, Zhao et al. [26]. determined the volatile compounds in the fruits of 4 apricot varieties in Xinjiang using HS-SPME-GC–MS technology and identified a total of 63 volatile compounds; these results likely occurred because of the use of different databases.

The aromas of fruits of different cultivars of the same tree species can differ. In this study, volatile metabolomics combined with the rOAV method was used to analyze the volatile components in the fruits of 24 apricot cultivars. Terpenoids were predominant in the fruits of the apricots and accounted for 17.95% of the total volatile components, followed by 70 esters; the esters accounted for 16.32% of the total volatile components. These percentages were consistent with the results from Zhao et al. and Aubert et al. [27,28]. However, Ayour et al. [29] identified volatile components in the fruits of 10 apricot clones from Morocco, and their results revealed that the dominant components in these fruits were aldehydes, alcohols, and acetic esters. Their results differed from those of this study, and the difference were likely caused by genetic factors or different environmental conditions. Similar findings have been reported for other tree species. For example, Giannetti et al. [30] detected and analyzed the volatile substances of 42 apple fruits in Italy and reported that ester compounds were their main volatile compounds. However, Liu et al. [31] reported that aldehyde volatile compounds were the main volatile substances in 'Ruixue' apples. These findings confirmed that the key aroma compounds in fruits of the same tree species and different cultivars greatly differed because of different genetic backgrounds.

The types and contents of volatile compounds may be related to cultivation conditions and climate factors, and substantial differences exist among the different varieties in the same producing region [32]. Wang et al. [33] studied the volatile compositions of the fruits of pear cultivars from different regions and reported considerable differences in the volatile components of these fruits. In Malatya, SPME-GC–MS was used to detect important aroma compounds in the fruits of 15 cultivated apricot varieties, and large differences were observed in the volatile compounds in the fruits of apricot varieties grown under the same conditions; their results were in agreement with those from this study [34]. Thus, under the same cultivation environment, the contents of volatile components in the fruits of 24 apricot cultivars greatly varied and ranged from 112.77 to 317.67 μg/g. Therefore, cultivation conditions and genetic background are likely important factors leading to differences in volatile compounds.

The aroma contribution of a volatile organic compound to fruits depends not only on the volatile species and its content but also on its odor threshold [35]. The rOAV method has been widely used to quantitatively evaluate the odor contribution of aromatic compounds [36–38]. In addition, the rOAV method was used to identify the volatile compounds of 24 apricot cultivars. Ninety-three key aroma compounds were identified, and terpenoids, esters, aldehydes, and heterocyclic compounds were the key aroma compounds. In addition, substances with rOAVs above 1000 were screened. Here, (2S,4R)-4-methyl-

2-(2-methylprop-1-enyl)oxane, (E)-non-2-enal, (3-methyl-3-sulfanylbutyl) formate, and thiophen-2-ylmethanethiol were identified as key aroma compounds. These results were inconsistent with those from Greger and Schieberle's study; these differences occurred because the upgraded plant metabolism database used in this study was more abundant and comprehensive [14]. In previous studies, key volatile compounds of mulberry [39], peach [27], and Gannan navel orange [40] were identified via the rOAV method. In our study, we confirmed the practicability and accuracy of this method from the perspective of the key aroma compounds in apricot fruit.

5. Conclusions

In summary, sensory evaluation revealed that the apricot fruits had typical fruity, sweet, and green aromas. A total of 429 volatile compounds were detected through qualitative and quantitative analyses of 24 apricot cultivars via the HS-SPME-GC–MS metabolomics method. These results indicated that terpenoids, heterocyclic compounds, aromatics, nitrogen compounds, aldehydes, alcohols, and hydrocarbons were the main volatile compounds in the apricot fruits. Forty-two key volatile compounds were screened according to their aroma activity, and the total volatile compound content in the fruits of 24 apricot cultivars ranged from 112.77 to 317.67 μg/g. The apricot cultivars with the highest and lowest total volatile compound contents were 'JNL' (317.67 μg/g) and 'ZSHYX' (112.77 μg/g), respectively. Forty-two key aroma compounds contributing to the aroma of apricot fruits were identified via the rOAV method, and (2S,4R)-4-methyl-2-(2-methylprop-1-enyl)oxane, (E)-non-2-enal, (3-methyl-3-sulfanylbutyl) formate, and thiophen-2-ylmethanethiol were determined to be the major active aroma compounds in apricot fruits. These results greatly contribute to the study of volatile metabolites in apricot fruits and reveal apricot germplasm resources with strong aromas; thus, our study provides a theoretical basis for apricot breeding and can aid in the improvement of apricot fruit quality.

Supplementary Materials: The following supporting information can be downloaded at https://www.mdpi.com/article/10.3390/foods13233912/s1: Supplementary Figure S1. Pictures of the mature fruit of the 24 apricot cultivars; Supplementary Figure S2. Representative total ion chromatogram of a quality control sample (a) and total ion flow diagram of the mixed QC sample (TIC, the sum of the strengths of all ions in the mass spectrometer at each time point) (b)—the horizontal coordinate is the retention time of metabolite detection (RT), and the vertical coordinate is the ion flow intensity of ion detection (CPS); Supplementary Table S1. Relative contents of the identified volatile compounds in the fruits of 24 apricot cultivars; Supplementary Figure S3. Column chart of the volatile compound contents in 24 apricot varieties. A, B, C, D, E, F, G, H, I, J, K, L, M, N, O, and P represent the terpenoids, esters, heterocyclic compounds, ketones, alcohols, aldehydes, aromatics, hydrocarbons, phenols, nitrogen compounds, halogenated hydrocarbons, amines, acids, sulfur compounds, ether, and other compound contents, respectively. Statistical analysis was performed via a one-way analysis of variance (ANOVA), and different letters indicate significant differences at the 0.05 level; and Supplementary Figure S4. Histogram of the aroma characteristics of fruits of different apricot germplasm resources.

Author Contributions: X.Z., data curation, methodology, visualization, software, formal analysis, investigation, and writing—original draft; S.Z., investigation, data curation, and resources; W.G., supervision; J.J., investigation and data curation; K.L., conceptualization and investigation; Z.T., resources; S.W., investigation; and W.Z., conceptualization, supervision, project administration, funding acquisition, and writing—review and editing. All authors have read and agreed to the published version of the manuscript.

Funding: This research was financially supported by the Major Science and Technology Projects in the Xinjiang Uygur Autonomous Region [2023A02008-1], the 2024 Special Fund for Forest and Grass of the Xinjiang Uygur Autonomous Region (Phenotypic Identification of Forest Fruit Germplasm Resources), the Xinjiang Endemic Fruit Germplasm Resources Conservation Project [19230836], and the National Horticultural Germplasm Xinjiang Special Fruit Trees and Rootstocks Subbank Operation Service [NHGRC2023-NH13].

Institutional Review Board Statement: The study was approved by the Institutional Review Board of College of Horticulture, Xinjiang Agricultural University (2 October 2024).

Informed Consent Statement: Informed consent was obtained from all subjects involved in the study.

Data Availability Statement: The original contributions presented in the study are included in the article and Supplementary Materials, further inquiries can be directed to the corresponding author.

Acknowledgments: We would like to express our gratitude to MWDB for providing the database.

Conflicts of Interest: The authors declare that they have no conflicts of interest.

References

1. Cheng, A.X.; Lou, Y.G.; Mao, Y.B.; Lu, S.; Wang, L.J.; Chen, X.Y. Plant terpenoids: Biosynthesis and ecological functions. *J. Integr. Plant Biol.* **2007**, *49*, 179–186. [CrossRef]
2. Dudareva, N.; Klempien, A.; Muhlemann, J.K.; Kaplan, I. Biosynthesis, function and metabolic engineering of plant volatile organic compounds. *New Phytol.* **2013**, *198*, 16–32. [CrossRef] [PubMed]
3. Hou, J.X.; Liang, L.; Wang, Y.X. Volatile composition changes in navel orange at different growth stages by HS-SPME–GC–MS. *Food Res. Int.* **2020**, *136*, 109333. [CrossRef] [PubMed]
4. Yuan, H.L.; Cao, G.P.; Hou, X.D.; Huang, M.L.; Du, P.M.; Tan, T.T.; Luo, J. Development of a widely targeted volatilomics method for profiling volatilomes in plants. *Mol. Plant* **2022**, *15*, 189–202. [CrossRef] [PubMed]
5. El Hadi, M.A.M.; Zhang, F.J.; Wu, F.F.; Zhou, C.H.; Tao, J. Advances in fruit aroma volatile research. *Molecules* **2013**, *18*, 8200–8229. [CrossRef] [PubMed]
6. Ma, X.W.; Su, M.Q.; Wu, H.X.; Zhou, Y.G.; Wang, S.B. Analysis of the volatile profile of core Chinese mango germplasm by headspace solid-phase microextraction coupled with gas chromatography–mass spectrometry. *Molecules* **2018**, *23*, 1480. [CrossRef]
7. Goff, S.A.; Klee, H.J. Plant volatile compounds: Sensory cues for health and nutritional value? *Science* **2006**, *311*, 815–819. [CrossRef]
8. Şeker, M.; Gündoğdu, M.A.; Ekinci, N.; Gür, E. Recent developments on aroma biochemistry in fresh fruits. *Int. J. Innov. Approaches Sci. Res.* **2021**, *5*, 84–103. [CrossRef]
9. Zhou, W.; Zhao, S.; Xu, M.; Niu, Y.; Nasier, M.; Fan, G.; Quan, S.; Zhang, S.; Wang, Y.; Liao, K. Identification of key genes controlling carotenoid metabolism during apricot fruit development by integrating metabolic phenotypes and gene expression profiles. *J. Agric. Food Chem.* **2021**, *69*, 9472–9483. [CrossRef] [PubMed]
10. Dondini, L.; Domenichini, C.; Dong, Y.; Gennari, F.; Bassi, D.; Foschi, S.; Tartarini, S. Quantitative trait loci mapping and identification of candidate genes linked to fruit acidity in apricot (*Prunus armeniaca* L.). *Front. Plant Sci.* **2022**, *13*, 838370. [CrossRef]
11. Tang, C.S.; Jennings, W.G. Volatile compound of apricot. *J. Agric. Food Chem.* **1967**, *15*, 24–28.
12. Solís-Solís, H.M.; Calderón-Santoyo, M.; Schorr-Galindo, S.; Luna-Solano, G.; Ragazzo-Sánchez, J.A. Characterization of aroma potential of apricot varieties using different extraction techniques. *Food Chem.* **2007**, *105*, 829–837. [CrossRef]
13. Zhang, H.H.; Liu, W.S.; Fang, J.B.; Chen, S.; Liu, Y.C.; Wu, B.; Li, S. Volatile profiles of apricot cultivars (*Prunus armeniaca* Lam.) evaluated by headspace solid phase microextraction gas chromatography–mass spectrometry. *Anal. Lett.* **2014**, *47*, 433–452. [CrossRef]
14. Greger, V.; Schieberle, P. Characterization of the key aroma compounds in apricots (*Prunus armeniaca*) by application of the molecular sensory science concept. *J. Agric. Food Chem.* **2007**, *55*, 5221–5228. [CrossRef] [PubMed]
15. Abdulra'uf, L.B.; Tan, G.H. Chemometric approach to the optimization of HS-SPME/GC–MS for the determination of multiclass pesticide residues in fruits and vegetables. *Food Chem.* **2015**, *177*, 267–273. [CrossRef]
16. Savelieva, E.I.; Gavrilova, O.P.; Gagkaeva, T.Y. Using solid-phase microextraction combined with gas chromatography–mass spectrometry for the study of the volatile products of biosynthesis released by plants and microorganisms. *J. Anal. Chem.* **2014**, *69*, 609–615. [CrossRef]
17. Fan, X.; Zhao, H.; Wang, X.; Cao, J.; Jiang, W. Sugar and organic acid composition of apricot and their contribution to sensory quality and consumer satisfaction. *Sci. Hortic.* **2017**, *225*, 553–560. [CrossRef]
18. Zhang, C.; Zhou, C.Z.; Xu, K.; Tian, C.Y.; Zhang, M.C.; Lu, L.; Guo, Y. A comprehensive investigation of macro-composition and volatile compounds in spring-picked and autumn-picked white tea. *Foods* **2022**, *11*, 3628. [CrossRef]
19. Yue, C.; Cao, H.L.; Zhang, S.R.; Hao, Z.L.; Wu, Z.J.; Luo, L.Y.; Zeng, L. Aroma characteristics of Wuyi rock tea prepared from 16 different tea plant varieties. *Food Chem. X* **2023**, *17*, 100586. [CrossRef]
20. Xue, J.J.; Liu, P.P.; Yin, J.F.; Wang, W.W.; Zhang, J.Y.; Wang, W.; Jiang, H. Dynamic changes in volatile compounds of shaken black tea during its manufacture by GC× GC–TOFMS and multivariate data analysis. *Foods* **2022**, *11*, 1228. [CrossRef]
21. Tan, F.; Wang, P.; Zhan, P.; Tian, H. Characterization of key aroma compounds in flat peach juice based on gas chromatography-mass spectrometry-olfactometry (GC-MS-O), odor activity value (OAV), aroma recombination, and omission experiments. *Food Chem.* **2022**, *366*, 130604. [CrossRef] [PubMed]
22. Klee, H.J. Improving the flavor of fresh fruits: Genomics, biochemistry, and biotechnology. *New Phytol.* **2010**, *187*, 44–56. [CrossRef] [PubMed]

23. Li, C.B.; Xin, M.; Li, L.; He, X.M.; Yi, P.; Tang, Y.Y.; Li, Z. Characterization of the aromatic profile of purple passion fruit (Passiflora edulis Sims) during ripening by HS-SPME-GC/MS and RNA sequencing. *Food Chem.* **2021**, *355*, 129685. [CrossRef] [PubMed]
24. Lan, T.; Gao, C.X.; Yuan, Q.Y.; Wang, J.Q.; Zhang, H.X.; Sun, X.Y.; Ma, T. Analysis of the aroma chemical composition of commonly planted kiwifruit cultivars in China. *Foods* **2021**, *10*, 1645. [CrossRef] [PubMed]
25. Xu, Y.; Yao, L.; Wang, Y.; Shen, J.; Chen, D.; Feng, T. Comparative analysis of the aromatic profiles of citri sarcodactylis fructus from various geographical regions using GC-IMS, GC-MS, and sensory evaluation. *Food Biosci.* **2024**, *58*, 103752. [CrossRef]
26. Zhao, C.; Sun, J.K.; Pu, X.L.; Shi, X.W.; Cheng, W.D.; Wang, B. Volatile compounds analysis and biomarkers identification of four native apricot (*Prunus armeniaca* L.) cultivars grown in Xinjiang region of China. *Foods* **2022**, *11*, 2297. [CrossRef]
27. Zhao, R.R.; Xu, Y.Y.; Li, C.Y.; Wang, X.W.; Du, J.; Wang, C.Q.; Wang, B. Analysis of physicochemical characteristics, antioxidant activity, and key aroma compounds of five flat peach cultivars grown in Xinjiang. *LWT* **2023**, *176*, 114550. [CrossRef]
28. Aubert, C.; Chanforan, C. Postharvest changes in physicochemical properties and volatile constituents of apricot (*Prunus armeniaca* L.). characterization of 28 cultivars. *J. Agric. Food Chem.* **2007**, *55*, 3074–3082. [CrossRef]
29. Ayour, J.; Benichou, M.; Alahyane, A.; Harrak, H. Relationships between biochemical criteria, volatile compounds, and sensory profiles of ten apricot clones at commercial and consumption ripening stages. *J. Food Qual.* **2020**, *2020*, 8873835. [CrossRef]
30. Giannetti, V.; Mariani, M.B.; Mannino, P.; Marini, F. Volatile fraction analysis by HS-SPME/GC–MS and chemometric modeling for traceability of apples cultivated in the Northeast Italy. *Food Control* **2017**, *78*, 215–221. [CrossRef]
31. Liu, X.J.; Feng, Y.F.; Li, S.S.; Li, D.M.; Yu, J.; Zhao, Z.Y. Jasmonate-induced MdMYC2 improves fruit aroma during storage of 'Ruixue' apple based on transcriptomic, metabolic and functional analyses. *LWT* **2023**, *185*, 115168. [CrossRef]
32. Zhang, M.Y.; Xue, C.; Xu, L.; Sun, H.; Qin, M.F.; Zhang, S.; Wu, J. Distinct transcriptome profiles reveal gene expression patterns during fruit development and maturation in five main cultivated species of pear (*Pyrus* L.). *Sci. Rep.* **2016**, *6*, 28130. [CrossRef] [PubMed]
33. Wang, X.H.; Chen, Y.Y.; Zhang, J.J.; Wang, Z.W.; Qi, K.J.; Li, H.X.; Yin, H. Comparative analysis of volatile aromatic compounds from a wide range of pear (*Pyrus* L.) germplasm resources based on HS-SPME with GC–MS. *Food Chem.* **2023**, *418*, 135963. [CrossRef]
34. Gokbulut, I.; Karabulut, I. SPME-GC–MS detection of volatile compounds in apricot varieties. *Food Chem.* **2012**, *132*, 1098–1102. [CrossRef]
35. Chen, Y.; Yin, H.; Wu, X.; Shi, X.; Qi, K.; Zhang, S. Comparative analysis of the volatile organic compounds in mature fruits of 12 Occidental pear (*Pyrus communis* L.) cultivars. *Sci. Hortic.* **2018**, *240*, 239–248. [CrossRef]
36. Gao, L.; Zhang, L.H.; Liu, J.; Zhang, X.; Lu, Y.H. Analysis of the volatile flavor compounds of pomegranate seeds at different processing temperatures by GC-IMS. *Molecules* **2023**, *28*, 2717. [CrossRef]
37. Xiao, Z.B.; Chen, H.T.; Niu, Y.W.; Zhu, J.C. Characterization of the aroma-active compounds in banana (Musa AAA Red green) and their contributions to the enhancement of sweetness perception. *J. Agric. Food Chem.* **2021**, *69*, 15301–15313. [CrossRef] [PubMed]
38. Yin, X.; Huang, J.A.; Huang, J.; Wu, W.L.; Tong, T.; Liu, S.J.; Zhang, S. Identification of volatile and odor-active compounds in Hunan black tea by SPME/GC–MS and multivariate analysis. *LWT* **2022**, *164*, 113656. [CrossRef]
39. Zhu, J.C.; Wang, L.Y.; Xiao, Z.B.; Niu, Y.W. Characterization of the key aroma compounds in mulberry fruits by gas chromatography-olfactometry (GC-O), odor activity value (OAV), gas chromatography–mass spectrometry (GC–MS) and flame photometric detection (FPD). *Food Chem.* **2018**, *245*, 775–785. [CrossRef]
40. Ye, Y.H.; Zheng, S.Y.; Wang, Y.X. Analysis of aroma components changes in Gannan navel orange at different growth stages by HS-SPME-GC–MS, OAV, and multivariate analysis. *Food Res. Int.* **2024**, *175*, 113622. [CrossRef]

Disclaimer/Publisher's Note: The statements, opinions and data contained in all publications are solely those of the individual author(s) and contributor(s) and not of MDPI and/or the editor(s). MDPI and/or the editor(s) disclaim responsibility for any injury to people or property resulting from any ideas, methods, instructions or products referred to in the content.

foods

Article

Flavor Characteristics of Ten Peanut Varieties from China

Bin Ding [1,†], Fei Wang [2,†], Bei Zhang [2], Mengshi Feng [1], Lei Chang [1], Yuyang Shao [2], Yan Sun [2], Ying Jiang [1], Rui Wang [1], Libin Wang [3], Jixian Xie [1,*] and Chunlu Qian [2,*]

[1] Taizhou Institute of Agricultural Sciences, Jiangsu Academy of Agricultural Sciences, Taizhou 210014, China; 20172201@jaas.ac.cn (B.D.); 20172202@jaas.ac.cn (M.F.); 20142207@jaas.ac.cn (L.C.); 20142204@jaas.ac.cn (Y.J.); wangruishiwei@163.com (R.W.)

[2] Department of Food Science and Engineering, School of Food Science and Engineering, Yangzhou University, Yangzhou 225012, China; wfei1101@163.com (F.W.); zhbeiang06@163.com (B.Z.); qq1046769248@163.com (Y.S.); 18705276730@163.com (Y.S.)

[3] College of Food Science and Technology, Nanjing Agricultural University, Nanjing 210095, China; wanglibin@njau.edu.cn

* Correspondence: 20162201@jaas.ac.cn (J.X.); clqian@yzu.edu.cn (C.Q.); Tel.: +86-514-87978050 (C.Q.)

† These authors contributed equally to this paper.

Abstract: To investigate the flavor characteristics of peanuts grown in Jiangsu, China, ten local varieties were selected. The amino acids, 5′-nucleotides and volatile substances were detected, and the flavor and odor characteristics of these varieties were estimated using an electronic tongue and nose. The results showed that the fat and protein contents of ten peanut varieties changed significantly ($p < 0.05$), and may have been negatively correlated with those of the Taihua 6 variety—in particular, having the highest protein content and the lowest fat content. The amino acid contents of the peanuts were 20.08 g/100 g (Taihua 4)–27.18 g/100 g (Taihua 6). Taihua 6 also contained the highest bitter (10.41 g/100 g) and sweet (6.06 g/100 g) amino acids, and Taihua 10 had the highest monosodium glutamate-like amino acids (7.61 g/100 g). The content of 5′-nucleotides ranged from 0.08 mg/g (Taihua 9725) to 0.14 mg/g (Taihua 0122–601). Additionally, 5′-cytidylate monophosphate (5′-CMP) and 5′-adenosine monophosphate (5′-AMP) were the major 5′-nucleotides detected in the peanuts. A total of 42 kinds of volatile flavor compounds were detected, with both Taihua 4 and 6 showing the most (18 kinds) and the highest content being in Taihua 4 (7.46%). Both Taihua 9725 and 9922 exhibited the fewest kinds (nine kinds) of volatile components, and the lowest content was in Taihua 9725 (3.15%). Formic acid hexyl ester was the most abundant volatile substance in peanuts, and the highest level (3.63%) was detected in Taihua 7506. The electronic tongue and nose indicated that the greatest taste difference among the ten varieties of peanuts was mainly related to sourness, and Taihua 4 and Taihua 9922 had special taste characteristics. On the other hand, the greatest smell difference among the ten varieties of peanuts was mostly for methane and sulfur organic substances, and Taihua 0605-2 had a special and strong smell characteristic. In conclusion, the content and composition differences of the flavor substances of ten peanut varieties were responsible for their divergences in taste and smell. These results will provide guidelines for the further use (freshly consumed or processed) of these ten peanut varieties.

Keywords: peanut; flavor; free amino acids; 5′-nucleotides; volatiles

1. Introduction

Peanuts (*Arachis hypogaea* L.) are among the most important economic crops. The seeds of peanuts can be eaten directly or used for oil, to make butter or in a variety of dishes [1]. China is the main producer and processor of peanuts, accounting for more than half of global peanut production; hence, peanuts are the main crop in many places across China. Because of the high consumption, China imports more peanuts than it exports, so the self-cultivation of peanuts is important in China. As the yield of peanuts increases, the evaluation of peanuts begins to shift from quantity to quality. The main ingredients of peanuts are fats and proteins, along with health-beneficial substances (oleic acid, vitamin E, folic acid, niacin, flavonoids, isoflavones) [2,3], and these components influence the flavor characteristics of peanuts. As a type of pulse food, peanuts can be freshly consumed and the flavor of peanuts directly influences their acceptance by consumers. As a raw material for other food production, peanuts can be consumed after processing or cooking, and the flavor of peanuts affects the flavor characteristics of processed or cooked foods, as well as their acceptance [4,5]. Therefore, the flavor of peanuts is one of the most vital commercial characteristics considered during the breeding, cultivation and processing of these pulses. The flavor of raw plant food materials, including peanuts, can be greatly affected by factors such as variety [6], location [7], harvest season and ripening stage [8], among which the variety has the greatest impact on the flavor characteristics of plant products, as the variety impacts the differences in genetic material.

Peanuts comprise over 500 varieties, each of which exhibits different planting characteristics and quality representations [9]. The flavor characteristics of certain peanut varieties are unique and heritable [10–12], so a comparative analysis of the flavors of similar varieties is necessary for planting and processing. Peanut varieties are discreetly and carefully selected for butter processing purposes based on their content and composition regarding color, flavor, nutritional substances and market acceptance [5,13]. Their flavor components comprise non-volatile and volatile substances, free amino acids and 5′-nucleotides as the main non-volatile flavor materials. These determine the taste characteristics of foods, and small-molecule volatile substances dominate the characteristic smells of foods. The flavor characteristics of foods are usually estimated via sensory evaluation, but high subjectivity and regional characteristics mean that the evaluation is not objective. The electronic tongue and nose equipped sensors can detect flavor substances and produce response values that express flavor characteristics in a quantitative manner, making this method much more objective and credible for evaluating flavor [14]. High-performance liquid chromatography (HPLC) and headspace solid-phase micro-extraction gas chromatography (HS-SPME–GC–MS) are efficient and well-established methods for separating various non-volatile and volatile flavor substances, and were previously successfully applied in the detection of flavor substances in coffee [15], meat [16] and lotus roots [14].

In this study, the flavors of ten Chinese peanut varieties were compared, their non-volatile and volatile flavor components were identified and their flavor characteristics were estimated using an electronic tongue and nose. The obtained results established the variety-specific flavor profiles of the peanuts from Jiangsu, China, and also revealed the flavor differences between ten varieties of peanuts, forming the basis for the further research and utilization (fresh, processed or cooked) of these ten peanut varieties.

2. Materials and Methods

2.1. Raw Materials

Ten varieties of fresh peanuts (*Arachis hypogaea* L.) were collected from the pilot farm of Taizhou Institute of Agricultural Science, Jiangsu Academy of Agricultural Sciences, Jiangsu, China. The ten varieties of peanuts were named 'Taihua0122-601', 'Taihua6', 'Taihua0311', 'Taihua0605-2', 'Taihua10', 'Taihua11', 'Taihua7506', 'Taihua9725', 'Taihua4' and 'Taihua9922'. All the varieties of peanuts are widely cultivated in China, and were collected and planted for a deep comparative analysis (production, resistance, flavor, processing, etc.) on the same pilot farm. All the fresh peanuts were dried under natural conditions (30–35 °C) and peeled, and the seeds were collected separately. Then, they were sealed in plastic bags (polyethylene, oxygen permeability, 12.5 $cm^3\ m^{-2}\ h^{-1}\ mpa^{-1}$) and stored at 4 °C with 50% relative humidity for a further analysis 1 week later. All the measurements were replicated in triplicate.

2.2. Physicochemical Properties of Peanuts

The water content of the peanuts was measured via further dehydration in a drying box (Yiheng DHG-9070A, Shanghai Yiheng Medical Equipment Co., Ltd., Shanghai, China) at 105 °C until a constant weight was obtained, and the difference between the weight before and weight after drying was recorded as the water content. The fat content of the peanuts was measured using the Soxhlet extraction method [17]. After the measurement of the water content, the dried sample (5 g) was crushed into powder (200 mesh pellets), wrapped in a filter paper cartridge and placed in a Soxhlet extractor for the total extraction of the fat. The protein content of the peanuts was measured using an automatic Kjeldahl nitrogen analyzer (Haineng K9840, Jinan Haineng Instrument Co., Ltd., Jinan, China). All operations followed the manufacturers' instructions for equipment use.

2.3. Determination of Non-Volatile and Volatile Components

The non-volatile (free amino acids and 5′-nucleotides) and volatile components of the peanuts were measured via HPLC and HS–SPME–GC–MS, and all the procedures were carried out employing the method that we previously developed [14]. A total of 1 g of the peanut sample was ground with distilled water, and incubated at 80 °C for 20 min (amino acids) or 100 °C for 5 min (5′-nucleotides). Then, the sample was topped up to a 10 mL volume after cooling to room temperature, and filtered using a 0.45 μm nylon membrane. During the extraction, both amino acids and 5′-nucleotides were separated on an inertsil–ODS–SP–C18 column (250 × 4.6 mm, Shimadzu, Kyoto, Japan), and determined via Waters E2695 series HPLC (Waters Technologies, Milford, MA, USA). A total of 3 g of the peanut sample was ground with 3 mL of sodium chloride (saturation) and 10 μL of 1pyr2-dichlorobenzene as the internal standard and put into a 20 mL glass bottle. The bottle was sealed, and the contents were mixed via magnetic stirring. Polydimethylsiloxane (PDMS) (65 μm) (preheated at 250 °C for 20 min) was inserted into the headspace (HS) at 55 °C for 30 min and then into the injection port of the GC–MS. The volatiles were separated and determined by employing the GC–MS apparatus (Trace TSQ, Themo Fisher Scientific, Waltham, MA, USA). The detected peaks and areas were compared against the National Institute of Standards and Technology (NIST) Standard Reference Database 78, and the calculations were performed using the peak area normalization method.

2.4. Determination of Taste and Smell Characteristics via Electronic Tongue and Nose

The taste and smell characteristics of the peanuts were estimated using the electronic tongue and nose, and all the procedures were carried out using the method we previously developed [14]. A total of 5 g of the peanut sample was ground with distilled water, topped up to a 50 mL volume, incubated at 55 °C and stirred for 10 min. After centrifugation at 3000× *g* for 10 min, the supernatant was collected and filtered, and then added to the electronic tongue (Shanghai Bosin Industrial Development Co., Ltd., Shanghai, China) to analyze the taste characteristics of the peanuts. A total of 5 g of the peanut sample was ground with distilled water (10 mL), and the homogenate was incubated in a sample bottle at 55 °C for 20 min. The headspace air was collected and analyzed using the electronic nose (Shanghai Bosin Industrial Development Co., Ltd.).

2.5. Statistical Analyses

All measurements were replicated in triplicate, and the data are expressed as the mean ± standard deviation (mean ± SD). A principal component analysis (PCA) was performed using RStudio (2023.07.0), a radar fingerprint analysis was performed using Origin and a variance (ANOVA) analysis was performed using SPSS Statistics 26.

3. Results

3.1. Physicochemical Properties of Ten Peanut Varieties

The water content of the peanuts was 4.752–5.483%, with the lowest being in Taihua 0605-2 and the highest in Taihua 11. The fat content of the peanuts was 48.24–53.83%, with the lowest in Taihua 6 and the highest in Taihua 0605-2. The protein content of the peanuts was 23.95–29.90%, the lowest being in Taihua 7506 and the highest in Taihua 6 (Table 1).

Table 1. Comparison of the water, fat and protein contents of the ten peanut varieties.

Relative Amount/%	Water Content	Fat Content	Protein Content
Taihua0122-601	5.152 ± 0.24 ab	50.87 ± 1.67 abc	27.54 ± 1.13 abc
Taihua6	5.386 ± 0.23 ab	48.24 ± 1.84 c	29.90 ± 1.46 a
Taihua0311	5.021 ± 0.16 ab	50.07 ± 1.13 abc	28.31 ± 1.28 ab
Taihua0605-2	4.752 ± 0.14 b	53.83 ± 1.53 a	28.58 ± 1.42 a
Taihua10	5.212 ± 0.27 ab	51.02 ± 1.45 abc	27.23 ± 1.18 abc
Taihua11	5.483 ± 0.35 a	48.47 ± 1.67 bc	28.67 ± 1.87 a
Taihua7506	5.274 ± 0.49 ab	53.59 ± 2.15 a	23.95 ± 1.45 c
Taihua9725	5.129 ± 0.25 ab	49.27 ± 1.57 bc	27.96 ± 1.26 ab
Taihua4	5.069 ± 0.18 ab	52.38 ± 2.05 ab	24.86 ± 1.06 bc
Taihua9922	5.269 ± 0.29 ab	50.43 ± 1.97 abc	28.71 ± 1.76 a

Different subscript letters in the same row for the same item indicate significant differences ($p < 0.05$).

3.2. *Free Amino Acid Contents of Ten Peanut Varieties*

As shown in Table 2, the total amino acid contents of the peanuts of ten varieties ranged from 20.08 g/100 g (Taihua 4) to 27.18 g/100 g (Taihua 6). The single amino acid contents were 0.39 g/100 g (L-tyrosine) to 3.50 g/100 g (L-aspartic). The bitter amino acid (histidine, arginine, valine, methionine, phenylalanine, isoleucine and leucine) contents were 5.90 g/100 g (Taihua 4) to 10.41 g/100 g (Taihua 6), accounting for 29.38% and 38.30% of the total amino acid contents in the peanuts, respectively. The sweet amino acid (glycine, serine, threonine and alanine) contents were 3.47 g/100 g (Taihua 0311) to 6.06 g/100 g (Taihua 6), accounting for 16.12% and 22.30% of the total amino acid contents in the peanuts, respectively. The monosodium glutamate (MSG)-like amino acid (aspartic, glutamic) contents were 4.77 g/100 g (Taihua 0605-2) to 7.61 g/100 g (Taihua 10), accounting for 20.99% and 28.78% of the total amino acid contents in the peanuts, respectively.

Table 2. Comparison of the free amino acid contents of the ten peanut varieties.

Relative Amount/g/100 g	Taihua0122-601	Taihua6	Taihua 0311	Taihua 0605-2	Taihua 10	Taihua 11	Taihua 7506	Taihua 9725	Taihua 4	Taihua 9922	Average
L-Aspartic	2.89 ± 0.31 d	4.14 ± 1.23 a	2.77 ± 0.33 d	2.81 ± 0.21 d	4.03 ± 1.18 ab	4.06 ± 1.97 a	3.55 ± 0.13 c	3.64 ± 0.16 c	3.71 ± 1.21 bc	3.42 ± 0.61 c	3.50
L-Glutamic	3.01 ± 0.98 bc	3.24 ± 0.76 ab	2.29 ± 0.88 d	1.96 ± 0.23 d	3.58 ± 1.22 a	3.22 ± 0.86 b	2.98 ± 0.98 bc	3.02 ± 0.45 bc	2.69 ± 0.11 c	3.11 ± 0.24 b	2.91
L-Serine	1.32 ± 0.87 de	1.95 ± 0.92 ab	1.15 ± 0.78 e	1.56 ± 0.31 cd	2.19 ± 0.56 a	1.84 ± 0.99 bc	1.23 ± 0.20 de	1.39 ± 0.35 de	1.41 ± 0.02 de	1.51 ± 0.11 cd	1.56
Glycine	1.22 ± 0.23 cd	1.95 ± 0.14 a	0.98 ± 0.02 d	1.71 ± 0.38 ab	1.86 ± 0.92 ab	1.94 ± 0.19 a	1.56 ± 0.29 bc	1.82 ± 0.12 ab	1.64 ± 0.18 ab	1.02 ± 0.09 d	1.57
L-Histidine	0.75 ± 0.09 abc	0.98 ± 0.23 a	0.64 ± 0.43 abcd	0.87 ± 0.05 ab	0.95 ± 0.19 a	0.89 ± 0.44 ab	0.48 ± 0.14 bd	0.29 ± 0.11 d	0.59 ± 0.32 bcd	0.86 ± 0.13 ab	0.73
L-Arginine	3.17 ± 1.43 bc	3.68 ± 0.93 a	3.09 ± 0.76 cd	3.29 ± 1.32 bcd	3.56 ± 1.22 ab	3.50 ± 0.23 abc	3.40 ± 0.76 abcd	2.78 ± 0.78 de	2.52 ± 0.98 e	2.67 ± 0.68 e	3.17
L-Threonine	0.31 ± 0.09 c	0.67 ± 0.23 abc	0.56 ± 0.45 abc	0.49 ± 0.29 bc	0.73 ± 0.45 ab	0.91 ± 0.13 a	0.64 ± 0.95 abc	0.77 ± 0.55 ab	0.82 ± 0.44 ab	0.65 ± 0.08 abc	0.66
L-Alanine	1.06 ± 0.34 cd	1.49 ± 0.13 b	0.78 ± 0.09 de	1.86 ± 0.17 a	1.08 ± 0.82 cd	1.33 ± 0.91 bc	0.88 ± 0.45 de	0.67 ± 0.29 e	1.27 ± 0.23 bc	0.75 ± 0.14 de	1.12
L-Proline	1.31 ± 0.93 a	1.23 ± 0.14 ab	0.88 ± 0.72 bcd	0.77 ± 0.24 d	1.14 ± 0.98 abc	1.01 ± 0.04 abcd	0.86 ± 0.12 cd	0.97 ± 0.78 bcd	0.76 ± 0.24 d	0.98 ± 0.5 bcd	0.99
L-Tyrosine	0.3 ± 0.09 abc	0.21 ± 0.12 bc	0.41 ± 0.23 abc	0.58 ± 0.08 a	0.59 ± 0.13 a	0.09 ± 0.03 c	0.28 ± 0.18 abc	0.31 ± 0.08 abc	0.54 ± 0.13 ab	0.56 ± 0.23 a	0.39
L-Valine	0.69 ± 0.34 cd	0.63 ± 0.21 cd	1.87 ± 0.37 a	0.84 ± 0.34 cd	0.97 ± 0.24 c	0.62 ± 0.23 d	1.54 ± 0.98 b	0.60 ± 0.56 d	0.61 ± 0.45 d	0.71 ± 0.09 cd	0.91
L-Methionine	0.11 ± 0.09 cd	0.44 ± 0.39 abc	0.28 ± 0.13 bcd	0.56 ± 0.23 ab	0.38 ± 0.16 bcd	0.72 ± 0.56 a	0.33 ± 0.23 bcd	0.38 ± 0.14 bcd	0.09 ± 0.06 d	0.21 ± 0.11 cd	0.35
L-Isoleucine	0.51 ± 0.34 bc	0.93 ± 0.45 a	0.82 ± 0.24 ab	0.65 ± 0.45 ab	0.55 ± 0.03 bc	0.71 ± 0.67 ab	0.86 ± 0.24 ab	0.61 ± 0.45 ab	0.21 ± 0.13 c	0.76 ± 0.34 ab	0.66
L-Leucine	0.67 ± 0.45 cd	1.89 ± 0.23 a	1.17 ± 0.56 b	1.04 ± 0.35 b	1.68 ± 0.45 a	1.57 ± 0.19 a	0.58 ± 0.34 d	0.98 ± 0.18 bc	0.85 ± 0.09 bcd	1.06 ± 0.45 b	1.15
L-Phenylalanine	1.37 ± 0.78 ef	1.86 ± 0.45 bcd	1.78 ± 0.12 bcd	1.98 ± 0.34 bc	2.10 ± 0.67 b	1.57 ± 0.36 de	1.64 ± 0.13 cde	1.80 ± 0.08 bcd	1.03 ± 0.15 f	2.54 ± 0.36 a	1.77
L-Lysine	1.89 ± 0.98 a	1.15 ± 0.74 b	1.23 ± 0.24 b	0.97 ± 0.54 bc	0.34 ± 0.06 e	1.06 ± 0.35 b	0.59 ± 0.21 de	0.71 ± 0.45 cd	0.43 ± 0.03 de	0.64 ± 0.24 cde	0.90
L-Tryptophan	0.88 ± 0.13 b	0.74 ± 0.21 b	0.82 ± 0.47 b	0.78 ± 0.09 b	0.71 ± 0.35 b	1.68 ± 0.56 a	0.29 ± 0.11 c	0.58 ± 0.23 bc	0.91 ± 0.43 b	0.59 ± 0.17 bc	0.80
Bitter	7.27 ± 3.52 bc	10.41 ± 2.89 a	9.65 ± 2.61 ab	9.23 ± 3.08 ab	10.19 ± 2.96 ab	9.58 ± 2.68 ab	8.83 ± 2.82 abc	7.44 ± 2.30 abc	5.9 ± 2.18 c	8.81 ± 2.16 abc	
Sweet	3.91 ± 1.53 b	6.06 ± 1.42 a	3.47 ± 1.34 b	5.62 ± 1.15 ab	5.86 ± 2.75 ab	6.02 ± 2.22 a	4.31 ± 1.89 ab	4.65 ± 1.31 ab	5.14 ± 0.87 ab	3.93 ± 0.42 b	
MSG-like	5.9 ± 1.29 ab	7.38 ± 1.99 a	5.06 ± 1.21 ab	4.77 ± 0.44 b	7.61 ± 2.40 a	7.28 ± 2.83 a	6.53 ± 1.11 ab	6.66 ± 0.61 ab	6.40 ± 1.32 ab	6.53 ± 0.85 ab	
Total	21.46 ± 4.23 b	27.18 ± 7.25 a	21.52 ± 5.45 b	22.72 ± 6.23 b	26.44 ± 7.44 a	26.72 ± 6.34 a	21.69 ± 5.25 b	21.32 ± 7.26 b	20.08 ± 5.48 b	22.04 ± 5.65 b	

Different subscript letters in the same row for the same item indicate significant differences ($p < 0.05$).

3.3. *5′-Nucleotide Contents of Ten Peanut Varieties*

As shown in Table 3, the 5′-nucleotide contents ranged from 0.08 mg/g (Taihua6, 10, 9725) to 0.14 mg/g (Taihua0122-601). The compounds 5′-cytidylate monophosphate (5′-CMP) and 5′-adenosine monophosphate (5′-AMP) comprised the main 5′-nucleotides in the peanuts, with Taihua7506 being the only variety with no 5′-AMP detected. On the other hand, 5′-uridine monophosphate (5′-UMP) was only detected in Taihua0311 and Taihua10; 5′-guanosine monophosphate (5′-GMP) was only detected in Taihua0122-601, Taihua0605-2, Taihua11, Taihua7506 and Taihua9922; and 5′-inosine monophosphate (5′-IMP) was not detected in any of the peanut varieties.

Table 3. Comparison of the 5′-nucleotide contents of the ten peanut varieties.

5′-Nucleotides/ mg/g	Taihua0122-601	Taihua6	Taihua0311	Taihua0605-2	Taihua10	Taihua11	Taihua7506	Taihua9725	Taihua4	Taihua9922
5-CMP	0.06 ± 0.03^{a}	0.06 ± 0.03^{a}	0.06 ± 0.04^{a}	0.06 ± 0.1^{a}	0.04 ± 0.04^{c}	0.05 ± 0.01^{b}	0.06 ± 0.07^{a}	0.06 ± 0.01^{a}	0.06 ± 0.03^{a}	0.06 ± 0.06^{a}
5′-UMP	ND	ND	0.05 ± 0.01^{a}	ND	0.02 ± 0.01^{b}	ND	ND	ND	ND	ND
5′-GMP	0.05 ± 0.03^{a}	ND	ND	0.05 ± 0.01^{a}	ND	0.05 ± 0.04^{a}	0.04 ± 0.01^{b}	ND	ND	0.05 ± 0.03^{a}
5′-IMP	ND	ND	ND	ND	ND	ND	ND	ND	ND	ND
5′-AMP	0.03 ± 0.03^{a}	0.02 ± 0.01^{b}	0.02 ± 0.01^{b}	0.02 ± 0.01^{b}	0.02 ± 0.00^{b}	0.02 ± 0.01^{b}	ND	0.02 ± 0.01^{b}	0.03 ± 0.01^{a}	0.02 ± 0.01^{b}
Total	0.14	0.08	0.13	0.13	0.08	0.12	0.10	0.08	0.09	0.13

Different subscript letters in the same row for the same item indicate significant differences ($p < 0.05$). ND indicates not detected.

3.4. Volatile Aroma Components of Ten Peanut Varieties

As shown in Table 4, a total of 42 kinds of volatile flavor compounds were identified within the ten peanut varieties, with 14, 18, 14, 15, 11, 9, 10, 9, 18 and 9 kinds of volatile compounds detected in Taihua0122-601, 6, 0311, 0605-2, 10, 11, 7506, 9725, 4 and 9922, respectively, and the total contents were 3.23, 4.94, 3.37, 6.39, 5.37, 6.54, 5.80, 3.15, 7.46 and 3.47%, respectively. Formic acid hexyl ester was the most abundant volatile substance in peanuts, and the highest content (3.63%) was identified in Taihua 7506. Only 1-nonanol, nonanal and 3-methyl-undecane were detected within all ten varieties of peanuts.

Table 4. Volatile compound contents of the ten peanut varieties.

Relative Amount/%	Taihua0122-601	Taihua6	Taihua0311	Taihua0605-2	Taihua10	Taihua11	Taihua7506	Taihua9725	Taihua4	Taihua9922
2-Octen-1-ol	ND	ND	ND	0.10 ± 0.04^{bc}	0.32 ± 0.11^{a}	ND	0.15 ± 0.07^{b}	0.33 ± 0.08^{a}	ND	ND
(E)-2-Octen-1-ol	0.10 ± 0.08^{bc}	0.24 ± 0.06^{b}	ND	ND	ND	ND	ND	ND	0.55 ± 0.28^{a}	ND
(Z)-2-Octen-1-ol	ND	ND	ND	ND	ND	ND	ND	ND	ND	0.08 ± 0.08^{a}
1-Nonanol	0.20 ± 0.08^{cd}	0.22 ± 0.04^{cd}	0.31 ± 0.13^{cd}	0.18 ± 0.03^{d}	0.62 ± 0.11^{ab}	0.45 ± 0.06^{abc}	0.26 ± 0.14^{cd}	0.41 ± 0.18^{bcd}	0.68 ± 0.06^{a}	0.34 ± 0.08^{cd}
2-Butyl-1-Octanol	ND	ND	ND	ND	ND	ND	ND	0.04 ± 0.04^{a}	ND	ND
Nonanal	0.67 ± 0.40^{bc}	1.28 ± 0.33^{bc}	0.77 ± 0.10^{bc}	0.37 ± 0.14^{c}	2.72 ± 1.00^{a}	1.59 ± 0.42^{b}	0.92 ± 0.17^{bc}	1.31 ± 0.44^{bc}	1.41 ± 0.28^{bc}	0.61 ± 0.11^{bc}
2-Nonenal	0.08 ± 0.05^{a}	ND	ND	ND	ND	ND	ND	ND	ND	0.06 ± 0.04^{a}
(E)-2-Nonenal	ND	ND	0.08 ± 0.03^{a}	ND	ND	ND	ND	ND	ND	ND
(Z)-2-Nonenal	ND	0.09 ± 0.07^{a}	ND	ND	0.11 ± 0.03^{a}	0.13 ± 0.07^{a}	ND	0.11 ± 0.04^{a}	0.10 ± 0.01^{a}	ND
Decanal	0.09 ± 0.07^{bc}	0.08 ± 0.04^{bc}	0.15 ± 0.07^{bc}	0.06 ± 0.06^{bc}	0.31 ± 0.20^{ab}	ND	ND	0.45 ± 0.24^{a}	0.13 ± 0.04^{bc}	ND
(E)-2-Decenal	ND	ND	ND	ND	ND	ND	0.10 ± 0.01^{a}	ND	0.14 ± 0.08^{a}	0.09 ± 0.08^{ab}
(Z)-2-Decenal	0.09 ± 0.04^{a}	ND	0.08 ± 0.08^{a}	ND	ND	ND	ND	ND	ND	ND
2,4-Decadienal	ND	ND	ND	ND	ND	0.39 ± 0.27^{a}	0.16 ± 0.08^{ab}	ND	0.11 ± 0.06^{b}	ND
(E,E)-2,4-Decadienal	ND	ND	0.08 ± 0.06^{a}	ND	ND	ND	ND	ND	ND	ND
2-Undecenal	ND	ND	0.05 ± 0.02^{a}	ND	ND	ND	ND	ND	ND	ND
2,4-Dimethyl-Benzaldehyde	ND	ND	ND	0.09 ± 0.04^{ab}	0.14 ± 0.08^{a}	ND	ND	ND	ND	0.15 ± 0.08^{a}
Undecanal	ND	ND	ND	ND	0.03 ± 0.02^{a}	ND	ND	ND	ND	ND
Formic Acid, Hexyl Ester	1.15 ± 0.92^{bc}	1.52 ± 0.74^{abc}	1.40 ± 0.24^{bc}	2.04 ± 1.47^{abc}	ND	3.10 ± 0.85^{ab}	3.63 ± 0.91^{a}	ND	2.01 ± 0.26^{abc}	1.98 ± 1.33^{abc}
Formic Acid, Heptyl Ester	0.08 ± 0.04^{a}	ND	ND	ND	ND	ND	ND	ND	ND	ND
Formic Acid, Octyl Ester	ND	0.35 ± 0.34^{a}	ND	ND	ND	0.37 ± 0.18^{a}	0.21 ± 0.04^{ab}	ND	ND	ND
Octanoic Acid, Methyl Ester	ND	0.15 ± 0.07^{c}	ND	0.40 ± 0.20^{bc}	0.68 ± 0.45^{ab}	0.30 ± 0.16^{bc}	ND	ND	0.89 ± 0.31^{a}	ND
Nonanoic Acid, Methyl Ester	ND	ND	ND	0.04 ± 0.03^{ab}	0.08 ± 0.01^{a}	ND	ND	ND	0.10 ± 0.07^{a}	ND
2-Methyl-Decane	0.11 ± 0.06^{a}	0.08 ± 0.03^{ab}	0.09 ± 0.08^{a}	0.04 ± 0.03^{ab}	ND	ND	ND	ND	ND	ND
3-Methyl-Decane	0.19 ± 0.08^{a}	0.14 ± 0.07^{a}	ND	0.17 ± 0.03^{a}	ND	ND	ND	ND	ND	ND
4-Methyl-Decane	0.18 ± 0.16^{a}	0.09 ± 0.06^{ab}	ND	ND	ND	ND	ND	ND	ND	ND
5-Methyl-Decane	ND	ND	0.12 ± 0.06^{a}	ND	ND	ND	ND	ND	ND	ND
2-Methyl-Undecane	ND	ND	ND	ND	ND	ND	ND	ND	0.18 ± 0.06^{a}	ND
3-Methyl-Undecane	0.12 ± 0.11^{ab}	0.05 ± 0.03^{b}	0.08 ± 0.06^{ab}	0.11 ± 0.06^{ab}	0.25 ± 0.20^{ab}	0.13 ± 0.04^{ab}	0.06 ± 0.04^{b}	0.10 ± 0.10^{ab}	0.31 ± 0.18^{a}	0.12 ± 0.07^{ab}
4-Methyl-Undecane	ND	ND	ND	ND	ND	ND	ND	ND	0.10 ± 0.04^{a}	ND
2,6-Dimethyl-Undecane	ND	0.19 ± 0.11^{ab}	ND	ND	ND	ND	0.19 ± 0.13^{ab}	0.38 ± 0.25^{a}	ND	ND
2,6,7-Trimethyl-Decane	ND	ND	ND	0.07 ± 0.03^{a}	ND	ND	ND	ND	ND	ND
2,6,8-Trimethyl-Decane	ND	0.11 ± 0.04^{a}	ND	ND	ND	ND	ND	ND	ND	ND
2,6,11-Trimethyl-Dodecane	ND	ND	0.04 ± 0.03^{a}	ND	ND	ND	ND	ND	ND	ND
3-Methyl-Tridecane	0.10 ± 0.07^{ab}	0.16 ± 0.06^{a}	0.09 ± 0.01^{ab}	0.05 ± 0.03^{ab}	0.11 ± 0.08^{ab}	0.08 ± 0.07^{ab}	ND	ND	0.14 ± 0.06^{a}	0.04 ± 0.03^{ab}
1-Dodecene	0.07 ± 0.04^{ab}	ND	ND	ND	ND	ND	ND	ND	0.10 ± 0.06^{a}	ND
(E)-2-Dodecene	ND	0.02 ± 0.01^{a}	ND	ND	ND	ND	ND	ND	ND	ND
1-Tridecene	ND	ND	ND	ND	ND	ND	ND	ND	0.03 ± 0.02^{a}	ND
6-Tridecene	ND	0.01 ± 0.01^{a}	ND	ND	ND	ND	ND	ND	ND	ND
(R)-1-Methyl-5-(1-Methylethenyl)-Cyclohexene	ND	ND	ND	2.52 ± 1.19^{a}	ND	ND	ND	0.02 ± 0.01^{b}	ND	ND
o-Cymene	ND	ND	0.03 ± 0.01^{c}	0.15 ± 0.09^{a}	ND	ND	0.12 ± 0.08^{b}	ND	ND	ND
Nonanoic Acid	ND	0.16 ± 0.11^{ab}	ND	ND	ND	ND	ND	ND	0.26 ± 0.19^{a}	ND
2-Pentyl-Furan	ND	ND	ND	ND	ND	ND	ND	ND	0.33 ± 0.12^{a}	ND

Different subscript letters in the same row for the same item indicate significant differences ($p < 0.05$). ND indicates not detected.

3.5. Electronic Tongue for the Taste Characteristics of Peanuts

As shown in Figure 1, the sourness, bitterness, astringency, umami and saltiness of the ten peanut varieties were estimated using the electronic tongue, and all the peanuts showed high umami, saltiness, astringency and bitterness response values, and low sourness response values. The greatest taste difference between the ten peanut varieties was found for the response values of sourness, whereby Taihua 0311 showed the lowest value and Taihua 4 the highest. There were few differences among the response values for bitterness, astringency and umami. The flavor characteristics of the ten peanut varieties were effectively separated via the principal component analysis (PCA); the contribution rate of PCA1 was 97.06%, and the contribution rate of PCA2 was 2.63%, totaling 99.69% (Figure 2), which indicated that the two-dimensional scatter diagram of the PCA could reflect the taste differences between the ten varieties of peanuts. There were overlaps between ten varieties of peanuts, indicating great similarity regarding the tastes of the ten varieties, but there were still differences between Taihua 4, Taihua 9922 and other varieties. The loading plot of the PCA also indicated that the taste contributions of the peanuts were in the order of sourness, astringency, bitterness, umami and saltiness (Figure 3), and the group differences were mostly related to sourness (Figures 2 and 3).

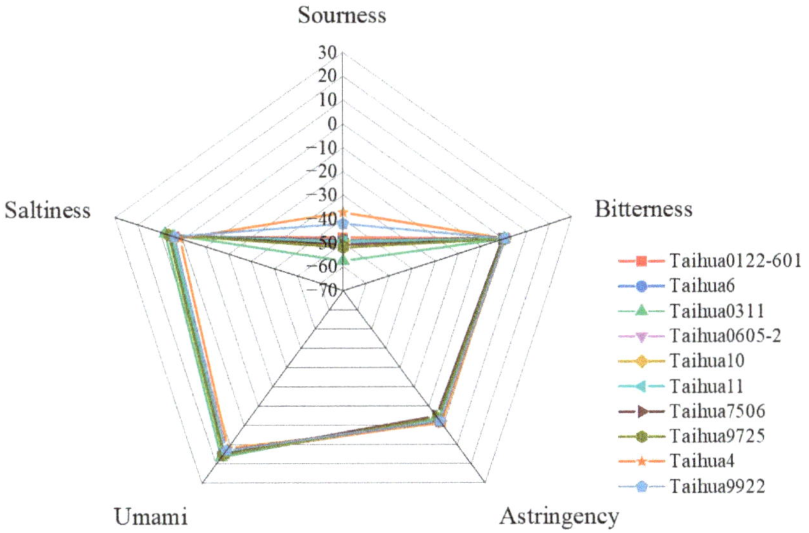

Figure 1. Radar map of the electronic tongue results for the ten peanut varieties.

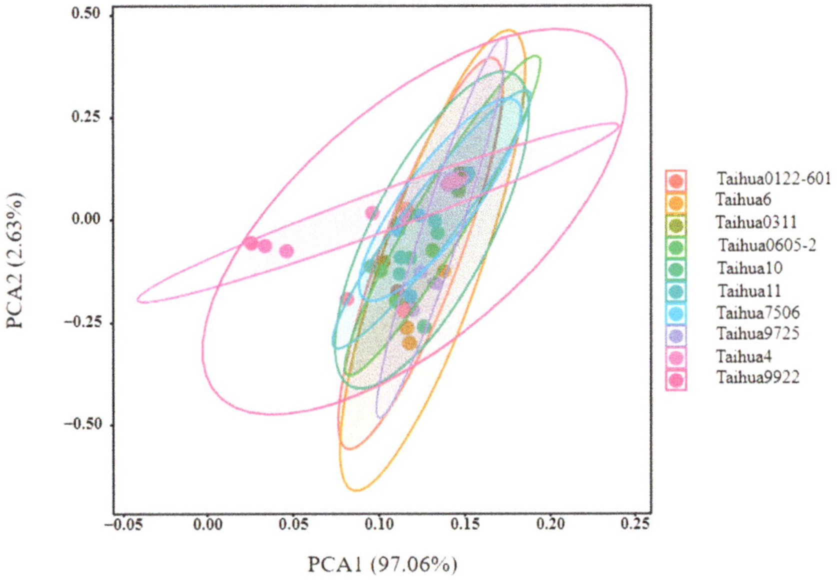

Figure 2. PCA of the electronic tongue results for the ten peanut varieties.

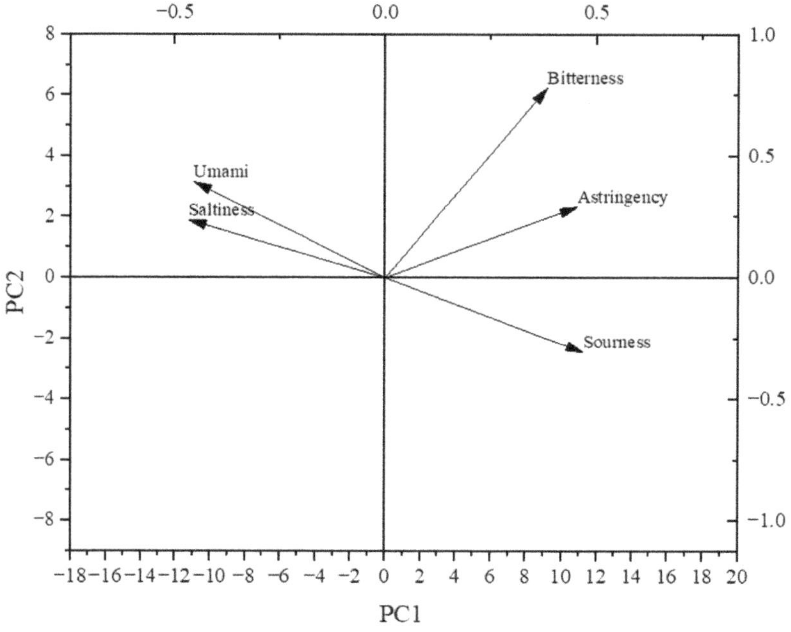

Figure 3. PCA loading plot of the electronic tongue results for the ten peanut varieties.

3.6. Electronic Nose for the Smell Characteristics of Peanuts

As shown in Figure 4, the smell characteristics of the ten peanut varieties were estimated using an electronic nose equipped with 10 kinds of sensors (Table 5), and the W1S, W1W, W2W and W5S sensors showed high response values, with Taihua0605-2 exhibiting the highest values. As shown in Figure 5, the contribution rate of PCA1 was 99.7% and the contribution rate of PCA2 was 0.01%, totaling 99.71%, which indicated that the two-dimensional scatter diagram of the PCA could effectively reflect the odor differences between ten varieties of peanuts. No overlaps appeared within PCA1 and PCA2, indicating that the smell characteristics differed across the ten varieties of peanuts, whereby Taihua 0605-2 exhibited significantly different smell characteristics compared to the other varieties, with a much smaller PCA1 value compared to that of the other varieties. The loading plot indicated that the smell contributions for the peanuts were in the order of W1S, W2S, W5S, W2W, W3S, W6S, W1W, W5C, W1C and W3C (Figure 6), and the reference substances were in the order of methane, carbon monoxide, nitrogen dioxide, hydrogen sulfide, methane, hydrogen, hydrogen sulfide, propane, toluene and benzene (Table 5).

Table 5. Volatile compounds responding to the 10 sensors of the electronic nose.

Sensor No.	Sensing Species	Reference Substance
W1C	aromatic	toluene
W1S	broad methane	methane
W1W	sulfur organic	hydrogen sulfide
W2S	broad alcohol	carbon monoxide
W2W	sulf–chlor	hydrogen sulfide
W3C	aromatic	benzene
W3S	methane–aliph	methane
W5C	arom–aliph	propane
W5S	broad-range	nitrogen dioxide
W6S	hydrogen	hydrogen

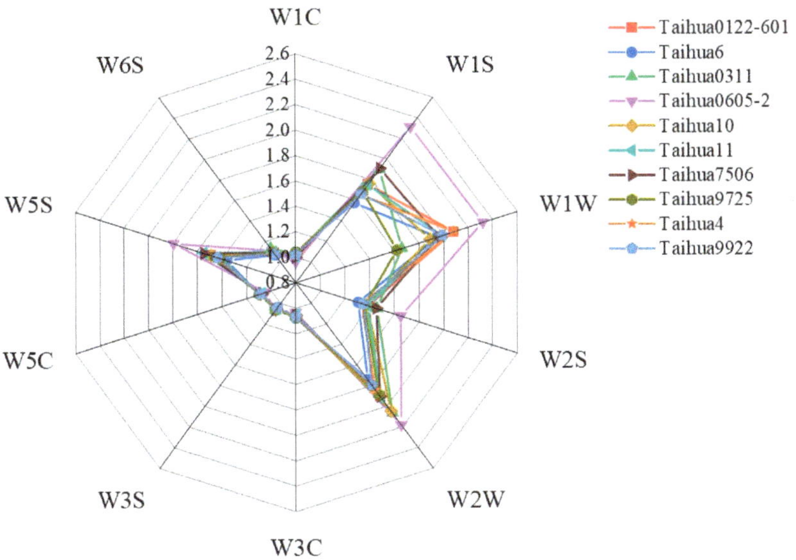

Figure 4. Radar map of the electronic nose results for the ten peanut varieties.

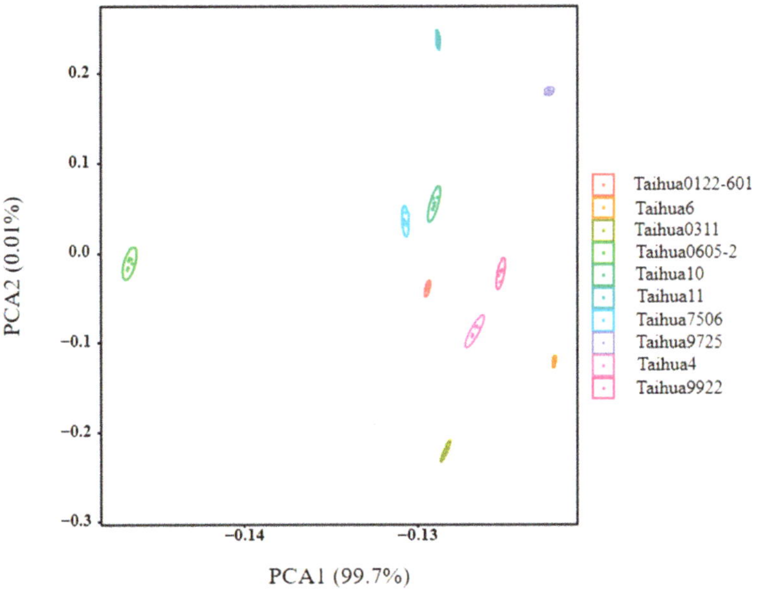

Figure 5. PCA of the electronic nose results for the ten peanut varieties.

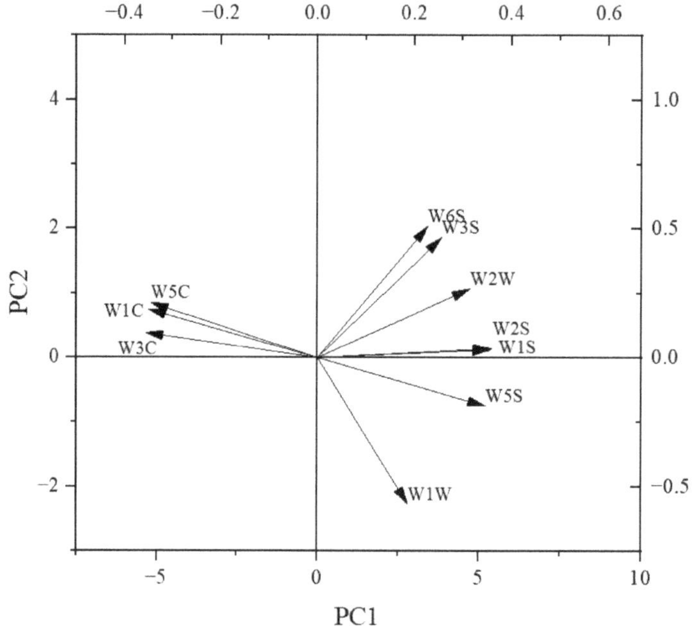

Figure 6. PCA loading plot of the electronic nose results for the ten peanut varieties.

4. Discussion

The water contents of the ten peanut varieties showed no significant ($p > 0.05$) differences. Taihua 0605-2 was the only variety exhibiting a significantly ($p < 0.05$) lower water content than Taihua 11 (Table 1), indicating that the degree of dehydration in peanuts before storage was similar. Moreover, Taihua 0605-2 peanuts may not require such stringent drying conditions (thin shell; small fruit) compared to the other varieties. The fat and protein contents were negatively correlated, especially for Taihua 6, which exhibited the highest protein content and lowest fat content. This is directly affected by hereditary properties. The protein and fat contents in this study are similar to those previously reported [18,19]. The non-volatile aroma compounds were mainly proteins and their degradation products (amino acids or polypeptides), and the volatile aroma compounds were fats (small-molecule lipid substances such as formic acid hexyl ester). Thus, the differences in the content and composition of proteins and fats in peanuts, largely influenced by the genetic material, are of crucial importance for their flavor [20].

Amino acids were the main flavor substances, manifesting different taste characteristics (bitter, sweet and MSG-like tastes) [21]. They also comprised the basic protein components [22]; hence, a higher protein content in peanuts also implied a higher amino acid content. Taihua6 peanuts showed the highest protein, amino acid and bitter and sweet amino acid contents. The differences in amino acid contents between the ten peanut varieties were affected by the protein component, directly determined according to genomic and expression differences [23]. Nucleotides also comprised flavor substances, with certain nucleotide components (5′–GMP) presenting stronger flavor characteristics than MSG [24,25]. The interaction between nucleotides and MSG can greatly enhance the flavor of each, producing a much stronger flavor than that of the nucleotides and MSG alone or together [26,27]. The ten varieties of peanuts showed relatively high nucleotide contents (see Table 5), but with very different nucleotide components, similar to previous reports [28,29]. The electronic tongue response values revealed that the greatest taste difference among the ten peanut varieties was related to sourness, whereby Taihua 0311 had a low sour taste and Taihua 4 had a highly sour taste.

PCA is a statistical process that uses orthogonal transformations to convert the observed values of a group of potentially related variables into the values of a group of linearly unrelated variables called the principal components [30]. The deviation distribution of PCA1 of Taihua 4 and Taihua 9922 implied their taste specificities. The low contents of amino acids and 5′-nucleotides in Taihua 4 may lead to its flavor specificity, while the high contents of amino acids and 5′-nucleotides in Taihua 9922 may also influence its special taste characteristics. The loading plot of the PCA also confirmed that sourness made an important contribution to the flavor of peanuts. Sourness is mostly attributed to H^+, and bitter materials can neutralize acids; therefore, the lowest bitter amino acid content in Taihua 4 may contribute to the highest sourness response value obtained via the electronic tongue for it.

The volatile components in the ten varieties ranged from 3.15% to 7.46%, with 9 to 18 kinds. Only 1-nonanol, nonanal and 3-methyl-undecane were detected in all ten peanut varieties. These results indicated great differences in the content and composition of volatile substances between the ten varieties of peanuts, and these results are similar to those previously reported [31]. The electronic nose response value revealed that the broad methane and broad alcohol were the main smell substances, and the smell of the peanuts was similar to that of the mixture of methane, carbon monoxide and nitrogen dioxide. The smell differences between the ten varieties of peanuts were mostly attributed to methane and sulfur organic substances. The PCA indicated that Taihua 0605-2 exhibited a particular odor characteristic based on the electronic nose signal, because its large deviation was distributed at PCA1, likely caused by its unique presence of 2,6,7-trimethyl-decane and high content of fats. More research on this special smell characteristic of Taihua 0605-2 should be carried out on variously processed and cooked peanuts.

5. Conclusions

The fat and protein contents of the ten peanut varieties were significantly ($p < 0.05$) different and negatively correlated. The amino acid contents of the peanuts were 20.08 g/100 g–27.18 g/100 g, and the 5′-nucleotide contents were 0.08 mg/g–0.14 mg/g. Forty-two kinds of volatile flavor compounds were detected. The electronic tongue and nose indicated that the dominant taste of peanuts was attributed to the mixture of sourness, astringency and bitterness, and the taste differences between the ten varieties of peanuts were due to sourness, with Taihua 4 and Taihua 9922 having special taste characteristics. The dominant smell of the peanuts was due to the mixture of methane, carbon monoxide and nitrogen dioxide, and the smell differences between the ten varieties of peanuts depended on methane and sulfur organic substances. Taihua 0605-2 exhibited special and strong smell characteristics. These results provide a theoretical basis for flavor research on peanuts, and guidelines for the utilization (fresh, processed or cooked) of these varieties.

Author Contributions: C.Q.: Conceptualization, Methodology and Funding Acquisition; B.D., F.W. and B.Z.: Investigation and Writing—Original Draft; M.F. and L.C.: Validation and Investigation; Y.S. (Yuyang Shao) and Y.S. (Yan Sun): Validation and Visualization; Y.J. and R.W.: Investigation; J.X.: Conceptualization; C.Q. and L.W.: Conceptualization and Writing—Review and Editing. All authors have read and agreed to the published version of the manuscript.

Funding: This research was funded by Jiangsu Modern Agricultural Industrial Technology System (grant number: JATS [2022] 298; Jiangsu Provincial Key Research and Development Project (Modern Agriculture) (grant number: BE2022411); Jiangsu Agricultural Science and Technology Independent Innovation Fund (grant number: CX (22)3086); Taizhou Science and Technology Support Program (Agriculture) Project (grant number: TN202128).

Data Availability Statement: The data used to support the findings of this study can be made available by the corresponding author upon request.

Conflicts of Interest: The authors declare that they have no conflict of interest.

References

1. Li, W.; Yoo, E.; Lee, S.; Sung, J.; Noh, H.J.; Hwang, S.J.; Desta, K.T.; Lee, G.-A. Seed Weight and Genotype Influence the Total Oil Content and Fatty Acid Composition of Peanut Seeds. *Foods* **2022**, *11*, 3463. [CrossRef] [PubMed]
2. Derbyshire, E.J. A review of the nutritional composition, organoleptic characteristics and biological effects of the high oleic peanut. *Int. J. Food Sci. Nutr.* **2014**, *65*, 781–790. [CrossRef]
3. Nawade, B.; Mishra, G.P.; Radhakrishnan, T.; Dodia, S.M.; Ahmad, S.; Kumar, A.; Kumar, A.; Kundu, R. High oleic peanut breeding: Achievements, perspectives, and prospects. *Trends Food Sci. Technol.* **2018**, *78*, 107–119. [CrossRef]
4. Nepote, V.; Mestrallet, M.G.; Grosso, N.R. Oxidative stability in fried-salted peanuts elaborated with high-oleic and regular peanuts from Argentina. *Int. J. Food Sci. Technol.* **2006**, *41*, 900–909. [CrossRef]
5. Sithole, T.R.; Ma, Y.-X.; Qin, Z.; Liu, H.-M.; Wang, X.-D. Influence of Peanut Varieties on the Sensory Quality of Peanut Butter. *Foods* **2022**, *11*, 3499. [CrossRef] [PubMed]
6. Pandit, S.S.; Chidley, H.G.; Kulkarni, R.S.; Pujari, K.H.; Giri, A.P.; Gupta, V.S. Cultivar relationships in mango based on fruit volatile profiles. *Food Chem.* **2009**, *114*, 363–372. [CrossRef]
7. Kelebek, H.; Canbas, A.; Jourdes, M.; Teissedre, P.-L. HPLC-DAD-MS Determination of Colored and Colorless Phenolic Compounds in Kalecik Karasi Wines: Effect of Different Vineyard Locations. *Anal. Lett.* **2011**, *44*, 991–1008. [CrossRef]
8. Hegmann, E.; Niether, W.; Phillips, W.; Rohsius, C.; Lieberei, R. Besides variety, also season and ripening stage have a major influence on fruit pulp aroma of cacao (*Theobroma cacao* L.). *J. Appl. Bot. Food Qual.* **2020**, *93*, 266.
9. Fritz, K.R.; Dean, L.L.; Hendrix, K.W.; Andres, R.J.; Newman, C.S.; Oakley, A.T.; Clevenger, J.P.; Dunne, J.C. Flavor quality and composition of accession resources in the North Carolina State University peanut breeding program. *Crop Sci.* **2022**, *62*, 1880–1890. [CrossRef]
10. Pattee, H.E.; Isleib, T.G.; Giesbrecht, F.G.; McFeeters, R.F. Investigation into genotypic variations of peanut carbohydrates. *J. Agric. Food Chem.* **2000**, *48*, 750–756. [CrossRef]
11. Pattee, H.E.; Isleib, T.G.; Gorbet, D.W.; Giesbrecht, F.G.; Cui, Z. Parent selection in breeding for roasted peanut flavor quality1. *Peanut Sci.* **2001**, *28*, 51–58. [CrossRef]
12. Wang, Z.; Yan, L.; Chen, Y.; Wang, X.; Huai, D.; Kang, Y.; Jiang, H.; Liu, K.; Lei, Y.; Liao, B. Detection of a major QTL and development of KASP markers for seed weight by combining QTL-seq, QTL-mapping and RNA-seq in peanut. *Theor. Appl. Genet.* **2022**, *135*, 1779–1795. [CrossRef]
13. Dhamsaniya, N.K.; Patel, N.C.; Dabhi, M.N. Selection of groundnut variety for making a good quality peanut butter. *J. Food Sci. Technol.* **2012**, *49*, 115–118. [CrossRef] [PubMed]
14. Qian, C.L.; Jiang, Y.Y.; Sun, Y.; Yin, X.D.; Zhang, M.; Kan, J.; Liu, J.; Xiao, L.X.; Jin, C.H.; Qi, X.; et al. Changes in the Texture and Flavor of Lotus Root after Different Cooking Methods. *Foods* **2023**, *12*, 2012. [CrossRef] [PubMed]
15. Zhang, K.; Cheng, J.H.; Hong, Q.D.; Dong, W.J.; Chen, X.A.; Wu, G.P.; Zhang, Z.Z. Identification of changes in the volatile substances of robusta coffee beans during drying based on HS—SPME/GC—MS and E—Nose analyses with the aid of chemometrics. *LWT Food Sci. Technol.* **2022**, *161*, 113317. [CrossRef]
16. Zhang, Z.Q.; Zang, M.W.; Zhang, K.H.; Wang, S.W.; Li, D.; Li, X.M. Effect of two types of thermal processing methods on the aroma and taste profile of three commercial plant—Based beef analogues and beef by GC—MS, E—Nose, E—Tongue, and sensory evaluation. *Food Control* **2023**, *146*, 109551. [CrossRef]
17. Singh, S.K.; Castell-Perez, M.E.; Moreira, R. Viscosity and Textural Attributes of Reduced-fat Peanut Pastes. *J. Food Sci.* **2000**, *65*, 849–853. [CrossRef]
18. Akhtar, S.; Khalid, N.; Ahmed, I.; Shahzad, A.; Suleria, H.A.R. Physicochemical characteristics, functional properties, and nutritional benefits of peanut oil: A review. *Crit. Rev. Food Sci. Nutr.* **2014**, *54*, 1562–1575. [CrossRef]
19. Wang, M.L.; Chen, C.Y.; Tonnis, B.; Pinnow, D.; Davis, J.; An, Y.Q.C.; Dang, P. Changes of seed weight, fatty acid composition, and oil and protein contents from different peanut FAD2 genotypes at different seed developmental and maturation stages. *J. Agric. Food Chem.* **2018**, *66*, 3658–3665. [CrossRef]
20. Du, L.; Ye, Y.; Lin, T.; Li, M.; Bi, Y.; Li, Q.; Liu, H. The evaluation of the nutritional quality of common wild edible fungi in Yunnan Province. *J. Food Sci.* **2023**, *88*, 1879–1889. [CrossRef]
21. Kong, Y.; Zhang, L.; Sun, Y.; Zhang, Y.; Sun, B.; Chen, H. Determination of the Free Amino Acid, Organic Acid, and Nucleotide in Commercial Vinegars. *J. Food Sci.* **2017**, *82*, 1116–1123. [CrossRef] [PubMed]
22. Kopper, R.; Van, T.; Kim, A.; Helm, R. Release of Soluble Protein from Peanut (*Arachis hypogaea*, Leguminosae) and Its Adsorption by Activated Charcoal. *J. Agric. Food Chem.* **2011**, *59*, 236–240. [CrossRef]
23. Lan, R.J.; Duan, M.Y.; Wu, F.D.; Lai, R.F.; Mo, Z.W.; Pan, S.G.; Tang, X.R. Enhancement of ultrasonic seed treatment on yield, grain quality characters, and 2-acetyl-l-pyrroline biosynthesis in different fragrant rice genotypes. *Phyton Int. J. Exp. Bot.* **2022**, *91*, 2461–2473. [CrossRef]
24. Wang, K.; Zhuang, H.N.; Bing, F.L.; Chen, D.; Feng, T.; Xu, Z.M. Evaluation of eight kinds of flavor enhancer of umami taste by an electronic tongue. *Food Sci. Nutr.* **2021**, *9*, 2095–2104. [CrossRef]
25. Yang, J.-H.; Lin, H.-C.; Mau, J.-L. Non-volatile taste components of several commercial mushrooms. *Food Chem.* **2001**, *72*, 465–471. [CrossRef]
26. Wifall, T.C.; Faes, T.M.; Taylor-Burds, C.C.; Mitzelfelt, J.D.; Delay, E.R. An Analysis of 5′-Inosine and 5′-Guanosine Monophosphate Taste in Rats. *Chem. Senses* **2007**, *32*, 161–172. [CrossRef]

27. Tsai, S.-Y.; Tsai, H.-L.; Mau, J.-L. Non-volatile taste components of *Agaricus blazei*, *Agrocybe cylindracea* and *Boletus edulis*. *Food Chem.* **2008**, *107*, 977–983. [CrossRef]
28. Kasuga, A.; Fujihara, S.; Aoyagi, Y. The relationship between the varieties of dried shiitake mushrooms (*Lentinus edodes* (Berk.) Sing.) and chemical composition: The effect of the varieties of dried shiitake mushrooms on their taste: Part I. *J. Jpn. Soc. Food Sci. Technol. Nippon. Shokuhin Kagaku Kogaku Kaishi* **1999**, *46*, 692–703. [CrossRef]
29. Kasuga, A.; Fujihara, S.; Aoyagi, Y. The behavior of different varieties of dried shiitake mushrooms (*Lentinus edodes* (Berk.) Sing.) in rehydration: The effect of different varieties of dried shiitake mushrooms on their taste, Part II. *J. Jpn. Soc. Food Sci. Technol. Nippon. Shokuhin Kagaku Kogaku Kaishi* **2000**, *47*, 347–354. [CrossRef]
30. Lu, L.; Hu, Z.; Hu, X.; Li, D.; Tian, S. Electronic tongue and electronic nose for food quality and safety. *Food Res. Int.* **2022**, *162*, 112214. [CrossRef]
31. Cherif, A.O.; Ben Messaouda, M.; Kaabi, B.; Boukhchina, S.; Pepe, C.; Kallel, H. Comparison of the concentrations of long-chain alcohols (Policosanol) in three tunisian peanut varieties (*Arachis hypogaea* L.). *J. Agric. Food Chem.* **2010**, *58*, 12143–12148. [CrossRef] [PubMed]

Disclaimer/Publisher's Note: The statements, opinions and data contained in all publications are solely those of the individual author(s) and contributor(s) and not of MDPI and/or the editor(s). MDPI and/or the editor(s) disclaim responsibility for any injury to people or property resulting from any ideas, methods, instructions or products referred to in the content.

foods

MDPI

Review

Based on CiteSpace Insights into *Illicium verum* Hook. f. Current Hotspots and Emerging Trends and China Resources Distribution

Zhoujian He [1,2,†], Jie Huan [3,†], Meng Ye [1,*], Dan Liang [4], Yongfei Wu [5], Wenjun Li [6], Xiao Gong [1] and Liqiong Jiang [6]

1 College of Forestry, Sichuan Agricultural University, Huimin Road 211, Wenjiang District, Chengdu 611130, China; hezhouj@163.com (Z.H.); gongxiao202209@163.com (X.G.)
2 School of Life Sciences, Sichuan University, Wangjiang Road 29, Wuhou District, Chengdu 610064, China
3 Enyang District Agriculture and Rural Bureau of Bazhong City, No. 6, Planning Road 40, Enyang District, Bazhong 636600, China; 15388135911@163.com
4 Baoxing County Natural Resources and Planning Bureau of Yaan City, Lingxiu Road 256, Baoxing County, Yaan 625700, China; 2020206034@stu.sicau.edu.cn
5 Animal Nutrition Institute, Sichuan Agricultural University, Huimin Road 211, Wenjiang District, Chengdu 611130, China; wyf061003@163.com
6 Institute of Forestry, Chengdu Academy of Agriculture and Forestry Sciences, Nongke Road 200, Wenjiang District, Chengdu 611130, China; lwj13980429970@163.com (W.L.); liqiong_jiang@163.com (L.J.)
* Correspondence: yemeng5581@163.com; Tel.: +86-028-8629-1456
† These authors contributed equally to this work.

Abstract: *Illicium verum* Hook. f. is a globally significant spice, which is recognized in China as a food-medicine homolog and extensively utilized across the pharmaceutical, food, and spice industries. China boasts the world's leading resources of *I. verum*, yet its comprehensive utilization remains relatively underexplored. Through a resource survey of *I. verum* and the application of bibliometric visualization using CiteSpace, this study analyzed 324 papers published in the Web of Science Core Collection (WOSCC) from 1962 to 2023 and 353 core documents from China's three major databases (CNKI, Wanfang Database, and VIP Database). *I. verum* from Guangxi province towards various southern provinces in China, with autumn fruits exhibited superior quality and market value over their spring fruits. Literature in WOSCC emerged earlier, with a research emphasis on food science technology and pharmacology pharmacy domains. WOSCC research on *I. verum* could be divided into two phases: an embryonic period (1962–2001) and a growth period (2002–2023), showing an overall upward trend in publication. The three major Chinese databases contain a larger number of publications, with a focus on the food sector, which could be categorized into three stages: an embryonic period (1990–1999), a growth period (2000–2010), and a stable period (2011–2023), with an overall downward trend in publication. Both Chinese and international research hotspots converge on the medical applications of *I. verum*, with antioxidant bioactivity research emerging as a prevailing trend. This study delineated the resource distribution of *I. verum* across China and identified the research hotspots and trends both in China and internationally. The findings are beneficial for guiding researchers in swiftly establishing their research focus and furnishing decision-makers with a comprehensive reference for industry information.

Keywords: *Illicium verum*; resource; CiteSpace; bibliometrics

1. Introduction

Star anise (*Illicium verum* Hook. f.), as known globally, generates an annual output value exceeding 10 billion RMB [1] and is extensively utilized in both the food and pharmaceutical industries [2]. This spice is a common ingredient in the culinary traditions of China and Southeast Asia, enhancing the flavor of meat dishes [3], and it is also employed in Western countries for the production of sweet wines, confections, and cakes [4]. Beyond its culinary applications, *I. verum* is recognized as a medicinal resource, possessing antimicrobial, antiviral, and anti-inflammatory properties [5]. It features in traditional Chinese medicine and the Indian healthcare system [4]. The plant is rich in bioactive compounds such as shikimic acid, essential oil, tannins, and anethole (85–90%), which can inhibit various viruses, including the avian influenza virus H5N1 and the human influenza virus H1N1 [6,7]. These compounds also find their way into daily chemical products [1]. In China, *I. verum* is classified as a food-medicinal plant and has been included in successive editions of the Chinese Pharmacopoeia [8]. The *I. verum* belongs to the Schisandraceae Blume family, within the *Illicium* L. genus [9], and is widely distributed across the southwestern part of the Asian continent [4], with the primary production areas located in the subtropical provinces of southern China [1]. China has a history of cultivating *I. verum* spanning over 300 years [10] and is the sole large-scale exporter of *I. verum* globally [11]. Understanding the distribution of this valuable resource is crucial for its optimal development and utilization. *I. verum* holds significant economic value on a global scale [6–8]. Gaining insights into its research hotspots and trends is anticipated to further stimulate the development and application of this valuable spice. However, the current landscape is replete with a vast body of literature on *I. verum*, both within China and internationally. Traditional literature review methods are often insufficient to extract the hotspot and trend information pertinent to *I. verum* research, underscoring the need for advanced analytical tools to navigate this complex field.

Literature review is instrumental in understanding the cutting-edge developments within a field. However, the sheer volume of publications in databases presents a challenge in achieving a comprehensive grasp of the domain's evolution [12,13]. Traditional reviews often focus on a specific domain, and it is seldom feasible to encompass the entirety of the literature to analyze the current state of research within a particular field [13], thus necessitating more convenient methods for literature analysis. Rapidly identifying the developmental hotspots and trends within a field can significantly save researchers time by swiftly guiding them toward promising research directions [10,14]. Moreover, such insights also serve to provide a more comprehensive industry information base, which is invaluable for strategic planning and decision-making [15,16]. Bibliometrics offers a solution by visualizing the literature, which can be employed to explore the dynamic changes in past research and predict future trends [14,15]. CiteSpace is a pivotal statistical software within the realm of bibliometrics, capable of rapidly mapping subject networks and assisting researchers in identifying trends and hotspots [16–18]. The application of the CiteSpace tool has facilitated the identification of research hotspots and trends across a spectrum of scientific disciplines, thereby significantly advancing scientific inquiry. Notably, this tool has been instrumental in fields such as food science [12,19], medicine [20], traditional Chinese medicine [21], and ecology [22], where it has helped to distill complex information and highlight areas ripe for further exploration. *I. verum*, as a globally significant research material, has been the subject of several literature reviews. However, these reviews often focus on food or medical aspects [2,23] and do not adequately reflect the dynamic evolution of research in this area.

The present study synthesizes the distribution of *I. verum* resources in China and employs CiteSpace to visualize journal articles from the three major Chinese databases and the Web of Science Core Collection (WOSCC). By doing so, we analyze the trends and hotspots in both Chinese and international research on *I. verum*. The findings of this study not only showcase the current state of *I. verum* resources in China but also serve as a reference for policymakers and growers. Furthermore, this research facilitates a better understanding for scientific researchers of the dynamic evolution and emerging trends within *I. verum* studies.

2. Materials and Methods

International literature databases such as Scopus and Google Scholar are available, but the Web of Science (WOS) is more suited for large-scale bibliometric analysis [24]. The WOSCC is an integral component of the WOS database, encompassing a wide array of influential literature from around the globe. The subject search was conducted using the search function of WOSCC, employing the following strategy: Topical Subject = Illicium verum OR Title = Illicium verum OR Author Keywords = Illicium verum OR Abstract = Illicium verum; the language was restricted to English, and the document type was limited to articles. The search was accessed on 25 March 2024. Literature published in 2024 was excluded, and the analysis focused on articles, resulting in a selection of 324 papers for analysis.

The literature search was conducted across three Chinese databases: China National Knowledge Infrastructure (CNKI https://www.cnki.net/, accessed on 8 February 2024), Weipu Database (VIP http://www.cqvip.com/, accessed on 8 February 2024), and the Wanfang Database (https://w.wanfangdata.com.cn/index.html?index=true, accessed on 8 February 2024). The search criteria included publications with "Illicium verum" in the subject or title from 1990 to 2023. The query was restricted to journals indexed as core by Peking University, and duplicates were removed. This process yielded a total of 353 papers for analysis. The specific analytical procedure is depicted in Figure 1.

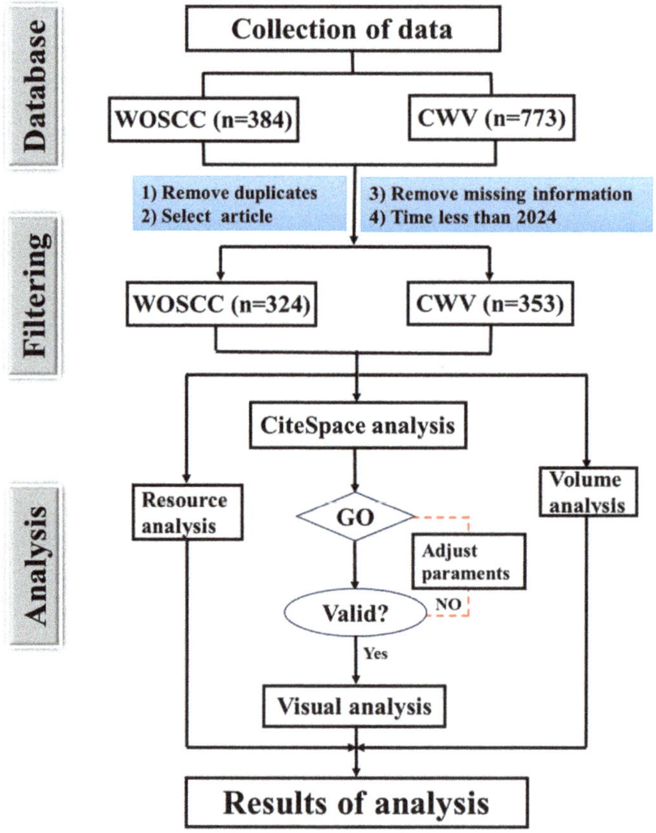

Figure 1. The process of literature analysis. CWV: CNKI (https://www.cnki.net/), Wanfan data (https://w.wanfangdata.com.cn/index.html?index=true), and VIP data (http://www.cqvip.com/) about publications of *I. verum*. WOSCC: Web of Science core collection about publications of *I. verum*.

Visualization of the literature was performed using CiteSpace V.6.2.R4 (Drexel University, Philadelphia, PA, USA), with the following parameter settings: (1) Time slicing was set from 1962 to 2023, with a slice width of 5 years; (2) The g-index was configured with a k-value of 25; (3) For network pruning, the Pathfinder and slicing network pruning algorithms were selected. Keyword co-occurrence maps were generated using the Log-Likelihood Ratio (LLR) algorithm to extract keywords. All clustering results had modularity above 0.3 and silhouette values above 0.7, indicating robust clustering outcomes [12].

3. Results

3.1. Resource of I. verum

In this study, we have mapped the primary distribution areas of the *I. verum* resources in China [1,9,25–28], as depicted in Figure 2. According to the literature, *I. verum* was predominantly found in the southern regions of the country [1], near the tropic of cancer [11]. The resource was mainly distributed across provinces such as Guangxi, Yunnan, Guangdong, Guizhou, Hunan, Jiangxi, Fujian, and Zhejiang, with small-scale cultivation in the Shigang Forest Farm in Lu County, Sichuan, and sporadic plantings in other provinces of the south [1]. Among the cultivating provinces, Guangxi had the largest planting area, covering 36,506 km^2 across 24 counties (cities or districts), with Baise and Ningming counties known for their superior quality and cultivation history exceeding 300 years [1,28]. Longzhou County, Fangcheng District, and Teng County are the top producers of *I. verum*, with an output exceeding 200,000 kg. The distribution of *I. verum* resources was centered in Guangxi and radiated to the adjacent southern provinces.

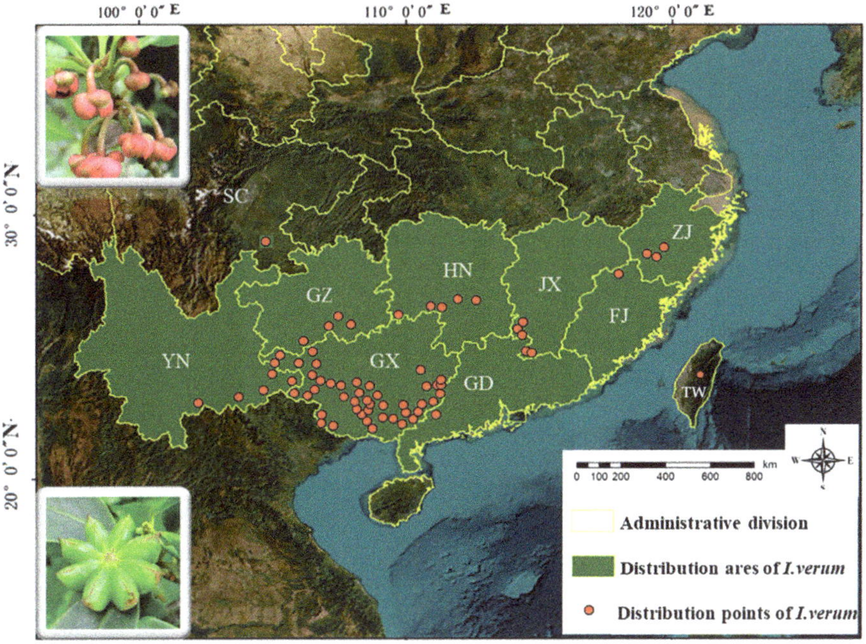

Figure 2. Distribution of *I. verum* in China. The green-shaded regions in the figure delineate the primary areas of *I. verum* cultivation in China, while the red dots indicate the distribution points of *I. verum* across various provinces. YN: Yunnan province; SC: Sichuan province; GZ: Guizhou province; GX: Guangxi Zhuang Autonomous Region; HN: Hunan province; GD: Guangdong province; JX: Jiangxi province; FJ: Fujian province; ZJ: Zhejiang province; TW: Taiwan province. Morphological images of *I. verum* fruits and flowers. The image was created using ArcGIS 10.5.

In the *Illicium*, certain species closely resemble *I. verum*, leading to the emergence of counterfeit *I. verum* that was challenging to distinguish from the authentic product. These counterfeits had not only disrupted the market for *I. verum* but had been responsible for incidents of poisoning due to their toxic properties. Here, we listed the common adulterants of *I. verum* and their toxic parts, as depicted in Table 1.

Table 1. Common adulterants and toxic parts of *I. verum*.

Species	Toxic of Location	Resources
Illicium anisatum	seeds and carpels	[22,27]
Illicium floridanum	fruits and leaves	[22,27]
Illicium merrillianum	pericarps	[22,27]
Illicium majus	fruits	[22,27]
Illicium verum	fruits	[22,27]
Illicium arborescens	fruits	[22,27]
Illicium brevystylum	fruits	[22,27]
Illicium henryi	fruits	[22,27]
Illicium macranthum	fruits	[22,27]
Illicium majus	fruits	[22,27]
Illicium minwanense	fruits	[22,27]
Illicium simonsii	fruits	[22,27]
Illicium ternstroemioides	fruits	[22,27]

In addition to the presence of morphologically similar counterfeits, *I. verum* was often classified into two varieties based on the harvest time: the spring fruit (horn flower star anise) and the autumn fruit (big red star anise). These two varieties differ not only in their chemical composition but also in their market value [29]. Data from the Chinese Medicinal Materials World Network (https://www.zyctd.com/jiage/xq8.html, accessed on 6 March 2024), which span from 2019 to 2023, provides a detailed analysis of the prices of different types of *I. verum* across various trading markets, as well as the morphological characteristics of the spring and autumn fruits, as depicted in Figure 3. The price trend of *I. verum* exhibited an overall "inverted V" pattern, reaching its peak in 2021 before stabilizing. Despite this stabilization, prices remained higher compared to those in 2019 and 2020. Interestingly, the price of the spring fruit is consistently lower than that of the autumn fruit across different regions, with minimal price fluctuations observed within the same type of fruit (Figure 3A). Within the same year, the price of autumn fruit was relatively constant across different trading markets (Figure 3C). Morphologically, the autumn fruit of *I. verum* was characterized by a better opening degree and a bright yellow fruit color, which gives it an overall superior appearance compared to the spring fruit (Figure 3B,D). The microscopic morphology of *I. verum* is illustrated in Figure S1. These findings underscore the importance of distinguishing between the different varieties of *I. verum* not only to combat counterfeit products but also to understand the economic implications of seasonal variations in the harvest.

3.2. Annual Number and Trend of Publications

The annual variation in the number of publications was a significant metric for gauging the academic research activity within a field [30], and it could reflect the pace of research development [31]. An analysis of articles related to *I. verum* in three Chinese databases and the WOSCC was presented in Figure 4. The initial publication of literature in WOSCC occurred in 1962, predating the first literature in three Chinese databases, which appeared in 1990. The annual publication trend of literature in three Chinese databases could be divided into three distinct phases: the embryonic phase (1990–1999), during which the number of publications was relatively low; the growth phase (2000–2010), characterized by an overall upward trend in publication numbers, peaking at 29 articles in 2010; and the stable phase (2011–2023), where the overall publication trend experienced a decline, yet the number of publications remained above seven per year. These findings provided valuable insights into the trajectory of academic interest and research output in the field of *I. verum*, highlighting the dynamic nature of research focus and the potential for future exploration.

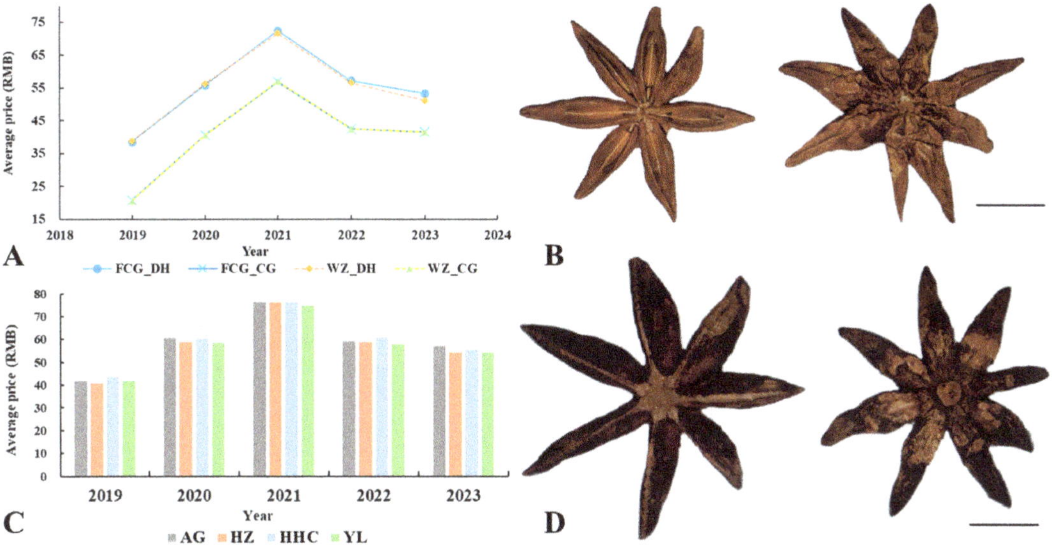

Figure 3. Price and Morphology of Two *I. verum* Varieties from 2019 to 2023. (**A**) Price of *I. verum* in Fangchenggang and Wuzhou cities in the Guangxi Zhuang Autonomous Region. FCG_CG: spring fruits of *I. verum* in Fangchenggang city; FCG_DH: autumn fruits of *I. verum* in Fangchenggang city; WZ_CG: spring fruits of *I. verum* in Wuzhou city; WZ_DH: autumn fruits of *I. verum* in Wuzhou city. (**C**) Annual price changes for autumn fruits of *I. verum* in four major traditional Chinese medicine markets in China. AG: Anguo Medicine Market in Baoding; HZ: Heze Medicine Market in Bozhou; HHC: Hehuachi Medicine Market; YL: Yulin Medicine Market in Yulin. (**B**) Morphological features of autumn fruits of *I. verum*. (**D**) Morphological features of spring fruits of *I. verum*. The prices depicted in the figure represent the average annual values for each variety. Scale = 1 cm.

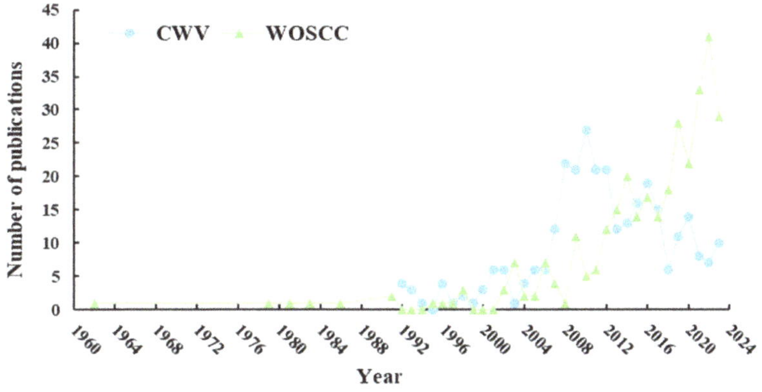

Figure 4. Trends of the number of publications issued by *I. verum*. CWV: CNKI (https://www.cnki.net/), Wanfan data (https://w.wanfangdata.com.cn/index.html?index=true), and VIP data (http://www.cqvip.com/) about publications of *I. verum*. WOSCC: Web of Science core collection about publications of *I. verum*.

The growth trend of literature in WOSCC on *I. verum* could be delineated into two distinct phases. The first phase, termed the embryonic phase (1962–2001), was characterized by an unstable number of publications. The second phase, known as the growth phase (2002–2023), witnessed an overall upward trend in publication volume, culminating in a peak of 41 articles in the year 2022.

3.3. Main Research Forces

3.3.1. Countries and Institutions

International scientific collaboration played a significant role in integrating into the global research community, reflecting the collaborative relationships between different countries and academic institutions [32]. The WOSCC, as a pivotal bibliographic database, effectively mirrors international cooperation. By employing CiteSpace for visualization, we depicted the leading institutions and countries in terms of publication output on *I. verum* within the WOSCC, as shown in Figure 5. Both the Chinese Academy of Sciences and Shandong Agricultural University have published over 10 articles related to *I. verum* (Figure 5A), with all these publications originating from China, indicating a close engagement of Chinese institutions in the research of *I. verum*. Among three Chinese database research institutions, those from Guangxi Zhuang Autonomous Prefecture had the highest number of publications on *I. verum* (Figure S2). The research on *I. verum* spanned over 50 countries and regions, demonstrating its global attention. The country with the most extensive research on *I. verum* was the People's Republic of China, with over 120 publications (Figure 5B). The size of the circles was proportional to the number of publications from each country [33]. Among the top 10 countries in *I. verum* research, the majority were located in Asia. In summary, the visualization analysis of research institutions and countries indicated a close correlation between the volume of scientific literature and the distribution of *I. verum* resources.

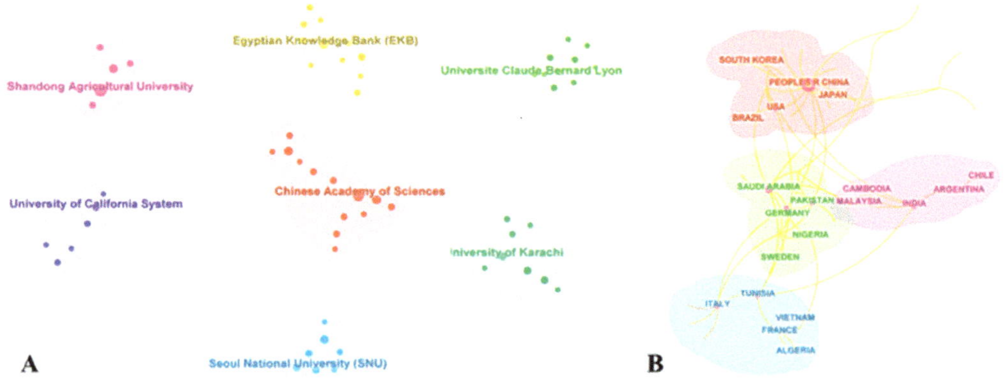

Figure 5. Network visualization map of cooperation between countries and institutions. (**A**) CiteSpace network map of institutions involved in *I. verum*, Modularity (Q = 0.9243), Silhouette (S = 0.9903). (**B**) CiteSpace network map of countries involved in *I. verum*, the size of the circle indicates the number of countries that had published papers, and the yellow line indicates the cooperation between different countries. Only the top 20 countries with the number of articles were shown. Modularity (Q = 0.5526), Silhouette (S = 0.7253).

The tier of research journals serves as a critical instrument for gauging the quality of research [13]. Subsequently, we conducted a statistical analysis of the top 10 journals by publication count, both Chinese and internationally, with the results presented in Table 2. Based on the data from WOSCC, the research field of *I. verum* abroad was primarily concentrated in "Food Science & Technology" (60 publications) and "Pharmacology &

Pharmacy" (47 publications). Among the top 10 journals with the highest publication count on *I. verum* in the WOSCC, "Molecules" (11 publications) published the most papers, showcasing a diverse range of research topics. The proportion of journals with a Journal Citation Report (JCR) rank of Q1 is 60%, indicating a high level of research quality. In the three Chinese databases, the research field of *I. verum* was predominantly focused on "Light Industrial Handicrafts" and "Traditional Chinese Medicine", with a relatively stable research domain. In the top 10 Chinese journals by publication count, "China Condiment" (60 publications) published the most papers. Only the journal "Food Science" was recognized as an outstanding Chinese journal, suggesting that the overall level of research on *I. verum* in Chinese journals was relatively low, and the influence was weak.

Table 2. Top 10 Journals by publication count for *I. verum* research.

Type	Journal	Impact Factor	Categories	Number
WOSCC	Molecules	4.6	Biochemistry & molecular biology	11
	Industrial crops and products	5.9	Agricultural engineering	7
	Food chemistry	8.8	Chemistry, applied	5
	Journal of agricultural and food chemistry	6.1	Agriculture, multidisciplinary	5
	Poultry science	4.4	Agriculture, dairy, and animal science	5
	Food and chemical toxicology	4.3	Toxicology	4
	Phytotherapy research	7.2	Chemistry, medicinal	4
	Animal science journal	2.0	Agriculture, dairy, and animal science	3
	Environmental science and pollution research	5.8	Environmental sciences	3
	Evidence-based complementary and alternative medicine	2.65	Integrative and complementary medicine	3
CWV	China condiment	2.749	Light industrial handicrafts	60
	Food science and technology	1.623	Light industrial handicrafts	15
	Lishizhen medicine and materia medica research	1.496	Traditional Chinese medicine	14
	Science and technology of the food industry	2.592	Light industrial handicrafts	14
	Journal of Anhui Agricultural Sciences	0.900	Comprehensive agriculture	11
	Food science	3.894	Light industrial handicrafts	9
	Food research and development	2.251	Light industrial handicrafts	9
	Chinese journal of pharmaceutical analysis	1.665	Medicine	9
	Chinese traditional and herbal drugs	4.694	Traditional Chinese medicine	9
	Journal of the Chinese cereals and oils association	2.278	Light industrial handicrafts	11

Note: The impact factors and subject categories of international journals were obtained through the WOSCC. The impact factors listed were as of the year 2022, and where multiple subject categories are available, the one with the highest ranking in the Journal Citation Reports (JCR) was selected. Journals highlighted in red were those classified in the Q1 zone of the JCR. The acronym "CWV" represents a composite of the China National Knowledge Infrastructure (CNKI) (https://www.cnki.net/), Wanfang (https://w.wanfangdata.com.cn/index.html?index=true), and VIP (http://www.cqvip.com/). The impact factors and subject categories for journals were retrieved via the CNKI platform.

3.3.2. Main Author

The co-citation network of literature represents the structure and development of a field, and identifying key researchers within this network could aid in discovering leading research teams [30,34]. We analyzed the primary authors and co-cited authors within the WOSCC, as depicted in Figure 6. The correlation between publication count and citation frequency in *I. verum* research was not significant. Both the most prolific author and the author with the highest citation count were scholars from China, indicating a deep level of research engagement in this field within the country. Yang Zaibin had the highest number of publications (12 papers), suggesting extensive research contributions. Wang GW had the highest number of citations (114), indicating a high quality of published work. Further analysis of *I. verum* researchers in China's three major databases was presented in Figure S3. The principal investigator in Chinese *I. verum* research is "Miu Jianhua", who was characterized by close collaborations with other researchers within China.

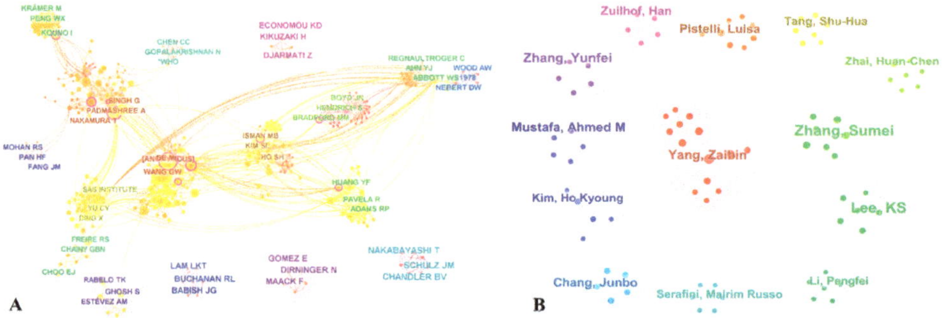

Figure 6. Citespace network of co-cited authorship and TOP 13 authors. (**A**) co-cited authorship in the field of *I. verum* research. The size of the circle is positively correlated with the cited counts of the authors, and the yellow lines represent a collaboration between two authors. Modularity (Q = 0.7921), Silhouette (S = 0.8684). (**B**) TOP 13 author in the field of *I. verum* research. The small number of points within the module reflects the number of articles published by the author. Modularity (Q = 0.9661), Silhouette (S = 0.9828).

3.4. Research Trend and Hotspot Analysis

3.4.1. Hotspot in *I. verum*

Understanding the relevance of other related topics could provide insights into a specific field of study [18]. The evolution of article keywords could reflect the structure and dynamics of a knowledge domain [35]. By utilizing CiteSpace to visualize the high-frequency keywords in the *I. verum* literature from the WOSCC, we could reveal the hotspots of research in this field, as shown in Figure 7. The effective clustering results encompass a total of 12 clusters (Figure 7A), which were primarily divided into two main categories. The first category pertains to the classification research of *I. verum* resources, including cluster #1, "star anise", cluster #2, "Illicium verum", and cluster #7, "foeniculum vulgare mill". The timelines of these clusters were notably extensive (Figure 7B), indicating that resource-related issues had become a significant foundation for research on *I. verum*.

The second category pertains to the application research of *I. verum*, which includes cluster #0, "antibacterial activity", cluster #3, "laying hen", cluster #4, "Aspergillus flavus", cluster #5, "natural fumigant", cluster #6, "Aedes aegypti", cluster #8, "shikimic acid", cluster #9, "quality control", cluster #10, "Blattella germanica", and cluster #11, "behavior". The clusters within the application domain of *I. verum* demonstrated a primary focus on the medical and pharmaceutical fields. The keyword clustering of *I. verum* in three Chinese databases (Figures S4 and S5) could likewise be divided into two main categories. However, the three Chinese databases' keyword clusters were more concentrated on the resources of *I. verum* rather than its applications. Both international and Chinese research on the application of *I. verum* emphasizes the medical field.

3.4.2. Trend in *I. verum*

Keyword clustering analysis enables the identification of emerging research trends [17]. By conducting an analysis of the strongest citation bursts for keywords in both Chinese and international databases, the results are depicted in Figure 8. Based on the top 10 keywords bursts from the WOSCC, the outcomes could be categorized into two groups (Figure 8A): the first group pertains to *I. verum* resources (10%), and the second to the application of *I. verum* (90%), indicating that research on *I. verum* was predominantly focused on its applications. Within the domain of *I. verum* applications, high-frequency keywords could be divided into two categories: the first category involves the chemical composition and extraction methods (60%), and the second category concerns the application of bioactive substances (30%). "Essential oil" and "Illicium verum Hook f" emerged as the earliest high-frequency keywords, suggesting that the initial research on *I. verum* centered on its chemical constituents and resources. While the chemical composition and extraction methods were once the primary high-frequency terms associated with *I. verum*, more recent research trends have shifted towards "antioxidant" and "nutrient digestibility". The emergence of "nutrient digestibility" as a new high-frequency keyword likely extends from the plant's antimicrobial and antioxidant properties.

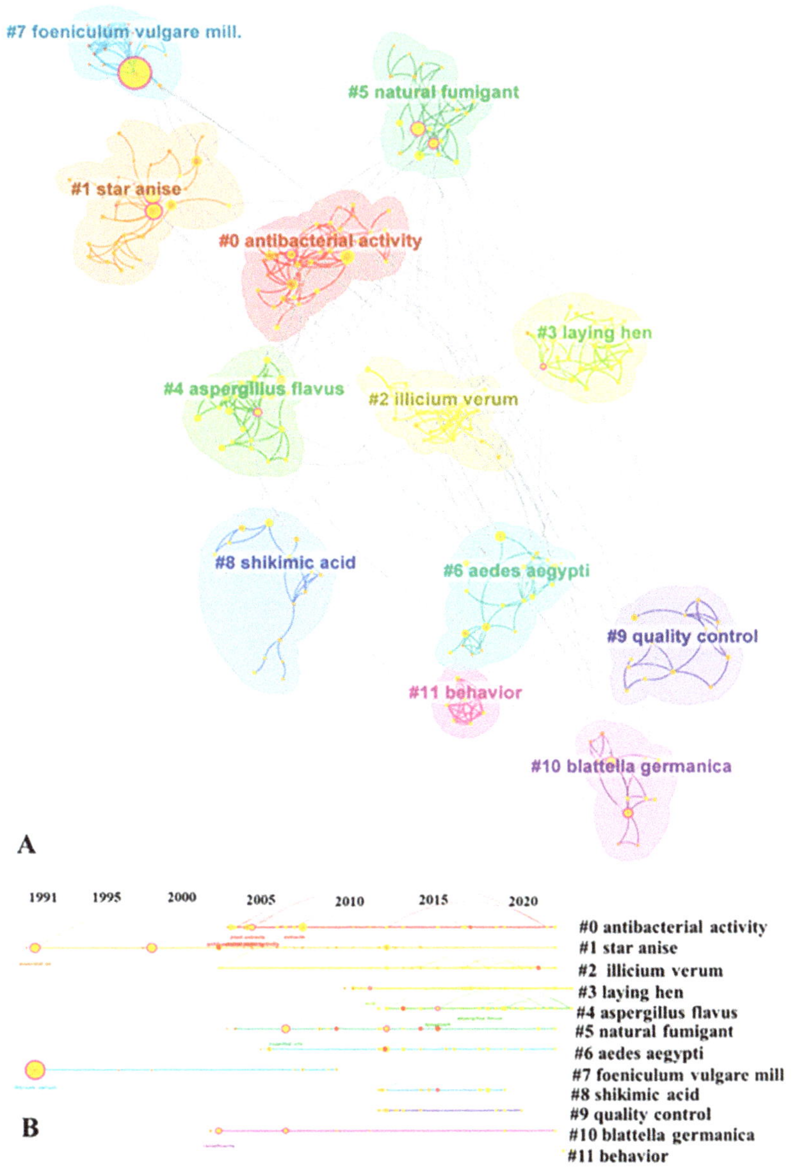

Figure 7. Keyword analysis in *I. verum*-related study. (**A**) Visual mapping of keywords in *I. verum* using CiteSpace; The black lines showed the connections between them. (**B**) The keyword time zones of *I. verum*. Modularity (Q = 0.6483), Silhouette (S = 0.8486).

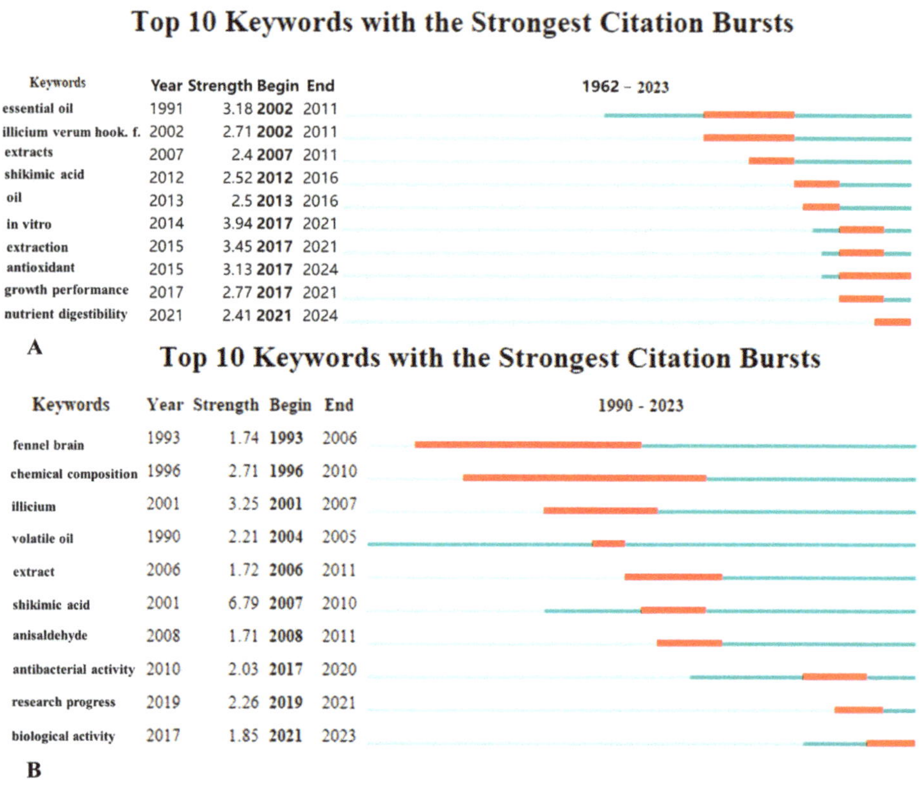

Figure 8. Keywords with the strongest citation bursts of *I. verum*. (**A**) Keywords from 1962 to 2023 appear in WOSCC. (**B**) The keywords of 1990~2023 in the three major Chinese databases appear. The red part indicated that the keyword was in a high-frequency state during the time period.

Based on the top 10 keyword bursts from three Chinese databases, the research results were categorized into two groups: resources and applications, with the latter still constituting the majority. The early study of the "fennel brain" might be attributed to its relevance to the medicinal quality of *I. verum* [36]. Subsequently, "Illicium" emerged as a research trend, potentially due to concerns regarding counterfeits in the resource pool of *I. verum*. In recent years, "biological activity" has become a prominent research trend, which might be associated with the antimicrobial and antioxidant characteristics of *I. verum* [4]. This shift in focus towards the biological activity of *I. verum* reflected the growing interest in understanding and harnessing the plant's therapeutic potential.

4. Discussion

4.1. Resource of I. verum

The plants within the *Illicium* exhibit similar morphological characteristics, yet they can have significant differences in their chemical compositions. These differences have led to the emergence of various counterfeits of *I. verum* (Table 1). The *I. verum* itself is non-toxic and is utilized across multiple domains due to its content of volatile oils and trans-anethole, among other active constituents [2,8]. However, these morphologically similar counterfeits pose a threat to human health because they contain toxins [23]. Reports from Europe and the United States have indicated that accidental consumption of these counterfeits can lead to adverse neurological reactions [37]. The distribution and habitat preferences of different counterfeits may vary. For instance, *Illicium simonsii* Maxim is found in the mountainous regions of Yunnan, Guizhou, and Sichuan provinces in China [38], while *Illicium henryi* Diels is distributed in the moist areas of Hubei and Sichuan provinces, among other locations in China [9]. Although these counterfeits can be harmful if consumed directly, they possess significant value as they can be extracted to yield shikimic acid, a compound that has been used to treat avian influenza, among other applications [23]. Therefore, the judicious utilization of these *I. verum* counterfeits could be beneficial to humanity. Understanding the distribution of *I. verum* resources is crucial for the identification and differentiation of counterfeits.

The *I. verum* is native to China and Vietnam [23] and thrives in subtropical regions of southern Asia [9]. However, China stands out as the country with the richest resources and the capacity for substantial export [11]. The plant is predominantly found in provinces near the Tropic of Cancer in China, such as Guangxi, Guangdong, and Guizhou provinces in China, where it is cultivated extensively. Although *I. verum* has been introduced to other provinces, it has not reached a significant scale of cultivation [1]. Economic development has prompted the introduction of *I. verum* in many regions, but often, the outcomes have been less than satisfactory, failing to yield substantial economic benefits. There are several potential reasons for the poor performance of *I. verum* in introduced settings. Firstly, while introduction to colder climates may reduce the incidence of pests and diseases [39], different species vary in their adaptability, and excessively low temperatures can limit plant growth [40]. Secondly, when temperatures exceed the optimal range for the plant, net photosynthesis may decline due to increased respiration and stomatal closure [41], potentially preventing the plant from completing reproductive tasks and merely maintaining normal growth. Lastly, when plants are introduced to new environments, they may encounter new herbivores or re-encounter local herbivores [42], leading to a decrease in their adaptive capabilities. Biogenic volatiles are crucial substances that plants produce in response to biotic or abiotic stress, enhancing their survival [43]. Variations in the content and composition of volatiles in *I. verum* from different regions [2,4,23,44] indicate that different environments can influence the accumulation of active components in *I. verum*, which also limits the effectiveness of its introduction to new areas. Consequently, these factors contribute to the current distribution pattern of *I. verum*.

4.2. Hotspot and Trend of *I. verum* in Pharmaceutical Field

The fruits and foliage of *I. verum* contain a variety of important active compounds. These include over 20 active substances such as cis- and trans-anethole and tannins, a small number of nitrogenous components, 14 types of hydrocarbon constituents, and 22 oxygenated hydrocarbon derivatives [2,45,46]. These compounds are known for their antimicrobial, antioxidant, and insecticidal properties [47,48]. As a food-medicinal plant, the *I. verum* is widely used in both food and pharmaceutical applications. In countries such as China, Vietnam, and India, *I. verum* is frequently used in the preparation of meat dishes, while in Western countries, it is employed in the making of alcoholic beverages, teas, and jams [4]. Beyond enhancing the flavor of food, *I verum* is also utilized as a pharmaceutical agent due to its numerous active ingredients. For instance, it is used in the manufacturing of the antiviral drug oseltamivir (Tamiflu®) [49]. Trans-anethole and volatile oil are critical indicators of the medicinal quality of *I. verum*, and since the 2010 edition of the Chinese Pharmacopoeia, a minimum trans-anethole content of 4% has been stipulated [36]. An analysis of keyword clustering trends and hotspots in recent years reveals a pronounced focus on the pharmaceutical applications of *I. verum* (Figures 7 and 8), highlighting the plant's significance in medical research.

The primary research on *I. verum* has predominantly focused on the pharmaceutical domain due to several key reasons. Initially, advancements in extraction technologies have facilitated the identification of a growing number of active constituents. *I. verum*, like many traditional Chinese medicinal herbs, is a rich source of bioactive compounds [38]. The evolution of extraction techniques has led to the uncovering of an increasing variety of these compounds, thereby enhancing research efforts in their medicinal applications [50].

With the discovery of the harmful effects of antibiotic misuse in both livestock and human medicine, the search for alternatives to antibiotics has become a research priority [51,52]. The *I. verum*, with its antiviral and antioxidant properties, has emerged as a significant candidate in this context. Its potential to serve as a substitute for conventional antibiotics has garnered considerable attention in the scientific community. In conclusion, the future trajectory of research on *I. verum* is poised to continue its emphasis on pharmaceutical applications. The exploration of its bioactive components and the development of novel therapeutic strategies leveraging these properties will likely remain at the forefront of scientific inquiry.

5. Conclusions

China stands as the country with the richest resources of *I. verum*, predominantly centered around the Guangxi province and extending to surrounding provinces. Chinese scholars have the highest publication count in WOSCC and engage in close collaboration with researchers. The abundance of *I. verum* resources facilitates the advancement of research in academic institutions. Currently, the focus of research interest in *I. verum* is primarily within the medicinal domain, and the investigation of its bioactive properties is poised to become a new direction for future studies.

Due to the presence of many morphologically similar toxic plants within the same genus, the application of *I. verum* in the food industry has been constrained. However, *I. verum* and its related species possess significant medicinal value, as documented in references [4–8], which has led to a focused research interest in the medical field for *I. verum*. Based on the analysis of current research hotspots and trends regarding *I. verum* presented in this article, potential future developments in star anise research may include the following areas.

(1) The proportion of research into the selection and breeding of superior resources of *I. verum* and its counterfeits is increasing. Current trends in research from both China and internationally indicate a growing focus on antioxidant activity, which is emerging as a prevailing research trend. This outcome is expected to stimulate an increased demand for the antioxidant components of *I. verum* in the future. Consequently, the development of superior varieties of *I. verum* rich in antioxidant constituents or the cultivation of high-quality resources from related species within the same genus may present new avenues for research;

(2) The research proportion dedicated to the application of *I. verum* in the food sector is anticipated to expand. A growing number of researchers are likely to follow the prevailing trends in *I. verum* research, delving into the study of its chemical constituents. The discovery of an increasing array of compounds may facilitate the development of *I. verum* within the food industry, potentially leading to the creation of a variety of foods with distinctive flavors. The promotion of superior sources of *I. verum* is anticipated to augment the yield of this spice. Concurrently, an increase in production is likely to enhance the circulation of *I. verum* in the market, thereby elevating its proportion within the culinary sector.

Supplementary Materials: The following supporting information can be downloaded at https://www.mdpi.com/article/10.3390/foods13101510/s1, Figure S1: Morphology of spring and autumn fruits of *I. verum*. Figure S2: Institution collaboration network in *I. verum* based on three Chinese databases. Figure S3: Researcher collaboration network in *I. verum* based on three Chinese databases. Figure S4: Visual mapping of keyword in *I. verum* using CiteSpace based on three Chinese databases. Figure S5: The keyword time zones of *I. verum* based on three Chinese databases.

Author Contributions: Z.H. and J.H.: Conceptualization, Methodology, Formal analysis, Data curation, Writing—original draft preparation; Y.W.: Conceptualization, Supervision, Software; W.L. and L.J.: Data curation, Funding acquisition; X.G.: Conceptualization, Supervision; M.Y.: Data curation, Visualization, Project administration, Funding acquisition; D.L.: Software. All authors have read and agreed to the published version of the manuscript.

Funding: Research was funded by the Sichuan Agricultural University Double Support Plan (035/03571828), Chengdu Functional Plant Germplasm Collection, and Tianfu Forest Granary Science and Technology Innovation Demonstration Base Construction Project.

Institutional Review Board Statement: Not applicable.

Informed Consent Statement: Not applicable.

Data Availability Statement: No new data were created or analyzed in this study. Data sharing is not applicable to this article.

Conflicts of Interest: The authors declare that the research was conducted in the absence of any commercial or financial relationships that could be construed as a potential conflict of interest.

References

1. Huang, K.S.; Li, G.Q.; An, J.C.; Li, K.X. Developing situation of processing and utilization industry of characteristic resources of *Star anise* (*Illicium verum*). *Biomass Chem. Eng.* **2020**, *54*, 6–12. Available online: https://www.cnki.com.cn/Article/CJFDTOTAL-LCHG202006002.htm (accessed on 25 March 2024).
2. Noumi, E.; Ahmad, I.; Adnan, M.; Patel, H.; Merghni, A.; Haddaji, N.; Bouali, N.; Alabbosh, K.F.; Kadri, A.; Caputo, L.; et al. *Illicium verum* L. (*Star anise*) essential oil: GC/MS profile, molecular docking study, in silico ADME profiling, quorum sensing, and biofilm-inhibiting effect on foodborne bacteria. *Molecules* **2023**, *28*, 7691. [CrossRef] [PubMed]
3. Li, W.; Wu, Z.; Xia, Y.; Tan, J.; Zhao, H.; Chen, S.; Li, Y.; Tang, H.; Wang, G.; Zhang, Y. Antiviral and antioxidant components from the fruits of *Illicium verum* Hook.f. (Chinese *Star anise*). *J. Agric. Food Chem.* **2022**, *70*, 3697–3707. [CrossRef] [PubMed]
4. Patra, J.K.; Das, G.; Bose, S.; Banerjee, S.; Vishnuprasad, C.N.; Del Pilar Rodriguez-Torres, M.; Shin, H.S. *Star anise* (*Illicium verum*): Chemical compounds, antiviral properties, and clinical relevance. *Phytother. Res.* **2020**, *34*, 1248–1267. [CrossRef] [PubMed]
5. George, C.K. 24—*Star anise*. In *Handbook of Herbs and Spices*, 2nd ed.; Peter, K.V., Ed.; Woodhead Publishing: Cambridge, UK, 2012; pp. 487–503. Available online: https://www.sciencedirect.com/book/9780857090393/handbook-of-herbs-and-spices (accessed on 25 March 2024).
6. Bradley, D. Star role for bacteria in controlling flu pandemic? *Nat. Rev. Drug Discov.* **2005**, *4*, 945–946. [CrossRef]
7. Yang, E.C.; Hsieh, Y.Y.; Chuang, L.Y. Comparison of the phytochemical composition and antibacterial activities of the various extracts from leaves and twigs of *Illicium verum*. *Molecules* **2021**, *26*, 3909. [CrossRef]
8. Zhou, H.; Li, Z.J.; Yang, Y.; Chen, X.L.; Zhou, H.; Xu, B.C.; Li, C. Research progress on flavor characteristics of Strong fragrance spices and their effects on flavor formation of meat products. *Sci. Technol. Food Ind.* **2024**, *1*, 1–16. [CrossRef]
9. Editorial Committee of Flora of China. *Flora of China*; Science Press: Beijing, China, 1996; pp. 228–231. Available online: http://www.iplant.cn/frps (accessed on 25 March 2024).
10. Hu, Y.; Duan, T.X.; Cao, F.; Yu, J.P.; Huang, J.M. Evaluation of LC-MS chromatograms of pericarps from *Illicium* species. *China J. Chin. Mater. Medica* **2010**, *35*, 1836–1839. Available online: https://www.cnki.com.cn/Article/CJFDTOTAL-ZGZY201014021.htm (accessed on 25 March 2024).
11. Bao, Z.Y.; Li, Y.W.; Li, Y.; Qian, H.F.; Zhang, H.; Wang, L. Biological activity and application of *Illicium verum*. *Food Ferment. Ind.* **2023**, *49*, 323–329. [CrossRef]
12. Zhang, X.L.; Zheng, Y.; Xia, M.L.; Wu, Y.N.; Liu, X.J.; Xie, S.K.; Wu, Y.F.; Wang, M. Knowledge domain and emerging trends in vinegar research: A bibliometric review of the literature from WoSCC. *Foods* **2020**, *9*, 166. [CrossRef]
13. Zhang, J.J.; Wei, Z.X.; Lu, T.; Qi, X.Z.; Xie, L.; Vincenzetti, S.; Polidori, P.; Li, L.J.; Liu, G.Q. The research field of meat preservation: A scientometric and visualization analysis based on the web of science. *Foods* **2023**, *12*, 4239. [CrossRef]
14. Liu, Z.; Wang, S.Z.; Zhang, Y.D.; Feng, Y.C.; Liu, J.J.; Zhu, H.D. Artificial intelligence in food safety: A decade review and bibliometric analysis. *Foods* **2023**, *12*, 1242. [CrossRef] [PubMed]
15. Wang, N.; Tang, G.; Jiang, B.; He, Z.; He, Q. The development of green enterprises: A literature review based on VOS viewer and Pajek. *Aust. J. Manag.* **2023**, *48*, 204–234. [CrossRef]
16. Chen, C. Searching for intellectual turning points: Progressive knowledge domain visualization. *Proc. Natl. Acad. Sci. USA* **2004**, *101*, 5303–5310. [CrossRef] [PubMed]
17. Chen, C.M. CiteSpace II: Detecting and visualizing emerging trends and transient patterns in scientific literature. *J. Am. Soc. Inf. Sci. Technol.* **2006**, *57*, 359–377. [CrossRef]
18. Chen, C. Science mapping: A systematic review of the literature. *J. Data Inf. Sci.* **2017**, *2*, 1–40. [CrossRef]
19. Wang, Q.; Zhao, Z.Y.; Wang, Z.Y. Data-driven analysis of risk-assessment methods for cold food chains. *Foods* **2023**, *12*, 1677. [CrossRef]
20. Shi, Y.M.; Luo, J.M.; Wang, X.Q.; Zhang, Y.Q.; Zhu, H.; Su, D.S.; Yu, W.F.; Tian, J. Emerging trends on the correlation between neurotransmitters and tumor progression in the last 20 Years: A bibliometric analysis via CiteSpace. *Front. Oncol.* **2022**, *12*, 800499. [CrossRef] [PubMed]
21. Wu, L.; Xiang, T.; Chen, C.; Isah, M.B.; Zhang, X.Y. Studies on *Cistanches herba*: A bibliometric analysis. *Plants* **2023**, *12*, 1098. [CrossRef]

22. Kao, X.; Wang, W.; Kao, Q.; Zhang, J. Visualization of energy-environment-economy system research characteristics and hotspots evolution trends based on CiteSpace. *Environ. Res. Commun.* **2022**, *4*, 055004. [CrossRef]
23. Howes, M.J.; Kite, G.C.; Simmonds, M.S. Distinguishing Chinese *Star anise* from Japanese *Star anise* using thermal desorption-gas chromatography-mass spectrometry. *J. Agric. Food Chem.* **2009**, *57*, 5783–5789. [CrossRef]
24. Pranckute, R. Web of Science (WoS) and Scopus: The titans of bibliographic information in today's academic world. *Publications* **2021**, *9*, 12. [CrossRef]
25. Lu, S.D. Analysis and forecast of production and sales of *Star anise* in Guangxi in 2000. *Chin. J. Inf. Tradit. Chin. Med.* **2001**, *2*, 70. Available online: https://www.cnki.com.cn/Article/CJFDTOTAL-XXYY200102050.htm (accessed on 25 March 2024).
26. Yan, J.H.; Xiao, X.X.; Huang, K.L. Component analysis of volatile oil from *Illicium Verum* Hook.f. *J. Cent. South. Univ. Technol.* **2002**, *9*, 173–176. Available online: http://qikan.cqvip.com/Qikan/Article/Detail?id=1000424003 (accessed on 25 March 2024). [CrossRef]
27. Lin, Y.L.; Huang, L.F.; Li, Q.; Wang, L.F.; Xie, C.X.; Suo, F.M.; Sun, C.; You, J.W.; Lu, J.W. Ecological suitability evaluation and diqital regionalization for *lllicium verum* Hook.f. *Mod. Tradit. Chin. Med. Mater. Medica-World Sci. Technol.* **2010**, *12*, 468–472. Available online: https://www.cnki.com.cn/Article/CJFDTOTAL-SJKX201003036.htm (accessed on 25 March 2024).
28. Zhao, L.B. Analysis on the current situation and countermeasures of *Star anise* industry development a case study of Ningming, Guangxi. *J. Guangxi Agric.* **2021**, *36*, 51–55+63. Available online: https://www.cnki.com.cn/Article/CJFDTOTAL-GXLB2021020 14.htm (accessed on 25 March 2024).
29. Chen, S.S.; Xu, M.; Wu, Q.; Yang, W.B.; Chen, H.; Zheng, L. Research progress in adulterants identification of *Star anise*. *J. Food Saf. Qual.* **2023**, *14*, 126–133. [CrossRef]
30. He, Z.J.; Ye, M.; Huan, J.; Li, Y.F.; Gong, X. CiteSpace and web of science of research status and trends of agarwood. *Chin. J. Trop. Agric.* **2024**, *10*, 1–10. Available online: http://kns.cnki.net/kcms/detail/46.1038.S.20240226.1518.004.html (accessed on 25 March 2024).
31. Huffman, M.D.; Baldridge, A.; Bloomfield, G.S.; Colantonio, L.D.; Prabhakaran, P.; Ajay, V.S.; Suh, S.; Lewison, G.; Prabhakaran, D. Global cardiovascular research output, citations, and collaborations: A time-trend, bibliometric analysis (1999–2008). *PLoS ONE* **2013**, *8*, e83440. [CrossRef]
32. Maina, M.B.; Ahmad, U.; Ibrahim, H.A.; Hamidu, S.K.; Nasr, F.E.; Salihu, A.T.; Abushouk, A.I.; Abdurrazak, M.; Awadelkareem, M.A.; Amin, A.; et al. Two decades of neuroscience publication trends in Africa. *Nat. Commun.* **2021**, *12*, 3429. [CrossRef]
33. Wu, H.Y.; Li, Y.Q.; Tong, L.J.; Wang, Y.L.; Sun, Z.M. Worldwide research tendency and hotspots on hip fracture: A 20-year bibliometric analysis. *Arch. Osteoporos.* **2021**, *16*, 73. [CrossRef] [PubMed]
34. Persson, O. The intellectual base and research fronts of JASIS 1986–1990. *J. Am. Soc. Inf. Sci.* **1994**, *45*, 31–38. [CrossRef]
35. Sun, Y.Q.; Wu, S.M.; Gong, G.Y. Trends of research on polycyclic aromatic hydrocarbons in food: A 20-year perspective from 1997 to 2017. *Trends Food Sci. Technol.* **2019**, *83*, 86–98. [CrossRef]
36. The Pharmacopoeia Commission of the People's Republic of China. *Pharmacopoeia of the People's Republic of China*; China Medical Science Press: Beijing, China, 2020; Volume 1, p. 5. Available online: https://ydz.chp.org.cn/ (accessed on 25 March 2024).
37. Schmidt, T.J.; Okuyama, E.; Fronczek, F.R. The molecular structure of 2 R-hydroxyneoanisatin and structure-activity relationships among convulsant sesquiterpenes of the secoprezizaane and picrotoxane types. *Bioorg. Med. Chem.* **1999**, *7*, 2857–2865. [CrossRef] [PubMed]
38. Zhang, Z.P. Market survey of *Star anise* and shikimic acid production. *Mod. Chin. Med.* **2006**, *4*, 41–42. Available online: https://www.cnki.com.cn/Article/CJFDTOTAL-YJXX200604015.htm (accessed on 25 March 2024).
39. Pellissier, L.; Roger, A.; Bilat, J.; Rasmann, S. High elevation *Plantago lanceolata* plants are less resistant to herbivory than their low elevation conspecifics: Is it just temperature? *Ecography* **2014**, *37*, 950–959. [CrossRef]
40. Anstett, D.N.; Ahern, J.R.; Glinos, J.; Nawar, N.; Salminen, J.P.; Johnson, M.T. Can genetically based clines in plant defence explain greater herbivory at higher latitudes? *Ecol. Lett.* **2015**, *18*, 1376–1386. [CrossRef] [PubMed]
41. Taylor, T.C.; Smith, M.N.; Slot, M.; Feeley, K.J. The capacity to emit isoprene differentiates the photosynthetic temperature responses of tropical plant species. *Plant Cell Environ.* **2019**, *42*, 2448–2457. [CrossRef] [PubMed]
42. Berkum, P.M.V.; Schmöckel, E.; Bischoff, A.; Carrasco-Farias, N.; Catford, J.A.; Feldmann, R.; Groten, K.; Henry, H.A.L.; Bucharova, A.; Hänniger, S.; et al. Plant geographic distribution influences chemical defences in native and introduced Plantago lanceolata populations. *Funct. Ecol.* **2024**, *38*, 883–898. [CrossRef]
43. Li, T.; Holst, T.; Michelsen, A.; Rinnan, R. Amplification of plant volatile defence against insect herbivory in a warming Arctic tundra. *Nat. Plants.* **2019**, *5*, 568–574. [CrossRef]
44. Kang, G.; Mishyna, M.; Appiah, K.S.; Yamada, M.; Takano, A.; Prokhorov, V.; Fujii, Y. Screening for plant volatile emissions with allelopathic activity and the identification of L-Fenchone and 1,8-Cineole from *Star anise* (*Illicium verum*) Leaves. *Plants* **2019**, *8*, 457. [CrossRef]
45. Sasidharan, S.; Chen, Y.; Saravanan, D.; Sundram, K.; Latha, L.Y. Extraction, isolation and characterization of bioactive compounds from plants' extracts. *Afr. J. Tradit. Complement. Altern. Med.* **2011**, *8*, 1–10. Available online: https://www.ncbi.nlm.nih.gov/pmc/articles/PMC3218439/pdf/AJT0801-0001.pdf (accessed on 25 March 2024). [CrossRef] [PubMed]
46. Aly, S.E.; Sabry, B.A.; Shaheen, M.S.; Hathout, A.S. Assessment of antimycotoxigenic and antioxidant activity of *Star anise* (*Illicium verum*) in vitro. *J. Saudi Soc. Agric Sci.* **2016**, *15*, 20–27. [CrossRef]

47. Wang, G.W.; Hu, W.T.; Huang, B.K.; Qin, L.P. *Illicium verum*: A review on its botany, traditional use, chemistry and pharmacology. *J. Ethnopharmacol.* **2011**, *136*, 10–20. [CrossRef] [PubMed]
48. Salem, M.A.; El-Shiekh, R.A.; Hashem, R.A.; Hassan, M. In vivo antibacterial activity of *Star anise* (*Illicium verum* Hook.) extract using murine MRSA skin infection model in relation to its metabolite profile. *Infect. Drug Resist.* **2021**, *14*, 33–48. [CrossRef] [PubMed]
49. Xu, S.; Hossain, M.M.; Lau, B.B.; To, T.Q.; Rawal, A.; Aldous, L. Total quantification and extraction of shikimic acid from *Star anise* (*llicium verum*) using solid-state NMR and cellulose-dissolving aqueous hydroxide solutions. *Sustain. Chem. Pharm.* **2017**, *5*, 115–121. [CrossRef]
50. Rocha, L.; Candido Tietbohl, L.A. Chapter 85—*Star anise* (*Illicium verum* Hook.) oils. In *Essential Oils in Food Preservation, Flavor and Safety*; Preedy, V.R., Ed.; Academic Press: San Diego, CA, USA, 2016; pp. 751–756. Available online: https://www.sciencedirect.com/book/9780124166417/essential-oils-in-food-preservation-flavor-and-safety (accessed on 25 March 2024).
51. Ghosh, C.; Sarkar, P.; Issa, R.; Haldar, J. Alternatives to Conventional Antibiotics in the Era of Antimicrobial Resistance. *Trends Microbiol.* **2019**, *27*, 323–338. [CrossRef]
52. Bahaddad, S.A.; Almalki, M.H.K.; Alghamdi, O.A.; Sohrab, S.S.; Yasir, M.; Azhar, E.I.; Chouayekh, H. Bacillus Species as Direct-Fed Microbial Antibiotic Alternatives for Monogastric Production. *Probiotics Antimicrob. Proteins* **2023**, *15*, 1–16. [CrossRef]

Disclaimer/Publisher's Note: The statements, opinions and data contained in all publications are solely those of the individual author(s) and contributor(s) and not of MDPI and/or the editor(s). MDPI and/or the editor(s) disclaim responsibility for any injury to people or property resulting from any ideas, methods, instructions or products referred to in the content.

foods

MDPI

Article

Insights into the Flavor Differentiation between Two Wild Edible *Boletus* Species through Metabolomic and Transcriptomic Analyses

Kaixiang Chao [1,†], Tuo Qi [2,†], Qionglian Wan [1] and Tao Li [1,*]

1 School of Chemistry Biology and Environment, Yuxi Normal University, Yuxi 653100, China; ckx@yxnu.edu.cn (K.C.); wanqionglian@yxnu.edu.cn (Q.W.)
2 Ecological Security and Protection Key Laboratory of Sichuan Province, Mianyang Normal University, Mianyang 621000, China; qituo777@163.com
* Correspondence: litao1976@yxnu.edu.cn
† These authors have contributed equally to this work and share first authorship.

Abstract: Despite the popularity of wild edible mushrooms due to their delectable flavor and nutritional value, the mechanisms involved in regulating and altering their taste remain underexplored. In this study, we analyzed the metabolome and transcriptome of *Boletus brunneissimus (B. brunneissimus) and Leccinum extremiorientale (L. extremiorientale)*, two *Boletus* species collected from different environments. Using UHPLC-MS, we annotated 644 peaks and identified 47 differential metabolites via OPLS-DA analysis. Eight of these were related to flavor, including L-Aspartic acid, Glycine, D-Serine, L-Serine, L-Histidine, Tryptophan, L-Isoleucine, Isoleucine, and alpha-D-Glucose. These differential metabolites were mainly concentrated in amino acid metabolism pathways. Transcriptome analysis revealed differential genes between *B. brunneissimus* and *L. extremiorientale*, which were enriched in protein processing in the endoplasmic reticulum, as well as differential genes of the same *Boletus* species in different environments that were enriched in the ribosome pathway. The combination of metabolome and transcriptome analyses highlighted Glycine, L-Serine, and L-Aspartic acid as the key compounds responsible for the differences between the two *Boletus* species. Using the O2PLS model and Pearson's coefficient, we identified key genes that modulate the differences in metabolites between the two species. These results have significant implications for the molecular breeding of flavor in edible mushrooms.

Keywords: mushroom flavor; wild edible *Boletus*; UHPLC-MS; metabolome; transcriptome

1. Introduction

Wild edible *boletus* species is highly valued for its delicious taste, rich nutrition, and various medicinal benefits and is a popular choice among consumers [1]. Research on the flavor compounds of this mushroom has become an important area of study. Non-volatile compounds, such as soluble sugars, organic acids, free amino acids, and 5′-nucleotides, are closely associated with mushroom flavor formation [2,3]. The distinct umami taste of mushrooms mainly originates from free amino acids, which can be classified into four categories: umami, sweet, bitter, and tasteless [4]. 5′-nucleotides found in mushrooms work by enhancing taste through synergistic reactions with flavoring amino acids, such as 5′-guanosine monophosphate and 5′-adenosine monophosphate [5,6]. However, research on the mechanism of flavor alteration in mushrooms remains limited, with most focusing only on changes in component content [6,7].

One of the reasons for the difficulty in studying the change mechanism of flavor substances in wild edible mushrooms may be that the artificial cultivation of most wild edible fungi is still difficult [8]. The flavor component will change significantly due to growth conditions, specifications, cooking, preservation, and storage conditions, which makes it difficult to accurately control experimental samples [9]. Researchers have attempted to overcome this obstacle by comparing different types of wild mushrooms. For example, Li et al. utilized widely targeted metabolomics to identify key taste components in two wild edible *Boletus* mushrooms [10].

Boletus brunneissimus (*B. brunneissimus*) and *Leccinum extremiorientale* (*L. extremiorientale*) are both wild edible *boletus* mushrooms commonly found in Yunnan markets. *B. brunneissimus* has a fruiting body with brown to dark brown color, a nearly flat cap, and dark blue flesh after injury. *L. extremiorientale*, on the other hand, has a thick and large fruiting body with an uneven cap surface that is nearly hemispherical. When it matures, it often has radial cracks, and the flesh of the fungus is light yellow, with no color change after injury [11,12]. These two *boletus* mushrooms differ in taste and flavor. Liang et al. (2016) isolated and purified three umami peptides from L. extremiorientale [13]. Wang Xiaoxiao (2017) used high-performance liquid chromatography (HPLC) to analyze the flavoring substances of *B. speciosus*, *B. auripes*, *B. edulis*, *L. extremiorientale*, and *B. brunneissimus*. The results showed that B. brunneissimus had the highest content of 5′-nucleotides [14]. However, rare studies have investigated the metabolic and molecular mechanisms underlying the differences in flavor components between these two *Boletus* mushrooms.

This study applied targeted metabolomics and transcriptome sequencing technology to investigate the differences in metabolites and gene expression between *L. extremiorientale* and *B. brunneissimus* from three distinct growing regions. The study aimed to identify the key flavor components responsible for the taste differences between the two *Boletus*, uncover the mechanisms behind these differences, and identify candidate genes that may play a role in regulating flavor components. By analyzing the molecular mechanisms of flavor component regulation in wild edible *boletus*, this research may ultimately contribute to the molecular breeding of artificially cultivated mushrooms through approaches such as genome editing and transgenic methods.

2. Materials and Methods

2.1. Mushroom Materials

To minimize the impact of environmental differences on experimental results, fruiting bodies of *B. brunneissimus* and *L. extremiorientale* were collected from three different locations in Yuxi City, Yunnan Province, at the same maturity stage [15]. The locations comprised Xiaoshiqiao Town, Hongta District, Lvzhi Town, Yimen County, and Liujie Street, Yimen County.

The experimental samples were purchased from the local market. Before purchasing, it was confirmed to the seller that the samples were collected on the local mountain and not transported from other places. Three biological replicates were collected for each *boletus* species at each site, resulting in a total of 18 experimental samples. Samples were immediately stored at −80 °C in a low-temperature refrigerator until testing. The experiment focused on the pileus, as it typically contains a higher concentration of the flavor component than the stipe [16].

2.2. Metabolites Extraction

To extract the samples after freeze-drying, they were first crushed in a mixer mill at 35 Hz for 45 s. The resulting material was weighed, then mixed with 700 L of extract solution, consisting of methanol and water in a ratio of 3:1, and pre-cooled to −40 °C. The internal standard was also added. Aliquots of 50 mg from each individual sample were added to Eppendorf tubes and homogenized at 35 Hz for 4 min, following a 30 s vortex, before being sonicated for 5 min in an ice water bath. This process was repeated three times. Next, the samples were shaken overnight at 4 °C and centrifuged for 15 min at 4 °C at 12,000 rpm (RCF = 13,800 (g), R = 8.6 cm). To prepare for ultrahigh-performance liquid chromatography-tandem mass spectrometry (UHPLC-MS), the supernatant was filtered through a 0.22 m microporous membrane and then diluted 20 times with an internal standard-containing solution of methanol and water (*v*:*v* = 3:1). The resulting supernatant was transferred to 2 mL glass vials. 50 L from each sample was taken and pooled as quality control(QC) samples before being stored at −80 °C [17].

2.3. UHPLC-MS Analysis

For UHPLC separation, an EXIONLC System from Sciex was used. The mobile phase A consisted of 1% formic acid in water, while acetonitrile served as the mobile phase B. The column temperature was fixed at 40 °C, and the autosampler was set at 4 °C. The injection volume was 2 L. For the test, a Sciex QTrap 6500+ was used, with typical ion source settings that included IonSpray Voltage at +5500/−4500 V, Curtain Gas at 35 psi, Temperature at 400 °C, Ion Source Gas 1 at 60 psi, Ion Source Gas 2 at 60 psi, and DP at 100 V. The data collected was processed using the SCIEX Analyst Work Station Software (version 1.6.3), and multiple reaction monitoring (MRM) was used. Custom R programming and database were used for peak detection and annotation based on the data matrix that comprised retention time (RT), the mass-to-charge ratio (m/z), and peak intensity. An in-house metabolite database was compared with the self-compiled R program for peak detection and annotation [18–20]. The metabolite profiling was conducted by Personal Biotechnology Cp. Ltd., based in Shanghai, China.

Based on the local metabolic database, the metabolites of the samples were qualitatively and quantitatively analyzed by mass spectrometry. The multi-reaction monitoring mode MRM metabolite detection Multi-peak map shows the substances that can be detected in the sample, with each color-coded mass spectrum peak representing one metabolite detected. The characteristic ions of each substance were screened by a triple four-level bar, and the signal strength (CPS) of the characteristic ions was obtained in the detector. The mass spectrometry file of the sample was opened with MultiQuant software, and the integration and correction of chromatographic peaks were carried out. The peak Area of each chromatographic peak represented the relative content of the corresponding substance. Finally, all chromatographic peak area integral data are derived and saved. In order to compare the difference in substance content of each metabolite in different samples among all detected metabolites, according to the information on metabolite retention time and peak type, the mass spectrum peak of each metabolite detected in different samples was corrected to ensure the qualitative and quantitative accuracy [21,22]. During the analysis, SIMCA software (version 16.0.2, Sartorius Stedim Data Analytics AB, Umea, Sweden) was used for principal component analysis (PCA) and orthogonal projections to latent structures—discriminant analysis (OPLS-DA) [23,24]. To evaluate the quality of the PCA, PLS-DA, and OPLS-DA models, R^2X or R^2Y and Q2 values were used (R^2 represents the interpretive rate of the model, and Q^2 represents the predictability of the model). The OPLS-DA method was used to visualize group separation and identify significantly differentially expressed metabolites (DEMs). DEMs with VIP (variable importance) > 1 and $p < 0.05$ (as determined by student t test) were considered to have undergone significant changes [25,26]. To further analyze the data, commercial databases, such as KEGG (http://www.genome.jp/kegg/ (accessed on 5 December 2022)) and MetaboAnalyst (http://metaboanalyst.ca/ (accessed on 5 December 2022)), were used for pathway enrichment analysis. For the metabolome KEGG pathway analysis, "*Agaricus bisporus* var. *bisporus* H97" was used as a comparative reference species [27]. The metabolite profiling and subsequent analyses were carried out by Personal Biotechnology Co., Ltd., based in Shanghai, China.

2.4. RNA Sequencing and De Novo Transcriptomic Analysis

The same batch of samples was sent separately for transcriptome and metabolome testing. To extract the total RNA from the pileus, a magnetic tissue total RNA kit was used (Product Series: DP762-T1A, TIANGEN, Beijing, China) following the manufacturer's instructions. The extracted RNA was then evaluated for concentration, purity, and integrity using a NanoDrop spectrophotometer from Thermo Scientific. Three micrograms of RNA were utilized as the starting material for RNA sample procedures. The TruSeq RNA Sample Preparation Kit from Illumina (San Diego, CA, USA) was used to produce sequencing libraries. mRNA was initially extracted from the whole RNA using poly-T oligo-attached magnetic beads, after which fragmentation was carried out at high temperature using divalent cations in a buffer developed by Illumina. A first strand of cDNA was then produced using random oligonucleotides and Super Script II. The second strand of cDNA was synthesized using DNA Polymerase I and RNase H. The remaining overhangs were transformed into blunt ends using exonuclease/polymerase activities, after which the enzymes were eliminated. Illumina PE adapter oligonucleotides were ligated to the 3′ ends of the DNA fragments following adenylation to prepare them for hybridization. The library fragments were purified using AMPure XP technology to identify cDNA fragments between 400–500 bp in length (Beckman Coulter, Beverly, CA, USA). A 15-cycle PCR process was used to preferentially select DNA fragments with ligated adapter molecules at both ends using the Illumina PCR Primer Cocktail. The products were then purified using the AMPure XP system, and the quantification was carried out using the Agilent high-sensitivity DNA assay on the Bioanalyzer 2 system (Agilent, Inc, Santa Clara, CA, USA). Shanghai Personal Biotechnology Co., Ltd. sequenced the sequencing library on the NovaSeq 6000 platform (Illumina, San Diego, CA, USA).

To filter the raw data, Cutadapt (version 1.16) was used [28]. This experiment employed the no-reference genome analysis approach. The Trinity software (version 2.5.1) was utilized to assemble clean reads into transcripts to support further analysis [29]. Once the montaging was complete, transcript sequence files in FASTA format were obtained. The longest transcript for each gene was extracted and used as the unigene or the representative sequence of the gene. To annotate the gene function, the databases used were NR (NCBI non-redundant protein sequences), GO (Gene Ontology), KEGG (Kyoto Encyclopedia of Genes and Genome), eggNOG (evolutionary genetics of genes: non-supervised Orthologous Groups) [30], Swiss-Prot, and Pfam. RSEM (version 1.2.31) statistics were employed to compare the Read Count values for each gene for its original expression. FPKM was then used to standardize the expression. DESeq (version 1.32.0) was used for the differential analysis of gene expression (DESeq). The screening conditions for differentially expressed genes were: multiple of the expression difference $|\log2FoldChange| > 1$, significance p-value < 0.05.

2.5. Correlation Analysis of Transcriptomic and Metabolomic Data

The analysis of the flavor metabolism mechanisms of the two edible fungi was carried out by combining the enrichment results of transcriptome differentially expressed genes and metabolome differentially expressed metabolites. To identify the differential genes that strongly correlated with the differential metabolites, the Pearson correlation coefficient ($|r| > 0.8$, $p < 0.05$) was used. The O2PLS model was then utilized to analyze the correlation of the two data matrices from transcriptomics and metabolomics and employed for two-way modeling and prediction in two data matrices. This approach was helpful in selecting candidate genes for future functional verification [31].

3. Results

3.1. Metabolic Profiling

The original data from the UHPLC-MS analysis consisted of 3 quality control (QC) samples and 18 experimental samples. In total, 644 peaks were extracted from both *Boletus* species, with isomers having the same peak being classified as the same class. Figure 1a shows the classification of all metabolites, with the highest number of metabolites belonging to the alkaloids (13.0%), followed by phenols (12.0%), flavonoids (10.1%), terpenoids (8.5%), and their derivative metabolite classes. Interestingly, no differences were found between the two *Boletus* species in the types of metabolites tested.

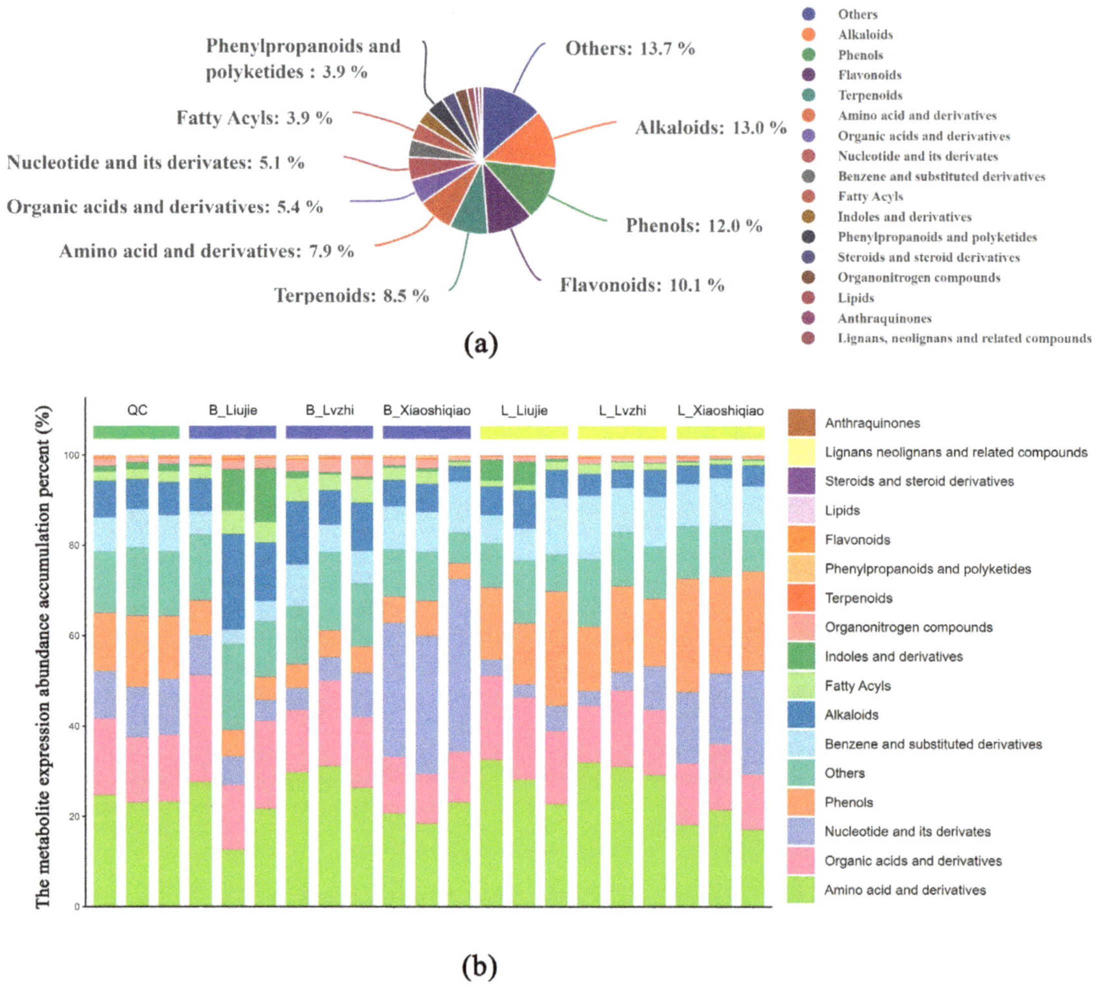

Figure 1. The classification of all metabolites and the metabolite expression abundance accumulation of the two *Boletus* species. (**a**) The classification of all metabolites of the two *Boletus* species. (**b**) The metabolite expression abundance accumulation of the two *Boletus* species. QC: Quality control samples. B_(Liujie, Lvzhi, Xiaoshiqiao): Collected from different locations of the *B. brunneissimus* sample. L_(Liujie, Lvzhi, Xiaoshiqiao): Collected from different locations of the *L. extremiorientale* sample.

From the perspective of metabolite expression abundance accumulation, amino acid and derivatives, organic acids and derivatives, nucleotide and its derivatives, phenols, benzene, and substituted derivatives accounted for a relatively high proportion of the total abundance accumulation of the whole sample expression (Figure 1b). The proportion of these metabolite classes varied substantially depending on sampling sites and species, with the proportion of amino acids and derivatives, phenols, alkaloids, and amino acids and their derivatives varying the most. However, the proportion of organic acids and their derivatives in the expression abundance of total metabolites was relatively stable, with a small range. Therefore, the key class that influences the flavor difference between the two *Boletus* species might not be organic acids and their derivatives.

3.2. Analysis of the Applicability of the OPLS-DA Model to This Study

In this study, nine samples from two *Boletus* species collected from three locations were compared as a group. The Pearson correlation coefficients of metabolites in each sample are shown in Figure S1. The principal component analysis (PCA) score scatter plot shows that all data are within the 95% confidence interval (Hotelling's T-squared ellipse). The *B. brunneissimus* and *L. extremiorientale* are located on both sides of the QC, and the *Bolete* from the same location are gathered, as can be seen in Figure 2a. These results suggest that there are significant differences in metabolites between *B. brunneissimus* and *L. extremiorientale* and that the metabolites in the fruiting bodies also differ between the different locations. Intra-group differences can help to highlight stable metabolites between each region and balance metabolites that have excessively high values in a sample, thereby optimizing the calculation of differential metabolites between groups.

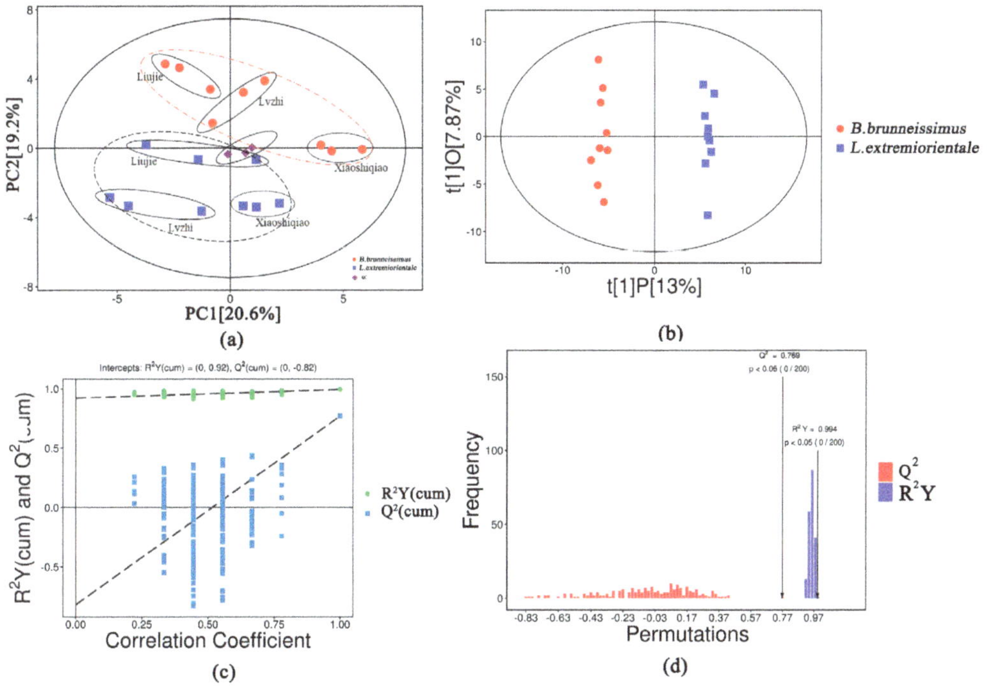

Figure 2. Analysis of the applicability of the OPLS-DA model to this study. (**a**) The principal component analysis (PCA) of samples from two *Boletus* species. (**b**) PCA of OPLS-DA model. (**c**) The correlation coefficient analysis of the OPLS-DA model. (**d**) The permutations analysis of the OPLS-DA model.

After logarithmic (LOG) conversion and UV formatting of the metabolite data, the orthogonal projections to late structures-discriminant analysis (OPLS-DA) model score scatter plot of the two groups of samples is highly significant, with all samples within the 95% confidence interval (Hotelling's T-squared ellipse, Figure 2b). The OPLS-DA modeling analysis shows that the modeling ability ($R^2Y(cum) = 0.994$) and predictive ability ($Q^2(cum) = 0.769$) were 99.4% and 76.9%, respectively (Figure 2c). In the OPLS-DA permutation plot (Figure 2d), all the blue Q^2 points are lower than the original blue Q^2 point on the right, and the regression line of the Q^2 point intersects with the ordinate or is less than zero, indicating that the original model is stable and there is no overfitting phenomenon ($R^2Y(cum) = (0,0.92)$, $Q^2(cum) = (0,-0.82)$). These results indicate that the OPLS-DA model is suitable for evaluating the differences between *B. brunneissimus* and *L. extremiorientale*.

3.3. *Difference of Flavor Metabolite between B. brunneissimus and L. extremiorientale*

Based on the OPLS-DA analysis, 47 metabolites were found to have significant differences (VIP > 1, $p < 0.05$) between *B. brunneissimus* and *L. extremiorientale*, which accounts for 7.3% of the total metabolites. The most significant class of differential expression metabolites (DEMs) included "amino acid and derivatives", "nucleotide and its derivatives", and "phenols" (Table S1, Figure 3). Among these DEMs, 18 metabolites in the fruiting body of *L. extremiorientale* were substantially greater than those in *B. brunneissimus*, while 29 metabolites were significantly lower.

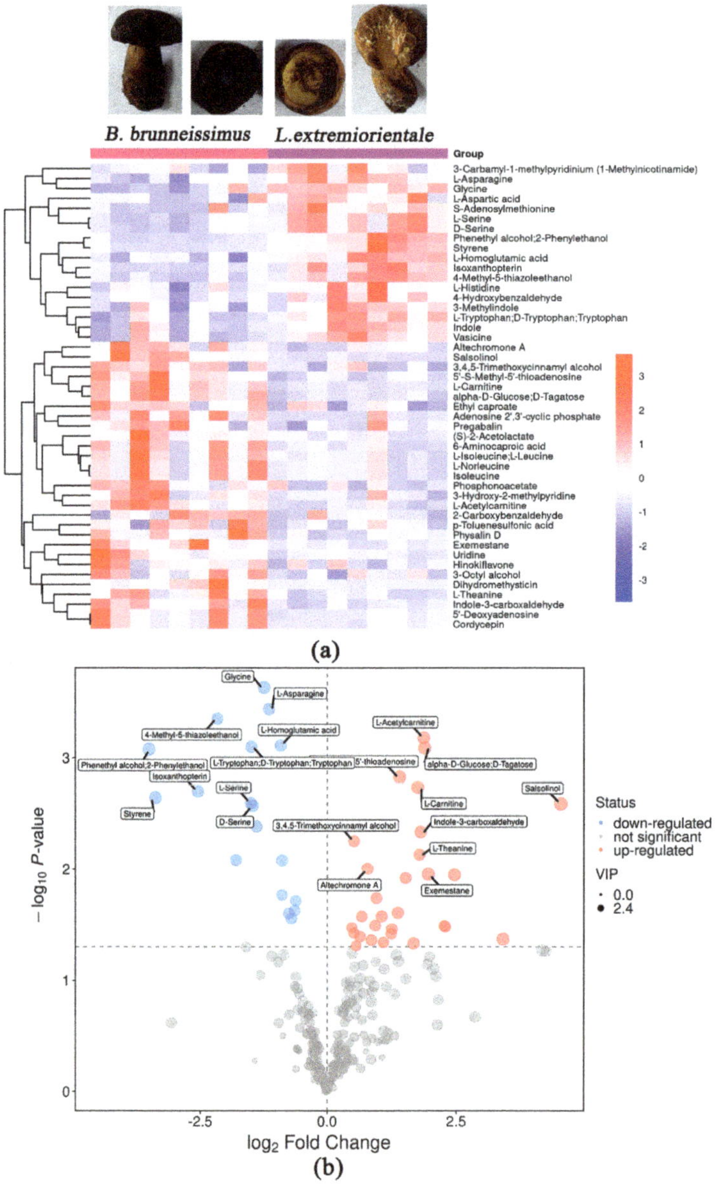

Figure 3. The heatmap plot (**a**) and the volcano plot (**b**) of differentially expressed metabolites.

Regarding mushroom flavor metabolites, 17 free amino acids, three soluble sugars, two organic acids, and one type of 5′ nucleotide were identified. Among them, seven free amino acids and one organic acid were found to be significant metabolites (Table 1). L-Aspartic acid (umami taste), Glycine (sweet taste), D-Serine (sweet taste), L-Serine (sweet taste), L-Histidine (bitter taste), and Tryptophan (L-Tryptophan, D-Tryptophan; bitter taste) were substantially greater in the free amino acids of *L. extremiorientale* compared to *B. brunneissimus*, whereas L-Isoleucine (L-Leucine, bitter taste) and Isoleucine (bitter taste) were higher in *B. brunneissimus*. The flavor-related soluble sugar alpha-D-Glucose (D-Tagatose, sweet taste) was also found to be higher in L. extremiorientale than in *B. brunneissimus*. Moreover, the content of ethyl acetate in *B. brunneissimus* was found to be significantly higher than that in *L. extremiorientale*. Ethyl acetate is an edible spice with a fruity smell and may be the reason why *B. brunneissimus* smells more fragrant than *L. extremiorientale*.

3.4. Transcriptomic Difference of the Fruiting Bodies of Two Wild Edible Boletus

The KEGG database was employed to analyze the differential metabolites and 47 differential metabolites were annotated, mostly enriched in metabolic pathways (abv01100). The most significantly enriched ($p < 0.001$) pathways were "Cysteine and methionine metabolism (abv00270)", "Glycine, serine and threonine metabolism (abv00260)", "Biosynthesis of amino acids (abv01230)", "ABC transporters (abv02010)", "D-Amino acid metabolism (abv00470)", "Cyanoamino acid metabolism (abv00460)", "Aminoacyl-tRNA biosynthesis (abv00970)", and "2-Oxocarboxylic acid metabolism (abv01210)". Among these pathways, "Tryptophan metabolism (abv00380)", "Valine, leucine and isoleucine biosynthesis (abv00280)", "Cysteine and methionine metabolism (abv00270)", "Glycine, serine and threonine metabolism (abv00260)", and "Alanine, aspartate and glutamate metabolism (abv00250)" were the top 5 enriched pathways in DEMs (Figure 4).

To better understand the regulation patterns of these DEMs, we conducted RNA-seq analysis of these wild edible *Boletus*. We obtained 80.67–93.79% clean reads, and a total of 232975214 Unigenes were predicted. Among them, 17,935 differentially expressed genes (DEGs) were identified between *B. brunneissimus* and *L. extremiorientale*. The DEGs clustered in clust1 and clust4 were highly expressed in *B. brunneissimus*, while clust 5, clust 6, clust 7, and clust 8 were highly expressed in *L. extremiorientale*(Figure 5). In addition, clust1 comprised genes that were upregulated in *B. brunneissimus* while surprisingly not expressed or very little expressed in *L. extremiorientale*. These genes were most enriched in the processes of peroxisome, mRNA surveillance, and nucleocytoplasmic transport. The same expression trend appeared on clust4, and these were linked to the processes of the cell cycle, MAPK signaling pathway, and nucleocytoplasmic transport. On the contrary, genes clustered in clust 5 that were upregulated in *L. extremiorientale* exhibited the lowest expression level in *B. brunneissimus*. For clust5, a large number of genes were enriched in protein processing in the endoplasmic reticulum, cell cycle, and MAPK signaling pathways. Clust 7 and clust 8 showed similar expression patterns to clust 5 and were involved in the processes of nucleocytoplasmic transport and biosynthesis of secondary metabolites. Clust 3 showed one peak in the *B. brunneissimus* collected from Xiaoshiqiao, and the genes were enriched in the ribosome, spliceosome, and protein processing in the endoplasmic reticulum pathway. Clust 2 showed similar expression patterns mostly involved in metabolic and secondary metabolites in samples collected in Liujie, indicating these genes may be related to the microenvironment (Figure S2). The heat map of each sample's DEGs clustering is presented in Figure S3. Further functional analysis of differential genes showed that the GO enrichment categories of protein folding and unfolded protein binding and the Protein processing in the endoplasmic reticulum pathway (ko04141) were significantly clustered (Figure 6). This result reflects the overall enrichment of transcriptome differential genes in the fruiting body of the two *Boletus* species.

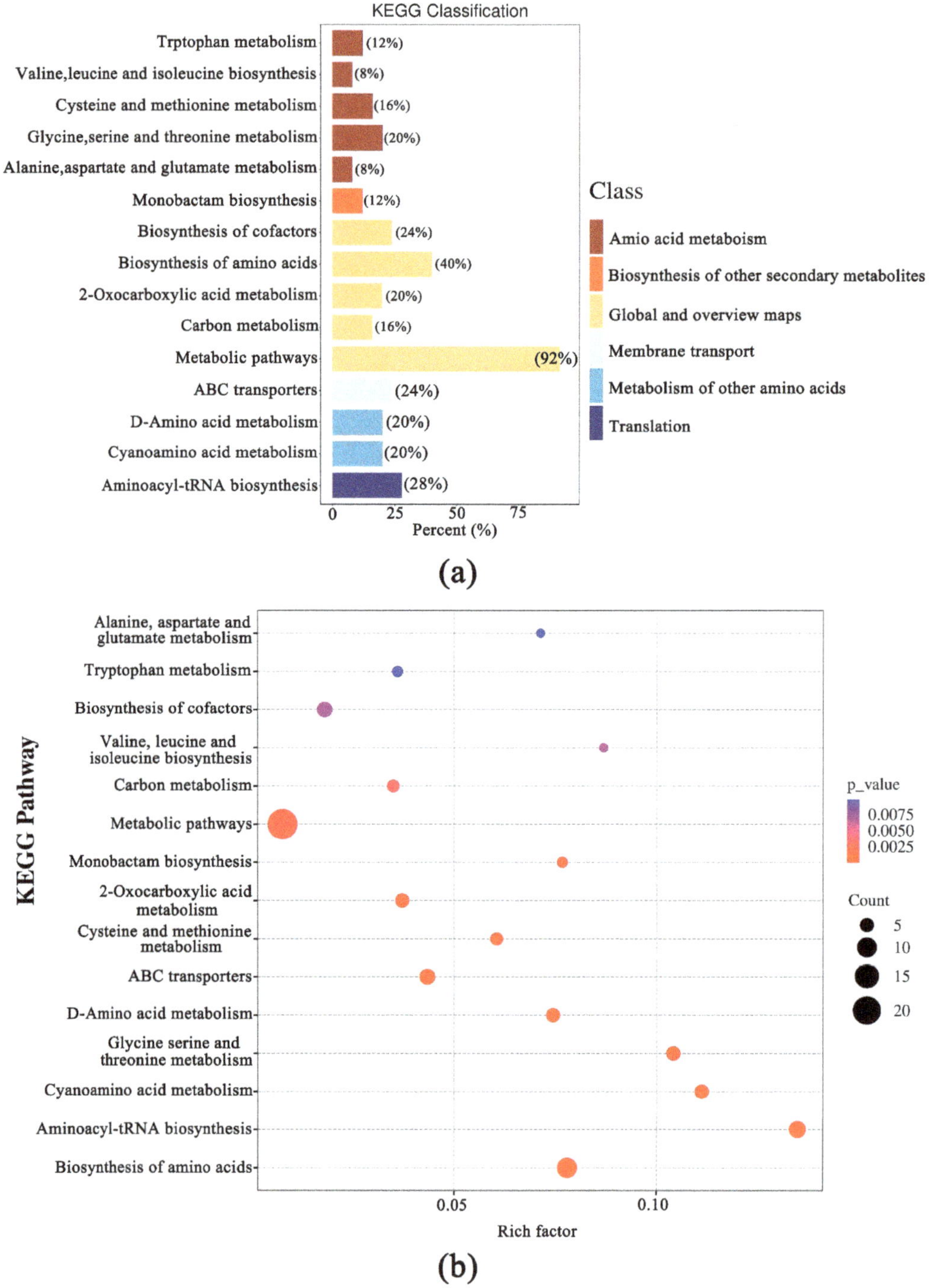

Figure 4. The KEGG pathway rich plot (**a**) and the bubble plot (**b**) of differentially expressed metabolites.

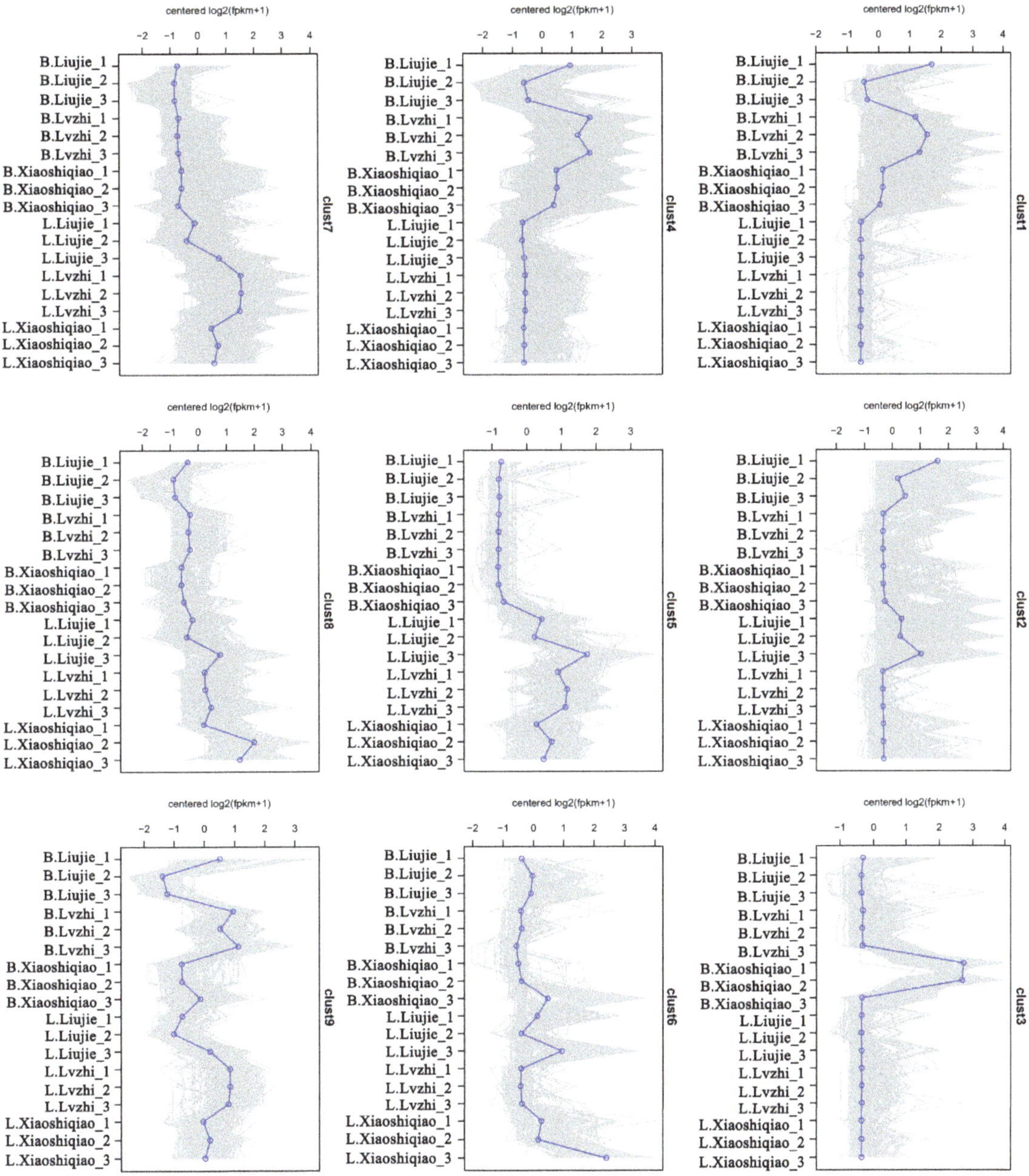

Figure 5. The cluster of DEGs of 18 wild edible *Boletus* samples. B.(Liujie, Lvzhi, Xiaoshiqiao): Collected from different locations of the *B.brunneissimus* sample. L.(Liujie, Lvzhi, Xiaoshiqiao): Collected from different locations of the *L. extremiorientale* sample.

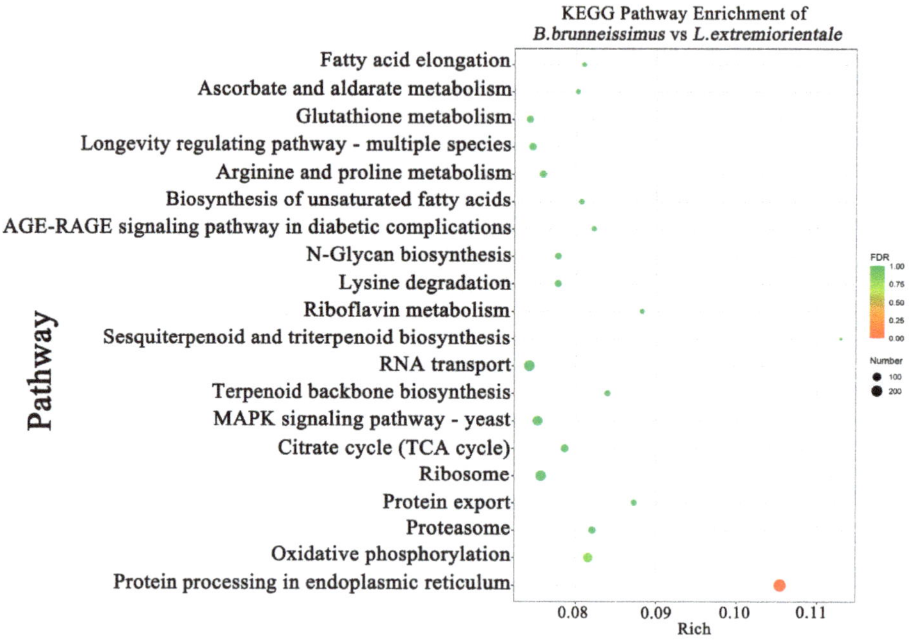

(a)

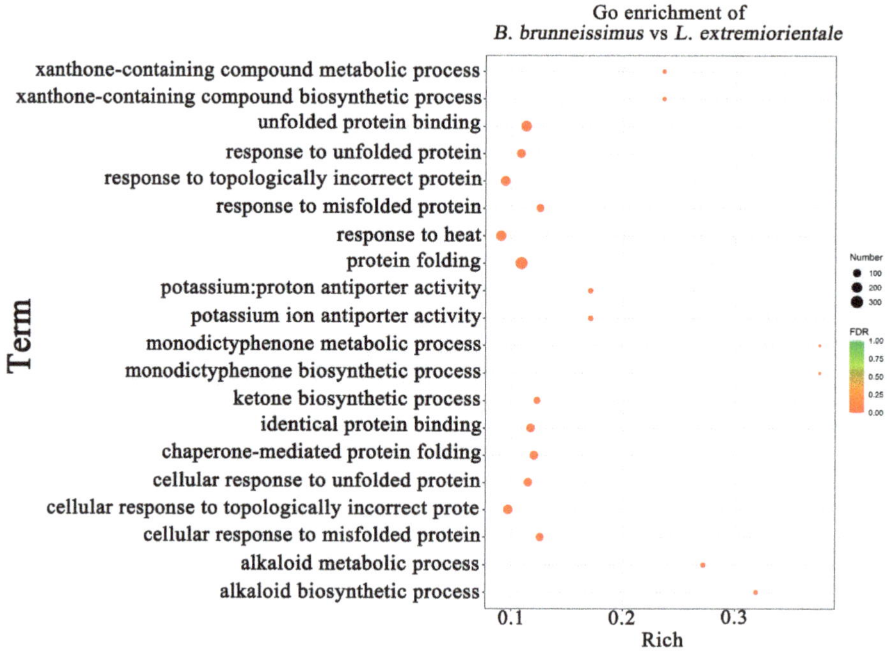

(b)

Figure 6. The KEGG pathway enrichment (**a**) and Go term (**b**) enrichment of *B. brunneissimus* vs. *L. extremiorientale*.

3.5. Analysis of DEMs and DEGs in the Two Boletus Fruiting Bodies in Different Locations

The 18 samples were divided into nine groups for DEMs and DEGs analysis based on comparisons between the two *Boletus* species in the same and different locations. All nine combinations were appropriate for OPLS-DA analysis to identify differential metabolites. The differential metabolites screened by different comparison combinations are listed in Table S2. The Venn plot (Figure 7a) and Upset plot (Figure 7b) comparing the two *Boletus* species from three regions and the same *Boletus* species from three regions showed that alpha-D-glucose (D-Tagatose) and L-carnitine were common differential metabolites between the two *Boletus* species in different regions, suggesting that the differences in these two substances between the *Boletus* species are not influenced by the environment. For the one *Boletus* species in different environment combinations, hypoxanthine, L-alanine (DL-alanine), and salsolinol were the common differential metabolites of *B. brunneissimus* and *L. extremiorientale*, respectively. These three peaks are most susceptible to environmental impact. The results of the differential metabolites analysis of *Boletus* fruiting bodies in different environments indicate that the differential metabolites of the two *Boletus* species' fruiting bodies vary between environments.

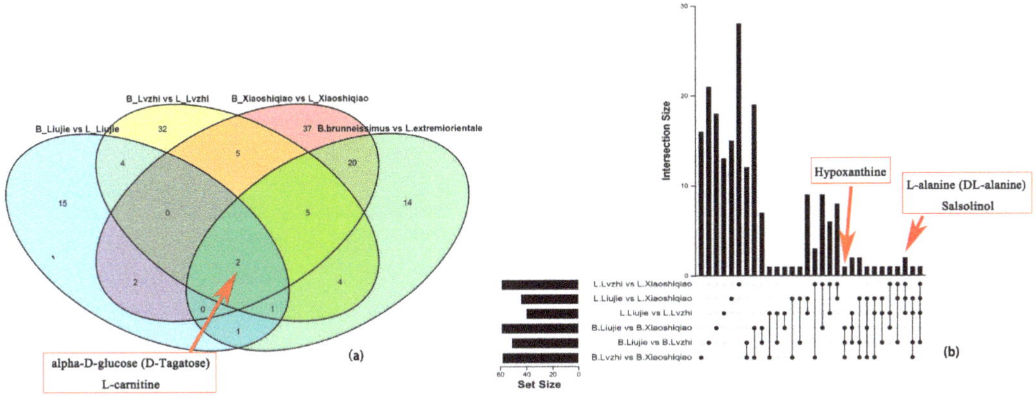

Figure 7. The Venn plot (**a**) and Upset(**b**) plot of different comparison combinations of *B. brunneissimus vs L. extremiorientale*. B_(Liujie, Lvzhi, Xiaoshiqiao): Collected from different locations of the *B. Brunneissimus sample*. L_(Liujie, Lvzhi, Xiaoshiqiao): Collected from different locations of the *L. extremiorientale* sample.

The DEGs screened by different comparison combinations are listed in Table S2. For the comparison of two *Boletus* species in the same region, the findings of enrichment analysis after KEGG annotation of DEGs were consistent with the prior comparative combination results, and the major enrichment was in the Protein processing in the endoplasmic reticulum (ko04141) pathway. However, the GO annotation did not have any obvious common enrichment terms. For the comparison of the same *Boletus* species in different environments, the majority of KEGG-annotated DEGs were enriched in the Ribosome pathway (ko03010), and there were no obvious common enrichment terms in the GO annotation (Table S2).

3.6. Correlation Analysis of Metabolomic and Transcriptomic Data

Correlation analysis between the metabolomics and transcriptomics data (comparing *B. brunneissimus* to *L. extremiorientale*, with all nine samples of each *Boletus* as a group) revealed that out of 17,935 DEGs, 4240 DEGs showed a strong correlation with 47 DEMs using the Pearson correlation coefficient ($|r| > 0.8$, $p < 0.05$). The KEGG annotation of these 4240 DEGs was significantly enriched for the protein processing in the endoplasmic reticulum (ko04141) pathway, the same as the result obtained from the analysis of all the DEGs. However, the GO annotation was significantly enriched for binding and protein binding terms, which differed from the result obtained from the analysis of all the DEGs (Figure S4).

The regulatory network analysis of differential metabolites revealed that the key differential metabolites of *B. brunneissimus* and *L. extremiorientale* are Glycine, L-Serine, and L-Aspartic acid. The differential metabolites produced by these three amino acid metabolism pathways account for 38% of the total differential metabolites. Enzymes related to DEMs in the pathway were used to compare DEGs, and it was observed that some enzymes that affect the formation of DEMs were not hit by any differential genes (Figure 8). Furthermore, through KEGG analysis, a large number of genes were annotated as "hypothetical protein", indicating that the genome and transcriptome of wild *boletus* require more in-depth research.

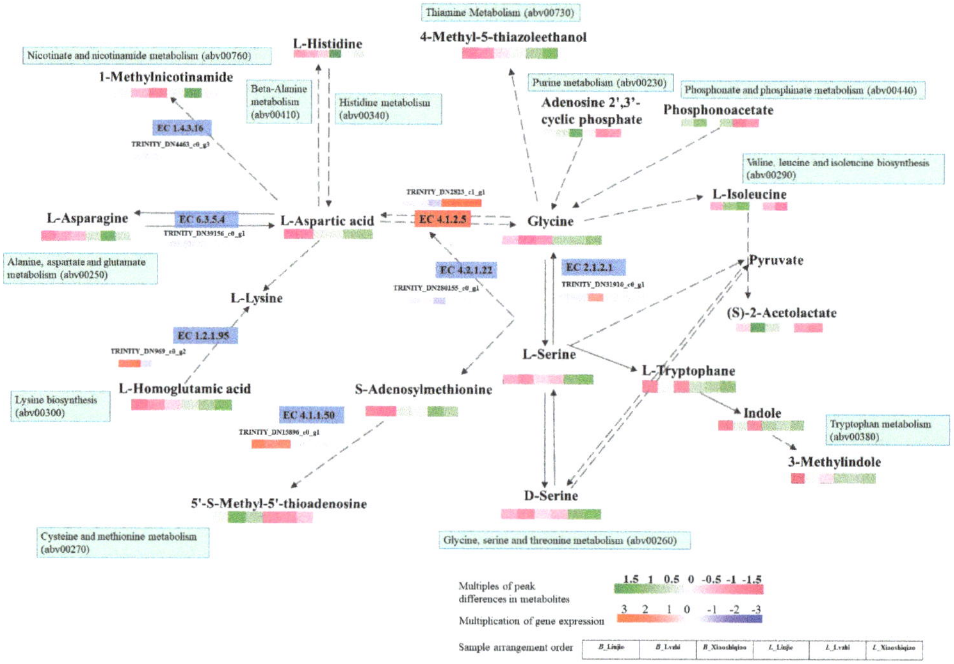

Figure 8. Correlation analysis between the metabolomics and transcriptomics data. The solid line represents the direct conversion of differential compounds; The dashed line represents the indirect transformation of differential compounds.

To better explore the differentially expressed genes that affect the metabolites of the two wild *Boletus* species, we used the O2PLS model to analyze the DEGs and metabolites strongly associated with the DEMs (with the logarithm of the peak value used for analysis). The results are presented in Figure 9, and the O2PLS model better explains the association between the metabolome and transcriptome (with $R^2X = 0.89$, $R^2Y = 0.692$, $R^2Xcorr = 0.75$, $R^2Ycorr = 0.574$). The top 10 genes, such as "TRINITY-DN343297_c0_g1," have the highest sum of loading squared values and can be used as candidate genes for further studying the interspecific differences between the two *Boletus* species (Table 2).

To investigate the key genes responsible for flavor differences between the two types of *Boletus* species, we conducted Pearson correlation analysis ($|r| > 0.8$, $p < 0.05$) using seven amino acid differential metabolites related to flavor (L-Aspartic acid, Glycine, D-Serine, L-Serine, L-Histidine, Tryptophan, and L-Isoleucine) and differential genes, and a total of 276 differential genes were identified. Subsequently, using the seven DEMs and 276 strongly associated DEGs for O2PLS analysis (with $R^2X = 0.95$, $R^2Y = 0.82$, $R^2Xcorr = 0.79$,

R^2Ycorr = 0.82), we selected ten genes, including "TRINITY_DN455_c1_g1," with the highest loading square value. These candidate genes can be used to further study the differences in the flavor of the two *Boletus* species (Table 2).

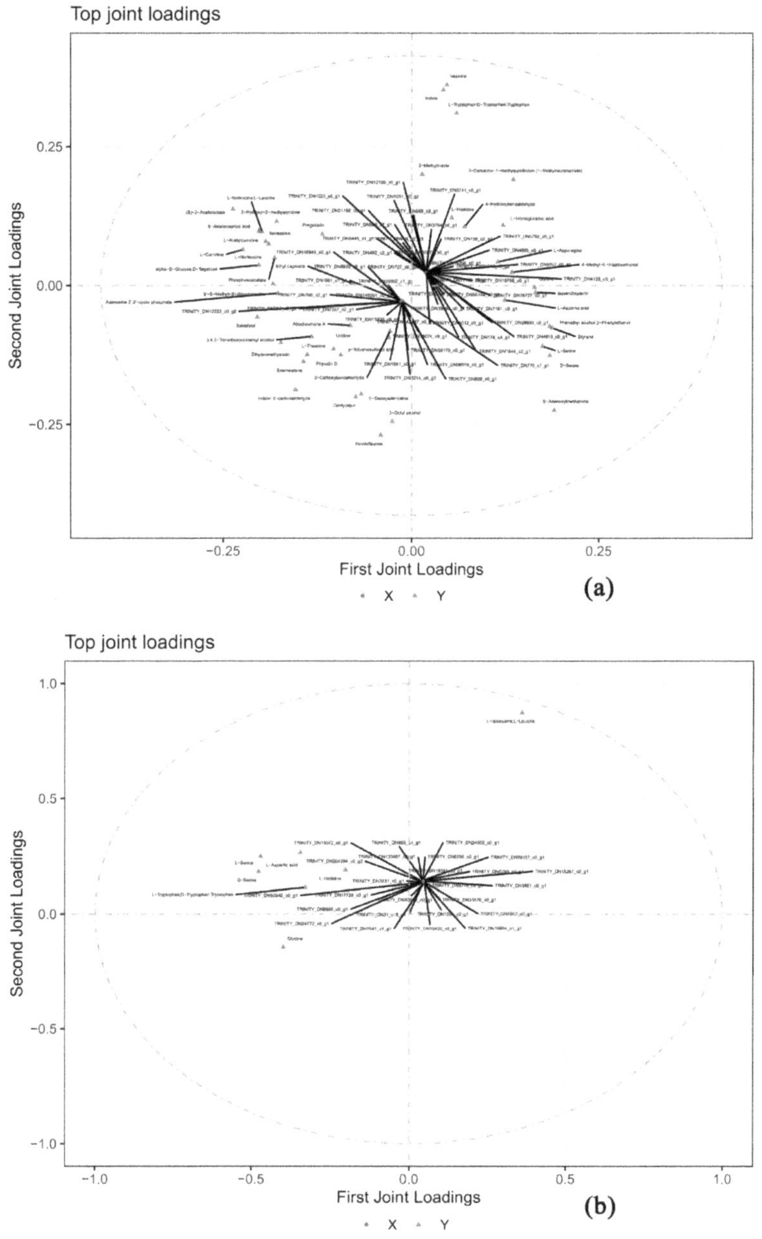

Figure 9. Analysis results of O2PLS model. (**a**) The relationship between all DEMs and all DEGs. (**b**) The relationship between flavor-related DEMs and all DEGs.

Table 1. Statistics of accumulating free amino acids, soluble sugars, organic acids, 5′-nucleotides, and miscellaneous in the fruiting body of. " *B. brunneissimus*"and "*L. extremiorientale*".

Index	Compound name	Class	Taste	Peak Area		VIP	*p*-Value	*B. brunneissimus* vs *L. extremiorientale*
				B. brunneissimus	*L. extremiorientale*			
1	L-Aspartic acid	Free amino acids	Umami taste (MSG-like)	394,035.09 ± 192,939.3	618,031.63 ± 180,516.1	1.51	0.02	UP
2	D-Aspartic acid	Free amino acids	Umami taste (MSG-like)	433,700.93 ± 232,571.85	632,223.46 ± 316,790.67	1.03	0.13	-
3	L-Glutamic acid	Free amino acids	Umami taste (MSG-like)	6,201,893.95 ± 3,509,410.93	6,921,258.43 ± 1,860,205.86	0.62	0.76	-
4	L-Alanine; DL-Alanine	Free amino acids	Sweet taste	658,758.14 ± 413,661.32	1,382,503.87 ± 922,614.21	0.06	0.22	-
5	Glycine	Free amino acids	Sweet taste	48,034.62 ± 15,782.87	114,075.09 ± 16,451.79	2.19	0	UP
6	D-Proline	Free amino acids	Sweet taste	8,407,880.54 ± 583,9726	11,620,670.17 ± 4,494,533.82	0.99	0.3	-
7	D-Serine	Free amino acids	Sweet taste	114,362.85 ± 68,449.37	316,275.73 ± 147,831.59	1.82	0	UP
8	L-Serine	Free amino acids	Sweet taste	111,179.47 ± 70,914.14	316,275.73 ± 147,831.59	1.81	0	UP
9	L-Threonine	Free amino acids	Sweet taste	576,328.69 ± 392,016.92	692,317.82 ± 153,705.11	0.88	0.44	-
10	L-Histidine	Free amino acids	Bitter taste	5,495,343.35 ± 2,751,086.31	9,208,274.72 ± 3,538,910.94	1.47	0.03	UP
11	L-Isoleucine; L-Leucine	Free amino acids	Bitter taste	378,672.76 ± 218,557.65	188,871.1 ± 108,314.07	1.3	0.46	DOWN
12	Isoleucine	Free amino acids	Bitter taste	400,448.58 ± 258,714.71	176,708.49 ± 108,579.05	1.43	0.03	DOWN
13	L-Methionine	Free amino acids	Bitter taste	106,123.63 ± 97,300.41	47,486.83 ± 42,604.14	1.28	0.13	-
14	L-Phenylalanine;D-(+)-Phenylalanine; DL-Phenylalanine	Free amino acids	Bitter taste	21,820,440.32 ± 12,846,566.66	19,022,942.97 ± 13,038,839.53	0.47	0.61	-
15	Tryptophan(L-Tryptophan; D-Tryptophan)	Free amino acids	Bitter taste	17,996.51 ± 16,094.24	53,685.79 ± 18,460.91	1.79	0	UP
16	L-Valine	Free amino acids	Bitter taste	409,058.87 ± 139,076.01	401,702.04 ± 71,429.27	0.05	0.69	-
17	L-Arginine	Free amino acids	Bitter taste	12,571,655.39 ± 1,273,7601.53	21,443,582.43 ± 13,314,464.99	1.2	0.2	-
18	L-Lysine;L-Glutamine	Free amino acids	Tasteless	18,074,414.78 ± 7,399,759.51	23,351,362.08 ± 8,585,564.65	0.9	0.2	-
19	L-Tyrosine	Free amino acids	Tasteless	1,958,434.87 ± 1,135,802.84	2,938,080.2 ± 1,187,109.94	1.16	0.19	-
20	Inositol	Soluble sugars; Polyols	Sweet taste	3100.23 ± 1151.09	2397.99 ± 1140.6	1	0.16	-
21	D-Xylulose	Soluble sugars; Reducing sugars	Sweet taste	73,430.52 ± 62,704.61	27,330.13 ± 25,050.39	-	-	-
22	alpha-D-Glucose; D-Tagatose	Soluble sugars; Reducing sugars	Sweet taste	301,634.92 ± 128,026.33	83,124.42 ± 34,406.66	2.25	0	UP
23	Beta-D-Glucose; alpha-D-Glucose	Soluble sugars; Reducing sugars	Sweet taste	3648.68 ± 1313.4	5118.91 ± 2292.21	-	-	-
24	Sucrose	Soluble sugars; Non-reducing sugars	Sweet taste	92,833.83 ± 62,857.55	349,897.24 ± 510,820.3	-	-	-

Table 1. *Cont.*

Index	Compound name	Class	Taste	Peak Area		VIP	*p*-Value	*B. brunneissimus* vs *L. extremiorientale*
				B. brunneissimus	*L. extremiorientale*			
25	L-Malic acid	Organic acids	sour taste	2013.11 ± 1300.98	1580.5 ± 1538.06	-	-	-
26	Ascorbic acid	Organic acids	sour taste	7771.71 ± 2522.87	6542.41 ± 2207.86	0.83	0.2	-
27	Adenosine 5′-monophosphate	5′-nucleotides	-	1,703,525.03 ± 710,544.37	2,947,050.82 ± 3,418,622.63	0.12	0.32	-
28	Ethyl caproate	Miscellaneous	-	64,844.26 ± 21,913.66	41,651.35 ± 22,101.1	1.48	0.03	DOWN

Table 2. The top 10 DEGs associated with DEMs by O2PLS analysis.

<table>
<tr><th colspan="5">The top 10 DEGs Associated with All DEMs by O2PLS Analysis</th></tr>
<tr><th>DEGs</th><th>Loading_1</th><th>Loading_2</th><th>Loading2 [1]</th><th>NR Annotation of DEGs</th></tr>
<tr><td>TRINITY_DN343297_c0_g1</td><td>−0.011662098</td><td>−0.031064851</td><td>0.00110103</td><td>gi | 1000864728 | gb | KXN90516.1 | Vesicle transport protein SFT2B [Leucoagaricus sp. SymC.cos]</td></tr>
<tr><td>TRINITY_DN15220_c0_g1</td><td>−0.013141608</td><td>−0.030371976</td><td>0.001095159</td><td>gi | 751055722 | gb | KIK90884.1 | hypothetical protein PAXRUDRAFT_831297 [Paxillus rubicundulus Ve08.2h10]</td></tr>
<tr><td>TRINITY_DN7287_c2_g1</td><td>−0.013351728</td><td>−0.030237734</td><td>0.001092589</td><td>gi | 914258129 | gb | KNZ73472.1 | WD repeat-containing protein 5 [Termitomyces sp. J132]</td></tr>
<tr><td>TRINITY_DN50170_c0_g1</td><td>−0.011257006</td><td>−0.030734214</td><td>0.001071312</td><td>gi | 750997013 | gb | KIK39305.1 | hypothetical protein CY34DRAFT_89493 [Suillus luteus UH-Slu-Lm8-n1]</td></tr>
<tr><td>TRINITY_DN18074_c0_g1</td><td>−0.011755046</td><td>−0.030544198</td><td>0.001071129</td><td>gi | 751064906 | gb | KIK99469.1 | hypothetical protein PAXRUDRAFT_131997 [Paxillus rubicundulus Ve08.2h10]</td></tr>
<tr><td>TRINITY_DN3215_c0_g3</td><td>−0.012219336</td><td>−0.03025251</td><td>0.001064527</td><td>gi | 749895491 | gb | KIJ63208.1 | hypothetical protein HYDPIDRAFT_29895 [Hydnomerulius pinastri MD-312]</td></tr>
<tr><td>TRINITY_DN262_c0_g2</td><td>−0.014473261</td><td>−0.029082473</td><td>0.001055266</td><td>- [2]</td></tr>
<tr><td>TRINITY_DN700_c2_g1</td><td>−0.014895131</td><td>−0.028608579</td><td>0.001040316</td><td>gi | 749834671 | gb | KIJ12323.1 | hypothetical protein PAXINDRAFT_156963 [Paxillus involutus ATCC 200175]</td></tr>
<tr><td>TRINITY_DN10222_c0_g2</td><td>−0.013377283</td><td>−0.028754748</td><td>0.001005787</td><td>gi | 751062861 | gb | KIK97499.1 | hypothetical protein PAXRUDRAFT_136196 [Paxillus rubicundulus Ve08.2h10]</td></tr>
<tr><td>TRINITY_DN5782_c0_g1</td><td>0.019328067</td><td>0.024823556</td><td>0.000989783</td><td>gi | 749846545 | gb | KIJ21256.1 | hypothetical protein PAXINDRAFT_64023 [Paxillus involutus ATCC 200175]</td></tr>
</table>

Table 2. *Cont.*

The top 10 DEGs Associated with All DEMs by O2PLS Analysis				
DEGs	**Loading_1**	**Loading_2**	**Loading^2 [1]**	**NR Annotation of DEGs**
The top 10 DEGs associated with flavor related DEMs by O2PLS analysis				
DEGs	Loading_1	Loading_2	Loading^2 [1]	NR annotation of DEGs
TRINITY_DN455_c1_g1	0.040883277	0.149953165	0.024157394	gi | 1035381610 | gb | OAX38220.1 | hypothetical protein K503DRAFT_850105 [*Rhizopogon vinicolor* AM-OR11-026]
TRINITY_DN24602_c0_g1	0.044125314	0.147860791	0.023809857	gi | 749890950 | gb | KIJ58942.1 | hypothetical protein HYDPIDRAFT_162871 [*Hydnomerulius pinastri* MD-312]
TRINITY_DN6766_c0_g1	0.043559178	0.147316192	0.023599462	gi | 628854451 | ref | XP_007774347.1 | hypothetical protein CONPUDRAFT_112209 [*Coniophora puteana* RWD-64-598 SS2]
TRINITY_DN19391_c0_g1	0.043673434	0.146674332	0.023420729	gi | 749903138 | gb | KIJ70603.1 | hypothetical protein HYDPIDRAFT_23695 [*Hydnomerulius pinastri* MD-312]
TRINITY_DN59157_c0_g1	0.04454385	0.146046036	0.023313599	gi | 763724697 | gb | KJA22132.1 | hypothetical protein HYPSUDRAFT_41267 [*Hypholoma sublateritium* FD-334 SS-4]
TRINITY_DN5293_c0_g1	0.043968374	0.145400401	0.023074494	gi | 749841699 | gb | KIJ16704.1 | acetolactate synthase [*Paxillus involutus* ATCC 200175]
TRINITY_DN15287_c0_g1	0.047024185	0.142239221	0.02244327	gi | 749892543 | gb | KIJ60404.1 | hypothetical protein HYDPIDRAFT_177471 [*Hydnomerulius pinastri* MD-312]
TRINITY_DN3921_c0_g1	0.046666069	0.141663398	0.02224624	gi | 749830483 | gb | KIJ08773.1 | glycoside hydrolase family 89 protein [*Paxillus involutus* ATCC 200175]
TRINITY_DN130487_c0_g1	0.04132318	0.142957832	0.022144547	gi | 749845521 | gb | KIJ20338.1 | hypothetical protein PAXINDRAFT_108316 [*Paxillus involutus* ATCC 200175]
TRINITY_DN9716_c0_g1	0.046452929	0.141217718	0.022100318	gi | 749901809 | gb | KIJ69300.1 | hypothetical protein HYDPIDRAFT_145192 [*Hydnomerulius pinastri* MD-312]

Note: [1]. Loading^2 = (Loading_1)^2 + (Loading_2)^2; [2]. "-"represents not being annotated.

4. Discussion

Edible mushrooms offer a delicious and nutritious addition to one's diet, thanks to their high levels of various secondary metabolites, such as phenolic compounds, polyketones, terpenoids, and steroids. These compounds have numerous beneficial health effects, such as immune regulation, lipid-lowering, and anti-tumor properties. Additionally, consumers enjoy their rich flavor. This experiment utilized UHPLC-MS technology to perform metabolomics analysis on the fruiting bodies of *B. brunneissimus* and *L. extremiorientale*, both of which possess strong flavor differences. Of 644 annotated compounds, alkaloids, phenols, flavonoids, and terpenoids were prominently identified in both wild *Boletus* species, indicating that they are nutrient-rich and have high utilization value.

Through OPLS-DA analysis, 47 differential metabolites were identified, primarily in amino acid metabolic pathways. Transcriptome analysis found differential genes in Protein processing in the endoplasmic reticulum (ko04141) pathway for *B. brunneissimus* and *L. extremiorientale*, while the differential genes for the same wild *Boletus* species were mainly enriched in the Ribosome pathway (ko03010). A total of 8 differential metabolites relating to soluble sugars, organic acids, free amino acids, and 5′ -nucleotides were identified between *B. brunneissimus* and *L. extremiorientale*. Joint analysis of the transcriptome and metabolome showed that Glycine, L-Serine, and L-Aspartic acid were key compounds accounting for the metabolic differences between *B. brunneissimus* and *L. extremiorientale*. Finally, the O2PLS model combined with the Pearson coefficient was used to conduct further research on the screening of DEGs associated with DEMs.

Mushroom flavor is complex and mainly formed by the aliphatic components, amino acids and flavorful nucleotides, such as 1-octen-3-ol,2-octen1-ol, 3-octanol, 1-octanol, 3-octanone, 5′-GMP and glutamic acid [32]. These compounds probably form a basal flavor and may be modified to create a more complex flavor in other species, such as Lentinus edodes [33]. Lipidtemns like terpenoids, 3-(methylthio)propanal, and N-heterocyclic compounds contribute to individual odors between different mushroom species. Ref. [34] is very common in most mushroom species. Aldehydes, esters, alcohols, acids, pyrazines, ketones, and phenols; 1-octen-3-ol and 2,5-dimethylpyrazine were enriched in boletus varieties, while 3-(methylthio) propionaldehyde and 2,6-dimethylpyrazine were the main compounds in B. edulis. Benzaldehyde, 3-methyl-1-butanol, Acetoin, 2-pentylfuran, 2-heptenal, 3-octanone and ethyl tiglate composite flavor and fragrance ingredient in B. pinophilus. [35], in this study, we focus on the flavored nucleotide and amino acid and found that the key differential metabolites of *B. brunneissimus* and *L. extremiorientale* are Glycine, L-Serine, and L-Aspartic acid. The differential metabolites produced by these three amino acid metabolism pathways account for 38% of the total differential metabolites. One of the key difficulties in studying the metabolites of wild *Boletus* is the lack of control. This is in contrast to cultivated edible mushrooms, where samples from different varieties of the same species can be more easily compared to studying differential metabolites and genes. In the case of wild *Boletus*, the composition and content of metabolites are influenced by various growth factors, making it challenging to collect samples and guarantee differences. To overcome this challenge, this study utilized three sets of samples from different growth environments to create intra-group differences, thereby mitigating errors and focusing on differences in species characteristics. The experimental results showed that the different metabolites screened from different locations varied and that expanding the sample size could yield more accurate results. By conducting research in this way, we can gain a better understanding of the unique characteristics of wild *Boletus* and their metabolites.

This study involved screening out the genes related to flavor differences between *B. brunneissimus* and *L. extremiorientale* through combined analysis of the metabolome and transcriptome. Gene annotation revealed that a significant number of genes were annotated as "hypothetical protein", suggesting that there is a wealth of untapped genetic resources in wild *Boletus*. From a breeding perspective, these genes from wild fungi can be used in the breeding of artificial fungi through large-scale fungal genetic transformation technology. The wide variety of wild *Boletus* provides an extensive gene resource pool for improving artificial mushroom varieties [36]. To study the molecular mechanism of *boletus* flavor substances regulation, it is possible to identify key regulatory genes, clone and study their functions through yeast genetic transformation, and ultimately employ genome editing, transgenic, and other technologies to improve the flavor of artificial mushrooms. Such research has important breeding implications for enhancing the flavor of artificial mushrooms.

Supplementary Materials: The following supporting information can be downloaded at: https://www.mdpi.com/article/10.3390/foods12142728/s1, Figure S1: The pearson correlation coefficients of metabolites in each sample are shown in Figure S1; Figure S2: The KEGG Enrichment of Clust 1–9; Figure S3: Heat map of DEGs clustering for all samples; Figure S4: The KEGG and GO enrichment of 4240 DEGs which strong correlation with 47 DEMs; Table S1: Difference of metabolite between Boletus brunneissimus and Leccinum extremiorientale; Table S2: Table 2 The DEMs and DEGs in the two boletus fruiting bodies in different compare combinations.

Author Contributions: Writing—original draft, review & editing, K.C. and T.Q.; formal analysis and resources, K.C.; project administration, Q.W.; funding acquisition, T.L. and K.C. All authors have read and agreed to the published version of the manuscript.

Funding: This research was funded by Yunnan Fundamental Research Projects (grant NO. 202001AU070086), the Special Basic Cooperative Research Programs of Yunnan Provincial Undergraduate Universities Association (grant NO. 202001BA0000-084), The National Natural Science Foundation of China (grant NO. 21667031).

Data Availability Statement: Data are contained within the article or Supplementary Material.

Acknowledgments: We appreciate the sequencing and technical consulting services provided by Yeren Long (a Personal Biotechnology company in Shanghai).

Conflicts of Interest: The authors declare no conflict of interest.

References

1. Rizzo, G.; Goggi, S.; Giampieri, F.; Baroni, L. A review of mushrooms in human nutrition and health. *Trends Food Sci. Technol.* **2021**, *117*, 60–73. [CrossRef]
2. Mau, J.L. The Umami taste of edible and medicinal mushrooms. *Int. J. Med. Mushrooms* **2005**, *7*, 119–126. [CrossRef]
3. Aisala, H.; Sola, J.; Hopia, A.; Linderborg, K.M.; Sandell, M. Odor-contributing volatile compounds of wild edible Nordic mushrooms analyzed with HS–SPME–GC–MS and HS–SPME–GC–O/FID. *Food Chem.* **2019**, *283*, 566–578. [CrossRef]
4. Komata. The taste and constituents of foods. *Nippon. Sho-Kuhin Kogyo Gakkaishi* **1969**, *3*, 26.
5. Poojary, M.M.; Orlien, V.; Passamonti, P.; Olsen, K. Enzyme-assisted extraction enhancing the umami taste amino acids recovery from several cultivated mushrooms. *Food Chem.* **2017**, *234*, 236–244. [CrossRef]
6. Samarasiri, M.; Chen, W.N. Variations of nonvolatile taste components of mushrooms with different operating conditions and parameters from farm to fork. *Crit. Rev. Food Sci.* **2022**, 1–20. [CrossRef]
7. Sun, L.-B.; Zhang, Z.-Y.; Xin, G.; Sun, B.-X.; Bao, X.-J.; Wei, Y.-Y.; Zhao, X.-M.; Xu, H.-R. Advances in umami taste and aroma of edible mushrooms. *Trends Food Sci. Technol.* **2020**, *96*, 176–187. [CrossRef]
8. Wen, X.; Geng, F.; Xu, Y.; Li, X.; Liu, D.; Liu, Z.; Luo, Z.; Wang, J. Quantitative transcriptomic and metabolomic analyses reveal the changes in Tricholoma matsutake fruiting bodies during cold storage. *Food Chem.* **2022**, *381*, 132292. [CrossRef]
9. Dong, J.E.; Zuo, Z.T.; Zhang, J.; Wang, Y.Z. Geographical discrimination of Boletus Edulis using two dimensional correlation spectral or integrative two dimensional correlation spectral image with ResNet. *Food Control* **2021**, *129*, 108132. [CrossRef]
10. Li, J.; Wu, H.; Wang, L.; Huang, Y.; Wang, L. Key taste components in two wild edible Boletus mushrooms using widely targeted metabolomics. *Biochem. Syst. Ecol.* **2021**, *96*, 104268. [CrossRef]
11. Feng, B.; Yang, Z. Studies on diversity of higher fungi in Yunnan, southwestern China: A review. *Plant Divers.* **2018**, *40*, 165–171. [CrossRef]
12. Wang, X.; Liu, P.; Yu, F. *Color Atlas of Wild Commercial Mushrooms in Yunnan*; Yunnan Science and Technology Press: Kunming, China, 2004. (In Chinese)
13. Liang, J.; Chen, L.; Li, Y.; Hu, X. Isolation and identification of umami-flavored peptides from *Leccinum extremiorientale* and their taste characteristic. *J. Food Process. Preserv.* **2021**, *45*, e15255. [CrossRef]
14. Wang, X. Study of Taste Active Components of Five Wild Edible Mushrooms in Yunnan. Master's Thesis, Kunming University of Science and Technology, Kunming, China, 2015. (In Chinese).
15. Falandysz, J.; Saba, M.; Strumińska-Parulska, D. 137Caesium, 40K and total K in Boletus edulis at different maturity stages: Effect of braising and estimated radiation dose intake. *Chemosphere* **2021**, *268*, 129336. [CrossRef]
16. Zhang, Y.; Venkitasamy, C.; Pan, Z.; Wang, W. Recent developments on umami ingredients of edible mushrooms—A review. *Trends Food Sci. Technol.* **2013**, *33*, 78–92. [CrossRef]

17. Mamat, A.; Tusong, K.; Xu, J. Identification of metabolic pathways related to rough-skinned fruit formation in Korla pear. *Sci. Hortic-Amst.* **2021**, *288*, 110414. [CrossRef]
18. Zhang, Z.-M.; Tong, X.; Peng, Y.; Ma, P.; Zhang, M.-J.; Lu, H.-M.; Chen, X.-Q.; Liang, Y.-Z. Multiscale peak detection in wavelet space. *Analyst* **2015**, *140*, 7955–7964. [CrossRef]
19. Kuhl, C.; Tautenhahn, R.; Böttcher, C.; Larson, T.R.; Neumann, S. CAMERA: An Integrated Strategy for Compound Spectra Extraction and Annotation of Liquid Chromatography/Mass Spectrometry Data Sets. *Anal. Chem.* **2012**, *84*, 283–289. [CrossRef]
20. Smith, C.A.; Want, E.J.; O'Maille, G.; Abagyan, R.; Siuzdak, G. XCMS: Processing Mass Spectrometry Data for Metabolite Profiling Using Nonlinear Peak Alignment, Matching, and Identification. *Anal. Chem.* **2006**, *78*, 779–787. [CrossRef]
21. Chen, W.; Gong, L.; Guo, Z.; Wang, W.; Zhang, H.; Liu, X.; Yu, S.; Xiong, L.; Luo, J. A Novel Integrated Method for Large-Scale Detection, Identification, and Quantification of Widely Targeted Metabolites: Application in the Study of Rice Metabolomics. *Mol. Plant* **2013**, *6*, 1769–1780. [CrossRef]
22. Zhang, H.; Lu, Q.; Liu, R. Widely targeted metabolomics analysis reveals the effect of fermentation on the chemical composition of bee pollen. *Food Chem.* **2022**, *375*, 131908. [CrossRef]
23. Dunn, W.B.; Broadhurst, D.; Begley, P.; Zelena, E.; Francis-McIntyre, S.; Anderson, N.; Brown, M.; Knowles, J.D.; Halsall, A.; Haselden, J.N.; et al. Procedures for large-scale metabolic profiling of serum and plasma using gas chromatography and liquid chromatography coupled to mass spectrometry. *Nat. Protoc.* **2011**, *6*, 1060–1083. [CrossRef]
24. Jolliffe, I. *Principal Component Analysis*; Wiley Online Library: Hoboken, NJ, USA, 2002.
25. Wiklund, S.; Johansson, E.; Sjöström, L.; Mellerowicz, E.J.; Edlund, U.; Shockcor, J.P.; Gottfries, J.; Moritz, T.; Trygg, J. Visualization of GC/TOF-MS-Based Metabolomics Data for Identification of Biochemically Interesting Compounds Using OPLS Class Models. *Anal. Chem.* **2008**, *80*, 115–122. [CrossRef]
26. Trygg, J.; Wold, S. Orthogonal projections to latent structures (O-PLS). *J. Chemom.* **2002**, *16*, 119–128. [CrossRef]
27. Saccenti, E.; Hoefsloot, H.C.J.; Smilde, A.K.; Westerhuis, J.A.; Hendriks, M.M.W.B. Reflections on univariate and multivariate analysis of metabolomics data. *Metabolomics* **2014**, *10*, 361–374. [CrossRef]
28. Martin, M. Cutadapt removes adapter sequences from high-throughput sequencing reads. *EMBnet J.* **2011**, *17*, 10–12. [CrossRef]
29. Grabherr, M.G.; Haas, B.J.; Yassour, M.; Levin, J.Z.; Thompson, D.A.; Amit, I.; Adiconis, X.; Fan, L.; Raychowdhury, R.; Zeng, Q.D.; et al. Full-length transcriptome assembly from RNA-Seq data without a reference genome. *Nat. Biotechnol.* **2011**, *29*, 644–652. [CrossRef]
30. Powell, S.; Szklarczyk, D.; Trachana, K.; Roth, A.; Kuhn, M.; Muller, J.; Arnold, R.; Rattei, T.; Letunic, I.; Doerks, T.; et al. eggNOG v3.0: Orthologous groups covering 1133 organisms at 41 different taxonomic ranges. *Nucleic. Acids Res.* **2011**, *40*, D284–D289. [CrossRef]
31. Bouhaddani, S.E.; Houwing-Duistermaat, J.; Salo, P.; Perola, M.; Jongbloed, G.; Uh, H.-W. Evaluation of O2PLS in Omics data integration. *BMC Bioinform.* **2016**, *17*, S11. [CrossRef]
32. Cho, I.H.; Kim, S.Y.; Choi, H.-K.; Kim, Y.-S. Characterization of Aroma-Active Compounds in Raw and Cooked Pine-Mushrooms (Tricholoma matsutake Sing.). *J. Agr. Food Chem.* **2006**, *54*, 6332–6335. [CrossRef]
33. Dijkstra, F.Y. Studies on mushroom flavours 3. Some flavour compounds in fresh, canned and dried edible mushrooms. *Eur. Food Res. Technol.* **1976**, *160*, 401–405. [CrossRef]
34. Aisala, H.; Manninen, H.; Laaksonen, T.; Linderborg, K.M.; Myoda, T.; Hopia, A.; Sandell, M. Linking volatile and non-volatile compounds to sensory profiles and consumer liking of wild edible Nordic mushrooms. *Food Chem.* **2020**, *304*, 125403. [CrossRef]
35. Bozok, F.; Zarifikhosroshahi, M.; Kafkas, E.; Taşkin, H.; Buyukalaca, S. Comparison of Volatile Compounds of Fresh Boletus edulis and B. pinophilus in Marmara Region of Turkey. *Not. Bot. Horti Agrobot.* **2015**, *43*, 192–195. [CrossRef]
36. Balabanova, L.A.; Shkryl, Y.N.; Slepchenko, L.V.; Yugay, Y.A.; Gorpenchenko, T.Y.; Kirichuk, N.N.; Khudyakova, Y.V.; Bakunina, I.Y.; Podvolotskaya, A.B.; Bulgakov, V.P.; et al. Development of host strains and vector system for an efficient genetic transformation of filamentous fungi. *Plasmid* **2019**, *101*, 1–9. [CrossRef]

Disclaimer/Publisher's Note: The statements, opinions and data contained in all publications are solely those of the individual author(s) and contributor(s) and not of MDPI and/or the editor(s). MDPI and/or the editor(s) disclaim responsibility for any injury to people or property resulting from any ideas, methods, instructions or products referred to in the content.

fermentation

MDPI

Article

Dynamics of Physicochemical Properties, Flavor, and Microbial Communities of Salt-Free Bamboo Shoots during Natural Fermentation: Correlation between Microorganisms and Metabolites

Xiaofeng Xu [1,†], Zhijian Long [1,2,†], Wanning Du [1], Qiyang Chen [1], Yu Zhang [1] and Shanglian Hu [1,2,*]

1 School of Life Science and Engineering, Southwest University of Science and Technology, Mianyang 621010, China; 18788981570m@sina.cn (X.X.); long20053182@swust.edu.cn (Z.L.); dwncie@sina.com (W.D.); chenqiyang@swust.edu.cn (Q.C.); vividz2010@126.com (Y.Z.)
2 Engineering Research Center for Biomass Resource Utilizaiton and Modification of Sichuan Province, Mianyang 621010, China
* Correspondence: hushanglian@swust.edu.cn
† These authors contributed equally to this work.

Abstract: Sour bamboo shoot is a Chinese fermented vegetable with unique flavors and is favored by local consumers. In this study, at different fermentation times, the texture of bamboo shoots and the changing rules of pH, titratable acid (TA), reduced sugar, and nitrite in bamboo shoot fermentation broth were explored. Headspace solid-phase microextraction (HS-SPME) combined with gas chromatography-mass spectrometry (GC-MS) and orthogonal partial least squares discriminant analysis (OPLS-DA) were used to investigate the dominant aroma compounds. 16S rRNA high-throughput sequencing technology (HTS) was employed to investigate the core microbial communities. The results show that the chewiness, fracturability, hardness, and pH decreased, while TA increased during the 60-day fermentation. The contents of reducing sugar and nitrite peaked on the 14th day of fermentation and then decreased. A total of 80 volatile compounds were detected during sour bamboo shoot fermentation, with 2,4-Di-tert-butylphenol having the highest concentration. Among them, 12 volatile compounds (VIP $\geq$ 1) were identified as characteristic aroma substances of sour bamboo shoots. The dominant bacterial phyla in sour bamboo shoots were Firmicutes and Proteobacteria, while *Bacillus* and *Acinetobacter* were the dominant genus. Correlation analysis showed that Firmicutes exhibited a positive correlation with 3,6-Nonadien-1-ol, (E,Z)-, Oxalic acid, isobutyl hexyl ester, and (-)-O-Acetylmalic anhydride, whereas *Bacillus* exhibited a negative correlation with Silanediol, dimethyl-, and Oxime-, methoxy-phenyl-. A detailed picture of the microbial community of fermented bamboo shoots has been provided by this study, and it may provide insight into the Chinese traditional fermented vegetable microbial structure.

Keywords: fermentation; sour bamboo shoots; volatile compounds; core microbiota

1. Introduction

Fermented vegetables originated during the Shang and Zhou periods, around 3100 years ago, and have become one of the important traditional fermented foods both domestically and internationally [1]. Fermented bamboo shoots are now commonly used in Southern Chinese households and are a traditional fermented food in Guangxi and surrounding regions. They are soaked in spring water or low-salt water and naturally fermented to acquire a sour taste. The distinctive aroma and flavor of sour bamboo shoots are beloved by locals [2].

The flavor of fermented vegetables is usually closely related to microorganisms. Different regions' fermented vegetables exhibit differences in dominant microbial communities and flavor-associated microbial communities. For example, *Lactobacillus* in Sichuan paocai (pickled vegetables), Jiangxi-jiancai (fermented vegetables), and Northeast-jiancai are considered as the core functional microbial groups influencing flavor [3]. The main microbial communities in vegetable fermentation were lactic acid bacteria (LABs), including *Leuconostoc*, *Weissella*, *Lactococcus*, *Lactobacillus*, etc., and these microbial communities contribute to the maturation of fermented vegetables [4–7]. Sour bamboo shoots contain a rich resource of lactic acid bacteria, including *Lactobacillus*, *Streptococcus*, *Lactococcus*, and *Weissella*. Microbial metabolism is crucial in the flavor of sour bamboo shoots. For example, in a study by Xu et al. (2020), the authors investigated the progression of bacterial and fungal communities and changes in key odorants during the fermentation of red pepper. The results revealed that species such as *Aspergillus*, *Bacillus*, *Brachybacterium*, *Microbacterium*, and *Staphylococcus* were responsible for producing the majority of the key odorants throughout the fermentation process [8]. *Lactobacillus* and *Streptococcus* showed a significant positive correlation with certain organic acids, low-threshold aldehydes, phenols, and high-threshold alcohols and esters. The complex microbial community provides the foundation for flavor development, and the two are closely intertwined [9]. Therefore, it is important to investigate the relationship between the flavor of sour bamboo shoots and microorganisms.

2. Materials and Methods

2.1. Preparation of Bamboo Shoots

Fresh Jipo bamboo shoots were collected from Leshan, Sichuan Province, China, and quickly transported to the laboratory within 24 h. Approximately 10 kg of bamboo shoots were peeled off, leaving the tender edible part; then, the shoots were washed and cut into 13 cm long, 0.5 cm wide, and 0.5 cm thick strips. These strips were placed in plastic bottles with a total volume of 0.5 L (50 g) and filled with sterile water [10,11]. The bottles were sealed and fermented at room temperature for 60 days. During the fermentation process, samples were collected on the 1st, 7th, 14th, 21st, 28th, and 60th day, resulting in a total of six samples labeled as JPW1, JPW2, JPW3, JPW4, JPW5, and JPW6; for sampling, three unopened bottles were selected at random each time. The collected bamboo shoot samples were immediately subjected to texture analysis, while the collected fermentation liquid samples were stored in a −80 °C freezer for subsequent analysis.

2.2. Texture Analysis

The texture of sour bamboo shoots during fermentation was evaluated using a TA-XT Plus texture analyzer. Uniform sour bamboo shoot strips (30 mm × 5 mm × 5 mm) were positioned beneath the P36R probe for texture measurement. The analysis was conducted with the following measurement parameters: pre-test speed of 2 mm/s, test speed of 1 mm/s, post-test speed of 1 mm/s, and a compression of 70% during testing, pause time of 5 s, data acquisition rate of 400 pps, and trigger force of 5 g.

2.3. Determination of pH and Total Acid (TA)

The pH of the samples was determined using a precise pH meter (PHS-25, Greifensee, Switzerland). The total acidity (TA) was measured according to the standard GB/T 12456-2021 [12].

2.4. Determination of Reducing Sugar Content

The reducing sugar content of the fermentation liquids was measured following GB 5009.7-2021 [13], and the direct titration method was used for determination.

2.5. Determination of Nitrite Content

The nitrite content in bamboo shoot fermentation liquid was measured using the spectrophotometric method according to GB 5009.33-2016 [14]. Place 18 g of bamboo shoot fermentation liquid in a 250 mL volumetric flask, add 2.5 mL of saturated borax solution, and then add 12 mL of 70 °C water. Mix well, heat in a boiling water bath for 15 min, remove and cool in cold water, and then let it cool to room temperature. Quantitatively transfer the above extraction solution to a beaker, add 1 mL of 106 g/L potassium ferrocyanide solution, shake well, and then add 5 mL of 220 g/L zinc acetate solution to precipitate proteins. Add water to make up the volume to 40 mL, shake well, let it stand for 30 min, remove the upper fat layer, filter the supernatant with filter paper, and reserve the filtrate. Add 8 mL of the filtered solution into a 10 mL stoppered colorimetric tube, then add 0.4 mL of 4 g/L sulfanilic acid solution, mix well, and after standing for 3–5 min, add 0.2 mL of 2 g/L naphthyl ethylenediamine hydrochloride solution. Add water to the mark, mix well, let it stand for 15 min, and measure the absorbance at 538 nm using a 1 cm cuvette. Each sample was measured three times. Based on the measured absorbance, find the corresponding mass of nitrite (micrograms) in the bamboo shoot fermentation liquid from the nitrite standard curve, and use the formula to calculate the nitrite content (mg/kg).

2.6. Determination of Volatile Compounds

To determine the volatile compounds in bamboo shoot fermentation liquid during different fermentation processes, this study appropriately modified the method reported by He et al. [15]. Analysis was performed using the headspace solid-phase microextraction (HS-SPME) [16] technique and gas chromatography-mass spectrometry (GC-MS). First, after adding 100 μL of an internal standard solution containing 38.12 μg of cyclohexanone to the samples for quantitative analysis, 8 mL of bamboo shoot fermentation liquid and 3 g of sodium chloride were placed in a 10 mL headspace vial for pretreatment. The vial was then preheated for 10 min in a 60 °C water bath. Next, a divinylbenzene/carboxylic acid/polydimethylsiloxane (CAR/PDMS/DVB, 65 μm) fiber was used for headspace adsorption of volatile compounds for 30 min, followed by desorption of the volatile compounds from the fiber onto a 250 °C heated GC injection port for 5 min. For gas chromatography analysis, a temperature program was set as follows: initial temperature 40 °C, increased at a rate of 12 °C/min to 100 °C and held for 2 min, increased at a rate of 5 °C/min to 120 °C, further increased at a rate of 8 °C/min to 180 °C and held for 5 min, then increased at a rate of 12 °C/min to 210 °C, and finally increased at a rate of 8 °C/min to 250°C for the final temperature. A flexible capillary column (HP-5MS, 30 m × 0.25 mm × 0.25 μm) was used, with the mass spectrometer set at a source temperature of 230 °C, an ionization potential of 70 eV, and a mass scan range of 40–400 m/z. Lastly, preliminary identification of the compounds was performed using the NIST 17 mass spectral library, and volatile compounds with a match quality greater than 85% were selected for qualitative analysis.

2.7. Microbial Analysis

The total genomic DNA from the bamboo shoot fermentation liquid was extracted according to the cetyltrimethylammonium bromide (CTAB) method. And triplicate samples of bamboo shoot fermentation liquid at different fermentation stages were collected, with 5 mL of DNA extracted from each sample. Next, PCR amplification was performed on the extracted DNA using the universal primers 338F and 806R, and only samples with brightness readings between 400 and 550 bp were selected for sequencing. To obtain more accurate and reliable results, the Illumina platform for PE250 bp sequencing was used to provide paired-end reads of 250 bp. Finally, using Trimatic v0.33 software for filtering, cutadapt 1.9.1 software to eliminate primer sequences, and the dada2 method in QIME2 2020.6 for denoising the analysis, the raw reads were obtained. An operational taxonomic unit (OTU) is a grouping of sequences with at least 97% similarity.

2.8. Statistical Analysis

The IBM SPSS Statistics 25.0 software was utilized for statistical analysis. Statistically significant differences between two groups of data were determined using one-way analysis of variance (ANOVA) and post hoc Tukey's honestly significant difference (HSD) tests.

Additionally, Pearson correlation networks and cell landscape (3.9.0) interaction networks were employed to explore the interactions between physicochemical properties, flavor, and microbiota. Throughout this process, mean values $\pm$ standard deviation (SD) were used to represent all measured values. A significance level of 0.05 was used to determine statistical significance, with a *p*-value below this threshold considered as statistically significant.

3. Results and Discussion

3.1. Physiochemical Property Analysis

3.1.1. Texture Analysis of Bamboo Shoots

Evaluating the quality of bamboo shoots involves considering various factors, including texture. Hardness, fracturability, and chewiness are particularly important aspects of texture assessment [10]. As shown in Figure 1A, the hardness of bamboo shoots significantly ($p < 0.05$) decreases within the first 1–14 days and slowly decreases from day 14 to day 60. Similarly, the fracturability and chewiness of bamboo shoots gradually decrease in a relatively stable and slow manner (Figure 1B,C). The softening of bamboo shoots may be caused by a decrease in cell membrane permeability due to a pH decrease during fermentation or by the hydrolysis of native pectin in bamboo shoots [17].

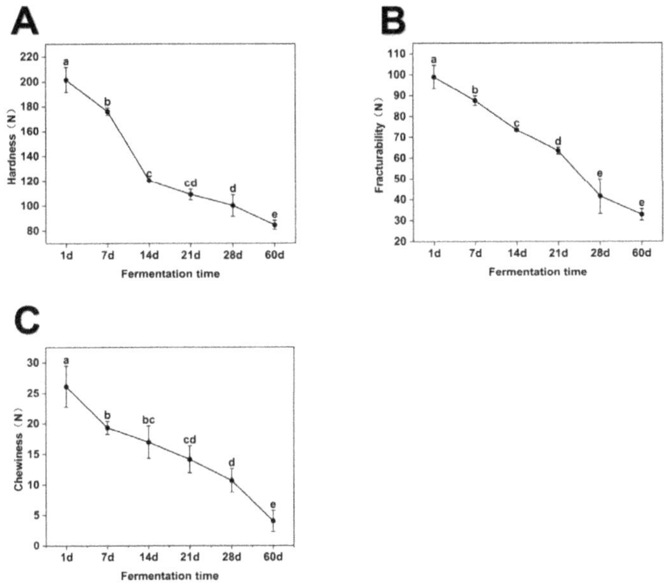

Figure 1. Texture properties of the sour bamboo shoot during the fermentation process. (**A**) Changes in hardness of sour bamboo shoots during fermentation. (**B**) Changes in fracturability of sour bamboo shoots during fermentation. (**C**) Changes in chewiness of sour bamboo shoots during fermentation. Statistically significant differences are indicated by different letters at $p < 0.05$.

3.1.2. Physicochemical Properties of Fermentation Liquids

The flavor quality of bamboo shoots is closely related to their acidity, and an appropriate increase in acidity helps enhance the flavor of the product [18]. During the fermentation process, lactic acid bacteria decompose and utilize carbohydrates in bamboo shoots to produce organic acids, thus forming the flavor of bamboo shoots. With the accumulation of organic acids, the pH and titratable acidity (TA) change [19]. As shown in Figure 2A, the pH of the fermentation broth remains stable throughout the fermentation process and then decreases significantly ($p < 0.05$), while the total acidity increases (Figure 2B). The decline in pH levels could potentially be attributed to the presence of *Bacillus* spores and lactic acid bacteria [20]. During the fermentation of bamboo shoots, the entry of lactic acid bacteria or airborne microorganisms into the fermentation system causes changes in acidity. As fermentation progresses, the ethanol content gradually decreases, indicating the generation of acidic components, leading to an increase in the acidity of the sample with prolonged fermentation time [21]. However, when the acidity reaches a certain level, the growth of lactic acid bacteria may be inhibited due to the reduction of nutrients in the fermentation environment, resulting in a slow increase in acidity in the later stages [22].

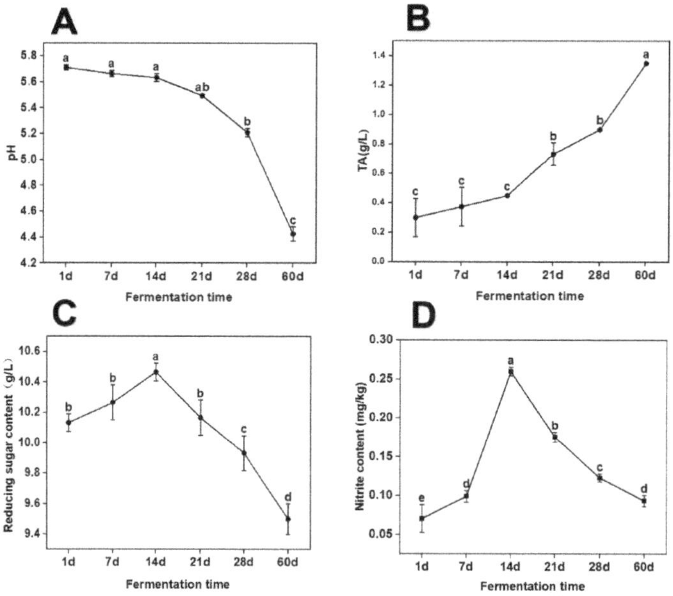

Figure 2. Changes in the pH (**A**), TA (**B**), reducing sugar content (**C**), and nitrite content (**D**) during different fermentation periods. Statistically significant differences are indicated by different letters at $p < 0.05$.

As shown in Figure 2C, the content of reducing sugars shows an upward trend from day 1 to day 14 during fermentation, possibly due to the flow of some bamboo shoot juice into the fermentation broth after cutting. Additionally, during the fermentation process, polysaccharides and disaccharides in bamboo shoots undergo hydrolysis, leading to the generation of reducing sugars and an increase in their content [23]. From day 14 to day 60, there is a significant downward trend in reducing sugar content, possibly because the growth metabolism of microorganisms during fermentation consumes the carbon source. With the rapid reproduction of microorganisms, the carbon source is continuously utilized and depleted, resulting in a significant decrease in reducing sugar content.

Nitrite is a recognized chemical carcinogen mainly produced by nitrite-reducing bacteria during vegetable fermentation [24]. The content of nitrites may be related to the growth of spoilage microorganisms [25]. As shown in Figure 2D, the nitrite content of bamboo shoots increases during the first 1–14 days of fermentation, possibly due to the rapid proliferation of harmful microorganisms such as *Enterobacteriaceae* in the initial stage of fermentation, leading to a significant increase in nitrite content [26]. A "nitrite peak" appears on day 14, followed by a significant decrease. On the one hand, this may be due to the presence of microorganisms with efficient nitrite degradation capabilities during fermentation, effectively reducing the production of nitrites in bamboo shoots [17]. On the other hand, as fermentation progresses, lactic acid bacteria gradually form dominant microbial populations, inhibiting the growth of harmful microorganisms and thereby suppressing nitrite production [26].

3.2. Volatile Compounds

3.2.1. Dynamic Changes in Volatile Compounds of Fermented Bamboo Shoots

Under specific detection conditions, headspace solid-phase microextraction-gas chromatography-mass spectrometry (HSSPME/GC-MS) spectroscopy was employed to analyze the volatile compounds in fermented bamboo shoot samples [27]. It detected and quantified various volatile compounds present in fermented bamboo shoot fermentation liquor. The volatile flavor compounds in fermented bamboo shoots primarily include phenols, aldehydes, alcohols, esters, and other ketones and furans, which are the major contributors to the unique flavor of fermented bamboo shoots [9]. As shown in Figure 3A, a total of 80 volatile compounds were detected and identified throughout the fermentation process of bamboo shoots. Among them, alcohols and esters had a greater variety of substances, which are likely important contributors to the pleasant aroma in fermented vegetables [28]. Alcohols accounted for the largest number of volatile compounds, with a total of 24, followed by aldehydes and ketones, accounting for 14 and 12 compounds, respectively. At day 1 of fermentation, a total of 53 volatile compounds were detected in the fermented bamboo shoot liquor, including 12 alcohols, 10 aldehydes, 10 ketones, 8 esters, 3 acids, 2 phenols, and 8 other volatile compounds. Silanediol, dimethyl-, 2,4-Di-tert-butylphenol, and 1,2-Benzenedicarboxylic acid, bis(2-methylpropyl) ester had significantly higher contents than other volatile compounds, making them the major volatile substances at day 1 of fermentation. At day 7, the content of volatile compounds decreased, and some volatile compounds even disappeared, such as 1-Hexanol, 2-ethyl-, Linalool, and 2-Pentadecanone. As the fermentation proceeded, both the variety and content of volatile compounds increased. At day 21, the volatile compounds reached their peak, with a total of 65 compounds identified, including 20 alcohols, 13 aldehydes, 11 ketones, 8 esters, 4 acids, 3 phenols, and 6 other volatile compounds. Among them, Oxime-, methoxy-phenyl-, 2,4-Di-tert-butylphenol, and Silanediol, dimethyl- had the highest contents.

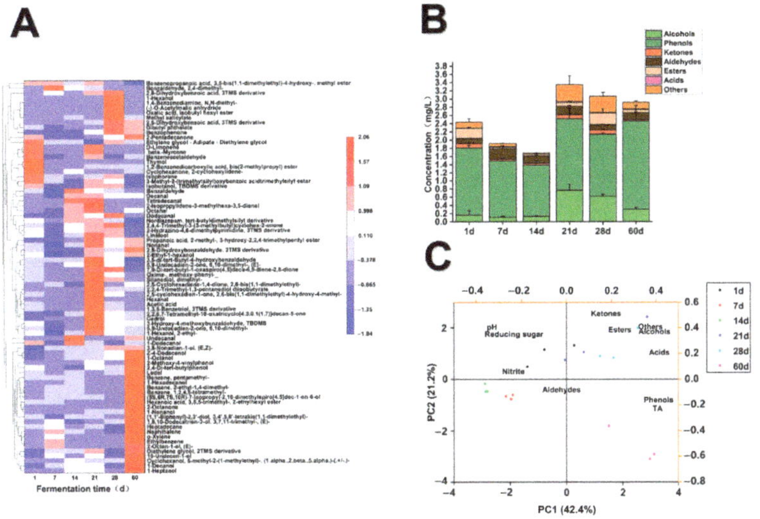

Figure 3. Hierarchical clustering heatmap of volatile compounds in fermented bamboo shoots (**A**), column chart of volatile compounds in fermented bamboo shoots (**B**), and PCA analysis of physicochemical properties and volatile compounds in fermented bamboo shoots (**C**).

From Figure 3B, it can be observed that the levels of volatile flavor compounds showed a trend of an initial decrease, followed by an increase, and then a gradual decrease again. The total content of volatile compounds reached its peak at day 21 of fermentation. Throughout the fermentation process, among the volatile compounds, phenols exhibited the highest content, followed by alcohols and aldehydes. The content of phenolic volatile compounds generally decreased at first and then increased. The variety of phenolic compounds increased during fermentation, with the detection of 2,4-Di-tert-butylphenol, 2-Methoxy-4-vinylphenol, and 1,3,5-Benzetriol, 3TMS derivative. Among them, 2,4-Di-tert-butylphenol had the highest content, and it has been found to be a candidate compound related to anti-diabetic enzymes [29]. The content of alcohol and aldehyde volatile compounds was next after phenols. Alcohols contribute pleasing aromas but have a higher odor threshold, and their impact on characteristic aromas in fermented vegetables is not significant [30]. As the fermentation progressed, the number of alcohol volatile compounds increased, including 1-Heptanol, 2-Octen-1-ol, (E)-, 1-Decanol, 1-Nonanol, Cedrol, among others. The content of aldehyde volatile compounds generally increased and then decreased, but it increased compared to day 1 of fermentation. The aldehyde compound with the highest content was 2,5-Dihydroxybenzaldehyde, 2TMS derivative, followed by Benzaldehyde, 2,4-dimethyl-. Most aldehyde compounds can provide a fresh, fruity, and nutty aroma to fermented vegetables [31]. Aldehyde compounds are mainly formed through the carbon chain oxidation or decarboxylation of unsaturated fatty acids, which are exothermic reactions during fermentation and easily reduced to alcoholic compounds. Although their content in bamboo shoots is not high, these compounds have a low threshold [32]. Ester substances are one of the characteristic flavor components in fermented foods [30]. Esters are primarily formed by microbial metabolism during fermentation through the combined action of acyl-coenzyme A and alcohol acyltransferases, as well as by the reaction of acids with alcohols catalyzed by lipases [9]. Ester compounds in bamboo shoots primarily serve to decrease strong odors, enhance aroma, and provide a balancing effect. They also contribute to increasing the content of health-promoting compounds [33]. Ketones contribute to aroma, but their content in bamboo shoots is relatively low, so their impact on the flavor of bamboo shoots may be limited [34]. Acid compounds are mainly formed by microbial biochemical reactions utilizing starch, fat, protein, and other organic substances [35]. The quantity of other volatile compounds remained relatively stable with a slight decrease. Due to variations in raw materials, processing methods, and environmental changes during fermentation, the content and variety of volatile flavor components in fermented bamboo shoots may differ.

As shown in Figure 3C, the analysis shows that PC1 and PC2 explain 42.2% and 21.2% of the total variance, respectively, with a cumulative value of 63.4%. The distance between day 60 of fermentation and other fermentation days is relatively large, indicating significant changes in the physicochemical properties and flavor compared to day 1 of fermentation. The distance between day 7 and day 14 is relatively close, suggesting similar flavors, which is consistent with the results of the column chart of volatile compounds (Figure 3B). At day 1 of fermentation, the content of volatile compounds was closely related to pH, reducing sugar content, and nitrite content. Representative volatile compounds at day 7 and day 14 of fermentation were aldehydes. At day 21 and day 28 of fermentation, the representative volatile compounds were ketones, esters, alcohols, acids, and other compounds. At day 60, the representative volatile compounds were phenols, strongly associated with total acid content.

3.2.2. Characteristic Volatile Compounds of Fermented Bamboo Shoots

By conducting an OPLS-DA analysis and obtaining VIP (variable importance in the projection) values, compounds with a VIP value of ≥ 1 were considered as contributing to the formation of the aroma in fermented bamboo shoots. According to our knowledge, 2,4-dimethylbenzaldehyde and 2,4-di-tert-butylphenol were detected as the main volatile compounds in fermented bamboo shoots for the first time, and these differences could be attributed to factors such as the type of raw materials, the growth environment of the raw materials, and fermentation conditions [36].

The table below presents the volatile compounds with VIP values ≥ 1, arranged according to their VIP values. As shown in Table 1, the compound with the highest VIP value is 1-Heptanol, which has a floral aroma [34] that contributes to the floral aroma of fermented bamboo shoots and is considered to have the greatest contribution to the flavor formation of fermented bamboo shoots. Phenolic volatile compounds have the highest content in bamboo shoots, and phenolic volatile substances typically produce off-flavors and irritating odors [34]. 2,4-Di-tert-butylphenol is the second compound after 1-Heptanol, considered to have a significant contribution to the flavor of fermented bamboo shoots. Benzaldehyde, 2,4-dimethyl- also has a VIP value greater than 1 and is considered to contribute to the flavor of fermented bamboo shoots with an almond/spicy flavor [37]. Alcohols have pleasant aromas. Silanediol, dimethyl-, and 3,6-Nonadien-1-ol, (E,Z)- have a relatively smaller contribution, but they provide a unique aroma to fermented bamboo shoots. Ester compounds with complex flavors have relatively less content throughout the fermentation process but have a variety of types. Although Oxalic acid, isobutyl hexyl ester, and 1,2-Benzenedicarboxylic acid, bis(2-methylpropyl) ester have VIP values greater than 1, their impact on the flavor is relatively minor, contributing slightly to the aroma of fermented bamboo shoots. O-Xylene has a faint aromatic odor, and Isophorone has a tobacco/woody aroma. D-Limonene contributes relatively less, has a lemon-like taste, and is a natural ingredient found in many citrus fruits, vegetables, and spices. It has a small content in bamboo shoots and may have a limited impact on the flavor of fermented bamboo shoots [34].

Table 1. Characteristic compounds in fermented bamboo shoots.

Compounds	VIP
1-Heptanol	4.64
2,4-Di-tert-butylphenol	3.50
Silanediol, dimethyl-	3.18
Oxime-, methoxy-phenyl-_	2.96
Benzaldehyde, 2,4-dimethyl-	2.84
Oxalic acid, isobutyl hexyl ester	2.63
(-)-O-Acetylmalic anhydride	2.57
1,2-Benzenedicarboxylic acid, bis(2-methylpropyl) ester	2.46
3,6-Nonadien-1-ol, (E,Z)-	1.71
o-Xylene	1.12
Isophorone	1.11
D-Limonene	1.05

3.3. Microorganisms

The microorganisms in bamboo shoots are mainly bacteria, including the genera *Lactobacillus*, *Weissella*, *Leuconostoc*, *Lactococcus*, *Enterobacter*, and *Serratia*, which are commonly enriched in fermented vegetables [9]. Throughout the bamboo shoot fermentation process, lactic acid bacteria mainly produce metabolites such as lactic acid, acetic acid, and ethanol, which help maintain the acidic environment necessary for fermentation [38,39]. These metabolites play an important role in giving bamboo shoots a certain flavor, as well as serving as the basis for the formation of acids and esters during the fermentation process.

Alpha diversity is a comprehensive index of microbial richness and evenness in the habitat [40]. Chao1 index and ACE index are used to compare community abundance [41], while Shannon index and Simpson index are used to compare community diversity [42]. As shown in Table S2, the coverage of all experimental samples is 1, indicating that the sequencing results can basically represent the actual situation of the samples [43]. Chao1, ACE, Shannon, and Simpson all showed a trend with an initial increase, then a decrease, and then an increase again. By comparing the Chao1, Simpson, and Shannon indices of the six fermentation stages, it can also be observed that the species diversity, species richness, and evenness of bamboo shoots at 21 days of fermentation were significantly higher than other fermentation stages. This phenomenon is due to the proliferation of microorganisms and increased species richness during the 1-day to 21-day fermentation stage. During the 21-day to 28-day fermentation stage, the bacterial community structure gradually became more singular, with some bacteria becoming dominant as the fermentation environment changed. During the 28–60 day fermentation stage, there may be the growth of miscellaneous bacteria in the fermentation bottles, leading to an increase in species richness [44]. PCoA and UPGMA analysis (Figures S2 and S3) revealed significant differences in microbial structure during the later stages of fermentation (21–60 days), with minimal fluctuation during the earlier stages.

3.3.1. Analysis of Beta Diversity during the Fermentation Process of Sour Bamboo Shoots

The following Figure 4A shows the top ten abundant phyla detected at the phylum level. At the phylum level, representatives include Cyanobacteria, Proteobacteria, Firmicutes, and Bacteroidetes [9]. Except for day 1 of fermentation, Firmicutes has the highest relative abundance in all other stages, indicating that Firmicutes is the dominant phylum during the 7–60 day stage of fermentation. At day 1 of fermentation, Proteobacteria has the highest relative abundance at 66.45%, followed by Firmicutes at 28.54%. Proteobacteria plays a dominant role during this stage. Except for day 1 of fermentation, Firmicutes has the highest relative abundance in all other stages, indicating that Firmicutes is the dominant phylum during the 7–60 day stage of fermentation. Therefore, Firmicutes plays a major role in the entire fermentation process, followed by Proteobacteria.

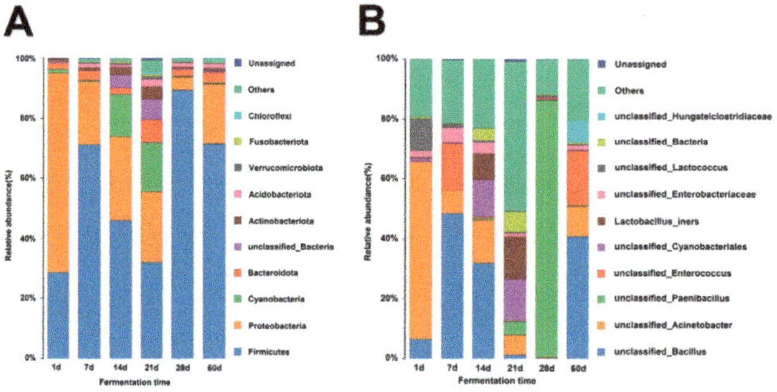

Figure 4. The phylum level (**A**) and genus level (**B**) of microbial community.

The following Figure 4B shows the top ten abundant genera detected at the genus level. It can be seen from the figure that the dominant bacterial genera vary under different fermentation conditions. At day 1 of fermentation, the dominant bacterial genus is *unclassified_Acinetobacter*. At day 7, 14, 60 of fermentation, the dominant bacterial genus is *unclassified_Bacillus*. At day 21 of fermentation, the dominant bacterial genus is *Lactobacillus_iners*. At day 28 of fermentation, the dominant bacterial genus is *unclassified_Paenibacillus*. Studies have shown that the microbial community in traditional fermented bamboo shoots has a high diversity, mainly including genera like *Lactobacillus*, *Lactococcus*, *Weissella*, *Shuttleworthia*, *Enterobacter*, and *Actinobacillus* [45]. The dominant genera identified in this sequencing are consistent with previous reports. The relative abundance of *unclassified_Enterobacteriaceae* gradually decreases with fermentation time, as lactic acid produced by lactic acid bacteria inhibits the growth of *unclassified_Enterobacteriaceae*. *Lactobacillus* is mostly facultative anaerobes with strong acid tolerance, making them more suitable for growth and metabolism in the low oxygen environment of the mid- to late-fermentation stages [46]. The relative abundance of *Enterococcus* shows a decreasing trend initially and then increases with fermentation. *Enterococcus* can produce short-chain fatty acids and organic acids in the human body and degrades sugars through glycolysis and pentose phosphate pathways [47].

3.3.2. Representative Bacteria in Sour Bamboo Shoots

LDA effect size (LEfSe) was employed as a statistical tool to assess dissimilarities among bacterial groups, discover distinctive features with diverse abundances and associated categories, and explore potential biomarkers [48]. Species with LDA values greater than four were selected for mapping [11]. As shown in Figure 5, on the first day of fermentation, sour bamboo shoots included two species biomarkers, two genus biomarkers, one family biomarker, one order biomarker, one class biomarker, and one phylum biomarker. On the 21st day of fermentation, sour bamboo shoots included five family biomarkers, four species biomarkers, four genus biomarkers, two order biomarkers, three class biomarkers, and three phylum biomarkers. On the 28th day of fermentation, sour bamboo shoots included one species biomarker, one genus biomarker, one family biomarker, one order biomarker, one class biomarker, and one phylum biomarker. On the 60th day of fermentation, sour bamboo shoots included one species biomarker, one genus biomarker, one family biomarker, and one domain biomarker. Combining the analysis of Alpha diversity and Beta diversity, as well as the succession between dominant bacterial phyla and genera, it can be inferred that the microbial succession process during the fermentation of bamboo shoots consists of Proteobacteria being gradually replaced by Firmicutes.

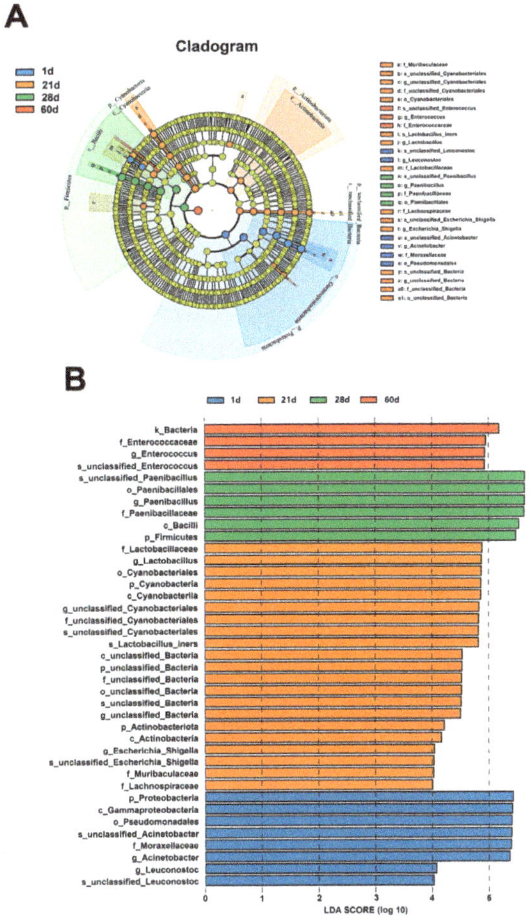

Figure 5. The composition of microbial community in LEfSe at the phylum (**A**) and genus (**B**) levels.

3.3.3. The Relationship between the Physicochemical Properties, Flavor, and Microbial Composition of Bamboo Shoots

The physicochemical properties and major volatile compounds during fermentation are closely related to the quantity and types of microbial communities. By using Pearson correlation analysis, the relationship between the physicochemical properties of bamboo shoots, characteristic volatile compounds (VIP ≥ 1), and dominant bacterial phyla (top ten) and genera (top ten) of the microbial community were studied.

The microbial population may be a key factor in the formation of volatile flavor substances in sour shoots [3,49–51]. As shown in Figure 6A, the vertical axis represents volatile flavor compounds with VIP values greater than one, as well as basic indicators such as pH, total acid, nitrite, and reducing sugar. The horizontal axis represents the top ten dominant bacterial phyla. $p \leq 0.05$ represents a significant correlation. When the correlation coefficient is 1, it indicates a positive correlation, while when the correlation coefficient is -1, it indicates a negative correlation. At the phylum level, Firmicutes and Proteobacteria are the dominant bacterial phyla. As shown in the figure, Firmicutes showed a significant positive correlation with 3,6-Nonadien-1-ol, (E,Z)-, and a significant negative correlation with D-Limonene. Proteobacteria showed a significant positive correlation with 1,2-Benzenedicarboxylic acid, bis(2-methylpropyl) ester, and a significant negative correlation with 3,6-Nonadien-1-ol, (E,Z)-. Alcohol compounds are related to the aroma of bamboo shoots, while ester compounds mainly reduce the irritating odor [30,33]. This indicates that Firmicutes is beneficial to the aroma production of bamboo shoots. Proteobacteria is beneficial for reducing the irritating odor of pickled bamboo shoots. Cyanobacteria showed a significant positive correlation with nitrite content, indicating that Cyanobacteria is beneficial for the nitrite content in bamboo shoots. Firmicutes showed a negative correlation with fracturability, hardness, and chewiness, indicating that it is not conducive to the texture of bamboo shoots, which will cause the bamboo shoots to become soft and decrease their taste. Proteobacteria showed the opposite trend, and it showed a positive correlation with the three texture indicators of bamboo shoots, indicating that it is conducive to the texture of bamboo shoots.

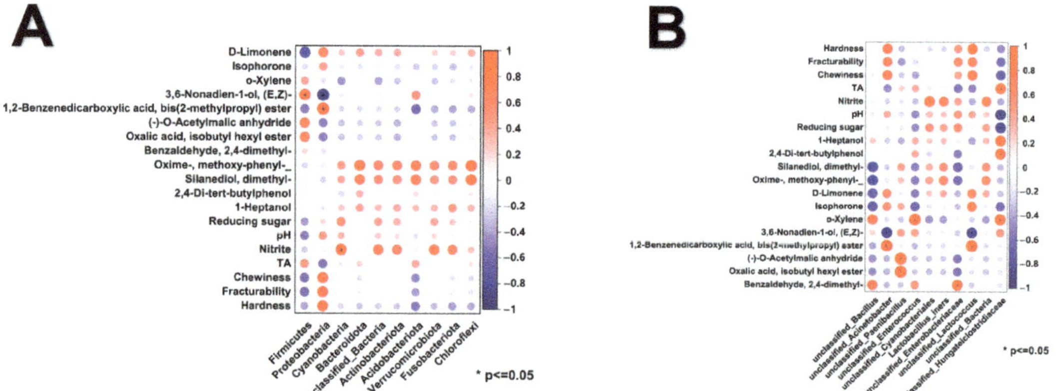

Figure 6. The relative abundance of the top 10 phyla (**A**) and genera (**B**) and the Pearson correlation analysis of volatile organic compounds (VIP ≥ 1).

As shown in Figure 6B, at the genus level, the vertical axis represents volatile flavor compounds with VIP values greater than one, as well as basic indicators such as pH, total acid, nitrite, and reducing sugar. The horizontal axis represents the top ten dominant bacterial genera. *Unclassified_Acinetobacter* and *unclassified_Lactococcus* showed a significant positive correlation with 1,2-Benzenedicarboxylic acid, bis(2-methylpropyl) ester, and a significant negative correlation with 3,6-Nonadien-1-ol, (E,Z)-, indicating that these two genera are not beneficial for the aroma of bamboo shoots [28]. *Unclassified_Paenibacillus* showed a significant positive correlation with (-)-O-Acetylmalic anhydride and Oxalic acid, isobutyl hexyl ester. *Unclassified_Enterococcus* showed a significant positive correlation with o-Xylene. *Unclassified_Cyanobacteriales* showed a significant positive correlation with nitrite. This indicates that *unclassified_Cyanobacteriales* will influence the safety of bamboo shoots. When its content increases, it will lead to an increase in nitrite content in bamboo shoots, which will have an impact on consumer's health when the nitrite content reaches a certain level [24]. *Unclassified_Enterobacteriaceae* showed a significant positive correlation with Benzaldehyde, 2,4-dimethyl-. *Unclassified_Hungateiclostridiaceae* showed a significant positive correlation with o-Xylene, 2,4-Di-tert-butylphenol, and total acid, and a significant negative correlation with pH and reducing sugar. This indicates that *unclassified_Hungateiclostridiaceae* is conducive to the formation of the flavor of bamboo shoots but not to the texture. *Unclassified_Acinetobacter* and *unclassified_Lactococcus* showed a positive correlation with the chewiness, hardness, and fracturability of bamboo shoots, indicating that both genera are beneficial to the texture of bamboo shoots.

4. Conclusions

Through the above experimental investigation, the relationship between the physicochemical properties of bamboo shoots, characteristic volatile compounds, and core microbial communities was studied. It was found in the study that there were significant changes in the physicochemical properties of sour bamboo shoots during the 60-day fermentation. 1-Heptanol was considered the most significant volatile compound contributing to the formation of bamboo shoots flavor. Firmicutes showed a significant positive correlation with 3,6-nonadien-1-ol (E, Z). Proteobacteria showed a significant positive correlation with 1,2-benzenedicarboxylic acid, bis (2-methylpropyl) ester and a significant negative correlation with 3,6-nonadien-1-ol (E, Z). Proteobacteria showed a positive correlation with the texture of sour bamboo shoots. *Unclassified_Bacillus* showed a negative correlation with Silanediol, dimethyl-, and oxime-, methoxy-phenyl-. *Classified_Acinetobacter* showed a significant positive correlation with 1,2-benzenedicarboxylic acid, bis (2-methylpropyl) ester and a significant negative correlation with 3,6-nonadien-1-ol (E, Z). The findings of this study provide valuable references for the advancement of industry-scale production of sour bamboo shoots.

Supplementary Materials: The following supporting information can be downloaded at: https://www.mdpi.com/article/10.3390/fermentation9080733/s1. Figure S1: Correlation among reducing sugar, hardness, fracturability, chewiness, TA, pH, and nitrite of sour bamboo shoots conducted using Pearson's correlation analysis; Figure S2: PCoA analysis of the microbial community in sour bamboo shoots during fermentation process; Figure S3: UPGMA analysis of the microbial community in sour bamboo shoots during fermentation process; Table S1: Volatile compounds of sour bamboo shoots during fermentation measured using GC-MS; Table S2: Alpha diversity of the microbial community in sour bamboo shoots during the fermentation process.

Author Contributions: Conceptualization, Q.C. and S.H.; investigation, Z.L., X.X. and W.D.; writing—original draft preparation, X.X. and Z.L.; writing—review and editing, Z.L. and Q.C.; funding acquisition, Y.Z. and S.H. All authors have read and agreed to the published version of the manuscript.

Funding: This research was funded by the National Key R&D Program of China (2021YFD2200504_1 and 2021YFD2200505_2), Science and Technology Project of Sichuan Province, China (2023JDRC0130, 2021YFYZ0006, and 2022NSFSC0093), and Ph.D. Foundation (no. 22zx7).

Institutional Review Board Statement: Not applicable.

Informed Consent Statement: Not applicable.

Data Availability Statement: Data is shown in the article.

Acknowledgments: The authors would like to acknowledge the National Key R&D Program of China, Sichuan Provincial Department of Science and Technology for the research project funds received for the project. The authors would like to thank the Southwest University of Science and Technology for the internal project fund to support the research.

Conflicts of Interest: The authors declare no conflict of interest.

References

1. Liu, A.; Li, X.Y.; Pu, B.; Ao, X.L.; Zhou, K.; He, L.; Chen, S.J.; Liu, S.L. Use of psychrotolerant lactic acid bacteria (*Lactobacillus* spp. and *Leuconostoc* spp.) isolated from chinese traditional paocai for the quality improvement of Paocai products. *J. Agric. Food Chem.* **2017**, *65*, 2580–2587. [CrossRef] [PubMed]
2. Qin, Y.L.; Zhao, X.Y.; Shen, Y.Y.; Yu, F.T.; Liu, X.L. Screening and identification of anti-inflammatory lactic acid bacteria in pickled bamboo shoots. *Food Sci.* **2022**, *43*, 121–127.
3. Xiao, Y.S.; Huang, T.; Huang, C.L.; Hardie, J.; Peng, Z.; Xie, M.Y.; Xiong, T. The microbial communities and flavour compounds of Jiangxi yancai, Sichuan paocai and Dongbei suancai: Three major types of traditional Chinese fermented vegetables. *LWT-Food Sci. Technol.* **2020**, *121*, 108865. [CrossRef]
4. Jeong, S.H.; Jung, J.Y.; Lee, S.H.; Jin, H.M.; Jeon, C.O. Microbial succession and metabolite changes during fermentation of dongchimi, traditional Korean watery kimchi. *Int. J. Food Microbiol.* **2013**, *164*, 46–53. [CrossRef] [PubMed]
5. Li, Q.Q.; Kang, J.M.; Ma, Z.; Li, X.P.; Liu, L.; Hu, X.Z. Microbial succession and metabolite changes during traditional serofluid dish fermentation. *LWT-Food Sci. Technol.* **2017**, *84*, 771–779. [CrossRef]
6. Liang, H.P.; Chen, H.Y.; Zhang, W.X.; Yu, C.X.; Ji, C.F.; Lin, X.P. Investigation on microbial diversity of industrial Zhacai paocai during fermentation using high-throughput sequencing and their functional characterization. *LWT-Food Sci. Technol.* **2018**, *91*, 460–466. [CrossRef]
7. Wang, Z.X.; Shao, Y.Y. Effects of microbial diversity on nitrite concentration in pao cai, a naturally fermented cabbage product from China. *Food Microbiol.* **2018**, *72*, 185–192. [CrossRef]
8. Xu, X.X.; Wu, B.B.; Zhao, W.T.; Pang, X.L.; Lao, F.; Liao, X.J.; Wu, J.H. Correlation between autochthonous microbial communities and key odorants during the fermentation of red pepper (*Capsicum annuum* L.). *Food Microbiol.* **2020**, *91*, 103510. [CrossRef]
9. Xue, B.J.; Hu, R.; Wu, L.R.; Yang, J.L.; Li, B.; Zhang, F.S.; Zheng, J. Advances in research on microbial flora succcession and flavor substance formation mechanism in Suansun. *Food Ferment. Ind.* **2023**, 1–7.
10. Chen, C.X.; Cheng, G.G.; Liu, Y.P.; Yi, Y.X.; Chen, D.J.; Zhang, L.; Wang, X.J.; Cao, J.X. Correlation between microorganisms and flavor of Chinese fermented sour bamboo shoot: Roles of *Lactococcus* and *Lactobacillus* in flavor formation. *Food Biosci.* **2022**, *50*, 101994. [CrossRef]
11. Guan, Q.Q.; Huang, T.; Peng, F.; Huang, J.Q.; Liu, Z.G.; Peng, Z.; Xie, M.Y.; Xiong, T. The microbial succession and their correlation with the dynamics of flavor compounds involved in the natural fermentation of suansun, a traditional Chinese fermented bamboo shoots. *Food Res. Int.* **2022**, *157*, 111216. [CrossRef] [PubMed]
12. *GB/T 12456-2021*; Determination of Total Acid in Foods. Standards Press of China: Beijing, China, 2021.
13. *GB 5009.7-2021*; Determination of Reducing Sugar in Foods. Standards Press of China: Beijing, China, 2021.
14. *GB 5009.33-2016*; Determination of Nitrite and Nitrate in Food. Standards Press of China: Beijing, China, 2016.
15. He, Z.; Chen, H.Y.; Wang, X.Y.; Lin, X.P.; Ji, C.F.; Li, S.J.; Liang, H.P. Effects of different temperatures on bacterial diversity and volatile flavor compounds during the fermentation of suancai, a traditional fermented vegetable food from northeastern China. *LWT-Food Sci. Technol.* **2020**, *118*, 108773. [CrossRef]
16. Drakula, S.; Mustač, N.Č.; Novotni, D.; Voučko, B.; Krpan, M.; Hruškar, M.; Ćurić, D. Optimization and validation of a HS-SPME/GC–MS method for the analysis of gluten-free bread volatile flavor compounds. *Food Anal. Methods* **2021**, *15*, 1155–1170. [CrossRef]
17. Liu, Y.Y.; Lin, H.; Yang, C.; Yang, W.Y.; Peng, B.Z. Development and quality analysis of Liuzhou sour bamboo shoots with low nitrite. *Food Ferment. Ind.* **2022**, *48*, 220–224.
18. Ren, T.; Liu, Y.L.; Peng, Y.M.; Zeng, S.; He, Y.C.; Yang, B.; Luan, X.X.; Luo, Y.L. Effect of mixed starter cultures on quality of tumorous stem mustard (*Brassica Juncea* Var. Tumida Tsen et Lee) pickle. *Food Sci. Technol.* **2020**, *45*, 29–35.
19. You, S.-Y.; Yang, J.-S.; Kim, S.H.; Hwang, I.M. Changes in the Physicochemical Quality Characteristics of Cabbage Kimchi with respect to Storage Conditions. *J. Food Qual.* **2017**, *2017*, 9562981. [CrossRef]
20. Jeyaram, K.; Romi, W.; Singh, T.A.; Devi, A.R.; Devi, S.S. Bacterial species associated with traditional starter cultures used for fermented bamboo shoot production in Manipur state of India. *Int. J. Food Microbiol.* **2010**, *143*, 1–8. [CrossRef] [PubMed]
21. Zhou, X.R.; Chen, X.Y.; Jiang, L.W.; Liu, C.G.; Huang, H. Changes of flavor substances in natural fermentation of bamboo shoots. *Chin. Brew.* **2019**, *38*, 20–24.
22. Zhuo, J.W.; Peng, S.Y.; Qin, L.; Yang, Q.; Mao, R.F. Effect of addition of red yeast rice on the properties of lactic acid fermentation agent. *Chin. Condiment* **2023**, *48*, 65–70.
23. Fan, L.Q.; Huang, Q.; Wang, Y.F. Effects of *Astragalus* extract on physicochemical properties, microbial flora, and sensory quality of low-salt naturally fermented pickles. *Food Ferment. Ind.* **2022**, *48*, 213–218.

24. Paik, H.D.; Lee, J.Y. Investigation of reduction and tolerance capability of lactic acid bacteria isolated from kimchi against nitrate and nitrite in fermented sausage condition. *Meat Sci.* **2014**, *97*, 609–614. [CrossRef]
25. Lu, H.H.; Huang, C.J.; Yu, K.; Liu, Z.M. Effects of mixed inoculation of Leuconostoc citreum and Lactobacillus plantarum on suansun (Sour bamboo shoot) fermentation. *Food Biosci.* **2022**, *47*, 101688. [CrossRef]
26. Cui, N.; Hui, X.; Wang, Y.X.; Wang, B.; Ao, Y.F.; Gong, X. Effects of different fermentation methods on the content of nitrite in sour bamboo shoots. *Food Saf. Guide* **2022**, *33*, 46–49.
27. Li, H.; Liu, F.; Kun-Farkas, G.; Kiss, Z. Quantitative analysis of flavor volatiles in beer using headspace solid-phase microextraction and gas chromatography–flame ionization detection (HS-SPME-GC-FID). *J. Am. Soc. Brew. Chem.* **2015**, *73*, 261–265. [CrossRef]
28. Corrëa Lelles Nogueira, M.; Lubachevsky, G.; Rankin, S.A. A study of the volatile composition of Minas cheese. *LWT-Food Sci. Technol.* **2005**, *38*, 555–563. [CrossRef]
29. Sompong, S.; Apirak, P.; Preecha, M. Biological activity and inhibition potential against α-glucosidase and α-amylase of 2,4-di-tert-butylphenol from bamboo shoot extract by in vitro and in silico studies. *Process Biochem.* **2022**, *126*, 15–22.
30. Cai, Y.; Jia, L.R.; Du, M.; Guo, H.Y. Analysis of volatile flavor components of Guangxi sour bamboo shoots. *Chin. Condiment* **2021**, *46*, 142–145+150.
31. Li, J.M.; Duan, H.; Zhao, R.H.; Xu, Y.; Wang, D. Study on main components of apple wine. *Food Sci.* **2007**, *28*, 362–365.
32. Liu, H.; Wang, Z.Y.; Zhang, D.Q.; Shen, Q.W.; Pan, T.; Hui, T.; Ma, J.R. Characterization of key aroma compounds in Beijing roasted duck by Gas Chromatography–Olfactometry–Mass Spectrometry, Odor-Activity Values, and Aroma-Recombination Experiments. *J. Agric. Food Chem.* **2019**, *67*, 5847–5856. [CrossRef]
33. Tian, Y.F.; Sun, M.H.; Xu, J.M.; Huang, Y.F.; Ma, Z.Z.; Ma, Y.C.; Chen, C.J.; Liu, X.L.; Li, S.B. Characterization of key aroma compounds in Guangxi fermented bamboo shoots. *Food Ferment. Ind.* **2023**, 1–10. [CrossRef]
34. Zhu, Z.H. Study on Fermented Bamboo Shoot Nutrition and Flavor Components. Master's Thesis, Guangxi University, Nanning, China, 2014.
35. Sun, Y.L.; Huang, Y.G.; Zhu, X.C.; Ma, Y.; Jiang, X.; Yin, S.M. Analysis of characteristic flavor compounds of Chinese light-soy sauce aroma type liquors. *Food Sci.* **2020**, *41*, 199–208.
36. Zhang, W.T.; Dong, P.; Lao, F.; Liu, J.Y.; Liao, X.J.; Wu, J.H. Characterization of the major aroma-active compounds in Keitt mango juice: Comparison among fresh, pasteurization and high hydrostatic pressure processing juices. *Food Chem.* **2019**, *289*, 215–222. [CrossRef] [PubMed]
37. Wu, Y. Study on Reaction Mechanism of Carbonylating Synthesis of 2,4-Dimethylbenzaldehyde from Meta-xylene. *Chem. Eng. Technol. Dev.* **2011**, *40*, 10–12+44.
38. Caporaso, J.G.; Kuczynski, J.; Stombaugh, J.; Bittinger, K.; Bushman, F.D.; Costello, E.K.; Fierer, N.; Peña, A.G.; Goodrich, J.K.; Gordon, J.I.; et al. QIIME allows analysis of high-throughput community sequencing data. *Nat. Methods* **2010**, *7*, 335–336. [CrossRef]
39. Sun, C.; Liu, Y. The distribution of lactic acid bacteria in traditional liquor production and main metabolites. *Chin. Brew.* **2012**, *31*, 1–4.
40. Chen, M.C.; Zhu, Y.J.; Liu, B.; Wang, J.P.; Liu, X.G.; Yang, W.J. Diversity research of the soil bacteria of jasminum sambac Ait based on metagenome. *J. Agric. Biotechnol.* **2018**, *26*, 1480–1493.
41. Bao, W.; Han, J.J.; Zhang, Z.X.; Lu, C.Y.; Zhou, J.; Ming, T.H.; Li, Y.; Su, X.R. Analysis of microbial diversity in traditional fermented vegetables in Zhejiang based on high-throughput sequencing technology. *Food Sci.* **2021**, *42*, 178–185.
42. Xia, F.; Chen, X.; Guo, M.Y.; Bai, X.H.; Liu, Y.; Shen, G.R.; Li, Y.L.; Lin, J.; Zhou, X.W. High-throughput sequencing-based analysis of endogenetic fungal communities inhabiting the Chinese Cordyceps reveals unexpectedly high fungal diversity. *Sci. Rep.* **2016**, *6*, 33437. [CrossRef]
43. Zhong, Y.; Zhang, Y.W.; Lv, J.M.; Chen, M.L.; Ma, H.W.; Guo, A.L.; Tang, D.P. Analysis of microbial diversity of Guangxi sour bamboo shoots and study on their lactic acid bacteria characteristics. *Chin. Condiments* **2023**, *48*, 1–9.
44. Li, X.Y. Study on the Variation Law of Microbial Community and Main Components in Natural Fermentation of Sour Bamboo Shoots. Master's Thesis, Hunan Agricultural University, Changsha, China, 2020.
45. Nirmala, C.; Bisht, M.S.; Laishram, M. Bioactive compounds in bamboo shoots: Health benefits and prospects for developing functional foods. *Int. J. Food Sci. Technol.* **2014**, *49*, 1425–1431. [CrossRef]
46. Wang, S.Q.; Bao, Y.; Dong, X.M.; Su, F.; Yao, G.Q.; Zhang, H.P. Physiological function and application of *Lactobacillus plantarum*. *China Agric. Sci. Technol. Bull.* **2010**, *12*, 49–55.
47. Liu, Y.Y. Characterization of Microflora and Their Functions on Flavor Compounds in Shaoxing Rice Wine. Master's Thesis, Jiangnan University, Wuxi, China, 2015.
48. Chen, C.; Li, J.; Cheng, G.G.; Liu, Y.P.; Yi, Y.X.; Chen, D.J.; Wang, X.J.; Cao, J.X. Flavor changes and microbial evolution in fermentation liquid of sour bamboo shoots. *J. Food Compos. Anal.* **2023**, *120*, 105273. [CrossRef]
49. Xiao, Y.S.; Xiong, T.; Peng, Z.; Liu, C.G.; Huang, T.; Yu, H.; Xie, M.Y. Correlation between microbiota and flavours in fermentation of Chinese Sichuan Paocai. *Food Res. Int.* **2018**, *114*, 123–132. [CrossRef] [PubMed]

50. Zheng, X.J.; Liu, F.; Li, K.X.; Shi, X.W.; Ni, Y.Q.; Li, B.K.; Bin, Z.G. Evaluating the microbial ecology and metabolite profile in Kazak artisanal cheeses from Xinjiang, China. *Food Res. Int.* **2018**, *111*, 130–136. [CrossRef]
51. Zhao, H.W.; Chen, Y.F.; Yan, P.M. Study on microorganisms in fermentation of pickles. *Chin. Condiments* **2022**, *47*, 211–216.

Disclaimer/Publisher's Note: The statements, opinions and data contained in all publications are solely those of the individual author(s) and contributor(s) and not of MDPI and/or the editor(s). MDPI and/or the editor(s) disclaim responsibility for any injury to people or property resulting from any ideas, methods, instructions or products referred to in the content.

foods

MDPI

Article

The Volatile Compounds Change during Fermentation of *Saccharina japonica* Seedling

Jingni Gong [1], Xiaolin Wang [2], Hui Ni [2,3,4,*] and Yonghua Wang [1,5,*]

1 School of Food Science and Engineering, South China University of Technology, Guangzhou 510640, China
2 College of Food and Biological Engineering, Jimei University, Xiamen 361021, China
3 Fujian Provincial Key Lab of Food Microbiology and Enzyme, Jimei University, Xiamen 361021, China
4 School of Marine Biology, Xiamen Ocean Vocational College, Xiamen 361000, China
5 Guangdong Youmei Institute of Intelligent Bio-Manufacturing, Foshan 528225, China
* Correspondence: nihui@jmu.edu.cn (H.N.); yonghw@scut.edu.cn (Y.W.); Tel.: +86-20-8711-3842 (H.N.)

Abstract: It is important to eliminate the fishy odor and improve the aroma quality of seafood. In this study, the *Saccharina japonica* (*S. japonica*) seedling, which is a new food material, was investigated for the effects of fermentation with *Saccharomyces cerevisiae* (*S. cerevisiae)* through sensory evaluation, GC–MS, and odor activity value (OAV) analysis. GC–MS analysis revealed the presence of 43 volatile compounds in the unfermented *S. japonica* seedling, with 1-octen-3-ol, hexanal, and trans-2,4-decadienal identified as the main contributors to its fishy odor. After fermentation with *S. cerevisiae*, 26 volatile compounds were identified in the *S. japonica* seedling. Notably, the major malodorous fish compounds, including 1-octen-3-ol, hexanal and trans-2,4-decadienal, were no longer present. The results indicate that fermentation with *S. cerevisiae* is an effective method for removing fishy malodor compounds and enhancing the volatile components with fruity, sweet, green, and floral notes in the *Saccharina japonica* seedling. This process facilitates the elimination of fishy malodor and enhance the fruity, sweet, green, and floral notes of *S. japonica* seeding and other seaweeds.

Keywords: *Saccharina japonica* seedling; aroma; volatile compounds; *Saccharomyces cerevisiae*; GC-MS

1. Introduction

Saccharina japonica (J.E. Areschoug) C.E. Lane, C. Mayes, Druehl & G.W. Saunders, also known as kelp and kombu, is a type of brown algae. *S. japonica* is a marine product that is rich in nutrients and contains a variety of functional substances, including protein, amino acids, polysaccharides, minerals, and vitamins [1,2]. *S. japonica* is particularly rich in iodine, and consuming this marine product could help prevent thyroid diseases [3,4]. In addition, *S. japonica* exhibits a variety of functional properties, including antioxidant, anti-inflammatory, hypoglycemic, and antihypertensive effects [5–9]. Therefore, *S. japonica* represents an important marine alga with the potential to provide health benefits to humans.

The aroma quality of food has a noticeable effect on the consumption preference of consumers. However, *S. japonica* products always have a strong fishy odor, which becomes the main barrier to human consumption [10,11]. In order to increase the consumption of *S. japonica*, it is essential to eliminate its fishy odor [12]. The methods of deodorization mainly include physical, chemical, and biological methods [13]. Compared with physical and chemical methods, biological methods are attracting more and more attention from researchers and manufacturers, due to their safety, environmental friendliness, and ease of implementation. Seo et al. [14] proposed fermenting *S. japonica* extract with *Aspergillus oryzae* to reduce its off-flavor. Although researchers have developed effective deodorization processes to remove the off-flavor from extracts and blends of *S. japonica*, few studies have been conducted to eliminate the fishy malodor from the foods made from *S. japonica* [14].

Recently, fermentation has been proposed to eliminate the off-flavors of *S. japonica* and improve the quality of *S. japonica* food products [15]. Nie et al. [16] demonstrated that the fermentation of *S. japonica* resulted in the reduction of fishy odorants, including hexanal and nonanal. Zhu et al. [17] demonstrated that fermentation with *Saccharomyces cerevisiae* could reduce 91% of unsaturated aldehydes from *S. japonica*. In addition, *Lactobacillus plantarum* and *Acetobacter pasteurianus* have also been used to ferment foods with desirable aromas [18,19]. In comparison, fermentation with *Saccharomyces cerevisiae* not only reduces fishy odors but also enhances fruit and floral aromas [17]. Therefore, researchers prefer to use *S. cerevisiae* to enhance the flavor of plant, seaweed, and fish products [20,21], which could be a practical method to enhance the flavor quality of food products made from *S. japonica*.

The *S. japonica* seedling is the *S. japonica* harvested at the early growth stage. It has higher nutritional and bioactivity values [22], making it increasingly important in food processing in recent years. However, *S. japonica* seedlings have a fishy odor, which limits consumer acceptance. Therefore, this study aimed to investigate the effectiveness of fermentation with *Saccharomyces cerevisiae* in removing the fishy odor from *S. japonica* seedlings. This study contributes to the understanding of the fishy odor of the novel food material *S. japonica* seedlings and provides a reference for enhancing the fruity, sweet, green, and floral notes of food products derived from *S. japonica* seedlings.

2. Materials and Methods

2.1. Chemicals

Standard 1-octen-3-ol (98%), 2-octen-1-ol (97%), 1-octanol (≥99%), anethol (≥99%), phenethyl alcohol (≥99%), 1-nonanol (≥98%), decyl aldehyde (analytical standard), undecanal (analytical standard), 1-octen-3-one (≥98%), 3-acetyl-2-octanone (≥98%), β-ionone (≥96%), 1,3-octadiene (≥95%), undecene (97%), 4-methyl-1-hendecene (97%), 8-heptadecene (98%), hexyl acetate (analytical standard), phenethyl acetate (analytical standard), hexyl butyrate (98%), ethyl caprate (≥99%), ethyl laurate (analytical standard), methyl palmitate (analytical standard), (E)-2-nonenal (≥95%), β-cyclocitral (≥95%), hexanal (95%), nonanal (96%), limonene (97%), and α-terpinene (≥95%) were purchased from Sigma Aldrich (St. Louis, MO, USA). Standard 2,4-dimethylbenzaldehyde (97%) and 2-pentylfuran (≥98%) were obtained from Alfa Aesar Co., Ltd. (Heysham, UK). A standard series of C_8–C_{20} alkanes (analytical standard) was used for retention index (RI) determination, and the internal standard cyclohexanone was purchased from Sigma Co., Ltd. (St. Louis, MO, USA).

2.2. Preparing the Algae and Yeast

A *Saccharina japonica* seedling was obtained from the FuJian province of China in 2020. It was subsequently divided into 2 cm × 2 cm pieces. *Saccharomyces cerevisiae* CICC 1464 was purchased from the China Food and Fermentation Industry Research Institute Co., Ltd. (Beijing, China). The *S. cerevisiae* strain was activated in a 250 mL flask containing 50 mL of malt extract medium (malt extract 130 g/L and chloramphenicol 0.1 g/L) using a ZQZY-CF shaker (Shanghai Chuzhi Biotechnology Co., Ltd. (Shanghai, China) at 180 rpm and 28 °C for 3 days. The *S. cerevisiae* cells were harvested by centrifugation at 1500 rpm and 4 °C for 5 min.

2.3. Fermentation Procedure

The fermentation was conducted following the methods described in the literature, with minor modifications [23]. In detail, the yeast seed was adjusted to an optical density at 600 nm (OD_{600}) of 0.8 with sterilized water. Ten grams of the *S. japonica* seedling was mixed with 100 mL of activated *S. cerevisiae* solution, followed by fermentation at 28 °C and 150 rpm for 120 min. The fermented *S. japonica* seedling was then removed and washed three times with 20 times the volume of distilled water. The seaweed was then dried at 40 °C to a moisture content of 9.5%.

2.4. Sensory Evaluation of Odor

According to the relevant literature [23] and ISO 13299:2016-05 [24], quantitative descriptive analysis (QDA) was used for the sensory evaluation of the odor. Ten panelists (5 males and 5 females) aged 22–26 years were trained to become familiar with the intensity of "fishy", "floral", "green", "fatty", "earthy", "fruity" and "sweet" notes. For the sensory evaluation of the samples, five grams of *S. japonica* seedling was placed in a 100 mL Erlenmeyer flask, and the sensory evaluation was carried out at 26 ± 2 °C in a clean environment. After each sniff, the evaluators needed a 20 s gap with fresh air to refresh their olfactory fatigue. The panelists rated the sample on a scale of 0–9, with 0 representing no perceived attribute intensity and 9 representing very strong attribute intensity. All analyses were conducted in triplicate, and the average value was considered the final score of the sample, which was then visualized on the radar plot. The rights and privacy of all participants were protected during the research. This included ensuring no coercion to participate, providing full disclosure of study requirements and risks, obtaining verbal consent from the participants, refraining from releasing participant data without their knowledge, and allowing the participants to withdraw from the study at any time.

2.5. GC-MS Analysis

Thirty milliliters of distilled water and 1 g of *S. japonica* seedling were added to an extraction vial containing 10 µL of the internal standard (cyclohexanone). The sample was equilibrated at 60 °C for 30 min, followed by extraction for 30 min with a 50/30 µm DVB/CAR/PDMS extraction fiber.

A QP2010 gas chromatography–mass spectrometry (GC-MS) system (Shimadzu, Kyoto, Japan) was equipped with an Rtx-5MS column (60 m × 0.32 mm × 0.25 µm, Restek Corporation, Bellefonte, PA, USA). Helium (99.999% purity) was used as the carrier gas. The column flow rate was 3.16 mL/min in splitless injection mode. The injection port temperature was 230 °C. The column temperature was 40 °C at the start, increased to 100 °C at a rate of 6 °C/min, and then further increased to 230 °C at a rate of 5 °C/min and held for 5 min. The temperatures of the ion source and the interface were 220 °C and 250 °C, respectively [23]. The MS was set to electron impact (EI) ionization mode with a mass scan range from 35 to 500 amu, and the electron energy was set at 70 eV. The solvent delay time was 3 min.

Mass spectra were compared to the NIST library (NIST11, NIST11s, FFNSC1.3) to identify relevant molecules. The compounds with a mass spectrometry match greater than 80% were screened out. The identification was then confirmed by combining the base peak, characteristic ion peak, and retention index (RI). The retention index was calculated based on a standard mixture of n-alkanes (C_8-C_{20}) using Equation (1):

$$RI_x = 100n + 100 \times \left(\frac{RT_x - RT_n}{RT_{n+1} - RT_n} \right) \tag{1}$$

In this equation, RI_x is the retention index of the component to be measured, n is the number of carbon atoms in the n-alkanes, RT_x is the retention time of each odorant (x), and RT_n and RT_{n+1} are the retention times of the n-alkanes eluted before and after the odorant (x) under identical chromatographic circumstances conditions.

The volatile compounds with chemical standards were quantitatively analyzed using their respective calibration curves. The volatile compounds without chemical standards were tentatively identified, and their concentrations were approximately estimated using the internal standard according to Equation (2):

$$C_x = \frac{P_x \times C_i}{P_i} \tag{2}$$

In this equation, C_x represents the concentration of the compound to be tested, P_x represents the peak area of the compound being tested, C_i represents the concentration of the internal standard, and P_i represents the peak area of the internal standard.

2.6. Analysis of odor Activity Value (OAV)

OAV was calculated by dividing the concentration of the volatile compounds by their odor threshold in water using Equation (3). The threshold value was obtained from the literature [25]. If OAV ≥ 1, the compound is considered to be an aroma compound that may be a major contributor.

$$\mathrm{OAV} = \mathrm{C_i}/\mathrm{OT_i} \tag{3}$$

In this equation, C_i is the concentration of volatile compounds and OT_i is the odor threshold.

2.7. Statistical Analysis

SPSS 19.0 (SPSS, Inc., Chicago, IL, USA) was used to calculate mean and errors and to conduct significance analysis ($p < 0.05$). Analysis of variance (ANOVA) was performed to determine differences in the amounts of the aroma compounds of the quantified volatile compounds before and after fermentation.

3. Results

3.1. Sensory Evaluation of *S. japonica* seedling before and after Fermentation

As shown in Figure 1, the *S. japonica* seedling primarily exhibited a fishy odor, accompanied by subtle notes of green, fatty, earthy, floral, and fruity scents. After fermentation with *S. cerevisiae*, the fishy odor diminished, while the green, floral, sweet, and fruity notes intensified. The results indicated that fermentation with *S. cerevisiae* could enhance the aroma of the *S. japonica* seedling. This finding is consistent with the study of Xu et al. [23].

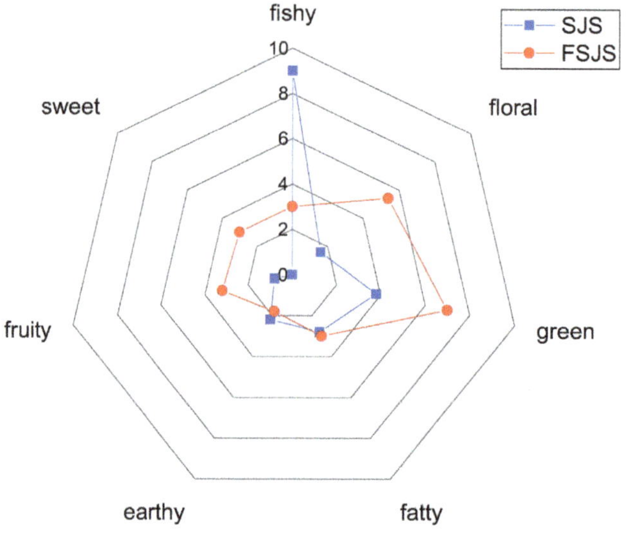

Figure 1. A spider diagram illustrating the sensory characteristics of the *S. japonica* seedling before and after fermentation (SJS refers to the *S. japonica* seedling without fermentation. FSJS is the *S. japonica* seedling after fermentation. 0 refers to unrecognized, 2 refers to can be recognized, 4 refers to weak, 6 refers to middle, 8 refers to strong, 10 refers to very strong).

3.2. Qualitative and Quantitative Analysis of Volatile Compounds before and after Fermentation

A total of 43 volatile compounds were detected in the unfermented *S. japonica* seedling, including 6 alcohols, 10 aldehydes, 10 ketones, 4 alkanes, 8 alkenes, 3 esters, and 2 other compounds (Table 1). As shown in Table 1, aldehydes and ketones are the most abundant compounds, followed by alkenes and alcohols. Among the aldehydes, 1-nonanal and hexanal are the most abundant in concentration. Ketones are the second most abundant chemical after aldehydes. Among the ketones, 2,2,6-trimethylcyclohexanone and β-ionone showed relatively high concentrations. Alkenes are the third most abundant chemical after ketones. Among the ketones, 3,5,5-trimethyl-2-hexene and 8-heptadecene exhibited the highest relative abundance in concentration. As for the alcohols, 1-octen-3-ol had the highest concentrations. These findings are consistent with those of Zhu et al. [17].

Table 1. Quantitative analysis and approximate estimates of volatile compounds (relative to cyclohexanone) in the *S. japonica* seedling before and after *S. cerevisiae* fermentation.

NO.	RT/min	Compounds	RI ^{a}	RI ^{b}	Ion Fragment	Identification	Standard Curves	R^2	Concentration (μg/g)	
									SJS	FSJS
		Alcohols								
1	9.79	1-Octen-3-ol	979	976	57 72	MS, RI, Std	$Y = 1.490988X - 0.071092$	0.999	62.4 ± 3.0	-
2	11.37	2-Octen-1-ol	1067	1069	57 68	MS, RI, Std	$Y = 6.551167X - 1.078406$	0.999	4.6 ± 1.3	-
3	12.22	1-Octanol	1198	1194	79 91	MS, RI, Std	$Y = 1.087064X - 7.655122$	0.989	2.6 ± 0.4	-
4	16.18	Cis-Anethol	1214	1213	117 147 148	MS, RI, Std	$Y = 1.325534X - 2.097888$	0.998	0.5 ± 0.3	-
5	16.31	Trans-2-decen-1-ol	1265	1273	57 67	MS, RI, Std	$Y = 3.258641X - 0.458222$	0.994	32.1 ± 3.4	-
6	19.03	Phenethyl alcohol	1299	1277	57 71	MS, RI, Std	$Y = 1.505271X - 1.243503$	0.999	-	3.8 ± 1.1
7	25.08	1-Nonanol	1568	1569	55 56 70	MS, RI, Std	$Y = 2.986852X - 1.702484$	0.993	0.6 ± 0.1	-
8	27.50	Cedrol	1601	1601	95 150 222	MS, RI, Std	$Y = 2.743444X - 1.758543$	0.994	-	8.2 ± 0.2
		Aldehydes								
1	5.81	Hexanal	801	800	56 72	MS, RI, Std	$Y = 15.83681X - 0.051104$	0.995	24.5 ± 0.5	-
2	10.55	1-Nonanal	1103	1102	57 71	MS, RI, Std	$Y = 6.034336X - 0.225634$	0.993	47.4 ± 1.7 ^{c}	11.6 ± 1.5 ^{d}
3	10.68	2-Nonenal	1157	1158	55 70	MS, RI, Std	$Y = 11.75698X - 0.063402$	0.995	1.3 ± 0.7	-
4	10.93	2,4-Dimethylbenzaldehyde	1168	1175	105 133 134	MS, RI, Std	$Y = 2.029053X - 3.532768$	0.994	0.8 ± 0.1	-
5	13.50	Oct-2-Enal	1192	1196	41 81 110	MS, RI, Std	$Y = 1.023356X - 5.775350$	0.999	2.8 ± 0.6 ^{c}	12.0 ± 2.2 ^{d}
6	14.93	Decyl aldehyde	1204	1204	57 70	MS, RI, Std	$Y = 1.649864X - 3.354954$	0.997	3.0 ± 0.8 ^{c}	7.1 ± 0.4 ^{d}
7	15.59	β-Cyclocitral	1212	1214	109 137 152	MS, RI, Std	$Y = 3.236918X + 0.228739$	0.993	3.8 ± 0.3 ^{c}	20.5 ± 2.9 ^{d}
8	18.94	2,6,6-Trimethyl-1-Cyclohexene-1-Acetaldehyde	1248	1251	107 151 166	MS, RI	Estimated	-	1.3 ± 0.4 ^{c}	8.1 ± 1.2 ^{d}
9	19.25	Undecanal	1305	1308	57 82	MS, RI, Std	$Y = 1.493987X - 0.218564$	0.996	0.9 ± 0.1	-
10	25.06	Trans-2,4-decadienal	1403	1402	81 95 152	MS, RI, Std	$Y = 2.045690X + 0.061021$	0.999	3.4 ± 0.3	-
11	26.03	Dodecyl aldehyde	1412	1412	57 82	MS, RI, Std	$Y = 0.743598X - 0.101732$	0.993	2.3 ± 0.6 ^{c}	4.1 ± 0.4 ^{c}
		Ketones								
1	10.13	1-Octen-3-one	973	974	30 41 43	MS, RI, Std	$Y = 0.396353X - 3.222123$	0.999	7.1 ± 0.1	-
2	10.15	2,2,6-Trimethylcyclohexanone	1030	1027	56 82 140	MS, RI, Std	$Y = 3.682041X - 1.250163$	0.993	49.3 ± 1.7	-
3	11.59	3-Methyl-2-cyclohexen-1-one	1053	1055	54 82 110	MS, RI, Std	$Y = 0.721058X - 3.702106$	0.999	19.3 ± 0.4	-
4	11.89	3,5-Octadien-2-one	1090	1081	81 95 124	MS, RI, Std	$Y = 4.146752X - 0.034237$	0.995	8.1 ± 2.2	-
5	12.15	3-Acetyl-2-octanone	1114	1118	54 82 138	MS, RI, Std	$Y = 2.084362X - 0.239612$	0.994	1.3 ± 0.5	-
6	12.74	2,6,6-Trimethyl-2-cyclohexanedione	1139	1139	68 96 152	MS, RI	Estimated	-	3.8 ± 0.4	-
7	21.85	2-Methyloctan-3-one	1400	1407	58 71 198	MS, RI, Std	$Y = 6.201378X - 0.920482$	0.998	3.0 ± 0.3 ^{d}	1.3 ± 0.6 ^{d}
8	21.92	β-Ionone	1415	1412	135 177 192	MS, RI, Std	$Y = 4.936785X - 0.014371$	0.997	42.8 ± 3.1 ^{c}	108.4 ± 5.6 ^{d}
9	22.14	4-(2,6,6-trimethylcyclohexen-1-yl) butan-2-one	1425	1433	121 161 194	MS, RI	Estimated	-	2.8 ± 0.7	-
10	23.23	5,9-Undecadien-2-one, 6,10-dimethyl-	1443	1451	69 136 194	MS, RI	Estimated	-	6.3 ± 0.8	-
		Alkanes						-		
1	21.64	n-Pentadecane	1499	1500	57 71 212	MS, RI	Estimated	-	2.0 ± 0.2 ^{c}	5.2 ± 1.7 ^{c}
2	21.82	n-Hexadecane	1600	1600	57 71 226	MS, RI	Estimated	-	0.9 ± 0.1 ^{c}	4.7 ± 1.5 ^{d}
3	21.87	n-Heptadecane	1700	1700	57 71 240	MS, RI	Estimated	-	37.8 ± 8.5 ^{c}	23.3 ± 4.2 ^{d}
4	24.97	n-Dodecane	1199	1200	57 71 170	MS, RI	Estimated	-	-	6.1 ± 2.0
5	26.55	Iodoheptane	1261	1260	57 71 85	MS, RI	Estimated	-	5.2 ± 1.0	-
6	27.13	2,6,11-Trimethyldodecane	1274	1275	57 71 212	MS, RI	Estimated	-	-	20.7 ± 4.2
7	27.94	Butyl-nonane	1282	-	57 71 184	MS, RI	Estimated	-	-	2.5 ± 1.4
8	28.68	4,6-Dimethyl dodecane	1320	1325	57 71 113	MS, RI	Estimated	-	-	7.6 ± 2.3
9	32.01	Tetradecane	1399	1400	57 71 198	MS, RI	Estimated	-	-	3.8 ± 1.4
		Alkenes								
1	4.05	5,5-Dimethyl-2-ethyl-1,3-cyclopentadiene	835	840	79 107 122	MS, RI	Estimated	-	5.4 ± 1.3	-
2	8.32	3,5,5-Trimethyl-1-hexene	967	968	57 70 126	MS, RI, Std	$Y = 0.985246X - 0.512014$	0.999	3.9 ± 0.6	-

Table 1. *Cont.*

NO.	RT/min	Compounds	RI [a]	RI [b]	Ion Fragment	Identification	Standard Curves	R^2	Concentration (μg/g)	
									SJS	FSJS
3	8.91	3,5,5-Trimethyl-2-hexene	971	985	57 72	MS, RI, Std	$Y = 2.150213X - 3.251024$	0.997	68.4 ± 10.4	-
4	10.13	Limonene	997	995	68 93 136	MS, RI, Std	$Y = 1.494374X - 0.117216$	0.998	7.4 ± 1.0	-
5	10.65	α-Terpinene	1012	1014	55 69	MS, RI, Std	$Y = 2.10673X - 0.0584598$	0.998	10.4 ± 0.4	-
6	29.87	3-Methyl-6-(1-methyl vinyl)-cyclohexene	1680	1680	55 69 238	MS, RI	Estimated	-	13.7 ± 20.7	-
7	30.28	1,3-Octadiene	1692	1692	55 97 238	MS, RI, Std	$Y = 6.231025X - 0.045231$	0.999	7.8 ± 0.2	-
8	34.69	D-limonene	1729	1728	68 93 136	MS, RI, Std	$Y = 6.131623X - 3.776805$	0.997	-	8.3 ± 3.5
9	37.83	1-Undecene	1791	1791	55 69 97	MS, RI, Std	Estimated	-	-	3.2 ± 0.3
10	38.23	4-Methyl-1-hendecene	1799	1785	57 71	MS, RI, Std	$Y = 0.453254X - 4.741452$	0.995	-	15.5 ± 2.6
11	40.28	Heptadecene	1821	1822	55 97 238	MS, RI, Std	$Y = 4.965865X - 0.552135$	0.998	31.0 ± 3.1	-
		Esters								
1	12.52	Hexyl acetate	1004	1004	43 56 84	MS, RI, Std	$Y = 0.886523X - 2.443578$	0.999	5.1 ± 1.3	-
2	16.87	Phenethyl acetate	1283	1281	43 91 105	MS, RI, Std	$Y = 3.412776X - 0.743251$	0.999	-	8.7 ± 2.4
3	18.12	Hexyl butyrate	1282	1284	33 61 92	MS, RI, Std	$Y = 5.120312X - 0.635227$	0.996	0.6 ± 1.5	-
4	18.43	Ethyl caprate	1289	1287	43 58 97	MS, RI, Std	$Y = 1.523712X - 0.641224$	0.998	-	2.3 ± 1.2
5	23.31	Ethyl laurate	1724	1725	45 63 112	MS, RI, Std	$Y = 3.265843X - 0.536241$	0.999	-	3.6 ± 1.4
6	32.03	Methyl palmitate	1827	1827	43 60 102	MS, RI, Std	$Y = 4.776543X - 3.156005$	0.993	65.3 ± 8.2 [c]	44.3 ± 10.7 [d]
		Others								
1	9.62	2-Pentylfuran	988	987	81 138	MS, RI, Std	$Y = 1.063317X - 0.033126$	0.994	18.9 ± 3.1	-
2	18.81	2-Methyl-2-allylphenol	1343	1349	105 133 148	MS, RI	Estimated	-	-	2.7 ± 0.6
3	22.16	1,6-Dimethylnaphthalene	1503	1502	141 155 156	MS, RI	Estimated	-	0.4 ± 0.5	-

Note: SJS refers to unfermented *S. japonica* seedling; FSJS refers to fermented *S. japonica* seedling; RT represents the retention time of aroma compounds on the Rtx-5MS column. RI [a] was calculated based on the retention time of C_8–C_{20}. RI [b] was obtained from the web: http://webbook.nist.gov/chemistry/ (accessed on 20 March 2020). Identification method: MS, mass spectrum comparison using (NIST11, NIST11s, FFNSC1.3) library; RI: retention index in agreement with literature value; Std: validated with authentic standards. Estimations suggest that the unquantified volatile compounds were treated as equivalent units of internal standard added to a sample matrix. [c,d] Signal intensities are shown as mean ± standard deviation (n = 3). Significant differences were performed using Tukey's test ($p < 0.05$) in SPSS 19.0.

The fermented *S. japonica* seedling was found to contain 26 volatile compounds (Table 1), including 2 alcohols, 6 aldehydes, 2 ketones, 8 alkanes, 3 alkenes, 4 esters, and 1 other compound. Fermentation affects the concentration of volatile compounds. As shown in Figure 2, the quantities of alcohols, aldehydes, alkanes, esters, ketones, and others decrease. Among these, alcohols and alkenes showed a significant decrease. Alkanes are the only volatile compounds that behaved differently, increasing by almost a third. A comparative analysis of the volatile compounds present in the *S. japonica* seedlings before and after fermentation revealed significant variations in the composition of alcohols, aldehydes, and ketones (Table 1). For example, the content of alcohols such as 1-octen-3-ol, 2-octen-1-ol, 1-octanol, cis-anethol, trans-2-decen-1-ol, and 1-nonanol completely disappeared after fermentation and was replaced by phenethyl alcohol and cedrol. The content of aldehydes, such as hexanal, 2-nonenal, 2,4-dimethylbenzaldehyde, undecanal, and trans-2,4-decadienal, was completely depleted, while the amount of 1-nonanal decreased from 47.4 ± 1.7 to 11.6 ± 1.5 μg/g after fermentation. Most of the ketones disappeared after fermentation, such as 1-octen-3-one, 2,2,6-trimethylcyclohexanone, 3-methyl-2-cyclohexen-1-one, 3,5-octadien-2-one, 3-acetyl-2-octanone, 2,6,6-trimethyl-2-cyclohexanedione, 4-(2,6,6-trimethylcyclohexen-1-yl) butan-2-one, and 5,9-undecadien-2-one, 6,10-dimethyl.

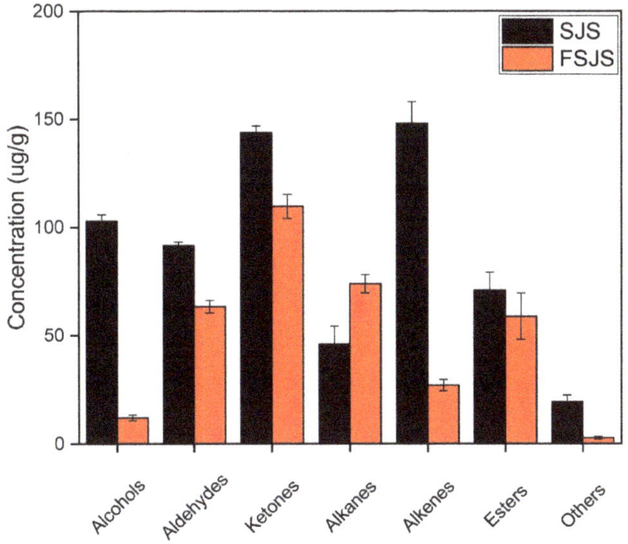

Figure 2. Volatile compound concentrations in *S. japonica* seedling before and after fermentation by *S. cerevisiae*. SJS means unfermented *S. japonica* seedling; FSJS means fermented *S. japonica* seedling.

3.3. OAV Analysis of Volatile Compounds of S. japonica Seedling before and after Fermentation

The odor activity value (OAV) is the ratio of the compound concentration to its flavor threshold [26]. An OAV $\geq$ 1 indicates that the compound is a significant contributor to the overall odor [27]. As shown in Table 2, the *S. japonica* seedling before fermentation had 14 compounds with OAV $\geq$ 1, including 1-octen-3-ol, 2-octen-1-ol, 1-octanol, 1-nonanol, hexanal, 1-nonanal, 2-nonenal, 2,4-dimethylbenzaldehyde, decyl aldehyde, β-cyclocitral, trans-2,4-decadienal, dodecyl aldehyde, β-ionone, and 2-pentylfuran. Among these compounds, 1-octen-3-ol, 1-nonanal, and trans-2,4-decadienal, which had fishy odors, had the highest OAV values. Obviously, these three compounds are the main contributors to the fishy odor of the *S. japonica* seedling. This result is consistent with the findings of Nie et al. [16] and Xu et al. [23].

After fermentation, the *S. japonica* seedling had 7 compounds with OAV $\geq$ 1, including 1-nonanal, decyl aldehyde, β-cyclocitral, dodecyl aldehyde, and β-ionone. Among these volatiles, 1-nonanal exhibited earthy and fatty odors; decyl aldehyde had ocean, cucumber, and herbal scents; β-cyclocitral emitted liquorice, fruity, and fresh aromas; dodecyl aldehyde presented herbal, fatty, and soap odors; and β-ionone showed floral and raspberry odors. Moreover, after fermentation, the OAVs of 1-octen-3-ol, hexanal, and trans-2,4-decadienal decreased to 0, and the OAV of 1-nonanal decreased from 43,093 to 10,541. This indicates that the fishy malodor of the fermented *S. japonica* seedling was dramatically reduced, while the floral and herbal notes were significantly enhanced.

Combining Figure 1 with Figure 3A, the OAV results are consistent with sensory properties. Unfermented *S. japonica* seedlings mainly present a fishy odor, which is reflected in OAV of trans-2,4-decadienal, 1-nonanal, and 1-octen-3-ol showing higher values. After fermentation, the sensory properties of the *S. japonica* seedling exhibited stronger notes of green, floral, sweet, and fruity. The significantly higher levels of β-ionone, β-cyclocitral, and decyl aldehyde in the OAV support their role as contributors to green, floral, sweet, and fruity notes.

Table 2. OAVs of the volatile compounds of the *S. japonica* seedling before and after fermentation.

No.	Volatile Compounds	Odor Description [a]	Threshold Value [b] (μg/kg)	OAV Value	
				SJS	FSJS
1	1-Octen-3-ol	Mushroom, fatty, fishy	1.5	41,603 ± 1.995	-
2	2-Octen-1-ol	Mushroom, fatty	840	5 ± 1.538	-
3	1-Octanol	Lemon, floral	0.13	20 ± 3.204	-
4	1-Nonanol	Earthy, fatty	280	2 ± 0.355	-
5	Hexanal	Herbal, fishy	5	4895 ± 0.104	-
6	1-Nonanal	Herbal, paint, fatty, fishy	1.1	43,093 ± 1.628 [c]	10,541 ± 1.355 [d]
7	2-Nonenal	Fishy, herbal	0.4	3253 ± 1.751	-
8	2,4-Dimethylbenzaldehyde	Bitter Almond	200	4 ± 0.500	-
9	Decyl aldehyde	Ocean, cucumber, herbal	1.97	1.11 ± 0.30 [c]	2.615 ± 0.138 [d]
10	β-Cyclocitral	Liquorice, fruity, fresh	3	1274 ± 0.103 [c]	6831 ± 0.974 [d]
11	Undecanal	Fruity	12.5	72 ± 0.004	-
12	Trans-2,4-decadienal	Fishy, fatty	10	340 ± 30.002	-
13	Dodecyl aldehyde	Herbal, fatty, soap	2	1150 ± 0.295 [c]	2050 ± 0.201 [d]
14	3,5-Octadien-2-one	Woody, sweet	150	54 ± 0.006	-
15	β-Ionone	Floral, raspberry	2	21,431 ± 0.441 [c]	54,243 ± 0.803 [d]
16	D-Limonene	Lemon, citrusy	200	-	42 ± 0.015
17	Methyl palmitate	Fatty	2000	33 ± 0.006 [c]	22 ± 0.013 [d]
18	2-Pentylfuran	Beany, spicy	5.8	3204 ± 0.521	-

Note: SJS refers to unfermented *S. japonica* seedling; FSJS refers to fermented *S. japonica* seedling. [a] Odor description refers to the Odor Database (http://www.odour.org.uk/). [b] Threshold values were according to the reported compilations of odor threshold values in air, water, and other media [25] [c,d] Signal intensities were shown in mean ± standard deviation (n = 3). Significant differences were performed using Tukey's test ($p < 0.05$) in SPSS 19.0.

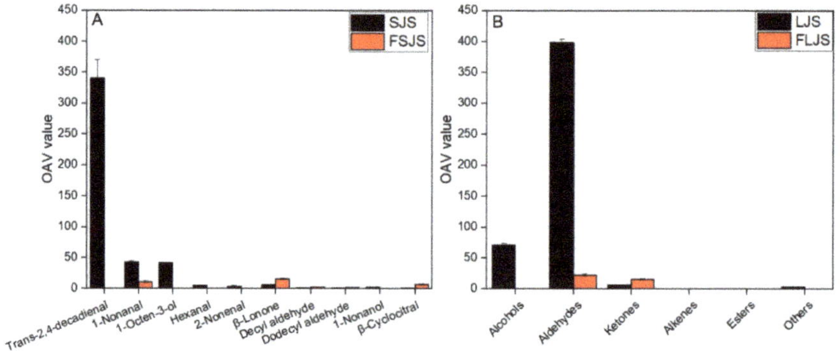

Figure 3. OAVs of the volatile compounds of the *S. japonica* seedling before and after *S. cerevisiae* fermentation. SJS means unfermented *S. japonica* seedling; FSJS means fermented *S. japonica* seedling (A refers to the OAV value of volatile compounds relate with sensory properties; B refers to the OAV value of alcohols, aldehydes, ketones, alkenes and esters).

Comparing the OAV before and after fermentation, as shown in Figure 3B, it is evident that alcohols, aldehydes, and ketones make significant contributions to the overall aroma. Aldehydes exhibit the highest OAV, indicating that aldehydes are the main contributors to the odor of the *S. japonica* seedling. After fermentation, the OAV of alcohols decreased to 0, while aldehydes decreased and ketones increased slightly. The change in the OAV of the *S. japonica* seedling after fermentation indicates that aldehydes and ketones are the primary contributors to the fermented *S. japonica* seedling odor. The OAV of alkanes, alkenes, and esters equals to 0, essentially indicating that these three compounds do not significantly contribute to the odor of the *S. japonica* seedling.

4. Discussion

It is important to eliminate the fishy odor in seafood, as this is a practical approach to enhance the flavor quality of seafood and encourage consumption [21,28–30]. The *S. japonica* seedling, which is the *S. japonica* harvested at the early growth stage and considered a new food material, was first investigated in terms of its aroma profile, aroma-active volatiles, and aroma enhancement through fermentation.

One of the objectives of this study was to identify the volatile compounds that contribute to the off-flavor of the *S. japonica* seedling. The result of this study, showing that the *S. japonica* seedling has 43 volatile compounds, is consistent with the findings of Zhu et al. [17]. Among these volatile compounds, 1-nonanal and hexanal were the most abundant in concentration. Furthermore, 1-nonanal, hexanal, and 1-octen-3-ol were the main contributors to the fishy odor of the *S. japonica* seedling. Nie et al. [16] reported that 1-octen-3-ol is a typical volatile compound contributing to the fishy odor in kelp. Peinado et al. [31] reported that hexanal and nonanal play an important role in determining the fishy odor of seaweed samples. Thus, this study identifies the presence of 1-octen-3-ol, trans-2,4-decadienal, 1-octen-3-one, and hexanal as responsible for the pronounced fishy odor in the *S. japonica* seedling, which is identical to the fishy odor reported in previous studies by Nie et al. [16] and Peinado et al. [31]. This is the first study on the volatile compounds of the *S. japonica* seedling that adds to our understanding of aroma volatile compounds in seafood.

The second aim of the study was to investigate the effectiveness of fermentation with *S. cerevisiae* in enhancing the aroma of the *S. japonica* seedling. Several groups of volatile compounds increased after fermentation, including β-ionone (108.4 μg/g), β-cyclocitral (20.5 μg/g), D-limonene (8.3 μg/g), dodecyl aldehyde (4.1 μg/g), and decyl aldehyde (7.1 μg/g). OAV results indicated that these volatile compounds contributed floral and fruity notes. The sensory characteristics of the fermented *S. japonica* seedling showed enhanced notes of green, floral, sweet, and fruity aromas. Thus, this study identifies β-ionone, β-cyclocitral, DL-limonene, dodecyl aldehyde, and decyl aldehyde as the key fragrance components in the *S. japonica* seedling. Liang et al. [32] reported that *Gracilaria lemaneiformis* produced floral and fruity aroma compounds after fermentation, with β-cyclocitral being the main aroma contributor. Xu et al. [23] reported that D-limonene is the main sweet contributor of *Bangia fusco-purpurea*. β-Ionone, β-cyclocitral, D-limonene, dodecyl aldehyde, and decyl aldehyde belong to terpenoids. The increase in terpenoids after fermentation may be due to the mevalonic acid (MVA) pathway in *S. cerevisiae*. The MVA pathway involves a series of enzymatic reactions starting with acetyl-coenzyme A (acetyl-CoA) and culminating in the production of isopentenyl diphosphate (IPP) and dimethylallyl diphosphate (DMAPP) [33,34]. Enzymes pivotal to this pathway include mevalonate kinase, phosphomevalonate kinase, and mevalonic acid dehydrogenase [35]. IPP and DMAPP are the basis for the synthesis of all terpenoid compounds [36].

The results showed that fishy odor contributors such as 1-octen-3-ol, trans-2,4-decadienal, 1-octen-3-one, and hexanal completely disappeared after fermentation with *S. cerevisiae*. Additionally, the concentration of 1-nonanal decreased from 47.4 $\pm$ 1.7 to 11.6 $\pm$ 1.5 μg/g. The results indicated that the fishy odor of the *S. japonica* seedling was dramatically weakened after fermentation with *S. cerevisiae*. This finding is consistent with the studies of Xu et al. [23] and Liang et al. [32], which demonstrated that fermentation with *Saccharomyces cerevisiae* effectively eliminated the fishy odor from *Bangia fusco-purpurea* and *Gracilaria lemaneiformis*, respectively, confirming the potential of *S. cerevisiae* fermentation to enhance the flavor profile of various seaweed species such as *S. japonica* seedlings. This finding is partially consistent with that of Zhu et al. [17], who found that the fishy odorant 1-octen-3-one disappeared from kelp after fermentation with *S. cerevisiae*, accompanied by the formation of 3-octanone. 1-Octen-3-ol, hexanal, heptanal, nonanal, and 2,4-heptadienal, which have been reported as the fishy odorants [16,23,31], are mainly derived from the oxidation of unsaturated fatty acids [37]. Unsaturated aldehydes and ketones, such as 1-octen-3-one, 2,6-nonadienal, 2,4-decadienal, and 2-nonenal, have been identified as the contributors to the fishy odor of kelp [17]. Microbial fermentation has been utilized to deodorize seaweed, fish, and other seafood [38–40], as enone reductase catalyzes the reduction of unsaturated bonds [17]. In addition, the reduction of fishy odor is related to the reactions of ester synthesis, dehydrogenation, reduction, and deformylation–oxygenation catalyzed by enone reductase, oxidases, aldehyde deformylating oxygenase, aldehyde dehydrogenase, ester synthase, aldoketo reductase, alcohol dehydrogenases, epoxide hydrolase, dehydrogenase, and acyltransferase [23]. Moreover, the disappearance of 1-octen-3-ol may be related to the catalysis of enzymes such as ADH, ALDH, ES, and ATF, which can convert 1-octen-3-ol to esters [41]. Enzymes could catalyze the hexanal to alcohol and esters through the reactions of reduction, dehydrogenation, and ester synthesis [42]. In addition, 1-nonanal could be converted to fatty acids and enol through ester synthesis, dehydrogenation, and reduction reactions in yeast [43]. Thus, the removal of the fishy odor and the enhancement of the overall aroma profile could be attributed to the bioreactions involving reduction, dehydrogenation, and ester synthesis during fermentation with *S. cerevisiae*. This will facilitate the understanding of how microorganism metabolism enhances the flavor of seaweeds, such as *S. japonica* seedlings. However, it is still unclear as to what the transformation pathways are for the eliminating or removing the fishy odor. Future research is required to explore how the fishy odors evolve during the fermentation process with *S. cerevisiae*.

5. Conclusions

The new food material *S. japonica* seedling was sensorily evaluated and found to have a strong fishy odor. Fermentation with *S. cerevisiae* is effective in eliminating the fishy note and enhancing the fruity, sweet, green, and floral notes of *S. japonica*. After fermentation, the number of volatile compounds decreased from 43 to 26; compounds such as 1-octen-3-ol, 2-octen-1-ol, hexanal, 2-nonenal, and 1-octen-3-one disappeared, while 1-nonanal decreased from 47.4 to 11.6 μg/g. On the contrary, some volatile compounds representing aroma increased after fermentation: β-ionone increased from 42.8 to 108.4 μg/g, β-cyclocitral from 3.8 to 20.5 μg/g, D-limonene from 0 to 8.3 μg/g, dodecyl aldehyde from 2.3 to 4.1 μg/g, and decyl aldehyde from 3.0 to 7.1 μg/g. The OAV revealed 14 major odor compounds (OAV $\geq$ 1), including 1-octen-3-ol, 1-nonanal, and trans-2,4-decadienal, which contribute to the strong fishy malodor; and β-ionone, β-cyclocitral, dodecyl aldehyde, and decyl aldehyde, which contribute to the floral and fruity aroma. After fermentation, the OAV of these fishy odorants decreases to 0, while the floral and fruity aroma increases. Fermentation with *S. cerevisiae* is a practical method to enhance the flavor quality of the new food material *S. japonica* seedling.

Author Contributions: J.G.: Conceptualization, data curation, formal analysis, investigation, methodology, validation, visualization, writing—original draft, writing—review and editing. H.N.: Conceptualization, methodology, project administration, writing—review and editing. X.W.: Data curation, investigation. Y.W.: Funding acquisition, supervision. All authors have read and agreed to the published version of the manuscript.

Funding: This research was funded by the Fujian Provinceal Project for Promoting High-Quality Development of Marine and Fishery Industries, funding number: FJHYF-L-2023-3.

Institutional Review Board Statement: Not applicable.

Informed Consent Statement: Not applicable.

Data Availability Statement: The original contributions presented in the study are included in the article, further inquiries can be directed to the corresponding author.

Conflicts of Interest: The authors declare no conflicts of interest.

References

1. Mouritsen, O.G.; Rhatigan, P.; Pérez-Lloréns, J.L. World cuisine of seaweeds: Science meets gastronomy. *Int. J. Gastron. Food S.* **2018**, *14*, 55–65. [CrossRef]
2. Park, E.; Yu, H.; Lim, J.-H.; Hee Choi, J.; Park, K.-J.; Lee, J. Seaweed metabolomics: A review on its nutrients, bioactive compounds and changes in climate change. *Food Res. Int.* **2023**, *163*, 112221. [CrossRef] [PubMed]

3. Aoe, S.; Yamanaka, C.; Ohtoshi, H.; Nakamura, F.; Fujiwara, S. Effects of Daily Kelp (*Laminaria japonica*) Intake on Body Composition, Serum Lipid Levels, and Thyroid Hormone Levels in Healthy Japanese Adults: A Randomized, Double-Blind Study. *Mar. Drugs* **2021**, *19*, 352. [CrossRef] [PubMed]
4. Miyai, K.; Tokushige, T.; Kondo, M.; Iodine Res, G. Suppression of Thyroid Function during Ingestion of Seaweed "Kombu" (*Laminaria japonoca*) in Normal Japanese Adults. *Endocr. J.* **2008**, *55*, 1103–1108. [CrossRef] [PubMed]
5. Wu, J.; Wang, H.; Liu, Y.; Xu, B.; Du, B.; Yang, Y. Effect of Ultrasonic Irradiation on the Physicochemical and Structural Properties of *Laminaria japonica Polysaccharides* and Their Performance in Biological Activities. *Molecules* **2023**, *28*, 8. [CrossRef] [PubMed]
6. Cho, J.-Y.; Khan, M.N.A.; Hong, Y.-K. Anti-inflammatory Activities of *Undaria pinnatifida* and *Laminaria japonica* (*Phaeophyta*). *Can. J. Fish. Aquat.Sci* **2007**, *10*, 127–132. [CrossRef]
7. Kim, K.-B.-W.-R.; Kim, M.-J.; Ahn, D.-H. Anti-inflammatory Activity of an Ethanol Extract of Laminaria japonica Root on *Lipopolysaccharide*-induced Inflammatory Responses in RAW 264.7 Cells. *Korean J. Food Sci. Technol.* **2014**, *46*, 729–733. [CrossRef]
8. Tong, A.; Li, Z.; Liu, X.; Ge, X.; Zhao, R.; Liu, B.; Zhao, L.; Zhao, C. Laminaria *japonica polysaccharide* alleviates type 2 diabetes by regulating the microbiota-gut-liver axis: A multi-omics mechanistic analysis. *Int. J. Biol. Macromol.* **2023**, *258*, 128853. [CrossRef] [PubMed]
9. Patra, K.; Das, G.; Baek, H. Chemical composition and antioxidant and antibacterial activities of an essential oil extracted from an edible seaweed, *Laminaria japonica* L. *Molecules* **2015**, *20*, 12093–12113. [CrossRef]
10. Li, S.; Hu, M.; Tong, Y.; Xia, Z.; Tong, Y.; Sun, Y.; Cao, J.; Zhang, J.; Liu, J.; Zhao, S.; et al. A review of volatile compounds in edible macroalgae. *Food Res. Int.* **2023**, *165*, 112559. [CrossRef]
11. Yang, Z.; Li, X.; Yu, M.; Jiang, S.; Qi, H. Effects of Different Processing Methods on the Quality and Physicochemical Characteristics of *Laminaria japonica*. *Foods* **2023**, *12*, 1619. [CrossRef] [PubMed]
12. Ferraces-Casais, P.; Lage-Yusty, M.A.; Rodríguez-Bernaldo de Quirós, A.; López-Hernández, J. Rapid identification of volatile compounds in fresh seaweed. *Talanta* **2013**, *115*, 798–800. [CrossRef] [PubMed]
13. Liu, Y.; Huang, Y.; Wang, Z.; Cai, S.; Zhu, B.; Dong, X. Recent advances in fishy odour in aquatic fish products, from formation to control. *Int. J. Food Sci. Tech.* **2021**, *56*, 4959–4969. [CrossRef]
14. Seo, Y.-S.; Bae, H.-N.; Eom, S.-H.; Lim, K.-S.; Yun, I.-H.; Chung, Y.-H.; Jeon, J.-M.; Kim, H.-W.; Lee, M.-S.; Lee, Y.-B.; et al. Removal of off-flavors from sea tangle (*Laminaria japonica*) extract by fermentation with Aspergillus oryzae. *Bioresour. Technol.* **2012**, *121*, 475–479. [CrossRef] [PubMed]
15. Yue, Q.; Wang, Z.; Yu, F.; Tang, X.; Su, L.; Zhang, S.; Sun, X.; Li, K.; Zhao, C.; Zhao, L. Changes in metabolite profiles and antioxidant and hypoglycemic activities of *Laminaria japonica* after fermentation. *LWT Food Sci. Technol.* **2022**, *158*, 113–122. [CrossRef]
16. Nie, J.; Fu, X.; Wang, L.; Xu, J.; Gao, X. Impact of *Monascus purpureus* fermentation on antioxidant activity, free amino acid profiles and flavor properties of kelp (*Saccharina japonica*). *Food Chem.* **2023**, *400*, 133990. [CrossRef] [PubMed]
17. Zhu, W.; Jiang, B.; Zhong, F.; Chen, J.; Zhang, T. Effect of Microbial Fermentation on the Fishy-Odor Compounds in Kelp (*Laminaria japonica*). *Foods* **2021**, *10*, 2532. [CrossRef] [PubMed]
18. Xu, H.; Xiao, N.; Xu, J.; Guo, Q.; Shi, W. Effect of *Lactobacillus plantarum* and flavourzyme on physicochemical and safety properties of grass carp during fermentation. *Food Chem. X* **2022**, *15*, 100392. [CrossRef] [PubMed]
19. Zheng, Y.; Liu, H.; Zhang, K.; Wang, M. Headspace Solid-Phase Microextraction Analysis of Volatile Compounds in Hawthorn Vinegars Fermented by two Strains of Acetobacter. In Proceedings of the 2010 3rd International Conference on Biomedical Engineering and Informatics (BMEI 2010), Yantai University, Yantai, China, 16–18 October 2010; pp. 2057–2061.
20. Du, X.; Xu, Y.; Jiang, Z.; Zhu, Y.; Li, Z.; Ni, H.; Chen, F. Removal of the fishy malodor from Bangia fusco-purpurea via fermentation of *Saccharomyces cerevisiae*, *Acetobacter pasteurianus*, and *Lactobacillus plantarum*. *J. Food Biochem.* **2021**, *45*, e13728. [CrossRef]
21. Hosoglu, M.I. Aroma characterization of five microalgae species using solid-phase microextraction and gas chromatography–mass spectrometry/olfactometry. *Food Chem.* **2018**, *240*, 1210–1218. [CrossRef]
22. Ni, H.; Wang, X.; Jiang, Z. Nutritional, taste and texture characteristics of juvenile and adult kelp. *Food Sci.* **2022**, *43*, 34–40. [CrossRef]
23. Xu, Y.-X.; Jiang, Z.-D.; Du, X.-P.; Zheng, M.-J.; Fan-Yang, Y.; Ni, H.; Chen, F. The identification of biotransformation pathways for removing fishy malodor from Bangia fusco-purpurea using fermentation with Saccharomyces cerevisiae. *Food Chem.* **2022**, *380*, 132103. [CrossRef] [PubMed]
24. *ISO PN-EN ISO 13299:2016-05*; Sensory Analysis-Methodology-General Guidelines for Determining the Sensory Profile. ISO: Geneva, Switzerland, 2016. Available online: https://www.iso.org/obp/ui/#iso:std:iso:13299:ed-2:v1:en (accessed on 15 June 2024).
25. Van Gemert, L.J. *OdourThesholds. Compilation of Odour Threshold Values in Air, Water and Other Media, 2011 Ed.*; Oliemans Punter & Partners BV: Zeist, The Netherlands, 2011.
26. Guo, X.; Schwab, W.; Ho, C.-T.; Song, C.; Wan, X. Characterization of the aroma profiles of oolong tea made from three tea cultivars by both GC-MS and GC-IMS. *Food Chem.* **2022**, *376*, 131933. [CrossRef] [PubMed]
27. Qiu, D.; Duan, R.; Wang, Y.; He, Y.; Li, C.; Shen, X.; Li, Y. Effects of different drying temperatures on the profile and sources of flavor in semi-dried golden pompano (*Trachinotus ovatus*). *Food Chem.* **2023**, *401*, 134112. [CrossRef] [PubMed]
28. Van Durme, J.; Goiris, K.; De Winne, A.; De Cooman, L.; Muylaert, K. Evaluation of the Volatile Composition and Sensory Properties of Five Species of Microalgae. *J. Agric. Food Chem.* **2013**, *61*, 10881–10890. [CrossRef] [PubMed]

29. Guo, Q.; Chen, X.; Yang, K.; Yu, J.; Liang, F.; Wang, C.; Yang, B.; Chen, T.; Li, Z.; Li, X.; et al. Identification and evaluation of fishy odorants produced by four algae separated from drinking water source during low temperature period: Insight into odor characteristics and odor contribution of fishy odor-producing algae. *Chemosphere* **2023**, *324*, 138328. [CrossRef] [PubMed]
30. Lomartire, S.; Gonçalves, A.M.M. Novel Technologies for Seaweed Polysaccharides Extraction and Their Use in Food with Therapeutically Applications—A Review. *Foods* **2022**, *11*, 2654. [CrossRef]
31. Peinado, I.; Girón, J.; Koutsidis, G.; Ames, J.M. Chemical composition, antioxidant activity and sensory evaluation of five different species of brown edible seaweeds. *Food Res. Int.* **2014**, *66*, 36–44. [CrossRef]
32. Liang, Z.; Yang, C.; He, Z.; Lin, X.; Chen, B.; Li, W. Changes in characteristic volatile aroma substances during fermentation and deodorization of *Gracilaria lemaneiformis* by lactic acid bacteria and yeast. *Food Chem.* **2023**, *405*, 134971. [CrossRef] [PubMed]
33. Kong, X.; Wu, Y.; Yu, W.; Liu, Y.; Li, J.; Du, G.; Lv, X.; Liu, L. Efficient synthesis of limonene in *Saccharomyces cerevisiae* using combinatorial metabolic engineering strategies. *J. Agr. Food Chem.* **2023**, *71*, 7752–7764. [CrossRef]
34. Tricarico, M.; Crovella, S.; Celsi, F. Mevalonate pathway blockade, mitochondrial dysfunction and autophagy: A possible link. *Int. J. Mol. Sci.* **2015**, *16*, 16067–16084. [CrossRef] [PubMed]
35. Miziorko, M. Enzymes of the mevalonate pathway of isoprenoid biosynthesis. *Arch. Biochem. Biophys.* **2011**, *505*, 131–143. [CrossRef] [PubMed]
36. Singh, S.; Chhatwal, H.; Pandey, A. Deciphering the Complexity of Terpenoid Biosynthesis and Its Multi-level Regulatory Mechanism in Plants. *J. Plant Growth Regul.* **2024**. [CrossRef]
37. Ma, R.; Liu, X.; Tian, H.; Han, B.; Li, Y.; Tang, C.; Zhu, K.; Li, C.; Meng, Y. Odor-active volatile compounds profile of triploid rainbow trout with different marketable sizes. *Aquacult. Rep.* **2020**, *17*, 100312. [CrossRef]
38. Zhang, X.; Gao, P.; Xia, W.; Jiang, Q.; Liu, S.; Xu, Y. Characterization of key aroma compounds in low-salt fermented sour fish by gas chromatography-mass spectrometry, odor activity values, aroma recombination and omission experiments. *Food Chem.* **2022**, *397*, 133773. [CrossRef] [PubMed]
39. Zhou, Y.; Wu, S.; Peng, Y.; Jin, Y.; Xu, D.; Xu, X. Effect of lactic acid bacteria on mackerel (*Pneumatophorus japonicus*) seasoning quality and flavor during fermentation. *Food Biosci.* **2021**, *41*, 100971. [CrossRef]
40. López-Pérez, O.; Picon, A.; Nuñez, M. Volatile compounds and odour characteristics of seven species of dehydrated edible seaweeds. *Food Res. Int.* **2017**, *99*, 1002–1010. [CrossRef] [PubMed]
41. Foo, J.L.; Susanto, A.V.; Keasling, J.D.; Leong, S.S.J.; Chang, M.W. Whole-Cell Biocatalytic and De Novo Production of Alkanes From Free Fatty Acids in Saccharomyces cerevisiae. *Biotechnol. Bioeng.* **2017**, *114*, 232–237. [CrossRef]
42. Park, Y.C.; Shaffer, C.E.H.; Bennett, G.N. Microbial formation of esters. *Appl. Microbiol. Biot.* **2009**, *85*, 13–25. [CrossRef]
43. Vermeulen, N.; Czerny, M.; Gaenzle, M.G.; Schieberle, P.; Vogel, R.F. Reduction of (E)-2-nonenal and (E,E)-2,4-decadienal during sourdough fermentation. *J. Cereal Sci.* **2007**, *45*, 78–87. [CrossRef]

Disclaimer/Publisher's Note: The statements, opinions and data contained in all publications are solely those of the individual author(s) and contributor(s) and not of MDPI and/or the editor(s). MDPI and/or the editor(s) disclaim responsibility for any injury to people or property resulting from any ideas, methods, instructions or products referred to in the content.

foods

MDPI

Article

Time-Series Sensory Analysis Provided Important TI Parameters for Masking the Beany Flavor of Soymilk

Miyu Masuda †, Yuko Terada †, Ryoki Tsuji, Shogo Nakano and Keisuke Ito *

Department of Food and Nutritional Sciences, Graduate School of Integrated Pharmaceutical and Nutritional Sciences, University of Shizuoka, 52-1 Yada, Suruga-ku, Shizuoka 422-8526, Japan; s22209@u-shizuoka-ken.ac.jp (M.M.); yukoterada@u-shizuoka-ken.ac.jp (Y.T.); f20017@u-shizuoka-ken.ac.jp (R.T.); snakano@u-shizuoka-ken.ac.jp (S.N.)

* Correspondence: sukeito@u-shizuoka-ken.ac.jp; Tel.: +81-54-264-5544

† These authors contributed equally to this work.

Abstract: The aim of this study is to provide a new perspective on the development of masking agents by examining the application of their time-series sensory profiles. The analysis of the relationship between 14 time-intensity (TI) parameters and the beany flavor masking ability of 100 flavoring materials indicate that the values of AreaInc, DurDec, and AreaDec, TI parameters related to the flavor release in the increasing and decreasing phases, were significantly higher in the top 10 masking score materials than in the bottom 10 materials. In addition to individual analysis, machine learning analysis, which can derive complex rules from large amounts of data, was performed. Machine learning-based principal component analysis and cluster analysis of the flavoring materials presented AreaInc and AreaDec as TI parameters contributing to the classification of flavor materials and their masking ability. AreaDec was suggested to be particularly important for the beany flavor masking in the two different analyses: an effective masking can be achieved by focusing on the TI profiles of flavor materials. This study proposed that time-series profiles, which are mainly used for the understanding of the sensory characteristics of foods, can be applied to the development of masking agents.

Keywords: time-series sensory evaluation; time-intensity method; masking; soymilk; flavor materials; machine learning

1. Introduction

Flavor, consisting of taste and aroma, is an important element of food that determines consumer preferences and purchase intentions. In the food industry, masking unpleasant flavors is a significant challenge. The addition of flavoring agents is one of the most commonly used approaches to masking unpleasant flavors in food [1]. For example, based on cooking experiences, flavoring agents related to spices and the Maillard reaction were added to soy-based meat to mask its beany flavor [2,3]. In other studies, flavoring materials that effectively mask the unpleasant odors of orange juice and Pu-erh tea were screened using gas chromatography-olfactometry [4,5]. As in the abovementioned examples, various attempts have been made to mask unpleasant flavors using flavoring agents. A detailed understanding of the sensory properties of the flavoring materials would lead to a more strategic masking of unpleasant flavors.

The characteristics of flavoring materials are typically analyzed using sensory evaluation methods [6–8]. Sensory evaluation methods are generally classified into two types: endpoint types and time-series types. The endpoint type is a method of statistically evaluating the sensory characteristics of food and is suitable for the comprehensive evaluation of flavor characteristics. Time-series type is a method of dynamically evaluating the sensory characteristics of food, which enables the analysis of the temporal changes in flavor that occur in the mouth [9]. The time-intensity (TI) method is one of the time-series sensory evaluation methods. It is a technique that analyzes the temporal changes in the intensity of one sensory attribute [10,11]. TI analysis yields dynamic parameters (time-series sensory profiles) (e.g., maximum intensity, time point of maximum intensity, and end point of sensation) [12,13]. Moreover, the TI method is used to obtain a detailed understanding of the temporal changes in flavor intensity that are difficult to capture with the endpoint method. For instance, the temporal changes in flavor intensity in fruit juices, wine, beer, and ice cream are analyzed using the TI method [14–18].

In this study, we aimed to provide a new perspective on the development of masking agents by assessing the application of time-series sensory profiles for analyzing the flavor materials. Soymilk was used as a model food with an unpleasant flavor because soy, which has a low environmental impact, is being anticipated as an alternative protein source [2,3]. TI profiles for the 100 flavoring materials were obtained, and then a detailed relationship analysis between these profiles and the beany flavor masking ability of the materials was conducted. As a result of the analysis, the TI parameters that are related to the masking ability were successfully identified. Time-series sensory profiles can be applied to the development of effective masking materials.

2. Experimental Section

2.1. Food-Flavoring Materials

One hundred commercially available food-flavoring materials (T&M Corporation, Chiba, Japan) were used for the sensory evaluation. These flavoring materials were classified into three types based on their chemical features: essence type (e), oil type (o), and flavor type (f) (Table 1). Each material name was denoted as flavor name_type (e, o, or f). For example, apple essence, cinnamon oil, and coffee flavor were denoted as apple_e, cinnamon_o, and coffee_f, respectively. The essence-type materials have water solubility and high volatility and are dissolved into alcohols. The oil-type materials are characterized by their oil solubility and low volatility and are dissolved in glycerol fatty acid esters and vegetable oils. The flavor-type materials have intermediate polarity and high volatility and are dissolved into propylene glycol and glycerin.

Table 1. Classification of the flavoring materials used in this study.

Material Type	Notation	Main Solvents	Chemical Features
Essence type	e	Alcohols	Water-soluble/High volatility
Oil type	o	Glycerol fatty acid esters/Vegetable oils	Oil-soluble/Low volatility
Flavor type	f	Propylene glycol/Glycerin	Intermediate polarity/ High volatility

The flavor materials were classified into three types based on their chemical features: essence type (e), oil type (o), and flavor type (f). Each material name was shown as flavor name_type (e, o, or f) (e.g., orange oil and maple flavor were noted as orange_o and maple_f, respectively). The type (e, o, or f) of the 100 flavoring materials is shown in Supplementary Table S1.

In this study, a total of 100 flavor materials were used, including 33 essence-type, 33 oil-type, and 34 flavor-type materials (Supplementary Table S1). The detailed flavor categories and numbers belonging to each category were as follows: 55 fruits (8 berries, 12 citrus, and 35 other flavors), 5 nuts, 7 herbs and spices, 3 flowers, 21 sweets, and 9 drink flavors. The flavor materials consisted of artificially blended materials and natural-derived materials, such as essential oils. Some of the materials have the same food name, such as orange essence and orange oil, but have different compositions.

2.2. Sample Preparation for Sensory Evaluation

In our preliminary study, the beany flavor of soymilk interfered with the TI analysis of the food-flavoring materials. To overcome this issue, a fake soymilk without beany flavor was prepared and used for the TI analysis. Considering that the fake soymilk used in the study was almost tasteless and odorless, its impact on the flavor TI evaluation would be minimal. This fake soymilk had the same content of carbohydrates, proteins, and fats as plain soymilk, but was prepared by mixing dextrin, soy protein RT-1, almond milk, and water to eliminate the beany flavor. Table 2 shows the composition of the fake soymilk. The almond milk was purchased from a local supermarket. Soy protein RT-1 and dextrin were provided by Fuji Oil Holdings, Inc. (Osaka, Japan). The food-flavoring materials were added to the fake soymilk at a concentration of 0.2% as per the manufacturer's recommendation. In our pilot study, we determined the appropriate concentration to yield the TI curve for each of the flavoring materials. The fake soymilk (10 mL) and plain soymilk (10 mL) were each mixed with one of the food-flavoring materials (20 µL) and used for the TI analysis and the beany flavor masking test, respectively. The plain soymilk was purchased from a local supermarket.

Table 2. Composition of the fake soymilk.

Ingredient	g
Almond milk	34
Soy protein RT-1	3.0
Dextrin	3.0
Water	60

2.3. Panelists and Training

The panelists were 10 adults (3 men and 7 women) between the ages of 21 and 26 (average age: 22.9 years) and were non-smokers. The outline of the research was explained to each panelist, and approval for data use was obtained from all panelists. The participants were asked to refrain from eating, brushing their teeth, or talking during and at least 1 h before testing. All evaluations were conducted in a sensory evaluation room at room temperature (25 °C). Each panelist received samples in their own booth. After achieving proficiency in (1) Training on TI analysis, the panelists proceeded to (2) Training on the beany flavor masking test.

(1) Training on TI analysis: The panelists were trained to evaluate changes in the flavor intensity of the flavoring materials over time using the TI method. The training utilized the TI program of the Fizz software (Version 2.0, Biosystems, Couternon, France). The panelists drank fake soymilk added to the flavoring materials and assessed the flavor TI. The panelists synchronized their breathing with a metronome set to 40 beats per minute (bpm), as flavor perception is known to be influenced by the rate of breathing [19–21]. During the training, the experimenters observed the upward and downward movement of the panelists' shoulders from behind and checked for synchronization with a metronome. The training was completed when the panelists were able to produce smooth TI curves with high reproducibility.

(2) Training on the beany flavor masking test: The panelists were trained to evaluate the masking ability of the flavor materials against the beany flavor of soymilk using the 100-mm visual analog scale (VAS) method. To ensure a shared perception of the beany flavor in the soymilk among all panelists, they were provided with plain soymilk during the training and were briefed about the sensation of beany flavor. The panelists drank the plain soymilk to note their sense of the beany flavor as a control sample. The beany flavor masking ability of all 100 flavoring materials was assessed in our preliminary study. Pineapple essence (pineapple_e) was selected as a moderate sample for the panelist training, because it exhibited a medium-level beany flavor masking ability, reducing the beany intensity from 10 to around 7.5. The value of 7.5 represents a challenging evaluation point, as it is difficult to determine whether the beany flavor is completely masked (around 0) or not masked at all (around 10). This challenging material was intentionally used in the training process to assess the panel's ability to accurately evaluate the beany flavor masking. The panelists tasted soymilk with pineapple_e or plain soymilk and rated the intensity of the beany flavor using a VAS sheet. The training was completed when the

panelists evaluated and rated the intensity of the beany flavor of plain soymilk as around 10 and that of soymilk with added pineapple_e as around 7.5.

2.4. TI Analysis of the 100 Food-Flavoring Materials

The TI method is a time-series sensory analysis method that analyzes the temporal changes in the intensity of one sensory attribute [10,11]. It allows for a detailed analysis of the temporal changes in flavor intensity by providing numerical data on various TI parameters derived from the obtained TI curves [12,13]. In this study, the TI method was employed to evaluate the temporal changes in flavor intensity of the flavor materials. Specifically, each flavor material was added to the fake soymilk, and the trained panel evaluated the temporal changes in flavor intensity. The detailed methods are described below. All samples were consumed at room temperature throughout all evaluations. The panelists evaluated intensity changes in the flavor of the 100 flavoring materials using the TI analysis. To avoid interference from the beany flavor of soymilk, fake soymilk without beany flavor was used as the test medium. The time intensity program of the Fizz software was used for the analysis. The panelists were provided with the fake soymilk (10 mL) added to one of the flavoring materials (20 μL) just before the test. According to the instructions on the computer screen, the panelists kept the samples in their mouths for 5 s, which were then swallowed 10 s after the test started. The panelists controlled the cursor on a scale of 0 (no sense of flavor) to 100 (same intensity with reference samples) on the computer screen until the flavor sensation disappeared. The reference samples used were peppermint oil (peppermint_o) and rose oil (rose_o), which had strong flavor intensity and were given a score of 100 on the scale. These reference samples were presented to the panelists before the TI test, and the panelists memorized their maximum flavor intensity as 100. Subsequently, the panelists evaluated the flavor TI of each flavoring material. To confirm that the panelists were able to evaluate the flavor intensity of the materials relative to the reference samples, the TI data of white peach essence (white peach_e) and vanilla essence (vanilla_e), which had low flavor intensity in the preliminary study, were examined to see if their Imax values were approximately half compared to the reference samples. The TI data were collected every second. During the evaluation, the panelists synchronized their breathing with a metronome set to 40 beats per minute (bpm). Each material was evaluated five to eight times, with an average of three evaluations being used for analysis. The three representative curves were selected based on the intermediate measurements of three important TI parameters: maximum intensity of the curve (Imax), plateau start time (TsPl), and end time (Tend). This approach allowed for the exclusion of outlying data and ensured the reproducibility of the results.

The TI data were collected every second using the Fizz software, and 14 TI parameters were extracted from the obtained TI curves. These parameters included the following: start time (Tstart), end time (Tend), maximum intensity of the curve (Imax), plateau start time (TsPl), plateau end time (TePl), duration of the plateau phase (DurPl), duration of the increasing phase (DurInc), duration of the decreasing phase (DurDec), maximum slope measured in the increasing phase (SlMInc), maximum slope measured in the decreasing phase (SlMDec), the total area under the curve (AreaTse), area under the curve during the increasing phase (AreaInc), area under the curve during the decreasing phase (AreaDec), and area under the curve during the plateau phase (AreaPl).

2.5. Evaluation of the Masking Ability of the Food-Flavoring Materials against the Beany Flavor of Soymilk

The masking ability of the 100 flavor materials against the beany flavor of soymilk was evaluated using the VAS method. All samples were consumed at room temperature throughout the evaluation. The soymilk (10 mL) added to one of the food-flavoring materials (20 μL) was given to the panelists. Each flavor material was added to soymilk just before drinking. Each sample was labeled with a random three-digit number. As previously described in Section 2.4, the panelists breathed in synchrony with a metronome set at 40 bpm during the evaluation. The panel swallowed the plain soymilk to note their sense of the beany flavor as a control sample. During a one-minute break, the panelists rinsed their mouths with distilled water until the flavor sensation disappeared. Five seconds after the start of the evaluation, the panelists put the sample into their mouths, which was then swallowed 5 s later. The panel continued to evaluate the beany flavor until the flavor sensation disappeared. The panelists rated their sense of beany flavor using the VAS, ranging from "no sense of beany flavor" at the left end to "equivalent to the beany flavor of plain soymilk" at the right end. The median intensity of the beany flavor in the samples added with each flavoring material was calculated based on the VAS data. The masking ability of each flavoring material against the beany flavor was scored on a 10-point scale using the following equation: 10 − (median intensity of beany flavor) = masking score. For example, if the median intensity of the beany flavor in the sample with a flavoring material was 4.0, its masking score would be 6.0. To avoid panelist fatigue, each panelist evaluated five samples per test for a total of 20 tests (a total of 100 samples).

2.6. Statistical Analysis

The obtained TI data were analyzed using Fizz Calculations (version 2.61, Biosystemes). After smoothing the raw data (Figure 1A,B), an average curve was generated (Figure 1C). The values for the 14 TI parameters shown in Figure 1D were calculated from the average curve. The Mann–Whitney U test (two-tailed, $\alpha = 0.05$) in GraphPad Prism 5 (version 5.03, GraphPad Software, CA, USA) was used to test for significant differences between the top 10 and bottom 10 masking materials (Figure 4). p-values of <0.05 were considered statistically significant.

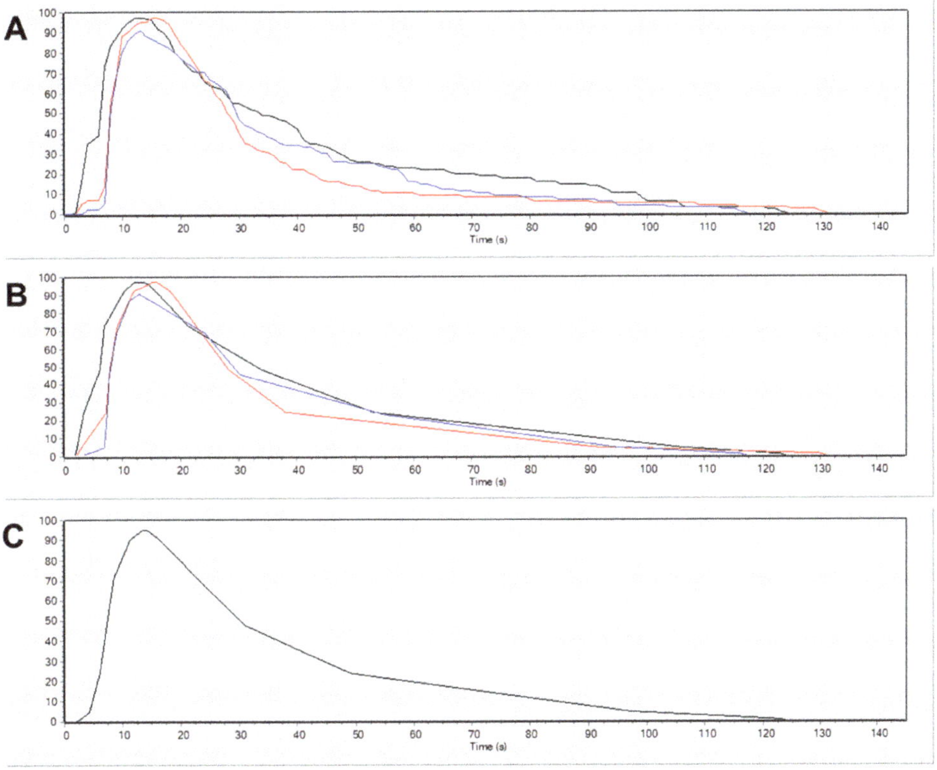

Figure 1. *Cont.*

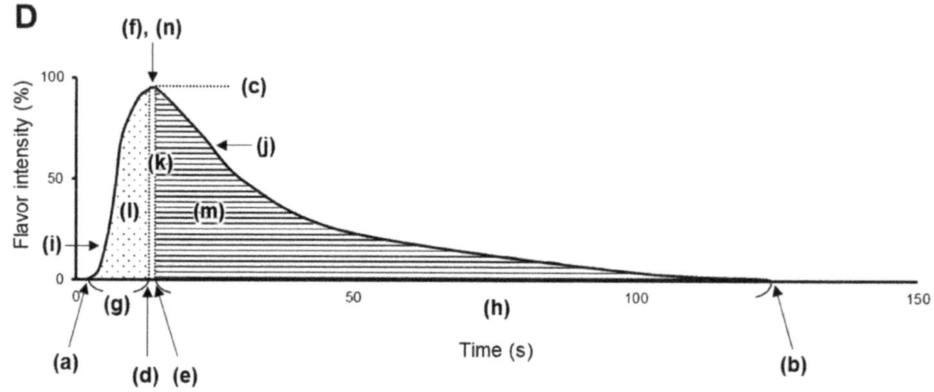

Figure 1. Example of time-intensity (TI) curve analysis and 14 TI parameters. Maple flavor (Maple_f) is shown as an example. (**A**) Raw data, (**B**) data after smoothing, (**C**) average curve, and (**D**) 14 TI parameters extracted from the average TI curve: (a) start time (Tstart), (b) end time (Tend), (c) maximum intensity of the curve (Imax), (d) plateau start time (TsPl), (e) plateau end time (TePl), (f) duration of the plateau phase (DurPl), (g) duration of the increasing phase (DurInc), (h) duration of the decreasing phase (DurDec), (i) maximum slope measured in the increasing phase (SlMInc), (j) maximum slope measured in the decreasing phase (SlMDec), (k) total area under the curve (AreaTse), (l) area under the curve during the increasing phase (AreaInc), (m) area under the curve during the decreasing phase (AreaDec), and (n) area under the curve during the plateau phase (AreaPl).

2.7. In Silico Analysis of the Time-Series Sensory Data for the Classification of the Food-Flavoring Materials

The obtained TI data were analyzed using unsupervised learning to classify the 100 flavoring materials based on their time-series profiles. The original Python script was used for the unsupervised learning. The data were first standardized to have a mean of 0 and a variance of 1 to avoid a strong influence from the parameters with large digits. Principal component analysis (PCA) was then performed to reduce the dimensionality of the data, and the calculated first and second principal components (PCA1 and PCA2) were used as the x- and y-axes, respectively. The number of clusters (k-value) was set to 3 because it resulted in a satisfactory separation of the data. Cluster analysis was performed using the k-means clustering algorithm from the scikit-learn library.

3. Results and Discussion

3.1. TI Analysis of the 100 Food-Flavoring Materials

To obtain fundamental sensory data for the 100 food-flavoring materials, the temporal changes in the flavor intensity of the materials were analyzed using the TI method. The TI curves were obtained for each flavor material, and then 14 parameter values were extracted from the curves (Supplementary Table S2). Four parameters, plateau start time (TsPl), plateau end time (TePl), end time (Tend), and the total area under the curve (AreaTse), were excluded from the data analysis since their values were similar to those of other parameters (Supplementary Figure S1A–D). As shown below, the values of some parameters were highly correlated. TsPl, TePl, and the duration of the increasing phase (DurInc) are parameters related to the increasing phase. The data values of these 3 parameters show a strong positive correlation with each other (correlation coefficient r = 0.99). Accordingly, DurInc was used for data analysis as the representative parameter related to the increasing phase. Four parameters, duration of the decreasing phase (DurDec), Tend, area under the curve during the decreasing phase (AreaDec), and AreaTse, are related to the decreasing phase. DurDec and Tend and AreaDec and AreaTse show high positive correlations, and their correlation coefficients were 0.99 and 0.98, respectively. Accordingly, DurDec and

AreaDec were used for the data analysis as parameters related to the decreasing phase. Two parameters related to the plateau phase, the area under the curve during the plateau phase (AreaPl) and the duration of the plateau phase (DurPl), were also excluded from the analysis because none of the flavor materials reached a plateau phase (Supplementary Figure S1E,F). The obtained TI curves were divided into four areas, and eight parameters were selected to represent each area for data analysis: starting time (Tstart), maximum intensity (Imax), increasing phase (AreaInc, DurInc, and SIMInc), and decreasing phase (AreaDec, DurDec, and SIMDec). The values for those eight parameters were arranged in descending order from left to right (Figure 2A–H). The names and values of the materials for each TI parameter were listed in descending order in Supplementary Table S3. The values of four parameters, namely, AreaInc, DurInc, AreaDec, and DurDec, exhibited distinct trends based on the type of materials, essence, oil, and flavor (Figure 2C,D,F,G).

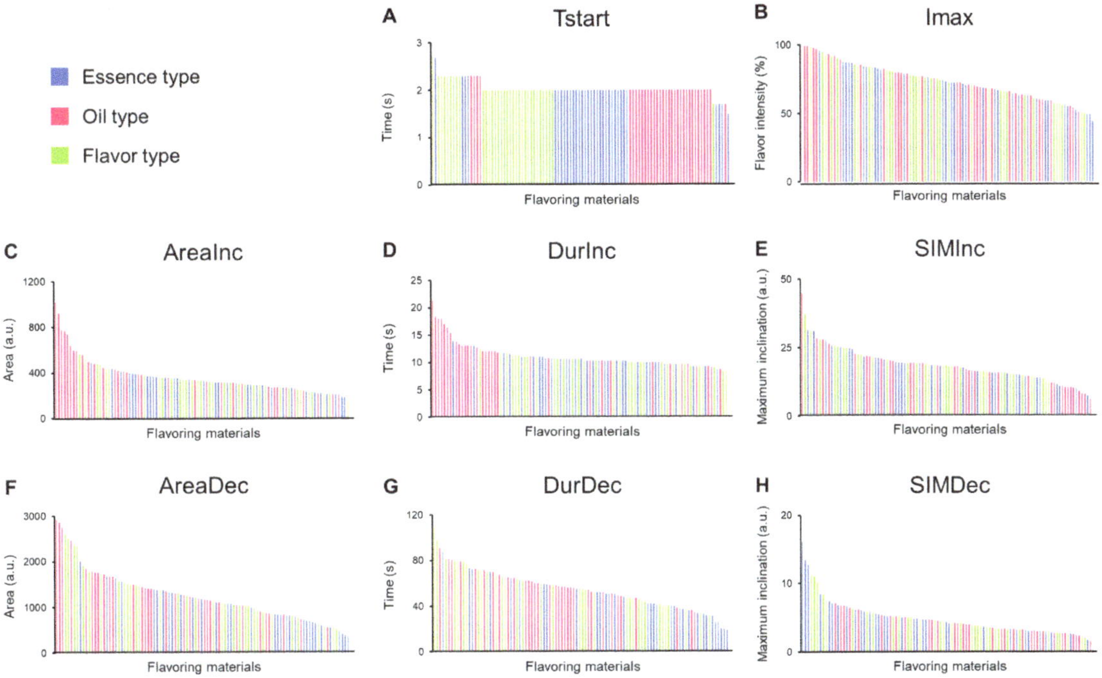

Figure 2. Comparison of 8 time-intensity (TI) parameter values among the 100 food-flavoring materials. The data values of the 100 flavor materials were arranged in descending order from left to right. The eight TI parameters are as follows: (**A**) Tstart, (**B**) Imax, (**C**) AreaInc, (**D**) DurInc, (**E**) SIMInc, (**F**) AreaDec, (**G**) DurDec, and (**H**) SIMDec. The colors of the bars indicate the material types: blue for essence type, magenta for oil type, and light green for flavor type. a.u.: arbitrary unit.

For a concise examination of the flavor materials' characteristics based on each TI parameter, the top 10 and bottom 10 materials were selected for analysis. In terms of AreaInc and DurInc, 9 out of the top 10 materials (top 10% of the 100 materials) belong to the oil type (Table 3). Particularly, in regard to DurInc, 20 out of the top 25 materials were classified as oil type. These results indicate that the oil-type materials tended to have higher values of AreaInc and DurInc compared with the essence- and flavor-type materials. AreaInc represents the area under the curve of the increasing phase, while DurInc represents the duration of the increasing phase. Therefore, larger values of AreaInc and DurInc indicate a milder increase in flavor intensity in the increasing phase, indicating a slower flavor release in the increasing phase. Consequently, the oil-type materials were suggested to have a slower flavor release in the increasing phase compared with the essence- and flavor-type materials.

Table 3. Top 10 AreaInc and DurInc materials.

Top	AreaInc	Area (a.u.)	DurInc	Time (s)
1	Orange essential_o	1016	Satsuma mandarin essential_o	21.3
2	Satsuma mandarin essential_o	919	Yuzu essential_o	18.3
3	Coconut_o	776	Coconut_o	18.0
4	Orange_o	765	Orange essential_o	18.0
5	Cherry_o	735	Orange_o	17.0
6	Yuzu essential_o	644	Apple_o	16.3
7	Rose_o	597	Banana_o	15.3
8	Mango_o	595	Orange_e	14.0
9	Maple_f	563	Cherry_o	13.7
10	Apple_o	558	Rose_o	13.3

The name of the flavoring materials was denoted as flavor name_type (e, o, or f). For example, orange essence, cherry oil, and maple flavor were shown as orange_e, cherry_o, and maple_f, respectively.

Regarding AreaDec and DurDec, 6 of the bottom 10 materials were essence-type materials, and for DurDec, 8 of the bottom 10 materials belonged to the essence type (Table 4). These results indicate that the essence-type materials tended to have lower values of AreaDec and DurDec compared with the oil- and flavor-type materials. AreaDec and DurDec represent the area under the curve during the decreasing phase and the duration of the decreasing phase, respectively. The smaller values of AreaDec and DurDec were associated with a sharper disappearance of flavor in the decreasing phase, indicating a lower flavor persistence in the decreasing phase. Therefore, the essence-type materials indicate lower flavor persistence in the decreasing phase compared with the oil- and flavor-type materials.

Table 4. Bottom 10 AreaDec and DurDec materials.

Bottom	AreaDec	Area (a.u.)	DurDec	Time (s)
1	White peach_e	352	Pineapple_e	18.7
2	Pineapple_e	380	White peach_e	19.7
3	Honey_e	414	Cheese_e	20.0
4	Caramel_e	484	Caramel_e	25.7
5	White peach_f	515	Cranberry_e	25.7
6	Honey_f	539	Orange_e	31.0
7	Cheese_e	552	Pineapple_f	31.3
8	Raspberry_o	554	Cinnamon_e	31.7
9	Brown sugar_f	565	Honey_e	33.0
10	Orange_e	594	White peach_f	33.3

The differences in the time-series sensory profiles among oil, essence, and flavor-type materials may be attributed to their different hydrophobicity and volatility. Philippe et al. suggested that the hydrophobicity of flavor compounds affects the release of compounds from food model solutions [22]. The hydrophobic interaction between flavor compounds and proteins or lipids is an important molecular mechanism controlling flavor release [23,24]. In this study, the oil materials showed a higher lipophilicity and slower flavor release in the increasing phase compared with the essence- and flavor-type materials. The slower flavor release may be related to the hydrophobic interactions between the oil materials and proteins or lipids in the test solution.

Buffo et al. indicated that vapor pressure is an important parameter affecting flavor persistence as it is an index of substance volatility [25]. Moreover, they also showed an inverse correlation between vapor pressure and flavor persistence. Consistent with this report, our study also found that the essence types with higher volatility exhibited lower flavor persistence in the decreasing phase. The differences in the chemical properties of each material type are likely responsible for the differences in their TI profiles, as discussed above.

3.2. Analysis of the Relationship between TI Profile and Beany Flavor Masking Ability of the Food-Flavoring Materials

Since soy is an important alternative protein source due to its low resource consumption, masking its beany flavor is a significant challenge. In this study, soymilk was used as a model food with an unpleasant flavor. The masking ability of the 100 flavor materials against the beany flavor of soymilk was evaluated and scored on a 10-point scale; the higher the score, the higher the masking ability. The average masking score of the 100 flavoring materials was 7.1 (Figure 3, Supplementary Table S2). Cinnamon oil (cinnamon_o) and peppermint oil (peppermint_o) show the highest masking score, 9.9 points, while cocoa essential (cocoa_e) shows the lowest masking score, 1.2 points. Table 5 shows the flavor materials that have the top 10 and bottom 10 masking scores. Seven of the top 10 materials were classified as oil types as follows: cinnamon_o, peppermint_o, orange essential_o, pineapple_o, rose_o, yuzu essential_o, and cheese_o. The bottom 10 materials were composed of five flavor-type materials (kabosu_f, sudachi_f, lemon_f, kyoho grape_f, and framboise_f), four essence-type materials (cocoa_e, white peach_e, lemon_e, and apple_e), and two oil-type materials (raspberry_o and apple_o).

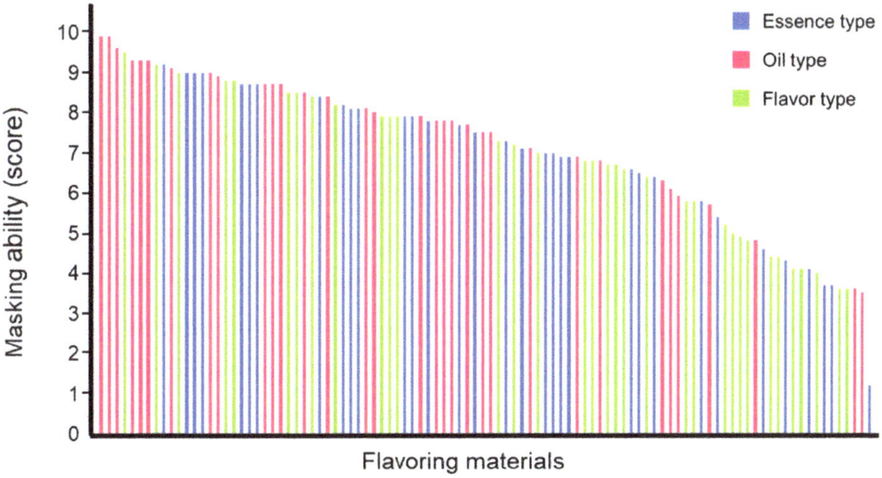

Figure 3. The masking score of the 100 flavor materials against beany flavor. The masking ability (score) of the 100 flavoring materials against the beany flavor of soymilk was arranged in descending order from left to right. The masking score of each flavoring material was calculated on a 10−point scale using the following equation: 10 − (median intensity of beany flavor) = masking score. The color of the bar indicates the material types: blue for essence type, magenta for oil type, and light green for flavor type.

To extract the important TI parameters related to the beany flavor masking, the values of the eight TI parameters were compared between the top 10 and bottom 10 materials (Figure 4). The values of five parameters, Tstart, Imax, DurInc, SIMInc, and SIMDec, were not significantly different between the top 10 and bottom 10 materials, whereas the values of three parameters, AreaInc, AreaDec, and DurDec, were significantly higher in the top 10 group than those in the bottom 10 group. These results suggested that these three TI parameters are important for the beany flavor masking by the flavor materials; an effective masking can be achieved by focusing on the TI profiles of flavoring materials.

Table 5. Top 10 and bottom 10 masking score materials.

Top	Material Name	Masking Score	Bottom	Material Name	Masking Score
1	Cinnamon_o	9.9	1	Cocoa_e	1.2
2	Peppermint_o	9.9	2	Raspberry_o	3.5
3	Orange essential_o	9.6	3	Apple_o	3.6
4	Herb_f	9.5	4	Kabosu_f	3.6
5	Pineapple_o	9.3	5	Sudachi_f	3.6
6	Rose_o	9.3	6	White peach_e	3.7
7	Yuzu essential_o	9.3	7	Lemon_e	3.7
8	Coffee_f	9.2	8	Lemon_f	4.0
9	Rose_e	9.2	9	Apple_e	4.1
10	Cheese_o	9.1	10	Kyoho grape_f	4.1
			11	Framboise_f	4.1

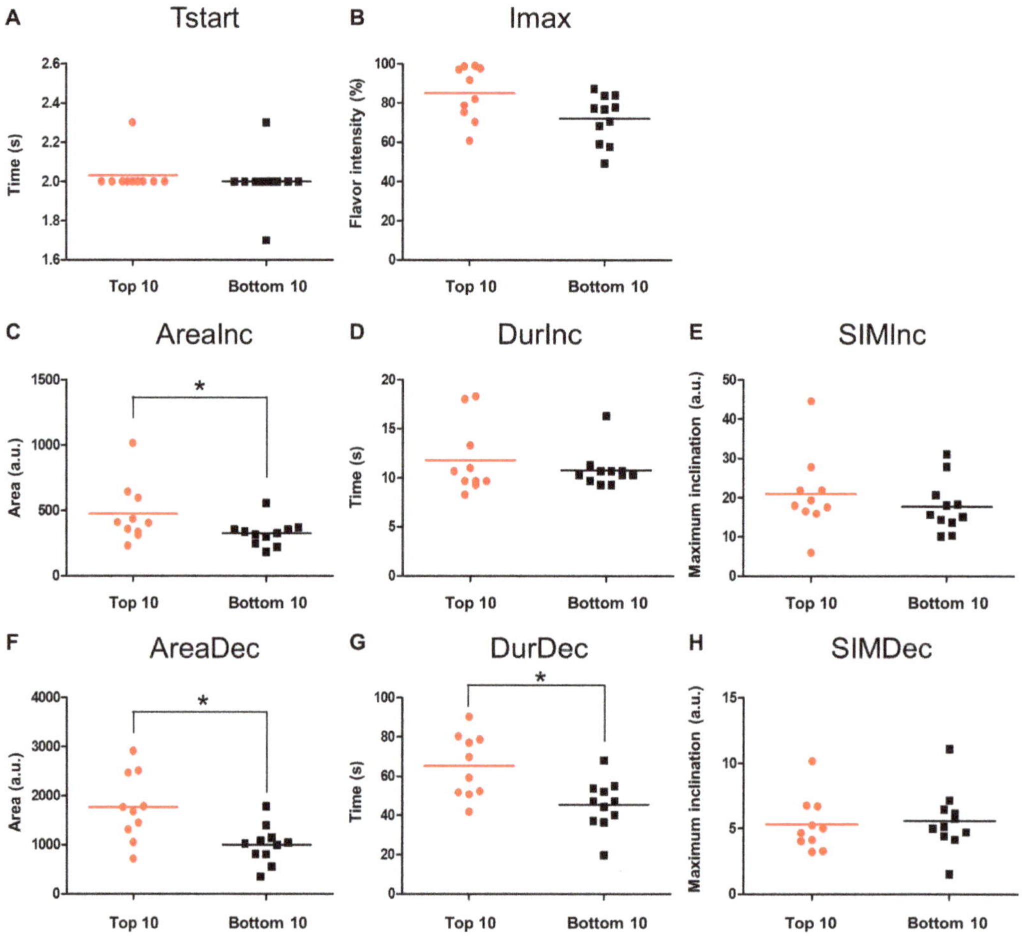

Figure 4. Comparison of TI parameter values between the top 10 and bottom 10 masking score materials. The eight TI parameter values, (**A**) Tstart, (**B**) Imax, (**C**) AreaInc, (**D**) DurInc, (**E**) SIMInc, (**F**) AreaDec, (**G**) DurDec, and (**H**) SIMDec were compared between the top 10 (red) and bottom 10 (black) masking score materials. The Mann–Whitney U test (two-tailed, $\alpha = 0.05$) was used to test the significance of the differences between the two groups. * indicates $p < 0.05$.

3.3. Classification of the Food-Flavoring Materials by Unsupervised Machine Learning Based on Time-Series Sensory Data

Machine learning is one of the powerful data analysis approaches that can derive rules from complex datasets. To analyze the relationships between the sensory profiles and the beany flavor masking ability of the 100 flavor materials, the materials were classified based on their sensory profiles. Their TI profile datasets were subjected to PCA and cluster analysis using unsupervised machine learning. The flavor materials were divided into three clusters: clusters 0, 1, and 2 (Figure 5). Cluster 2 was clearly separated from clusters 0 and 1. The contribution of PCA1 and PCA2 to the clustering analysis was 98% and 2%, respectively. This means that the contribution of PCA2 was very small, while PCA1 accounted for the majority of the clustering. Accordingly, the flavor materials can be classified by the PCA1 components. Table 6 and Supplementary Table S4 show the sample name and masking score of the materials that compose cluster 2 and clusters 0 and 1, respectively. The average masking score of the 8 materials was 8.2 points, which was significantly higher than that of clusters 0 and 1, i.e., 6.9 points (Mann–Whitney U test, two-tailed, $\alpha = 0.05$, $p < 0.05$). The 8 materials that compose cluster 2 contain 2 of the top 10 materials, but not the bottom 10 materials. Accordingly, the flavor materials can be classified by the PCA1 components. Moreover, the high values of PCA1 can be associated with the high masking ability of the materials. PCA1 was composed of the eight TI parameters: Imax, Tstart, DurInc, DurDec, SIMInc, SIMDec, AreaInc, and AreaDec. Since the data were standardized in the initial stages of the analysis, the coefficients of the explanatory variables can be used for the quantitative comparison of their contributions. Among the eight explanatory variables, the coefficients of AreaDec and AreaInc were higher than those of the others, indicating that they had a greater influence on PCA1 (Table 7). Therefore, it can be inferred that the components of PCA1, specifically AreaDec and AreaInc, are important for the classification and the high masking ability of the flavoring materials. AreaInc and AreaDec are parameters related to the flavor release in the increasing phase and that in the decreasing phase, respectively. Because the coefficient of AreaDec was approximately six times larger than that of AreaInc for PCA1, the flavor release in the decreasing phase is a more important factor than that in the increasing phase for the classification and the high masking ability of the flavor materials. As a result of the PCA and the cluster analysis using unsupervised machine learning, the flavor materials were classified according to their TI profiles. Moreover, TI parameters related to the classification and the beany flavor masking ability were extracted for the flavor materials. The application of machine learning to TI data analysis was considered a valuable approach capable of deriving complex rules from large datasets.

Table 6. Eight materials belonging to cluster 2.

Material Name	Masking Score
Satsuma mandarin essential_o	8.4
Rose_o	9.3
Cherry_o	9.0
Maple_f	7.0
Herb_f	9.5
Plum_f	8.8
Custard_f	4.8
Banana_e	9.0
Average	8.2

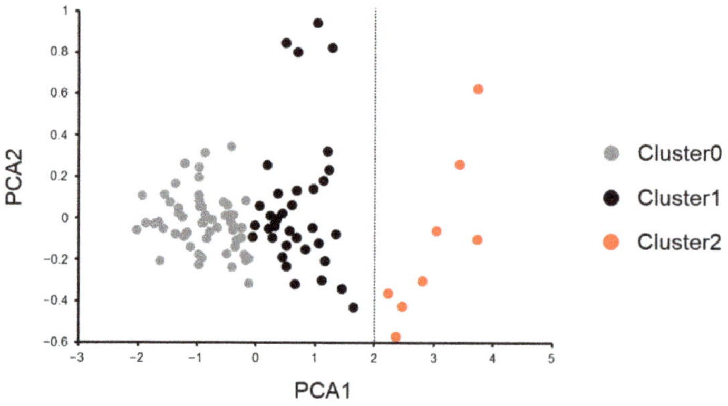

Figure 5. Classification of the 100 flavoring materials based on their TI profiles: principal component analysis and cluster analysis using unsupervised learning. Cluster analysis using the k-means method (k = 3) was performed for the flavor materials based on their TI profiles, with the first principal component (PCA1) and the second principal component (PCA2) set as the x- and y-axes, respectively. Gray, cluster 0; black, cluster 1; and red, cluster 2.

Table 7. Components of principal component analysis 1.

Imax	Tstart	DurInc	DurDec	SIMInc	SIMDec	AreaInc	AreaDec
0.015	-6.00×10^{-5}	0.0015	0.026	0	0	0.167	0.985

AreaInc and AreaDec, particularly AreaDec, were identified as important TI parameters that are related to the beany flavor masking in machine learning analysis and individual analysis. The in silico analysis successfully determined the contribution ratio of each TI parameter to the classification of the flavor materials, which is related to their masking ability. Machine learning derives certain rules from complex phenomena. In our previous studies, the rules of ligands on proton-coupled oligopeptide transporters and transient receptor potential (TRP) channels were clarified [26,27]. This study showed the rules for masking by food-flavoring materials.

AreaDec, a TI parameter related to the flavor release in the decreasing phase, was suggested to be important for the beany flavor masking by flavor materials. Effective masking can be achieved by focusing on the flavor material's AreaDec. In the present study, the AreaDec value of soymilk was 2074, indicating a strong flavor intensity in the decreasing phase. The mean value of AreaDec for masking the top 10 materials was 1763, which was close to that of soymilk, suggesting that the top 10 materials effectively masked the beany flavor of soymilk in the decreasing phase. Numerous studies have focused on controlling the flavor release from foods, e.g., glycerol, polysaccharides, and protein addition, and flavor encapsulation is one of the approaches used to modify the release of flavor compounds [28–31]. Jones et al. reported that the addition of glycerol and polysaccharides enhanced and weakened the flavor intensity of white wine, respectively [28]. Santos et al. showed that the encapsulation of mint flavor using gum arabic increased the persistence of the flavor in chewing gum [32]. Accordingly, controlling the release of flavor materials would enable more effective masking.

4. Conclusions

To provide a new perspective on the development of masking agents, this study examined the application of the time-series sensory profiles of food-flavoring materials. The individual analysis between the TI parameters and the masking ability indicated that the values of AreaInc, DurDec, and AreaDec were significantly higher in the top 10 materials with high masking scores compared to the bottom 10 materials. Machine learning-based PCA and cluster analysis of the flavoring materials presented AreaInc and AreaDec as TI parameters contributing to the classification of flavor materials and their masking ability. Both the individual analysis and the in silico analysis consistently indicate the importance of AreaDec, a TI parameter associated with flavor release during the decreasing phase, in the masking of beany flavor by flavoring materials. Time-series sensory evaluation is a useful method for understanding the sensory characteristics of foods. In addition to individual analysis, machine learning, which can derive complex rules from large amounts of data, will be a powerful tool for developing masking agents. In the future, an approach combining time-series sensory profiles and machine learning will lead to a more strategic masking process.

Supplementary Materials: The following supporting information can be downloaded at: https://www.mdpi.com/article/10.3390/foods12142752/s1, Figure S1: Six TI parameter values of the food-flavoring materials; Table S1: List of the 100 food-flavoring materials; Table S2: Masking scores and TI profiles of the flavor materials; Table S3: List of the 100 food-flavoring materials arranged in descending order for each of the eight TI parameters; Table S4: Materials belonging to clusters 0 and 1.

Author Contributions: Conceptualization, M.M., Y.T. and K.I.; data curation, M.M. and Y.T.; Formal analysis, M.M., R.T. and S.N.; funding acquisition, Y.T. and K.I.; investigation, M.M. and Y.T.; methodology, K.I.; project administration, K.I.; software, S.N. and R.T.; supervision, Y.T. and K.I.; validation, M.M. and Y.T.; visualization, M.M., Y.T. and K.I.; writing—original draft, M.M. and Y.T.; writing—review and editing, S.N. and K.I. All authors have read and agreed to the published version of the manuscript.

Funding: This research was funded by the Japan Society for the Promotion of Science (KAKENHI grant JP 21H02144 and 21K19092 to K.I., and 22K14852 to Y.T.) and the Fuji Foundation for Protein Research (K.I.).

Institutional Review Board Statement: Ethical review and approval were waived for this study because samples used in the sensory evaluation were soymilk and food-flavoring materials that are eaten in daily life.

Informed Consent Statement: Informed consent was obtained from all subjects involved in the study.

Data Availability Statement: The data underlying this article are available in the article and in its online Supplementary Materials.

Acknowledgments: We acknowledge the generous support from the Japan Society for the Promotion of Science and the Fuji Foundation for Protein Research to K.I. and Y.T.

Conflicts of Interest: The authors declare no conflict of interest.

References

1. Hu, S.; Liu, X.; Zhang, S.; Quan, D. An Overview of Taste-Masking Technologies: Approaches, Application, and Assessment Methods. *AAPS PharmSciTech* **2023**, *24*, 67. [CrossRef] [PubMed]
2. Wang, Y.; Tuccillo, F.; Lampi, A.M.; Knaapila, A.; Pulkkinen, M.; Kariluoto, S.; Coda, R.; Edelmann, M.; Jouppila, K.; Sandell, M.; et al. Flavor challenges in extruded plant-based meat alternatives: A review. *Compr. Rev. Food Sci. Food Saf.* **2022**, *21*, 2898–2929. [CrossRef] [PubMed]
3. Li, X.; Li, J. The Flavor of Plant-Based Meat Analogues. *Cereal Foods World* **2020**, *65*, 40.
4. Burseg, K.; de Jong, C. Application of the Olfactoscan method to study the ability of saturated aldehydes in masking the odor of methional. *J. Agric. Food Chem.* **2009**, *57*, 9086–9090. [CrossRef]
5. Zhang, T.; Ni, H.; Qiu, X.J.; Li, T.; Zhang, L.Z.; Li, L.J.; Jiang, Z.D.; Li, Q.B.; Chen, F.; Zheng, F.P. Suppressive Interaction Approach for Masking Stale Note of Instant Ripened Pu-Erh Tea Products. *Molecules* **2019**, *24*, 4473. [CrossRef]
6. Montero-Fernandez, I.; Marcia-Fuentes, J.A.; Cascos, G.; Saravia-Maldonado, S.A.; Lozano, J.; Martin-Vertedor, D. Masking Effect of Cassia grandis Sensory Defect with Flavoured Stuffed Olives. *Foods* **2022**, *11*, 2305. [CrossRef]
7. Fiorentini, M.; Kinchla, A.J.; Nolden, A.A. Role of Sensory Evaluation in Consumer Acceptance of Plant-Based Meat Analogs and Meat Extenders: A Scoping Review. *Foods* **2020**, *9*, 1334. [CrossRef]
8. Tempere, S.; Schaaper, M.H.; Cuzange, E.; de Lescar, R.; de Revel, G.; Sicard, G. The olfactory masking effect of ethylphenols: Characterization and elucidation of its origin. *Food Qual. Prefer.* **2016**, *50*, 135–144. [CrossRef]

9. Marques, C.; Correia, E.; Dinis, L.T.; Vilela, A. An Overview of Sensory Characterization Techniques: From Classical Descriptive Analysis to the Emergence of Novel Profiling Methods. *Foods* **2022**, *11*, 255. [CrossRef]
10. Visalli, M.; Galmarini, M.V. Multi-attribute temporal descriptive methods in sensory analysis applied in food science: Protocol for a scoping review. *PLoS ONE* **2022**, *17*, e0270969. [CrossRef]
11. Lawless, H.T.; Heymann, H. Time-Intensity Methods. In *Sensory Evaluation of Food: Principles and Practices*; Lawless, H.T., Heymann, H., Eds.; Springer: Boston, MA, USA, 1999; pp. 265–300.
12. Cliff, M.; Heymann, H. Development and use of time-intensity methodology for sensory evaluation: A review. *Food Res. Int.* **1993**, *26*, 375–385. [CrossRef]
13. Dijksterhuis, G.B.; Piggott, J.R. Dynamic methods of sensory analysis. *Trends Food Sci. Technol.* **2001**, *11*, 284–290. [CrossRef]
14. Bolini, H.M.A.; Lima, R.S.; Freitas, R.L.; Medeiros, A.C. Preference Drivers for Blackberry Nectar (*Rubus* spp., Rosaceae) with Different Sweeteners. *Foods* **2023**, *12*, 549. [CrossRef]
15. Medeiros, A.; Tavares, E.; Bolini, H.M.A. Descriptive Sensory Profile and Consumer Study Impact of Different Nutritive and Non-Nutritive Sweeteners on the Descriptive, Temporal Profile, and Consumer Acceptance in a Peach Juice Matrix. *Foods* **2022**, *11*, 244. [CrossRef] [PubMed]
16. Goodstein, E.S.; Bohlscheid, J.C.; Evans, M.; Ross, C.F. Perception of flavor finish in model white wine: A time-intensity study. *Food Qual. Prefer.* **2014**, *36*, 50–60. [CrossRef]
17. Vázquez-Araújo, L.; Parker, D.; Woods, E. Temporal Sensory Methods for Beer. *J. Sens. Stud.* **2013**, *28*, 387–395. [CrossRef]
18. de Medeiros, A.C.; Tavares Filho, E.R.; Bolini, H.M.A. Temporal profile of low calorie lactose-free ice cream chocolate flavor: Temporal dominance sensation and multiple time-intensity analysis. *J. Food Sci. Technol.* **2021**, *58*, 3164–3173. [CrossRef]
19. Pellegrino, R.; Atchley, A.; Ali, S.; Shingleton, J.; Luckett, C.R. Retronasal Habituation: Characterization and Impact on Flavor Perception Using Time-Intensity. *Chemosens. Percept.* **2018**, *13*, 1–10. [CrossRef]
20. Hanaoka, K.; Vallet, N.; Giampaoli, P.; Heyd, B.; MacLeod, P. Possible influence of breathing on detection frequency and intensity rating in gas chromatography-olfactometry. *Food Chem.* **2001**, *72*, 97–103. [CrossRef]
21. Pionnier, E.; Chabanet, C.; Mioche, L.; Le Quere, J.L.; Salles, C. 1. In Vivo Aroma Release during Eating of a Model Cheese: Relationships with Oral Parameters. *J. Agric. Food Chem.* **2004**, *52*, 557–564. [CrossRef]
22. Philippe, E.; Seuvre, A.M.; Colas, B.; Langendorff, V.; Schippa, C.; Voilley, A. Behavior of flavor compounds in model food systems: A thermodynamic study. *J. Agric. Food Chem.* **2003**, *51*, 1393–1398. [CrossRef] [PubMed]
23. Zhang, J.; Kang, D.; Zhang, W.; Lorenzo, J.M. Recent advantage of interactions of protein-flavor in foods: Perspective of theoretical models, protein properties and extrinsic factors. *Trends Food Sci. Technol.* **2021**, *111*, 405–425. [CrossRef]
24. Ammari, A.; Schroen, K. Flavor Retention and Release from Beverages: A Kinetic and Thermodynamic Perspective. *J. Agric. Food Chem.* **2018**, *66*, 9869–9881. [CrossRef] [PubMed]
25. Buffo, R. Persistence of aroma compounds in human breath after consuming an aqueous model aroma mixture. *Food Chem.* **2005**, *89*, 103–108. [CrossRef]
26. Ito, K.; Hikida, A.; Kawai, S.; Lan, V.T.; Motoyama, T.; Kitagawa, S.; Yoshikawa, Y.; Kato, R.; Kawarasaki, Y. Analysing the substrate multispecificity of a proton-coupled oligopeptide transporter using a dipeptide library. *Nat. Commun.* **2013**, *4*, 2502. [CrossRef]
27. Terada, Y.; Tanaka, K.; Matsuyama, M.; Fujitani, M.; Shibuya, M.; Yamamoto, Y.; Kato, R.; Ito, K. Collection of Data Variation Using a High-Throughput Image-Based Assay Platform Facilitates Data-Driven Understanding of TRPA1 Agonist Diversity. *Appl. Sci.* **2022**, *12*, 1622. [CrossRef]
28. Jones, P.R.; Gawel, R.; Francis, I.L.; Waters, E.J. The influence of interactions between major white wine components on the aroma, flavour and texture of model white wine. *Food Qual. Prefer.* **2008**, *19*, 596–607. [CrossRef]
29. Lesme, H.; Rannou, C.; Famelart, M.-H.; Bouhallab, S.; Prost, C. Yogurts enriched with milk proteins: Texture properties, aroma release and sensory perception. *Trends Food Sci. Technol.* **2020**, *98*, 140–149. [CrossRef]
30. Premjit, Y.; Pandhi, S.; Kumar, A.; Rai, D.C.; Duary, R.K.; Mahato, D.K. Current trends in flavor encapsulation: A comprehensive review of emerging encapsulation techniques, flavour release, and mathematical modelling. *Food Res. Int.* **2022**, *151*, 110879. [CrossRef]
31. Saifullah, M.; Shishir, M.R.I.; Ferdowsi, R.; Tanver Rahman, M.R.; Van Vuong, Q. Micro and nano encapsulation, retention and controlled release of flavor and aroma compounds: A critical review. *Trends Food Sci. Technol.* **2019**, *86*, 230–251. [CrossRef]
32. Santos, M.G.; Carpinteiro, D.A.; Thomazini, M.; Rocha-Selmi, G.A.; da Cruz, A.G.; Rodrigues, C.E.C.; Favaro-Trindade, C.S. Coencapsulation of xylitol and menthol by double emulsion followed by complex coacervation and microcapsule application in chewing gum. *Food Res. Int.* **2014**, *66*, 454–462. [CrossRef]

Disclaimer/Publisher's Note: The statements, opinions and data contained in all publications are solely those of the individual author(s) and contributor(s) and not of MDPI and/or the editor(s). MDPI and/or the editor(s) disclaim responsibility for any injury to people or property resulting from any ideas, methods, instructions or products referred to in the content.

fermentation

MDPI

Article

Characterization of the Key Aroma Compounds of Soybean Flavor in Fermented Soybeans with *Bacillus subtilis* BJ3-2 by Gene Knockout, Gas Chromatography–Olfactometry–Mass Spectrometry, and Aroma Addition Experiments

Zhaofeng Chen, Yongjun Wu *, Shuoqiu Tong, Jing Jin, Lincheng Zhang, Chen Li, Qibo Tan, Feng Wen and Yi Tao

Key Laboratory of Plant Resource Conservation and Germplasm Innovation in Mountainous Region (Ministry of Education), College of Life Sciences/Institute of Agro-Bioengineering, Guizhou University, Guiyang 550025, China; zfchen2024@outlook.com (Z.C.); m18786120419@163.com (S.T.); jjin3@gzu.edu.cn (J.J.); zhanglc201312@163.com (L.Z.); cenli@gzu.edu.cn (C.L.); qibotan@outlook.com (Q.T.); fengwen0211@outlook.com (F.W.); yitaomsu@outlook.com (Y.T.)

* Correspondence: wyjbio@163.com

Abstract: Soybean flavor is considered to be essential for the aroma quality of fermented soybeans (FS) with *Bacillus subtilis* BJ3-2 (BJ3-2) at 37 °C. However, the key aroma compounds of the soybean flavor must be further elucidated. In this study, two candidate genes (*sdaAA* and *katX*) of BJ3-2 involved in the control of soybean flavor production were screened using prior multi-omics data. FS samples with BJ3-2, BJ3-2Δ*sdaAA*, BJ3-2Δ*katX*, and BJ3-2Δ*sdaAA*Δ*katX* were analyzed by quantitative descriptive sensory analysis (QDA), gas chromatography–olfactometry–mass spectrometry (GC-O-MS), relative odor activity values (ROAV), and aroma addition experiments. The QDA revealed that the aroma profile of the soybean flavor in FS consisted of "sweaty", "smoky", "beany", "roasted", and "sweet" attributes. A total of 20 aroma-active compounds were detected, and 13 of them with ROAV > 1 were identified as key aroma compounds. Moreover, aroma addition experiments were conducted to further confirm the key aroma compounds of soybean flavor. Among them, 2-methylbutyric acid, 2,3,5-trimethylpyrazine, and guaiacol contributed higher aroma intensity values and ROAV, resulting in "sweaty", "roasted", and "smoky" attributes of soybean flavor in FS, respectively, while 1-octen-3-ol was associated with the "beany" attribute. These findings provide novel insights into the aroma attributes of soybean flavor in FS and a new strategy for revealing the key aroma compounds in fermented foods.

Keywords: *Bacillus subtilis*; fermented soybeans; soybean flavor; soybean flavor candidate gene; aroma compounds

1. Introduction

Fermented soybean foods are widely popular in East and Southeast Asian countries due to their unique flavor and nutritional value. *Bacillus subtilis* (*B. subtilis*) is a predominant functional bacterium in these foods [1]. During the fermentation of soybeans by *B. subtilis*, the exogenous enzymes secreted by *B. subtilis* (mainly proteases and lipases) can decompose the proteins and lipids into amino acids, fatty acids, and organic acids, which then form volatile compounds [2,3]. The soybean flavor is a special flavor formed by fermenting the raw material by microorganisms (namely Chi-flavor, a special flavor with an ammonia scent distinct from natto) [4]. Previous research has reported that the characteristics of *B. subtilis* BJ3-2 fermented soybean differ at different temperatures. For example, fermented soybeans at 37 °C enhance soybean flavor compared to those at 45 °C and 53 °C, while fermented soybeans at 45 °C and 53 °C enhance soy sauce-like flavor compared to those at 37 °C [5]. Although traditional soybean flavor products such as Douchi and

Chi-flavor-type Baijiu are popular and accepted worldwide due to their prominent soybean flavor [6,7], there are only a few reports on the study of soybean flavor. It is speculated that the synergistic effects of multiple metabolic pathways and multiple genes may produce soybean flavor [5]. Meanwhile, the key aroma compounds contributing to soybean flavor in fermented soybeans are poorly known.

Systematic characterization of the aroma-active compounds therein is a prerequisite for improving the flavor quality of fermented soybeans [8]. Food aroma compounds are usually studied using an integrated approach encompassing isolation, identification, and quantification of volatile compounds, a deep analysis of aroma compounds, and reconstitution the aroma using the key aroma compounds [9,10]. Gas chromatography–olfactometry–mass spectrometry (GC-O-MS), a combination of gas chromatography–olfactometry (GC-O) and gas chromatography–mass spectrometry (GC-MS), separates the volatile compounds in a complex sample through gas chromatography, enables the determination of which of these volatiles are aromatically active through olfactometric detection, and allows the identification of the compounds eluting from the column through mass spectrometry [11]. Thus far, the aroma compounds in several *B. subtilis* fermented soybean foods have been examined using GC-O-MS. Previous studies have demonstrated that pyrazines play a crucial role in the flavor of natto (a traditional Japanese food made from fermented soybeans), among which 2,3,5-trimethylpyrazine, 2,5-dimethylpyrazine, 2,3,5,6-tetramethylpyrazine, and 2,3,5-methyl-6-ethylpyrazine substantially contribute to natto flavor [12]. Furthermore, Mathatheeranan et al. identified the dominant aroma compounds of thua nao (indigenous Thai fermented soybeans) fermented by commercial *B. subtilis* under controlled conditions including 2-methylbutyric acid, phenylethyl alcohol, benzaldehyde, 2,5-dimethylpyrazine, 2,6-dimethylpyrazine, and 2-methoxyphenol, which contribute to the fermented/sour, floral, almond-like, and nutty aromas of fermented soybeans [13]. More recently, GC-O-MS combined with relative odor activity values (ROAV) were applied to examine the aroma compounds of sufu [14], Yongchuan Douchi [6], typical fermented soy foods [15], and so on. In these studies, e-nose analysis and quantitative description sensory analysis (QDA) were used to analyze the aroma profiles of different fermented soy products. Differences in the aroma profiles of the various products have been identified and associated with key aroma compounds.

In general, microbial metabolism is crucial for producing flavor in fermented foods [16]. To elucidate the mechanism of soybean flavor and soy sauce-like flavor production by *B. subtilis* BJ3-2, Wu et al. measured and analyzed the transcriptome, proteome, and metabolome of *B. subtilis* BJ3-2 at three temperatures (37 °C, 45 °C, and 53 °C). The functions of differentially expressed genes (DEGs) have been analyzed by KEGG and GO enrichment analyses. These DEGs are enriched in the relevant metabolic pathways, including the glycine, serine, threonine, cysteine, and methionine metabolite pathways, among others [5]. Meanwhile, amino acids are substrates for the formation of aroma compounds [17]. The production of soybean flavor may be closely linked to these metabolic pathways. Nevertheless, the candidate genes that may regulate the production of soybean flavor need to be further investigated.

Therefore, this study aimed to (a) screen soybean flavor candidate genes from the analysis of multi-omics data; (b) obtain mutants with significantly altered soybean flavor by knocking out the soybean flavor candidate gene to better analyze the relationship between sensory attributes and aroma compounds of soybean flavor in fermented soybeans (FS); (c) characterize key aroma compounds of soybean flavor in FS using GC-O and ROAV; and (d) verify the correlation between key aroma compounds and sensory attributes by partial least squares regression (PLSR) analysis. In the present study, by employing multi-omics data, gene knockout, and the headspace solid-phase microextraction (HS-SPME) method, combined with GC-O-MS, ROAV, and aroma addition experiments, we have become the first to characterize the aroma of soybean flavor in FS samples with *B. subtilis* BJ3-2 and its different mutants. These data can help gain a deeper understanding of the key aroma compounds in the production of soybean flavor in fermented soybeans with *B. subtilis*.

They also provide a new strategy for the screening of soybean flavor candidate genes and the modification of fermentation flavor by genetic engineering.

2. Materials and Methods

2.1. Strains and Culture Conditions

B. subtilis BJ3-2 (GI: CP025941) strains were cultured on liquid Luria–Bertani (LB) medium or solid LB at 37 °C. Chloramphenicol (5 μg/mL) or erythromycin (1 μg/mL) were used for screening when necessary. *E. coli* DH5α was cultured on liquid or solid LB medium with ampicillin (100 μg/mL).

2.2. Chemicals

Commercial standard n-alkanes (C7-C30), 2-octaol (≥99.0%), 2,3-butanedione (≥99.0%), 3-octanone (≥99.5%), and 3-ethyl-2,5-dimethylprazine (98%) were obtained from Macklin Biochemical Technology Co. (Shanghai, China). 2,3,5-Trimethylprazine (>98.0%), 2-ethylfuran (>98.0%), 2-pentylfuran (>98.0%), methyl 2-methylbutyrate (>98.0%), 2,4,5-trimethyloxazole (>98.0%), 2-heptaone (>98.0%), 2,5-dimethylprazine (>98.0%), 1-octen-3-ol (>98.0%), 2-methylbutyric acid (>98.0%), and guaiacol (>98.0%) were obtained from TCI Co., Ltd. (Shanghai, China). Propylene glycol (99%) and anhydrous ethanal (≥99.7%) were purchased from Solarbio Science and Technology Co., Ltd. (Beijing, China).

2.3. Reverse Transcription Quantitative Real-Time PCR (RT-qPCR)

To validate the expression of *sdaAA* and *katX* in the multi-omics data, the total RNA of BJ3-2 was extracted, and the first strand of cDNA was synthesized using the FastKing method as a template (TIANGEN, Beijing, China). Then, the cDNA was detected by the CFX96 Touch PCR instrument (Bio-Rad, Hercules, CA, USA). The reaction system and the settings applied followed a previously described approach [18]. 16S rRNA was used as the internal reference gene. The primers used are listed in Table S1.

2.4. Construction and Transformation of Homologous Recombinant Knockout Vectors

HLarm is the homologous left arm of the target gene; HRarm is the homologous right arm; *cm* is the chloramphenicol gene; and *erm* is the erythromycin gene. DEDP is a double-exchange detection primer (Table S1) used to assess the success of the knockout. The genome of BJ3-2 was used as a template for PCR amplification of the HLarm and HRarm of *sdaAA* and *katX*, respectively. The *cm* was amplified by PCR using pHT01cas9-p43 as the template. The fragment and pUC18 were digested by the corresponding restriction endonuclease *Eco*R I, *Kpn* I, *Bam*H I, *Sal* I, and *Pst* I and then linked with T4 DNA ligase. The connected vector was then transformed into *E. coli* DH5α-competent cells. Finally, knockout vectors pUC18-(HLarm-*cm*-HRarm)/*sdaAA* and pUC18-(HLarm-*cm*-HRarm)/*katX* were constructed. Then, vectors were transformed into BJ3-2 and screened with chloramphenicol; the method used for the transformation was chemical transformation [19,20]. PCR validation of positively transformed colonies was performed using DEDP (Figures S1 and S2). Sequencing was then performed to validate the knockout strains (Figures S4 and S5). BJ3-2Δ*sdaAA* and BJ3-2Δ*katX* were obtained. In addition, the genome of BJ3-2 was used as a template to amplify the *sdaAA*, and the pMarA was used as a template for PCR amplification of the *erm*.

Another knockout vector, pUC18-*sdaAA*::*erm*, was constructed using the same method. pUC18-*sdaAA*::erm was transformed into BJ3-2 and screened with erythromycin. BJ3-2Δ*sdaAA* was obtained after validation by PCR using DEDP, and then pUC18-(HLarm-*cm*-HRarm)/*katX* was transferred into BJ3-2Δ*sdaAA* to obtain BJ3-2Δ*sdaAA*Δ*katX*. The double-knockout strain was validated by PCR using DEDP and sequencing (Figures S3 and S6). All primer synthesis and sequencing were performed by Tsingke Biotechnology Co., Ltd. (Beijing, China).

2.5. Preparation and Aroma Profile Evaluation of Fermented Soybeans (FS)

The preparation of FS samples was performed following the method previously described by Liu et al. [4], with slight modifications as follows: single colonies of BJ3-2, BJ3-2Δ*sdaAA*, BJ3-2Δ*katX*, and BJ3-2Δ*sdaAA*Δ*katX* were inoculated into 5 mL of LB medium and shaken at 180 rpm for 14 h at 37 °C. Soybeans were then autoclaved at 121 °C for 20 min after soaking in water for 12-13 h. Appealed bacterial suspensions of BJ3-2, BJ3-2Δ*sdaAA*, BJ3-2Δ*katX*, and BJ3-2Δ*sdaAA*Δ*katX* were inoculated onto cooled autoclaved soybeans (1%). Next, the soybeans and the bacterial suspensions were shaken and placed in the incubator at 37 °C for 72 h.

The aroma profile evaluation of FSs was assessed by quantitative descriptive analysis (QDA), and the sensory evaluation team consisted of 10 panelists from the School of Life Sciences, Guizhou University (4 males and 6 females, 20–27 years old). Each team member underwent a one-week sensory training course on fermented soybeans. Firstly, a descriptive test was conducted to elicit the sensory descriptor. Each FSs (5 g each) was put into 50 mL Teflon vessels, and each member smelled them to freely record the sensory descriptors. The panel then discussed and reached a unanimous agreement based on the frequency and intensity of the sensory attributes of soybean flavor. After that, five sensory attributes were selected to describe the soybean flavor, including "sweaty", "smoky", "beany", "roasted", and "sweet". Finally, for FS samples coded with 3-digit labels, the intensity of the aroma contribution was evaluated using a 6-point scale (0 = no smell, 1 = very weak, 2 = weak, 3 = moderate, 4 = strong, and 5 = very strong with significant irritation) [21], and each sample was analyzed three times. The flavor radar chart was plotted based on the obtained five-dimensional aroma wheel.

2.6. Extraction of Volatile Compounds in FS Samples by Using HS-SPME

HS-SPME extraction of FSs was performed according to Peng et al. [22] by putting 2.5 mL of distilled water, 0.625 g of NaCl, and 1.5 g of the fermented soybean sample into a 15 mL headspace vial. Samples were then pre-equilibrated in a 50 °C water bath (Dk-s24, Shanghai, China) for 20 min, after which the aged SPME (50/30 μm DVB/CAR/PDMS (2 cm); Bellefonte, PA, USA) was placed 1 cm above the liquid sample and adsorbed for 40 min. Next, SPME fiber was immediately inserted into the GC injection port (250 °C) of the GC-system and desorbed for 5 min. Each sample was analyzed three times.

2.7. Identification of Aroma-Active Compounds in FS Samples by GC-O-MS

GC-O-MS analysis was performed using an Agilent 8890B GC (Agilent Technologies, Santa Clara, CA, USA) equipped with a polar capillary column DB-WAX (60 m × 0.25 mm × 0.25 μm film thickness, Agilent Technologies) and coupled to an olfactory detector (ODP4; Gerstel, Mülheim an der Ruhr, Germany) and an Agilent 5977B MS (Agilent Technologies, Santa Clara, CA, USA). The mass spectrometer and the sniffer detection port received the volatile compounds separated by the capillary column in a 1:1 ratio. Humidified air was pumped into the sniffer arm at 40 mL/min. The following temperature program was applied: the column temperature was initially set at 40 °C and held for 3 min, ramped up to 130 °C at a rate of 3 °C/min and held for 5 min, and finally ramped up to 200 °C at a rate of 5 °C/min and held for 5 min. The GC inlet temperature was 250 °C, and high-purity helium was used as the carrier gas (He > 99.999%) and transported through the capillary column at a constant flow rate of 1.0 L/min. GC-O: The aroma intensity (AI) was evaluated using the approach of Peng et al. with minor adjustments (from a 5-point scale to a 6-point scale) [23]. Four panelists (2 males and 2 females, an average of 25 years old) were asked to record the AI values, as well as odor characteristics and retention time. A 6-point scale ranging from 0 (none) to 5 (extreme) was used to evaluate the aroma intensity. A higher number means that the compound has a stronger odor. Analyses was repeated three times by each panelist. Finally, the AI values were the average of four panelists.

For qualitative and quantitative analysis, volatile compounds were identified using mass spectra from the database (NIST 20.0) with reference to retention index (RI) and odor characteristics from the literature. Some important compounds were further identified using standard compounds. Retention indices were calculated for individual compounds based on n-alkane standards from C7 to C30.

For quantification, 8 μL of 2-octanol (0.18 mg/mL) was added to fermented soybeans as an internal standard, and all compounds were analyzed semi-quantitatively, and the relative concentration of each compound was calculated following the method of Zhao et al. [24]. The concentration of the volatile compounds is calculated from the peak areas of the volatiles and the concentration of the internal standard.

2.8. Calculation of Relative Odor Activity Values (ROAVs)

ROAVs were calculated based on the ratio of the relative content of volatile compounds to their absolute thresholds in water using the following formulae [25]:

$$\mathrm{ROAV} = \mathrm{Cx}/\mathrm{OTx}$$

where Cx is the average concentration of the compounds and OTx is the odor threshold of the compound. The thresholds used in this study were those in water obtained from published literature [26,27].

2.9. Aroma Addition Experiment

Aroma addition experiments were conducted to compare the aroma profile changes after the key aroma compounds screened from GC-O and ROAV were added to 1.5 g of FSs. Two FSs with BJ3-2 and BJ3-2Δ*sdaAA*Δ*katX*, which have greater differences in soybean flavor, were selected. The differences in concentrations of key aroma compounds in the two FSs were calculated from the semi-quantitative results (Table S5). All chemical solutions were prepared in water or propylene glycol solution and diluted to the corresponding concentrations [28] (Table S5). Then, 1.5 g of each of the two FSs matrices were placed into 15 mL Teflon vessels; 11 compounds were added to FSs with BJ3-2, and 2 compounds were added to FSs with BJ3-2Δ*sdaAA*Δ*katX* to achieve the same concentration in both FSs (Table S5). The two FSs were then equilibrated at 25 °C for 30 min. The aroma profiles of the two FSs were then evaluated by the ten panelists using QDA, and the flavor radar chart was drawn based on average.

2.10. Statistical Analysis

At least three replications of each experiment and sample analysis were conducted, and the results were reported as mean ± standard deviation. Data were statistically analyzed by SPSS software (version 26.0, Inc., Chicago, IL, USA) using a *t*-test to ascertain a significant difference between BJ3-2, BJ3-2Δ*sdaAA*, BJ3-2Δ*katX*, and BJ3-2Δ*sdaAA*Δ*katX* at $p \leq 0.05$. Origin 2022 software (version 9.8.0.200) was used for plotting. The correlation between sensory attributes and key aroma compounds was analyzed by PLSR using the Unscrambler 9.7 version software (CAMO ASA, Oslo, Norway). TBtools v2.042 was used to make the heat map.

3. Results and Discussion

3.1. Screening of Soybean Flavor Candidate Genes of BJ3-2

To screen genes that may significantly impact the formation of soybean flavor, we conducted further analysis and screening of the prior multi-omics data from BJ3-2 incubated at different temperatures, as described in the literature [29]. Briefly, differential genes, proteins, and metabolites with $|\mathrm{Log_2FC}$ (FOLDCHANGE)$| \geq 2$, $|\mathrm{Log_2FC}| \geq 1.5$, and $|\mathrm{Log_2FC}| \geq 1$ were screened according to p-value < 0.05 and from the transcriptome, proteome, and metabolome for KEGG pathway mapping. Co-analyzed genes and metabolites are mainly involved in the glycine, serine, and threonine metabolism; ABC transporter, arginine and proline metabolism; cysteine and methionine metabolism; and tryptophan metabolism metabolic pathways. Meanwhile, differential genes and proteins with $|\mathrm{Log_2FC}| \geq 2.5$ were screened from the joint analysis results, and a total of *sdaAA* and *katX* were studied as candidate genes. In the transcriptome data, *sdaAA* was down-regulated by 2.50~fold (log_2FC) after incubating at 53 °C, and *katX* was down-regulated by 4.40~fold (log_2FC) after incubating at 53 °C (Table S2). RT-qPCR was further performed to confirm the gene expression of BJ3-2 (Table S3). As shown in Figure 1, the results show that the expression of *sdaAA* and *katX* was low at 53 °C and high at 37 °C, indicating the transcriptome data and the RT-qPCR results were consistent.

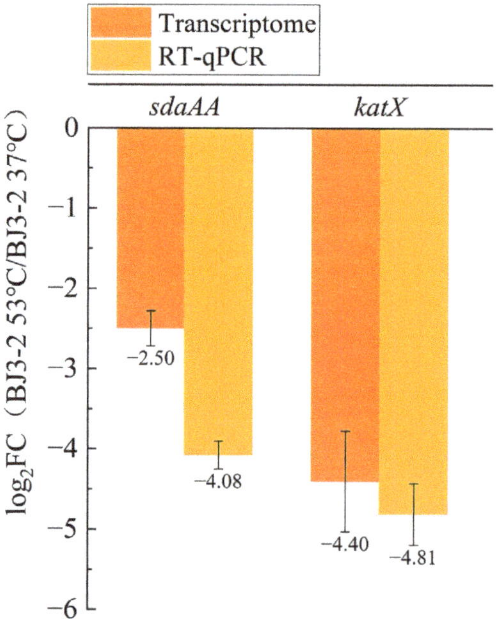

Figure 1. RT-qPCR verification of the expressions *sdaAA* and *katX*.

3.2. *Knockout of sdaAA and katX Had No Significant Effect on the Morphology and Growth of BJ3-2*

In this study, we constructed two single-knockout strains and one double-knockout strain by homologous recombination. After obtaining the knockout strain, it was determined whether the knockout affected the colonies' phenotypic characteristics and growth ability. The morphology of single colonies of BJ3-2 on solid plates was first identified. Colonies appeared white, dry, irregular in morphology, and internally wrinkled. The morphology of BJ3-2Δ*sdaAA*, BJ3-2Δ*katX*, and BJ3-2Δ*sdaAA*Δ*katX* colonies was comparable to that of BJ3-2 (Figure 2A,C,E,G).

Furthermore, the four strains (BJ3-2, BJ3-2Δ*sdaAA*, BJ3-2Δ*katX*, and BJ3-2Δ*sdaAA*Δ*katX*) were stained with a blue-purple tint and a short stick form by Gram staining, as shown in Figure 2B,D,F,H. The growth curves were then used to determine the growth ability of the strains. The results are shown in Figure 2I. The OD values of BJ3-2Δ*sdaAA*, BJ3-2Δ*katX*, and BJ3-2Δ*sdaAA*Δ*katX* were similar to those of BJ3-2 at different time points. These findings showed that the morphology and growth of BJ3-2 were not significantly affected by the knockout of *sdaAA* and *katX*.

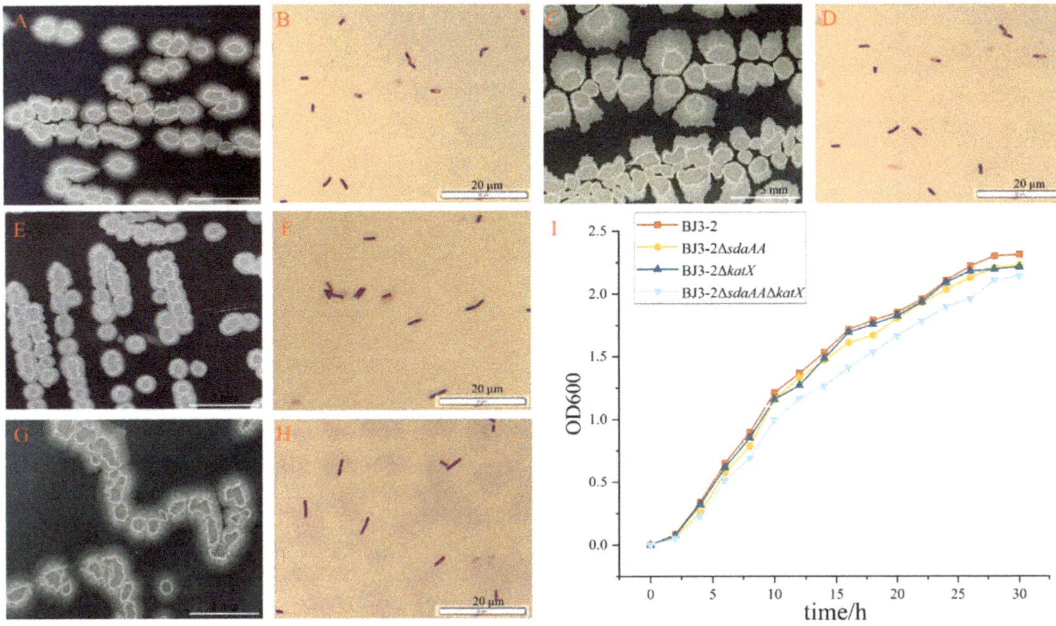

Figure 2. Colony morphology and growth curves of BJ3-2, BJ3-2Δ*sdaAA*, BJ3-2Δ*katX*, and BJ3-2Δ*sdaAA*Δ*katX*. (**A**) Colony morphology of BJ3-2. (**B**) Gram staining of BJ3-2. (**C**) Colony morphology of BJ3-2Δ*sdaAA*Δ*katX*. (**D**) Gram staining of BJ3-2Δ*sdaAA*Δ*katX*. (**E**) Colony morphology of BJ3-2Δ*sdaAA*. (**F**) Gram staining of BJ3-2Δ*sdaAA*. (**G**) Colony morphology of BJ3-2Δ*katX*. (**H**) Gram staining of BJ3-2Δ*katX*. (**I**) Growth curves of BJ3-2, BJ3-2Δ*sdaAA*, BJ3-2Δ*katX*, and BJ3-2Δ*sdaAA*Δ*katX*.

3.3. Sensory Analysis of Soybean Flavor in FSs with BJ3-2, BJ3-2ΔsdaAA, BJ3-2ΔkatX, and BJ3-2ΔsdaAAΔkatX at 37 °C

Quantitative descriptive analyses (QDAs) were applied to assess the differences in aroma profiles of four FSs after fermentation of soybeans with BJ3-2, BJ3-2Δ*sdaAA*, BJ3-2Δ*katX*, and BJ3-2Δ*sdaAA*Δ*katX* at 37 °C (Figure 3A). As shown in Figure 3B, the aroma profiles of soybean flavor from FSs were described by five sensory attributes: "sweaty", "smoky", "beany", "roasted", and "sweet". BJ3-2 was more frequently associated with the "sweaty (4.5 ± 0.45)" attribute, whereas the other attributes were at low levels, probably because acids obscure other attributes. BJ3-2Δ*sdaAA* and BJ3-2Δ*katX* had diminished "sweaty" attributes and enhanced "smoky", "beany", and "sweet" attributes compared to BJ3-2. BJ3-2Δ*sdaAA*Δ*katX* presented the lowest intensity of the "sweaty (2.3 ± 0.32)" attribute and the highest intensity of the "beany (2.6 ± 0.12)" and "roasted (3.1 ± 0.33)" attributes, with higher differences from BJ3-2. In addition, the "sweet" attribute had the lowest intensity of soybean flavor in FSs, and knockout of *sdaAA* and *katX* resulted in an increase in the intensity of the "sweet" attribute. The QDA results suggest that knockout of *sdaAA* and *katX* has a substantial effect on the flavor formation of soybean flavor and also validate the analysis of the mechanism of soybean flavor formation using multi-omics data.

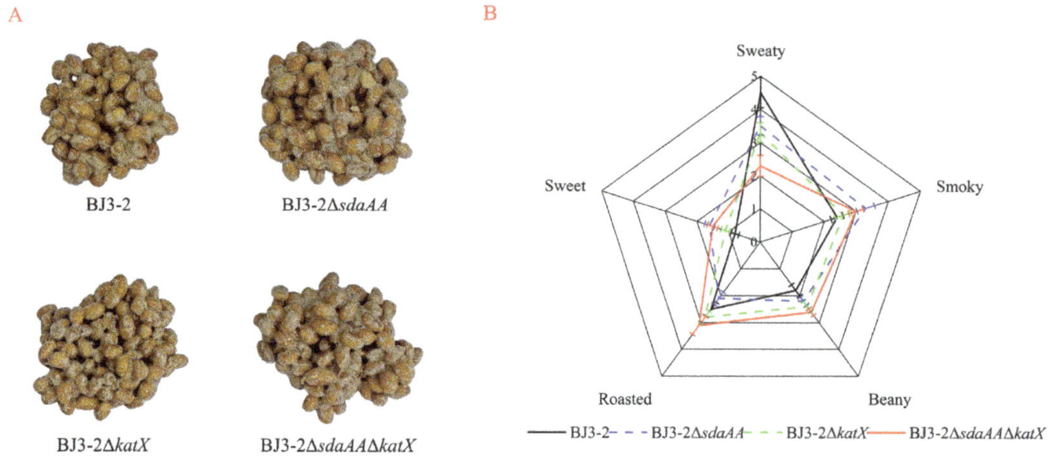

Figure 3. Aroma profile analysis of soybean flavor in fermented soybean samples with BJ3-2, BJ3-2Δ*sdaAA*, BJ3-2Δ*katX*, and BJ3-2Δ*sdaAA*Δ*katX*. (**A**) Fermented soybean samples with BJ3-2, BJ3-2Δ*sdaAA*, BJ3-2Δ*katX*, and BJ3-2Δ*sdaAA*Δ*katX* at 37 °C. (**B**) Aroma profile analysis of soybean flavor in fermented soybean samples with BJ3-2, BJ3-2Δ*sdaAA*, BJ3-2Δ*katX*, and BJ3-2Δ*sdaAA*Δ*katX* via QDA (panel number = 10).

3.4. Identification of Aroma-Active Compounds of Soybean Flavor in FSs with BJ3-2, BJ3-2ΔsdaAA, BJ3-2ΔkatX, and BJ3-2ΔsdaAAΔkatX by GC-MS-OSME

GC-MS-OSME was conducted to investigate the aroma-active compounds in four FSs. OSME (odor-specific magnitude estimation) is often used to determine the contribution of aroma-active compounds to the flavor based on aroma intensity (AI); higher AI values indicate a greater contribution to the flavor [25,30]. A total of 20 aroma-active compounds were detected, including 6 ketones, 5 pyrazines, 3 alcohols, 2 furans, 1 ester, 1 phenol, 1 acid, and 1 other (Table S4).

A total of 19, 20, 20, and 18 aroma-active compounds were found in BJ3-2, BJ3-2Δ*sdaAA*, BJ3-2Δ*katX*, and BJ3-2Δ*sdaAA*Δ*katX* FS samples, respectively (Table 1). Eight compounds with relatively high AI values coexisted in all FSs, including 2,3-butanedione (1.74 ± 0.20–2.58 ± 0.16), methyl 2-methylbutyrate (2.32 ± 0.14–2.98 ± 0.20), 2,4,5-trimethyloxazole (1.54 ± 0.46–2.48 ± 0.11), 2,5-dimethylpyrazine (2.89 ± 0.27–3.04 ± 0.12), 2,3,5-trimethylpyrazine (3.02 ± 0.17–3.46 ± 0.09), 1-octen-3-ol (2.65 ± 0.18–3.04 ± 0.16), 2-methylbutyric acid (3.34 ± 0.27–3.85 ± 0.24), and guaiacol (2.91 ± 0.37–3.12 ± 0.28). Therefore, these compounds are regarded as important contributors to soybean flavor. The compound 2-methylbutyric acid, known as an off-flavor compound in *Bacillus*-fermented foods [31,32], showed the highest AI (3.85 ± 0.24) in BJ3-2, followed by BJ3-2Δ*sdaAA* (3.58 ± 0.16), BJ3-2Δ*katX* (3.46 ± 0.14), and BJ3-2Δ*sdaAA*Δ*katX* (3.34 ± 0.27). In this study, through the aroma characterization of 2-methylbutyric acid in GC-O and its standard compounds, we concluded that 2-methylbutyric acid is responsible for the "sweaty" attribute.

The aroma compounds 2,5-dimethylpyrazine, 2,3,5-trimethylpyrazine, and 3-ethyl-2,5-dimethylpyrazine have previously been reported in natto [12]. These pyrazines typically provide a "roasted" or "nutty" attribute. In this study, these three pyrazines showed different AI values among the four FSs. Guaiacol also showed a relatively higher AI value in four FSs, and it has a smoky flavor, which should be related to the "smoky" attributes of soybean flavor. 1-Octen-3-ol imparts a mushroom flavor, which is here related to the "beany" attribute. In addition, 2,4,5-trimethyloxazole contributed to the "roasted" attribute and 2-pentylfuran to the "beany" attribute.

Table 1. GC-O/AI and ROAV analysis of aroma-active compounds of soybean flavor in fermented soybean samples with BJ3-2, BJ3-2Δ*sdaAA*, BJ3-2Δ*katX*, and BJ3-2Δ*sdaAA*Δ*katX*.

	GC-O (AI)				ROAV			
Compounds	**BJ3-2**	**BJ3-2Δ*sdaAA***	**BJ3-Δ*katX***	**BJ3-2Δ*sdaAA*Δ*katX***	**BJ3-2**	**BJ3-2Δ*sdaAA***	**BJ3-2Δ*katX***	**BJ3-2Δ*sdaAA*Δ*katX***
3-Methyl-2-butanone	0.16 ± 0.03	0.1 ± 0.02	0.13 ± 0.03	0.17 ± 0.05	0.1 ± 0.00	0.01 ± 0.00	0.01 ± 0.00	0.01 ± 0.00
2-Ethylfuran	1.25 ± 0.09	1.37 ± 0.12	1.85 ± 0.06	1.45 ± 0.26	3.30 ± 0.04	5.38 ± 1.09	6.70 ± 2.72	4.68 ± 0.27
2,3-Butanedione	1.74 ± 0.20	2.1 ± 0.18	2.35 ± 0.23	2.58 ± 0.16	13.72 ± 2.96	10.06 ± 0.38	29.19 ± 7.18	54.08 ± 5.94
Methyl 2-methylbutyrate	2.72 ± 0.17	2.98 ± 0.20	2.32 ± 0.14	2.49 ± 0.09	56.80 ± 8.96	109.30 ± 18.00	55.16 ± 7.72	48.45 ± 8.24
2-Heptanone	1.14 ± 0.08	1.05 ± 0.04	1.26 ± 0.09	1.31 ± 0.08	1.12 ± 0.59	1.69 ± 0.10	2.64 ± 0.53	2.06 ± 0.16
2,4,5-Trimethyloxazole	2.17 ± 0.12	1.54 ± 0.46	2.32 ± 0.22	2.48 ± 0.11	17.33 ± 0.94	4.20 ± 0.32	17.98 ± 1.78	18.27 ± 1.08
2-Pentylfuran	1.06 ± 0.04	1.12 ± 0.13	1.53 ± 0.08	1.35 ± 0.11	1.63 ± 0.12	2.39 ± 0.49	3.41 ± 1.24	2.90 ± 0.88
3-Octanone	1.5 ± 0.09	1.54 ± 0.10	1.38 ± 0.16	1.62 ± 0.12	2.93 ± 0.83	1.87 ± 1.60	4.29 ± 0.27	2.96 ± 0.52
2-Octanone	0.95 ± 0.07	0.37 ± 0.09	0.54 ± 0.02	0.00 ± 0.00	1.35 ± 0.39	0.72 ± 0.14	0.64 ± 0.25	0.00 ± 0.00
2,5-Dimethylpyrazine	2.89 ± 0.27	2.95 ± 0.35	2.93 ± 0.14	3.04 ± 0.12	67.54 ± 2.72	88.58 ± 0.46	76.63 ± 1.35	79.50 ± 0.55
2-Nonanone	1.05 ± 0.15	0.19 ± 0.07	0.84 ± 0.17	0.32 ± 0.28	1.01 ± 0.09	0.64 ± 0.05	1.15 ± 0.10	0.79 ± 0.08
1-Hexanol	0.43 ± 0.04	0.45 ± 0.05	0.54 ± 0.12	0.57 ± 0.16	0.13 ± 0.01	0.15 ± 0.02	0.17 ± 0.01	0.20 ± 0.01
2-Ethyl-5-methylpyrazine	0.00 ± 0.00	0.24 ± 0.06	0.19 ± 0.12	0.00 ± 0.00	0.01 ± 0.00	0.02 ± 0.00	0.01 ± 0.00	0.02 ± 0.00
2,3,5-Trimethylpyrazine	3.02 ± 0.17	3.27 ± 0.14	3.35 ± 0.19	3.46 ± 0.09	185.94 ± 9.72	147.72 ± 1.56	208.56 ± 6.54	212.19 ± 3.40
1-Octen-3-ol	2.65 ± 0.18	2.74 ± 0.24	2.95 ± 0.14	3.04 ± 0.16	45.85 ± 1.43	52.28 ± 3.59	53.02 ± 1.49	55.10 ± 2.82
3-Ethyl-2,5-dimethylpyrazine	1.76 ± 0.23	1.83 ± 0.19	1.94 ± 0.35	1.68 ± 0.17	2.19 ± 0.30	2.16 ± 0.01	3.06 ± 0.23	3.93 ± 0.11
Tetramethylpyrazine	0.42 ± 0.04	0.58 ± 0.06	0.45 ± 0.02	0.38 ± 0.09	0.01 ± 0.00	0.00 ± 0.00	0.01 ± 0.00	0.01 ± 0.00
3-Octanol	0.56 ± 0.06	0.73 ± 0.10	0.94 ± 0.05	1.23 ± 0.07	0.03 ± 0.00	0.02 ± 0.00	0.02 ± 0.00	0.04 ± 0.00
2-Methylbutyric acid	3.85 ± 0.24	3.58 ± 0.16	3.46 ± 0.14	3.34 ± 0.27	591.46 ± 9.51	341.41 ± 3.40	327.74 ± 11.86	241.16 ± 10.44
Guaiacol	2.91 ± 0.37	3.08 ± 0.18	2.88 ± 0.24	3.12 ± 0.28	164.85 ± 19.76	248.76 ± 31.59	179.23 ± 22.06	236.32 ± 5.65

Note: Orange color in the table indicates compounds that are common to both GC-O and ROAV > 1. Data are presented as mean ± SD.

Several other compounds, including 2-ethylfuran, 2,3-butanedione, methyl 2-methylbutyrate, 2-heptanone, and 3-octanone, were associated with the "sweet" attribute, as these compounds were perceived as having a pleasant flavor, neutralizing the overall flavor of soybean. Concentration differences mainly caused differences in AI values in fermented soybean samples from different strains. Therefore, these compounds need to be quantified and relative odor activity values calculated.

3.5. Key Aroma Compounds of Soybean Flavor in FSs with BJ3-2, BJ3-2ΔsdaAA, BJ3-2ΔkatX, and BJ3-2ΔsdaAAΔkatX Identified via ROAV

The contribution of aroma-active compounds to overall flavor is related to concentration and odor threshold values. After semi-quantification using the internal standard method, ROAV was further calculated to determine the contribution of aroma-active compounds to soybean flavor. Aroma-active compounds with ROAV > 1 were considered to be the contributors to the soybean flavor of FSs. As shown in Table 1, there were 15, 13, 14, and 13 compounds with ROAV > 1 in FSs with BJ3-2, BJ3-2Δ*sdaAA*, BJ3-2Δ*katX*, and BJ3-2Δ*sdaAA*Δ*katX*, respectively.

Among these compounds, the highest values of ROAV were for 2-methylbutyric acid (ROAV 241.16 ± 10.44–591.46 ± 9.51), 2,3,5-trimethylpyrazine (ROAV 147.72 ± 1.56–212.19 ± 3.40), and guaiacol (ROAV 164.85 ± 19.76–248.76 ± 31.59), implying they contribute significantly to the aroma of soybean flavor in FSs. Relatively lower ROAV for 2,3-butanedione (ROAV 10.06 ± 0.38–54.08 ± 5.94), methyl 2-methylbutyrate (ROAV 48.45 ± 8.24–109.3 ± 18.00), 2,4,5-trimethyloxazole (ROAV 4.2 ± 0.32–18.27 ± 1.08), 2,5-dimethylpyrazine (ROAV 67.54 ± 2.72–88.58 ± 0.46), and 1-octen-3-ol (ROAV 45.85 ± 1.43–55.1 ± 2.82). The lowest ROAV were found for 2-ethylfuran (ROAV 3.3 ± 0.04–6.7 ± 2.72), 2-heptanone (ROAV 1.12 ± 0.59–2.64 ± 0.53), 2-pentylfuran (ROAV 1.63 ± 0.12–3.41 ± 1.24), 3-octanone (ROAV 1.87 ± 1.60–4.29 ± 0.27), and 3-ethyl-2,5-dimethylpyrazine (ROAV 2.16 ± 0.01–3.93 ± 0.11). These compounds are also closely related to the aroma formation of soybean flavor in FSs. Taken together, the ROAV calculations for most of the compounds were in agreement with the AI values of the GC-O, suggesting a high degree of agreement between the two methods. Of these 20 aroma-active compounds from GC-O, 13 were detected with ROAV > 1 in all FSs, indicating their crucial influence on the formation of soybean flavor in FSs. The 13 key aroma compounds are listed in Figure 4. The chemical structures of these compounds were further identified by comparing the retention index (RI) and aroma profile with those of the standard compounds and used for further analysis.

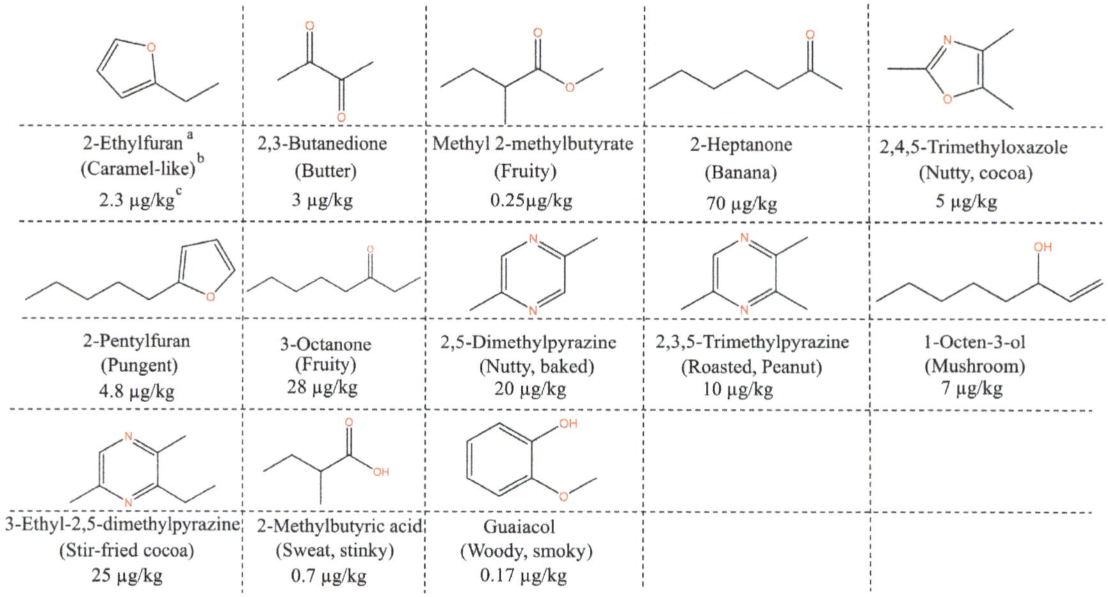

Figure 4. Information on thirteen key aroma compounds from GC-O (AI) combined with ROAV. [a] Thirteen key aroma compounds from GC-O combined with ROAV; [b] aroma characteristics of key aroma compounds; [c] odor thresholds of compounds in water reported in the literature.

3.6. Relationship between Key Aroma Compounds and Sensory Attributes of Soybean Flavor in FSs

Partial least squares regression (PLSR) was performed to investigate the relationship between sensory attributes and key aroma compounds in the four FS samples. The X-matrix was set as the thirteen key aroma compounds, and the Y-matrix was set as five sensory attributes. The PLSR consisted of two significant PCs explaining 75% of the X-variance and 92% of the Y-variance (Figure 5), while the inner ellipse and outer ellipse explained 50% and 100% of the variance, respectively. All variables are located outside the small ellipse except 2-ethylfuran (1) and 3-octanone (7), which shows that these variables were well explained by the PLSR model. The low values of RMSEC (0.11) and RMSRP (0.19) also indicate that the regression model performs well. As shown in Figure 5, the closer to the variable Y, the higher the correlation between X and Y [33]. The compound 2-Methylbutyric acid (12) was highly associated with the "sweaty" attribute. The "smoky" attribute was strongly influenced by guaiacol (13). In addition, 2,3,5-trimethylpyrazine (9) and 1-octen-3-ol (10) showed robust correlation with "roasted" and "beany" attributes, respectively. In addition, the "sweet" attribute was weakly associated with certain aroma compounds. This might be because more compounds are implicated in the formation of this sensory attribute. These results are consistent with the GC-O and ROAV analyses (Table 1).

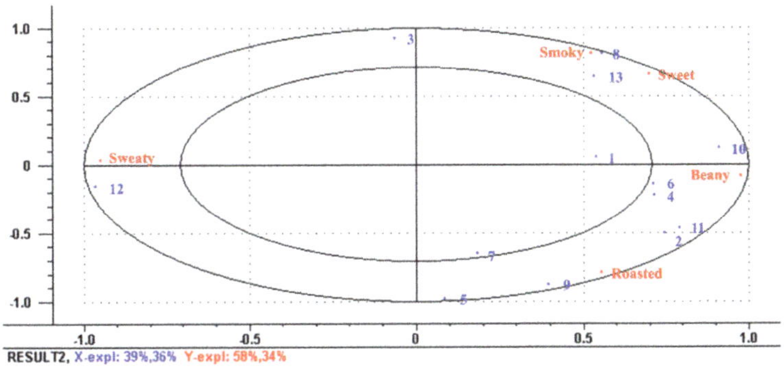

Figure 5. The correlation loading plots between the concentration of the thirteen key aroma compounds (X matrix) and scores of sensory attributes (Y matrix) in FSs with BJ3-2, BJ3-2Δ*sdaAA*, BJ3-2Δ*katX*, and BJ3-2Δ*sdaAA*Δ*katX*; the validation method is full cross-validation; ellipses indicate $r^2 = 0.5$ and 1.0, respectively. Numbers indicate the following compounds: 1, 2-ethylfuran; 2, 2,3-butanedione; 3, methyl 2-methylbutyrate; 4, 2-heptanone; 5, 2,4,5-trimethyloxazole; 6, 2-pentylfuran; 7, 3-octanone; 8, 2,5-dimethylpyrazine; 9, 2,3,5-trimethylpyrazine; 10, 1-octen-3-ol; 11, 3-ethyl-2,5-dimethylpyrazine; 12, 2-methylbutyric acid; 13, guaiacol.

3.7. Changes in Key Aroma Compounds in FSs between BJ3-2, BJ3-2ΔsdaAA, BJ3-2ΔkatX, and BJ3-2ΔsdaAAΔkatX

Based on semi-quantitative results, significant difference analyses ($p \leq 0.05$) were performed on the content of 13 key aroma compounds between the four FSs with BJ3-2, BJ3-2Δ*sdaAA*, BJ3-2Δ*katX*, and BJ3-2Δ*sdaAA*Δ*katX*.

As shown in Figure 6, in comparison to FSs with BJ3-2, the content of 2-methylbutyric acid in FSs with BJ3-2Δ*katX*, BJ3-2Δ*sdaAA*, and BJ3-2Δ*sdaAA*Δ*katX* was significantly lower, with BJ3-2Δ*sdaAA*Δ*katX* having the lowest content. The lowest content of 2-methylbutyric acid in BJ3-2Δ*sdaAA*Δ*katX* also explains well why it possesses the lowest intensity of "sweaty" attributes (Figure 3B). This compound may have a positive correlation with the intensity of "sweaty" attributes. 2,3,5-Trimethylpyrazine was decreased in BJ3-2Δ*sdaAA* and significantly increased in both BJ3-2Δ*katX* and BJ3-2Δ*sdaAA*Δ*katX*. Meanwhile, in contrast to BJ3-2, the "roasted" attribute of BJ3-2*sdaAA* was weakened, whereas the "roasted" attribute of BJ3-2Δ*katX* and BJ3-2Δ*sdaAA*Δ*katX* was enhanced (Figure 3B), indicating that 2,3,5-trimethylpyrazine was associated with the "roasted" attribute. Guaiacol and 1-octen-3-ol were increased in BJ3-2Δ*katX*, BJ3-2Δ*sdaAA*, and BJ3-2Δ*sdaAA*Δ*katX*, which is consistent with the QDA results (Figure 3B) of enhanced "smoky" and "beany" attributes of soybean flavor in FSs. Several other compounds also showed some variations in the content of the four FSs. In summary, knockout of *sdaAA* and *katX* resulted in a significant increase or decrease in the content of key aroma compounds, which mainly resulted in a significant decrease in 2-methylbutyric acid and a significant increase in 2,5-dimethylpyrazine, 2-ethylfuran, 2-pentylfuran, 2-heptanone, 1-octen-3-ol, and guaiacol, thus affecting the soybean flavor.

Additionally, KEGG pathway analysis shows that *sdaAA* (which encodes serine dehydratase) is involved in metabolic pathways, mainly including glycine, serine, and threonine metabolism and cysteine and methionine metabolism. *katX* (encodes catalase) is involved in metabolic pathways, mainly tryptophan metabolism, glyoxylate metabolism, and dicarboxylate metabolism. This implies that the knockout of *sdaAA* and *katX* may impact the production of aroma compounds by affecting these metabolic pathways, leading to altered sensory attributes in FS with its mutants. However, the roles and specific regulatory mechanisms of *sdaAA* and *katX* in the production of soybean flavor still require further investigation.

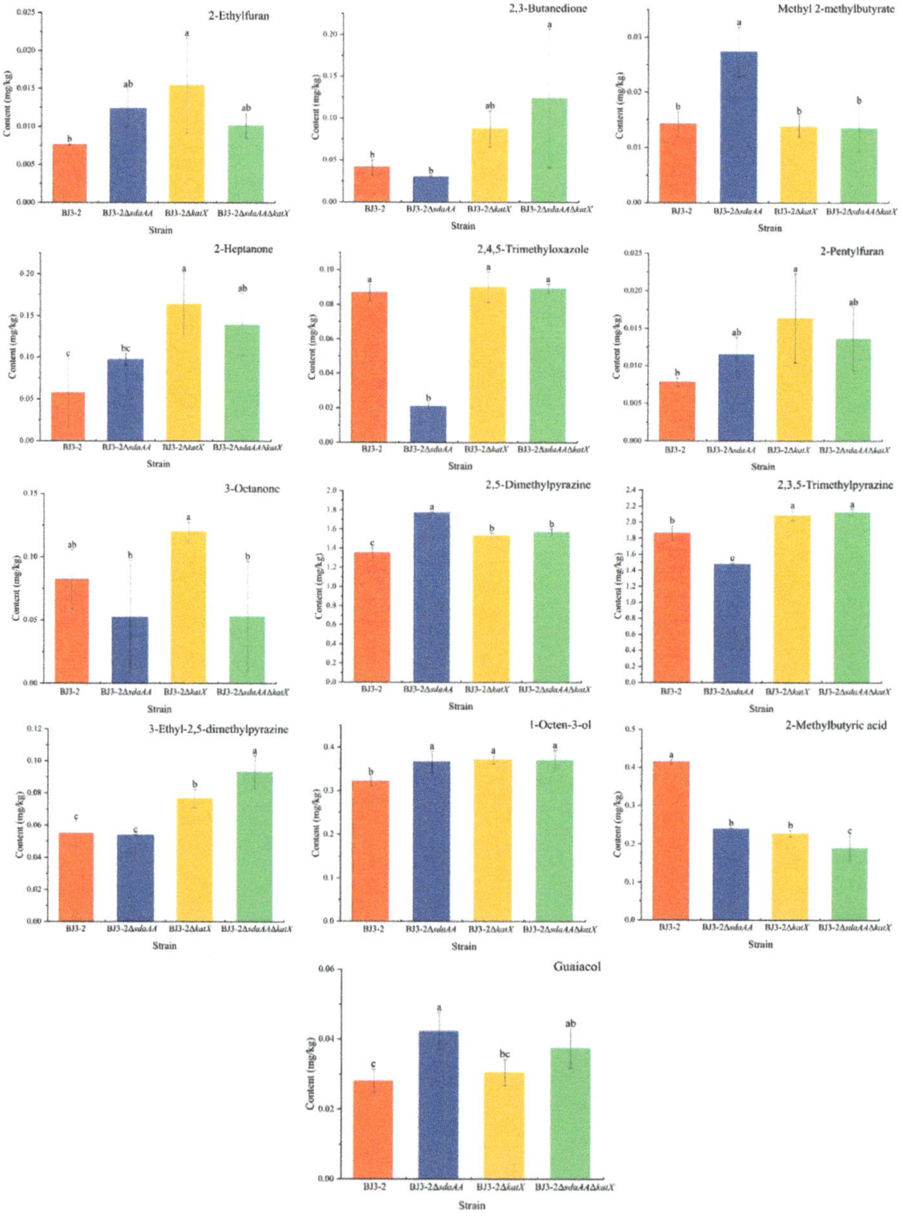

Figure 6. Changes in content of key aroma compounds among the four fermented soybean samples with BJ3-2, BJ3-2Δ*sdaAA*, BJ3-2Δ*katX*, and BJ3-2Δ*sdaAA*Δ*katX*. Significant differences between groups are indicated by letters a, b, c.

3.8. Aroma Addition Experiment

Based on our analyses and further validation of key aroma compounds, BJ3-2 and BJ3-2Δ*sdaAA*Δ*katX*, which have greater differences in aroma profiles, were selected for an aroma addition experiment. Aroma addition experimental models were prepared by adjusting the contents of key aroma compounds in the two samples to be consistent based on the difference in concentrations of the 13 key aroma compounds in the two FSs. After that, it was determined whether there was a change in the aroma profiles of soybean flavor in the two FSs or whether the aroma profiles converged to verify the effect of these key aroma compounds. As shown in Figure 7, QDA was applied to access the aroma profiles of the two FSs after adding the aroma compounds. The aroma profiles of both samples changed after the addition of the differential aroma compounds, and both samples converged in "sweaty", "smoky", "roasted", and "beany" attributes but still differed in "sweet". This may be due to the fact that the "sweet" attribute of BJ3-2 is masked by the strong intensity of the "sweaty" attribute. The result of the aroma addition experiment suggests that these compounds are closely related to the aroma profiles of soybean flavor and also validates these substances as key aroma compounds of soybean flavor in FS.

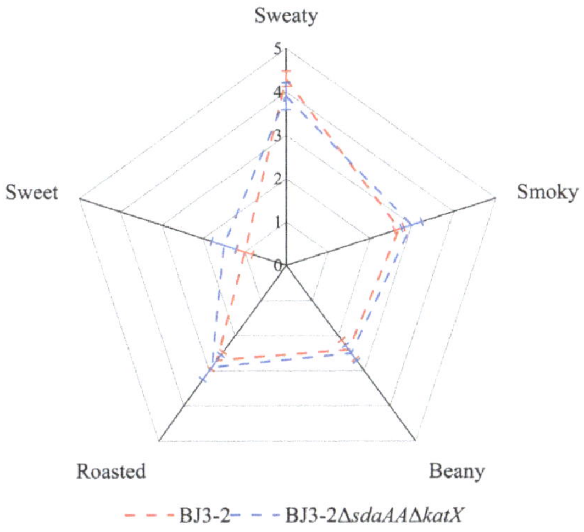

Figure 7. Aroma profile of soybean flavor in fermented soybean samples with BJ3-2 and BJ3-2Δ*sdaAA*Δ*katX* after addition of aroma compounds (panel number = 10).

3.9. Source of Key Aroma Compounds of Soybean Flavor in Fermented Soybeans

The results of the key aroma compounds of soybean flavor in FS showed that 2-methylbutyric acid produced the main compounds contributing to "sweaty" attributes. Moreover, 2-methylbutyric acid was identified as a short-chain fatty acid. The source of 2-methylbutyric acid may be due to the ability of *B. subtilis* to metabolically convert isoleucine, leucine, and valine to 2-methylbutyric acid during fermentation [34]. Meanwhile, 2-methylbutyric acid can be converted to methyl 2-methylbutyrate [35].

Furthermore, 2,5-dimethylpyrazine, 2,3,5-trimethylpyrazine, and 3-ethyl-2,5-dimethylpyrazine have been detected in a number of fermented foods, such as white wine, curd, and natto [12,36,37]. Pyrazines can be produced by non-enzymatic Maillard reactions and enzymatic reactions caused by *B. subtilis* [13]. During fermentation, exogenous enzymes secreted by *B. subtilis* hydrolyze proteins and carbohydrates into monosaccharides and free amino acids, respectively. These pyrazines can be produced from glucose and L-threonine [3].

Guaiacol is also found to be an important aroma compound and can be produced by the degradation of vanillin by the enzyme vanillin dehydrogenase in *B. subtilis* [38]. Moreover, 2,3-butanedione usually has a pleasant, creamy, or buttery taste and can be metabolized by *B. subtilis* to produce α-acetolactate (which can be converted to acetoin) and 2,3-butanedione (which can also be converted to each other under the catalysis of acetoin/butanediol dehydrogenase) [39]. Cysteine/alanine has been reported to react with 2,3-butanedione to form 2,4,5-trimethyloxazole [40], and 2-heptanone can also be produced by the Maillard reaction or biosynthesized from acid [41]. The metabolic pathways of

3-octanone and 1-octen-3-ol in fungi are well defined, whereas the biosynthetic pathway in *B. subtilis* has not been elucidated. The fungal pathway is mainly based on the oxidative degradation of unsaturated fatty acids [42,43].

4. Conclusions

In this study, the aroma profiles and key aroma compounds of soybean flavor in FS with *B. subtilis* BJ3-2 and its knockout mutants at 37 °C were analyzed using QDA, GC-O-MS, ROAV calculation, and aroma addition experiments. QDA results revealed that the overall aroma of soybean flavor in FSs consists of "sweaty", "smoky", "roasted", "beany", and "sweet" attributes. In addition, the knockout of soybean flavor candidate genes (*sdaAA* and *katX*) did not significantly affect the morphology and growth of BJ3-2 but had a greater effect on the aroma profiles of soybean flavor in FSs with BJ3-2. A total of 20 aroma-active compounds were detected by GC-O-MS from four FSs with BJ3-2, BJ3-2Δ*sdaAA*, BJ3-2Δ*katX*, and BJ3-2Δ*sdaAA*Δ*katX*, and 13 key aroma compounds were identified as key aroma compounds according to their high ROAVs. Furthermore, PLSR revealed a good correlation between the key aroma compounds and sensory attributes. Knockout of *sdaAA* and *katX* resulted in a significant increase or decrease in the content of key aroma compounds. Finally, combined with the aroma addition experiments, 2-methylbutyric acid, 2,3,5-trimethylpyrazine, guaiacol, and 1-octen-3-ol were identified as the aroma compounds responsible for the main aroma profiles of soybean flavor. The findings of this study provide new insights into the analysis of the soybean flavor of fermented soybeans produced by *B. subtilis* BJ3-2, which is important for quality control and improvement of the aroma of fermented soybeans.

Supplementary Materials: The following supporting information can be downloaded at: https://www.mdpi.com/article/10.3390/fermentation10080409/s1, Table S1: Primers used in this study; Table S2: Transcriptome data at 37 °C and 53 °C; Table S3: Reverse transcription–quantitative real-time PCR at 37 °C and 53 °C; Table S4: Relative content (ug/kg) of volatile compounds in FSs by HS-SPME-GC-O-MS (μg/kg); Table S5: Data used in the aroma addition experiment; Figure S1: Construction and transformation of *sdaAA* homologous recombination knockout vector; Figure S2: Construction and transformation of *katX* homologous recombination knockout vector; Figure S3: Construction and transformation of *sdaAA* and *katX* homologous recombination knockout vector; Figure S4: Sequencing of BJ3-2Δ*sdaAA*; Figure S5: Sequencing of BJ3-2Δ*katX*; Figure S6: Sequencing of BJ3-2Δ*sdaAA*Δ*katX*; Figure S7: Fermented soybeans with BJ3-2, BJ3-2Δ*sdaAA*, BJ3-2Δ*katX*, and BJ3-2Δ*sdaAA*Δ*katX* at 37 °C.

Author Contributions: Data curation, Z.C., F.W. and Y.T.; Formal analysis, Z.C., Q.T. and F.W.; Funding acquisition, Y.W.; Methodology, Z.C.; Project administration, S.T. and C.L.; Validation, J.J.; Writing—original draft, Z.C.; Writing—review and editing, Y.W. and L.Z. All authors have read and agreed to the published version of the manuscript.

Funding: This work was supported by the National Natural Science Foundation of China (Grant No. 32060585).

Institutional Review Board Statement: As the experimental samples used in the study were consumed in daily life, the study was exempted from ethical review and approval.

Informed Consent Statement: Informed consent was obtained from all panelists involved in the study. Participants agreed to the use of their data/answers. The study strictly adhered to a protocol designed to protect the rights and privacy of all participants.

Data Availability Statement: The original contributions presented in the study are included in the article/Supplementary Material, and further inquiries can be directed to the corresponding author.

Conflicts of Interest: The authors declare no conflicts of interest.

References

1. Gopikrishna, T.; Suresh, K.H.; Perumal, K.; Elangovan, E. Impact of *Bacillus* in fermented soybean foods on human health. *Ann. Microbiol.* **2021**, *71*, 30. [CrossRef] [PubMed]
2. Ardö, Y. Flavour formation by amino acid catabolism. *Biotechnol. Adv.* **2006**, *24*, 238–242. [CrossRef] [PubMed]
3. Visessanguan, W.; Benjakul, S.; Potachareon, W.; Panya, A.; Riebroy, S. Accelerated proteolysis of soy proteins during fermentation of Thua-nao inoculated with *Bacillus subtilis*. *J. Food Biochem.* **2005**, *29*, 349–366. [CrossRef]
4. Liu, Z.; Wu, Y.; Zhang, L.; Tong, S.; Jin, J.; Gong, X.; Zhong, J. *rocF* affects the production of tetramethylpyrazine in fermented soybeans with *Bacillus subtilis* BJ3-2. *BMC Biotechnol.* **2022**, *22*, 1–18. [CrossRef] [PubMed]
5. Wu, Y.; Tao, Y.; Jin, J.; Tong, S.; Li, S.; Zhang, L. Multi-omics analyses of the mechanism for the formation of soy sauce-like and soybean flavor in *Bacillus subtilis* BJ3-2. *BMC Microbiol.* **2022**, *22*, 1–13. [CrossRef] [PubMed]
6. Liu, Y.; Li, W.; Tan, Q.; Li, Z.; Wang, H.; Suo, H. Characterization and comparison of key aroma compounds in traditional and rapid Yongchuan douchi by molecular sensory science. *LWT* **2024**, *191*, 115607. [CrossRef]
7. Zhao, W.; Liang, Z.; Qian, M.; Li, X.; Dong, H.; Bai, W.; Wei, Y.; He, S. Evolution of microbial communities during fermentation of Chi-flavor type Baijiu as determined by high-throughput sequencing. *LWT* **2022**, *170*, 114102. [CrossRef]
8. Wang, S.; Chang, Y.; Liu, B.; Chen, H.; Sun, B.; Zhang, N. Characterization of the Key Aroma-Active Compounds in Yongchuan Douchi (Fermented Soybean) by Application of the Sensomics Approach. *Molecules* **2021**, *26*, 3048. [CrossRef]
9. Tan, F.; Wang, P.; Zhan, P.; Tian, H. Characterization of key aroma compounds in flat peach juice based on gas chromatography-mass spectrometry-olfactometry (GC-MS-O), odor activity value (OAV), aroma recombination, and omission experiments. *Food Chem.* **2022**, *366*, 130604. [CrossRef]
10. Frauendorfer, F.; Schieberle, P. Identification of the Key Aroma Compounds in Cocoa Powder Based on Molecular Sensory Correlations. *J. Agric. Food Chem.* **2006**, *54*, 5521–5529. [CrossRef]
11. Zellner, B.D.A.; Dugo, P.; Dugo, G.; Mondello, L. Gas chromatography–olfactometry in food flavour analysis. *J. Chromatogr. A* **2008**, *1186*, 123–143. [CrossRef] [PubMed]
12. Liu, Y.; Song, H.; Luo, H. Correlation between the key aroma compounds and gDNA copies of *Bacillus* during fermentation and maturation of natto. *Food Res. Int.* **2018**, *112*, 175–183. [CrossRef] [PubMed]
13. Mathatheeranan, P.; Wongprasert, T.; Na Nan, T.; Suwannakul, E.; Wang, Y.; Lu, T.; Fang, M.; Suppavorasatit, I. Volatile profiles and aroma-active components of Northern Thai fermented soybean (thua nao) produced from controlled and uncontrolled fermentations. *Int. J. Gastron. Food Sci.* **2023**, *33*, 100763. [CrossRef]
14. Xie, C.; Zeng, H.; Wang, C.; Xu, Z.; Qin, L. Volatile flavour components, microbiota and their correlations in different sufu, a Chinese fermented soybean food. *J. Appl. Microbiol.* **2018**, *125*, 1761–1773. [CrossRef] [PubMed]
15. Guo, Q.; Peng, J.; He, Y. A Systematic Comparative Study on the Physicochemical Properties, Volatile Compounds, and Biological Activity of Typical Fermented Soy Foods. *Foods* **2024**, *13*, 415. [CrossRef] [PubMed]
16. McFeeters, R.F. Fermentation Microorganisms and Flavor Changes in Fermented Foods. *J. Food Sci.* **2004**, *69*, FMS35–FMS37. [CrossRef]
17. Maoz, I.; Lewinsohn, E.; Gonda, I. Amino acids metabolism as a source for aroma volatiles biosynthesis. *Curr. Opin. Plant Biol.* **2022**, *67*, 102221. [CrossRef]
18. Zhang, L.; Zhu, M.; Ren, L.; Li, A.; Chen, G.; Hu, Z. The *SlFSR* gene controls fruit shelf-life in tomato. *J. Exp. Bot.* **2018**, *69*, 2897–2909. [CrossRef] [PubMed]
19. Zhang, X.Z.; Cui, Z.L.; Hong, Q.; Li, S.P. High-level expression and secretion of methyl parathion hydrolase in *Bacillus subtilis* WB800. *Appl. Environ. Microbiol.* **2005**, *71*, 4101–4103. [CrossRef]
20. Phan, T.T.P.; Nguyen, H.D.; Schumann, W. Novel plasmid-based expression vectors for intra- and extracellular production of recombinant proteins in *Bacillus subtilis*. *Protein Expr. Purif.* **2006**, *46*, 189–195. [CrossRef]
21. Zhang, L.; Qin, Z.; Zhang, L.; Jiang, Y.; Zhu, J. Dynamic changes of quality and flavor characterization of Zhejiang rosy vinegar during fermentation and aging based on untargeted metabolomics. *Food Chem.* **2023**, *404*, 134702. [CrossRef]
22. Peng, X.; Li, X.; Shi, X.; Guo, S. Evaluation of the aroma quality of Chinese traditional soy paste during storage based on principal component analysis. *Food Chem.* **2014**, *151*, 532–538. [CrossRef]
23. Peng, Y.; Bishop, K.S.; Zhang, J.; Chen, D.; Quek, S.Y. Characterization of phenolic compounds and aroma active compounds in feijoa juice from four New Zealand grown cultivars by LC-MS and HS-SPME-GC-O-MS. *Food Res. Int.* **2020**, *129*, 108873. [CrossRef]
24. Zhao, M.; Li, T.; Yang, F.; Cui, X.; Zou, T.; Song, H.; Liu, Y. Characterization of key aroma-active compounds in Hanyuan Zanthoxylum bungeanum by GC-O-MS and switchable GC × GC-O-MS. *Food Chem.* **2022**, *385*, 132659. [CrossRef]
25. Yue, Y.; Wang, C.; Chen, Y.; Zheng, M.; Zhang, Y.; Deng, Q.; Zhou, Q. Aroma characteristics of flaxseed milk via GC-MS-O and odor activity value calculation: Imparts and selection of different flaxseed varieties. *Food Chem.* **2024**, *432*, 137095. [CrossRef]
26. Leo, V. *Compilations of Flavour Threshold Values in Water and Other Media*; Science Press: Beijing, China, 2015.
27. Giri, A.; Osako, K.; Okamoto, A.; Ohshima, T. Olfactometric characterization of aroma active compounds in fermented fish paste in comparison with fish sauce, fermented soy paste and sauce products. *Food Res. Int.* **2010**, *43*, 1027–1040. [CrossRef]
28. Chen, J.; Pu, D.; Shi, Y.; Sun, B.; Guo, H.; Li, K.; Zhang, Y. Characterization of the Key Aroma Compounds in Different Yeast Proteins by GC-MS/O, Sensory Evaluation, and E-Nose. *Foods* **2023**, *12*, 3136. [CrossRef]

29. Tao, Y. Multi-Omics Screening of Key Genes of Soybean Flavor Produced by *Bacillus subtilis* and Their Functional Studies. Master's Thesis, Guizhou University, Guiyang, China, 2022.
30. Qian, M.; Nelson, C.; Bloomer, S. Evaluation of fat-derived aroma compounds in blue cheese by dynamic headspace GC/Olfactometry-MS. *J. Am. Oil Chem. Soc.* **2002**, *79*, 663–667. [CrossRef]
31. Leejeerajumnean, A.; Duckham, S.C.; Owens, J.D.; Ames, J.M. Volatile compounds in *Bacillus*-fermented soybeans. *J. Sci. Food Agric.* **2001**, *81*, 525–529. [CrossRef]
32. Hong, Y.; Jung, H.; Kim, H. Aroma characteristics of fermented Korean soybean paste (Doenjang) produced by *Bacillus amyloliquefaciens*. *Food Sci. Biotechnol.* **2012**, *21*, 1163–1172. [CrossRef]
33. Zheng, Y.; Li, Y.; Pan, L.; Guan, M.; Yuan, X.; Li, S.; Ren, D.; Gu, Y.; Liang, M.; Yi, L. Aroma and taste analysis of pickled tea from spontaneous and yeast-enhanced fermentation by mass spectrometry and sensory evaluation. *Food Chem.* **2024**, *442*, 138472. [CrossRef] [PubMed]
34. Kai, M. Diversity and Distribution of Volatile Secondary Metabolites Throughout *Bacillus subtilis* Isolates. *Front. Microbiol.* **2020**, *11*, 559. [CrossRef]
35. Zhang, Y.; Fraatz, M.A.; Birk, F.; Rigling, M.; Hammer, A.; Zorn, H. Enantiomeric ratios of 2-methylbutanoic acid and its methyl ester: Elucidation of novel biogenetic pathways towards (R)-methyl 2-methylbutanoate in a beverage fermented with shiitake. *Food Chem.* **2018**, *266*, 475–482. [CrossRef]
36. Yoon, K.H.; Chang, Y.I.; Lee, G.H. Characteristic aroma compounds of cooked and fermented soybean (Chungkook-Jang) inoculated with various *Bacilli*. *J. Sci. Food Agric.* **2013**, *93*, 85–92. [CrossRef]
37. Fan, W.; Xu, Y.; Zhang, Y. Characterization of pyrazines in some Chinese liquors and their approximate concentrations. *J. Agric. Food Chem.* **2007**, *55*, 9956–9962. [CrossRef]
38. Graf, N.; Wenzel, M.; Altenbuchner, J. Identification and characterization of the vanillin dehydrogenase YfmT in *Bacillus subtilis* 3NA. *Appl. Microbiol. Biotechnol.* **2016**, *100*, 3511–3521. [CrossRef]
39. Xiao, Z.; Lu, J.R. Generation of acetoin and its derivatives in foods. *J. Agric. Food Chem.* **2014**, *62*, 6487–6497. [CrossRef]
40. Marchand, S.; Almy, J.; de Revel, G. The cysteine reaction with diacetyl under wine-like conditions: Proposed mechanisms for mixed origins of 2-methylthiazole, 2-methyl-3-thiazoline, 2-methylthiazolidine, and 2,4,5-trimethyloxazole. *J. Food Sci.* **2011**, *76*, C861–C868. [CrossRef]
41. GEHRIG, R.F.; KNIGHT, S.G. Formation of 2-Heptanone from Caprylic Acid by Spores of Various Filamentous Fungi. *Nature* **1961**, *192*, 1185. [CrossRef]
42. Selli, S.; Guclu, G.; Sevindik, O.; Kelebek, H. Variations in the key aroma and phenolic compounds of champignon (*Agaricus bisporus*) and oyster (*Pleurotus ostreatus*) mushrooms after two cooking treatments as elucidated by GC-MS-O and LC-DAD-ESI-MS/MS. *Food Chem.* **2021**, *354*, 129576. [CrossRef] [PubMed]
43. Jo, Y.J.; Cho, I.H.; Song, C.K.; Shin, H.W.; Kim, Y.S. Comparison of fermented soybean paste (Doenjang) prepared by different methods based on profiling of volatile compounds. *J. Food Sci.* **2011**, *76*, C368–C379. [CrossRef]

Disclaimer/Publisher's Note: The statements, opinions and data contained in all publications are solely those of the individual author(s) and contributor(s) and not of MDPI and/or the editor(s). MDPI and/or the editor(s) disclaim responsibility for any injury to people or property resulting from any ideas, methods, instructions or products referred to in the content.

foods

MDPI

Article

Characterization of Key Aroma Compounds of Soy Sauce-like Aroma Produced in Ferment of Soybeans by *Bacillus subtilis* BJ3-2

Qibo Tan [1], Yongjun Wu [1,*], Cen Li [1], Jing Jin [1], Lincheng Zhang [1], Shuoqiu Tong [1], Zhaofeng Chen [1], Li Ran [1], Lu Huang [1] and Zeyan Zuo [2]

1 Key Laboratory of Plant Resource Conservation and Germplasm Innovation in Mountainous Region (Ministry of Education), College of Life Sciences/Institute of Agro-Bioengineering, Guizhou University, Guiyang 550025, China; qibotan@outlook.com (Q.T.); cenli@gzu.edu.cn (C.L.); jjin3@gzu.edu.cn (J.J.); zhanglc201312@163.com (L.Z.); m18786120419@163.com (S.T.); zfchen2024@outlook.com (Z.C.); rainnie0722@outlook.com (L.R.); hl.2025@outlook.com (L.H.)

2 Guizhou Institute of Products Quality Inspection & Testing, Guiyang 550016, China; 15285996397@163.com

* Correspondence: wyjbio@163.com

Abstract: Fermented soybeans are popular among many for their rich soy sauce-like aroma. However, the precise composition of this aroma remains elusive, with key aroma compounds unidentified. In this study, we screened the candidate genes *ilvA* and *serA* in BJ3-2 based on previous multi-omics data, and we constructed three mutant strains, BJ3-2-Δ*serA*, BJ3-2-Δ*ilvA*, and BJ3-2-Δ*serA*Δ*ilvA*, using homologous recombination to fermented soybeans with varying intensities of soy sauce-like aroma. Our objective was to analyze samples that exhibited different aroma intensities resulting from the fermented soybeans of BJ3-2 and its mutant strains, thereby exploring the key flavor compounds influencing soy sauce-like aroma as well analyzing the effects of *ilvA* and *serA* on soy sauce-like aroma. We employed quantitative descriptive sensory analysis (QDA), gas chromatography–olfactometry–mass spectrometry (GC-O-MS), relative odor activity value analysis (rOAV), principal component analysis (PCA), orthogonal partial least squares-discriminant analysis (OPLS-DA), and partial least squares regression analysis (PLSR). QDA revealed the predominant soy sauce-like aroma profile of roasted and smoky aromas. GC-MS detected 99 volatile components, predominantly pyrazines and ketones, across the four samples, each showing varying concentrations. Based on rOAV (>1) and GC-O, 12 compounds emerged as primary contributors to soy sauce-like aroma. PCA and OPLS-DA were instrumental in discerning aroma differences among the samples, identifying five compounds with VIP > 1 as key marker compounds influencing soy sauce-like aroma intensity levels. Differential analyses of key aroma compounds indicated that the mutant strains of *ilvA* and *serA* affected soy sauce-like aroma mainly by affecting pyrazines. PLSR analysis indicated that roasted and smoky aromas were the two most important sensory attributes of soy sauce-like aroma, with pyrazines associated with roasted aroma and guaiacol associated with smoky aroma. In addition, substances positively correlated with the intensity of soy sauce-like aroma were verified by additional experiments. This study enhances our understanding of the characteristic flavor compounds in soy sauce-like aroma ferments, provides new perspectives for analyzing the molecular mechanisms of soy sauce-like aroma formation, and provides a theoretical framework for the targeted enhancement of soy sauce-like aroma in various foods.

Keywords: *Bacillus subtilis*; soy sauce-like aroma; GC-O-MS; key aroma compounds; fermentation

1. Introduction

Soy sauce-like aroma is widely appreciated in various foods for its lightness, delicacy, richness, and longevity [1,2]. It plays a crucial role in traditional foods such as soy sauce, soybean paste, tempeh, and saucy white wine, influencing their acceptability [3,4]. To understand its formation, researchers have analyzed soy sauce-like aroma using various analytical tools with these foods as subjects.

Fan et al. investigated the volatile components of soy sauce-like aroma in Lang and Maotai sauce-type liquors, highlighting compounds like 2,3,5-trimethyl-6-ethylpyrazine, 3-methylbutanoic acid, 2,3,5-trimethylpyrazine, 2,3,5,6-tetramethylpyrazine, 4-methylguaiacol, and others [5–7]. Lin et al. identified 14 compounds that accurately mimicked the aroma of Pixian bean sauce, including 2,3,5,6-tetramethylpyrazine, 2,5-dimethylpyrazine, furfural, and 4-ethyl-2-methoxyphenol, using SPME-GC-O/MS and aroma recombination, emphasizing pyrazines, furans, and aldehydes as key contributors to soy sauce-like aroma [8]. In 2024, Duan et al. demonstrated that 2,3,5-trimethylpyrazine, 3-hydroxy-2-butanone, and several other substances may be critical flavor compounds in Langjiu's soy sauce-like aroma through recombination and omission tests [9]. Pyrazines, 3-methylbutyric acid, maltol, and methoxyphenol are found in higher concentrations in cheonggukjang, contributing to its characteristic sauce aroma [1]. Recently, Zhao et al. proposed a new theory based on their study of volatiles in three Japanese soy sauces, linking amino acids closely to the production of volatile compounds like pyrazines [10]. These studies collectively indicate that soy sauce-like aroma is not attributed to a single substance but arises from the synergistic action of various aroma compounds [11,12]. Despite these findings, the complete composition of soy sauce-like aroma remains incompletely elucidated.

It has been proposed that beans and flour can undergo fermentation by microorganisms to yield a rich soy sauce-like aroma [13]. *Bacillus subtilis* is closely associated with the production of soy sauce-like aroma [14], with wide-ranging applications in the fermentation industry and beyond [15,16]. Transcriptome, proteome, and metabolome analyses have further elucidated the differential expression of genes and metabolic pathways, particularly in amino acid metabolism under varying temperatures, suggesting their role in soy sauce-like aroma accumulation [17]. Based on BJ3-2's capacity to generate soy sauce-like aroma from fermented soybeans at 53 °C and supported by multi-omics data, *Bacillus subtilis* BJ3-2 serves as an ideal model for investigating the molecular mechanisms and key aroma compounds of soy sauce-like aroma. The flavor candidate genes *ilvA* and *serA* were screened by combined transcriptomic, proteomic, and metabolomic analyses of BJ3-2 in relation to those that are likely to be associated with the soy sauce-like aroma. In the KEGG pathway, *ilvA* is the gene encoding threonine dehydratase, and *serA* is the gene encoding D-3-phosphoglycerate dehydrogenase. The metabolic pathways involved in both *ilvA* and *serA* are the glycine, serine, and threonine metabolic pathways [18,19]. It has been shown that the metabolism of amino acids is closely related to the formation of the soy sauce-like aroma [20]. And it has been reported that the metabolism of serine and threonine is highly related to the formation of pyrazines [21], so the knockout of *ilvA* and *serA* may lead to pyrazines and other aroma compounds, thus affecting the soy sauce-like aroma. Thus, we could systematically analyze aroma-active compounds that produce soy sauce-like aroma in soybeans fermented at 53 °C using wild-type strains of Bacillus subtilis BJ3-2 and mutant strains of *ilvA* and *serA* to elucidate the formation of soy sauce aroma in solid-state fermentation. For flavor analysis, GC-MS combined with a headspace solid-phase microextraction technique (HS-SPME technique) can be used to concentrate analytes in samples by extracting phase-encapsulated fiber heads and then enable the effective separation and identification of volatile compounds across a wide range of foods in GC-MS [22]. Additionally, the integration of GC-O, quantitative analysis, and various data analysis methods offers insights into the intricate relationships between sensory perceptions and differential aroma substances among different samples [23].

In this study, the aim of this study was (1) to screen soy sauce-like aroma candidate genes through multi-omics data analysis and (2) to obtain mutants with significantly altered soy sauce-like aroma by knocking out soy sauce-like aroma candidate genes, which produce a more pronounced soy sauce aroma, so as to better analyze the relationship between sensory attributes and soy sauce flavor aroma compounds; (3) Simulated fermentations were conducted using these mutant strains and BJ3-2 to produce soy sauce-like aroma, followed by sensory evaluations. (4) Volatile compounds from the fermentations exhibiting soy sauce-like aroma were analyzed using GC-MS with a semi-quantitative approach. (5) Characteristic aroma compounds of soy sauce-like aroma were identified using rOAV in conjunction with GC-O. (6) The variability among samples with different soy sauce-like aroma intensities was further analyzed using PCA and OPLS-DA. (7) The correlation between key aroma compounds and sensory attributes was analyzed and validated by partial least squares regression analysis (PLSR) and additive experiments. (8) The effect of knockout candidate genes on the formation of soy sauce-like aroma compounds was analyzed by the results of the KEGG metabolic pathway analysis of candidate genes and the statistical difference analysis of key aroma substances. In this study, we characterized for the first time the aroma profile of soy sauce-like aroma in fermented samples containing *Bacillus subtilis* BJ3-2 and its different mutants using multi-omics data, gene knockout, and headspace solid-phase microextraction (HS-SPME) methods combined with gas chromatography–mass spectrometry (GC-O-MS), rOAV, and an aroma addition experiment. These investigations focused on key volatiles enhancing soy sauce-like aroma, deepening our understanding of the characteristic aroma compounds in foods with soy sauce-like aroma, gaining insights into the key aroma compounds that produce a soy sauce-like aroma during the fermentation of soybeans by Bacillus subtilis, and providing theoretical insights for the targeted enhancement of soy sauce-like aroma in food products. These data also provide new strategies for screening soy sauce-like aroma candidate genes and modifying fermentation flavors through genetic engineering.

2. Materials and Methods

2.1. Strains and Vectors

Bacillus subtilis BJ3-2 is a patented strain from our laboratory (Patent No. 201110023795.4; link) [24]. Escherichia coli DH5α and pUC18 vectors were purchased from TaKaRa (Dalian, China). The pHT-01cas9-p43 vector was obtained from Biomics Biotechnology (Nantong, China), and the pMarA vector was a generous gift from China Agricultural University.

2.2. Chemicals

Retention indices (RI) were calculated using n-alkane standards (C7-C30) (Supelco, Bellefonte, PA, USA). The internal standard sec-octanol (99%) was purchased from Aladdin Industries Corporation (Shanghai, China). Reference standards were sourced as follows: 2,3-butanedione ($\geq$99.0%), 3-ethyl-2,5-dimethylpyrazine (98%), and malic acid from Macklin Biochemical Technology Corporation (Shanghai, China); 3-hydroxy-2-butanone (98%) and 2,3,5,6-tetramethylpyrazine from Aladdin Industries Corporation (Shanghai, China); and 2,3,5-trimethylpyrazine (>98.0%), 2,3-dimethylpyrazine (>98.0%), 2,4,5-trimethyloxazole (>98.0%), 2,5-dimethylpyrazine (>98.0%), 1-octen-3-ol (>98.0%), guaiacol (>98.0%), and 2-methylbutanoic acid (>98.0%) from TCI Co., Ltd. (Shanghai, China). Propylene glycol (99%) and anhydrous ethanol ($\geq$99.7%) were purchased from Solarbio Science and Technology Co., Ltd. (Beijing, China).

2.3. Reverse Transcription-Quantitative Real-Time PCR (RT-qPCR)

Total RNA was extracted using the Column Bacterial Total RNA Extraction and Purification Kit (Sangon Biotech, Shanghai, China), following the manufacturer's instructions. First-strand cDNA synthesis from total RNA was performed using StarScript III RT Mix with a gDNA Remover (Genstar, Beijing, China). Quantitative real-time PCR (qRT-PCR) was carried out using 2 $\times$ RealStar Green Fast Mixture (Genstar, Beijing, China). Expression levels were determined based on Ct values, and the data were normalized using the reference gene 16S rRNA. Primers for both the internal reference and target genes are listed in Table S1. Relative gene expression was normalized using the $2^{-\Delta\Delta CT}$ method. Each RT-qPCR experiment was performed in triplicate [25].

2.4. Construction and Transformation of Homologous Recombinant Knockout Vectors

HLarm stands for the homologous left arm of the target gene, while HRarm represents the homologous right arm. The cm gene refers to chloramphenicol resistance, and erm denotes erythromycin resistance. DEDP is a double-exchange assay primer (Table S1) used to verify successful knockouts. To construct the knockdown vectors, the HLarm and HRarm sequences of *ilvA* and *serA* were PCR-amplified from the *Bacillus subtilis* BJ3-2 genome. The cm gene was amplified from the PHT01cas9-p43 vector. HLarm, cm, and HRarm fragments were sequentially double-digested with EcoRI/*Kpn*I, *Kpn*I/SalI, and SalI/HindIII, respectively. These fragments were then ligated into the pUC18 vector to create the homologous recombinant knockdown vectors pUC18-HLarm(*ilvA*)-cm-HRarm(*ilvA*) and pUC18-HLarm(*serA*)-cm-HRarm(*serA*). For the *serA* and erm knockdown vectors, *serA* was PCR-amplified from the BJ3-2 genome, and erm was amplified from the pMarA plasmid. These fragments were double-digested with *Kpn*I/BamHI and *Pst*I, respectively, and sequentially ligated into the pUC18 vector to generate the homologous recombinant knockdown vector pUC18-*serA*::Em^r. Following construction, the recombinant plasmids were transformed into *Bacillus subtilis* BJ3-2, as described previously [26,27]. The transformants were confirmed by PCR using DEDP primers (Table S1). Genomic DNA from the transformants was extracted using the Bacterial DNA Kit (OMEGA, Woburn, MA, USA) according to the manufacturer's instructions and sequenced by Tsingke Biotechnology Co., Ltd. (Beijing, China).

2.5. Soy Sauce-like Aroma Fermentation Experiment

BJ3-2, BJ3-2Δ*serA*, BJ3-2Δ*ilvA*, and BJ3-2Δ*serA*Δ*ilvA* were inoculated into 5 mL of liquid LB medium and incubated at 37 °C for 12 h with agitation at 180 rpm. The soybeans were autoclaved at 121 °C for 20 min. A bacterial suspension of the four strains (OD600 = 0.465) was inoculated into sterilized soybeans (the bacterial inoculation was 1% of the soybean medium) and fermented for 72 h at 53 °C.

2.6. Sensory Evaluation

The soybeans' fermentation medium was employed for simulating the fermentation of soy sauce-like aroma [28]. Upon the completion of the fermentation, sensory evaluation was conducted. To ensure the accurate assessment of fermentation characteristics, a panel of 10 trained sensory evaluators participated in quantitative descriptive analysis (QDA). The panelists, consisting of 5 males and 5 females aged 20–30 years, underwent rigorous three-day training to familiarize themselves with the odors associated with soy sauce-like aroma fermentation. Firstly, a descriptive test was conducted to elicit the sensory descriptor. Each sample from the fermentation with soy sauce-like aroma (20 g each) was put into 50 mL Teflon vessels, and each member smelled them to freely record the sensory descriptors. The panel then discussed and reached a unanimous agreement based on the frequency and intensity of the sensory attributes of the soy sauce-like aroma of fermented soybeans. Through the panelists' discussion and after reaching a unanimous opinion, six aroma attributes were selected for evaluation: soy sauce-like aroma, roasted, smoky, sweet, buttery, and grain aroma. Roasted aroma evokes scents reminiscent of aromas of chocolate and roasted nuts [9]; smoky aroma resembles the fresh scent of wood and grass [9]; buttery aroma is akin to the aroma of cream and yogurt [9]; sweet aroma resembles caramelized popcorn; and grain aroma is similar to the scent of steamed bean grains [9]. Soy sauce-like aroma is the overall aroma of fermented soybeans and is described as a smooth, lingering aroma akin to that found in saucy fermented products [17]. Samples from the fermentation with soy sauce-like aroma were placed in 50 mL odorless, clear PET bottles, each labeled with a random three-digit number and assigned randomly to assessors for scoring. For each odor attribute, the intensity scores ranged from 1 (faint odor) to 5 (strong odor) in increments of 1 [29]. All experiments were conducted in triplicate, and average scores were calculated for each attribute.

2.7. Extraction of Volatile Compounds from Fermentations by HS-SPME

We referred to the HS-SPME technique of Peng et al. for the extraction of volatile compounds in fermentation [30]. Volatiles from fermented soybeans were extracted using a 2 cm 50/30 µm DVB/CAR/PDMS extraction head combined with an autosampler. The extraction head was preconditioned at 270 °C for 30 min prior to the experiment to eliminate residual substances. To prepare the sample, 3 g of fermented soybeans were accurately weighed using a 0.001 g electronic balance and placed into a 20 mL headspace vial. A 5 µL aliquot of 180 µg/mL sec-octanol solution was added as the internal standard, and the vial was sealed. The headspace vial was placed in an incubator-equipped autosampler with a vibration frequency set at 250 rpm. After pre-treatment at 50 °C for 10 min, the extraction head was inserted into the headspace vial for sampling. The sample underwent headspace extraction for 30 min, followed by injection into the GC pre-sample port for analysis, which was conducted over a 3 min period.

2.8. GC-MS Analysis

The analytical method of GC-MS was evaluated based on the method of Peng et al., with slight modifications [30]. The measurements were conducted using a 7890-5977B GC-MS system equipped with a polar WB-WAX column under specific experimental conditions. Helium served as the carrier gas at a flow rate of 2 mL/min. The column temperature program was initiated at 40 °C and held for 3 min, followed by an increase to 130 °C at a rate of 3 °C/min. It was maintained at 130 °C for 5 min and then ramped at 4 °C/min to 200 °C and held for 5 min. An electron ionization source was employed with an energy of 70 eV, scanning from m/z 35 to 500 at a rate of 1.8 scans/s. The inlet temperature was set at 250 °C, the ion source temperature was set at 230 °C, and the quadrupole rod temperature was set at 150 °C. The injection was non-split, with a solvent delay of 3 min.

2.9. GC–O Analysis

C-O analysis was conducted on an Agilent 7890B-5977B GC-MS equipped with an ODP4 olfactory detection port. The GC effluent was evenly split between the MS and sniffing port at a 1:1 ratio. Helium (99.99%) served as the carrier gas for continuously separating the volatiles in the column, delivered to the sniffing port at a flow rate of 1.5 mL/min. The delivery line temperature was maintained at 250 °C. To ensure comfort during sniffing, moist air was introduced into the sniffing port. Fermented soybean samples simulating sauces were analyzed by three trained and experienced experimental team members (two males and one female), who conducted an olfactory detection of odor descriptors through sniffing. Each sample was evaluated in triplicate. The panelists rated the intensity of the detected odors on a scale from 1 to 5, with higher ratings indicating stronger odors of the compounds detected by sniffing [31].

2.10. Qualitative and Quantitative Analyses

The volatile compounds contributing to the soy sauce-like aroma ferment were identified by comparing their mass spectra with those in the NIST 20.0 database and by referencing retention indices (RI) and odor profiles from relevant literature. RI values were calculated using a homologous series of n-alkane standards (C7–C30) under identical chromatographic conditions. The formulas were derived from comprehensive studies on major volatile compounds. Semi-quantitative analysis of volatile compounds was conducted using 5 µL of sec-octanol (180 µg/mL) as an internal standard. Concentrations of aromatic compounds were determined based on peak areas of the volatiles and the known concentration of the internal standard. The relative abundance of each compound was assessed following the methodology described by Zhao et al. [32]. The concentration of aromatic compounds was calculated using the following formula [32]:

$$C_x = \frac{A_x \times C_s \times V_s}{A_s \times m_x}$$

where C_x is the content of compound x (ug/g); A_x and A_s are the peak areas of compound x and internal standard s, respectively; m_x is the mass of the sample (g); and C_s and vs. are the concentration (μg/mL) and volume (mL) of the added internal standard, respectively.

2.11. Relative Odor Activity Values (rOAV) Determination

The rOAV for each volatile substance was calculated using the formula rOAV = Ci/OTi, where Ci represents the relative concentration of the compound and OTi denotes its odor threshold in water. The odor thresholds for the aroma-active compounds were sourced from literature references [33].

2.12. Adding Experiments

Based on the sensory evaluation and analysis of variance, it is known that S4 is the sample with the strongest soy sauce-like aroma, and S1 is the sample with the weakest soy sauce-like aroma. Sample S1 was used as the addition matrix. Based on the semi-quantitative results, the difference values of substances positively correlated with the soy sauce-like aroma were calculated, and these substances were added to sample S1. The odors were equilibrated for 10 min, and the evaluators rated both S1 and the added sample S1-A for each odor attribute and overall soy sauce-like aroma.

2.13. Data Analysis

At least three replications of each experiment and sample analysis were conducted. The GC-MS results were analyzed and processed using Excel 2019, with experimental data expressed as the mean ± SD. The data were statistically analyzed by SPSS software (version 26.0, Inc., Chicago, IL, USA) and Origin 2021 (OriginLab Corporation, Northampton, MA, USA) using a Fisher-LSD to analyze a significant difference between BJ3-2(S1), BJ3-2*ΔilvA*(S2), BJ3-2*ΔserA*(S3), and BJ3-2*ΔserAΔilvA*(S4) at $p \leq 0.05$. Sensory evaluation results were visualized using radar plots, while combined rOAV and GC-O analyses were depicted using Venn diagrams. Bar, radar, and Venn diagrams were generated using Origin 2021 (OriginLab Corporation, Northampton, MA, USA), and clustered heatmaps were created using TBtools version 2019 (Heatmap Illustrator, Wuhan, China). Principal Component Analysis (PCA) and Orthogonal Partial Least Squares Discriminant Analysis (OPLS-DA) were performed using SIMCA-P 14.1 (Umetrics, Umeå, Sweden). Correlations between sensory attributes and key aroma compounds were analyzed using PLSR in The Unscrambler version 9.7 software (CAMO ASA, Oslo, Norway).

3. Results and Discussion

3.1. Expression of *ilvA* and *serA* in BJ3-2 Incubated at 37 °C and 53 °C

The transcriptome, proteome, and metabolome of *Bacillus subtilis* BJ3-2 at 37 °C and 53 °C were previously analyzed [17]. In the current study, we further analyzed and screened the multi-omics data. Screening conditions for differentially expressed genes, proteins, and metabolites were set at |log2FoldChange| ≥ 2, |log2FoldChange| ≥ 1.4, and |log2FoldChange| ≥ 1, respectively, with a significance level of $p < 0.05$. The enriched differentially expressed genes, proteins, and metabolites, along with their associated KEGG metabolic pathways, were jointly analyzed. We found that genes meeting these screening criteria were enriched in the glycine, serine, and threonine metabolic pathways, which have been previously implicated in the literature in relation to the formation of soy sauce-like aroma in *Bacillus subtilis* [21]. Within the glycine, serine, and threonine metabolic pathway, *gbsA*, *ilvA*, and *serA* were highly expressed at 53 °C. The knockdown of *gbsA* also significantly impaired the growth of *Bacillus subtilis* [34]. The |log2FoldChange| values for *ilvA* and *serA* at 53 °C versus 37 °C were 2.00 and 2.16 (Table S2), respectively, identifying *serA* and *ilvA* as candidate genes for further study. We validated the expression of *ilvA* and *serA* in BJ3-2 at both temperatures using RT-qPCR (Table S3), as depicted in Figure 1. The expression of *ilvA* and *serA* was notably higher at 53 °C, consistent with the RNA-seq findings.

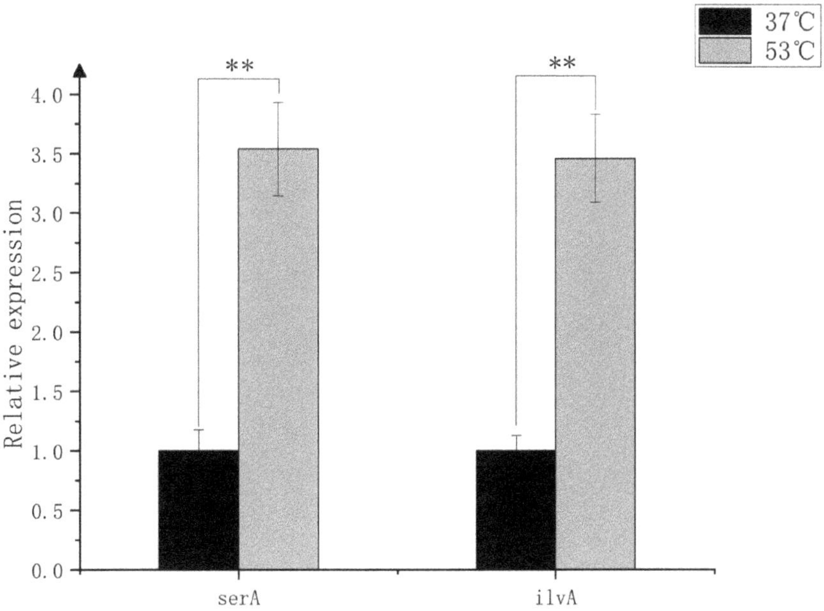

Figure 1. RT-qPCR verified the target gene expression. Using SPSS version 26.0, the data were analyzed using one-way ANOVA and then least significant difference (LSD) to assess the significance of mean differences. ** shows significant differences ($p < 0.05$). The results were calculated by the $2^{-\Delta\Delta CT}$ method.

3.2. Generation of Knockout Strains

Knockout vectors, pUC18-*serA*::*Em*^{r} and pUC18-HLarm-cm-HRarm, targeting *serA* and *ilvA*, respectively, were constructed and introduced into *Bacillus subtilis* BJ3-2. Positive clones were confirmed via PCR (Figures S1 and S2). Subsequent sequencing of the positive clones (Figures S3 and S4) revealed the successful replacement of *ilvA* with cm in the BJ3-2 genome using the pUC18-HLarm-cm-HRarm vector and the insertion of erm into *serA* in pUC18-*serA*::*Em*^{r}. This confirmed the knockdown of *serA* and *ilvA* in *Bacillus subtilis* BJ3-2, resulting in strains designated as BJ3-2Δ*serA* and BJ3-2Δ*ilvA*, respectively. The pUC18-HLarm-cm-HRarm vector was subsequently transformed into BJ3-2Δ*serA*, and positive clones were validated by PCR (Figure S5) and sequenced (Figure S6). These results confirmed the successful knockdown of both *serA* and *ilvA* in *Bacillus subtilis* BJ3-2, generating the double knockdown strain designated as BJ3-2Δ*serA*Δ*ilvA*.

3.3. Characterisation of Strains

To assess the growth characteristics of BJ3-2Δ*ilvA*, BJ3-2Δ*serA*, and BJ3-2Δ*serA*Δ*ilvA*, their appearance on plates showed similar colors and shapes, characterized by rough surfaces and irregular folds (Figure 2A–D). Additionally, Gram staining revealed that BJ3-2Δ*ilvA*, BJ3-2Δ*serA*, and BJ3-2Δ*serA*Δ*ilvA* exhibited a microscopic morphology of short purple rods, consistent with BJ3-2 (Figure 2E–H). There were no significant differences in morphology or Gram-staining characteristics compared to BJ3-2. Similarly, their growth curves (Figure 2I) demonstrated nearly identical patterns to BJ3-2, indicating comparable growth capacities among BJ3-2Δ*ilvA*, BJ3-2Δ*serA*, BJ3-2Δ*serA*Δ*ilvA*, and the wild-type strain.

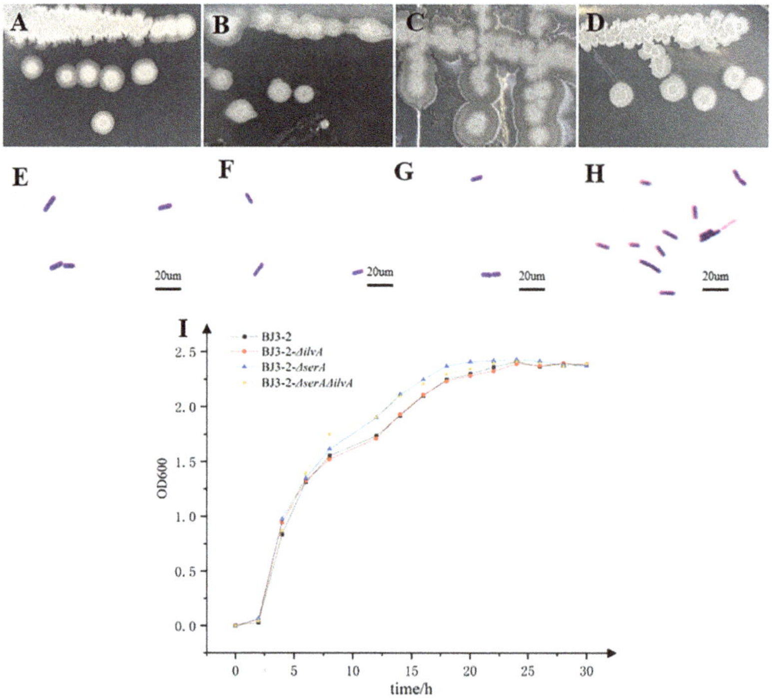

Figure 2. Colony morphology and growth curves. (**A–D**) Colony morphology of BJ3-2, BJ3-2*ΔilvA*, BJ3-2*ΔserA*, and BJ3-2*ΔserAΔilvA*; (**E–H**) Gram staining of BJ3-2, BJ3-2*ΔilvA*, BJ3-2*ΔserA*, and BJ3-2*ΔserAΔilvA*; (**I**) BJ3-2 Growth curves of BJ3-2*ΔilvA*, BJ3-2*ΔserA*, and BJ3-2*ΔserAΔilvA*.

3.4. Sensory Evaluation

Fermented samples (S1, S2, S3, and S4) from four strains (BJ3-2, BJ3-2*ΔilvA*, BJ3-2*ΔserA*, and BJ3-2*ΔserAΔilvA*) were analyzed for their aroma profiles. A QDA analysis based on six sensory attributes was conducted to evaluate the overall aroma profile (Figure 3). In sample S4 of BJ3-2*ΔserAΔilvA* fermentation, both roasted and smoky aromas scored high, suggesting these attributes dominate the soy sauce-like aroma, while sweet, buttery, and grain aromas scored relatively low, indicating lesser influence on the soy sauce-like profile. Differences were observed in the overall soy sauce-like aroma among samples S1, S2, S3, and S4. Samples S2 and S3 exhibited slightly stronger soy sauce-like aromas compared to S1, whereas sample S4 showed a significantly stronger overall soy sauce-like aroma than S1, S2, and S3. In order to analyze whether the differences in soy sauce-like aroma among the four samples were statistically significant, we performed statistical analysis using SPSS. The results of the analysis showed that the sensory differences in soy sauce-like aroma produced by the four samples, S1, S2, S3, and S4, were statistically significant ($p \leq 0.05$). The soy sauce-like aroma of S2, S3, and S4 was significantly enhanced compared to that of the S1 sample. Meanwhile, the soy sauce-like aroma of S4 was significantly enhanced compared to that of the S2 and S3 samples. Sample S4 specifically displayed intensified roasted and smoky aromas, contributing to its overall more intense soy sauce-like profile. The panelists unanimously agreed that sample S4 had the strongest soy sauce-like aroma. These findings suggest that the knockdown of *ilvA* and *serA* enhances the soy sauce-like aroma during BJ3-2 fermentation. This result prompts further investigation into the specific substances responsible for enhancing this aroma profile.

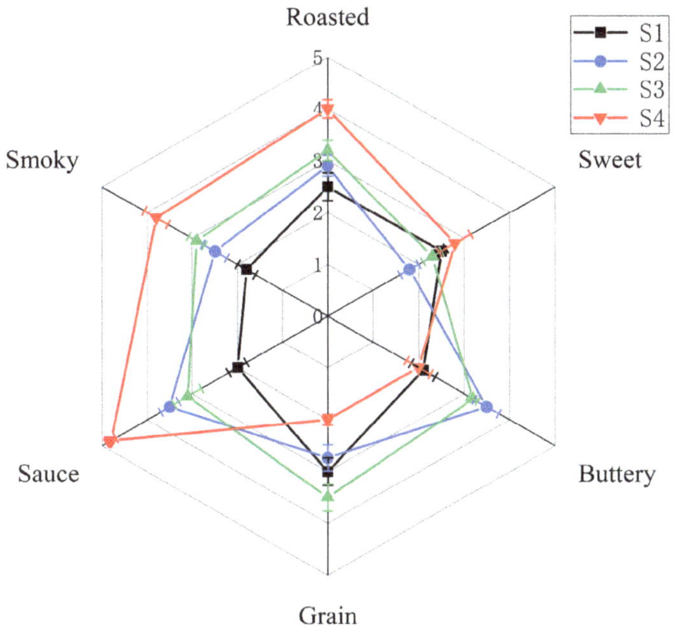

Figure 3. Sensory characteristic of samples S1, S2, S3, and S4 by QDA (S1, S2, S3, and S4 are samples fermented by *Bacillus subtilis* BJ3-2, BJ3-2Δ*ilvA*, BJ3-2Δ*serA*, and BJ3-2Δ*serA*Δ*ilvA*, respectively).

3.5. Qualification and Quantification of Volatile Compounds

SPME and GC-MS detected the volatiles of soy sauce-like aroma fermentation samples fermented by different knockout strains. An internal standard, sec-octanol, was added for quantification, with the results shown in Table S4. A total of 99 compounds were identified and categorized into pyrazines (13), ketones (19), acids (6), alcohols (22), aldehydes (16), phenols (3), and others. Compared to S1 (10338.15 μg/kg), a sample of soy sauce-like aroma fermentation using *Bacillus subtilis* BJ3-2, other samples fermented by knockout strains (S2: 11061.10 μg/kg, S3: 11048.72 μg/kg, and S4: 12366.8530 μg/kg) showed an overall increase in the total amount. Possibly due to the knockout of *serA* and *ilvA*, altering metabolic pathways involving serine, glycine, and threonine thereby increased pyrazines and other compounds, leading to a higher total amount.

To facilitate a clearer comparison among the four samples of soy sauce-like fermentations, volatile compounds were analyzed using a clustered heat map (Figure 4). This heat map illustrates the average content distribution of each volatile through color shades, employing hierarchical cluster analysis (HCA). The color intensity ranges from maximum (red) to minimum (blue), indicating consistency from high to low volatile contents [35]. The HCA results categorized the fermentation samples into two main groups: one comprising S1, S2, and S3 and the other consisting of S4, aligning with sensory evaluation outcomes. This observation suggests that the simultaneous knockdown of *serA* and *ilvA* significantly impacts the soy sauce-like aroma.

In addition, volatile compounds were classified into two main groups according to the tree diagram. Group A comprised 60 compounds, encompassing various classes such as acids and pyrazines. Tetramethylpyrazine and 2,3,5-trimethylpyrazine were notably higher in Group A, potentially contributing significantly to the soy sauce-like aroma. Previous studies indicated that pyrazines such as tetramethylpyrazine and 2,3,5-trimethylpyrazine impart predominantly nutty, cocoa, and chocolate aromas, thereby influencing the roasted composition in the soy sauce-like aroma of the samples [36]. These compounds are considered crucial flavor contributors to the soy sauce-like aroma profile of Moutai, a type of sauce baijiu [7]. Guaiacol, known for its smoky and herbaceous odor, was predominantly detected in foods with a soy sauce-like aroma [7]. Samples from the experimental group exhibited significantly higher levels of guaiacol, correlating with a notable enhancement of the soy sauce-like aroma. Compared to S1 acids, S4 exhibited a 25.85% increase. Research has demonstrated that acidic compounds are particularly conducive to enhancing the soy sauce-like aroma, playing a pivotal role in enriching the overall aromatic profile [2]. This finding underscores their importance in enhancing the richness of the soy sauce-like aroma in S4 samples.

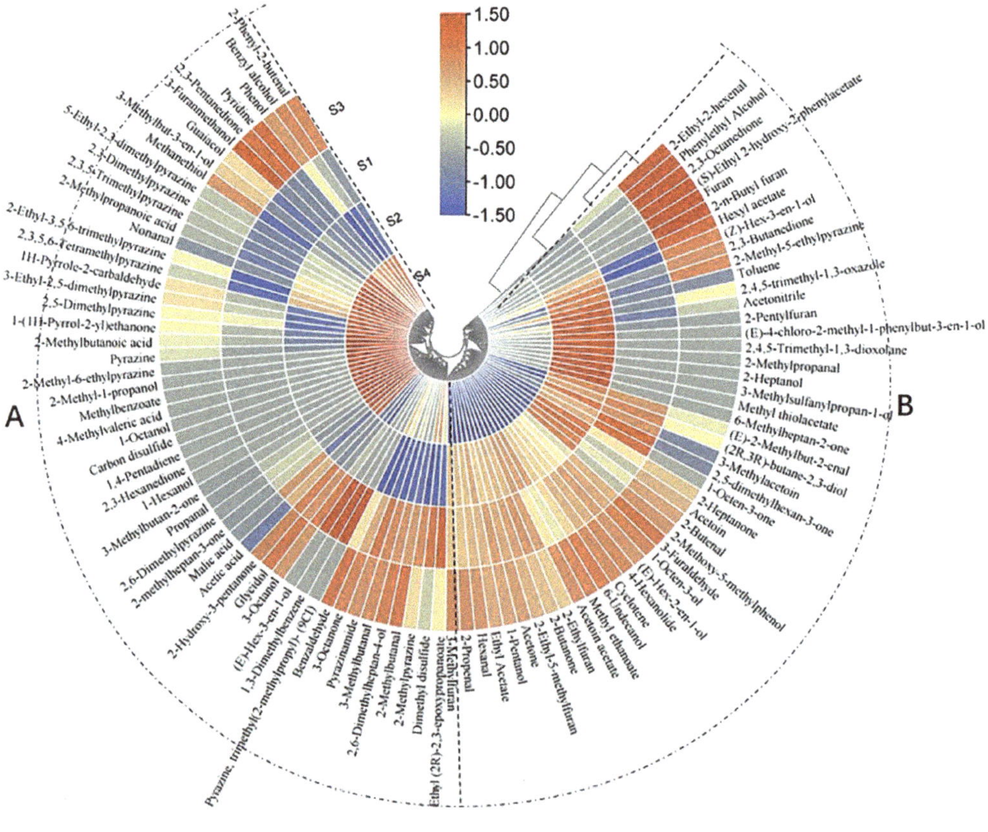

Figure 4. Hierarchical clustering and thermogram visualization of volatile compounds for four samples. High concentrations are represented in red, while low concentrations are represented in blue. A and B: According to the cluster analysis of the heat map, all the compounds detected in the four samples were clustered into two groups, group A and group B respectively.

Group B comprised 47 compounds, predominantly ketones and others. Compared to S1, ketones in S2 increased by 18.92% and in S3 by 12.56%. It has been shown that the increase in ketones can enrich the soy sauce-like aroma [2]. Notably, 2,4,5-trimethyloxazole may be closely related to the formation of soy sauce-like aroma [7]. In the present experiments, 2,4,5-trimethyloxazole was also detected, with relatively higher levels in S2 and S3 compared to S1 samples. These findings suggest that the increased presence of 2,4,5-trimethyloxazole and ketones may significantly contribute to enhancing the soy sauce-like aroma in S2 and S3 samples compared to S1.

3.6. Characteristic Flavor Substances of Soy Sauce-like Aroma by rOAV and GC-O

Numerous studies have demonstrated that many volatile compounds in food matrices may not significantly contribute to the aroma profile, with only a small fraction being involved in aroma perception [37]. Odor Specific Magnitude Estimation (OSME) is an effective method for evaluating compounds responsible for aroma perception based on Aroma Intensity (AI) [38], where compounds with stronger aromas exhibit higher AI values [39]. The contribution of aroma compounds to the soy sauce-like aroma depends not only on their concentration but also on their odor thresholds and interactions [35]. To further assess the contribution of aroma-active compounds to the aroma characteristics of sauces, their rOAV are often used. rOAV is the ratio of the concentration of volatile compounds to their odor thresholds, with compounds having rOAV $\geq$ 1 considered significant contributors to the aroma profile [40]. Therefore, identifying characteristic aroma-active compounds among the 99 compounds in the four samples is crucial for understanding soy sauce-like aroma formation, often analyzed through a combination of rOAV and GC-O.

Nineteen aroma actives were identified through rOAV analysis (Table S5), with 15, 15, 16, and 18 compounds in S1 to S4 having rOAV values above 1, respectively. Fifteen of these compounds exhibited rOAV > 1 across all four samples. Among them, guaiacol (rOAV 320.4–693.4), 2,3,5-trimethylpyrazine (rOAV 123.6–141.6), 2,3-butanedione (rOAV 364.3–449.7), and 2,5-dimethylpyrazine (rOAV 41.7–71.3) showed particularly high rOAV values, indicating their significant contribution to the soy sauce-like aroma. Additionally, 1-octen-3-ol (rOAV 28.9–36.1), 2,4,5-trimethyloxazole (rOAV 10.9–26.3), dimethyl disulfide (rOAV 24.8–29.7), 2,3,5,6-tetramethylpyrazine (rOAV 6.0–9.54), 3-hydroxy-2-butanone (rOAV 3.09–5.24), 3-ethyl-2,5-dimethylpyrazine (rOAV 1.8–3.2), malic acid (rOAV 0.96–2.71), 3-methylbutyric acid (rOAV 0.97–1.36), 2-methylbutyraldehyde (rOAV 9.4–13.4), 2-ethylfuran (rOAV 6.6–7.4), and 2-n-pentylfuran (rOAV 2.7–5.3) exhibited relatively lower rOAV values but still contributed significantly to the overall aroma.

Only a small fraction of compounds in the volatile components of food matrices have an aroma effect. GC-O is commonly used to assess the odor intensity of compounds detectable by experimenters [41]. GC-O intensity is rated by experimenters based on the perceived level of odor intensity of the compound, with aromatic compounds detected by more than two experimenters being recorded. As detailed in Table S6, a total of 17 volatile active compounds were identified from the four samples through GC-O experiments, including eight pyrazines, two alcohols, three acids, two ketones, and one each of oxazoles and phenols. These volatiles are presumed to be key flavor substances in fermented sauces. Specifically, 2,3,5-trimethylpyrazine, 2,3-butanedione, guaiacol, 2,5-dimethylpyrazine, malic acid, 2-methylbutanoic acid, and 3-hydroxy-2-butanone exhibited high GC-O intensities ($\geq$3) across samples, indicating their significant contribution to the sauce's aroma profile.

The results of rOAV analysis and GC-O detection were further analyzed using a Venn diagram (Figure 5). This combined analysis revealed 12 characteristic flavor compounds detected across all four samples, including 2,3,5-trimethylpyrazine, 2,5-dimethylpyrazine, 2,3,5,6-tetramethylpyrazine, 2,3-dimethylpyrazine, 3-ethyl-2,5-dimethylpyrazine, guaiacol, malic acid, 2-methylbutanoic acid, 2,3-butanedione, 3-hydroxy-2-butanone, 1-octen-3-ol, and 2,4,5-trimethyloxazole. These compounds are identified as key aroma contributors influencing the soy sauce-like aroma.

They were categorized into four groups based on aroma characteristics, detailed in Table 1. Group 1 (2,3,5-Trimethylpyrazine, 2,5-Dimethylpyrazine, 2,3,5,6-tetramethylpyrazine, 2,3-Dimethylpyrazine, and 3-Ethyl-2,5-Dimethylpyrazine) exhibits chocolatey, nutty, and baked-potato aromas, predominantly contributing roasted notes. Group 2 (2,3-butanedione, 3-hydroxy-2-butanone, and malic acid) presents pleasant aromas reminiscent of yogurt, buttery, and burnt sugar. Group 3 (guaiacol and 2,4,5-trimethyloxazole) imparts grassy, woody, and cucumber aromas, offering a refreshing scent. Group 4 (2-methylbutyric acid and 1-octen-3-ol) is characterized by a slightly pungent odor resembling sweat and peas. These key aroma compounds form the foundation of the soy sauce-like aroma. Additionally, four compounds were exclusively detected by GC-O, and seven compounds showed rOAV values > 1 only. These aromatic compounds likely play a crucial role in harmonizing the overall soy sauce-like aroma.

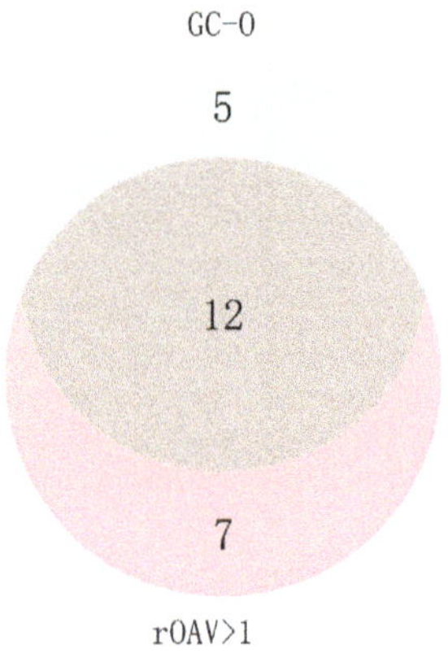

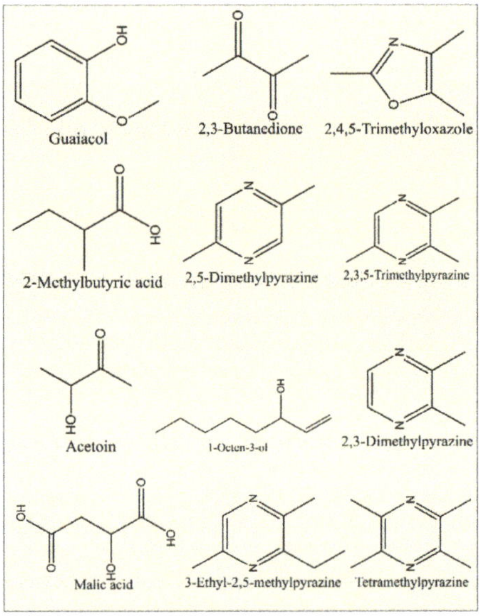

Figure 5. Analysis of aroma-active compounds in soy sauce by rOAV and GC-O.

Table 1. Odor descriptions of the main aroma compounds in the four samples.

No.	Compounds	Odor Description [a]	Related Sample [b]
Group 1			
	2,3,5-trimethylpyrazine	roasted peanut, cocoa	all
	2,5-dimethylpyrazine	woody grass medical	all
	2,3,5,6-tetramethylpyrazine	chocolate, nutty	all
	3-ethyl-2,5-methylpyrazine	roasted potato	all
	2,3-dimethylpyrazine	nutty, coffee	S4
Group 2			
	2.3-butanedione	strong butter, yoghurt	all
	3-hydroxy-2-butanone	buttery, creamy	all
	malic acid	caramel, popcorn	all
Group 3			
	guaiacol	smoky, woody	all
	2,4,5-trimethyloxazole	earthy, cucumbers	all
Group 4			
	2-methylbutyric acid	stinky feet, cheese	all
	1-octen-3-ol	mushroom.	all

[a]: Description of the odor detected by the GC-O sniffer port. [b] indicates the sample where the compound was detected by GC-O and rOAV > 1.

3.7. PCA and OPLS-DA Analyses

Based on the content of key flavor compounds with rOAV > 1 or detected by GC-O, and to better visualize the variability between samples with different degrees of sauce intensity, multivariate analysis (PCA and OPLS-DA) was conducted in this study. PCA is an unsupervised data analysis method for visualizing differences in complex information between samples without prior knowledge of the dataset [42]. PCA analysis of the 12 aforementioned flavor compounds in each sample revealed that the two principal components, PC1 and PC2, accounted for over 80% of the total variance, indicating that the model effectively explains sample variability. As depicted in Figure 6A, the 12 samples from the 4-sample group formed four distinct clusters. Sample S1 was located at the positive end of PC2, while S2, S3, and S4 were located at the negative end of PC2, clearly separated from each other. At the same time, sample S4, characterized by the strongest soy sauce-like aroma, was positioned at the negative end of PC1, while S1, S2, and S3 were positioned at the positive end and were separable from each other along PC2. And we also analyzed the PC1 loadings plot. By analyzing these loadings, it is possible to reveal the correlations between the variables and their influence on the formation of the principal component [43]. In Figure 6B, it can be observed that 2,3-butanedione, 2,4,5-trimethyloxazole, 3-hydroxy-2-butanone, and 1-octen-3-ol are positive signals, and 2,3,5-trimethylpyrazine, 2,3-dimethylpyrazine, 3-ethyl-2,5-methylpyrazine, tetramethylpyrazine, 2,5-dimethylpyrazine, malic acid, guaiacol, and 2-methylbutanoic acid are negative signals. Whereas the height of the column reflects the degree of contribution to the principal components, in Figure 6B, it can be observed that pyrazines and guaiacol contribute to the principal components to a greater extent. Notably, in the PCA score plot, the distance between S4 and S1, S2, and S3 was significant, consistent with sensory evaluation results indicating a pronounced difference between S4 and the other three samples. This suggests that the simultaneous knockdown of *serA* and *ilvA* significantly influences the soy sauce-like aroma in fermentation, enhancing the overall soy sauce-like aroma, as observed in sensory experiments.

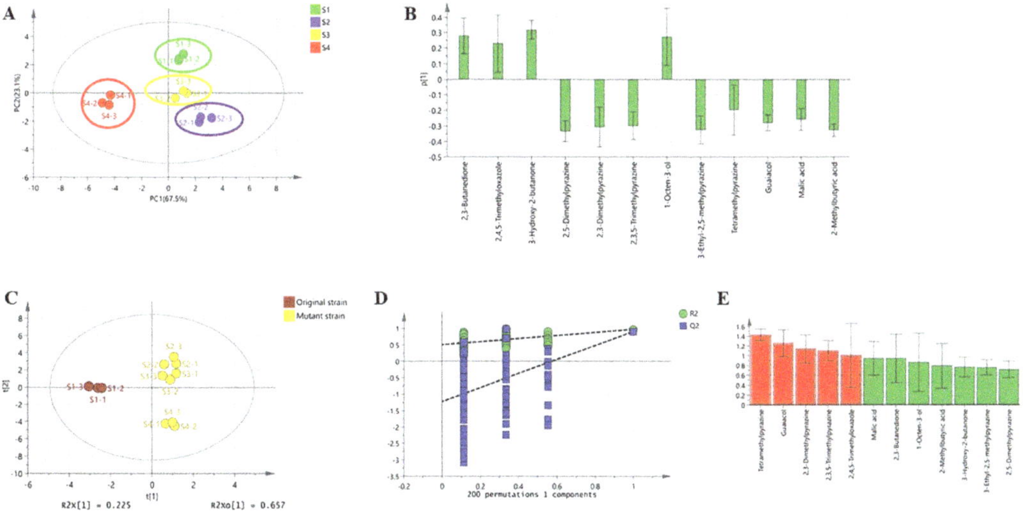

Figure 6. PCA and OPLS-DA analyses were performed on four samples S1–S4 with different sauce intensities. (**A**) PCA score plots; (**B**) PCA loading column plot; (**C**) OPLS-DA score plots; (**D**) 200 permutation test cross-validation plots; and (**E**) VIP values, where compounds with a VIP > 1 are highlighted in red, indicating their significance.

Based on PCA analysis, to further investigate differences between samples exhibiting varying levels of soy sauce-like aroma intensity, subsequently, OPLS-DA analysis was performed for 12 aroma actives. OPLS-DA, a discriminant analysis method, maximizes the interpretation of between-group differences using sample grouping information [44]. The samples were classified based on fermentation strains: one group included sample S1 fermented by the original strain BJ3-2, and the other comprised samples S2, S3, and S4 fermented by mutant strains. These groups were analyzed using OPLS-DA. Figure 6C depicts the OPLS-DA scores plot showing two well-separated clusters (R2 Y = 0.975), indicating high predictive efficacy (Q2 = 0.907). In order to assess whether the OPLS-DA model is overfitting, we performed alignment tests (n = 200) on the OPLS-DA model separately, as shown in Figure 6D. Figure 6D illustrates that RY2 exceeds QY2, affirming the good predictability and reliability of the OPLS-DA model without overfitting, as verified by 200 permutation tests. At the same time, we performed a CV-ANOVA test on the model of OPLS-DA. In the CV-ANOVA test result, the p-value should be as low as possible (less than 0.05) and the F-value should be as high as possible (greater than 3.85). The F-value of the CV-ANOVA test result is 9.4, which is greater than 3.85, and the p-value of the CV-ANOVA test result is 0.006, which is less than 0.05; this also indicates that there is no overfitting of the OPLS-DA analytical model. VIP was used to identify the most significant variables, and it is generally accepted that variables with a VIP greater than 1 are considered to have a significant impact on the discrimination of samples. Figure 6E demonstrates the VIP values for varying compounds. Variables with VIP values exceeding 1, including 2,3,5,6-tetramethylpyrazine, 2,3,5-trimethylpyrazine, guaiacol, 2,3-dimethylpyrazine, and 2,4,5-trimethyloxazole (VIP > 1), are significant for sample discrimination. The detection of these five differential flavor volatiles can serve as markers for distinguishing enhanced soy sauce-like aroma intensity.

3.8. Correlation between Key Aroma Compounds and Sensory Attributes

Partial Least Squares Regression (PLSR) was utilized to establish correlations between identified key aroma compounds and sensory attributes. PLSR comprised two principal components explaining 90% of the variance in X (12 key aroma compounds) and 89% in Y (six sensory attributes) (Figure 7). The inner and outer ellipses represent 50% and 100% of the explained variance, respectively, intersecting at the variances' location, indicating a well-explained PLSR model. The soy sauce-like aroma exhibited positive correlations with smoky and roasted attributes. These findings align with sensory evaluations (Figure 1) and GC-O analysis. Pyrazines such as 2,3,5-trimethylpyrazine, 2,5-dimethylpyrazine, 2,3-dimethylpyrazine, 3-ethyl-2,5-dimethylpyrazine, and 2,3,5,6-tetramethylpyrazine showed strong correlations with roasted aromas; guaiacol was closely associated with smoky aromas; malic acid correlated strongly with sweet aromas; and 3-hydroxy-2-butanone and 2,3-butanedione were highly correlated with creamy aromas. These aroma-active compounds, identified as key aroma contributors in soy sauce-like foods, are consistent with findings in previous studies [2,3,7,11].

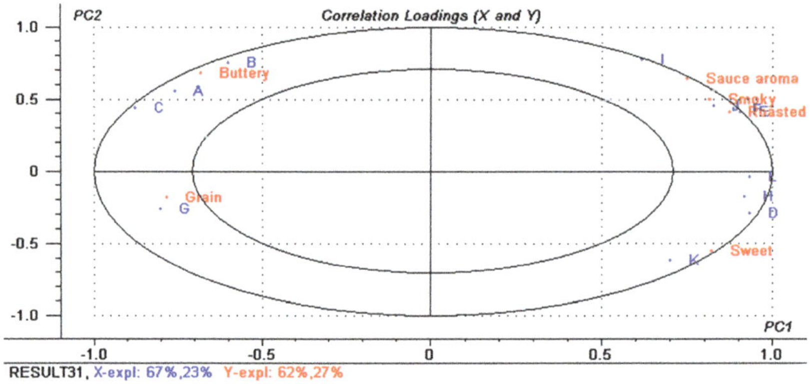

Figure 7. An overview of the variation found in the mean data from the partial least-squares regression (PLSR) correlation loading plot of sensory attributes and key aroma compounds. Ellipses represent r^2 = 0.5 and 1.0 ((A) 2,3-butanedione; (B) 2,3,5-trimethylpyrazine; (C) 3-hydroxy-2-butanone; (D) 2,5-dimethylpyrazine; (E) 2,3-dimethylpyrazine; (F) 2,3,5-trimethylpyrazine; (G) 1-octen-3-ol; (H) 3-ethyl-2,5-dimethylpyrazine; (I) 2,3,5,6-tetramethylpyrazine; (J) guaiacol; (K) malic acid).

To further explore which volatile compounds significantly contributed to sensory attributes, PLS1 regression analyses were conducted. Additionally, significant attribute variables were examined by calculating estimated regression coefficients using the jack-knife uncertainty test [45]. Figure 8 illustrates that 2,3,5-trimethylpyrazine, 2,5-dimethylpyrazine, 2,3-dimethylpyrazine, 3-ethyl-2,5-dimethylpyrazine, 2,3,5,6-tetramethylpyrazine, guaiacol, 3-hydroxy-2-butanone, 2,3-butanedione, 2,4,5-trimethyloxazole, and 2-methylbutyric acid were positively correlated with the soy sauce-like aroma. Notably, pyrazines and guaiacol were the most influential contributors to the soy sauce-like aroma, while ketones and acids also made significant contributions.

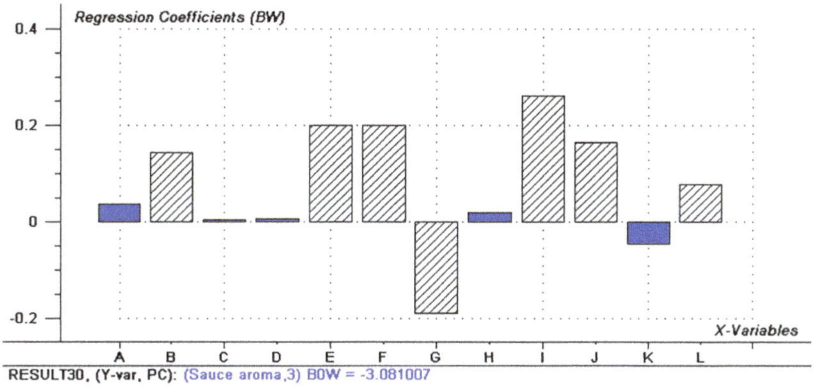

Figure 8. Regression coefficients and significance indicators of sensory attributes derived from PLS1: sauce aroma.

Combining the above data and results reveals that pyrazines, phenolic compounds, and ketones notably enhance the soy sauce-like aroma. This aroma profile comprises a complex system of baked, smoky, sweet, creamy, and cereal notes that harmoniously interact. Based on these characteristics and changes in the compound content during sensory evaluation, it is evident that pyrazines (2,3,5,6-tetramethylpyrazine, 2,3,5-trimethylpyrazine, 2,5-dimethylpyrazine, 2,3-dimethylpyrazine, 3-ethyl-2,5-dimethylpyrazine, and 2,4,5-trimethyloxazole) and phenols (guaiacol) make substantial contributions to the soy sauce-like aroma. Ketones (2,3-butanedione, 3-hydroxy-2-butanone), alcohols, and acids also contribute, forming a complex soy sauce-like aroma system.

3.9. Changes in Key Aroma Compounds of the Soy Sauce-like Aroma in Fermented Samples between BJ3-2, BJ3-2ΔilvA, BJ3-2ΔserA, and BJ3-2ΔserAΔilvA

A total of 12 key aroma compounds from four samples were analyzed for significance of differences. The results of the analysis showed statistically significant ($p \leq 0.05$) differences in the content of the 12 key aroma compounds among the four fermented samples (S1, S2, S3, and S4) produced from BJ3-2, BJ3-2*ΔilvA*, BJ3-2*ΔserA*, and BJ3-2*ΔserAΔilvA* fermented soybeans. As shown in Figure 9, the contents of tetramethylpyrazine, 2,3,5-trimethylpyrazine, and 2,3-dimethylpyrazine were significantly increased in the S2, S3, and S4 samples added with BJ3-2*ΔilvA*, BJ3-2*ΔserA*, and BJ3-2*ΔserAΔilvA*, as compared to the S1 samples fermented from BJ3-2. Among them, the fermented samples of BJ3-2*ΔserAΔilvA* had the highest content of tetramethylpyrazine, 2,3,5-trimethylpyrazine, 2,3-dimethylpyrazine, and other pyrazines. The highest pyrazines content in the BJ3-2*ΔserAΔilvA* fermented samples also explains why it has the highest intensity of "roasted" properties (Figure 3). A significant increase in guaiacol was also seen in the fermentation samples of BJ3-2*ΔilvA*, BJ3-2*ΔserA*, and BJ3-2*ΔilvAΔserA*, which is in agreement with the results of the QDA of the "smoky" enhancement in S2, S3, and S4 (Figure 3). On the other hand, 1-octen-3-ol and 3-hydroxy-2-butanone were significantly reduced in BJ3-

2*ΔilvAΔserA*, which is in line with the QDA results of sample S4, with reduced "grain" and "buttery" aromas (Figure 3). These results also show consistency with the PLSR results (Figure 7) as well. Several other compounds also showed some variation in content between the four samples.

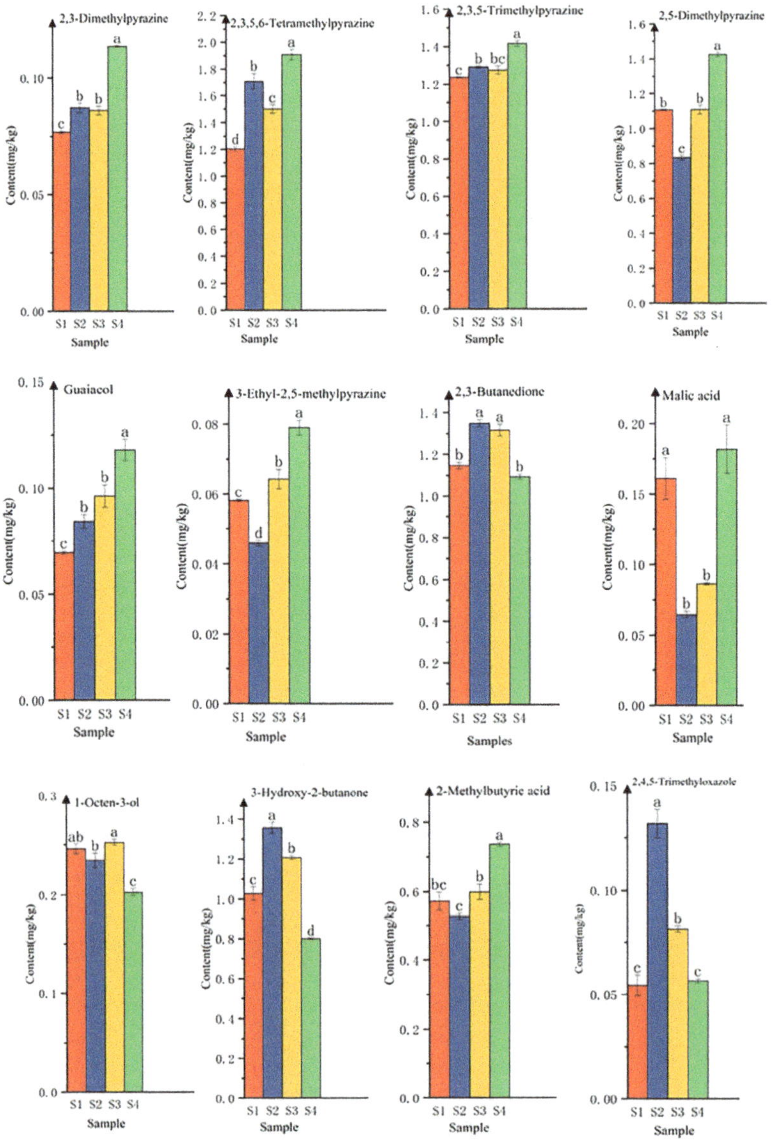

Figure 9. Changes in the content of key aroma compounds among the four fermented soybean samples with BJ3-2, BJ3-2*ΔilvA*, BJ3-2*ΔserA*, and BJ3-2*ΔilvAΔserA*. Significant differences between groups are indicated by letters a, b, d, c.

In addition, KEGG pathway analysis showed that *ilvA* (encoding threonine dehydratase) and *serA* (encoding D-3-phosphoglycerate dehydrogenase) participate in metabolic pathways that are glycine-, serine-, and threonine-metabolizing routes [18,19]. This suggests that the knockout of *ilvA* and *serA* may affect the production of aroma compounds by influencing this metabolic pathway, resulting in altered sensory attributes of BJ3-2 and its mutants that produce soy sauce-like aroma. The effects of the knockdown of *ilvA* and *serA* on the formation of soy sauce-like aroma compounds can therefore be further analyzed based on the results of KEGG pathway and differential analysis. First, the knockout of *serA* leads to the deregulation of the feedback inhibition of serine [46], which may lead to an increase in the serine and 3P-glycerate content, which in turn leads to an increase in the pyruvate content. And the increase in the pyruvate content may lead to an increase in the 2,3-butanedione and 3-hydroxy-2-butanone content. It has been documented that when oxygen is present, 2,3-butanedione is destabilized by the reduction to 3-hydroxy-2-butanone [47], which potentially contributes to the increase in the tetramethylpyrazine content. If *ilvA* were knocked out, there would theoretically be a relative increase in the content of L-threonine, which could lead to an increase in the content of aminoacetone. Aminoacetone is a precursor substance for the synthesis of dimethylpyrazine and trimethylpyrazine [21]. In turn, aminopyruvic acid was generated by amine oxidase, aldo-keto transferase, and D-lactate dehydrogenase [21], and NH4+ was also generated in this process, which resulted in an increase in trimethylpyrazine and tetramethylpyrazine levels. Theoretically, the content of 2,5-dimethylpyrazine should have also increased, but the results showed a decrease, presumably due to the preferential synthesis of pyruvic acid by aminopyruvic acetone without the spontaneous reaction to produce 2,5-dimethylpyrazine. The simultaneous knockout of *ilvA* and *serA* may had led to an increase in the pyruvate and aminopyruvate content, which may have then led to an increase in the dimethylpyrazine and trimethylpyrazine content. At the same time, the metabolism of amino acids, such as a threonine and a serine, provides large amounts of NH4+ [20], which may lead to an increase in the tetramethylpyrazine content. The BJ3-2Δ*serA*Δ*ilvA* fermentation samples exhibited reduced levels of 3-hydroxy-2-butanone and 2,3-butanedione, which was hypothesized to be due to the large amount of 3-hydroxy-2-butanone and 2,3-butanedione involved in the massive synthesis of 2,3,5-trimethylpyrazine and tetramethylpyrazine leading to the reduction in their levels. From the above analysis, it can be suggested that *serA* and *ilvA* may be the main factors affecting the formation of soy sauce-like aroma by influencing the production of key aroma substances such as pyrazines. However, the roles and specific regulatory mechanisms of *ilvA* and *serA* in the production of the soy sauce-like aroma still need to be further investigated.

3.10. Adding Experiments

To verify whether 2,3,5-trimethylpyrazine, 2,4,5-trimethyloxazole, 2,5-dimethylpyrazine, 2,3,5,6-tetramethylpyrazine, 2,3-dimethylpyrazine, 3-ethyl-2,5-methylpyrazine, guaiacol, and 2-methylbutyric acid were positively correlated with soy sauce-like aroma, based on the results of the semi-quantitative study, these compounds were added to the S1 samples. After 10 min of odor equilibration, sensory evaluation was performed on the added sample (S1-A) and the original sample (S1). As shown in Figure 10, the blue line is the aroma profile of the S1 sample, and the red line is the aroma profile of the addition model. Compared with the S1 sample, the overall aroma profile of the additive model showed a significant increase in the intensity of roasted, smoky, and soy sauce-like aromas, which was closer to the profile of the S4 sample. This result indicates that the addition of these seven substances is indeed positively correlated with soy sauce-like aroma. The data suggest that the addition model can well represent that the saucy aroma is enhanced by the addition of these substances, but further considerations are needed to fully establish this. In summary, the soy sauce-like aroma of S1-A was significantly enhanced. The overall sensory evaluation scores suggest that these substances have an important contribution to the soy sauce-like aroma and can enhance the soy sauce aroma.

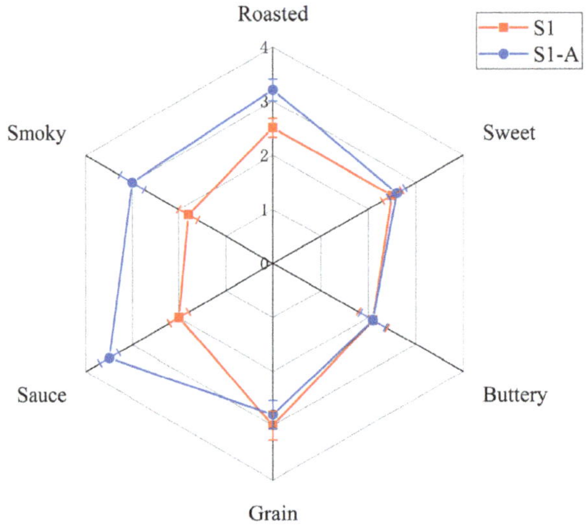

Figure 10. Aroma characteristics of S1 and additive models. (The blue line is the aroma profile of the S1 sample, and the red line is the aroma profile of the addition model).

4. Conclusions

In this study, we analyzed the soy sauce-like aroma characteristics and major aroma compounds of fermented samples of Bacillus subtilis BJ3-2 and its knockout mutant at 53 °C by using QDA, GC-O-MS, rOAV, PCA, OPLS-DA, PLSR, and an aroma addition experiment. The QDA results showed that the overall aroma of the soy sauce-like aroma in the fermented samples consisted of "roasted", "smoky", "buttery", "grain", and "sweet" aromas. In addition, the knockdown of the candidate genes for the soy sauce-like aroma (*serA* and *ilvA*) did not significantly affect the morphology and growth of BJ3-2 but had a greater effect on the aroma profile of soy sauce-like aroma in the fermented samples of BJ3-2. In summary, *serA* and *ilvA* genes of the *B. subtilis* BJ3-2 strain, unfavorable for producing soy sauce-like aroma at 53 °C, were knocked out via homologous recombination. This led to the generation of the mutant strains BJ3-2-*ΔserA*, BJ3-2-*ΔilvA*, and BJ3-2-*ΔserAΔilvA*. These mutant strains, along with the wild strain, were inoculated into the soybean fermentation medium. Sensory evaluations indicated that fermentation by the mutant strains resulted in a stronger soy sauce-like aroma, with BJ3-2-*ΔserAΔilvA* (Sample S4) producing the most pronounced aroma, significantly enhanced compared to that of BJ3-2 (Sample S1). Knockouts of *serA* and *ilvA* result in significant increases and decreases in the levels of key aroma compounds such as pyrazines. Combining sensory evaluations with analyses of differential volatiles and PLSR revealed that pyrazines (2,3,5,6-tetramethylpyrazine, 2,3,5-trimethylpyrazine, 2,5-dimethylpyrazine, 2,3-dimethylpyrazine, and 3-ethyl-2,5-dimethylpyrazine) and phenols (guaiacol) were the primary contributors to the soy sauce-like aroma. These compounds, along with 2,3-butanedione, 2,4,5-trimethyloxazole, 1-octen-3-ol, 2-methylbutyric acid, and other ketones, alcohols, and acids, collectively formed a complex soy sauce-like aroma system. The results of this study provide new insights for analyzing the soy sauce-like aroma of fermented soybeans produced by Bacillus subtilis BJ3-2, which provides a theoretical basis for the quality control and aroma improvement of the soy sauce aroma in fermented foods.

Supplementary Materials: The following supporting information can be downloaded at: https://www.mdpi.com/article/10.3390/foods13172731/s1, Table S1: Primers used in this study; Table S2: Transcriptome data at 37 °C and 53 °C; Table S3: Reverse transcription-quantitative real-time PCR at 37 °C and 53 °C; Table S4: Relative content (ug/kg) of volatile compounds in FSs by

HS-SPME-GC-O-MS (μg/kg); Table S5: rOAVs of the odor-active substances (rOAV > 1) in samples; Table S6: Odor intensity of the odorants in four samples identified using GC-O; Figure S1: *ilvA* homologous recombination knockout vector construction and transformation of the vector; Figure S2: *serA* homologous recombination knockout vector construction and transformation of the vector; Figure S3: Sequencing of BJ3-2Δ*ilvA*; Figure S4: Sequencing of BJ3-2Δ*serA*; Figure S5: PCR validation of BJ3-2Δ*serA*Δ*ilvA*; Figure S6: Sequencing of BJ3-2Δ*serA*Δ*ilvA*.

Author Contributions: Writing—Original Draft Preparation, Q.T.; Methodology, Q.T. and Z.C.; Formal Analysis, Q.T.; Investigation, Q.T., L.R. and Z.Z.; Writing—Review and Editing, Y.W., L.Z. and C.L.; Project Administration, J.J. and S.T.; Visualization, Q.T. and Y.W.; Supervision, Y.W., J.J., L.Z. and C.L.; Data Curation, Q.T., Z.C. and L.H.; Funding Acquisition, Y.W. All authors have read and agreed to the published version of the manuscript.

Funding: This work was supported by the National Natural Science Foundation of China (Funding number No. 32060585). The National Natural Science Foundation of China is the funder of the project.

Institutional Review Board Statement: In this study, participants in the sensory evaluations were selected from a pool of highly trained evaluators who volunteered to perform the sensory test. All participants were informed of the purpose and the procedures of the experiment in advance and gave written consent to take part in the sensory analysis for voluntary. People could withdraw from the test at any time. Panelists were in good health and had no known allergy to the components.

Informed Consent Statement: Informed consent was obtained from all panelists involved in the study. The participants agreed to the use of their data/answers. The study strictly adhered to a protocol designed to protect the rights and privacy of all participants.

Data Availability Statement: The original contributions presented in the study are included in the article, further inquiries can be directed to the corresponding author.

Conflicts of Interest: All authors declare that there are no conflicts of interest associated with the content of this article.

References

1. Lee, S.H.; Lee, S.; Lee, S.H.; Kim, H.J.; Singh, D.; Lee, C.H. Integrated Metabolomics and Volatolomics for Comparative Evaluation of Fermented Soy Products. *Foods* **2021**, *10*, 2516. [CrossRef] [PubMed]
2. Li, J.; Sun, C.; Zhang, M.; He, J.; Shen, Z.; Geng, L.; Yuan, L.; Wang, S.; Wang, C. Study on the factors affecting the sauce-flavor compounds of soybean paste in fermentation. *Food Biosci.* **2024**, *59*, 103872. [CrossRef]
3. Gao, L.H.; Liu, T.; An, X.J.; Zhang, J.L.; Ma, X.R.; Cui, J.M. Analysis of volatile flavor compounds influencing Chinese-type soy sauces using GC-MS combined with HS-SPME and discrimination with electronic nose. *J. Food Sci. Technol. Mysore* **2017**, *54*, 130–143. [CrossRef] [PubMed]
4. An, F.Y.; Wu, J.R.; Feng, Y.Z.; Pan, G.Y.; Ma, Y.Y.; Jiang, J.H.; Yang, X.M.; Xue, R.X.; Wu, R.A.; Zhao, M.M. A systematic review on the flavor of soy-based fermented foods: Core fermentation microbiome, multisensory flavor substances, key enzymes, and metabolic pathways. *Compr. Rev. Food Sci. Food Saf.* **2023**, *22*, 2773–2801. [CrossRef]
5. Fan, W.; Qian, M.C. Headspace solid phase microextraction and gas chromatography-olfactometry dilution analysis of young and aged Chinese "Yanghe Daqu" liquors. *J. Agric. Food Chem.* **2005**, *53*, 7931–7938. [CrossRef]
6. Fan, W.; Qian, M.C. Characterization of aroma compounds of chinese "Wuliangye" and "Jiannanchun" liquors by aroma extract dilution analysis. *J. Agric. Food Chem.* **2006**, *54*, 2695–2704. [CrossRef] [PubMed]
7. Fan, W.; Xu, Y.; Qian, M.C. Identification of Aroma Compounds in Chinese "Moutai" and "Langjiu" Liquors by Normal Phase Liquid Chromatography Fractionation Followed by Gas Chromatography/Olfactometry. In *Flavor Chemistry of Wine and Other Alcoholic Beverages*; American Chemical Society: Washington, DC, USA, 2012.
8. Lin, H.B.; Liu, Y.; He, Q.; Liu, P.; Che, Z.M.; Wang, X.M.; Huang, J.Q. Characterization of odor components of Pixian Douban (broad bean paste) by aroma extract dilute analysis and odor activity values. *Int. J. Food Prop.* **2019**, *22*, 1223–1234. [CrossRef]
9. Duan, J.W.; Cheng, W.; Lv, S.L.; Deng, W.Y.; Hu, X.J.; Li, H.H.; Sun, J.Y.; Zheng, F.P.; Sun, B.G. Characterization of key aroma compounds in soy sauce flavor baijiu by molecular sensory science combined with aroma active compounds reverse verification method. *Food Chem.* **2024**, *443*, 138487. [CrossRef]
10. Zhao, G.Z.; Feng, Y.X.; Hadiatullah, H.; Zheng, F.P.; Yao, Y.P. Chemical Characteristics of Three Kinds of Japanese Soy Sauce Based on Electronic Senses and GC-MS Analyses. *Front. Microbiol.* **2021**, *11*, 579808. [CrossRef]
11. Cai, W.C.; Wang, Y.R.; Wang, W.P.; Shu, N.; Hou, Q.C.; Tang, F.X.; Shan, C.H.; Yang, X.Q.; Guo, Z. Insights into the Aroma Profile of Sauce-Flavor Baijiu by GC-IMS Combined with Multivariate Statistical Analysis. *J. Anal. Methods Chem.* **2022**, *2022*, 4614330. [CrossRef]

12. Duan, J.W.; Yang, S.Q.; Li, H.H.; Qin, D.; Shen, Y.; Li, H.H.; Sun, J.Y.; Zheng, F.P.; Sun, B.G. Why the key aroma compound of soy sauce aroma type baijiu has not been revealed yet? *Lwt-Food Sci. Technol.* **2022**, *154*, 112735. [CrossRef]
13. Song, D.H.; Chun, B.H.; Lee, S.; Son, S.Y.; Reddy, C.K.; Mun, H.I.; Jeon, C.O.; Lee, C.H. Comprehensive Metabolite Profiling and Microbial Communities of Doenjang (Fermented Soy Paste) and Ganjang (Fermented Soy Sauce): A Comparative Study. *Foods* **2021**, *10*, 641. [CrossRef]
14. Gao, Y.X.; Hu, M.; Meng, W.M.; Wen, W.; Zhang, P.F.; Fan, B.; Wang, F.Z.; Li, S.Y. Study on the quality of soybean proteins fermented by *Bacillus subtilis* BSNK-5: Insights into nutritional, functional, safety, and flavor properties. *Food Chem.* **2024**, *443*, 138523. [CrossRef] [PubMed]
15. Van Dijl, J.M.; Hecker, M. *Bacillus subtilis*: From soil bacterium to super-secreting cell factory. *Microb. Cell Factories* **2013**, *12*, 3. [CrossRef] [PubMed]
16. Schallmey, M.; Singh, A.; Ward, O.P. Developments in the use of Bacillus species for industrial production. *Can. J. Microbiol.* **2004**, *50*, 1–17. [CrossRef] [PubMed]
17. Wu, Y.; Tao, Y.; Jin, J.; Tong, S.; Li, S.; Zhang, L. Multi-omics analyses of the mechanism for the formation of soy sauce-like and soybean flavor in *Bacillus subtilis* BJ3-2. *BMC Microbiol.* **2022**, *22*, 142. [CrossRef]
18. Rosenberg, J.; Müller, P.; Lentes, S.; Thiele, M.J.; Zeigler, D.R.; Tödter, D.; Paulus, H.; Brantl, S.; Stülke, J.; Commichau, F.M. ThrR, a DNA-binding transcription factor involved in controlling threonine biosynthesis in *Bacillus subtilis*. *Mol. Microbiol.* **2016**, *101*, 879–893. [CrossRef]
19. Révora, V.; Marchesini, M.I.; Comerci, D.J. Brucella abortus Depends on l-Serine Biosynthesis for Intracellular Proliferation. *Infect. Immun.* **2020**, *88*, 10–1128. [CrossRef]
20. Liu, Z.L.; Wu, Y.J.; Zhang, L.C.; Tong, S.Q.; Jin, J.; Gong, X.; Zhong, J. rocF affects the production of tetramethylpyrazine in fermented soybeans with *Bacillus subtilis* BJ3-2. *BMC Biotechnol.* **2022**, *22*, 18. [CrossRef]
21. Zhang, L.J.; Cao, Y.L.; Tong, J.N.; Xu, Y. An Alkylpyrazine Synthesis Mechanism Involving L-Threonine-3-Dehydrogenase Describes the Production of 2,5-Dimethylpyrazine and 2,3,5-Trimethylpyrazine by *Bacillus subtilis*. *Appl. Environ. Microbiol.* **2019**, *85*, e01807-19. [CrossRef]
22. Zhao, S.S.; Sai, Y.; Liu, W.T.; Zhao, H.W.; Bai, X.; Song, W.Y.; Zheng, Y.; Yue, X.Q. Flavor Characterization of Traditional Fermented Soybean Pastes from Northeast China and Korea. *Foods* **2023**, *12*, 3294. [CrossRef] [PubMed]
23. Zhang, J.X.; Xia, D.Z.; Li, T.H.; Wei, Y.M.; Feng, W.Z.; Xiong, Z.C.; Huang, J.L.; Deng, W.W.; Ning, J.M. Effects of different over-fired drying methods on the aroma of Lu'an Guapian tea. *Food Res. Int.* **2023**, *173*, 113224. [CrossRef] [PubMed]
24. Jia, D.; Wu, Y.; Li, Y.; Xu, W. Screening and Identification of Bacillus for Lobster Sauce Fermentation. *Food Science* **2009**, *30*, 217–221.
25. Liu, L.; Zhang, L.C.; Yang, L.H.; Zheng, J.H.; Jin, J.; Tong, S.Q.; Wu, Y.J. Genome-wide characterization and expression analysis of the HD-Zip II gene family in response to drought and GA3 stresses in *Nicotiana tabacum*. *Mol. Biol. Rep.* **2024**, *51*, 581. [CrossRef] [PubMed]
26. Zhang, X.-Z.; Cui, Z.-L.; Hong, Q.; Li, S.-P. High-level expression and secretion of methyl parathion hydrolase in *Bacillus subtilis* WB800. *Appl. Environ. Microbiol.* **2005**, *71*, 4101–4103. [CrossRef]
27. Phan, T.T.P.; Nguyen, H.D.; Schumann, W. Novel plasmid-based expression vectors for intra- and extracellular production of recombinant proteins in *Bacillus subtilis*. *Protein Expr. Purif.* **2006**, *46*, 189–195. [CrossRef]
28. Li, W.; Xie, H. Effect of tdh gene of Bacillus subtilis on Maotai-flavor. *Food Ferment. Ind.* **2021**, *47*, 116–123. [CrossRef]
29. Zhang, L.; Qin, Z.; Zhang, L.; Jiang, Y.; Zhu, J. Dynamic changes of quality and flavor characterization of Zhejiang rosy vinegar during fermentation and aging based on untargeted metabolomics. *Food Chem.* **2023**, *404*, 134702. [CrossRef]
30. Peng, X.; Li, X.; Shi, X.; Guo, S. Evaluation of the aroma quality of Chinese traditional soy paste during storage based on principal component analysis. *Food Chem.* **2014**, *151*, 532–538. [CrossRef]
31. Nuzzi, M.; Lo Scalzo, R.; Testoni, A.; Rizzolo, A. Evaluation of Fruit Aroma Quality: Comparison Between Gas Chromatography-Olfactometry (GC-O) and Odour Activity Value (OAV) Aroma Patterns of Strawberries. *Food Anal. Methods* **2008**, *1*, 270–282. [CrossRef]
32. Zhao, M.; Li, T.; Yang, F.; Cui, X.Y.; Zou, T.T.; Song, H.L.; Liu, Y. Characterization of key aroma-active compounds in Hanyuan *Zanthoxylum bungeanum* by GC-O-MS and switchable GC x GC-O-MS. *Food Chem.* **2022**, *385*, 132659. [CrossRef]
33. Leo, V. *Compilations of Odour Threshold Values in Air, Water and Other Media*; Science Press Ltd.: Beijing, China, 2015.
34. Boch, J.; Kempf, B.; Schmid, R.; Bremer, E. Synthesis of the osmoprotectant glycine betaine in *Bacillus subtilis*: Characterization of the gbsAB genes. *J. Bacteriol.* **1996**, *178*, 5121–5129. [CrossRef] [PubMed]
35. Ni, R.J.; Wang, P.; Zhan, P.; Tian, H.L.; Li, T. Effects of different frying temperatures on the aroma profiles of fried mountain pepper (*Litsea cubeba* (Lour.) Pers.) oils and characterization of their key odorants. *Food Chem.* **2021**, *357*, 129786. [CrossRef]
36. Fan, W.; Xu, Y.; Zhang, Y. Characterization of pyrazines in some Chinese liquors and their approximate concentrations. *J. Agric. Food Chem.* **2007**, *55*, 9956–9962. [CrossRef] [PubMed]
37. Vidal, A.M.; Alcalá, S.; De Torres, A.; Moya, M.; Espínola, J.M.; Espínola, F. Fresh and Aromatic Virgin Olive Oil Obtained from Arbequina, Koroneiki, and Arbosana Cultivars. *Molecules* **2019**, *24*, 3587. [CrossRef] [PubMed]
38. Tan, F.L.; Wang, P.; Zhan, P.; Tian, H.L. Characterization of key aroma compounds in flat peach juice based on gas chromatography-mass spectrometry-olfactometry (GC-MS-O), odor activity value (OAV), aroma recombination, and omission experiments. *Food Chem.* **2022**, *366*, 130604. [CrossRef]

39. Niu, Y.W.; Yao, Z.M.; Xiao, Z.B.; Zhu, G.Y.; Zhu, J.C.; Chen, J.Y. Sensory evaluation of the synergism among ester odorants in light aroma type liquor by odor threshold, aroma intensity and flash GC electronic nose. *Food Res. Int.* **2018**, *113*, 102–114. [CrossRef]
40. Xie, J.L.; Wang, L.L.; Deng, Y.L.; Yuan, H.B.; Zhu, J.Y.; Jiang, Y.W.; Yang, Y.Q. Characterization of the key odorants in floral aroma green tea based on GC-E-Nose, GC-IMS, GC-MS and aroma recombination and investigation of the dynamic changes and aroma formation during processing. *Food Chem.* **2023**, *427*, 136641. [CrossRef]
41. Feng, Y.; Cai, Y.; Fu, X.; Zheng, L.; Xiao, Z.; Zhao, M. Comparison of aroma-active compounds in broiler broth and native chicken broth by aroma extract dilution analysis (AEDA), odor activity value (OAV) and omission experiment. *Food Chem.* **2018**, *265*, 274–280. [CrossRef]
42. Yang, Y.; Wang, B.; Fu, Y.; Shi, Y.G.; Chen, F.L.; Guan, H.N.; Liu, L.L.; Zhang, C.Y.; Zhu, P.Y.; Liu, Y.; et al. HS-GC-IMS with PCA to analyze volatile flavor compounds across different production stages of fermented soybean whey tofu. *Food Chem.* **2021**, *346*, 128880. [CrossRef]
43. Krstić, G.; Anđelković, B.; Choi, Y.H.; Vajs, V.; Stević, T.; Tešević, V.; Gođevac, D. Metabolic changes in Euphorbia palusrtis latex after fungal infection. *Phytochemistry* **2016**, *131*, 17–25. [CrossRef]
44. Khalil, M.N.A.; Fekry, M.I.; Farag, M.A. Metabolome based volatiles profiling in 13 date palm fruit varieties from Egypt via SPME GC-MS and chemometrics. *Food Chem.* **2017**, *217*, 171–181. [CrossRef] [PubMed]
45. Liu, F.; Song, S.Q.; Zhang, X.M.; Tan, C.; Karangwa, E. Effect of sterilization methods on ginger flavor beverage assessed by partial least squares regression of descriptive sensory analysis and gas chromatography-mass spectrometry. *Eur. Food Res. Technol.* **2014**, *238*, 247–257. [CrossRef]
46. Grant, G.A.; Xu, X.L.; Hu, Z. Role of an interdomain Gly-Gly sequence at the regulatory-substrate domain interface in the regulation of Escherichia coli. D-3-phosphoglycerate dehydrogenase. *Biochemistry* **2000**, *39*, 7316–7319. [CrossRef]
47. Zalán, Z.; Hudáček, J.; Tóth-Markus, M.; Husová, E.; Solichová, K.; Hegyi, F.; Plocková, M.; Chumchalová, J.; Halász, A. Sensorically and antimicrobially active metabolite production of Lactobacillus strains on Jerusalem artichoke juice. *J. Sci. Food Agric.* **2011**, *91*, 672–679. [CrossRef] [PubMed]

Disclaimer/Publisher's Note: The statements, opinions and data contained in all publications are solely those of the individual author(s) and contributor(s) and not of MDPI and/or the editor(s). MDPI and/or the editor(s) disclaim responsibility for any injury to people or property resulting from any ideas, methods, instructions or products referred to in the content.

foods

MDPI

Article

Revealing the Mechanism of Aroma Production Driven by High Salt Stress in *Trichomonascus ciferrii* WLW

Fangying Xian [1,†], Lin Yang [1,2,†], Huaqing Ye [1], Jinlin Xu [1], Xiaoping Yue [2,*] and Xiaolan Wang [1,*]

[1] School of Life Science (Health), Jiangxi Normal University, Nanchang 330022, China; vifyx@foxmail.com (F.X.); linyang2023@jxnu.edu.cn (L.Y.); 202240100759@jxnu.edu.cn (H.Y.); 202341500029@jxnu.edu.cn (J.X.)

[2] College of Chemistry and Chemical Engineering, Jiangxi Normal University, Nanchang 330022, China

* Correspondence: xiaoping@hbu.edu.cn (X.Y.); xlwang@jxnu.edu.cn (X.W.)

† These authors contributed equally to this work.

Abstract: Douchi is a Chinese traditional fermented food with a unique flavor. Methyl anthranilate (MA) plays an important role in formation of this flavor. However, the complicated relationship between the MA formation and the metabolic mechanism of the key functional microorganisms remains unclear. Here, we elucidated the response mechanism of aroma production driven by high salt stress in *Trichomonascus ciferrii* WLW (*T. ciferrii* WLW), which originates from the douchi fermentation process. The highest production of MA was obtained in a 10% NaCl environment. The enhanced expression of the key enzyme genes of the pentose phosphate pathway and shikimic acid pathway directed carbon flow toward aromatic amino acid synthesis and helped sustain an increased expression of *metK* to synthesize a large amount of the methyl donor *S*-adenosylmethionine, which promoted methyl anthranilate yield. This provides a theoretical basis for in-depth research on the applications of the flavor formation mechanisms of fermented foods.

Keywords: douchi; methyl anthranilate; salt stress responses; transcriptome; food microbiology

1. Introduction

Douchi is a traditional Chinese fermented condiment that is widely consumed in China due to its unique flavor and texture [1]. A wide variety of volatile substances, such as esters, alcohols, and pyrazines, are involved in composing the flavor of douchi [2]; among them is methyl anthranilate (MA), which is an aromatic ester with a grapey, often sweet, warming aroma. It has several applications and is of economic value [3], with interest growing in the medical research field [4]. In addition, MA has been shown to bind to receptor proteins while inhibiting the development of inter-microbial signaling systems and reducing the formation of biofilms; thus, it has very promising applications in water treatment technology [3,5]. These applications demonstrate the important role of MA in the food industry and biomedicine.

Although MA plays an important role in many fields, it is mostly of plant origin [4]. However, due to the low content in plants and the complex and costly extraction process [4,6], there is no effective report on the commercial production of plant-derived MA, so the current commercial production of MA is dominated by chemical synthesis. Wang et al. [7] optimized the production of MA using a response surface methodology based on a chemical method. Unfortunately, the chemical method involves chemical principles such as Hoffman rearrangement, and has the disadvantages of a complex reaction mechanism, many by-products, and difficulty in product purification [7,8]. Therefore, it is not an environmentally friendly synthesis method. In contrast, microbial synthesis, as a more competitive and emerging approach, is a hot research topic nowadays, and this method has good sustainability, high productivity, low production cost, and high scalability. It can also effectively address the concerns about rapid global population growth and climate change [9,10]. Therefore, the use of microbial methods for MA synthesis will contribute to the green development of the industry, increase the sustainability of the synthesis process, reduce the cost of MA production, and improve production efficiency [4,11]. However, to the best of our knowledge, little research has been reported on MA synthesis using microorganisms. Luo et al. [4] constructed an MA biosynthetic pathway by introducing a maize-derived anthranilic acid, methyltransferase, to *Escherichia coli*. Kuivanen et al. [11] used metabolic engineering to modify *Saccharomyces cerevisiae*, which addressed the gap in the literature of MA synthesis by eukaryotes. In addition, a method for microbial-mediated (wood-rotting fungi, etc.) dimethyl anthranilate demethylation was published as a patent. This method can be used for demethylation to synthesize MA, but the need to add the exogenous substrate dimethyl anthranilate, which is toxic to most cells, and the requirement for the strict optimization and control of the reaction conditions have hampered the popularity of this method [12]. Interestingly, in the present study, we identified a yeast strain, *Trichomonascus ciferrii* WLW (*T. ciferrii* WLW), which produced high yields of MA in high-salt environments without the addition of exogenous substrates. Therefore, it is interesting to investigate the relationship between MA and environmental factors (salt stress). To the best of our knowledge, the regulatory mechanism of MA synthesis by microorganisms under high-salt conditions has not been elucidated. Therefore, unraveling the resistance mechanism of cells could provide theoretical support for improving the MA production capacity of cells, which could further promote industrial applications.

Thus, in this study, we aimed to reveal the mechanism of aroma production in *T. ciferrii* WLW driven by high salt stress. We examined the resilience of the aroma-producing yeast *T. ciferrii* WLW using a salinity tolerance test and detected the change in MA production capacity of this yeast with increasing salinity, and finally, we used transcriptome sequencing to further reveal the regulatory mechanism of the synthesis of MA by this yeast in response to salt stress. This study provides a theoretical basis for further understanding the potential application of *T. ciferrii* WLW yeast in the industry to provide effective strategies for subsequent metabolic regulation studies and the expansion of the production process.

2. Materials and Methods

2.1. Strain Source

The strain *T. ciferrii* WLW used in this experiment was isolated from *Aspergillus*-type douchi (Nanchang Taoxiangyuan Seasoned Foods Co., Ltd., Nanchang, Jiangxi, China) in the previous work of our group. It was obtained from douchi samples by the dilution coating method, isolation purification, and ITS sequence identification. The strain was stored in a refrigerator at −80 °C in the laboratory at Guangdong Provincial Mycological Strain Conservation Center (GDMCC), with the conservation number GDMCC No.61319.

2.2. Salt Tolerance and Aroma Production Capacity in Salt Conditions Test

The strain was pre-cultured in yeast–peptone–dextrose (YPD) solid medium at 30 °C for 48 h. A sufficiently activated bacterial solution was serially diluted 10^5-fold, and 100 mL each of solutions 10^3–10^5 was spread on YPD plates with salinities of 0% (control), 5%, 10%, 15%, and 20% (w/v, three biological replicates). The number of single colonies on the plates was counted (cfu/L) after 7 d. Subsequently, suitable single colonies were selected and cultured in the YPD liquid medium for 24 h, and this process was repeated once for proper activation. In order to study the trend of *T. ciferrii* WLW aroma production under salt stress and to determine the optimal salt concentration for aroma production, active seed inoculums were inoculated at 1% [13] (v/v) in YPD liquid medium (250/500 mL) at different salt concentrations (0%, 2%, 4%, 6%, 8%, 10%, 12%, 14%, and 16%, w/v) and assessed for MA production by high-performance liquid chromatography (HPLC).

2.3. Chemical Analysis

Supernatants were collected from centrifuged yeast liquid cultures and analyzed for MA titers using HPLC [4,14]. Prior to HPLC analysis, the samples were filtered through 0.22 μm syringe filters and the culture samples were diluted 1:1 with methanol (Aladdin Co., Ltd., Shanghai, China). The methanol used here to dissolve the samples contained

3 mmol L^{-1} of benzyl alcohol (Aladdin Co., Ltd., Shanghai, China) as an internal standard. The concentration of MA was determined using HPLC (Primaide, Hitachi, Co., Ltd., Tokyo, Japan) with a C18 column (5 μm, 4.6 × 150 mm, LaChrom Ultra, Hitachi, Co., Ltd., Tokyo, Japan) operated at 30 °C. The sample volume was 20 μL. The injection volume was 20 μL. The ratio of mobile phase A (aqueous solution containing 0.1% trifluoroacetic acid) to mobile phase B (methanol solution containing 0.1% trifluoroacetic acid) was 4:6, and the flow rate was set to 1 mL min^{-1}. Detection was performed at 220 nm using a photodiode array detector in isocratic elution with a re-equilibration time of 2 min between each run.

2.4. Library Construction and Sequencing

The precipitate from the centrifuged yeast fermentation liquid was collected for the extraction of total RNA using TRIzol® Reagent with three biological replicates, following the manufacturer's instructions. RNA was purified using an RNA Purification Kit (Majorbio, Shanghai, China). Subsequent assessments of concentration and purity were performed using a Nanodrop2000 spectrophotometer. Agarose gel electrophoresis was utilized for integrity evaluation, and the RIN value was ascertained with an Agilent5300 Fragment Analyzer. Messenger RNA was isolated using oligo (dT) beads via poly(A) selection. Fragmentation was performed using a fragment buffer, followed by the synthesis of double-stranded cDNA using a SuperScript Double-Stranded cDNA Synthesis Kit (Invitrogen, Carlsbad, CA, USA) and random hexamer primers (Illumina, San Diego, CA, USA). End repair, phosphorylation, and A-tailing of the synthesized cDNA were performed according to the Illumina library construction protocol. The library size was selected for 300 bp cDNA target fragments on 2% low-range ultra-agarose. This was followed by 15 cycles of PCR amplification using Phusion DNA Polymerase (NEB). Sequencing was performed by Majorbio Bio-pharm Biotechnology Co., Ltd. (Shanghai, China) using a NovaSeq 6000 sequencer.

2.5. Transcriptome Data Analysis

Using default parameters, we trimmed and quality-controlled raw paired-end reads using fastp [15]. Next, we aligned the clean reads to the reference genome, which was obtained from our genome sequencing of *T. ciferrii* WLW, using HISAT2 software (http://ccb.jhu.edu/software/hisat2/index.shtml, v2.2.2.1, accessed on 14 December 2020) in the directional mode [16]. The mapped reads of each sample were assembled using StringTie3 with a reference-based method. To identify the differentially expressed genes (DEGs) between two different samples, we calculated the expression level of each transcript by the transcripts per million method. We used RSEM to quantify gene abundance [17]. We performed differential expression analysis using DESeq2 [18]. The DEGs with $|Log_2FC| \geq 1$ and Padj value < 0.05 (DESeq2) were considered as significant DEGs. In addition, we performed enrichment analysis based on the Gene Ontology (GO, https://geneontology.org/ accessed on 14 December 2020) and Kyoto Encyclopedia of Genes and Genomes (KEGG, https://www.kegg.jp/ accessed on 14 December 2020) databases to determine which DEGs were significantly enriched in GO terms and in terms of metabolic pathways compared to the transcriptome background at a Bonferroni-corrected p-value (Padj) < 0.05. Analyses of GO functional enrichment and KEGG pathways were conducted using GoATools and KOBAS [19], respectively.

2.6. Data Analysis

Experimental data are expressed as mean ± standard deviation. Data statistics were established using IBM SPSS Statistics 24 and Prism GraphPad 8.0.2. One-way analysis of variance (ANOVA) was employed to analyze the effect of change in salt concentration on survival, and generalized linear models were run to analyze the effects of independent variables (time and salt concentration) on MA production and the interaction effects of independent variables. Results were statistically significant when $p < 0.05$.

3. Results and Discussion

3.1. Characterization of Salt Tolerance and MA Production in *T. ciferrii* WLW

In general, microorganisms capable of growing in saline conditions above 15% (*w*/*v*) NaCl are defined as extremely halotolerant microorganisms [20]. The black yeast *Hortaea werneckii*, for example, can grow in NaCl concentrations ranging from 0% (*w*/*v*) to 32% saturation (*w*/*v*), and thus is classified as an extremely halotolerant fungus [21]. An aroma-producing strain, *Trichomonascus ciferrii* WLW (*T. ciferrii* WLW), previously screened by our group from high-salt fermented food environments, was tested for salt tolerance, and the results showed that *T. ciferrii* WLW had a fairly high survival rate and entered a growth plateau in 20% NaCl (*w*/*v*) aqueous saline environments on day 6 (Figure 1A), suggesting that it has the ability to grow stably. Using a salt-free condition as a control, the survival rate of solid medium plates in the 10% NaCl (*w*/*v*) range decreased, but not significantly ($p > 0.05$). In contrast, the change in the survival rate of the solid medium plates in the range of 15% (*w*/*v*) NaCl reached a significant level ($p < 0.05$), but the survival rate was still as high as 50.13% (Figure 1B). The above results indicated that *T. ciferrii* WLW has high salinity tolerance and is an extremely halotolerant microorganism. In addition, the high-salt conditions promoted an enhanced production of methyl anthranilate (MA) in *T. ciferrii* WLW, which led us to speculate that a correlation exists between the production of MA by *T. ciferrii* WLW and environmental factors (salt stress).

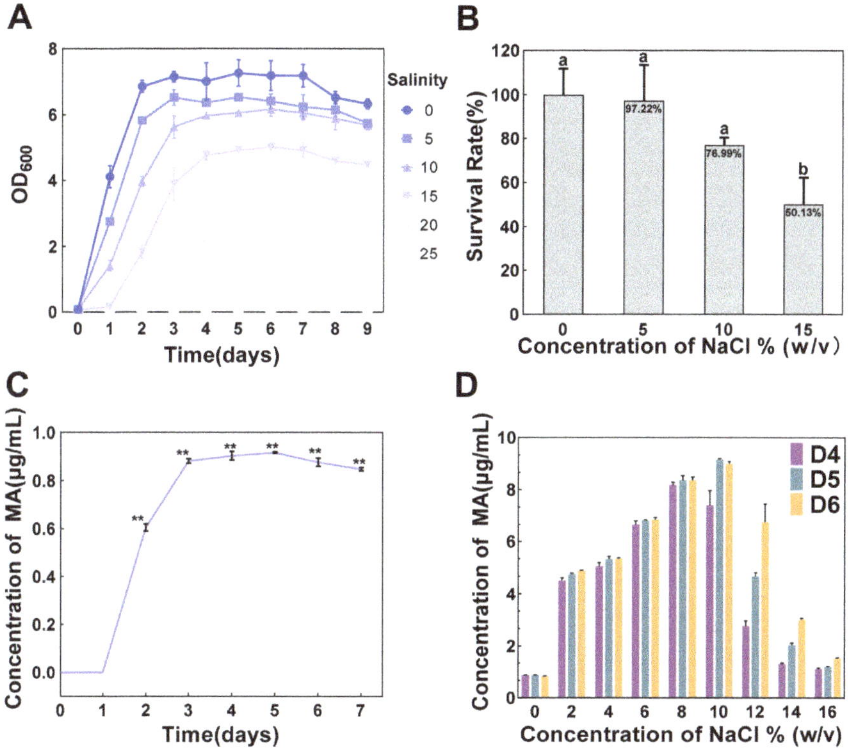

Figure 1. Effect of salt concentration on the ability of *T. ciferrii* WLW to survive and synthesize MA. (**A**) Curves of OD_{600} of *T. ciferrii* WLW in liquid medium at 0–25% (*w*/*v*) salinity; (**B**) changes in *T. ciferrii* WLW survival rate at 0–15% (*w*/*v*) salinity on solid medium plates (different letters above the columns indicate significant differences between groups, $p < 0.05$); (**C**) MA production curve of MA synthesized by *T. ciferrii* WLW in a salt-free state; (**D**) the change in aroma production of *T. ciferrii* WLW is in the range of 0%–16% (*w*/*v*) salinity. (D4–D6 represent Day 4–Day 6, respectively, and the value is mean standard deviation, $n = 3$. ** represents the level of significant difference at $p < 0.01$).

High-performance liquid chromatography (HPLC) is a technique capable of effectively detecting MA concentrations [4]. To further explore the optimal salt concentration for the synthesis of MA by *T. ciferrii* WLW, MA yield variations in a range of salt concentrations (0%–16% NaCl) were determined. The results showed that under salt-free conditions, the MA concentration reached a significant level on day 2 ($p < 0.01$) and entered a stabilization period on day 3 (Figure 1C), indicating that this strain has the ability to accumulate MA. A previous paper reported a method to generate MA using the natural demethylation of *Bacillus megaterium* [22]. However, this method requires the addition of exogenous substrates, which not only increases the cost but also the toxicity of the substrate to the microorganism. This means that the production process needs to be strictly controlled in terms of substrate concentration, thus limiting the production efficiency and discouraging the scale-up to commercial production. In contrast, *T. ciferrii* WLW-synthesized MA is clearly a more natural and less costly microbial pathway. In addition, MA production increased with increasing salinity (0–10% NaCl) compared to a salt-free environment (Figure 1D), and this effect of salt pressure driving the synthesis of MA reached a significant level (Table S1). The results showed that, on day 6, MA production at 10% (w/v) NaCl was approximately 9-fold higher than that at 0% salinity (Figure 1D). Interestingly, MA production decreased with increasing salinity when salinity exceeded 10% (Figure 1D). As a result, 10% (w/v) NaCl played the most important role in the optimal salt concentration for MA production by *T. ciferrii* WLW. In conclusion, salt pressure promoted MA synthesis in *T. ciferrii* WLW yeasts and MA production reached a maximum on day 6 at 10% salinity.

In fact, food microbial growth and metabolism are often influenced by the interaction of multiple external factors [23]. Generalized linear models are often used to reveal the interaction effects of independent variables on dependent variables [24,25]. Based on this model, in our study, there was a significant effect of salt concentration and time on MA yield effects ($p < 0.001$) (Table S1), and more importantly, there was still a significant effect on MA yield when time and salt concentration interacted ($p < 0.001$) (Table S1). Previous studies have demonstrated a significant effect of salt type and concentration on the resulting microbial community, as well as on the type and concentration of metabolites present in fermented foods [26,27]. This is similar to our findings, but the salt-tolerance mechanisms and molecular mechanisms of salt-pressure-driven MA synthesis in *T. ciferrii* WLW remain unclear and need to be further analysis.

3.2. Transcriptome Assembly and Difference Analysis

The transcriptome sequencing of *T. ciferrii* WLW at 0% and 10% (w/v) salinity was performed to determine the intrinsic relationship between MA production and salt pressure at the transcriptomic level. Based on the aroma production curve (250/500 mL) at 10% (w/v) salinity (Figure 2A), we selected samples from the 0% (w/v) NaCl control and 10% (w/v) NaCl-treated groups on Days 2, 4, and 6, and extracted RNA for the library construction and subsequent analysis. The reference genome was derived from the previously obtained whole-genome sequencing data of *T. ciferrii* WLW. Notably, after data quality control, the Q30 values of all sequencing data met the criteria (>90%) and the total mapping rates of the reference genomes were higher than 95% (Table S2). This result indicated that the quality of the sequencing data reached a qualified level and provided a reliable basis for subsequent analyses [28,29].

Subsequently, we further used the difference analysis mediated by DESeq2 [18] to define significant DEGs according to the rule of Padj value ≤ 0.05, $|Log_2FC| \geq 1$ [30]. As shown in Figure 2B, the comparison of the treatment group with the control group on Day 2 indicated a total of 1340 DEGs, of which 509 genes were upregulated and 831 genes were downregulated. On Day 4, a total of 2061 DEGs were obtained, of which 1022 genes were upregulated and 1039 genes were downregulated; on Day 6, a total of 1654 DEGs, of which 951 genes were upregulated and 703 genes were downregulated, were found. In summary, the number of differential genes under salt stress (10% NaCl) tended to fluctuate, suggesting that, compared to the controls, *T. ciferrii* WLW showed genetic differences at the transcriptome level under high-salt conditions, and the mechanism of salt-stress-driven MA synthesis may be a complex and dynamic process. Therefore, it is necessary to perform further differential gene analyses to reveal the molecular mechanisms by which salt stress drives *T. ciferrii* WLW to synthesize MA.

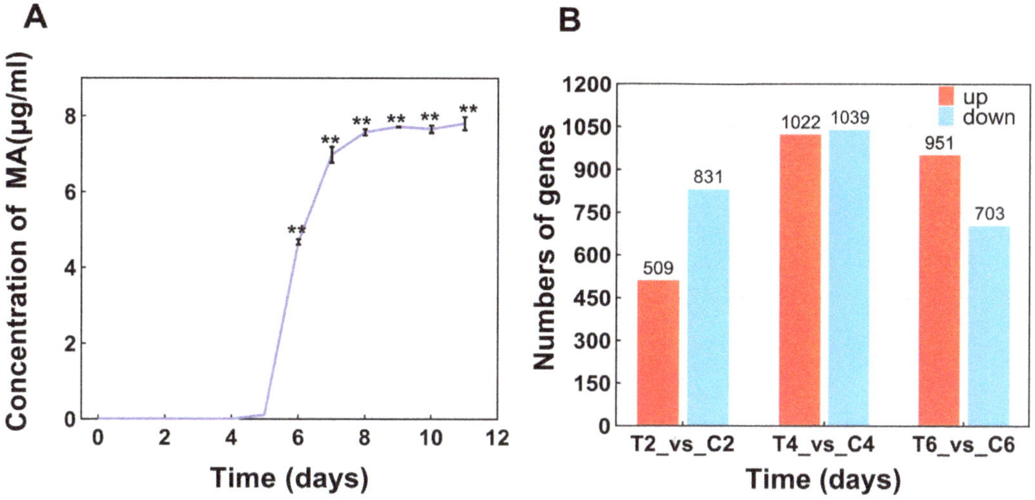

Figure 2. Trends in aroma production and genetic changes in *T. ciferrii* WLW under 10% salinity. (**A**) Aroma production curve of *T. ciferrii* WLW under 10% salinity; (**B**) number of significant DEGs for different days under salt stress. (** represents the level of significant difference at $p < 0.01$. "T2/C2", "T4/D4", and "T6/C6" represent data on days 2/4/6 for the treatment and control groups, respectively).

3.3. Mechanisms of T. ciferrii WLW Response to Salt Stress

In addition to the ability of *T. ciferrii* WLW to produce MA while tolerating high salt, the underlying molecular mechanism of the response to salt stress also attracted our attention. A series of studies related to the mechanism of salt tolerance have shown that membrane proteins, ion homeostasis, and transcription factors (TFs) are closely related to the strong salt tolerance of microorganisms under high-salt conditions [31–33]. Therefore, we used transcriptomic techniques to explore the reason for survival despite high-salt conditions from multiple perspectives, including the expression of key genes in terms of membrane proteins, ionic homeostasis, and salt-stress-related transcription factors.

Important genes related to the salt stress response in salt environments tend to show significant changes at the transcriptional level [34,35]. Therefore, we differentially analyzed the transcription of genes in conjunction with the Gene Ontology (GO, https://geneontology.org/ accessed on 14 December 2020) database. As shown in the Venn diagram (Figure 3A), more period-specific DEGs than common genes were present in *T. ciferrii* WLW at Days 2–6, suggesting that substantial metabolic changes may occur in *T. ciferrii* WLW under salt stress at different stages. Then, GO annotation and enrichment analysis were performed based on common genes at Days 2–6 to elucidate the persistent changes in *T. ciferrii* WLW transcript levels under high-salt conditions. The annotation results showed a high proportion of entries for cellular processes, metabolic processes, and localization activities (Figure 3B). These items are related to membrane transport, enzyme activity, ion localization, intracellular metabolism, and cell cycle changes, indicating that *T. ciferrii* WLW may alter its physiological and metabolic status by modulating genes related to transport, catalysis, bioregulation, localization, and organelle and cell membrane fractions to improve its tolerance to high-salt environments. In addition, the enrichment analysis of the common genes based on the GO database showed that four items related to membrane activity were significantly enriched (Padj < 0.05) (transmembrane transporter activity, transporter activity, integral component of membrane, and intrinsic component of membrane), suggesting that membrane-associated genes play an important role in the salt stress response (Figure 3C).

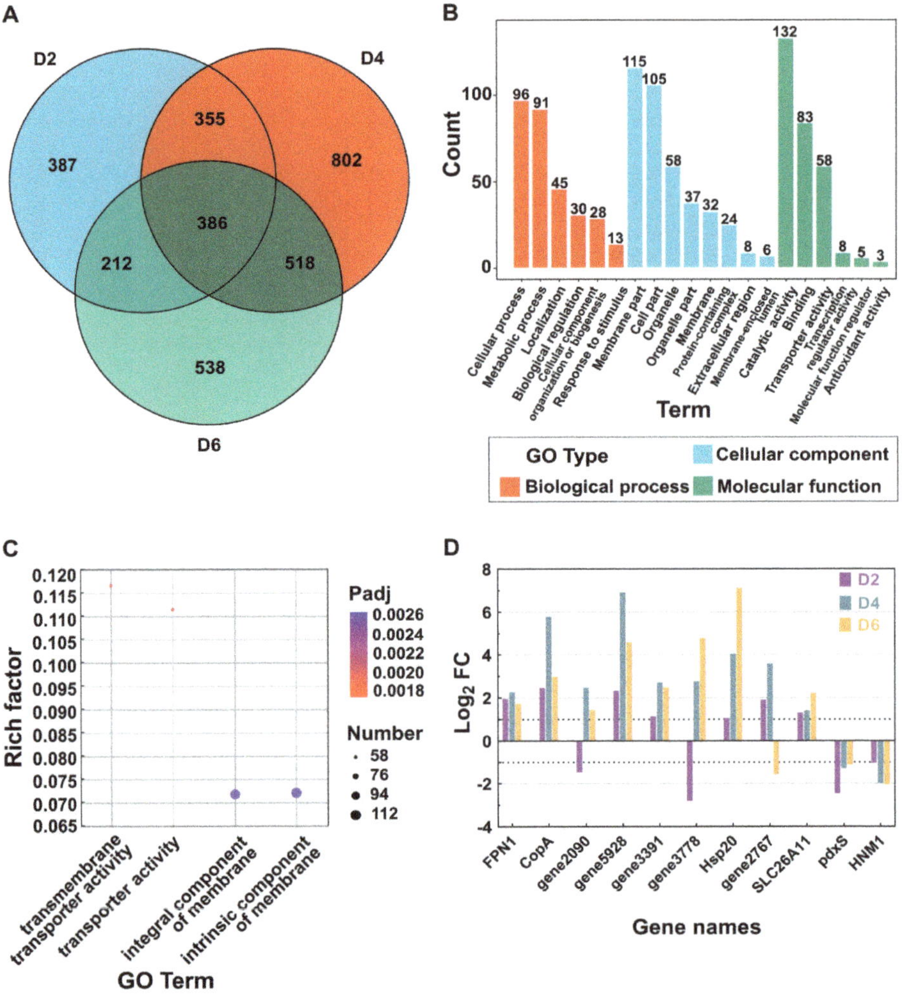

Figure 3. Days 2–6 of DEG counting and results of common DEG annotations and enrichment based on GO database. (**A**) Venn diagram of DEGs at Days 2–6. (**B**) Results of common significant DEGs at Days 2–6, annotated based on the GO database. (**C**) Bubble diagram of enrichment analysis of common DEGs based on GO database. Only terms with significant enrichment (Padj < 0.05) are shown. (**D**) Log_2FC for 10 representative genes at Days 2–6.

Because membrane proteins have an important effect on the environmental tolerance of cells [36], we investigated the transcription of relevant membrane protein genes using transcriptomic techniques. Notably, among the many DEGs identified in this study, some membrane protein genes continued to be significantly downregulated on Days 2–6 (Figure 3D). Amino acids are known to be important for resistance to salt stress [37]. One of these membrane protein genes is the pyridoxal phosphate synthase subunit gene *pdxS* (Figure 3D), which is a cofactor for amino acid catabolic enzymes in the pyridoxal phosphate metabolism and is involved in amino acid catabolic reactions, such as transamination and decarboxylation [38]. In the present study, *pdxS* was significantly downregulated at the transcriptional level at Days 2–6, which is similar to previously reported results [39]. In the present study, the persistent and significant downregulation of *pdxS* at the transcriptional level may be a result of the harsh high-salt environment, and to a certain extent, it may prevent amino acid degradation, thus helping to maintain the salt tolerance of *T. ciferrii* WLW. In addition, glycine betaine is involved in maintaining intracellular osmotic homeostasis and avoiding the toxicity of intracellular enzymes induced by high salt and inorganic ion concentrations [40]. The glycine betaine transmembrane transporter gene *HNM1* was significantly downregulated within Days 2–6, which may be due to the harsh, high-salt environment to which *T. ciferrii* WLW was subjected.

Inorganic ion homeostasis plays an important role in plant salt tolerance [41]. Thus, we performed further GO enrichment analysis of the genes involved in ion homeostasis. The top 20 significant entries were selected for further analysis, with the most significant ones being related to Zn, Mn, Cu, and Fe ion homeostasis (Figure 4). The genes for the ion homeostasis-related proteins *FPN1*, *CopA*, and *SLC26A11* were significantly upregulated in response to Fe, Cu, and Cl ions, respectively (Figure 3D). *SLC26A11* is usually reported to be an anion exchange channel with Cl channel effects [42,43]. The latter regulates the uptake and efflux of Cl and other anions from the cell membrane, thereby affecting cellular osmotic homeostasis, ionic balance, and pH regulation, which, in turn, enhances cellular salt tolerance [44]. Thus, the upregulation of Cl channels in *T. ciferrii* WLW may play a positive role in the salt stress response [45,46]. In addition, Ca ions are cofactors for a variety of metabolic enzymes, and calmodulin plays an important role in the salt stress response [47]. Therefore, as shown in Figure S1C, the regulation of Ca ion homeostasis and the overexpression of calmodulin-related genes contributed to salt tolerance in the yeast. To the best of our knowledge, the involvement of Cu/Fe/Mn ions in the response of fungi to salt stress has not yet been reported. Their significant enrichment in our study improves our understanding of the mechanisms by which yeast responds to abiotic stress. Calcium ions modulate the abundance of plasma membranes ZIP8 and ZIP14, thereby regulating the cellular uptake of Mn(II) [48], which acts as a coenzyme in metabolism. Iron and Cu ions have been reported to have strong redox activity and protein-binding capacity and participate in antioxidant processes, such as acting as antioxidant enzyme cofactors and inactivating free radicals [49,50]. Salt stress usually leads to oxidative stress [51]. In our study, Cu and Fe ion import-related genes (*gene5665* and *gene5666*) remained significantly upregulated (Figure S1A,B), suggesting that Fe and Cu ion homeostasis plays an important role in the response of *T. ciferrii* WLW to salt stress. In summary, the expression levels of ionic homeostasis-related genes were significantly altered under salt stress, indicating that ionic homeostasis plays an important role in the salt resistance of *T. ciferrii* WLW.

In terms of TFs, which are important for regulating stress response [52], we analyzed the TF data obtained from DEGs under salt stress conditions based on the JASPAR database (http://jaspar.genereg.net/, accessed on 12 October 2019). Sixteen TFs were identified in our results, of which four families were known: C4-GATA-related factors, heterodimeric CCAAT-binding factors, other factors with up to three neighboring zinc fingers, and TBP-related factors (Table S3). The GATA transcription factors *Gln3* and *Gat1* regulate the transcription of permease and catabolite genes and inhibit nitrogen catabolism, playing an important role in the salt response of yeast [53,54]. Their significant upregulation during the pre-salt stress period (Days 2–4) (Figure 5A,B) might contribute to the high salt tolerance of *T. ciferrii* WLW. *YOX1/YHP1* are co-repressors of DNA transcription, and when DNA replication is stressed, high *YOX1* expression (Figure 5C) blocks cell cycle progression until it is able to overcome replication defects [55], thereby increasing cellular resilience to environmental stress [56], which might be an important reason for the high salt tolerance of *T. ciferrii* WLW. However, the downregulation of *CAT8* accelerates cell growth and glucose consumption, and alters the type of energy metabolism in yeast cells, in which many genes related to the mitochondrial respiratory chain are downregulated, leading to a decrease in aerobic respiration and the tricarboxylic acid cycle [57]. Our results were similar to this (Figure 5A,B), suggesting that salt stress might alter the energy metabolic composition of *T. ciferrii* WLW. Overall, TFs are strongly associated with salt resistance in *T. ciferrii* WLW.

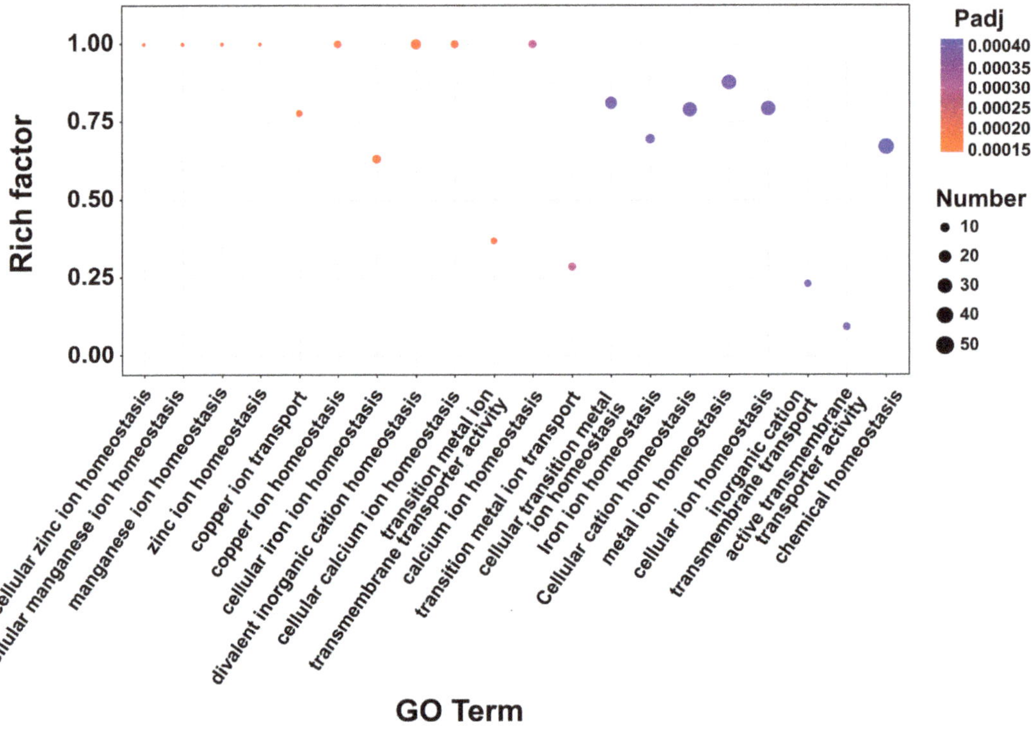

Figure 4. GO-enriched bubble plots of ion-homeostasis-related genes (only the first 20 significantly enriched terms are shown).

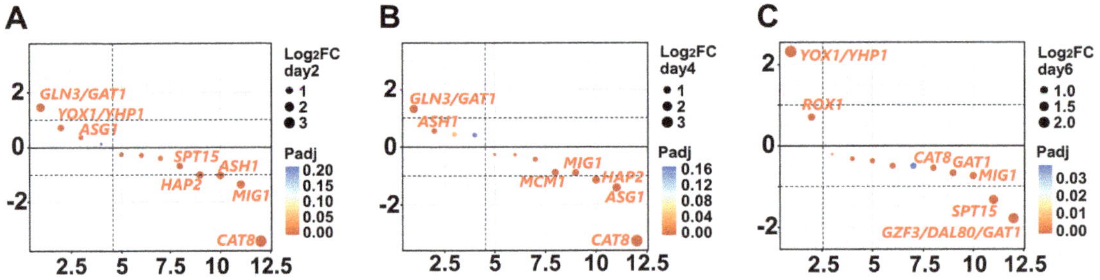

Figure 5. Bubble map of fold-change order of transcription factors. (**A**–**C**) correspond to the differential gene ordering of transcription factors on Days 2, 4, and 6, respectively. (Note: Horizontal coordinates indicate gene sequencing based on differential ploidy, and vertical coordinates indicate Log_2FC).

In summary, *T. ciferrii* WLW regulates the expression of key genes in terms of membrane proteins, ionic homeostasis, and salt stress-related TFs to enhance salt stress capacity, thus laying the foundation for high MA production despite the high salt environment.

3.4. Timing Analysis of the MA-Synthesis-Pathway-Associated DEGs

To reveal the mechanism of MA production driven by salt stress, a timing analysis of the MA-pathway-associated DEGs was performed based on the difference analysis. Several previous studies have shown that the MA de novo synthesis pathway flows through the glucose through glycolysis pathway (EMP), pentose phosphate pathway (PPP), shikimate pathway, and chorismate synthesis tryptophan pathway (trp biosynthesis) [4,11].

As shown in Figure 6, in the case of glucose through EMP, normally, glucose is broken down by glycolysis and flows to the citric acid cycle (TCA cycle), the main pathway of energy synthesis, to synthesize energy with the aid of oxidative phosphorylation [58]. However, in the present study, three key regulatory enzyme genes of the TCA cycle, *gltA* (citrate synthase), *IDH3* (isocitrate dehydrogenase), and *SDH1* (succinate dehydrogenase), were significantly downregulated on Days 2, 4, and 6, respectively (Figures 6 and S3, and Table S4). These results imply that the TCA cycle may be impaired and oxidative phosphorylation is reduced. Notably, the downregulation of the TCA cycle may help carbon flow toward the PPP and, thus, toward MA synthesis.

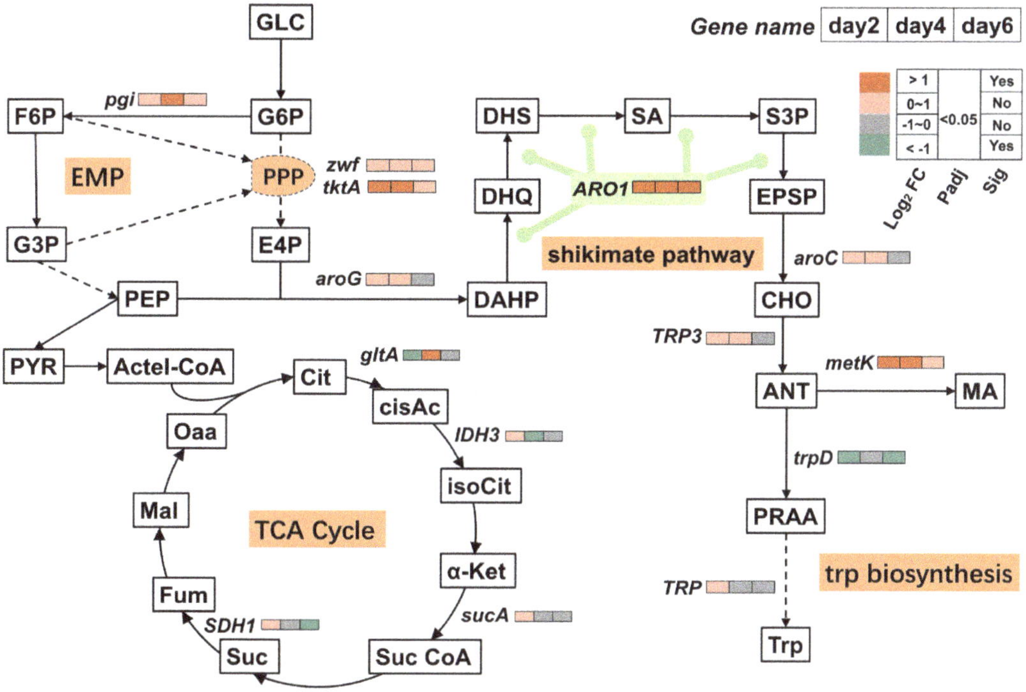

Figure 6. Expression of pathways related to MA synthesis. GLC: glucose; G6P: glucose-6-phosphate; F6P: fructose-6-phosphate; G3P: glyceraldehyde-3-phosphate; PEP: phosphoenolpyruvate; PYR: py-ruvate; TCA cycle: tricarboxylic acid cycle; E4P: Erythritol-4-phosphate; DAHP: 3deoxy-α-arabinoheptulosonate-7-phosphate; DHQ: 3-dehydroquinate; DHS: 3- dehydroshikimic acid; SA: shikimic acid; S3P: shikimate-3-phosphate; EPSP: 3-enol pyruvylshhikimate-5-phosohate; CHO: cho-rismite; ANT: anthranilate; MA: methyl anthranilate; PRAA: N-(5′-phosphoribosyl) anthranilate; Trp: tryptophan; Cit: citrate; cisAc: cis-Aconitate; isoCit: isocitrate; α-Ket: α-ketoglutarate; Suc CoA: succinyl CoA; Suc: succinate; Fum: fumarate; Mal: malate; Oaa: oxalo-acetate. (Note: Solid lines indicate one-step reactions, while dashed lines indicate multi-step reactions; the abbreviation 'Sig' is used to indicate the significance of genes).

Therefore, we found that the expression of the pathway genes in both the control and salt-treated groups and key gene of the PPP, *tktA*, which encodes ketolase, were significantly upregulated on Days 2 and 4 under salt conditions (Figure 6). The results suggest that salt pressure may be an important factor driving the overexpression of key genes.

The shikimate pathway, which plays an important role in MA synthesis, is the third pathway through which carbon flows to synthesize MA [4]. Notably, *ARO1*, a key enzyme gene in the shikimate pathway, encodes a pentafunctional AROM polypeptide that is responsible for positively catalyzing the five key steps of the shikimate pathway [59]. In our text, the expression of *ARO1*maintained a high differential fold value ($Log_2FC > 10$) on Days 2–4 of salt stress (Figure 6 and Figure S2). This means that the high expression of the *ARO1* gene in *T. ciferrii* WLW in a saline environment is helpful for the forward progress of the shikimic acid pathway and may further promote carbon flow towards MA synthesis. The shikimate pathway is directly involved in the synthesis of aromatic amino acids (phenylalanine, tyrosine, and tryptophan) [60]. In addition to its important role in salt tolerance [61–63], more importantly, tryptophan synthesis is directly related to MA synthesis [4]. In our study, the upregulation of the shikimate pathway not only contributed to an improvement in cellular salt tolerance, but also aided in the direction of carbon flow towards MA synthesis.

According to some previous reports, in the chorismate synthesis tryptophan pathway, the synthetic reaction of anthranilic acid (ANT) to MA is catalyzed by methyltransferase, and *S*-adenosylmethionine (SAM) is required as a methyl donor for the reaction [64,65]. Therefore, methyltransferases play an important role in MA synthesis. As a common coenzyme involved in methyl transfer reactions, SAM is used to synthesize MA from ANT by donating a methyl group to the ANT molecule, leading to MA formation [4]. In our study, we noticed that the upregulation of the shikimate pathway continued up to ANT and into the synthesis of MA, and the SAM synthase gene (*metK*) was significantly upregulated on Days 2 and 4 ($Log_2FC > 1$, $Padj < 0.05$) (Figure 6), suggesting that *T. ciferrii* WLW has sufficient SAM at this stage, and that the methyltransferases on this metabolic branch from ANT to MA may be highly active, thus contributing to MA synthesis. This may be an important reason why *T. ciferrii* WLW produces a high amount of MA under salt stress. Noteworthily, the related genes in tryptophan biosynthesis located after ANT, *trpD* (anthranilate phosphoribosyl transferase) and *TRP* (tryptophan synthase), were consistently downregulated (Figure 6), suggesting that *T. ciferrii* WLW chooses to downregulate tryptophan synthesis under salt stress. As a result, this could help in the synthesis and accumulation of MA. In addition, in the tryptophan catabolism–kynurenine pathway, the gene encoding the enzyme for catalyzing the conversion of formyl anthranilate to ANT was consistently upregulated on Days 2 and 4, which may also contribute to the accumulation of ANT and further aid in the synthesis of MA (Figure 6).

In summary, the above-mentioned genes in the MA synthesis pathway play a key role in increasing MA production and can be activated under salt pressure. *T. ciferrii* WLW regulated the expression of key genes (e.g., *metK*, *tkt*, and *ARO1*) in the TCA cycle, the PPP, the shikimate pathway, and tryptophan biosynthesis in response to salt stress, which in turn helped increase the yield of MA.

4. Conclusions

In conclusion, this report not only characterizes the high salt tolerance and optimal salinity for MA synthesis in the fragrance-producing yeast *T. ciferrii* WLW, but also reveals the molecular mechanisms underlying its salt tolerance and salt-driven fragrance production. *T. ciferrii* WLW exhibited a survival rate of 50.13% at 15% salinity, demonstrating that *T. ciferrii* WLW is an extremely halotolerant microorganism. The best performance of MA production by *T. ciferrii* WLW was achieved at 10% salinity, which emphasizes the favorable effect of an appropriate salt concentration environment on MA synthesis by *T. ciferrii* WLW. Statistical analyses based on generalized linear models showed that the time–salt concentration interaction effect significantly affected the ability of *T. ciferrii*

WLW to synthesize MA. Transcriptome analysis revealed that membrane proteins, ion homeostasis, and transcription factors are important to salt tolerance in *T. ciferrii* WLW. Salt-stressed environments can affect cell survival and aroma production capacity by inhibiting the TCA cycle. Metabolic pathways such as the PPP, the shikimate pathway, and tryptophan biosynthesis formed important components of the molecular mechanism of salt-pressure-driven aroma production in *T. ciferrii* WLW, and key genes in these pathways acted together to help the flow of carbon towards MA synthesis, which resulted in an increase in MA production in *T. ciferrii* WLW. These findings facilitate future enhancements in the fermentation process of *T. ciferrii* WLW for MA production, particularly when integrated with response surface metabolomics and other advanced techniques. This research not only offers valuable insights for optimizing production methods and expanding the application of critical functional microorganisms in the manufacturing of traditional food products, but also lays a solid theoretical foundation for exploring the mechanisms behind flavor development in fermented foods and their broader applications.

Supplementary Materials: The following supporting information can be downloaded at: https://www.mdpi.com/article/10.3390/foods13111593/s1, Figure S1: Difference in expression of genes related to iron and copper ion homeostasis; Figure S2: Differences in the timing of expression of key genes for MA synthesis; Figure S3: Bubble map of fold-change order of the main different genes included citrate cycle; Table S1. Effect testing of generalized linear model based on data from salt tolerance test; Table S2: Summary of trimming and read mapping results of the sequences generated from *T. ciferrii* WLW yeast with or without salt condition; Table S3: Transcription factors predicted by JASPAR database; Table S4: The main different genes included in citrate cycle.

Author Contributions: Conceptualization, F.X. and L.Y.; methodology, F.X. and L.Y.; software, X.Y.; validation, F.X.; formal analysis, X.Y.; investigation, H.Y.; resources, X.W.; data curation, F.X. and H.Y.; writing—original draft preparation, F.X. and L.Y.; writing—review and editing, X.W. and X.Y.; visualization, J.X.; supervision, X.W. and X.Y.; project administration, X.W. and X.Y.; funding acquisition, X.W. and L.Y. All authors have read and agreed to the published version of the manuscript.

Funding: This work was supported by the National Natural Science Foundation of China [No. 22308137] and Natural Science Foundation of Jiangxi [20232ACB203012, 20232BAB213041].

Institutional Review Board Statement: Not applicable.

Informed Consent Statement: Not applicable.

Data Availability Statement: The original contributions presented in the study are included in the article/Supplementary Material, further inquiries can be directed to the corresponding authors.

Conflicts of Interest: The authors declare no conflicts of interest.

References

1. He, B.; Li, H.R.; Hu, Z.H.; Zhang, Y.H.; Sun, M.; Qiu, S.K.; Zeng, B. Difference in microbial community and taste compounds between *Mucor*-type and *Aspergillus*-type Douchi during koji-making. *Food Res. Int.* **2019**, *121*, 136–143. [CrossRef] [PubMed]
2. Luo, Y.Y.; Guo, Y.; Hu, X.Y.; Liu, W.H.; Liu, B.Q.; Yang, J.; Tu, Z.C.; Huang, Y.H. Flavor improvement of fermented soybean foods by co-fermentation with *Bacillus velezensis* and *Lactiplantibacillus plantarum*. *LWT-Food Sci. Technol.* **2023**, *186*, 9. [CrossRef]
3. Wang, J.; De Luca, V. The biosynthesis and regulation of biosynthesis of Concord grape fruit esters, including 'foxy' methylanthranilate. *Plant J.* **2005**, *44*, 606–619. [CrossRef] [PubMed]
4. Luo, Z.W.; Cho, J.S.; Lee, S.Y. Microbial production of methyl anthranilate, a grape flavor compound. *PNAS* **2019**, *116*, 10749–10756. [CrossRef] [PubMed]
5. Feng, Y.R.; Liang, J.; Liu, X.H.; Gao, K.X.; Zhang, Y.; Li, A.; Chen, C.; Hou, L.A.; Yang, Y. Graphene oxide/methyl anthranilate modified anti-biofouling membrane possesses dual functions of anti-adhesion and quorum quenching. *J. Membr. Sci.* **2023**, *668*, 10. [CrossRef]
6. Sulzbach, M.; da Silva, M.A.S.; Gonzatto, M.P.; Marques, M.M.O.; Böettcher, G.N.; Silvestre, W.P.; Silvad, J.; Pauletti, G.F.; Schwarz, S.F. Effect of distillation methods on the leaf essential oil of some *Citrus* cultivars. *J. Essent. Oil Res.* **2021**, *33*, 452–463. [CrossRef]
7. Wang, B.; Diao, Y.W.; Yuan, J.Q.; Zhang, F.Y.; Zhou, H.Y.; Du, L. Optimization of Methyl Anthranilate Synthesis Process by Response Surface Methodology and Its Reaction Mechanism. *Synthesis* **2022**, *54*, 5261–5272. [CrossRef]

8. Wang, B.; Wang, C.; Du, L.; Zhao, Q.; Yin, J. Optimization of Process and Improvement in Continuous Production of Synthesis of Methyl Anthranilate. *Chem. Eng. Trans.* **2020**, *81*, 139–144.
9. Choi, K.R.; Lee, S.Y. Systems metabolic engineering of microorganisms for food and cosmetics production. *Nat. Rev. Bioeng.* **2023**, *1*, 832–857. [CrossRef]
10. Johnson, A.; Deshmukh, P.; Kaushik, S.; Sharma, V. Microbial Bio-production of Proteins and Valuable Metabolites. In *Microbial Interventions in Agriculture and Environment: Volume 1: Research Trends, Priorities and Prospects*; Singh, D.P., Gupta, V.K., Prabha, R., Eds.; Springer: Singapore, 2019; pp. 381–418.
11. Kuivanen, J.; Kannisto, M.; Mojzita, D.; Rischer, H.; Toivari, M.; Jäntti, J. Engineering of *Saccharomyces cerevisiae* for anthranilate and methyl anthranilate production. *Microb. Cell Factories* **2021**, *20*, 34. [CrossRef]
12. Gregory, P.; Bonita, S.; Mohamad, F. A method for the preparation of "natural" methyl anthranilate. WO8900203A, 1 December 1989.
13. Chandrasekaran, S.D.; Vaithilingam, M.; Shanker, R.; Kumar, S.; Thiyur, S.; Babu, V.; Selvakumar, J.N.; Prakash, S. Exploring the *In Vitro* Thrombolytic Activity of Nattokinase From a New Strain *Pseudomonas aeruginosa* CMSS. *Jundishapur J. Microbiol.* **2015**, *8*, 8. [CrossRef] [PubMed]
14. Nozal, M.J.; Bernal, J.L.; Toribio, L.; Jiménez, J.J.; Martín, M.T. High-performance liquid chromatographic determination of methyl anthranilate, hydroxymethylfurfural and related compounds in honey. *J. Chromatogr. A* **2001**, *917*, 95–103. [CrossRef] [PubMed]
15. Chen, S.F.; Zhou, Y.Q.; Chen, Y.R.; Gu, J. fastp: An ultra-fast all-in-one FASTQ preprocessor. *Bioinformatics* **2018**, *34*, 884–890. [CrossRef] [PubMed]
16. Kim, D.; Landmead, B.; Salzberg, S.L. HISAT: A fast spliced aligner with low memory requirements. *Nat. Methods* **2015**, *12*, 357-U121. [CrossRef] [PubMed]
17. Li, B.; Dewey, C.N. RSEM: Accurate transcript quantification from RNA-Seq data with or without a reference genome. *BMC Bioinform.* **2011**, *12*, 16. [CrossRef] [PubMed]
18. Love, M.I.; Huber, W.; Anders, S. Moderated estimation of fold change and dispersion for RNA-seq data with DESeq2. *Genome Biol.* **2014**, *15*, 38. [CrossRef] [PubMed]
19. Xie, C.; Mao, X.Z.; Huang, J.J.; Ding, Y.; Wu, J.M.; Dong, S.; Kong, L.; Gao, G.; Li, C.Y.; Wei, L.P. KOBAS 2.0: A web server for annotation and identification of enriched pathways and diseases. *Nucleic Acids Res.* **2011**, *39*, W316–W322. [CrossRef] [PubMed]
20. Chung, D.; Kim, H.; Choi, H.S. Fungi in salterns. *J. Microbiol.* **2019**, *57*, 717–724. [CrossRef] [PubMed]
21. Czachura, P.; Owczarek-Koscielniak, M.; Piatek, M. *Salinomyces polonicus:* A moderately halophilic kin of the most extremely halotolerant fungus *Hortaea werneckii*. *Fungal Biol.* **2021**, *125*, 459–468. [CrossRef] [PubMed]
22. Taupp, M.; Harmsen, D.; Heckel, F.; Schreier, P. Production of Natural Methyl Anthranilate by Microbial N-Demethylation of N-Methyl Methyl Anthranilate by the Topsoil-Isolated Bacterium Bacillus megaterium. *J. Agric. Food Chem.* **2005**, *53*, 9586–9589. [CrossRef]
23. Zhu, M.; Zheng, J.; Xie, J.; Zhao, D.; Qiao, Z.W.; Huang, D.; Luo, H.B. Effects of environmental factors on the microbial community changes during medium-high temperature Daqu manufacturing. *Food Res. Int.* **2022**, *153*, 11. [CrossRef] [PubMed]
24. Long, J.; Huang, H.; Tang, P.; Liang, J.; Liao, Q.; Chen, J.; Pang, L.; Yang, K.; Wei, H.; Chen, M.; et al. Associations between maternal exposure to multiple metals and metalloids and blood pressure in preschool children: A mixture-based approach. *J. Trace Elem. Med. Biol. Organ Soc. Miner. Trace Elem. (GMS)* **2024**, *84*, 127460. [CrossRef] [PubMed]
25. Geng, S.; Zhang, L. Decorrelated empirical likelihood for generalized linear models with high-dimensional longitudinal data. *Stat. Probab. Lett.* **2024**, *211*, 110135. [CrossRef]
26. Lee, M.A.; Choi, Y.J.; Kim, Y.S.; Chon, S.Y.; Chung, Y.B.; Park, S.H.; Yun, Y.R.; Min, S.G.; Yang, H.C.; Seo, H.Y. Effects of salt type on the metabolites and microbial community in kimchi fermentation. *Heliyon* **2022**, *8*, 9. [CrossRef] [PubMed]
27. Mi, T.; Wang, D.K.; Yao, S.J.; Yang, H.; Che, Y.L.; Wu, C.D. Effects of salt concentration on the quality and microbial diversity of spontaneously fermented radish paocai. *Food Res. Int.* **2022**, *160*, 14. [CrossRef] [PubMed]
28. Hao, S.; Li, S.; Wang, J.; Zhao, L.; Yan, Y.; Cao, Q.; Wu, T.T.; Liu, L.Y.; Wang, C.T. Transcriptome Analysis of Phycocyanin-Mediated Inhibitory Functions on Non-Small Cell Lung Cancer A549 Cell Growth. *Mar. Drugs* **2018**, *16*, 511. [CrossRef]
29. Yao, Z.Y.; Gong, J.S.; Liu, Y.R.; Jiang, J.Y.; Zhang, Y.S.; Su, C.; Li, H.; Kang, C.L.; Liu, L.; Xu, Z.H.; et al. Genetic variation reveals the enhanced microbial hyaluronan biosynthesis via atmospheric and room temperature plasma. *Carbohydr. Polym.* **2023**, *312*, 13. [CrossRef]
30. Zeng, Z.; Liu, Y.; Feng, X.Y.; Li, S.X.; Jiang, X.M.; Chen, J.Q.; Shao, Z.Q. The RNAome landscape of tomato during arbuscular mycorrhizal symbiosis reveals an evolving RNA layer symbiotic regulatory network. *Plant Commun.* **2023**, *4*, 19. [CrossRef] [PubMed]
31. Guo, H.; Nie, C.Y.; Li, Z.; Kang, J.; Wang, X.L.; Cui, Y.N. Physiological and Transcriptional Analyses Provide Insight into Maintaining Ion Homeostasis of Sweet Sorghum under Salt Stress. *Int. J. Mol. Sci.* **2023**, *24*, 11045. [CrossRef]
32. Banik, S.; Dutta, D. Membrane Proteins in Plant Salinity Stress Perception, Sensing, and Response. *J. Membr. Biol.* **2023**, *256*, 109–124. [CrossRef]
33. Dabravolski, S.A.; Isayenkov, S.V.J.H. The Role of the γ-Aminobutyric Acid (GABA) in Plant Salt Stress Tolerance. *Horticulturae* **2023**, *9*, 230. [CrossRef]
34. Li, J.L.; Pang, Q.Y.; Yan, X.F. Unique Features of the m^6A Methylome and Its Response to Salt Stress in the Roots of Sugar Beet (*Beta vulgaris*). *Int. J. Mol. Sci.* **2023**, *24*, 11659. [CrossRef]

35. Liu, J.G.; Han, X.; Yang, T.; Cui, W.H.; Wu, A.M.; Fu, C.X.; Wang, B.C.; Liu, L.J. Genome-wide transcriptional adaptation to salt stress in Populus. *BMC Plant Biol.* **2019**, *19*, 367. [CrossRef] [PubMed]
36. Gill, R.A.; Ahmar, S.; Ali, B.; Saleem, M.H.; Khan, M.U.; Zhou, W.; Liu, S. The Role of Membrane Transporters in Plant Growth and Development, and Abiotic Stress Tolerance. *Int. J. Mol. Sci.* **2021**, *22*, 12792. [CrossRef] [PubMed]
37. Danaeifar, A.; Khaleghi, E.; Zivdar, S.; Mehdikhanlou, K. Physiological, biochemical, and gene expression of Sour Orange (*Citrus aurantium* L.) to Iron(II)-Arginine Chelate under salinity, alkalinity, and salt-alkali combined stresses. *Sci. Hortic.* **2023**, *319*, 112146. [CrossRef]
38. Heo, L.; Cho, Y.M.; Choi, J.; Lee, J.; Han, Y.; Han, S.W. Proteomic and Phenotypic Analyses of a Putative YggS Family Pyridoxal Phosphate-Dependent Enzyme in Acidovorax citrulli. *Plant Pathol. J.* **2023**, *39*, 235–244. [PubMed]
39. Chen, H.; Xiong, L.M. Pyridoxine is required for post-embryonic root development and tolerance to osmotic and oxidative stresses. *Plant J.* **2005**, *44*, 396–408. [CrossRef] [PubMed]
40. Zhu, M.Y.; Li, Q.X.; Zhang, Y.S.; Zhang, M.C.; Li, Z.H. Glycine betaine increases salt tolerance in maize (*Zea mays* L.) by regulating Na^+ homeostasis. *Front. Plant Sci.* **2022**, *13*, 978304. [CrossRef] [PubMed]
41. Daraz, U.; Ahmad, I.; Li, Q.S.; Zhu, B.; Saeed, M.F.; Li, Y.; Ma, J.; Wang, X.-B. Plant growth promoting rhizobacteria induced metal and salt stress tolerance in *Brassica juncea* through ion homeostasis. *Ecotox. Environ. Safe.* **2023**, *267*, 115657. [CrossRef]
42. Mount, D.B.; Romero, M.F. The *SLC26* gene family of multifunctional anion exchangers. *Pflügers Arch. Eur. J. Physiol.* **2004**, *447*, 710–721. [CrossRef]
43. Rahmati, N.; Kunzelmann, K.; Xu, J.; Barone, S.; Sirianant, L.; De Zeeuw, C.I.; Soleimani, M. Slc26a11 is prominently expressed in the brain and functions as a chloride channel: Expression in Purkinje cells and stimulation of V H^+-ATPase. *Pflügers Arch. Eur. J. Physiol.* **2013**, *465*, 1583–1597. [CrossRef] [PubMed]
44. Malakar, P.; Chattopadhyay, D. Adaptation of plants to salt stress: The role of the ion transporters. *J. Plant Biochem. Biotechnol.* **2021**, *30*, 668–683. [CrossRef]
45. Diédhiou, C.; Golldack, D. Salt-dependent regulation of chloride channel transcripts in rice. *Plant Sci.* **2006**, *170*, 793–800. [CrossRef]
46. Nakamura, A.; Fukuda, A.; Sakai, S.; Tanaka, Y. Molecular Cloning, Functional Expression and Subcellular Localization of Two Putative Vacuolar Voltage-gated Chloride Channels in Rice (*Oryza sativa* L.). *Plant Cell Physiol.* **2006**, *47*, 32–42. [CrossRef] [PubMed]
47. Li, Y.Q.; Zhang, H.D.; Dong, F.Y.; Zou, J.; Gao, C.B.; Zhu, Z.W.; Liu, Y.K. Multiple roles of wheat calmodulin genes during stress treatment and TaCAM2-D as a positive regulator in response to drought and salt tolerance. *Int. J. Biol. Macromol.* **2022**, *220*, 985–997. [CrossRef] [PubMed]
48. Steimle, B.L.; Bailey, D.K.; Smith, F.M.; Rosenblum, S.L.; Kosman, D.J.; Hanson, P. Calcium and the Ca-ATPase SPCA1 modulate plasma membrane abundance of ZIP8 and ZIP14 to regulate Mn(II) uptake in brain microvascular endothelial cells. *J. Biol. Chem.* **2022**, *298*, 102211. [CrossRef] [PubMed]
49. Wojtunik-Kulesza, K.; Oniszczuk, A.; Waksmundzka-Hajnos, M. An attempt to elucidate the role of iron and zinc ions in development of Alzheimer's and Parkinson's diseases. *Biomed. Pharmacother.* **2019**, *111*, 1277–1289. [CrossRef] [PubMed]
50. Gao, L.; Zhang, A.Q. Copper-instigated modulatory cell mortality mechanisms and progress in oncological treatment investigations. *Front. Immunol.* **2023**, *14*, 1236063. [CrossRef]
51. Yin, Y.L.; Xu, Y.N.; Li, X.N.; Fan, S.G.; Wang, G.Y.; Fu, J.M. Physiological integration between Bermudagrass ramets improves overall salt resistance under heterogeneous salt stress. *Physiol. Plant.* **2022**, *174*, e13655. [CrossRef]
52. Ma, R.; Liu, B.W.; Geng, X.; Ding, X.; Yan, N.; Sun, X.; Wang, W.L.; Sun, X.Z.; Zheng, C.S. Biological Function and Stress Response Mechanism of *MYB* Transcription Factor Family Genes. *J. Plant Growth Regul.* **2023**, *42*, 83–95. [CrossRef]
53. Tate, J.J.; Marsikova, J.; Vachova, L.; Palkova, Z.; Cooper, T.G. Effects of Abolishing Whi2 on Nitrogen Catabolite Repression-Sensitive GATA-Factor Localization and Protein Production. *Faseb J.* **2022**, *36*. [CrossRef]
54. Crespo, J.L.; Daicho, K.; Ushimaru, T.; Hall, M.N. The GATA transcription factors *GLN3* and *GAT1* link TOR to salt stress in *Saccharomyces cerevisiae*. *J. Biol. Chem.* **2001**, *276*, 34441–34444. [CrossRef] [PubMed]
55. Gómez-Escoda, B.; Ivanova, T.; Calvo, I.A. *Yox1* links MBF-dependent transcription to completion of DNA synthesis. *EMBO Rep.* **2011**, *12*, 84–89. [CrossRef] [PubMed]
56. Conti, M.M.; Ghizzoni, J.M.; Gil-Bona, A.; Wang, W.; Costanzo, M.; Li, R.; Flynn, M.J.; Zhu, L.J.; Myers, C.L.; Boone, C.; et al. Repression of essential cell cycle genes increases cellular fitness. *PLoS Genet.* **2022**, *18*, e1010349. [CrossRef] [PubMed]
57. Deng, H.; Du, Z.D.; Lu, S.R.; Wang, Z.; He, X. Regulation of Cat8 in energy metabolic balance and glucose tolerance in *Saccharomyces cerevisiae*. *Appl. Microbiol. Biotechnol.* **2023**, *107*, 4605–4619. [CrossRef]
58. MacLean, A.; Legendre, F.; Appanna, V.D. The tricarboxylic acid (TCA) cycle: A malleable metabolic network to counter cellular stress. *Crit. Rev. Biochem. Mol. Biol.* **2023**, *17*, 81–97. [CrossRef] [PubMed]
59. Duncan, K.; Edwards, R.M.; Coggins, J.R. The pentafunctional arom enzyme of *Saccharomyces cerevisiae* is a mosaic of monofunctional domains. *Biochem. J.* **1987**, *246*, 375–386. [CrossRef] [PubMed]
60. Maeda, H.; Dudareva, N. The Shikimate Pathway and Aromatic Amino Acid Biosynthesis in Plants. *Annu. Rev. PlantBiol.* **2012**, *63*, 73–105. [CrossRef] [PubMed]
61. Ingrisano, R.; Tosato, E.; Trost, P.; Gurrieri, L.; Sparla, F. Proline, Cysteine and Branched-Chain Amino Acids in Abiotic Stress Response of Land Plants and Microalgae. *Plants* **2023**, *12*, 3410. [CrossRef]

62. Chen, Y.L.; Huang, W.X.; Zhang, F.T. Metabolomic Profiling of Dongxiang Wild Rice Under Salinity Demonstrates the Significant Role of Amino Acids in Rice Salt Stress. *Front. Plant Sci.* **2021**, *12*, 729004. [CrossRef]
63. Xie, E.; Wei, X.J.; Ding, A.Z.; Zheng, L.; Wu, X.N.; Anderson, B. Short-Term Effects of Salt Stress on the Amino Acids of *Phragmites australis* Root Exudates in Constructed Wetlands. *Water* **2020**, *12*, 569. [CrossRef]
64. Blaszczyk, A.J.; Wang, B.; Silakov, A.; Ho, J.V.; Booker, S.J. Efficient methylation of C2 in L-tryptophan by the cobalamin-dependent radical *S*-adenosylmethionine methylase TsrM requires an unmodified N1 amine. *J. Biol. Chem.* **2017**, *292*, 15456–15467. [CrossRef] [PubMed]
65. Köllner, T.G.; Lenk, C.; Zhao, N.; Seidl-Adams, I.; Gershenzon, J.; Chen, F.; Degenhardt, J. Herbivore-Induced SABATH Methyltransferases of Maize That Methylate Anthranilic Acid Using *S*-Adenosyl-L-Methionine. *Plant Physiol.* **2010**, *153*, 1795–1807. [CrossRef] [PubMed]

Disclaimer/Publisher's Note: The statements, opinions and data contained in all publications are solely those of the individual author(s) and contributor(s) and not of MDPI and/or the editor(s). MDPI and/or the editor(s) disclaim responsibility for any injury to people or property resulting from any ideas, methods, instructions or products referred to in the content.

MDPI AG
Grosspeteranlage 5
4052 Basel
Switzerland
Tel.: +41 61 683 77 34

MDPI Books Editorial Office
E-mail: books@mdpi.com
www.mdpi.com/books

Disclaimer/Publisher's Note: The title and front matter of this reprint are at the discretion of the Topic Editors. The publisher is not responsible for their content or any associated concerns. The statements, opinions and data contained in all individual articles are solely those of the individual Editors and contributors and not of MDPI. MDPI disclaims responsibility for any injury to people or property resulting from any ideas, methods, instructions or products referred to in the content.

www.ingramcontent.com/pod-product-compliance
Ingram Content Group UK Ltd.
Pitfield, Milton Keynes, MK11 3LW, UK
UKHW060224061026
15625UKWH00008B/484

9 783725 844999